Handbook of Geochemistry

Vol. II/5

Elements La (57) to U (92)

Executive Editor: K. H. Wedepohl

Springer-Verlag Berlin Heidelberg New York

Printed in Germany

ISBN 3-540-09022-3 Springer-Verlag Berlin · Heidelberg · New York
ISBN 0-387-09022-3 Springer-Verlag New York · Heidelberg · Berlin

Library of Congress Catalog Card Number: 78-85402

Typesetting, printing, and binding: Universitätsdruckerei H. Stürtz AG, Würzburg

2121/3140-5 4 3 2 1 0

LIST OF CHAPTERS

Chemical Elements Occurring in Nature

Atomic numbers (in heavy print) are identical with chapter numbers in Volume II of this handbook. Elements with very close relations, such as the noble gases, the lanthanides and the platinum group elements, are combined and covered under one element of the respective group. Elements printed in italics will not be included in this book.

STANDARD SECTIONS OF CHAPTERS

(Chapters have the atomic numbers of the elements)

A. Crystal chemistry
B. Isotopes in nature
C. Abundance in cosmos, meteorites, lunar materials and tektites
D. Abundance in rock-forming minerals; phase equilibria; minerals of the respective element
E. Abundance in common igneous rock types
F. Behavior in magmatogenic processes (pegmatites, gas transport etc.) and ore deposition
G. Behavior during weathering and alteration of rocks; abundance in soils
H. Solubilities of compounds which control concentrations of the element in natural waters; adsorption processes; valence states in natural environments
I. Abundance in natural waters and in the atmosphere
K. Abundance in common sediments and sedimentary rock types
L. Biogeochemistry; abundance in coal, crude oil ect.
M. Abundance in common metamorphic rock types
N. Behavior in metamorphic processes
O. Relations to other elements, economic importance etc.

References (Common abbreviations: SB = Strukturbericht; SR = Structure Reports)

LIST OF SYMBOLS DESIGNATING ANALYTICAL METHODS

S	Optical emission sprectrography or spectrometry: arc, spark etc.
A	Atomic absorption spectrometry
E	Electrolysis
F	Optical emission spectrometry: flame photometry
X	X-ray spectrometry and X-ray fluorescence spectrometry
W	Chemical (wet conventional and rapid) methods etc.
C	Colorimetric, spectrophotometric methods
P	Polarographic methods
N/R	Neutron activation, radiometric measurements
I	Isotope dilution, mass spectrometric measurements
M	Microprobe analysis
L	Fluorimetric measurements
Mass	Spark source mass spectrometry etc.

Or combinations of these symbols

Yttrium and Lanthanides 39, **57**–71

A	J. Felsche	(Fachbereich Chemie, Universität Konstanz, Germany)
B—M, O	A. G. Herrmann	(Geochemisches Institut der Universität Göttingen, Germany)

Introduction

The element yttrium ($Z=39$) and the lanthanides: lanthanum ($Z=57$), cerium ($Z=58$), praseodymium ($Z=59$), neodymium ($Z=60$), samarium ($Z=62$), europium ($Z=63$), gadolinium ($Z=64$), terbium ($Z=65$), dysprosium ($Z=66$), holmium ($Z=67$), erbium ($Z=68$), thulium ($Z=69$), ytterbium ($Z=70$), and lutetium ($Z=71$), are associated in all natural matter, because the chemical behavior of Y is quite similar to that of members of the group La—Lu. For this reason it is suitable to discuss the geochemical data of Y and La—Lu in one chapter. Because of the lack of similarities, scandium is treated in a separate chapter.

The members of group III B of the periodic table are called in English "rare earth elements" (REE), in French "terres rares", in German "Seltene Erden", and in Russian "Redkie zemli". However, several of these elements are not really rare compared with numerous metals. For example, Ce, La, Y, and Nd are more abundant in the igneous rocks of the upper earth's crust than Pb, Co, Sn, Mo, and W. Hence in this chapter the rare earth elements are either denoted by the chemical symbols Y, La—Lu or by the name lanthanides (including yttrium).

The Y, La—Lu distribution patterns of meteorites and terrestrial matter correspond to the Oddo-Harkins rule. The simple plots ppm Y, La—Lu versus atomic number show the striking differences in abundance between successive lanthanides with odd and even atomic numbers (Fig. 39, 57—71-C-1). For a clear graphical presentation of natural fractionations among the lanthanides, relative abundances are usually plotted. In the following paper the abundance of each element in the minerals or rocks under investigation is divided by the abundance of the corresponding element in chondrites. Another technique is to standardize all the lanthanide concentrations to the value of one member of the series, e.g. La or Yb, taken as 1.00 (SCHMITT *et al.*, 1963; HASKIN, FREY *et al.*, 1966, for instance). These ratios are plotted on a logarithmic ordinate versus an abscissa scale either with the ionic radii (for example listed by AHRENS, 1952, and TEMPLETON and DAUBEN, 1954) or with the atomic numbers of Y, La—Lu. In this chapter the division based on atomic numbers is preferred (see HASKIN *et al.*, 1968). Many authors (CORYELL *et al.*, 1963; CHASE, WINCHESTER, and CORYELL, 1963; SCHMITT *et al.*, 1963; HASKIN, FREY *et al.*, 1966) prefer the chondrites for comparison in the distribution of the lanthanides, because no fractionation between the lighter and the heaver lanthanides in these meteorites has been observed. In this chapter the average absolute contents of Y and La—Lu data from 26 determinations on 22 individual chondrites and a mixed sample of 9 chondrites are used for the graphical comparison plots (see Section 39, 57—71-C).

Another method of comparing data on lanthanides in different rocks is to report the ratios

$$\frac{\Sigma \text{ ppm La—Eu}}{\Sigma \text{ ppm Gd—Lu, Y}}$$

for the respective minerals and rocks (see for instance BALASHOV and TURANSKAYA,

1960, 1962; LEONOVA and BALASHOV, 1963; BALASHOV, 1963; BALASHOV *et al.*, 1964). The present author also uses these ratios to demonstrate the fractionation of the lanthanides in natural matter.

The following papers present compilations of data on the Y, La—Lu distribution in cosmic and terrestrial matter: BALASHOV (1963), SEMENOV (1963), AHRENS (1964), FLEISCHER (1965), JENSEN and BRUNFELT (1965), HASKIN, FREY *et al.* (1966), HASKIN and FREY (1966), NEUMANN *et al.* (1966), chapters Y and Lanthanides in VLASOV (1966), BORODIN (1967), HASKIN *et al.* (1968).

Acknowledgements

The writer wishes to thank Prof. Dr. K. H. WEDEPOHL for spirited discussions and Dr. D. BAUMGARTEN and Dr. G. S. PUNDEER who kindly corrected the English of the manuscript.

Revised manuscript received: February 1969

39, 57—71-A. Crystal Chemistry

I. Electronic Configuration and Chemical Bonding

Of the rare earth elements, lanthanum has a natural place in Group III A of the Periodic Table. Therefore the fourteen succeeding elements cerium to lutetium are often referred to as lanthanoids[1]. Above it in the Periodic Table are scandium and yttrium and below it is the element actinium. These elements have many chemical similarities and therefore rare earth minerals are commonly found to contain some of these elements as major or minor constituents. Yttrium is a most abundant element in natural rare earth compounds because of the 0.89 Å ionic radius of the Y^{3+}-ion. This fits perfectly into the lanthanoid series of RE^{3+}-ions, which exhibit ionic radii of 1.06 Å (La^{3+}) to 0.85 Å (Lu^{3+}) for octahedral coordination. This is the reason for including yttrium in the discussion of crystal chemical properties of the lanthanoids The four elements scandium, yttrium, lanthanum, and actinium are the first members of the four series of transition elements. These series consist of elements with unfilled lower orbitals. The first two transition element series have unfilled d-orbitals and the last two, which comprise the rare earths or lanthanoids and the actinoids[1] respectively, have unfilled f-orbitals.

With increasing atomic number in each of the transition element series the nuclear charge increases in proton units, and with this increase successive electrons are added to the least energetic orbital available. Because of the imperfect shielding from the nuclear charge of one electron followed by another, shrinkage of the atomic volume is normally observed in all four transition element series. This contraction of the atomic volume is extremely regular within the 4f-transition element series of RE^{3+}-ions and is therefore commonly referred to as the "lanthanoid contraction". It will be noted that the largest contraction occurs between lanthanum and cerium, and also that the contraction from lanthanum to gadolinium is greater than from gadolinium to lutetium. A similar decrease in radius occurs in lanthanoid ions with charges other than three.

One of the most striking effects of the lanthanoid contraction occurs in the behaviour of yttrium. This element, which has a much smaller nuclear charge ($Z = 39$), has a radius similar to that of holmium ($Z = 67$). This is the reason why yttrium appeared among the heavy RE^{3+}-ions in many of the older separations; these depend on fractional crystallization and basicity separations, in which the ionic radius and charge/radius ratio are important. The striking similarities between some other elements not belonging to the rare earth group, such as that between zirconium-hafnium and niobium-tantalum, are also due to the small size of the ion of the second element of these pairs arising from the lanthanoid contraction.

[1] The names "lanthanoids" and "actinoids" are recommended by the 1970—I.U.P.A.C. rules instead of "lanthanides" and "actinides" commonly used. "Rare earth" elements may be used for Sc, Y and La to Lu inclusive.

Contrary to the two short series of 3d- and 4d-transition elements respectively, in which the d-electrons may participate in the formation of chemical bonds, in the third long period of the lanthanoids the 4f-electrons are shielded by the upper 5s- and 5p-orbitals which have already been filled at xenon (Z = 54). Commencing with cerium, 14 electrons are added successively to the 4f shell to give the sequence of lanthanoid elements. The peculiarities of the lanthanoid elements depend very largely on the 4f electron levels which project to the periphery of the atoms and thereby determine the possible valency states. The 4f electrons are shielded by eight electrons, $5s^2$ and $5p^6$, so that their chemical significance is so slight as to allow the remarkable similarity of the lanthanoids.

The 4f electrons are not always added in single steps from cerium to lutetium, although there is a tendency for the electron to enter the 4f shell rather than the 5d shell. The configurations of cerium and terbium are uncertain, but it is established that the 4f shell is half-filled at europium and filled at ytterbium. The normal charge state of the lanthanoid elements is three, and the RE^{3+}-ions are formed by losing the $6s^2$ and generally a 4f electron, but sometimes the $5d^1$ electron. Thus the RE^{3+}-ion has the xenon core with 4f electrons. It is significant that no charge other than three is known for lanthanum, gadolinium and lutetium with one exception available for compounds REX (see Table 39, 57—71-A-12). Some of the lanthanoid elements also have charges of two (Sm^{2+}, Eu^{2+}, Yb^{2+}) or four (Ce^{4+}, Pr^{4+}, Tb^{4+}), the ionic configurations being (Xe) $4f^{n+1}$ and (Xe) $4f^{n-1}$ respectively. It has been argued for some time that these unusual valency states are associated with the extra stability of the empty, the half-filled, and the filled 4f configurations, but it is now known that these oxidation states are more widely distributed and cannot be accounted for by this explanation alone.

The character of the chemical bond of the rare earths with the most common anions of the halogens and chalcogens is largely ionic, decreasing with increasing atomic numbers of the VIA and VIIA elements. Thus with values of the electronegativity χ(RE) varying between 1.1 and 1.2 within the series of rare earth elements, the percentage in ionic bonding character in the lanthanoid–halogen bond is 85%, for example with fluorine ($\chi = 4.0$), decreasing to 40% in the lanthanoid–iodine bond. Corresponding values exist for the lanthanoid–chalcogenides. The proportion of the ionic character in the lanthanoid–oxygen bond is 75%. It decreases in the sulfur-, selenium- (both 30%) and tellurium (20%)-rare earth bond (all values of electronegativity according to PAULING, 1960).

Covalency might well be expected to be most likely in complexes with strongly chelating ligands and in particular with ligands that can participate in back-bonding through π-interactions. As yet, the available evidence is insufficient to allow definite conclusions as to the degree of covalency or the exact nature of the participating orbitals. Minor proportions of covalent bonding in certain solid chalcogenides, pnictides, borides, and non-stoichiometric halides and oxides have been suggested from theoretical considerations (BURNS, 1967; BANDURKIN *et al.*, 1969; JØRGENSON *et al.*, 1965; JØRGENSON, 1973). The experimental data from magnetic resonance, photo-electron spectroscopy and thermodynamical measurements do not however allow definite conclusions. In a model for the electronic configuration of the lanthanoid metals, which suggests two kinds of 4f electrons (the "atomic" 4f-electrons and the 4f-"bond" electrons), GSCHNEIDNER (1971) has briefly discussed the possible influence of the "bonding" 4f-electrons on effects of covalency in compounds of the lanthanoids.

Magnetic susceptibility measurements and spectroscopic methods are being increasingly employed to determine the state of valence of the rare earth ions in a given compound and thus its stoichiometry. Knowledge of the electronic configuration in the solid rare earth compounds is extremely important for understanding their chemical and physical properties. Modern technology has made recent use of the unique magnetic and optical properties of these "electronic materials".

The determination of the electronic configuration of a rare earth element (as well as of an actinoid element) is difficult. The classical method depends upon a term analysis of the emission spectrum of the element but the spectra of the rare earths are exceedingly complex. Results deduced from spectroscopy have been confirmed and supplemented by deductions from atomic beam resonance (DIEKE, 1968). This method depends upon direct measurement of the electronic angular momentum, and the atomic g-values of the atoms, from which the ground states can be deduced. Present views on the electronic structures of the elements RE^0 and the terpositive ions RE^{3+} are given in Table 39, 57—71-A-1.

Table 39, 57-71-A-1. *Electronic configurations of 'outer' electronic shells with 'inner' Xe- core ($1s^2\ 2s^2\ 2p^6\ 3s^2\ 3p^6\ 3d^{10}\ 4s^2\ 4p^6\ 4d^{10}$) and permanent magnetic moments μ_B of the lanthanoid elements and their terpositive ions* (MOELLER, 1973)

Atomic No.	Element	Configuration (spectroscopic term)		Magnetic moment μ_B (RE^{3+})
		RE^0	RE^{3+}	
57	La	$5d^16s^2$ ($^2D_{3/2}$)	$4f^0$ ($^1S_0$)	0.00
58	Ce	$4f^15d^16s^2$ ($^1G_4$)	$4f^1$ ($^2F_{5/2}$)	2.56
59	Pr	$4f^37s^2$ ($^4I_{9/2}$)	$4f^2$ (H_4)	3.62
60	Nd	$4f^46s^2$ ($^5I_4$)	$4f^3$ ($^4I_{9/2}$)	3.68
61	Pm	$4f^56s^2$ ($^6H_{5/2}$)	$4f^4$ ($^5I_4$)	2.83
62	Sm	$4f^66s^2$ ($^7F_0$)	$4f^5$ ($^6H_{5/2}$)	1.55—1.65
63	Eu	$4f^76s^2$ ($^8S_{7/2}$)	$4f^6$ ($^7F_0$)	3.40—3.51
64	Gd	$4f^75d^16s^2$ ($^9D_2$)	$4f^7$ ($^8S_{7/2}$)	7.94
65	Tb	$4f^96s^2$ ($^6H_{15/2}$)	$4f^8$ ($^7F_6$)	9.7
66	Dy	$4f^{10}6s^2$ ($^5I_8$)	$4f^9$ ($^6H_{15/2}$)	10.6
67	Ho	$4f^{11}6s^2$ ($^4I_{15/2}$)	$4f^{10}$ ($^5I_8$)	10.6
68	Er	$4f^{12}6s^2$ ($^3H_6$)	$4f^{11}$ ($^4I_{15/2}$)	9.6
69	Tm	$4f^{13}6s^2$ ($^2F_{7/2}$)	$4f^{12}$ ($^3H_6$)	7.6
70	Yb	$4f^{14}6s^2$ ($^1S_0$)	$4f^{13}$ ($^2F_{7/2}$)	4.5
71	Lu	$4f^{14}5d^16s^2$ ($^2D_{3/2}$)	$4f^{14}$ ($^1S_0$)	0.00

The spectroscopic data are in agreement with the results of magnetic measurements on the electronic configuration of the rare earths. In common with many of the transition elements, most of the lanthanoid metals and their compounds are strongly paramagnetic. Paramagnetism is always associated with the presence of unpaired electrons, which make spin and orbital contributions to the magnetic moment. In the first and second transition series, the unpaired electrons lie in the 3d or 4d shells, but in the lanthanoid elements they are in the 4f shell. It has been stated already that the 4f electrons do not participate in chemical bonding, since they are effectively shielded by the overlying 5s and 5p electrons. For this reason, there is an important difference between the magnetic properties of the lanthanoids and the first two series of transition elements. Since the 3d and 4d electrons of the latter elements can participate in chemical bonding, their magnetic moments lose the orbital contribution when complexing occurs. Such reactivity is not displayed by the lanthanoid elements. This fact is of considerable practical consequence, as it means that the measurement of the magnetic susceptibility of a lanthanoid compound gives a very accurate assessment of its charge state.

Another important test for electronic structure can be made by examining lanthanoids in other charge states. Cerium, praseodymium and terbium have common tetravalent states. Similarly Sm^{2+}, Eu^{2+}, and Yb^{2+} are isoelectronic with the trivalent Eu^{3+}, Gd^{3+}, and Lu^{3+} respectively. In nearly all cases, the moments of the trivalent lanthanoids are sufficiently well separated for magnetic measurements to give an unequivocal assessment of the charge state.

In naturally occuring rare earth compounds, the similarity in chemical properties of the rare earth ions leads to a combination of several lanthanoids in one mineral. No known mineral contains only one lanthanoid as the major constituent. Therefore in an early approach to the crystal chemistry of the rare earths Goldschmidt *et al.* (1926) grouped the lanthanoid minerals into three categories. To the first group belong pegmatitic and magmatic minerals which contain major quantities of the rare earths (i.e. monazite, which contains mainly the light RE-ions, and xenotime, which contains yttrium and the heavy RE-ions). Other minerals in this category are euxenite and gadolinite which have a more uniform distribution of lanthanoids. The second category includes minerals in which the lanthanoid elements form minor constituents but nevertheless are still in the trivalent state. In the most common minerals, partial substitution of trivalent lanthanoids for large bivalent ions, such as those of calcium, strontium, or lead, occurs. Many calcium minerals, for example apatite, contain minor quantities of lanthanoids. Cases are also known where the lanthanoid substitution can vary between wide limits. The extremes occur in mixed crystals between a lanthanoid and a non-lanthanoid mineral respectively. Examples of this are found in the fluorspar-yttrofluorite and epidote-orthite series. In the third category, small quantities of bivalent lanthanoids replace large uni- or bivalent ions; this replacement is usually confined to the two ions Eu^{2+} and Sm^{2+} (small quantities of europium are found in some minerals of the strontianite, fluorite, and feldspar types).

Experiments to prepare rare earth feldspar minerals by solid state reaction have not been successful so far. Early results published by Eitel *et al.* (1930) suggested isomorphous replacement of Al^{3+} by La^{3+}, Nd^{3+}, Y^{3+} in nepheline type structures. These results could not be reproduced, however, by modern analytical techniques using solid state materials. There is some reason to assume that the unique X-ray powder diffraction pattern shown by samples of different composition is due to the apatite-like phases, which play a dominant role in these systems (Felsche, 1972).[1]

[1] Bettermann, P., and Liebau, F., recently (16. Annual Session "Arbeitsgemeinschaft Kristallographie", Abstract in Z. Kristallogr. **144**, 1976) reported on solid solution series in the system $CaAl_2Si_2O_8$—$Na_{0.5}La_{0.5}Al_2Si_2O_8$. Based on powder diffraction data e.g. the phase —$Na_{0.5}La_{0.5}Al_2Si_2O_8$ shows orthorhombic symmetry.

II. Coordination and Ionic Radii

The early concepts of correlations between ionic size and stability of certain crystal structure types as developed by GOLDSCHMIDT *et al.* (1926) and PAULING (1927, 1929, 1960) did not lose their importance in a modern view of crystal chemistry. On the basis of reliable crystal structure data which give empirical bond lengths on a level of confidence of at least ± 0.01 Å, some correction terms to the early model of charged solid spheres have been proposed recently. The early concept, which is mainly based on the two parameters known as radius ratio r (cation) : r (anion) and electrostatic valence principle, corresponds to a rather simple model of electrostatic interaction of spherical ions with formal charges. This concept also provides a reasonable approach to rare earth compounds as far as the electrostatic bonding character is concerned. The largely ionic bonding character, for example 75% and 90%, in the RE—O and RE—F bonds respectively favours this approach as in other typical ionic compounds, e.g. NaCl and CaF_2, with comparable values of 90% after PAULINGS (1960) scale of electronegativity. In a large number of rare earth compounds however, this simple concept of crystal structures and prediction of isotypic compounds becomes less meaningful.

The peculiarities of crystal structures containing rare earth elements as major constituents can be understood from the following features:

a) The rare earth cations show a size varying between 1.32 Å to 0.68 Å along the lanthanoid series La^{3+} to Lu^{3+} corresponding to coordination numbers 6 to 9 towards O^{2-} and F^-. Thus their compounds have to be assigned to the group of large-cation structures.

b) As compared to their regular valence state (3+), the rare earths represent a unique set of elements within the Periodic Table with regard to the ratio of size to valence. Rare earth compounds generally have a complicated stoichiometry. Even for the simple binary halogenide or chalcogenide compounds this leads to 1:3 or 2:3 compositions. This has the consequence that the nature of coordination (geometry of coordination polyhedra and their linkage) cannot be simply due to the ratio of the radii of the cation to anion and the electrostatic charge balance.

c) Strong polarisation forces of RE^{3+} cations are especially effective against the large and "soft" anions of halogen and chalcogen elements because of the poor screening of the nuclear charge by the f-electrons. This might cause a selective shortening of the bonding lengths.

d) A different bonding character of a second type of cation B present in a structure with the common anion X might govern the RE-X-B bonds and therefore control the structural features of a compound. As shown by many RE-O-Si bonds in the rare earth silicates, the rather rigid [SiO_4] complex anions have a major influence on the structural features of these compounds.

In most crystal structures 6-, 7-, 8-, or 9-fold coordination around the rare earth cations does not occur in regular coordination but in strongly distorted polyhedra. The cation-anion distances within a given coordination vary over a wide range. This also corresponds with variations in the bonding angles. In most cases it is not possible to describe the coordination in terms of common polyhedra such as octahedron, cube, square antiprism, dodecahedron or cube-octahedron.

With regard to the complicated coordination geometry around the rare earth cations obtained from various precise structure determinations, one in general, should consider the assignment of a Cartesian polyhedra as an approximation. The assignment of a coordination number to each constituent in a crystal structure is always possible to a large degree of probability, however. BRUNNER and SCHWARZENBACH (1971) have shown by evaluation of some hundred crystal structures that the coordination number 'CN' might clearly be defined in terms of the coordination sphere of the first order which is significantly separated from the sphere of the second order around each given lattice point (atom or ion). Thus throughout this article the coordination numbers given for the rare earth cations generally denote the number of next neighbours in an arbitrary position around the cation unless otherwise stated. In its coordination sphere, each ion tends to achieve spherical shielding for itself and minimize repulsion between the coordination ions of opposite charge. Therefore in the ionic rare earth crystal structures the values for the interatomic distances of a coordination shell represent the equilibrium distance r_0 between ions. More quantitatively this value is given by the well-known Born-Mayer treatment (BORN *et al.*, 1932) of the lattice energy U:

$$U = \left(1 - \frac{1}{n}\right) \frac{NAe^2Z^2}{r_0}.$$

Many crystal structures have been successfully discussed in terms of the Madelung part of the lattice energy (e.g. MERTENS *et al.*, 1966, and HOPPE, 1970) in order to determine the equilibrium positions of ions in alternative structure models.[1]

In the discussion of bond length variations within a given coordination sphere several empirical approaches have been published recently, evaluating the most accurate crystal structure data available from all kinds of inorganic compounds. The starting point of all these modern approaches is Paulings concept of the electrostatic valence principle (PAULING, 1929) by which the sums, p, of mean bond strengths, $\bar{s}$, between the cations with coordination number CN and anions are approximately equal to their valence z:

$$\bar{s} = \frac{z}{CN};\ p = \Sigma \bar{s}.$$

Since this model had been selected for ionic compounds built up of rather regular coordination polyhedra with nearly equal cation-anion distances, large deviations for the bond strengths values s up to 40% have been observed in recent studies. BAUR (1970, 1971) uses the bond strength deviations to predict which bonds will be longer or shorter than the average in a particular crystal. Using a linear expression, $R_{ij} = a_i + b_i p_j$, he relates the individual bond lengths, R_{ij}, to the sum, p_j, of the mean bond strength around the anions j. For several cations A_i in different coordinations of 130 ternary oxides he determined the empirical constants a_i and b_i. However, this approach failed in a number of rare earth crystal structures because this model does not take into account the influence of RE-O-B bonds of cations B on RE-O distances. The correspondence between variations in RE-O and Si-O distances in rare earth silicates was recently explained (FELSCHE, 1973) by application of a similar

[1] Recent advance in this field is given by a semi-empirical potential-field model with atomic constants as parameters for calculation of crystal energies as an aid in structural chemistry: SKORCZYK, R.: Acta Cryst. **A 32**, 447 (1976).

function as that proposed by DONNAY *et al.* (1970) and BROWN *et al.* (1973). It could be shown that the charge balance p_j on the individual oxygen j coordinated by CN cations

$$p_j = \sum^{cn} s_i$$

s strongly correlated with the individual bond strength s_i

$$s_i = \frac{z_i}{CN_i}\left(\frac{\bar{R}}{R_{ij}}\right)^N$$

(i: individual cations of valence z and coordination number CN having a particular distance R_{ij}, which differs from the mean bond length $\bar{R}$ of this coordination shell). As a result of the small variation in the degree of covalency in the individual rare earth-oxygen bonds, it could be shown that the exponent N is equal to 2 for all known rare earths silicate structures of different RE^{3+} ions with different coordination numbers. In their similar relationship of bond-strength and bond-length

$$s = s_0\left(\frac{R}{R^0}\right)^{-N}$$

BROWN *et al.* (1973) derived single pairs of parameters N and R_0 for 25 different cation-oxygen pair bonds. For sets of isoelectronic ions the values of N and R_0 are independent of the bond character and coordination number. Although the values of N and R_0 represent just numbers of a proper mathematical function without any chemical relation, the value of the bond strength s is directly related to the covalent character of the bond. The sum of s-values around each atom is within about 5% of the valence value.

Up to the present time this powerful type of correlation has been applied only to silicates of the rare earths. Additional studies should be concentrate on other types of rare earth compounds for a better understanding of the complex geometrical relations in these structures in terms of bonding!

In spite of the difficulty in describing large cation crystal structures in terms of cation–anion polyhedra some approaches have been successfully applied on oxygen–rare earth coordination polyhedra. The oxygens in structures of RE_2O_3, $RE_2[MO_4]O$ and some rare earth oxysalts are tetrahedrally coordinated by the rare earth cations. $[O\,RE_4]^{+n}$ tetrahedra are forming 2-dimensional or 3-dimensional networks with common edges (CARO, 1968—1973; FELSCHE, 1971; BAERINGHAUSEN *et al.*, 1971).

Most of the rare earth compounds show extensive polymorphism. This means that there are small differences in the Gibbs free energy between different structural arrangements. Two or more structures of a single compound can occur at different temperatures (or pressures). The study of polymorphism in rare earth compounds reveals the unique dependency of interatomic equilibrium distances on temperature or on the ionic radius for a given structure type. Increase of the interatomic distances occurs either from increasing thermal motion of the ions or substitution of cations of different size. To demonstrate this correspondence between temperature and ion size effects, polymorphism of the compounds RE_2O_3 and $RE_2Si_2O_7$ is shown in Fig. 39, 57—71-A-1. The fields of stability of the 5 modifications A, B, C, H, X of the rare earth sesquioxides are shown in Fig. 39, 57—71-A-1a and the 7 structure types of $RE_2Si_2O_7$-compounds in Fig. 39, 57—71-A-1b.

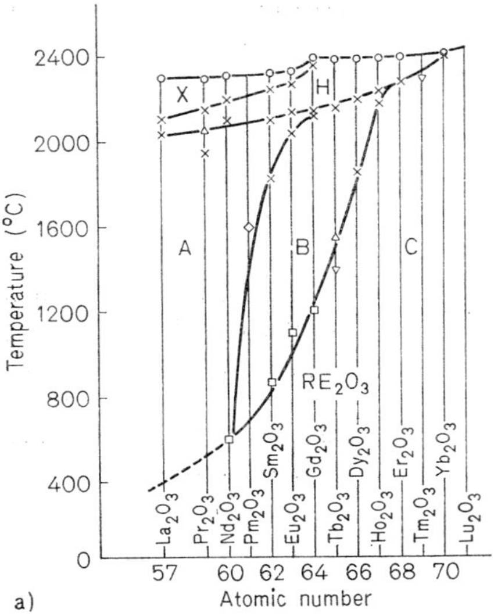

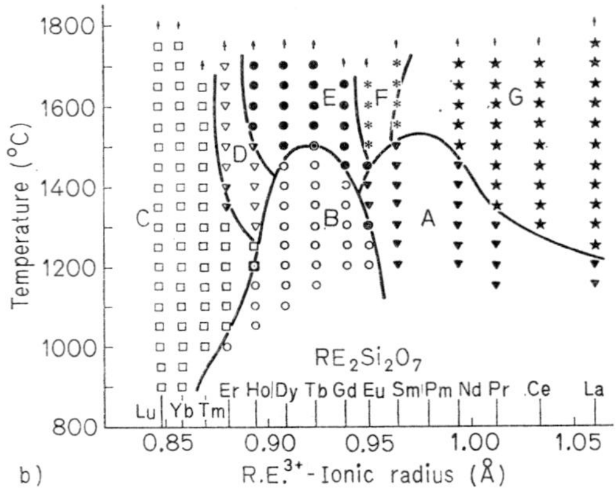

Fig. 39, 57-71-A-1 a and b. Phase diagrams of compounds RE_2O_3 (a) and $RE_2Si_2O_7$ (b) demonstrate extensive polymorphism, which is shown by most of the binary and ternary rare earth compounds. The fields of stability of the individual structure types are denoted by capital letters. Structure types A, B, C of compounds RE_2O_3 are illustrated in Fig. 39, 57-71-A-2a, b, c, crystal data are listed in Table 39, 57-71-A-3. Structures A to G of compounds $RE_2Si_2O_7$ are characterized in Table 39, 57-71-A-8. X/T diagrams of compounds RE_2O_3 are taken from Foex *et al.* (1966) of compounds $RE_2Si_2O_7$ from Felsche (1970)

Following the sequence of ionic radii in the series $La^{3+} \ldots Lu^{3+}$, e.g. as determined for 6-fold coordination from C-type RE_2O_3 (Shannon *et al.*, 1970a), one finds the radius ratio of $r(RE^{3+}) : r(O^{2-})$ extending from 0.60 for $r(Lu^{3+}) : r(O^{2-})$ to 0.76 for $r(La^{3+}) : r(O^{2-})$. Since the value is considerably larger than the radius ratio of 0.41 for atoms in octahedral coordination (e.g. in the corundum structure of Al_2O_3), the C-type structure occurs as the most stable form of the oxides of the smaller rare earths. The structure has a cubic space group symmetry, but a rather distorted type of 6-fold coordination as shown in Fig. 39, 57—71-A-2. The $[REO_6]$-coordination polyhedra show some resemblance to the $[REO_8]$-cube like the coordination polyhedron of the fluorite structure type of the REO_2 compounds (RE: Ce, Pr, Tb) or RE_nO_{2n-2} (RE: Ce, Pr, Tb $n = 4 \ldots \infty$) but with exactly two of the corner oxygens missing. As a consequence of the high degree of distortion, which corresponds to a strong deviation from the minimum point symmetry $m3/T_h$ of the ideal octahedron, this structure changes to the A-, B-, or H-form at elevated temperatures depending on the size of the rare earths. The number of polymorphic forms increases from Lu_2O_3 to La_2O_3.

In the hexagonal A-type structure the lanthanoid ion is coordinated with six oxygens of an octahedron-like polyhedron and a seventh oxygen through one face of the latter. In Fig. 39, 57—71-A-2 the A-type structure is shown as proposed by Pauling (1928) and confirmed by Koehler *et al.* (1958). Mueller-Buschbaum *et al.* (1965) presented an advanced type of a statistical model for the A-structure which

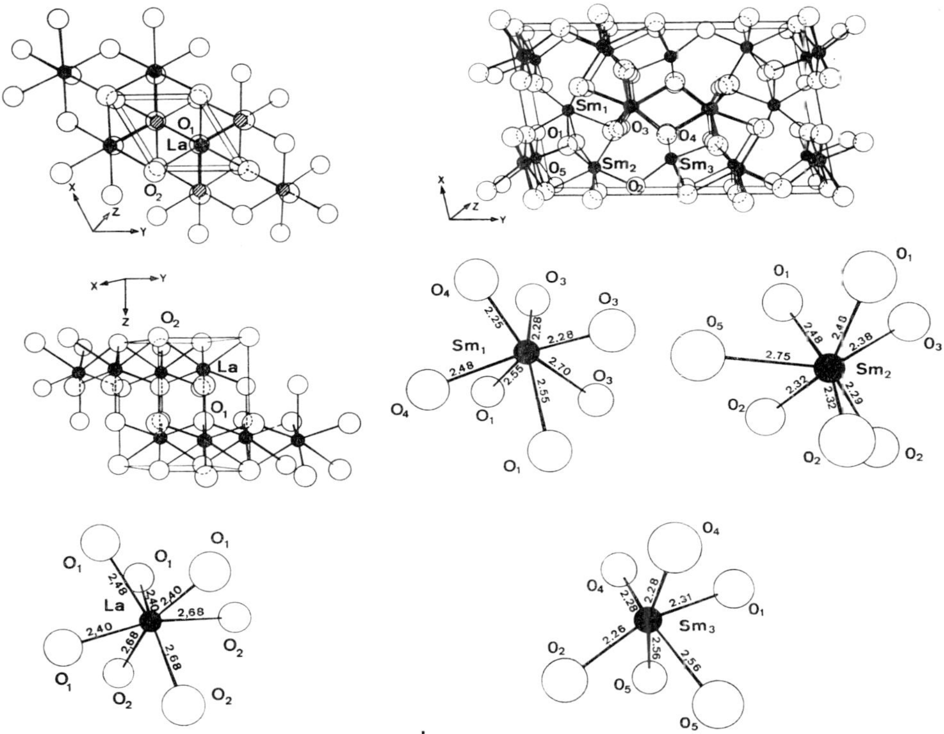

a b

Fig. 39, 57-71-A-2a—d. Crystal structures of La_2O_3, Sm_2O_3, Dy_2O_3, Eu_3O_4 demonstrating the problem in rare earth compounds of describing cation-anion coordination in terms of common cartesian coordination polyhedra. Data for A-type La_2O_3 are taken from PAULING (1928), KOEHLER *et al.* (1953)), for B-type Sm_2O_3 from CROMER (1957), for C-type Dy_2O_3 from PATON *et al.* (1965) and for $[CaFe_2O_4]$-type Eu_3O_4 from RAU (1966). For further crystal data of compounds RE_2O_3 see Table 39, 57-71-A-3 and Fig. 39, 57-71-A-1 (X/T-diagram). For compound Eu_3O_4 see Table 39, 57-71-A-13

satisfies the systematic extinction of the reflexion 001 which had been neglected by PAULING (1928) and KOEHLER (1958). According to MUELLER-BUSCHBAUM *et al.* the macroscopic A—La_2O_3 crystal consists of microscopic domains of arbitrary size, each of which shows cations on either of the following positions 1/3, 2/3, z; 2/3,1/3, $\bar{z}$; 2/3, 1/3, 1 2+z; 1/3, 2/3, 1/2—z; with z = 0,234 and the oxygens on positions 0, 0, 0; 0, 0, 1/2; 1/3, 2/3, z; 2/3, 1/3, $\bar{z}$; 2/3, 1/3, 1/2+z; 1/3, 2/3, 1/2—z with z = 0,639 in terms of symmetry restrictions of space group $P6_3/mmc$. It has been argued that this micro-twinning in La_2O_3 crystals could have been caused by the high temperature phase transitions X→H→A of the La_2O_3 crystal (CARO, 1968).

The monoclinic B-type structure contains different coordination numbers: 7, 7, 6. The sevenfold coordinated cations have six oxygens on the corners of a trigonal-like prism and the seventh coordinating through one prism face. The sixfold coordination of the third type of cation position in the B-type structure shows a rather regular octahedron with one additional oxygen directed at one face of it, but too far away

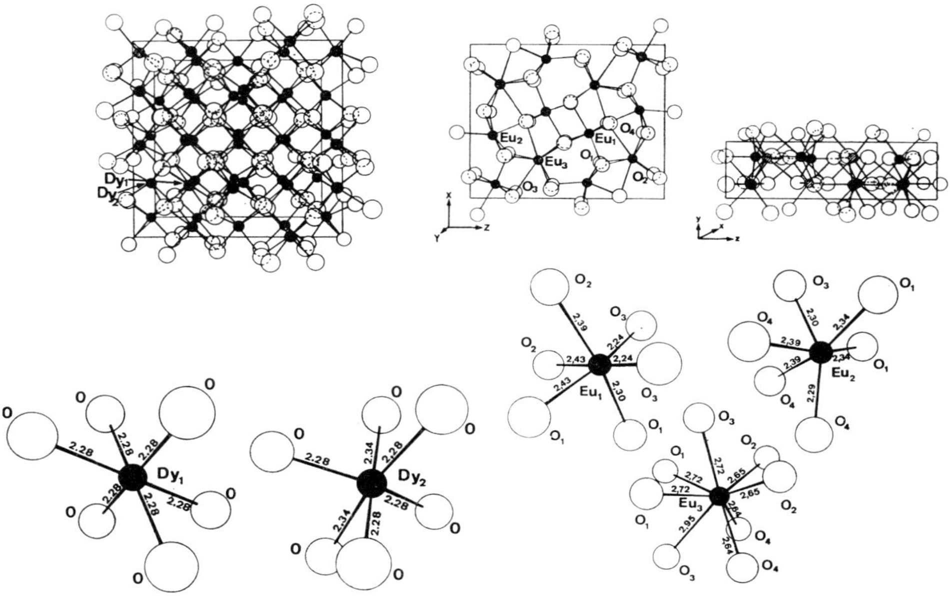

c

d

Fig. 39, 57-71-A-2c and d (For caption see 39, 57-71-A-9)

to be involved in the first coordination sphere of the cation. Structure types H and X represent high temperature modifications which occur in the temperature range 2000—2400° C. From the similar powder pattern the hexagonal H-type form is assumed to be close to the A-type structure (Foex *et al.*, 1966). However, structure determinations for this high temperature form, as well as for the cubic X-modification, have not been reported as yet.

The largest, intermediate, and smallest lanthanoid ions prefer RE_2O_3 structures with average coordination numbers of (7), (6-2/3) and (6) for the lanthanoid cations found as the A-type, B-type and C-type forms respectively. This is a logical consequence of the decreasing ionic size or the decreasing ratio of the radii. The shape of the rare earth coordination polyhedra of coordination numbers 6 and 7 in the RE_2O_3 structures of Fig. 39, 57—71-A-2 clearly demonstrates that the assignment of a Cartesian polyhedron involves a considerable degree of approximation. This accounts, to an even greater extent, for the larger coordination numbers of compounds with more complex crystal structures as given in Tables 39, 57—71-A-2 to A-14.

An empirical set of effective ionic radii in oxides and fluorides has been evaluated by Shannon and Prewitt (1969, 1970) from modern X-ray data. In those tables data are also given for the rare earth ions of RE^{3+}(RE^{4+}, RE^{2+}). The effective radii were determined from the average cation-anion distances in coordination polyhedra of crystal structures which had been determined and refined to e.s.d.'s $\leqq 0.01$ Å for regular interatomic distance varying between 2.20—2.80 Å for RE-O and RE-F bonds. Since both the cation- and the anion radii are known to vary with their coordination number, appropriate anion radii were subtracted from the average

Table 39, 57-71-A-2. *Binary and ternary halide structures REX_3 and $REMX_4$*

Structure type	Compounds, composition	Coordination, formula	Symmetry, cryst. system, space group	Unit cell data					Reference
				RE	a_0 [Å] α^0	b_0 [Å] β^0	c_0 [Å] γ^0	Z	
Tysonite	RE F_3 RE: La…Lu	$RE^{[5+4+2]}F_3^{[3]}$	hexagonal $P6_3/mmc$	La	7.186		7.352	6	THOMA (1966) ZALKIN *et al.* (1966)
YF_3	RE F_3 RE: Nd…Lu	$RE^{[8+1]}F_3^{[3]}$	orthorhombic Pnma	Y	6.353	6.850	4.393	4	ZALKIN *et al.* (1953)
UCl_3	RE Cl_3 RE: La…Gd	$RE^{[9]}Cl_3^{[3]}$	hexagonal $P6_3/m$	La	7.483		4.364	2	ZACHARIASEN (1948a) TEMPLETON *et al.* (1954a)
	RE Br_3 RE: La…Nd	$RE^{[9]}Br_3^{[3]}$	hexagonal $P6_3/m$	La	7.967		4.510	2	ZACHARIASEN (1948a)
$AlCl_3$	RE Cl_3 RE: Dy…Lu	$RE^{[6]}Cl_3^{[2]}$	monoclinic C2/m	Er	6.80	11.79 110° 42′	6.39	4	TEMPLETON *et al.* (1954b)
$PuBr_3$	RE Br_3 RE: La…Nd	$RE^{[8]}Br_3^{[3]}$	orthorhombic Ccmm	Nd	4.11	12.65	9.16	4	ZACHARIASEN (1948a)
	RE I_3	$RE^{[8]}I_3^{[3]}$	orthorhombic Ccmm	Nd	4.284	13.979	9.948	4	ZACHARIASEN (1948a)
BiI_3	RE I_3 RE: Sm…Lu	$RE^{[6]}I_3^{[2]}$	hexagonal $R\bar{3}$	Gd	7.539		20.83	6	THOMA (1966)
Scheelite	RE Li F_4 RE: Eu…Lu	$RE^{[8]}Li^{[4]}F_4^{[3]}$	tetragonal $I\,4_1/a$	Yb	5.134		10.588	4	KELLER *et al.* (1965a), THOMA (1966), THOMA *et al.* (1970)
	K RE $(WO_4)_2$ RE: La…Nd	$K^{[8]}RE^{[8]}[W^{[4]}O_4^{[3]}]$	tetragonal $I\,4_1/a$	La	5.42		12.08	2	KLEVTSOV *et al.* (1969)

Table 39, 57-71-A-2 (continued)

Nd Na F_4	$RENaF_4$ RE: La ... Lu	$RE^{[9]}(RE, Na)^{[9]}F_4$	hexagonal P6	La	6.157		3.822	2	Thoma *et al.* (1966), Burns (1965)
Fluorite	$RENaF_4$ RE: Pr ... Lu (HT-form)	$RE^{[8]}Na^{[8]}F_4^{[4]}$	cubic Fm3m	Pr	5.695			2	Thoma (1966)
	$REKF_4$ RE: La ... Ce	$RE^{[8]}K^{[8]}F_4^{[4]}$	cubic Fm3m	La	5.93			2	Zachariasen (1949)
K Ce F_4		$K^{[4]}Ce^{[9]}F_2^{[5]}F_2^{[4]}$	orthorhombic Pnma		6.289	3.804	15.596	4	Brunton (1969)
Y_3 K F_{10}		$Y_3^{[8]}K^{[16]}F_{10}^{[3,4]}$	cubic Fm3m		11.542			8	Pierce *et al.* (1973)

Table 39, 57-71-A-3. *Binary and ternary chalcogenide structures RE_2X_3, $REMX_2$*

Structure type	Compounds, composition	Coordination, formula	Symmetry, cryst. system, space group	Unit cell data RE	a_0[Å] α^0	b_0[Å] β^0	c_0[Å] γ^0	Z	Reference
A-RE_2O_3	RE_2O_3 RE: La…Nd	$RE_2^{[7]}O_3^{[6]}$	hexagonal $P6_3/mmc$	La	3.939		6.128	1	PAULING (1928), KOEHLER *et al.* (1953), MÜLLER-BUSCHBAUM *et al.* (1965), HOEKSTRA (1966)
	RE_2O_2X RE: La…Lu X: S, Se	$RE_2^{[7]}O_2^{[6]}S^{[6]}$	hexagonal $P6_3/mmc$	La / S	4.051		6.943	1	EICK (1958, 1960), PORTIER *et al.* (1973)
B-RE_2O_3	RE_2O_3 RE: Sm…Lu	$RE_2^{[7,6]}O_3^{[6,5]}$	monoclinic C2/m	Sm	14.177	3.633 99.96	8.847	6	CROMER (1957), WILLER *et al.* (1969)
C-RE_2O_3	RE_2O_3 RE: Eu…Lu	$RE_2^{[6]}O_3^{[4]}$	cubic Ia3	Y	10.664			16	O'CONNOR *et al.* (1969), PAULING *et al.* (1932), PATON *et al.* (1965), MERTENS *et al.* (1966), BRAUER *et al.* (1972), BOULESTEIX *et al.* (1974)
X-RE_2O_3 HT-form	RE_2O_3 RE: La…Gd	structure unknown	cubic	Dy	4.25 (at 2370 °C)			1	FOEX *et al.* (1966), BADIE (1971)
H-RE_2O_3 HT-form	RE_2O_3 RE: La…Tm	similar to A-RE_2O_3	hexagonal	La	4.063 (at 2100° C)		6.43	1	FOEX *et al.* (1966), BADIE (1971)
A-RE_2S_3	RE_2S_3 RE: La…Dy	$RE^{[7]}RE^{[8]}S_3^{[5]}$	orthorhombic Pnma	Nd	7.442	4.029	15.519	4	KLEMM *et al.* (1937), ELISEEV *et al.* (1972), PREWITT *et al.* (1968)

Table 39, 57-71-A-3 (continued)

Structure type	Compounds, composition	Coordination, formula	Symmetry, cryst. system, space group	Unit cell data					Reference
				RE	a_0[Å] α^0	b_0[Å] β^0	c_0[Å] γ^0	Z	
B-RE_2S_3	RE_2S_3 RE: La...Nd	unknown structure							FLAHAUT *et al.* (1965)
C-RE_2S_3 (defect Th_3P_4)	RE_2X_3 RE: La...Lu X: S, Se, Te in part: h.p.-form	$[RE_{2.67}\square_{0.33}]^{[8]}X^{[4]}$	cubic I $\bar{4}$3d	La/S	9.055			4	FLAHAUT *et al.* (1965a), WEBB *et al.* (1969), SLEIGHT *et al.* (1968), CARTER (1972), ATOJI (1971), CLARK *et al* (1975)
D-RE_2S_3	RE_2S_3 RE: Dy...Tm	$RE^{[6]}RE^{[7]}S_2^{[4]}S^{[5]}$	monoclinic $P2_1/m$	Ho	17.50	4.002 99.4	10.15	6	WHITE *et al.* (1967), RANGE *et al.* (1974)
E-RE_2S_3 (α-Al_2O_3)	RE_2S_3 RE: Yb, Lu	$RE^{[6]}S^{[4]}$	trigonal R $\bar{3}$c	Yb	6.748		18.191	6	SLEIGHT *et al.* (1968), RANGE *et al.* (1975b)
U_2S_3	RE_2Se_3 RE: Gd...Dy RE_2Te_3 RE: Nd...Gd RE_2S_3 RE: Y, Ho...Lu h.p.-form	$RE_2^{[7]}X_3^{[5]}$	orthorhombic Pnma	Dy/Se	11.077	4.007	10.912	4	FLAHAUT *et al.* (1965b), RANGE *et al.* (1975), CLARK *et al.* (1975, 1976)
Sc_2S_3 (NaCl-like)	RE_2Se_3 RE: Ho...Lu RE Te_3 RE: Tb...Lu	$RE_2^{[6]}X_3^{[4]}$	orthorhombic Fddd	Ho/Se	11.412	8.133	24.27	16	DIMUSKES *et al.* (1965)
Er_3ScS_6	RE_2S_3 RE: Lu...Tm h.p.-form		monoclinic $P2_1/m$	Lu	11.049	3.845 109.07	10.813	4	RANGE *et al.* (1976)
Ce_2O_2S (A-RE_2O_3-like)	RE_2O_2X RE: La...Lu X: S, Se	$RE_2^{[4O,3S]}O_2^{[6]}S^{[6]}$	trigonal P 3m	La/S	4.052		6.941	1	FLAHAUT (1968)

La_2O_2Te	RE_2O_2Te RE: La...Dy	unknown structure	tetragonal	La	4.12		13.10	2	Flahaut (1968)
α-$RELiO_2$ (α-$FeLiO_2$)	$RELiO_2$ RE: Y, Er...Lu RE Na O_2 RE: La...Gd	$RE^{[6]}Li^{[6]}O_2^{[6]}$	tetragonal $I4_1/amd$	Lu	4.37		9.995		Bertaut *et al.* (1962), Waintal *et al.* (1967)
β-$RELiO_2$	$RELiO_2$ RE: Dy, Ho, Y RE Na O_2 RE: Tb...Er		monoclinic $P2_1/c$	Dy	6.21	6.17 119.20	6.30	2	Bertaut *et al.* (1962), Waintal *et al.* (1967)
γ-$RELiO_2$ (diaspore)	RE Li O_2 RE: Sm...Tb	$RE^{[6]}Li^{[6]}O_2^{[4]}$	orthorhombic Pbnm	Eu	5.335	11.045	3.471	4	Gondrand *et al.* (1963), Baernighausen (1963)
δ-$RELiO_2$ (ZrO_2)	RE Li O_2 RE: La...Eu	$RE^{[7]}Li^{[4]}O^{[3]}O^{[4]}$	monoclinic $P2_1/c$	Eu	5.682	5.989 103° 10′	5.662	2	Baernighausen (1965), Waintal *et al.* (1967), Gondrand (1967)
α-$NaFeO_2$ (NaCl-like)	Li RE S_2 RE: Y, Ho...Lu Na RE S_2 RE: Sm...Er, Y	$M^{[6]}RE^{[6]}X_2^{[6]}$	rhombohedric $R\bar{3}m$	$\frac{Er}{Li}$	3.875		18.63	3	Flahaut (1968)
	K RE S_2 RE: La...Yb								
	Na RE Se_2 RE: La...Er, Y M RE O_2 RE: La...Lu M: K, Rb Na RE O_2 RE: Y, Tm...Lu								Clos *et al.* (1967)
Na Cl	Li RE S_2 RE: Pr...Dy, Y Na RE S_2 RE: La...Nd	$M^{[6]}RE^{[6]}S_2^{[6]}$	cubic Fm3m	$\frac{Pr}{Li}$	5.686			2	Flahaut (1968)

Table 39, 57-71-A-4. *Trihydroxid, oxyhydroxid and oxyhalide structures* $RE(OH)_3$, $REO(OH)$, $REOX$, *and other halide and hydrate structures*

Structure type	Compounds, composition	Coordination, formula	Symmetry, cryst. system, space group	Unit cell data RE	a_0 [Å] α^0	b_0 [Å] β^0	c_0 [Å] γ^0	Z	Reference
REOOH	REOOH RE: La...Lu	$RE^{[7]}O^{[4]}(OH)^{[3]}$	monoclinic $P2_1/m$	Eu	6.109	3.748 108.6	4.347	2	DEXPERT *et al.* (1972), KLEVTSOVA *et al.* (1964), CHRISTENSEN (1965), BAERNIGHAUSEN (1965)
UCl_3	RE $(OH)_3$ Re: La... Yb $RE[(OH)_{1-x}F_x]_3$	$RE^{[9]}(OH)_3^{[3]}$	hexagonal $P6_3/m$	La	6.523		3.850	2	MARBEUF *et al.* (1971), ZACHARIASEN (1948a), KLEVTSOV *et al.* (1965), CHRISTENSEN (1966)
LaOF (fluorite-super-structure)	REOF RE: La...T, Y	$RE^{[40,4F]}O^{[4]}F^{[4]}$	rhombohedral $R\bar{3}m$	La	7.132 32.39			2	KLEMM *et al.* (1941), ZACHARIASEN (1951), BAENZIGER *et al.* (1954), POPOV *et al.* (1954)
Fluorite	EuOF	$RE^{[40,4F]}O^{[4]}F^{[4]}$	cubic Fm3m	Eu	5.805			4	BEDFORD (1970)
PbFCl	REOCl RE: La...Lu	$[RE^{[40,5Cl]}O^{[4]}]Cl^{[4]}$	tetragonal P4/nmm	La	4.109		6.865	2	TEMPLETON *et al.* (1953), ZACHARIASEN (1949), BAERNIGHAUSEN *et al.* (1965)
	REOBr RE: La...Lu	$[RE^{[40,5Cl]}O^{[4]}]Br^{[4]}$	tetragonal P4/nmm	La	4.145		7.359	2	TEMPLETON *et al.* (1953), ZACHARIASEN (1949), BAERNIGHAUSEN *et al.* (1965)
	REOI RE: La...Lu	$[RE^{[40,5Cl]}O^{[4]}]I^{[4]}$	tetragonal P4/nmm	La	4.144		9.126	2	TEMPLETON *et al.* (1953), ZACHARIASEN (1949), BAERNIGHAUSEN *et al.* (1965)

Bastnaesite	RE F CO_3 RE: La...Nd		hexagonal P $\bar{6}$2c	La	7.16		9.787	6	Donnay *et al.* (1953), Sawyer *et al.* (1973)
$LaOHCO_3$	RE OH CO_3 RE: La...Nd		orthorhombic C mmm (?)	La	21.891	12.639	10.047	36	Haschke (1975)
PbFCl	RESF RE: La...Eu	$RE^{[9]}S^{[4]}F^{[4]}$	tetragonal P4/nmm	La	4.02		6.99	2	Flahaut (1968)
CeSI	RESX RE: La, Ce (Pr) X: Cl, Br, I	$RE^{[4S,3X]}S^{[4]}X$	orthorhombic Pcab	La ― I	7.38	14.57	7.10	8	Flahaut (1698)
SmSI	RESI RE: Ce...Sm		monoclinic B2/m	Sm	7.86	11.23 103.8	4.53	4	Flahaut (1968)
Eu_3O_4Br	RE_3 O_4 Br RE: Nd, Sm, Eu, Yb	$RE_2^{[50,2Br]}RE^{[50]}O_3^{[4]}$ $—0^{[3]}Br^{[4]}$	orthorhombic	Eu	11.986	11.854	4.124	4	Baernighausen *et al.* (1967)
La $(OH)_2Cl$	$RE(OH)_2X$ RE: La...Gd, Y X: Cl, NO_3	$RE^{[6OH,2Cl]}[OH]_2^{[4]}Cl^{[1]}$	monoclinic $P2_1$, $P2_1/m$	La	6.325	3.992 113.90	6.934	2	Carter *et al.* (1969), Tarchova *et al.* (1970), Klevtsova et *al.* (1966), Dornberger *et al.* (1967)
$Y(OH)_2Cl$	RE $(OH)_2X$ RE: La...Lu; X: Cl, NO_3	$RE^{[6OH,2Cl]}[OH]_2Cl$	orthorhombic Pbnm/$Pbn2_1$	Y ― Cl	6.244	12.045	3.605	4	Carter *et al.* (1969), Klevtsova *et al.* (1965), Haschke *et al.* (1971, 1974)
$Y_3(OH)_5OCl_2$	RE: Y, Yb	$Y_3^{[8]}[OH]_5O$ Cl_2	orthorhombic Pnmm/$Pnm2_1$	Yb	2.41	13.108	3.610	2	Carter *et al.* (1969)
$Yb_3[SiO_4]_2Cl$		$Yb_2^{[7O+1Cl]}Yb^{[6O+2Cl]}$ $[SiO_4]_2$ $Cl^{[2]}$	orthorhombic Pnma		6.731	17.556	6.129	4	Ayasse *et al.* (1973)

Table 39, 57-71-A-4 (continued)

Structure type	Compounds, composition	Coordination, formula	Symmetry, cryst. system, space group	Unit cell data					Reference
				RE	a_0 [Å] α^0	b_0 [Å] β^0	c_0 [Å] γ^0	Z	
$Nd_2O_2CO_3$	$RE_2\ O_2\ CO_3$	$RE_2^{[8]}\ O_2\ CO_3$	hexagonal $P6_3/mmc$	Nd	3.974		15.703	2	Christensen (1970, 1973) Sawyer *et al.* (1971)
$Ce_2[SO_4]_3\cdot 4H_2O$	$RE_2[SO_4]_3\cdot 4H_2O$	$RE_2^{[9]}[S^{[4]}O_4]_3\cdot 4H_2O$	monoclinic $P2_1/c$	Ce	13.390	7.247 134.20	18.328	8	Dereigne *et al.* (1972)
$La_2[SO_4]_3\cdot 9H_2O$	$RE_2[SO_4]_3\cdot 9H_2O$ RE: La...Pr	$RE^{[12]}RE^{[9]}[SO_4]_3\cdot 9H_2O$	hexagonal $P6_3/m$	La	10.98		8.13	2	Hunt *et al.* (1954)
$Nd_2[SO_4]_3\cdot 8H_2O$	$RE_2[SO_4]_3\cdot 8H_2O$ RE: Pr...Lu	$RE_2^{[9]}[SO_4]_3\cdot 8H_2O$	monoclinic C2/c		13.66	6.837	18.430	4	Iverovna *et al.* (1951)
$Gd\ Cl_3\cdot 6H_2O$	$RE\ Cl_3\cdot 6H_2O$ RE: Nd...Lu	$RE^{[8]}Cl_3\cdot 6H_2O$	monoclinic P2/n		9.651	6.525 93.65°	7.923	2	Marezio *et al.* (1951)
$Nd[BrO_3]_3\cdot 9H_2O$		$Nd^{[9]}[BrO_3]_3\cdot 9H_2O$	hexagonal		11.73		6.76	2	Helmholz (1939)

Table 39, 57-71-A-5. *Ternary chalcogenide structures* $REMX_3$

Structure type	Compounds, composition	Coordination, formula	Symmetry, cryst. system	Unit cell dada					Reference
				RE/M	a_0 [Å] α°	b_0 [Å] β°	c_0 [Å] γ°	Z	
Perovskite (ideal)	$LaAlO_3$ $LaYbO_3$	$RE^{[12]}M^{[6]}O_3^{[6]}$ $RE^{[12]}M^{[6]}O_3^{[6]}$	cubic Pm3m	Al	3.818			1	Geller *et al.* (1956)
$LaAlO_3$ (perovskite-like)	$REAlO_3$ RE: La … Sm	$RE^{[12]}Al^{[6]}O_3^{[6]}$	rhombohedral R3c	La	5.357 60°6′			2	Geller *et al.* (1956), Kim (1968), Burbank (1970)
	$LaMO_3$ M: Ga, Ni, Co, Mn	$RE^{[12]}M^{[6]}O_3^{[6]}$	rhombohedral R3c	Ni	5.461 60°49′			2	Geller (1956), Koehler *et al.* (1957), Wood *et al.* (1962)
$YAlO_3$ (perovskite-like)	$REAlO_3$ RE: Sm … Er, Y	$RE^{[8]}Al^{[5]}O_3^{[6]}$	hexagonal $P6_3/mmc$	Dy	3.700		10.50	3	Bertaut *et al.* (1963)
	$REGaO_3$ RE: La … Eu	$RE^{[8]}Ga^{[5]}O_3^{[6]}$	hexagonal $P6_3/mmc$	Nd	3.870		10.72	3	Sallavuard *et al.* (1969)
$LuMnO_3$ (perovskite-like)	$REMnO_3$	$RE^{[7]}Mn^{[5]}O_3^{[6]}$	hexagonal $P6_3cm$	Er	6.115		11.41	6	Yakel *et al.* (1963)
$GdFeO_3$ (perovskite like)	$REAlO_3$ RE: Sm … Yb	$RE^{[8]}Al^{[6]}O_3^{[4,5]}$	orthorhombic Pbnm	Yb	5.12	5.33	7.31	4	Geller *et al.* (1956), Garion *et al.* (1967)
	$ReMO_3$ RE: La … Lu M: Cr, Ga, Fe, Mn, Rh, Ti	$RE^{[8]}M^{[6]}O_3^{[4,5]}$	orthorhombic Pbnm	Gd/Cr	5.312	5.525	7.606	4	Quezel-Ambrunaz *et al.* (1963), Marezin *et al.* (1966, 1968, 1970), Quezel-Ambrunaz (1968) Eibschütz (1965), Shannon (1970c), McCarthy *et al.* (1969a), McCarthy *et al.* (1972b)

Table 39, 57-71-A-5 (continued)

Structure type	Compounds, composition	Coordination, formula	Symmetry, cryst. system	Unit cell data					Reference
				RE M	a_0 [Å] α°	b_0 [Å] β°	c_0 [Å] γ°	Z	
	$REMO_3$ RE: La…Er, Y M: V, Sc, In	$RE^{[8]}M^{[6]}O_3^{[4,5]}$	orthorhombic $Pbn2_1$	Gd Sc	5.487	5.756	7.925	4	REUTER (1965), WEIGEL (1969), SHANNON (1967)
	$RECoO_3$ RE: Pr…Tb	$RE^{[8]}Co^{[6]}O_3^{[4,5]}$	orthorhombic $Pbn2_1$	Nd	5.336	5.336	7.547	4	BERTAUT *et al.* (1956)
	$REMO_3$ RE: La(Pr) M: Y, Ho…Lu	$RE^{[8]}M^{[6]}O_3^{[4,5]}$	orthorhombic $Pbn2_1$	La Lu	5.824	6.009	8.385	4	MOREAU (1968)
	$NdGaO_3$ $La\,Yb\,O_3$	$RE^{[8]}M^{[6]}O_3^{[4,5]}$ $RE^{[8]}M^{[6]}O_3^{[4,5]}$	orthorhmobic $Pbn2_1$		5.426 5.838	5.502 6.028	7.706 8.402	4 4	BRUSSET *et al.* (1967), MÜLLER-BUSCHBAUM *et al.* (1969)
$PrAlO_3$ perovskite-like)	$PrAlO_3$	$Pr^{[8]}Al^{[6]}O_3^{[6]}$	monoclinic 12/m		5.347	5.322 90.0	7.481	4	BURBANK (1970)
Aragonite	$REBO_3$ RE: La…Eu	$RE^{[9]}[B^{[3]}O_3]$	orthorhombic Pmcn	Nd	5.041	8.080	5.729	4	WEIDELT *et al.* (1968), WEIGEL (1969), MEYER (1969)
$LaBO_3$ (HT-form)	$REBO_3$ RE: La, Ce	$RE^{[9]}[B^{[3]}O_3]$	monoclinic $P2_1/m$	La	6.348	5.084 107.89	4.186	2	WEIDELT (1970), BÖHLHOFF *et al.* (1971)
Calcite	$LuBO_3$	$Lu^{[6]}[B^{[3]}O_3]$	hexagonal $R\bar{3}c$		4.913		16.214	12	LEVIN *et al.* (1961)
$YbBO_3$ (vaterite-like)	$REBO_3$ RE: Sm…Lu	$Yb^{[6,12]}B^{[4]}O_3$	hexagonal $P\bar{6}c2$	Yb	3.735		8.747	2	BRADLEY *et al.* (1966)
NH_4CdCl_3	$RECrSe_3$ RE: La…Nd	$RE^{[9]}Cr^{[6]}Se_3^{[5]}$	orthorhombic Pnam	La	8.11	13.79	3.96	4	HUY-DUNG *et al.* (1969) HUY-DUNG *et al.* (1971)
$YScS_3$	$REScS_3$ RE: La…Er, Y	$RE^{[6]}Sc^{[6]}S_3^{[4]}$	orthorhombic $Pna2_1$	La	7.18	6.53	9.60	4	RODIER *et al.* (1969)
$LaAlS_3$	$REMX_3$ RE: La…Er, Y M: Al, Ga X: S, Se	unknown structure	hexagonal		10.17		6.08		FLAHAUT (1968)

Table 39, [illegible]

Structure type	Compounds, composition	Coordination, formula	Symmetry, cryst. system, space group	Unit cell data: RE	a_0 [Å] $\alpha°$	b_0 [Å] $\beta°$	c_0 [Å] $\gamma°$	Z	Reference
Zircon	$REPO_4$ RE: Tb…Lu	$RE^{[8]}[P^{[4]}O_4^{[3]}]$	tetragonal $I4_1/amd$	Dy	6.917		6.053	4	SCHWARZ (1963a)
	$REAsO_4$ RE: Sm…Lu	$RE^{[8]}[As^{[4]}O_4^{[3]}]$	tetragonal $I4_1/amd$	Dy	7.074		6.307	4	SCHWARZ (1963a)
	$REVO_4$ RE: Ce…Lu	$RE^{[8]}[V^{[4]}O_4^{[3]}]$	tetragonal $I4_1/amd$	Dy	7.145		6.310	4	SCHWARZ (1963a)
	$RECrO_4$ RE: Pr…Lu	$RE^{[8]}[Cr^{[4]}O_4^{[3]}]$	tetragonal $I4_1/amd$	Dy	7.141		6.269	4	SCHWARZ (1963b)
Scheelite	$RENbO_4$ RE: La…Yb	$RE^{[8]}[Nb^{[4]}O_4^{[3]}]$	tetragonal $I4_1/a$	Y	5.186		11.075	4	
	$CeGeO_4$	$Ce^{[8]}[Ge^{[4]}O_4^{[3]}]$	tetragonal $I4_1/a$		5.045		11.167	4	DURIF-VARAMBON (1959)
M-fergusonite	$RENbO_4$ RE: La…Lu	$RE^{[8]}Nb^{[4]}O_4^{[3]}$	monoclinic I2	Y	5.293	10.943 94°32′	5.065	4	KELLER *et al.* (1965b)
	$RETaO_4$ RE: Nd…Lu	$RE^{[8]}Ta^{[4]}O_4^{[3]}$	monoclinic I2	Y	5.239	10.893 95°31′	5.056	4	KELLER (1962)
M′-fergusonite	$RETaO_4$ RE: Sm…Yb	$RE^{[8]}Ta^{[4]}O_4^{[3]}$	monoclinic P2/a	Y	5.292	5.451 96°44′	5.110	2	WOLTEN (1967), WOLTEN *et al.* (1967)
Monazite	$REPO_4$ RE: La…Gd	$RE^{[8]}[P^{[4]}O_4]$	monoclinic $P2_1/n$	La	6.83	7.05 103°34′	6.48	4	WEIGEL *et al.* (1965), JAUHUES (1972)
	$REAsO_4$ RE: La…Nd	$RE^{[8]}[As^{[4]}O_4]$	monoclinic $P2_1/n$	La	7.008	7.212 104°29′	6.767	4	KELLER *et al.* (1965b)
Fluorite	$REPaO_4$ RE: La…Lu	$RE^{[8]}Pa^{[8]}O_4^{[4]}$	cubic Fm3m	Gd	5.403			4	KELLER (1964), KELLER (1965a)
	$REUO_4$ RE: Nd…Lu	$RE^{[8]}U^{[8]}O_4^{[4]}$	cubic Fm3m	Gd	5.385			4	KELLER *et al.* (1969a)
	$RENpO_4$ RE: La…Lu	$RE^{[8]}Np^{[8]}O_4^{[4]}$	cubic Fm3m	Gd	5.403			4	KELLER *et al.* (1969b)
Fluorite (disordered)	$La_2Ce_2O_7$	$La_2^{[8]}Ce_2^{[8]}O_7^{[4]}$	cubic Fm3m(?)		5.570			2	BRISSE *et al.* (1967)

Table 39, 57-71-A-7. *Ternary chalcogenide structures RE_2MX_4*

Structure type	Compounds, composition	Coordination, formula	Symmetry, cryst. system, space group	Unit cell data					Reference
				RE M	a_0 [Å] α°	b_0 [Å] β°	c_0 [Å] γ°	Z	
Y_2BeO_4	RE_2BeO_4 RE: Y	$Y_2^{[6]}Be^{[3]}O_4$	orthorhombic Pmcn	Y	3.532	9.899	10.400	4	HARRIS *et al.* (1967)
Spinell (normal)	RE_2MS_4 RE: Tm, Yb, Lu M: Cd, Mg, Mn	$RE_2^{[6]}M^{[4]}S^{[4]}$	cubic Fd3m	Yb Mg	10.957			8	FLAHAUT (1968)
	RE_2CdS_4 RE: Y, Ho...Lu	$RE_2^{[6]}M^{[4]}S^{[4]}$	cubic Fd3m	Ho	11.168			8	FLAHAUT (1968)
	RE_2MSe_4 RE: Y, Ho...Lu M: Cd, Mg, Mn	$RE_2^{[6]}M^{[4]}S^{[4]}$	cubic Fd3m	Ho Mg	11.48			8	FLAHAUT (1968)
$CaFe_2O_4$	$BaRE_2O_4$ RE: La...Ho, Y	$Ba^{[8]}RE_2^{[6]}O_4^{[5]}$	orthorhombic Pnam	Nd	10.592	12.454	3.602	4	BKHARGAVA *et al.* (1966)
	$SrRE_2O_4$ RE: Nd...Lu	$Sr^{[8]}RE_2^{[6]}O_4^{[5]}$	orthorhombic Pnam	Nd	10.157	12.211	3.571	4	BKHARGAVA *et al.* (1965), BAERNIGHAUSEN *et al.* (1962)
	$BaRE_2X_4$ RE: Na...Lu X: S, Se	$Ba^{[8]}RE_2^{[6]}X_4^{[5]}$	orthorhombic Pnam	Ho	12.07	14.47	4.02	4	FLAHAUT (1968), PATRIE *et al.* (1969)
	$SrRE_2X_4$ RE: Tb...Lu	$Sr^{[8]}RE_2^{[6]}X_4^{[5]}$	orthorhombic Pnam	Ho	11.89	14.27	3.98	4	FLAHAUT (1968), PATRIE *et al.* (1969)
	$PbRE_2X_4$ RE: Ho...Lu X: S, Se	$Pb^{[8]}RE_2^{[6]}X_4^{[5]}$	orthorhombic Pnam	Ho	11.89	14.25	4.01	4	FLAHAUT (1968), PATRIE *et al.* (1969)

MnY_2S_4	MRE_2S_4 RE: Tb ... Tm M: Mn, Mg	$M^{[6]}RE_2^{[6]}S_4$	orthorhombic $Bb2_1m$	Ho	12.54	12.66	3.76	4	Flahaut (1968), Patrie *et al.* (1966)
	MRE_2S_4 RE: Y, Er ... Lu M: Ca, Cr	$M^{[6]}RE_2^{[6]}S_4$	orthorhombic $Bb2_1m$	Ho / Cr	12.51	12.51	3.76	4	Patrie *et al.* (1967)
Th_3P_4	MRE_2S_4 RE: La ... Dy (Ho) M: Ca, Pb, Cd	$M^{[8]}RE_2^{[8]}S^{[6]}$	cubic I 4 3d	Nd / Ca	8.533			4	Meisel (1932), Flahaut (1968)
	MRE_2X_4 RE: La ... Nd (Sm) M: Sr, Ba, Pb X: S, Se	$M^{[8]}RE_2^{[8]}S^{[6]}$	cubic I 4 3d	Nd / Ba	8.793			4	Flahaut (1968)
$CaHo_2Se_4$ (NaCl like)	$CaRE_2Se_4$ RE: Y, Ho ... Lu	$Ca^{[6]}RE_2^{[6]}Se^{[6]}$	rhombohedral R 32	Ho	7.193 33°08′			0.5	Souleau *et al.* (1969), Golabi *et al.* (1965), Guittard *et al.* (1968)

Table 39, 57-71-A-8. *Ternary oxide structures* $RE_2M_2O_7$

Structure type	Compounds, composition	Coordination, formula	Symmetry, cryst. system, space group	Unit cell data					Reference
				RE M	a_0 [Å] $\alpha°$	b_0 [Å] $\beta°$	c_0 [Å] $\gamma°$	Z	
Thortveitite C-$RE_2Si_2O_7$	$RE_2Si_2O_7$ RE: Y, Ho ... Lu	$RE_2^{[6]}[Si_2O_7]$	monoclinic C2/m	Yb	6.789	9.068 101° 50′	4.682	2	Smolin *et al.* (1970b), Felsche (1970b), Shannon *et al.* (1970), Ito *et al.* (1968), Felsche (1973)
D-$RE_2Si_2O_7$	$RE_2Si_2O_7$ RE: Y, Ho, Er	$RE_2^{[6]}[Si_2O_7]$	monoclinic $P2_1/a$	Er	5.588	10.794 95.82	4.688	2	Bocquillon *et al.* (1976), Smolin *et al.* (1970b), Felsche (1973), Ito *et al.* (1968)
E-$RE_2Si_2O_7$	$RE_2Si_2O_7$ RE: Eu ... Ho, Y	$RE_2^{[7]}[Si_2O_7]$	orthorhombic Pnam	Gd	12.87	5.073	8.33	4	Ito *et al.* (1968), Felsche (1970b), Smolin *et al.* (1970b)
G-$RE_2Si_2O_7$ (β-$Ca_2P_2O_7$)	$RE_2Si_2O_7$ RE: La ... Sm	$RE_2^{[8]}[Si_2O_7]$	monoclinic $P2_1/m$	Pr	5.405	8.674 90.00	12.996	4	Smolin (1970b), Felsche (1971c)
F-$RE_2Si_2O_7$	$RE_2Si_2O_7$ RE: Sm, Eu	similar to G-$RE_2Si_2O_7$	triclinic P1, $P\bar{1}$	Eu	8.517 91.65	12.848 92.24	5.385 90.44	4	Felsche (1969), Felsche (1970b), Felsche (1973)
A-$RE_2Si_2O_7$ (α-$Ca_2P_2O_7$)	$RE_2Si_2O_7$ RE: La ... Eu	$RE_2^{[7,8,9]}[Si_2O_7]$	tetragonal $P4_1$	Pr	6.766		24.608	8	Smolin *et al.* (1970c), Felsche (1971c)
B-$RE_2Si_2O_7$ ($La_2Ge_2O_7$)	$RE_2Si_2O_7$ RE: Eu ... Er, Y)	$RE_4^{[8]}[SiO_4][Si_3O_{10}]$	triclinic $P\bar{1}$	Ho	6.612 85.81	6.669 89.38	12.085 88.57	8	Ito *et al.* (1968), Felsche (1972, 1973), Bocquillon *et al.* (1976)

$La_2Ge_2O_7$	$RE_2Ge_2O_7$ RE: La … Er	$La_4^{[8]}[Ge^{[4]}O_4]_3$ $[Ge_3^{[4]}O_{10}]$	triclinic $P\bar{1}$	La	12.76 90° 57′	7.070 90° 21′	7.006 94° 06′	8	Smolin *et al.* (1970a)
$Er_2Ge_2O_7$	$RE_2Ge_2O_7$ RE: Tb … Yb	$RE_2^{[7]}Ge_2^{[4]}O_7$	tetragonal $P4_12_12$	Er	6.778		12.34	8	Toropov *et al.* (1971)
Pyrochlore	$RE_2M_2O_7$ RE: (La, Ce)·Lu M: Sn, Ru, Pt, Ir	$RE_2^{[8]}M_2^{[6]}O_6^{[4]}O^{[4]}$	cubic Fd3m	Nd Sn	10.568			8	Brisse *et al.* (1968), Shannon *et al.* (1968), Bayer (1969)
	$RE_2Ti_2O_7$ RE: Sm … Lu	$RE_2^{[8]}M_2^{[6]}O_6^{[4]}O^{[4]}$	cubic Fd3m	Gd	10.182			8	Knop *et al.* (1969), Collongues *et al.* (1965)
	$RE_2M_2O_7$ RE: La … Gd M: Zr, Hf	$RE_2^{[8]}M_2^{[6]}O_6^{[4]}O^{[4]}$	cubic Fd3m	Ce Zr	10.701			8	Collongues *et al.* (1965), Knop *et al.* (1970)
	$RE_2Ge_2O_7$ RE: Gd … Lu (high press. form)			Gd	9.995			8	Shannon *et al.* (1968)

Table 39, 57-71-A-9. *Ternary oxide structures RE_2MO_5 and $RE_2M_3O_{12}$*

Structure type	Compounds, composition	Coordination, formula	Symmetry, cryst. system, space group	Unit cell data					Reference
				RE	a_0 [Å] $\alpha°$	b_0 [Å] $\beta°$	c_0 [Å] $\gamma°$	Z	
A-RE_2SiO_5	RE_2SiO_5 RE: La ... Tb	$RE^{[4+3]}RE^{[8+1]}$ $[SiO_4]O^{[4]}$	monoclinic $P2_1/c$	Gd	9.131	7.045 107.52	6.479	4	SMOLIN *et al.* (1969b), FELSCHE (1971a, 1973)
B-RE_2SiO_5	RE_2MO_5 RE: Y, Tb ... Lu M: Si, Ge	$RE^{[4+2]}RE^{[5+2]}$ $[SiO_4]O^{[4]}$	monoclinic B2/b	Yb	14.28	10.28	6.653	8	MICHEL *et al.* (1967), BUISSON *et al.* (1968), SMOLIN (1970)
La_2TiO_5	La_2TiO_5	$La_2^{[7]}Ti^{[5]}O_5$	orthorhombic Pnam		10.97	11.37	3.937	4	GUILLEN *et al.* (1966)
Gd_2TiO_5 (defect fluorite)	RE_2TiO_5 RE: Eu ... Dy, Y	$RE_2^{[?]}Ti^{[?]}O_5^{[4]}$	monoclinic	Gd	15.49	5.25 87° 39′	15.28	12	COLLONGUES *et al.* (1965), MCCARTHY *et al.* (1969b)
La_2SnS_5	RE_2SnS_5 RE: La ... Nd	$RE_2^{[9]}Sn^{[6]}S_5$	orthorhombic Pbam	La	11.22	7.915	3.96	2	JAULMES (1974)
$Eu_2W_3O_{12}$ (defect scheelite)	$RE_2W_3O_{12}$ RE: La ... Ho $RE_2Mo_3O_{12}$ RE: Sm ... Dy	$RE_2^{[8]}[W^{[4,5]}O_4]_3$ $RE_2^{[8]}[W^{[4,5]}O_4]_3$	monoclinic C2/c monoclinic C2/c	Eu	7.676	11.463 109.63	11.396	4	TEMPLETON *et al.* (1963), BORCHERT *et al.* (1967), SLEIGHT *et al.* (1973), SLEIGHT *et al.* (1973)
$La_2Mo_3O_{12}$ (defect scheelite)	$RE_2Mo_3O_{12}$ RE: La ... Sm	$RE_2^{[8]}[W^{[4]}O_4^{[1+1]}]_2$- $[W^{[4]}O_4^{[1+2]}]$	monoclinic C2/c	La	17.006	11.952 108.44	16.093	12	JEITSCHKO (1973), BRIXNER *et al.* (1972)
$Gd_2Mo_3O_{12}$ (LT-form)	$RE_2Mo_3O_{12}$ RE: Pr ... Ho	$RE_2^{[7]}$- $[Mo^{[4]}O_2^{[1+1]}O_2^{[1+2]}]_3$	orthorhombic Pba2	Gd	10.388	10.419	10.700	4	KEVE *et al.* (1971), JEITSCHKO (1972)
$Gd_2Mo_3O_{12}$ (HT-form)	$RE_2Mo_3O_{12}$ RE: Pr ... Ho	$RE_2^{[7]}$- $[Mo^{[4]}O_3^{[1+1]}O^{[1+2]}]_3$	tetragonal $P\bar{4}2_1m$	Gd	7.393		10.670	2	JEITSCHKO (1972)
$Sc_2W_3O_{12}$ RE: (Gd, Tb) ... Lu	$RE_2M_3O_{12}$ M: Mo, W		monoclinic						ABRAHAMS (1966)

Table 39, 57-71-A-10. *Miscellaneous ternary oxide structures*

Structure type	Compounds, composition	Coordination, formula	Symmetry, cryst. system, space group	Unit cell data					Reference
				RE	a_0 [Å] $\alpha°$	b_0 [Å] $\beta°$	c_0 [Å] $\gamma°$	Z	
Apatite	$RE_{10}Si_6O_{26}$ RE: La ... (Tm, Yb, Lu)	$RE_6^{[6+1]}[RE_{3.33}\square_{0.67}]^{[9]}[SiO_4]_6O_2^{[3]}$	hexagonal $P6_3/m$	Gd	9.431	6.873		1	Kuzmin *et al.* (1966), Smolin *et al.* (1969), Felsche (1971c)
Garnet	$RE_3M_5O_{12}$ RE: La ... Lu M: Ga, Fe	$RE_3^{[8]}M_2^{[6]}[M^{[4]}O_4]_3$	cubic Ia3d	Gd/Fe	12.471			8	Geller (1967)
	$RE_3Al_5O_{12}$ RE: Gd ... Lu	$RE_3^{[8]}M_2^{[6]}[M^{[4]}O_4]_3$	cubic Ia3d	Gd	12.123			8	Rubinstein *et al.* (1965), Bondar *et al.* (1967)
$RE_4Zr_3O_{12}$ (defect fluorite)	$RE_4Zr_3O_{12}$ RE: Yb, Lu	$RE_4^{[6,7]}Zr_3^{[7]}O_{12}^{[4]}$	rhombohedral ?		8.259 72°40′			3	Collongues *et al.* (1965), Thornberg *et al.* (1968), Thornberg *et al.* (1970)
$La_4(WO_4)_3O_3$	$RE_4(WO_4)_3O_3$ RE: La ... Nd	structure unknown	tetragonal $P4_2/nmc$	La	10.06		12.63	4	Timchenko *et al.* (1969)
$La_4Ge_3S_{12}$	$RE_4Ge_3S_{12}$ RE: La ... Gd	unknown structure	trigonal $R\bar{3}c$, R3c	La	19.40		8.10	2	Flahaut (1968)
LaB_3O_6	REB_3O_6 RE: La ... Tb	$RE^{[10]}B^{[3]}B_2^{[4]}O_6$	monoclinic I 2/a	La	7.965	8.161 93.63	6.499	4	Kindermann (1974), Bambauer *et al.* (1969), Ysker *et al.* (1970), Pakhomov *et al.* (1969), Weidelt (1970)
$La_{26}B_8O_{51}$	$RE_{26}B_8O_{51}$ RE: La, (Y, Ho ... Lu)?	$La_{26}^{[6,7,8]}[B^{[3]}O_3]_8O_{27}^{[4La]}$	monoclinic $P2_1/c$	La	6.926	12.926 99.41	14.542	1	Breit (1975)

Table 39, 57-71-A-10 (continued)

Structure type	Compounds, composition	Coordination, formula	Symmetry, cryst. system, space group	Unit cell data RE	a_0 [Å] $\alpha°$	b_0 [Å] $\beta°$	c_0 [Å] $\gamma°$	Z	Reference
Y_5S_7	RE_4MS_7 RE: Ho, Y M: Cd, Mn, Cr, Mg	$RE_4^{[6]}M^{[6]}S_7$	monoclinic C2/m	Ho	12.79	3.82 105.70	11.46	2	ADOLPHE *et al.* (1971), ADOLPHE (1965)
	$Sc_2RE_3S_7$	$Sc_2^{[6]}RE^{[6]}S_7$	monoclinic C2/m	Gd	12.29	3.741 104.9	11.26	2	RODIER *et al.* (1972), CARRE *et al.* (1974)
HoP_5O_{14}	REP_5O_{14} RE: La ... Yb	$RE^{[8]}P_5^{[4]}O_{14}$	orthorhombic Pnma	Ho	8.726	12.710	8.926	4	BAGIEN-BEUCHER *et al.* (1970), TRANQUI *et al.* (1972)
$Eu_4Al_2O_9$	$RE_4Al_2O_9$ RE: Sm ... Lu $RE_4Ga_2O_9$ RE: La ... Gd	$RE_2^{[6]}RE_2^{[7]}Al_2^{[4]}O^{[3,4]}$	monoclinic $P2_1/c$	$\frac{Eu}{Al}$	7.608	10.626	11.101	4	BRANDLE *et al.* (1969)

Table 39, 57-71-A-11. *Miscellaneous quaternary oxide structures*

Structure type	Compounds, composition	Coordination, formula	Symmetry, cryst. system, space group	Unit cell data RE	a_0 [Å] $\alpha°$	b_0 [Å] $\beta°$	c_0 [Å] $\gamma°$	Z	Reference
$NaNdSiO_4$	$NaRESiO_4$ RE: Ce…Nd	$Na^{[7]}RE^{[8]}[SiO_4]$	orthorhombic $Pna2_1$	Nd	20.00	9.28	5.45	12	CHICHAGOV *et al.* (1968a), MAKSIMOV *et al.* (1969a)
$NaSmSiO_4$	Na RE Si O_4 RE: Nd…Ho	$Na^{[7]}RE^{[6]}[SiO_4]$	tetragonal I4/m	Sm	11.84		5.45	8	CHICHAGOV *et al.* (1968b), AVETISYAN *et al.* (1971)
$NaYSiO_4$ (olivine-like)	$MRESiO_4$ RE: Y, Er…Lu M: Na, Li	$Na^{[6]}RE^{[6]}[SiO_4]$	orthorhombic $Pbn2_1$/Pbnm	Y Na	4.95	11.30	6.30	4	MAKSIMOV *et al.* (1967), BLASSE *et al.* (1967)
	$NaREGeO_4$ RE: Sm…Lu	$Na^{[6]}[RE^{6]}[GeO_4]$	orthorhombic $Pbn2_1$/Pbnm	Sm	5.27	11.70	6.50	4	CHICHAGOV *et al.* (1967) CHENAVAS *et al.* (1969)
$CeBSiO_5$	$REBSiO_5$ RE: La…Nd		hexagonal $P3_1$	Ce	6.85		6.70	3	VORONKOV *et al.* (1967), NEUMANN *et al.* (1966)
$Na_3ScSi_2O_7$	$Na_3RESi_2O_7$ RE: Y, Ho…Lu	$Na_3^{[5,4]}RE^{[6]}[Si_2O_7]$	orthorhombic Pbnm	Sc	5.354	9.347	13.089	4	SKSHAT *et al.* (1969)
Pyrochlore	RE_2FeSbO_7 RE: La…Lu	$RE^{[8]}Fe^{[6]}Sb^{[6]}O_6^{[4]}O^{[4]}$	cubic	Eu	10.305			8	KNOP *et al.* (1968)
$K_3YSi_3O_8(OH)_2$	$K_3RESi_3O_8(OH)_2$ RE: Y, (Ho, Er) (?)	$K^{[9]}K_2^{[7]}RE^{[4+2]}$- $[Si_3O_8](OH)_2$	orthorhombic Pmnb	Y	13.53	13.17	5.867	4	MAKSIMOV *et al.* (1969b)
Garnet	$RE_3M_2Li_3O_{12}$ RE: La…Lu M: Te, W	$RE_3^{[8]}M_2^{[6]}[Li^{[4]}O_4]_3$	cubic Ia3d	Gd Te	12.390			8	KASPER (1969)

Table 39, 57-71-A-11 (continued)

Structure type	Compounds, composition	Coordination, formula	Symmetry, cryst. system, space group	Unit cell data					Reference
				RE	a_0 [Å] $\alpha°$	b_0 [Å] $\beta°$	c_0 [Å] $\gamma°$	Z	
α-$Li_2Eu_5O_8$	$Li_2BaEu_4O_8$ α-Li_2 Sr Eu_4 O_8	$Li_2^{[4]}Ba^{[8]}Eu_4^{[7]}$ $O_2^{[4]}O_6^{[5]}$	monoclinic B2/m	Ba	13.429	9.725	3.593 119.57	2	BAERNIGHAUSEN *et al.* (1973)
β-$Li_2Eu_5O_8$	β-$Li_2SrEu_4O_8$	$Li^{[4]}Li^{[5]}Eu^{[6]}Eu_3^{[7]}$ $Sr^{[7]}O_2^{[4]}O_6^{[5]}$	orthorhombic $A2_1am$	Sr	13.156	17.131	3.543	4	BAERNIGHAUSEN *et al.* (1973)
Garnet	$RE_3M_2Ge_3O_{12}$ RE: Y, Ho … Lu M: Sr, Ca	$RE_3^{[8]}M_2^{[6]}[Ge^{[4]}O_4]_3$	cubic Ia3d	Y/Ca	12.805			8	GELLER (1967)
Apatite	$MRE_9[SiO_4]_6O_2$ RE: La … Lu M: Li, Na	$M^{[6+1]}RE_5^{[6+1]}RE_4^{[9]}$- $[SiO_4]_6O_2^{[3]}$	hexagonal $P6_3/m$	La/Li	9.681		7.160	1	FELSCHE (1972b)
	$M_2RE_8[SiO_4]_6O_2$ RE: La … Er, Y M: Mg, Ca, Sr, Pb, Mn, (Ba)	$M_2^{[6+1]}RE_4^{[6+1]}RE_4^{[9]}$- $[SiO_4]_6O_2^{[3]}$	hexagonal $P6_3/m$	La/Mn	9.63		7.08	1	ITO (1968), SUWA *et al.* (1968)
	$Ba_2La_8[GeO_4]_6O_2$	$M_2^{[6+1]}RE_4^{[6+1]}RE_4^{[9]}$-	hexagonal $P6_3/m$	La/Ca	9.93		7.37	1	COCKBAIN (1968)
	$Ca_8RE_2[PO_4]_6O_2$	$[SiO_4]_6O_2^{[3]}$	hexagonal $P6_3/m$						
	RE: La, Y				9.48		6.95	1	ITO (1968)
	$M_4RE_6[SiO_4]_6$-$(OH)_2$ RE: La … Lu M: Ca, Sr, Pb, (Mn, Ba)	$M_4^{[9]}RE_6^{[6+1]}[SiO_4]_6$- $(OH)_2$		La/Ca	9.66		7.12	1	ITO (1968), COCKBAIN (1968)

$LiEu_3O_4$	LiM_2EuO_4 M: Sr, Ba	$Li^{[4]}Eu^{[6]}M^{[6+1]}M^{[7]}O_4$	orthorhombic Pbnm	Ba	11.883	12.130	3.601	4	BAERNIGHAUSEN *et al.* BAERNIGHAUSEN (1970)
Perovskite	$BaREMoO_6$ RE: Eu…Lu	$Ba^{[12]}RE^{[6]}Mo^{[6]}O_6^{[6]}$	cubic Fm3m	Eu	8.466			4	BRANDLE *et al.* (1971)
La_3CuSiS_7	RE_3BCX_7 RE: La…Dy B: Cu (Ag) C: Si, Ge X: S, Se	$RE^{[8]}B^{[6]}C^{[4]}X_7$	hexagonal $P6_3$		10.310		5.794	2	COLLIN *et al.* (1971), FLAHAUT *et al.* (1970)
	RE_3BAlS_7 RE: La…Gd B: Cr, Mn, Fe		hexagonal $P6_3$						

Table 39, 57-71-A-12. *Binary halide and chalcogenide structures REX and REX_2 of bivalent rare earths RE^{2+}*

Structure type	Compounds, composition	Coordination, formula	Symmetry, cryst. system, space group	Unit cell data RE/X	a_0 [Å] $\alpha°$	b_0 [Å] $\beta°$	c_0 [Å] $\gamma°$	Z	Reference
NaCl	EuO (SmO, YbO not existent!)	$RE^{[6]}O^{[6]}$	cubic Fm3m		5.1439			4	Klemm *et al.* (1939), Brauer (1964), Brauer *et. al.* (1967a, 1967b), McCarthy *et al.* (1970), Flahaut *et al.* (1968), Bedford *et al.* (1971), Bärnighausen (1966a) Kaldis (1971)
	RE X RE: La … Lu X: S, Se, Te	$RE^{[6]}X^{[6]}$	cubic Fm3m	La/S	5.854			4	
CaF_2	REF_2 RE: Sm, Eu, Yb	$RE^{[8]}F_2^{[4]}$	cubic Fm3m	Eu	5.840			4	Döll *et al.* (9939), Petzel *et al.* (1973), Bärnighausen *et al.* (1971)
$PbCl_2$	$RECl_2$ RE: Nd, Sm, Eu	$RE^{[6+3]}Cl_2^{[3]}$	orthorhombic Pbnm	Eu	8.958	7.534	4.504	4	Döll *et al.* (1939), Petzel *et al.* (1973),
	EuF_2 (h.p-form) $REBr_2$, RE:Nd, Sm		orthorhombic Pbnm		7.435	6.324	3.803	4	Bärnighausen *et al.* (1971)
SrI_2	$RECl_2$ RE: (Dy, Ho), Tm, Yb EuI_2 $REBr_2$, RE: Dy, Tm (Yb)	$RE^{[7]}Cl_2^{[3]}$	orthorhombic Pbca	Yb	13.150	6.942	6.693	8	Baernighausen *et al.* (1974)

$SrBr_2$	$REBr_2$ RE: Sm, Eu NdI_2 $DyCl_2$	$RE^{[7+1]}RE^{[8]}Br_4^{[4]}$	tetragonal P 4/n	Eu	11.574		7.098	10	BAERNIGHAUSEN *et al.* (1971)
$CaCl_2$	$YbBr_2$	$RE^{[6]}Br_2^{[3]}$	orthorhombic Pnnm		6.63	6.93	4.37	2	BAERNIGHAUSEN *et al.* (1971
m-EuI_2	REI_2 RE: Sm, Eu	$RE^{[7]}I_2^{[3,4]}$	monoclinic P 2_1/c	Eu	7.64	8.62 98.0	7.88	4	BAERNIGHAUSEN *et al.* (1971)
CdI_2	REI_2 RE: Tm, Yb	$RE^{[6]}I_2^{[3]}$	trigonal P $\bar{3}$ml	Yb	4.503		6.972	1	BAERNIGHAUSEN *et al.* (1971)

Table 39, 57-71-A-13. *Miscellaneous ternary structures of bivalent rare earths RE^{2+}*

Structure type	Compounds, composition	Coordination, formula	Symmetry, cryst. system, space group	Unit cell data					Reference
				RE	a_0 [Å] α°	b_0 [Å] β°	c_0 [Å] γ°	Z	
$Sr(OH)_2 \cdot H_2O$	$Eu(OH)_2 \cdot H_2O$	$Eu^{[6(OH)]+2H_2O]}(OH)_2 \cdot H_2O$	orthorhombic $P2_1am$	Eu	6.701	6.197	3.652	2	BAERNIGHHAUSEN *et al.* (1967a), BAERNIGHAUSEN (1966b)
Perovskite	$EuMO_3$ M: Ti, Zr $Eu\,Cs\,X_3$ X: F, Cl	$Eu^{[12]}Zr^{[6]}O_3^{[6]}$	cubic Pm3m		4.18			1	SHAFER (1965), MCCARTHY *et al.* (1973)
Scheelite	$EuMO_4$ M: W, Mo	$Eu^{[8]}M^{[4]}O^{[3]}$	tetragonal $I\,4_1/a$	W	5.411		11.936	4	MCCARTHY (1971)
$K_2\,Ni\,F_4$	Eu_2MO_4 M: Ti, Zr	$Eu_2^{[9]}Zr^{[6]}O_4^{[6]}$	tetragonal I 4/mmm	Ti	3.883		12.532	2	MCCARTHY *et al.* (1969a, 1969b), MCCARTHY (1971)
$CaFe_2O_4$	Eu_3O_4	$Eu^{[8]}Eu_2^{[6]}O_4^{[5]}$	orthorhombic Pnam	Nd	10.085	12.054	3.502	4	BAERNIGHAUSEN *et al.* (1962),
	$EuRE_2O_4$ RE: Nd, Gd	$Eu^{[8]}RE_2^{[6]}O_4^{[5]}$	orthorhombic Pnam		10.14	12.16	3.15	4	LOPATO *et al.* (1970)
Yb_3S_4	Yb_3S_4	$Yb^{[7]}Yb_2^{[6]}S_3^{[5]}S^{[4]}$	orthorhombic Pnam	Ho	12.82	12.95	3.82	4	CHEVALIER *et al* (1967)
	$YbRE_2X_4$ RE: Y, Ho...Lu X: S, Se	$Yb^{[7]}RE_2^{[6]}S^{[4]}S_3^{[5]}$	orthorhombic Pnam		12.90	13.04	3.86	4	CHEVALIER *et al.* (1967)
$CaHo_2Se_4$ (NaCl-like)	$YbRE_2S_4$ RE: Y, Ho...Lu	$(Yb, \square)^{[6]}RE_2^{[6]}Se_4^{[6]}$	rhombohedral R 32	Y	4.12		20.40	1	SOULEAU *et al.* (1969)
Aragonite	$EuCO_3$	$Eu^{[9]}CO_3$	orthorhombic Pmcn	Eu	5.102	8.422	6.030	4	MAYER *et al.* (1964)

Baryte	$EuSO_4$	$Eu^{[12]}SO_4$	orthorhombic Pnma	Eu	8.333	5.326	6.862	4	MAYER *et al.* (1964)
$LiEu_3O_4$	$LiEu_3O_4$	$Li^{[4]}Eu^{[3+6]}Eu^{[6+1]}$ $Eu^{[7]}O_4$	orthorhombic Pbnm		11.565	11.535	3.480	4	BAERNIGHAUSEN (1970, 1967)
α-$Li_2Eu_5O_8$	$Li_2Eu_5O_8$	$Li_2^{[4]}Eu_4^{[7]}Eu^{[8]}O_2^{[4]}O_6^{[5]}$	monoclinic B 2/m		13.378	9.666	3.539 119.52	2	BAERNIGHAUSEN *et al.* (1973)
β-$Li_2Eu_5O_8$	$Li_2Eu_5O_8$	$Li^{[4]}Li^{[5]}Eu^{[6]}Eu_3^{[7]}$ $Eu^{[7]}\,O_2^{[4]}\,O_6^{[5]}$	orthorhombic $A2_1am$		13.152	17.137	3.545	4	BAERNIGHAUSEN *et al.* (1973)
$SrTi_2O_4$ (MnY_2S_4-like)	$EuTi_2O_4$	$Eu^{[8]}Ti_2^{[6]}O_4^{[5]}$	orthorhombic $Bm2_1b$		3.87	10.10	9.95	4	SIELER *et al.* (1970)
β-Ca_2SiO_4	Eu_2SiO_4 LT-form	$Eu^{[8]}Eu^{[10]}[SiO_4]$	monoclinic $P\,2_1/c$		5.661	7.101 122.01	11.518	4	SHAFER (1965), BUSCH *et al.* (1970), KALDIS *et al.* (1970), FELSCHE (1971b) FELSCHE *et al.* (1972)
α_H-Ca_2SiO_4	Eu_2SiO_4 HT-form	$Eu^{[9]}Eu^{[10]}[SiO_4]$	orthorhombic $Pmnb/P2_1nb$		5.656	7.137	9.767	4	BUSCH *et al.* (1970), FELSCHE (1973)
Sr_3SiO_5	Eu_3SiO_5	$Eu_3^{[6]}[SiO_4]O^{[6]}$	tetragonal $I\,\bar{4}c2$		6.004		10.760	4	FELSCHE (1973)
$Sr_3Ti_2O_7$	$Eu_3Ti_2O_7$	(K_2NiF_4-perovskite-like?)	tetragonal I 4/mmm (?)		3.899		20.280	2	MCCARTHY *et al.* (1969a)
$Sr_4Ti_3O_{10}$	$Eu_4Ti_3O_{10}$				3.90		28.1	2	MCCARTHY *et al.* (1969a)
Pyrochlore	$RE_2Mo_2O_7$ RE: Eu, Sm	$RE_2^{[8]}Mo_2^{[7]}O_6^{[4]}O^{[4]}$	cubic F d3m	Eu	10.380			8	MCCARTHY (1971)
Eulytine	$REM_3(PO_4)_3$ RE: La … Lu M: Sr, Pb, Eu^{2+}		cubic $I\,\bar{4}\,3\,d$	$\frac{La}{Eu^{2+}}$	10.819			4	ENGEL *et al.* (1975)

Table 39, 57-71-A-14. *Fluorite-like binary and some ternary oxide structures containing tetravalent rare earths* RE^{4+}

Structure type	Compounds, composition	Coordination, formula	Symmetry, cryst. system, space group	Unit cell data					Reference
				RE	a_0 [Å] α°	b_0 [Å] β°	c_0 [Å] γ°	Z	
Fluorite	RE O_2 RE: Ce, Pr, Tb $O_{1.95}$	$RE^{[8]}O_2^{[4]}$	cubic Fm3m	Ce Pr Tb	5.409 5.394 5.220			4	BRAUER (1964, 1966, 1968)
RE_nO_{2n-2} (fluorite-related)	RE $O_{1.5-2.0}$ RE: Ce, Pr, Tb							4	BEVAN *et al.* (1964), HYDE *et al.* (1966a), HYDE *et al.* (1966b)
α-phase n: 7 → ∞	RE $O_{1.7-2.0}$ RE: Ce, Pr, Tb	$RE^{[7-8]}[O_{2-x}\square_x{}^{[4]}]$ (fluorite/C-RE_2O_3-like: disorder)	cubic Fm3m (?)	Ce	5.411			4	EYRING (1967), JENKINS *et al.* (1970), TURCOTTE *et al.* (1971)
β-phase n: 12	RE $O_{1.833}$ RE: Pr, Tb		monoclinic P n	Tb	6.7	23.2 125.2°	15.5		TURCOTTE *et al.* (1973), KUNZMANN *et al.* (1975)
δ-phase n: 11	RE $O_{1.818}$ RE: Ce, Pe, Tb		triclinic	Tb	6.5 90.0°	9.9 99.6°	6.5 96.3°		
ε-phase n: 10	RE $O_{1.800}$ RE: Ce, Pr		monoclinic P n	Pr	6.7	19.3 125.2°	15.5		
ζ-phase n: 9	RE $O_{1.778}$ RE: Ce, Pr		triclinic	Pr	6.5 97.3°	8.4 99.6°	6.5 75.0°		
i-phase n: 7	RE $O_{1.714}$ RE: Ce, Pr, Tb	$RE^{[6,7]}O_{1,714}^{[3,4]}$ (ordered)	rhombohedral R $\bar{3}$	Pr	6.750 99°23′				
σ-phase	RE $O_{1.5-1.7}$ RE: Ce, Pr, Tb	C-RE_2O_3-like (disordered)	cubic body-centered	Ce	11.10				
A-RE_2O_3 n: 4	RE $O_{1.5}$ RE: Ce, Pr, Tb	see Table 39, 57-71-A-3							
B-RE_2O_3 n: 4	RE $O_{1.5}$ RE: Tb								
C-RE_2O_3 n: 4	RE $O_{1.5}$ RE: Pr, Tb								

Table 39, 57-71-A-14 (continued)

Structure type	Compounds, composition	Coordination, formula	Symmetry, cryst. system, space group	Unit cell data					Reference
				RE	a_0 [Å] $\alpha°$	b_0 [Å] $\beta°$	c_0 [Å] $\gamma°$	Z	
NaCl	α-Na_2REO_3 RE: Ce, Pr, Tb K_2TbO_3	$Na_2^{[6]}RE^{[6]}O_3^{[6]}$	cubic Fm3m	Ce	4.82			4/3	Hoppe (1965)
Li_2SnO_3	β-Na_2REO_3 RE: Ce, Pr		monoclinic C2/c	Ce	3.448 pseudo-hexagonal		16.49		Hoppe (1965)
Scheelite	$CeGeO_4$	$Ce^{[8]}Ge^{[4]}O_4^{[3]}$	tetragonal I 4_1/a		5.045	11.167		4	Durif-Varambon (1959)
$Ce[IO_3]_4$		$Ce^{[8]}[IO_3]_4$	tetragonal P 4_2/n		9.90		5.32	2	Cromer *et al.* (1956)
$Ce[IO_3]_4 \cdot H_2O$		$Ce^{[8]}[I^{[3+3]}O_3]_4 \cdot H_2O$	monoclinic P 2_1/n		9.57	14.92 97.58	8.00	4	Cromer *et al.* (1956)

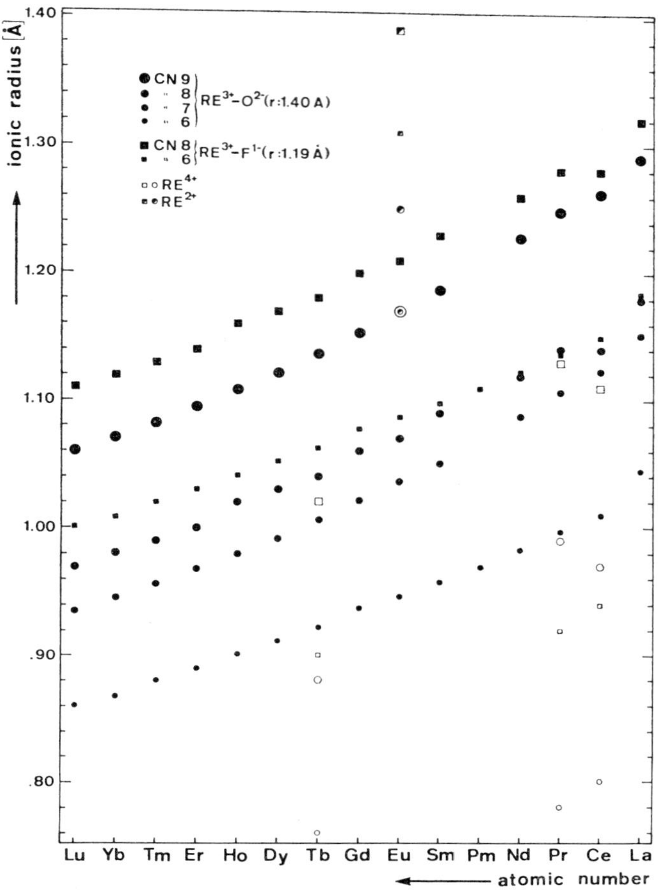

Fig. 39, 57-71-A-3. Effective ionic radii r[Å] of the rare earth ions RE^{3+} (RE^{2+}, RE^{4+}) in different coordination after SHANNON *et al.* (1969, 1970) and FELSCHE (1973)

cation-anion distances. The values of interest for the rare earth oxides and fluorides are

$$O^{[2]} = 1.35\ Å, \quad O^{[3]} = 1.36\ Å \quad O^{[4]} = 1.38\ Å, \quad O^{[6]} = 1.40\ Å$$
$$F^{[2]} = 1.29\ Å, \quad F^{[3]} = 1.30\ Å, \quad F^{[4]} = 1.31\ Å, \quad F^{[6]} = 1.33\ Å$$

(the values of the radii of the oxygen and the fluorine anions are given as a function of their coordination number). Based on the assumption that with a constant anion, the volume of the unit cell V of isostructural series of compounds is proportional to the volume of the cations, the ionic radii r for the complete series of trivalent rare earths showing CN's 6 and 8, (partly 7 and 9), were derived from plots $r^3[Å^3]$ vs. $V[Å^3]$. Recently RE^{3+} ionic radii for silicate structures from isotypic series of compounds $RE_2[Si_2O_7]$, $RE_2[SiO_4]O$, $RE_{9.33}[SiO_4]_6O_2$, were determined in the same way (FELSCHE, 1973).

The most reliable values of rare earth ionic radii from the recent evaluations (SHANNON *et al.*, 1969, 1970; FELSCHE, 1973) are listed in Fig. 39, 57—71-A-3. Corresponding to the accuracy of the evaluated data, the error for each given value should be limited to ±0.01 Å. It should be mentioned, however, that this relatively high level of confidence for the given values is actually restricted to the isostructural series of compounds of which the particular members have been determined. As the discussion of bond-length to bond-strength variation has shown these values of ionic radii cannot be used for the prediction or determination of other crystal structures, to the same degree of accuracy, since each radius $r(RE^{3+})$ value is "effective" only for a particular RE—X set of bonds of special bond character empirically evaluated.

III. The Major Binary and Ternary Structure Families of the Rare Earth Compounds

The number of new rare earth compounds has rapidly increased during the last two decades. There are mainly two reasons for this development. First there are the unique physical properties of the lanthanoid elements which are related to the $4f^n$-electronic configuration and the shielding of these orbitals by the 5s and 5p electrons. Thus from the field of material science there is a large interest in preparing new compounds for application of, e.g., magnetic and optical materials. Secondly, and related to the purpose of preparing compounds containing just one element removed from the series of lanthanoids, new methods of fractional separation have been developed. By methods of ion exchange chromatography rare earth compounds are commercially available nowadays up to a purity of 99.999% at reasonable prices. From this knowledge many rare earths containing compounds known from the early investigations, and from the variety of minerals, could be clearly assigned to a limited number of structure types. The problems of polymorphism, which is a general feature in most of the rare earth compounds, were successfully investigated emplying phases of a high degree of purity.

Several phase diagrams of rare earth halide, chalcogenide and pnictide systems have been investigated and provided a large number of new compounds. Most of these new compounds have been since investigated by X-ray crystal structure analysis. It is the high precision crystal structure data available for most of the binary and ternary compounds which has led to a more thorough understanding of the crystal chemical behaviour of the rare earth ions and the physical properties of those compounds. The unique size of the rare earth ions related to the high valency state of the ions (2^+) 3^+ (4^+) and their gradual monotonic variation of chemical properties makes this series of transition elements a unifying concept in basic crystal chemistry studies with special attention to the field of large cation structures.

Crystal chemical data of a selection of inorganic rare earth compounds belonging to the major structure families are listed in the following tables. Rare earth compounds containing elements of groups III A, IV A and V A and intermetallic compounds will not be considered. As far as possible, certain groups of compounds have been evaluated from recent reviews to avoid a large extension of the list of references. For the purpose of better transparency we will use the term of binary or ternary structure types with symbols $A_x^{[m]}B_y^{[m]}X_n^{[m]}$, in which the letters A, B denote different

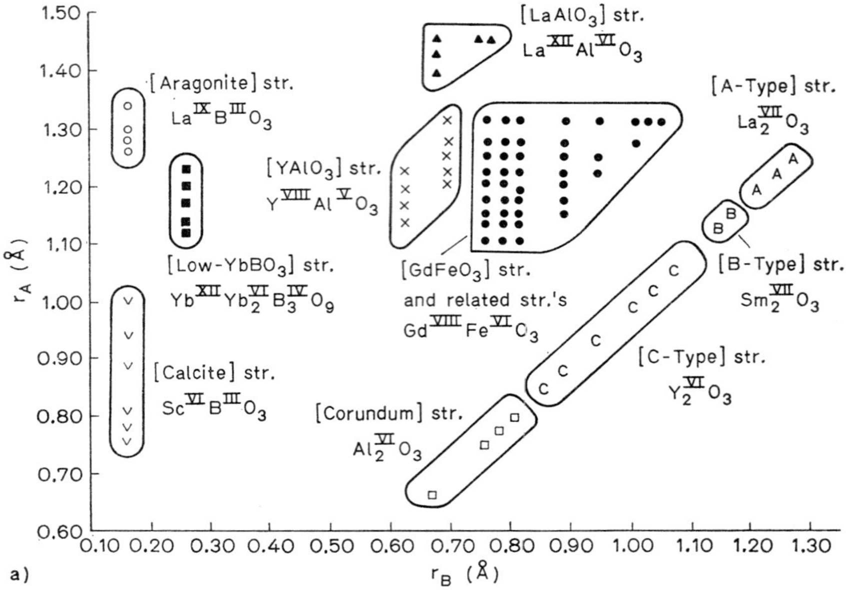

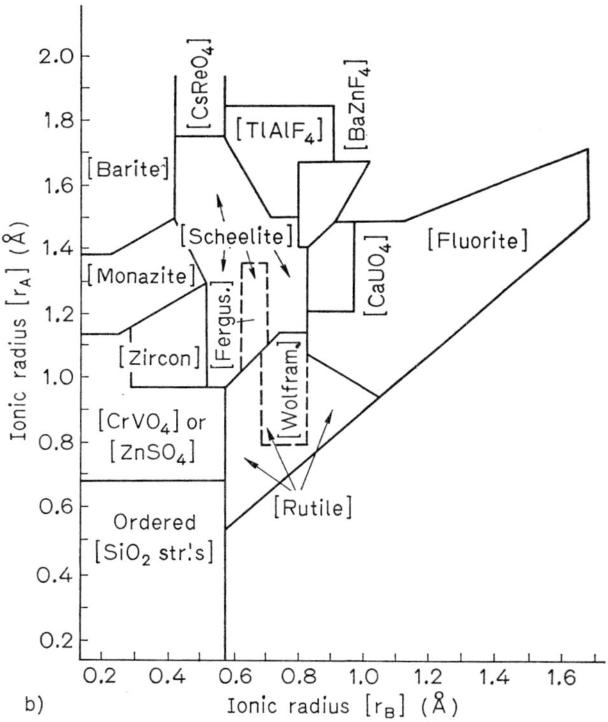

Fig. 39, 57-71-A-4a—d. Structure field maps of ternary structures ABO_3 (a), ABX_4 (b), A_2BX_4 (c) and $A_2B_2O_7$ (d) show the regions of stability of the individual structure types in terms of ionic radii r(A) vs. r(B) for the individual structure types. For crystal data of particular compounds crystallizing in one of the given structure types see Tables 39, 57-71-A-5, 39, 57-71-A-6, 39, 57-71-A-7, 39, 57-71-A-8. Diagrams of structures ABO_3, ABX_4, A_2BX_4 have been taken from Muller and Roy (1974), of structures $A_2B_2O_7$ from Felsche (1973)

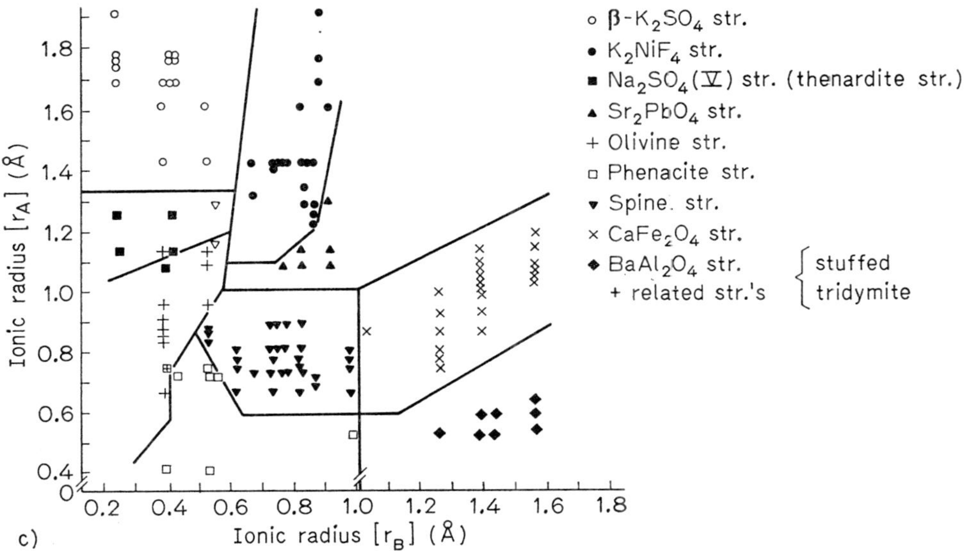

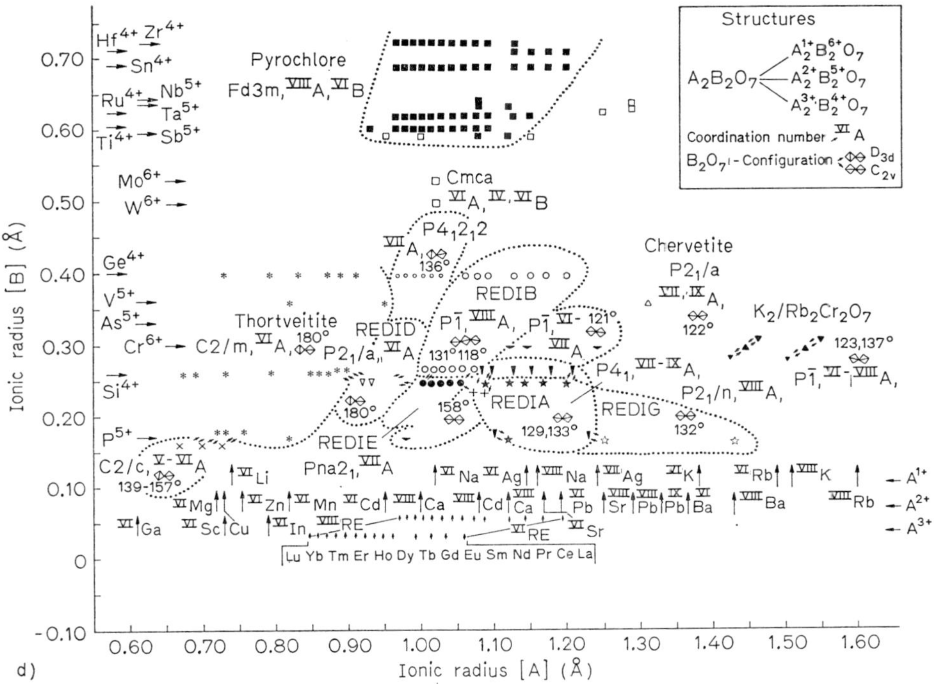

Fig. 39, 57-71-A-4c and d

cations located on different lattice sites. The cation symbol A is often replaced by the symbol RE which is used throughout this paper as an abbreviation of rare earths with the particular elements denoted for each compounds composition. Rare earth always denotes the trivalent state of oxydation (RE^{3+}) unless otherwise stated. In cases where the cation symbol B could be mixed with the element boron it is replaced by the symbol M. X presents the anion in a given compound and the numbers x, y, n give information on the stoichiometry related to the state of valency for the individual ions. [m] denotes the coordination number which presents the number of atoms belonging to the first shell of coordination around a given ion as outlined in Subsection 57-A-II.

In Tables 39, 57—71-A-2 to A-14 the major structure families are listed.

The striking usefulness of the empirical ionic radii for describing the stability of a given structure type is clearly shown in the 2-dimensional diagrams of Fig. 39, 57—71-A-4. In plots of the ionic radii r(A) vs. r(B) of compounds e.g. $A_xB_yX_n$ with fixed numbers x, y, n the fields of stability of individual structure types readily appear within certain boundaries. These marked by the individual pairs of ionic radii denoting cations which might be substituted in the given lattice sites in a particular crystal structure. With some precaution to the reliability of the boundaries with respect to the "type" values of the ionic radii, these plots of structure types are of great help in understanding interstructural relationships and structural properties of compounds which look to have a complicated chemical composition. In Fig. 39, 57—71-A-4 the stability fields of some ternary structures, ABO_3, ABX_4, A_2BX_4 and $A_2B_2O_7$ have been plotted, the crystal chemical data of which have been listed in Tables 39, 57—71-A-5, 39, 57—71-A-6, 39, 57—71-A-7 and 39, 57—71-A-8.

Revised manuscript received: April 1976

References: Section 39, 57—71-A

ABRAHAMS, S. C., BERNSTEIN, J. L.: Crystal structures of the transition metal molybdates and tungstates II. $Sc_2(WO_4)_3$. J. Chem. Phys. **45**, 2745 (1966).

ADOLPHE, C.: Contribution a l'etude d'un groupe de sulfures isostructuraux de terres rares et d'yttrium de type: Y_5S_7 et FeY_4S_7. Ann. Chim. Fr. **10**, 271 (1965).

ADOLPHE, C., LARUELLE, P.: Structure cristalline de $SeHo_4S_7$ et des certains composés isotypes. Bull. Soc. Franc. Mineral. Crist. **91**, 219 (1968).

ATOJI, M.: Neutron diffraction study of Ce_2S_3 at 300–5° K. Chem. Phys. **54**, 3226 (1971).

AVETISYAN, E. I., CHICHAGOV, A. V., BELOV, N. V.: Crystal structure of sodium gadolinium orthosilicate. Kristallografiya **15**, 926 (1971).

AYASSE, C., EICK, H. A.: The synthesis and crystal structure of triytterbium chloro-orthosilicate: $Yb_3(SiO_4)_2Cl$. Proc. 10th Rare Earth Res. Conf., 1973, p. 510 (1973).

BADIE, J. M.: Etude à haute température de la transformation H—X des sesquioxydes de terbium et de dysprosium. Compt. Rend. C **273**, 932 (1971).

BAENZIGER, N. C., HOLDEN, J. R., KNUDSEN, L.E., POPOV, A. I.: Unit cell dimensions of some rare earth oxyfluorides. J. Am. Chem. Soc. **76**, 4734 (1954).

BAGIEN-BEUCHER, M., TRANQUI, D.: Les ultraphosphates des terres rares et d'yttrium du types TP_5O_{14}. Bull. Soc. Franc. Mineral. Crist. **93**, 505 (1970).

BAERNIGHAUSEN, H.: Darstellung, Gitterkonstanten und Raumgruppe von $LiEuO_2$. Acta Cryst. **16**, 1073 (1963).

BAERNIGHAUSEN, H.: Darstellung und Kristalldaten von Alpha-$LiEuO_2$. Acta Cryst. **19**, 1047 (1965).

BAERNIGHAUSEN, H.: Untersuchungen am System EuO—Eu_2O_3. J. Prakt. Chem. **34**, 1 (1966a).

BAERNIGHAUSEN, H.: Gitterkonst. u. Raumgruppe der isotypen Verbindungen $Eu(OH)_2 \cdot H_2O$, $Sr(OH)_2 \cdot H_2O$, $Ba(OH)_2 \cdot H_2O$. Z. Anorg. Allgem. Chem. **342**, 233 (1966b).

BAERNIGHAUSEN, H.: $LiEu_3O_4$, eine neue, mit $LiSr_2EuO_4$ isotype Europium (II, III)-Verbindung. Z. Anorg. Allgem. Chem. **349**, 280 (1967).

BAERNIGHAUSEN, H.: The crystal structure of $LiEu_3O_4$. Z. Anorg. Allgem. Chem. **374**, 201 (1970).

BAERNIGHAUSEN, H., BECK, H. P., GRÜNINGER, H.-W.: The crystal chemistry of the salt-like rare earth dihalides. Proc. 9th Rare Earth Res. Conf. **1971**, 74 (1971).

BAERNIGHAUSEN, H., BRAUER, G.: Ein neues Europiumoxid Eu_3O_4 und die isotype Verbindung Eu_2SrO_4. Acta Cryst. **15**, 1059 (1962).

BAERNIGHAUSEN, H., BRAUER, G., SCHULTZ, N.: Darstellung und Kristallstruktur der Samarium-, Europium- u. Ytterbium-oxibromide LnOBr und Ln_3O_4Br. Z. Anorg. Allgem. Chem. **338**, 250 (1965).

BAERNIGHAUSEN, H., WEIDLEIN, J.: Die Kristallstruktur von Strontiumhydroxid-Monohydrat. Acta Cryst. **22**, 252 (1967a).

BAERNIGHAUSEN, H., PÄTOW, H., BECK, H. P.: Die Kristallstruktur von $YbCl_2$. Z. Anorg. Allgem. Chem. **403**, 45 (1974).

BAERNIGHAUSEN, H., SCHMID, R. D.: Darstellung und Kristalldaten von $LiBa_2EuO_4$. Z. Anorg. Allgem. Chem. **354**, 27 (1967b).

BAERNIGHAUSEN, H., SCHUSTER, R., VOGT, K.: The crystal structures of the new Eu(II)—Eu(III)-compounds $Li_2Eu_5O_8$, $Li_2SrEu_4O_8$ and $Li_2BaEu_4O_8$. Proc. 10th Rare Earth Res. Conf. **1973**, 490 (1973).

BAMBAUER, H. U., WEIDELT, J., YSKER, J.-ST.: Röntgenographische und optische Untersuchungen an Boraten Seltener Erden $SE(BO_2)_3$. Z. Krist. **130**, 207 (1969).

BANDURKIN, G. A., DZHURINSKII, B. F., TANANAEV, I. V.: f-degeneration in the rare earth elements. Dokl. Akad. Nauk SSSR **189**, 94 (1969).

BAUR, W. H.: Bond length variation and distorted coordination polyhedra in inorganic crystals. Trans. Am. Cryst. Assoc. 6, 125 (1970).

BAUR, W. H.: The prediction of bond length variations in silicon-oxygen bonds. Am. Mineralogist **56**, 1573 (1971).

BAYER, G.: Zur Kristallchemie des Tellurs. Fortschr. Mineral. **46**, 1 (1969).

BAYER, G.: New rare earth germanates with garnet structure. Naturwissenschaften **58**, 622 (1971).

BEDFORD, R. G., CATALANO, E.: Studies of the europium-oxygen-fluorine and samarium-oxygen-fluorine systems. J. Sol. State Chem. **2**, 585 (1970).

BEDFORD, R. G., CATALANO, E.: Investigations into the Eu—Eu_2O_3, the Eu—Pt—Eu_2O_3, the Sm—Sm_2O_3 and the Yb—Yb_2O_3 systems. J. Sol. State Chem. **3**, 112 (1971).

BERTAUT, F., FORRAT, F.: Sur le deformation è dans le perovskite a base de terres rares et d'elements de transition trivalents. J. Phys. Radium **17**, 129 (1956).

BERTAUT, F., GONDRAND, M.: Etude de combinaisons d'oxydes de terres rares et de lithium du type $TLiO_2$. Compt. Rend. **255**, 1135 (1962).

BERTAUT, F., MARESCHAL, J.: Un nouveau type de structure hexagonale: $AlTO_3$(T=Y, Eu, Gd, Tb, Dy, Ho, Er). Compt. Rend. **257**, 867 (1963).

BEVAN, D. J. M., KORDIS, J.: Mixed oxides of the type MO_2(fluorite)-M_2O_3—I. J. Inorg. Nucl. Chem. **26**, 1509 (1964).

BKHARGAVA, K. D., KOVBA, L. M., MARTYNENKO, L. I., SPITSYN, V. I.: New compounds of rare earths, strontium and barium. Dokl. Akad. Nauk SSSR **161**, 594 (1965).

BKHARGAVA, K. D., KOVBA, L. M., MARTYNENKO, L. I., SPITSYN, V. I.: Reaction of barium oxide with oxides of the rare earth elements. Zh. Neorgan. Khim. **11**, 1965 (1966).

BLASSE, G., DEVRIES, J.: A new family of lanthanide compounds: lithium lanthanide silicates and germanates. J. Inorg. Nucl. Chem. **29**, 1541 (1967).

BOCQUILLON, G., CHATEAU, C., LORIERS, C., LORIERS, J.: Polymorphism and pressure of the disilicates of the heavier lanthanides $Ln_2Si_2O_7$ (Ln=Tm, Yb, Lu). J. Sol. State Chem. (1976) (in press).

BÖHLHOFF, R., BAUMBAUER, H. U., HOFFMANN, W.: Die Kristallstruktur von Hoch-$LaBO_3$. Z. Krist. **133**, 386 (1971).

BONDAR, I. A., TOROPOV, N. A.: Preparation and properties of rare earth silicates and aluminates. Mat. Res. Bull. **2**, 479 (1967).

BORCHERT, H. J., BIERSTEDT, P. E.: Ferroelectric rare-earth molybdates. J. Appl. Phys. **38**, 2057 (1967).

BOULESTEIX, C., CARO, P. E., GASGNIER, M., LA BLANCHETALS, B., PARDO, B., SCHIFFMACHER, G.: La structure cubic C des sequioxides de terres rares. Phys. Stat. Sol. **23**, 597 (1974).

BRADLEY, W. F., GRAF, D. L., ROTH, R. S.: The vaterite-type ABO_3 rare earth borates. Acta Cryst. **20**, 283 (1966).

BRANDLE, C. D., STEINFINK, H.: The crystal structure of $Eu_4Al_2O_9$. Inorg. Chem. **8**, 1320 (1969).

BRANDLE, C. D., STEINFINK, H.: Phase transition in ordered barium-rare earth-molybdenum perovskites. Inorg. Chem. **10**, 922 (1971).

BRAUER, G.: In EYRING, L. (ed.): Progress in the Science and Technology of the Rare Earths. Oxford: Pergamon Press 1964, **1**, 152; 1966, **2**, 312; 1968, **3**, 434.

BRAUER, G., BAERNIGHAUSEN, H., SCHULTZ, N.: Versuche zur Darstellung niederer Oxide von Neodym, Samarium und Ytterbium. Z. Anorg. Allgem. Chem. **356**, 46 (1967a).

BRAUER, G., GINGERICH, K., HOLTSCHMIDT, U.: Über die Oxyde des Cers-IV. Die Sauerstoffzersetzungsdrucke im System der Ceroxyde. J. Inorg. Nucl. Chem. **16**, 77 (1960).

BRAUER, G., MOHR-ROSENBAUM, E.: Zur Darstellung der C-Modifikation von La_2O_3. Z. Anorg. Allgem. Chem. **394**, 301 (1972).

BRAUER, G., SCHULTZ, N.: Die Mischkristallreihe EuO—SrO und ein neues Verfahren zur Darstellung von EuO. J. Less-Common. Metals **13**, 213 (1967b).

BREIT, U.: Kristallstrukturverfeinerung und chemische Zusammensetzung des Lanthanborats $La_{26}B_8O_{51}$. Diplomarbeit Universität Münster (1976).

BRISSE, F., KNOP, O.: Pyrochlores II. An investigation of $La_2Ce_2O_7$ by neutron diffraction. Can. J. Chem. **45**, 609 (1967).

Brisse, F., Knop, O.: Pyrochlores III. X-ray, neutron, infrared and dielectric studies of $A_2Sn_2O_7$. Can. J. Chem. **46**, 859 (1968).
Brixner, L. H., Sleight, A. W., Licis, M. S.: Cell dimensions of the molybdates $La_2(MoO_4)_3$, $Ce_2(MoO_4)_3$, and $Nd_2(MoO_4)_3$. J. Sol. State Chem. **5**, 247 (1972).
Brown, I. D., Shannon, R. D.: Empirical bond-strength- bond-length curves for oxides. Acta Cryst. A **29**, 266 (1973).
Brunner, G., Schwarzenbach, D.: Zur Abgrenzung der Koordinationssphäre und Ermittlung der Koordinationszahl in Kristallstrukturen. Z. Krist. **133**, 127 (1971).
Brunton, G.: The crystal structure of β-$KCeF_4$. Acta Cryst. B **25**, 600 (1969).
Bruisset, H., Gillier-Pandraud, H., Berdot, J. L.: Structure cristalline du gallarte de néodyme $NdGaO_3$. Bull. Soc. Chim. France **1967**, 2886 (1967).
Buisson, G., Michel, C.: Isomorphous orthosilicates and orthogermanates of the rare earths. Mat. Res. Bull. **3**, 193 (1968).
Burbank, J. D.: A qualitative X-ray diffraction study of the 205° K and 151° K phase transformations in praseodymium aluminate ($PrAlO_3$). J. Appl. Cryst. **3**, 112 (1970).
Burns, G.: Covalent effects for rare earth ions with 6, 8 and 12 neighbours. Phys. Letters **25** A, 15 (1967).
Burns, J. H.: Crystal structure of hexagonal sodium neodymium fluorite and related compounds. Inorg. Chem. **4**, 881 (1965).
Busch, G., Kaldis, E., Verreault, R., Felsche, J.: A phase transition in Eu_2SiO_4. Mat. Res. Bull. **5**, 9 (1970).
Caro, P. E.: OM_4 tetrahedra linkages and the cationic group $(MO)_n^{n+}$ in rare earth oxides and oxysalts. J. Less-Common. Metals **16**, 367 (1968).
Caro, P.: Structure lamellaires: syntaxie, polytypie et non-stoechiometrie. J. Sol. State Chem. **6**, 396 (1973).
Carter, F. L.: Vacancy and charge ordering in the Th_3P_4 related structures. J. Sol. State Chem. **5**, 300 (1972).
Carter, F. L., Levinson, S.: The hydrothermal preparation and data for some lanthanide dihydroxy chlorides. Inorg. Chem. **8**, 2788 (1969).
Chenavas, J., Waintal, A., Capponi, J. J., Gondrand, M.: Etude sous haute pression et haute temperature des composés $NaTSiO_4$ et $NaTGeO_4$. Mat. Res. Bull. **4**, 425 (1969).
Chevalier, R., Laruelle, P., Flahaut, J.: Structure cristallographique Yb_3S_4. Bull. Soc. Franc. Mineral. Crist. **90**, 564 (1967).
Chichagov, A. V., Belov, N. V.: Crystal structure of $NaNdSiO_4$. Geochem. **12**, 1456 (1968a).
Chichagov, A. V., Ilyukhin, V. V., Belov, N. V.: Crystal structure of $NaSmSiO_4$. Sov. Phys. Dokl. Cryst. **12**, 997 (1968b).
Chichagov, A. V., Litvin, B. N., Belov, N. V.: Synthetic series of sodium rare earth silicates and germanates. Sov. Phys. Cryst. **14**, 95 (1969).
Christensen, A. N.: On the crystal structures of HoOOH, ErOOH, and YbOOH. Acta Chem. Scand. **19**, 1391 (1965).
Christensen, A. N.: Hydrothermal preparation of some Rare Earth trihydroxides and Rare Earth oxide hydroxides. Acta Chem. Scand. **20**, 896 (1966).
Christensen, A. N.: Hydrothermal preparation of neodymium oxide carbonate. Acta Chem. Scand. **24**, 2440 (1970).
Christensen, A. N.: Hydrothermal preparation and magnetic properties of $RE_2O_2CO_2$ (RE: Dy, Ho, Er, Yb). Acta Chem. Scand. **27**, 1835 (1973).
Clark, J. B., Range, K. J.: Quenchable high pressure phases of $GdYbS_3$. Z. Naturforsch. **30** b, 896 (1975).
Clark, J. B., Range, K. J.: Quenchable high pressure phases of $DyYbS_3$. Z. Naturforsch. **31** b, 275 (1976).
Clos, R., Devalette, M., Hagenmüller, P., Hoppe, R., Paletta, E.: Sur quelques oxydes doubles de potassium et de lanthanides de formule $KLnO_2$, à structure $NaFeO_2$. Compt. Rend. **265** C, 801 (1967).
Cockbain, A. G., Smith, G. V.: Alkaline-earth rare earth silicate and germanate apatites. Mineral. Mag. **36**, 411, 654 (1968).

COLLIN, G., LARUELLE, P.: Structure de $La_6Cu_2Si_2S_{14}$. Bull. Soc. Franc. Mineral. Crist. **94**, 175 (1971).

COLLOGUES, R., QUEYROUX, F., PEREZ, M., JORBA, Y., GILIES, J.-C.: Structures et propriétés des composés formés par les oxydes de terres rares avec les oxydes des éléments du groupe IVA. Bull. Soc. Chim. **1965**, 1141 (1965).

CROMER, D. I.: The crystal structure of monoclinic Sm_2O_3. J. Phys. Chem. **61**, 753 (1957).

CROMER, D. I., LARSON, A. C.: The crystal structure of $Ce(IO_3)_4$. Acta Cryst. **9**, 1015 (1956).

DEREIGNE, A., MANOLI, J. M., PANNETIER, G., HERPIN, P.: Cristallochimie des sulphates hydrates des terres rares: structure cristalline du $Ce_2(SO_4)_3 \cdot 4H_2O$. Bull. Soc. Franc. Mineral. Crist. **95**, 269 (1972).

DEXPERT, H., LEMAITRE-BLAISE, M., CARO, P.: Chemistry and reactivity of compounds with layered structures in the ternary heterogeneous systems rare earth oxides-carbon-dioxide-water. In: ANDERSON, J. S., ROBERTS, F. S., STONE. F. S. (eds.), Reactivity of Solids. London 1972.

DIEKE, G. H.: Spectra and Energy Levels of Rare Earth Ions in Crystals. New York: John Wiley and Sons, 1968.

DIMUSKES, J. P., WHITE, J. P.: Rare earth sesquioxides and sesquitellurides with the Sc_2S_3 structure. Inorg. Chem. **4**, 970 (1965).

DÖLL, W., KLEMM, W.: Über die Struktur einiger Dihalogenide. Z. Anorg. Allgem. Chem. **241**, 239 (1939).

DONNAY, G., ALLMANN, R.: How to recognize O^{2-}, OH^-, and H_2O in crystal structures determined by X-rays. Am. Mineralogist **55**, 1003 (1970).

DONNAY, G., DONNAY, J. D. H.: The crystal structure of bastnaesite. Am. Mineralogist **38**, 932 (1953).

DORNBERGER-SCHIFF, K., KLETSOVA, R. F.: On the relation between the monoclinic and the orthorhombic form of yttrium. Acta Cryst. **22**, 435 (1967).

DURIF-VARAMBON, A.: Etude de la substitution du silicium dans quelques types d'orthosilicates, hydroxychloride. Bull. Soc. Franc. Mineral. Crist. **82**, 285 (1959).

EIBSCHÜTZ, M.: Lattice constants of orthoferrites. Acta Cryst. **19**, 337 (1965).

EIBSCHÜTZ, M., COPPENS, P.: Determination of the crystal structure of yttrium orthoferrite and refinement of gadolinium orthoferrite. Acta Cryst. **19**, 524 (1965).

EICK, H. A.: The preparation, lattice parameters and some chemical properties of the rare earth mono-thiooxides. J. Am. Chem. Soc. **80**, 43 (1958).

EICK, H. A.: The crystal structure and lattice parameters of some rare earth mono-seleno oxides. Acta Cryst. **13**, 161 (1960).

EITEL, W., HERLINGER, E., TRÖMEL, G.: Zur Kristallchemie der Alumosilicate. Beziehungen der Anorthitgruppe zur Gruppe des Nephelins. Naturwissenschaften **18**, 469 (1930).

ELISEEV, A. A., USPENSKAJA, S. I., FEDOROV, A. A., TOLSTOVA, V. A.: Crystal structure of α—M_2S_3 (M=La, Nd, Sm). Zh. Strukt. Khim. **13**, 77 (1972).

ENGEL, G., KIRCHBERGER, W.: Neue Phosphate mit Eulytinstruktur, insbesondere Europium(II)-Verbindungen. Z. Anorg. Allgem. Chem. **417**, 81 (1975).

EYRING, L.: Fluorite-related oxide phases of the rare-earth and actinide elements. Advanc. Chem. Ser. **71**, 67 (1967).

FELSCHE, J.: Crystal data on two polymorphs of europium disilicate $Eu_2Si_2O_7$. J. Appl. Cryst. **2**, 1380 (1969).

FELSCHE, J.: Crystal data on the polymorphic disilicate $Y_2Si_2O_7$. Naturwissenschaften **57**, 127 (1970).

FELSCHE, J.: Polymorphism and crystal data of the rare-earth disilicates of type $R.E._2Si_2O_7$. J. Less-Common. Metals **21**, 1 (1970).

FELSCHE, J.: Rare earth silicates of the type $RE_2(SiO_4)O$. Naturwissenschaften **58**, 565 (1971a).

FELSCHE, J.: The crystal structure of the low form of Eu_2SiO_4. Naturwissenschaften **58**, 218 (1971b).

FELSCHE, J.: The crystal structures of the dimorphic rare earth disilicate $Pr_2Si_2O_7$. Z. Krist. **133**, 364 (1971c).

FELSCHE, J.: A new silicate structure containing linear (Si_3O_{10}) groups. Naturwissenschaften **59**, 35 (1972a).

FELSCHE, J.: Rare earth silicates with the apatite structure. J. Sol. State Chem. **5**, 266 (1972b).
FELSCHE, J.: The crystal chemistry of the rare earth silicates. Struct. Bonding **13**, 99 (1973).
FELSCHE, J., KALDIS, E.: Thermal oxidation of Eu_2SiO_4—a topoactic solid state reaction. J. Sol. State Chem. **5**, 49 (1972).
FLAHAUT, J.: In: EYRING, L. (ed.), Progress in the Science and Technology of the Rare Earths. Vol. 3, p. 209. Oxford: Pergamon Press 1968.
FLAHAUT, J., DOMANGE, L., GUITTARD, M., PARDO, M. P.: Etude crist des séléniures et des tellurures des terres rares de type U_2S_3. Bull. Soc. Chim. France **1965**, 326 (1965b).
FLAHAUT, J., GUITTARD, M., PATRIE, M., PARDO, M. P., GOLABI, S. M., DOMANGE, L.: Phases cubiques type Th_3P_4 dans les sulfures, les séléniures et les tellurures L_2X_3 et L_3X_4 des terres rares. Acta Cryst. **19**, 14 (1965a).
FLAHAUT, J., LARUELLE, P.: In: The Chemistry of Extended Defects in Non-metallic Solids, p. 109. Amsterdam: North Holland Publ. 1970.
FOEX, M., TRAVERSE, J.-P.: Remarques sur les transformations cristallines présentée à haute température par les sesquioxides de terres rares. Rev. Int. Hautes Temper. et Refract. **3**, 429 (1966).
GARTON, G., WANKLYN, B. M.: The rare earth aluminates. J. Cryst. Growth **1**, 164 (1976).
GELLER, S., BALA, V. B.: Crystallographic studies of perovskite-like compounds. II. Rare earth aluminates. Acta Cryst. **9**, 1019 (1956).
GELLER, S.: Crystal chemistry of the garnets. Z. Krist. **125**, 1 (1967).
GOLABI, S. M., FLAHAUT, J., DOMANGE, L.: Systemès formés entre le séléniure de calcium et les séléniure de terres rares L_2Se_3. Compt. Rend. **260**, 6385 (1965).
GOLDSCHMIDT, V. M., BARTH, T., LUNDE, G., ZACHARIASEN, W. H.: Geochemische Verteilungsgesetze der Elemente. Skrifter Norske Videnskaps Akad. Oslo I, Mat. Naturv. Kl. No. 2 (1926).
GONDRAND, M.: Données crystallographic sur de nouveaux composés α—$SmLiO_2$; $NdLiO_2$; $PrLiO_2$; $LaLiO_2$. Bull. Soc. Franc. Mineral. Crist. **90**, 107 (1967).
GONDRAND, M., BERTAUT, E. F.: Une nouvelle classe de composés $LiTO_2$ (T=Sm, Eu, Gd). Bull. Soc. Franc. Mineral. Crist. **86**, 301 (1963)
GSCHNEIDER, K. A.: On the nature of 4f bonding in the lanthanide elements and their compounds. J. Less-Common. Met. **25**, 405 (1971).
GUILLEN, M., BERTAUT, E. F.: Structure de La_2TiO_5. Bull. Soc. Franc. Ceram. **72**, 57 (1966).
GUITTARD, M., FLAHAUT, J.: Les séléniures inferieurs des élément des terres rares. Bull. Soc. Chim. France **1968**, 4759 (1968).
HARRIS, L. A., YAKEL, H. L.: Crystal structure of Y_2BeO_4. Acta Cryst. **22**, 354 (1967).
HASCHKE, J. M.: Preparation, phase equilibria, crystal chemistry and some properties of lanthanide hydroxide nitrates. Inorg. Chem. **13**, 1812 (1974).
HASCHKE, J. M.: The lanthanum hydroxide fluoride carbonate system: the preparation of synthetic bastnaesite. J. Sol. State Chem. **12**, 115 (1975).
HELMHOLZ, L.: The crystal structure of neodymium bromate enneahydrate $Nd(BrO_3)_3 \cdot 9H_2O$. J. Am. Chem. Soc. **61**, 1544 (1939).
HOEKSTRA, R. H.: Phase relationship in the rare earth sesquioxides at high pressure. Inorg. Chem. **5**, 754 (1966).
HOPPE, R.: Ternäre Oxide der Alkalimetalle. Bull. Soc. Chim. France **1965**, 1115 (1965).
HOPPE, R.: Madelung constants as a new guide to the structural chemistry of solids. In: STACEY, M., TATLOW, C. T., SHARPE, A. G. (eds.), Advances in Fluorine Chemistry, p. 387. London: Butterworths 1970.
HUNT, E. B., RUNDLE, R. E., STOSICK, A. J.: The structure of lanthanum sulfate enneahydrate. Acta Cryst. **7**, 106 (1954).
HUY-DUNG, N., ETTIENNE, J., FLAHAUT, J.: Sur une nouvelle famille de séléniures doubles formes par le Cr^{3+} et La à Nd. Compt. Rend. **269**C, 120 (1969).
HUY-DUNG, N., ETTIENNE, J., LARUELLE, P.: Isoelement et structure cristalline d'une de composés des terres rares. Bull. Soc. Chim. France **1971**, 2433 (1971).
HYDE, B. G., BEVAN, D. J. M., EYRING, L.: On the praseodymium oxygen system. Phil. Trans. Roy. Soc. London, Ser. A, No. 1106, **259**, 583 (1966a).
HYDE, B. G., EYRING, L.: In: EYRING, L. (ed.), Rare Earth Research III. New York: Gordon Breach 1966b.

IBERS, J. A., CROMER, D. T.: The least-squares refinement of the crystal structure of $Ce(IO_3)_4 \cdot H_2O$ (Ceric iodate monohydrate). Acta Cryst. **11**, 794 (1958).
ITO, J.: Silicate apatites and oxyapatites. Am. Mineralogist **53**, 890 (1968).
ITO, J., JOHNSON, H.: Synthesis and study of yttrialite. Am. Mineralogist **53**, 1940 (1968).
IVEROVNA, V. I., TARASOVA, V. P., UMANSKII, M. M.: Preparation and crystal data of $RE(SO_4) \cdot nH_2O$. Izv. Akad. Nauk SSSR, Ser. Fiz. **15**, 164 (1951).
JAULMES, S.: Affinement de la structure de $LaPO_4$. Bull. Soc. Franc. Mineral. Crist. **95**, 42 (1972).
JAULMES, S.: Structure crystalline du sulfure d'etain et de lanthane, La_2SnS_5. Acta Cryst. B **30**, 2283 (1974).
JEITSCHKO, W.: Crystal structure of $La_2 (MoO_4)_3$, a new ordered defect scheelite type. Acta Cryst. B **29**, 2074 (1973).
JEITSCHKO, W.: A comprehensive X-ray study of the phases of $Gd_2(MoO_4)_3$. Acta Cryst. B **28**, 60 (1972).
JENKINS, M. S., THURCOPPE, R. P., EYRING, L.: In: The Chemistry of Extended Defects in Non-metallic Solids. Amsterdam: North Holland Publ. 1970.
JØRGENSON, C. K.: The inner mechanism of rare earths elucidated by photo-electron spectra. Structure and Bonding **13**, 199 (1973).
JØRGENSON, C. K., PAPPALARDO, R., FLAHAUT, J.: L'effect néphelauxetique dans sulfures des terres rares. J. Chim. Phys. **1965**, 444 (1965).
KALDIS, E.: High-temperature chemical transport. J. Cryst. Growth. **9**, 281 (1971).
KALDIS, E., VERREAULT, R.: Crystal growth and characterization of europium (II) orthosilicate. J. Less. Common. Metals **20**, 177 (1970).
KASPER, H. M.: A new series of rare earth garnets $Ln_3M_2Li_3O_{12}$. Inorg. Chem. **8**, 1000 (1969).
KELLER, C.: Über ternäre Oxide des Niobs und Tantals vom Typ ABO_4. Anorg. Allg. Chem. **318**, 83 (1962).
KELLER, C.: Über die Reaktion von Protactiniumpentoxid mit den Oxiden einiger dreiwertiger Metalle. Inorg. Nucl. Chem. **26**, 2069 (1964).
KELLER, C., ENGERER, H., LEIPNER, L., SRIYOTHA, U.: Phasengleichgewicht in den Systemen Uranoxid-Oxide der seltenen Erden (Ho, Er, Tm, Yb, Lu). J. Inorg. Nucl. Chem. **31**, 965 (1969a).
KELLER, C., ENGERER, H., SEIFERT, H.: Ternäre Oxide des fünf- und sechswertigen Neptuniums mit seltenen Erden. J. Inorg. Nucl. Chem. **31**, 2727 (1969b).
KELLER, C., SCHMUTZ, H.: Die Reaktion von Lithiumfluorid mit den Trifluoriden der Lanthaniden und einiger Actiniden. J. Inorg. Nucl. Chem. **27**, 900 (1965a).
KELLER, C., WALTER, K. H.: Einige ternäre Oxide des Plutoniums, Americiums und Curiums vom Typ $Me^{III}X^{V}O_4$. J. Inorg. Nucl. Chem. **27**, 1233 (1965b).
KEVE, E. T., ABRAHAMS, S. C., BERNSTEIN, J. L.: Ferroelectric, ferroelastic paramagnetic beta-$Gd_2(MoO_4)_3$ crystal structure of the transition metal molybdates and tungstates. J. Chem. Phys. **54**, 3185 (1971).
KIM, Y. S.: Crystallographic study of cerium aluminate ($CeAlO_3$). Acta Cryst. B **24**, 295 (1968).
KINDERMANN, B.: Untersuchungen über Kristallarten und Gläser in den Systemen SE_2O_3—CaO—B_2O_3 (SE: La, Gd, Yb). Dissertation Universität Münster (1974).
KLEMM, W., KLEIN, H. A.: Lanthanoxyfluorid. Z. Anorg. Allgem. Chem. **248**, 137 (1941).
KLEMM, W., KOCZY, A.: Über die Selenide der seltenen Erden. Z. Anorg. Allgem. Chem. **233**, 84 (1937).
KLEMM, W., SENFF, H.: Chalkogenide des zweiwertigen Europiums. Z. Anorg. Allgem. Chem. **241**, 259 (1939).
KLEVTSOV, P. V., KOZEEVA, L. P.: Synthesis and X-ray and thermal studies of potassium rare earth tungstates $KLn(WO_4)_2$, Ln=rare earth elements. Sov. Phys. Cryst. **14**, 185 (1969).
KLEVTSOV, P. V., SHEINA, L. P.: Hydrothermal synthesis and crystal structure of earth hydroxides. Izv. Akad. Nauk SSSR, Neorgan. Mat. **1**, 912 (1965).
KLEVTSOVA, R. F., KLEVTSOV, P. V.: The crystal structure of YOOH. Zh. Strukt. Khim. **5**, 860 (1964).

KLEVTSOVA, R. F., KLEVTSOV, P. V.: Shapes of crystal growth and crystal structure of $YCl(OH_2)$. Dokl. Akad. Nauk SSSR **162**, 1049 (1965).
KLEVTSOVA, R. F., KLEVTSOV, P. V.: Zh. Strukt. Khim. **7**, 556 (1966).
KNOP, O., BRISSE, F., BARKER, W. W., GRAHAM, J.: The Chemistry of Extended Defects in Non-metallic Solids. Amsterdam: North Holland Publ. 1970.
KNOP, O., BRISSE, F., CASTELLITZ, L.: Pyrochlores V. Studies of $A_2Ti_2O_7$ titanates. Can. J. Chem. **47**, 971 (1969).
KNOP, O., BRISSE, F., MEADS, R. E., BAINBRIDGE, J.: Pyrochlores IV. Crystallographic and Mössbauer studies of A_2FeSbO_7 pyrochlores. Can. J. Chem. **46**, 3829 (1968).
KOEHLER, W. C., WOLLAN, E. O.: Neutron-diffraction study of the structure of the A-form of the rare earth sesquioxides. Acta Cryst. **6**, 741 (1953).
KOEHLER, W. C., WOLLAN, E. O.: Neutron-diffraction study of the magnetic properties of perovskite-like compounds $LaBO_3$. J. Phys. Chem. Solids **2**, 100 (1957).
KUNZMANN, P., EYRING, L.: On the crystal structures of the fluorite-related intermediate rare earth oxides. J. Sol. State Chem. **14**, 229 (1975).
KUZMIN, E. A., BELOV, N. V.: Crystal structure of the simplest silicates of La and Sm. Sov. Phys. Dokl. **10**, 1009 (1966).
LEVIN, E. M., ROTH, R. S., MARTIN, J. B.: Polymorphism of ABO_3 type rare earth borates. Am. Mineralogist **46**, 1030 (1961).
LOPATO, L. M., SHEVCHENKO, A. V.: Reaction of EuO with rare earth neutral sesquioxides. Zh. Neorg. Khim. **15**, 1027 (1970).
MAKSIMOV, B. A., ILYUKHIN, V. V., BELOV, N. V.: Crystal structure of the Na-Y orthosilicate $NaY(SiO)_4$. Sov. Phys. Cryst. **11**, 583 (1967).
MAKSIMOV, B. A., ILYUKHIN, V. V., BELOV, N. V.: Crystal structure of K, Y-triorthosilicate $K_3Y(Si_3O_8(OH)_2)$. Sov. Phys. Dokl. Cryst. **13**, 638 (1969b).
MAKSIMOV, B. A., ILYUKHIN, V. V., LITVIN, B. N., BELOV, N. V.: Hydrothermal crystallization in the system A_2O-TR_2O_3—SiO_2—H_2O. Sov. Phys. Cryst. **14**, 407 (1969a).
MARBEUF, A., DEMAZEAU, G., TURELL, S., HAGENMÜLLER, P., DEROUET, J., CARO, P.: Les hydroxyfluorures de terres rares. J. Sol. Chem. **3**, 637 (1971).
MAREZIO, M., REMEIKA, J. P., DERNIER, P. D.: High pressure synthesis of $YGaO_3$, $GdGaO_3$, and $YbGaO_3$. Mat. Res. Bull. **1**, 247 (1966).
MAREZIO, M., REMEIKA, J. P., DERNIER, P. D.: Rare earth orthogallates. Inorg. Chem. **7**, 1337 (1968).
MAREZIO, M., REMEIKA, J. P., DERNIER, P. D.: The crystal chemistry of the rare earth orthoferrites. Acta Cryst. B **26**, 2008 (1970).
MAYER, I., LEVY, E., GLASNER, A.: The crystal structure of $EuSO_4$ and $EuCO_3$. Acta Cryst. **17**, 1071 (1964).
MCCARTHY, G. J.: Divalent europium compounds in the system Eu—Mo—O, Eu—W—O. Mat. Res. Bull. **6**, 31 (1971).
MCCARTHY, G. J., FISCHER, R. D., SANZGIRI, J.: Crystal chemistry and subsolidus phase reactions in the system Eu—W—O. J. Sol. State Chem. **5**, 200 (1972a).
MCCARTHY, G. J., FISCHER, R. D.: The system Eu—Fe—O. J. Sol. State Chem. **4**, 340 (1972b).
MCCARTHY, G. J., GREEDAN, J. E.: Crystal chemistry and magnetic properties of Eu_2TiO_4 and $Eu_3Ti_2O_7$. Mat. Res. Bull. **7**, 531 (1972c).
MCCARTHY, G. J., JOHNSTON, R. G., KOULIHAN, J. F.: Crystal chemistry and magnetic properties of Eu^{2+} ternary oxides. Proc. 10th Rare Earth Res. Conf. **1973**, 989 (1973).
MCCARTHY, G. J., WHITE, W. B.: On the stabilities of the lower oxides of the rare earths. J. Less-Common. Metals **22**, 409 (1970).
MCCARTHY, G. J., WHITE, W. B., ROY, R.: Preparation and structure of the rare earth titanates. Mat. Res. Bull. **4**, 251 (1969a).
MCCARTHY, G. J., WHITE, W. B., ROY, R.: The system Eu—Ti—O: Phase relations in a portion of the 1400 °C isotherm. J. Inorg. Nucl. Chem. **31**, 329 (1969b).
MEISEL, K.: Über die Kristallstruktur des Rheniumtrioxids. Z. Anorg. Allgem. Chem. **207**, 121 (1932).
MEREZIO, M., PLETTINGER, H. A., ZACHARIASEN, W. H.: The crystal structure of gadolinium trichloride hexahydrate. Acta Cryst. **14**, 234 (1961).

MERTENS, H. E., ZEMANN, J.: Elektrostatische Gitterenergien für den Bixbyit-Typ. Acta Cryst. **21**, 467 (1966).
MEYER, H. J.: Orthoborate der seltenen Erden mit Aragonit-Struktur. Naturwissenschaften **56**, 458 (1969).
MICHEL, C., BUISSON, G., BERTRAUT, E. F.: Structure de Y_2SiO_5. Compt. Rend. **264**, 397 (1967).
MOELLER, I.: Comprehensive Inorganic Chemistry, Vol. 4, 1. Oxford: Pergamon Press 1973.
MOREAU, J. M.: Etude cristallographique et magnetique des oxydes doubles de terre rare de type $LaTO_3$. Mat. Res. Bull. **3**, 427 (1968).
MÜLLER-BUSCHBAUM, H. K., VON SCHNERING, H. G.: Strukturuntersuchungen an La_2O_3. Z. Anorg. Allgem. Chem. **340**, 232 (1965).
MÜLLER-BUSCHBAUM, H. K., TESKE, C.: Zur Kenntnis der Kristallstruktur von $LaYbO_3$. Z. Anorg. Allgem. Chem. **369**, 255 (1969).
MULLER, O., ROY, R.: The Major Ternary Structural Families. Berlin-Heidelberg-New York: Springer 1974.
NAUMANN, H., BERGSTOL, S., NILSSEN, B: Stillwellite in den Langesundfiord nepheline syenite pegmatite dykes. Norske Geol. Tidskr. **46**, 327 (1966).
O'CONNOR, B. H., VALENTINE, P. M.: A neutron diffraction study of the crystal structure of the C-form of yttrium sesquioxide. Acta Cryst. B **25**, 2140 (1969).
PAKHOMOV, V. I.: The structure of rare earth metaborates. Izv. Akad. Nauk SSSR, Neorg. Mat. **5**, 1661 (1969).
PATON, M. G., MASLEN, E. N.: A refinement of the crystal structure of yttria. Acta Cryst. **19**, 307 (1965).
PATRIE, M., CHEVALIER, R.: Sur une famille de composés orthorhombiques du type MnY_2S_4. Compt. Rend. **263** C, 1061 (1966).
PATRIE, M., FLAHAUT, J.: Sur une famille de composés CaL_2S_4 formés par les éléments des terres rares. Compt. Rend. **264** C, 395 (1967).
PATRIE, M., GUITTARD, M., PARDO, M.-P.: Systèmes L_2X_3—PbX (L=Lanthanides, X=S, Se, Te). Bull. Soc. Chim. France **1969**, 3832 (1969).
PAULING, L.: The sizes of ions and the structure of ionic crystals. J. Am. Chem. Soc. **49**, 765 (1927).
PAULING, L.: The crystal structure of the A-modification of the rare earth sesquioxides. Z. Krist. **69**, 514 (1928).
PAULING, L.: The principles determining the structure of complex ionic crystals. J. Am. Chem. Soc. **51**, 1010 (1929).
PAULING, L.: The Nature of the Chemical Bond. New York: Cornell University Press 1960.
PAULING, L., SHAPPEL, M. D.: The crystal structure of bixbyite and the C-modification of the sesquioxides. Z. Krist. **75**, 128 (1932).
PETZEL, T., GREIS, O.: Über die Anwendung eines neuartigen Reduktionsverfahrens zur Reindarstellung von SmF_2, EuF_2, YbF_2, SmS, YbS u. EuO. Z. Anorg. Allgem. Chem. **396**, 95 (1973).
PIERCE, J. W., HONG, Y.-P.: Structural Studies in the System KF-YF_3. Proc. 10th Rare Earth Res. Conf. **1973**, 527 (1973).
POPOV, A. I., KNUDSON, G. E.: Preparation and properties of rare earth fluorides and oxyfluorides. J. Am. Chem. Soc. **76**, 3921 (1954).
PORTIER, R., FAYARD, M., BOULESTEIX, C., CARO, P.: Mise en evidence par microscopie electronique de marches a la surface de cristaux minces de sesquioxides des terres rares. Mat. Res. Bull. **8**, 693 (1973).
PREWITT, C. T., SLEIGHT, A. W.: Structure of gadolinium sesquisulfide. Inorg. Chem. **7**, 1090 (1968).
QUEZEL-AMBRUNAZ, S., MARESCHAL, M.: Parametre cristalline de chromites de terres rares. Bull. Soc. France Mineral. Crist. **86**, 204 (1963).
QUEZEL-AMBRUNAZ, S., MARESCHAL, M.: Parameters de manganide de terres rares perovskites. Bull. Soc. France Mineral. Crist. **91**, 339 (1968).
RANGE, K.-J., LEEB, R.: Hochdruckmodifikationen der Lanthanoiden (III) sulfide Ln_2S_3 (Ln=Lu—Ho, Y) mit U_2S_3-Struktur. Z. Naturforsch. **30**b, 889 (1974).

RANGE, K. J., LEEB, R.: Die Kristallstruktur von Lu_2S_3. Z. Naturforsch. **30**b, 637 (1975).

RANGE, K. J., LEEB, R.: Darstellung und Kristallstruktur der Hochdruckphase Tm_2S_3—II. Z. Naturforsch. **31**b, (1976).

RAU, R. C.: Thy crystal structure of Eu_3O_4. Acta Cryst. **20**, 716 (1966).

REUTER, B.: Über Doppeloxidsysteme des Vanadiums. Bull. Soc. Chim. France **1965**, 1053 (1965).

RODIER, N., LARUELLE, P.: Sur un nouveau sulphure mixte d'erbium et de scandium de formula Sc_2(Gd ... Ho, $Er)_3S_7$ Bull. Soc. France Mineral. Crist. **95**, 548 (1972).

RODIER, N., LARUELLE, P., FLAHAUT, J.: Sur une nouvelle série de combinaisons des sesquisulfures de terres rares et d'yttrium avec le sesquisulfure de scandium. Compt. Rend. **269**C, 1391 (1969).

RUBINSTEIN, C. B., BARNS, R. L.: Crystal data for rare earth aluminium garnets II. Am. Mineralogist **50**, 782 (1965).

SALLAVUARD, G., SZABO, G., PARIS, R. A.: Sur les monogallates lanthanidiques $LnGaO_3$. Compt. Rend. **286**C, 1050 (1969).

SAWYER, J. O., CARO, P., EYRING, L.: Herstellung, Analyse und röntgenographische Identifizierung der Dioxymonocarbonate des Lanthans und der Lanthanidenelemente. Monatsh. Chem. **102**, 333 (1971).

SAWYER, J. O., CARO, P., EYRING, L.: Hydroxy-carbonates of the lanthanide elements. Rev. Chim. minerale **10**, 93 (1973).

SCHWARZ, H.: Die Phosphate, Arsenate und Vanadate der Seltenen Erden. Z. Anorg. Allgem. Chem. **323**, 44 (1963a).

SCHWARZ, H.: Die Chromate (V) der Seltenen Erden, $SECrO_4$, von Samarium bis Lutetium. Z. Anorg. Allgem. Chem. **323**, 275 (1963b).

SHAFER, M. W.: Preparation and crystal chemistry of divalent europium compounds. J. Appl. Phys. **36**, 1145 (1965).

SHANNON, R. D.: Synthesis of some new perovskites containing indium and thallium. Inorg. Chem. **6**, **1474** (1967).

SHANNON, R. D.: Cell dimensions of rare earth orthorhodites. Acta Cryst. B**26**, 447 (1970c).

SHANNON, R. D., PREWITT, C. T.: Effective ionic radii in oxides and fluorides. Acta Cryst. B **25**, 925 (1969).

SHANNON, R. D., PREWITT, C .T.: Synthesis of pyrosilicates and pyrogermanates having the thortveitite structure. J. Sol. State Chem. **2**, 199 (1970a).

SHANNON, R. D., PREWITT, C. T.: Revised values of effective ionic radii. Acta Cryst. B **26**, 1046 (1970b).

SHANNON, R. D., PREWITT, C. T.: Effective ionic radii and crystal chemistry. J. Inorg. Nucl. Chem. **32**, 1427 (1970).

SHANNON, R. D., SLEIGHT, A. W.: Synthesis of new high-pressure pyrochlore phases. Inorg. Chem. **7**, 1649 (1968).

SHINN, D. B., EICK, H. A.: Phase analyses of lanthanide oxide fluorides. Inorg. Chem. **8**, 232 (1969).

SIELER, J., KAISER, J.: Struktur und Eigenschaften von SE Übergangsmetalloxiden von K_2NiF_4 und $CaFe_2O_4$ Typen. Z. Anorg. Allgem. Chem. **377**, 316 (1970).

SKSHAT, S. M., SIMONOV, V. T., BELOV, N. V.: Crystal structure of the synthetic sodium-scandium silicate $Na_3Sc(Si_2O_7)$. Sov. Phys. Dokl. Cryst. **14**, 9 (1969).

SLEIGHT, A. W., BRIXNER, L. H.: Polymorphism in rare earth molbdates and tungstates of the types $R_2(MoO_4)_3$ and R_2MO_6. Proceed. 10th Rare Earth Res. Conf. **1973**, 570 (1973).

SLEIGHT, A. W., PREWITT, C. I.: Crystal chemistry of the rare earth sesquisulfides. Inorg. Chem. **7**, 2282 (1968).

SMOLIN, YU. I.: The crystal structure of ytterbium oxyorthosilicate Yb_2SiO_5. Sov. Phys. Cryst. **14**, 854 (1970).

SMOLIN, YU. I., SHEPELEV, YU. F.: Crystal structure of $7Gd_2O_3 \cdot 9SiO_2$. Izv. Akad. Nauk SSSR, Neorg. Mater. **5**, 1823 (1969a).

SMOLIN, YU. I., SHEPELEV, YU. F.: The crystal structures of the rare earth pyrosilicates. Acta Cryst. B **26**, 484 (1970b).

SMOLIN, YU. I., SHEPELEV, YU. F., BUTIKOVA, I. K.: The crystal structure of the low temperature form of $Sm_2Si_2O_7$. Sov. Phys. Cryst. **15**, 214 (1970c).

SMOLIN, YU. I., SHEPELEV, YU. F., UPATOVA, T. V.: The (Ge_3O_{10}) radical in crystals of composition $La_2Ge_2O_7$. Sov. Phys. Cryst. **14**, 630 (1970a).

SMOLIN, YU. I., TKACHEV, S. P.: Determination of the structure of gadolinium oxyortho-silicate $Gd_2O_3 \cdot SiO_2$. Sov. Phys. Cryst. **14**, 15 (1969b).

SOULEAU, C., GUITTARD, M., LARUELLE, P.: Systèmes YbSe—L_2Se_3. Bull. Soc. Chim. France **1969**, 9 (1969).

SUWA, Y., NAKA, S., NODA, T.: Preparation and properties of yttrium magnesium silicate with apatite structure. Mat. Res. Bull. **3**, 139 (1968).

TEMPLETON, D. H., CARTER, G. F.: The crystal structure of yttrium trichloride and similar compounds. J. Phys. Chem. **58**, 940 (1954a).

TEMPLETON, D. H., DAUBEN, C. H.: Crystal structures of rare earth oxychlorides. J. Am. Chem. Soc. **75**, 6069 (1953).

TEMPLETON, D. H., DAUBEN, C. H.: Lattice parameters of some rare earth compounds and a set of crystal radii. J. Am. Chem. Soc. **76**, 5237 (1954b).

TEMPLETON, D. H., ZALKIN, A.: Crystal structure of europium tungstate. Acta Cryst. **16**. 762 (1963).

THARCHOVA, T. N., GRISM, J. A., MIRONOV, N. N.: The crystal structure of $La(OH)_2Cl$. Zh. Neorg. Chem **15**, 2584 (1970).

THOMA, R. E.: In: Progress in the Science and Technology of the Rare Earths, Vol. 2, p. 90. Oxford: Pergamon Press 1966.

THOMA, R. E., BRUNTON, G. D., PENNEMANN, R. A., KEENAN, F. K.: Equilibrium relations and crystal structure of lithium fluorolanthanate phases. Inorg. Chem. **9**, 1096 (1970).

THOMA, R. E., INSLEY, H., HERBERT, G. M.: The sodium fluoride-lanthanide trifluoride systems. Inorg. Chem. **5**, 1222 (1966).

THORNBERG, M. G., BEVAN, D. K. M.: Crystal structures of the high and low temperature forms of $Zr_3Yb_4O_{12}$. J. Sol. State Chem. **1**, 536 (1970).

THORNBERG, M. R., BEVAN, D. J. M., GRAHAM, J.: Mixed oxides of the type MO_2 (fluorite)-M_2O_3. III. Crystal structures of the intermediate phases $Zr_5Sc_2O_{13}$ and $Zr_3Sc_4O_{12}$. Acta Cryst. B **24**, 1183 (1968).

TIMCHENKO, T. I., PETUSHKO, L. V., POBEDIMSKAYA, E. A., PASHKOVA, A. V.: A new lanthanum oxytungstate $2\,La_2O_3 \cdot 3\,WO_3$. Sov. Phys. Cryst. **14**, 188 (1969).

TOROPOV, N. A., BONDAR, I. A., LAZAREV, A. N., SMOLIN, YU. I.: In: Rare Earths Silicates and their Analogues. Leningrad: Akad. Nauk. SSSR (NAUKA) 1971.

TRANQUI, D., BAGIEN-BEUCHER, M., DURIF, A.: Structure cristalline de l'ultraphosphates d'holmium HoP_5O_{14}. Bull. Soc. France Mineral. Crist. **95**, 437 (1972).

TURCOTTE, R. P., JENKINS, M. S., EYRING, L.: Pseudophase behavior in the epsilon and iota regions of the praseodymium oxide oxygen system. J. Sol. State Chem. **7**, 454 (1973).

TURCOTTE, R. P., WARMKESSEL, J. M., TILLEY, R. J. D., EYRING, L.: On the phase interval $PrO_{1.5}$ to $PrO_{1.71}$ in the praseodymium oxide-oxygen system. J. Sol. State Chem. **3**, 265 (1971).

VORONKOV, A. A., PYATENKO, Y. A.: X-ray diffraction study of the atomic structure of stillwellite $CeBO(SiO_4)$. Sov. Phys. Cryst. **12**, 214 (1967).

WAINTAL, A., GONDRAND, M.: Transformation sous très haute pression et à très haute temperature des composes $SmLiO_2$, $EuLiO_2$, $TbLiO_2$. Mat. Res. Bull. **2**, 889 (1967).

WEBB, A. W., EATOUGH, N. L., HALL, H. T.: High pressure Th_3P_4-type polymorphs of rare earth sesquichalcogenides. Inorg. Chem. **8**, 2069 (1969).

WEIDELT, J.: Die Verbindungen $SE(BO_2)_3$. Z. Anorg. Allgem. Chem. **374**, 26 (1970).

WEIDELT, J., BAMBAUER, H. U.: Ein mit Aragonit isotypes Cerborat. Naturwissenschaften **55**, 342 (1968).

WEIGEL, F.: Die Chemie des Promethiums. Fortschr. Chem. Forsch. **12**, 539 (1969).

WEIGEL, F., SCHERER, V., HENSCHEL, H.: Unit cells of the monazite type-rare earth phosphates. J. Am. Ceram. Soc. **48**, 383 (1965).

WHITE, J. G., YOCOM, P. N., LERNER, S.: Structure determination and crystal preparation of monoclinic rare earth sesquioxides. Inorg. Chem. **6**, 1872 (1967).

WILLER, B., DAIRE, M.: Données cristalographiques sur la variété B des sesquioxides La_2O_3 et Pr_2O_3. Bull. Soc. France Mineral. Crist. **92**, 33 (1969).

WILTEN, G. M., CHASE, A. B.: Synthetic fergusonite and a new polymorph of yttrium tantalate. Am. Mineralogist **52**, 1536 (1967).
WOLTEN, G. M.: The structure of the M'-phase of $YTaO_4$, a third fergusonite polymorph. Acta Cryst. B **23**, 939 (1967).
WOOD, E. A., MILLER, R. C., REMEIKA, J.-P.: The field-induced ferroelectric phase of sodium niobate. Acta Cryst. **15**, 1273 (1962).
YAKEL, H. L., KOEHLER, W. C., BERTAUT, E. F., FORRAT, E. F.: On the crystal structure of the manganese (III) trioxides of the heavy lanthanides and yttrium. Acta Cryst. **16**, 957 (1963).
YSKER, J. ST., HOFFMANN, W.: Die Kristallstruktur des $La(B_3O_6)$. Naturwissenschaften **57**, 129 (1970).
ZACHARIASEN, W. H.: The UCl_3 type of crystal structure. J. Chem. Phys. **16**, 254 (1948a).
ZACHARIASEN, W. H.: Crystal chemical studies of the 5f-series of elements. I. Acta Cryst. **1**, 265 (1948b).
ZACHARIASEN, W. H.: Crystal chemical studies of the 5f-series of elements. XII. Acta Cryst. **2**, 388 (1951a).
ZACHARIASEN, W. H.: Crystal chemical studies of the 5f-series of elements. XIV. Acta Cryst. **4**, 231 (1951b).
ZALKIN, A., TEMPLETON, D. H.: The crystal structures of YF_3 and related compounds. J. Am. Chem. Soc. **75**, 2453 (1953).
ZALKIN, A., TEMPLETON, D. H., HOPKINS, T. E.: The atomic parameters in the lanthanum trifluoride structure. Inorg. Chem. **5**, 1466 (1966).

Revised manuscript received: April 1976

39, 57—71-B. Isotopes in Nature

Only one stable isotope is known of the following elements: Y, Pr, Tb, Ho, and Tm, whereas from two to seven isotopes are found of natural La, Ce, Nd, Sm, Eu, Gd, Dy, Er, Yb, and Lu (Table 39, 57—71-B-1). The element promethium ($Z=61$) is not detected in natural matter because of the fast decay of its isotopes. The half-lives of ^{145}Pm and ^{147}Pm with the slowest decay rate are in the range of a few years. Four naturally occurring unstable isotopes ^{138}La, ^{144}Nd, ^{147}Sm, and ^{176}Lu decay with half-lives ranging from 10^{15} to 10^{10} years (see Table 39, 57—71-B-1). Because of their rarity and their slow decay these isotopes are of little or no use in determining geological age. Their production of heat is unimportant in the total radiogenic heat production of the earth. It should be mentioned that the α-type of radioactive disintegration occurring in the isotopes listed in Table 39, 57—71-B-1 is unusual for low or medium atomic number isotopes.

Table 39, 57—71-B-1

Element	Isotopes in nature	Abundance %	Reference
$_{39}$Y	^{89}Y	100	Cameron and Wichers (1962)
$_{57}$La	^{138}La [a]	0.089	Cameron and Wichers (1962)
	^{139}La	99.911	
$_{58}$Ce	^{136}Ce	0.193	Inghram *et al.* (1947)
	^{138}Ce	0.250	
	^{140}Ce	88.48	
	^{142}Ce	11.07	
$_{59}$Pr	^{141}Pr	100	Cameron and Wichers (1962)
$_{60}$Nd	^{142}Nd	27.11	Cameron and Wichers (1962)
	^{143}Nd	12.17	
	^{144}Nd [b]	23.85	
	^{145}Nd	8.30	
	^{146}Nd	17.22	
	^{148}Nd	5.73	
	^{150}Nd	5.62	
$_{61}$Pm	There are no stable isotopes of Pm in nature		
$_{62}$Sm	^{144}Sm	3.09	Heath (1964/65)
	^{147}Sm [c]	14.97	
	^{148}Sm	11.24	
	^{149}Sm	13.83	
	^{150}Sm	7.44	
	^{152}Sm	26.72	
	^{154}Sm	22.71	
$_{63}$Eu	^{151}Eu	47.82	Cameron and Wichers (1962)
	^{153}Eu	52.18	

Table 39, 57—71-B-1 (Continued)

Element	Isotopes in nature	Abundance %	Reference
$_{64}$Gd	^{152}Gd	0.20	HEATH (1964/65)
	^{154}Gd	2.15	
	^{155}Gd	14.73	
	^{156}Gd	20.47	
	^{157}Gd	15.68	
	^{158}Gd	24.87	
	^{160}Gd	21.90	
$_{65}$Tb	^{159}Tb	100	CAMERON and WICHERS (1962)
$_{66}$Dy	^{156}Dy	0.052	HEATH (1964/65)
	^{158}Dy	0.090	
	^{160}Dy	2.29	
	^{161}Dy	18.88	
	^{162}Dy	25.53	
	^{163}Dy	24.97	
	^{164}Dy	28.18	
$_{67}$Ho	^{165}Ho	100	CAMERON and WICHERS (1962)
$_{68}$Er	^{162}Er	0.136	HEATH (1964/65)
	^{164}Er	1.56	
	^{166}Er	33.41	
	^{167}Er	22.94	
	^{168}Er	27.07	
	^{170}Er	14.88	
$_{69}$Tm	^{169}Tm	100	CAMERON and WICHERS (1962)
$_{70}$Yb	^{168}Yb	0.135	HEATH (1964/65)
	^{170}Yb	3.03	
	^{171}Yb	14.31	
	^{172}Yb	21.82	
	^{173}Yb	16.13	
	^{174}Yb	31.84	
	^{176}Yb	12.73	
$_{71}$Lu	^{175}Lu	97.41	CAMERON and WICHERS (1962)
	^{176}Lu [d]	2.59	

For data on the abundance of isotopes in nature, see also the Nuclear Data Sheets 1959—1965 and later.

(a) ^{138}La $\xrightarrow{EC}$ ^{138}Ba 70% $t_{1/2}$ 1.6×10^{11} y; ^{138}La $\xrightarrow{\beta^-}$ ^{138}Ce 30% — GLOVER and WATT (1957).

(b) ^{144}Nd $\xrightarrow{\alpha}$ ^{140}Ce $t_{1/2}$ 2.1×10^{15} y — Nuclear Data, sect. B, vol. 2 (1967)

(c) ^{147}Sm $\xrightarrow{\alpha}$ ^{143}Nd $t_{1/2}$ $1.15—1.28 \times 10^{11}$ y — BEARD and WIEDENBECK (1954), LESLIE (1956), BEARD and KELLEY (1958)

(d) ^{176}Lu $\xrightarrow{\beta^-}$ ^{176}Hf $t_{1/2}$ 2.2×10^{10} y — Nuclear Data Sheets (1959—1965)

EC = orbital electron capture.

Revised manuscript received: February 1969

39, 57—71-C. Abundance in Cosmos, Meteorites, Tektites and Lunar Specimens

I. Cosmos

Very little is known about the abundance of the lanthanides in the sun and other stars. ALLER (1961, 1962; quoted from HASKIN, FREY *et al.*, 1966; see also ALLER, 1967) determined the following ratios: 5.7 Y atoms/10^6 Si atoms and 1.1 Yb atoms/10^6 Si atoms in the solar atmosphere. Later the same author (ALLER, 1962) published higher abundance ratios: 50 Y atoms/10^6 Si atoms and 5.9 Yb atoms/10^6 Si atoms. Probably the high Y and both Yb ratios are dubious assessments. The high Y value is 14 times and the two Yb values are 6 and 33 times the corresponding Y and Yb ratios in 22 individual chondrites and a composite of 9 (Table 39, 57—71-C-1).

Table 39, 57—71-C-1. *Atomic abundances, relative to 10^6 Si atoms, of Y and several lanthanides in the sun, in 6 F-stars, and in the average of chondrites (average absolute concentrations of Y and La-Lu in chondrites, see Table 39, 57—71-C-3). The data for the chondrites were calculated using an average Si content of 17 wt.-% in these meteorites* (UREY and CRAIG, 1953; MASON, 1962). WALLERSTEIN's *data have been renormalized by* HASKIN, FREY *et al.* (1966) *to a Sc meteoritic value of 32 (atomic abundance relative to 10^6 Si atoms.)*

Element	Sun	Average of 6 F-stars	Average of 22 individual chondrites and a composite mixture of 9 chondrites	Reference for sun and stars	
$_{39}$Y	5.7	—	3.6	quoted from HASKIN, FREY *et al.* (1966)	ALLER (1961)
	50	—			ALLER (1962)
	6.4	3.2			WALLERSTEIN (1966 and unpublished)
$_{57}$La	2.6	1.3	0.38		
$_{58}$Ce	1.3	1.3	1.11		
$_{59}$Pr	(0.8) (possibly incorrect)	—	0.14		
$_{60}$Nd	2.1	(2.1)	0.69		
$_{62}$Sm	0.6	0.5	0.22		
$_{63}$Eu	(0.3) weak line	—	0.079		
$_{64}$Gd	(0.3) weak line	—	0.33		
$_{70}$Yb	1.1	—	0.18		ALLER (1961)
	5.9	—			ALLER (1962)

WALLERSTEIN (1966) measured the abundance of Y and La—Gd relative to Sc in the sun and in six F-stars (see also HASKIN, FREY *et al.*, 1966). Abundances of only the elements Y, Ce, Sm, and Gd in stars and the sun agree within a factor of 2 with meteoritic values. Because of the uncertainties and incompleteness mentioned, the best source of information on the cosmic abundance of the lanthanide elements is meteorites, especially chondrites.

II. Meteorites

The first analyses of lanthanides in iron and stone meteorites were obtained from X-ray spectroscopy by I. NODDACK (1935). These data demonstrated that Y, La—Lu are much more concentrated in chondrites than in iron meteorites, and that the abundance of neighboring even and odd atomic numbers corresponds to the Oddo-Harkins rule. Some twenty years later SCHMITT *et al.* (1963, 1964), SCHMITT and SMITH (1963), HASKIN and GEHL (1962), HASKIN, FREY *et al.* (1966) determined the abundances of the lanthanides in meteorites by neutron activation analysis. Later on PHILPOTTS *et al.* (1967a) and SHIMA and HONDA (1967) made mass spectrometric measurements.

Absolute concentrations of $\sum$ Y, La—Lu in 26 determinations on 22 individual chondrites and in a composite sample of 9 chondrites range from 2.7 to $\approx$9.9 ppm (Table 39, 57—71-C-2). The arithmetic mean for the Y, La—Lu concentrations in these specimens is given in Table 39, 57—71-C-3 (see also Fig. 39, 57-71-C-1).

HASKIN, FREY *et al.* (1966) observed negligible internal fractionation of the lanthanides among the five principal groups of chondrites including the carbonaceous chondrites which are thought to represent the closest approach to primitive matter. Atomic ratios of La/10^6 Si and Yb/10^6 Si indicate a descending order of the lanthanides relative to Si in carbonaceous, ordinary and enstatite chondrites.

The abundances of Y and La—Lu have been determined in 7 specimens of Ca-rich achondrites and 3 specimens of Ca-poor achondrites: SCHMITT *et al.* (1963, 1964), SCHMITT and SMITH (1963), PHILPOTTS *et al.* (1967a), see also Table 39, 57—71-C-2. The absolute concentrations of Y, La—Lu in 4 Ca-rich achondrites of the class Ap (eucritic of "basaltic" achondrites) are about 11 times higher than the corresponding means of 22 chondrites. No appreciable fractionation of the lighter and the heavier lanthanides has occurred in these Ap-achondrites relative to the chondrites (HASKIN, FREY *et al.*, 1966, Fig. 39, 57—71-C-2). The same authors discuss a potential differentiation of basaltic achondrites from chondrites in the light of chemical relations. In two specimens of Ca-rich achondrites of the class Ado (nakhlites), a slight enrichment of the lighter lanthanides can be observed (Fig. 39, 57—71-C-2). The absolute Y, La—Lu contents of these specimens are only about 4 times higher than those of the chondrites (Table 39, 57—71-C-2).

The relative lanthanide distribution of the only specimen of howardite is comparable to that of Ap-achondrites and chondrites (PHILPOTTS *et al.*, 1967a).

In Ca-poor achondrites the lighter lanthanides, especially Eu, are depleted relative to the chondrites (Fig. 39, 57—71-C-2). The absolute abundances of Y and La—Lu in one specimen are almost identical with chondrites whereas in two others they are lower (Table 39, 57—71-C-2). The Eu depletion probably indicates mineral fractionation under different redox conditions between Ca-poor achondrites and chondrites.

Table 39, 57—71-C-2. *Absolute abundances of Y and La—Lu in meteorites in ppm (dash: no value reported). Σ Y, La—Lu includes estimates for missing analytical values*

Group	Name	Class symbol	Method	$_{39}$Y	$_{57}$La	$_{58}$Ce	$_{59}$Pr	$_{60}$Nd	$_{62}$Sm	$_{63}$Eu	$_{64}$Gd	$_{65}$Tb	$_{66}$Dy	$_{67}$Ho	$_{68}$Er	$_{69}$Tm	$_{70}$Yb	$_{71}$Lu	Σ Y, La—Lu	Reference
Ca-poor achondrites	Norton County	Ae	N/R	2.09	0.21	0.81	0.11	0.63	0.22	0.022	0.38	0.061	0.40	0.100	0.25	0.036	0.22	0.039	5.6	1
	Johnstown	Ab	N/R	1.22	0.044	—	—	—	0.080	0.0089a	—	—	0.14	0.036	0.14	0.021	0.15	0.033	≈2.6	1
	Shalka	Ab	N/R	0.22	0.015	—	0.0062	—	0.0103	0.0040	—	—	—	0.0046	0.021	0.0037	0.024	0.0058	≈0.43	1
Ca-rich achondrites	Lafayette	Ado	N/R	4.4	1.76	5.48	0.80	3.35	0.85	0.24	0.92	0.120	0.89	0.146	0.44	0.057	0.32	0.051	19.8	2, 3
	Nakhla	Ado	N/R	3.17	1.57	6.19	0.67	3.18	0.73	0.20	0.94	0.109	—	0.140	0.34	0.047	0.23	0.044	18.3	2, 3
	Bununu	Aor	I	—	—	4.02	—	3.31	1.064	0.357	1.289	—	1.558	0.92	—	0.984b	—	—	—	4
	Pasamonte	Ap	N/R	—	3.21	8.09	1.26	5.10	1.90	0.68	2.69	—	3.06	0.69	1.65	0.30	1.67	0.50	51.1	1
	Nuevo Laredo	Ap	N/R	—	4.03	10.7	1.47	8.0	2.19	0.75c	2.46	0.59	4.14	0.84	2.87	0.48	2.34	0.30	66.4	1
	Juvinas	Ap	N/R	17.1	2.53	7.25	0.98	5.01	1.74	0.62	2.55	0.40	3.03	0.59	1.78	0.28	1.37	0.231	45.5	3
	Stannern	Ap	N/R	28	4.89	12.9	1.96	10.0	3.17	0.83	4.47	0.71	4.89	1.07	3.05	0.47	2.37	0.38	79.2	3
Chondrites	Abee	Ce_1	N/R	1.02	0.15	0.48	0.054	0.24	0.095	0.050	0.16	0.025	0.16	0.040	0.131	0.014	0.094	0.019	2.7	1
	Abee, 4—6 samples	Ce_1	N/R, I	—	0.33	0.71	—	0.57	0.18	0.058	0.31	—	0.31	—	0.18	—	0.20	0.023	≈5.7	5
	Indarch	Ce_1	N/R	1.46	0.21	0.59	0.11	0.37	0.14	0.042	0.21	0.031	0.18	0.053	0.15	0.023	0.123	0.023	3.7	1
	St. Marks	Ce_1	N/R	—	0.25	0.66	0.13	0.36	0.14	0.045	0.21	0.036	0.21	0.046	0.14	0.024	0.135	0.027	3.8	1
	St. Marks	Ce_1	N/R	1.74	—	0.79	0.094	0.478	0.157	0.049	0.179	0.041	—	0.060	0.172	0.025	0.166	0.029	≈4.3	6
	Forest City	CH	N/R	2.15	0.33	—	0.12	0.62	0.22	0.088d	0.40	0.057	—	0.068e	0.25	0.053	0.19	—	≈9.9	7
	Richardton	CH	N/R	—	0.32	0.48	0.12	0.61	0.20	0.080f	0.34	0.053	0.34	0.068	0.21	0.033	0.19	0.033	5.1	8, 1
	Richardton, Troilite phase		N/R	—	0.74g	—	0.058	—	0.080	—	—	—	—	0.034	—	—	0.052	0.010	—	3
	Allegan	CH	N/R	—	0.33	0.55	0.12	0.65	0.24	0.087	0.34	0.049	0.39	0.082	0.22	0.043	0.20	0.038	5.8	8, 1
	Miller	CH	N/R	2.09	0.29	0.88	0.13	0.52	0.20	0.066	0.29	0.052	0.26	0.074	0.22	0.027	0.170	0.030	5.3	1
	Holbrook	CL	N/R	—	0.45	2.12	0.16	0.79	0.28	0.093h	0.39	0.068	0.42	0.103	0.31	0.043	0.233	0.037	8.6	1
	Kyushu	CL	N/R	1.97	0.30	0.92	0.13	0.57	0.22	0.000	0.36	0.044	0.26	0.068	0.22	0.030	0.209	0.035	5.1	1
	Bruderheim 2—3 samples	CL	N/R, I	—	0.61	1.3	—	0.78	0.27	0.090	0.36	0.064	0.37	—	—	—	0.24	0.036	≈6.9	5
	Ramsdorf	CL	N/R	2.60	0.332	1.19	0.130	—	0.198	0.074	0.299	0.072	—	0.089	0.267	0.044	0.281	0.043	≈6.2	6
	Modoc	CL	N/R, I	2.07	0.34	1.51	0.13	0.63	0.20	0.080i	0.33	0.048	0.30	0.084	0.23	0.033	0.150	0.036	6.2	3
	Modoc	CL		—	0.374	0.835	—	0.702	0.216	0.070	0.277	—	0.329	—	0.208	—	0.222	0.035	—	9
	Manbhoom	CLL	N/R	1.97	0.31	0.91	0.123	0.65	0.244	0.079	0.26	0.059	0.35	0.089	0.24	0.036	0.20	0.033	5.6	3
	Vavilovka	CLL	N/R	1.9	0.25	0.86	0.075	0.76	0.190	0.067	0.31	0.046	0.32	0.066	0.166	0.032	0.16	0.029	4.5	10
	Felix	CHL	N/R	2.42	0.34	1.08	0.14	0.81	0.28	0.101	0.40	0.061	—	0.096	0.28	0.037	0.195	0.037	6.8	1
	Mokoia	CHL	N/R	2.44	0.46	1.06	0.16	0.84	0.31	0.103	0.40	0.073	0.42	0.098	0.26	0.047	0.246	0.038	7.0	1
	Grosnaja	CHL	N/R	2.4	0.38	1.24	0.13	0.80	0.27	0.097	0.46	0.057	0.39	0.086	0.24	0.038	0.22	0.034	6.8	10
	Murray	Cc_2	N/R	—	0.39	1.05	0.15	0.62	0.21	0.072	0.28	0.049	0.32	0.079	0.23	0.037	0.184	0.030	6.0	1
	Mighei	Cc_2	N/R	1.77	0.29	1.24	0.13	0.61	0.20	0.078	0.38	0.047	0.36	0.076	0.21	0.030	0.17	0.032	5.6	3
	Ivuna	Cc_2	N/R	1.69	0.19	0.59	0.090	0.40	0.134	0.055	0.24	0.066	—	0.058	0.149	0.022	0.113	0.024	4.1	3
	Orgueil	Cc_2	N/R	1.44	0.19	0.67	0.097	0.44	0.131	0.050	—	0.022	0.22	0.054	0.132	0.022	0.16	0.022	3.8	3

Table 39, 57—71-C-2 (Continued)

Group	Name	Class symbol	Method	$_{39}$Y	$_{57}$La	$_{58}$Ce	$_{59}$Pr	$_{60}$Nd	$_{62}$Sm	$_{63}$Eu	$_{64}$Gd	$_{65}$Tb	$_{66}$Dy	$_{67}$Ho	$_{68}$Er	$_{69}$Tm	$_{70}$Yb	$_{71}$Lu	Σ Y, La—Lu	Reference
Chondrites	Boriskino	Cc_2	N/R	2.1	0.25	0.94	0.094	0.57	0.19	0.072	0.30	0.041	—	0.062	0.26	0.031	0.18	0.027	5.4	10
Chondrites	Composite of 9 chondrites	—	N/R	1.96	0.330	0.88	0.112	0.60	0.181	0.069	0.249	0.047	—	0.070	0.200	0.030	0.200	0.034	5.3	6
Stony-Irons	Brenham (Olivine phase)	P	N/R	1.57	—	≦0.17	0.032	0.16	0.112	0.018	—	0.038	—	0.051	0.094	0.0096	0.043	0.0059	≈2.8	3
Stony-Irons	Brenham (Olivine phase)	P	I	—	0.0177	0.0293	—	0.0104	0.0016	—	0.0012	—	0.0009	—	0.0007	—	0.0026	0.0008	—	9
Stony-Irons	Thiel Mts. (Olivine phase)	P	N/R	—	0.12[k]	0.28	0.022	0.12	0.022	0.0031	0.020	—	0.016	0.0034	—	0.0010	0.0086	0.0017	≈0.73	3
Stony-Irons	Estherville (Silicate phase, non magnetic portion)	M	N/R	5.6	1.12	2.98	0.37	2.08	0.68	0.22	1.10	0.14	0.97	0.23	0.63	0.096	0.50	0.083	16.8	1
Stony-Irons	Veramin (Silicate phase)	M	N/R	2.47	0.29	0.87	0.114	0.57	0.23	0.118	0.44	0.054	0.37	0.097	0.26	0.041	0.21	0.044	6.2	3
Irons	Yardymly (Aroos)	Og	N/R	—	4.1×10^{-4}	—	1.9×10^{-4}	—	79×10^{-6}	27×10^{-6}	—	—	1.5×10^{-4}	$<4.2\times10^{-4}$	—	—	—	—	—	1
Irons	Odessa	Og	N/R	—	—	—	—	—	36×10^{-6}	12×10^{-6}	—	—	—	$<10\times10^{-6}$	—	—	—	—	—	1

[a] Bate *et al.* (1960) report 0.0098 ppm Eu. [b] Doubtful value (Herrmann). [c] Bate *et al.* (1960) report 0.76 ppm Eu. [d] Bate *et al.* (1960) report 0.073 ppm Eu. [e] Corrected after Haskin, Frey *et al.* (1966). [f] Bate *et al.* (1960) report 0.081 ppm Eu. [g] The high La value may possibly be due to contamination (Schmitt *et al.*, 1964). [h] Bate *et al.* (1960) report 0.081 ppm Eu. [i] Bate *et al.* (1960) report 0.080 ppm Eu. [k] The La value for another sample in Thiel Mts. was 0.022 ppm (Schmitt *et al.*, 1964).

Reference: 1. Schmitt *et al.* (1963); 2. Schmitt and Smith (1963); 3. Schmitt *et al.* (1964); 4. Philpotts *et al.* (1967a); 5. Shima and Honda (1967); 6. Haskin *et al.* (1968); 7. Haskin and Gehl (1962); 8. Schmitt *et al.* (1960); 9. Masuda (1968); 10. Schmitt *et al.*, quoted from Haskin, Frey *et al.* (1966);

Table 39, 57—71-C-3. *Absolute concentrations from Y and La—Lu from 26 determinations on 22 individual chondrites and a composite mixture of 9 chondrites. Data for chondrites were taken from Table 39, 57—71-C-2*

Element	ppm
$_{39}$Y	1.96
$_{57}$La	0.32
$_{58}$Ce	0.94
$_{59}$Pr	0.12
$_{60}$Nd	0.60
$_{62}$Sm	0.20
$_{63}$Eu	0.073
$_{64}$Gd	0.31
$_{65}$Tb	0.050
$_{66}$Dy	0.31
$_{67}$Ho	0.073
$_{68}$Er	0.21
$_{69}$Tm	0.033
$_{70}$Yb	0.19
$_{71}$Lu	0.031
Σ Y, La—Lu	5.42
$\frac{\text{ppm La}}{\text{ppm Nd}}$	0.53
$\frac{\text{ppm Pr}}{\text{ppm Sm}}$	0.60
$\frac{\Sigma \text{ ppm La—Eu}}{\Sigma \text{ ppm Gd—Lu, Y}}$	0.71

The absolute concentrations of Y and La—Lu in olivines of two pallasites (class P) are lower than in the chondrites (SCHMITT *et al.*, 1964; MASUDA, 1968b; Table 39, 57—71-C-2), because olivine allows only a minimum substitution of Y and La—Lu (see Section 39, 57—71-D). The relative enrichment of the lighter lanthanides in the olivine phase of the pallasites "Thiel" (SCHMITT *et al.*, 1964) and "Brenham" (MASUDA, 1968b) resembles the fractionation of the terrestrial lanthanide distribution (Fig. 39,57—71-C-3). On the other hand the olivine phase of the pallasite "Brenham" analysed by SCHMITT *et al.* (1964) shows higher absolute concentrations of Y and La—Lu and a relative enrichment of the heavier lanthanides. It seems likely that one of the patterns obtained for "Brenham" does not represent a genuine olivine phase. No significant lanthanide fractionation was observed in the silicate phase of two M stone-iron meteorites (Fig. 39, 57—71-C-3).

The absolute lanthanide concentrations observed in two iron meteorites are approximately four orders of magnitude lower than in chondrites (SCHMITT *et al.*, 1963; Table 39, 57—71-C-2). They can probably be attributed to non-iron inclusions present in the iron meteorites (HASKIN, FREY *et al.*, 1966).

A detailed theoretical interpretation of the abundances of the isotopes of Y and La—Lu relative to Si in meteoritic matter with respect to nucleosynthesis and stability is given by HASKIN, FREY *et al.* (1966).

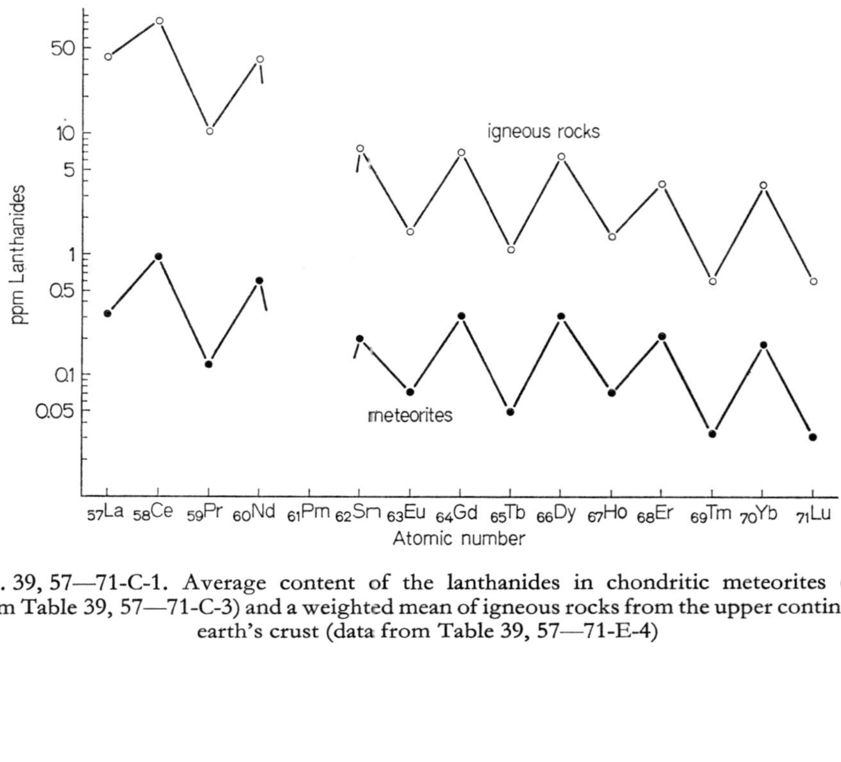

Fig. 39, 57—71-C-1. Average content of the lanthanides in chondritic meteorites (data from Table 39, 57—71-C-3) and a weighted mean of igneous rocks from the upper continental earth's crust (data from Table 39, 57—71-E-4)

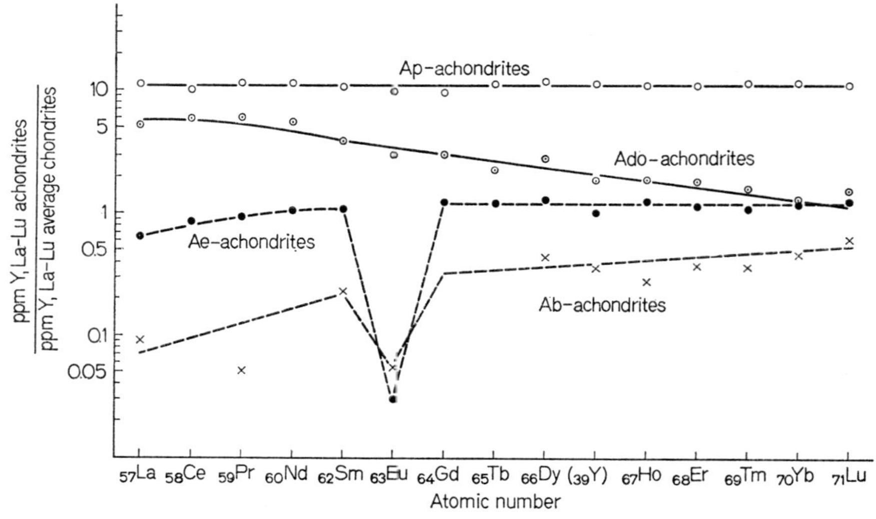

Fig. 39, 57—71-C-2. Abundances of Y and La—Lu in 4 Ap and 2 Ado Ca-rich achondrites and 1 Ae and 2 Ab Ca-poor achondrites relative to chondrites. Data from Schmitt *et al.* (1963, 1964); see also Table 39, 57—71-C-2

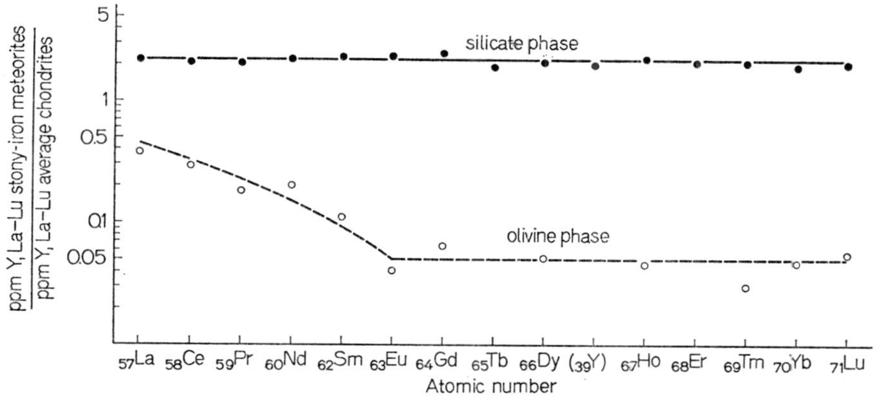

Fig. 39, 57—71-C-3. Abundances of Y and La—Lu in the silicate phase of two M stony-iron meteorites and olivine phase of the P stony-iron meteorite "Thiel" relative to chondrites. Data from SCHMITT *et al.* (1963, 1964); see also Table 39, 57—71-C-2

III. Tektites

The lanthanide abundance in one bediasite, one philippinite, one australite, one indochinite, one moldavite, one tektite from the Ivory Coast, and impact glasses from the Bosumtwi crater in Ghana were determined by HASKIN and GEHL (1963a), CHASE, SCHNETZLER *et al.* (1963) and SCHNETZLER *et al.* (1967), see Table 39, 57—71-C-4. In contrast to the meteorites, the Y, La—Lu concentrations of the tektites, for which nearly complete analyses are available, are similar to the average of sedimentary rock types and consequently to the upper continental earth's crust (see also Fig. 39, 57—71-C-4, and Table 39, 57—71-K-2). SCHNETZLER *et al.* (1967) reports very similar absolute and relative lanthanide abundances in impact glasses from the Bosumtwi crater and in a sample of graywacke occuring close to this crater.

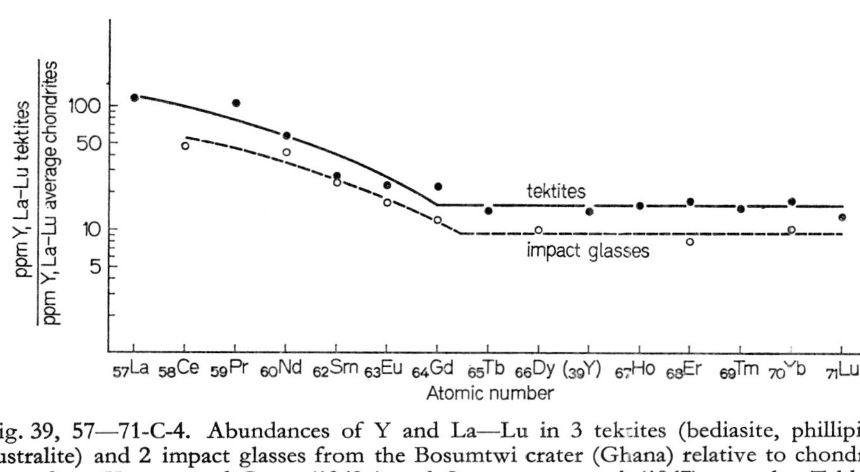

Fig. 39, 57—71-C-4. Abundances of Y and La—Lu in 3 tektites (bediasite, phillipinite, australite) and 2 impact glasses from the Bosumtwi crater (Ghana) relative to chondrites. Data from HASKIN and GEHL (1963a) and SCHNETZLER *et al.* (1967); see also Table 39, 57—71-C-4

Table 39, 57—71-C-4. *Absolute abundances of Y and La—Lu in tektites in ppm (dash: no value reported). Σ Y, La—Lu includes estimates for missing analytical values*

Type and locality	Method	$_{39}$Y	$_{57}$La	$_{58}$Ce	$_{59}$Pr	$_{60}$Nd	$_{62}$Sm	$_{63}$Eu	$_{64}$Gd	$_{65}$Tb	$_{66}$Dy	$_{67}$Ho	$_{68}$Er	$_{69}$Tm	$_{70}$Yb	$_{71}$Lu	Σ Y, La—Lu	Reference
Bediasite	N/R	31	42	—	14.8	35	4.4	2.3	8.2	0.61	—	1.36	3.6	0.55	4.01	0.47	≈230	HASKIN and GEHL (1963a)
Philippinite	N/R	31	33	—	10.2	31	6.85	1.4	5.6	0.95	—	0.94	3.6	0.57	3.4	—	≈200	HASKIN and GEHL (1963a)
Australite	N/R	27	38	—	12.6	34.5	5.2	1.5	7.7	0.59	—	1.15	4.0	0.40	2.56	0.36[a]	≈210	HASKIN and GEHL (1963a)
Indochinite, Kouang-Tcheou-Wan	N/R	—	47	—	—	—	—	1.5	—	—	5.3	—	—	—	—	—	—	CHASE, SCHNETZLER *et al.* (1963)
Moldavite, Radomilice ČSSR (Bohemia)	N/R	—	29	—	—	—	—	0.83	—	—	2.8	—	—	—	—	—	—	CHASE, SCHNETZLER *et al.* (1963)
Ivory Coast tektite, 2 km south of Adzopé (GSFC No. 4)	I	—	—	49.5	—	25.5	4.83	1.14	3.68	—	2.95	—	1.54	—	1.72	—	—	SCHNETZLER *et al.* (1967)
Composite. Ivory Coast tektites (GSFC No. 49), 10 samples	I	—	—	45.2	—	24.9	4.59	1.13	3.57	—	2.76	—	1.36	—	1.68	—	—	SCHNETZLER *et al.* (1967)
Bosumtwi Crater, impact glass (GSFC No. 10), Ghana	I	—	—	47.9	—	26.7	4.97	1.22	—	—	2.90	—	—	—	1.51	—	—	SCHNETZLER *et al.* (1967)
Bosumtwi Crater, impact glass (GSFC No. 158), Ghana	I	—	—	44.7	—	25.0	4.90	1.20	3.81	—	3.42	—	1.77	—	1.94	—	—	SCHNETZLER *et al.* (1967)

[a] Corrected by HASKIN, FREY *et al.* (1966).

IV. Lunar Specimens

Y and La—Lu determinations of the lunar samples from Apollo 11 (around Tranquillity Base) were made by the lunar sample preliminary examination team (1969)[1], Goles *et al.* (1970), Haskin *et al.* (1970), Philpotts and Schnetzler (1970), and Schmitt *et al.* (1970); see also table 39, 57—71-C-5. The lanthanides are enriched by one and two orders of magnitude, as compared to the chondritic abundances of these elements. The calculated absolute concentrations of ΣY, La—Lu increase from ≈195 ppm in the type B lunar rocks (medium grained igneous rocks) to ≈310—400 ppm in the types D, C and A (lunar "soil", "breccias"; "basalts") (Haskin *et al.*, 1970). Schmitt *et al.* (1970) found in two chips from the type B rocks (10050 and 10017) 279 and 514 ppm ΣY, La—Lu, respectively.

Relative lanthanide abundance patterns resemble those of chondrites, calcium-rich achondrites, and some oceanic basalts. The difference between lunar samples and terrestrial continental rocks, where high total lanthanide contents are associated with relative accumulation of light lanthanides, is striking. Whole-rock lunar specimens are abnormally depleted in Eu (Fig. 39, 57—71-C-5).

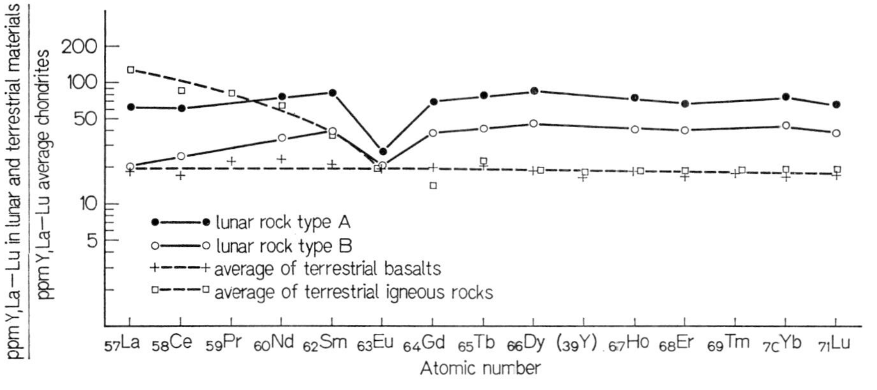

Fig. 39,57-71-C-5. Abundances of Y and La—Lu in lunar rocks type A and B (Haskin *et al.*, 1970) and in average basalts and igneous rocks of the upper earth crust, relative to chondrites. Data from Table 39,57-71-C-5 and 39,57-71-E-4

Plagioclase, pyroxenes, and opaques were separated from the type B rock (samples number 10044 E, F, G) by Goles *et al.* (1970). The plagioclase is enriched in Eu, and both the pyroxenes and opaques are Eu depleted. The bulk of the rest of the lanthanides is preferentially concentrated in the pyroxenes and in opaques.

[1] The lunar sample preliminary examination team: Preliminary examination of lunar samples from Apollo 11. Science **165**, 1211—1227 (1969).

Table 39,57-71-C-5. *Absolute abundances of Y and La—Lu in lunar materials from Apollo 11 in ppm (dash: no value reported).* Σ Y, *La—Lu includes estimates for missing analytical values*

Sample number	Type	Method	$_{39}$Y	$_{57}$La	$_{58}$Ce	$_{59}$Pr	$_{60}$Nd	$_{62}$Sm	$_{63}$Eu	$_{64}$Gd	$_{65}$Tb	$_{66}$Dy	$_{67}$Ho	$_{68}$Er	$_{69}$Tm	$_{70}$Yb	$_{71}$Lu	Σ Y, La—Lu	Reference
22, 72, 57, 20	Type A rocks ("basalts")	S	185—310	—	—	—	—	—	—	—	—	—	—	—	—	2.5—7	—	—	1
Average 10022, 10071 A, 10071 B		N/R	—	23	79	—	—	18.4	2.1	—	4.2	—	—	—	—	16.4	2.5	—	2
Average 10003, 10020, 10022, 10057, 10072		N/R	—	20	56	—	52	16.6	1.96	22	4.0	27.5	4.6—6.4	14.4	—	15.0	2.11	≈400	3
17, 58, 45, 50	Type B rocks (medium grained igneous rocks)	S	100—310	—	—	—	—	—	—	—	—	—	—	—	—	1.3—5	—	—	1
Average 10017, 10045, 10044 A		N/R	—	14.5	49	—	—	15.2	2.1	—	3.2	—	—	—	—	13.8	2.0	—	2
10045		N/R	—	6.7	22.5	—	21	7.9	1.52	12.3	2.11	14.4	2.2—3.5	8.7	—	8.6	1.17	≈195	3
21, 61	Type C rocks (breccias)	S	115—300	—	—	—	—	—	—	—	—	—	—	—	—	1.8—4.5	—	—	1
Average 10018, 10019, 10021, 10060 A, 10060 B		N/R	—	17.1	55	—	—	14	1.93	—	2.8	—	—	—	—	12.1	1.7	—	2
Average 10048, 10060		N/R	—	20.5	57	—	43	15	2.0	22	3.6	25.7	4.3—5.3	15	—	12.9	2.1	≈375	3
37	Type D (fine material from the lunar "soil")	S	130	—	—	—	—	—	—	—	—	—	—	—	—	2.5	—	—	1
Average 10084 A, 10084 B		N/R	—	14.4	52	—	—	12.6	1.78	—	2.6	—	—	—	—	11.1	1.58	—	2
10084		N/R	—	16.6	47	—	43	13.7	1.77	15	3.2	20	4.2—4.7	12	—	11.3	1.66	≈310	3
10044 E	Plagioclase from Type B rocks	N/R	—	4.4	14	—	—	5.75	6.15	—	1.1	—	—	—	—	2.6	0.35	—	2
10044 F	Pyroxenes from Type B rocks	N/R	—	2.52	24	—	—	7.60	0.57	—	—	—	—	—	—	10.1	1.47	—	2
10044 G	Opaques from Type B rocks	N/R	—	4.85	21	—	—	7.3	0.32	—	0.8	—	—	—	—	11.3	1.60	—	2

Reference: 1. The lunar sample preliminary examination team: Preliminary examination of lunar samples from Apollo 11. Science *165*, 1211—1227 (1969); 2. Goles *et al.* (1970); 3. Haskin *et al.* (1970).

Revised manuscript received: February 1969

39, 57—71-D. Minerals of Yttrium and the Lanthanides (I); Abundance in Rock-Forming Minerals (II)

I.

The mineralogy of yttrium and the lanthanides is characterized by more effective selectivity in the separation of individual elements or element groups than is known from reactions in wet analytical chemistry. Similarities among elements 57—71 are a consequence of the addition of electrons to the inner 4*f* orbitals rather than the outer shells. The predominance of the trivalent state (with the possible exception of: Ce^{4+}, Eu^{2+} [Sm^{2+}, Yb^{2+}]) and the contraction of ion size with increasing atomic number are two major consequences of the specific shell structure. The lanthanide contraction is the origin of gradual change in the crystal chemical behavior of the elements, forming at least two major groups which have long been distinguished as yttrium-dominant (heavy lanthanides) and cerium-dominant (lighter lanthanides) minerals. The fact that separate lanthanide minerals are formed from rather low concentrations of lanthanides in natural melts indicates the high "stability" of the latter and the resistance of the lanthanides against being camouflaged to a higher rate by major elements.

Systematic studies of the abundance and distribution of Y and La—Lu in minerals were started by GOLDSCHMIDT and THOMASSEN (1924). Further information about the composition of lanthanide minerals in more than a thousand specimens have been published by MURATA *et al.* (1953, 1957, 1959), VAINSHTEIN *et al.* (1956), SEMENOV (1957, 1958, 1963), SEMENOV and BARINSKII (1958), BORODIN (1960, 1961), FRONDELL (1964), IVANTIŠIN *et al.* (1964), CHASE, WINCHESTER, and CORYELL (1963), ZAYATS and KUTS (1964), FLEISCHER (1965, 1966), NEUMANN *et al.* (1966), HASKIN, FREY *et al.* (1966), chapter Y and Lanthanides in VLASOV (1966), JENSEN (1967a, 1967b), LEE and BASTRON (1967), SVERDRUP (1968) etc.

At present nearly 200 minerals containing > 0.01 wt.-% lanthanides are known. Table 39, 57—71-D-1 lists important minerals of Y and the lanthanides. For a more complete compilation see SEMENOV (1963).

The highest concentrations of Σ Y, La—Lu (in brackets: Σ $(Y, La—Lu)_2O_3$) can be found in the economically important minerals: bastnaesite with amounts up to 64 wt.-% ($\approx$75 wt.-%), monazite with 60 wt.-% ($\approx$70 wt.-%), and cerite with 59 wt.-% (69 wt.-%).

The distribution of Y and La—Lu in minerals is variable. GOLDSCHMIDT and THOMASSEN (1924), MURATA *et al.* (1953, 1957), SEMENOV and BARINSKII (1958), BORODIN (1960), and SEMENOV (1963) divided these minerals into several groups containing preferentially either the lighter or the heavier lanthanides or having compositions with relatively uniform proportions of both. SEMENOV and BARINSKII (1958) and SEMENOV (1963) distinguish between selective lanthanum minerals like davidite, selective cerium minerals like tynsonite, loparite, burbankite, parisite, monazite, cerite, and rhabdophane, selective cerium-neodymium minerals like

Table 39, 57—71-D-1. *Minerals of Y and the lanthanides, the most common species are printed in italics. The minerals listed follow a selection by* SEMENOV (1963). *In general the formulae are taken from* STRUNZ (1966), *some from* SEMENOV (1963), *and the chapter Y and Lanthanides in* VLASOV (1966)

Mineral	Formula
	Fluorides
Yttrofluorite	$(Ca, Y)F_{2-2.33}$
Gagarinite	$NaCaYF_6$
Tynsonite (fluocerite)	$(Ce, La)F_3$
	Oxides
Loparite	$(Na, Ce, Ca)TiO_3$
Davidite	$(Fe^{\cdot\cdot}, Fe^{\cdot\cdot\cdot}, U, Ce, La)_2(Ti, Fe^{\cdot\cdot\cdot}, Cr, V)_5O_{12}$
Marignacite	$(Ce, Ca)_2(Nb, Ti)_2O_7$
Obruchevite	$(Y, Na, Ca)_2(Nb, Ti, Ta)_2O_6(O, OH, F)$
Euxenite	$(Y, Er, Ce, U, Pb, Ca)(Nb, Ta, Ti)_2(O, OH)_6$
Samarskite	$(Y, Er, Fe, Mn, Ca, U, Th, Zr)(Nb, Ta)_2(O, OH)_6$
Kobeite[a]	$YFeZrTi_3O_{11}$
Yttrocrasite	Y, Ti_2O_5OH
Aeschynite	$(Ce, Th, Ca, \ldots)[(Ti, Nb, Ta)_2O_6]$
Fergusonite	$YNbO_4$
Cerianite	$(Ce, Th)O_2$
	Carbonates
Sahamalite	$(Ce, La, Nd)_2(Mg, Fe)[CO_3]_4$
Burbankite	$Na_2(Ca, Sr, Ba, Ce, La)_4[CO_3]_5$
Bastnaesite	$Ce[F \mid CO_3]$
Parisite	$CaCe_2[F_2 \mid (CO_3)_3]$
Roentgenite	$Ca_2Ce_3[F_3 \mid (CO_3)_5]$
Synchisite	$CaCe[F \mid (CO_3)_2]$
Huanghoite	$BaCe[F \mid (CO_3)_2]$
Cordylite	$Ba(Ce, La, Nd)_2[F_2 \mid (CO_3)_3]$
Calkinsite	$(Ce, La,\ldots)_2[CO_3]_3 \cdot 4H_2O$
Lanthanite	$(La, Dy, Ce)_2[CO_3]_3 \cdot 8H_2O$
Tengerite	$Y_2(CO_3)_3 \cdot nH_2O$
Ancylite	$Sr_3(Ce, La, Dy)_4[(OH)_4 \mid (CO_3)_7] \cdot 3H_2O$
	Borates
Braitschite	$7(Ca, Na_2)O \cdot RE_2O_3 \cdot 11B_2O_3 \cdot 7H_2O$ (RAUP *et al.* 1968)
	Sulfates
Chuchrovite	$Y, Ca_3Al_2(SO_4)F_{13} \cdot 10H_2O$
	Phosphates
Xenotime	$Y[PO]_4$
Monazite	$Ce[PO_4]$
Rhabdophane	$Ce[PO_4] \cdot {}_{0-0.5}H_2O$
Florencite	$CeAl_3[(OH)_6 \mid (PO_4)_2]$
Belovite	$CeNaSr_3[OH \mid (PO_4)_3]$
Britholite	$(Na, Ce, Ca)_5[F \mid (SiO_4, PO_4)_3]$
Weinschenkite	$(Y, Er)[PO_4] \cdot 2H_2O$

Table 39, 57—71-D-1 (Continued)

Mineral		Formula
		Silicates
Keilhauite		$(Ca, Y, Ce)(Ti, Al, Fe^{\cdots})[O \mid SiO_4]$
Steenstrupine		$Na_2Ce(Mn, Ta, Fe,\ldots)H_2[(Si, P)O_4]_3$
Cerite		$(Ca, Mg)_2(Ce)_8[SiO_4]_7 \cdot 3H_2O$
Toernebohmite		$(Ce, La, Al)_3[OH \mid (SiO_4)_3]$
Spencite		$(Y, Ca, Ce, La, Th, Al\ldots)_{2.8-3.8}(Si, B)_{3.4-3.9}(O, OH, F)_{13}$
Gadolinite		$Y_2FeBe_2[O \mid SiO_4]_2$
Melanocerite	by chance	$Na_4Ca_{16}(Y, La)_3(Zr, Ce)_6[F_{12} \mid (BO_3)_3 \mid (SiO_4)_{12}]$
Thalenite		$Y_2[Si_2O_7]$
Hellandite		$(Ca, Y, Er, Mn)_{<3}(Al, Fe)[(OH)_2 \mid Si_2O_7]$
Mosandrite (rinkolite)		$(Ca, Na, Y)_3(Ti, Zr, Ce)[(F, OH, O)_2 \mid Si_2O_7]$
Allanite (orthite)		$(Ca, Ce)_2(Fe^{\cdot\cdot}, Fe^{\cdots})Al_2[O \mid OH \mid SiO_4 \mid Si_2O_7]$
Tschevkinite		$(Ce, La)_2Ti_2[O_4 \mid Si_2O_7]$(?)
Nordite	by chance	$Na_4(Ce, La, \ldots)_{1.5}(Sr, Mn, Ca, Mg)_3Si_8O_{23}$
Cappelenite		$(Ba, Ca, Ce, Na)_3(Y, Ce, La)_6[(BO_3)_6 \mid Si_3O_9]$
Stillwellite		$(Ce, La)_3[B_3O_6 \mid Si_3O_9]$
Kainosite		$Ca_2Y_2[CO_3 \mid Si_4O_{12}] \cdot H_2O$
Tundrite		$Ce_2Ti(Si, P)O_7 \cdot 4H_2O$
Carnasurtite[b]		$(La, Ce, Th)(Ti, Nb)(Al, Fe)[(OH)_4 \mid (Si, P)_2O_7] \cdot 3H_2O$
Wudjawrite[c]		$Ce_4(Ti_2O_6)_3 \cdot n SiO_2 \cdot m H_2O$

[a] Probably a metamict hydration product of polycrase, Strunz (1966).

[b] Perhaps a mixture with lessingite-britholite or a La-rich variety of rhabdophane (Strunz, 1966).

[c] Amorphous yellow conversion product of "lovtschorrite", Kola Peninsula. The latter is a metamict rinkite (Strunz, 1966).

aeschynite, selective gadolinium-yttrium minerals like samarskite, selective yttrium minerals like euxenite, xenotime and fergusonite, and complex minerals with several peaks in the distribution curves like allanite and gadolinite (see also Fig. 39, 57—71-D-1). If the ratio between maximum and minimum concentrations of even lanthanides (including Y and La) exceeds 50, the composition may be referred to as selective, if it is lower, as complex (Semenov, 1963).

A comparison of the distribution of Y and La—Lu in a number of minerals of different origin shows that in many geochemical processes the lanthanides undergo differentiation. For instance, Lyakhovich and Barinskii (1961), and Lyakhovich (1962, 1967) have shown that there are notable changes in the concentrations of Y, La—Lu and the relative distribution of the lanthanides in accessory minerals in the sequence from granites to hydrothermal veins formed during the emplacement of granitic massifs and also from unaltered and altered granitic rocks. The already mentioned distinction of two groups containing the "more basic" (lighter) and "less basic" (heavier) lanthanides (Borodin, 1960, Borodin and Barinskii, 1960) is one of the results of differentiation. Fleischer (1965) pointed out that any such grouping is rather arbitrary and that most properties show gradual rather than abrupt changes among these groups.

The distribution of lanthanides is assumed to be influenced by the geological environment and the crystal structures of the minerals. Fleischer (1965) has shown

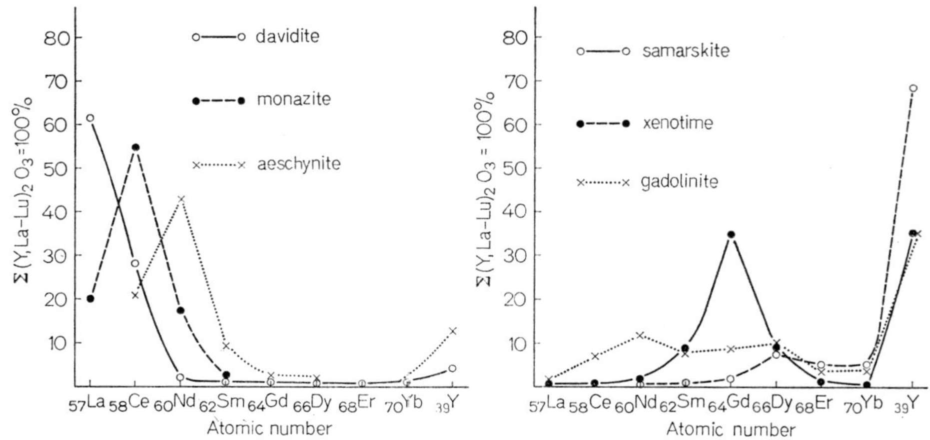

Fig. 39, 57—71-D-1. Distribution of Y and La—Lu in some lanthanide minerals proper (Semenov, 1963). Davidite (selective lanthanum mineral), monazite (selective cerium mineral), aeschynite (selective cerium-neodymium mineral), samarskite (selective gadolinium-yttrium mineral), xenotime (selective yttrium mineral), gadolinite (complex mineral)

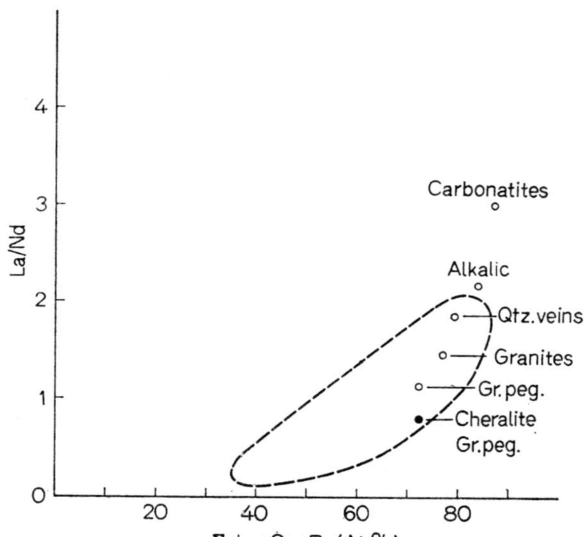

Fig. 39, 57—71-D-2. Plot of $\sum$ La+Ce+Pr against La/Nd of monazites from different parageneses on the basis of more than 200 analyses. Quoted from Fleischer (1965)

that monazites from granitic pegmatites are more enriched in heavier lanthanides as compared to those from granites, while monazites from alkalic rocks and carbonatites contain more lighter lanthanides (Fig. 39, 57—71-D-2). He also mentions that any discussion of ratios of lanthanide pairs in minerals must be restricted to samples of similar paragenesis.

II.

At present only a few data are known as to the abundance and distribution of Y and La—Lu in the individual co-existing rock-forming minerals from different rocks (Table 39, 57—71-D-2). The absolute concentration and the relative proportions of lanthanides in these minerals vary widely. For example, the absolute content of Y and La—Lu in pyroxenes of basalts and gabbros is higher than that in plagioclases of the same rocks. The lanthanide contents are very low in the olivines of basaltic rocks (SCHNETZLER and PHILPOTTS, 1968). Biotites from granites contain higher amounts of lanthanides than the co-existing feldspars and quartz. The abundances of lathanides in accessory minerals from different rocks may be some orders of magnitude higher than those in the main rock-forming minerals, but the lanthanide pattern of a rock is not always controlled by the distribution of these elements in the accessory minerals (apatite in basaltic rocks, for instance). A special study of the distribution of the lanthanides in pyroxenes and pyroxenites from peridotitic and alkaline rocks as from carbonatite massifs was published by BORODIN (1965).

A marked enrichment of lighter lanthanides relative to the abundance in chondrites was observed in feldspars from basalts (SCHNETZLER and PHILPOTTS, 1968; HERRMANN, 1968), from one gabbro (TOWELL, WINCHESTER and VOLFOVSKY SPIRN, 1965), one norite (felsic minerals; FREY *et al.*, 1968), and from one granite (TOWELL, WINCHESTER, and VOLFOVSKY SPIRN, 1965). The same happens in biotites from one granite, and in apatites from one gabbro. Contrary to this a relative increase of heavier lanthanides has been reported in the pyroxene and hornblende of the San Marcos gabbro (TOWELL, WINCHESTER, and VOLFOVSKY SPIRN, 1965; Fig. 39, 57—71-D-3). In other rocks pyroxene can be characterized by a relative accumulation of lighter lanthanides: pyroxenes from some eclogites (BAKUN-CZUBAROW, 1965; FREY and HASKIN, quoted from HASKIN, FREY *et al.*, 1966; PHILPOTTS *et al.*, 1967b), and diopsidic phenocrysts from the Recent Stromboli, Italy (HERRMANN, 1968). However, unlike the pyroxene, the garnet in eclogites shows a relative enrichment of heavier lanthanides (FREY and HASKIN, quoted from HASKIN, FREY *et al.*, 1966; Fig. 39, 57—71-D-4). MASUDA (1967b) reported a very remarkable linear relationship between the lighter lanthanides in garnet and those in pyroxene of eclogite. The different lanthanide pattern of the major minerals from potential mantle rocks, as well as that of basaltic phenocrysts, may influence the lanthanide abundances of basaltic melts formed either from partial melting or from differentiation (HERRMANN, 1968; etc.).

For some feldspar specimens an anomalously high Eu content is remarkable, while Eu is relatively depleted in the co-existing pyroxene and hornblende of the San Marcos gabbro and in the biotite of the Rubidoux leucogranite (TOWELL, WINCHESTER, and VOLFOVSKY SPIRN, 1965; Fig. 39, 57—71-D-3).

For the understanding of the distribution of Y and La—Lu in minerals and rocks and of their behavior during magmatic differentiation etc., the analytical determination of partition coefficients between co-existing minerals from different rocks (KHOMYAKOV, 1967), and phenocrysts and residual melts (SCHNETZLER and PHILPOTTS, 1968; ONUMA *et al.*, 1968) is very important. The latter authors have published data on amphibole, clinopyroxene, orthopyroxene, olivine, and plagioclase phenocrysts and their respective basaltic matrix and the results are given in Table 39, 57—71-D-3.

Table 39, 57—71-D-2. *Absolute abundances of Y and La—Lu in rock-forming minerals in ppm (dash: no value reported). Σ Y, La—Lu includes estimates for missing analytical values*

Mineral	Source	Method	$_{39}$Y	$_{57}$La	$_{58}$Ce	$_{59}$Pr	$_{60}$Nd	$_{62}$Sm	$_{63}$Eu	$_{64}$Gd	$_{65}$Tb	$_{66}$Dy	$_{67}$Ho	$_{68}$Er	$_{69}$Tm	$_{70}$Yb	$_{71}$Lu	Σ Y, La—Lu	Reference
Basalt matrix	Ankaramite-oceanite, from Fantana Creek, Tahiti (GSFC No. 25). Large-zoned phenocrysts of olivine (20%) and augite (15%) make up about 35% of the rock	I	—	—	87	—	54	11.6	3.1	—	—	7	—	2.8	—	2.2	—	—	1
Olivine		I	—	—	0.8	—	0.36	0.07	0.02	—	—	0.06	—	0.026	—	—	—	—	1
Intermediate augite		I	—	—	15.4	—	17	5	1.46	5	—	3.9	—	1.5	—	1.06	—	—	1
Core augite		I	—	—	12	—	12.5	3.8	1.12	4	—	2.9	—	1.1	—	0.8	—	—	1
Rim augite		I	—	—	48	—	23.5	5.9	1.7	5	—	3.5	—	1.4	—	1.06	—	—	1
Basalt matrix	Comptonite, from Canada cement quarry, Montreal, Canada, with large phenocrysts of amphibole (GSFC No. 29)	I	—	—	173	—	98	16.6	4.6	12	—	8	—	3.7	—	2.84	—	—	1
Amphibole		I	—	—	20	—	16	4	1.2	—	—	2.5	—	0.9	—	0.66	—	—	1
Basalt matrix	Nodule, consisting of an amphibole phenocryst and altered "limburgite" trap-rock. Deborah Volcanic Formation, Kakanui, New Zealand (GSFC No. 45)	I	—	—	70	—	47	9.1	2.6	6.8	—	4.5	—	1.65	—	1.24	—	—	1
Amphibole		I	—	—	18.2	—	18.7	5.1	1.63	4.5	—	2.8	—	0.78	—	0.36	—	—	1
Basalt matrix	Ankaramite-oceanite, from route 19, near Poopoo Creek, west flank of Mauna Kea, Hawaii (GSFC) No. HHP-66-19)	I	—	—	—	—	35	8.1	2.5	7.3	—	5.7	—	2.7	—	2.1	—	—	1
Plagioclase An_{79}		I	—	—	4.9	—	4	0.8	0.85	—	—	0.51	—	—	—	—	—	—	1
Augite		I	—	—	7	—	8.9	3.1	0.97	3.4	—	2.9	—	1.23	—	0.88	—	—	1
Basalt matrix	Alkaline olivine basalt or hawaiite from Takashima, North Kyushu, Japan	N/R	—	33.8	66.1	—	30.4	6.4	2.15	6.1	0.86	—	—	—	0.36	2.33	0.33	—	2
Augite		N/R	—	2.85	11.0	—	11.6	5.0	1.62	5.0	0.83	—	—	—	0.40	2.34	0.32	—	2
Bronzite		N/R	—	—	0.17	—	—	0.09	0.05	—	0.04	—	—	—	—	0.25	0.037	—	2
Plagioclase (An_{55}), 59.4%	San Marcos Gabbro. Batholith of southern California (for whole rock see Table 39, 57—71-E-3)	N/R	1.1	2.3	3.6	—	—	0.26	0.81	—	—	0.17	0.05	—	0.022	0.15	0.023	≈ 9.5	3
Hornblende, 17.2%		N/R	73	2.4	—	—	—	9.0	2.1	—	1.8	13	2.5	—	1.2	7.7	1.07	≈130	3

Augite, 6.9%	San Marcos Gabbro. Batholith of southern California (for whole rock see Table 93, 57—71-E-3	N/R	44	0.41	—	—	—	4.4	0.92	—	1.3	7.8	1.5	—	0.69	4.0	0.74	≈85	3
Apatite 0.29%		N/R	327	392	—	172	534	97	17	—	12	67	14	—	3.3	20	3.1	≈2,600	3
Felsic minerals	Norite, Bushveld (for whole rock see Table 39, 57—71-E-3)	N/R	0.69	6.8	9.1	1.1	4.9	0.59	1.18	0.48	0.09	—	0.05	0.15	0.02	0.05	0.035	≈26	4
Mafic minerals (ortho- and clinopyroxene)		N/R	7.7	6.1	5.3	0.6	—	0.92	0.14	1.07	0.35	—	0.29	0.95	0.17	2.1	0.34	≈29	4
Diopsidic pyroxene	Phenocrysts from the Recent Stromboli, Italy	X, S	41	14	36	6	31	7.2	2.2	9	1	6.1	—	4.4	0.6	3.2	0.6	≈163	5
Feldspar	Syenite pegmatite, Wausau	N/R	16	34	—	10.1	36	7	2.9	6.4	1.04	—	0.84	1.6	—	0.98	0.17	≈180	6
Arfvedsonite		N/R	170	270	—	109	380	127	21	88	—	—	14.6	40	6.0	35	6.8	≈1,800	6
Plagioclase (An_{20}), 30.3%	Rubidoux Mountain Leucogranite. Batholith of southern California (for whole rock see Table 39, 57—71-E 5)	N/R	8.5	10.7	—	—	—	1.7	0.97	—	—	1.7	0.45	—	—	0.90	0.13	≈60	3
Potassium feldspar, 27.9%		N/R	4.7	9.7	—	—	—	1.1	0.94	—	—	0.74	0.17	—	0.08	0.51	0.07	—	3
Biotite, 3.03%		N/R	226	225	—	68	222	48	0.36	—	7.3	43	9.6	—	3.2	22	2.7	≈1,300	3
Feldspar and quartz	Granite, Kirovograd district (Ukraine, USSR)	X	—	23	50	4.8	19	3.4	—	—	—	—	—	—	—	—	—	>100[a]	7
Biotite and chlorite		X	—	125	244	31	121	26	—	19	—	11	—	—	—	—	—	>600[a]	7
Garnet		X	27	—	14	—	10	4.4	—	27	—	73	13	52	—	25	4	>250[a]	7
Apatite		X	1,440	275	1,000	198	902	350	—	450	115	400	98	143	—	—	—	>5,400[a]	7
Monazite		X	—	133,000	230,000	22,000	88,000	15,000	—	12,000	—	—	—	—	—	—	—	>500,000[a]	7
Felsic portion	Gneiss, Neillsville	N/R	10.5	52	—	9.3	23	3.6	0.96	2.5	0.25	—	0.35	1.06	0.16	0.82	0.17	≈150	6
Mafic portion		N/R	12.4	83	—	14.7	52	6.4	0.73	—	0.31	—	0.57	1.17	0.28	1.01	0.17	≈280	6

Table 39, 57—71-D-2 (Continued)

Mineral	Source	Meth-od	$_{39}$Y	$_{57}$La	$_{58}$Ce	$_{59}$Pr	$_{60}$Nd	$_{62}$Sm	$_{63}$Eu	$_{64}$Gd	$_{65}$Tb	$_{66}$Dy	$_{67}$Ho	$_{68}$Er	$_{69}$Tm	$_{70}$Yb	$_{71}$Lu	Σ Y, La—Lu	Refer-ence
Garnet	Eclogite, large inclusion in dunite, Shikoku, Japan (for whole rock see Table 39, 57—71-M-1)	N/R	20	0.82	2.2	0.20	1.5	1.7	1.04	3.3	0.57	—	0.85	2.6	0.36	3.0	0.49	≈43	8
Pyroxene		N/R	0.54	5.1	12.6	1.6	7.2	1.7	0.50	0.74	0.065	—	0.024	0.051	0.0056	0.041	0.012	≈30	8
Garnet (G_1)	Eclogite (G_1) Nowa Wieś, Poland. Intercalations in Gierattów gneisses (for whole rock see Table 39, 57—71-M-1)	W	see Tb	0.25	0.40	0.08	0.77	0.43	Eu+Gd 1.74		Y+Tb 15.7	2.35	0.26	0.44	0.09	Yb+Lu 0.26		23	9
Omphacite (O_1)		W	see Tb	0.25	0.89	0.08	0.95	0.09	Eu+Gd 0.09		Y+Tb+Dy 0.16		—	—	—	Yb+Lu —		≈3	9
Garnet (G_2)	Eclogite (G_2), Nowa Wieś, Poland. Intercalations in Gierattów gneisses (for whole rock see Table 39, 57—71-M-1)	W	see Tb	0.20	0.39	0.08	0.20	0.41	Eu+Gd 0.73		Y+Tb 4.24	0.84	—	0.31	—	Yb+Lu —		≈8	9
Omphacite (O_2)		W	see Tb	1.45	3.08	0.50	2.15	0.52	Eu+Gd 0.35		Y+Tb+Dy 0.36		—	—	—	Yb+Lu —		≈9	9

[a] For this table the relative values of the original paper have been converted into ppm.

Reference: 1. Schnetzler and Philpotts (1968); 2. Onuma *et al.* (1968); 3. Towell, Winchester and Volfoysky Spirn (1965); 4. Frey *et al.* (1968); 5. Herrmann (1968); 6. Gehl, quoted from Haskin, Frey *et al.* (1966); 7. Gavrilova and Turanskaya (1958); 8. Frey and Haskin, quoted from Haskin, Frey *et al.* (1966); 9. Bakun-Czubarow (1965).

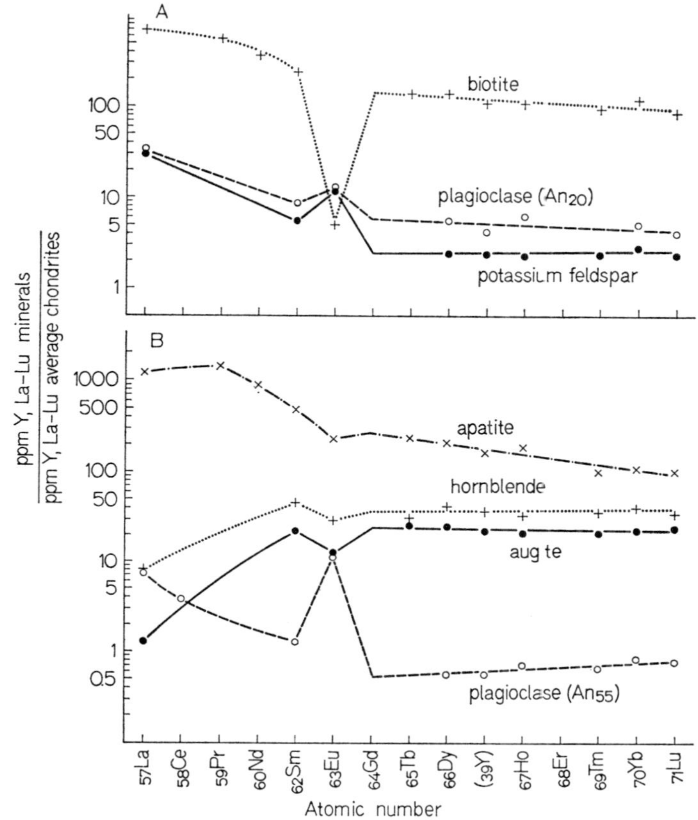

Fig. 39, 57—71-D-3. Abundances of Y and La—Lu in minerals from the Rubidoux Mt. leucogranite (A) and the San Marcos gabbro (B) relative to chondrites. Data from Towell, Winchester and Volfovsky Spirn (1965); see also Table 39, 57—71-D-2

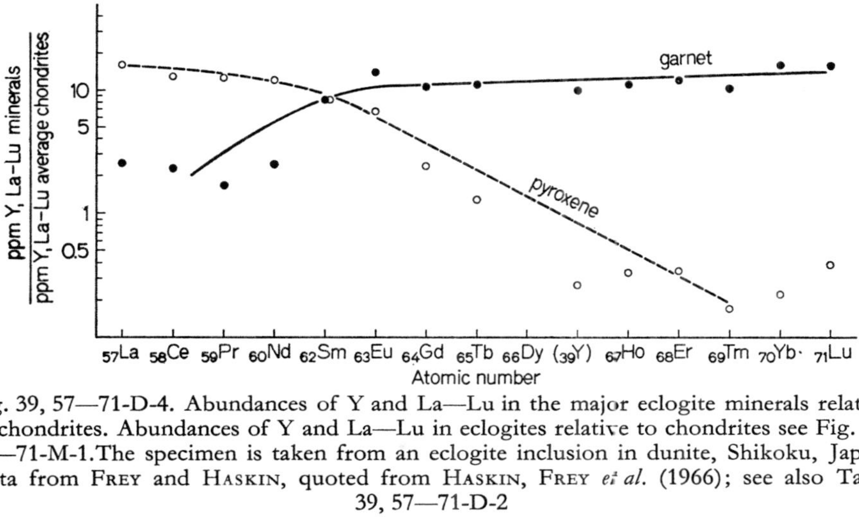

Fig. 39, 57—71-D-4. Abundances of Y and La—Lu in the major eclogite minerals relative to chondrites. Abundances of Y and La—Lu in eclogites relative to chondrites see Fig. 39, 57—71-M-1. The specimen is taken from an eclogite inclusion in dunite, Shikoku, Japan. Data from Frey and Haskin, quoted from Haskin, Frey *et al.* (1966); see also Table 39, 57—71-D-2

Table 39, 57—71-D-3. *Partition coefficients:* $\frac{\text{mineral}}{\text{rock matrix}}$

Schnetzler and Philpotts (1968)[1,2]

Mineral	Ce	Nd	Sm	Eu	Gd	Dy	Er	Yb
Amphibole	0.12 to 0.26	0.16 to 0.40	0.24 to 0.56	0.26 to 0.63	0.28 to 0.66	0.31 to 0.62	0.24 to 0.47	0.23 to 0.30
Clinopyroxene	0.18	0.26 to 0.32	0.38 to 0.43	0.39 to 0.48	0.46 to 0.57	0.50 to 0.56	0.46 to 0.53	0.43 to 0.48
Olivine	0.009	0.007	0.006	0.006	—	0.009	0.009	—
Plagioclase	—	0.11	0.10	0.35	0.09	0.09	—	—

Onuma *et al.* (1968)

Mineral	La	Ce	Nd	Sm	Eu	Gd	Tb	Tm	Yb	Lu
Clinopyroxene	0.084	0.166	0.382	0.736	0.753	0.82	0.97	1.09	1.01	0.95
Orthopyroxene	—	0.0026	—	0.014	0.023	—	0.046	—	0.107	0.111

[1] Schnetzler, C. C., and J. A. Philpotts: Partition coefficients of rare-earth elements between igneous matrix material and rock-forming mineral phenocrysts — II. Geochim. Cosmochim. Acta *34*, 331—340 (1970).

[2] Masuda, A., and I. Kushiro: Experimental determination of partition coefficients of ten rare earth elements and barium between clinopyroxene and liquid in the synthetic silicate system at 20 kilobar pressure. Contr. Mineral. and Petrol. *26*, 42—49 (1970).

Revised manuscript received: February 1969

39, 57—71-E. Abundance in Common Igneous Rock Types

I. Peridotites

BALASHOV and TURANSKAYA (1962), SCHMITT *et al.* (1963), SOBOLEV (1965), FREY and HASKIN, quoted from HASKIN, FREY *et al.* (1966), HERRMANN and WEDEPOHL (1966), and NAGASAWA *et al.* (1969), have studied the abundance of Y and La—Lu in some peridotites of different origin. The samples included peridotites from the Urals, one peridotite nodule from a kimberlite, mylonitized peridotites from St. Paul's Island, 3 peridotites with "high temperature" contacts, and 6 lherzolite inclusions from alkali olivine basalts and nephelinite tuffs in N.W. Germany and Hawaii, respectively (Table 39, 57—71-E-1). The total content of the lanthanides in some peridotites is of the same order of magnitude as that observed in chondrites, but in other samples the absolute Y, La—Lu concentrations are 3—10 times (≈ 15 to ≈ 47 ppm Σ Y, La—Lu) higher than those of chondrites. The absolute concentrations of these elements are apparently a function of the olivine to pyroxene ratio of the rocks. As stated in Section 39, 57—71-D, olivines are very low in the lanthanides whereas diopsidic pyroxenes are usually very much higher in them than olivines (and orthopyroxenes). A high olivine content of more than 50% is characteristic for peridotites. A comparable olivine content has been assumed for rocks forming the upper mantle of the earth (see Chapter 6 in Volume I of this handbook). It is probable that the absolute lanthanide contents of the mantle rocks are similar to those observed in peridotites and some eclogites (Section 39, 57—71-M), consequently ranging from about 5 to about 50 ppm Σ Y, La—Lu.

Of the specimens listed in Table 39, 57—71-E-1, only the peridotite of the Polar Urals, the sample H-2 from St. Paul's Island and the lherzolite inclusion from the alkali olivine basalt in N.W. Germany have a relative lanthanide pattern similar to that of chondrites. The "high temperature" peridotites are relatively depleted and some specimens of St. Paul's Island are relatively enriched in lighter lanthanides as compared to chondrites (Fig. 39, 57—71-E-1). No definite conclusions regarding the relative lanthanide distribution pattern of the upper mantle can be reached until more samples have been analyzed.

The absolute lanthanide contents of three South African kimberlites and some specimens from kimberlite pipes in the Yakutian A.S.S.R. (FREY and HASKIN, quoted from HASKIN, FREY *et al.*, 1966; BURKOV and PODPORINA, 1966) are about one order of magnitude higher than those of the analyzed peridotite samples. These kimberlites have relative lanthanide patterns strongly dominated by light lanthanides whereas a peridotite inclusion in kimberlite is much lower in these elements (Table 39, 57—71-E-1, Fig. 39, 57—71-E-1).

Table 39, 57—71-E-1. *Absolute abundances of Y and La—Lu in peridotites and kimberlite rocks in ppm (dash: no value reported). Σ Y, La—Lu includes estimates for missing analytical values*

Rock	Locality	Method	$_{39}$Y	$_{57}$La	$_{58}$Ce	$_{59}$Pr	$_{60}$Nd	$_{62}$Sm	$_{63}$Eu	$_{64}$Gd	$_{65}$Tb	$_{66}$Dy	$_{67}$Ho	$_{68}$Er	$_{69}$Tm	$_{70}$Yb	$_{71}$Lu	Σ Y, La—Lu	Reference
Peridotites	Polar Urals. Region of Mt. Slyudyanaya, southern part of the Marun-Keu Range	X	4.8	1.8	3.3	0.60	2.4	0.88	—	0.90	0.19	0.95	0.24	0.47	—	0.57	—	≈ 18	1
	St. Paul's Island. Mylonitized peridotite																		
	PH—i	N/R	5.0	4.5	9.3	1.76	6.2	1.11	0.32	1.16	0.31	—	0.18	0.53	0.068	0.50	0.072	≈ 32	2
	Pl—i	N/R	5.0	4.6	18.7	2.6	8.0	1.7	0.58	2.5	0.33	—	0.23	0.70	0.086	0.57	0.047	≈ 47	2
	H—2	N/R	1.24	0.58	0.57	0.078	0.31	0.095	0.53	0.19	0.047	—	0.045	0.13	0.027	0.16	0.022	≈ 4.2	2
	"High temperature" intrusions																		
	Tinaquillo, Venezuela	N/R	2.1	0.0041	0.107	0.017	0.20	0.113	0.068	0.27	0.079	—	0.095	0.27	0.049	0.45	0.031	≈ 4.4	2
	Lizard, Cornwall	N/R	1.21	0.0046	0.059	0.00111	0.042	0.033	0.017	0.071	0.081	—	0.052	0.17	0.030	0.27	0.021	≈ 2.6	2
	Mt. Albert, Quebec	N/R	0.54	0.0067	0.054	0.0031	0.042	0.0090	0.0059	0.047	0.038	—	0.022	0.081	0.0149	0.17	0.0148	≈ 1.1	2
	Peridotite nodule from Kimberlite of the Wesselton mine in southern Africa	N/R	—	3.3	8.0	1.02	3.1	0.57	0.16	0.65	0.088	0.59	0.14	0.36	0.053	0.43	0.064	≈ 23	3
	Lherzolite inclusions from alkaline olivine basalt. Northern Hessia (Hirzstein), Germany. 2 samples	X, S	≈1.8	0.9	3.6	0.3	2.2	0.4	—	—	—	—	—	—	—	—	—	≈ 15	4
Kimberlite rocks	Premier, southern Africa	N/R	4.9	26	—	9.0	24	7.0	2.0	5.2	0.58	—	0.67	0.87	—	0.96	—	≈140	2
	Mwadui, southern Africa	N/R	7.3	26	72	6.6	40	5.6	1.39	8.9	—	—	—	—	0.22	1.18	0.19	≈176	2
	Dutoitspan, southern Africa	N/R	12.9	200	510	36	134	11.6	0.28	5.3	—	—	0.63	1.38	0.22	1.20	0.16	≈920	2
	Kimberlite pipes in the Yakut ASSR																		
	Mir. Kimberlite breccias (6 samples)	X	—	39	102	11	33	5	0.9	4	0.5	1.6	—	1.2	—	1.1	—	≈212[a]	5
	Zarnitsa, Kimberlite breccias (8 samples)	X	—	40	94	9	33	4.3	0.3	2.8	0.3	1.5	0.3	0.8	—	1	—	≈203[a]	5
	Udachnaya-western Kimberlite breccias (3 samples)	X	—	67	158	19	65	6.4	0.9	4.2	0.5	1.7	0.5	—	—	2.0	—	≈340[a]	5
	Udachnaya-eastern basaltoid Kimberlites (7 samples)	X	—	48	119	12	46	6.9	0.6	3.7	0.2	1.7	0.2	0.3	—	1.1	—	≈246[a]	5

[a] For this table the relative values of the original paper have been converted into ppm.

Reference: 1. Balashov and Turanskaya (1962); 2. Frey and Haskin, quoted from Haskin, Frey *et al.* (1966); 3. Schmitt *et al.* (1963); 4. Herrmann and Wedepohl (1966); 5. Burkov and Podporina (1966).

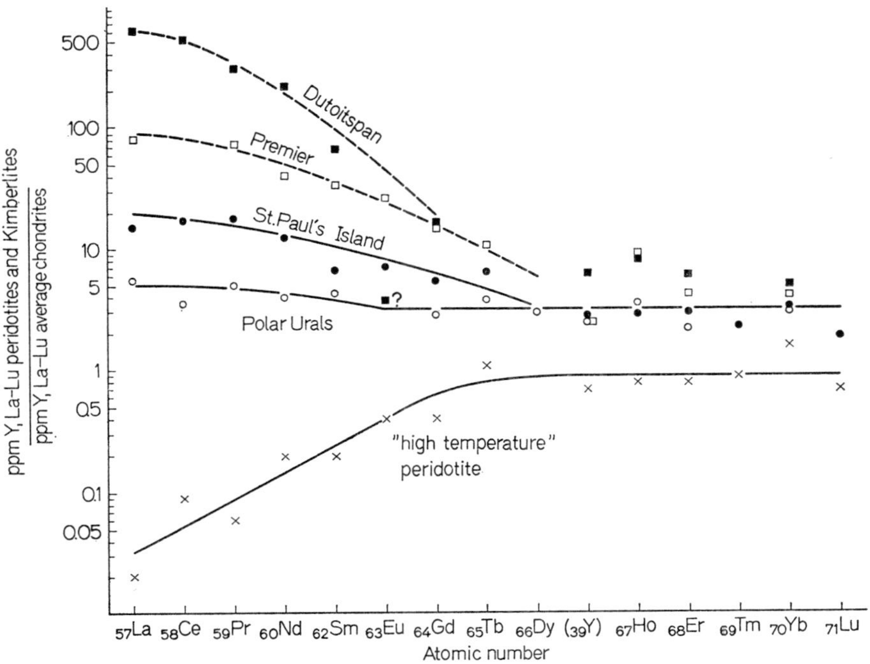

Fig. 39, 57—71-E-1. Abundances of Y and La—Lu in several peridotites (2 specimens of peridotites PH-i and Pl-i from St. Paul's island; one specimen from the Polar Urals; 3 specimens of high-temperature peridotite intrusions) and kimberlite rocks (Dutoitspan and Premier, South Africa) relative to chondrites. Data from BALASHOV and TURANSKAYA (1962); FREY and HASKIN, quoted from HASKIN, FREY *et al.* (1966); see also Table 39, 57—71-E-1

II. Basalts

The results of lanthanide analyses on different basalt types are listed in Table 39, 57—71-E-2a and 2b. In terms of their relative lanthanide patterns, these basalt types can be clearly divided into two distinct groups: 1. oceanic and continental tholeiitic basalts (including the coarser-grained dolerites and diabases in American nomenclature), and 2. alkali basalts. Ten samples of oceanic tholeiitic basalts from the mid-Atlantic ridge and the east Pacific rise have relative lanthanide distribution patterns nearly the same as those of chondrites and Ca-rich achondrites. In these samples and in tholeiitic basalts from Japan the chondritic pattern is disturbed only by a slight depletion of La, Ce, and Pr (Fig. 39, 57—71-E-2). But the average concentration of the total lanthanides in oceanic tholeiitic basalts ($\approx$96 ppm Σ Y, La—Lu) is 20 times that of chondrites (FREY and HASKIN, 1964; FREY *et al.*, 1968; MASUDA, 1968a). Recently GRAHAM and NICHOLLS (1969) have published different total contents of the lanthanides in tholeiitic basalts from different parts of the Atlantic Ocean floor (for instance, 8.3 and 45 ppm Σ Y, La—Lu).The continental tholeiitic basalts from the Siberian platform, from the Deccan plateau (BALASHOV, 1965; BALASHOV and NESTERENKO, 1966), from the large Triassic sill in the USA (HASKIN and GEHL, 1963b; TOWELL, VOLFOVSKY, and WINCHESTER, 1965; ALEKSIEV and BOYADJIEVA,

Table 39, 57—71-E-2a. *Absolute abundances of Y and La—Lu in basalts in ppm (dash: no value reported). Σ Y, La—Lu includes estimates for missing analytical values*

Rock	Locality	Method	$_{39}$Y	$_{57}$La	$_{58}$Ce	$_{59}$Pr	$_{60}$Nd	$_{62}$Sm	$_{63}$Eu	$_{64}$Gd	$_{65}$Tb	$_{66}$Dy	$_{67}$Ho	$_{68}$Er	$_{69}$Tm	$_{70}$Yb	$_{71}$Lu	Σ Y, La—Lu	Reference
Tholeiitic basalts from oceanic ridges and rises	Mid-Atlantic Ridge. Station 7 (GE-159)	N/R	34	4.5	18	2.1	13.2	4.4	1.8	6.5	1.25	—	1.70	4.7	0.79	4.0	—	≈ 97	1
	Mid-Atlantic Ridge. Station 8 (GE-160)	N/R	47	3.0	18	2.5	13.5	4.8	2.1	8.4	1.27	—	2.10	5.8	0.83	4.0	0.92	≈114	1
	Mid-Atlantic Ridge. Station 20 (GE-260). Altered rock	N/R	34	5.0	—	2.4	14.0	4.4	1.9	7.3	1.70	—	1.40	—	1.18	4.2	0.79	≈100	1
	Mid-Atlantic Ridge. D-2-1	N/R	33	5.0	11.8	2.1	9.7	4.3	1.7	5.5	1.19	—	1.50	3.9	0.56	3.0	0.52	≈ 84	2
	Mid-Atlantic Ridge. D-3	N/R	47	6.2	—	2.7	20	6.6	1.9	7.1	1.47	—	1.8	5.3	0.76	4.5	—	≈117	2
	Mid-Atlantic Ridge. D-4	N/R	28	2.8	9.2	1.7	8.7	3.2	1.2	4.9	0.93	—	1.15	3.4	0.48	3.0	0.56	≈ 69	2
	East Pacific Rise. D-1	N/R	15.8	1.91	5.6	0.96	5.6	1.97	0.83	2.6	0.50	—	0.68	1.76	0.25	1.5	0.25	≈ 40	2
	East Pacific Rise. D-7 (Mendocino)	N/R	50	4.7	13	4.0	19	5.6	2.1	10.4	1.6	—	2.1	6.1	1.0	5.2	—	≈126	2
	East Pacific Rise. Molokai	N/R	33.6	2.84	8.6	1.83	9.9	3.5	1.3	5.0	1.0	—	1.3	3.7	0.51	3.4	0.56	≈ 77	2
	East Pacific Rise. Guadelupe, Experimental Mohole basalt	N/R	30	1.9	14	1.9	8.6	2.8	1.35	4.6	0.83	—	1.06	3.3	0.41	2.8	0.57	≈ 74	1
Tholeiitic basalts from oceanic islands and continents	Kilauea Iki-22, Hawaii	N/R	—	10.5	35	3.9	17.8	4.2	1.31	4.7	0.66	3.0	0.64	1.69	0.21	1.11	0.20	≈104	3
	Kohala Mt., Hawaii	N/R	24	9.7	22	4.3	18.8	5.2	2.1	6.2	1.2	—	0.96	2.8	0.39	2.3	0.35	≈110	2
	Columbia Plateau	N/R	53	25.3	66	7.9	37	10.0	2.3	10.2	1.66	9.6	1.94	5.43	0.75	3.9	0.64	≈235	4
	Deccan Plateau, India	N/R	52	45	143	11.7	42	8.2	2.3	8.1	1.08	—	1.46	4.3	0.65	4.4	0.59	≈325	2
	Deccan Plateau, India, 2 samples	X	24	21.5	40	5.5	27.5	6.5	2.2	6.0	1.3	5.5	1.4	2.4	—	2.3	0.4	≈147	5
	Siberian Platform, Noril'sk region. Permian-Triassic volcanic cycles; 1,268 m in thickness. 163 samples	X	23	13	29	4.3	16.5	4.2	1.3	4.7	1.3	4.3	1.3	2.5	—	2.0	0.4a	≈108	5
	Siberian Platform. Western, central, and north-eastern part of the platform. Permian-Triassic age. 81 samples. Rocks quoted as diabases	X	30	6.9	18.8	3.1	12.3	4.2	1.5	5.0	1.2	4.9	1.4	2.9	—	2.5	0.7	≈ 96	5
	Averages of vertical sections of differentiated intrusives of the Siberian Platform. Rocks quoted as diabases																		
	Chernogorka sill (Noril'sk region) 9 samples	X	28	10.5	23	3.2	15.2	5.0	2.0	5.2	1.1	5.7	1.4	2.9	—	2.5	0.4	≈107	5
	Padunsk sill (Angara river) 7 samples	X	35	11	25	4.1	19	6.4	2.2	5.7	1.4	6.7	1.5	3.9	—	3.4	0.55	≈126	5
	Alamdzhakh sill (Vilui river basin) 8 samples	X	30	9.9	23.2	3.7	17.2	4.8	1.5	5.8	1.3	5.2	1.2	3.2	—	3.0	0.5	≈111	5
	Gottsbüren-Gieselwerder; Sababurg; Kottenberg in Northern Hessia, N.W. Germany. Tertiary volcanism. 3 samples	X, S	19	8.8	27	3.6	19.1	5.3	1.1	5.4	0.9	3.4	1.0	2.1	0.5	1.8	0.5	100	6
	Rocks quoted as dolerites																		
	Ferrar, Antarctica	I	—	—	25.8	—	15.0	3.5	0.98	—	—	3.96	—	2.36	—	2.35	—	—	7

Tholeiitic basalts from oceanic islands and continents	Red Hill dike, Tasmania	I	—	—	25.4	—	14.6	3.4	0.83	3.4	—	3.7	—	2.22	—	2.33	—	—	7
	Karroo, South Africa	I	—	—	19.6	—	13.1	3.4	1.05	—	—	4.1	—	2.38	—	2.34	—	—	7
	5 km S.S.W. of Atsumi, southwest coast of Yamagata Prefecture (Miocene). Rock quoted as olivine dolerite	I	—	21.7—26.2	35.7	—	16.0	4.5	—	4.4	—	4	—	2.36	—	2.05	—	—	8
	Tanna Basin; Ô-shima; Hachijô-jima; Central Japan (Tertiary volcanism). 5 samples	I	—	3.3	8.0	—	9.6	3.4	1.1	4.6	—	5.6	—	3.4	—	3.5	0.53	—	9
	Bull Run Quarry, on U.S. Highway 211, S.W. Centerville, Fairfax County, Va., U.S.A. (Triassic). Rock quoted as diabase W-1	N/R	23.8	11.7	24.3	3.68	15.1	3.8	1.09	4.2	0.75	—	1.35	2.57	0.36	2.1	0.33	≈101	10
	W-1	N/R	28	9.31	15.1	3.51	20.2	3.46	1.3	—	0.81	4.4	0.86	—	0.33	2.23	0.35	≈ 97	11
	W-1	X	21	11.2	20.0	3.5	13.3	3.8	0.9	4.5	2.0	3.8	1.6	1.7	0.5	3.0	0.5	91.3	12
Spilitic rocks	Sauerland region, N.W. Germany. 3 samples. Middle Devonian	X, S	22	11.6	31	4	19	4.6	—	6	—	4.8	—	3.3	—	2.8	—	≈113	13
Transitional tholeiitic basalts from oceanic ridges and from the continent	Mid-Atlantic Ridge (4519-65)	N/R	26	8.8	17.8	2.8	14.4	3.4	1.14	4.3	1.00	—	1.03	3.1	0.44	2.9	0.49	≈ 92	2
	Mid-Atlantic Ridge (4519-56)	N/R	19.3	13.6	25	3.4	14.6	3.6	1.20	4.1	0.84	—	0.87	2.4	0.33	2.2	0.34	≈ 96	2
	Bühl, near Weimar. Northern Hessia, N.W. Germany	X, S	19	32	55	6.5	24	4.1	1.2	4.3	0.7	3.4	0.8	2.2	0.5	1.7	0.5	156	6
	Composite of 282 continental basalts	N/R	27.3	32	59	7.8	34	6.6	1.87	6.2	1.0	—	1.05	2.9	0.42	2.6	0.42	≈189	2

[a] Corrected after personal communication with the authors.

Reference: 1. Frey and Haskin (1964); Frey *et al.* (1968); 2. Frey *et al.* (1968); 3. Schmitt *et al.* (1963); 4. Schmitt *et al.* (1964); 5. Balashov and Nesterenko (1966); 6. Herrmann (1968); 7. Philpotts and Schnetzler (1968a); 8. Masuda (1966); 9. Masuda (1968a); 10. Haskin and Gehl (1963b); 11. Towell, Volfosky, and Winchester (1965); 12. Aleksiev and Boyadjieva (1966); 13. Herrmann (1969), unpublished.

Table 39, 57—71-E-2b. *Absolute abundances of Y and La—Lu in alkali olivine basalts, olivine nephelinites, trachytes, and phonolites in ppm (dash: no value reported). Σ Y, La—Lu includes estimates for missing analytical values*

Rock	Locality	Method	$_{39}$Y	$_{57}$La	$_{58}$Ce	$_{59}$Pr	$_{60}$Nd	$_{62}$Sm	$_{63}$Eu	$_{64}$Gd	$_{65}$Tb	$_{66}$Dy	$_{67}$Ho	$_{68}$Er	$_{69}$Tm	$_{70}$Yb	$_{71}$Lu	Σ Y, La—Lu	Reference
Alkali olivine basalts	500 m N.N.W. of Misaki, south to Saigo Bay, Oki Dogo Island, Japan (Pliocene)	I	—	40—48	59	—	30.5	6.3	—	6.7	—	5.2	—	2.34	—	1.6	—	—	1
	East Coast of the island of Takashima, N.W. of Karatsu, northern Kyushu, Japan (Pleistocene)	I	—	33—42	63	—	29.5	6.1	—	6.4	—	4.1	—	1.92	—	1.16	—	—	1
	Kiso district, Japan	I	—	23.6	42	—	26	5.5	1.87	5.2	—	4.7	—	2.33	—	2.1	0.30	—	2
	Jefferson Co., Colorado, U.S.A.	N/R	40	79	210	40	105	20	6.6	14.5	2.3	—	2.3	—	0.81	2.6	0.57	≈ 530	3
	Tertiary volcanism in Northern Hessia and Lower Saxony, N.W.-Germany (13 samples)	X, S	20	87	149	16.2	55	7.9	2.1	7.9	1.3	4.7	1.4	2.9	0.6	2.2	0.6	359	4
Nepheline basanite	Kuhberg, Druseltal. Northern Hessia, N.W.-Germany. Tertiary volcanism	X, S	16	99	178	19	71	11.7	1.9	7.4	1.5	6.2	2.0	4.1	0.8	3.4	0.6	423	4
Olivine nephelinites	Northern Hessia, N.W.-Germany. Tertiary volcanism (5 samples)	X, S	25	163	278	30	103	13.9	2.9	11.5	1.8	6.8	2.2	3.8	1.0	3.1	0.8	647	4
Nepheline leucite tephrites	Eifel, W.-Germany (3 samples)	X, S	28	131	225	26	84	10	2.9	10.5	1.6	5.2	1.9	3.1	0.6	2.7	0.6	533	4
Trachytes	Westerwald area, W.-Germany. (3 samples)	X, S	30	197	295	30	95	11	1.9	8.7	1.2	5.8	1.3	3.1	0.7	2.9	0.7	684	4
	Gough Island	N/R	81.2	223	503	46.5	149	23	0.50	17.8	3.0	—	3.2	8.8	1.28	7.0	1.21	≈1,069	3
	Ascension Island	N/R	34	83	—	11.8	—	11	1.7	6.3	1.13	—	1.12	4.6	0.52	4.0	—	≈ 400	3
Phonolites	Malberg near Leuterod, Westerwald area, W.-Germany	X, S	18	120	185	15	37	3.6	0.5	4.8	0.8	3.5	—	2.5	0.5	2.0	0.5	395	4
	Kaiserstuhl, S.W.-Germany (5 samples)	X	11	120	290	≈40	55	≈25	—	≈15	≈20	≈15	—	≈10	—	—	—	—	5

Reference: 1. Masuda (1966); 2. Masuda (1968a); 3. Frey *et al.* (1968); 4. Herrmann (1968); 5. van Wambeke (1964).

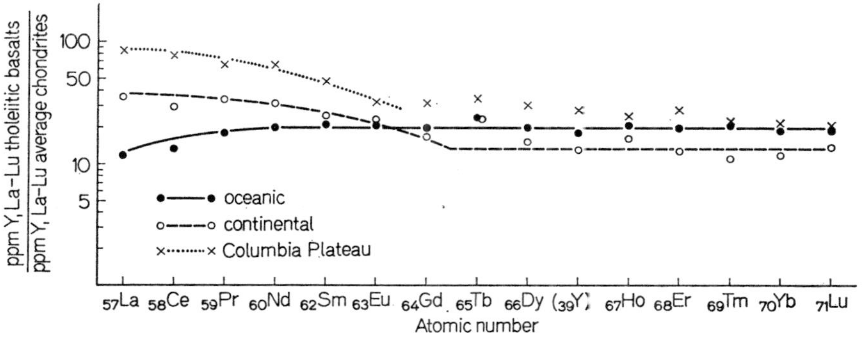

Fig. 39, 57—71-E-2. Abundances of Y and La—Lu in oceanic and continental tholeiitic basalts relative to chondrites. The average of 10 oceanic tholeiitic basalts based on the analyses of 6 specimens from the Mid-Atlantic Ridge and 4 specimens from the East Pacific Rise (FREY and HASKIN, 1964; FREY *et al.*, 1968). The average of continental tholeiitic basalts based on the analyses of 1 specimen from the Kilauea Iki-22 (SCHMITT *et al.*, 1963), from the Kohala Mt., Hawaii (FREY *et al.*, 1968), 6 composites of basalts from the Siberian platform and the Deccan plateau (BALASHOV and NESTERENKO, 1966), and 3 specimens of basalts of the late Tertiary volcanism in Northern Hessia and Lower Saxony, N.W. Germany (HERRMANN, 1968). See also Table 39, 57—71-E-2a

1966), from Tasmania, South Africa (PHILPOTTS and SCHNETZLER, 1968a), and from N.W. Germany (HERRMANN, 1968), have relative lanthanide distribution patterns showing only a slight enrichment of the lighter lanthanides as compared to that of chondrites (Fig. 39, 57—71-E-2). Paleozoic extrusive and intrusive spilitic rocks in greenschist facies (Devonian — Carboniferous) from different regions in N.W. and W. Germany contain the same absolute and relative lanthanide concentrations as typical continental tholeiitic basalts (HERRMANN and WEDEPOHL, 1969; Table 39, 57—71-E-2a). The absolute concentrations of the lanthanides in the continental theoleiitic basalts are similar to those from the oceans.

It must be mentioned that the lanthanide distributions in some peridotites, eclogites, and tholeiitic basalts are more similar to those of chondrites and Ca-rich Apachondrites than any other rocks of the earth's crust.

The behavior of the lanthanides in different rock types from one volcanic province has been studied in detail on samples from Hawaii and N.W. Germany by SCHILLING and WINCHESTER (1966) and HERRMANN (1968) respectively. The absolute Σ Y, La—Lu (in brackets) and the ratio

$$\frac{\Sigma \text{ ppm La—Eu}}{\Sigma \text{ ppm Gd—Lu, Y}} \text{ [in brackets]}$$

increase in the following order: tholeiitic basalts (≈ 100 ppm) [1.9], "intermediate" basalt (≈ 156 ppm) [3.7], alkali olivine basalts (≈ 360 ppm) [7.6], nepheline basanite (≈ 420 ppm) [9.1], and olivine nephelinites (≈ 647 ppm) [10] from the Tertiary in N.W. Germany. Fig. 39,57—71-E-3 demonstrates the increasing enrichment of the lightest lanthanides relative to chondritic meteorites.

A comparable relative enrichment of the lighter lanthanides was also found in the Hawaiian basalts in the following order: tholeiitic basalts, alkali olivine basalts, basanites, and nepheline melilite basalts (SCHILLING and WINCHESTER, 1966).

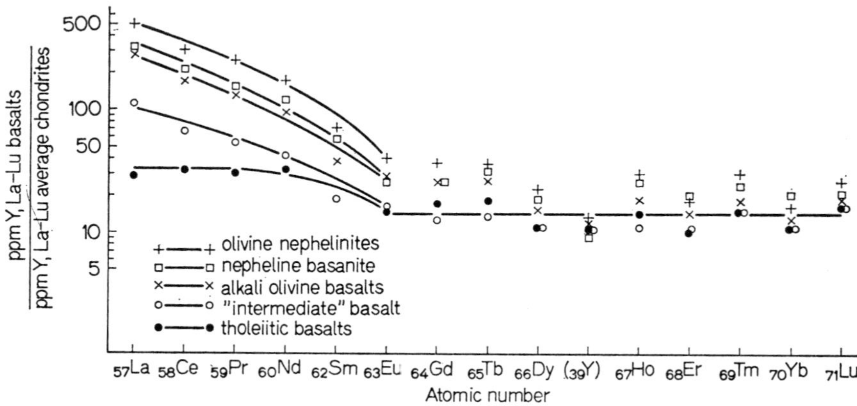

Fig. 39, 57—71-E-3. Abundances of Y and La—Lu in several types of basalts of the late Tertiary volcanism in Northern Hessia and Lower Saxony, N.W. Germany, relative to chondrites. The "intermediate" basalt is a rock between tholeiitic basalts and alkali olivine basalts in composition. Data from HERRMANN (1968); see also Table 39, 57—71-E-2a

The increasing lanthanide fractionation in this sequence of basalts is associated with a higher K, Ba content and a higher

$$\frac{Fe_2O_3}{FeO}$$

ratio, as well as a decrease of SiO_2. These chemical data contribute significantly to considerations about the origin of the different basaltic rock types (for detailed discussions see, for example, HASKIN, FREY *et al.*, 1966; SCHILLING and WINCHESTER, 1966; HERRMANN, 1968). MASUDA (for example 1962, 1966, 1967a, 1968a), MASUDA and MATSUI (1965), and MATSUI (1966) estimated partition coefficients of the lanthanides, calculated depths of magma sources, reported about areal relationships in the patterns of lanthanides of basalts from the Pacific Ocean and the Japan Sea, and about parallelism between fractionation of the lanthanides and the $^{87}Sr/^{86}Sr$ ratios in basalts. The latter could not be confirmed for the Tertiary basalts from N.W. Germany (HOEFS and WEDEPOHL, 1968).

It is important to note that tholeiitic basalts are some orders of magnitude more abundant than alkali olivine basalts. Regarding the origin of different magma types, it is interesting that fractional crystallization of the Skaergaard magma in Greenland did not give rise to any strong change of the original Y, La—Lu pattern (as evidenced by the contact gabbro) in the course of solidification. Only the absolute lanthanide concentrations increased steadily during solidification of the layered series of the Skaergaard intrusion (L. A. HASKIN and M. A. HASKIN, 1968).

The high absolute Y and La—Lu contents and the relative enrichment of the lighter lanthanides in the olivine nephelinites from N. W. Germany as compared to alkali olivine basalts cannot be explained by assimilation of sedimentary carbonates through the alkali olivine basalt magma (Fig. 39, 57—71-E-3).

The lanthanide fractionation in olivine nephelinites, leucite tephrites, essexites, phonolites from Kaiserstuhl (Southern Germany) and in nepheline leucite tephrites from the Eifel area, as well as in trachytes and one phonolite from the Westerwald in Western Germany is as strong as that of the olivine nephelinites already mentioned (HERRMANN, 1968; see also Table 39, 57—71-E-2b, Fig. 39, 57—71-E-4).

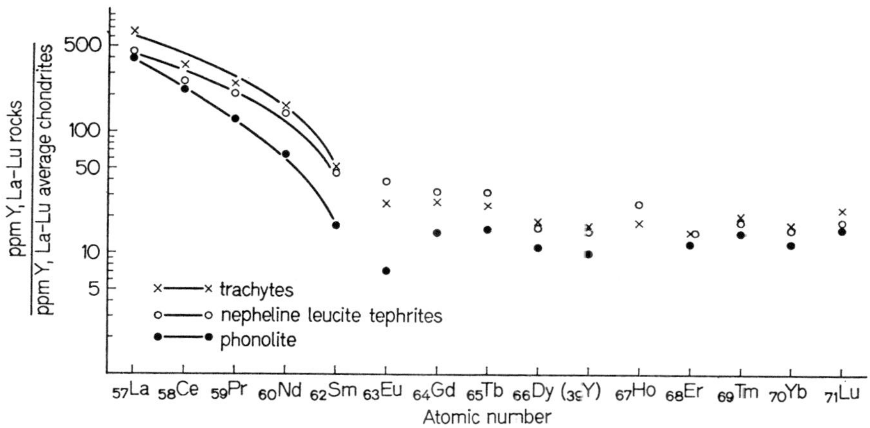

Fig. 39, 57—71-E-4. Abundances of Y and La—Lu in trachytes and phonolite from the Westerwald area (Tertiary) and nepheline leucite tephrites from the Eifel area (Pleistocene) of Germany relative to chondrites. Data from HERRMANN (1968); see also Table 39, 57—71-E-2b

III. Gabbros, Norites and Anorthosites

Information about the lanthanide abundances in gabbros is limited to specimens from Lapland (SAHAMA, 1945), the batholith of southern California (TOWELL, WINCHESTER, and VOLFOVSKY SPIRN, 1965), Ironton, Missouri (FREY *et al.*, 1968), Central Armenia (PAVLENKO *et al.*, 1966), Zekarsk, Georgia (BALASHOV and KEKELIYA, 1965), Skaergaard (L. A. HASKIN and M. A. HASKIN, 1968), and Bushveld (FREY *et al.*, 1968), see also Table 39, 57—71-E-3. The absolute lanthanide concentrations (from about 20 to about 120 ppm Σ Y, La—Lu) and their relative patterns are almost comparable with continental tholeiitic basalts as might be expected (Fig. 39, 57—71-E-5). The ratios of

$$\frac{\Sigma \text{ ppm La—Eu}}{\Sigma \text{ ppm Gd—Lu, Y}} = 2.3$$

for the Zekarsk gabbro, 2 for the norite, 1.5 for the Ironton and the Skaergaard gabbro, 1.4 for the San Marcos gabbro, and 1.2 for the sample from Armenia, are very similar to those observed for various peridotites, eclogites, and tholeiitic basalts.

PHILPOTTS *et al.* (1966) and ANDERSON (1966) have reported abundance data of some lanthanides in anorthosites and a mangerite (Table 39, 57—71-E-3). The former authors have published a very peculiar fractionation pattern of the plagioclase accumulation, again with an Eu anomaly (Fig. 39, 57—71-E-5).

Average Y, La—Lu contents of basaltic rocks representing the most common types in the upper earth's crust can best be obtained from a composite of oceanic and continental tholeiitic basalts (see Chapter 7 in Volume I of this handbook). Accordingly, from a mixture of 70% oceanic tholeiitic basalts and 30% continental tholeiitic basalts (under special consideration of W-1 [1]), an average value ≈ 99 ppm Σ Y, La—Lu is calculated for basalts (Table 39, 57—71-E-4). The lanthanide distribution pattern of this average basalt is nearly the same as that of chondrites and Ca-rich Ap-achondrites (Fig. 39, 57—71-E-6).

[1] Special determinations of Ce and Eu in some standard rocks are published by BRUNFELT and STEINNES (1967).

Table 39, 57—71-E-3. *Absolute abundances of Y and La—Lu in gabbros, one norite and one anorthosite in ppm (dash: no value reported). Σ Y, La—Lu includes estimates for missing analytical values*

Rock	Locality	Method	$_{39}$Y	$_{57}$La	$_{58}$Ce	$_{59}$Pr	$_{60}$Nd	$_{62}$Sm	$_{63}$Eu	$_{64}$Gd	$_{65}$Tb	$_{66}$Dy	$_{67}$Ho	$_{68}$Er	$_{69}$Tm	$_{70}$Yb	$_{71}$Lu	Σ Y, La—Lu	Reference
Gabbro and dolerite	Southern Lapland, Finland (24 samples)	X	4.7	1.7	3.4	0.9	4.3	0.9	—	1.6	—	0.9	—	0.9	—	≈0.8	—	≈ 22	1
Gabbros	San Marcos	N/R	15.5	4.01	14.5	2.15	7.7	2.17	1.05	—	0.48	2.8	0.57	—	0.25	1.7	0.26	≈ 57	2
	Ironton, Missouri	N/R	26	12.8	—	4.7	20	5.9	3.7	5.1	0.50	—	1.21	4.4	0.72	2.7	0.60	≈ 123	3
	Skaergaard intrusion. Chilled olivine gabbro from the marginal border group	N/R	12.7	5.6	12.1	1.88	9.2	2.51	1.04	2.88	0.44	—	0.54	1.45	0.20	1.2	0.19	≈ 55	4
Norite	Bushveld	N/R	6.2	4.8	6.4	1.4	4.3	1.1	0.59	1.0	—	—	0.23	0.81	0.10	0.54	—	≈ 28	3
Anorthosite	Tchitogama Lake, Quebec	I	—	—	4.3	—	2.16	0.35	0.65	0.29	—	0.16	—	—	—	0.023	—	—	5

Reference: 1. Sahama (1945); 2. Towell, Winchester, and Volfovsky Spirn (1965); 3. Frey *et al.* (1968); 4. L. A. Haskin and M. A. Haskin (1968); 5. Philpotts *et al.* (1966).

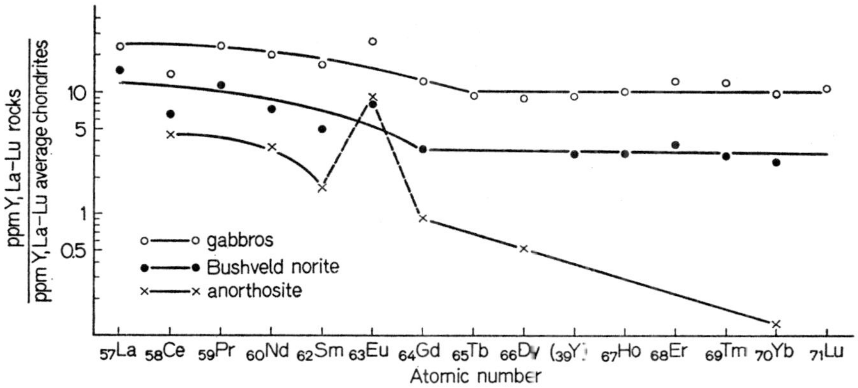

Fig. 39, 57—71-E-5. Abundances of Y and La—Lu in gabbros (San Marcos; Ironton, Missouri; Skaergaard), the Bushveld norite and one anorthosite relative to chondrites. Data from Towell, Winchester and Volfovsky Spirn (1965), Frey *et al.* (1968), L. A. Haskin and M. A. Haskin (1968) and Philpotts *et al.* (1966); see also Table 39, 57—71-E-3

Table 39, 57—71-E-4. *Average absolute concentrations of Y and La—Lu in basalts, intermediate rocks, granitic rocks, and a weighted mean of the igneous rocks of the upper continental earth's crust. For details on averaging see text of Section 39, 57—71-E*

Element	Average of basalts ppm	Average of intermediate rocks ppm	Average of granitic rocks ppm	Weighted average of igneous rocks ppm
$_{39}$Y	32	35	38	35
$_{57}$La	6.1	31	55	42
$_{58}$Ce	16	60	104	81
$_{59}$Pr	2.7	7.4	12	10
$_{60}$Nd	14	31	47	40
$_{62}$Sm	4.3	6.2	8	7.6
$_{63}$Eu	1.5	1.3	1.1	1.5
$_{64}$Gd	6.2	6.8	7.4	7.2
$_{65}$Tb	1.1	1.1	1.1	1.1
$_{66}$Dy	5.9	6.1	6.2	5.8
$_{67}$Ho	1.4	1.5	1.5	1.4
$_{68}$Er	3.6	3.9	4.2	3.9
$_{69}$Tm	0.60	0.65	0.69	0.62
$_{70}$Yb	3.2	3.8	4.3	3.7
$_{71}$Lu	0.55	0.62	0.68	0.61
ΣY, La—Lu	≈99	≈196	≈290	≈241
$\frac{\text{ppm La}}{\text{ppm Nd}}$	0.4	1	1.2	1.1
$\frac{\text{ppm Pr}}{\text{ppm Sm}}$	0.6	1.2	1.5	1.3
$\frac{\Sigma\text{ ppm La—Eu}}{\Sigma\text{ ppm Gd—Lu, Y}}$	0.8	2.3	3.5	3.1

IV. Granitic Rocks

The average Y, La—Lu content of granitic rocks in the upper continental earth's crust is computed from a proportion of 80% of a composite sample which contains 213 granites with >70% SiO_2, 10% of a sample of the Precambrian granites from the Western USA (HASKIN *et al.*, 1968) and 10% of the standard granite G-1 (HASKIN and GEHL, 1963b; TOWELL, VOLFOVSKY, and WINCHESTER, 1965; ALEKSIEV and BOYADJIEVA, 1966; see also Table 39, 57—71-E-5).

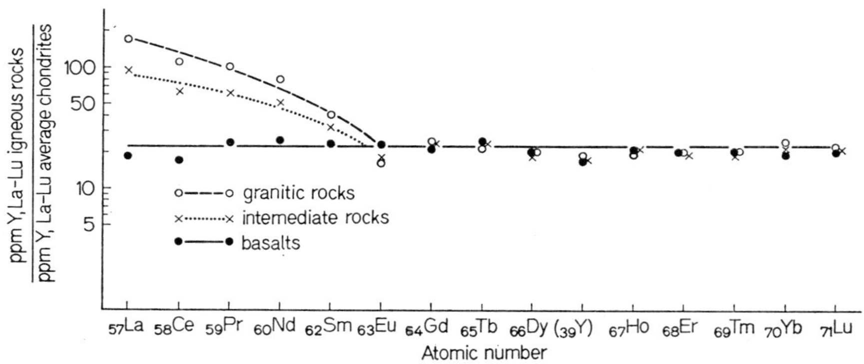

Fig. 39, 57—71-E-6. Abundances of Y and La—Lu in average basalts, intermediate rocks and granitic rocks of the upper earth's crust, relative to chondrites. Data from Table 39, 57—71-E-4

Giving a total of about 290 ppm Σ Y, La—Lu (Table 39, 57—71-E-4), the content of the lanthanides in granitic rocks is nearly 3 times that of the average basaltic rocks. A large relative increase in the lighter lanthanides, as compared to chondrites, basalts, and gabbros, characterizes the granitic rocks of the upper earth's crust (Fig. 39, 57—71-E-6). With regard to Y, La-Lu abundances in granitic rocks, the following papers contribute data (Table 39, 57—71-E-5): SAHAMA and VÄHÄTALO (1941), SAHAMA (1945), GAVRILOVA and TURANSKAYA (1958), BALASHOV (1963), KOLBE and TAYLOR (1966), TAYLOR *et al.* (1968), BALASHOV *et al.* (1968). Rhyolites have been investigated by HASKIN *et al.* (1968) and EWART, TAYLOR and CAPP (1968), and obsidians by HASKIN, FREY and POETZ, quoted from HASKIN, FREY *et al.* (1966).

The composite sample with 60—70% SiO_2, listed in Table 39, 57—71-E-5, may be taken to represent the average lanthanide composition of granodiorites.

At least two major types of magmatic differentiation by crystal settling have been effective in intrusions. They depend on oxygen fugacity in the melt and are characterized by an increase of either silica or ferrous iron. In the batholith of Southern California the exposed proportions of granitic to gabbroic rocks are not small enough to explain the former as differentiation products of the latter. Here the absolute concentrations of Σ Y, La—Lu increase from 57 to 170 ppm in the sequence: gabbro, tonalite, granodiorite, and leucogranite (quartz monzonite) (TOWELL, WINCHESTER, and VOLFOVSKY SPIRN, 1965, Table 39, 57—71-E-4 and E-5). In the same sequence these authors observed an increasing relative enrichment of the lighter lanthanides. In the Susamyr batholith (Central Tien Shan) the absolute concentrations Σ Y, La—Lu and the degree of fractionation within the lanthanide groups passes a maximum in

Table 39, 57—71-E-5. *Absolute abundances of Y and La—Lu in granitic rocks in ppm (dash: no value reported). Σ Y, La—Lu includes estimates for missing analytical values*

Rock	Locality	Method	$_{39}$Y	$_{57}$La	$_{58}$Ce	$_{59}$Pr	$_{60}$Nd	$_{62}$Sm	$_{63}$Eu	$_{64}$Gd	$_{65}$Tb	$_{66}$Dy	$_{67}$Ho	$_{68}$Er	$_{69}$Tm	$_{70}$Yb	$_{71}$Lu	Σ Y, La—Lu	Reference
Granitic rocks	Westerly, Rhode Island, U.S.A.																		
	G-1	N/R	12.5	102	134	20.9	54.6	8.6	1.04	4.88	0.50	—	0.50	1.40	0.20	0.625	0.17	≈345	1
	G-1	N/R	13.3	92.5	150	12.9	55.2	8.25	1.36	—	0.637	2.52	0.408	—	0.156	0.937	0.123	≈345	2
	G-1	X	10.0	105	150	21	75	5	1	10.2	1.5	5.5	1.3	1.0	0.8	1.32	0.5	389	3
	Composite of 213 samples, >70% SiO_2	N/R	42	50	100	11.4	46	8.3	1.10	7.6	1.12	—	1.62	4.7	0.74	4.8	0.78	≈288[a]	4
	Composite of 191 samples, 60—70% SiO_2	N/R	33.5	43	83	11.0	44	8.5	2.0	7.4	—	—	1.27	3.7	0.54	3.25	0.54	≈249[a]	4
	Composite of 85 samples, <60% SiO_2	N/R	31	37	97	10.0	45	8.3	2.5	7.6	1.03	—	1.21	3.4	0.50	3.0	0.48	≈238[a]	4
	Wisconsin Ruby; Wisconsin Red, (2 samples)	N/R	29	34	—	9.8	37	7.7	0.60	6.2	1.06	—	1.2	2.7	0.52	3.3	0.62	≈205	5
	Composite Western U.S.A., Precambrian granites	N/R	31	49	97	11.0	42	7.2	1.25	5.8	0.94	—	1.22	3.2	0.53	3.5	0.52	≈260	4
	Saganaga granite formation	N/R	2.17	7.0	15.0	1.58	6.2	0.95	0.37	0.72	0.08	—	0.07	0.2	0.03	0.19	0.02	≈ 35	4
	Batholith of Southern California, Rubidoux Mountain, U.S.A.	N/R	17.3	24.5	88	5.6	23.5	3.8	0.63	—	0.63	3.1	0.74	—	0.26	1.30	0.20	≈170	12
	Kirovograd district, Ukraine, U.S.S.R.	X	2.5	91	170	17	68	12.5	—	7.4	0.17	1.9	0.26	0.69	—	0.23	0.04	≈370	6
	E. Tuva, Agash, U.S.S.R.	X	72	22	45	8.7	36	9.0	—	9.7	—	7.9	1.1	5.4	—	5.5	—	≈230	7
	E. Tuva, Nizhne-Kadraus, U.S.S.R.	X	88	53	100	12	48	11	—	11	4.3	12	3.0	8.8	—	10	—	≈363	7
	E. Tuva, Radyros, U.S.S.R.	X	29	130	210	19	58	9.3	—	6.4	—	4.6	—	2.3	—	1.7	—	≈473	7
	Ustinovskii, Ukraine, U.S.S.R. (Rapakivi granite)	X	20	90	190	23	95	18	—	12	—	9.5	—	2.0	—	1.9	—	≈468	7
	Composite of Finnish granites	N/R	36	47	77	13	64	6.1	1.00	5.7	1.6	—	1.13	3.0	0.57	3.0	—	≈230	8
	Onas, Finland	X	63	34	86	17	68	17	2.6	17	3.5	13	3.5	8.7	1.8	8.8	3.5	347	9
	Obbnäs, Finland	X	160	260	430	60	170	52	8.6	69	8.7	26	8.7	26	<8.8	18	8.8	1,315	9
	Nattanen, Finland	X	24	26	77	17	43	8.6	1.7	8.7	1.7	4.4	1.7	2.6	<1.7	2.6	1.7	222	9
	Unonen in Kalvola, Finland	X	38	34	60	13	43	15	2.6	17	2.6	13	2.6	8.7	0.9	6.1	2.6	259	9
	Astelsholm, Bällarby in Karjaa, Finland	X	24	26	43	8.6	26	8.6	1.7	8.7	1.7	5.2	1.7	3.5	0.9	3.5	1.7	165	9
	Säppi, Finland	X	100	17	60	8.6	26	8.6	0.9	8.7	0.9	6	1.7	3.5	0.9	4.4	1.8	249	9
	Sorkka, Finland	X	63	34	77	17	43	8.6	1.7	17	1.7	8.7	2.6	6.2	0.9	5.3	1.8	289	9
Granite pegmatite	Luikonlahti, Maarianvaara massif, Northern Karelia (2 samples)	X	39	8.5	17	4.3	8.6	4.8	0.9	6.5	1.7	8.2	1.7	6.5	0.9	5.3	1.7	116	9
Gneiss granite and granite gneiss	Southern Lapland, Finland (42 samples)	X	14	19	30	7.7	24	6	3	7	3	4.3		3.5	—	3	—	≈126	10
Ferrodiorite	Skaergaard intrusion. Upper zone-(UZa) of the layered series. No. EG 5181	N/R	9.6	2.8	7.2	1.1	6.0	1.88	1.59	2.27	0.32	—	0.40	1.1	0.17	0.96	0.16	≈ 38	11

Table 39, 57—71-E-5 (Continued)

Rock	Locality	Method	$_{39}$Y	$_{57}$La	$_{58}$Ce	$_{59}$Pr	$_{60}$Nd	$_{62}$Sm	$_{63}$Eu	$_{64}$Gd	$_{65}$Tb	$_{66}$Dy	$_{67}$Ho	$_{68}$Er	$_{69}$Tm	$_{70}$Yb	$_{71}$Lu	Σ Y, La—Lu	Reference
Ferrodiorite	Skaergaard intrusion. Upper zone (UZc) of the layered series, the last differentiate of the intrusion. No. EG 4330	N/R	118	76.5	171	26	113	35	14.0	35.5	5.8	—	6.2	15.4	2.4	13.7	2.7	≈702	11
Grano-diorite	Woodson Mountain, Batholith of Southern California, U.S.A.	N/R	24.6	25.3	34.2	7	22.3	3.96	0.80	—	0.83	3.0	0.90	—	0.37	2.93	0.39	≈127	12
Tonalite	Bonsall, Batholith of Southern California, U.S.A.	N/R	21.5	13.4	22.5	5.7	17.2	3.65	1.06	—	0.84	2.6	0.75	—	0.39	2.0	0.30	≈ 92	12
Quartz-diorite	Del Fustiarbu, Sardinia, Italy	I	10.6	15.5	20.7	2.84	8.1	2.8	0.38	3.5	0.36	2.23	0.35	0.77	0.10	0.16	0.07	68	13
Rhyolites	Not reported	N/R	44	56	91	12.2	46	8.5	1.4	7.3	1.09	—	1.66	4.6	0.72	4.6	0.72	≈287	4
	Taupo Volcanic Zone, North Island, New Zealand (26 samples) pleistocene — Recent age.	I	28	28	44	6.4	18	5.5	1.0	4.3	1.2	—	1.4	3.9	0.5	3.6	—	≈150	14
Obsidians	Rotorua, New Zealand	N/R	31	28	42	6.7	—	5.4	0.96	4.5	1.08	—	1.07	3.2	0.53	3.7	—	≈160	15
	Ascension Island	N/R	52	63	—	13.5	—	14.1	4.8	9.2	1.4	—	1.7	4.5	0.89	5.0	—	≈350	15
	Easter Island	N/R	160	60	—	34	80	29	7.6	31	5.4	—	7.1	20	3.1	15	2.6	≈710	15

[a] Data for the same composite specimens also reported by Haskin, Frey *et al.* (1966). Note differences for several lanthanides between the publications Haskin, Frey *et al.* (1966), and Haskin *et al.* (1968). The present table considers data from the latter publication only.

Reference: 1. Haskin and Gehl (1963b); 2. Towell, Volfovsky, and Winchester (1965); 3. Aleksiev and Boyadjieva (1966); 4. Haskin *et al.* (1968); 5. Gehl (1964), quoted from Haskin, Frey *et al.* (1966); 6. Gavrilova and Turanskaya (1958); the converted values quoted from Haskin, Frey *et al.* (1966); 7. Balashov (1963); 8. L. A. Haskin, Frey, Poetz, and M. A. Haskin, quoted from Haskin, Frey *et al.* (1966); 9. Sahama and Vähätalo (1941); 10. Sahama (1945); 11. L. A. Haskin and M. A. Haskin (1968); 12. Towell, Winchester, and Volfovsky Spirn (1965); 13. Matteucci (1967); 14. Ewart, Taylor, and Capp (1968); 15. Frey, Haskin, and Poetz, quoted from Haskin, Frey *et al.* (1966).

the sequence of differentiation from gabbros to aplites (BALASHOV, 1966; Table 39, 57—71-E-6). It is of importance to note that the other type of fractionation shows no relative lanthanide fractionation in the order: gabbro, ferrodiorite, hedenbergite granophyre, as sampled at the Skaergaard intrusion (L. A. HASKIN and M. A. HASKIN, 1968).

The total quantity of lanthanides in the igneous rocks of the upper continental earth's crust can be roughly estimated to be 241 ppm, using the following proportions of common rock types: 45% granite, 35% granodiorite, 5% diorite (intermediate rocks), and 15% tholeiitic basalt instead of gabbro (WEDEPOHL, 1967, and Chapter 7 in Volume I of this handbook, see also Table 39, 57—71-E-4).

V. Intermediate Igneous Rocks

A composite of 85 igneous rocks with $<60\%$ SiO_2 and most probably made up of rocks of intermediate composition has been analyzed by HASKIN *et al.* (1968). Its lanthanide content of about 238 ppm Σ Y, La—Lu probably resembles that of average intermediate rocks (Table 39, 57—71-E-5). The total amount of Y, La—Lu in these rocks can also be estimated from a 1:1 mixture of basaltic and granitic rocks to be about 196 ppm Σ Y, La—Lu (Table 39, 57—71-E-4).

VI. Alkalic Rocks, Carbonatites

Some systematic studies of the lanthanide abundances in composite alkali rock massifs have been made by BALASHOV and TURANSKAYA (1960, 1961), ZLOBIN and BALASHOV (1961), BALASHOV (1962, 1963, 1966), LEONOVA and BALASHOV (1963), YES'KOVA and GANZEYEV (1964), and chapters Y and lanthanides in VLASOV (1966). These authors analyzed nepheline syenites from the Lovozero massif (Kola Peninsula), the alkali rock massifs of Ilimausuk (Southern Greenland), of Sandyk (Central Tien Shan), of the Vishnevye Mountains (the Urals), of Khibiny (Kola Peninsula), and of Kzyl-Ompul (eastern spurs of the Kirghiz Range, Northern Tien Shan).

Variations in the distribution of the lanthanides were observed to depend on changes in mineral associations and in the ratio

$$\frac{Na+K}{Al}$$

during magmatic differentiation. The ratio

$$\frac{\Sigma \text{ ppm La—Eu}}{\Sigma \text{ ppm Gd—Lu, Y}}$$

decreases and the absolute lanthanide concentrations increase in the sequence of rock formation in the Lovozero massif, following an increase of mafic minerals at a ratio

$$\frac{Na+K}{Al}$$

greater than one. Contrary to the above observation, an increase in the ratio

$$\frac{\Sigma \text{ ppm La—Eu}}{\Sigma \text{ ppm Gd—Lu, Y}}$$

and in the absolute lanthanide content was observed in the sequence from the early (alkalic gabbroids) to the late (nepheline syenites) phases of the Sandyk massif characterized by a ratio

$$\frac{Na+K}{Al}$$

less than one (Fig. 39, 57—71-E-7). This again suggests that the accumulation of the heavier lanthanides is obviously higher in the more melanocratic rocks from

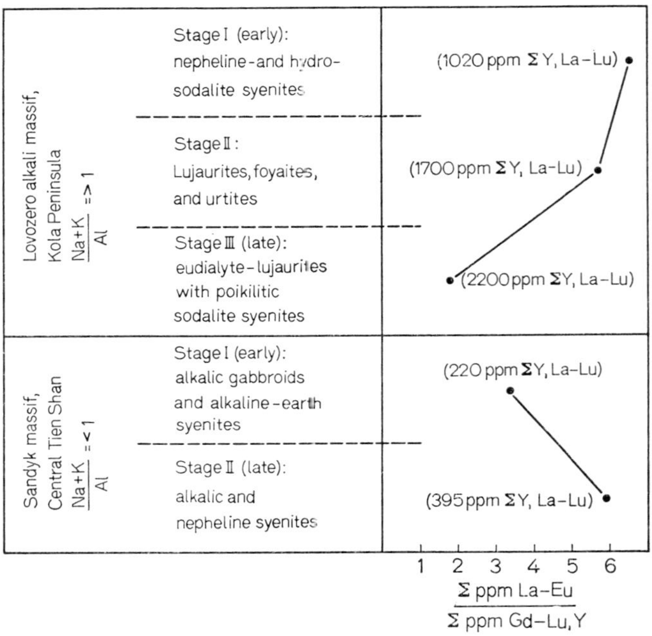

Fig. 39, 57—71-E-7. Fractionation of Y and La—Lu in composite alkalic rock massifs where different stages of evolution can be separated, plotted as

$$\frac{\Sigma \text{ ppm La—Eu}}{\Sigma \text{ ppm Gd—Lu, Y}}$$

against progress of evolution. Data from BALASHOV (1966); see also Table 39, 57—71-E-6

the massifs examined. From a general comparison of different intrusions, it can be stated that the fractionation of Y and La—Lu during this specific magmatic differentiation is significantly influenced by the ratio

$$\frac{Na+K}{Al}$$

of the melt (BALASHOV, 1966; Fig. 39, 57—71-E-8). Probably the presence of alkalic pyroxenes in the rocks of the above-mentioned massifs gives a ratio of

$$\frac{Na+K}{Al}$$

greater than one, while feldspathoids are characterized by a ratio of one or less than one.

All igneous rocks from composite intrusives are relatively enriched in lighter lanthanides as compared to the pattern of chondrites. The absolute Y, La—Lu abundances in alkalic rocks are generally greater than those of the average upper continental earth's crust (Table 39, 57—71-E-4 and E-6).

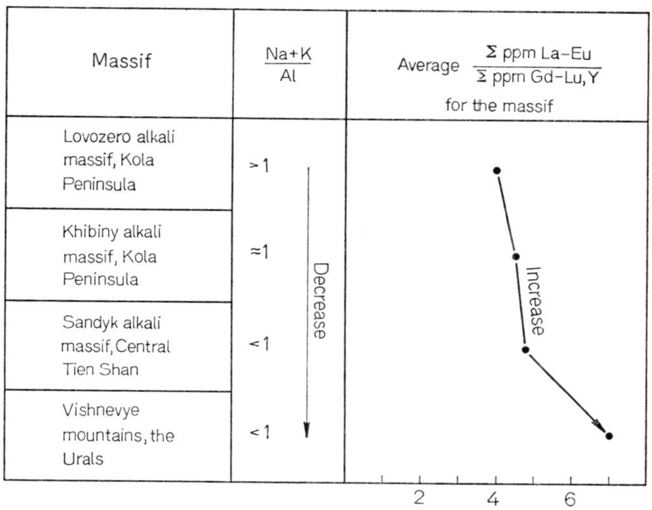

Fig. 39, 57—71-E-8. Fractionation of

$$\frac{\Sigma \text{ ppm La—Eu}}{\Sigma \text{ ppm Gd—Lu, Y}}$$

in alkalic rocks (nepheline syenites) of different massifs in relation to the alkali-to-aluminium ratio. Data from BALASHOV (1966)

SCHOFIELD and HASKIN (1964), VAN WAMBEKE (1964), and BALASHOV and POZHARITSKAYA (1968) have published absolute Y, La—Lu concentrations in carbonatites (Table 39, 57—71-E-6); VAINSHTEIN, POZHARITSKAYA, and TURANSKAYA (1961) and especially KAPUSTIN (1966) mentioned that in the early phases of carbonatite forming processes the lanthanides are almost entirely scattered in minerals like calcite, pyrochlore, apatite, amphibole, and ankerite, while in later phases more than 90% of them are fixed in lanthanide minerals proper (monazite, parisite). VAINSHTEIN, POZHARITSKAYA, and TURANSKAYA (1961) observed an increase in the absolute Y, La—Lu abundances in the total rocks toward the later phases of carbonatite formation. The light lanthanides are predominant in all the carbonatite rocks as a whole, as well as in their individual minerals (Fig. 39, 57—71-E-9). Carbonatite lavas can contain high sodium concentrations; the soda-rich carbonate lavas from Oldoinyo Lengai, Tanganyika (DAWSON, 1962) contain 29.5—29.7% Na_2O. We also know of carbonatitic lavae low in Na_2O, however, such as that from Fort Portal Area in western Uganda containing 0.7—1% Na_2O, 600 ppm La and 75 ppm Y on the average (VON KNORRING and DU BOIS, 1961).

Table 39, 57—71-E-6. *Absolute abundances of Y and La—Lu in several intrusive stages of different igneous rock massifs, alkalic rocks, and carbonatites in ppm (dash: no value reported). Σ Y, La—Lu includes estimates for missing analytical values*

Locality	Intrusive phases and rocks	Method	$_{39}$Y	$_{57}$La	$_{58}$Ce	$_{59}$Pr	$_{60}$Nd	$_{62}$Sm	$_{63}$Eu	$_{64}$Gd	$_{65}$Tb	$_{66}$Dy	$_{67}$Ho	$_{68}$Er	$_{69}$Tm	$_{70}$Yb	$_{71}$Lu	Σ Y, La—Lu	Reference
Susamyr batholith (Central Tien Shan)	Stage I (early): stocks and xenoliths of gabbrodiorites and diorites	X	64[a]	45	90	13	47	12	—	11	3.1	11	3.8	5.7	—	6.5	—	≈ 330	1
	Stage II: Granodiorites, tonalites, granites and adamellites	X	36	92	170	20	66	11	—	7	—	6.7	—	3.4	—	3.2	—	≈ 420	1
	Stage III (late): Leucocratic granites	X	75	63	120	14	53	13	—	12	3.2	11	4.2	7.4	—	7.4	—	≈ 390	1
	Aplites	X	15	10	220	—	20	10	—	13	—	22	—	15	—	17	—	≈ 300	1
	Average for the batholith[b]	X	42	87	160	19	64	11	—	7.8	—	7.3	—	4.0	—	3.8	—	≈ 410	1
Lovozero alkalic massif, Kola Peninsula	Stage I (early): nepheline- and hydrosodalite — syenites (14 samples)	X	76	230	390	50	180	37	—	24	5.5	14	3.7	7.4	—	5.5	—	≈1,020	2
	Stage II: Lujaurites, foyaites, and urtites (45 samples)	X	150	380	690	73	264	39	≈10	34	6.5	29	5.0	15	2.7	12	2.0	≈1.708	2
	Stage III (late); eudialyte-lujaurites with poikilitic sodalite — syenites (53 samples)	X	480	290	600	79	330	92	13	79	22	90	26	48	8	46	10	2,213	2
	Average for the massif[c] (116 samples)	X	210	362	671	74	274	48	10	42	9.3	40	8.8	21	3.5	18	3.4	1.788	2
Nepheline syenites	Turpi massif, Alai range	X	36	56	110	14	51	12	—	8.7	2.0	9.2	2.5	3.6	—	—	—	≈ 310	3
	Khibiny massif, Kola Peninsula	X	62	120	220	28	100	16	—	12	3.0	10	3.1	4.1	—	5.2	—	≈ 590	3
Syenite	Burpala pluton, North Baikal region (8 samples)	X, S	—	90	200	12	86	9.5	1.2	5.8	0.9	3.3	1.7	0.9	—	0.9	—	≈ 460[d]	4
Carbonatites	Panda Hill, Mbeya, Tanganyika	N/R	94.5	181	272	66.5	282	46	14.3	34	11	—	5.1	14.5	4.5	9.0	2.4	≈1,060	5
	Sangu complex, Tanganyika	N/R	38	76	391	25	94	18	6.7	16.7	2.5	—	2.05	4.5	0.53	1.50	—	≈ 687	5
	Kaiserstuhl, S.W.-Germany (25 samples) [sövite]	X	49	528	832	96	170	41	—	25	<30	<30	—	<25	—	—	—	—	6
	Kaiserstuhl, S.W.-Germany (16 samples) [alvikite]	X	35	860	1,050	60	188	38	—	20	<20	<20	—	<20	—	—	—	—	6

[a] Corrected value; probable error in decimal point used by BALASHOV (1966). [b] Considering the relative proportions of the stages. [c] In the calculation of the average contents the volume of the stages has been considered. [d] For this table the relative values of the original paper have been converted into ppm.

Reference: 1. BALASHOV (1966); 2. BALASHOV and TURANSKAYA (1960, 1961), BALASHOV (1962), quoted from BALASHOV (1966); 3. BALASHOV (1963); 4. PORTNOV *et al.* (1967); 5. SCHOFIELD and HASKIN (1964); 6. VAN WAMBEKE (1964).

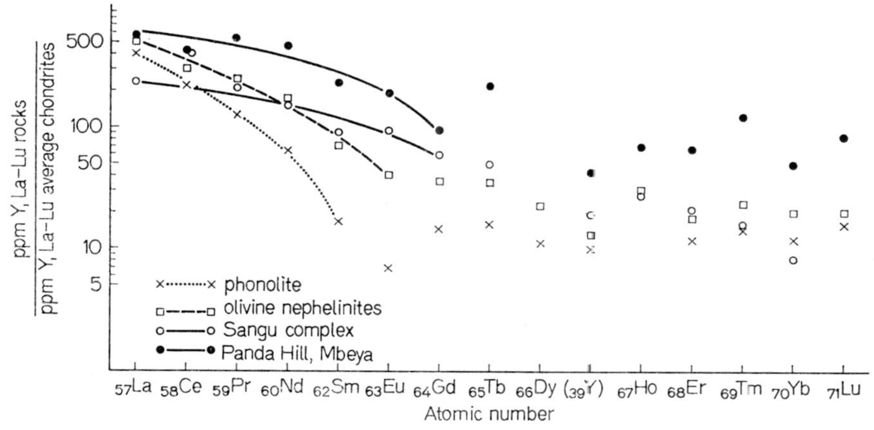

Fig. 39, 57—71-E-9. Abundances of Y and La—Lu in two carbonatites from Tanganyika and for olivine nephelinites and phonolite relative to chondrites. Data from SCHOFIELD and HASKIN (1964) and HERRMANN (1968); see also Table 39,57—71-E-2b and E-6

Revised manuscript received: February 1969

39, 57-71-F. Behavior in Magmatogenic Processes (Pegmatites, Hydrothermal Transport etc.)

I. Pegmatites

SLEPNEV (1957) studied the abundance and the distribution of the lanthanides in the nepheline syenite pegmatites of the Khibina alkalic massif, Kola Peninsula (in this connection see also the studies of the lanthanide abundances in the Lovozero massif, made by BALASHOV and TURANSKAYA, 1960, 1961). SLEPNEV showed that the Si, Al, Ti, Fe, Ca, K, and Na contents of the pegmatites do not differ from those of the rocks of the massif. However, the pegmatites contain 75 times more lanthanides than are found in the rocks of the massif. The lanthanide contents of the pegmatites as a whole (Σ Y, La—Lu $\approx$ 7,500 ppm) are about 30 times higher than the average igneous rocks of the upper earth's crust (Σ Y, La—Lu $\approx$ 241 ppm). The lanthanides are predominantly concentrated in dark-colored, calcium-bearing minerals like lovchorrite, loparite, pyrochlore, sphene etc., which crystallized during the later stages of pegmatite formation. During the stage of initial crystallization of a pegmatitic melt the lanthanides are trapped in small amounts in potash feldspar, nepheline and arfvedsonite, but are mostly concentrated in mosandrite. Pegmatites show a relative enrichment of the lighter lanthanides as compared to the pattern in chondrites.

Data on the distribution of lanthanides in minerals of the granite-pegmatites of North-western, South-western and Northern Karelia were reported by KALITA (1959) and ZHIROV *et al.* (1961). KALITA found that in the pegmatites of North-western and South-western Karelia the early-formed lanthanide minerals, allanite and monazite contain mainly lighter lanthanides. Later minerals crystallized with higher fractions of Y and heavier lanthanides, as for example blomstrandine and euxenite. Reactions by late, "metasomatic" processes have produced secondary minerals which are characterized by a low content of heavier lanthanides. KALITA (1959) concluded that this distribution apparently is the result of the considerably greater mobility of the complex compounds of heavier lanthanides in solutions of higher alkalinity.

ZHIROV *et al.* (1961) reported that the distribution of lanthanides in monazites and xenotimes from pegmatites in Northern Karelia is quite constant, and no variation with the paragenetic characteristics or geologic environment of these pegmatites is apparent. However, the distribution of lanthanides in allanites and uraninites depends on the paragenesis of these minerals. If the paragenesis contains phosphates, allanite has a normal composition, rich in cerium, and uraninite is characterized by a Gd, Dy, and Er maximum. In the absence of phosphates, allanite is Y-rich and uraninite has a Dy, Er, and Yb maximum. Later alteration processes may influence the abundance of the lanthanides in some minerals.

II. Hydrothermal Transport, Metasomatism

Kosterin (1959), Bandurkin (1961), and Mineyev (1963) investigated the problem of the transport of Y and La—Lu in hydrothermal solutions. Kosterin (1959) suggests that most of the lanthanides in hydrothermal deposits were probably transported in alkaline solutions in the form of carbonate, fluorine, or sulfate complexes. As reasons for the breakdown of these complexes and the precipitation of lanthanide minerals, he mentions decrease of pressure, fixation of CO_3^{2-}, or a change in alkalinity. As heavy lanthanides form more stable complexes than the light ones, they stay in solution until precipitated in late-stage deposits. Bandurkin (1961) and Mineyev (1963) assume that the transport of lanthanides in fluoride complexes should be possible. The stability of $NaYF_4$ under hydrothermal conditions has been proved experimentally. Transport of the respective La and Ce complexes together with that of Y in a temperature gradient from 500 to 550° C was experimentally demonstrated by Mineyev *et al.* (1966).

The origin of lanthanide minerals in greisen deposits can possibly be explained this way (Bandurkin, 1961). Mineyev (1963) again mentions the assumption of other authors that the differentiation of the lanthanides in minerals and rocks from light to heavy lanthanides (plus yttrium) may to some extent be explained by the different stability of complexes as a function of decreasing alkalinity in postmagmatic solutions.

An example of different lanthanide distribution in the Ognitsk Middle Paleozoic intrusive complex with alaskites and leukocratic alkalic granites from the central part of East Sayan, genetically connected with high temperature metasomatic processes, was reported by Kovalenko *et al.* (1966). In this complex the rise in SiO_2 of the mineral parageneses is accompanied by an increase in the relative contribution of heavier lanthanides. This observation is characterized by a decrease in the ratio

$$\frac{\Sigma \text{ ppm La—Eu}}{\Sigma \text{ ppm Gd—Lu, Y}}.$$

It is important to point out, however, that all mineral associations of this complex are relatively enriched in light lanthanides as compared to the lanthanide pattern of chondrites.

Fractionations between the lighter and the heavier lanthanides are observed in fluorites of hydrothermal origin (Herget, 1965; Huber-Schausberger and Schroll, 1967). The latter authors distinguish between three groups of lanthanide distribution in fluorites (with Σ Y, La—Lu from 330—1,600 ppm, see also Fig. 39, 57—71-F-1). No genetic explanation is given for the different fractionation. Portnov *et al.* (1967) found an enrichment of heavy lanthanides in fluorites from the fluorite-apatite association in hydrothermal veins of the Burpala pluton in the North Baikal region.

The data of Hegemann and Wilk (1964) on lanthanides in fluorites from Wölsendorf/Opf. (Germany) are only semi-quantitative. The total concentration of lanthanides in fluorites reported by Herget (1965) is much lower than those reported from English, French, Italian, and S. W. African deposits.

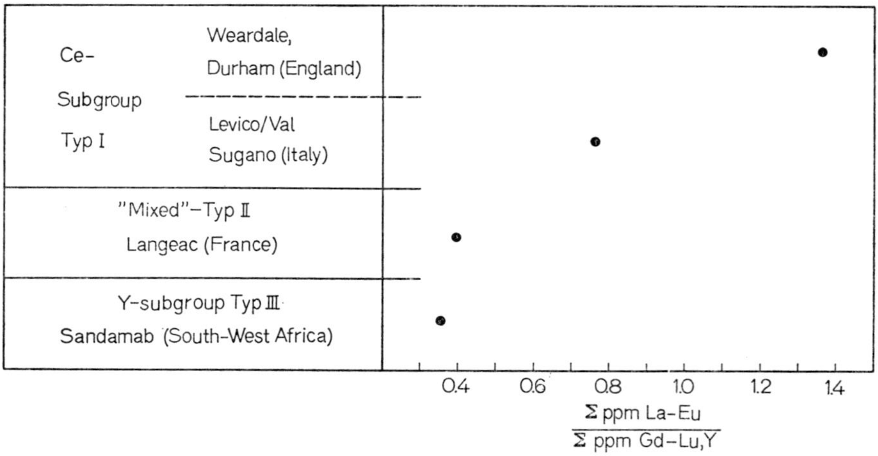

Fig. 39, 57—71-F-1. Fractionation of

$$\frac{\Sigma \text{ ppm La—Eu}}{\Sigma \text{ ppm Gd—Lu, Y}}$$

in three types of green fluorites. Data from HUBER-SCHAUSBERGER and SCHROLL (1967)

Revised manuscript received: February 1969

39, 57—71-G. Behavior during Weathering and Alteration of Rocks

BALASHOV *et al.* (1964) and RONOV *et al.* (1967) determined the Y and La—Lu concentrations in various sediments from the Russian platform which were deposited under either arid or humid conditions during the Middle Paleozoic time. RONOV *et al.* (1967) make the statement, which is generally applicable to weathering and sediment accumulation, that weathering products depend on the mineralogical and chemical composition of the primary rocks as well as on the physicochemical conditions in the weathering environment. The shales and sandstones of the Famennian stage (Upper Devonian) were accumulated under conditions of increasing aridity and salinity in the Moskovian basin. The shales and sandstones do not differ much in average lanthanide concentrations, containing about 270 ppm and about 310 ppm Σ Y, La—Lu respectively. The lanthanide concentrations of the sandstone samples from the Russian platform are high, as compared with other specimens of this type (see Table 39, 57—71-K-1). The ratios

$$\frac{\Sigma \text{ ppm La—Eu}}{\Sigma \text{ ppm Gd—Lu, Y}}$$

of the shales (continental, arid = 2.6, saline, lagoonal = 2.5) and of the sandstones (continental, arid = 2.6, saline, lagoonal = 1.8) are also rather similar (BALASHOV *et al.*, 1964). Therefore BALASHOV *et al.* (1964) and RONOV *et al.* (1967) reach the general conclusion on the basis of a great number of samples that there is no significant difference and fractionation in the lanthanides among clastic rocks of the arid environment of erosion and sedimentation.

Humid conditions apparently can cause lanthanide concentrations and distribution patterns in detrital sediments (Yasnopolyansk substage, Lower Carboniferous) which differ from the above-mentioned arid formations. On average the shales contain about 370 ppm Σ Y, La—Lu and the sandstones about 225 ppm Σ Y, La—Lu (Table 39, 57—71-K-1), with the highest abundances of the lanthanides adjacent to the source areas. Towards the center of the basin and towards the coal-bearing zone, the lanthanides in both the shales (Σ Y, La—Lu about 310 ppm) and sandstones (Σ Y, La—Lu about 198 ppm) decrease to a minimum.

There is a considerable change in the ratio

$$\frac{\Sigma \text{ ppm La—Eu}}{\Sigma \text{ ppm Gd—Lu, Y}}$$

from continental (including coal-bearing deposits) to near-shore shales and sandstones (see Fig. 39, 57—71-G-1). In this direction a relative increase in the heavy lanthanides is to be observed (see also RONOV *et al.*, 1967). The marine carbonate rocks also indicate an enrichment in heavy lanthanides.

BALASHOV *et al.* (1964) assume that with the increase in alkalinity of weathering solutions the lanthanides are precipitated and fixed like Al, Ga and Ti. A low pH

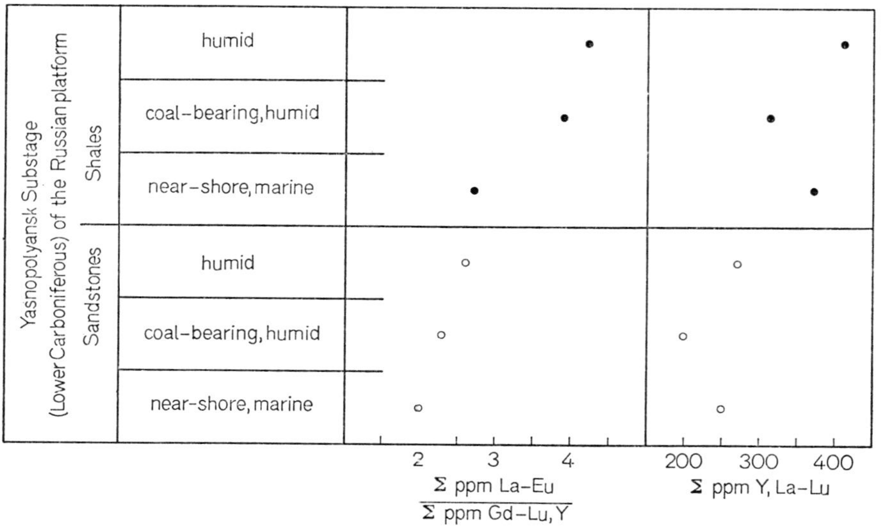

Fig. 39, 57—71-G-1. Fractionation of

$$\frac{\Sigma \text{ ppm La—Eu}}{\Sigma \text{ ppm Gd—Lu, Y}}$$

in shales and sandstones from different environments. Total Σ ppm Y, La—Lu are plotted for comparison. Data from BALASHOV (1964); see also Table 39, 57—71-K-1

favors the solution and removal of the lanthanides. This process may explain the minimum of the lanthanide abundances in shales and sandstones from coal-bearing deposits (BALASHOV *et al.*, 1964).

Investigations of the lanthanides carried by the Volga river confirm a preferential removal of the heavy lanthanides by river waters in the humid zones (BALASHOV and KHITROV, 1967; see also Section 39, 57—71-H and 39, 57—71-I). The studies by RONOV *et al.* (1967) of the changes in the lanthanides between source rocks and their weathering products lack a reference element or mineral for comparing mass equivalences in primary rocks and weathered matter (for details on balances of this type see KÖSTER, 1961; etc.).

The behavior of some lanthanides during the weathering of eudialyte from nepheline syenites and granitic rocks, of calcite and parisite from carbonatites and of the whole granitic rock was studied by BALASHOV *et al.* (1965), BALASHOV and KUDINOV (1966), and BURKOV and PODPORINA (1967). The latter authors observed a significant decrease of lanthanides in the residium (kaolin zone) of granitic rock compared with the parent rock. It was suggested that the lighter lanthanides are particularly mobile in an acid environment and are leached during the development of the kaolinite zone.

Revised manuscript received: February 1969

39, 57—71-H. Solubilities of Compounds which Control Concentrations of the Lanthanides in Natural Waters (I); Valence States in Natural Environments (II)

I.

In natural waters the soluble fraction of the lanthanides can exist in different states, for example as ions, complexes or colloids (BALASHOV *et al.*, 1964; BALASHOV and KHITROV, 1967). The solubility products of lanthanide hydroxides are given in Table 39, 57—71-H-1 (VICKERY, 1953). Y and La—Lu form the most stable complexes with some organic compounds. The stability of complexes with organic acids depends upon the dissociation and structure of the acid (RYABCHIKOV and TERENTYEVA, 1964). During weathering and alteration of rocks most of the dissolved lanthanides can be transported in the form of carbonate and perhaps organic complexes (BALASHOV *et al.*, 1964). The stability of complexes with alkali carbonates and organic compounds increases from La—Lu. Hence a higher content of HCO_3^- in natural waters in a humid environment causes a higher solubility of the heavy lanthanides as compared to the light elements of this group (BALASHOV *et al.*, 1964).

Table 39, 57-71-H-1. *Solubility product K at 25° C of trivalent hydroxides* (VICKERY, 1953)

La	1.0×10^{-19}
Ce	1.5×10^{-20}
Pr	2.7×10^{-20}
Nd	1.9×10^{-21}
Sm	6.8×10^{-22}
Eu	3.4×10^{-22}
Gd	2.1×10^{-22}
Er	1.3×10^{-23}
Tm	3.3×10^{-24}
Yb	2.9×10^{-24}
Lu	2.5×10^{-24}
for comparison:	
$Al(OH)_3$	3.7×10^{-15} (25° C)[a]
$Ca(OH)_2$	5.47×10^{-6} (18° C)[a]
$Fe(OH)_2$	4.8×10^{-16} (18° C)[a]
$Fe(OH)_3$	3.8×10^{-38} (18° C)[a]

[a] D'ANS and LAX (1967); for additional data see MARTELL and SILLÉN (1964).

ROBINSON (1943) and ROBINSON *et al.* (1958) observed a relatively high portion of lanthanides to be exchangeable in soils. HASKIN, WILDEMAN *et al.* (1966) removed a concentration of x ppm radioactive Eu^{3+} by precipitation of only a small amount of $CaCO_3$. The solubility product of $Eu_2(CO_3)_3$ was not reached.

II.

Krumholz (1964) has published a compilation of experimental data on the valence changes of the lanthanides Ce, Sm, Eu, and Yb. Most abundant under natural conditions is the oxidation state +3 for Y and La—Lu, but Ce, Tb, and Pr can be expected to be oxidized to the +4 state and Eu, Yb, and Sm can be expected to be reduced to the +2 state. Some lanthanide distribution patterns of minerals, rocks and stone meteorites indicate an "anomalous" behavior of Ce and Eu with respect to the neighboring lanthanides. No similar "anomalies" are observed for the elements Tb, Pr, Yb, and Sm in naturally formed minerals and rocks.

Some lanthanide oxidation-reduction potentials are given in Table 39, 57—71-H-2 (Vickery, 1953).

Table 39, 57—71-H-2. *Lanthanide element electrode potentials* ε° *(hydrogen scale)* (Vickery, 1953)

$La = La^{3+} + 3e^-$	2.37 v	
$Y = Y^{3+} + 3e^-$	ca. 2.1 v	
$Pr = Pr^{3+} + 3e^-$	ca. 2.0 v	
$Sm^{2+} = Sm^{3+} + e^-$	ca. 2.0 v	(ethanolic chloride solution)
	1.72 v	(polarographic determination)
$Yb^{2+} = Yb^{3+} + e^-$	1.15 v	
$Eu^{2+} = Eu^{3+} + e^-$	0.43 v	
$Pr^{3+} = Pr^{4+} + e^-$	< -1.60 v	
$Ce^{3+} = Ce^{4+} + e^-$	-1.61 v	

Haskin *et al.* (1968) reported a relative depletion of Ce in some individual basalts (peridotites) and in some granitic rocks. Goldberg *et al.* (1963) have observed the remarkably low Ce concentration in sea water compensated by a relative accumulation of Ce in marine ferromanganese nodules, probably after oxidation. Ronov *et al.* (1967) assume that the oxidation of Ce^{3+} to Ce^{4+} takes place in a weakly alkaline environment under surface conditions, usually in zones of weathering and sedimentation. They also mention that in its chemical properties the tetravalent cerium is closer to the heavy lanthanides than to the light ones.

The anomalous behavior of Eu, deviating from a smooth distribution curve, has been frequently observed in minerals and in rocks (see also Section 39, 57—71-D). It can most clearly be observed in Ca-minerals, being controlled by the similar crystal chemical properties of Ca and Eu^{2+}. This effect has been explained by the reduction of Eu^{3+} to Eu^{2+}.

Philpotts and Schnetzler (1968b) have used the Eu anomaly for an interpretation of the genesis of basalt. Haskin *et al.* (1968) have compared Eu concentration in a few samples of geologically old sedimentary rocks with that in North American shales. In these cases the Eu concentration in material 3.5 billion years old was twice that of sediments accumulated less than 1.5 billion years ago. According to Ronov *et al.* (1967) there has been no change in the lanthanides of shales during the past 1.5 billion years.

Revised manuscript received: February 1969

39, 57—71-I. Abundance in Natural Waters

The concentration of the lanthanides in sea water (Indian Ocean, Pacific Ocean, Atlantic Ocean) is very low. 3 groups of authors have published complete sets of data on Σ Y, La—Lu: BALASHOV and KHITROV (1961), see also BALASHOV and LISITSIN (1968); GOLDBERG *et al.* (1963); HØGDAHL (1967), and HØGDAHL *et al.* (1968). HAYES, HOOD and SLOWEY (1964; quoted from HASKIN, FREY *et al.*, 1966) have contributed values for single elements of this group in surface water from the Gulf of Mexico. BALASHOV and KHITROV (1961) found a gradual increase of the Σ Y, La—Lu with depth in sea water from 10 stations in the Indian Ocean. In agreement with this observation, water samples from the Pacific Ocean approximately 15 km west of San Diego show a concentration gradient of four in Σ Y, La—Lu between a water depth of 4,000 meters and the surface (GOLDBERG *et al.*, 1963; Table 39, 57—71-I-1a). In absolute concentrations, there is a discrepancy of one to two orders of magnitude between BALASHOV and KHITROV (1961) and the other groups of authors. The present author has quoted the results from the most complete sampling analysed by neutron activation (Table 39, 57—71-I-1a; Fig. 39, 57—71-I-1).

The residence time for lanthanides in ocean water was calculated by GOLDBERG *et al.* (1963) and, on the basis of more recent data by WILDEMAN and HASKIN (1965), was found to be in the range of 200 years for La, Pr, Nd, Sm, Eu, Gd, in the range of 400 years and more for Ho—Lu, but as low as 50 years for Ce only. BALASHOV and KHITROV (1961) mentioned that no more than 50% of the Σ Y, La—Lu in these waters is associated with suspended particles > 0.7 μ in diameter and not more than 0.1—0.3% with the plankton. They also noted that the content of Σ Y, La—Lu in suspended matter increases with depth and decreases with shore distance. HAYES, HOOD and SLOWEY (1964; quoted from HASKIN, FREY *et al.*, 1966) and HAYES *et al.* (1965) again mention that approximately from one-half to two-thirds of the total lanthanide concentrations were fixed to fine-grained matter (> 0.45 μ) in surface water from an offshore station in the Gulf of Mexico (Table 39, 57—71-I-1a). The soluble fraction of Y, La—Lu in sea water is less than that required to precipitate the least soluble hydroxides of the lanthanides (BALASHOV and KHITROV, 1961). BALASHOV and KHITROV (1961) assume that terrigenous material is a major source of the lanthanides in the ocean (see also WILDEMAN and HASKIN, 1965).

The lanthanide pattern of ocean water certainly differs from that of the average crustal rocks. Ocean waters are relatively enriched in heavy lanthanides as compared to the chondritic pattern. This observation may be explained by a higher solubility or a lower sorption of the heavy lanthanides already observable in the difference in residence time between light and heavy lanthanides (for additional information see Section 39, 57—71-H).

In various ocean sediments and in sediments deposited in shallow seas, WILDEMAN and HASKIN (1965) have found distribution patterns of the lanthanides similar

Table 39, 57—71-I-1a and 1b. *Absolute abundances of Y and La—Lu in ocean water in ppb and in ocean sediments in ppm (dash: no value reported). Σ Y, La—Lu includes estimates for missing analytical values*

Table 39, 57—71-I-1a

Water type	Locality	Method	$_{39}$Y	$_{57}$La	$_{58}$Ce	$_{59}$Pr	$_{60}$Nd	$_{62}$Sm	$_{63}$Eu	$_{64}$Gd	$_{65}$Tb	$_{66}$Dy	$_{67}$Ho	$_{68}$Er	$_{69}$Tm	$_{70}$Yb	$_{71}$Lu	Σ Y, La—Lu	Reference
Ocean water	100 m depth in Pacific Ocean approximately 15 km west of San Diego	N/R	—	0.0029	0.0013	0.00064	0.0023	0.00042	0.000114	0.0006	—	0.00073	0.00022	0.00061	0.00013	0.00052	0.00012	≈0.013	1
	Sample from 4,000 m depth, same locality	N/R	—	—	—	—	—	—	—	—	—	—	0.00072	—	—	0.00280	0.00064	—	1
	North Atlantic deep water. 11 samples	N/R	0.0133	0.0034	0.0012	0.00064	0.0028	0.00045	0.00013	0.00070	0.00014	0.00091	0.00022	0.00087	0.00017	0.00082	0.00015	0.026	2
	Gulf of Mexico, surface water																		
	Particulate phase	N/R	—	—	0.026	0.010	—	0.0091	0.0024	—	—	0.0072	(0.015)[a]	—	—	0.0024	0.0041	—	3
	Soluble phase	N/R	—	—	—	0.0031	—	0.0021	0.0004	—	—	0.0053	0.0015	—	—	0.0027	0.0005	—	3
	Total	N/R	—	—	0.026	0.0131	—	0.0113	0.0028	—	—	0.0125	(0.016)[a]	—	—	0.0051	0.0046	—	3

[a] Doubtful value (HERRMANN).

Reference: 1. GOLDBERG *et al.* (1963); 2. HØGDAHL (1967), HØGDAHL *et al.* (1968); 3. HAYES, HOOD, and SLOWEY (1964), quoted from HASKIN, FREY *et al.* (1966).

Table 39, 57—71-I-1b

Sample	Locality	Method	$_{39}$Y	$_{57}$La	$_{58}$Ce	$_{59}$Pr	$_{60}$Nd	$_{62}$Sm	$_{63}$Eu	$_{64}$Gd	$_{65}$Tb	$_{66}$Dy	$_{67}$Ho	$_{68}$Er	$_{69}$Tm	$_{70}$Yb	$_{71}$Lu	Σ Y, La—Lu	Reference
Brown clay	Argentine and Brazil basin, 3 samples	N/R	32	31	56	9.6	35	6.2	1.8	7.4	1.1	—	1.2	3.0	0.56	2.9	—	≈194	1
Glauconite	Pacific Ocean, coast near California	W	see Dy	1.05	4.15	0.26	1.05	0.26	Eu+Gd 0.26		—	Dy+Y 0.68	—	—	—	—	—	—	2
Manganese nodule	"Horizon" from 5,000 m depth about 20° N of Hawaii	N/R	—	115	518	36	138	64	10.4	46	7	30	5.8	13.8	<1.7	4.6	<1.2	≈1,090[a]	3

[a] The listed values are corrected by two orders of magnitude because of a typographical error in the original publication (GOLDBERG, private communication).

Reference: 1. WILDEMAN and HASKIN (1965); 2. BALASHOV and KAZAKOV (1968); 3. GOLDBERG *et al.* (1963).

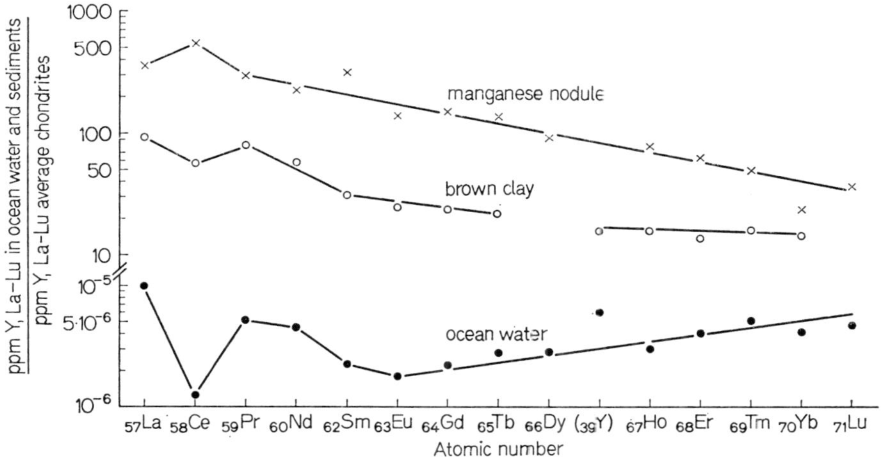

Fig. 39, 57—71-I-1. Abundances of Y and La—Lu in ocean water (Høgdahl, 1967; Høgdahl *et al.*, 1968) and ocean sediments (one manganese nodule, 3 brown clays) (Goldberg *et al.*, 1963; Wildeman and Haskin, 1965) relative to chondrites. See also Table 39, 57—71-I-1a and I-1b

to those in shales, except for abnormally low Ce concentrations (Table 39, 57—71-I-1b). This Ce depletion is compensated by a Ce peak in the lanthanide distribution of manganese oxides from the deep ocean (Fig. 39, 57—71-I-1). A relatively high rate of Ce precipitation has been indicated by the abnormally low concentration and residence time of this element in ocean water too. Likewise no Ce depletion was found by Balashov and Kazakov (1968) in glauconites from the Pacific Ocean. Haskin, Frey *et al.* (1966) have calculated that the total amount of Y and La—Lu in the oceans is the same as that in 0.2 mm sediment thickness.

Fresh waters from wells in crystalline schists of the Virginia Piedmont area west of Washington, D.C., contained about 2 and 60 ppb Σ Y, La—Lu (Robinson *et al.*, 1958). Seasonal changes between 0.001 and 1.76 ppb La in the Columbia River water were measured during 1962 by Silker (1964). Landström and Wenner (1965) found 0.20 ppb La in streams in Sweden. Turekian (this handbook, Volume I, p. 315) has estimated the concentrations for the other elements Y and Ce—Lu in river waters.

Revised manuscript received: February 1969

39, 57—71-K. Abundance in Common Sediments and Sedimentary Rock Types

The first determinations of Y and La—Lu in sedimentary rocks were made by MINAMI (1935), who analyzed three composite samples of Paleozoic and Mesozoic shales from Europe and Japan. Since 1935 until recently his values have been used for compilations of crustal abundances of the lanthanides. M. A. HASKIN and L. A. HASKIN (1966), HASKIN *et al.* (1968), and HERRMANN and WEDEPOHL (1967) likewise determined the lanthanide abundances in the original MINAMI sample "36 European Paleozoic shales" (see Tables 39, 57—71-K-1 and K-2). The absolute concentrations of La—Pr obtained by MINAMI (1935) were 2 to 3 times lower than those of the analyses published in 1966—1968. The absolute and relative lanthanide abundance patterns of the "36 European Paleozoic shales" from the 1966—1968 publications are very similar to those of a composite of 40 North American shales (M. A. HASKIN and L. A. HASKIN, 1966; Ce and Nd corrected by HASKIN *et al.*, 1968), of North American metamorphic shales (HASKIN *et al.*, 1968) and of shales from the Russian platform (BALASHOV *et al.*, 1964; RONOV *et al.*, 1967; see also Tables 39, 57—71-K-1 and K-2).

The absolute concentrations of Σ Y, La—Lu obtained for most of the shales range from ≈200 to 300 ppm. A lower Σ Y, La—Lu content as that of about 61 ppm in the shale from Trenton, Arbuckle Mountains (HASKIN, WILDEMAN *et al.*, 1966) can be caused by an abnormal portion of the sand fraction in the specimen.

The absolute Σ Y, La—Lu concentrations and the ratios

$$\frac{\text{ppm La}}{\text{ppm Nd}}, \quad \frac{\text{ppm Pr}}{\text{ppm Sm}}, \quad \text{and} \quad \frac{\Sigma \text{ ppm La—Eu}}{\Sigma \text{ ppm Gd—Lu, Y}}$$

of the average of igneous rocks of the upper continental earth's crust (Table 39, 57—71-E-4) are close to those of the "36 European Paleozoic shales" (Table 39, 57—71-K-2), as is to be expected if in the shales large units of the earth's crust have been sampled. Furthermore this seems to confirm the proportions of igneous rock types to be used for the "average". The sample "composite of 17 graywackes" contains samples from Germany as the rocks from the type locality Harz mountains (Table 39, 57—71-K-2). This specimen shows Σ Y, La—Lu = 190 ppm which corresponds to a ratio

$$\frac{\Sigma \text{ ppm Y, La—Lu average of igneous rocks of the upper earth's crust}}{\Sigma \text{ ppm Y, La—Lu composite of 17 graywackes}}$$

of 0.8. This factor must be due to a dilution of graywackes by 20 per cent quartz as compared with average igneous rocks (WEDEPOHL, 1968, unpublished[1]). Because a value of about 0.8 for this ratio is obtained for the individual elements, too, the relative lanthanide distribution patterns in the average of igneous rocks and in the

[1] Personal communication.

composite of 17 graywackes are almost identical (see Fig. 39, 57—71-K-1). Additional data for the lanthanide abundances in graywackes are contributed by HASKIN, WILDEMAN *et al.* (1966) and SCHNETZLER *et al.* (1967). The absolute lanthanide contents of some of these samples agree with the composite of 17 graywackes, whereas others have lower Y, La—Lu values. This may be due to higher proportions of quartz in the respective graywackes.

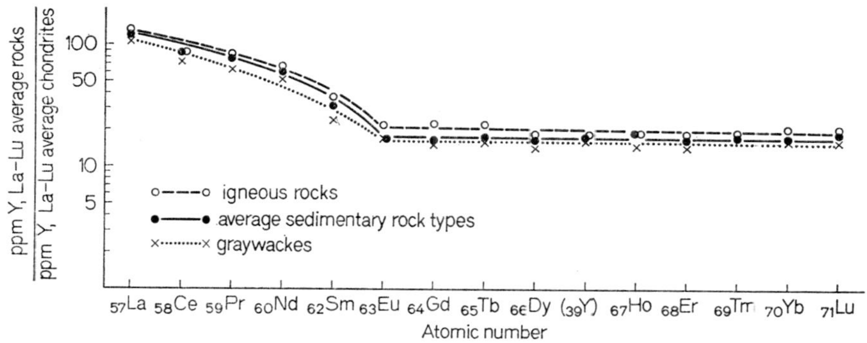

Fig. 39, 57—71-K-1. Abundances of Y and La—Lu in a weighted mean of igneous rocks from upper continental earth's crust (Table 39,57—71-E-4), average of sedimentary rock types as fraction of sediment mass on the continents (Σ Y, La—Lu ≈ 230 ppm) and a composite of 17 typical graywackes (Table 39,57—71-K-2) relative to chondrites

An arbitrary average of sedimentary rock types as a fraction of the sediment mass on the continents can be computed from the following proportions: 77% shales and graywackes, 15% sandstones and 8% limestones (WEDEPOHL, Chapter 8, Volume I of this handbook). On this basis we get a total of Σ Y, La—Lu ≈ 170 ppm or Σ Y, La—Lu ≈ 230 ppm for the sedimentary rocks (Table 39, 57—71-K-2). The latter value may be plausible as compared with the average ≈ 241 ppm Σ Y, La—Lu in igneous rocks as the ultimate source of the sediments. BALASHOV *et al.* (1964) give an average of 282 ppm Σ Y, La—Lu for the Middle Paleozoic sedimentary rocks of the Russian platform. This is high compared with 145 ppm Σ Y, La—Lu for the total sedimentary rocks from the same area (60% shales, 20% sandstones, 20% carbonate rocks) reported by RONOV *et al.* (1967).

Until now no distinct relationship between the lanthanide abundances in sedimentary rocks and their geological age could be observed (SCHNETZLER and PHILPOTTS, 1967).

The lanthanide contents in 4 samples of pelagic clay from the Atlantic Ocean range from about 160 to 260 ppm Σ Y, La—Lu (WILDEMAN and HASKIN, 1965; see also Table 39, 57—71-I-1b). Similar values were reported by OSTROUMOV (1953) for Recent clays from the Black Sea. The distribution of the lanthanides in these samples is similar to that in shales and graywackes, but has a distinct depletion of Ce (Fig. 39, 57—71-I-1). The lanthanides in manganese nodules from the bottom of the Pacific Ocean and the Indian Ocean have been determined by GOLDBERG *et al.* (1963) and PACHADZHANOV *et al.* (1963). The relative lanthanide pattern of the manganese nodule from the Pacific Ocean represents the well-known accumulation

Table 39, 57—71-K-1. *Absolute abundances of Y and La—Lu in sedimentary rock types in ppm (dash: no value reported). Σ Y, La—Lu includes estimates for missing analytical values*

Rock type	Locality	Method	$_{39}$Y	$_{57}$La	$_{58}$Ce	$_{59}$Pr	$_{60}$Nd	$_{62}$Sm	$_{63}$Eu	$_{64}$Gd	$_{65}$Tb	$_{66}$Dy	$_{67}$Ho	$_{68}$Er	$_{69}$Tm	$_{70}$Yb	$_{71}$Lu	Σ Y, La—Lu	Reference
Shales	Composite of 36 European Paleozoic shales	N/R	31.8	41.1	85.9	10.4	41.7	7.3	1.52	6.03	1.05	—	1.20	3.55	0.56	3.29	0.58	≈242	1
		X, S	45[a]	48.8	96	10.9	41	7	1.2	6.0	1.0	5.8	1.8	4.0	0.6	3.9	0.8	≈274	2
	Composite of 40 North American shales	N/R	27	32	73	7.9	33	5.7	1.24	5.2	0.85	—	1.04	3.4	0.50	3.1	0.48	≈200	1
	Composite of North American metamorphic shales	N/R	27.6	34.0	73.8	8.2	35	6.05	1.51	5.70	0.88	—	1.12	3.30	0.48	2.71	0.41	≈206	3
	U.S.A. Precambrian, Ordovician, Mississippian Pennsylvanian (4 samples)	N/R	16.7	28	49	5.8	17.6	3.9	0.85	3.5	0.37	—	0.59	1.96	0.30	1.82	0.42	≈133	4
	Sillery, Quebec, Canada	N/R	38	73	81	22	47	10.2	2.4	9.6	1.2	—	1.8	—	0.66	5.3	0.93	≈304	5
	Fig tree shale from the Swaziland System of Southern Africa. Age ≈ 3.0 billion years	I	—	—	62	—	34	6.7	1.54	—	—	3.4	—	1.8	—	1.9	—	—	6
	Famennian stage of the Russian Platform (Upper Devonian)																		
	D_1fm. Continental arid (71 samples)	X	66	48	89	12	44	11	—	9	—	7	—	5.8	—	5.4	—	≈300	7
	D_1fm. Saline, lagoonal (70 samples)	X	45	41	81	11	38	8	—	6.6	—	6.4	—	4.4	—	4.3	—	≈250	7
	D_3fm. Average for the basin (141 samples)	X	54	44	84	11	40	9.2	—	7.7	—	6.9	—	4.9	—	4.8	—	≈270	7
	Yasnopolyansk substage of the Russian platform (lower Carboniferous)																		
	C_1jasn. Humid (69 samples)	X	51	79	152	19	63	12	—	10	—	7	—	4.7	—	4.2	—	≈410	7
	C_1jasn. Coal-bearing, humid (45 samples)	X	40	58	113	14	50	10	—	7.6	2	5.6	1.5	3.5	—	3.1	—	≈310	7
	C_1jasn. Near-shore, marine (34 samples)	X	65	58	118	15	60	14	—	12	—	9	—	6	—	6	—	≈370	7
	C_1jasn. Average for the basin (148 samples)	X	53	67	130	16	58	12	—	9.5	2.7	7	2	4.6	—	4.4	—	≈370	7
	Average for D_3fm. and C_1jasn. (289 samples)	X	53	61	120	15	54	11	—	9	—	6.9	—	4.6	—	4.5	—	≈345	7
Gray-wackes	U.S.A. Precambrian, Devonian or Carboniferous, Jurassic, Cretaceous (5 samples)	N/R	20	21.6	31.6	6	22.2	4.7	1.0	4.1	0.61	—	0.79	2.1	0.37	2.1	0.37	≈120	5
	Sample vicinity Bosumtwi Crater area of Ghana with an age of about 2 billion years	I	—	—	54	—	42	6.7	1.7	5.5	—	3.72	—	1.8	—	1.56	—	≈190	8
Sand-stones	U.S.A. (4 samples), Libyan Desert sand (1 sample)	N/R	10.8	5.6	—	2.2	8.5	2.4	0.43	2.0	0.39	—	0.40	0.67	0.29	0.93	—	≈ 50	4
	Famennian stage of the Russian Platform (Upper Devonian)																		
	D_3fm. Continental arid (38 samples)	X	61	57	105	14	47	10	—	10	—	8.4	—	5.9	—	5.5	—	≈330	7
	D_3fm. Saline, lagoonal (38 samples)	X	69	45	81	12	39	10	—	8	1.9	10	2	6	—	5.3	—	≈290	7
	D_3fm. Average for the basin (76 samples)	X	65	51	93	12	43	9.8	—	9	2	9.2	2.2	6	—	5.6	—	≈310	7
	Yasnopolyansk substage of the Russian platform (lower Carboniferous)																		
	C_1jasn. (3 samples)	X	—	49	86	12	39	8	—	—	—	—	—	—	—	—	—	≈270	7
	C_1jasn. Coal-bearing (16 samples)	X	37	33	60	9	30	7	—	6.2	—	4.5	—	3.5	—	3.1	—	≈198	7
	near-shore, marine (11 samples)	X	57	35	74	10	39	8.2	—	8.2	1.5	6	—	4.8	—	4.4	—	≈250	7

Sand-stones	C_1jasn. Average for the basin (30 samples)	X	45	36	69	9.7	34	7.6	—	7.1	—	5.2	—	4	—	3.7	—	≈225	7
	Average for D_3fm. and C_1jasn. (106 samples)	X	54	42	78	11	38	8.4	—	7.1	—	6.9	—	4.9	—	4.4	—	≈260	7
Lime-stones	7 samples from England and U.S.A. Pre-cambrian, Silurian, Carboniferous, Jurassic	N/R, S	9	8	—	2	6.5	1.4	0.4	1.7	0.44	—	0.28	0.7	0.16	0.9	0.16	≈49	9
	Solnhofen (Germany), Jurassic	I	—	—	2.6	—	1.42	0.28	0.06	0.24	—	0.22	—	0.12	—	0.104	—	—	6
	Ave. for Yasnopolyansk substage (lower Carboniferous) of the Russian platform (116 samples)	X	23	10	20	2.5	9	2.1	—	2.7	—	2.1	—	1.7	—	1.6	—	≈ 77	7
	Florida Bay, Carbonate sediment (Recent)	N/R	4.15	3.55	—	0.99	3.05	0.58	0.10	0.43	0.058	—	0.21	0.21	0.054	0.42	0.063	≈ 22	10
Corals	Florida Bay (Pleistocene)	N/R	0.099	0.052	—	0.028	—	0.013	0.0034	0.013	—	—	0.006	—	—	0.0084	—	≈ 0.4	11
	Florida Bay (Recent). Porites porites	N/R	0.124	0.08	2.7	—	0.71	0.079	0.058	0.072	0.0058	—	0.0048	0.019	0.0032	0.018	—	≈ 4	11
Sea shells	New Jersey, coast near Ocean City (composite)	N/R	0.22	0.13	0.78	0.053	0.26	0.038	0.012	0.055	0.010	—	0.010	0.036	0.005	0.019	0.003	≈ 1.7	11
Marble	Marble Point, Surko Creek, Antarctica, and Beldens Formation, West Rutland, Vermont (2 samples)	N/R	5.7	5.3	5.5	1.2	4.1	0.94	0.16	0.64	0.19	—	0.18	0.6	0.09	0.55	0.09	≈ 26	4
Soils	Average contents in various soils of the U.S.S.R. (5 samples)	X	—	28	53	6.7	26	6.4	<0.7	4.2	<0.7	3	<0.7	<0.7	<0.7	<0.7	<0.7	≈150	12
Average of 511 Middle Paleozoic sedimentary rocks of the Russian platform		X	49	50	98	12	44	9.5	—	7.8	1.8	6.2	1.8	4.3	—	4.2	—	≈282	7

[a] Probably somewhat high.

Reference: 1. M. A. Haskin and L. A. Haskin (1966), Haskin *et al.* (1968); 2. Herrmann and Wedepohl (1967); 3. Haskin *et al.* (1968); 4. Haskin and Gehl (1962), Haskin, Wildeman *et al.* (1966); 5. Haskin, Wildeman *et al.* (1966); 6. Schnetzler and Philpotts (1967); 7. Balashov *et al.* (1964); 8. Schnetzler *et al.* (1967); 9. Haskin and Gehl (1962), Haskin, Wildeman *et al.* (1966), H. Levine and I. May, quoted from Graf (1960); 10. Haskin and Gehl (1962); 11. Schofield and Haskin (1964); 12. Vinogradov (1959).

Table 39, 57—71-K-2. *Absolute abundances of Y, La—Lu in an average of shales, graywackes, sandstones and limestones*

Element	Composite of 36 European Paleozoic shales. Average from analyses M. A. Haskin and L. A. Haskin (1966), Herrmann and Wedepohl (1967), Haskin *et al.* (1968) ppm	8,616 shales from the Russian platform (Paleozoic, Mesozoic, Cenozoic). Ronov *et al.* (1967). Method: X ppm	Composite of 17 typical graywackes. Herrmann (1967), unpublished. Methods: X, S ppm	6,051 sandstones from the Russian platform. Ronov *et al.* (1967). Method: W, X ppm	11,205 limestones from the Russian platform. Ronov *et al.* (1967). Method: W ppm	Average sedimentary rock types as fraction of sediment mass on the continents (77% shales, 15% sandstones, 8% limestones [a]) ppm	Average of sedimentary rock types as fraction of sediment mass on the continents (77% shales, 15% sandstones, 8% limestones [b]) ppm
$_{39}$Y	38.4	30	31	15	3.8	26	35
$_{57}$La	45	34	33	17	4.1	29	40
$_{58}$Ce	91	67	62	33	6.5	57	80
$_{59}$Pr	10.7	8.8	7.7	4	1.7	7.5	9.5
$_{60}$Nd	41.4	30.5	31	16.5	4.6	26.4	37
$_{62}$Sm	7.2	6.9	5	3.7	1.4	6.0	6.4
$_{63}$Eu	1.4	1.4	1.3	0.7	Σ Eu + Gd	1.2	1.3
$_{64}$Gd	6.0	6.2	5.1	3.1	1.2	5.4	5.5
$_{65}$Tb	1.0	—	—	—	—	0.7[c]	0.9
$_{66}$Dy	5.8	4.5	4.5	2.6	0.8	4.0	5.2
$_{67}$Ho	1.6	—	1.1	—	—	0.9[c]	1.4
$_{68}$Er	3.8	3.3	3.1	1.6	0.4	2.8	3.4
$_{69}$Tm	0.6	—	—	—	—	0.4[c]	0.6
$_{70}$Yb	3.6	2.6	3.0	1.2	0.3	2.2	3.3
$_{71}$Lu	0.7	—	0.5	—	—	0.4[c]	0.6
Σ Y, La—Lu	258	≈200[d]	≈190[d]	≈100[d]	≈25[d]	≈170	≈230
ppm La / ppm Nd	1.1	1.1	1.1	1.0	0.9	1.1	1.1
ppm Pr / ppm Sm	1.5	1.5	1.5	1.1	1.2	1.3	1.5
Σ ppm La—Eu / Σ ppm Gd—Lu, Y	3.2	3.0	2.8	3.0	2.8	3.0	3.1

[a] The average lanthanide concentrations are calculated on the basis of 8,616 shales, 6,050 sandstones and 11,205 limestones from the Russian platform (Ronov *et al.*, 1967). [b] The average lanthanide concentrations are calculated on the basis of the composite of 36 European Paleozoic shales (77% for shales plus 1% shales instead of the clay residue of 8% limestones) and the composite of 17 typical graywackes (for sandstones). [c] Values estimated through comparison with the neighboring lanthanides. [d] Including estimates for elements not determined.

of the lighter lanthanides but the elements Ce and Sm show distinct maximas (Fig. 39, 57—71-I-1; see also Table 39, 57—71-I-1 b). For a discussion of the peculiar distribution of the lanthanides in ocean matter, see Section 39, 57—71-I.

SCHOFIELD and HASKIN (1964) determined the distribution of lanthanides in two corals and a composite of sea shells, WILDEMAN and HASKIN (1965) in globigerina ooze and diatomaceous ooze. In the corals and shells the absolute lanthanide content is very low (Σ Y, La—Lu $\approx$ 0.5—2 ppm, see also Table 39, 57—71-K-1). The source of the higher lanthanide contents found in the globigerina and diatomaceous oozes (Σ Y, La—Lu $\approx$ 120—160 ppm) is most probably due to the detrital clay proportions in these samples. The average of Y, La—Lu in soils of the U.S.S.R. is $\approx$ 150 ppm (VINOGRADOV, 1959), ranging from 6.2 ppm Σ Y, La—Lu in a B-horizon on basic rocks to 270 ppm in mountain tundra soils (see also Table 39, 57—71-K-1).

Revised manuscript received: February 1969

39, 57—71-L. Biogeochemistry

I. Plants and Coals

ROBINSON *et al.* (1938, 1958), ROBINSON (1943), and ROBINSON and EDGINGTON (1945) reported data on Y and La—Lu in dry hickory leaves ranging from 3 to ≈2,300 ppm Σ Y, La—Lu and in dry crop plants ranging from 4 to ≈170 ppm Σ Y, La—Lu. In small twigs, trunk wood and taproot wood of a hickory tree 40 to ≈60 ppm Σ Y, La—Lu were found, and in bark of taproot 140 ppm Σ Y, La—Lu. Nut meats, nut shells and nut husks from another hickory tree contained 5—17 ppm and the leaves 981 ppm Σ Y, La—Lu (ROBINSON and EDGINGTON, 1945). Leaves from hickory on soils originating from granite and gneiss have the highest quantities of lanthanides, compared with those grown on soils from basic rocks and sandstones (ROBINSON, 1943). In comparison with the soil the hickory trees do not appreciably fractionate the lanthanides.

PAPPAS *et al.* (1963) found ≈30—150 ppm Σ Y, La—Lu in samples of opium ash of different origin. The variations in the distribution patterns of the lanthanides in opium ashes have tentatively been used to identify the origin of the specimens.

Early determinations of Y, Ce, Nd, Sm, Gd, Dy, Er, and Yb in the ash of a number of German and British coals were made by GOLDSCHMIDT and PETERS (1933), who obtained an average of 100 ppm Y and a maximum of 800 ppm Y. YERSHOV (1961) reported on coals from the Kizelovskii coal basin that Y and Yb are fixed predominantly in the organic matter of the coal and that the total of the Y, La—Lu oxides contains almost 53% Y. KOSTERIN *et al.* (1963) and SCHOFIELD and HASKIN (1964) analysed the lanthanides in coals from the Chekez lignite deposit and from Western Pennsylvania and Southern Illinois respectively (see also Table 39, 57—71-L-1). The average ash content of the sampled North American coals is about 7%. From this a content of ≈500 ppm Σ Y, La—Lu in the ash can be computed which was considered to be too high for a purely detrital origin of the lanthanides (SCHOFIELD and HASKIN, 1964; HASKIN, FREY *et al.*, 1966). The coals are relatively enriched in lighter lanthanides as compared to the pattern of chondrites, but not as compared to the pattern of graywackes.

II. Phosphates

Few determinations have been made of lanthanides in animal tissues. Bone ash of cattle contains ≦0.01 ppm lanthanides (LUX, 1938) and the ash of a mouse ≈1.8 ppm Σ Y, La—Lu (POETZ and HASKIN, quoted from HASKIN, FREY *et al.*, 1966; see also Table 39, 57—71-L-1). These two specimens are relatively enriched in lighter lanthanides as compared to the lanthanide patterns in chondrites.

ARRHENIUS and BONATTI (1965) describe high concentrations of Y and La—Lu in pelagic fish bone apatite. KOCHENOV and ZINOV'EV (1960) studied the abundance of Y, La—Lu in phosphatic fish remains (with a few tenths of a percent up to 5% of organic matter) of various Mesozoic and Cenozoic sediments from the Maikop beds

Table 39, 57—71-L-1. *Absolute abundances of Y and La—Lu in phosphatic material and in coal in ppm (dash: no value reported). Σ Y, La—Lu includes estimates for missing analytical values*

Material	Origin	Method	$_{39}$Y	$_{57}$La	$_{58}$Ce	$_{59}$Pr	$_{60}$Nd	$_{62}$Sm	$_{63}$Eu	$_{64}$Gd	$_{65}$Tb	$_{66}$Dy	$_{67}$Ho	$_{68}$Er	$_{69}$Tm	$_{70}$Yb	$_{71}$Lu	Σ Y, La—Lu	Reference
Phosphatic material	Average in bone detritus of Mesozoic and Cenozoic fishes (67 samples)	X	1.750	1.500	2.700	500	1.500	430	20	330	30[a]	300	50[a]	220	—	100	—	≈9.500	1
	Average in the remains of Paleozoic fishes of the Russian platform (12 samples)	X	1.362	1.083	2.263	538	1.817	448	—	468	61	305	52	122	35	53	18	≈8.700[b]	2
	Cattle (bone ash)	X	0.02	0.01	0.0185	—	0.0091	—	—	0.0042	—	0.0031	—	—	—	—	—	—	3
	Mouse (ash)	N/R	0.24	0.33	0.58	0.073	0.33	0.043	0.016	0.049	0.012	—	0.017	0.035	0.010	0.026	0.0075	≈1.8	4
	Concretions, nodules. 21 phosphorite samples from the Russian platform (Paleozoic; Mesozoic)	X	—	95	192	36	133	29	4	31	4	21	5	14	3	13	3	≈720[b]	5
	Phosphorite, taken from a depth of 100 m off the continental shelf north of Magdalena Bay	N/R	—	56	102	14.5	57	11.2	2.8	15	2.1	12.3	3.3	9	1.06	6.7	1.2	≈ 380	6
	Phosphorite, Phosphoria formation	N/R	196	160	—	34	96	24	5.8	—	3.5	—	3.9	—	2.1	11	—	≈ 900	7
Coal	Western Pennsylvania and Southern Illinois (2 samples)	N/R	4.7	4.1	11.5	2.2	4.7	1.6	0.7[c]	1.6	0.3	—	0.3	0.6	0.1	0.5	0.07	≈ 34	7

[a] Kochenov and Zinov'ev (1960) report 330 ppm Tb and 500 ppm Ho. These values are probably ten times too high.
[b] For this table the relative values of the original paper have been converted into ppm.
[c] Calculated figure using a Eu-value reported by Haskin, Frey *et al.* (1966).

Reference: 1. Kochenov and Zinov'ev (1960); 2. Blokh (1961); 3. Lux (1938); 4. Poetz and Haskin (1966), quoted from Haskin, Frey *et al.* (1966); 5. Semenov *et al.* (1962); 6. Goldberg *et al.* (1963); 7. Schofield and Haskin (1964);

of the Caucasus and from some other deposits of the U.S.S.R. A direct linear relationship between the lanthanides and the P_2O_5 content is noted in bone detritus from sulfide-bearing layers and from other sediments. The lanthanide concentration and the lanthanide-to-P_2O_5 ratio increase from the large bones to the fine fractions of bone detritus separated from the sample. The solid, massive bones, teeth and fin rays of large fish, ear bones of whales etc., have the lowest concentrations of lanthanides. A low lanthanide-to-P_2O_5 ratio was also found in mineralized bones the pores of which are filled with calcite, celestite or barite. These minerals separate the phosphates from the surrounding matrix and prevent accumulation of the lanthanides in them. The highest amounts of the lanthanides occur in the surface parts of bones. KOCHENOV and ZINOV'EV (1960) state that in general the lowest concentrations of the lanthanides occur mainly in phosphatic remains of continental sediments deposited in freshwater basins. The highest concentrations in bones are characteristic of normal marine argillites, while somewhat lower concentrations occur in carbonate rocks and glauconitic sandstones. The absolute abundance of the lanthanides in fossil bones is somehow related to organic matter. KOCHENOV and ZINOV'EV (1960) stated that the lanthanide content in the phosphatic remains involves a coupled replacement of Ca^{++} by Na^{+} and the trivalent lanthanides. 67 specimens of bone residues of fish contain on average 9,500 ppm Σ Y, La—Lu (KOCHENOV and ZINOV'EV, 1960), and 12 samples of Paleozoic fish remains in sediments of the Russian platform 8,700 ppm Σ Y, La—Lu (BLOKH, 1961; see also Table 39, 57—71-L-1). BLOKH (1961) assumes that the accumulation of these high lanthanide concentrations in the phosphates during the lifetime of fish is improbable. The bones of recent fish do not contain concentrations of the lanthanides of more than 100 ppm (VINOGRADOV, quoted from BLOKH, 1961). The author concludes that the lanthanides accumulated in the bone phosphates during diagenesis of the sediments. The phosphate of phosphorite deposits has not undergone such changes. SEMENOV *et al.* (1962), GOLDBERG *et al.* (1963), and SCHOFIELD and HASKIN (1964) reported analyses of phosphorites and Recent marine phosphates ranging from 400 to 900 ppm Σ Y, La—Lu (see also Table 39, 57—71-L-1).

The fish remains are relatively enriched in lighter lanthanides as compared to the pattern of chondrites.

Revised manuscript received: February 1969

39, 57—71-M. Abundances in Common Metamorphic Rock Types

At present only a few analyses of Y and La—Lu in metamorphic rocks are available. A group of metamorphic shales (HASKIN *et al.* 1968) is listed under sedimentary rocks in the Table 39, 57—71-K-1. This average is very close to that of non-metamorphic shales. The similarity seems to be due to a lack of influence of metamorphism on the composition of the lanthanides. Investigations of 6 eclogites must be mentioned: two from kimberlite pipes in South Africa (Roberts Victor mine and Dutoitspan), one from a basic breccia pipe near Delegate, Australia, one from a large inclusion in dunite, Shikoku, Japan, and two eclogites from intercalations in Gierałtów gneisses, Nowa Wieś, Poland (SCHMITT *et al.* 1963; FREY, HASKIN, and POETZ, quoted from HASKIN, FREY *et al.*, 1966; SCHMITT *et al.*, 1964; BAKUN-CZUBAROW, 1965). The absolute Y, La—Lu concentrations in these samples range from about 25 to 120 ppm (Table 39,57—71-M-1). When compared with chondrites, the distribution patterns and the ratios

$$\frac{\Sigma \text{ ppm La—Eu}}{\Sigma \text{ ppm Gd—Lu, Y}} \approx 1$$

of five out of six eclogites analysed so far show no significant fractionation between lighter and heavier lanthanides (Fig. 39, 57—71-M-1) comparable with Ca-rich Ap achondrites, oceanic tholeiitic basalts and some continental tholeiitic basalts (from N.W. Germany for instance, see Section 39, 57—71-E). Only the Dutoitspan specimen containing about 117 ppm Σ Y, La—Lu shows an enrichment in lighter lanthanides as compared with chondritic patterns (Fig. 39, 57—71-M-1). It must be considered that the Dutoitspan specimen was picked from a kimberlite with a content of nearly 1,000 ppm Σ Y, La—Lu, strongly dominated by the light lanthanides. Even a slight contamination from its host material would seriously affect the absolute concentrations of the lanthanides. Unfortunately no petrographic characteristics are given for some of the eclogite specimens used for the analyses reported.

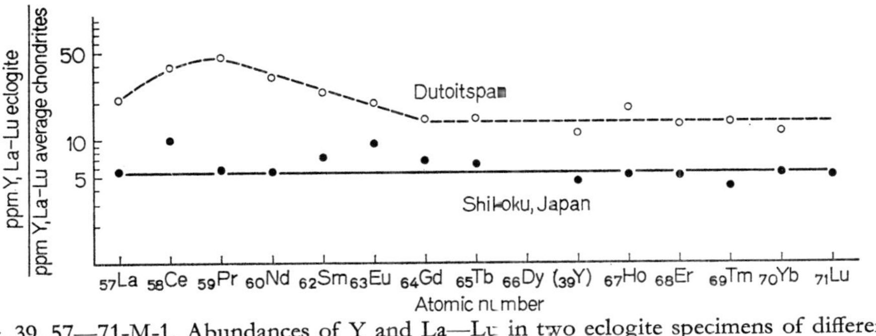

Fig. 39, 57—71-M-1. Abundances of Y and La—Lu in two eclogite specimens of different origin (Dutoitspan, kimberlite pipe, South Africa; eclogite inclusion in dunite, Shikoku, Japan) relative to chondrites. Data from FREY, HASKIN and POETZ, quoted from HASKIN, FREY *et al.* (1966); see also Table 39, 57—71-M-1

Table 39, 57—71-M-1. *Absolute abundances of Y and La—Lu in eclogites in ppm (dash: no value reported). Σ Y, La—Lu includes estimates for missing analytical values*

Locality	Method	$_{39}$Y	$_{57}$La	$_{58}$Ce	$_{59}$Pr	$_{60}$Nd	$_{62}$Sm	$_{63}$Eu	$_{64}$Gd	$_{65}$Tb	$_{66}$Dy	$_{67}$Ho	$_{68}$Er	$_{69}$Tm	$_{70}$Yb	$_{71}$Lu	Σ Y, La—Lu	Reference
Roberts Victor mine, Kimberlite pipe. South Africa	N/R	8.8	4.2	9.7	1.55	5.6	1.5	0.46	1.71	0.25	≤2.31	0.45	1.64	0.22	1.55	0.26	40	1
Basic breccia pipe near Delegate, N.S.W. Australia	N/R	27	3.6	14.1	2.5	11.7	3.8	1.25	5.2	0.75	5.7	1.00	2.75	0.41	2.00	0.32	82	2
Dutoitspan, Kimberlite pipe. South Africa	N/R	23	7.4	37	5.4	20	4.9	1.44	4.7	0.75	—	1.33	2.9	0.46	2.3	—	≈117	3
Large inclusion in dunite, Shikoku, Japan	N/R	9.2	1.8	9.5	0.70	3.4	1.45	0.70	2.1	0.32	—	0.39	1.11	0.14	1.13	0.16	≈ 34	3
Nowa Wieś, in the region of Śnieznik Kłodzki, Poland. Average samples E_1 and E_2	W	see Tb	2.0	3.1	0.8	2.5	1.4		1.9	Tb+Y 9.0	1.7	0.24	1.0	0.09		0.5	24	4

Reference: 1. Schmitt *et al.* (1963); 2. Schmitt *et al.* (1964); 3. Frey, Haskin, and Poetz, quoted from Haskin, Frey *et al.* (1966); 4. Bakun-Czubarow (1965).

Revised manuscript received: February 1969

39, 57—71-O. Relations to other Elements, Economic Importance etc.

Economic minerals used for the industrial production of some lanthanide oxides and metals are monazite, bastnaesite, davidite, gadolinite, euxenite, fergusonite, uraninite. These and other independent lanthanide minerals are primarily associated with alkalic and granitic rocks and their pegmatites and with hydrothermal mineralization stages. Many economically important monazite deposits are secondary accumulations as, for example, are the beach placers in India.

According to calculations by KOGAN (1961), the total reserves of lanthanide oxides in non-Soviet countries amount to 12—20 million tons (containing 99.9% of the lighter lanthanides), the U.S.A. having 48% of this mass, followed by India, Canada, Kenya, Brazil, and the Republic of South Africa. KOGAN (1961) calculated that during 1957 the lanthanide production in the western countries including Africa and India was 35,000 tons when based on a monazite concentrate containing 60% $(Y, La—Lu)_2O_3$.

R. M. MANDLE and H. H. MANDLE (1964) compiled information about the technical, industrial or commercial applications of lanthanides as follows: I. Metals and alloys: application in iron metallurgy as mischmetall to remove C, O, S, N, P from the melt, production of heat and corrosion-resistant stainless steels, additions to cast iron in order to increase its mechanical properties etc. II. Glass and ceramic industry: cerium oxide as excellent glass decolorizer for ultraviolet absorbing glasses, special and ceramic paints. III. Cored arc light carbons (high light output). IV. Miscellaneous. This last category includes catalysts, promoters, phosphors, and activators; the most recent use is in solid-state physics (semiconductors etc.), electronics, and nucleonics. However, as indicated by MANDLE and MANDLE, the highest consumption of relatively low priced lanthanides in the United States continues to be in the first three categories.

Y and La—Lu are of importance in industry because of their physical rather than their chemical properties.

Revised manuscript received: February 1969

References: Sections 39, 57—71-B to 39, 57—71-O

AHRENS, L. H.: The use of ionization potentials. Part 1. Ionic radii of the elements. Geochim. Cosmochim. Acta **2**, 155—169 (1952).

— Aspects of the geochemistry of the rare earths. Progress in the Science and Technology of the Rare Earths, vol. I, p. 1—29, edit. by LEROY EYRING. Oxford-London-New York-Paris: Pergamon Press 1964.

ALEKSIEV, E., and R. BOYADJIEVA: Content of rare earths in the standard igneous rocks G-1, W-1, and G-B. Geochim. Cosmochim. Acta **30**, 511—513 (1966).

ALLER, L. H.: Earth, chemical composition of, and its comparison with that of the sun, moon and planets. In: International Dictionary of Geophysics; general editor S. K. RUNCORN, p. 285—291. Oxford-London-Edinburgh-New York-Toronto-Sydney-Paris-Braunschweig: Pergamon Press 1967.

ANDERSON, A. T., JR.: Mineralogy of the Labrieville anorthosite, Quebec. Am. Mineralogist **51**, 1671—1711 (1966).

D'ANS, J., u. E. LAX: Taschenbuch für Chemiker und Physiker, 3. Aufl., Bd. I. Berlin-Heidelberg-New York: Springer 1967.

ARRHENIUS, G., and E BONATTI: Neptunism and vulcanism in the ocean. Progress in Oceanography, vol. 3, p.7— 22, editor M. Sears. Oxford-London-Edinburgh-New York-Paris-Frankfurt: Pergamon Press 1965.

BAKUN-CZUBAROW, N.: Rare-earth elements content in eclogites from Nowa Wieś in the region of Śnieżnik Kłodzki. Bull. Acad. Polon. Sci., Sér. Sci. Géol. Géograph. **13**, 187—194 (1965).

BALASHOV, YU. A.: Evolution of rare earth composition and content in the intrusive phases of the Lovozero alkalic massif (Kola Peninsula). Geochemistry (USSR), English Transl., 233—247 (1962).

— Regularities in the distribution of the rare earths in the earth's crust. Geochemistry (USSR), English Transl., 107—124 (1963).

— Analogies in the distribution of the rare earths in meteorites and in rocks of the earth's crust. (Ob analogii razdelenija redkozemel'nych elementov v meteoritach i porodach zemnoj kory.) Probl. Geochimii, 81—90 (1965). Izdatel'stvo "Nauka", Moskva, [in Russian].

— Differentiation of rare-earth elements during magmatic processes. Chap. in A. P. VINOGRADOV, Chemistry of the Earth's Crust, vol. I, p. 372—387 (1963). Izdatel'stvo Akademii Nauk SSSR. Translated from Russian, Israel Program for Scientific Translations, Jerusalem (1966).

—, M. D. DORFMAN, and N. V. TURANSKAYA: Separation of Ce from the lanthanides during the weathering of eudialite. (Otdelenie cerija of redkozemel'nych elementov pri vyvetrivanii evdialita.) Novye dannye o mineralach SSSR, mineralogičeskij muzej im. A. E. FERSMANA, trudy, vyp. 16, p. 205—208 (1965). Izdatel'stvo "Nauka', Moskva [in Russian].

—, and G .A. KAZAKOV: Source of the rare earths in Pacific ocean glauconite. Dokl. Acad. Nauk SSSR, Earth Sci. Sect. **179**, 181—183 (1968).

—, and M. A. KEKELIYA: The effect of facies changes on the distribution of the rare earths in the rocks of the Zekarsk gabbro-diorite intrusive. Geochemistry Internat. **2**, 814—821 (1965).

—, and L. M. KHITROV: Distribution of the rare earths in the waters of the Indian Ocean. Geochemistry (USSR), English Transl., 877—890 (1961).

— — Composition of the rare earths in the clastic material carried by the Volga river. Geochemistry Internat. **4**, 404—407 (1967).

BALASHOV, YU. A., and YU. A. KUDINOV: Separation of Ce from the lanthanides during the weathering of the minerals calcite and parisite in carbonatites. (Otdelenie cerija of redkozemel'nych elementov pri vyvetrivanii kal'zita i parizita iz karbonatitov.) Novye dannye o mineralach SSSR, mineralogičeskij muzey im. A. E. FERSMANA, trudy, vyp. 17, p. 176—179 (1966). Izdatel'stvo "Nauka", Moskva [in Russian].

—, and A. P. LISITSIN: Migration of rare-earth elements in the ocean. (Migrazija redkozemel'nych elementov i okeane.) Rezul'taty issledovanij po meždunarodnym geofizičeskim proektam. Okeanologičeskie issledovanija, No. 18, 213—282 (1968). Izdatel'stvo "Nauka", Moskva, [in Russian].

—, and G. V. NESTERENKO: Distribution of the rare earths in the traps of the Siberian platform. Geochemistry Internat. **3**, 672—679 (1966).

—, and L. K. POZHARITSKAYA: Factors governing the behavior of rare-earth elements in the carbonatite process. Geochemistry Internat. **5**, 271—288 (1968).

—, A. B. RONOV, A. A. MIGDISOV, and N. V. TURANSKAYA: The effect of climate and facies environment on the fractionation of the rare earths during sedimentation. Geochemistry (USSR), English Transl., 951—969 (1964).

—, R. N. SOBOLEV, and M. M. CHEKHOVSKIKH: Variation in the ratios of the rare earth elements as a geochemical indicator of hybridization of magma. Geochemistry Internat. **5**, 38—47 (1968).

—, and N. V. TURANSKAYA: Pattern of distribution of the rare earths in the rocks of the differentiated complex of the Lovozero alkalic massif and certain problems of the genesis of the complex. Geochemistry (USSR), English Transl., 845—859 (1960).

— — Rare earths in the eudialyte complex of the Lovozero alkalic massif. Geochemistry (USSR), English Transl., 1213—1226 (1961).

— — Rare earth elements in the peridotite of the Polar Urals. Geochemistry (USSR), English Transl., 433—435 (1962).

BANDURKIN, G. A.: Behaviour of the rare earths in fluorine-bearing media. Geochemistry (USSR), English Transl., 159—167 (1961).

BATE, G. L., H. A. POTRATZ, and J. R. HUIZENGA: Scandium, chromium and europium in stone meteorites by simultaneous neutron activation analysis. Geochim. Cosmochim. Acta **18**, 101—107 (1960).

BEARD, G. B., and W. H. KELLEY: The use of a samarium loaded liquid scintillator for the determination of the half-life of Sm^{147}. Nucl. Phys. **8**, 207—209 (1958).

—, and M. L. WIEDENBECK: Nuclear radioactivity of Sm^{147}. Phys. Rev. **95**, 1245—1246 (1954)

BLOKH, A. M.: Rare earths in the remains of Paleozoic fishes of the Russian platform. Geochemistry (USSR), English Transl., 404—415 (1961).

BORODIN, L. S.: Correlations among rare-earth elements and some characteristics of their fractionation under endogenic conditions. Geochemistry (USSR), English Transl., 604—616 (1960).

— Distribution and correlation of rare earth elements in specific cerium minerals. (Korreljaciionnye svjazi redkozemel'nych elementov i nekotorye osobennosti ich raspredelenija i cerievych i cerijsoderžaščich mineralach.) Voprosy mineralogii i geochimii redkich elementov. Trudy, vyp. **7**, 3—25 (1961). Akademija nauk SSSR, Institut mineralogii, geochimii i kristallochimii redkich elementov [in Russian].

— Rare earth elements in pyroxenes and pyroxenites from ultrabasites, alakaline rocks and carbonatite complex massifs. (Redkozemel'nye elementy v piroksenach i piroksenitach iz kompleksnych massivov ul'trabazitov, ščeločnych porod i karbonatitov.) Petrologija i geochimičeskie osobennosti kompleksa ul'trabazitov ščeločnych porod i karbonatitov, 263—281 (1965). Izdatel'stvo "Nauka", Moskva [in Russian].

— Die Seltenen Erden als geochemische Indikatoren bei der Lösung von Fragen der endogenen Mineralbildung. Z. Angew. Geol. **13**, 9—16 (1967). The same paper in Russian: Isv. Akad. Nauk SSSR, Ser. Geol. 3—20 (1965).

—, and R. L. BARINSKII: Rare earths in perovskites (knopites) from massifs of ultrabasic alkalic rocks. Geochemistry (USSR), English Transl., 343—351 (1960).

BRUNFELT, A. O., and E. STEINNES: Cerium and europium content of some standard rocks. Chem. Geology **2**, 199—207 (1967).

BURKOV, V. V., and YE. K. PODPORINA: First data on rare earths in kimberlite. Dokl. Acad. Sci. USSR, Earth Sci. Sect. **171**, 215—219 (1966).
— — Rare earths in granitoid residuum. Dokl. Acad. Sci. USSR, Earth Sci. Sect. **177**, 214—216 (1967).
CAMERON, A. E., and E. WICHERS: Report of the International Commission on Atomic Weights (1961). J. Am. Chem. Soc. **84**, 4175—4197 (1962).
CHASE, J. W., CH. C. SCHNETZLER, G. K. CZAMANSKE, and J. W. WINCHESTER: The lanthanum, europium, and dysprosium contents of two tektites. J. Geophys. Res. **68**, 577—579 (1963).
—, J. W. WINCHESTER, and CH. D. CORYELL: Lanthanum, europium, and dysprosium distributions in igneous rocks and minerals. J. Geophys. Res. **68**, 567—575 (1963).
CORYELL, CH. D., J. W. CHASE, and J. W. WINCHESTER: A procedure for geochemical interpretation of terrestrial rare-earth abundances patterns. J. Geophys. Res. **68**, 559—566 (1963).
DAWSON, J. B.: Sodium carbonate lavas from Oldoinyo Lengai, Tanganyika. Nature **195**, 1075—1076 (1962).
EWART, A., S. R. TAYLOR, and A. C. CAPP: Trace and minor element geochemistry of the rhyolitic volcanic rocks, Central North Island, New Zealand. Contr. Mineral. and Petrol. **18**, 76—104 (1968).
FLEISCHER, M.: Some aspects of the geochemistry of yttrium and the lanthanides. Geochim. Cosmochim. Acta **29**, 755—772 (1965).
— Rare earths in the aeschynite-priorite series. The status of lyndochite. Mineral. Mag. **35**, 801—809 (1966).
FREY, F. A., and L. HASKIN: Rare earths in oceanic basalts. J. Geophys. Res. **69**, 775—780 (1964).
—, M. A. HASKIN, J. A. POETZ, and L. A. HASKIN: Rare earth abundances in some basic rocks. J. Geophys. Res. **73**, 6085—6098 (1968).
FRONDEL, J. W.: Variation of some rare earths in allanite. Am. Mineralogist **49**, 1159—1177 (1964).
GAVRILOVA, L. K., and N. V. TURANSKAYA: Distribution of rare earths in rock-forming and accessory minerals of certain granites. Geochemistry (USSR), English Transl., 163—170 (1958).
GLOVER, R. N., and D. E. WATT: The natural radioactivity of lanthanum. Phil. Mag. **2**, 49—56 (1957).
GOLDBERG, E. D., M. KOIDE, R. A. SCHMITT, and R. H. SMITH: Rare-earth distributions in the marine environment. J. Geophys. Res. **68**, 4209—4217 (1963).
GOLDSCHMIDT, V. M., u. Cl. PETERS: Über die Anreicherung seltener Elemente in Steinkohlen. Nachr. Ges. Wiss. Goettingen, Math.-Physik. Kl., Fachgruppe IV, Geologie u. Mineralogie, Nr. 40, 371—386 (1933).
— u. L. THOMASSEN: Geochemische Verteilungsgesetze der Elemente. III. Röntgenspektrographische Untersuchungen über die Verteilung der seltenen Erdmetalle in Mineralen. Videnskapsselskapets Skrifter, I. Mat.-naturv. Kl. **1**, No. 5, 1—58 (1924).
GOLES, G. G., M. OSAWA, K. RANDLE, R. L. BEYER, D. Y. JÉROME, D. J. LINDSTROM, M. R. MARTIN, S. M. MCKAY, and T. L. STEINBORN: Instrumental neutron activationan alyses of Apollo XI specimens. Lunar Science Conference, Houston, Texas, January 1970. Preprint.
GRAHAM, A. L., and G. D. NICHOLLS: Mass spectrographic determinations of lanthanide element contents in basalts. Geochim. Cosmochim. Acta **33**, 555—568 (1969).
HASKIN, L. A., and F. A. FREY: Dispersed and not-so-rare earths. Science **152**, 299—314 (1966).
— —, R. A. SCHMITT, and R. H. SMITH: Meteoritic, solar and terrestrial rare earth distributions. Physics and Chemistry of the Earth, vol. 7, p. 167—321. Oxford and New York: Pergamon Press 1966.
HASKIN, L., and M. A. GEHL: The rare-earth distribution in sediments. J. Geophys. Res. **67**, 2537—2541 (1962).
— — Rare-earth elements in tektites. Science **139**, 1056—1058 (1963a).
— — The rare-earth contents of standard rocks G-1 and W-1 and their comparison with other rare-earth distribution patterns. J. Geophys. Res. **68**, 2037—2043 (1963b).

HASKIN, L. A., and M. A. HASKIN: Rare-earth elements in the Skaergaard intrusion. Geochim. Cosmochim. Acta **32**, 433—447 (1968).
— —, F. A. FREY, and T. R. WILDEMAN: Relative and absolute terrestrial abundances of the rare earths. Origin and distribution of the elements, edit. by L. H. AHRENS. Intern. Ser. Monographs Earth Sci. **30**, 889—912 (1968).
—, P. A. HELMKE, and R. O. ALLEN: Rare-earth elements in Apollo XI returned lunar samples. Lunar Science Conference, Houston, Texas, January 1970. Preprint.
—, T. R. WILDEMAN, F. A. FREY, K. A. COLLINS, C. R. KEEDY, and M. A. HASKIN: Rare earths in sediments. J. Geophys. Res. **71**, 6091—6105 (1966).
HASKIN, M. A., and L. A. HASKIN: Rare earths in European shales: a redetermination. Science **154**, 507—509 (1966).
HAYES, D. W., J. F. SLOWEY, and D. W. HOOD: Rare earth distribution in waters of Gulf of Mexico. Trans. Am. Geophys. Union **46**, 548—549 (1965).
HEATH, R. L.: Table of isotopes. Handbook of Chemistry and Physics, 45th edit. B-4—B-92 (1964/1965).
HEGEMANN, F., and G. W. WILK: Emissionsspektrographische Bestimmung der Seltenen Erdmetalle im Flußspat von Wölsendorf/Opf. Neues Jahrb. Mineral., Monatsh., 147—159 (1964).
HERGET, G.: Quantitative emissionsspektrographische Bestimmung der seltenen Erdmetalle im Flußspat der Grube Cäcilia bei Nabburg/Opf. Neues Jahrb. Mineral,. Monatsh., 115—126 (1965).
HERRMANN, A. G.: Die Verteilung der Lanthaniden in basaltischen Gesteinen. Contr. Mineral. and Petrol. **17**, 275—314 (1968).
—, u. K. H. WEDEPOHL: Die Verteilung des Yttriums und der Lanthaniden in einem Olivin-Alkali-Basalt mit Peridotit-Einschlüssen. Contr. Mineral. and Petrol. **13**, 366—373 (1966).
— — Die quantitative Bestimmung der Seltenen Erden (La—Lu) und des Yttriums in silicatischen Gesteinen. Z. Anal. Chem. **225**, 1—13 (1967).
— — Die Verteilung der Lanthaniden in paläozoischen spilitischen Gesteinen („Diabasen"). 47. Jahrestag. der Deutschen Mineralogischen Ges. in Bern, 1969. Vortragsreferat.
HOEFS, J., and K. H. WEDEPOHL: Strontium isotope studies on young volcanic rocks from Germany and Italy. Contr. Mineral. and Petrol. **19**, 328—338 (1968).
HØGDAHL, O.: Distribution of the rare-earth elements in sea water. Central Institute for Ind. Res., Oslo, NATO Research Grant 203, Semiannual Progr. Rept. 4 (1967).
HØGDAHL, O. T., S. MELSOM, and V. T. BOWEN: Neutron activation analysis of lanthanide elements in sea water. Advances in Chemistry Series 73; Trace inorganics in water, p. 308—325 (1968).
HUBER-SCHAUSBERGER, I., u. E. SCHROLL: UV-Lumineszenz und Seltenerdgehalte in Flußspaten. Geochim. Cosmochim. Acta **31**, 1333—1341 (1967).
INGHRAM, M. G., R. J. HAYDEN, and D. C. HESS JR.: The isotopic constitution of lanthanum and cerium. Phys. Rev. **72**, 967—970 (1947).
IVANTIŠIN, M. N., O. A. KYL'SKAJA, G. R. GORNYJ, and G. D. ELISEEVA: Geochemistry and analytical chemistry of the rare earths. (Geochimija i analitičeskaja chimija redkozemel'nych elementov.) Izdatel'stvo "Naukova dumka", Kiev, p. 167 (1964 [in Russian].
JENSEN, B. B. Distribution patterns of rare earth elements in cerianite. Norsk Geol. Tidsskr. **47**, 1—8 (1967a).
— Distribution patterns of rare earth elements in cerium rich minerals. Norsk Geol. Tidsskr. **47**, 9—19 (1967b).
—, and A. O. BRUNFELT: Distribution patterns of rare earth elements in terrestrial rocks. Norsk Geol. Tidsskr. **45**, 249—283 (1965).
KALITA, A. P.: Distribution of rare earths in the pegmatite minerals of north western and south western Karelia. Geochemistry (USSR), English Transl., 171—177 (1959).
KAPUSTIN, YU. L.: Geochemistry of rare earth elements in carbonatites. Geochemistry Internat. **3**, 1054—1064 (1966).
KHOMYAKOV, A. P.: Chemical and crystallochemical factors in the distribution of rare earths. Geochemistry Internat. **4**, 127—135 (1967).

KNORRING, O. VON, and C. G. B. DU BOIS: Carbonatitic lava from Fort Portal area in western Uganda. Nature **192**, 1064—1065 (1961).
KOCHENOV, A. V., and V. V. ZINOV'EV: Distribution of rare earth elements in phosphatic remains of fish from the maikop deposits. Geochemistry (USSR), English Transl., 860—873 (1960).
KÖSTER, H. M.: Vergleich einiger Methoden zur Untersuchung von geochemischen Vorgängen bei der Verwitterung. Beitr. Mineral. Petrog. **8**, 69—83 (1961).
KOGAN, B. I.: Economic data about the rare earth elements. (Ékonomičeskie očerki po redkim semljam). Izdatel'stvo Akademii Nauk SSSR, Moskva, p. 440 (1961). [In Russian].
KOLBE, P., and S. R. TAYLOR: Geochemical investigation of the granitic rocks of the Snowy Mountains area, New South Wales. J. Geol. Soc. Australia **13**, 1—25 (1966).
KOSTERIN, A. V.: The possible modes of transport of the rare earths by hydrothermal solutions. Geochemistry (USSR), English Transl., 381—387 (1959).
—, D. F. KOROLEV, and V. E. KIZYURA: Rare earths in the Chekhez lignite deposit. Geochemistry (USSR), English Transl., 721—723 (1963).
KOVALENKO, V. I., A. S. ZNAMENSKAYA, V. P. AFONIN, G. V. PAVLINSKIY, and V. M. MAKOV: Behavior of rare-earth elements and yttrium in metasomatically altered alkalic granites of the Ognitsk complex (east Sayan). Geochemistry Internat. **3**, 406—418 (1966).
KRUMHOLZ, P.: Solution chemistry. Progress in the Science and Technology of the Rare Earths, vol. I, edit. by LEROY EYRING, p. 110—138. Oxford-London-New York-Paris: Pergamon Press 1964.
LANDSTRÖM, O., and C. G. WENNER: Neutron-activation analysis of natural water applied to hydrogeology. Aktiebolaget Atomenergi (Sweden) AE-204 (1965).
LEE, D. E., and H. BASTRON: Fractionation of rare-earth elements in allanite and monazite as related to geology of the Mt. Wheeler mine area, Nevada. Geochim. Cosmochim. Acta **31**, 339—356 (1967).
LEONOVA, L. L., and YU. A. BALASHOV: Distribution of uranium, thorium, and the rare earths in the granitoids of the Susamyr batholith (Central Tien-Shan). Geochemistry (USSR), English Transl., 1047—1055 (1963).
LESLIE, G. E.: The radioactivity of samarium. Thesis, North Carolina State College (1954). Nucl. Sci. Abstr. No. 1099 (1956).
LEVINE, H., and I. MAY; quoted from D. L. GRAF: Geochemistry of carbonate sediments and sedimentary carbonate rocks. III. Minor element distribution. Illinois State Geol. Surv., Circ. **301**, 1—71 (1960).
LUX, H.: Die Häufigkeit der seltenen Erden in Tierknochen. Z. Anorg. Allgem. Chem. **240**, 21—30 (1939).
LYAKHOVICH, V. V.: Rare earth elements in the accessory minerals of granitoids. Geochemistry (USSR), English Transl., 39—55 (1962).
— Distribution of rare earths among the accessory minerals of granites. Geochemistry Internat. **4**, 691—696 (1967).
—, and R. L. BARINSKII: Characteristics of the rare earth assemblages in the accessory minerals of granitoids. Geochemistry (USSR), English Transl., 495—509 (1961).
MANDLE, R. M., and H. H. MANDLE: Uses and applications. Progress in the Science and Technology of the Rare Earths, vol. I, edit. by LEROY EYRING, p. 416—500. Oxford-London-New York-Paris: Pergamon Press 1964.
MARTELL, A. E., and L. G. SILLÉN: Stability constants of metal-ion complexes. Special publication No. 17. The Chemical Society, Burlington House, London, W. 1 (1964).
MASON, B.: Meteorites. New York and London: John Wiley & Sons, Inc. 1962.
MASUDA, A.: Simple regularity in the variation of relative abundances of rare earth elements. J. Earth Sci., Nagoya Univ., 125—134 (1957).
— Regularities in variation of relative abundances of lanthanide elements and an attempt to analyse separation index patterns of some minerals. J. Earth Sci., Nagoya Univ., 173—187 (1962).
— Lanthanides in basalts of Japan with three distinct types. Geochem. J. **1**, 11—26 (1966).
— An estimation of lanthanide partition coefficient for the "calcium effect". Earth Planet. Sci. Lett. **2**, 277—279 (1967a).

MASUDA, A.: Lanthanide concentration ratios between pyroxene and garnet. Earth Planet. Sci. Lett. **3**, 25—28 (1967b).
— Geochemistry of lanthanides in basalts of Central Japan. Earth Planet. Sci. Lett. **4**, 284—292 (1968a).
— Lanthanide concentrations in the olivine phase of the Brenham pallasite. Earth Planet. Sci. Lett. **5**, 59—62 (1968b).
—, and Y. MATSUI: Depths of magma sources of an eclogite and a peridotite, tentatively estimated from the pattern of lanthanides. Japan. J. Geol. Geography, Trans. **36**, 163—169 (1965).
MATSUI, Y.: Parallelism between the rare-earth fractionation and the rubidium to strontium ratio in the earth's mantle. Japan. J. Geol. Geography, Trans. **37**, 175—185 (1966).
MATTEUCCI, E.: Distribuzione degli elementi delle terre rare nella "quarzdiorite del fustiarbu" (Sardegna) e suo confronto con altre distribuzioni. Rend. Soc. Mineral. Ital. **23**, 275—313 (1967).
MINAMI, E.: Gehalte an seltenen Erden in europäischen und japanischen Tonschiefern. Nachr. Ges. Wiss. Goettingen. Math.-Physik. Kl. Fachgruppe IV, **1**, Nr. 14, 155—170 (1935).
MINEYEV, D. A.: Geochemical differentiation of the rare earths. Geochemistry (USSR), English Transl., 1129—1149 (1963).
—, YU. P. DIKOV, B. P. SOBOLEV, and V. L. BORUTSKAYA: Differentiation of rare earth elements under supercritical conditions. Geochemistry Internat. **3**, 357—359 (1966).
MURATA, K. J., C. V. DUTRA, M. TEIXEIRA DA COSTA, and J. J. R. BRANCO: Composition of monazites from pegmatites in eastern Minas Gerais, Brazil. Geochim. Cosmochim. Acta **16**, 1—14 (1959).
—, H. J. ROSE JR., and M. K. CARRON: Systematic variation of rare earths in monazite. Geochim. Cosmochim. Acta **4**, 292—300 (1953).
— — —, and J. J. GLASS: Systematic variation of rare-earth elements in cerium-earth minerals. Geochim. Cosmochim. Acta **11**, 141—161 (1957).
NAGASAWA, H., H. WAKITA, H. HIGUCHI, and N. ONUMA: Rare earths in peridotite nodules: An explanation of the genetic relationship between basalt and peridotite nodules. Earth Planet. Sci. Lett. **5**, 377—381 (1969).
NEUMANN, H., B. B. JENSEN, and A. O. BRUNFELT: Distribution patterns of rare earth elements in minerals. Norsk Geol. Tidsskr. **46**, 141—179 (1966).
NODDAK, I.: Die Häufigkeit der seltenen Erden in Meteoriten. Z. Anorg. Allgem. Chem. **225**, 337—364 (1935).
Nuclear Data Sheets, 1959—1965. New York and London: Academic Press.
Nuclear Data, B 2-1-67. Vol. 2, No. 1, 1967. Section B, Nuclear Data Sheets. New York and London: Academic Press.
ONUMA, N., H. HIGUCHI, H. WAKITA, and H. NAGASAWA: Trace element partition between two pyroxenes and the host lava. Earth Planet. Sci. Lett. **5**, 47—51 (1968).
OSTROUMOV, E. A.: Rare earths in sediments of the Black Sea. (Redkie zemli v glubokovodnch otloženijach černogo morja). Dokl. Akad. Nauk SSSR **91**, 1175—1178 (1953). [In Russian.]
PACHADZHANOV, D. N., G. A. BANDURKIN, A. A. MIGDISOV, and YU. P. GIRIN: Data on the geochemistry of manganese nodules from the Indian Ocean. Geochemistry (USSR), English Transl., 520—527 (1963).
PAPPAS, A. C., J. ALSTAD, and G. LUNDE: Determination of trace elements in opium by means of activation analysis. I. Methods of analysis and studies referring to the determination of the origin of opium. Radiochim. Acta **1**, 109—117 (1963).
PAVLENKO, A. S., YU. A. BALASHOV, R. G. GEVORKYAN, and N. V. TURANSKAYA: Behavior of the rare earths during the formation of calc-alkalic and alkalic comagmatic basaltoids of Central Armenia. Geochemistry Internat. **3**, 124—125 (1966).
PHILPOTTS, J. A., and C. C. SCHNETZLER: Genesis of continental diabases and oceanic tholeiites considered in light of rare-earth and barium abundances and partition coefficients. Origin and Distribution of the Elements, edit. by L. H. AHRENS. Intern. Ser. Monographs Earth Sci. **30**, 939—947 (1968a).
— — Europium anomalies and the genesis of basalt. Chem. Geol. **3**, 5—13 (1968b).

PHILPOTTS, J. A., and C. C. SCHNETZLER: Apollo XI returned lunar samples: K, Rb, Sr, Ba, and rare-earth concentrations in some rocks and separated phases. Lunar Science Conference, Houston, Texas, January 1970. Preprint.
— —, and H. H. THOMAS: Rare earth abundances in an anorthosite and a mangerite. Nature **212**, 805—806 (1966).
— — — Rare-earth and barium abundances in the Bununu howardite. Earth Planet. Sci. Lett. **2**, 19—22 (1967a).
— — — Meteorite-eclogite-basalt: A possible genetic sequence considered in terms of the rare earths. The Geological Society of America, Annual Meetings, New Orleans, Louisiana. Program with Abstracts, p. 177—178 (1967b).
PORTNOV, A. M., A. A. GANZEYEV, and K. V. BURSUK: Geochemistry of rare earth elements in the Burpala pluton. Dokl. Acad. Sci. USSR, Earth Sci. Sect. **174**, 196—198 (1967).
RAUP, O. B., A. J. GUDE, E. J. DWORNIK, F. CUTTITTA, and H. ROSE JR.: Braitschite, a new hydrous calcium rare-earth borate mineral from the Paradox basin, Grand County, Utah. Am. Mineralogist **53**, 1081—1095 (1968).
ROBINSON, W. O.: The occurence of rare earths in plants and soils. Soil Sci. **56**, 1—6 (1943).
—, H. BASTRON, and K. J. MURATA: Biogeochemistry of the rare-earth elements with particular reference to hickory trees. Geochim. Cosmochim. Acta **14**, 55—67 (1958).
—, and G. EDGINGTON: Minor elements in plants, and some accumulator plants. Soil Sci. **60**, 17—28 (1945).
—, R. WHETSTONE, and B. F. SCRIBNER: The presence of rare earths in hickory leaves. Science **87**, 470 (1938).
RONOV, A. B., YU. A. BALASHOV, and A. A. MIGDISOV: Geochemistry of the rare earths in the sedimentary cycle. Geochemistry Internat. **4**, 1—17 (1967).
RYABCHIKOV, D. I., and E. A. TERENTYEVA: Recent Soviet research on the chemistry of rare earth complexes. Progress in the Science and Technology of the Rare Earths, vol. I, edit. by LEROY EYRING, p. 139—151. Oxford-London-New York-Paris: Pergamon Press 1964.
SAHAMA, TH. G.: Spurenelemente der Gesteine im südlichen Finnisch-Lappland. Bull. Comm. Geol. Finlande **135**, 1—86 (1945).
—, and V. VÄHÄTALO: The rare earth content of wilkite. Bull. Comm. Geol. Finlande **125**, 97—109 (1939).
— — X-ray spectrographic study of the rare earths in some Finnish eruptive rocks and minerals. Bull. Comm. Geol. Finlande **126**, 50—83 (1941).
SCHILLING, J. G., and J. W. WINCHESTER: Rare earths in Hawaiian basalts. Science **153**, 867—869 (1966).
SCHMITT, R. A., A. W. MOSEN, C. S. SUFFREDINI, J. E. LASCH, R. A. SHARP, and D. A. OLEHY: Abundances of the rare-earth elements, lanthanum to lutetium, in chondritic meteorites. Nature **186**, 863—866 (1960).
—, and R. H. SMITH: Implications of similarity in rare-earth fractionation of nakhlitic meteorites and terrestrial basalts. Nature **199**, 550—551 (1963).
— —, J. E. LASCH, A. W. MOSEN, D. A. OLEHY, and J. VASILEVSKIS: Abundances of the fourteen rare-earth elements, scandium, and yttrium in meteoritic and terrestrial matter. Geochim. Cosmochim. Acta **27**, 577—622 (1963).
— —, and D. A. OLEHY: Rare-earth, yttrium and scandium abundances in meteoritic and terrestrial matter — II. Geochim. Cosmochim. Acta **28**, 67—86 (1964).
—, H. WAKITA, and P. REY: Abundances of 30 elements in Apollo XI lunar rocks, soil, and core samples. Lunar Science Conference, Houston, Texas, January 1970. Preprint.
SCHNETZLER, C. C., and J. A. PHILPOTTS: Has the earth's crust changed with time? Rare-earth abundances in ancient sediments. Unpublished manuscript (1967).
— — Partition coefficients of rare-earth elements and barium between igneous matrix material and rock-forming mineral phenocrysts-I. Origin and distribution of the elements, edit. by L. H. AHRENS. Intern. Ser. Monographs Earth Sci. **30**, 929—938 (1968).
— —, and H. H. THOMAS: Rare-earth and barium abundances in Ivory Coast tektites and rocks from the Bosumtwi crater area, Ghana. Geochim. Cosmochim. Acta **31**, 1987—1993 (1967).

SCHOFIELD, A., and L. HASKIN: Rare-earth distribution patterns in eight terrestrial materials. Geochim. Cosmochim. Acta **28**, 437—446 (1964).

SEMENOV, E. I.: Isomorphism and camouflage of rare earths. Geochemistry (USSR), English Transl., 735—748 (1957).

— Relationship between composition of rare earths and composition and structures of minerals. Geochemistry (USSR), English Transl., 574—586 (1958).

— Mineralogy of the rare earths. (Mineralogija redkich semel'.) Izdatel'stvo Akademii Nauk SSSR, Moskva, p. 412 (1963) [in Russian].

—, and R. L. BARINSKII: The composition characteristics of the rare earths in minerals. Geochemistry (USSR), English Transl., 398—419 (1958).

—, V. N. KHOLODOV, and R. L. BARINSKII: Rare earths in phosphorites. Geochemistry (USSR), English Transl., 501—507 (1962).

SHIMA, M., and M. HONDA: Distributions of alkali, alkaline earth and rare earth elements in component minerals of chondrites. Geochim. Cosmochim. Acta **31**, 1995—2006 (1967).

SILKER, W. B.: Variations in elemental concentrations in the Columbia River. Limnol. Oceanog. **9**, 540—545 (1964).

SLEPNEV, Y. S.: Geochemical peculiarities of the lovochorrite-rinkolite pegmatites of the Khibina alkaline massif. Geochmistry (USSR), English Transl., 480—489 (1957).

SOBOLEV, S. F.: Rare earth elements in the ultramafic and mafic rocks of the Urals. Geochemistry Internat. **2**, 281—283 (1965).

STRUNZ, H.: Mineralogische Tabellen, 4. Aufl. Leipzig: Akademische Verlagsgesellschaft Geest & Portig K.-G. 1966.

SVERDRUP, TH. L.: Yttrofluorite — yttrocerite — cerfluorite in Norwegian pegmatites. Norsk Geol. Tidsskr. **48**, 245—252 (1968).

TAYLOR, S. R., A. EWART, and A. C. CAPP: Leucogranites and rhyolites: trace element evidence for fractional crystallisation and partial melting. Lithos **1**, 179—186 (1968).

TEMPLETON, D. H., and C. H. DAUBEN: Lattice parameters of some rare earth compounds and a set of crystal radii. J. Am. Chem. Soc. **76**, 5237—5239 (1954).

TOWELL, D. G., R. VOLFOVSKY, and J. W. WINCHESTER: Rare earth abundances in the standard granite G-1 and standard diabase W-1. Geochim. Cosmochim. Acta **29**, 569—572 (1965).

—, J. W. WINCHESTER, and R. V. VOLFOVSKY SPIRN: Rare earth distributions in some rocks and associated minerals of the batholith of Southern California. J. Geophys. Res. **70**, 3485—3496 (1965).

UREY, H. C., and H. CRAIG: The composition of the stone meteorites and the origin of the meteorites. Geochim. Cosmochim. Acta **4**, 36—82 (1953).

VAINSHTEIN, E. E., L. K. POZHARITSKAYA, and N. V. TURANSKAYA: Behavior of the rare earths in the process of formation of carbonatites. Geochemistry (USSR), English Transl., 1151—1154 (1961).

—, A. I. TUGARINOV, and N. V. TURANSKAYA: Regularities in the distribution of rare earths in certain minerals. Geochemistry (USSR), English Transl., 159—178 (1956).

VICKERY, R. C.: Chemistry of the lanthanons, pp. 296. London: Butterworths Sci. Publ. 1953.

VINOGRADOV, A. P.: The geochemistry of rare and dispersed chemical elements in soils. Translated from Russian. 2nd edit., pp. 209. New York: Consultants Bureau, Inc.; London: Chapman & Hall Ltd. 1959.

VLASOV, K. A. (editor): Geochemistry and mineralogy of rare elements and genetic types of their deposits, vol. I, Geochemistry of Rare Elements. Translated from Russian by Z. LERMAN. Israel Program for Scientific Translations, Jerusalem, 188—277 (1966).

WALLERSTEIN, G.: A preliminary analysis of the abundances of the rare earths in the sun. Icarus **5**, 75—78 (1966).

WAMBEKE, L. VAN: La géochimie des roches du Kaiserstuhl. Dans Monographie: Les roches alcalines et les carbonatites du Kaiserstuhl. 232 p. (1964). Communauté Européenne de l'Énergie Atomique-Euratom. EUR 1827. d, f, e. Direction Générale Recherche et Enseignement Section Minéralogie-Géochimie. Bruxelles.

WEDEPOHL, K. H.: Geochemie. Sammlung Göschen, Bd. 1224/1224a/1224b; pp. 220. Berlin: Walter de Gruyter & Co. 1967.

WILDEMAN, T. R., and L. HASKIN: Rare-earth elements in ocean sediments. J. Geophys. Res. **70**, 2905—2910 (1965).
YERSHOV, V. M.: Rare earth elements in the coals of the Kizelovskii coal basin. Geochemistry (USSR), English Transl., 306—308 (1961).
YES'KOVA, YE. M., and A. A. GANZEYEV: Rare earth elements in accessory minerals of the Vishnev mountains. Geochemistry Internat. 1152—1163 (1964).
ZAYATS, A. P., and V. P. KUTS: Rare earth elements in the accessory minerals of gneisses of the Ukrainian crystalline shield. Geochemistry Internat. 1126—1128 (1964).
ZHIROV, K. K., G. A. BANDURKIN, and YU. G. LAVRENT'YEV: Geochemistry of rare earth elements in pegmatites of northern Karelia. Geochemistry (USSR), English Transl. 1107—1118 (1961).
ZLOBIN, B. L., and YU. A. BALASHOV: Distribution and ratios of the rare earth elements in the alkalic plumasitic series: essexite-nepheline syenite. Geochemistry (USSR), English Transl., 862—866 (1961).

Revised manuscript received: February 1969

72-A. Crystal Chemistry

The crystal structure and the chemical properties of hafnium compounds are identical or at least very similar to those of corresponding zirconium compounds. This close relationship is due to the similarity of the ionic and atomic radii of Hf and Zr as a consequence of the lanthanide contraction. All zirconium minerals contain hafnium; no separation of Hf from Zr is found and therefore pure hafnium minerals are not known. The content of hafnium in minerals rarely exceeds that of zirconium; exceptions are certain thortveitites (LEVINSON and BORUP, 1960a). Zircon, $ZrSiO_4$, and baddeleyite, ZrO_2, are the most important sources of hafnium and these minerals usually contain upto 2% hafnium. Some of the Norwegian zircons were found with Hf-concentrations in the order of 20% (LEVINSON and BORUP, 1960b).

Metallic hafnium crystallizes in two modifications like zirconium, in the hexagonal close-packed α-form (Mg-type), and in the body-centered cubic β-form (W-type) at high temperatures. The reversible $\alpha \rightleftharpoons \beta$ transformation occurs at a much higher temperature (1950° C) than in the case of Zr. Hafnium like zirconium forms interstitial compounds with H, B, C, N and O. HfB_2 (AlB_2-type), HfC and HfN (NaCl-type) are very hard and have high melting temperatures (KIEFFER and BENESOVSKY, 1963). The oxygen atoms occupy the interstitial positions of α-Hf at random with a steady increase in the lattice parameters up to the composition Hf_4O (DAGERHAM, SR 1961, 463).

Hafnium oxide, HfO_2, is one of the most stable refractory oxides with a melting point of 2,900° C. It is known in three crystalline modifications, monoclinic, tetragonal and cubic, which are structurally identical to the corresponding forms of ZrO_2. The average Hf—O distances are slightly shorter than the average Zr—O distances (RUH and CORFIELD, 1970). In addition an orthorhombic high pressure form of HfO_2 has been reported (BOCQUILLON *et al.*, 1968). The reversible transformation of monoclinic to tetragonal HfO_2 was found to occur at a higher temperature (1,500—1,650° C) and with a smaller hysteresis than in the case of zirconia (WOLTEN, 1963). Stabilization of HfO_2 in the fluorite structure can be achieved as for ZrO_2 by addition of CaO, Y_2O_3 or rare earth oxides, except that higher temperatures are necessary to form these solid solutions (ALPER, 1970). The phase 1 $CaO \cdot 4\,HfO_2$ has a distorted fluorite structure (DELAMARRE, 1967). Hafnium silicate ("hafnon") which is not found as a mineral is structurally identical to zircon (LEFEVRE and COLLONGUES, SR 1960, 457). A compilation of Hf-oxide compounds is given in Table 72-A-1; in all cases isostructural Zr-oxide compounds are known. The oxygen coordination of Hf is predominantly 6-fold and 8-fold.

As well as the oxide compounds, the halides, oxyhalides, and chalkogenides etc. of Hf are structurally identical to the corresponding zirconium compounds (WELLS, 1962; KREBS, 1968); examples are HfI_3($MoBr_3$-type), HfI_4(SnI_4-type), K_2HfIF_6 (K_2ThF_6-type), Na_3HfF_7(α-Na_3UF_7-type), HfS_2, $HfSe_2$(CdI_2-type). The coordination

Table 72-A-1. *Oxide compounds of hafnium*

Compound	Coordination number of Hf	Structural type	Reference
$HfMo_2O_8$	6, 8	—	RIMSKY *et al.* (1968)
$Hf[P_2O_7]$	6	$Me^{4+}[P_2O_7]$	VOELLENKLE *et al.* (1963)
HfO_2	7	baddeleyite	RUH and CORFIELD (1970)
HfO_2	8	fluorite, distorted	CURTIS, DONEY and JOHNSON (SR 1955, 362)
$Hf[SiO_4]$	8	zircon	LEFEVRE and COLLONGUES (SR 1960, 457)
$Hf[GeO_4]$	8	scheelite	LEFEVRE and COLLONGUES (SR 1960, 457)
$HfTiO_4$	6	α-PbO_2	HARARI *et al.* (1968)
$Li_2Hf(WO_4)_3$	6	wolframite	CHANG (1967)
$HfTe_3O_8$	6	$Me^{4+}Te_3O_8$	MEUNIER (1970)
$Gd_2Hf_2O_7$, $Sm_2Hf_2O_7$	6	pyrochlore	DUCLOT *et al.* (1970)
$CaHfO_3$, $SrHfO_3$, $BaHfO_3$	6	perovskite	GODINA and KELLER (SR 1959 374)
Sr_2HfO_4, Ba_2HfO_4	6	K_2NiF_4	SCHOLDER *et al.* (1968)
$NaScHfO_4$	6	$CaFe_2O_4$	REID *et al.* (1968)

number of hafnium in such compounds varies from 4 in HfI_4 (KRAUSE *et al.*, SR 1960, 298), to 7 in Na_3HfF_7 (HARRIS, SR 1959, 291), and 8 in HfF_4 (ZACHARIASEN, SR 1949, 168) and K_2HfF_6 (BODE and TEUFER, SR 1956, 224).

Revised manuscript received: July 1972

References: Section 72-A

ALPER, A. M.: High Temperature Oxides. Part II: Oxides of Rare Earths, Titanium, Zirconium, Hafnium, Niobium and Tantalum. New York: Academic Press Inc. 1970.

BOCQUILLON, G., SUSE, C., VODAR, B.: Allotropy of hafnium oxide under high pressure. Rev. Int. Hautes Temp. Refract. **5** (4), 247 (1968).

CHANG, L. L. Y.: $Li_2Zr(WO_4)_3$, a wolframite-type compound. Mineral Mag. **36**, 436 (1967).

DELAMARRE, C.: Le système HfO_2—CaO. Silicates Ind. **32**, 345 (1967).

DUCLOT, M., VICAT, J., DEPORTES, C.: Mise en evidence et étude de la phase ordonné $Y_2Hf_7O_{17}$ dans le système HfO_2—Y_2O_3. J. Solid State Chem. **2**, 236 (1970).

HARARI, A., BOCQUET, J.-P., HUBER, M., COLLONGUES, R.: Structure cristalline de l'oxyde mixte $HfTiO_4$. Compt. Rend. **267**, 1316 (1968).

KIEFER, R., BENESOVSKY, F.: Hartstoffe. Wien: Springer 1963.

KREBS, H.: Grundzüge der anorganischen Kristallchemie. Stuttgart: Enke 1968.

LEVINSON, A. A., BORUP, R. A.: New data on the hafnium, zirconium and yttrium content of thortveitite. Am. Mineralogist **45**, 712 (1960a).

LEVINSON, A. A., BORUP, R. A.: High hafnium zircon from Norway. Am. Mineralogist **45**, 562 (1960b).

MEUNIER, G.: Les systèmes MO_2—TeO_2 (M = Ti, Sn, Hf, Zr). Thèse, Université de Bordeaux, 1970.

REID, A. F., WADSLEY, A. D., SIENKO, M. J.: Crystal chemistry of sodium scandium titanate, $NaScTiO_4$, and its isomorphs. Inorg. Chem. **7**, 112 (1968).

RIMSKY, A., THORET, J., FREUNDLICH, W.: Structure cristalline de la phase hexagonale $HfMo_2O_8$. Compt. Rend. C **267**, 1468 (1968).

RUH, R., CORFIELD, P. W. R.: Crystal structure of monoclinic hafnia and comparison with monoclinic zirconia. J. Am. Ceram. Soc. **53** (3), 126 (1970).

SCHOLDER, R., RAEDE, D., SCHWARZ, H.: Über Zirkonate, Hafnate und Thorate von Barium, Strontium, Lithium und Natrium. Z. Anorg. Allgem. Chemie **362**, 149 (1968).

VOELLENKLE, H., WITTMANN, A., NOWOTNY, H.: Über Diphosphate vom Typ Me(IV)P_2O_7. Monatsh. Chem. **94**, 956 (1963).

WELLS, A. F.: Structural Inorganic Chemistry, 3rd ed. Oxford: Clarendon Press 1962.

WOLTEN, G. M.: Diffusionless phase transformations in zirconia and hafnia. J. Am. Ceram. Soc. **46**, 418 (1963).

Revised manuscript received: July 1972

72-B. Isotopes in Nature

The conventional abundance of the stable isotopes of hafnium is as follows:

Isotope:	^{174}Hf	^{176}Hf	^{177}Hf	^{178}Hf	^{179}Hf	^{180}Hf
%:	0.18	5.20	18.50	27.14	13.75	35.24

The heavier isotopes are more abundant since they are predominantly formed by the r-process during nucleosynthesis (SEEGER *et al.*, 1965). No natural isotopic fractionation has been reported, and none is expected at these mass numbers, except for the possibility of ^{176}Hf forming by β^- decay of ^{176}Lu ($t_{\frac{1}{2}} = 2.4 \times 10^{10}$ years; RANKAMA, 1963). Hafnium radioactive isotopes with mass numbers 168, 169, 170, 171, 172, 173, 175, 178, 179, 180, 181, 182, 183 have been prepared in the laboratory.

The atomic weight of hafnium is 178.49.

Manuscript received: August 1977

72-C. Abundance in Cosmos, Meteorites, Tektites and Lunar Samples

I. Abundance in Meteorites

It is convenient to first discuss the abundance of hafnium in meteorites, before the section dealing with cosmic abundances is presented. The abundance of hafnium in meteorites is not only important for defining cosmic abundances, but also with regard to possible fractionation between Zr and Hf in the earth and meteorites (ERLANK and WILLIS, 1964).

Although the precise distribution of hafnium within meteorites is not accurately known, available data indicate, as expected, that it follows zirconium in being predominantly lithophile in its geochemical behaviour. MASON and GRAHAM (1970) and MACDOUGALL and KERRIDGE (1976) report the presence of zircon in mesosiderites and chondrites, and only relatively insignificant amounts of hafnium appear to be present in the metal and troilite phases of meteorites (SETSER and EHMANN, 1964).

The first determinations of hafnium in chondrites (average 1.4 ppm Hf) made by MERZ (1962) yielded high cosmic abundances and suggested Zr-Hf fractionation between the earth and meteorites (ERLANK and WILLIS, 1964). However, subsequent analyses have shown that the hafnium (and zirconium) data of MERZ (1962) are in error. All available subsequent data for chondrites are shown in Table 72-C-1. The high value of 2.28 ppm Hf in the Cc1 chondrite Alais is clearly anomalous, and possibly related to the relative abundance of zircon in this meteorite (MACDOUGALL and KERRIDGE, 1976). However it is interesting that this meteorite is also enriched in other refractory elements such as Zr, Ta and the rare earths (GANAPATHY *et al.*, 1976). Apart from this, it is difficult to detect any real fractionation between the various chondrite classes, as shown by analyses for any particular meteorite by different analysts and by the averages shown in Table 72-C-1. Analytical error may be partly the cause, although the small sample size used by many of the workers may also be responsible (in the latter respect SHIMA and HINTENBERGER, 1976, report 0.22 and 0.32 ppm Hf for two different samples of Allende, yet both have Zr/Hf ratios of 34).

Data for achondrites are also given in Table 72-C-1 and show a distinct fractionation between the Ca-rich and Ca-poor achondrites. Some of the former (i.e. the eucrites) have abundance levels comparable to that observed in some terrestrial tholeiitic basalts. The only data for iron meteorites are given by SETSER and EHMANN (1964) who reported from 4–9 ppb Hf in four samples.

The distribution of hafnium within meteorites is shown in Table 72-C-2. Apart from the previously mentioned possible presence of zircon, which will undoubtedly contain hafnium, it appears that much, or the bulk of, hafnium in meteorites is contained in clinopyroxenes. This is well illustrated by the value of 2.8 ppm Hf in an Al-rich pyroxene (fassaite) separated from a pyroxene-melilite chondrule ex-

Table 72-C-1. *Hafnium abundances and Zr/Hf ratios in chondrites and achondrites.* Method N/R exept for CHURCH *et al.* (I), SHIMA and HINTENBERGER (I) and MARTIN and MASON (Mass)

Meteorite	ppm Hf	Zr/Hf	Reference
Carbonaceous Chondrites Type 1 (Cc1)			
Orgueil	0.30	31	EHMANN and REBAGAY (1970)
	0.10	52	EHMANN and CHYI (1974)
	0.11	28	GANAPATHY *et al.* (1976)
Ivuna	0.34	25	EHMANN and REBAGAY (1970)
Alais	2.28	76	GANAPATHY *et al.* (1976)
Average (excluding) (Alais)	0.21	34	
Carbonaceous Chondrites Type 2 (Cc2)			
Murray	0.26	—	SETSER and EHMANN (1964)
	0.21	25	EHMANN and REBAGAY (1970)
	0.17	35	SHIMA and HINTENBERGER (1976)
Mighei	0.26	—	SETSER and EHMANN (1964)
	0.14	41	EHMANN and REBAGAY (1970)
Boriskino	0.22	30	EHMANN and REBAGAY (1970)
Haripura	0.15	49	EHMANN and CHYI (1974)
Murchison	0.13	48	EHMANN and CHYI (1974)
	0.14	33	GANAPATHY *et al.* (1976)
	0.29	34	SHIMA and HINTENBERGER (1976)
Cold Bokkeveld	0.11	48	EHMANN *et al.* (1975)
Nawapali	0.11	62	EHMANN *et al.* (1975)
Average	0.21	41	
High iron—low metal group chondrites (CHL or Cc3)			
Ornans	0.25	48	EHMANN and REBAGAY (1970)
Karoonda	0.29	33	EHMANN and REBAGAY (1970)
	0.20	40	EHMANN and CHYI (1974)
Allende	0.16	81	EHMANN and REBAGAY (1970)
	0.23	29	CHURCH *et al.* (1972)
	0.2	50	MARTIN and MASON (1974)
	0.20	32	EHMANN *et al.* (1975)
	0.19	31	GANAPATHY *et al.* (1976)
	0.22	34	SHIMA and HINTENBERGER (1976)
	0.32	34	SHIMA and HINTENBERGER (1976)
Leoville	0.22	34	SHIMA and HINTENBERGER (1976)
Yamato C	0.20	35	SHIMA and HINTENBERGER (1976)
Yamato 74–662	0.17	34	SHIMA and HINTENBERGER (1976)
Average	0.22	40	
High iron group chondrites (CH)			
Morland	0.46	32	MERZ and SCHRAGE (1964)
Potter	0.30	31	MERZ and SCHRAGE (1964)
Plainview	0.29	41	MERZ and SCHRAGE (1964)
	0.22	—	SETSER and EHMANN (1964)
	0.38	18	EHMANN and REBAGAY (1970)
Pultusk	0.16	—	SETSER and EHMANN (1964)
Forest City	0.21	—	SETSER and EHMANN (1964)
	0.12	47	EHMANN and REBAGAY (1970)
Ochansk	0.10	57	EHMANN and REBAGAY (1970)

Table 72-C-1 (continued)

Meteorite	ppm Hf	Zr/Hf	Reference
Richardton	0.16	38	Ehmann and Rebagay (1970)
	0.20	36	Shima and Hintenberger (1976)
Mount Browne	0.14	51	Ehmann and Rebagay (1970)
Coolidge	0.26	46	Ehmann and Chyi (1974)
Canyon City	0.11	36	Ehmann *et al.* (1975)
Breitscheid	0.25	35	Shima and Hintenberger (1976)
Pantar	0.17	36	Shima and Hintenberger (1976)
Yamato d	0.15	34	Shima and Hintenberger (1976)
Yamato 74–371	0.16	32	Shima and Hintenberger (1976)
Yamato 74–647	0.15	33	Shima and Hintenberger (1976)
Yamato 75–258	0.20	34	Shima and Hintenberger (1976)
Average	0.21	37	
Low iron group chondrites (CL)			
Elenovka	0.19	—	Setser and Ehmann (1964)
Lake Labyrinth	0.11	—	Setser and Ehmann (1964)
Harleton	0.26	—	Setser and Ehmann (1964)
Mocs	0.13	—	Setser and Ehmann (1964)
	0.20	25	Ehmann and Rebagay (1970)
Bruderheim	0.19	—	Setser and Ehmann (1964)
	0.17	30	Ehmann and Rebagay (1970)
	0.20	33	Ganapathy *et al.* (1976)
	0.17	35	Shima and Hintenberger (1976)
Kiel	0.19	35	Shima and Hintenberger (1976)
Yamato 74–362	0.19	36	Shima and Hintenberger (1976)
Ioka	0.15	42	Ehmann and Rebagay (1970)
Saratov	0.18	38	Ehmann and Rebagay (1970)
Beenham, N. M.	0.24	28	Ehmann and Rebagay (1970)
Kunashak	0.10	43	Ehmann and Rebagay (1970)
Peace River	0.09	50	Ehmann and Rebagay (1970)
Barwell	0.26	30	Ehmann and Rebagay (1970)
Chainpur	0.16	44	Ehmann and Rebagay (1970)
	0.14	46	Ehmann and Chyi (1974)
Wethersfield	0.12	57	Ehmann and Chyi (1974)
Leedey	0.19	30	Church *et al.* (1972)
Upper Volta	0.12	56	Ehmann *et al.* (1975)
Average	0.17	39	
Low iron—low metal group chondrites (CLL)			
Olivenza	0.21	44	Ehmann and Rebagay (1970)
Jelica	0.14	49	Ehmann and Rebagay (1970)
Dhurmsala	0.15	35	Ehmann and Chyi (1974)
St. Severin	0.29	37	Shima and Hintenberger (1976)
Average	0.20	41	
Enstatite chondrites (Ce)			
Abee (Ce 1)	0.17	—	Setser and Ehmann (1964)
	0.10	38	Ehmann and Rebagay (1970)
Atlanta (Ce 2)	0.15	—	Setser and Ehmann (1964)
	0.12	34	Ehmann and Rebagay (1970)
Daniel's Kuil (Ce 2)	0.21	37	Ehmann and Rebagay (1970)
Average	0.16	35	

Table 72-C-1 (continued)

Meteorite	ppm Hf	Zr/Hf	Reference
Overall average for chondrites	0.20	39	
Ca poor achondrites			
Norton County (Ae)	0.10	—	SETSER and EHMANN (1964)
	0.02	65	EHMANN and REBAGAY (1970)
Cumberland Falls (Ae)	0.12	—	SETSER and EHMANN (1964)
	0.01	77	EHMANN and REBAGAY (1970)
Johnstown (Ab)	0.16	—	SETSER and EHMANN (1964)
	0.04	48	EHMANN and REBAGAY (1970)
	0.18	32	SHIMA and HINTENBERGER (1976)
Shalka (Ab)	0.01	100	EHMANN and REBAGAY (1970)
Yamato b (Ab)	0.021	30	SHIMA and HINTENBERGER (1976)
North Haig (Aop)	0.055	71	EHMANN and CHYI (1974)
Average	0.072	60	
Ca rich achondrites			
Pasamonte (Ap)	0.21	—	SETSER and EHMANN (1964)
	0.97	63	EHMANN and REBAGAY (1970)
Binda (Ap)	0.61	28	EHMANN and REBAGAY (1970)
Juvinas (Ap)	1.3	—	WÄNKE *et al.* (1972)
Stannern (Ap)	1.23	32	SHIMA and HINTENBERGER (1976)
Kapoeta (Aor)	0.75	79	EHMANN and REBAGAY (1970)
	0.60	—	WÄNKE *et al.* (1972)
Yamato 1 (Aor)	0.15	33	SHIMA and HINTENBERGER (1976)
Average	0.73	47	

tracted from the Allende (CHL) chondrite, attesting to the refractory nature of hafnium (and zirconium).

Zr/Hf ratios: The similar chemical properties of zirconium and hafnium lead to their close geochemical association, and any fractionation between these elements within different meteorite types, and between meteorites (particularly chondrites), the moon and the earth as a whole, would have important consequences for any models dealing with the origin of these bodies.

Condensation models for meteorites would not suggest any fractionation of the Zr/Hf ratio during their formation. Zr/Hf ratios for chondrites in Table 72-C-1 show some variation, but this is probably mainly due to analytical error. Average Zr/Hf ratios (excluding Alais) for the different chondrite groups are: Cc1 = 34, Cc2 = 41, CHL = 40, CH = 37, CL = 39, CLL = 41, Ce = 35. The overall average for 70 chondrites yields a Zr/Hf ratio of 39, which is similar to that of lunar and terrestrial surface rocks. However, closer inspection of Table 72-C-1 indicates that the correct chondrite average may be somewhat lower. If only the data of CHURCH *et al.* (1972), GANAPATHY *et al.* (1976) and SHIMA and HINTENBENGER (1976) are considered, then it can be seen that the calculated Zr/Hf ratios vary between the range 28–37, with an average of 34 (26 samples). It appears that more accurate zirconium data are required to substantiate these lower Zr/Hf ratios, since the disagreement between the various analyses cited in Table 72-C-1 is larger for zirconium than for hafnium.

Table 72-C-2. *Hafnium abundances and Zr/Hf ratios in separated phases and minerals from meteorites* (brackets denote number of samples).

Phase or Mineral	Meteorite	Class	ppm Hf	Zr/Hf	Method	Reference
Total meteorite	Allende	CHL	0.2	50	Mass	1
Melilite chondrules (10)	Allende	CHL	1.3	43	Mass	1
Ca, Al-rich aggregates (5)[a]	Allende	CHL	<0.3	—	Mass	1
Ca, Al-rich aggregates (2)[b]	Allende	CHL	2.4	44	Mass	1
Olivine-rich chondrules (3)	Allende	CHL	<0.3	—	Mass	1
Total meteorite	Allende	CHL	0.20	32	N/R	2
Fusion crust	Allende	CHL	0.21	32	N/R	2
Black matrix	Allende	CHL	0.20	31	N/R	2
Chondrules (2)	Allende	CHL	0.12	33	N/R	2
White inclusions (2)	Allende	CHL	0.14	36	N/R	2
Clinopyroxene	Winona	Chondrite[c]	4.0[d]	23	Mass	3
	Moore County	Ap	0.9	16	Mass	4
	Haraiya	Ap	1.3	22	Mass	4
	Modoc	CL	2.9[d]	38	Mass	4
	Allende	CHL	2.8	35	Mass	5
Melilite	Allende	CHL	<0.3	—	Mass	5
Phosphate	Modoc	CL	18[d]	36	Mass	3
Troilite	Canyon Diablo	Og	0.03	—	N/R	6

[a] fractionated REE patterns; [b] unfractionated REE patterns; [c] precise classification unknown; [d] possible zircon impurity.

References: 1. Martin and Mason (1974). 2. Ehmann *et al.* (1975). 3. Mason and Graham (1970). 4. Allen and Mason (1973). 5. Mason and Martin (1974). 6. Setser and Ehmann (1964).

The achondrite data in Table 72-C-1 show a larger variation in Zr/Hf ratios and have higher averages than found in the chondrites. However, this is again probably due to analytical error because the data of Shima and Hintenberger (1976) imply no Zr/Hf fractionation between chondrites and achondrites.

II. Cosmos

Three types of approach may be used to establish the "cosmic" or solar system abundance of hafnium. The first and most commonly used involves the use of chondrites. Since it is generally accepted that the Cc1 chondrites may best represent primitive meteoritic material, with the specimen Orgueil being particularly important in this regard, only the most recent abundances for this meteorite have been used for obtaining the atomic abundances shown in Table 72-C-3.

The second approach involves the use of modern theories of nucleosynthesis to predict cosmic abundances. Although these predictions are not entirely independent, since they relied initially to some extent on meteorite abundances, it is noteworthy that the hafnium data in the compilation by Seeger *et al.* (1965) is in agreement with the subsequent meteoritic estimates.

Table 72-C-3. *Atomic abundance of hafnium* (Si = 10^6)

Reference	Source	Atomic abundance
Ehmann and Chyi (1974)	Orgueil	0.15
Ganapathy *et al.* (1976)	Orgueil	0.16
Seeger *et al.* (1965)	Theoretical	0.15
Ross and Aller (1976)	Solar	0.14

Abundance measurements in the solar atmosphere have also been made for hafnium, the most recent being those incorporated in the compilation presented by Ross and Aller (1976). It is encouraging that the atomic abundances derived by these three types of approach, as summarized in Table 72-C-3, are in good agreement.

III. Tektites

The few hafnium data given for tektites in Table 72-C-4 do not allow any meaningful conclusions to be drawn but suggest that a variation in hafnium concentration occurs between the different tektite types, probably indicative of differences in parent material composition. However, the Zr/Hf ratios show a smaller variation and are generally similar to those discussed for chondrites.

Since the Apollo missions to the moon, it is now generally accepted that tektites originate by cometary or meteoritic impact on terrestrial sedimentary rocks. Thus the australite data in Table 72-C-4 is similar to that shown for Henbury glass pro-

Table 72-C-4. *Hafnium abundances and Zr/Hf ratios in tektites, impact glasses and possible parent materials*

Sample type	No. of samples	ppm Hf		Zr/Hf	Method	Reference
		Range	Average			
Bediasite	1		4.2	31	N/R	1
Billitonite	1		8.7	30	N/R	1
Indochinite	1		2.6	35	N/R	1
Javaite	1		3.1	30	N/R	1
Moldavite	1		7.6	38	N/R	1
Philippinite	1		5.9	34	N/R	1
Thailand tektite	1		10.9	35	N/R	1
Australite	1		9.6	42	N/R	1
Australite	5	3.7–5.0	4.2	66	Mass	2
Libyan Desert glass	1		2.5	48	N/R	1
Darwin impact glass	1		2.9	46	N/R	1
Henbury impact glass	1		2.5	38	N/R	1
Henbury impact glass	1		3.7	108	Mass	2
Henbury subgreywacke	1		8.5	46	Mass	2
Wanapitei Lake impactite	4	0.63–1.9	1.4	25	I	3

References: 1. Setser and Ehmann (1964). 2. Taylor (1966). 3. Winzer *et al.* (1976).

duced from Henbury sub-greywacke, although the concentration ranges indicated are rather large. Similar comments apply for the respective Zr/Hf ratios, indicating a need for more reliable data.

IV. Lunar Materials

It is an irony of our times that some 1,000 hafnium analyses (mostly by the neutron activation technique) are available for 340 lunar samples or sub-samples which have been studied by various investigators. A summary of this data is given in Table 72-C-5, arranged in order of the various Apollo missions (see TAYLOR, 1975, for a description of the landing sites) and in terms of the simple classification basalt, breccia and fines, in keeping with other chapters in this handbook. Only ranges in concentration are given, because such a classification has no real genetic significance, and because in many cases the same samples have been analysed by different investigators. As an example, note that the Apollo 16 rocks show a range in hafnium concentration from 0.01–22 ppm. This large range is due to the fact that these rocks include cataclastic anorthosites (very low Hf) and a variety of breccias containing several types of glass and rock, most of which are probably not locally derived. Thus Table 72-C-5 should be treated simply as a reference source in order to obtain hafnium data for any particular landing site.

A more meaningful interpretation is presented in Table 72-C-6. The first set of data are taken directly from the useful book by TAYLOR (1975) while the compilation by GARG and EHMANN (1976) contains data obtained in one laboratory only. Slightly different terminologies should not prevent comparison of the two sets of data. Thus, with the exception of the high-K Apollo 11 basalts, it may be seen that the maria (or mare) basalts, together with KREEP-poor non-mare basalts (the acronym KREEP has been designated by some workers for rocks rich in K, REE, P), have hafnium contents comparable to terrestrial basalts (see Table 72-E-3). Although the highland rocks are mixtures of complex breccias and difficult to interpret, careful study has revealed the presence of several different rock types. The anorthositic and troctolitic clan have, predictably, low hafnium concentrations, but the Fra Mauro or KREEP basalts (and breccias) have very high hafnium contents for *basaltic* rocks, and thus have special importance.

Hafnium, together with zirconium and other so-called "incompatible" elements, is concentrated in the residual or last crystallising fraction (mesostasis) of many lunar rocks. A variety of Zr-rich minerals occur in this fraction including zircon, baddeleyite, tranquillityite, zirconolite (zirkelite?) and some Fe-Ti-Zr rich minerals which are too small to characterize. Substantial concentrations of hafnium occur in these phases (see Table 72-D-2) which are documented in greater detail in Subsection 40-C-IV. Mention should, however, also be made of the presence of hafnium in lunar ilmenites (Table-72 D-2) in view of the greater abundance of this mineral in lunar rocks and the discussion given below.

Zr/Hf and other inter-element ratios: Several authors have noted that the Zr/Hf ratio is fractionated in lunar materials and this is apparent from the data given in Table 72-C-6. Specifically, if the data of GARG and EHMANN (1976) are used, then the anorthositic clan and the maria basalts have lower ratios (Zr/Hf $<$ 40) than the non-maria and KREEP-rich basalts (Zr/Hf $>$ 40). However, if lunar soils are considered

Table 72-C-5. *Hafnium abundances in lunar materials* (Number of samples in brackets)

Sample type		Mission	ppm Hf	Method	Reference
Basalts	(18)	Apollo 11	6.5–24.5	Mass, N/R	1, 2, 3, 4, 5, 7, 8, 9, 18, 22, 24, 80
	(17)	Apollo 12	1.0–10.0	Mass, N/R	13, 14, 15, 16, 17, 18, 20, 21, 22, 23, 24, 28, 80
	(3)	Apollo 14	3.0–28.4	Mass, N/R	27, 31, 37, 43, 53
	(20)	Apollo 15	0.90–32.7	Mass, N/R	35, 36, 40, 44, 45, 53, 54, 56, 58, 59, 80, 83
	(35)	Apollo 17	5.4–14.0	N/R	69, 72, 73, 76, 77, 79, 81, 82, 84
Breccias	(14)	Apollo 11	8.9–16.3	Mass, N/R	1, 2, 3, 5, 6, 7, 8, 9, 22
	(4)	Apollo 12	12.0–25.2	N/R	13, 15, 16, 25
	(11)	Apollo 14	6.8–40.6	Mass, N/R	27, 29, 30, 31, 32, 33, 34, 43, 50, 53, 81
	(18)	Apollo 15	0.10–18.5	Mass, N/R	39, 40, 53, 54, 56, 58, 77, 79, 83
	(41)	Apollo 16	0.01–22.0	I, Mass, N/R	45, 46, 52, 55, 56, 57, 59, 61, 62, 63, 65, 66, 67, 69, 70, 72, 74, 77, 79, 80, 81
	(30)	Apollo 17	0.10–21.5	I, Mass, N/R	56, 61, 64, 66, 69, 70, 72, 73, 74, 75, 76, 77, 78, 79, 80, 81 82, 83
Fines (soils)	(1)	Apollo 11	7.6–12.9	Mass, N/R	1, 3, 4, 5, 6, 7, 8, 9, 17, 19, 21, 22, 24, 53, 81
	(7)	Apollo 12	9.8–17.2	Mass, N/R	13, 14, 15, 16, 17, 18, 19, 20, 21, 22, 24, 28, 82
	(11)	Apollo 14	19.0–30.0	I, Mass, N/R	10, 11, 12, 29, 30, 31, 32, 33, 34, 35, 50, 82
	(22)	Apollo 15	3.2–13.0	I, Mass, N/R	27, 29, 32, 35, 38, 40, 41, 45, 51, 54, 56, 58, 59, 70, 79, 82
	(31)	Apollo 16	1.5–6.9	Mass, N/R	25, 40, 42, 45, 46, 47, 48, 49, 51, 52, 53, 55, 56, 57, 59, 61, 67, 68, 71, 77, 79, 81
	(29)	Apollo 17	4.2–7.8	N/R	55, 59, 60, 61, 67, 69, 72, 73, 74, 75, 80, 82

Note: All Apollo 16 rocks have been classed as breccias for the purposes of this Table.

References: 1. GOLES *et al.* [P/1], p. 1177. 2. GOLES *et al.* [P/1], p. 1165. 3. HASKIN *et al.* [P/1]. 4. MAXWELL *et al.* [P/1]. 5. MORRISON *et al.* [P/1]. 6. SMALES *et al.* [P/1]. 7. TUREKIAN *et al.* [P/1]. 8. WAKITA *et al.* [P/1]. 9. WÄNKE *et al.* [P/1]. 10. SCHNETZLER and NAVA (1971). 11. BRUNFELT *et al.* (1971). 12. TAYLOR *et al.* (1971). 13. GOLES *et al.* [P/2]. 14. TAYLOR *et al.* [P/2]. 15. MORRISON *et al.* [P/2]. 16. WÄNKE *et al.* [P/2]. 17. BOUCHET *et al.* [P/2]. 18. SMALES *et al.* [P/2]. 19. TRAVESI *et al.* [P/2]. 20. BRUNFELT *et al.* [P/2]. 21. VOBECKY *et al.* [P/2]. 22. KHARKAR *et al.* [P/2]. 23. WAKITA *et al.* [P/2]. 24. GOLES (1971). 25. SHOWALTER *et al.* (1972). 26. TAYLOR *et al.* (1972a). 27. BRUNFELT *et al.* [P/3]. 28. EHMANN *et al.* [P/3]. 29. LAUL *et al.* [P/3]. 30. LINDSTROM *et al.* [P/3]. 31. TAYLOR *et al.* [P/3]. 32. WÄNKE *et al.* [P/3]. 33. HELMKE *et al.* [P/3]. 34. PHILPOTTS *et al.* [P/3]. 35. BRUNFELT [A/15]. 36. CHURCH *et al.* [A/15]. 37. EHMANN *et al.* [A/15]. 38. HELMKE and HASKIN [A/15]. 39. LAUL *et al.* [A/15]. 40. TAYLOR *et al.* [A/15]. 41 WÄNKE *et al.* [A/15]. 42. BRUNFELT *et al.* [L/4]. 43. EHMANN *et al.* [L/4]. 44. HELMKE *et al.* (1973a). 45. TAYLOR *et al.* (1973a). 46. MORRISON *et al.* [L/4]. 47. TAYLOR *et al.* [L/4]. 48. WÄNKE *et al.* [L/4]. 49. HELMKE *et al.* (1973b). 50. WIIK *et al.* (1973). 51. LAUL and SCHMITT (1973). 52. BRUNFELT *et al.* [P/4]. 53. CHYI and EHMANN [P/4]. 54. FRUCHTER *et al.* [P/4]. 55. HASKIN *et al.* [P/4]. 56. LAUL and SCHMITT [P/4]. 57. MORRISON *et al.* [P/4]. 58. TAYLOR *et al.* [P/4]. 59. WÄNKE *et al.* [P/4]. 60. BRUNFELT *et al.* [L/5]. 61. CHYI and EHMANN [L/5]. 62. HASKIN *et al.* [L/5]. 63. LAUL *et al.* [L/5]. 64. LAUL and SCHMITT [L/5]. 65. FRUCHTER *et al.* [L/5]. 66. TAYLOR *et al.* [L/5]. 67. EHMANN and CHYI [P/5]. 68. FRUCH-

TER *et al.* [P/5]. 69. LAUL *et al.* [P/5]. 70. HUBBARD *et al.* [P/5]. 71. PHILPOTTS *et al.* [P/5]. 72. WÄNKE *et al.* [P/5]. 73. BAEDECKER *et al.* [P/5]. 74. EHMANN *et al.* [L/6]. 75. LAUL and SCHMITT [L/6]. 76. LAUL *et al.* [L/6]. 77. WÄNKE *et al.* [L/6]. 78. LAUL and SCHMITT [P/6]. 79. WÄNKE *et al.* [P/6]. 80. EHMANN *et al.* [P/6]. 81. BOYNTON *et al.* [P/6]. 82. GARG and EHMANN [P/7]. 83. WÄNKE *et al.* [P/7]. 84. RHODES *et al.* [P/7].

References designated [P/1] to [P/7] are from the Proceedings of the First to the Seventh Lunar Science Conferences, published annually as Supplements 1 (1970) to 7 (1976) of Geochimica et Cosmochimica Acta. References designated [L/4] to [L/6] comprise the abstracts presented at the Fourth (1973), Fifth (1974) and Sixth (1975) Lunar Science Conferences, and reproduced and distributed by the Lunar Science Institute, 3303 NASA Road 1, Houston, Texas 77058. References designated [A/15] are from a special issue of abstracts devoted to the Apollo 15 samples, and also reproduced by the Lunar Science Institute (1972). Other references not designated as indicated above may be found in the bibliography at the end of this chapter.

Table 72-C-6. *Average hafnium abundances and Zr/Hf ratios in lunar rocks and soils* from TAYLOR (1975)

Maria basalts	ppm Hf	Zr/Hf	Highland rocks	ppm Hf	Zr/Hf
Green glass, Apollo 15	0.42	52	Anorthositic gabbro	0.17	65
Olivine basalt, Apollo 12	4.0	28	Gabbroic anorthosite	1.4	35
Olivine basalt, Apollo 15	2.0	47	Troctolite	1.5	56
Quartz basalt, Apollo 12	4.2	29	Low-K Fra Mauro basalt	9.8	49
High-K basalt, Apollo 11	19	29	Medium-K Fra Mauro basalt	17	55
Low-K basalt, Apollo 11	11	32			
High-Ti basalt, Apollo 17	5.7	32			
Al-rich basalt, Apollo 12	6.5	31			

from GARG and EHMANN (1976)[a]

Rocks and breccias	ppm Hf	Zr/Hf	Soils	ppm Hf	Zr/Hf
High-Ti mare basalt	10.2	32	Apollo 11	9.4	36
Low-Ti mare basalt	3.6	36	Apollo 12	13.6	41
KREEP-poor non-mare basalt	5.4	46	Apollo 14	25	40
			Apollo 15	4.8	37
Anorthosite	0.012	21	Apollo 16	4.3	44
Troctolite	0.36	37	Apollo 17	6.4	35
KREEP-rich basalt and breccia	23	45	Orange soil, 74420	6.1	32
KREEP-poor breccia	6.8	44			

[a] Method N/R; all data from same laboratory.

to represent well-mixed samples of lunar surface rocks, then it is significant that the average Zr/Hf ratio of 39 for Apollo 11 to 17 soils shown in Table 72-C-6 is identical to the average previously calculated for chondrites. For the moon as a whole, TAYLOR (1976) has computed the following data: Zr = 24 ppm, Hf = 0.54 ppm, Zr/Hf = 44.

The fractionation of the Zr/Hf ratio in lunar materials has led to suggestions that this is due to partial reduction of Zr^{4+} to Zr^{3+} (but not Hf^{4+}), with exclusion of Zr^{3+} from early crystallising Fe-Ti minerals such as ilmenite or because zirconium is present in different phases from hafnium in the lunar interior (TAYLOR *et al.*, 1972b; CHYI and EHMANN, 1973; GARG and EHMANN, 1976). This interesting suggestion has not been confirmed and has been questioned by LAUL *et al.* (1974).

The importance of the inter-element relationships existing between the refractory elements Zr, Hf, Nb and Ta with respect to the origin of lunar rocks and the moon as a whole has been discussed in some detail by WÄNKE *et al.* (1975).

Manuscript received: August 1977

72-D. Abundance in Rock-Forming Minerals, Phase Equilibria, Hafnium Minerals

I. Rock-Forming Minerals

The discussion of the hafnium content of rock-forming minerals has been divided into two sections: (a) the distribution of hafnium in zircons and (b) other silicate and oxide minerals. The data for hafnium in minerals has similarly been divided into two sections, viz, zircons in Table 72-D-1 and other minerals in Table 72-D-2. Additional information on hafnium in minerals can be found in FLEISCHER (1955) and VLASOV (1966).

Hafnium and zirconium are very closely associated, perhaps more so than any other pair of elements (RANKAMA and SAHAMA, 1950). Hafnium is lithophile in its behaviour and occurs in oxides and silicates as Hf^{4+}, with the Hf-O bond being dominantly ionic in character (see Section 40-A). The ionic radii, ionization potentials and electronegativities of these two elements may be found in Section 40-A, 40-D, 72-A and in Chapter 12 of Vol. 1 of this handbook. The most recent ionic radii for 6-fold coordination and 4+ oxidation states given by WHITTAKER and MUNTUS (1970) are 0.80 Å for zirconium and 0.79 Å for hafnium. The reason for a near equivalence of radii is evidently due to the effect of the entry of the 4f electrons, often termed the lanthanide contraction. The elements that zirconium and hafnium are most likely to replace in common rock-forming minerals have already been discussed for zirconium in Section 40-D.

a) Hafnium and the Zr/Hf Ratio in Zircons

Hafnium, for reasons discussed above, is readily admitted into the structure of zirconium minerals and with a terrestrial abundance roughly one-fiftieth that of zirconium (FLEISCHER, 1955), discrete hafnium minerals are extremely rare. VAINSHTEIN *et al.* (1959) have proposed that zircons containing 10 wt.% or more HfO_2 be regarded as hafnium minerals. CURTIS *et al.* (1954) have applied the name hafnon to pure synthetic $HfSiO_4$ and CORREIA NEVES *et al.* (1974) have suggested a nomenclature to be used for the solid solution series (Zr, Hf) SiO_4. Zircons commonly contain up to 3 wt.% Hf although concentrations of Hf from 20 to 37 wt.% (see Table 72-D-1) have been recorded from South-West Africa, Rhodesia, Norway, and the U.S.S.R. Recently, true hafnons have been reported from the tantalite pegmatites in Mozambique (CORREIA NEVES *et al.*, 1974), that contain >90 mole % of $HfSiO_4$.

Before the advent of suitable analytical methods (e.g. N/R, Mass, I) to detect the low concentrations of hafnium in rocks and sediments, the behaviour and distribution of hafnium and the Zr/Hf ratio in igneous and metasomatic processes was studied by analysing zircons extracted from these rocks. Many workers (e.g. LIPOVA *et al.*, 1957; KOSTERIN *et al.*, 1960; KOSTERIN *et al.*, 1964) have determined the Zr/Hf ratio in zircons by X-ray fluorescence methods and do not report the absolute hafnium contents. For this reason, and because of the unknown purity and degree of alteration

Table 72-D-1. *Hafnium abundances and Zr/Hf ratios of hafnons, zircons and varieties of zircon*

Mineral	No. of samples	Rock type and locality	Wt% Hf Range	Wt% Hf Mean	Zr/Hf	Method	Reference
Hafnons and hafnian zircons							
Hafnon	2	Pegmatite, Zambézia (Mozambique)	59.12–61.47	60.29	0.03	M	1
Hafnian cyrtolite	1	Pegmatite, N. Karelia (U.S.S.R.)		37.00	0.9	X	2
Hafnian zircon	2	Pegmatites, S. W. Africa and Rhodesia	17.30–27.00	22.20		X	3
Hafnian zircon	2	Iveland (Norway)	18.70–20.40	19.60	1.7	X	4
Mafic rocks							
Zircon	2	Diabase porphyries, Kuramin Range (U.S.S.R.)			47	X	5
Zircon	2	Gabbros, California batholith, (U.S.A.)		1.00	50	S	6
Zircon	2	Leucocratic gabbros, Kuramin Range (U.S.S.R.)			55	X	5
Zircon	2	Gabbros, N. Kirgizia (U.S.S.R.)			71	X	7
Diorites and granodiorites							
Zircon	16	Diorites, Yeoval diorite complex, N. S. W. (Australia)	1.59–2.09	1.80	27	M	8
Zircon	9	Quartz diorites, various locations, U.S.S.R.	0.70–1.47	1.08	38	X	9
Zircon	2	Quartz diorites, Ural Mountains (U.S.S.R.)			41	X	10
Zircon	2	Diorite porphyries, Kuramin Range (U.S.S.R.)			42	X	5
Zircon	1	Quartz diorite, N. Kazakhstan (U.S.S.R.)		0.80	63	X	11
Zircon	5	Diorite and lamprophyres, Kuramin Range (U.S.S.R.)			44	X	5
Zircon	6	Diorites and granodiorites, N. Kazakhstan (U.S.S.R.)			45	X	12

Table 72-D-1 (continued)

Mineral	No. of samples	Rock type and locality	Wt% Hf Range	Wt% Hf Mean	Zr/Hf	Method	Reference
Zircon	5	Diorites and granodiorites, N. Kirgizia (U.S.S.R.)			59	X	7
Zircon	25	Granodiorites, California batholith (U.S.A.)	1.12–1.80	1.42	34	S	6
Zircon	7	Granodiorites, various locations, U.S.S.R.	0.80–1.36	1.08	35	X	9
Zircon	3	Granodiorites, Kuramin Range (U.S.S.R.)			37	X	5
Zircon	7	Granodiorites, Ural Mountains (U.S.S.R.)			40	X	10
Zircon	2	Granodiorites, various locations, U.S.S.R.			41	X	13
Zircon	1	Granodiorite, Mojave Desert (U.S.A.)		1.15	43	S	6
Zircon	7	Granodiorite porphyries, Kuramin Range (U.S.S.R.)			50	X	5
Zircon	9	Granodiorites and granites, N. Kazakhstan (U.S.S.R.)	1.20–1.60	1.39	36	X	11
Acidic rocks							
Zircon	1	Pink granite, E. Desert, Egypt			13	X	14
Cyrtolite	6	Granites, N. Kazakhstan (U.S.S.R.)	1.40–4.30	2.50	18	X	11
Cyrtolite	12	Granites and alaskites, N. Kazakhstan (U.S.S.R.)	1.20–3.30	2.07	19	X	11
Zircon	19	Granites, Yeoval diorite complex, N. S. W. (Australia)	1.29–3.57	1.86	27	M	8
Zircon	9	Granites, various locations, U.S.S.R.	0.74–1.74	1.09	30	X	15
Zircon	4	Granites and alaskites, N. Kazakhstan (U.S.S.R.)	1.40–1.80	1.60	31	X	11
Zircon	10	Granites, Ukraine (U.S.S.R.)	0.59–1.91	1.07	31	X	16
Zircon	34	Granites, various locations, U.S.S.R.	0.77–2.50	1.31	32	X	9

Zircon	15	Granites, various locations, U.S.S.R.			32	X	13
Cyrtolite	3	Granites, various locations, U.S.S.R.	0.93–1.38	1.10	33	X	9
Zircon	5	Granites, Mojave Desert (U.S.A.)	1.35–2.02	1.60	33	S	6
Zircon	2	Leucocratic granites, Kuramin Range (U.S.S.R.)			34	X	5
Zircon	10	Alaskite granites, N. Kirgizia (U.S.S.R.)			36	X	7
Zircon	4	Granites, Ural Mountains (U.S.S.R.)			37	X	10
Zircon	11	Granites, N. Kazakhstan (U.S.S.R.)	1.20–1.60	1.35	37	X	11
Zircon	1	Granite, California batholith (U.S.A.)		1.32	38	S	6
Zircon	15	Granites, various locations, U.S.A.			40	X	17
Zircon	3	Granites, Kuramin Range (U.S.S.R.)			44	X	5
Zircon	6	Biotite granites, N. Kirgizia (U.S.S.R.)			45	X	7
Zircon	35	Granites, S. E. Ontario (Canada)			46	X	18
Zircon	7	Granites and qtz. syenites, N. Kazakhstan (U.S.S.R.)	1.00–1.40	1.19	46	X	11
Cyrtolite	3	Granites, N. Kazakhstan (U.S.S.R.)			46	X	12
Zircon	7	Granites, Stanovoi Range (U.S.S.R.)			47	X	19
Zircon	10	Granites, various locations, S. Africa	0.71–1.07	0.94	48	X	20
Zircon	28	Granites, N. Italy	0.95–4.93	2.23	49	M	21
Zircon	16	Granites, N. Kazakhstan (U.S.S.R.)			52	X	12
Zircon	9	Quartz and granite porphyries, Kuramin (U.S.S.R.)			54	X	5
Zircon	2	Adamellites, Ural Mountains (U.S.S.R.)			42	X	10
Zircon	8	Tonalites, California batholith (U.S.A.)	1.01–1.20	1.10	46	S	6
Zircon	3	Monzonites, California batholith (U.S.A.)	1.32–1.73	1.55	32	S	6
Zircon	1	Quartz monzonite, Mojave Desert (U.S.A.)		1.40	36	S	6
Zircon	3	Quartz monzonites, Kuramin Range (U.S.S.R.)			44	X	5
Zircon	2	Monzonites, N. Kazakhstan (U.S.S.R.)			62	X	12
Cyrtolite	3	Aplites, N. Kazakhstan (U.S.S.R.)	1.60–2.60	2.13	23	X	11

Table 72-D-1 (continued)

Mineral	No. of samples	Rock type and locality	Wt% Hf		Zr/Hf	Method	Reference
			Range	Mean			
Zircon	6	Aplites, various locations, U.S.S.R.	1.00–2.71	1.81	26	X	9
Zircon	5	Aplites, S. Stanovoi Range (U.S.S.R.)			37	X	19
Zircon	3	Sodic felsites, E. Desert, Egypt			22	X	14
Zircon	6	Felsites and rhyolites, Kuramin Range (U.S.S.R.)			49	X	5
Pegmatites							
Naegite (Metamict)	1	Pegmatite, Naegi (Japan)		5.10	7	X	2
Cyrtolite	6	Uranium pegmatites, various locations	2.27–5.19	3.25	10	X	22
Malacon (semi-metamict)	2	Pegmatites, various locations, U.S.S.R.	2.20–6.90	4.55	12	X	2
Cyrtolite	14	Pegmatites, various locations, U.S.S.R.	0.93–5.00	2.60	13	X	23
Zircon	1	Pegmatite, N. Kazakhstan (U.S.S.R.)			14	X	12
Zircon	1	Pegmatite, Gordonia (S. Africa)		2.12	17	X	20
Cyrtolite	4	Pegmatites, N. Kazakhstan (U.S.S.R.)	1.20–2.60	2.10	19	X	11
Zircon	6	Pegmatites, various locations, U.S.S.R.	1.25–6.02	2.56	20	X	9
Zircon	47	Pegmatites, S. Stanovoi Range (U.S.S.R.)			23	X	19
Zircon	4	Pegmatites, Ukraine, (U.S.S.R.)	0.81–1.06	0.97	31	X	16
Zircon	9	Pegmatites, various locations, U.S.S.R.	0.93–2.08	0.94	34	X	23
Zircon	7	Pegmatites, various locations	0.61–0.20	0.94	50	X	22
Zircon	2	Pegmatites, various locations, U.S.A.			61	X	17
Zircon (fresh and metamict)	3	Pegmatites, Wichita Mts., Oklahoma (U.S.A.)	0.50–0.90	0.70	68	S	24
Zircon	1	Subalkalic pegmatite, Karg River (U.S.S.R.)		1.61	20	X	15
Zircon	7	Syenite pegmatites, Il'men Mts. (U.S.S.R.)	0.85–1.02		49	X	23

Zircon	12	Pegmatites, Lovozero Massif (U.S.S.R.)	0.76–1.10			X	25
Zircon and malacon	12	Alkali pegmatites, various locations, U.S.S.R.	0.34–1.14	0.69	55	X	15
Zircon	6	Miaskite pegmatites, Il'men Mts. (U.S.S.R.)	0.85–1.40	1.17	37	X	23
Miscellaneous igneous rock types							
Zircon	8	Alkaline ultrabasic rock, Kola Peninsula (U.S.S.R.)	0.55–1.31	1.06	44	X	26
Zircon	1	Kimberlite, Yakutia (U.S.S.R.)		1.40	33	X	2
Zircon	2	Kimberlites, various locations, S. Africa	0.86–1.24	1.05	45	X	20
Zircon	13	Kimberlites, various locations, U.S.S.R.	0.76–1.36	0.99	48	X	27
Zircon	1	Carbonatite, Madras (India)		1.35	36	X	2
Zircon	14	Mainly syenites, various locations, U.S.S.R.			42	X	13
Zircon	4	Syenites, N. Kirgizia (U.S.S.R.)			44	X	7
Zircon	2	Nepheline Syenites, Ukraine (U.S.S.R.)	0.85–0.98	0.92	51	X	16
Zircon	4	Nepheline syenites, various locations, U.S.S.R.	0.25–0.57	0.46	104	X	15
Zircon	5	Miaskites, various locations, U.S.S.R.			82	X	13
Zircon	7	Foyaites, various locations, U.S.S.R.			56	X	13
Zircon	2	Andesites, Kuramin Range (U.S.S.R.)			52	X	5
Cyrtolite (metamict)	9	Various rock types and locations	1.10–37.0	6.30	16	X	2
Cyrtolite (semi-metamict)	5	Various rock types and locations	1.00–7.00	3.78	18	X	2
Zircon and malacon	6	Subalkalic rocks, various locations, U.S.S.R.	0.76–2.54	1.39	34	X	15
Cyrtolite, facetted (metamict)	2	Various rock types and locations	1.0–1.20	1.10	39	X	2
Zircon, facetted (crystalline)	6	Various rock types and locations	0.85–1.70	1.31	38	X	2
Zircon, facetted (metamict)	2	Variuos rock types and locations	0.95–1.35	1.15	39	X	2
Cyrtolite, facetted (semi-metamict)	3	Various rock types and locations	0.60–1.6	1.17	41	X	2
Zircon (various types)	32	Granite, pegmatite and syenite, E. Ontario (Canada)			65	X	28
Lunar zircon	1	Rock 12013, Apollo 12		2.88	16	M	29

Table 72-D-1 (continued)

Mineral	No. of samples	Rock type and locality	Wt% Hf		Zr/Hf	Method	Reference
			Range	Mean			
Sedimentary materials							
Zircon	13	Alluvial sediments, Lena River (U.S.S.R.)	0.68–1.40	1.10	38	X	27
Zircon	12	Black sand, Mediterranean coast, Rosetta (Egypt)			39	X	14
Zircon	9	Alluvial deposits, various locations			41	X	17
Zircon	2	Beach sands, various locations, S. Africa	1.08–1.11	1.09	43	X	20
Zircon	2	Placer deposits, various locations	0.95–1.10	1.03	47	X	2
Zircon	13	Zirconium ores, various locations			54	S	30
Zircon	3	Alluvial sediments, Suordaak River (U.S.S.R.)	0.76–0.85	0.79	57	X	27
Zircon	2	Sediments, various locations, S. Africa	1.14–1.17	1.15	40	X	20
Metamorphic and metasomatic rocks							
Zircon	4	Migmatites and gneisses, Ukraine (U.S.S.R.)			26	X	16
Zircon	22	Pyroxene gneisses, S. E. Ontario (Canada)			29	X	18
Zircon	57	Biotite gneisses, S. E. Ontario (Canada)			35	X	18
Zircon	2	Gneisses, various locations, S. Africa	1.23–1.40	1.31	34	X	20
Zircon	10	Gneisses, various locations, U.S.A.			41	X	17
Zircon	7	Gneisses, S. Stanovoi Range (U.S.S.R.)			41	X	19
Zircon (various types)	6	Granite gneisses, E. Ontario (Canada)			50	X	28
Zircon and malacon (autometasomatic)	3	Subalkalic rocks, various locations, U.S.S.R.	1.57–3.82	2.63	15	X	15
Zircon and malacon (autometasomatic)	5	Nepheline syenites, various locations, U.S.S.R.	0.55–1.78	1.02	44	X	15
Zircon	7	Albitites, Lovozero Massif (U.S.S.R.)	1.00–1.22	1.12	42	X	31

Zircon	3	Albitites, Ukraine (U.S.S.R.)	0.17–0.47	0.34	62	X	16
Cyrtolite, alvite and naegite	6	Replacement pegmatites, various locations	5.60–11.70	8.40	5.2	X	22
Cyrtolite	4	Albitized pegmatites, various locations	4.49–6.28	5.15	8.9	X	22
Zircon	5	Albitized miaskite pegmatites, various locations	0.75–1.70	1.12	37	X	23
Zircon	3	Albitized pegmatites, Vishnevyye Mts. (U.S.S.R.)	0.98–1.57	1.25	41	X	23
Zircon	3	Hydrothermal veins, N. Kirgizia (U.S.S.R.)			29	X	7
Zircon (metasomatic)	7	Dolomites, Ukraine, (U.S.S.R.)	0.38–0.81	0.54	67	X	16
Zircon and malacon	7	Metasomatic veins, various locations, U.S.S.R.	0.21–0.85	0.30	143	X	15

References: 1. Correia Neves *et al.* (1974); 2. Lipova and Mayeva (1971); 3. Von Knorring and Hornung (1961); 4. Levinson and Borup (1960a); 5. Kosterin *et al.* (1960); 6. Gottfried and Waring (1964); 7. Kosterin *et al.* (1958); 8. Gulson (1970); 9. Lyakhovich and Shevaleevskii (1962); 10. Lipova *et al.* (1957); 11. Yeliseyeva *et al.* (1968); 12. Kosterin *et al.* (1963); 13. Shevaleevskii *et al.* (1960); 14. Abdel-Gawad (1966); 15. Pavlenko *et al.* (1957); 16. Tugarinov *et al.* (1956); 17. Mertie *et al.* (1958); 18. Currie (1968); 19. Kosterin *et al.* (1964); 20. Schutte (1966); 21. Pigorini and Veniale (1966); 22. Vainshtein *et al.* (1959); 23. Lipova and Shevaleevskii (1961); 24. Larsen *et al.* (1953); 25. Gerasimovskii *et al.* (1962); 26. Kukharenko *et al.* (1960), 27. Nekrasova and Rozhdestvenskaya (1970); 28. Currie (1962); 29. Haines *et al.* (1971); 30. Cooley *et al.* (1953); 31. Gerasimovskii and Shevaleevskii (1957).

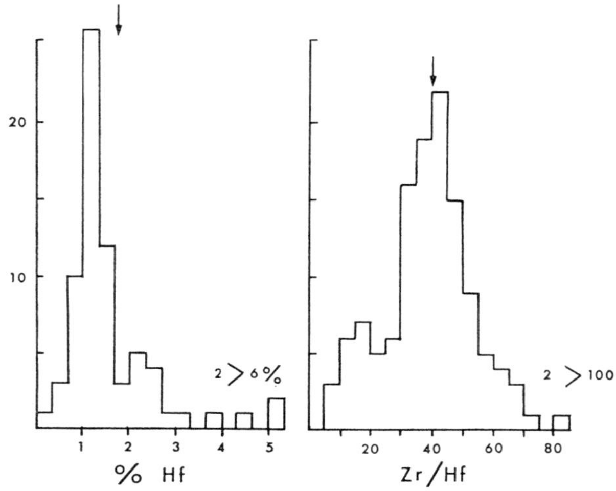

Fig. 72-D-1. Frequency distribution diagram for hafnium and Zr/Hf ratios in zircons, using mean values for the individual entries in Table 72-D-1. The arrows indicate the mean concentration for hafnium of 1.71% Hf for 72 entries in Table 72-D-1, and a mean Zr/Hf ratio of 40 for 123 entries

of many of the minerals separates in the literature, the distribution of hafnium in zircons will be discussed mainly with respect to changes in the Zr/Hf ratio.

The hafnium and Zr/Hf data which have been used in Table 72-D-1 is shown in Fig. 72-D-1 in order to provide an overall assessment of the ranges encountered. Note that only the means for each entry have been used in Fig. 72-D-1 as we consider that this introduces less sample bias.

(i) *Zircons from igneous rocks:* von Hevesy and Jantzen (1925) noted that zircons from granitic pegmatites have lower Zr/Hf ratios than zircons from granites, with those from nepheline syenites having the highest Zr/Hf ratios. These observations have been confirmed in several later papers (Fleischer, 1955; Pavlenko *et al.*, 1957; Lipova and Shevaleevskii, 1961; Currie, 1962), while Tugarinov *et al.* (1956) found virtually identical Zr/Hf ratios in zircons from granites and pegmatites but higher Zr/Hf ratios in zircons from nepheline syenites.

The range of hafnium contents in zircons from unaltered nepheline syenites is lower than that of granites (Pavlenko *et al.*, 1957, Table 72-D-1). Pavlenko *et al.* (1957) point out that although there is in general an overlapping range of Zr/Hf ratios in zircons from nepheline syenites (Zr/Hf 73 to 135) and their associated pegmatites (Zr/Hf 35 to 83), the zircons in pegmatites have lower Zr/Hf ratios than zircons from the corresponding host rocks.

Kosterin *et al.* (1958, 1960), Shevaleevskii *et al.* (1960), Gottfried and Waring (1964) and Gottfried *et al.* (1968) have observed a general increase in hafnium and a decrease in the Zr/Hf ratio in zircons from basic to acid rocks. Gottfried *et al.* (1968) record a twofold enrichment of hafnium in zircons from granites over that found in gabbroic rocks, in samples from Pennsylvania and Tasmania. Other workers who have analysed zircons from granite complexes and pegmatites have noted a similar trend of increasing hafnium and decreasing Zr/Hf from the early to the more evolved rocks (LIPOVA and SHEVALEEVSKII, 1961; LYAKHOVICH *et al.*, 1962; KOSTERIN *et al.*, 1964; YELISEYEVA *et al.*, 1968; CONDIE and LO, 1971). In contrast to the above investigations GULSON (1970) found no systematic change in the Zr/Hf ratio in zircons from gabbros, diorites and granites from the Yeoval Diorite Complex, Australia. Similarly KOSTERIN *et al.* (1960) found no systematic change in the Zr/Hf ratio in zircons from relatively rapidly cooled hypabyssal rocks, ranging in composition from basalts to quartz felsites.

The average Zr/Hf ratios computed for zircons from various rock types, using the data listed in Table 72-D-1 give the following values: gabbros (8 samples) 56, diorites and granodiorites (107 samples) 38, granites (327 samples) 39 and oversaturated pegmatites (107 samples) 25. The general decrease from gabbros to pegmatites of the Zr/Hf ratio should be interpreted with caution due partly to the small number of zircons from gabbros, the possible over-representation of unusual zircon types from the pegmatites and analytical error. The similar Zr/Hf ratio in zircon from diorites, granodiorites, granites and the average of all the entries in Table 72-D-1 (see Fig. 72-D-1) of ~39 is significantly lower than the abundance ratio given by RANKAMA and SAHAMA (1950) of 49 but is in good agreement with the estimate given by BROOKS (1970) of 37 for crustal rocks and by SHAW *et al.* (1976) of 41 for the Canadian Shield.

The assumption has often been made that the Zr/Hf ratio of zircons reflects the Zr/Hf ratio of the bulk rock (e.g. KOSTERIN *et al.*, 1958; CURRIE, 1962). The early data of LIPOVA *et al.* (1957) tended to confirm this assumption, as the average Zr/Hf ratio of 38 from zircons was within experimental error of the Zr/Hf ratio of 40 in two granite samples. This assumption has been seriously questioned more recently, particularly for rocks containing abundant melanocratic minerals (VLASOV, 1966). SHEVALEEVSKII *et al.* (1960) and BUTLER and THOMPSON (1965) point out that pyroxenes and alkali amphiboles may contain appreciable amounts of the hafnium present in the whole rock. In addition hafnium is often inhomogeneously distributed within and between zircon grains (GOTTFRIED and WARING, 1964; SILVER and CHODOS, 1966; GULSON, 1970). YELISEYEVA *et al.* (1968) suggest the separation of zircons from at least 10 kg of rock sample in order to obtain reliable geochemical data from the zircons.

The processes occurring in magmas that cause the fractionation between hafnium and zirconium are clearly complex. RINGWOOD (1955) has proposed that Hf^{4+} and Zr^{4+} behave predominantly as complex formers in silicate magmas although some free 4+ ions may exist. From ionic potential considerations given by RINGWOOD (1955) the complex HfO_4^{4-} is very slightly more stable than the ZrO_4^{4-} complex in the melt. AHRENS (1964) has outlined the slight differences in physical and chemical properties of simple zirconium and hafnium compounds. From the small differences in the fourth ionization potentials between zirconium and hafnium he suggests that the Zr-ligand bond should be slightly more covalent than the Hf-ligand bond. Support for this conclusion is found in the fact that HfO_2 is a weak base while ZrO_2 is amphoteric (e.g. CAMPBELL, 1970). CAMPBELL (1970) has pointed out the difficulties in measuring the acid-base properties of these very slightly soluble compounds. This more basic nature of Hf has been proposed as the reason for the fractionation of zirconium and hafnium (e.g. SHEVALEEVSKII *et al.*, 1960; VLASOV, 1966; GERASI-

MOVSKII *et al.*, 1972), particularly in alkali rocks. VLASOV (1966) has proposed that zirconium and hafnium replace magnesium and iron in the structure of melanocratic minerals, and that the more basic hafnium will more easily replace these basic ions, than will zirconium, the latter becoming concentrated in the accessory minerals.

BURNS and FYFE (1966, 1967) have pointed out the necessity for considering all the thermodynamic quantities involved when crystals form from magmas, to account for the geochemical behaviour of the elements, although the application of thermodynamics is hampered by the lack of suitable data (RANDLE, 1974). ESSON *et al.* (1968) have documented the unit cell volumes of a number of corresponding zirconium and hafnium compounds, and in all cases the hafnium compound has the smaller unit cell volume, indicating that Hf^{4+} occupies less space than Zr^{4+} in these structures. They propose that this property of Hf^{4+} leads to a slightly greater stabilization of hafnium in a silicate melt relative to zirconium as the proportion of octahedral sites in the melt decreases during differentiation. This in turn should lead to a decrease of the Zr/Hf ratio in the melt as differentiation proceeds.

The variety of possible causes that have been suggested for fractionation of zirconium from hafnium, testify to the complexity of the problem and there is no clear answer at this stage.

The early data of hafnium, uranium and thorium content for zircons suggested an increase in the radioactive elements in zircon as hafnium increased (RANKAMA and SAHAMA, 1950; MARTIN and PIZZOLATO, 1954). More recent investigations, however, do not confirm this (LIPOVA *et al.*, 1957; LEVINSON and BORUP, 1960a; SCHUTTE, 1966). LIPOVA and MAYEVA (1971) found no correlation of the Zr/Hf ratio and the metamict state of the zircon, although there is a tendency for hafnium to increase in cyrtolites in the more metamict zircons. They did, however, observe a correlation between the Zr/Hf ratio and morphology of the zircon crystals. Most zircons with Zr/Hf ratios less than 20 in the samples they studied were imperfect cystals or spherulites.

(ii) *Zircons from metasomatic and metamorphic rocks:* PAVLENKO *et al.* (1957) have analysed corroded poikilitic zircons from autometasomatic rocks, and found them to have very low Zr/Hf ratios compared to zircons from the equivalent unaltered rocks. The Zr/Hf ratio in altered subalkali rocks ranged from 12 to 17 while the equivalent range from altered alkali rocks was from 27 to 68 (see Table 72-D-1).

Lower Zr/Hf ratios appear to be a common feature (KUKHARENKO *et al.*, 1960; LYAKHOVICH *et al.*, 1962; VLASOV, 1966) for zircons from autometasomatic rocks. These authors have attributed the lowering of the Zr/Hf ratio to the removal of zirconium during autometasomatism, rather than to the addition of hafnium. On the other hand zircons in veins deposited from hydrothermal solutions have a Zr/Hf ratio range of 48 to 190 (PAVLENKO *et al.*, 1957) and malacons from metasomatic bodies in dolomite (TUGARINOV *et al.*, 1956) have a Zr/Hf range of 52–78. The Zr/Hf ratio is generally higher in these zircons than those in common rock types. The formation and relative stability of zirconium and hafnium fluoride complexes has been proposed to account for this distribution (TUGARINOV *et al.*, 1956; GERASIMOVSKII *et al.*, 1972), the compound Na_2HfF_6 being less stable than the equivalent zirconium compound (TUGARINOV *et al.*, 1956). Zirconium would therefore be more stable and more abundant than hafnium in hydrothermal solutions, and zircons crystallising from such solutions would have higher Zr/Hf ratios. The relative stability of the zirconium complex would enable zirconium to be leached relative to hafnium from zircons during autometasomatism (VLASOV, 1966). BUTLER and THOMPSON (1965) have opposed this point of view, as they consider the differing rates of diffusion of the fluoride complexes (rate of diffusion $\propto 1/\sqrt{M}$) would result in a hafnium enriched aqueous fluoride system.

Data for hafnium in zircons from the common metamorphic rocks are scarce (see Table 72-D-1). The few data given by TUGARINOV *et al.* (1956) indicate similar to slightly lower Zr/Hf ratios in zircons from migmatites and gneisses compared to zircons from granites.

(iii) *Zircons in sedimentary materials:* MERTIE (1958) has determined the Zr/Hf ratio in zircons from alluvial deposits, granites and metamorphic rocks, mainly from the United States. Nine samples of zircons from alluvial deposits gave an average Zr/Hf ratio of 41 (method: X). The average for thirty samples of sedimentary, metamorphic and igneous zircons is 42. He computed the average Zr/Hf ratio of zircons from 40 samples listed in FLEISCHER (1955) that are probably of alluvial origin and obtained the value of 37.

NEKROSOVA and ROZHDESTVENSKAYA (1970) have analysed zircons from alluvial sediments in areas of kimberlite and carbonatite magmatism; the average Zr/Hf ratio 57 is relatively high, while zircons from placer deposits on the Lena river in the U.S.S.R. gave an average Zr/Hf ratio of 38, typical of granitic rocks.

b) Hafnium and the Zr/Hf ratio in Oxide and Silicate Minerals except Zircon

Data for terrestrial and lunar oxide and silicate minerals, other than zircon, are given in Table 72-D-2. It should be noted in this table that, in some cases, Zr and Hf have been determined by different analytical methods. The reader should take this into account when evaluating the Zr/Hf ratios, as it may often be the case that the zirconium data may be less reliable than that for hafnium, for example where optical spectrographic techniques have been used for the estimation of zirconium. We have indicated in Table 72-D-2 where this situation occurs; these comments also apply to all subsequent tables. The comments in Section 40-D-1 pertaining to the problems of zircon contamination of mineral separates should be borne in mind when considering the data, particularly for the non-zirconium minerals. Zircon contains the highest absolute concentrations of hafnium, while other zirconium minerals listed in Table 72-D-2 generally have hafnium contents in the range of 0.1 to 2 wt.% Hf. Hafnium almost certainly replaces zirconium in these minerals, and in the non-zirconium minerals it probably replaces similar elements to zirconium, the relevant information being given in Section 40-D-1.

Fractionation of hafnium relative to zirconium has been demonstrated by EL GORESY *et al.* (1973) between co-existing baddeleyite (Zr/Hf, 44–58) and zirconolite (Zr/Hf, 27–31) from Apollo 16 breccias. KUKHARENKO *et al.* (1960) observed the opposite trend of higher Zr/Hf ratios in zirkelite than baddeleyite in rocks from the Kola Peninsula; however equilibrium between the minerals was not shown. Note that there is considerable confusion in the literature about the zirconolite-zirkelite terminology (BUSCHE *et al.*, 1972; WARK *et al.*, 1973), although WARK *et al.* (1973) favour the zirconolite formula for lunar and terrestrial "zirconolite-zirkelite" phases.

Table 72-D-2. *Hafnium abundances and Zr/Hf ratios of oxide and silicate minerals*

Mineral	No. of samples	Rock type and locality	Wt% Hf		Zr/Hf	Method[b]	Reference
			Range	Mean			
Oxides							
Baddeleyite	3	Zirconium ores, various locations			32	S	1
Baddeleyite	10	Alkaline ultrabasic rock, Kola Peninsula (U.S.S.R.)	1.61–2.12	1.33	36	X	2
Baddeleyite	2	Troctolite, Apollo 17		1.86	37	M	3
Baddeleyite	2	Breccias, Apollo 16	1.20–1.30	1.25	51	M	4
Baddeleyite	2	Carbonatites, various locations, S. Africa			56	X	5
Belyankinite	2	Pegmatites, Lovozero Massif (U.S.S.R.)	0.09–0.13	0.11	47	X	6
Zirkelite	5	Alkaline ultrabasic rock, Kola Peninsula, (U.S.S.R.)	0.21–0.70	0.47	70	X	2
Zirkelite (zirconolite?)	3	Breccias, Apollo 12 and 14	0.24–0.74	0.38	89	M	7
Zirconolite	4	Basalts and breccias, Apollo 11, 14 and 15	0.73–1.10	0.91	27	M	8
Zirconolite	1	Breccia, Apollo 16		0.87	29	M	4
Zirconolite	2	Arbarastkh and Afrikanda Massifs (U.S.S.R.)	0.33–0.47	0.40	64	M	8
Ilmenite	1	Basalt, Apollo 11		16.5[a]	37	N/R	9
Ilmenite	1	Ferrodiorite, Skaergaard intrusion (E. Greenland)		12.5[a]	37	N/R, X	10
Ilmenite	13	Kimberlites, southern Africa	13–34	18[a]		N/R	11
Ilmenite	10	Megacrysts, kimberlite, southern Africa	17–38	29[a]	33	X	12
Chronium spinel	1	Kimberlites, southern Africa		10.5[a]		N/R	11
Silicates							
Tranquillityite	4	Basalts, Apollo 11 and 12	0.04–0.60	0.17	88	M	13
Catapleite	2	Pegmatites, Lovozero Massif (U.S.S.R.)	0.52–0.57	0.55	40	X	6

Catapleite	3	Pegmatites, Lovozero Massif (U.S.S.R.)	0.30–0.36			X	14
Eudialyte	9	Lujavrites and foyaites, Lovozero Massif (U.S.S.R.)	0.14–0.24	0.21	47	X	14
Eudialyte	4	Pegmatites, Lovozero Massif (U.S.S.R.)	0.14–0.23	0.17	48	X	14
Eudialyte	19	Various rock types, Lovozero Massif (U.S.S.R.)	0.11–0.19	0.16	51	X	6
Eudialyte	1	Kangerdlugssuaq alkaline intrusion (Greenland)		0.12	128	N/R, X	15
Keldyshite	2	Foyaites, Lovozero Massif (U.S.S.R.)	0.61–0.69	0.65		X	14
Lavenite	1	Pegmatite, Lovozero Massif (U.S.S.R.)		0.24	48	X	6
Lovozerite	2	Lovozerite-lujavrites, Lovozero Massif (U.S.S.R.)	0.12–0.17	0.15	67	X	6
Lovozerite	3	Lujavrites, Lovozero Massif (U.S.S.R.)	0.16–0.21			X	14
Zirfesite	2	Pegmatites, Lovozero Massif (U.S.S.R.)	0.76–0.79	0.78	32	X	6
Zirfesite	3	Pegmatites, Lovozero Massif (U.S.S.R.)	0.46–0.48			X	14
Thortveitite	3	Iveland (Norway) and Befanamo (Madagascar)	0.51–1.44	1.05	1.2	X	16
Thorite	1	Alkalic granite, Nigeria		0.09	18	X	17
Astrophyllite	1	Kangerdlugssuaq alkaline intrusion (E. Greenland)		0.20	58	N/R, X	15
Arfvedsonite	4	Kangerdlugssuaq akaline intrusion (E. Greenland)	8–74	33[a]	37	N/R, X	15
Hornblende	1	Standard HB-1		0.61[a]		N/R	18
Aegirine	6	Kangerdlugssuaq alkaline intrusion (E. Greenland)	52–264	128[a]	34	N/R, X	15
Aegirine	2	Lujavrites, Lovozero Massif (U.S.S.R.)	80–100	90[a]	35	X	14
Clinopyroxene	9	Alkaline ultrabasic rock, Kola Peninsula (U.S.S.R.)	0.03–0.28	0.12	69	X	2
Chromium diopside	16	Kimberlites, southern Africa	0.5–5.5	2.9[a]		N/R	11
Pyroxene	1	Basalt, Apollo 15		1.4[a]	41	MASS	19

Table 72-D-2 (continued)

Mineral	No. of samples	Rock type and locality	Wt%Hf Range	Wt%Hf Mean	Zr/Hf	Method[b]	Reference
Pyroxene	1	Basalt, Apollo 11		5.9[a]	32	N/R	9
Pyroxene	1	Ferrodiorite, Skaergaard intrusion (E. Greenland)		3.6[a]	30	N/R, X	10
Sphene	1	Kangerdlugssauq alkaline intrusion (E. Greenland)		12.4[a]		N/R	15
Plagioclase	5	Anorthosites, Fiskenaesset complex (Greenland)	0.003–0.035	0.016[a]		N/R	20
Plagioclase	6	Pumices, various locations	0.30–0.91	0.56[a]		N/R	21
Plagioclase	1	Basalt, Apollo 15		0.21[a]		MASS	19
Biotite	2	Kangerdlugssuaq alkaline intrusion (E. Greenland)	0.9–1.1	1.0[a]	110	N/R, X	15
Biotite	1	Standard BI-1		1.0[a]		N/R	18
Garnet (red)	3	Kimberlites, southern Africa	2.2–3.2	2.8[a]		N/R	11
Garnet (orange)	4	Kimberlites, southern Africa	1.2–2.9	2.0[a]		N/R	11

[a] Values as ppm

[b] Where two analytical methods are given, the second refers to the method used for the zirconium determination.

References: 1. Cooley *et al.* (1953); 2. Kukharenko *et al.* (1960); 3. Dymek *et al.* (1975); 4. El Goresy *et al.* (1973); 5. Schutte (1966); 6. Gerasimovskii and Shevaleevskii (1957); 7. Busche *et al.* (1972); 8. Wark *et al.* (1973); 9. Ehmann *et al.* (1975); 10. Brooks (1969); 11. Fesq *et al.* (1976); 12. Mitchell *et al.* (1973); 13. Lovering *et al.* (1971); 14. Gerasimovskii *et al.* (1962); 15. Brooks (1970); 16. Levinson and Borup (1960b); 17. Heinrich (1963); 18. Randle (1974); 19. Taylor *et al.* (1973b); 20. Henderson *et al.* (1976); 21. Dudas *et al.* (1971).

GERASIMOVSKII *et al.* (1962) have noted that the Zr/Hf ratios of other zirconium silicate minerals such as eudialyte and lovozerite from the Lovozero Massif are higher than the ratios for the total rock, the latter being lowered by the incorporation of appreciable amounts of hafnium into the more abundant minerals such as aegirine and arfvedsonite, that have low Zr/Hf ratios (see Table 72-D-2).

Thortveitite has attracted attention due to the very low Zr/Hf ratios reported for this mineral (VON HEVESY and JANTZEN, 1924; LEVINSON and BORUP, 1960b) of around 1, with hafnium commonly exceeding the zirconium content. Thortveitite is, however, a very rare mineral and LEVINSON and BORUP (1960b) have pointed out that the zircons associated with it also have exceptionally high hafnium concentrations ($\sim$20% Hf) and low Zr/Hf ratios ($\sim$2).

Ilmenites from Apollo 11 basalt (EHMANN *et al.*, 1975) have virtually identical hafnium contents and Zr/Hf ratios as those extracted from ferrodiorite in the Skaergaard intrusion (BROOKS, 1969) and ilmenites extracted from heavy mineral concentrates, and found as megacrysts, in kimberlites (FESQ *et al.*, 1976; MITCHELL *et al.*, 1973). EHMANN *et al.* (1975) have noted that early crystallising lunar ilmenite does not appear to be anomalous in terms of the Zr/Hf ratio, as might be expected if Zr^{3+} ions were common in lunar magmas, although they have stressed the difficulties of obtaining pure ilmenite separates. The high zirconium concentrations in lunar ilmenites reported by ARRHENIUS *et al.* (1971) will probably contain high hafnium contents as well.

Pyroxenes from lunar and terrestrial basaltic rocks, and garnets from kimberlites contain varying but significant amounts of hafnium. Biotites from the Kangerdlugssuaq alkaline intrusion have low Hf contents and significantly higher Zr/Hf ratios than most other minerals from this intrusion, this possibly being related to a crystallochemical property of the biotite structure (BROOKS, 1970).

The hafnium content of plagioclase is very low, with the data reported by HENDERSON *et al.* (1976) being significantly lower than the rest of the plagioclase data given in Table 72-D-2.

II. Phase Equilibria

Experimental studies by CURTIS *et al.* (1954) show that ZrO_2 and HfO_2 form a continuous series of solid solutions and that HfO_2 and SiO_2 react to form only one compound $HfSiO_4$, which these authors called hafnon. This synthetic product was accepted as a mineral name by the Commission of New Minerals and Mineral Names of the International Mineralogical Association. CURTIS *et al.* (1954) did not present a phase diagram for the system HfO_2 and SiO_2, and we are not aware of any other experimental studies dealing with this system, nor with the system $ZrSiO_4$—$HfSiO_4$. Hydrothermal synthesis of zircon and hafnon using various zirconium and hafnium salts was achieved by CARUBA *et al.* (1975) who also showed that hafnium could substitute for zirconium. Further information relating to the phase equilibria between hafnium and other elements may be found in the compilation provided by LEVIN *et al.* (1964). The phase relationships for the system ZrO_2—SiO_2 (i.e. for zircon) are discussed in Subsection 40-D-II.

III. Hafnium Minerals

Although many workers (e.g. BAYER, Section 72-A) have considered that hafnium does not form a mineral of its own, this can no longer be substantiated in the light of evidence provided earlier in this section. Thus VAINSHTEIN *et al.* (1959) have proposed that zircons containing 10 wt.% or more HfO_2 be regarded as hafnium minerals. The discovery of natural hafnons (using the name given by CURTIS *et al.*, 1954, to synthetic $HfSiO_4$) by CORREIA NEVES *et al.* (1974) demonstrates conclusively that hafnium minerals do occur, and the terminology proposed by CORREIA NEVES *et al.* (1974) should be adopted, viz:

Name	Zircon	Hafnian zircon	Zirconian hafnon	Hafnon
Mole % $HfSiO_4$	0–10	10–50	50–90	90–100

Analytical data are provided in Table 72-D-1.

Manuscript received: August 1977

72-E. Abundance in Common Igneous Rocks; Crustal Abundance

The relatively low abundance of hafnium in common rocks (VINOGRADOV, 1962, estimated the average igneous rock to contain 1 ppm Hf), coupled with the propensity of hafnium to substitute for zirconium in zirconium minerals led to the relatively late discovery of hafnium in 1923 as described by CORRENS in Volume 1 of this handbook. Additionally, the lack of suitably sensitive analytical techniques resulted in the fact that much of our early knowledge on the distribution and abundance of hafnium was derived from analysis of zirconium-rich minerals, notably zircon, as indicated in the previous section (72-D-1). Thus, little information on the distribution of hafnium is provided in the reviews by RANKAMA and SAHAMA (1950) and GOLDSCHMIDT (1954), and the subsequent compilations by FLEISCHER (1955) and VLASOV (1966) rely heavily on the distribution of hafnium in zircons and other zirconium-rich minerals. Nevertheless, these earlier workers recognized some of the important features of the distribution of hafnium, such as the general increase in concentration in the sequence ultramafic-mafic-intermediate-granitic rocks, and the striking concentration of hafnium in agpaitic and syenitic igneous rocks.

Since the introduction of modern techniques such as neutron activation analysis and, to a lesser extent, spark-source and isotope-dilution mass spectrometry, a great many hafnium analyses have become available for geological materials. For some types (see, for example, the data for meteorites and lunar materials in Section 72-C) the position is satisfactory, but there is still a lack of hafnium data for many common rocks. In those cases where hafnium data are not available then, to a first approximation, reference may be made to zirconium data in Chapter 40, in view of the close geochemical coherence existing between these two elements, as discussed in Section 72-D. FLEISCHER (1955) estimated a Zr/Hf ratio of 50 for the earth's crust, although our compilations (see later) show that a Zr/Hf ratio of approximately 40 may be more appropriate.

I. Standard Reference Rocks

It has become common practice to use standard reference rocks for calibrating or checking the accuracy of an analytical method. The first reference rocks to be used internationally were G-1, a granite, and W-1 a "diabase" or dolerite. Compilations of data for these have been issued at periodic intervals, and the most recent data, together with other standards which are more generally used, are listed in Table 72-E-1 in terms of the average values and magnitudes proposed by FLANAGAN (1973). This reference, together with RANDLE (1974) and ABBEY (1975) may be consulted for the source and locality of these reference standards; the latter two references also list additional hafnium data for these and other reference standards.

Table 72-E-1. *Hafnium values and Zr/Hf ratios for standard reference rocks.* (Data from FLANAGAN (1973))

Average values				Magnitudes			
Sample		ppm Hf	Zr/Hf	Sample		ppm Hf	Zr/Hf
Basalt	BCR-1	4.7	40	Dunite	DTS-1	0.01	300
Spilite	BM	3.1	34[a]	Peridotite	PCC-1	0.06	117
Diabase	W-1	2.67	39[a]	Basalt	JB-1	3.5	86
Andesite	AGV-1	5.2	43[a]	Norite	NIM-N	5	5
Granodiorite	GSP-1	15.9	31[a]	Granodiorite	JG-1	3.5	46
Granite	G-1	5.2	40[a]	Granite	NIM-G	12	25
Granite	G-2	7.35	41[a]	Syenite	NIM-S	0.6	50
Granite	GM	4.7	31[a]	Lujavrite	NIM-L	250	—
Slate	TB	4.8	36[a]	Limestone	KH	0.8	38[a]

[a] Recommended Zr values used to calculate the Zr/Hf ratios; other Zr values are magnitudes (see Table 40-E-1).

II. Ultramafic and Ultrabasic Rocks

The hafnium contents of ultramafic and ultrabasic rocks can be expected to be low on the basis of their mineralogy and low zirconium contents, and this is borne out by the few data available (Table 72-E-2). Concentrations of <1 ppm Hf are generally indicated, except for some pyroxenites analysed by SOBACHENKO *et al.* (1974). Details for the latter are lacking, but these rocks appear to be from ultrabasic alkali rock massifs which contain high zirconium contents and which may have been metasomatised (Section 40-E). It may be noted that very low levels of hafnium are present in the reference standards peridotite PCC-1 (0.06 ppm) and dunite DTS-1 (0.01 ppm) as shown in Table 72-E-1, as compared with the abundances in Table 72-E-2 and the figure of 0.6 ppm Hf estimated by TUREKIAN and WEDEPOHL (1961) for the average ultramafic rock.

The three Zr/Hf ratios given in Table 72-E-2 do not allow any meaningful comment to be made with regard to Zr/Hf variations in ultramafic rocks.

III. Mafic Rocks

Data for hafnium in mafic rocks (basalts, gabbros) and some of their differentiates are presented in Table 72-E-3. Spilites and metabasalts are also included since we can expect hafnium, like zirconium (Subsection 40-E-III), to be immobile during the subsequent processes that have affected these rocks. The abundances given in Table 72-E-3 may be compared with the average of 2 ppm Hf used by TUREKIAN and WEDEPOHL (1961) and TAYLOR (1964) for basaltic rocks; this figure appears to be low as the average of all the individual averages in Table 72-E-3 is 3.4 ppm Hf. The following generalisations are suggested by the data in Table 72-E-3 and the literature.

i) The dominant rock type in the oceanic crust is tholeiitic (variously referred to by such terms as oceanic tholeiite, abyssal tholeiite, MORB) with a mean concentration around 2.8 ppm Hf and a mean Zr/Hf ratio of 42. Regional variations, and

Table 72-E-2. *Hafnium abundances and Zr/Hf ratios of ultramafic and ultrabasic rocks*

Rock type	No. of samples	Locality	ppm Hf		Zr/Hf	Method[a]	Reference
			Range	Mean			
Peridotite	1	Garabal Hill-Glen Fyne igneous complex (Scotland)		1.0	26	N/R, X	1
Peridotitic komatiite (greenstone)	1	Archaean, Barberton (South Africa)		0.5		N/R	2
Orthopyroxenite	2	Archaean, Barberton (South Africa)	0.2–0.4	0.3		N/R	2
Pyroxenites	?	Gabbro-pyroxenite formation, Sayan-Baykal (U.S.S.R.)		3.6	33		3
Eclogite	1	Roberts Victor Mine (South Africa)		0.81	33	N/R	4
Garnet anorthosite	2	Precambrian, Fiskenaesset complex (West Greenland)	0.2–0.3	0.25		N/R	5
Anorthosite	6	Precambrian, Fiskenaesset complex (West Greenland)	0.07–1.2	0.63		N/R	6

[a] Where two analytical methods are given, the second refers to the method used for the zirconium determination.

References: 1. Brooks (1970); 2. Herrmann *et al.* (1976); 3. Sobachenko *et al.* (1974); 4. Rebagay (1969); 5. Morgan *et al.* (1976); 6. Henderson *et al.* (1976).

Table 72-E-3. *Hafnium abundances and Zr/Hf ratios of mafic rocks*

Rock type	No. of samples	Locality	ppm Hf		Zr/Hf	Method[a]	Reference
			Range	Mean			
Oceanic volcanic rocks							
Tholeiitic basalt	4	DSDP[b] Leg 2, Sites 10, 11	2.5–3.2	3.0	32	N/R, S	1
Tholeiitic basalt	16	DSDP Leg 3, Sites 14, 15, 18, 19, 20	0.90–3.9	2.1	42	N/R, S	1
Tholeiitic basalt	7	DSDP Leg 11, Sites 100, 105	0.94–1.5	1.3	37	MASS	2
Tholeiitic basalt	2	DSDP Leg 22, Sites 214, 216	3.9–4.1	4.0	37	N/R, S	3
Tholeiitic basalt	4	DSDP Leg 26, Sites 253, 254	0.94–3.9	2.9	56	N/R, S	3
Tholeiitic basalt	11	DSDP Leg 27, Sites 259, 260, 261	1.4–4.7	2.3	32	MASS	4
Tholeiitic basalt	21	DSDP Leg 28, Sites 265, 266, 267, 274	1.8–4.0	2.9	40	N/R, S	5
Transitional basalt	9	DSDP Leg 33, Site 315A	3.7–4.9	4.4	44	N/R, S	6
Tholeiitic basalt	21	DSDP Leg 33, Site 317A	0.9–1.7	1.3	55	N/R, S	6
Tholeiitic basalt	12	DSDP Leg 34, Sites 319, 320, 321	1.7–4.9	3.1	42	N/R, S	7
Tholeiitic basalt	23	DSDP Leg 34, Sites 319, 320, 321	1.8–4.8	3.5	37	N/R, X	8
Tholeiitic basalt	30	DSDP Leg 37, Sites 332, 334, 335	0.61–2.6	1.3		N/R	9
Tholeiitic basalt	2	Carlsberg Ridge, Indian Ocean	1.9–2.9	2.4	40	N/R, X	10
Tholeiitic basalt	2	Mid-Atlantic Ridge	1.8–2.1	2.0	46	N/R, X	10
Alkali olivine basalt	4	Jan Mayen, Mid-Atlantic Ridge		4.9		N/R	11
Tholeiitic basalt	6	Tertiary to Recent, eastern Iceland	2.3–5.9	4.2	32	N/R, X	12
Tholeiitic basalt	5	Maio complex, Pre-Cenozoic, Cape Verde	1.8–2.3	2.0		N/R	13
Tholeiitic basalt	1	Mauna Loa, Hawaii		3.3	57	N/R, X	10
Alkali olivine basalt	1	Keauhou, Hualalai		2.9	61	N/R, X	10
Nepheline basalt	1	Nakuku, Honolulu		3.5	67	N/R, X	10
Melilite nepheline basalt	1	Moiliili, Honolulu		6.6	65	N/R, X	10
Hawaiite	1	Mauna Kea, Hawaii		10.1	72	N/R, X	10
Ankaramite	1	Mauna Kea, Hawaii		4.1	59	N/R, X	10
Continental volcanic rocks							
Basalt	17	Steens Mountain, Oregon (U.S.A.)		4.8		N/R	14
Transitional basalt	4	Afar rift, Ethiopia	4.0–8.0	6.0	32	N/R, X	15
Lava	1	Ventersdorp (South Africa)		2.3		N/R	16
Basaltic komatiite (greenstone)	3	Archaean, Barberton (South Africa)	0.7–1.3	1.1		N/R	17

Hypabyssal (doleritic) rocks							
Dolerite lower zone	7	Great Lake intrusion, Tasmania	0.50–1.5	0.80	69	N/R, S	18
Dolerite, central zone	6	Great Lake intrusion, Tasmania	0.48–3.1	2.2	53	N/R, S	18
Dolerite chilled contact	1	Great Lake intrusion, Tasmania		2.1	24	N/R, S	18
"Diabase"	5	Dillsburg, Pennsylvania (USA)	1.1–3.2	1.7	54	N/R, S	18
Quartz dolerite	1	Palolo Quarry, Honolulu		2.7	53	N/R, X	10
"Diabase"	1	Simmer Dyke (South Africa)		2.6		N/R	16
Metadolerite	1	Archaean, Barberton (South Africa)		1.2	44	MASS	19
Plutonic rocks							
Gabbro		Gabbro-Pyroxenite formation, Sayan-Baykal (U.S.S.R.)		3.9	33		20
Gabbro		Gabbro-Syenite formation Sayan region (U.S.S.R.)	0.4–6.0		30		20
Gabbro	2	Garabal Hill-Glen Fyne igneous complex (Scotland)		1.3	41	N/R, X	10
Gabbro	3	Skaergaard intrusion (East Greenland)	1.8–6.6	4.4	37	N/R, X	21
Chilled margin	1	Skaergaard intrusion (East Greenland)		2.6	36	N/R, X	21
Picrite	1	Skaergaard intrusion (East Greenland)		1.4	42	N/R, X	21
Gabbro-dolerite	12	Various intrusions, Siberia (U.S.S.R.)	2.2–8.2	4.3	23	S	22
Gabbro	1	Archaean, Barberton (South Africa)		1.4		N/R	17

[a] Where two analytical methods are given, the second refers to the method used for the zirconium determination.
[b] DSDP=Deep Sea Drilling Project.

References: 1. Frey *et al.* (1974); 2. Ayuso *et al.* (1976); 3. Frey and Sung (1974); 4. Robinson and Whitford (1974); 5. Ford (1975b); 6. Jackson *et al.* (1976); 7. Thompson *et al.* (1976); 8. Rhodes *et al.* (1976); 9. Blanchard *et al.* (1976); 10. Brooks (1970); 11. Weigand *et al.* (1972); 12. Wood (1976); 13. De Paepe *et al.* (1974); 14. Helmke and Haskin (1973); 15. Barberi *et al.* (1975); 16. Uken and Erasmus (1970); 17. Herrmann *et al.* (1976); 18. Gottfried *et al.* (1968); 19. Glikson (1976); 20. Sobachenko *et al.* (1974); 21. Brooks (1969); 22. Nesterenko *et al.* (1971).

variations with distance from active mid-ocean ridges, in hafnium content and Zr/H ratio have not been noted.

ii) Data for tholeiitic rocks from island arcs are lacking, but on the basis of their zirconium contents (Table 40-E-3), and the data given for andesites (Table 72-E-5), these rocks can be expected to have lower hafnium contents than tholeiites associated with mid-oceanic ridges. Oceanic island basalts in turn appear to have higher hafnium contents when compared to mid-ocean ridge tholeiites, with hafnium increasing in the sequence island tholeiite—alkali basalt —nephelinite. The variations are presumably related to source rock (mantle) compositions, tectonic environment, degree of partial melting and/or fractional crystallisation, and the preference for hafnium to concentrate in the liquid phase during the latter processes, hence the use of the terms "incompatible" and "residual" element by many authors. This is further exemplified by higher concentrations of hafnium in basaltic differentiates (e.g. hawaiites) and the general tendency for hafnium to increase with increasing differentiation will become evident in subsequent discussion and tables in this section. Hafnium in mafic rocks can be expected to be mainly present in clinopyroxenes and the mesostasis, since zircons generally appear to be absent.

It is noteworthy that the Hawaiian basalts analysed by BROOKS (1970) have distinctly higher Zr/Hf ratios (about 60–70) than the other oceanic basalts quoted in Table 72-E-3, but the reason for this is not clear.

iii) Continental mafic rocks (averages of the individual averages for the volcanic, hypabyssal and plutonic rocks in Table 72-E-3 are 3.3 ppm Hf and Zr/Hf, 41) do not appear to show any systematic or appreciable differences when compared to the oceanic basalts previously discussed. Some varieties, such as the Archaean tholeiites from South Africa (HERRMANN *et al.*, 1976) and some dolerite intrusions (GOTTFRIED *et al.*, 1968) do, however, appear to have lower hafnium abundances, which can also be expected in cumulate rocks when little intercumulus liquid is present (viz. some gabbros in Table 72-E-3). Finally, there is no real evidence for a secular change in mantle (or basaltic) composition with geologic age, nor for contamination by the continental crust, for hafnium.

IV. Kimberlites, Carbonatites and Alkali-rich Lavas

The tabulation used for Table 72-E-4 is not genetic but based on uniqueness and generally undersaturated and highly alkaline character of the rocks concerned. Some of the rock types are volumetrically unimportant, but geologically significant, since their origin is not well understood and various mechanisms have been considered in the literature to account for their production (see discussion in Subsection 40-E-IV). As may be seen from Table 72-E-4 there is a general tendency for kimberlites to be enriched in hafnium when compared to mafic rocks, with the former having a mean concentration around 8 ppm Hf. The same comment applies to the single carbonatite (10 ppm Hf) and the alkali-rich lavas from the African rift system, the latter having a mean content of about 9 ppm Hf.

Zr/Hf ratios, using the individual means in Table 72-E-4, average 44 and 43 for kimberlites and alkali-rich lavas respectively. Zircons from kimberlites appear to have similar Zr/Hf ratios to those of the host rock (Table 72-D-1).

Table 72-E-4. *Hafnium abundances and Zr/Hf ratios of kimberlites, carbonatites and alkali rich lavas*

Rock type	No. of samples	Locality	ppm Hf		Zr/Hf	Method[a]	Reference
			Range	Mean			
Kimberlite	15	Premier Mine (South Africa)	1.6–3.4	2.5	43	N/R, X	1
Kimberlite	10	Koffiefontein (South Africa)	2.5–2.9	2.7	43	N/R, X	1
Kimberlite	5	Ebenhaezer East (South Africa)	2.8–3.3	3.1	46	N/R, X	1
Kimberlite	5	Ebenhaezer West (South Africa)	2.6–3.3	3.1	42	N/R, X	1
Kimberlite	10	Water fissure, Bellsbank (South Africa)	2.1–4 1	3.0	47	N/R, X	1
Kimberlite	6	Bobbejaan fissure, Bellsbank (South Africa)	5.2–7.7	6.4	44	N/R, X	1
Kimberlite	4	Main fissure, Bellsbank (South Africa)	6.7–8.9	7.5	41	N/R, X	1
Kimberlite	6	Wesselton (South Africa)	8–16	12		N/R	2
Kimberlite	4	Swartruggens (South Africa)	9–13	11		N/R	2
Kimberlite	1	Monastery (South Africa)		16		N/R	2
Micaceous kimberlite	7	Central India	17–30	24		N/R	3
Basaltic kimberlite	5	South India	2.8–5.4	3.9		N/R	3
Carbonatite	2	Western rift zone, Uganda and Ruanda	9.4–10.5	10.0	82	X	4
Picrite basalt	5	Eastern rift zone, Tanzania	4.5–15	7.9	31	X	4
Andesite basalt	2	Eastern rift zone, East Africa	9.4–15	12	35	X	4
Leucite basalt	6	Western rift zone, Uganda and Ruanda	8.3–11.4	10.0	41	X	4
Basalt (various types)	4	Western rift zone, Uganda and Ruanda	3.7–7.2	5.6	41	X	4
Ugandite	3	Western rift zone, Uganda and Ruanda	6.6–8.5	7.7	49	X	4
Leucite-nephelinite	1	Western rift zone, Uganda and Ruanda		10.5	38	X	4
Mafurite	1	Western rift zone, Uganda and Ruanda		5.3	48	X	4
Katungite	1	Western rift zone, Uganda and Ruanda		6.4	55	X	4
Nephelinite	6	Eastern rift zone, Tanzania	10–17	13	54	X	4
Melanephelinite	2	Eastern rift zone, Tanzania	8.0–9.8	8.9	35	X	4

[a] Where two analytical methods are given, the second refers to the method used for the zirconium determination.

References: 1. KABLE *et al.* (1975); 2. MITCHELL and BRUNFELT (1975); 3. PAUL *et al.* (1975); 4. GERASIMOVSKII *et al.* (1972).

Table 72-E-5. *Hafnium abundances and Zr/Hf ratios of intermediate rocks*

Rock type	No. of samples	Locality	ppm Hf		Zr/Hf	Method[a]	Reference
			Range	Mean			
Oceanic volcanic rocks							
Andesite	3	DSDP[b] leg 28, site 264	3.6–3.8	3.7	35	N/R, S	1
Andesitic rock	5	Eolian Islands	1.7–4.1	3.0		N/R	2
Andesite, low Si	5	Bougainville	1.3–2.5	2.0	50	Mass, S	3
Andesite, high K	1	Bougainville		4.5	38	Mass, S	3
Andesite	2	Bougainville	2.4–2.7	2.6	44	Mass, S	3
Andesite, low K	2	Saipan (Mariana Islands)	1.9–2.0	2.0	45	Mass, S	3
Andesite	3	Central North Island (New Zealand)	1.6–2.1	1.9	49	Mass, S	4
Andesite, tholeiitic suite	7	Viti Levu (Fiji)	0.82–4.5	2.1	53	Mass, X	5
Andesite, calc-alkali suite	2	Viti Levu (Fiji)	1.9–3.0	2.5	44	Mass, X	5
Shoshonitic suite	2	Viti Levu (Fiji)	0.8–1.1	0.95	54	Mass, X	5
Trachyte	1	Puu Anahulu, Hualalai		17.7	85	N/R, X	6
Continental rocks							
Diorite	2	Yeoval complex, New South Wales (Australia)	1.4–2.7	1.6	44	Mass, X	7
Basaltic andesite	7	Eocene, Kastamonu area, northern Turkey	2.3–3.1	2.6	44	Mass, X	8
Andesite	7	Eocene, Kastamonu area, northern Turkey	3.0–6.2	4.4	43	Mass, X	8
Trachyte	3	Afar rift, Ethiopia	14–17	16	32	N/R, X	9
Trachyte	3	Western rift zone, Uganda and Ruanda	11–14	12	54	X	10
Quartz trachyte	2	Eastern rift zone, East Africa	20–21	21	36	X	10
Pyroxene-mica diorite	1	Garabal Hill-Glen Fyne igneous complex (Scotland)		3.5	50	N/R, X	6
Ferrodiorite	1	Isle of Skye (Scotland)		12.1	38	N/R, X	6
Quartz diorite and granodiorite	3	Hercynian, Verkhissetsk, Ural Mountains (U.S.S.R.)	2.7–3.0	2.8	42	S	11
Calc-alkalic rocks	3	Various locations, U.S.S.R.			47	X	12

Diorite and quartz diorite	8	Tannuolsky complex, Tuva (U.S.S.R.)		3.5	27		13
Diorite and quartz diorite	9	Dzhidinsky complex, West Zabaikalje (U.S.S.R.)		1.7	50		13
Diorite and quartz diorite	7	Balsheporozhsky complex, West Sajan (U.S.S.R.)		2.5	47		13
Granophyre	4	Skaergaard intrusion (East Greenland)	22–44	29	54	N/R, X	14
Granophyre	2	Dillsburg, Pennsylvania (USA)	5.4–6.3	5.9	48	N/R, S	15
Granophyre	2	Great Lake intrusion, Tasmania	4.9–8.0	6.5	33	N/R, S	15
Granophyre	5	Alamdzhakh intrusion, Siberia (U.S.S.R.)	12–18	15	38	S?	16

[a] Where two analytical methods are given, the second refers to the method used for the zirconium determination.
[b] DSDP=Deep Sea Drilling Project.

References: 1. Ford (1975a); 2. Klerkx *et al.* (1974); 3. Taylor *et al.* (1969); 4. Ewart *et al.* (1968b); 5. Gill (1970); 6. Brooks (1970); 7. Gulson *et al.* (1972); 8. Peccerillo and Taylor (1976); 9. Barberi *et al.* (1975); 10. Gerasimovskii *et al.* (1972); 11. Lipova *et al.* (1957); 12. Shevaleevskii *et al.* (1960); 13. Znamenskiy *et al.* (1972); 14. Brooks (1969); 15. Gottfried *et al.* (1968); 16. Nesterenko *et al.* (1971).

V. Intermediate Rocks

Hafnium abundances for intermediate rocks are given in Table 72-E-5. It is clear that calc-alkaline rocks (island arc andesites and continental diorites) are not noticeably enriched in hafnium, since they generally contain 2–4 ppm Hf, in keeping with their low zirconium concentrations (Subsection 40-E-V). In contrast, alkali and tholeiitic basalt differentiates of similar bulk composition contain higher hafnium contents, and the rate of increase of hafnium concentration with differentiation in the latter two series is also more accentuated. Thus, high concentrations of hafnium may be encountered in trachytes derived from alkali basalts and granophyres derived from tholeiitic dolerites, as shown for some examples in Table 72-E-5.

There do not appear to be any systematic variations of the Zr/Hf ratio for the different magma types represented in Table 72-E-5 with an average Zr/Hf ratio of 45 being obtained from the individual values shown.

VI. Syenites

Data for syenites and related rocks are given in Table 72-E-6 and may be compared with the estimate of 11 ppm Hf given by TUREKIAN and WEDEPOHL (1961). It is customary to consider syenitic rocks as being of miaskitic or agpaitic type, and discussion is given on this basis.

i) *Miaskitic rocks* with molecular $(Na_2O + K_2O)/Al_2O_3 < 1$ are more abundant than the agpaitic varieties, but contain less hafnium probably because zircon crystallises at a relatively early stage, preventing the build-up of hafnium in residual magmas. Thus, rocks from the Kangerdlugssuaq intrusion and the Dzhidinsky complex (Table 72-E-6) do not have noticeably high hafnium contents (~4–7 ppm Hf) or abnormal Zr/Hf ratios.

ii) *Agpaitic syenites* are characterized not only by molecular $(Na_2O + K_2O)/Al_2O_3$ > 1, but also by their high contents of Zr, H_2O, F and Cl, and their exotic mineralogy (notably the zirconium mineral eudialyte). Some of these rocks contain the highest hafnium concentrations of all rock clans as may be seen from Table 72-E-6 for rocks from the famous Lovozero massif, where hafnium varies from 25–480 ppm, with the increase in concentration occurring towards the end of the magmatic process. These rocks contain most hafnium in zirconium minerals such as eudialyte, but soda-rich amphiboles and pyroxenes (arfvedsonite and aegirine) may also contain substantial proportions of hafnium (Table 72-D-2).

The extreme enrichment of hafnium in the Lovozero rocks is presumably related to the reasons for the extreme enrichment of zirconium, viz. the suppression of zircon crystallisation due to the alkaline and volatile-rich nature of these magmas (DIETRICH, 1968) and the role of stable, soluble complexes in allowing the build-up of hafnium in residual magmas. Further discussion on this point relating to zirconium is given in Chapter 40 (Sections D and E).

GERASIMOVSKII *et al.* (1962) noted that Zr/Hf ratios in the Lovozero massif increased (35 to 52) towards the end of the magmatic stage, but this is not apparent from the data given in Table 72-E-6. The latest summary by GERASIMOVSKII (1966) does suggest a slight increase considering the Zr/Hf averages of 35, 41 and 43 for the first, second and third intrusive phases respectively.

Table 72-E-6. *Hafnium abundances and Zr/Hf ratios of syenites and related rocks*

Rock type	No. of samples	Locality	ppm Hf		Zr/Hf	Method[a]	Reference
			Range	Mean			
Nepheline syenite (first complex)	5	Lovozero Massif, Kola Peninsula (U.S.S.R.)	25–66	39	34	X	1
Urtite (second complex)	3	Lovozero Massif, Kola Peninsula (U.S.S.R.)	28–95	53	32	X	1
Foyaite (second complex)	10	Lovozero Massif, Kola Peninsula (U.S.S.R.)	13–416	101	37	X	1
Lujavrite (second complex)	11	Lovozero Massif, Kola Peninsula (U.S.S.R.)	31–153	87	32	X	1
Sodalite syenite (third complex)	3	Lovozero Massif, Kola Peninsula (U.S.S.R.)	26–75	44	31	X	1
Lujavrite (third complex)	12	Lovozero Massif, Kola Peninsula (U.S.S.R.)	212–483	285	37	X	1
Nepheline syenite	2	Lovozero Massif, Kola Peninsula (U.S.S.R.)			36	X	2
Lujavrite	9	Lovozero Massif, Kola Peninsula (U.S.S.R.)			42	X	2
Foyaite	5	Lovozero Massif, Kola Peninsula (U.S.S.R.)			46	X	2
Urtite	4	Lovozero Massif, Kola Peninsula (U.S.S.R.)			41	X	2
Foyaite	1	Kangerdlugssuaq alkaline intrusion (East Greenland)		7.5	47	N/R, X	3
Nordmarkite	2	Kangerdlugssuaq alkaline intrusion (East Greenland)	3.9–6.8	5.4	37	N/R, X	3
Pulaskite	1	Kangerdlugssuaq alkaline intrusion (East Greenland)		4.5	37	N/R, X	3
Ijolite	1	Eastern rift zone, Tanzania		22	30	X	4
Monzonite, syenite and quartz syenite	12	Dzhidinsky complex, West Zabaikalje (U.S.S.R.)		6.0	63		5

[a] Where two analytical methods are given, the second refers to the method used for the zirconium determination.

References: 1. Gerasimovskii *et al.* (1958); 2. Gerasimovskii *et al.* (1962); 3. Brooks (1970); 4. Gerasimovskii *et al.* (1972); 5. Znamenskiy *et al.* (1972).

Table 72-E-7. *Hafnium abundances and Zr/Hf ratios of granitic rocks*

Rock type	No. of samples	Locality	ppm Hf		Zr/Hf	Method[a]	Reference
			Range	Mean			
Tonalite	1	Viti Levu (Fiji)		3.8	32	Mass, X	1
Granite	1	Green Mountain, Ascension		29	64	N/R, X	2
Granite	12	Carboniferous, Schluchsee, Black Forest (Germany)	5.2–16	8.4	20	N/R, X	3
Granite	9	Carboniferous, Barhalde, Black Forest (Germany)	3.4–5.8	4.6	13	N/R, X	3
Granite	10	Carboniferous, Malsburg, Black Forest (Germany)	4.7–9.1	7.0	29	N/R, X	3
Biotite granite	3	Younger granite suite, Liruei (Nigeria)	13–34	23	17	N/R, X	4
Amphibole granite	3	Younger granite suite, Liruei (Nigeria)	21–43	31	27	N/R, X	4
Granite	6	Younger granite suite, Amo (Nigeria)	9–66	23	24	N/R, X	4
Granite	6	Basement granite, Nigeria	10–20	13	60	S	5
Amphibole granite	6	Younger granite suite, Nigeria	8–14	11	60	S	5
Biotite granite	6	Younger granite suite, Nigeria	5–18	12	27	S	5
Alkaline granite	7	Younger granite suite, Nigeria	9–110	50	48	S	5
Pegmatite	4	Younger granite suite, Nigeria	0.9–1.6	1.2	41	S	5
Granite, sub-alkaline	6	Rhode Island, New England (U.S.A.)		7.5	42	N/R	6
Granite, alkaline	9	Massachusetts, New England (U.S.A.)		18	39	N/R	6
Quartz monzonite	3	Precambrian, Louis Lake batholith, Wyoming (U.S.A.)	4.2–9.4	6.1	46	N/R, X	7
Granite	3	Precambrian, Louis Lake batholith, Wyoming (U.S.A.)	3.7–5.8	4.7	44	N/R, X	7
Aplite	6	Precambrian, Louis Lake batholith, Wyoming (U.S.A.)	1.1–3.6	2.6	46	N/R, X	7
Granodiorite	17	Precambrian, Louis Lake batholith, Wyoming (U.S.A.)	3.9–7.5	5.9	57	N/R, X	7

Biotite granite	4	Perak (Malaya)	2.5–12	6.1	24	N/R, X	4
Granodiorite	1	Yeoval diorite complex (Australia)		3.0	47	Mass, X	8
Granodiorite	2	Garabal Hill-Glen Fyne igneous complex (Scotland)	4.2–4.4	4.3	36	N/R, X	2
Pegmatite	1	Garabal Hill-Glen Fyne igneous complex (Scotland)		0.8	32	N/R, X	2
Epigranite	4	Isle of Skye (Scotland)	10–22	17	37	N/R, X	2
Granite	1	Caledonian, Verkhissetsk Ural Mountains (U.S.S.R.)		1.0	33	X	9
Granite	3	Hercynian, Verkhissetsk, Ural Mountains (U.S.S.R.)	1.3–2.9	2.0	43	X	9
Muscovite granite	2	Hercynian, Verkhissetsk, Ural Mountains (U.S.S.R.)	0.8–1.8	1.3	33	X	9
Adamellite	2	Hercynian, Verkhissetsk, Ural Mountians (U.S.S.R.)	0.9–2.3	1.6	42	X	9
Aplite	1	Hercynian, Verkhissetsk, Ural Mountains (U.S.S.R.)		1.5	32	X	9
Pegmatite	1	Hercynian, Verkhissetsk, Ural Mountains (U.S.S.R.)		0.5	39	X	9
Granite	65	Paleozoic, East Tuva and North Mongolia (U.S.S.R.)			32		10
Granitoid	9	Middle Proterozoic, West Baykalia (U.S.S.R.)	2–15	10	35	S	11
Granite	9	Brensk complex, Tuva (U.S.S.R.)	1.0–1.4	1.2	30		12
Alkalic granite	3	Alkalic complex, Tuva (U.S.S.R.)	0.9–2.5	1.5	31		12
Altered alkalic granite	3	Alkalic complex, Tuva (U.S.S.R.)	1.6–3.8	2.6	15		12
Granite	4	Archaean, Barberton (South Africa)	2.6–9.1	5.9	38	Mass	13
Tonalite	3	Archaean, Barberton (South Africa)	2.1–3.1	2.6	32	Mass	13
Plagiogranite	25	Mainsky complex, West Sajan (U.S.S.R.)		2.4	35		14
Plagiogranite	8	Mujsky complex, Pribaykalje (U.S.S.R.)		3.5	36		14
Tonalite and plagiogranite	10	Tannuolsky complex, Tuva (U.S.S.R.)		2.0	34		14
Granodiorite and granite	10	Tannuolsky complex, Tuva (U.S.S.R.)		4.0	37		14
Granodiorite	15	Dzhidinsky complex, West Zabaikalje (U.S.S.R.)		6.2	40		14

Table 72-E-7 (continued)

Rock type	No. of samples	Locality	ppm Hf		Zr/Hf	Method[a]	Reference
			Range	Mean			
Granodiorite	8	Balsheporozhsky complex, West Sajan (U.S.S.R.)		3.6	37		14
Granite (two mica)	15	Balsheporozhsky complex, West Sajan (U.S.S.R.)		3.5	32		14
Quartz diorite and granodiorite	36	Shakhtaminsky complex, East Zabaikalje (U.S.S.R.)		3.5	80		14
Granodiorite and granite	8	Irelsky complex, East Pribaikalje (U.S.S.R.)		8.7	53		14
Tonalite, granodiorite and granite	116	Susamyrsky complex, Tjan-Shan (U.S.S.R.)		3.0	67		14
Porphyry granite	10	Dzhojsky complex, West Sajan (U.S.S.R.)		4.0	41		14
Biotite and alaskite granite	14	Sjutkholsky complex, Tuva (U.S.S.R.)		3.4	21		14
Porphry granite	8	Primorsky complex, Pribakalje (U.S.S.R.)		9.0	31		14
Porphyry granite and granodiorite	93	Amudzhikano-Sretensky complex, Zabaikalje (U.S.S.R.)		4.0	75		14
Biotite granite	103	Kukulbejsky complex, East Zabaikalje (U.S.S.R.)		4.8	44		14
Leucocratic granite	26	Nerchugansky complex, East Zabaikalje (U.S.S.R.)		5.4	42		14

[a] Where two analytical methods are given, the second refers to the method used for the zirconium determination.

References: 1. Gill (1970); 2. Brooks (1970); 3. Esson *et al.* (1968); 4. Butler and Thompson (1965); 5. Tugarinov *et al.* (1968a); 6. Buma *et al.* (1971); 7. Condie and Lo (1971); 8. Gulson *et al.* (1972); 9. Lipova *et al.* (1957); 10. Tugarinov *et al.* (1968b); 11. Sobachenko *et al.* (1974); 12. Pavlenko *et al.* (1957); 13. Glikson (1976); 14. Znamenskiy *et al.* (1972).

VII. Granitic Rocks

TUREKIAN and WEDEPOHL (1961) estimated averages of 2.3 and 3.9 ppm Hf for low calcium and high calcium granitic rocks respectively while TAYLOR (1964) derived an average of 4 ppm Hf for granites. These figures can be seen to serve as reasonable magnitudes when compared with the data assembled for granitic rocks in Table 72-E-7, in view of the wide ranges in concentration encountered (~1–100 ppm Hf).

In the typical and more abundant (calc-alkaline) granitic rocks the hafnium concentration does not rise to high levels and most samples have <10 ppm Hf. In some cases, for example the Verkhissetsk intrusions, U.S.S.R. (LIPOVA *et al.*, 1957), the hafnium content decreases with age of intrusion. In contrast, hafnium does not show the same behaviour in more alkaline granites, and increases markedly in concentration with increasing differentiation, resulting in higher abundance levels. This may be seen for the younger granites of Nigeria, where concentrations of 10–20 ppm Hf are common, and with some varieties containing up to 100 ppm Hf (Table 72-E-7). Zircon is relatively rare in the latter granites and hafnium is probably mostly present in the alkali amphiboles.

The average Zr/Hf ratio of 39 for granitic rocks calculated from the individual entries in Table 72-E-7 is identical to the average calculated for granitic zircons in Table 72-D-1. This is hardly surprising since, at least for the common calc-alkaline granitic rocks, the bulk of the hafnium in these rocks can be expected to be present in zircon.

VIII. Felsic Lavas

Data for felsic lavas are given in Table 72-E-8 and it will be recognized that the abundances of hafnium are related to associated parental magma types. Thus, calc-alkaline rocks, such as dacites and rhyolites from island arcs, contain less hafnium than do those which have a tholeiitic parentage which in turn contain less hafnium than the types associated with the alkali basalt clan. Highest concentrations (up to 60 ppm Hf) are found in peralkaline comendites, pantellerites and rhyolites associated with the African rift systems and the Liruei suite in Nigeria (Table 72-E-8). Previous comments made in this section with regard to the solubility of hafnium in alkaline melts are again relevant. Acid lavas possibly derived by fusion of crustal materials have hafnium abundances similar to the common granitic rocks.

The Zr/Hf ratios shown in Table 72-E-8 do not reveal any systematic variations with parental type or tectonic setting; an average Zr/Hf ratio of 36 is derived from the individual entries provided.

IX. Crustal Abundance of Hafnium

Estimates of the abundance of elements in the earth's continental crust have traditionally involved assumptions as to the relative volumes of igneous rocks present (granitic, intermediate, mafic etc.). This type of approach was used by TAYLOR (1964) and HORN and ADAMS (1966) who estimated 3.0 and 3.9 ppm Hf respectively for average igneous rocks. Such an approach, using the tables presented in this section, would indicate that the values given above are too low. For example, a 1/1 granitic-mafic rock average yields a figure of 5.5 ppm Hf. However, we consider it more

Table 72-E-8. *Hafnium abundances and Zr/Hf ratios of acid lavas*

Rock type	No. of samples	Locality	ppm Hf		Zr/Hf	Method[a]	Reference
			Range	Mean			
Oceanic volcanic rocks							
Rhyodacite, rhyolite	4	DSDP[b] Leg 28, Site 264	5.8–12.6	9.9	24	N/R, S	1
Dacite	1	Bougainville		3.8	26	Mass, S	2
Rhyolite, tholeiitic suite	1	Viti Levu (Fiji)		4.5	16	Mass, X	3
Acid lavas (various)	6	Eolian Islands	2.3–7.1	4.3		N/R	4
Dacite	1	Saipan (Mariana Island)		2.6	48	Mass, S	2
Dacite	1	Central North Island (New Zealand)		4.6	47	Mass, S	5
Rhyolitic lava	4	Central North Island (New Zealand)	3.8–4.9	4.5	37	Mass, S	5
Pantellerite	2	Mayor Island (New Zealand)	15.5–23.5	19.5	68	Mass, S	6
Continental volcanic rocks							
Pumice	1	Kyushu (Japan)		9.2		N/R	7
Pumice	5	Washington and Oregon (U.S.A.)	2.4–5.9	4.2		N/R	7
Rhyolitic rocks	14	Rocky Mountains (U.S.A.)	3.8–7.8	5.4	38	N/R	8
Pearlette-like ash	2	Pleistocene, Nebraska (U.S.A.)	7.6–7.9	7.8	23	N/R	9
Rhyolitic ash	9	Pleistocene, California, Utah and Nebraska (U.S.A.)	3.2–3.7	3.5	23	N/R	9
Ash	2	Pleistocene, Colorado (U.S.A.)	4.5–4.6	4.6	17	N/R	9
Alkali rhyolite	1	Afar rift, Ethiopia		27	33	N/R, X	10
Pantellerite	3	Afar rift, Ethiopia	30–32	31	36	N/R, X	10
Latite	2	Western rift Zone, Uganda and Ruanda	11–12	11	43	X	11
Phonolite	3	Eastern rift Zone, East Africa	14–20	18	44	X	11
Comendite	1	Eastern rift Zone, Tanzania		48	44	X	11
Comendite	1	Liruei, northern Nigeria		63	37	N/R, X	12
Rhyolite	2	Liruei, northern Nigeria	42–86	64	35	N/R, X	12
Felsite	1	Isle of Skye (Scotland)		11.3	33	N/R, X	13
Rhyolite and porphyry rocks	8	Nigeria	9–47	19	37	S	14
Rhyolite and albite porphyries	6	Archaean, Barberton (South Africa)	3.0–5.3	3.8	47	Mass	15

[a] Where two analytical methods are given, the second refers to the method used for the zirconium determination
[b] DSDP=Deep Sea Drilling Project

References: 1. Ford (1975a); 2. Taylor *et al.* (1969); 3. Gill (1970); 4. Klerkx *et al.* (1974); 5. Ewart *et al.* (1968b); 6. Ewart *et al.* (1968a); 7. Dudas *et al.* (1971); 8. Zielinski *et al.* (1977); 9. Borchardt *et al.* (1972); 10. Barberi *et al.* (1975); 11. Gerasimovskii *et al.* (1972); 12. Butler and Thompson (1965); 13. Brooks (1970); 14. Tugarinov *et al.* (1968a); 15. Glikson (1976).

Table 72-E-9. *Summary of Zr/Hf ratios in chondrites and lunar and terrestrial materials*

Material	Zr/Hf[a]	Source
Chondrites	39	Table 72-C-1
Soils, Apollo 11–17	39	Table 72-C-6
Whole moon	44	TAYLOR (1976)
Zircons (diorites, granodiorites)	38	Table 72-D-1
Zircons (granites)	39	Table 72-D-1
Zircons (pegmatites)	25	Table 72-D-1
Mafic rocks	45	Table 72-E-3
Kimberlites	44	Table 72-E-4
Alkali-rich lavas	43	Table 72-E-4
Intermediate rocks	45	Table 72-E-5
Syenites	39	Table 72-E-6
Granitic rocks	39	Table 72-E-7
Acidic lavas	36	Table 72-E-8
Sedimentary materials	35	Table 72-K-2
Metamorphic rocks	44	Table 72-M-1
Canadian Shield	41	SHAW *et al.* (1976)

[a] The Zr/Hf ratios quoted from tables in this chapter, except for Table 72-D-1, have been derived by averaging the average values given for individual entries, and do not represent weighted averages according to the numbers of samples involved.

instructive to consider values obtained for metamorphic rocks from the old shield areas (cratons) because they may provide a better sampling in view of the presence of granulite facies rocks commonly thought to be representative of deeper crustal levels. Thus the figure of 5.8 ppm Hf derived from 5512 rocks from the Canadian shield (SHAW *et al.*, 1976) is especially important. On the other hand, as shown in Section 40-E and Table 40-E-9, the zirconium value of 240 ppm obtained by SHAW *et al.* (1976) may be too high for a crustal estimate, and a figure of 190 ppm Zr was suggested as a more reasonable magnitude. If this figure is used with a Zr/Hf ratio of 40 for the continental crust, as discussed in the following section, then an abundance of 4.8 ppm Hf is obtained for the continental crust.

The oceanic crust is thought to be dominantly composed of abyssal tholeiitic basalt and, if so, then a figure of 2.8 ppm Hf is suggested by the data given in Table 72-E-3.

X. Zr/Hf Variations

Reference has been made in preceding sections to the variation of the Zr/Hf ratio in various materials. These may be briefly summarized as follows:

i) Although fractionation between Zr and Hf does occur in specific materials, there does not appear to be any significant fractionation between Zr and Hf in the earth as a whole, the moon and the chondritic meteorites, as shown by the data summarized in Table 72-E-9, which indicates a Zr/Hf ratio of about 40 for the solar system. The earlier estimates of Zr/Hf = 50 by FLEISCHER (1955) and VLASOV (1966) for the continental crust are thus slightly high.

ii) In granitic and other igneous rocks where the bulk of the zirconium and hafnium are likely to be present in zircons, there is often an enrichment of hafnium

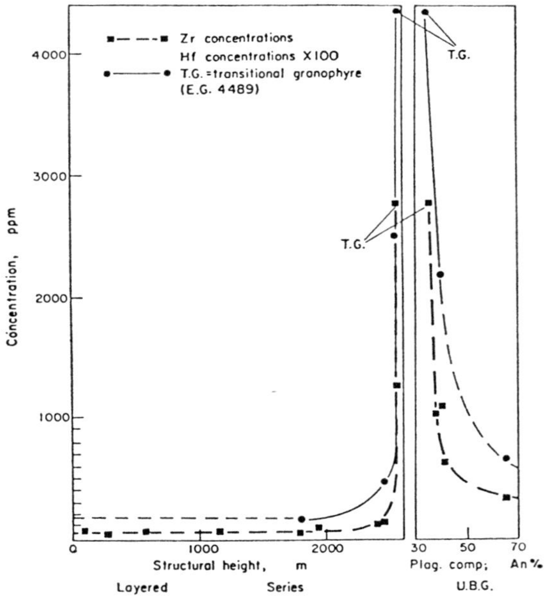

Fig. 72-E-1. Distribution of hafnium in the rocks of the exposed parts of the Layered Series and Upper Border Group from the Skaergaard intrusion. The curve for Zr is also shown for comparison. (From BROOKS, 1969)

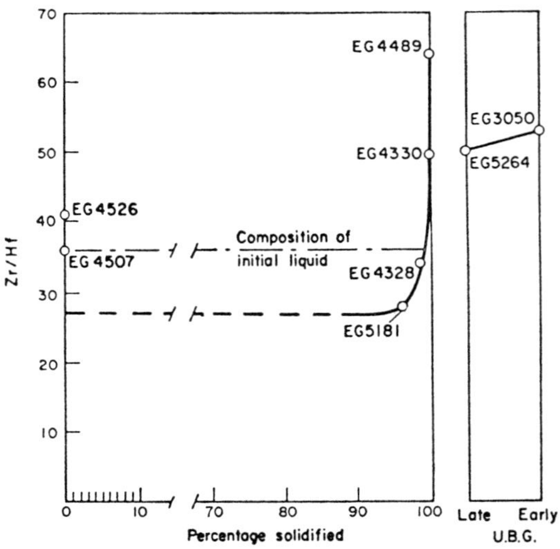

Fig. 72-E-2. Variation of the Zr/Hf ratio in the Skaergaard intrusion as a function of the percentage of magma solidified. (The average Zr/Hf ratio of the intrusion as given by the composition of the chilled margin as shown.) (From BROOKS, 1969)

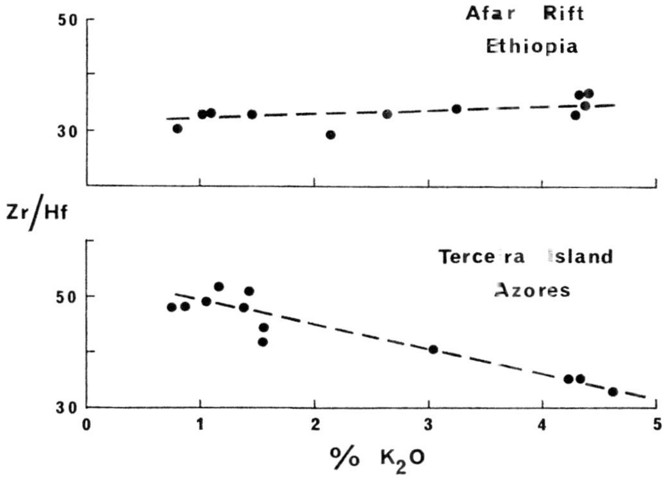

Fig. 72-E-3. Variation of Zr/Hf ratios in volcanic lavas from the Afar Rift, Ethiopia (BARBERI *et al.*, 1975) and Terceira island, Azores (ALLÈGRE *et al.*, 1977) with differentiation, as shown by K_2O content

relative to zirconium, as shown for example by the zircon data for pegmatites in Table 72-E-9, and the occurrence of hafnons and hafnian zircons in pegmatites (Table 72-D-1). This enrichment of hafnium is usually correlated with increasing differentiation, with the more evolved rocks, or their zircons, having lower Zr/Hf ratios. Full discussion of the factors that may be responsible for this has been given in Section 72-D, but it should be noted that there are no clear answers at this stage.

iii) There has been much discussion regarding the variation of Zr/Hf ratios during magmatic differentiation of mafic rocks. For example, in the Skaergaard intrusion (BROOKS, 1969), the Zr/Hf ratio appears to increase (Figs. 72-E-1 and 72-E-2). In two basalt-granophyre sequences (GOTTFRIED *et al.*, 1968) the Zr/Hf ratio apparently decreases, although the data are variable. In volcanic rocks no fractionation would be predicted if both zirconium and hafnium behaved as truly "incompatible" elements (i.e. if bulk solid/liquid partition coefficients are of the order of 0.1 or less). As shown in Fig. 72-E-3, lavas ranging in composition from basalt-rhyolite from the Afar rift, Ethiopia (BARBERI *et al.*, 1975) show only a slight increase in Zr/Hf with differentiation; in contrast Zr/Hf ratios in the basalt-trachyte series from Terceira island, Azores (ALLÈGRE *et al.*, 1977) clearly decrease with increasing differentiation. Other examples may be cited but it will not be possible to interpret these variations quantitatively until partition coefficient data, which are not yet available, are obtained for potential liquidus mineral phases that may be removed during fractional crystallisation processes.

Manuscript received: August 1977

72-F. Behavior in Magmatogenic Processes

From the discussion in Section 72-E, it will be appreciated that, for those rock suites where hafnium is incorporated in early crystallising zircon, the ability for it to become concentrated in late stage fluids is reduced. Thus, *pegmatites* and *aplites* from typical granitic rocks have low concentrations of hafnium (1 ppm or less in many cases) as may be seen from Table 72-E-7 (e.g. data of LIPOVA *et al.* (1957) for the Verkhissetsk intrusives). Extensive precipitation of zirconium minerals is confined to alkaline-rich pegmatites or to the pegmatite replacement stage (albitisation) where zircon or its varieties (cyrtolite, malacon) may occur (VLASOV, 1966). However, it should be noted that whereas in the early stage of the pegmatitic process the Zr/Hf ratios of the zirconium minerals is close to that of the parent rock, in the later stages of the pegmatitic process the Zr/Hf ratios of zircons and its varieties may decrease sharply. VLASOV (1966) documents several examples where this occurs. In other words, although both zirconium and hafnium may be reduced in absolute concentration during the course of pegmatite forming processes (i.e. reduction in the total amount of zircon crystallising), the rate of depletion of zirconium is faster than that for hafnium, leading to a decrease of the Zr/Hf ratio in zircon and its varieties. Hafnons and hafnian zircons presumably form under extreme conditions of differentiation which allow substantial enrichment of hafnium (and other elements) relative to zirconium (Section 40-D).

Pegmatites associated with agpaitic nepheline syenites contain a variety of zirconium minerals and are obviously enriched in zirconium (VLASOV, 1966). Hafnium data are not available, but relatively high concentrations are to be expected. Similarly, zirconium is known to concentrate in hydrothermal veins, generally associated with alkali rocks (VLASOV, 1966), and relatively high contents of hafnium may again be anticipated.

In summary, the behavior of hafnium is such that it becomes concentrated in high temperature magmatic processes, and its refractory nature dictates that it is not geochemically active or prominent in gas-transport or ore-deposition processes.

The behavior of hafnium in metasomatic processes is discussed in Section 72-N.

Manuscript received: August 1977

72-G. Behavior during Weathering and Alteration of Rocks

Apart from the question of abundances, the geochemical characteristics of hafnium are very much like those of zirconium (see Section 40-G) and the present discussion will seek to emphasize the differences between the behavior of these two elements rather than repeat similarities. Small differences in the chemical properties of the two elements are shown largely by the solubilities of compounds in water, in ion-exchange and solvent extraction behavior (COTTON and WILKINSON, 1966), in hydrolytic behavior (BILINSKI and TYREE, 1971), in acid-base reactions and in biochemical complex formation (RONOV *et al.*, 1961). In general, hafnium is more basic, precipitates at higher pH values in most given anionic media, hydrolyses less readily and forms less stable complexes (HERNDON, 1955). Differences in the geochemical behavior of the two elements are accentuated in those natural systems where sharp changes in pH and other physiochemical factors occur (RONOV *et al.*, 1961; VLASOV, 1966) and this probably accounts for the fact that hafnium does not follow zirconium quite as relentlessly in the supergene cycle as it does in magmatic processes.

When most of the zirconium and hafnium in an eroding profile is preserved intact in detrital zircon there can be no great change in the Zr/Hf ratio of the alluvial products, which will reflect the ratio of the two elements in the parent material (NEKRASOVA and ROZHDESTVENSKAYA, 1970). It should be noted, however, that hafnium-rich zircons seem to be relatively rapidly broken down (NEKRASOVA and ROZHDESTVENSKAYA, 1970; GOLDSCHMIDT, 1954). This phenomenon has been attributed to two possible causes: (i) a correlation between elevated hafnium and uranium concentrations in many zircons, with attendant additional radiation damage, and (ii) lattice strain or point defects (PYATENKO, 1965) caused by the presence of hafnium atoms in place of zirconium (see Section 40-G). The destruction of these hafnium-enriched zircons may, in part, account for the rather elevated Zr/Hf ratio observed for sandstones and some shales (Table 72-K-1).

Conversely, any marked separation of zirconium and hafnium will probably be confined to the hydrolysate fraction of the weathering products. This principle is supported in some measure by the data of Table 72-K-1 and 72-K-2, which show, without putting too fine a point on it, that it is the carbonates and pelagic sediments which most frequently exhibit the greatest deviations from the mean crustal Zr/Hf ratio of about forty (Section 72-D). BOGDANOV *et al.* (1976), in particular, reported a very low mean Zr/Hf ratio for pelagic sediments (20) and stated explicity that the zirconium and hafnium in these materials occurred as hydrolysates. Nevertheless, not all of the available information is consistent with this simplistic model. For example, factor analysis by DYMOND *et al.* (1977) of geochemical data from a deep sea core did not indicate that hafnium was specifically associated with either a biological, hydrothermal or detrital/hydrogenous fraction in these materials.

The work of RONOV *et al.* (1961) on the clays of the Russian Platform provides examples of the kinds of influences which may alter the proportion of zirconium to hafnium in weathering rock material. They found that Zr/Hf was slightly higher in clays formed in humid (generally acid, deeply weathered) areas than in clays originating in arid (generally alkaline and less deeply weathered) regions. Other observations (see Section 72-K) led them to postulate that zirconium forms soluble humic complexes more readily than does hafnium and is therefore more mobile than hafnium in the presence of dissolved organic acids.

CONDIE and SNANSIENG (1971) noted that although the zirconium content of certain sedimentary formations in northern California sometimes served as a useful "provenance indicator", in that it could be used to identify the source rocks from which the sediment had been derived, the hafnium content could not be used in this manner. This suggests that significant separation of the two elements may have occurred during weathering and transportation.

Manuscript received: August 1977

72-H. Solubilities of Compounds which Control Concentrations of Hafnium in Natural Waters; Adsorption Processes

The forms of hafnium in natural solutions are expected to be similar to those of zirconium in almost all respects (Section 40-H). It is generally agreed that the free aqueous hafnium ion is not favoured and that the metal may exist, for example, as $Hf(OH)_4^0$ (LECKIE and JAMES, 1974).

There is but little information about the absorption characteristics of hafnium in natural environments. Experiments by DOSHI *et al.* (1973) indicate that hafnium, like zirconium, is quantitatively adsorbed onto manganese dioxide which is precipitated from seawater. Hafnium in layered manganese nodules from the South Pacific appeared to be associated with the outer layer of manganese oxides and not with the manganese-poor core. The metal was irregularly distributed throughout the mass of homogeneous nodules composed mainly of manganese oxides (FRANA *et al.*, 1973).

Manuscript received: August 1977

72-I. Abundance in Natural Waters and in the Atmosphere

SCHUTZ and TUREKIAN (1965) found the abundance of hafnium in *sea water* to be below the detection limit of 8 ng/l. According to BREWER (1975), sea water contains about 7 ng Hf per litre, and thus apparently has a Zr/Hf ratio of 4.3. If this is true, it may be of significance in explaining certain aspects of the sedimentary cycle of hafnium.

COWGILL (1973) studied a pondwater profile for over a year, but could at no stage detect hafnium in the water. Hafnium was not found in concentrations above natural levels in the Yodo *river* in Japan, although several other metal pollutants were present (MAMURO *et al.*, 1971b).

There has been some work on hafnium in the *atmosphere*. The metal can be readily detected in city dust (MAMURO, 1973) and over some cities, for example Osaka and Sakai (MAMURO *et al.*, 1971a), the distribution of the element has a strong area dependancy. SHUM and LOVELAND (1974) showed that certain industries can cause slight hafnium air pollution. However, hafnium in the air in urban areas is not related to automobile emission (PSZENNY and RUSSELL, 1974) and the metal could not be detected in jet fuels contributing to air pollution at airports (FORDYCE and SHEIBLEY, 1974). In most instances the level of concentration of hafnium and the values for interelement ratios (e.g. Hf/Sc) in air are compatible with an origin from crustal rocks (RANCITELLI *et al.*, 1974).

TANNER *et al.* (1974), for example, found that hafnium was not particularly enriched in urban (St. Louis, U.S.A.) aerosols (0.49 ng/m^3) when compared to samples taken from 20 to 60 miles downwind of the city (0.31 to 0.94 ng/m^3). If the data were expressed as enrichment factors, relative to the concentration of iron, then the concentration of hafnium in the St. Louis aerosols was only about one half of that of the downwind samples, indicating a natural geochemical origin for hafnium in the atmosphere.

Hafnium is amongst the heavy metals which were detected in particles occurring in noctilucent clouds formed at altitudes between 79 and 113.5 km (HEMENWAY *et al.*, 1972). It is strongly argued that this material originates in the sun and rains down on the earth.

Manuscript received: August 1977

72-K. Abundance in Common Sediments and Sedimentary Rock Types

Apparently not a single hafnium analysis of any sedimentary rock had been published before 1955 (FLEISCHER, 1955) and in all probability published values for fewer than 1,000 samples exist. Estimates of the mean hafnium content of various sediments and sedimentary rocks are given in Table 72-K-1. Relatively recent analyses of hafnium in sedimentary materials are compiled in Table 72-K-2. There are a number of discrepancies in these data, but this is not unexpected in view of the small sample population. Nevertheless, the following general conclusions may be drawn: (i) The data support the view of VLASOV (1966) that there is no marked accumulation of hafnium in sedimentary rocks. (ii) The absolute abundance of hafnium is roughly similar (2.5 to 6.5 ppm) in psammites, greywackes and pelites, but is apparently lower in carbonates. GRAF (1960) noted that the hafnium content of carbonate materials is low and often below the detection limits of routine conventional instrumental analysis. Statistical treatment of analytical data for twenty marine limestones from Germany (PAREKH *et al.*, 1977) showed that the hafnium content of the carbonate fraction of the limestones is extremely low, but that the element is nevertheless not enriched in the clastic fraction relative to the average crustal abundance. (iii) The Zr/Hf ratios (see Table 40-K-2) of the various types of sedimentary materials apparently increase in the following order: deep-sea pelites, other pelites, psammites, carbonates. On the other hand, VLASOV (1966) claimed that "non-terriginous sedimentary materials" generally have low Zr/Hf ratios.

Some pertinent observations made by RONOV *et al.* (1961) about the distribution of hafnium in various types of Russian claystones have been mentioned in Section 72-G. In addition, they noted a regular relationship between the Zr/Hf ratio and facies changes in these claystones (see Fig. 72-K-1). The clays from the Loshaki area

Table 72-K-1. *Estimates of the mean hafnium contents and Zr/Hf ratios of sedimentary materials*

Material	ppm Hf	Zr/Hf	Reference
Sandstones	3.9	56	TUREKIAN and WEDEPOHL (1961)
Sandstones	3.0	68	HORN and ADAMS (1966)
Shales	2.8	57	TUREKIAN and WEDEPOHL (1961)
Shales	3.1	45	HORN and ADAMS (1966)
Clays and shales	6	33	VINOGRADOV (1962)
Deep sea clays	4.1	37	TUREKIAN and WEDEPOHL (1961)
Oceanic clays	4.6	32	HORN and ADAMS (1966)
Carbonates	0.3	63	TUREKIAN and WEDEPOHL (1961)
Carbonates	0.23	77	HORN and ADAMS (1966)
Deep sea carbonates	0.41	49	TUREKIAN and WEDEPOHL (1961)
Oceanic carbonates	1.1	75	HORN and ADAMS (1966)
Average sediment	0.1—18	—	JOHNSON (1976)

Table 72-K-2. *Hafnium abundances and Zr/Hf ratios of sediments and sedimentary rocks*

Rock type	No. of samples	Locality	ppm Hf		Zr/Hf	Method[a]	Reference
			Range	Mean			
Marine placers		Recent, Baltic Sea	30–780				1
Conglomerate	82	Precambrian, Kimberley Reef (South Africa)	0.29–52	6.5		N/R	2
Quartzite	22	Precambrian, Kimberley Reef (South Africa)	1.2–7.7	3.0		N/R	2
Quartzite	4	Precambrian, Witwatersrand (South Africa)	1.3–3.6	2.5		N/R	3
Sandstone and siltstone	5	Siberian Platform (U.S.S.R.)	3.9–9.1	6.1	32	S	4
Arkose	1	Triassic, Perth Basin (Australia)		4.7		MASS	5
Subgreywacke	3	Cambrian, Amadeus Basin (Australia)	4.2–7.4	6.0		MASS	5
Subgreywacke	3	Permian, Perth Basin (Australia)	3.5–17	11.5		MASS	5
Subgreywacke	2	Henbury (Australia)	6.6–8.5	7.6	46	MASS, S	6
Fresh subgreywacke	7	Ordovician, Duzel formation California (U.S.A.)		5.0	31	N/R, X	7
Altered subgreywacke	2	Ordovician, Duzel formation California (U.S.A.)		5.5	21	N/R, X	7
Greywacke	12	Silurian, Gazelle formation Califonria (U.S.A.)	1.8–4.9	3.1	28	N/R, X	7
Siltstone	1	Carboniferous, Canning Basin (Australia)		5.4		MASS	5
Shale	3	Precambrian, Witwatersrand (South Africa)	1.9–4.6	3.4		N/R	3
Shale	4	Proterozoic, Mt. Isa group (Australia)	2.1–4.1	2.8		MASS	5
Shale	6	Upper Proterozoic, Amadeus Basin (Australia)	3.9–5.9	4.5		MASS	5
Shale	7	Silurian, Camp Hill, Canberra (Australia)	2.3–4.0	3.1		MASS	5
Shale	3	Triassic, Perth Basin (Australia)	4.3–5.3	4.9		MASS	5
Carbonaceous limey shale	1	Krivoi Rog Suite (U.S.S.R.)		9.3	35	X	8

Claystone (arid provenance)	187	Famennian, Russian platform (U.S.S.R.)		3.8	44	X	9
Claystone (humid provenance)	90	Carboniferous, Moscow Basin (U.S.S.R.)		7.1	47	X	9
Limestone, marine	20	Southern Germany	0.05–1.6	0.4		N/R	10
Arenaceous limestone	1	Carboniferous, Canning Basin (W. Australia)		1.9		Mass	5
Dolomite	1	Krivoi Rog Suite (U.S.S.R.)		3.0	17	X	8
Deep sea sediment	1	Pacific Ocean		4.4	34	N/R	11
Deep sea sediment	1	Atlantic Ocean		1.4	51	N/R	11
Metalliferous deep sea sediment	18	Nazca Plate, Pacific Ocean	0.81–5.0	2.5		N/R	12
Pelagic ooze ($CaCO_3$ poor)	90	Southeast Pacific Ocean	0.3–35	4.8	20	N/R, S	13
Argillaceous deep sea sediment	6	Various locations	2–5	3	42	S	14
Calcareous deep sea sediment	2	Atlantic Ocean	<1–5	3	47	S	14

[a] Where two analytical methods are given, the second refers to the method used for the zirconium determination.

References: 1. Martynov *et al.* (1972); 2. Rasmussen and Fesq (1973); 3. Uken and Erasmus (1970); 4. Nesterenko *et al.* (1971); 5. Nance and Taylor (1976); 6. Taylor (1966); 7. Condie and Snansienc (1971); 8. Vainshtein *et al.* (1958); 9. Ronov *et al.* (1961); 10. Parekh *et al.* (1977); 11. Setser and Ehmann (1964); 12. Dymond *et al.* (1976, 1977); 13. Bogdanov *et al.* (1976); 14. El Wakeel and Riley (1961).

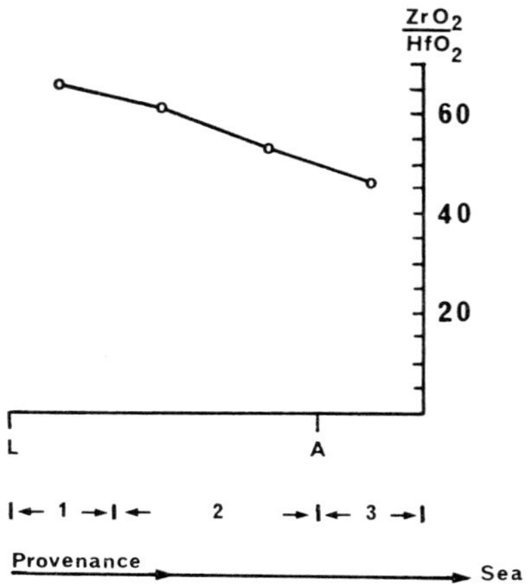

Fig. 72-K-1. Variation in ZrO_2/HfO_2 ratio in a section across the facies of the Yasnaya Polyana humid basin of the Russian platform (after RONOV *et al.*, 1961). Locations: *L* Loshaki, *A* Abakumovo; Facies: *1* Variegated, *2* Coal-bearing, *3* Near-shore marine. See text for full explanation

were deposited under acidic oxidizing conditions, but towards the Abakumovo region the depositional environment was increasingly more reducing. The Zr/Hf ratio decreases systematically along this traverse even though the absolute values of the concentrations of the two elements fluctuate considerably. The interpretation of this phenomenon may be complex, but the authors suggest that the relative depletion of zirconium in reducing, organic-rich environments may, in part, be due to the preferential mobilisation and dispersion of zirconium in the form of soluble humic complexes. They also noted that fine fractions (0.1 to 0.01 and also <0.001 mm) separated from these claystones had hafnium contents and Zr/Hf ratios similar to those of the bulk rock.

Hafnium is one of the few rare metals which are sharply enriched in sedimentary rocks immediately below the main gold-bearing reef in the ERPM mine of the Witwatersrand goldfield (UKEN and ERASMUS, 1970). The authors suggested that this fact might be of use in stratigraphic correlation and identification of the pay band.

Manuscript received: August 1977

72-L. Biogeochemisty

Naturally occurring compounds of hafnium are unlikely to have any significant influence on organisms. Metallic hafnium and HfO_2 are not toxic to mice in doses up to 10 g per kg. Doses of 2 g per kg and day for nine months produced only minor effects in guinea pigs (ZHILOVA, 1967). Hafnium is not regarded as a serious hazard to industrial workers but it is considered desirable to limit the concentrations of hafnium compounds—especially the halides—in factory air. The permissible limit in the U.S.A. is 0.5 mg/m³ (ZHILOVA, 1967).

FURR *et al.* (1974) found between < 0.13 and 4.5 ppm hafnium in twelve samples of protein-rich food materials. The highest values occurred in seaweed or its derivatives, or in commercial dogfood. The authors concluded that hafnium is a non-essential trace contaminant in organisms. VELANDIA and PERKONS (1973, 1974) found the concentration of hafnium in the heart tissue of rats to be too low to be detectable by the method they used (N/R) and they could find only traces of the metal in the kidneys of rabbits. According to JOHNSON (1976) the mean hafnium content of benthic organisms is 0.4 ppm (dry weight basis).

MOROSHKINA and VANEEVA (1972) found that certain plant ashes contained between about 0.1 and 1 ppm hafnium. The ashes of six species of lichen and eleven species of moss from Finland contained 1–19 ppm Hf (ERÄMETSÄ and YLIRUOKANENI, 1971). COWGILL (1973) reported that the pondweed *Pontedaria cordataa* contined 0.06 ppm Hf (dry weight basis), and apparently has a Zr/Hf ratio of 387. Hafnium could not be detected in the pond water.

Manuscript received: August 1977

72-M/N. Abundance in Common Metamorphic Rock Types, Behavior in Metamorphic Processes

There is a considerable range of the Hf contents and Zr/Hf ratios in the data given in Table 72-M-1 for metamorphic rocks, regardless of analytical technique. Note that data for greenstones of basaltic composition are included in Table 72-E-3 for reasons discussed in Section 72-E.

No attempt has been made to calculate the average Zr/Hf ratio of metamorphic rocks in Table 72-M-1 due to the presumed variable quality of the analytical data and limited number of samples. The average given by SHAW *et al.* (1976) for the Canadian Shield (Zr/Hf = 41) is in good agreement with the average Zr/Hf ratio in zircons from intermediate and granitic rocks of 39 (see Section 72-D). There is wide regional variation in the average values used by SHAW *et al.* (1976) to compile the average for the Canadian Shield. The composite samples from Southern Quebec have significantly higher hafnium contents (8.27 ppm) and Zr/Hf ratios (46) than the other areas where hafnium is in the range of 4.3 to 5.8 ppm and the Zr/Hf ratio 36 to 41.

The fused xenolith of gneiss in the Skaergaard intrusion has higher hafnium and zirconium and lower Zr/Hf ratio than the surrounding gneiss (see Table 72-M-1). BROOKS (1969) has attributed this to the precipitation of zirconium (probably hafnium as well) in the fused acid xenolith from the mafic magma. VLASOV (1966) and YES'KOVA and YEFIMOV (1970) have noted an increase in the hafnium and zirconium contents (see Section 40-M) of metasomatic rocks, compared to the parent rock, coupled with a decrease in the Zr/Hf ratio. EHMANN *et al.* (1975) do not quote any data but note that they have found no change in the Zr/Hf ratio (Zr/Hf = 47.3 ± 0.8) with metamorphic grade in metamorphic pelitic rocks from New Hampshire, although the individual zirconium and hafnium abundances varied by a factor of two.

Further information an the behavior of hafnium in metamorphic reactions has been discussed in Section 72-D.

Table 72-M-1. *Hafnium abundances and Zr/Hf ratios of metamorphic and metasomatic rocks*

Rock type	No. of samples	Locality	ppm Hf		Zr/Hf	Method[a]	Reference
			Range	Mean			
Metamorphic rocks							
Quartzite	2	N. Krivoi Rog (U.S.S.R.)	1.3–5.7	3.5	62	X	1
Serpentinite		Gabbro-Pyroxenite formation, Sayan-Baykal (U.S.S.R.)		2.8	35		2
Amphibolite	3	Kan-Biryusa area (U.S.S.R.)		4.0	58		3
Schist	3	Middle Krivoi Rog (U.S.S.R.)	0.9–5.3	2.5	27	X	1
Schist	4	N. Krivoi Rog (U.S.S.R.)	2.3–12.0	6.0	40	X	1
Schist	9	Verkhissetsk area (U.S.S.R.)			95	X	4
Gneiss and schist	4	Nigeria	1.1–8.1	5.0	69	S	5
Gneiss	12	Sayan region (U.S.S.R.)		14	16		3
Gneiss	4	Kan-Biryusa area (U.S.S.R.)		4.0	55		3
Acid gneiss (fused)	1	Skaergaard area (E. Greenland)		22.4	23	N/R, X	6
Grey gneiss	1	Skaergaard area (E. Greenland)		3.2	78	N/R, X	6
Granulite (basaltic)	1	Loch Inver area (N. W. Scotland)		2.6	25	N/R, X	7
Granulite (tonalitic)	1	Scourie area (N. W. Scotland)		2.8	47	N/R, X	7
Migmatite	2	Archaean, Barberton (South Africa)	1.8–6.2	4.0	43	Mass	8
Metasomatic rocks							
Albitite	1	N. Krivoi Rog (U.S.S.R.)		20	28	X	1
Apoeffusive alkali metasomatite	12	Urals (U.S.S.R.)		50	80		9
Albitized granite	104	Zabaikalje (U.S.S.R.)		8.5	12		10
Greisenized granite	34	Zabaikalje (U.S.S.R.)		2.8	23		10
Greisenized granite	4	Younger granite suite (Nigeria)	2.3–65	23	16	S	5
Crustal average							
Average Precambrian Shield	5512	Canada		5.8	41	N/R, X	11

[a] Where two analytical methods are given, the second refers to the method used for the zirconium determination.

References: 1. Vainshtein *et al.* (1958); 2. Sobachenko *et al.* (1974); 3. Znamenskiy *et al.* (1974); 4. Lipova *et al.* (1957); 5. Tugarinov *et al.* (1968a); 6. Brooks (1969); 7. Brooks (1970); 8. Glikson (1976); 9. Yes'kova and Yeefimov (1970); 10. Koval (1972); 11. Shaw *et al.* (1976).

Manuscript received: August 1977

72-O. Economic Importance

Zirconium is more widely used in industry than the geochemically closely associated element hafnium. Many of the applications of zirconium and its compounds (Section 40-O) are unaffected by minor amounts of hafnium. Hafnium, is however, extracted from zirconium ores by the Kroll Process and particularly suitable deposits of granitic and alluvial zircons, which contain 3–5% HfO_2, occur in Nigeria.

Hafnium is used for the production of filaments for electric light bulbs, X-ray cathode tubes (VLASOV, 1966), as alloys with Ti, Nb, Ta, Fe etc. (Handbook of Physics and Chemistry, 1975) and in the ceramics industry. Very pure hafnium which has a good absorption cross section for thermal neutrons and excellent mechanical properties is used for the production of reactor control rods.

The occurrence, uses, production and outlook of zirconium and hafnium have been discussed recently by KLEMIC (1975).

Manuscript received: August 1977

References: Sections 72-B to 72-O

ABBEY, S.: Studies in standard samples of silicate rocks and minerals. Part 4, 1974. Edition of usable values. Geol. Surv. Canada Paper **74–41** (1975).

ABDEL-GAWAD, A. M.: X-ray spectrographic determination of hafnium-zirconium ratio in zirconium minerals. Am. Mineralogist **51**, 464 (1966).

AHRENS, L. H.: The significance of the chemical bond for controlling the geochemical distribution of the elements. Part I. Phys. Chem. Earth **5**, 1 (1964).

ALLÈGRE, C. J., TREUIL, M., MINSTER, J.-F., MINSTER, B., ALBARÈDE, F.: Systematic use of trace element in igneous processes. Contrib. Mineral. Petrol. **60**, 57 (1977).

ALLEN, R. O. (Jr.), MASON, B.: Minor and trace elements in some meteoritic minerals. Geochim. Cosmochim. Acta **37**, 1435 (1973).

ARRHENIUS, G., EVERSON, J. E., FITZGERALD, R. W., FUJITA, H.: Zirconium fractionation in Apollo 11 and Apollo 12 rocks. Proc. Second Lunar Sci. Conf., Geochim. Cosmochim. Acta, Suppl. **2**, 169 (1971).

AYUSO, R. A., BENCE, A. E., TAYLOR, S. R.: Upper Jurassic tholeiitic basalts from DSDP Leg 11. J. Geophys. Res. **81**, 4305 (1976).

BARBERI, F., SANTACROCE, R., FERRARA, G., TREUIL, M., VARET, J.: A transitional basalt-pantellerite sequence of fractional crystallisation, the Boina Centre (Afar Rift, Ethiopia). J. Petrol. **16**, 22 (1975).

BILINSKI, H., TYREE, S. Y. (Jr.): The rate of hydrolysis of hafnium in 1 M NaCl. In: HEM, J. D. (ed.): Nonequilibrium Systems in Natural Water Chemistry. Advances in Chemistry Series, Vol. **106**. Washington: Amer. Chem. Soc. 1971.

BLANCHARD, D. P., RHODES, J. M., DUNGAN, M. A., RODGERS, K. V., DONALDSON, C. H., BRANNON, J. C., JACOBS, J. W., GIBSON, E. K.: The chemistry and petrology of basalts from Leg 37 of the DSDP. J. Geophys. Res. **81**, 4231 (1976).

BOGDANOV, YU. A., LISITZIN, A. P., LUKASHIN, V. N., GORDEEV, V. V., ZVERINSKAYA, I. B., ZELJDINA, B. B., KUZINOV, A. D., ZHIVAGON, V. N.: Chemical composition of sediments. In: HOLLISTER, C. D., CRADDOCK, C. *et al.* Initial Reports of the Deep Sea Drilling Project, Vol. 35 Washington: U.S. Govt. Printing Office 1976.

BORCHARDT, G. A., ARUSCAVAGE, P. J., MILLARD, T.: Correlation of the Bishop ash, a Pleistocene marker bed, using instrumental neutron activation analysis. J. Sediment. Petrol. **42**, 301 (1972).

BREWER, P. G.: Minor elements in sea water. In: RILEY, J. P., SKIRROW, G. (eds.): Chemical Oceanography, 2nd. edit., Vol. I. London: Academic Press 1975.

BROOKS, C. K.: On the distribution of zirconium and hafnium in the Skaergaard intrusion, East Greenland. Geochim. Cosmochim. Acta **33**, 357 (1969).

BROOKS, C. K.: The concentrations of zirconium and hafnium in some igneous and metamorphic rocks and minerals. Geochim. Cosmochim. Acta **34**, 411 (1970).

BRUNFELT, A. O., HEIER, K. S., STEINNES, E., SUNDVOLL, B.: Determination of 36 elements in Apollo 14 bulk fines 14163 by activation analysis. Earth Planet. Sci. Lett. **11**, 351 (1971).

BUMA, G., FREY, F. A., WONES, D. R.: New England granites: trace element evidence regarding their origin and differentiation. Contrib. Mineral. Petrol. **31**, 300 (1971).

BURNS, R. G., FYFE, W. S.: Distribution of elements in geological processes. Chem. Geol. **1**, 49 (1966).

BURNS, R. G., FYFE, W. S.: Trace element distribution rules and their significance. Chem. Geol. **2**, 89 (1967).

BUSCHE, F. D., PRINZ, M., KEIL, K., KURAT, G.: Lunar zirkelite: a uranium-bearing phase. Earth Planet. Sci. Lett. **14**, 313 (1972).

BUTLER, J. R., THOMPSON, A. J.: Zirconium: hafnium ratios in some igneous rocks. Geochim. Cosmochim. Acta **29**, 167 (1965).

BUTTERMAN, W. C., FOSTER, W. R.: Zircon stability and the ZrO_2—SiO_2 phase diagram. Am. Mineralogist **52**, 880 (1967).

CAMPBELL, J. A.: Chemical Systems. San Francisco: Freeman and Co. 1970.

CARUBA, R., BAUMER, A., TURCO, G.: Nouvelles synthèses hydrothermales du zircon: substitutions isomorphiques; relation morphologie-milieu de croissance. Geochim. Cosmochim. Acta **39**, 11 (1975).

CHURCH, S. E., BANSAL, B. M., WIESMANN, H.: The distribution of K, Ti, Zr, U and Hf in Apollo 14 and 15 materials. In: CHAMBERLAIN, J. W., WATKINS, C. (eds.): The Apollo 15 Lunar Samples. Houston: Lunar Science Institute 1972.

CHYI, L. L., EHMANN, W. D.: Zirconium and hafnium abundances in some lunar materials and implications of their ratios. Proc. Fourth Lunar Sci. Conf., Geochim. Cosmochim. Acta, Suppl. **4**, 1219 (1973).

CONDIE, K. C., LO, H. H.: Trace element geochemistry of the Louis Lake batholith of early Precambrian age, Wyoming. Geochim. Cosmochim. Acta **35**, 1099 (1971).

CONDIE, K. C., SNANSIENG, S.: Petrology and geochemistry of the Duzel (Ordovician) and Gazelle (Silurian) Formations, Northern California. J. Sediment. Petrol. **41**, 741 (1971).

COOLEY, R. A., MARTIN, A. V., FELDMAN, C., GILLESPIE, J.: The hafnium to zirconium abundance ratio and specific radioactivity of some ores. Geochim. Cosmochim. Acta **3**, 30 (1953).

CORREIA NEVES, J. M., LOPES NUNES, J. E., SAHAMA, TH. G.: High hafnium members of the zircon-hafnon series from the granite pegmatites of Zambézia, Mozambique. Contrib. Mineral. Petrol. **48**, 73 (1974).

COTTON, F. A., WILKINSON, G.: Advanced Inorganic Chemistry. New York: J. Wiley and Son 1966.

COWGILL, U. M.: The determination of all detectable elements in the aquatic plants of Linsley Pond and Cedar Lake (North Branford, Connecticut) by X-ray emission and optical emission spectroscopy. Appl. Spectroscopy **27**, 5 (1973).

CURRIE, K. L.: Variations of the hafnium-zirconium ratio of granitic rocks from Eastern Ontario. Can. Mineralogist **7**, 264 (1962).

CURRIE, K. L.: Variations in the hafnium zirconium ratio of high grade metamorphic rocks from southeastern Ontario, Canada. Earth Planet. Sci. Lett. **4**, 299 (1968).

CURTIS, C. E., DONEY, L. M., JOHNSON, J. R.: Some properties of hafnium oxide, hafnium silicate, calcium hafnate, and hafnium carbide. J. Am. Ceram. Soc. **37**, 458 (1954).

DIETRICH, R. V.: Behaviour of zirconium in certain artificial magmas under diverse P–T conditions. Lithos **1**, 20 (1968).

DOSHI, G. R., KRISHNAMOORTHY, T. H., SASTRY, V. N., SARMA, T. P.: Sorption of trace nuclides in sea water on manganese dioxide. Indian J. Chem. **11**, 158 (1973).

DUDAS, M. J., SCHMITT, R. A., HARWARD, M. E.: Trace element partitioning between volcanic plagioclase and dacitic pyroclastic matrix. Earth Planet. Sci. Lett. **11**, 440 (1971).

DYMEK, R. F., ALBEE, A. L., CHODOS, A. A.: Comparative petrology of lunar cumulate rocks of possible primary origin: Dunite 72415, troctolite 76535, norite 78235, and anorthosite 62237. Proc. Sixth Lunar Sci. Conf., Geochim. Cosmochim. Acta, Suppl. **6**, 301 (1975).

DYMOND, J., CORLISS, J. B., HEATH, G. R.: History of metalliferous sedimentation at deep sea drilling site 319, in the south eastern Pacific. Geochim. Cosmochim. Acta **41**, 741 (1977).

DYMOND, J., CORLISS, J. B., STILLINGER, R.: Chemical composition and metal accumulation rates of metalliferous sediments from sites 319, 320 and 321. In: YEATS, R. S., HART, S. R. *et al.* Initial Reports of the Deep Sea Drilling Project, Vol. 34. Washington: U.S. Govt. Printing Office 1976.

EHMANN, W. D., CHYI, L. L.: Zirconium and hafnium in meteorites. Earth Planet. Sci. Lett. **21**, 230 (1974).

EHMANN, W. D., CHYI, L. L., GARG, A. N., HAWKE, B. R., MA, M.-S., MILLER, M. D., JAMES, W. D., Jr., PACER, R. A.: Chemical studies of the lunar regolith with emphasis on zirconium and hafnium. Proc. Sixth Lunar Sci. Conf., Geochim. Cosmochim. Acta, Suppl. **6**, 1351 (1975).

EHMANN, W. D., REBAGAY, T. V.: Zirconium and hafnium in meteorites by activation analysis. Geochim. Cosmochim. Acta **34**, 649 (1970).

EL GORESY, A., RAMDOHR, P., MEDENBACH, O.: Lunar samples from the Descartes site: mineralogy and geochemistry of the opaques. In: CHAMBERLAIN, J. W., WATKINS, C. (eds.), Lunar Science IV. Houston: Lunar Science Institute 1973.

EL WAKEEL, S. K., RILEY, J. P.: Chemical and mineralogical studies of deep-sea sediments. Geochim. Cosmochim. Acta **25**, 110 (1961).

ERÄMETSÄ, O., YLIRUOKANENI, I.: Niobium, molybdenum, hafnium, tungsten, thorium and uranium in lichens and mosses. Suomen Kemistilehti B. **44**, 327 (1971).

ERLANK, A. J., WILLIS, J. P.: The zirconium content of chondrites and the zirconium-hafnium dilemma. Geochim. Cosmochim. Acta **28**, 1715 (1964).

ESSON, J., HAHN-WEINHEIMER, P., JOHANNING, H.: Determination of hafnium by neutron activation, and variation in the Zr/Hf ratio of some granite masses. Talanta **15**, 1111 (1968).

EWART, A., TAYLOR, S. R., CAPP, A. C.: Geochemistry of the pantellerites of Mayor Island, New Zealand. Contrib. Mineral. Petrol. **17**, 116 (1968a).

EWART, A., TAYLOR, S. R., CAPP, A. C.: Trace and minor element geochemistry of the rhyolitic volcanic rocks, central North Island, New Zealand. Total rock and residual liquid data. Contrib. Mineral. Petrol. **18**, 76 (1968b).

FESQ, H. W., KABLE, E. J. D., GURNEY, J. J.: The geochemistry of some selected South African kimberlites and associated heavy minerals. Report **1703**, National Institute for Metallurgy, Johannesburg (1976).

FLANAGAN, F. J.: 1972 values for international geochemical reference samples. Geochim. Cosmochim. Acta **37**, 1189 (1973).

FLEISCHER, M.: Hafnium content and hafnium-zirconium ratio in minerals and rocks. U.S. Geol. Surv. Bull. **1021-A** (1955).

FORD, A. B.: Volcanic rocks of Naturaliste Plateau, eastern Indian Ocean, site 264, DSDP Leg 28. In: HAYES, D. E., FRAKES, L. A. *et al.* Initial Reports of the Deep Sea Drilling Project, Vol. 28. Washington: U.S. Govt. Printing Office 1975a.

FORD, A. B.: Antarctic deep-sea basalt, southeast Indian Ocean and Balleny Basin, DSDP Leg 28. In: HAYES, D. E., FRAKES, L. A. *et al.* Initial reports of the Deep Sea Drilling Project, Vol. 28. Washington: U.S. Govt. Printing Office 1975b.

FORDYCE, J. S., SHEIBLEY, D. W.: Estimate of contribution of jet aircraft operations to trace element concentrations at or near airports. Tech. Memo. NASA TMX-3054, NASA (1974).

FRANA, J., VOBECKY, M., BAUER, J.: Instrumental neutron activation analysis of manganese nodules. Sb. Vysoké Školy Chem.-Technol. Praze, Mineral. G. **15**, 77 (1973).

FREY, F. A., BRYAN, W. B., THOMPSON, G.: Atlantic ocean floor: geochemistry and petrology of basalts from Legs 2 and 3 of the DSDP. J. Geophys. Res. **79**, 5507 (1974).

FREY, F. A., SUNG, C. M.: Geochemical results for basalts from sites 253 and 254. In: DAVIES, T. A., LUYENDYK, B. P. *et al.* Initial Reports of the Deep Sea Drilling Project, Vol. 36. Washington: U.S. Govt. Printing Office 1974.

FURR, A. K., KOSIKOWSKI, F. V., BACHE, C. A., LISK, D. J.: Elemental analysis of protein-containing food materials from various sources. J. Food Sci. **39**, 887 (1974).

GANAPATHY, R., PAPIA, G. M., GROSSMAN, L.: The abundances of zirconium and hafnium in the solar system. Earth Planet. Sci. Lett. **29**, 302 (1976).

GARG, A. N., EHMANN, W. D.: Zr-Hf fractionation in chemically defined lunar rock groups. Proc. Seventh Lunar Sci. Conf., Geochim. Cosmochim. Acta, Suppl. **7**, 3397 (1976).

GERASIMOVSKII, V. I.: Geochemical features of agpaitic nepheline syenites. In: VINOGRADOV, A. P. (ed.). Chemistry of the Earth's Crust, Vol. I. Jerusalem: Israel Program for Scientific Translations 1966.

GERASIMOVSKII, V. I., NESMEYANOVA, L. I., KAKHANA, M. M., KHAZIZOVA, V. D.: On regularities of zirconium and hafnium distribution in effusive rocks of East African rift zones. [In Russian.] Geochimiya **12**, 1595 (1972).

GERASIMOVSKII, V. I., SHEVALEEVSKII, I. D.: The zirconium-hafnium ratio in zirconium minerals from the Lovozero Massif. Geochemistry **8**, 816 (1957).

GERASIMOVSKII, V. I., TUZOVA, A. M., SHEVALEEVSKII, I. D.: The zirconium hafnium (celtium) ratio in the Lovozero Massif rocks. Geochemistry **8**, 926 (1958).

GERASIMOVSKII, V. I., TUZOVA, A. M., SHEVALEEVSKII, I. D.: Zirconia-hafnia ratio in minerals and rocks of the Lovozero Massif. Geochemistry **6**, 585 (1962).

GILL, J. B.: Geochemistry of Viti Levu, Fiji, and its evolution as an island arc. Contrib. Mineral. Petrol. **27**, 179 (1970).

GLIKSON, A. Y.: Trace element geochemistry and origin of early Precambrian acid igneous series, Barberton Mountain Land, Transvaal. Geochim. Cosmochim. Acta **40**, 1261 (1976).

GOLDSCHMIDT, V. M.: Geochemistry. MUIR, A. (ed.). Oxford: Clarendon Press 1954.

GOLES, G. G.: Instrumental activation analysis of Columbia River basalts and of lunar rocks. In: BRUNFELT, A. O., STEINNES, E. D. (eds.). Activation Analysis in Geochemistry and Cosmochemistry. Oslo: Universitetsforlaget 1971.

GOTTFRIED, D., GREENLAND, L. P., CAMPBELL, E. Y.: Variation of Nb—Ta, Zr—Hf, Th—U and K—Cs in two diabase-granophyre suites. Geochim. Cosmochim. Acta **32**, 925 (1968).

GOTTFRIED, D., WARING, C. L.: Hf content and Hf/Zr ratio in zircon from the southern Californian batholith. U.S. Geol. Surv. Prof. Papers **501-B**, B 88 (1964).

GRAF, D. L.: Geochemistry of carbonate sediments and sedimentary carbonate rocks. Part III. Minor element distribution. Illinois State Geol. Surv. Circ. **301** (1960).

GULSON, B. L.: Electron microprobe determination of Zr/Hf ratios in zircons from the Yeoval Diorite Complex, N.S.W., Australia. Lithos **3**, 17 (1970).

GULSON, B. L., LOVERING, J. F., TAYLOR, S. R., WHITE, A. J. R.: High-K diorites, their place in the calc-alkaline association and relationship to andesites. Lithos **5**, 269 (1972).

HAINES, E. L., ALBEE, A. L., CHODOS, A. A., WASSERBURG, G. J.: Uranium-bearing minerals of lunar rock 12013. Earth Planet. Sci. Lett. **12**, 145 (1971).

HEINRICH, E. W.: Xenotime and thorite from Nigeria. Am. Mineralogist **48**, 206 (1963).

HELMKE, P. A., BLANCHARD, D. P., HASKIN, L. A., TELANDER, K., WEISS, C., JACOBS, J. W.: Major and trace elements in igneous rocks from Apollo 15. The Moon **8**, 129 (1973).

HELMKE, P. A., BLANCHARD, D. P., JACOBS, J. W., HASKIN, L. A.: Rare earths, other trace elements and iron in Luna 20 samples. Geochim. Cosmochim. Acta **37**, 869 (1973).

HELMKE, P. A., HASKIN, L. A.: Rare-earth elements, Co, Sc and Hf in the Steens Mountain basalts. Geochim. Cosmochim. Acta **37**, 1513 (1973).

HEMENWAY, C. L., HALLGREN, D. S., SCHMALBERGER, D. C.: Stardust. Nature **238**, 256 (1972).

HENDERSON, P., FISHLOCK, S. J., LAUL, J. C., COOPER, T. D., CONARD, R. L., BAYNTON, W. V., SCHMITT, R. A.: Rare earth element abundances in rocks and minerals from the Fiskenaesset Complex, West Greenland. Earth Planet. Sci. Lett. **30**, 37 (1976).

HERNDON, J. M.: Hafnium removal. In: LUSTMAN, B., KERZE, F., Jr. (eds.): The Metallurgy of Zirconium. New York: McGraw-Hill 1955.

HERRMANN, A. G., BLANCHARD, D. P., HASKIN, L. A., JACOBS, J. W., KNAKE, D., KOROTEV, R. L., BRANNON, J. C.: Major, minor, and trace element compositions of peridotitic and basaltic komatiites from the Precambrian crust of Southern Africa. Contrib. Mineral. Petrol. **59**, 1 (1976).

HORN, M. K., ADAMS, J. A. S.: Computer-derived geochemical balances and element abundances. Geochim. Cosmochim. Acta **30**, 279 (1966).

HEVESY, G. VON, JANTZEN, V. T.: Hafnium content of zirconium ores. Chem. News **128**, 341 (1924).

HEVESY, G. VON, JANTZEN, V. T.: The hafnium content of zirconium ores. Chem. News **130**, 179 (1925).

JACKSON, E. D., BARGAR, K. E., FABBI, B. P., HEROPOULOS, C.: Petrology of the basaltic rocks drilled on Leg 33 of the DSDP. In: SCHLANGER, S. O., JACKSON, E. D. *et al.* Initial Reports of the Deep Sea Drilling Project, Vol. 33. Washington: U.S. Govt. Printing Office 1976.

JOHNSON, R.: Mechanisms and problems of marine pollution in relation to commercial fisheries. In: JOHNSON, R. (ed.): Marine Pollution. London and New York: Academic Press 1976.

KABLE, E. J. D., FESQ, H. W., GURNEY, J. J.: The significance of the inter-element relationships of some minor and trace elements in South African kimberlites. Phys. Chem. Earth **9**, 709 (1975).

KLEMIC, H.: Zirconium and hafnium minerals. In: LEFOND, S. J. (ed.). Industrial Minerals and Rocks, 4th. ed. New York: A.I.M.E. 1975.

KLERKX, J., DEUTSCH, S., HERTOGEN, J., DE WINETR, J., GIJBELS, R., PICHLER, H.: Comments on "Evolution of Eolian arc volcansim (Southern Tyrrhenian Sea)" by BARBERI, F., FERRARA, G., INNOCENTI, F., KELLER, J., VILLARI, L. Earth Planet. Sci. Lett. **23**, 297 (1974).

KNORRING, O. VON, HORNUNG, G.: Hafnian zircons. Nature **190**, 1098 (1961).

KOSTERIN, A. V., SHEVALEEVSKII, I. D., KIZYURA, V. E.: Behaviour of zirconium and hafnium in the pegmatites of the southern Stanovoi Range. Geochem. Intern. **5**, 989 (1964).

KOSTERIN, A. V., SHEVALEEVSKII, I. D., RYBALOVA, K. E.: The Zr/Hf ratio in the zircons of some igneous rocks of the northern slope of the Kuramin Mountain Range. Geochemistry **5**, 541 (1960).

KOSTERIN, A. V., SHEVALEEVSKII, I. D., RYBALOVA, K. E., TOLOK, K. P.: Zr/Hf ratio in zircons from some igneous rocks of northern Kazakhstan. Geochemistry **10**, 1007 (1963).

KOSTERIN, A. V., ZUEV, V. N., SHEVALEEVSKII, I. D.: Zr/Hf ratio in zircons in some igneous rocks of Northern Kirgizia. Geochemistry **1**, 116 (1958).

KOVAL, P. V.: Geochemistry of high temperature post-magmatic processes in granites. In: Granitisation, Granites and Pegmatites. [In Russian.] First Internat. Geochem. Congress, Moscow 1971, **2**, 371 (1972).

KUKHARENKO, A. A., VAINSHTEIN, E. E., SHEVALEEVSKII, I. D.: Zirconium and hafnium ratios in rock-forming pyroxenes and in zirconium minerals of the Paleozoic complex of the ultrabasic and alkaline rocks of the Kola Peninsula. Geochemistry **7**, 730 (1960).

LARSEN, E. S., WARING, C. L., BERMAN, J.: Zoned zircon from Oklahoma. Am. Mineralogist **38**, 1118 (1953).

LAUL, J. C., HILL, D. W., SCHMITT, R. A.: Chemical studies of Apollo 16 and 17 samples. Proc. Fifth Lunar Sci. Conf., Geochim. Cosmochim. Acta, Suppl. **5**, 1047 (1974).

LAUL, J. C., SCHMITT, R. A.: Chemical composition of Luna 20 rocks and soil and Apollo 16 soils. Geochim. Cosmochim. Acta **37**, 927 (1973).

LECKIE, J. O., JAMES, R. O.: Control mechanisms for trace metals in natural waters. In: RUBIN, A. J. (ed.): Aqueous Environmental Chemistry of Metals, Vol. I. Michigan: Ann Arbor Science 1974.

LEVIN, E. M., ROBINS, C. R., MCMURDIE, H. F.: Phase Diagrams for Ceramists. Ohio: The Amer. Ceramic Soc. 1964.

LEVINSON, A. A., BORUP, R. A.: High hafnium zircon from Norway. Am. Mineralogist **45**, 562 (1960a).

LEVINSON, A. A., BORUP, R. A.: New data on the Hf, Zr and Y content of thortveitite. Am. Minerlaogist **45**, 712 (1960b).

LIPOVA, I. M., MAYEVA, M. M.: The relation of Zr/Hf ratio in zircon to crystal morphology. Geochem. Intern. **8**, 785 (1971).

LIPOVA, I. M., SHEVALEEVSKII, I. D.: On the Zr:Hf ratio in zircons from pegmatites of different compositions. Geochemistry **7**, 686 (1961).

LIPOVA, I. M., SHEVALEEVSKII, I. D., TUZOVA, A. M.: On the ratio of zirconium and hafnium in granitoids of the Verkhissetsk intrusion. Geochemistry **2**, 159 (1957).

LOVERING, J. F., WARK, D. A., REID, A. F., WARE, N. G., KEIL, K., PRINZ, M., BUNCH, T. E., EL GORESY, A., RAMDOHR, P., BROWN, G. M., PECKETT, A., PHILLIPS, R., CAMERON, E. N., DOUGLAS, J. A. V., PLANT, A. G.: Tranquillityite: A new silicate mineral from Apollo 11 and Apollo 12 basaltic rocks. Proc. Second Lunar Sci. Conf., Geochim. Cosmochim. Acta, Suppl. **2**, 39 (1971).

LYAKOVICH, V. V., SHEVALEEVSKII, I. D.: Zr:Hf ratio in accessory zircon of granitoids. Geochemistry **5**, 508 (1962).

MACDOUGALL, J. D., KERRIDGE, J. F.: Unusual anhydrous mineral assemblage in the Alais (C 1) meteorite. Meteoritics **11**, 326 (1976).

MAMURO, T.: Nondestructive multielement analysis using radiation. Genshiryoku Kogyo **19**, 69 (1973).
MAMURO, T., MATSUDA, Y., MIZOBATA, A., MATSUNAMI, T.: Activation analysis of airborne dust. Ann. Rep. Radiat. Cent. Osaka Prefect. **12**, 1 (1971a).
MAMURO, T., MATSUDA, Y., MIZOBATA, A., MATSUNAMI, T., TAKEUCHI, T.: Radiometric analysis of Yodo river water. Mizu Shori Gijutsu **12**, 33 (1971b).
MARTIN, D.R., PIZZOLATO, P. J.: Hafnium. In: Rare Metals Handbook. New York: Reinhold Publishing Corp. 1954.
MARTIN, P. M., MASON, B.: Major and trace elements in the Allende meteorite. Nature **249**, 333 (1974).
MARTYNOV, YU. T., SEMENOV, G. S., SULIN, V. V., SHATSOV, A. N., SHORIKOV, S. I.: Application of nuclear physical methods for the study of rock, ore, and mineral samples. Estimation of the titanium, zirconium, and hafnium contents in samples of recent placers of the Baltic Sea by gamma-activation methods. Tr. Vses. Nauch.-Issled. Inst. Yad. Geofiz. Geokhim. **13**, 32 (1972).
MASON, B., GRAHAM, A. L.: Minor and trace elements in meteoritic minerals. Smithsonian Contrib. Earth Sci. No. **3**, 1 (1970).
MASON, B., MARTIN, P. M.: Minor and trace element distribution in melilite and pyroxene from the Allende meteorite. Earth Planet. Sci. Lett. **22**, 141 (1974).
MERTIE, J. B.: Zirconium and hafnium in the southeastern Atlantic States. U.S. Geol. Surv. Bull. **1082-A** (1958).
MERZ, E.: The determination of hafnium and zirconium in meteorites by neutron activation analysis. Geochim. Cosmochim. Acta **26**, 347 (1962).
MERZ, E., SCHRAGE, E.: Activation analysis determination of the zirconium-hafnium ratios in stony meteorites and minerals. Geochim. Cosmochim. Acta **28**, 1873 (1964).
MITCHELL, R. H., BRUNFELT, A. O.: Rare earth element geochemistry of kimberlite. Phys. Chem. Earth **9**, 671 (1975).
MITCHELL, R. H., BRUNFELT, A. O., NIXON, P. H.: Ilmenite association trace element studies. In: NIXON, P. H. (ed.). Lesotho Kimberlites. Maseru: Lesotho National Development Corporation 1973.
MORGAN, J. W., GANAPATHY, R., HIGUCHI, H., KRÄHENBÜHL, U.: Volatile and siderophile trace elements in anorthositic rocks from Fiskenaesset, West Greenland: comparison with lunar and meteoritic analogues. Geochim. Cosmochim. Acta **40**, 861 (1976).
MOROSHKINA, T. M., VANEEVA, L. V.: Concentration of microamounts of rare elements in new chelating resins and their spectrographic determination in natural materials. Anal. Metody. Geokhim. Issled. Mater. Geokhim. Conf. 4th Gorn Inst. Leningrad **96** (1972).
NANCE, W. B., TAYLOR, S. R.: Rare earth element patterns and crustal evolution—I. Australian post-Archaean sedimentary rocks. Geochim. Cosmochim. Acta **40**, 1539 (1976).
NEKRASOVA, R. A., ROZHDESTVENSKAYA, I. V.: The ZrO_2/HfO_2 ratio in zircons from kimberlites and alluvial sediments. Geochim. Intern. **7**, 536 (1970).
NESTERENKO, G. V., ZNAMENSKIY, YE. B., AL'MUKHAMEDOV, A. I., TSYKHANSKIY, V. D.: Nb, Ta, Zr, and Hf in trap magma differentiation. Geochim. Intern. **8**, 725 (1971).
PAEPE, P. DE, KLERKX, J., HERTOGEN, J., PLINKE, P.: Oceanic tholeiites on the Cape Verde Islands: petrochemical and geochemical evidence. Earth Planet. Sci. Lett. **22**, 347 (1974).
PAREKH, P. P., MÖLLER, P., DULSKI, P.: Distribution of trace elements between carbonate and non-carbonate phases of limestone. Earth Planet. Sci. Lett. **34**, 39 (1977).
PAUL, D. K., POTTS, P. J., GIBSON, I. L., HARRIS, P. G.: Rare-earth abundances in Indian kimberlites. Earth Planet. Sci. Lett. **25**, 151 (1975).
PAVLENKO, A. S., VAINSHTEIN, E. E., SHEVALEEVSKII, I. D.: On the hafnium-zirconium ratio in zircons of igneous and metasomatic rocks. Geochemistry **5**, 411 (1957).
PECCERILLO, A., TAYLOR, S. R.: Geochemistry of Eocene calc-alkaline volcanic rocks from the Kastamonu area, northern Turkey. Contrib. Mineral. Petrol. **58**, 63 (1976).
PIGORINI, B., VENIALE, F.: Studio mediante microsonda elettronica dei diversi tipi di zircone accessorio nei graniti di Baveno, M. Orfano e Alzo. Atti Soc. Ital. Sci. Nat. Museo Civico Storia Nat. Milano **105**, 207 (1966).
PSZENNY, A. A. P., RUSSELL, I. J.: Analysis of air particulates generated in the Boston Sumner tunnel and environs. Report COO-3015-10, Dep. Chem. Boston Coll., Mass. (1974).

PYATENKO, YU. A.: Isomorphism of atoms and its mineralogical consequences. Geochem. Intern. **2**, 268 (1965).
RANCITELLI, L. A., COOPER, J. A., PERKINS, R. W.: Multielement characterization of atmospheric aerosols by neutron activation and direct γ-ray analysis, and X-ray fluorescence analysis. Proc. Comp. Stud. Food. Environ. Contam. Symp., 431. Vienna: I.A.E.A. (1974).
RANDLE, K.: Some trace element data and their interpretation for several new reference samples obtained by neutron activation analysis. Chem. Geol. **13**, 237 (1974).
RANKAMA, K.: Progress in Isotope Geology. New York, London and Sydney: Interscience Publ. 1963.
RANKAMA, K., SAHAMA, T. G.: Geochemistry. Chicago: University of Chicago Press 1950.
RASMUSSEN, S. E., FESQ, H. W.: Neutron activation analysis of samples from the Kimberley Reef conglomerate. Report 1563, National Institute for Metallurgy, Johannesburg (1973).
REBAGAY, T. V.: The determination of zirconium and hafnium in meteorites and terrestrial materials by activation analysis. Ph.D. Dissertation, Dept. of Chemistry, University of Kentucky (1969).
RHODES, J. M., BLANCHARD, D. P., RODGERS, K. V., JACOBS, J. W., BRANNON, J. C.: Petrology and chemistry of basalts from the Nazca Plate: Part 2— Major and trace element chemistry. In: YEATS, R. S., HART, S. R., *et al.* Initial Reports of the Deep Sea Drilling Project, Vol. 34. Washington: U.S. Govt. Printing Office 1976.
RINGWOOD, A. E.: The principles governing trace-element behaviour during magmatic crystallization. Part II. The role of complex formation. Geochim. Cosmochim. Acta **7**, 242 (1955).
ROBINSON, P. T., WHITFORD, D. J.: Basalts from the eastern Indian Ocean, DSDP Leg 27. In: VEEVERS, J. V., HEIRTZLER, J. R., *et al.* Initial Reports of the Deep Sea Drilling Project, Vol. 27. Washington: U.S. Govt. Printing Office 1974.
RONOV, A. B., VAINSHTEIN, E. E., TUZOVA, A. M.: Geochemistry of hafnium, zirconium and some other hydrolyzate elements in clays. Geochemistry **4**, 343 (1961).
ROSS, J. E., ALLER, L. H.: The chemical composition of the sun. Science **191**, 1223 (1976).
SCHNETZLER, C. C., NAVA, D. F.: Chemical composition of Apollo 14 soils 14163 and 14259. Earth Planet. Sci. Lett. **11**, 345 (1971).
SCHUTTE, C. E. G.: Die chemiese samestelling en hafnium/sirkonium verhouding van Suid-Afrikaanse sirkone. Geol. Surv. S. Africa, Bull. **46** (1966).
SCHUTZ, D. F., TUREKIAN, K. K.: The investigation of the geographical and vertical distribution of several trace elements in sea water using neutron activation analysis. Geochim. Cosmochim. Acta **29**, 259 (1965).
SEEGER, P. A., FOWLER, W. A., CLAYTON, D. D.: Nucleosynthesis of heavy elements by neutron capture. Astrophys. J., Suppl. Series **11**, 121 (1965).
SETSER, J. L., EHMANN, W. D.: Zirconium and hafnium abundances in meteorites, tektites, and terrestrial materials. Geochim. Cosmochim. Acta **28**, 769 (1964).
SHAW, D. M., DOSTAL, J., KEAYS, R.: Additional estimates of continental surface Precambrian shield composition in Canada. Geochim. Cosmochim. Acta **40**, 73 (1976).
SHEVALEEVSKII, I. D., PAVLENKO, A. S., VAINSHTEIN, E. E.: Dependance of the behaviour of zirconium and hafnium on the petrochemical characteristics of igneous and alkalic metasomatic rocks. Geochemistry **3**, 262 (1960).
SHIMA, M., HINTENBERGER, H.: Elemental and isotopic abundances of Ti, Zr and Hf in stony meteorites. Meteoritics **11**, 364 (1976).
SHOWALTER, D. L., WAKITA, H., SMITH, R. H., SCHMITT, R. A.: Chemical composition of sawdust from lunar rock 12013 and comparison of a Java tektite with the rock. Science **175**, 170 (1972).
SHUM, Y. S., LOVELAND, W. D.: Atmospheric trace element abundances and their application in air pollution tracing. Report RLO-2227-T7-27, Oregon State Univ. Dep. Chem. (1974).
SILVER, L. T., CHODOS, A. A.: Petrological and geochronological implications of composition variations in igneous zircons. Trans. Am. Geophys. Union **47**, 495 (1966).
SOBACHENKO, V. N., KOSTYNKOVA, YE. S., KUZNETSOVA, A. I., LEGEYDO, V. A., NOVIKOV, V. M., PETROV, L. L., SMIRNOVA, YE. V., TSYKHANSKIY, V. D.: Geological and geo-

chemical characterisation of the granitoids in the Davan and Primorya deep-fault zones in Baykalia. Geochem. Intern. **11**, 168 (1974).

Tanner, T. M., Young, J. A., Cooper, J. A.: Multielement analysis of St. Louis aerosols by nondestructive techniques. Chemosphere **5**, 211 (1974).

Taylor, S. R.: Abundance of chemical elements in the continental crust: a new table. Geochim. Cosmochim. Acta **28**, 1273 (1964).

Taylor, S. R.: Australites, Henbury impact glass and subgreywacke: a comparison of the abundances of 51 elements. Geochim. Cosmochim. Acta **30**, 1121 (1966).

Taylor, S. R.: Lunar Science: A Post-Apollo View. Oxford: Pergamon Press 1975.

Taylor, S. R.: Geochemical constraints on the composition of the moon. Proc. Seventh Lunar Sci. Conf., Geochim. Cosmochim. Acta, Suppl. **7**, 3461 (1976).

Taylor, S. R., Capp, A. C., Graham, A. L., Blake, D. H.: Trace element abundances in andesites. II. Saipan, Bougainville and Fiji. Contrib. Mineral. Petrol. **23**, 1 (1969).

Taylor, S. R., Gorton, M. P., Muir, P., Nance, W., Rudowski, R., Ware, N.: Trace element geochemistry of Apollo 16 soil 68501. Nature **239**, 205 (1972a).

Taylor, S. R., Gorton, M. P., Muir, P., Nance, W., Rudowski, R., Ware, N.: Composition of the Descartes region, lunar highlands. Geochim. Cosmochim. Acta **37**, 2665 (1973a).

Taylor, S. R., Gorton, M. P., Muir, P., Nance, W., Rudowski, R., Ware, N.: Lunar highlands composition: Apennine Front. Proc. Fourth Lunar Sci. Conf., Geochim. Cosmochim. Acta, Suppl. **4**, 1445 (1973b).

Taylor, S. R., Kaye, M., Muir, P., Nance, W., Rudowski, R., Ware, N.: Composition of the lunar uplands: Chemistry of Apollo 14 samples from Fra Mauro. Proc. Third Lunar Sci. Conf., Geochim. Cosmochim. Acta, Suppl. **3**, 1231 (1972b).

Taylor, S. R., Muir, P., Kaye, M.: Trace element chemistry of Apollo 14 lunar soil from Fra Mauro. Geochim. Cosmochim. Acta **35**, 975 (1971).

Thompson, G., Bryan, W. B., Frey, F. A., Dickey, J. S., Suen, C. J.: Petrology and geochemistry of basalts from DSDP Leg 34, Nazca Plate. In: Yeats, R. S., Hart S. R. *et al*: Reports of the Deep Sea Drilling Project, Vol. 34. Washington: U.S. Govt. Printing Office 1976

Tugarinov, A. I., Kovalenko, V. I., Znamensky, E. B., Legeido, V. A., Sobatovich, E. V., Brandt, S. B., Tsyhansky, V. D.: Distribution of Pb-isotopes, Sn, Nb, Ta, Zr and Hf in granitoids of Nigeria. In: Ahrens, L. H. (ed.): Origin and Distribution of the Elements. Oxford: Pergamon Press 1968.

Tugarinov, A. I., Pavlenko, A. S., Kovalenko, V. I.: Geochemical evidence on the origin of aplogranites. Geochim. Intern. **5**, 1156 (1968).

Tugarinov, A. I., Vainshtein, E. E., Shevaleevskii, I. D.: Hafnium-zirconium ratio in the zircons of igneous and metasomatic rocks. Geochemistry **4**, 361 (1956).

Turekian, K. K., Wedepohl, K. H.: Distribution of the elements in some major units of the earth's crust. Bull. Geol. Soc. Am. **72**, 175 (1961).

Uken, E.-A., Erasmus, C. S.: The instrumental neutron-activation analysis of samples taken at different levels of the E.R.P.M. Mine. Report 1013, National Institute for Metallurgy, Johannesburg (1970).

Vainshtein, E. E., Ginzburg, A. I., Shevaleevskii, I. D.: The Hf/Zr ratio in zircons from granite pegmatites. Geochemistry **2**, 151 (1959).

Vainshtein, E. E., Tugarinov, A. I., Tuzova, A. M., Shevaleevskii, I. D.: Hafnium-zirconium ratios in metamorphic and metasomatic rocks. Geochemistry **3**, 305 (1958).

Velandia, J. A., Perkons, A. K.: Survey of 33 constituent elements in heart tissue by instumental activation analysis. J. Radioanal. Chem. **14**, 171 (1973).

Velandia, J. A., Perkons, A. K.: Survey of 43 constituent elements in kidney samples by instrumental activation analysis. J. Radioanal. Chem. **20**, 715 (1974).

Vinogradov, A. P.: Average contents of chemical elements in the principal types of igneous rocks of the earth's crust. Geochemistry **7**, 641 (1962).

Vlasov, K. A.: Geochemistry and mineralogy of rare elements and genetic types of their deposits. Vol. I. Geochemistry of Rare Elements. Jerusalem: Israel Program for Scientific Translations 1966.

Wänke, H., Baddenhausen, H., Balacescu, A., Teschke, F., Spettel, B., Dreibus, G., Palme, H., Quijano-Rico, M., Kruse, H., Wlotzka, F., Begemann, F.: Multielement analyses of lunar samples and some implications of the results. Proc. Third Lunar Sci. Conf., Geochim. Cosmochim. Acta, Suppl. **3**, 1251 (1972).

WÄNKE, H., PALME, H., BADDENHAUSEN, H., DREIBUS, G., JAGOUTZ, E., KRUSE, H., PALME, C., SPETTEL, B., TESCHKE, F., THACKER, R.: New data on the chemistry of lunar samples: Primary matter in the lunar highlands and the bulk composition of the moon. Proc. Sixth Lunar Sci. Conf., Geochim. Cosmochim. Acta, Suppl. **6**, 1313 (1975).

WARK, D. A., REID, A. F., LOVERING, J. F., EL GORESY, A.: Zirconolite (versus zirkelite) in lunar rocks. In: CHAMBERLAIN, J. W., WATKINS, C. (eds.): Lunar Science IV. Houston: Lunar Science Institute 1973.

WEIGAND, P. W., BRUNFELT, A. O., HEIER, K. S., SUNDVOLL, S., STEINNES, E.: Geochemistry of alkali olivine basalts from an eruption on Jan Mayen. Nature **235**, 31 (1972).

WHITTAKER, E. J. W., MUNTUS, R.: Ionic radii for use in geochemistry. Geochim. Cosmochim. Acta **34**, 945 (1970).

WIIK, H. B., MAXWELL, J. A., BOUVIER, J.-L.: Chemical composition of some Apollo 14 lunar samples. Earth Planet. Sci. Lett. **17**, 365 (1973).

WINZER, S. R., LUM, R. K. L., SCHUMANN, S.: Rb, Sr and strontium isotopic composition, K/Ar age and large ion lithophile trace element abundances in rocks and glasses from the Wanapitei Lake impact structure. Geochim. Cosmochim. Acta **40**, 51 (1976).

WOOD, D. A.: Spatial and temporal variation in the trace element geochemistry of the eastern Iceland flood basalt succession. J. Geophys. Res. **81**, 4353 (1976).

YELISEYEVA, O. P., MURAVITSKAYA, G. N., SEMENOVA, T. T.: Distribution of radioactive elements, yttrium and hafnium in granitoids of northern Kazakhstan. Geochim. Intern. **5**, 628 (1968).

YES'KOVA, E. M., YEFIMOV, A. F.: Distribution of rare elements in the apoeffusive alkalic metasomatites of the Urals. Geochim. Intern. **7**, 721 (1970).

ZHILOVA, N. A.: Toxic effect of hafnium and its compounds in mice and guinea pigs. Nov. Dannye Toksikol. Redk. Metal. Ikh. Soedin. **51** (1967).

ZIELINSKI, R. A., LIPMAN, P. W., MILLARD, H. T.: Minor-element abundances in obsidian, perlite and felsite of calc-alkalic rhyolites. Amer. Mineral. **62**, 426 (1977).

ZNAMENSKIY, E. B., GLAZUNOV, O. M., POPOLITOV, E. I., TSYKHANSKIY, V. D.: Distribution of Nb, Ta, Zr and Hf in basic and ultrabasic rocks of the Syan-Baykal folded region. Geochem. Intern. **11**, 606 (1974).

ZNAMENSKIY, E. B., POPOLITOV, E. I., TSYKHANSKIY, V. D.: The behaviour of Nb, Ta, Zr and Hf in granitoids of various stages of the formation of fold belts. In: Granitisation, Granites and Pegmatites. [In Russian.] First Internat. Geochem. Congress, Moscow 1971, **2**, 69 (1972).

Manuscript received: August 1977

Tantalum 73

A	O. W. FLÖRKE and W. GEBERT	(Institut für Mineralogie, Ruhr-Universität, Bochum, Germany)
B—G I—M, O	K. H. WEDEPOHL	(Geochemisches Institut der Universität, Göttingen, Germany)

73-A. Crystal Chemistry

The electron configuration of Ta in the ground state is Xe $4f^{14}\,5d^3\,6s^2$. Due to the lanthanide contraction Ta shows almost the same ionic radius as Nb in spite of its much higher atomic number. This causes close relationships in crystal chemistry between Nb and Ta (Goldschmidt, 1958), observable in extensive solid solution and isotypism.

Table 73-A-1. *Valency, coordination number and ionic radii of Ta*

Valency	Coordination number	Ionic radius [Å] according to:	
		Whittaker and Muntus (1970)	Shannon and Prewitt (1969)[a]
3+	VI	0.75	0.67
4+	VI	0.74	0.66
5+	VI	0.72	0.64
5+	VIII	0.77	0.69

[a] Based on $r(^{VI}O^{2-}) = 1.40$ Å.

Therefore most of the structure types and solid solution series discussed in the chapter on crystal chemistry of niobium are also valid for tantalum and need not be discussed here again in detail.

The crystal chemistry of Nb and Ta is not the same in all cases. For example no tantalum analogs of the well investigated nonmineral halides and oxyhalides of Nb have yet been found.

Tantalum, due to its only partially filled 5d level, shows transition metal character. The element crystallizes in a body-centered structure and has a very high melting point (ca. 3,300° K), which is higher than that of Nb despite tantalum's higher atomic mass.

The five-valent state is the most stable oxidation state of Ta. Ta^{5+} in oxide compounds prefers octahedral coordination (Table 73-A-1). It occurs in only one structure-type in tetrahedral coordination with oxygen. The tantalum minerals are almost exclusively oxygen compounds (Table 73-A-2), the only known non-oxide mineral, tantalum carbide TaC, crystallizing in the NaCl type structure (Frondel, 1962).

I. Oxide Minerals with Ta in Octahedral Coordination

a) Derivatives of the α-PbO_2 Type

Tantalite (Fe, Mn)$(Ta, Nb)_2O_6$ (Sturdivant, SR 1928—32, 337; Laves *et al.*, 1963), manganotantalite (Mn, Fe)Ta_2O_6 (Brandt, SR 1942—44, 179), wodginite

Table 73-A-2. *Structural properties of selected Ta minerals*

Name and composition	Structure type	Coordination number	Reference
Tantalite (Fe, Mn)(Ta, Nb)$_2$O$_6$	α-PbO$_2$	6	STURDIVANT (SR 1928—32, 337)
Stibiotantalite Sb(Ta, Nb)O$_4$	α-PbO$_2$	6	DIHLSTRÖM (SR 1938, 113)
Manganotantalite (Mn, Fe)Ta$_2$O$_6$	α-PbO$_2$	6	BRANDT (SR 1942—44, 179)
Wodginite (Ta, Nb, Sn, Mn, Fe)$_2$O$_4$	α-PbO$_2$	6	NICKEL *et al.* (1963)
Tapiolite (Fe, Mn)(Ta, Nb)$_2$O$_6$	trirutile	6	HUTTON (SR 1958, 345)
Mikrolithe (Ca, Na)$_2$Ta$_2$O$_6$(O, OH, F)	pyrochlore	6	KERR and HOLMES (SR 1945—46, 118)
Rijkeboerite (Ba,...)(Ta, Nb, Ti)$_2$O$_6$(O, OH, F)	pyrochlore	6	V.D. VEEN (1963)
Westgrenite (Bi, Ca)(Ta, Nb)$_2$O$_6$(OH)	pyrochlore	6	V. KNORRING and MROSE (SR 1962, 599)
Suculaite Sn$_2$Ta$_2$O$_7$	pyrochlore	6	VORMA and SIIVOLA (1967)
Thoreaulite Sn(Ta, Nb)$_2$O$_7$	—	6	MAKSIMOVA and ILYUKHIN (1967)
Simpsonite Al$_4$Ta$_3$O$_{13}$(OH)	—	6	KERR and HOLMES (SR 1945—46, 118)
Behierite (Ta, Nb)[BO$_4$]	zircon	8	ZASLAVSKII and ZVINCHUK (SR 1953, 525)
Formanite YTaO$_4$	scheelite	4	WOLTEN (1967)

(Ta, Nb, Sn, Mn, Fe)$_2$O$_4$ (NICKEL *et al.*, 1963). Tantalite formed at medium temperatures in nature has a disordered distribution of the A- and B-cations, thus resulting in a division of the a_0-dimension into three. No discussion is given of what kind of structure is produced by this disorder (FRENZEL, 1955; PANAGOS, 1964).

b) Derivatives of the Rutile Type

Tapiolite, (Fe, Mn) (Ta, Nb)$_2$O$_6$ (HUTTON, SR 1958, 345), crystallizes in the trirutile structure. The nonmineral GaTaO$_4$ in its high form existing above 1,803° K shows disordered A-B distribution and hence simple rutile structure (BAYER, SR 1962, 526). It is striking that in the ordered trirutiles AB$_2$O$_6$, the valency difference between A and B is always 3 whereas it is 2 in the disordered ABO$_4$ rutiles. Ta prefers the trirutile type and ordered Nb compounds of the composition AB$_2$O$_6$ prefer to crystallize in the niobite type which derives from the α-PbO$_2$ type (PANAGOS, 1964; LAVES *et al.*, 1963). Here again differences in the crystal chemistry of tantalum and niobium are obvious.

c) Derivatives of the Pyrochlore Type

Microlite (Ca, Na)$_2$Ta$_2$O$_6$(O, OH, F). Substitution in the A and B positions leads to rijkeboerite (Ba,...)$_2$(Ta, Nb, Ti)$_2$O$_6$(O, OH, F) (van der Veen, 1963), westgrenite (Bi, Ca)(Ta, Nb)$_2$O$_6$OH (v. Knorring and Mrose, SR 1962, 599), and suculaite Sn$_2$Ta$_2$O$_7$ (Vorma and Siivola, 1967). The nonmineral BiTaO$_4$, depending on temperature of formation, shows complicated polymorphism (Panagos, 1964).

d) Thoreaulite

Thoreaulite, Sn(Ta, Nb)$_2$O$_7$ (Maksimova and Ilyukin, 1967), has a structure based on two-dimensional infinite double sheets of octahedra parallel to (100). These sheets consist of pairs of (Ta, Nb)O$_6$ octahedra which are joined together by a common edge. The octahedra pairs are joined to neighboring pairs by common corners (Fig. 73-A-1). Six of the seven oxygen atoms of the formula unit build these double sheets, lying at the positions of a hexagonal close-packing of spheres with the packing axis parallel [010]. The Ta and Nb atoms occupy half of the octahedral holes in this arrangement in such a way that pairs of filled holes within the close-packed double steets of oxygen alternate with pairs of empty holes along [010] and [001]. The double sheets of octahedra are joined to each other by Sn atoms to which the 7th oxygen is coordinated, forming nearly planar [OSn$_4$] groups parallel (100). Control calculations lead to abnormal short Sn—O and O—O distances, and therefore this structure may need some refinement.

e) Stibiotantalite

Stibiotantalite, Sb(Ta, Nb)O$_4$ (Dihlström, SR 1938, 113), and bismutotantalite, Bi(Ta, Nb)O$_4$, crystallize in related structures (Frenzel, 1955; Hurlbut, SR 1957, 289; Panagos, 1964) with octahedral coordination of (Ta, Nb) and a one-sided coordination of Sb and Bi.

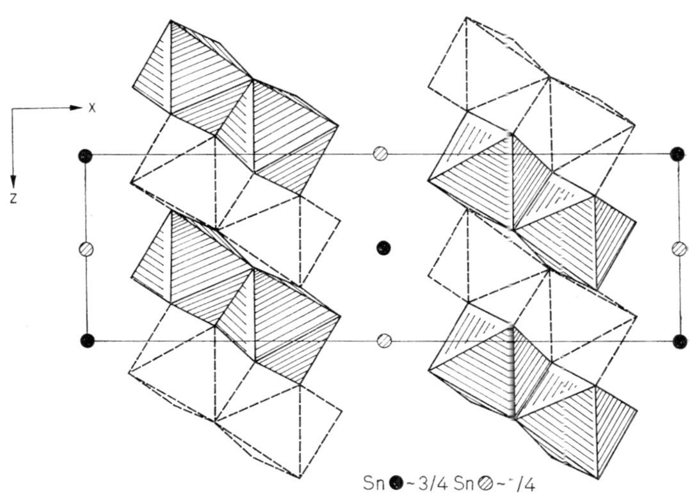

Fig. 73-A-1. Structure model of thoreaulite (according to Maksimova and Ilyukin, 1967). Only octahedra filled by (Ta, Nb) are drawn

II. Oxide Minerals with Ta in 8-fold Coordination

Behierite, (Ta, Nb)[BO_4] (ZASLAVSKIJ and ZVINCHUK, SR 1953, 525; MROSE and ROSE, 1961), crystallizes in the zircon structure.

III. Oxide Minerals with Ta in Tetrahedral Coordination

Formanite, $YTaO_4$, is isostructural with fergusonite (scheelite-type) (BUTLER and HALL, 1960). Polymorphism of this substance is described by WOLTEN (1967) and WOLTEN and CHASE (1967).

Revised manuscript received: January 1974

References: Section 73-A

BUTLER, J. R., HALL, R.: Chemical variations in members of the fergusonite-formanite series Mineral. Mag. **32**, 392 (1960).
FRENZEL, G.: Zur Kenntnis von Bismutotantalit. Neues Jahrb. Mineral., Monatsh. **1955**, 241 (1955).
FRONDEL, C.: Non-existence of native tantalum. Am. Mineralogist **47**, 786 (1962).
GOLDSCHMIDT, V. M.: Geochemistry. Oxford: Clarendon Press 1958.
LAVES, F., BAYER, G., PANAGOS, A.: Strukturelle Beziehungen zwischen den Typen α-PbO_2, $FeWO_4$ (Wolframit) und $FeNb_2O_6$ (Columbit) und über die Polymorphie des $FeNbO_4$. Schweiz. Mineral. Petrog. Mitt. **43**, 217 (1963).
MAKSIMOVA, N. V., ILYUKIN, V. V.: Crystal structure of thorolite, $Sn(Ta, Nb)_2O_7$. Soviet Phys. Cryst. **12**, 105 (1967).
MROSE, M. E., ROSE, H. J.: Behierite $(Ta, Nb)BO_4$, a new mineral from Manjaka, Madagascar. Mineral. Soc. Am. Prog. **1961**, 111 A (1961).
NICKEL, E. H., ROWLAND, J. F., MCADAM, R. C.: Wodginite — A new tin-manganese tantalate from Wodgina, Australia, and Bernic Lace, Manitoba. Canad. Mineralogist **7**, 390 (1963).
PANAGOS, A.: Vergleich des kristallchemischen Verhaltens von Nb und Ta in chemisch analogen Oxydverbindungen. Dissertation Nr. 3346 ETH Zürich (1964).
SHANNON, R. D., PREWITT, C. T.: Effective ionic radii in oxides and fluorides. Acta Cryst. **25 B**, 925 (1969).
VEEN, A. H. VAN DER: A study of pyrochlore. Verh. Kon. Nederl. Geol. Mignb. Genootschap, **22**, 1 (1963).
VORMA, A., SIIVOLA, J.: Sukulaite — $Ta_2Sn_2O_7$ — and wodginite as inclusions in cassiterite in the granite pegmatite in Sukula, Tammela in SW Finland. Bull. Comm. Geol. Finlande **229**, 173 (1967).
WHITTAKER, E. J., MUNTUS, R.: Ionic radii for use in geochemistry. Geochim. Cosmochim. Acta **34**, 945 (1970).
WOLTEN, C. M.: The structure of the M′-phase of $YTaO_4$, a third fergusonite polymorph. Acta Cryst. **23**, 939 (1967).
WOLTEN, C. M., CHASE, A. B.: Synthetic fergusonites and a new polymorph of yttrium tantalate. Am. Mineralogist **52**, 1536 (1967).

Revised manuscript received: January 1974

73-B. Isotopes in Nature

The relative abundances of the two tantalum isotopes are reported by WHITE *et al.* (1955) as ^{180}Ta 0.0123%, ^{181}Ta 99.9877%.

Manuscript received: April 1977

73-C. Abundance in Cosmos, Meteorites, Tektites and Lunar Materials

I. Cosmos

SUESS and UREY (1956) and CAMERON (1968) report a cosmic abundance of 0.065 and 0.022 atoms Ta per 10^6 atoms Si respectively.

II. Meteorites

Results from neutron activation analyses of stone and iron meteorites are listed in Table 73-C-1. The cosmic abundance published by CAMERON (1968) is slightly higher than the averages computed for 9 CL-, 4 CH- and 1 Cc_2 chondrites (0.018; 0.021; 0.019 atoms Ta/10^6 atoms Si), listed in Table 73-C-1.

In chondrites Ta is rather uniformly distributed. There is no systematic difference visible between the two sets of data published by different authors although they used different counting techniques. On the base of the averages from Tables 41-C-1 (0.49 ppm Nb) and 73-C-1 (0.022 ppm Ta) a mean Nb/Ta ratio of 23 can be computed which is at least twice that of common terrestrial crustal rocks. The mean chondritic Nb/Ta ratio decreases to 17 if only data from ordinary chondrites of Table 41-C-1 are used.

The Ta concentration in achondrites is much less uniform than the Nb abundance.

The lithophilic tendency of Ta is clearly demonstrated by the information compiled in Table 73-C-1.

III. Tektites

EHMANN (1965) has analyzed two philippinites (0.73; 3.3 ppm Ta), two indochinites from Thailand (0.61; 0.56 ppm Ta) and one Darwin glass from Tasmania (0.63 ppm Ta) by neutron activation. The average Nb concentration of tektites, reported in Subsection 41-C-III, is 18 ppm. If we include the high Ta figure for one philippinite in computing an average Nb/Ta ratio (14) we approach terrestrial Nb/Ta ratios. Otherwise the Nb/Ta ratio (29) is even higher than for chondrites.

IV. Lunar Materials

More than 400 pieces of data exist on tantalum in lunar rocks and fines (Table 73-C-2) obtained from neutron activation procedures.

The analytical information considers basaltic rocks, breccias and fines in about equal proportions. The variability of Ta values within and between rock types is not large. Data from different laboratories are consistant.

The average Ta concentrations in *basaltic rocks* from the Apollo 11, 12, 14, 15 and 17 missions are: 1.8, 0.48, 1.6, 0.48 and 1.7 ppm Ta respectively. Nb/Ta ratios

Table 73-C-1. *Tantalum abundance in meteorites* (Analytical method: N/R)

Name	Class	ppm Ta	Reference
Chondrites			
Ochansk	CH	0.018	1
Forest City	CH	0.024	1, 2
Plainview	CH	0.022	2
Penokee	CH	0.029	2
Bjurböle	CL	0.021	1
Cynthiana	CL	0.021	2
Chandakapur	CL	0.020	1
Bruderheim	CL	0.017	2
Chateau Renard	CL	0.027	1
Holbrook	CL	0.025	1
Long Island	CL	0.028	1, 2
Ness County	CL	0.022	1
Elenovka	CL	0.012	2
Murray	Cc2	0.017	2
Average ordinary chondrites		0.022	
Achondrites			
Cumberland Falls	Ae	<0.01	2
Norton County	Ae	~0.004	2
Johnstown	Ab	0.008	1
Pasamonte	Ap	0.12	2
Juvinas	Ap	0.12	3
Kapoeta	Aor	0.10	3
Siderites			
Canyon Diablo	Og	0.002	1
Henbury	Om	0.005	1
Odessa	Og	~0.0005	2
Sikhote-Alin	Ogg	~0.001	2
Williamstown	Om	~0.003	2
San Martin	H	0.0008	1
Troilite			
Canyon Diablo		~0.004	2

References: 1. Atkins and Smales (1960); 2. Ehmann (1965); 3. Wänke *et al.* (1972).

of the same rock groups (Tables 41-C-2 and 73-C-2) do not scatter as much as the Ta averages. They range from 12 (basalts of Apollo 11) to 13.6 (basalts of Apollo 14) only. These ratios are intermediate between those of terrestrial rocks (about 10; see Sect. 41-E) and an average Nb/Ta ratio of ordinary chondrites (17).

KREEP rich basalt is higher (2.3 to 3.9 ppm Ta) and anorthosite (≤ 0.07 ppm Ta) much lower in Ta than common lunar basalt. Ta in lunar feldspars is lower than in pyroxenes, olivine and ilmenite (Vobecky *et al.* [P2]; Brunfelt *et al.* [P3, P4])[1].

The following average Ta concentrations in *breccia* from the Apollo 11, 12, 14 to 17 missions can be computed: 1.8, 2.9, 4.1, 1.0, 0.74, 1.3 ppm Ta. The average

[1] For abbreviations of references see Table 73-C-2.

Table 73-C-2. *Tantalum concentrations in lunar materials.* (Number of samples in brackets; Most references are from Volume 2 of the 7 Proceedings of the Lunar Science Conferences. These are marked as [P1] to [P7]. The Proceedings are Supplements 1 to 7 of Geochim. Cosmochim. Acta, 1970 to 1976. Analytical method: N/R)

Rock Type	Range ppm Ta	Average ppm Ta	Reference
Apollo 11			
Basalt (8)	1.0 –2.7	2.0	Goles *et al.* [P1]
Basalt (2)	1.0 –1.8	1.4	Haskin *et al.* [P1]
Basalt (2)	—	5	Maxwell *et al.* [P1]
Basalt (5)	1.0 –1.8	1.4	Morrison *et al.* [P1]
Basalt (12)	1.1 –2.7	1.8	Turekian and Kharkar [P1] [P2]
Basalt (3)	2.0 –2.2	2.1	Wänke *et al.* [P1]
Breccia (16)	1.0 –2.1	1.7	Goles *et al.* [P1]
Breccia (2)	1.4 –3.0	2.2	Morrison *et al.* [P1]
Breccia (3)	1.4 –1.8	1.7	Turekian and Kharkar [P1] [P2]
Breccia (3)	1.7 –2.0	1.9	Wänke *et al.* [P1] [P2]
Fines (6)	1.2 –1.4	1.3	Goles *et al.* [P1]
Fines (1)	—	1.4	Haskin *et al.* [P1]
Fines (1)	—	3	Maxwell *et al.* [P1]
Fines (1)	—	2.3	Morrison *et al.* [P1]
Fines (3)	1.3 –1.8	1.5	Turekian and Kharkar [P1] [P2]
Fines (1)	—	1.3	Wänke *et al.* [P1]
Apollo 12			
Basalt (8)	0.28–0.48	0.39	Goles *et al.* [P2]
Basalt (4)	0.5 –0.7	0.6	Morrison *et al.* [P2]
Basalt (9)	0.33–0.54	0.43	Wänke *et al.* [P2]
Basalt (5)	0.35–0.42	0.38	Brunfelt *et al.* [P2]
Basalt (5)	0.55–0.88	0.68	Kharkar and Turekian [P2]
Basalt (1)	—	0.7	Ehmann *et al.* [P3]
Breccia (2)	3.09–3.72	3.4	Goles *et al.* [P2]
Breccia (1)	—	2.7	Morrison *et al.* [P2]
Breccia (1)	—	2.1	Wänke *et al.* [P2]
Fines (3)	1.4 –1.8	1.6	Ehmann *et al.* [P3]
Fines (3)	1.04–1.46	1.3	Wänke *et al.* [P2]
Fines (2)	1.9 –2.05	2.0	Goles *et al.* [P2]
Fines (3)	1.4 –3.3	2.2	Morrison *et al.* [P2]
Fines (1)	—	1.1	Travesi *et al.* [P2]
Fines (2)	0.61–1.4	1.0	Brunfelt *et al.* [P2]
Fines (1)	—	1.16	Vobecky *et al.* [P2]
Fines (3)	1.8 –2.0	1.9	Kharkar and Turekian [P2]
Apollo 14			
Basalt (1)	—	2.3	Brunfelt *et al.* [P3]
Basalt (1)	—	2.4	Laul *et al.* [P3]
Basalt (1)	—	1.2	Lindstrom *et al.* [P3]
Basalt fragment (2)	1 –1.2	1.1	Wänke *et al.* [P3]
Breccia (5)	4.8 –7.3	5.9	Lindstrom *et al.* [P3]
Breccia (2)	2.1 –3.2	2.7	Brunfelt *et al.* [P3]
Breccia (2)	2.3 –3.2	2.8	Laul *et al.* [P3]
Breccia (3)	3.2 –4.0	3.6	Wänke *et al.* [P3]
Breccia (3)	3.0 –3.9	3.4	Boynton *et al.* [P6]

Table 73-C-2 (continued)

Rock type	Range ppm Ta	Average ppm Ta	Reference
Fines (1)	—	2.7	Brunfelt *et al.* [P3]
Fines (5)	3.9 –5.7	4.7	Lindstrom *et al.* [P3]
Fines (2)	2.8 –3.2	3.0	Wänke *et al.* [P3]
Apollo 15			
KREEP-basalt	—	3.9	Laul *et al.* [P3]
Basalt (10)	0.35–0.77	0.46	Fruchter *et al.* [P4]
Basalt (3)	0.4 –0.5	0.43	Laul and Schmitt [P4]
Basalt (3)	0.31–0.45	0.37	Wänke *et al.* [P6]
Basalt (12)	0.31–(1.84)	0.53	Ma *et al.* [P7]
Norite (2)	2.1 –3	2.6	Fruchter *et al.* [P4]
Breccia (2)	0.56–1.4	1.0	Fruchter *et al.* [P4]
Breccia (6)	0.09–2.0	1.0	Laul and Schmitt [P4]
Fines (6)	1.1 –1.7	1.3	Fruchter *et al.* [P4]
Fines (4)	0.46–1.2	0.81	Wänke *et al.* [P3] [P6]
Fines (4)	0.60–1.2	0.85	Laul *et al.* [P3]
Fines (9)	0.75–1.2	1.1	Laul and Schmitt [P4]
Fines (6)	0.46–1.2	0.79	Wänke *et al.* [P4]
Apollo 16			
KREEP-basalt (1)	—	2.28	Brunfelt *et al.* [P4]
Gabbroic anorthosite (1)	—	0.29	Boynton *et al.* [P6]
Anorthos. gabbro (1)	—	1.02	Wänke *et al.* [P4]
Anorthosite (5)	<0.01–0.42	<0.09	Laul and Schmitt [P4]
Anorthosite (1)	—	0.011	Wänke *et al.* [P6]
Anorthosite (1)	—	<0.02	Laul *et al.* [P5]
Breccia (3)	0.14–0.45	0.34	Brunfelt *et al.* [P4]
Breccia (5)	0.14–0.73	0.35	Laul and Schmitt [P4]
Breccia (3)	0.4 –0.6	0.5	Fruchter *et al.* [P5]
Breccia (3)	0.83–2.3	1.7	Laul *et al.* [P5]
Breccia (9)	0.03–2.37	0.95	Wänke *et al.* [P5] [P6]
Breccia (5)	0.12–0.88	0.58	Warner *et al.* [P7]
Fines (2?)	—	0.47	Brunfelt *et al.* [P4] [P5]
Fines (11)	0.50–0.64	0.58	Laul and Schmitt [P4]
Fines (9)	0.22–0.55	0.45	Wänke *et al.* [P4] [P6]
Fines (11)	0.3 –0.6	0.47	Fruchter *et al.* [P5]
Fines (4)	0.5 –0.8	0.7	Finkelman *et al.* [6]
Fines (7)	0.14–0.61	0.47	Boynton *et al.* [P6]
Fines (2)	0.4 –0.6	0.5	Rhodes *et al.* [P6]
Apollo 17			
Basalt (high-Ti) (5)	1.54–1.81	1.6	Brunfelt *et al.* [P5]
Basalt (2)	1.6 –2.2	1.9	Boynton *et al.* [P6]
Basalt (3)	1.6 –2.6	1.9	Laul *et al.* [P5]
Basalt (1)	—	1.3	Baedecker *et al.* [P5]
Basalt (6)	1.55–2.01	1.7	Wänke *et al.* [P5] [P6]
Gabbro (1)	—	2.0	Boynton *et al.* [P6]
Gabbro (1)	—	1.1	Baedecker *et al.* [P5]
Breccia (1)	—	0.87	Brunfelt *et al.* [P5]
Breccia (8)	0.22–1.6	1.2	Laul *et al.* [P5]
Breccia (6)	0.94–1.82	1.3	Wänke *et al.* [P5] [P3]
Breccia (1)	—	0.79	Baedecker *et al.* [P5]

Table 73-C-2 (continued)

Rock type	Range ppm Ta	Average ppm Ta	Reference
Breccia (1)	—	1.8	Boynton *et al.* [P6]
Breccia matrix (1)	1.0 –1.8	1.4	Blanchard *et al.* [P7]
Orange glas (1)	—	0.9	Wasson *et al.* [P7]
Fines (5)	3 –10	5.2	Laul and Schmitt [P4]
Fines (< 1 mm) (12)	0.75–1.3	0.95	Laul *et al.* [P5]
Fines (9)	0.68–1.2	0.82	Wänke *et al.* [P5]
Fines (10)	0.57–1.4	0.84	Baedecker *et al.* [P5]
Fines (< 1 mm) (4)	0.7 –1.9	1.1	Blanchard *et al.* [P6]

Ta abundances in *fines* for the same sequence of missions are: 1.4, 1.4, 4.0, 1.0, 0.51, 1.4 ppm Ta respectively. These two series of data indicate some correlation between breccias and fines in their Ta content. A comparable correlation with the Ta concentration in basalts from different sampling areas is not observable. In some areas the tantalum concentration in fines and breccias exceeds that of the local magmatic rocks, especially in Apollo 14 samples. The latter might be due to a higher KREEP proportion. The Nb/Ta ratios of breccias and fines scatter over an appreciably larger range than those for basalts. Apollo 11 fines and Apollo 14 breccias are as low as 13.9 in their Nb/Ta ratio and Apollo 12 fines as high as 27.7. The average Nb/Ta ratio of fines from all missions is 19.9 and exceeds the Nb/Ta ratio of lunar basalts (12.5). A close correlation between Nb and RE elements has been demonstrated by Taylor (1975; Fig. 5.19).

Manuscript received: April 1977

73-D. Abundance in Rock-Forming Minerals, Tantalum Minerals, Phase-Equilibria

I. Rock-Forming Minerals

Tantalum occurs in igneous rocks in two ways: i) commonly it is dispersed as a camouflaged element in oxide and silicate minerals of iron and iron-titanium (less commonly in zirconium silicates). ii) In rare rocks, particularly complex pegmatites and certain alkalic, aluminous, or metasomatized granites ("apogranites") it may appear in its own mineral species (tantalite, euxenite, fergusonite, microlite).

The most common host minerals for vicarious tantalum include pyroxene (Table 41-D-4), amphibole (Tables 41-D-1/4), biotite (Tables 41-D-1/2), muscovite (Table 41-D-3), magnetite (Table 41-D-4), ilmenite (Tables 41-D-1/4), sphene (Tables 41-D-1/4), rutile, and perovskite (Table 41-D-4). Another accessory mineral that carries tantalum is cassiterite (Table 41-D-5). The various schemes of isomorphous substitution in niobium-tantalum minerals are listed in Table 41-D-9. RANKAMA (1944) has contributed numerous data on Ta in rare minerals.

Table 73-D-1. *Tantalum averages and Nb/Ta ratios in rock-forming minerals from magmatic rocks* (Number of samples in brackets)

Mineral	Average ppm Ta	Range Nb/Ta	Average Nb/Ta	Reference
Biotite (163)	10.1	1.7–37	11.3	Tables 41-D-1/2, SITNIN (1966)[b]
Muscovite (249)	14.2	0.6–30	5.6	Table 41-D-3, SITNIN (1966)[b], BEUS (1966)[c]
Pyroxene (29)	6.5	(2.5–12.9)[a]	6.5	Table 41-D-4
Hornblende (29)	7.2	(2.6–12.8)[a]	9.8	Table 41-D-4, BEUS and SITNIN (1968)
Sphene (28)	330	(1.1–21.3)[a]	11.5	Tables 41-D-1/4, BEUS and SITNIN (1968)
Titanomagnetite (9)	100	(5.3–10)[a]	6.8	Table 41-D-4, BEUS and SITNIN (1968)
Ilmenite (18)	250	(3.3–6.6)[a]	6.4	Tables 41-D-1/4, BEUS and SITNIN (1968)
Perovskite (62)	477	(8.5–16.8)[a]	11.3	Table 41-D-4
Zircon (6)	38	(24 –110)[a]	1.0	BEUS and SITNIN (1968)

[a] Range of group averages (partly).
[b] Unaltered granites, granodiorites and two-mica granites from the U.S.S.R.
[c] 150 muscovites from granitic pegmatites; concentrations $\geq$20 ppm Ta in muscovites from rare-metal pegmatites containing tantalite-columbite.

Table 73-D-2. *Tantalum in coexisting hornblende and biotite from magmatic rocks of the southern California batholith* (After GOTTFRIED and DINNIN, 1965)

Mineral	Parent rock	Tantalum (ppm)		Distribution coefficient[a] Ta in biotite/ Ta in hornblende
		W/C	N/R	
Hornblende	Gabbro (hornblende norite), San Marcos		0.17	
Hornblende	Tonalite, Green Valley	0.7		2.3
Biotite	Tonalite, Green Valley	1.6		2.3
Hornblende	Tonalite, Lakeview Mountain	0.8	0.88	2.1
Biotite	Tonalite, Lakeview Mountain	1.7	1.64	2.1
Hornblende	Tonalite, Bonsall	0.9		1.7
Biotite	Tonalite, Bonsall	1.5		1.7
Hornblende	Granodiorite, Woodson Mountain	2		1
Biotite	Granodiorite, Woodson Mountain	2	1.62	1
Hornblende	Quartz monzonite, coarse phase of Rubidoux Mountain	18	8.8	0.4
Biotite	Quartz monzonite, coarse phase of Rubidoux Mountain	6.6	6.4	0.4
Hornblende	Quartz monzonite, coarse phase of Rubidoux Mountain	10		0.5
Biotite	Quartz monzonite, coarse phase of Rubidoux Mountain	4.8		0.5
Hornblende	Quartz monzonite, fine phase of Rubidoux Mountain	12		0.5
Biotite	Quartz monzonite, fine phase of Rubidoux Mountain	6		0.5

[a] Calculated from chemical analyses.

ZNAMENSKII (1957), ZNAMENSKII *et al.* (1962), ZNAMENSKII and POPOLUTOV (1964), and ZNAMENSKII *et al.* (1968) in investigating the geochemistry of niobium and tantalum in granitoids distinguished between i) monazite-bearing granites with accessory ilmenite and rutile and ii) allanite-bearing granites with magnetite and sphene. In the first type the bulk of niobium and tantalum is concentrated in biotite, even though the ilmenite contains higher percentages of these elements. In the second type most of the niobium and tantalum is in sphene. For each of the most important host minerals of Nb and Ta in magmatic rocks a large range of different

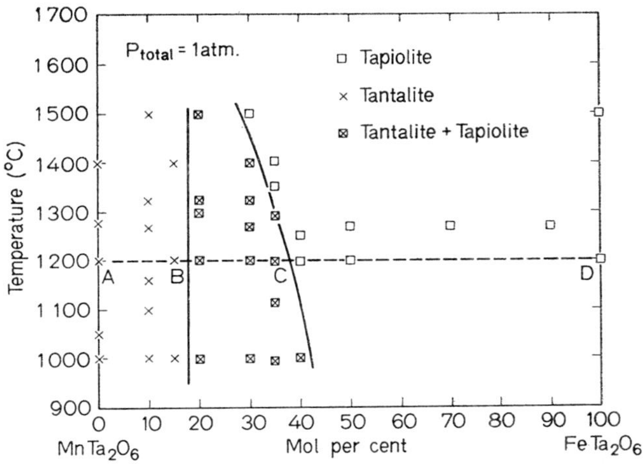

Fig. 73-D-1. The system $MnTa_2O_6$-$FeTa_2O_6$. The data points present the results of synthesis for various lengths of time, plus reversals of the equilibrium along the tapiolite field boundary (TURNOCK, 1965)

Nb/Ta ratios has been observed. Therefore the specific tendency of certain crystal structures to fractionate Nb and Ta is not yet clearly visible (Table 73-D-1).

In rocks of the composite Southern California batholith most of the tantalum occurs in biotite and coexisting hornblende (80% of the total Ta in tonalite and quartz monzonite). Sphene has the greatest tantalum content (36–100 ppm Ta) of any mineral of the batholith. There is a considerable increase in the tantalum content of the least mafic rock series, the Rubidoux Mountain quartz monzonite, over those of the older batholithic units of granodiorite, tonalite, and gabbro. The distribution coefficient of Ta between biotite and hornblende decreases systematically from tonalite to quartz monzonite (Table 73-D-2). BEUS and SITNIN (1968) state that biotites and muscovites of postmagmatically altered rocks (two-mica and muscovite granites) from the USSR contain 2 to 3 times as much tantalum as biotite from unaltered granites.

Feldspars are low in tantalum. BUMA *et al.* (1971) list 0.01 ppm Ta in plagioclase and 0.03 ppm Ta in microcline of granite from Narragansett Pier and 0.01 ppm Ta in microcline of granite from Westerly, New England (USA) (analytical method: N/R). KOSALS and MAZUROV (1970) report 0.4, 0.4 and 1.6 ppm Ta and Nb/Ta ratios 16, 16 and 8 in three samples of oligoclase from granites of the Bitu-Dzhida intrusion in SW Baykalia (USSR) (method: W/C). Potash feldspars of these rocks are comparable in their Ta content but have a Nb/Ta ratio of 8. It cannot be excluded that even these low Ta concentrations are due to intergrowth of feldspars with minerals higher in Ta.

II. Tantalum Minerals

Tantalum minerals are represented mainly by complex oxides and hydroxides (Tables 41-D-7, 41-D-8). There exists one borate. Tantalites are the most common Ta minerals. Stibiotantalite, euxenite, microlite, formanite, ixiolite and thoreaulite

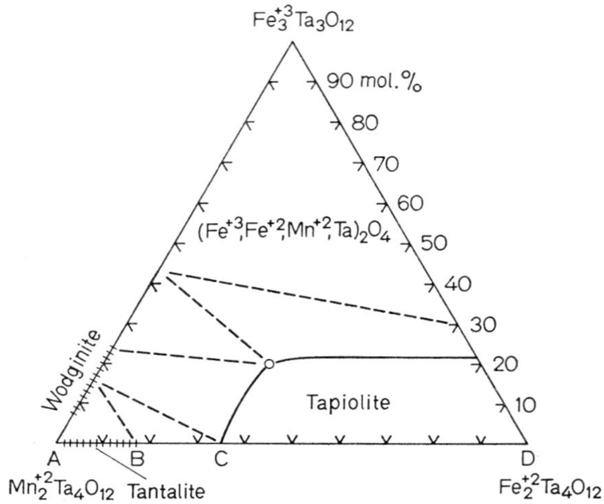

Fig. 73-D-2. The system $Fe_3Ta_3O_{12}$-$Mn_2Ta_4O_{12}$-$Fe_2Ta_4O_{12}$ at 1,200° C, total pressure 1 atm. Field boundaries in the ternary are drawn by extrapolation from the sides, plus projection from the system $MnTa_2O_6$-$FeTa_2O_6$-O for the tapiolite field. Points A, B, C, D are at the field boundaries of tantalite and tapiolite as in Fig. 73-D-1 (TURNOCK, 1965)

occur occasionally in certain rare metal pegmatites etc. The number of tantalum minerals is substantially less than the total of dominantly niobium minerals.

The different schemes of isomorphous substitution in niobium-tantalum bearing minerals are listed in Table 41-D-9.

III. Phase Equilibria

The system $MnTa_2O_6$-$FeTa_2O_6$ has been investigated by TURNOCK (1965). The conditions of stability of tapiolite and tantalite are plotted in Fig. 73-D-1. TURNOCK (1965) has additionally studied the oxidation of tantalites and tapiolites at various partial pressures of oxygen. The tantalite end-member, $MnTa_2O_6$, remained stable under the most oxidizing conditions (1 atm P_{O_2}). The iron compounds, in contrast, oxidized:

$$2\,Fe^{2+}Ta_2O_6 + 1/2\,O_2 \rightleftharpoons 2\,Fe^{3+}TaO_4 + Ta_2O_5$$

At medium partial pressures of oxygen tapiolites were partly oxidized, with $FeTaO_4$ remaining in solid solution via the substitution:

$$Fe_2Ta_4O_{12} \leftrightarrow Fe_3Ta_3O_{12}$$

Complete oxidation of the iron in materials containing both iron and manganese produced wodginite and (Fe, Mn) TaO_4. Fig. 73-D-2 shows a triangular join of three crystalline phases, wodginite + tantalite + tapiolite, which may exist at univariant equilibrium. The formation of wodginite requires an oxygen partial pressure greater than that of gas with CO: $CO_2 = 1:10$ as well as a chemical composition enriched in Mn, Ta, and Fe. The presence of wodginite in pegmatites is an indication of an oxidizing environment.

Phase equilibria within the columbite tapiolite series and of this group with euxenite [$Y(Ti, Nb, Ta)_2O_6$] have been studied by SCHRÖCKE (1966).

Manuscript received: April 1977

73-E. Abundance in Common Igneous Rock Types, Crustal Abundance

Biotite, amphibole and Ti minerals are the major host phases of tantalum in magmatic rocks (Sect. 73-D). Magmatic biotite contains about 10 ppm Ta (Table 41-D-2). Granitic rocks, the most abundant crustal constituents, often have 5 to 13% biotite. 10 percent biotite can contribute 1 ppm Ta which is about half the tantalum concentration of an average granitic rock (2.3 ppm Ta according to Table 41-E-3).

A major problem is that of the reliability of the analytical data on Ta in magmatic rocks. The relatively large spread of Nb/Ta ratios within rock classes might partly be due to some analytical uncertainty.

I. Standard Reference Rocks

Because of the low Ta concentration in common magmatic rocks, being often close to 1 ppm, routine methods of optical spectrography and X-ray spectrometry lack sensitivity for this type of analysis. Chemical preconcentration in combination with spectrum analysis has been applied by several authors. One of the earlier extensive investigations on this base is the paper by RANKAMA (1944). More sensitive and reliable are neutron activation procedures. The following authors list neutron activation values on the reference basalt BCR-1 in connection with their data on lunar rocks: HASKIN *et al.* [P1][1] (0.9 ppm Ta), GOLES *et al.* [P2][1] (0.99 ppm Ta), BRUNFELT *et al.* [P2][1] (0.74 ppm Ta), LAUL *et al.* [P3, P5][1] (0.74; 0.85 ppm Ta), BLANCHARD *et al.* [P6][1] (0.91 ppm Ta).

Table 73-E-1. *Recommended values, averages and magnitudes for selected international reference rocks* (FLANAGAN, 1973)

Symbol	Rock type	ppm Ta[a]	Symbol	Rock type	ppm Ta[a]
G-1	Granite (USGS)	1.5	GM	Granite (ZGI)	2.1[b]
G-2	Granite (USGS)	0.91	G*	Granite (NIM)	*4*[c]
GSP-1	Granodiorite (USGS)	1.0	JG-1	Granodiorite (GSJ)	*3*[c]
AGV-1	Andesite (USGS)	0.9	S*	Syenite (NIM)	*0.3*[c]
BCR-1	Basalt (USGS)	0.91	JB-1	Basalt (GSJ)	*4*[c]
W-1	"Diabase" (USGS)	0.50	BM	Spilite (ZGI)	*0.4*[c]
			N*	Norite (NIM)	*0.4*[c]
			PCC-1	Peridotite (USGS)	<0.1[b]
			DTS-1	Dunite (USGS)	<0.1[b]

[a] Recommended values; [b] Average; [c] *Magnitude* (italics).

[1] For abbreviations of references see Table 73-C-2.

A detailed paper on the instrumental activation analysis of standard rocks with results rather close to the recommended values of Table 73-E-1 has been published by GORDON *et al.* (1968). Only for W-1 (0.67 ppm Ta) and GSP-1 (1.4 ppm Ta) these authors deviate more than 10 percent from the recommended values.

FLANAGAN (1973) has compiled recommended values. averages and magnitudes for magmatic reference rocks listed in Table 73-E-1.

II. Common Magmatic Rocks, Excluding Alkalic Types

A reasonably large number of analytical data on the Ta concentration in non-alkalic granites has been listed in Tables 41-E-2 and 41-E-3. Information of Table 41-E-2 was partly compiled by PARKER and FLEISCHER (1968). Averages from these and additional data are reported in Table 73-E-2.

Several authors like GERASIMOVSKIY and KARPUSHINA (1965) have checked the analytical determinations and mentioned the existence of inaccurate procedures which might partly influence the large range of variations in absolute values and

Table 73-E-2. *Group averages of Ta concentrations and Nb/Ta ratios in common non-alkalic magmatic rocks*

Rock type, origin	Number of analyses (Analytical method)	ppm Ta	Nb/Ta	Reference
Calc alkalic granitic rocks, worldwide sampling	299 (W/S, W/C, S)	2.1	10.5	Tables 41-E-2/3
Granites, U.S.S.R.	180 (W/C)	2.5 (±0.2)	14.5	BEUS and SITNIN (1968)
Granites, New England (U.S.A.)	5 (N/R)	2.6	—	BUMA *et al.* (1971)
Gabbros, Greenland, U.S.A., U.S.S.R.	30 (N/R, W/C, S)	0.95	7.2	Table 41-E-9
Oceanic tholeiitic basalts, Iceland	3 (N/R)	2.0	6.9	WOOD *et al.* (1976)
Tholeiitic basalts, U.S.A., Australia	18 (N/R)	0.55	13.4	GOTTFRIED *et al.* (1968)
Tholeiitic basalts, U.S.A.	2 (N/R)	0.71	16.2	Tables 41-E-1, 73-E-1
Ferrobasalts, Snake River (U.S.A.)	7 (N/R)	6.3	—	LEEMAN *et al.* (1976)
Ultramafic rocks, U.S.A.	2 (N/R)	<0.1	—	Table 73-E-1
Ultramafic rocks, Kola Peninsula (U.S.S.R.)	3 (W/C)	0.55	10.8	GERASIMOVSKIY and KARPUSHINA (1965)
Dunites	2 (N/R)	0.018	—	ATKINS and SMALES (1960)
Pyroxenites, Yakutiya (U.S.S.R.)	4 (W, S)	0.8	5	NEKRASOV (1970)

in Nb/Ta ratios to be observed in several sets of data which have been used for averages.

Regional investigations as that by Beus and Sitnin (1968) on Ta in *granites* from different areas of the USSR have demonstrated that comparable rock associations in different provinces can contain specific tantalum abundances. In some examples the Nb and Ta contents in the same rock type between massifs show greater differences than the average contents of different rock types within one massif.

For some common rock types as *rhyolites*, *andesites* and different *basaltic varieties* the data for tantalum are too few to compute meaningful averages and average Nb/Ta ratios.

A comparable lack of information on tantalum in upper mantle rocks must be mentioned. The few data listed in Table 73-E-2 indicate that the average Ta content of *peridotites* is probably lower than 0.1% Ta. Pyroxene, phlogopite and ilmenite are expected to be the major host minerals of Ta in peridotites. Fesq *et al.* (1975) have reported a large range from 50 to 500 ppm Ta with abundant values at 100 ppm Ta in ilmenite from mantle rocks. Rankama (1944) has listed 12 out of 15 ultramafic rocks to contain less than 0.8 ppm Ta. His 3 samples with more than 0.8 ppm Ta are pyroxenites and a titanomagnetite rich peridotite from the Kola alkalic massif. Ordinary chondrites are as low as 0.02 ppm Ta (Table 73-C-1). Data on Ta concentrations in kimberlites are listed in Subsect. 73-E-III.

A few investigations exist on the behavior of tantalum within *rock series* with potential relations from differentiation. Gottfried *et al.* (1965) have studied the series from gabbro, tonalite, granodiorite to quartz monzonite of the southern California batholith. They observed an increase from 0.19 ppm Ta in gabbro to 0.45 ppm Ta in tonalite. Granodiorite and quartz monzonite contain tantalum at about the same level as tonalite (40 ppm Ta). Ta in hornblende from this rock sequence increases from gabbro to quartz monzonite, in biotite from tonalite to quartz monzonite and in sphene from tonalite to granodiorite. In the Skaergaard gabbro intrusion in East Greenland Ta increases from the chilled marginal gabbro (0.46 ppm Ta) to ferrogabbro (6 samples: 1.05 ppm Ta) to hedenbergite granophyre (3.53 ppm Ta; analytical method: N/R) (Atkins and Smales, 1960).

III. Alkalic Rocks

Both niobium and tantalum are enriched in alkalic rocks. This is already observable in a comparison of average calc alkalic and *alkalic granitic rocks*. The former contain 2.25 ppm Ta (484 samples of Table 73-E-2) and for the latter a mean of 3.9 ppm Ta has been computed (58 samples of Table 41-E-4). The average Nb/Ta ratio of calc alkalic granitic rocks is 10 to 12 being lower than that of alkalic species which have a ratio of 15. A representative Ta value for alkali rhyolites cannot be computed yet. Syenites and trachytes are probably close to 13 ppm Ta and a Nb/Ta ratio of 11 (Table 41-E-8); but these are averages of a few data.

Alkalic rocks containing nepheline are highly variable in Ta within and between rock types (Table 41-E-5 and 7) and from one to the other locality (Tables 41-E-5, 6 and Table 73-E-3). Table 41-E-6 and a compilation of Ta in alkalic rocks of the USSR by Borodin *et al.* (1969) (Table 73-E-3) proves nepheline syenites from the Lovozero complex as well as miaskitic alkalic complexes of the Kola Peninsula

Table 73-E-3. *Tantalum content in the principal nepheline syenite massifs of the U.S.S.R.* (According to BORODIN *et al.*, 1969; Analytical method: W/C)

Massif	Number of samples	Range ppm Ta	Average ppm Ta
I. Alkalic-granitoid group			
Kola	1	—	35.2
Ukraine	2	16.4–32.8	24.8
Urals	8	1.3–21.2	7.7
Kazakhstan	1	—	5.9
Middle Asia	10	0.7– 6.6	3.2
Tuva	61	0.3– 6.6	2.7
Yenisey Range	1	—	11.5
Northern Baykal	12	0.6–16.4	8.7
Vitim Plateau	14	0.8– 7.4	2.2
Chukotsk Peninsula	1	—	1.5
II. *Potassium-basaltoid group*			
Caucasus	21	1.7– 1.9	1.8
Kazakhstan	2	0.5– 1.8	1.1
Middle Asia	>3	1.1– 1.7	1.4
Northern Baykal	54	0.3– 4.1	1.9

(USSR) being dominant in high Ta concentrations. The average tantalum content generally seems to be higher in agpaitic than in miaskitic nepheline syenites. But Nb-Ta minerals (pyrochlore) are typical of the latter varieties (GERASIMOVSKIY and KARPUSHINA, 1966). Within a nepheline syenite massif as Khibiny and Lovozero (USSR) ijolite-urtites (BORODIN *et al.*, 1969), lujavrites and ijolite-melteigites (GERASIMOVSKIY *et al.*, 1966) are highest in Ta and reasonably low in Nb/Ta ratios.

Meaningful Ta averages of nepheline bearing alkalic rocks are difficult to obtain because of the large spread of values. Giving equal statistical weight to every value the average of all nepheline syenites, phonolites and other nepheline-bearing rocks of Table 41-E-5 would be 50 ppm Ta. But rare rock species, a high-Ta area (Kola) and the availability of Ta data which favour rocks relatively rich in Ta, have a dominant influence on this high average. BORODIN *et al.* (1969) in their survey of 80 nepheline syenite and ijolite massifs as the principal alkalic provinces of the Soviet Union (Table 73-E-3) compute an average tantalum content in these alkalic rocks of 8 ppm. An average Nb/Ta ratio, computed for 120 nepheline-bearing rocks of Table 41-E-5 is 12.4. This ratio must be too low according to the above mentioned parameters influencing a high Ta average for this table. Therefore it is to be assumed that tantalum is specifically accumulated in alkalic rocks relative to niobium.

Several Ta data on *kimberlites*, which are alkalic peridotites, are available. MITCHELL and BRUNFELT (1975) observed a range from 6 to 19 ppm Ta in 15 kimberlites, mainly from well known South African localites. Their average is 11.2 ppm Ta (analytical method: N/R). KABLE *et al.* (1975) report a Ta range from 1.4 to 21.3 ppm for 59 kimberlite samples from the Premier Mine and the Kimberley fissure and pipe materials (S. Africa). These contain on average 5.9 ppm Ta (analytical method: N/R) and a Nb/Ta ratio of 13.8. The Nb/Ta ratio varies from 7 to 23 for groups of the latter kimberlites whereas the Ti/Ta ratio ranges from 410 to 1,900. Relative large ranges of Ta concentrations might be due to the variable Ta content in ilmenites from the source (see Section 73-D) and the inhomogeneous distribution of ilmenite.

IV. Crustal Abundance

The average tantalum concentration in magmatic rocks of the upper continental crust and a related Nb/Ta ratio have been tentatively computed in Table 73-E-4 on the basis of figures from Table 73-E-2, Subsections 73-E-II and III, Table 41-E-11 and Table 7–8 (Vol. I). The final average of 2.1 ppm Ta is the same as that published by RANKAMA (1944) on a much smaller data base.

Table 73-E-4. *Upper crustal abundance of tantalum in magmatic rocks*

Magmatic rocks	Rock abundance in vol. %	ppm Ta	(Nb/Ta)[a]	(Nb/Ta)[b]
Granites, quartz monzonites, granodiorites, quartz diorites	86	2.3	9.8	12.0
Diorites	1	(1.5)		
Gabbros, basalts (tholeiitic)	13	0.9	11.	9.6
Syenites, alkalic rocks	<0.5	~10.	~10.	>11.7
Peridotites	<0.5	< 0.1	>15.	(10.8)
Average	~100	2.1	10	11.6

[a] Using all the available Nb data (Table 41-E-11).
[b] Using only samples for which both Nb and Ta data are reported (Table 73-E-2).

Manuscript received: April 1977

73-F. Behavior in Magmatogenic Processes

Tantalum generally tends to become more abundant with increasing SiO_2 content during magmatic differentiation within series which can be consanguineous (s. Subsect. 73-E-II). Biotite concentrates Nb relative to Ta. Therefore a descrease of biotite in leucocratic rocks can be partly responsible for the increase in Ta relative to Nb in late-stage differentiation products (PODOL'SKIY *et al.*, 1965). BEUS and SITNIN (1968) have demonstrated with data on granitic rocks and their minerals from the USSR that Ta contents of granites can be related to effects of high-temperature postmagmatic metasomatism.

The highest Ta concentrations appear in pegmatites, aplites, greisens and albitized granitoids, in which independent tantalum minerals may be formed.

Only *granite pegmatites* are known to contain independent tantalum minerals and deposits. In these pegmatites the concentration of tantalum can be comparable with that of niobium and may occasionally exceed that of the latter. Their content and the Nb/Ta ratio depends on both the original composition of the melt and a potential secondary alteration (autometasomatism). Highest Ta and Nb concentrations are found in effectively replaced pegmatites preferably with albite as prominent constituent. Generally the Nb/Ta ratio of rare-metal granite pegmatites varies from 0.3 to 3 and the Ta concentration may approach a level of several hundred ppm (VLASOV, 1966). In low-temperature fully differentiated rare-metal pegmatites Nb-Ta minerals are represented mainly by tantalocolumbite and niobotantalite. In highly replaced pegmatites Ta frequently predominates over Nb and beside tantalite (manganotantalite etc.) the following minerals might occur: microlite, djalmaite, thoreaulite, ixiolite, stibiotantalite, simpsonite. The Nb/Ta ratio is generally lowest for the central zones of replaced lithium-bearing complexes (with lepidolite and Li-bearing micas). In the latest mica replacements of granite pegmatites the Ta concentration may occasionally approach 3,000 ppm Ta (VLASOV, 1966). So of all comparable materials only granite pegmetites and granite greisens, mostly of lithium provinces, can here and there be of economic importance as a source of tantalum (beside Be, Li, Cs, Nb). The principal economic Ta minerals of these pegmatites are tantalite and its varieties microlite and (with lesser importance) simpsonite.

BEUS (1966) has contributed 150 analyses of granitic pegmatite muscovites to confirm some of the former statements. Muscovites with more than 20 ppm Ta are restricted to tantalite-columbite containing rare-metal pegmatites. For additional details on the Ta distribution in pegmatites see GMELIN (1969).

Independent Ta minerals are not formed in *alkali pegmatites* genetically associated with nepheline syenites etc. Here tantalum is incorporated in Ti, Zr, Fe, and Nb minerals. The Ta concentration in these pegmatites generally does not exceed 8 ppm (VLASOV, 1966). In *carbonatites* tantalum mainly occurs in the minerals of the pyrochlore and perovskite group. Their Ta content varies from <0.1 to 6.5% and their Nb/Ta ratio is commonly high (in pyrochlore often 60; references of Table 212 in VLASOV, 1966).

In *greisen* zones of small granite intrusions with an extensive occurrence of lithium bearing micas (lepidolite, lithian muscovite, zinnwaldite) and topaz economic Ta-mineral deposits can be formed. SITNIN (in VLASOV, 1965) reports 8 to 250 ppm Ta (Nb/Ta ~2) in greisenized amazonite granites from Siberia. According to RANKAMA (1944) greisens from Zinnwald contain 16 ppm Ta.

For additional information on the paragenesis of pegmatitic niobium-tantalum minerals see Sect. 41-F.

Manuscript received: April 1977

73-G. Behavior during Weathering and Rock Alteration

Tantalum occurs as Ta^{5+} in natural compounds mainly replacing Fe, Ti, Nb etc. in mica, hornblende, Ti minerals, Nb-Ta minerals etc. (see Sect. 73-D). During the partial destruction of magmatic and metamorphic minerals in the weathering cycle Ta will be distributed between newly formed and residual minerals.

Abundant *placer* minerals which have incorporated some Ta (see Tables 41-D-1, 4, 5 and 73-D-1) are: rutile, ilmenite, cassiterite, wolframite. Alluvial placers neighboring Ta rich rocks (pegmatites, greisens etc.) can approach economic proportions of tantalite (columbite, euxenite, pyrochlore-microlite).

RONOV and MIGDISOV (1965) in their investigation of the behavior of Ti, Zr, Hf, Nb, Ta and Th during weathering have observed the following Ta gradients from unaltered to weathered rocks (Ta concentrations in ppm, Nb/Ta ratios in brackets): pyroxenite 0.1 (7) to 1.1 (29); amphibolite 0.2 (11.5) to 0.7 (9.4); anorthosite 1.3 (7.4) to 2 (8.5); gabbro 5 (4) to 5 (15.4); granite 0.4 (7.3) to 7.9 (1.4). Ta together with the above mentioned elements increases from the fresh rock to the weathering products indicating low solubility relative to alkali, earth alkali and several other major elements. In granite the Nb/Ta ratio decreases during weathering, therefore Nb must have been more mobile than Ta. For pyroxenite and gabbro an opposite trend is to be observed. The different behavior must be due to the different host minerals in mafic and leucocratic rocks.

PACHADZHANOV (1963), who compared the behavior of tantalum under humid environment with that under arid conditions in clays of the Russian platform has observed a higher average Ta concentration of 2.4 ppm (Nb/Ta: 8.4) in clays from the humid environment. The products of arid conditions contain 0.9 ppm Ta and the higher Nb/Ta ratio of 14.8.

35 varieties of bauxites from the USSR contain on average 4.5 ppm Ta (Nb/Ta: 6.2) which is twice the crustal Ta abundance (PACHADZANOV, 1964). The Ta content varies from 0.2 to 28 ppm Ta; the latter value is related to bauxite derived from shale. 4 bauxites from Hungary, analyzed by the same author, contain on average 9.1 ppm Ta (Nb/Ta: 6.8).

Manuscript received: April 1977

73-I. Abundance in Natural Waters

Schutz and Turekian (1968) have proved that the concentration of tantalum in *sea water* samples from the 3 oceans is below the limit of detection of their neutron activation method, which means < 0.0025 ppb Ta. Perel'man and Batulin (1962: in Gmelin, 1969) report that the Ta abundance in natural waters is generally below 0.001 to 0.01 ppb.

Manuscript received: April 1977

73-K. Abundance in Common Sediments and Sedimentary Rock Types

There are a few data on the distribution of Ta in sediments available. In Table 41-K-1 354 shales and clays from the Russian Platform (PACHADZHANOV, 1963) are listed as containing on average 2 ppm Ta. This value and the Nb/Ta ratio of 9.1 is very close to the mean of magmatic rocks from the upper continental earth's crust (Table 73-E-4: 2.1 ppmTa; Nb/Ta 10–11.6). Pelitic materials from a humid environment are apparently higher in tantalum (2.4 ppm Ta) than those from arid conditions (0.9 ppm Ta). Nb and Ta concentrations correlate with Al, Ti and Zr. PACHADZHANOV (1963) assumes that half of the Ta content of clays and shales is contained in Ti minerals of the silt fraction and the other half in clay minerals.

HAMAGUCHI *et al.* (1963) report an average of 1.2 ppm for 3 clays from the Japan Sea (analytical method: N/R). 3 globigerina oozes contain 0.2 ppm Ta (0.1–0.4 ppm Ta). One sample of pelagic clay from the Mid-Pacific, analyzed by EHMANN (1965) with neutron activation, is lower than the above reported average, it has 0.58 ppm Ta. The only manganese nodule with information on its Ta content (0.17 to 2.3 ppm Ta, see Table 41-K-1) indicates a small Ta accumulation in the manganese oxide phases of deep sea deposits. PAREKH *et al.* (1977) report an average of 0.22 ppm Ta (range: 0.043 to 0.73 ppm Ta, method: N/R) for 24 limestones of Upper Jurassic age from southern Germany. A reasonably large proportion of this Ta is expected to occur in the non-carbonate residue (average: 8.7%).

Highest Ta concentrations are attained in placer deposits which have been supported from close-by greisens, alkalic rocks or granite pegmatites (see Sect. 41-D and K, 73-D).

Manuscript received: April 1977

73-L. Biogeochemistry

Some ascidians (*Molgula manhattensis*, *Styela plicata*) actively incorporate tantalum (and niobium). KOKUBU and HIDAKA (1965) report an average of 230 ppm Ta for dry material (method: W/C) which is about the same level as for Nb. Except this abnormal accumulation almost nothing is known about the Ta abundance in common organic matter.

RANKAMA (1944) reports 60 and 80 ppb Ta for two coal samples from England (gas coal) and East Karelia (shungite = anthracolite) respectively in addition to estimates of <200 and <20 for two coals from England and Germany. This gives an order of magnitude. HEADLEE and HUNTER (1955 in GMELIN, 1969) published an average of 49 ppm Nb for 590 coals from West Virginia (USA). A combination of the Nb and Ta data indicates an abnormal high Nb/Ta ratio of >500, which might be influenced by analytical errors.

Manuscript received: April 1977

73-M. Abundance in Common Metamorphic Rock Types

Two gneisses from western Keyvy (USSR) with 1.7 ppm Ta and a Nb/Ta ratio of 10.3 to 11.8 (GERASIMOVSKIY and KARPUSHINA, 1966) occur in a range of tantalum concentrations and Nb/Ta ratios which is expected to be typical for abundant crustal rocks (Table 73-E-4; Sect. 73-K). Additional information on Ta in this class of rock types is highly needed.

Manuscript received: April 1977

73-O. Economic Importance

The melting point of metallic Ta (3,015° C) is 650 degrees higher than that of niobium. Both metals have a high strength and resistancy against chemical attack and therefore are used as containers in the chemical industry. The main consumer of this metal is North America. In 1972 the United States produced 95% of their consumption as metal or alloy and 5% as carbide; 55% went into the electronics industry. The tantalum metal is chiefly used for condensors, electrodes, electronic tubes, heat exchangers, coolers, furnaces, coils, pumping devices, reaction vessels, medical instruments. Because of the stability of anodic oxide films on Ta it is used for electrolytic rectifiers and other electrical instruments. Tantalum carbide is extremely hard to be utilized in certain tools. The reactor industry and the manufacture of aircraft engines should be mentioned in connection with modern applications of tantalum or tantalum alloys.

Ta mineral deposits with relatively high Ta/Nb ratios are mainly connected with Sn-Li-bearing greisens and rare metal granite pegmatites (Zaire, Canada, Nigeria etc.) and their decomposition products in eluvial and alluvial placers (Nigeria, Zaire, Mozambique) (GOCHT, 1974). Otherwise Ta is the by-product of the processing of Nb-minerals from alkalic rocks. The main producers are Canada (1972: 425 sh.t concentrates with 50–60% Ta_2O_5, which is 46% of the world production; GOCHT, 1974), Brazil (235 sh.t) and Mozambique (120 sh.t). The largest reserves are expected to occur in Zaire, Nigeria, Canada and Brasil. A large proportion will be contributed from placers (60% as estimated by DE KUN, 1962).

Revised manuscript received: March 1977

References: Sections 73-B to G, I to M, O

ATKINS, D. H. F., SMALES, A. A.: The determination of tantalum and tungsten in rocks and meteorites by neutron activation analysis. Anal. Chim. Acta **22**, 462 (1960).
BEUS, A. A.: Distribution of tantalum and niobium in muscovites of granitic pegmatites. Geochem. Intern. **3**, 974 (1966).
BEUS, A. A., SITNIN, A. A.: Geochemical features of the distribution of tantalum in granitoids. Geochem. Intern. **5**, 490 (1968).
BORODIN, L. S., OSOKIN, E. D., BLYUM, I. A.: Average content and distribution of tantalum in alkalic rocks. Geochem. Intern. **6**, 1039 (1969).
BUMA, G., FREY, F. A., WONES, D. R.: New England granites: Trace element evidence regarding their origin and differentiation. Contrib. Mineral. Petrol. **31**, 300 (1971).
CAMERON, A. G. W.: A new table of abundances of the elements in the solar system. In: AHRENS, L. H. (ed.), Origin and Distribution of the Elements. Oxford: Pergamon Press 1968.
DE KUN, N.: The economic geology of columbium (niobium) and of tantalum. Econ. Geol. **57**, 377 (1962).
EHMANN, W.D.: On some tantalum abundances in meteorites and tektites. Geochim. Cosmochim. Acta **29**, 43 (1965).
FESQ, H. W., KABLE, E. J. D., GURNEY, J. J.: Aspects of the geochemistry of kimberlites from the Premier Mine and other selected South African occurrences with particular reference to the rare earth elements. Phys. Chem. Earth **9**, 687 (1975).
FLANAGAN, F. J.: 1972 values for international geochemical reference samples. Geochim. Cosmochim. Acta **37**, 1189 (1973).
GERASIMOVSKIY, V. I., KARPUSHINA, V. A.: On the relationship of Nb and Ta in igneous rocks. Geochem. Intern. **2**, 575 (1965).
GERASIMOVSKIY, V. I., KARPUSHINA, V. A.: Geochemistry of niobium and tantalum in nepheline syenites. Geochem. Intern. **7**, 609 (1966).
GERASIMOVSKIY, V. I., VOLKOV, V. P., KOGARKO, L. N., POLYAKOV, A. I., SAPRIYKINA, T. V. BALASHOV, Y. A.: Geochemistry of the Lovozero Massif. Akad. Nauk. S.S.S.R., Moscow 1966.
GMELIN, L.: Handbuch der Anorganischen Chemie (8. ed.) Tantal. Part A, 1. Weinheim: Verlag Chemie 1969.
GOCHT, W.: Handbuch der Metallmärkte. Berlin-Heidelberg-New York: Springer 1974.
GORDON, G. E., RANDLE, K., GOLES, G. G., CORLISS, J. B., BEESON, M. H., OXLEY, S. S.: Instrumental activation analysis of standard rocks with high-resolution γ-ray detectors. Geochim. Cosmochim. Acta **32**, 369 (1968).
GOTTFRIED, D., DINNIN, J. I.: Distribution of tantalum in some igneous rocks and coexisting minerals of the southern California batholith. U.S. Geol. Surv. Profess. Papers **525-B**, B96 (1965).
GOTTFRIED, D., GREENLAND, L. P., CAMPBELL, E. Y.: Variation of Nb-Ta, Zr-Hf, Th-U and K-Cs in two diabase-granophyre suites. Geochim. Cosmochim. Acta **32**, 925 (1968).
HAMAGUCHI, H., KURODA, R., WATANABE, Y.: Chemical investigations of deep sea sediments 29. Tantalum content of deep sea sediments (in Japanese). J. Chem. Soc. Japan, Pure Chem. Sect. **84**, 723 (1963).
KABLE, E. J. D., FESQ, H. W., GURNEY, J. J.: The significance of the inter-element relationships of some minor and trace elements in South African kimberlites. Phys. Chem. Earth **9**, 709 (1975).
KOKUBU, N., HIDAKA, T.: Tantalum and niobium in ascidians. Nature **205**, 1028 (1965).
KOSALS, Y. A., MAZUROV, M. P.: Behavior of molybdenum, tungsten, tin, niobium, and tantalum in the emplacement of the Bitu-Dzhida granite intrusion in southwestern Baykalia. Geochem. Intern. **7**, 506 (1970).

LEEMAN, W. P., VITALIANO, C. J., PRINZ, M.: Evolved lavas from the Snake River Plain: Craters of the Moon National Monument, Idaho. Contrib. Mineral. Petrol. **56**, 35 (1976).

MITCHELL, R. H., BRUNFELT, A. O.: Rare earth element geochemistry of kimberlite. Phys. Chem. Earth **9**, 671 (1975).

NEKRASOV, I. Y.: Distribution of tantalum and niobium in magmatic and postmagmatic rocks and minerals of the northeastern USSR. Geochem. Intern. **7**, 378 (1970).

PACHADZHANOV, D. N.: Geochemistry of niobium and tantalum in clays. Geochemistry **10**, 963 (1963).

PACHADZHANOV, D. N.: Geochemical bond between tantalum, niobium, and titanium in bauxites. Geochem. Intern **1**, 889 (1964).

PAREKH, P. P., MÖLLER, P., DULSKI, P., BAUSCH, W. M.: Distrubution of trace elements between carbonate and noncarbonate phases of limestone. Earth Planet. Sci. L. **34**, 39 (1977).

PARKER, R. L., FLEISCHER, M.: Geochemistry of niobium and tantalum. U.S. Geol. Surv. Profess. Papers **612** (1968).

PODOL'SKIY, A. M., NUMVEROV, S. V., GOLIKOV-ZAVOLZHENSKIY, I. V., MINTS, M. V., LARIN, V. N.: Tantalum in alaskites and subalkalic granites of eastern part of Central Kazakhstan. Geochem. Intern. **2**, 437 (1965).

RANKAMA, K.: On the geochemistry of tantalum. Bull. Géol. Comm. Finlande **133**, 1 (1944).

RONOV, A. B., MIGDISOV, A. A.: Principal features of the geochemistry of hydrolyzate elements in weathering and sedimentation. Geochem. Intern. **2**, 92 (1965).

SCHRÖCKE, H.: Über Festkörpergleichgewichte innerhalb der Columbit-Tapiolitgruppe sowie der Columbit-Tapiolitgruppe mit YTi(Nb, Ta)O_6, Euxenit und mit $FeNbO_4$. Neues Jahrb. Mineral. Abhandl. **106**, 1 (1966).

SCHUTZ, D. F., TUREKIAN, K. K.: The investigation of the geographical and vertical distribution of several trace elements in sea water using neutron activation analysis. Geochim. Cosmochim. Acta **29**, 259 (1965).

SITNIN, A. A.: Tantalum and niobium concentration in granitoid micas of the USSR. Geochem. Intern. **3**, 843 (1966).

SUESS, H. E., UREY, H. C.: Abundances of the elements. Rev. Mod. Phys. **28**, 53 (1956).

TAYLOR, S. R.: Lunar Science: a Post-Apollo View. New York: Pergamon Press Inc. 1975.

TURNOCK, A. C.: Synthetic wodginite, tapiolite and tantalite. Canada Dept. Mines, Tech. Surveys, Min. Sci. Divis. International Rept. MS-65-31, 1 (1965).

VLASOV, K. A. (ed.): Geochemistry of Rare Elements, Vol. 1, p. 383. Israel Program for Scientific Translations, Jerusalem 1966.

WÄNKE, H., BADDENHAUSEN, H., BALACESCU, A., TESCHKE, F., SPETTEL, B., DREIBUS, G., PALME, H., QUIJANO-RICO, M., KRUSE, H., WLOTZKA, F., BEGEMANN, F.: Multielement analyses of lunar samples and some implications of the results. Proc. Third Lunar Sci. Conf. Geochim. Cosmochim. Acta, Suppl. **3**, 2, 1251 (1972).

WHITE, F. A., COLLINS, T. L., Jr., ROURKE, F. M.: New naturally occurring isotope of tantalum. Phys. Rev. **97**, 566 (1955).

WOOD, D. A., GIBSON, I. L., THOMPSON, R. N.: Elemental mobility during zeolite facies metamorphism of the Tertiary basalts of eastern Iceland. Contrib. Mineral Petrol. **55**, 241 (1976).

ZNAMENSKII, E. B.: On the average contents of niobium and tantalum in igneous rocks and the crust of the earth. Geochem. **1957** (8), 855 (1957).

ZNAMENSKII, E. B., KONYSOVA, V. V., KRINBERG, I. A., POPOLITOV, E. I., FLEROVA, K. V., TSYKHANSKII, V. D.: Distribution of titanium, niobium, and tantalum in granitic rocks containing sphene. Geochemistry **1962**, 919 (1962).

ZNAMENSKII, E. B., POPOLITOV, E. I.: Dependence of the geochemical behavior of niobium and tantalum on the paragenesis of titanium and titanium containing minerals in granitic rocks. Akad. Nauk. S.S.S.R. Siberskoye Otdeleniye, Geokhim. Redkikh Elementov, 57 (1964).

ZNAMENSKIY, E. B., KUZ'MIN, M. I., KOVAL', P. V., TSYKHANSKIY, V. D., KRINBERG, I. A.: Tantalum and niobium in granitic rocks of different ages, eastern Transbaykaliya. Geochem. Intern. **5**, 85 (1968).

Manuscript received: April 1977

Tungsten (Wolfram) 74

A	H. T. EVANS, JR.	(U.S. Geological Survey, Reston, Virginia, U.S.A.)
B-O	K. B. KRAUSKOPF	(Department of Geology, Stanford University, Stanford, Calif., U.S.A.)

74-A. Crystal Chemistry

Tungsten and molybdenum are transition metal elements in the sixth and fifth rows, respectively, of group VIB of the periodic table. Thus, their chemistry is complex and encompasses a wide range of oxidation states. In nature, however, they are almost exclusively hexavalent in their oxygen compounds and are otherwise found only in the two sulfide species tungstenite and molybdenite. The two elements are separated by the lanthanum series of elements, the result being that through the lanthanide contraction phenomenon they have nearly identical atomic size. With respect to oxygen the radii of W^{6+} and Mo^{6+} are 0.41 and 0.42 Å, respectively, for tetrahedral coordination (SHANNON and PREWITT, 1969). Consequently, the chemical behavior of the two elements is very similar, and isomorphism between corresponding compounds is common, sometimes leading to solid solution systems. The most characteristic crystal chemical properties of tungsten are treated here, but reference is also made to the chapter on molybdenum (Section 42-A) for further information.

I. Tungsten Sulfides

The only natural tungsten sulfide is the rare species tungstenite $W^{[6]}S_2$. It was shown by VAN ARKEL (SB 1913—28, 215) to be isostructural with molybdenite, but no structure refinement has yet been made. The unusual structure in which W has trigonal pyramidal coordination with sulfur in a hexagonal layer structure is described in more detail in Section 42-A. Some evidence for the existence of a rhombohedral tungstenite (3R) has been presented by GAIT and MANDARINO (1970). Both tungsten and molybdenum readily form sulfosalt complexes such as K_2WS_4, but these are highly soluble with all cations and are not found in nature.

II. Oxotungstate Compounds

Two principal groups of oxo compounds are known as minerals, typified by wolframite, $(Fe, Mn)^{[6]}[W^{[6]}O_4]$, and scheelite, $Ca^{[8]}[W^{[4]}O_4]$. The latter has tetrahedral $[WO_4]^{2-}$ groups analogous to the usual $[MoO_4]^{2-}$ groups in wulfenite, $Pb^{[8]}[Mo^{[4]}O_4]$, but in the wolframite structure tungsten has been shown to be octahedrally coordinated (KEELING, SR 1957, 305) through a structure study of the isomorphous, synthetic $NiWO_4$. The monoclinic structure is based on a hexagonal close-packed arrangement of oxygen atoms, with all cations located in octahedral sites. As shown in Fig. 74-A-1, these octahedra form zigzag edge-shared chains, the tungsten chains alternating with parallel light-metal chains. Only Fe^{2+} (in ferberite, $Fe^{[6]}W^{[6]}O_4$), Mn^{2+} (in hübnerite, $Mn^{[6]}W^{[6]}O_4$), and Zn^{2+} (in sanmartinite, $(Zn, Mn)^{[6]}W^{[6]}O_4$), are represented in major amounts in this structure as minerals, but these are commonly replaced by small amounts of Ca and Mg. W may also be replaced by varying amounts of Nb and Ta. Analogous synthetic MWO_4 species with M = Zn, Ni, Co, and Mg have been reported. The structure of ferberite has been

Fig. 74-A-1. The crystal structure of wolframite, (Fe, Mn)[6]W[6]O_4. The view is nearly along the octahedral chain direction parallel to the monoclinic *c* axis (the *a* axis is vertical). Shaded octahedra contain W, unshaded octahedra contain (Fe, Mn). The monoclinic unit cell is shown in light lines

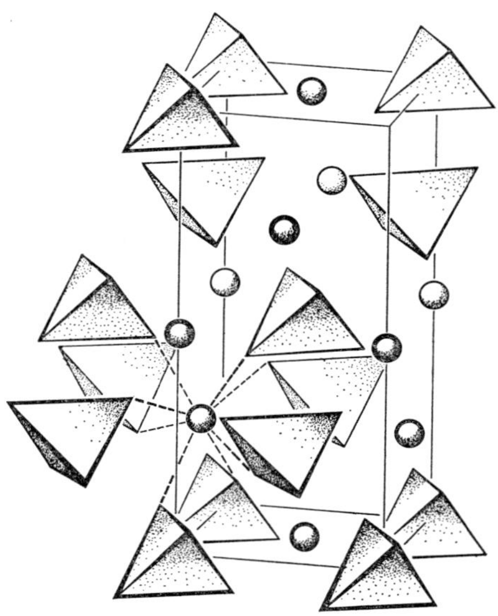

Fig. 74-A-2. The crystal structure of scheelite, Ca[8]W[4]O_4. The WO_4 groups are indicated by tetrahedra, and the Ca^{2+} ions by spheres. The eightfold coordination of one Ca^{2+} is shown by dashed lines; the tetragonal unit cell is outlined in full lines

refined using both X-ray and neutron diffraction by Ülkü (1967), giving W—O bond lengths in the distorted octahedron (in pairs) of 1.792 ± 0.005 Å, 1.905 ± 0.006 Å, and 2.124 ± 0.005 Å (the specimen from Tirpersdorf, Germany, contained 7 mol. percent Mn). Hübnerite was studied similarly by Dachs *et al.* (1967), who found corresponding W—O bond lengths of 1.81 ± 0.02 Å, 1.91 ± 0.02 Å, and 2.14 ± 0.02 Å (the specimen from Henderson, North Carolina, contained 3 mol. percent Fe). The antiferromagnetic ordering of the magnetic moments was determined in both these studies. A complete solid solution series between these two species is known in nature, comprising the wolframite series. Molybdenum is not known to enter this structure except under pressure (YOUNG and SCHWARTZ, 1963). The wolframite structure is very similar to that of columbite, (Fe, Mn)$^{[6]}$(Nb, Ta)$_2^{[6]}$O$_6$, the principal difference being in the ordering of the two types of cations.

A second important oxotungstate structural species is exemplified by scheelite, $Ca^{[8]}W^{[4]}O_4$. The earlier prevailing confusion about the structure of this species and its analogs was removed by an extensive study by SILLÉN and NYLANDER (SR 1942—44, 171), who showed conclusively that it contains tungsten in tetrahedral coordination. The cation is in eightfold coordination with triangular dodecahedral configuration (see Fig. 74-A-2). The structure of scheelite has been refined, using X-rays, by ZALKIN and TEMPLETON (SR 1964, 331) and, using neutrons, by KAY *et al.* (SR 1964, 331); these results have been critically analyzed by BURBANK (1965). The W—O bond length is found to be 1.785 ± 0.003 Å.

Scheelite is a member of a large homologous isostructural series of the type $M^{[8]}X^{[4]}Y_4$ where Y may be O or F; X may be W, Mo, Nb, I, Ge, Ru, Re, As, or Cr; and M may be Ca, Pb, Ba, Sr, Cd, Ag, Na, K, Rb, Bi, Ce, Y, Th, U, or Zr. Of these, only scheelite ($Ca^{[8]}W^{[4]}O_4$), stolzite ($Pb^{[8]}W^{[4]}O_4$), powellite ($Ca^{[8]}Mo^{[4]}O_4$), and wulfenite ($Pb^{[8]}Mo^{[4]}O_4$) are definitely known to occur in nature. The structure of stolzite has been refined by PLAKOV *et al.* (1970), who found a W—O bond length of 1.804 Å. Evidence is abundant of extensive solid solution between tungstate and molybdate species, and of calcium in the lead species.

The structure of raspite, $PbWO_4$, a monoclinic polymorph of stolzite, is not yet known, but it probably also contains tetrahedrally coordinated tungsten. The very rare russellite, Bi_2WO_6, has been little studied, but its crystallography and composition suggests a relationship to koechlinite, $Bi_2^{[3+4]}Mo^{[6]}O_6$, in which molybdenum has been reported to be octahedrally coordinated, as described in Section 42-A. A unique crystal chemical occurrence of tungsten has been found by MOORE (1967) in welinite, $(Mn^{2+}, W)_{2.6}^{[6]}Si^{[4]}(O, OH)_7$, in which W replaces octahedral Mn to as much as 15 mol. percent.

Three hydrated tungsten oxide minerals from weathering zones are known, having the composition $WO_3 \cdot nH_2O$: tungstite with $n \sim 1$, hydrotungstite with $n = 2$, and meymacite with $n \geqq 2$. The crystal chemistry of these poorly defined species is unknown, but they are probably related to the condensed phases that appear in acid environments in which tungsten approaches octahedral coordination with extensive edge-sharing. No tendency to become chemically reduced, with the appearance of color as with the molybdenum oxide hydrate ilsemannite, is observed for the tungsten oxides in nature.

Manuscript received: November 1973

References: Section 74-A

BURBANK, R. D.: An absolute integrated intensity measurement: applications to $CaWO_4$ and comparison of several refinements. Acta Cryst. **18**, 88 (1965).

DACHS, H., STOLL, E., WEITZEL, H.: Kristallstruktur und magnetische Ordnung des Hübnerits, $MnWO_4$. Z. Krist. **125**, 120 (1967).

GAIT, R. I., MANDARINO, J. A.: Polytypes of tungstenite. Can. Mineralogist **10**, 729 (1970).

MOORE, P. B.: The crystal structure of welinite. Arkiv Mineral. Geol. **4**, 459 (1968).

PLAKOV, G. F., POBEDIMSKAYA, E. A., SIMONOV, M. A., BELOV, N. V.: The crystal structure of $PbWO_4$. Kristallografiya **15**, 1067 (1970); Soviet Phys. -Cryst. **15**, 928 (1971).

SHANNON, R. D., PREWITT, C. T.: Effective radii in oxides and fluorides. Acta Cryst. **B 25**, 925 (1969).

ÜLKÜ, D.: Untersuchungen zur Kristallstruktur und magnetischen Struktur des Ferberits. Z. Krist. **124**, 192 (1967).

YOUNG, A. P., SCHWARTZ, C. M.: High-pressure synthesis of molybdates with the wolframite structure. Science **141**, 348 (1963).

Manuscript received: November 1973

74-B. Isotopes in Nature

Tungsten has the following stable isotopes (CLARK, 1966):

Atomic mass	180	182	183	184	186
Abundance, weight-%	0.1	26.4	14.4	30.6	28.4

No appreciable separation of the isotopes in natural processes would be expected for an element with so slight a percentage difference in atomic masses, and none has been reported.

Manuscript received: July 1969

74-C. Abundance in Meteorites, Tektites, and Cosmos

Recent analytical results, mostly by the technique of neutron activation, are markedly lower than older abundance figures (Table 74-C-1). These results have been verified in enough different laboratories to establish their reliability.

Table 74-C-1. *Abundance of tungsten in meteorites and tektites*

Samples	Number of analyses[a]	Abundance, ppm W[b]	Analytical method	Reference
Iron meteorites:				
San Martin	2	2.64 (2.60—2.68)	N/R	Atkins and Smales (1960)
Cañon Diablo	3	1.61 (1.42—1.87)	N/R	Atkins and Smales (1960)
Henbury	2	0.76 (0.75—0.77)	N/R	Atkins and Smales (1960)
Henbury	5	0.78 (0.32—1.27)	N/R	Hamaguchi *et al.* (1961)
Cañon Diablo	3	1.17 (0.70—1.28)	N/R	Hamaguchi *et al.* (1961)
Various	5	1.16 (0.78—1.45)	N/R	Amiruddin and Ehmann (1962)
Stony meteorites:				
Average	5, 12	0.136 (0.07—0.19)	N/R	Atkins and Smales (1961)
Average of chondrites		0.15	N/R	Vinogradov (1962a)
Average of chondrites	16, 29	0.14 (0.080—0.19)	N/R	Amiruddin and Ehmann (1962)
Average of achondrites	6, 13	0.067 (0.005—0.12)	N/R	Amiruddin and Ehmann (1962)
Olivine from Springwater pallasite		1.88	N/R	Amiruddin and Ehmann (1962)
Forest City (high-Fe chondrite)	4	0.26 (0.14—0.37)	N/R	Hamaguchi *et al.* (1961)
Modoc (low-Fe chondrite)	2	0.13 (0.11—0.15)	N/R	Hamaguchi *et al.* (1961)
Nuevo Laredo (chondrite)	1	0.11	N/R	Hamaguchi *et al.* (1961)
Admire (pallasite)	1	0.028	N/R	Hamaguchi *et al.* (1961)

Table 74-C-1 (continued)

Samples	Number of analyses[a]	Abundance, ppm W[b]	Analytical method	Reference
Troilite from meteorites:				
Cañon Diablo, chunk samples	2	0.013	N/R	AMIRUDDIN and EHMANN (1962)
Cañon Diablo, powdered sample	1	0.02	N/R	AMIRUDDIN and EHMANN (1962)
Tektites:				
Australite	1	1.84	N/R	AMIRUDDIN and EHMANN (1962)
Australite	4	0.65 (0.6—0.8)	I	TAYLOR (1966)
Indochinite	1	1.22	N/R	AMIRUDDIN and EHMANN (1962)
Javaite	1	1.12	N/R	AMIRUDDIN and EHMANN (1962)
Philippinite	1	2.35	N/R	AMIRUDDIN and EHMANN (1962)
Libyan desert glass	1	0.12	N/R	AMIRUDDIN and EHMANN (1962)

[a] Where two numbers are given, the first is the number of meteorites sampled and the second is the total number of analyses.

[b] Most of the abundance figures are averages of two or more analyses. Where the original reference gives a range, maximum and minimum figures are given in parenthesis.

For iron meteorites a reasonable average of recent figures is 1.24 ppm (WIENDL, 1968, p. 91) and for chondrites 0.14 ppm. The few figures for troilite are an order of magnitude lower. These abundances indicate that in the extreme reducing conditions of meteorite formation, tungsten behaves as a dominantly siderophile element with some lithophile character. In confirmation of this conclusion, AMIRUDDIN and EHMANN (1962) point out that the great variation in analytical results for achondrites can be at least partly explained by assuming that nearly all of the tungsten is in the metal phase.

The few tektite analyses permit only the generalizations that the tungsten content is of the order of 1 ppm and that there are small but real differences between tektites from different localities.

The cosmic abundance of tungsten (Table 74-C-2) has been estimated in two ways: by calculating rates of nucleosynthesis of tungsten atoms compared with rates for atoms of similar atomic weights, and by assuming that chondritic meteorites are a fair sample of the relatively nonvolatile matter in the universe. Recent estimates by the two methods are in good agreement: 0.1—0.2 atoms of W per 10^6 atoms of Si.

Table 74-C-2. *Cosmic abundance of tungsten*

Atoms of W per 10^6 atoms of Si	Method of estimation	Reference
0.49	Interpolation between abundances of Hf and Re	SUESS and UREY (1956)
0.105	Calculation from rate of nucleosynthesis	CAMERON (1959)
0.20	Calculation from rate of nucleosynthesis	CLAYTON and FOWLER (1961)
0.11 (0.136 ppm)	Average of 5 chondrites	ATKINS and SMALES (1960)
0.11 (0.14 ppm)	Average of 16 chondrites	AMIRUDDIN and EHMANN (1962)

Manuscript received: July 1969

II. Lunar Materials

The scanty data summarized in Table 74-C-3 and 74-C-4 were all obtained by neutron activation analysis. They provide little basis for generalizations. Reported abundances for lunar rocks cover the range 0.035—0.43 ppm, for breccias 0.19—1.21 ppm, and for fine-grained material ("soil") 0.19—0.90 ppm. Metal particles separated from lunar materials contain large amounts of tungsten, ranging from 25 to 223 ppm. WÄNKE *et al.* (1970a; 1971) point out that W, in contrast to the siderophile elements Au, Ir and Ge, is not depleted in lunar igneous rocks, which probably means that tungsten was not siderophile at the time of major differentiation of the moon. Tungsten is now strongly enriched in lunar metal particles, but it is probably not derived from meteorites because most meteoritic metal does not have such high tungsten concentrations. WÄNKE *et al.* suggest that tungsten enters the metal phase when lunar material is heated strongly, hence that conditions on the moon's surface now are more reducing than at the time of the moon's formation.

Table 74-C-3. *Tungsten in lunar rocks*

Material	Number of analyses	Abundance, ppm W		Reference
		range	average	
Apollo 11				
Rocks, type A	3	0.13—0.42	0.32	1 [P1]
Rocks, type B	2	0.35—0.36	0.36	1 [P1]
Rocks, type C (breccia)	2	0.15—0.35	0.25	1 [P1]
Fines (soil)	1		0.25	1 [P1]
Rocks, type A	1		0.43	2 [P1]
Rocks, type B	2	0.24—40	0.32	2 [P1]
Fines (soil)	1		0.22	2 [P1]
Apollo 12				
Rocks, type A	2	0.07—0.16	0.12	3 [P2]
Rocks, type AB	1		0.09	3 [P2]
Rocks, type B	1		0.15	3 [P2]
Rocks, type C (breccia)	1		0.86	3 [P2]
Fines (soil)	3	0.73—1.0	0.90	3 [P2]

Table 74-C-3 (continued)

Material	Number of analyses	Abundance, ppm W		Reference
		range	average	
Apollo 12				
Rocks	5	0.12—0.15	0.14	4 [P2]
Breccia	1		1.21	4 [P2]
Rocks	9	0.092—0.29	0.17	5 [P2]
Fines (soil)	2	0.64, 0.64	0.64	5 [P2]
Apollo 15				
Green glass	1		0.14	6 [P4]
Fines (soil)	2	0.34—0.52	0.43	6 [P4]
Breccia	10	0.08—0.31	0.19	6 [P4]
Apollo 16				
Fines	2	0.074—0.30	0.19	7 [P4]
Gabbro anorthosite	1		0.035	7 [P4]

References: (1) Morrison, G. H., Gerard, J. T., Kashuba, A. T., Gangadharam, E. V., Rothenberg, A. M., Potter, N. M., Miller, G. B. (2) Wänke, H., Rieder, R., Baddenhausen, H., Spettel, B., Teschke, F., Quijano-Rico, M., Balacescu, A. (3) Morrison, G. H., Gerard, J. T., Potter, N. M., Gangadharam, E. V., Rothenberg, A. M., Burdo, R. A. (4) Wänke, H., Wlotzka, F., Baddenhausen, H., Balacescu, A., Spettel, B., Teschke, F., Jagoutz, E., Kruse, H., Quijano-Rico, M., Rieder, R. (5) Brunfelt, A. O., Heier, K. S., Steinnes, E. (6) Taylor, S. R., Gorton, M. P., Muir, P., Nance, W., Rudowski, R., Ware, N. (7) Wänke, H., Baddenhausen, H., Dreibus, G., Jagoutz, E., Kruse, H., Palme, H., Spettel, B., Teschke, F.

From: Proceedings of the Apollo 11 Lunar Science Conference, Geochim. Cosmochim. Acta Suppl. 1, **2** (1970) [P1]; Proceedings of the Second Lunar Science Conference, Geochim. Cosmochim. Acta Suppl. 2, **2** (1971) [P2], Proceedings of the Fourth Lunar Science Conference, Geochim. Cosmochim. Acta, Suppl. 4; **2** (1973) [P4].

Table 74-C-4. *Tungsten in metal particles and minerals from lunar rocks*

Material	Number of analyses	Abundance, ppm W	Reference
Metal particles			
Apollo 11 fines	1	24	1 [P1]
Apollo 12 rocks	1	33	2 [P2]
Apollo 12 fines	1	110	2 [P2]
Apollo 14 rocks	1	46	3 [P3]
Apollo 14 fines	1	223	3 [P3]
Apollo 15 fines	1	25	3 [P3]
Apollo 16 fines	2	56 (32, 80)	4 [P4]
Apollo 11 fines			
Feldspar	1	<6	5 [P2]
Ilmenite	1	<10	5 [P2]
Clinopyroxene	1	27.5	5 [P2]
Green glass	1	24	5 [P2]

Table 74-C-4 (continued)

Material	Number of analyses	Abundance, ppm W	Reference
Apollo 12 fines			
Olivine	1	<2	5 [P2]
Pyroxene	1	<2	5 [P2]
Glass	4	<6	5 [P2]

References: (1) Wänke, H., Wlotzka, F., Jagoutz, E., Begemann, F. (2) Wänke, H., Wlotzka, F., Baddenhausen, H., Balacescu, A., Spettel, B., Teschke, F., Jagoute, E., Kruse, H., Quijano-Rico, M., Rieder, R. (3) Wlotzka, F., Jagoutz, E., Spettel, B., Baddenhausen, H., Balacescu, A., Wänke, H. (4) Wlotzka, F., Spettel, B., Wänke, H., (5) Vobecky, M., Fràna, J., Bauer, J., Randa, Z., Benada, J., Kurcir, J.
From: Proceedings of the Apollo 11 Lunar Science Conference, Geochim. Cosmochim. Acta Suppl. 1, **2** (1970) [P1]; Proceedings of the Second Lunar Science Conference, Geochim. Cosmochim. Acta, Suppl. 2, **2** (1971) [P2]; Proceedings of the Third Lunar Science Conference, Geochim. Cosmochim. Acta, Suppl. 3, **2** (1972) [P3]; Proceedings of the Fourth Lunar Science Conference, Geochim. Cosmochim. Acta, Suppl. 4, **2** (1973) [P4].)

Manuscript received: December 1973

D. Abundance in Rock-Forming Minerals (I), Tungsten Minerals (II)

I. Rock-Forming Minerals

The scanty available data are summarized in Table 74-D-1. One can generalize that quartz and feldspar seldom contain more than 2 ppm W and often have less than 1 ppm, although values up to 100 ppm are reported near tungsten deposits. Figures are not available for amphiboles, pyroxenes, or olivines, but the small total concentrations in mafic rocks indicate that the tungsten content of these minerals is generally less than 1 ppm. Of the abundant minerals in igneous rocks, mica is richest in tungsten; concentrations of 5—50 ppm are common, and muscovite in altered granites near tungsten deposits contains locally as much as 500 ppm. Among the accessory minerals, the iron oxides and minerals of titanium often show tungsten concentrations over 10 ppm. Niobium-tantalum minerals may contain one per cent or more of tungsten, presumably because W^{6+} can substitute for Nb^{5+} and Ta^{5+}.

Manganese oxide minerals locally contain extraordinary amounts of tungsten, especially oxides associated with hot-spring deposits. Hewett *et al.* (1963) list tungsten among the rare metals whose abundance makes possible a distinction between manganese oxides of supergene origin and those from a hypogene source.

II. The Minerals of Tungsten

The mineralogy of tungsten is fairly simple. The twenty minerals in Table 74-D-2 include all those whose existence is established beyond reasonable doubt. Except for the very rare sulfide, all are tungstates or forms of the oxide. Only scheelite and the three minerals of the wolframite group have any quantitative importance; tungstite and stolzite are minor constituents of many oxidized zones, but all the others are little more than mineralogical curiosities. Of the twenty minerals listed, the first eleven are usually primary minerals and the last nine are limited to zones of weathering.

Solid-solution relations among the tungsten minerals have been worked out only in part. The series huebnerite-wolframite-ferberite, long thought to be a classic example of complete solid solution, has been shown by Schröcke (1960) to exhibit complete mixing only at temperatures over about 400° C. Solid solution between scheelite and wolframite is limited to a fraction of a percent at ordinary temperatures, but becomes extensive at temperatures over 600° C (Grubb, 1967). Some tungstates show extensive solid solution with the corresponding molybdates. Amounts of molybdenum reported in the wolframite group are limited to the merest traces, but in scheelite molybdenum may be present as up to 24 per cent MoO_3 (seyrigite). A complete solid-solution series may exist between scheelite and powellite ($CaMoO_4$), but it has not been verified. Likewise stolzite and wulfenite ($PbMoO_4$) may form a complete series, but recorded analyses of mixtures (chillagite) are limited to the ranges 21.1—29.5 per cent WO_3 and 16.3—22.2 per cent MoO_3.

Table 74-D-1. *Tungsten in rock-forming minerals etc. (Method in brackets)*

Mineral	Number of analyses	Abundance[a], ppm W	Locality	Reference
Silicate minerals:				
Quartz from granite	1	0.1 (C)	Uganda	JEFFERY (1959)
Quartz from mica granite	6	1.0—1.6 (S)	E. Transbaikal	IVANOVA (1969)
Quartz from greisen	1	5.6 (S)	E. Transbaikal	IVANOVA (1969)
Feldspar from granite	1	0.1 (C)	Uganda	JEFFERY (1959)
Feldspar from mica granite	10	0.9—2.6 (S)	E. Transbaikal	IVANOVA (1969)
Feldspar from granite in tungsten ore zone	2	6.5—10 (S)	E. Transbaikal	IVANOVA (1969)
Feldspar from granite and pegmatite of W-Mo deposit		10—100 (S)	Kazakhstan	SHCHERBA *et al.* (1964)
Muscovite from mica schist	1	0.4 (C)	Uganda	JEFFERY (1959)
Muscovite from mica granite	11	14—70 (S)	E. Transbaikal	IVANOVA (1969)
Muscovite from mica granite	1	150 (S)	E. Transbaikal	IVANOVA (1969)
Muscovite from granite in tungsten ore zone	8	41—120 (S)	E. Transbaikal	IVANOVA (1969)
Muscovite from greisen	2	43—53 (S)	E. Transbaikal	IVANOVA (1969)
Muscovite from greisen	1	510 (S)	E. Transbaikal	IVANOVA (1969)
Biotite from biotite gneiss	1	6.0 (C)	Uganda	JEFFERY (1959)
Biotite from mica granite	4	~5 (S)	E. Transbaikal	IVANOVA (1969)
Green biotite from carbonatite	1	2.8 (C)	Uganda	JEFFERY (1959)
Phlogopite from carbonatite	1	22.4 (C)	Uganda	JEFFERY (1959)
Tourmaline	1	15.4 (C)	Uganda	JEFFERY (1959)
Niobium-tantalum minerals:				
Columbite	5	6,000—11,300 (?)	Sierra Leone	POLLETT (1957)
Columbite	2	200—3,200 (C)	Uganda	JEFFERY (1959)
Tantalite	1	93 (C)	Uganda	JEFFERY (1959)
Microlite	1	610 (C)	Uganda	JEFFERY (1959)
Pyrochlore	1	44 (C)	Uganda	JEFFERY (1959)

Table 74-D-1 (Continued)

Mineral	Number of analyses	Abundance[a], ppm W	Locality	Reference
Iron and titanium minerals:				
Magnetite	5	2.7—21.9 (C)	Uganda	JEFFERY (1959)
Hematite	5	3.7—35.7 (C)	Uganda	JEFFERY (1959)
Ilmenite	7	0.7—19.5 (C)	Uganda	JEFFERY (1959)
Ilmenite from pegmatite	1	56 (C)	Uganda	JEFFERY (1959)
Ilmenite from biotite granite	8	67.8 (?)	U.S.S.R.	LYAKHOVICH and BALANOVA (1969)
Ilmenite from greisenized and albitized granite	4	7.5 (?)	U.S.S.R.	LYAKHOVICH and BALANOVA (1969)
Ilmenite from gabbro	1	1 (?)	U.S.S.R.	LYAKHOVICH and BALANOVA (1969)
Ilmenite from alkaline syenite	1	13 (?)	U.S.S.R.	LYAKHOVICH and BALANOVA (1969)
Rutile	3	74—2,000 (C)	Uganda	JEFFERY (1959)
Perovskite	1	26.4 (C)	Uganda	JEFFERY (1959)
Sphene from biotite granite	5	12 (?)	U.S.S.R.	LYAKHOVICH and BALANOVA (1969)
Sphene from alaskitic granite	1	9 (?)	U.S.S.R.	LYAKHOVICH and BALANOVA (1969)
Sphene from diorite	4	33.8 (?)	U.S.S.R.	LYAKHOVICH and BALANOVA (1969)
Manganese oxide minerals:				
Hot-spring deposit		maximum 55,000 (W)	Golconda, Nevada	WHITE (1955)
Hot-spring deposit		maximum 17,000 (W)	Uncia, Bolivia	WHITE (1955)
Hot-spring deposit		maximum 17,500 (W)	Schwarzwald, Germany	KIRCHHEIMER (1953)
Hot-spring deposit		maximum 11,900 (W)	Chihuahua, Mexico	HEWETT and FLEISCHER (1960)
Pyrolusite from tungsten mine		maximum 8,000 (C)	Uganda	JEFFERY (1959)
Pyrolusite		4—12 (C)	Uganda	JEFFERY (1959)
Psilomelane		5.8 (C)	Uganda	JEFFERY (1959)
Supergene manganese oxides		<150—300 (S)		HEWETT *et al.* (1963)
Hypogene manganese oxides		<150—70,000 (S)		HEWETT *et al.* (1963)

[a] Letters in parenthesis designate analytical method.

Table 74-D-2. *Minerals of tungsten*

Mineral	Composition
Wolframite group:	
Huebnerite	$MnWO_4$ (0—20 atomic-% Fe)
Wolframite	(Fe, Mn)WO_4 (20—80 atomic-% Fe)
Ferberite[a]	$FeWO_4$ (80—100 atomic-% Fe)
Scheelite[b]	$CaWO_4$
Seyrigite[b]	Ca(W, Mo)O_4 (WO_3 ca. 28%)
Cuproscheelite	(Cu, Ca)WO_4
Sanmartinite	(Zn, Fe, Ca, Mn)WO_4
Tungstenite	WS_2
Russellite	$(Bi_2W)O_3$
Scheteligite	(Ca, Fe, Mn, Sb, B, Y)(Ti, Ta, Nb, W)$_2$(O, OH)$_7$
Yttrocrasite	(Y, Th, U, Ca)$_2$(Ti, Fe_3^+, W)$_4O_{11}$(?)
Stolzite	$PbWO_4$
Chillagite	Pb(W, Mo)O_4 (WO_3 21.1—29.5%)
Raspite	$PbWO_4$
Tungstite	$WO_3 \cdot H_2O$(?)
Hydrotungstite[c]	$WO_3 \cdot 2H_2O$
Cuprotungstite	$Cu_2WO_4(OH)_2$
Ferritungstite	$Ca_2Fe^{2+}{}_2Fe^{3+}{}_2(WO_4)_7 \cdot 9H_2O$
Anthoinite	$AlWO_4OH \cdot H_2O$
Yttrotungstite	Hydrated Y tungstate

Source of data: Wiendl (1968) and Dana (1944, 1951).

[a] Ferberite pseudomorphous after scheelite is sometimes called *reinite*.

[b] Scheelite with more than 3% MoO_3 is very rare. Substitution up to 24% MoO_3 (*seyrigite or molybdoscheelite*) is known, however, and also substitution of W for Mo in $CaMoO_4$ up to 14% WO_3.

[c] Amorphous hydrotungstite is called *meymacite*.

Analyses of wolframite indicate that considerable deviation from the stoichiometric formula is possible [ratio (Fe + Mn): W up to 1.21]. Schröcke (1960) finds experimentally that mix-crystals of the oxides form at high temperatures over wide ranges of composition:

	FeO—WO_3	MnO—WO_3
at 800° C	35—55% WO_3	25—65% WO_3
at 1,000° C	25—65% WO_3	20—70% WO_3

Schröcke notes also that $FeWO_4$ at 1,000° C can form mix-crystals with Fe_2WO_6 ($Fe_2O_3 \cdot WO_3$) and $FeWO_3$ ($FeO \cdot WO_2$), but finds no evidence for such mixtures in natural material.

Besides Mo, the principal elements that may substitute in minor amounts for tungsten are Ta and Nb. The valence difference is commonly made up by simultaneous substitution of Sc and V for Fe and Mn, and of Y and rare-earth metals for Ca.

Manuscript received: July 1969

74-E. Abundance in Common Igneous Rock Types

VON HEVESY and HOBBIE started to determine tungsten in rocks as early as 1933. The analytical data are not entirely consistent, but indicate as a general average for granitic rocks about 1.5 ppm W and for mafic rocks somewhat less, 0.5—1.0 ppm (Table 74-E-1). The few data for ultramafic rocks suggest still lower values, 0.1—0.8 ppm; the data for alkalic rocks indicate some enrichment, but are too scattered to give a meaningful average. The suggestion from these figures of a rough correlation between tungsten and silica content is borne out by analyses of five granites from the Llano region of Texas: tungsten decreases regularly from 2.4 ppm in a granite with 76.8% SiO_2 to 1.1 ppm in a granite with 68.2% SiO_2 (SANDELL, 1946). WIENDL (1968) hazards a guess that volcanic rocks have slightly more tungsten than plutonic rocks of similar composition, but the evidence for this generalization is slender.

Regional differences in tungsten abundance in igneous rocks are suggested by the generally high figures obtained by JEFFERY (1959) for a variety of rocks in Uganda, by DEKATE (1962) for rocks in India, and by IVANOVA (1969) for granites from eastern Transbaikal. At least to some extent these high figures reflect abnormal concentrations in rocks directly associated with tungsten deposits. IVANOVA (1963, 1969), however, generalizes from her extensive work in eastern Siberia that normal granites in regions of tungsten mineralization are not enriched in the metal; only rocks that have been altered (muscovitized), presumably by the ore-forming solutions, show anomalous concentrations. A similar conclusion is reached by SOTNIKOV and IZYUMOVA (1965) regarding granites associated with tungsten deposits in the Altai Mountains.

Older analyses (pre-1946) of the tungsten content of normal igneous rocks, indicating concentrations in the range 3—80 ppm, are almost certainly in error.

Table 74-E-1. *Tungsten in igneous rocks*

Rock	Number of analyses	Abundance, ppm W		Locality	Reference
		range	average		
Felsic rocks:					
G-1 granite	7	0.36—0.47	0.40 (N/R)	Rhode Island	ATKINS and SMALES (1960)
G-1 granite	2	0.37—0.38	0.375 (N/R)	Rhode Island	AMIRUDDIN and EHMANN (1962)
G-1 granite	4	0.36—0.60	0.50 (N/R)	Rhode Island	HAMAGUCHI *et al.* (1962)
Average granite			1.55	World	WIENDL (1968)
Average granite			1.42	World	HORN and ADAMS (1966)
Average granite			2.0	World	TAYLOR (1964)

Table 74-E-1 (Continued)

Rock	Number of analyses	Abundance, ppm W range	average	Locality	Reference
Felsic rocks, continued					
Granite and granodiorite, average			1.5	World	VINOGRADOV (1962b)
Low-calcium granite			2.2	World	TUREKIAN and WEDEPOHL (1961)
High-calcium granite			1.3	World	TUREKIAN and WEDEPOHL (1961)
Average granite			1.5	U.S.S.R.	LYAKHOVICH and BALANOVA (1969)
Average granite		1.3—2.4	(S)	U.S.S.R.	IVANOVA (1963)
Average granite	8	1.9—3.3	2.4 (S)	U.S.S.R.	VINOGRADOV *et al.* (1958)
Average granite			3.7 (C)	India	DEKATE (1967)
Average granite	52	0.1—3.7	1.4 (C)	Uganda	JEFFERY (1959)
Average granite	5	1.3—2.0	1.6 (W)	U.S. and Canada	KURODA and SANDELL (1954)
Average granite	13	1.0—2.6	1.5 (W)	mostly N. America	SANDELL (1946)
Granites of mineralized areas	6	2.6—12.0	8.0 (C)	Uganda	JEFFERY (1959)
Biotite granite	52	1—2.95	1.60 (S)	E.Transbaikal	IVANOVA (1969)
Two-mica granite	57	1.0—4.4	2.1 (S)	E.Transbaikal	IVANOVA (1969)
Muscovite granite	12	2.0—19.6	6.4 (S)	E.Transbaikal	IVANOVA (1969)
Muscovite granite	20	1.5—126	12.6 (S)	E.Transbaikal	IVANOVA (1959)
Muscovite granite	12	2.0—7,100	1,308 (S)	E.Transbaikal	IVANOVA (1959)
Greisen	26	6.5—5,000	568 (S)	E.Transbaikal	IVANOVA (1959)
Muscovite granite of ore zone	20	46—2,270	574 (S)	E.Transbaikal	IVANOVA (1959)
Muscovite granite of ore zone	19	1.4—8.9	4.4 (S)	E.Transbaikal	IVANOVA (1959)
Granophyre sill	5	0.41—0.50	0.46 (N/R)	Skaergaard	ATKINS and SMALES (1960)
Hedenbergite granophyre	7	1.02—1.17	1.11 (N/R)	Skaergaard	ATKINS and SMALES (1960)
Rhyolite	1		2.4 (W)	Minnesota	SANDELL (1946)
Obsidian	3	0.83—1.60	1.12 (N/R)	U.S., Iceland	AMIRUDDIN and EHMANN(1962)

Table 74-E-1 (Continued)

Rock	Number of analyses	Abundance, ppm W range	Abundance, ppm W average	Locality	Reference
Intermediate rocks:					
Granodiorite	3	2.1—3.3	2.7 (SW)	Caucasus, Kirghiz	VINOGRADOV *et al.* (1958)
Diorite	3	1.4—2.0	1.6 (SW)	Caucasus, Kirghiz	VINOGRADOV *et al.* (1958)
Diorite	3		1.2 (W)	California	KURODA and SANDELL (1954)
Diorite	7		1.2 (W)	California, Oregon	SANDELL (1946)
Diorite and andesite			1.1	World	VINOGRADOV (1962b)
Mafic rocks:					
W-1 basalt ("diabase")	4	0.41—0.47	0.43 (N/R)	Virginia	ATKINS and SMALES (1960)
W-1 basalt ("diabase")	2	0.26—0.27	0.27 (N/R)	Virginia	AMIRUDDIN and EHMANN (1962)
W-1 basalt ("diabase")	5	0.48—0.63	0.58 (N/R)	Virginia	HAMAGUCHI *et al.* (1962)
Plutonic mafic rocks, average			0.53	World	WIENDL (1968)
Mafic rocks, average			1.0	World	VINOGRADOV (1962b)
Gabbro and basalt	4	0.5—0.8	0.6 (SW)	Norilsk, E. Siberia	VINOGRADOV *et al.* (1958)
Mafic and ultramafic rocks			1.6 (C)	India	DEKATE (1967)
Basalts (dolerites)	6	0.9—14.5	6.3 (C)	Uganda	JEFFERY (1959)
Gabbro		0.09—0.14	0.12 (N/R)	Skaergaard	ATKINS and SMALES (1960)
Chilled marginal gabbro			0.22 (N/R)	Skaergaard	ATKINS and SMALES (1960)
Ferrogabbro		0.18—0.52	0.43 (N/R)	Skaergaard	ATKINS and SMALES (1960)
Basalt, average			1.17	World	WIENDL (1968)
Basalt, average			0.7	World	TUREKIAN and WEDEPOHL (1961)
Basalt, average			1.0 (C)	India	DEKATE (1967)
Basalt	3	0.6—1.6	1.0 (SW)	Kamchatka	VINOGRADOV *et al.* (1958)

Table 74-E-1 (Continued)

Rock	Number of analyses	Abundance, ppm W range	Abundance, ppm W average	Locality	Reference
Ultramafic rocks:					
Ultramafic rocks, average			0.1	World	VINOGRADOV (1962b)
Ultramafic rocks, average			0.77	World	TUREKIAN and WEDEPOHL (1961)
Dunite, average			0.1	World	VINOGRADOV (1962b)
Dunite, average	6		0.33 (SW)	N. Ural	VINOGRADOV *et al.* (1958)
Peridotite	1		0.35 (SW)	Siberia	VINOGRADOV *et al.* (1958)
Pyroxenite	1		0.6 (SW)	Siberia	VINOGRADOV *et al.* (1958)
Pyroxenite	1		10.4 (C)	Uganda	JEFFERY (1959)
Dunite	4	0.14—0.48	0.3 (N/R)	?	ATKINS and SMALES (1960)
Serpentine	3	0.44—0.65	0.5 (SW)	?	VINOGRADOV *et al.* (1958)
Alkalic rocks:					
Syenite, average			1.3	World	TUREKIAN and WEDEPOHL (1961)
Syenite	6		1.9 (W)	New York, Minnesota	SANDELL (1946)
Alkalic rocks, average			5.1 (C)	India	DEKATE (1967)
Nepheline syenite	1		0.7 (C)	Uganda	JEFFERY (1959)
Tinguaite	4	0.4—14.8	4.7 (C)	Uganda	JEFFERY (1959)
Trachyte	1		5.8 (C)	Uganda	JEFFERY (1959)
Phonolite	5	3.7—29.9	16.3 (C)	Uganda	JEFFERY (1959)
Nephelinite	5	5.8—17.0	10.7 (C)	Uganda	JEFFERY (1959)
Carbonatite	6	0.6—17.7	5.2 (C)	Uganda	JEFFERY (1959)
Alkalic rocks	6 of 27	2.0—4.0	3.0 (S)	E. Tuva	STUDENIKOVA and PAVLENKO (1960)
Alkalic rocks	13 of 13	1.7—3.1	(S)	N. Caucasus	STUDENIKOVA and PAVLENKO (1960)

Manuscript received: July 1969

74-F. Behavior in Magmatogenic Processes

In a crystallizing magma that contains appreciable tungsten, the tungsten should be concentrated in the water-rich residual solutions. Although this statement has not been verified by direct experiment, the marked difference in properties between W^{6+} and the major ions of silicate minerals, together with the demonstrated scarcity of tungsten in the common silicates, indicates that the element will favor a water-rich phase rather than the crystallizing silicate minerals.

Volatile tungsten compounds should not be present in the water-rich phase in appreciable amounts, as can be demonstrated by simple calculations from thermodynamic data. Many tungsten halides and oxyhalides (e.g., WF_6, $WOCl_4$) are indeed volatile, and in older speculations these are often called upon to transport tungsten, the mechanism seeming plausible because tungsten minerals are often associated with fluorine-bearing minerals like fluorite, mica, and topaz. But equilibria involving these compounds and water (e.g., $WF_6 + 3H_2O \rightleftharpoons WO_3 + 6HF$) are displaced so far toward the oxide that their presence in a gas phase would demand completely unreasonable concentrations of the halogen acids (Ivanova, 1966; Krauskopf, 1964). The thermochemical data are by no means complete, but the necessary extrapolations are not extreme. Tungstic oxide is very slightly volatile in the presence of water alone, presumably by forming a hydrate, but the vapor pressure of this compound is only about 10^{-5} mm Hg even at a temperature of 800°C, far too small to account for movement of tungsten under ore-forming conditions (Glemser and von Haeseler, 1956; Milner and Neugebauer, 1949; Gundlach and Thormann, 1960).

In contrast to volatility, the solubility of tungsten compounds in a magmatic water-rich phase should be large enough to easily account for transportation of the metal in ore-forming processes. In pure water the solubility of WO_3, $CaWO_4$, and $FeWO_4$ is low, even at high temperatures, but other constituents of magmatic fluids enhance the solubility enough to make transport possible over the entire temperature range from magmatic conditions down to ordinary temperatures. The solubility has been demonstrated experimentally up to 500° C, and extrapolation to higher temperatures seems reasonable. Uncertainty remains as to the particular ions or compounds of tungsten that play the principal roles in different ranges of temperature, pressure, and composition. In slightly acid solutions tungsten can be carried in small but appreciable amounts in the presence of Ca^{2+}, Fe^{2+}, and Mn^{2+}, perhaps as the simple tungstate ion WO_4^{2-}, or at higher temperatures as the relatively undissociated compounds H_2WO_4 and Na_2WO_4. Compounds of the divalent cations are precipitated, of course, when the acidity is neutralized. A more efficient and more likely form of transportation in natural silica-containing solutions is provided by the heteropoly-tungsten acids, particularly $H_8Si(W_2O_7)_6$. Gundlach and Thormann (1960) have shown that neutralization of acid solutions of this compound at ordinary temperatures (concentrations: 1.4% WO_3 and 0.12 M Ca^{2+}, Fe^{2+}, or Mn^{2+}) gives a precipitate of $FeWO_4$ at pH 5.9, $MnWO_4$ at 6.7, and $CaWO_4$ at 7.3; their experiments show also that the compound is stable at least to temperatures of 300° C. Possibly, as pointed out by IVANOVA and KHODAKOVSKII (1968), an oxyfluoride complex may contribute to the solubility in solutions where the concentration of F^- is exceptionally high. For tungsten, in contrast to many other metals, solubility relations are not complicated by the possibility of reduction to a form of lower valence, because under natural conditions redox potentials only very rarely fall low enough to reduce the valence below +6.

Deposition of tungsten from ore-forming fluids likewise does not pose serious problems. In the experiments of GUNDLACH and THORMANN, the three most common compounds of tungsten are shown to precipitate in the order usually found in nature. At higher temperatures, BRYZGALIN (1958) has shown that scheelite forms readily by reaction between Na_2WO_4 solutions and calcic plagioclase, provided the solution is no more than weakly alkaline; this fact is used to account for the observed common association of scheelite and fairly sodic plagioclase in skarn deposits. BRYZGALIN notes also that a faintly alkaline solution in the presence of calcite at 250 and 500° C will decompose wolframite and deposit tungsten as scheelite.

From these observations it seems clear that tungsten is carried in ore-forming solutions chiefly as tungstate ion, tungstic acid, sodium tungstate, or a heteropoly-acid, the relative amounts of these depending on temperature, pH, and the silica content of the solutions. From such fluids tungsten is deposited as wolframite if limestone is absent and Ca^{2+} is low, or as scheelite if limestone is present, probably usually in response to increasing pH of the solutions. The commonly observed rims of scheelite on wolframite crystals are explainable by the fact that, when all three of the ions Fe^{2+}, Mn^{2+}, and Ca^{2+} are present, scheelite requires a slightly higher pH for its precipitation.

Magmatogenic tungsten deposits are reported in all temperature zones, although the largest and most abundant are those formed at high temperatures. Where a clear association with plutonic rocks can be established, the rocks are nearly always granitic. An occurrence in serpentinized peridotite is reported in Azerbaijan (SITKOVSKII *et al.*, 1960), but the authors regard siliceous rocks nearby as the probable source of metal. Among the different varieties of granitic rock there seems to be no preference; tungsten ores are found with all kinds, from diorite to alaskitic granite. Wolframite occurs as a constituent of pegmatite, exceptionally even of granite itself. Both wolframite and scheelite are found in high-temperature quartz veins that locally are transitional into pegmatites. Scheelite is characteristic of skarn deposits at granite-limestone contacts. In the high-temperature deposits, usual associates of the tungsten minerals are cassiterite (especially with wolframite), molybdenite, arsenopyrite, and bismuth minerals; the more common sulfide minerals are generally present in minor amounts. Scheelite is found also in mesothermal and epithermal veins, sometimes with stibnite and cinnabar in addition to the more common sulfides. Where tin and tungsten minerals occur together, the ratio tungsten/tin increases toward the intrusive source.

The ratio Mn/Fe in wolframite has been the subject of much study. Evidence for correlation with temperature, depth, and environment of ore deposition is so contradictory, however, that no generalization seems warranted (POLKINGHORNE, 1953—1954; HEIER, 1955; BOLDUAN, 1954; OELSNER, 1954; KHASIN, 1949).

It is sometimes postulated that tungsten associated with intrusive rocks is ultimately of syngenetic sedimentary origin, the metal being incorporated into the melt formed by anatexis of the sedimentary rocks (DE MAGNÉE and ADERCA, 1960). Another possible ultimate relation of tungsten to associated intrusive rock is proposed by ŠTEMPROK and ŠULCEK (1969); on the basis of careful geologic and analytical work on a granitic cupola in the Erzgebirge, they demonstrate that tungsten cannot have come from exposed levels of the intrusive but must have ascended from an unknown source below, its occurrence with the granite being explained on structural rather than genetic grounds.

Evidence is fairly good that tungsten (unlike tin) does not show abnormal concentrations in granites associated with ore deposits, except where the granites have been altered (generally muscovitized) by solutions presumably related to the ore-forming process. Evidently, if the concentration of tungsten is an episode in the crystallization of felsic magma, the metal is excluded from the growing crystals (except for the minor amounts found in all igneous minerals) until the very end of the magmatic history, even if initial concentrations in the magma were abnormally high. This fact leaves open the question whether tungsten was actually a part of the magma originally or whether the magma served only to activate an original concentration that had been formed in some other way. Since only a very few granites have associated tungsten deposits, and since these few are seemingly in no way different from the great majority that show no concentrations of tungsten at all, it can be reasonably argued that the role of the magma is secondary and that the metal must have been at least moderately enriched by some earlier process before the magma started to form.

Whatever is the answer to this question, the actual emplacement of tungsten minerals is clearly by solutions of the sort generally supposed to be residual from the cooling of a water-rich magma. Evidence both from experiment and from observation of tungsten deposits indicates that precipitation of the tungstates can take place over a wide temperature range, from magmatic conditions down practically to ordinary temperatures, although a majority of large deposits have formed in the higher part of the normal range of ore formation.

Wolframite characteristically occurs in quartz veins cutting granite or rocks adjacent to granite that do not contain limestone. It is often, but by no means always, accompanied by cassiterite and the usual gangue minerals of tin veins. Scheelite is particularly abundant in skarn deposits at granite-limestone contacts. High temperature of formation of both minerals is often suggested by the character of the gangue minerals and the alteration of adjacent rocks, and also by the striking transitions between pegmatite dikes and feldspar-free quartz veins containing one or other of the tungstates. Both wolframite and scheelite occur also in quartz veins (scheelite in carbonate veins as well) farther from granite and evidently formed at lower temperatures. The composition of wolframite ranges from nearly iron-free huebnerite to nearly manganese-free ferberite, but reasons for this variation are not understood.

Manuscript received: July 1969

74-G. Behavior during Weathering and Rock Alteration

Observations on the weathering of tungsten minerals are not entirely consistent. It is recognized, of course, that wolframite and scheelite are fairly insoluble in the pH range of surface waters, but neither experiments nor observations of natural occurrences permit safe generalizations about the effects of different kinds of solutions or different weathering conditions.

The insolubility and high density of wolframite and scheelite suggest that they should accumulate in placers, but large concentrations of this sort are not common. Eluvial placers are more important than alluvial placers; much of the wolframite obtained from the enormous deposits of southern China, for example, has consisted of fragments left near the surface by erosion of more easily weathered material. The scheelite deposit at Atolia in southern California is in part an alluvial placer, representing transport by streams for distances up to a few kilometers (LEMMON and DORR, 1940). A few other placer deposits are well authenticated, but occasional claims (e.g., ZESCHKE, 1961) that scheelite has moved several hundred kilometers in stream gravels are probably invalid (DAVIDSON, 1962). The scarcity of tungsten placers is commonly explained by the brittleness of scheelite and wolframite, which leads to their breakdown into very fine particles that become widely dispersed in fine-grained sediments.

Tungsten minerals are slowly attacked by acid surface waters, especially by the strongly acid waters of weathering sulfide deposits. Scheelite is commonly stated to be more soluble in acid than wolframite, although decay of the latter is aided by the susceptibility of its iron and manganese to oxidation. Tungsten released by acid attack is probably partly dissolved as HWO_4^- (or a heteropoly-silicon acid) and partly converted to some form of tungstic oxide (tungstite, hydrotungstite, ferritungstite). These minerals are not often reported, perhaps in part because they are inconspicuous when mixed with limonite and jarosite, and in part because under usual conditions most of the tungsten is dissolved. If lead or copper sulfides are present the tungsten may form stolzite or cupritungstite, but more commonly it appears as a minor constituent of wulfenite or vanadinite (NEWHOUSE, 1934). DEKATE (1962), from an extensive study of deposits in India, concludes that the fate of tungsten released by acid attack depends in large measure on climate; under semi-arid conditions it forms anthoinite, cupritungstite, and ferritungstite, but in a warm humid climate it goes into tungstite and hydrotungstite if sulfides are present and is adsorbed on iron and manganese oxides if sulfides are absent.

Alkaline solutions in nature probably have little effect on tungsten minerals, although wolframite is somewhat dissolved by strongly alkaline solutions in the laboratory (GANNETT, 1919). The influence of organic matter is uncertain; GANNETT reports that peat extract has no effect on the solubility of tungsten compounds, but DOUCET (1966) claims that organic matter will complex tungsten at pHs in the range 6—7 and hence will aid its solution.

Movement of tungsten in near-surface solutions is shown by the existence of "haloes" in soils around ore deposits, but the tungsten haloes are much narrower than associated molybdenum haloes (TAKHOMIROV and MILLER, 1946; KRAINOV *et al.*, 1965). Secondary enrichment of tungsten is minor and local; a few examples are cited by LEVINA (1954) and by KRAINOV *et al.* (1965). The latter authors explain the lack of secondary enrichment by the trapping of most dissolved tungsten by adsorption on iron and manganese oxides.

Hydrothermal alteration of tungsten minerals is not often reported. HEIER (1955) observed the apparent change of wolframite to scheelite at a deposit in Norway, and showed experimentally that the change could be brought about by heating wolframite and CaO in a bomb at 200—800° C and 70—700 bars. BRYZGALIN (1958) describes similar experiments in which wolframite was decomposed by solutions of 0.005 N NaOH and 0.025 N $NaHCO_3$ at 350—500° C and its tungsten was deposited as scheelite on labradorite or limestone. OELSNER (1954) notes that wolframite in a deposit in southern Germany gives scheelite and huebnerite as a result of hydrothermal alteration, but no ferberite. A change in the opposite direction, from ferberite to scheelite, is described by DE MAGNÉE and ADERCA in Ruanda (1960).

Manuscript received: July 1969

74-H. Solubilities of Compounds which Control Concentrations in Natural Waters; Valence States

Tungsten under most natural conditions is stable in its highest valence, +6. Its oxide, WO_3, is nearly insoluble in neutral and weakly acid solutions but readily soluble in alkaline solutions. When acid is added to an alkaline solution the oxide is precipitated, generally as a hydrate, or it may form a colloid. Tungstates of most divalent metals are insoluble in neutral and alkaline solutions; in acid they are decomposed with formation of the oxide. The chemistry of tungsten is complicated in neutral and slightly acid solutions by its ability to form large complex ions and molecules. Significant characteristics not obviously predictable from its position in the periodic table are the extreme reducing conditions needed to stabilize the disulfide, the dominance of Ca, Fe, and Mn over all other divalent metals in tungstate minerals, and the particular stability of complexes with Si, P, and Sb in acid solutions.

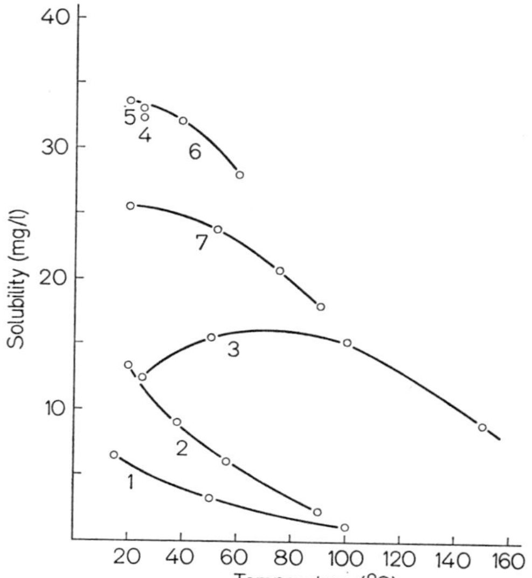

Fig. 74-H-1. Solubility of scheelite. 1: Gmelin (1933); 2: Bokii and Anikin (1956); 3: Khodakovskii and Mishin (1958); 4: Urusov *et al.* (1967); 5: Berkem (1946); 6: Bokii and Anikin (1956) (1 M LiCl); 7: Bokii and Anikin (1956) (1 M NaCl)

Quantitative solubility data for the tungstates are fragmentary and conflicting. The data are most complete for scheelite (Fig. 74-H-1 and Table 74-H-1), but even for this compound the highest and lowest values reported at ordinary temperatures differ by a factor of six. Two sets of measurements agree that the solubility of scheelite decreases as temperature rises, at least to 100° C, but the calculations of solubility product by Khodakovskii and Mishin (1968) indicate that the decrease begins only at higher temperatures. Solubilities of other tungstates have been compiled by Urusov *et al.* (1967) (Table 74-H-2).

Table 74-H-1. *Solubility products of $CaWO_4$, $FeWO_4$, and $MnWO_4$ to* 350° *C* (*—log K calculated from thermochemical data*)

Temperature, ° C	25°	50°	100°	150°	200°	250°	300°	350°
$-\log K_{CaWO_4}$	8.74	8.54	8.55	9.02	9.30	10.80	11.95	13.21
$-\log K_{FeWO_4}$	15.00	14.72	14.69	15.06	15.75	16.67	17.76	19.00
$-\log K_{MnWO_4}$	9.06	9.07	9.48	10.23	11.27	12.51	13.90	15.33

Sources: ($CaWO_4$): KHODAKOVSKII and MISHIN (1968); ($FeWO_4$ and $MnWO_4$): CHERNYSHEV and IVANOVA (1969).

Table 74-H-2. *Solubilities of tungstates*

Compound	Temperature, ° C	Solubility, g/liter	Logarithm of solubility product
$MgWO_4$		high	
$SrWO_4$	25	0.03	−8.17
$BaWO_4$	25	0.0096	−9.25
$MnWO_4$	25	0.0166	−8.85
$FeWO_4$	25		−9.2
$CuWO_4$	25		−12.8
$PbWO_4$	25		−12.9, −15.0 (?)

Source of data: URUSOV *et al.* (1967). The last three entries are calculated by the authors from thermochemical data. The figure for the solubility product of $FeWO_4$ differs by almost six orders of magnitude from that calculated by CHERNYSHEV and IVANOVA (1969) in Table 74-H-1.

Table 74-H-3. *Free energies of formation of tungsten compounds* (*in kcal/mole at* 25° *C*)

	LATIMER (1952)	KHODAKOVSKII and MISHIN (1968)	ROBIE *et al.* (1968)	ROSSINI *et al.* (1950)
WO_4^{2-}	−220	−222.5 ± 1.0		
$CaWO_4$	−368.7	−366.7	−376.906 ± 0.510	
$FeWO_4$				
$MnWO_4$				
WO_3	−181.2	−182.63	−182.631 ± 0.210	−182.47
WS_2	−47.5		−71.21 ± 0.64	−46.2

	KUBASCHEWSKI, EVANS, ALCOCK (1967)	BULAKH (1968)	URUSOV *et al.* (1967)
WO_4^{2-}			
$CaWO_4$		−218.2	−364.7 ± 0.5
$FeWO_4$			−259.8 ± 2.0
$MnWO_4$			−287.8 ± 1.0
WO_3	−183.0 ± 1.3		
WO_2	−47.4 ± 5.0		

Thermochemical data on which calculations of solubility might be based are incomplete, particularly the data for tungstate ion, WO_4^{2-}, and its hydrolysis products. The free energies of formation in Table 74-H-3 illustrate the kind of information available. The discrepancies among the free energies are so large, and the value for the ion is so uncertain, that further calculations of solubility seem pointless. At least the *order* of solubilities of scheelite, ferberite, and huebnerite as calculated from the data of URUSOV *et al.* agrees with the relative solubilities of the three minerals noted by several observers (e.g., GANNETT, 1919; GUNDLACH and THORMANN, 1960): scheelite most soluble, huebnerite slightly less, ferberite considerably less. The uncertain but very low solubility product calculated for $PbWO_4$ is consistent with the occurrence of stolzite and raspite as secondary tungsten minerals.

For tungstite and tungstenite no solubility data have been reported. Qualitatively tungstite is known to dissolve readily in alkali, presumably forming WO_4^{2-}. For the possible reaction

$$WO_3 + H_2O \rightleftharpoons WO_4^{2-} + 2H^+$$

a rough calculation from the data of Table 74-H-3 indicates that tungstite should be appreciably soluble (concentration of WO_4^{2-} greater than 10^{-5} M) at pHs greater than 4.5, and that the solubility increases rapidly as the pH rises (1 M at pH 7), but the limiting figure is very uncertain. Tungstenite should dissolve in slightly acid solutions according to the reaction

$$WS_2 + 4H_2O \rightleftharpoons WO_4^{2-} + 2H_2S + 4H^+ + 2e^-$$

and another rough calculation shows that the sulfide would be soluble under usual conditions of pH and Eh. This rare mineral, the only one in which tungsten has a valence less than +6, can be stable at ordinary temperatures only when conditions are unusually reducing.

The free-energy data permit a somewhat more reliable calculation for the reaction

$$MeWO_4 + 2H^+ \rightleftharpoons Me^{2+} + WO_3 + H_2O$$

where Me^{2+} may be Ca^{2+}, Mn^{2+}, or Fe^{2+}. The calculation indicates that appreciable alteration of scheelite to tungstite should take place in solutions near neutrality, and of huebnerite in faintly acid solutions, but that alteration of ferberite should require a pH of 2 or 3. This conclusion is in general accord with observations on the solubility of the three compounds under both natural and laboratory conditions, although in nature the reaction for ferberite and huebnerite would usually be aided by oxidation of the divalent ions.

Tungsten is known to form many complex ions and molecules which must influence the solubility of its compounds, but the only quantitative data are those reported by GUNDLACH and THORMANN (1960); when an acid solution of the heteropolysilicic acid $H_8Si(W_2O_7)_6$ containing 1.4% WO_3 and 0.12 M Ca^{2+}, Mn^{2+}, or Fe^{2+} is titrated with NaOH, the precipitation of $FeWO_4$ begins at pH 5.9, of $MnWO_4$ at 6.7, and of $CaWO_4$ at 7.3.

Adsorption of tungsten out of solution, especially on the oxides of iron and manganese, is mentioned by many authors. A few quantitative experiments are reported by KRAINOV *et al.* (1965); if a solution containing 100 ppb W in the form of WO_4^{2+} is maintained at a pH between 6 and 7, 60—80% of the tungsten is adsorbed on either ferric oxide or manganese dioxide.

Manuscript received: July 1969

74-I. Abundance in Natural Waters

The concentration of tungsten in natural waters is very low, seldom rising above 1 ppb except in a few hot springs and in groundwater directly associated with ore deposits.

In sea water, ISHIBASHI *et al.* (1953, 1954) give 0.00009—0.00012 ppb W in samples from the Pacific Ocean near Japan. GOLDBERG (1965) suggests a general average of 0.0001 ppb, and a residence time for tungsten in the ocean of about 1,000 years.

For surface waters, UDODOV and PARILOV (1961) report that of 2,223 samples from Siberian streams and lakes only 18 had detectable tungsten (greater than 0.8 ppb), and that the range in these 18 samples was 0.8—4.0 ppb W. The same authors measured 8 ppb in mine water from a wolframite-quartz deposit. LANDSTRÖM and WENNER (1965) report 0.03 ppb W for streams in Sweden. KRAINOV *et al.* (1965) conclude from an extensive study of groundwater associated with tungsten deposits in many parts of the U.S.S.R. that acid waters directly associated with ore can range up to 100 ppb W, and waters of the dispersion halo around a tungsten deposit up to 15 ppb. The tungsten content of mine waters shows a general increase with pH, but the relation is not simple; on the basis of a few analyses KRAINOV *et al.* suggest that "deep alkaline water" may contain 100—300 ppb W. The same authors find by experiment that contact with iron and manganese oxides reduces the tungsten content of slightly acid water from 100 ppb to 20—40 ppb.

The extraordinary concentration of 40—64 ppm W is reported in the brines of Searles Lake in California (CARPENTER and GARRETT, 1959). Among salt lakes Searles Lake is highly exceptional, however, for it lies in a region of many small tungsten deposits, and the tungsten in its water has accumulated over a period of at least several hundreds of thousands of years.

In hot-spring waters concentrations of tungsten up to a few hundred parts per billion are not uncommon. CARPENTER and GARRETT (1959) note that one of the springs feeding into Searles Lake contains 240 ppb W. For hot springs of a particular type (pH 9—10, temperature 28—100° C, SiO_2 at least 100 ppm, chief dissolved salts Na_2SO_4 and $NaHCO_3$, total mineralization generally less than 1 g/liter, chief accompanying gas N_2), KRAINOV (1965) reports tungsten concentrations of 15—300 ppb in the widely scattered localities of Bulgaria, Pamir, Tienshan, eastern Sayan, western Transbaikal, and the Maritime Provinces of eastern Siberia. In general the tungsten content increases with temperature and with total dissolved solids, and is highest in areas of tungsten mineralization.

Manuscript received: July 1969

74-K. Abundance in Common Sediments and Sedimentary Rocks

The analytical data in Table 74-K-1 indicate that tungsten has about the same average abundance in sedimentary rocks as in igneous rocks, 1—2 ppm. As with most of the heavy metals, the concentration in carbonate rocks is notably less than that in rocks derived from clastic sediments. WIENDL (1968) has called attention to the fact that tungsten is at least as abundant in sandstones as in shales; since quartz and feldspar are usually low in tungsten, he explains this fact by supposing that the metal in sandstones is concentrated in the heavy minerals or in clay cement. Data are too few to give a clear picture of the behavior of tungsten in deep-sea sediments or in sediments containing abundant organic matter.

Most of the scattered high figures in Table 74-K-1 can be plausibly explained by local sources of abnormal amounts of tungsten during deposition of the sediment. The large concentrations in marine sediments from the Okhotsk Sea (ISAYEVA, 1960) and from the Pacific near Kamchatka (PETELIN and OSTROUMOV, 1961) probably result from the activity of nearby volcanoes. A similar explanation may account for the high values obtained by AMIRUDDIN and EHMANN (1962) in a core from the mid-Pacific; these authors note that their four analyses indicate a steady downward increase in tungsten concentration, from 3.2 ppm within 16 cm of the sediment-water interface to 5.4 ppm at a depth of 3.7 m, and guess that this relation may indicate a decline in volcanic activity during deposition of the sediment. The high average values for sedimentary rocks in Uganda (JEFFERY, 1959) probably result from inclusion of rocks from a mineralized area with a regionally high concentration of tungsten (as noted by PETTIJOHN, 1963). The abnormal concentration found by DEKATE (1967) in ferruginous sandstones and shales may be correlated with the ease of adsorption of tungsten on ferric and manganese oxides. A similar correlation is suggested by ISAYEVA's (1960) observation that tungsten is particularly abundant in Okhotsk Sea sediments that contain abnormal amounts of iron and manganese, and of course also by the startling concentrations sometimes found in manganese oxide sediments (Table 74-D-1).

For soils no systematic study of tungsten abundance has been made, but a few scattered analyses have been published in reports on geochemical prospecting for tungsten ores. DEKATE (1967) notes that tungsten is preferentially concentrated in the B horizon of Indian soils, and that tungsten concentrations in soils of mineralized areas can reach 29 times background values. HOLMAN and WEBB (1957), using samples collected at a depth of 45 cm in thick tropical soils in Uganda, report a regional background over unmineralized rock of 4—6 ppm W, an increase to 6—12 ppm near tungsten deposits, and values directly over an ore body generally between 20 and 40 ppm but ranging up to 100 ppm. GRANIER (1958), working with the A_1 horizon of a soil near a quartz vein containing wolframite, finds a background value of 10 ppm and anomalies up to 80 ppm near the vein. These few data permit

Table 74-K-1. *Tungsten in sediments and sedimentary rocks*

Rock	Number of analyses	Abundance, ppm W range	average	Locality	Reference
Shales and clays:					
Shale, average			1.8	World	TUREKIAN and WEDEPOHL (1961)
Clays and shales, average			2.0	World	VINOGRADOV (1962b)
Shale, average			1.9	World	HORN and ADAMS (1966)
Clay	Composite of 7,614 samples		1.8 (SW)	Russian platform	VINOGRADOV *et al.* (1958)
Shales and clays			1.5 (C)	India	DEKATE (1967)
Shale, siltstone, mudstone	23	1.2—9.0[a]	3.8 (C)	Uganda	JEFFERY (1959)
Sandstones:					
Sandstone, average			1.6	World	TUREKIAN and WEDEPOHL (1961)
Sandstone, average			1.6	World	HORN and ADAMS (1966)
Sandstone and quartzite			1—2	World	PETTIJOHN (1963)
Sand	Composite of 6,107 samples		1.6 (SW)	Russian platform	VINOGRADOV *et al.* (1958)
Sandstone			1.5 (C)	India	DEKATE (1967)
Ferruginous sandstone and clay			4.3 (C)	India	DEKATE (1967)
Sandstone	5	1.2—5.5[b]	3.6 (C)	Uganda	JEFFERY (1959)
Carbonate rocks:					
Carbonate rocks, average			0.6	World	TUREKIAN and WEDEPOHL (1961)
Carbonate rocks, average			0.56	World	HORN and ADAMS (1966)
Limestone and dolomite			0.4 (C)	India	DEKATE (1967)
Limestone	4	0.2—0.7	0.48 (C)	Uganda	JEFFERY (1959)
Organic sedimentary rocks:					
Coal ash	596	40—440	90 (S)	West Virginia	HEADLEE and HUNTER (1953)
Anthracite ash		8—90	(?)	Pennsylvania	NUNN *et al.* (1953)
Carbonaceous shale and lignite			2.5 (C)	India	DEKATE (1967)

Table 74-K-1 (Continued)

Rock	Number of analyses	Abundance, ppm W		Locality	Reference
		range	average		
Modern marine sediments:					
Deep-sea clay, average		1—10		World	TUREKIAN and WEDEPOHL (1961)
Ocean clay, average			1.1	World	HORN and ADAMS (1966)
Deep-sea sediments	4	3.18—5.35	4.5 (N/R)	Mid-Pacific	AMIRUDDIN and EHMANN (1962)
Red clay	5	16—18	17 (C)	NE Pacific	ISAYEVA (1960)
Bottom sediments	145	5—70	12—20 (C)	Okhotsk Sea	ISAYEVA (1960)
Clay and sand		15—40	(?)	Near Kamchatka	PETELIN and OSTROUMOV (1961)
Arenaceous and argillaceous sediments		10—130		Black Sea	PILIPCHUK, VOLKOV (1966)
Deep-sea carbonate		0.1—1.0		World	TUREKIAN and WEDEPOHL (1961)
Ocean carbonate			0.11	World	HORN and ADAMS (1966)
Acid-soluble part of Mn nodules		50—100			RILEY and SINHASENI (1958)

[a] 2 samples with high concentrations omitted: 15.1 and 50.7 ppm.
[b] 1 sample with high concentration omitted: 33.8 ppm.

a tentative conclusion that tungsten is slightly enriched in the B horizon or lower A horizon of normal soils and that it forms well defined haloes in soils around ore deposits.

Manuscript received: July 1969

74-L. Biogeochemistry

Almost nothing is known about the biogeochemistry of tungsten. Some concentration by plants is indicated by DEKATE's (1967) observation that plants growing in a mineralized zone contained 2 to 18 times as much tungsten as the background value of 2.7 ppm. High values in some samples of coal ash (Table 74-K-1) may indicate concentration by the original coal-swamp vegetation, but they are equally well explained by later adsorption of tungsten on the dead organic material.

A major gap in the analytical data on tungsten is the lack of figures for black shales and other organic-rich sedimentary rocks. To some extent the deficiency is made up by abundance values for graphitic schists and phyllites, the probable metamorphic equivalents of black shales (Table 74-M-1). JEFFERY (1959) suggests that his high figures for graphitic phyllites in Uganda (8—86 ppm W) may mean an original concentration of the metal by organic processes, but that this conclusion is not beyond question because the phyllites occur in areas of tungsten mineralization. PARGETER (1956) and DE MAGNÉE and ADERCA (1960) are less cautious in their interpretation of similar phyllites in southern Uganda and adjacent parts of Ruanda; they feel that the high content of tungsten (50—192 ppm, according to analyses by JEDWAB, 1958), the correlation of tungsten with amount of carbon, the presence of nodular ferberite, and the stratigraphic control of tungsten concentration, all point to a syngenetic sedimentary origin of the tungsten enrichment. DEKATE (1967) records an abundance figure for graphitic phyllites and schists in India that is very similar to the figures obtained in Uganda and Ruanda.

Like other heavy metals, tungsten in more than traces is poisonous to most organisms. An important biogeochemical role for the element is therefore hardly likely, but minor concentration in organic processes is certainly possible and is strongly suggested by the fragmentary data on carbonaceous metamorphic rocks.

Manuscript received: July 1969

74-M. Abundance in Common Metamorphic Rock Types

The abundance figures in Table 74-M-1 suggest that metamorphic rocks contain roughly twice as much tungsten as their igneous and sedimentary equivalents. This seems an unlikely conclusion, and probably reflects the fact that the data come almost exclusively from two sources: JEFFERY's (1959) work in Uganda and DEKATE's (1967) in India. As noted previously for igneous and sedimentary abundances, JEFFERY's results indicate that rocks from parts of Uganda in general contain somewhat more tungsten than crustal averages. The very high figures that JEFFERY obtained for a few metamorphic rocks in mineralized areas have been excluded from the table, but apparently slight mineralization is so widespread that averages are still somewhat abnormal. DEKATE does not describe the source of his samples, but since his work was primarily a study of tungsten ore deposits in India it seems possible that his averages also are influenced by samples collected in mineralized areas. Perhaps it is significant that the rocks studied in both India and Uganda are almost exclusively Precambrian.

Table 74-M-1. *Tungsten in metamorphic rocks*

Rock	Number of analyses	Abundance, ppm W		Locality	Reference
		range	average		
Metamorphic rocks, general			3.7 (C)	India	DEKATE (1967)
Metasedimentary rocks			2.3 (C)	India	DEKATE (1967)
Phyllite	19	0.3—15.7[a]	4.7 (C)	Uganda	JEFFERY (1959)
Phyllite		4—400		Uganda	PARGETER (1956)
Mica schist	5	1.2—4.8	2.5 (C)	Uganda	JEFFERY (1959)
Other schists	5	0.5—6.2[b]	2.7 (C)	Uganda	JEFFERY (1959)
Pelitic schists			2.6 (C)	India	DEKATE (1967)
Schists	33	1—25	1—2 (WS)	New Zealand	WILSON and FIELDES (1944)
Graphitic phyllite	40	8—86	22.8 (C)	Uganda	JEFFERY (1959)
Graphitic phyllite		50—192	(C)	Ruanda	JEDWAB (1958)
Graphitic phyllite and schist			21.2 (C)	India	DEKATE (1967)
Quartzite	8	0.2—10.4[c]	3.4 (C)	Uganda	JEFFERY (1959)
Quartzite			0.5 (C)	India	DEKATE (1967)
Marble	3	0.5—0.8	0.7 (C)	Uganda	JEFFERY (1959)
Marble			0.4 (C)	India	DEKATE (1967)
Amphibolite	5	2.7—6.3	4.0 (C)	Uganda	JEFFERY (1959)
Metabasic rocks			6.8 (C)	India	DEKATE (1967)
Charnockite			0.6 (C)	India	DEKATE (1967)

[a] 3 samples with high concentrations omitted: 28.5, 89.9, and 128.0 ppm.
[b] 2 samples with high concentrations omitted: 12.5 and 25.0 ppm.
[c] 2 samples with high concentrations omitted: 33.8 and 38.6 ppm.

For metasedimentary rocks Table 74-M-1 bears out the general relationship noted for the original sedimentary types (Table 74-K-1); tungsten is conspicuously higher in rocks derived from clastic sediments than in those from carbonates (except that DEKATE's figure for quartzites, 0.5 ppm, does not fit the pattern). Particularly striking are the high values for phyllites and schists containing graphite. As noted in Section L, these concentrations suggest an initial enrichment of tungsten in the organic matter from which the graphite was derived.

The high figures for metamorphic equivalents of mafic igneous rocks (JEFFERY's "amphibolites" and DEKATE's "metabasic rocks") are unexplained. JEFFERY's figure is consistent with the high tungsten content found in Uganda dolerites (Table 74-E-1), but DEKATE's average value for unmetamorphosed mafic rocks in India is much lower, as is the general average for such rocks in other parts of the world.

Manuscript received: July 1969

74-N. Behavior in Metamorphic Reactions

The replacement of one tungstate mineral by another during hydrothermal alteration has been mentioned in Section 74-G. The change from wolframite to scheelite is well documented both by observations in nature and by experiment (e.g., Bryzgalin, 1958; Heier, 1955), and evidence for the reverse reaction from scheelite to ferberite has been described by de Magnée and Aderca (1960). Presumably the direction of such reactions depends primarily on ratios of the divalent cations in contact with the original minerals. If the hydrothermal solutions contain an abundance of sodium rather than divalent cations the tungsten minerals may simply be dissolved, particularly at high temperatures where Na_2WO_4 would be largely undissociated; this is suggested by an observation of Studenikova and Pavlenko (1960) that tungsten is often removed from rocks that have undergone sodium metasomatism but is unaffected by potassium metasomatism. A more complicated replacement reaction is needed to explain the observation of Rimskaya-Korsakova and Troyanov (1956) that tungstenite forms pseudomorphs after scheelite in a contact-metamorphic deposit in Uzbekistan.

Except for such local replacement reactions, the metamorphic behavior of tungsten is unknown. As a speculation, it is often suggested that the minute amounts of tungsten in ordinary rocks may be partly dissolved by fluids generated during metamorphism, transported long distances, and concentrated in places where the solutions change character. Tweto (1960), for example, suggests that minor concentrations of scheelite in lime-rich parts of Precambrian gneisses in Colorado may owe their tungsten to metamorphism of large volumes of shale containing only average amounts of the metal. De Magnée and Aderca (1960) propose a similar origin for the quartz-ferberite veins cutting the tungsten-rich graphitic phyllites in northern Ruanda, and carry the idea further by suggesting that most tungsten deposits, even those intimately associated with granites, owe their initial metal concentration to accumulation of tungsten in sediments which are subsequently metamorphosed and partly melted. From a purely chemical viewpoint the mechanism seems feasible, but conclusive evidence for its large-scale operation in nature is difficult to obtain.

Manuscript received: July 1969

74-O. General Considerations, Geochemical Relations to Other Elements, Economic Importance

I. General Considerations

As befits a heavy metal of the sixth subgroup of the Periodic Table, tungsten is a lithophile element in terrestrial environments, although in the extreme reducing conditions of meteorite formation it becomes chiefly siderophile. Its average abundance in crustal material lies certainly within the limits 0.5 and 2.0 ppm, and probably within the range 1.0–1.5. Evidence is fairly conclusive for slight inrichment in felsic and alkalic igneous rocks as compared with more mafic rocks. This enrichment, together with the very low abundance figures for chondritic meteorites, indicates that tungsten has been concentrated in crustal rocks rather than the mantle during the large-scale differentiation of the earth's material. Sedimentary processes only locally lead to much concentration of the element; shales and sandstones in general are not far from the crustal average, limestones two or three times less, and rocks with abundant organic matter sometimes considerably more.

More clearly than for most of the heavy metals, notable concentrations of tungsten are unambiguously linked to igneous processes. Most major ore deposits are associated with granitic rocks, but the metal shows no preference for particular rock compositions in the range from diorite to alkali granite.

Figures for average abundances of tungsten in major rock types and mineral groups are probably correct within a factor of two or three, but the values have too much scatter to establish with certainty trends in rock series or partition ratios between mineral pairs. Information about tungsten ore deposits is still almost entirely descriptive. The element would well repay further study, but a major difficulty is the lack of sensitivity of two common analytical techniques, optical spectrography and X-ray fluorescence spectrography. Chemical methods of great sensitivity have been developed, but are fairly laborious. In recent years neutron activation as an analytical technique has shown great promise.

Recent abundance figures are almost uniformly much smaller than older figures, often by more than an order of magnitude. The consistency of recent results obtained with a variety of analytical techniques indicates that most values dating from before 1946 must be in error.

The geochemistry of tungsten has been exhaustively reviewed in a recent dissertation by WIENDL (1968). This Chapter 74 is taken largely from WIEND's work and the references he cites, with a few additions from more recent literature.

II. Geochemical Relations to Other Elements

The closest relative of tungsten is molybdenum. The two metals appear in the same column of the Periodic Table, have similar ionic radii and electro-negativities, and exhibit the same valences of +4 and +6 in natural materials. Isomorphous substitution is extensive in some compounds; the limits of possible substitution have not been explored, but are known to be much wider for scheelite than for wolframite. Molybdates are uniformly more soluble than the corresponding tungstates, a fact correlated with the slightly greater electronegativity of molybdenum and hence with the greater covalent character of its bond with oxygen (POVARENNYKH, 1961; URUSOV *et al.*, 1967).

The separation of the two elements in nature depends on the tendency of molybdenum to form the sulfide MoS_2 in hypogene deposits and of tungsten to form tungstates. The difference is apparent in the free energies of the two reactions

$$MoS_2 + 3/2\ O_2 \rightleftharpoons MoO_3 + S_2(gas) \qquad \Delta G_{700^\circ K} = -80.6 \text{ kcal}$$
$$WS_2 + 3/2\ O_2 \rightleftharpoons WO_3 + S_2(gas) \qquad \Delta G_{700^\circ K} = -107.5 \text{ kcal.}$$

At concentrations of O_2 and S_2 to be expected in hypogene environments, the first equilibrium would favor the sulfide and the second the oxide (KRAUSKOPF, 1957; IVANOVA and KHODAKOVSKII, 1969). (The behavior of the oxides would reflect the behavior of the tungstates and molybdates found in nature, since free energies for reactions of the type

$$MeO + WO_3 \rightleftharpoons MeWO_4 \text{ and } MeO + MoO_3 \rightleftharpoons MeMoO_4$$

are relatively small.) In accordance with the free energies, molybdates and tungsten sulfide appear only very rarely in hypogene deposits. Molybdates can be stable only where O_2 is abundant, as in the zone of weathering. Although free energies help to quantify the difference in behavior of the two elements, the difference is not easily interpreted in terms of more fundamental atomic properties.

Molybdenum and tungsten resemble each other also in surface environments. Both form hydrated oxides on weathering, and both form insoluble compounds with lead, copper, and iron. The greater solubility of most of the compounds of molybdenum gives it a greater mobility and wider dispersion in surface waters. Molybdenum is an essential trace element in the life processes of some plants, and perhaps for this reason it is more concentrated than tungsten in the organic matter of sediments.

Two other elements whose geochemistry somewhat resembles that of tungsten are its close neighbors in the fifth group of the Periodic Table, niobium and tantalum. These metals are often reported in amounts of a few hundredths of a percent in analyses of scheelite and wolframite.

Manuscript received: July 1969

III. Economic Importance[1]

Tungsten is economically important because of its unusual properties and the properties it imparts to its alloys and compounds. The pure metal has a high density ($19.35\ g/cm^2$), the highest melting point (3410° C) and tensile strength ($41{,}500\ kg/cm^2$) of all metals, a high heat of sublimation, a low compressibility, and a low coefficient of thermal expansion. Many of its alloys have a remarkable ability to retain strength, hardness, and wear resistance at high temperatures. Tungsten carbides are valued for their extreme hardness and wear resistance. The major uses of tungsten follow

1 References of this subsection are listed at its end

directly from these properties: for the production of metal-working machinery, construction and mining machinery, tool steel, drill bits, turbine blades, rocket nozzles, armor-piercing shells, valves, tips for ball-point pens, dies for wire-drawing, and structural material for nuclear and space applications. An important use of tungsten, although the total amount needed is small in comparison with that for other uses, is in electrical equipment, particularly filaments for light bulbs, contact points, electrodes, and high-temperature thermocouples. Tungsten compounds are used in paints and ceramics and as catalysts; sodium tungstate is an important fireproofing agent; and tungsten sulfide and selenide are used as lubricants. Of the total quantity used commercially, in recent years about 68 percent has gone into tungsten carbide and 15 percent into alloy steels and nonferrous alloys; 15 percent has been used as tungsten metal and tungsten-base alloys, and the remaining 1 or 2 percent for various chemical purposes.

The two important ore minerals of tungsten are wolframite ($(Fe,Mn)WO_4$) and scheelite ($CaWO_4$). These are easily separated mechanically from most gangue minerals because of their high densities, and further purification is accomplished by magnetic separation (for wolframite) and flotation (for scheelite). Tungsten is obtained from the ore minerals or other compounds by reduction at high temperatures with carbon or hydrogen. For the production of alloy steels, tungsten may be added in the form of scheelite or as ferrotungsten prepared by reduction of tungsten ore with carbon. Tungsten carbides (WC and W_2C) are produced by heating tungsten powder with charcoal to temperatures of 1400–1600° C; the powdered carbides are commonly compacted and sintered with cobalt as a binder. Tungsten metal is prepared in the form of wire, rods, and sheets by the techniques of powder metallurgy.

A major problem in the use of tungsten and its alloys is the rapid oxidation of the metal at temperatures over 1000° C. Research on means of controlling the oxidation rate by coating or alloying the metal has so far met with little success. An alloy with rhenium has the necessary resistance to oxidation, but rhenium is too expensive for the alloy to be widely used.

Tungsten is considered a strategically important metal because in some of its industrial and military applications it has no satisfactory substitute. Molybdenum can be used in place of tungsten in tool steels and high-temperature nonferrous alloys, but the molybdenum alloys are inferior in hardness and wear resistance. Titanium carbide, alumina, and industrial diamonds can to some extent take the place of tungsten carbide in cutting tools. For tungsten in electrical apparatus, particularly for filaments in incandescent lights, no even partly satisfactory substitutes are known. Tungsten will presumably remain an essential material for industrial civilization during the foreseeable future.

Recent estimates of *world reserves* of tungsten ore range from 1.2 million to 2.2 million tons of contained tungsten. Between 60 and 75 percent of the total reserves are estimated to be in southern China; other important producers are Australia, Bolivia, Burma, Canada, Korea, USA, and USSR. The total *world demand* for tungsten in 1968 was about 25 million kilograms, and the estimated demand for the year 2000 is predicted to be between 50 and 65 million kilograms. In recent years the USA and USSR have each accounted for about 20 percent of the total demand; other major users are West Germany, Great Britain, Japan, and Poland. Comparison of figures for estimated future requirements and world reserves indicates that reserves will be seriously depleted within a few decades, possibly before 2000. Exploitation of lower-grade deposits and improved recovery of tungsten scrap, however, can augment supplies considerably.

Prices of tungsten have fluctuated widely during this century, reaching high levels that stimulated much mining activity during the two world wars and during the 1950's when the United States was stockpiling tungsten ore. From 1965 to 1976 prices were stabilized by the decision of the United States government to sell from its stockpile at a fixed price. Beginning in 1976 the price rose precipitately, and the rise has prompted the reopening of older mines and development of new ones.

References: Gibbs, E. C.: Tungsten—rising demand, new production for '77. Engineering and Mining Journal, **178**, 110 (1977). Hobbs, S. W., Elliott, J. E.: Tungsten. In: United States Mineral Resources, U.S. Geological Survey Professional Paper **820**, 667 (1973). Roethe, G.: Wolfram. In: Gocht, W. (ed.) Handbuch der Metallmärkte. Springer-Verlag, Berlin-Heidelberg-New York, 1974. Stevens, R. F.: Tungsten. In: Minerals Yearbook for 1973. U.S. Bureau of Mines, **1**, 1245 (1973). Stevens, R. F.: Tungsten. In: Mineral Facts and Problems. U.S. Bureau of Mines Bull. **667**, 1161 (1975).

Manuscript received: December 1977

References: Sections 74-B to 74-O

AMIRUDDIN, A., and W. D. EHMANN: Tungsten abundances in meteoritic and terrestrial materials. Geochim. Cosmochim. Acta **26**, 1011—1022 (1962).

ATKINS, D. H. F., and A. A. SMALES: The determination of tantalum and tungsten in rocks and meteorites by neutron activation analysis. Anal. Chim. Acta **22**, 462—478 (1960).

BERKEM, A. R.: Determination of the solubility of normal alkaline earth tungstates and lead tungstate. Rev. Fac. Sci. Univ. Istanbul **8**, 332—8 (1943).

BOKII, G. B., and I. N. ANIKIN: Determination of the solubility of scheelite in water and aqueous solutions. Zh. Neorgan. Khim. **1**, 1926—1928 (1956).

BOLDUAN, H.: Genetische Untersuchungen der Wolframitlagerstätte Pechtelsgrün/V. unter besonderer Berücksichtigung der Verteilung des H/F-Koeffizienten und der Spurenelemente Niob und Tantal im Wolframit. Freiberger Forschungsh. C **10**, 46—61 (1954).

BRYZGALIN, O. V.: The origin of scheelite in skarn ore deposits. Geochemistry **1958**, 297—304.

BULAKH, A. G.: Metody Termodinamiki v Mineralogii. (Thermodynamic methods in mineralogy.) Leningrad: Nedra 1968.

CAMERON, A. G. W.: A revised table of abundances of the elements. Astrophys. J. **129**, 676—699 (1959).

CARPENTER, L. C., and D. E. GARRET: Tungsten in Searles Lake. Mining Eng. **11**, 301—303 (1959).

CHERNYSHEV, L. V., and G. F. IVANOVA: Proizvedeniya rastvorimosti tverdykh faz peremennovo sostava i ikh geokhimicheskoye znacheniya. (Solubility products of solid phases with varying compositions and their geochemical importance.) Geokhimiya **1969**, 565—570

CLARK, S. P.: Isotopic abundances. Handbook of Physical Constants (S. P. CLARK, ed.). Geol. Soc. Am. Mem. **97**, 11—17 (1966).

CLAYTON, D. C., and W. A. FOWLER: Abundances of heavy nuclides. Preprint (1961). Cit. by AMIRUDDIN and EHMANN (1962).

DANA, J. D.: The system of mineralogy, 7th ed., vol. I. New York: Wiley 1944.

— The system of mineralogy, 7th ed., vol. II. New York: Wiley 1951.

DAVIDSON, C. F.: Uraninite-scheelite placers of the River Indus. Econ. Geol. **57**, 456—457 (1962).

DEKATE, Y. G.: Tungsten occurrences in India and their genesis. Econ. Geol. **62**, 556—561 (1967).

DOUCET, S.: Sur les conditions géochimiques d'altération, de migration et d'accumulation de quelques minéraux détritiques. Bull. Soc. Franç. Mineral. Crist. **89**, 120—127 (1966).

GANNETT, R. W.: Experiments relating to the enrichment of tungsten ore. Econ. Geol. **14**, 68—78 (1919).

GLEMSER, O., u. R. VON HAESELER: Über gasförmiges $WO_2(OH)_2$. Naturwissenschaften **47**, 467—468 (1960).

GOLDBERG, E. D.: Minor elements in sea water. Chemical Oceanography (J. P. RILEY and G. SKIRROW, eds.), vol. 1, p. 163—196 (1965).

GRANIER, C.: Dispersion secondaire du tungstène et de l'arsenic en sol résiduel. Bull. Soc. Franç. Mineral. Crist. **81**, 194—200 (1958).

GRUBB, P. L. C.: Solid solution relationships between wolframite and scheelite. Am. Mineralogist **52**, 418—426 (1967).

GUNDLACH, H., u. W. THORMANN: Versuch einer Deutung der Entstehung von Wolfram- und Zinnlagerstätten. Z. Deut. Geol. Ges. **112** (I), 1—35 (1960).

HAMAGUCHI, H., R. KURODA, T. SHIMIZU, I. TSUKAHARA, and R. YAMAMOTO: Values for trace elements in G-1 and W-1 with neutron activation analysis. Part II. Mo, Sn, Ta, W. Geochim. Cosmochim. Acta **26**, 503—505 (1962).
—, T. NAKAI, and E. IDENO: Determination of tungsten in rocks and meteorites by neutron activation. J. Chem. Soc. Japan **82**, 1493—1498 (1961).
HEADLEE, A. J. W., and R. G. HUNTER: Elements in coal ash and their industrial significance. Ind. Eng. Chem. **45**, 548—551 (1953).
HEIER, K.: The Ørsdalen tungsten deposit. Norsk Geol. Tidsskr. **35**, 69—85 (1955).
VON HEVESY G., u. R. HOBBIE: Die Ermittlung des Molybdän- und Wolframgehaltes von Gesteinen. Z. Anorg. Allgem. Chem. **212**, 134—144 (1933).
HEWETT, D. F., and M. FLEISCHER: Deposits of the manganese oxides. Econ. Geol. **55**, 1—55 (1960).
— —, and N. CONKLIN: Deposits of the manganese oxides: supplement. Econ. Geol. **58**, 1—51 (1963).
HOLMAN, R. H. C., and J. S. WEBB: Exploratory geochemical soil survey at Ruhiza ferberite mine, Uganda. 6th Commonwealth Mining Metallurg. Congr., Montreal: Methods and case histories in mining geophysics, 353—357 (1957).
HORN, M. K., and J. A. S. ADAMS: Computer-derived geochemical balances and element abundances. Geochim. Cosmochim. Acta **30**, 279—297 (1966).
ISAYEVA, A. B.: Volfram v donnikh otlozheniyakh Okhotskovo Morya. (Tungsten in bottom deposits of the Sea of Okhotsk.) Dokl. Akad. Nauk SSSR **131**, 416—419 (1960).
ISHIBASHI, M.: Studies on minute elements in sea water. Records Oceanog. Works Japan **1**, 88—92 (1953).
—, T. SHIGEMATSU, and Y. NAKAGAWA: Determination of tungsten and molybdenum in sea water. Bull. Inst. Chem. Res. Kyoto Univ. **32**, 199—202 (1954).
IVANOVA, G. F.: Content of tin, tungsten and molybdenum in granites enclosing tin-tungsten deposits. Geochemistry **1963**, 492—500.
— Thermodynamic evaluation of the possibility of tungsten transport as halogen compounds. Geochemistry **1966**, 964—973.
— Ob usloviyakh kontsentratsii volframa v greizenovom protsesse. (On the conditions of concentration of tungsten during greisenization.) Geokhimiya **1969**, 22—32.
—, and I. L. KHODAKOVSKII: Formy migratsii volframa v gidrotermalnykh rastvorakh. Geokhimiya **1968**, 930—940.
— — Ob usloviyakh obrazovaniya mineralov volframi i molibdena v gidrotermalnom protsesse. (Conditions of formation of tungsten and molybdenum minerals in hydrothermal processes.) Geokhimiya **1969**, 411—415.
JEFFERY, P. G.: The geochemistry of tungsten, with special reference to the rocks of the Uganda Protectorate. Geochim. Cosmochim. Acta **16**, 278—295 (1959).
KHASIN, R. A.: O zonalnosti izomorfnovo ryada ferberit-gyubnerit volframovykh mestorozhdenii (Zonal distribution of the isomorphous series ferberite-huebnerite in wolframite occurrences.) Dokl. Akad. Nauk SSSR **64**, 117—119 (1949).
— Über die zonale Verteilung der isomorphen Serie Ferberit-Hübnerit in den Wolframvorkommen. Dokl. Akad. Nauk SSSR **64**, 117—119 (1949). [Russ.] Abstract in: Zentr. Min. II, 423 (1953).
KHODAKOVSKII, I. L., and I. V. MISHIN: O proizvedenii rastvorimosti molibdata i volframata kaltsiya. (On the solubility products of calcium molybdate and calcium tungstate.) Geol. Rudn. Mestorozhd. **10**, No. 5, 51—62 (1968).
KIRCHHEIMER, F.: Weitere Untersuchungen über das Vorkommen von Uran im Schwarzwald. Abhandl. Geol. Landesamtes Baden-Wuerttemberg **1**, 1—60 (1953).
KRAINOV, S. R.: Geochemistry of fluorine, tungsten and germanium in nitrogenous thermal waters of crystalline rocks. Geochemistry **1965**, 1001—1011 (1965).
KRAUSKOPF, K. B.: The heavy metal content of magmatic vapor at 600° C. Econ. Geol. **52**, 786—807 (1957).
— The possible role of volatile metal compounds in ore genesis. Econ. Geol. **59**, 1—45 (1964).
KUBASCHEWSKI, O. E., E. LL. EVANS, and C. B. ALCOCK: Metallurgical Thermochemistry. Ed. 4. London: Pergamon 1967.

Kuroda, P. K., and E. B. Sandell: Geochemistry of molybdenum. Geochim. Cosmochim. Acta **6**, 35—63 (1954).

Landström, O., and C. G. Wenner: Neutron-activation analysis of natural water applied to hydrogeology. Aktiebolaget Atomenergi (Sweden) AE-204 1965.

Latimer, W. M.: The oxidation states of the elements and their potentials in aqueous solution. New York: Prentice Hall 1961.

Lemmon, D. M., and J. V. N. Dorr: Tungsten deposits of the Atolia District, San Bernadino and Kern Counties, California. U.S. Geol. Surv. Bull. **922**-H, 205—245 (1940).

Levina, S. D.: Tungsten minerals in the oxidized skarns of Central Asia. [Russ.] Zap. Vses. Mineralog. Obshchestva **83**, 106—116 (1954).

Lyakhovich, V. V., and T. T. Balanova: O srednem soderzhanii W, Mo, Sn, Ta, Nb, Zn v sfene i ilmenite iz granitoidov. (On the average content of W, Mo, Sn, Ta, Nb, Zn in sphene and ilmenite from granitic rocks.) Geokhimiya **1969**, 322—327.

Magnée, I. de, et B. Aderca: Contribution à la connaissance du tungsten-belt ruandais. Acad. Roy. Sci. Outre-Mer, Classe Sci. Nat. et Med., Mem. 8, N.S., **11**, No. 7, 55 p. (1960).

Millner, T., and J. Neugebauer: Volatility of the oxides of tungsten and molybdenum in the presence of water vapor. Nature **163**, 601—602 (1949).

Moenke, H.: Spurenelemente in variskischen und prävariskischen deutschen Graniten, eine spektrochemische Analyse altersverschiedener granitischer Gesteine. Chem. Erde **20**, 227—301 (1960).

Newhouse, W. H.: The source of V, Mo, W and Cr in oxidized lead deposits. Am. Mineralogist **19**, 209—220 (1934).

Nunn, R. C., H. L. Lovell, and C. C. Wright: Spectrographic analyses of trace elements in anthracite. Trans. Ann. Anthracite Conf., Lehigh Univ. **11**, 51—65 (1953).

Oelsner, O.: Bemerkungen über die Anwendbarkeit des H/F-Koeffizienten zur Deutung der Genese von Wolframiten. Freiberger Forschungsh. C **10**, 62—67 (1954b).

Pargeter, R. C.: The Ruhizha ferberite deposit, Kigezi. Records. Geol. Surv. Uganda 27—46 (1954).

Petelin, W. P., and E. A. Ostroumow: Geochemistry of sediments of Okhotsk Sea. [Russ.] Sovremen. Osadki Moreyi Okeanov 380—403 (1961).

Pettijohn, F. J.: Chemical composition of sandstones — excluding carbonate and volcanic sand. U.S. Geol. Surv. Profess. Papers **440**-S, 19 p. (1963).

Pilipchuk, M. F., and I. I. Volkov: Tungsten in Recent Black Sea sediments. Dokl. Acad. Nauk SSSR **167**, 430—433, 1966.

Polkinghorne, J. P. R.: Significance of variable iron content of wolframite in the west of England. Trans. Roy. Geol. Soc. (Cornwall) **18**, 526—534 (1953—54).

Pollett, J. D.: Annual report, Geological Survey Department for 1955, Freetown, Sierra Leone (1957).

Povarennykh, A. S.: The cause for different behavior of molybdenum and tungsten in endogenetic processes. [Russ.]. Geol. Zh. Akad. Nauk Ukrain. SSR **21**, No. 5, 39—47 (1961).

Riley, J. P., and P. Sinhaseni: Chemical composition of three manganese nodules from the Pacific Ocean. J. Marine Res. **17**, 466—482 (1958).

Rimskaya-Korsakova, O. M., and M. D. Troyanov: New data on tungstenite. [Russ.] Zap. Vses. Mineralog. Obshchestva **85**, 277—285 (1956).

Robie, R. A.: Thermodynamic properties of minerals. Handbook of Physical Constants (S. P. Clark, ed.). Geol. Soc. Am. Mem. **97**, 437—458 (1966).

Rossini, F. D., D. D. Wagman, W. H. Evans, S. Levine, and I. Jaffe: Selected values of chemical thermodynamic properties. U.S. Natl. Bur. Standards Circ. **500** (1952).

Sandell, E. B.: Abundance of tungsten in igneous rocks. Am. J. Sci. **244**, 643—648 (1946).

Schröcke, H.: Isomorphiebeziehungen in der Wolframitgruppe. Beitr. Mineral. Petrog. **7**, 166—206 (1960).

Shcherba, G. N., V. D. Gukova, A. V. Kudryashov, and N. P. Senchilo: Alkali feldspars from feldspar-quartz veinlets in molybdenum-tungsten deposits, Kazakhstan. Chapt. IV, quartz and feldspar-quartz veins and veinlets. Geochemistry **1964**, 141—191 (?).

SITKOVSKII, I. N., M. A. MUSTAFABELLI, and A. Z. VEZIR-ZADE. Uchenye Zapiski, Azerbaidzhan Gosudarst. Univ., Ser. Geol.-Geog. Nauk **1960**, No. 2, 29—35 (1960).

SOTNIKOV, V. I., and L. G. IZYUMOVA: O soderzhanii volframa v granitakh intrusivnikh massivov Gornova Altaya. (On the tungsten content of granite intrusives in the Altai Mountains.) Geokhimiya **1965**, 175—179 (1965).

ŠTEMPROK, M., and Z. ŠULCEK: Geochemical profile through an ore-bearing lithium granite. Econ. Geol. **64**, 392—404 (1969).

STUDENIKOVA, Z. V., and L. J. PAVLENKO: Contents of tungsten and molybdenum in alkaline rocks of the east Tuva and the northern Caucasus. Geochemistry **1960**, 709—717 (1960).

SUESS, H. E., and H. C. UREY: Abundances of the elements. Rev. Mod. Phys. **28**, 53—74 (1956).

TAKHOMIROV, N. I., and S. D. MILLER: On a physicochemical method of prospecting for molybdenum. [Russ.] Razvedka Nedr **12**, No. 2, 34—39 (1946).

TAYLOR, S. R.: Australites, Henbury impact glass and subgreywacke: a comparison of the abundance of 51 elements. Geochim. Cosmochim. Acta **30**, 1121—1136 (1966).

TUREKIAN, K. K., and K. H. WEDEPOHL: Distribution of the elements in some major units of the earth's crust. Bull. Geol. Soc. Am. **72**, 175—192 (1961).

TWETO, O.: Scheelite in the Precambrian gneisses of Colorado. Econ. Geol. **55**, 1406—1428 (1960).

UDODOV, P. A., and Y. S. PARILOV: Certain regularities of migration of metals in natural waters. Geochemistry **1961**, 763—771 (1961).

URUSOV, V. S., G. S. IVANOVA, and I. L. KHODAKOVSKIY: Energetic and thermodynamic properties of molybdates and tungstates and some features of their geochemistry. Geochemistry **1967**, 950—963 (1967).

VINOGRADOV, A. P.: Atomic abundance of elements in the sun and stony meteorites. Geochemistry **1962**a, 329—336 (1962).

— Average contents of chemical elements in the principal types of igneous rocks of the earth's crust. Geochemistry **1962**b, 641—664 (1962).

—, E. E. VAINSHTEIN, and L. J. PAVLENKO: Tungsten and molybdenum in igneous rocks (as related to the geochemistry of tungsten). Geochemistry **1958**, 497—509 (1958).

WHITE, D. E.: Thermal springs and epithermal ore deposits. Econ. Geol. 50th Anniv. vol., 90—154 (1955).

WIENDL, U.: Zur Geochemie und Lagerstättenkunde des Wolframs. Diss. Tech. Univ. Clausthal 1968.

WILSON, S. H., and M. FIELDES: A combined chemical and spectrographic method for the determination of traces of tungsten in rocks. Analyst **69**, 12—14 (1944).

ZESCHKE, G.: Prospecting for ore deposits by panning heavy minerals from river sands. Econ. Geol. **56**, 1250—1257 (1961).

Manuscript received: July 1969

Rhenium 75

A—C, D. F. C. Morris (Nuclear Science Centre,
E—G, and Brunel University, Uxbridge, U.K.)
I—L, E.L. Short
O

Introduction

As a result of its rarity the element of atomic number 75 was not discovered until 1925, although its existence had been predicted much earlier (MENDELEEV'S dvi-manganese) from considerations of periodicity[1]. The German investigators NODDACK, TACKE and BERG (1925)[2] are generally credited with the first detection of element 75 by its X-ray spectrum; they named the new element rhenium, after the West German province Rhineland. Other announcements claiming the isolation of this congener of manganese in periodic group VII A were made a few months later by LORING and DRUCE (1925) and by DOLEJŠEK and HEYROVSKÝ (1925). The history of the discovery has been summarized by DRUCE (1948).

[1] It is possible that KERN discovered rhenium in platinum ores in 1877 when he claimed the finding of a new element, "davyum" (NEWTON FRIEND and DRUCE, 1950).

[2] Fräulein IDA TACKE married Dr. WALTER NODDACK in 1926.

Revised manuscript received: October 1965

75-A. Chemistry (I), Crystal Chemistry (II) and Minerals (III)

Rhenium may be classed as a dispersed element in the sense of VERNADSKY (1924), i.e. it occurs almost entirely in minerals of other elements. Minerals in which rhenium is a major constituent are rare and have only been discovered recently.

I. The chemistry of rhenium resembles that of technetium in Group VII A but differs noticeably from that of manganese, although there are similarities in stoichiometries, e.g. MnO_4^- and ReO_4^-. Rhenium has been identified in all oxidation states from —1 to +7. As might be expected from its position in the third transition group, the element tends in most of its compounds and complexes to have low spin. Moreover, there appears to be no evidence for any simple +2 cationic species and little evidence for the formation of any cationic species. In these respects rhenium differs considerably from manganese for which the +2 state is the most stable and characteristic. Some oxidation-reduction potential data for rhenium are summarized in Fig. 75-A-1; these are of limited reliability. The oxidation states IV and VII are most important from a geochemical point of view.

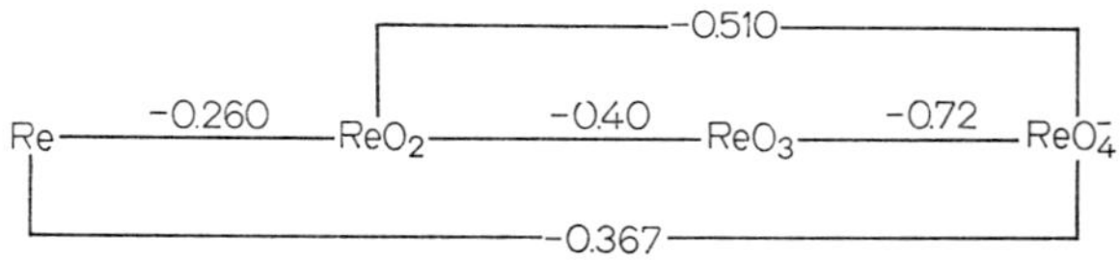

Fig. 75-A-1. Oxidation-reduction potential data for rhenium (BOYD *et al.*, 1953)

II. A close geochemical coherence exists between rhenium and molybdenum. The ionic radius according to AHRENS (1952) of Re^{4+} (0.74 Å) is similar to that of Mo^{4+} (0.70 Å) and the electronegativities of rhenium and molybdenum calculated using the Allred-Rochow formula, 1.46 and 1.30 respectively, are of like order (COTTON and WILKINSON, 1962). Thus the genetic link between the two elements is not unexpected.

The major source of rhenium at the moment is molybdenite. The phase relations and crystal structures of the systems Mo—S and Re—S have been discussed by JELLINEK (1963). It appears that MoS_2 and ReS_2 possess very similar structures and dimensions (Table 75-A-1) and that rhenium is carried by molybdenite isomor-

Table 75-A-1. *Structure types and unit-cell dimensions of disulphides of molybdenum and rhenium* (JELLINEK, 1963)

Phase	Structure type	Symmetry	a_0 in Å = M-6M	c_0 (Å)
$2s$ — MoS_2	$2s$ — MoS_2	hexagonal	3.1604	12.295
$3s$ — MoS_2	$3s$ — NbS_2	rhombohedral	3.161	18.36
$2s$ — ReS_2	$2s$ — MoS_2?	hexagonal	3.14	12.20
$3s$ — ReS_2	$3s$ — NbS_2	rhombohedral	3.18	18.2

phously (Kagle, 1960; Faramazyan and Khurshudyan, 1963; Ponomareva and Lebedev, 1962). Faramazyan and Khurshudyan (1963) have shown by X-ray methods that in rhenium-rich molybdenites the c_0 parameter of the crystal is reduced as a result of the substitution of rhenium, whilst the value of a_0 remains constant.

III. Although molybdenite is the major host mineral for rhenium, there have been cases reported where the rhenium content has exceeded the molybdenum content of sulphide ores and where the formation of a definite mineral of rhenium may arise. Moreover, leaching experiments by Pokalov (1963) have indicated that part of the rhenium in molybdenite may be present in a non-isomorphic form, distributed in intercrystalline voids or in defect layers of the molybdenite crystal lattice.

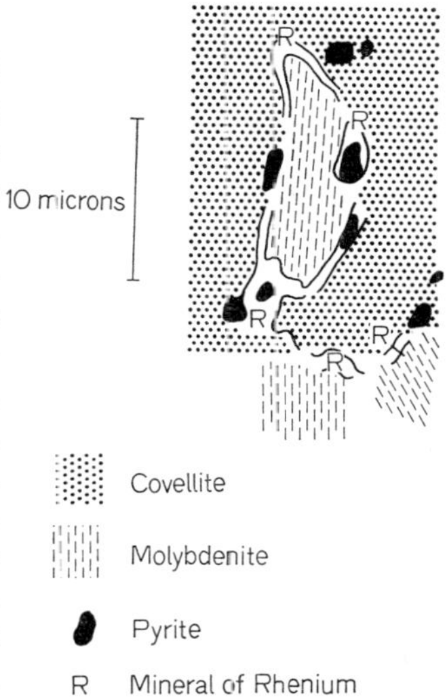

Fig. 75-A-2. Distribution of rhenium mineral in complex ore from Kipushi

A new type of rhenium accumulation has been found in the cupriferous sandstones of the Dzhezkazgan deposit of Kazakhstan, SSR. From a detailed mineralographic and chemical study of the dispersed ores, extremely fine aggregates of a new mineral, dzhezkazganite, were discovered in the sulphidic conglomerations, and less frequently in the dispersed sulphidic disseminations, in close contact with bornite, chalcocite and galena (Poplavko, Marchukova and Zak, 1962; Poplavko, 1962, 1964; Satpaeva *et al.*, 1962). According to Poplavko, the new mineral, which occurs in the form of very thin platelets (rarely >0.1 mm), contains Re 40—50% and Cu 15—20%. The properties of dzhezkazganite suggest that it is a sulphide, and the formula $CuReS_4$ has been proposed. Although Poplavko (1962) has pointed out that the molybdenum content did not exceed the rhenium content in any adjacent sulphide studied, Dedeshko *et al.* (1964) have suggested from the results of leaching experiments that the rhenium mineral is a rhenium molybdenum sulphide (Re: Mo $\sim$1) in which copper is not a mineral-forming component. Polyakova (1965) reports on the basis of X-ray patterns that the Re-minerals from the Dzhezkazgan deposits are solid solutions of ReS_2 and MoS_2.

Capitant *et al.* (1963) have studied the distribution of rhenium by means of an electron microanalyzer in a sample of mineral complex from the Prince Leopold Mine, Kipushi, Katanga, Congo. The major minerals in the sample were molybdenite, tennantite and bornite, in that order of abundance, with accessory minerals including chalcocite, covellite, pyrite and sphalerite. Distribution of rhenium through the molybdenite was found to be variable and the presence of local concentrations suggested the existence of a specific rhenium mineral. The species was located as some small areas on a grey-brown mineral found adjacent to molybdenite in contact with covellite — Fig. 75-A-2. The inclusions were too small for more than a semiquantitative analysis by the microprobe but contained "several tens of per cent of rhenium".

Revised manuscript received: October 1965

75-B. Isotopes in Nature

Naturally occurring rhenium consists of two isotopes, ^{185}Re (per cent abundance $\theta = 37.07$) and ^{187}Re ($\theta = 62.93\%$). In 1948 NALDRETT and LIBBY (1948) found from counting experiments that the isotope of mass number 187 was radioactive. Subsequently it has been shown that the β^--decay of the nuclide can be applied to age determinations for molybdenites and iron meteorites (HIRT, HERR and HOFFMEISTER, 1962).

For the application of the rhenium-osmium method of geochronology, it is necessary to have a direct knowledge of the half-life (T) of ^{187}Re. The β^--particles

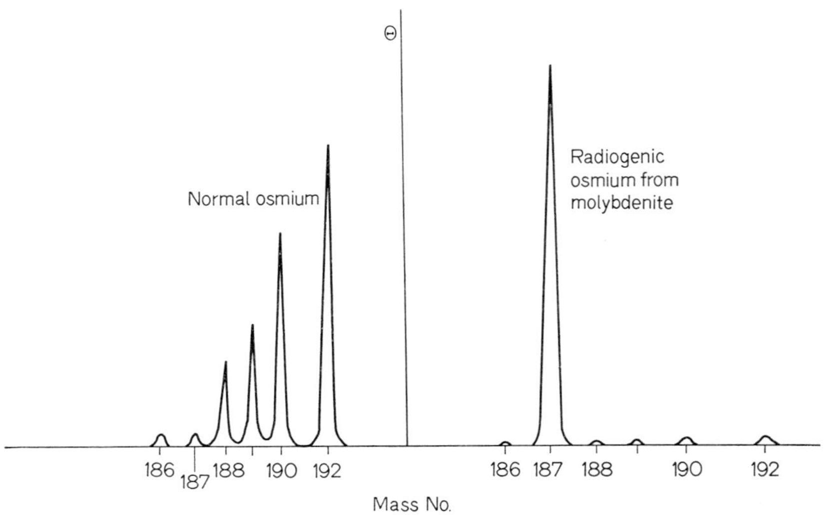

Fig. 75-B-1. Mass spectra of osmium. (After HINTERBERGER *et al.*, 1954)

Table 75-B-1. *Re/Os ages of molybdenites* (HIRT, HERR and HOFFMEISTER, 1962)

Molybdenite	Re content ppm	Radiogenic ^{187}Os content ppm	Age[a] 10^6 yr.
1. Rautio, Finland	55.7	0.995	1730 ± 100
2. Knaben Mine, Norway	30.6	0.384	940 ± 50
3. Vatterfjord, N. Norway	230.0	6.05	2540 ± 200
4. Haga Sigdal, Norway	14.2	0.148	1010 ± 100
5. Topland, Norway	2.02	0.0163	780 ± 100
6. Yetholme, Australia	71.4	0.239	330 ± 40
7. Glen Eden, Australia	3.45	0.0037	105 ± 40
8. Carpenters Gully, Australia	94.4	0.217	225 ± 40
9. Wombah, Mt. Perry, Australia	73.0	0.156	210 ± 40

[a] The errors indicated do not include the error of the half-life of ^{187}Re.

Table 75-B-2. *Rhenium and osmium contents of some meteorites* (Herr *et al.*, 1961; Hirt *et al.*, 1962)

Meteorites	Class	Average ppm Os	Average ppm Re	Ratio Os/Re
Iron meteorites:				
Central Missouri	Ogg	0.025	0.0024	10.4
Linwood	Ogg	4.52 ± 0.18	0.271 ± 0.026	16.7 ± 1.7
Mount Joy	Ogg	0.51 ± 0.15	0.026 ± 0.0055	19.6 ± 7.5
Odessa	Og	2.60 ± 0.10	0.251 ± 0.007	10.4 ± 0.5
Odessa Troilite			0.031 ± 0.011	
Treysa	Om	0.58 ± 0.04	0.093 ± 0.014	6.2 ± 1.0
Casas Grandes	Om	3.22 ± 0.27	0.45 ± 0.07	7.2 ± 1.2
Trenton	Om	1.60	0.200	8.0
Cañon Diablo	Om	2.07 ± 0.04	0.252 ± 0.011	8.21 ± 0.37
Henbury a	Om	2.25 ± 0.16	0.215 ± 0.019	10.5 ± 1.2
Henbury from Brown and Goldberg		13.25	1.31	10.1
Xiquipilco	Om	2.55 ± 0.15	0.219 ± 0.028	11.6 ± 1.6
Carbo	Om	15.7 ± 0.66	1.28 ± 0.08	12.3 ± 0.9
Gibeon/Mucerop	Of	2.52 ± 0.27	0.237 ± 0.013	10.6 ± 1.3
Bethany (Krantz)	Of	3.20 ± 0.19	0.271 ± 0.008	11.8 ± 0.8
Grant	Of	0.1	0.0048	
Tocopilla	H	1.29 ± 0.11	0.252 ± 0.023	5.12 ± 0.62
San Martin	H	1.23	0.205	6.0
Negrillos	H	50.4 ± 4.4	4.80 ± 0.24	10.5 ± 1.1
Colomera	H	12.46	0.784	15.9
Sikhote Alin	H, Ogg	0.025	0.01	
Monahans	D	12.7 ± 0.2	1.13 ± 0.05	11.3 ± 0.5
Tlacotepec	D	39.6 ± 2.2	2.92 ± 0.13	13.6 ± 1.0
Piñon	D	28.3 ± 2.2	1.83 ± 0.09	15.5 ± 1.6
Santa Catharina	D	0.16	0.03	
Otumpa	A	3.71	0.345	10.7
Lombard		1.14 ± 0.20	0.160 ± 0.007	7.1 ± 1.3
Dayton Ohio		0.047 ± 0.007	0.0050 ± 0.0019	9.4 ± 3.3
Loreto Baja		3.91	0.371	10.5
Aswam		0.225	0.016	13.7
Brenham (iron phase)	P	0.07	0.004	
Stone meteorites:				
Ramsdorf (iron Phase)		2.57	0.188	13.7
Mocs (iron phase)	CL	3.67	0.291	12.6
Mocs (stone phase)	CL	0.35	0.031	11.3

The error is 3 times standard error calculated on the basis of 3—6 independent measurements.

emitted by the radionuclide have such low energies ($E_{max} \sim 0.002$ Mev) however, that measurements by counting are difficult and have yielded values for T over a wide range (see Hirt, Tilton, Herr and Hoffmeister, 1963 and Table 12-5 of this Handbook (Vol. I).

In geologically old molybdenite containing rhenium, osmium is found, although from geochemical considerations it would not be expected to occur in the mineral.

Results of mass spectrometric investigation are shown in Fig. 75-B-1, and indicate that ^{187}Os is practically the only isotope of osmium present, i.e. the osmium in the mineral is almost entirely radiogenic. This observation permits an estimate of the half-life of ^{187}Re to be made. By determination of the traces of rhenium by neutron-activation and the ^{187}Os by isotope dilution in ten independently dated molybdenites, HIRT *et al.* (1963) have derived a value $(4.3 \pm 0.5) \times 10^{10}$ years, for the half-life of ^{187}Re. The error would be lowered if the molybdenites could have been independently dated directly, but unfortunately they contain no radioelements other than rhenium. However, their ages were assessed from measurements on rocks from which the mineral was obtained, or on neighbouring rocks, by use of the uranium-lead, lead-lead, rubidium-strontium or potassium-argon methods.

Among terrestrial materials studied until now, only molybdenites have contained enough rhenium to permit age determinations by the Re–Os method. Nevertheless this is useful in that it provides a direct means of dating veins and mineral deposits (SATPAEVA *et al.*, 1960). Some results obtained for the ages of molybdenites are shown in Table 75-B-1. Other specimens have been dated by ALLÈGRE *et al.* (1964); these workers used a spectrophotometric procedure to determine the osmium and this requires much larger samples than the isotope dilution method used by HIRT *et al.*

HIRT, HERR and HOFFMEISTER (1962) have shown that iron phases of meteorites contain sufficient rhenium and have large enough Re:Os ratios (Table 75-B-2) for application of the rhenium-osmium age method. For such samples the content of rhenium and common osmium was determined by neutron-activation analysis and the enrichment of radiogenic ^{187}Os was determined with a mass spectrometer (HERR *et al.*, 1961). The measurements and results of the study are completely compatible with the assumption that iron meteorites, stone meteorites, and also the Earth's crust were formed at approximately the same time, *ca.* $(4—5) \times 10^9$ years ago.

AHRENS (1957) has shown that a plot of θ for the heavier naturally occurring isotopes of rubidium, antimony and rhenium (these are the odd Z, even N nuclides ^{87}Rb, ^{123}Sb and ^{187}Re) against their neutron excess (N—Z) yields a curve that is convex upwards. If allowance is made for the decay of ^{87}Rb and ^{187}Re since the time of formation of the elements t, the relationship is changed significantly. The curve tends to straighten and a rectilinear plot is obtained with $t \sim 5 \times 10^9$ years.

Revised manuscript received: October 1965

75-C. Abundance in Meteorites (I), Tektites (II), Cosmos (III) and Lunar Materials (IV)

I. Meteorites

The occurrence of rhenium in meteorites was first reported by NODDACK and NODDACK (1931). They estimated the following abundances: iron phase 0.008 ppm Re (av. of 16), sulfide phase 0.001 ppm (av. of 7), silicate phase 0.0008 ppm (composite). More recent results have been obtained by neutron-activation analysis (BROWN and GOLDBERG, 1949; GOLDBERG und BROWN, 1950; HERR *et al.*, 1960, 1961; MORGAN and LOVERING, 1964; PEREZHOGIN, 1965) and must be regarded as much more reliable than those of the NODDACKS. The ranges of rhenium content in siderites and stony meteorites are summarized in Table 75-C-1.

Table 75-C-1. *Rhenium contents of meteorites*

Classification	Range (ppm) from —	to	Average (ppm)	References
I Siderites	0.25	1.4	0.62	BROWN and GOLDBERG (1949)
	0.0024	4.80	0.58	HERR *et al.* (1961)
Overall average			0.58	
II Stones				
A: Chondrites	0.036	0.068	0.060	MORGAN and LOVERING (1964)
	0.041	0.076	0.053	PEREZHOGIN (1965)
	0.046			HERR *et al.* (1961)
Overall average			0.055	
B: Achondrites	0.00025			MORGAN and LOVERING (1964)
	0.071			PEREZHOGIN (1965)

HERR *et al.* (1961) found 0.031 ppm Re in the silicate portion of the Mocs hypersthene chondrite and 0.291 ppm in the metallic portion (see Table 75-B-2).

The variation in concentration of various elements in iron meteorites in relation to gallium content and classification has been discussed by GOLDBERG *et al.* (1951) and by LOVERING *et al.* (1957), but from the results presently available no consistent correlation between rhenium and gallium contents can be observed. The measurements of MORGAN and LOVERING (1964) on five chondrites indicate no fractionation between the various chondrite types.

Rhenium may be classed as a decidedly siderophile element as data from meteorites indicate. It also shows pronounced chalcophile character in its geochemical distribution. NODDACK *et al.* (1940) have measured distribution coefficients of heavy metals, including rhenium, between iron and iron (II) sulfide. The metals

Table 75-C-2. *Estimates of the cosmic abundance of rhenium (atoms per million atoms Si)*

BROWN (1949)[a]	SUESS and UREY (1956)[b]	CAMERON (1959)[b]	ALLER (1961)[b]
0.41	0.155[c]	0.054	0.25

[a] Data from meteorites and sun.
[b] Values obtained by adjusting empirical solar and meteoritic abundances to conform to various theories of nucleogenesis.
[c] The value in the table in the reference is 0.135 but the text shows that it should be 0.155.

The average chondritic abundance of rhenium is 0.0512 (Si $= 10^6$ atoms) (MORGAN and LOVERING, 1964).

showed a tendency to concentrate in the iron, the inclination decreasing in the order Pd > Mo > Rh > Pt > Ru > Ir > Re > Os. The similarity in the single bond metallic radii (PAULING, 1960) of rhenium (1.283 Å), iron (1.165 Å), osmium (1.260 Å), and iridium (1.265 Å), as well as the relatively small (negative) free energy of oxidation, can be correlated with the tendency for rhenium to concentrate in meteoritic iron and possibly in minerals of the precious metals (NODDACK and NODDACK, 1931; GOLDSCHMIDT, 1954).

II. Tektites

Rhenium has been determined in tektites by LOVERING and MORGAN (1964) using neutron-activation. The specimens investigated were two australites, two javaites, one philippinite, two indochinites, two moldavites and one bediasite. The rhenium contents range from 0.014 to 0.123 ppb. Comparison of the results with data for terrestrial igneous and sedimantary rocks suggests that either type of rock could be a parent material for tektites.

III. Cosmos

Recent estimates of the cosmic abundance of rhenium are summarized in Table 75-C-2. All estimates are given relative to Si $= 10^6$ atoms rather than in ppm. For details of the assumptions involved in the selection of the various magnitudes the reader is directed to the original references.

Revised manuscript received: October 1965

IV. Lunar Materials

Because of the very low concentrations of rhenium in lunar samples, activation using irradiation with neutrons in a nuclear reactor has been the general method of analysis used. This has been based on either both of the nuclear reactions (i) and (ü) or on reaction (i) only

(i) $\quad {}^{185}\mathrm{Re}\,(n, \gamma)\,{}^{186}\mathrm{Re} \xrightarrow[t_{\frac{1}{2}} = 90.6\text{ h}]{\beta^-} {}^{186}\mathrm{Os}\,(\text{stable})$

(ii) $$^{187}\text{Re}(n,\gamma)\,^{188}\text{Re} \xrightarrow[t_{\frac{1}{2}}=17.0\text{ h}]{\beta^-} {}^{188}\text{Os (stable)}.$$

A number of complexities arise. MICHEL *et al.* (1972, 1973a, b) have applied the technique for the isotopic analysis of lunar samples for rhenium in the subnanogram range. They found by comparing the 137 keV photopeak of ^{186}Re with the 155 keV photopeak of ^{188}Re, measured with a Ge(Li)-detector and a multichannel analyzer, that ^{187}Re is enriched in lunar material relative to the terrestrial value by 1.4 to 1.8% (in 14 regoliths) and by as much as 29% in one of the Apollo 14 breccias. The (tungsten/rhenium) ratios varied between 1,100 (soils) and 41,000 (breccia 14,321), and this suggested that the isotopic anomaly is mainly caused through neutron capture by ^{186}W giving rise to ^{187}Re

(iii) $$^{186}\text{W}(n,\gamma)\,^{187}\text{W} \xrightarrow[t_{\frac{1}{2}}=24\text{ h}]{\beta^-} {}^{187}\text{Re}.$$

This is considered to be primarily due to epithermal neutrons of energies up to ~100 eV in the lunar flux. A further complication is that interference with the determination of rhenium may arise from the sequence (iii) + (ii) occurring on irradiation of the sample in a nuclear reactor (MORRIS and FIFIELD, 1961a; MICHEL *et al.* 1972, 1973a, b). This effect cannot be avoided, because it occurs inherently in the analytical method, but it may be allowed for by calculation and by simultaneous irradiation of suitable Re and W comparison standards.

Nuclear data show that the reaction

(iv) $$^{184}\text{W}(n,\gamma)\,^{185}\text{W} \xrightarrow[t_{\frac{1}{2}}=75\text{ d}]{\beta^-} {}^{185}\text{Re}$$

which leads to the formation of ^{185}Re is of much less consequence than (iii). Another possible cause of enrichment of ^{185}Re relative to ^{187}Re may be due to depletion of the latter in "solar wind" rhenium by a temperature dependent "faster" decay in the sun (CLAYTON, 1969). However, experiment indicates (HERPERS *et al.*, 1971; MICHEL *et al.*, 1972) that the effect of the captured "solar wind" rhenium on the lunar abundance ratio ^{185}Re/^{187}Re is of much less significance than the consequence of the nuclear reaction (iii) brought about by the cosmic ray produced neutron flux at the moon's surface.

It follows that a proper analysis for the rhenium content of materials on the lunar surface by neutron activation should involve the determination of both the individual isotopes ^{185}Re and ^{187}Re, together with simultaneous analysis for tungsten in order to correct for interference from reaction (iii). This proper procedure has not been employed in the majority of cases (KEAYS *et al.*, 1974).

Analyses of lunar materials are reported according to mission in Table 75-C-3.

Despite careful precautions taken by the investigators concerned, there is evidence that some of the samples may have become contaminated by rhenium before analysis. However, in general it appears that the lunar regolith is enriched in rhenium compared with the rocks themselves and this has been attributed to meteoritic influx (GANAPATHY *et al.*, 1973).

Table 75-C-3. *Rhenium concentration in lunar materials (Number of samples in brackets)*

Material	Re (ppb)	Reference
Apollo 11 (Mare Tranquillitatis)		
Basalt (A) (2) (grain size <0.5 mm)	1.18–1.73	1
Basalt (B) (1) (grain size >1 mm)	0.0016–0.03	1
Breccia (C) (2)	0.216–5.20	1, 2
Soil (D) (1)	8.68–10.62[a]	1
Soil (D) (1)	0.85[a]	2
Soil (D) (1)	7.25–7.85[a, b]	3
Soil (D) (1)	11.20[a, b]	4
Soil (D) (1)	0.71	5
Apollo 12 (Oceanus Procellarum)		
Basalt (A, AB, B) (6)	0.014–0.026	2
Basalt (A, AB, B) (1)	0.077–0.078	3
Basalt (A, AB, B) (3)	0.026–0.094[c]	3
Breccia (C) (1)	0.41–0.52	2
KREEP[d] (6)	0.097–0.19	6
Magnetic fraction (1) (30–100 mesh)	0.57	6
Soil (D) (4)	0.33–0.61	2
Soil (D) (1)	0.25	6
Soil (D) (1)	0.66–0.67[c]	3
Apollo 14 (Fra Mauro)		
Magmatic rock (1)	0.007	7
Basalt (1)	0.0066	8
Norite (3)	0.43–0.92	7
Magnetic fraction (1)	0.99	7
Magnetic fraction (3)	1.03–1.47	8
Glass (1)	0.91	7
Glass (9)	0.32–0.53	8
Microbreccia (4)	0.55–0.86	8
Breccia (1)	0.62–0.83	9, 10
Breccia (1)	0.64	11
Breccia (1)	0.021–0.037	9
Soil (7)	0.93–1.34	8
Soil (2)	0.96–1.41	9
Soil (1)	1.00	7
Soil (1)	1.07	5
Apollo 15 (Hadley-Apennine)		
Coarse basalt (1)	0.0080	12
Basalt (3)	0.0013–0.0081	11
Gabbro (2)	0.0015–0.0094	11
Anorthosite (1)	0.00084	11
Breccia (7)	0.13–0.63	11
Soil Breccia (2)	0.51–0.71	12
Glass (1)	0.87	11
Glass (1)	0.90	12
Breccia glass (2)	0.02–0.03	11
Soils (6)	0.19–0.76	11

Table 75-C-3 (continued)

Material	Re (ppb)	Reference
Apollo 16 (Descartes-Cayley)		
Crystalline magmatic rock (1)	0.0278	13
Basalt (4)	0.336–2.13	13
Anorthosite (3)	0.0016–0.0022	13
Gabbroic anorthosite (1)	0.2	14
Breccia (4)	0.103–0.79	13
Soil breccia (7)	0.0594–3.26	13
Soil, magnetic separate (2)	10.4–12.6	13
Soil (9)	0.527–2.34	13
Soil (2)	0.5–2.6	14
Apollo 17 (Mare Serenitatis)		
Soil (1)	≈0.2	14
Luna 16 (Mare Fecunditatis)		
Soils (2)	3.4–3.7 [e]	5
Luna 20 (between Mare Fecunditatis and Mare Crisium)		
Anorthosite (1)	1.19	11
Breccia (1)	1.04	6
Soil (<0.125 mm) (1)	5.34 [f]	6

[a] WÄNKE *et al.* (1970) examined a large sample (80 g) of soil 10084–18 and found it to be very heterogeneous, containing a significant amount of iron particles with a tungsten content of 24 ppm. LOVERING and BUTTERFIELD (1970) and LOVERING and HUGHES (1971) analyzed lunar soil 10084–53 and included all particles with grain sizes <1 mm. The analyses carried out by HERR *et al.* (1970, 1971) were on sample fractions with particle sizes of <100 μm. The results of GANAPATHY *et al.* (1970) indicate that siderophiles are enriched in the finest fractions of lunar soils and this may partially account for the large discrepancies in these soil analyses.

[b] Rhenium determined using the ^{185}Re (n, γ) ^{186}Re reaction and γ-ray spectrometry assuming a "normal" terrestrial isotopic abundance. Most other values involved β-counting of ^{186}Re and the assumption of a constant abundance ratio.

[c] Results take account of the lunar isotopic abundance and were based on reactions (a) and (b) and γ-ray spectrometry.

[d] KREEP is a component contained in certain lunar soils which is rich in potassium (K), the rare earth elements (REE), and phosphorus (P) (HUBBARD *et al.*, 1971).

[e] These samples had been passed through a 0.125 mm sieve. The rhenium figures are thought to be spurious and contamination is suspected.

[f] This sample is thought to have been contaminated prior to receipt.

References: 1. LOVERING and BUTTERFIELD (1970); 2. LOVERING and HUGHES (1971); 3. HERR *et al.* (1971); 4. HERR *et al.* (1970); 5. LAUL *et al.* (1972); 6. MORGAN *et al.* (1973a); 7. GANAPATHY *et al.* (1972); 8. MORGAN *et al.* (1972); 9. MICHEL *et al.* (1972); 10. HERPERS *et al.* (1973); 11. GANAPATHY *et al.* (1973); 12. MORGAN *et al.* (1973b); 13. KRÄHENBÜHL *et al.* (1973); 14. WÄNKE *et al.* (1973).

Revised manuscript received: September 1975

75-E. Abundance in Common Igneous Rocks

Data available are insufficient for a precise knowledge of the pattern of distribution of rhenium in rocks to be achieved. The most reliable analyses are probably those obtained by neutron-activation — Table 75-E-1. Although this method of analysis has unique advantages from the point of view of avoidance of contamination, sampling difficulties may arise. Thus the precision of analyses for rhenium in samples of the standard rocks G-1 and W-1 found experimentally is poorer than that which should be provided intrinsically by the method (MORRIS and FIFIELD, 1961a); this suggests a possible nonhomogeneity of distribution of rhenium in the rocks. The problem of contamination is illustrated by the results of analyses of Specpure precipitated silica. However, the available results suggest a magnitude of *ca.* 0.5 ppb for the abundance of rhenium in igneous rocks. This figure may be compared with that quoted in earlier reference works, namely 1 ppb, based on a single determination by NODDACK and NODDACK (1931) on a composite sample of 110 igneous rocks. According to BASITOVA (1958) the mean abundance of rhenium in igneous rocks is 5—7 ppb.

Table 75-E-1. *Rhenium content of some igneous rocks*

Rocks	Re, ppb	Reference
Standard Granite G 1,	(0.52—0.84) 0.67	MORRIS and FIFIELD (1961b)
Westerley, Rhode Island, U.S.A.	(0.33—0.80) 0.56	MORGAN (1965)
Standard Diabase W 1,	(0.56—0.84) 0.71	MORRIS and FIFIELD (1961b)
Centerville, Virginia, U.S.A.	(0.40—0.45) 0.42	MORGAN (1965)
Gabbroid, Insch Mass, Aberdeenshire, Scotland		MORRIS and FIFIELD (1961b)
Syenogabbro	0.36	
Syenite	0.18	
Troctolite	0.31	
Hybridised diorite	0.16	
Granitic, Liruei, Nigeria		MORRIS and FIFIELD (1961b)
Albite-riebeckite-granite	1.12	
Biotite-granite	0.22	
Quartz-pyroxene-fayalite-porphyry	0.29	
Rhyolite	1.14	
Riebeckite-aegirine-granite	0.25	
Comendite	0.33	
Specpure silica (Johnson, Matthey)		
Batch No. 19408	11.0	MORRIS and FIFIELD (1961b)
Batch No. 23453	42.6	MORGAN (1965)

Revised manuscript received: October 1965

75-F. Behaviour during Processes Connected with Magmatism

According to NODDACK and NODDACK (1931) rhenium separates from the crystallizing magma in the early sulphide fraction. Consequently the early-separated oxide and silicate minerals are poor in rhenium. The only exception to this rule appears to be chromite, Table 75-F-1. The early magmatic sulphides of the pyrrhotite-pentlandite paragenesis contain on average *ca.* 20 ppb Re.

A notable concentration of rhenium occurs in pegmatites and partly also in pneumatolytic rocks, the maximum being found in niobates, tantalates, thortveitite, zircon, and particularly gadolinite — Table 75-F-1.

Table 75-F-1. *Minerals other than sulphides containing enhanced contents of rhenium*

Mineral	Reference	Re, ppm
Alvite	NODDACK and NODDACK (1931)	0.2—0.6
Gadolinite	NODDACK and NODDACK (1931)	0.03—1.1
Thortveitite	NODDACK and NODDACK (1931)	0.6
Zircon	NODDACK and NODDACK (1931)	up to 0.05
Columbite	NODDACK and NODDACK (1931)	0.05—0.2
	SEN (1958)	0.26
	MESHRI, SHAH and HALDAR (1963)	0.4
Fergusonite	NODDACK and NODDACK (1931)	0.02
Samarskite	NODDACK and NODDACK (1931)	0.02
Scheelite	NODDACK and NODDACK (1931)	0.02
Tantalite	NODDACK and NODDACK (1931)	0.03—0.08
Wolframite	NODDACK and NODDACK (1931)	0.02
Chromite	NODDACK and NODDACK (1931)	0.01—0.02
Cassiterite	NODDACK and NODDACK (1931)	0.01
Precious metal mineral-alloys	NODDACK and NODDACK (1931)	up to 1

Rhenium constantly accompanies molybdenum through the whole of the pneumatolytic-hydrothermal and hydrothermal stages, forming appreciable concentrations in molybdenum deposits of various genetic types. The concentration of rhenium in molybdenite may vary from traces up to the value of 1.88 per cent noted in the deposits of Armenia (MAGAK'YAN *et al.*, 1963). The great tendency for rhenium to concentrate specifically in molybdenite is illustrated by the fact that spectrographic analyses by KAISER *et al.* (1954) of nearly 1000 samples of smelter products showed rhenium only in the molybdenum sulphide.

The data available give strong indications that the richest sources of rhenium in molybdenite are those closely associated with copper sulphide minerals. It appears also that the rhenium content may be correlated with the temperature of formation of the molybdenite (PAGANELLI, 1963; BADALOV, 1958; BADALOV, BASITOVA and GODUNOVA, 1962). A notable relation between the rhenium and selenium contents of molybdenites has been observed (KRUSHCHOV *et al.*, 1961; FARAMAZYAN, 1961; KARAMYAN, 1962; MOGAROVSKII, 1963), cf. Fig. 75-F-1.

Numerous authors refer to rhenium in molybdenites. The determinations published up to 1959 have been listed by FLEISCHER (1959, 1960). GOROKHOVA (1963, 1964a) has reviewed in detail the occurrence and distribution of rhenium in the various types of molybdenite deposits.

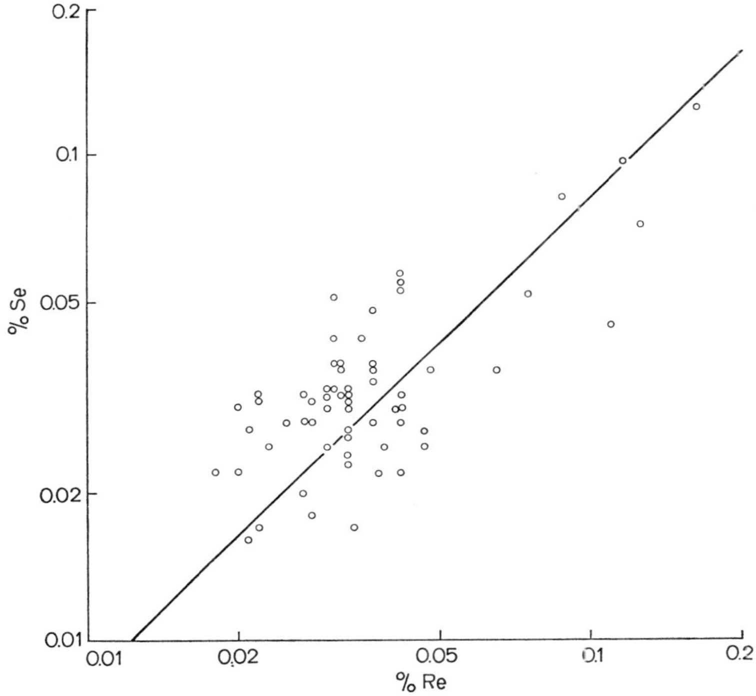

Fig. 75-F-1. Rhenium and selenium content of various samples of molybdenite from the Kadzharan deposit, Armenia SSR. (Data from FARAMAZYAN, 1961)

Revised manuscript received: October 1965

75-G and I. Behavior during Weathering and Occurrence in Natural Waters

It appears that Re^{VII} is the stable valency state in the zone of oxidation (see 75-A and Fig. 75-A-1). Rhenium (VII) oxide is soluble in water forming the strong perrhenic acid $HReO_4$, and perrhenates are stable in both acid and alkaline solutions.

Since molybdenite is the major host mineral for rhenium and the radii of Re^{7+} (0.56 Å) and Mo^{6+} (0.62 Å) are not widely different, the finding of rhenium in minerals resulting from the oxidation of molybdenite is not surprising. Thus FLEISCHER (1959) records up to 1.25 ppm Re in wulfenites. BADALOV (1958) in a study of the copper-molybdenum deposits of Almalyk has reported the occurrence of rhenium in powellite (0.015%) originating from oxidation of molybdenite (0.020% Re). In this case it appears that an insignificant separation of rhenium and molybdenum has occurred in the earlier stages of molybdenite decomposition. However, prolonged migration invariably causes a separation of the elements; thus wulfenite from the deposit was found to contain only 0.001% Re. That rhenium shows a greater migration capacity than molybdenum has been demonstrated by the experiments of MORACHEVSKII and NECHAEVA (1960). The fractionation of the two elements during weathering is discussed in more detail than can be given here by GOROKHOVA (1964a).

Rhenium, as perrhenate, is evidently present in sea water, although the concentration must be very low (RANKAMA and SAHAMA, 1950).

Revised manuscript received: October 1965

75-K. Occurrence in Common Sedimentary Rocks

Sedimentary deposits provide important sources of rhenium and the form of occurrence of the element in such deposits varies. At low concentrations of molybdenum, rhenium may precipitate in the form of an independent complex sulphide, dzhezkazganite, and possibly as the heptoxide. At high concentrations of molybdenum and sulphur, the rhenium is usually captured by molybdenite or jordisite. The sedimentary deposits rich in rhenium habitually contain multicomponent ore minerals (Gorokhova, 1964b).

Great interest from the point of view of the reserves of rhenium in the USSR lies in the cupriferous sandstone deposits, particularly those of the Dzhezkazgan group in Kazakhstan. As a rule, the accumulation of rhenium in such cupriferous sandstones is localized in zones of bornite and chalcopyrite-pyrite ores. Even at a low concentration of rhenium in the aqueous basin, the copper-bearing sulphide ores always have an elevated content of the element. Data for rhenium in such deposits are given in Table 75-K-1.

Table 75-K-1. *Rhenium in ores of the sedimentary deposits* (Gorokhova and Pokrovskaya, 1962)

Type of deposit	Locality	Re content ppm	Characteristic ore formers
Copper sandstones	Dzhezkazgan,	0.5—85	
	Kazakhstan, USSR	5—100	Copper and rhenium
	Kop-Kazgan, Kazakhstan, USSR	0.8—35	Copper
	Dalenskoye, Kirghiziya, USSR	2	Insignificant ore development
	Gul'chinskoye, Kirghiziya, USSR	2—5	Insignificant ore development
	Spasskoye, Kazakhstan, USSR	2	Insignificant ore development
	Udokan, East Siberia, USSR	Traces — 3	Copper
Coal-bearing deposits	Central Asia, USSR	0.095—0.327	Brown coals with high molybdenum content and other rare elements
Bituminous copper-bearing shales	Kupferschiefer, Mansfeld, E. Germany	23—235	Copper, lead, zinc, vanadium, nickel,
	Kupferschiefer, W. Germany (Ziegler, 1965)	0.75—1.1	molybdenum, rhenium and other rare elements
Uranium-copper deposits in sandstones	Sun Valley, N. Arizona, USA	50—700	The richest uranium-vanadium region of USA. Supplementary source of rhenium and other rare metals.

Concentrations of rhenium and other heavy metals also occur in bituminous sediments as, for example, are found in the well-known *Kupferschiefer* of Mansfeld in E. Germany (Schüller, 1958; Knitzschke, 1961) — Table 75-K-1. A local accumulation of rhenium is due to the action of later stage hydrothermal solutions from which deposits are precipitated in the already formed copper shales. During this process rhenium is paragenetically closely associated with molybdenum and uranium.

An occurrence of rhenium associated with uraninite in Coconino County, Arizona, USA has been reported by Petersen *et al.* (1959). Twenty-nine samples of sedimentary rocks of Triassic age from the Sun Valley uranium mine were analyzed spectrographically and 11 of the samples contained 0.005 to 0.1 per cent of rhenium. The rhenium is associated with uranium and molybdenum and is probably in the form of the heptoxide Re_2O_7. It may have been contained originally in jordisite; this is indicated by the association of the rhenium oxide with the water-soluble molybdenum mineral islemannite. Myers *et al.* (1960) have reported a concentration of rhenium up to 70 ppm in samples of uranium ore from the Runge Mine, South Dakota, USA.

The deposits mentioned in this section represent cases of rhenium enrichment; the abundance of the element in an "average sedimentary rock" is probably much lower (Lovering and Morgan, 1964).

Revised manuscript received: October 1965

75-L. Biogeochemistry

Little is known of the place of rhenium in the organic cycle. MYERS and HAMILTON (1961) have found up to 300 ppm Re in ashes of plants from soils rich in uranium and molybdenum. In marine biological organisms the rhenium content, ppb dry matter, varies from <0.31 in *Pandalus sp.* (prawn) up to 16 in *Ulva sp.* (seaweed) (FUKAI and MEINKE, 1962).

Rhenium is concentrated in certain carbonaceous deposits. Results obtained for the brown coals of Central Asia, USSR, are summarized in Table 75-K-1. The molybdenum content of the coals varies from 66 ppm to 0.521%. Investigations have shown that the rhenium and molybdenum occur in the lignites in the form of organometallic compounds and in micro-inclusions of sulphides disseminated in the coals. Moreover, the two elements are present in solution in waters circulating in the coal-bearing strata (KUZNETSOVA and SAUKOV, 1961; RAZENKOVA and KUZNETSOVA, 1963).

Revised manuscript received: October 1965

75-O. Concluding Notes on Relations to Other Elements, Economics, etc.

The fact that there is but a slight geochemical coherence between rhenium and manganese is illustrated by the work of HURD and HISKEY (1938). These investigators analyzed 80 samples of pyrolusite from many localities and found 0.1 ppm Re in 7 samples and 0.2 ppm in one sample; the element could not be detected in the remainder.

There are several applications of rhenium but the relatively high cost and rarity of the element have limited its widespread commercial use. Rhenium is useful as a constituent of filaments, thermocouple elements and sparking plug tips. It can also be of value in the fields of catalysis, electrical engineering, and rocketry.

GOROKHOVA and POKROVSKAYA (1962) have assessed the potential rhenium reserves of the western countries as *ca.* 1,000—1,043 tons Re. Of this amount, 454 to 680 tons were reckoned to lie in deposits in the USA. Countries such as the UK, France and W. Germany are importers from Chile of molybdenite concentrates which can have a high rhenium content (HERR and MERZ, 1955; CUSICANQUI and PAREDES, 1960).

Revised manuscript received: October 1965

References: Sections 75-A to 75-C, 75-E to 75-G, 75-I to 75-L, 75-O

AHRENS, L. H.: Use of ionization potentials. I. Ionic radii of the elements. Geochim. Cosmochim. Acta **2**, 155–169 (1952).

AHRENS, L. H.: Studies on the relative abundances of isotopes. Geochim. Cosmochim. Acta **11**, 1–27 (1957).

ALLÈGRE, C., DESCHAMPS, N., FAUCHERRE, J., HANOTEL-MONSTRAND, M.: Contribution à la géochronologie par la méthode rhenium-osmium. Compt. Rend. Soc. Géol. France **1964**, 297–299

ALLER, L. H.: The Abundance of the Elements. New York: Interscience Publ. 1961.

BADALOV, S. T.: The geochemistry of rhenium in the copper-molybdenum deposits of Almalyk. Renii, Tr. Vses. Soveshch. po Probl. Reniya, Akad Nauk S.S.R. Inst. Met. 1958, [Russian. Publ. 1961].

BADALOV, S. T., BASITOVA, S. M., GODUNOVA, L. I.: Distribution of rhenium in Central Asian molybdenites. Geokhimiya 813 (1962) [Russian].

BASITOVA, S. M.: Methods of extraction and determination of rhenium in rocks. Renii, Trudy Vses. Soveshch. po Probl. Reniya, Akad. Nauk S.S.R. Inst. Met. 1958 [Russian. Publ. 1961].

BOYD, G. E., COBBLE, J. W., SMITH, W. T.: Thermodynamic properties of technetium and rhenium compounds. III. Heats of formation of rhenium heptoxide and trioxide and a revised potential diagram for rhenium. J. Am. Chem. Soc. **75**, 5783 (1953).

BROWN, H.: A table of relative abundances of nuclear species. Rev. Mod. Phys. **21**, 625 (1949).

BROWN, H., GOLDBERG, E.: A new determination of the relative abundance of rhenium in nature. Phys. Rev. **76**, 1260 (1949).

CAMERON, A. G. W.: A revised table of abundance of the elements. Astrophys. J. **129**, 676 (1959).

CAPITANT, M., FRANCOTTE, J., PICOT, P., TROLY, G.: Hautes teneurs en rhenium dans une molybdénite de Kipushi. Compt. Rend. **257**, 3443 (1963).

CLAYTON, D. D.: Isotopic composition of cosmic importance. Nature **224**, 56 (1969).

COTTON, F. A., WILKINSON, G.: Advanced Inorganic Chemistry. New York: Interscience Publ. 1962.

CUSICANQUI, H., PAREDES, E.: Determination of rhenium in molybdenite. Univ. Chile, Fac. Cienc. Fis. Mat. Anales **17**, 133 (1960).

DEDESHKO, M. P., SATPAEVA, T. A., FAIN, E. E.: Chemical composition of a rhenium mineral from the Dzhezkazgan ore. Vestnik Akad. Nauk Kazakh. S.S.R., **20** (11), 47 (1964) [Russian].

DOLEJŠEK, V., HEYROVSKÝ, J.: The occurrence of dvi-manganese in manganese salts. Nature **116**, 782 (1925).

DRUCE, J. G. F.: Rhenium. Cambridge: Cambridge University Press 1948.

FARAMAZYAN, A. S.: Regularities in the distribution of rhenium in ores of the Kadzharan deposit. Izvest. Akad. Nauk Armyan S.S.R., Ser. Geol. i Geograf. Nauk **14**, No. 1, 39 (1961) [Russian].

FARAMAZYAN, A. S., KHURSHUDYAN, E. KH.: The isomorphism of molybdenum with rhenium in molybdenites. Doklady Akad. Nauk Armyan. S.S.R. **37**, (4), 211 (1963).

FLEISCHER, M.: The geochemistry of rhenium with special reference to its occurrence in molybdenite. Econ. Geol. **54**, 1406 (1959).

FLEISCHER, M.: The geochemistry of rhenium-addendum. Econ. Geol. **55**, 607 (1960).

FUKAI, R., MEINKE, W. W.: Activation analysis of vanadium, arsenic, molybdenum, tungsten, rhenium and gold in marine organisms. Limnol. Oceanog. **7**, 186 (1962).

GANAPATHY, R., KEAYS, R. R., LAUL, J. C., ANDERS, E.: Trace elements in Apollo 11 lunar rocks: Implications for meteoritic influx and origin of moon. Proceeding of the Apollo 11 Lunar Science Conference. Geochim. Cosmochim. Acta Supplement 1, **2**, 1117 (1970).

GANAPATHY, R., LAUL, J. G., MORGAN, J. W., ANDERS, E.: MOON: possible nature of the body that produced the Imbrian Basin from the composition of Apollo 11 samples. Science **175**, 55 (1972).

GANAPATHY, R., MORGAN, J. W., KRÄHENBÜHL, U., ANDERS, E.: Ancient meteoritic components in lunar highland rocks: Clues from trace elements in Apollo 15 and 16 samples. Proceedings of the Fourth Lunar Science Conference. Geochim. Cosmochim. Acta Supplement 4, **2**, 1239 (1973).

GOLDBERG, E., BROWN, H.: Radiometric determination of gold and rhenium. Anal. Chem. **22**, 308 (1950).

GOLDBERG, E., UCHIYAMA, A., BROWN, H.: The distribution of nickel, cobalt, gallium, palladium and gold in iron meteorites. Geochim. Cosmochim. Acta **2**, 1 (1951).

GOLDSCHMIDT, V. M.: Geochemistry. Oxford: Clarendon Press 1954.

GOROKHOVA, V. N.: Rhenium in molybdenum and copper-molybdenum deposits. Trudy Inst. Mineral. Geokhim. i Kristallokhim. Redkikh Elementov **10**, 171 (1963) [Russian].

GOROKHOVA, V. N.: Rhenium. In: VLASOV, K.A. (ed.) Rare Elements in the Earth's Crust. Mineral. i Genet. Tipy Mestorozhd. Redkikh Elementov, Akad. Nauk S.S.S.R. Gos. Geol. Kom. SSSR, Inst. Mineralog., Geokhim. i Kristallokhim. Redkikh Elementov.. Vol. 1, p. 612. Moscow: Izd. Nauka 1964a [Russian].

GOROKHOVA, V. N.: Rhenium in sedimentary deposits. Redkie Element v Osad. i Metamorficheskikh Porodakh, Akad. Nauk S.S.S.R. Inst. Mineralog. Geokhim. i Kristallokhim. Redkikh Elementov 183 (1964b) [Russian].

GOROKHOVA, V. N.: POKROVSKAYA, V. L.: Genetic types of rhenium deposits. Renii, Akad. Nauk S.S.S.R., Inst. Met., Tr. 2-go Vses. Soveshch., Moscow 1962 [Russian, Publ. 1964].

HERPERS, U., HERR, W., KULUS, H., MICHEL, R., THIEL, K., WOELFLE, R.: Manganese-53 profile, particle track studies and the rhenium-187 isotopic anomaly of breccia 14305. Proceedings of the Fourth Lunar Science Conference. Geochim. Cosmochim. Acta Supplement 4, **2**, 2157 (1973).

HERPERS, U., MICHEL, R., HERR, W.: Search for "solar wind" rhenium in moon samples. Angew. Chem. **10**, 818 (1971).

HERR, W., HERPES, U., HESS, B., SKERRA, B., WOELFLE, R.: Determination of manganese-53 by neutron activiation and other miscellaneous studies on lunar dust. Science **167**, 747 (1970).

HERR, W., HERPERS, U., MICHEL, R., ABDEL RASSOUL, A. A., WOELFLE, R.: Search for rhenium isotopic anomalies in lunar surface material by neutron bombardment. Proceedings of the Second Lunar Science Conference. Geochim. Cosmochim. Acta Supplement 2, **2**, 1337 (1971).

HERR, W., HOFFMEISTER, W., HIRT, B., GEISS, J., HOUTERMANS, F. G.: Versuch zur Datierung von Eisenmeteoriten nach der Rhenium-Osmium-Methode. Z. Naturforsch. **16**a, 1053 (1961).

HERR, W., HOFFMEISTER, W., LANGHOFF, J.: Die Bestimmung von Rhenium und Osmium in Eisenmeteoriten durch Neutronaktivierung. Z. Naturforsch. **15**a, 99 (1960).

HERR, W., MERZ, E.: Eine neue Methode zur Altersbestimmung von Rhenium-haltigen Mineralien mittels Neutronaktivierung. Z. Naturforsch. **10**a, 613 (1955).

HINTENBERGER, H., HERR, W., VOSHAGE, H.: Radiogenic osmium from rhenium-containing molybdenite. Phys. Rev. **95**, 1690 (1954).

HIRT, B., HERR, W., HOFFMEISTER, W.: Age determinations by the rhenium-osmium method. Radioactive Dating, p. 35–43. Proc. Symp. Athens 1962,.

HIRT, B., TILTON, G. R., HERR, W., HOFFMEISTER, W.: The half-life of ^{187}Re. In: GEISS, J., GOLDBERG, E.D. (eds) Earth Science and Meteoritics p. 273–280. Amsterdam: N. Holland Publ. Co. 1963.

HUBBARD, N. J., MEYER, C. JR., GAST, P. W., WIESMANN, H.: The composition and derivation of Apollo 12 soils. Earth Planet Sci. Letters **10**, 341 (1971).

HURD, L. C., HISKEY, C. F.: Determination of rhenium. Estimation in pyrolusite. Ind. Eng. Chem. Anal. Ed. **10**, 623 (1938).

JELLINEK, F.: Sulphides of the transition metals of Groups IV, V and VI. Arkiv Kemi **20**, 447 (1963).

KAGLE (1960). Unpublished paper, Univ. Utah, quoted in: MALOUF E. E., ZIMMERLEY S. R. 1962.

KAISER, E. P., HERRING, B. F., RABBITT, J. C.: Minor elements in some rocks, ores, and mill and smelter products. USAEC Report TEI-415, 1954.

KARAMYAN, K. A.: The correlation of rhenium, selenium and tellurium in molybdenites of the Kadzharan copper-molybdenum deposit. Geokhimiya 171 (1962).

KEAYS, R. R., GANAPATHY, R., LAUL, J. C., KRÄHENBÜHL, U., MORGAN, J. W.: The simultaneous determination of 20 trace elements in terrestrial, lunar and meteoritic material by radiochemical neutron activation analysis. Analyt. Chim. Acta **72**, 1 (1974).

KHRUSHCHOV, N. A., KRUGLOVA, V. G., PENSIONEROVA, V. M., PANKOVA, V. E., ROZOVSKAYA, G. V.: Distribution of rhenium, selenium, and tellurium in molybdenum deposits of the Soviet Union. Mineral. Syr'e, Moscow Sb. No. 1, 86 (1961).

KNITZSCHKE, G.: Vererzung, Hauptmetalle und Spurenelemente des Kupferschiefers in der Sangerhäuser und Mansfelder Mulde. Z. angew. Geol. **7**, 349 (1961).

KRÄHENBÜHL, U., GANAPATHY, R., MORGAN, J. W., ANDERS, E.: Volatile elements in Apollo 16 samples: implications for highland volcanism and accretion history of the moon. Proceedings of the Fourth Lunar Science Conference. Geochim. Cosmochim. Acta Supplement 4, **2**, 1325 (1973).

KUZNETSOVA, V. V., SAUKOV, A. A.: Possible forms of occurrence of molybdenum and rhenium in coals of Central Asia. Geokhimiya 750; English Transl. Geochemistry 822 (1961).

LAUL, J. C., GANAPATHY, R., MORGAN, J. W., ANDERS, E.: Meteoritic and non-meteoritic trace elements in Luna 16 samples. Earth Planet. Sci. Letters **13**, 450 (1972).

LORING, F. H., DRUCE, J. G. F.: Examination of crude dvi-manganese. Chem. News **131**, 337 (1925).

LOVERING, J. F., BUTTERFIELD, D.: Neutron activation analysis of rhenium and osmium in Apollo 11 lunar material. Proceedings of the Apollo 11 Lunar Science Conference. Geochim. Cosmochim. Acta Supplement 1, **2**, 1351 (1970).

LOVERING, J. F., HUGHES, T. C.: Rhenium and osmium abundance determinations and meteoritic contamination levels in Apollo 11 and Apollo 12 lunar samples. Proceedings of the Second Lunar Science Conference. Geochim. Cosmochim. Acta Supplement 2, **2**, 1331 (1971).

LOVERING, J. F., MORGAN, J. W.: Rhenium and osmium abundances in tektites. Geochim. Cosmochim. Acta **28**, 761 (1964)).

LOVERING, J. F., NICHIPORUK, W., CHODOS, A., BROWN, H.: The distribution of gallium, germanium, cobalt, chromium and copper in iron and stony-iron meteorites in relation to nickel content and structure. Geochim. Cosmochim. Acta **11**, 263 (1957).

MAGAK'YAN, I. G., PIDZHYAN, G. O., FARAMAZYAN, A. S.: Rhenium in Armenian copper-molybdenum deposits. Doklady Akad. Nauk Armyans. S.S.R. **37** (2), 77 (1963) [Russian].

MALOUF, E. E., ZIMMERLEY, S. R.: Mineralogical and geological occurrence of rhenium. in: GONSER, B. W. (ed.). Rhenium Papers presented at the Symp. on Re of the Electrothermics and Metallurgy Div. of the Electrochemical Soc. Chigaco, p. 3–5. Amsterdam and New York: Elsevier Publ. Co. 1962.

MESHRI, D. T., SHAH, S. D., HALDAR, B. C.: Determination of rhenium in columbite by neutron-activation analysis. Indian J. Appl. Chem. **26**, 1 (1963).

MICHEL, R., HERPERS, U., KULUS, H., HERR, W.: Isotopic anomalies in lunar rhenium. Proceedings of the Third Lunar Science Conference. Geochim. Cosmochim. Acta Supplement 3, **2**, 1917 (1972).

MICHEL, R., HERPERS, U., KULUS, H., HERR, W.: Isotopic abundance determination of submicro-amounts of rhenium by neutron activation. Am. Soc. Testing Mater., Spec. Tech. Publ. **539**, 140 (1973a).

MICHEL, R., HERPERS, U., KULUS, H., HERR, W.: Neutron-activation technique for isotopic analysis of rhenium in the sub-nanogram region. J. Radioanalyt. Chem. **17**, 177 (1973b).

MOGAROVSKII, V. V.: On the bond between rhenium and selenium in molybdenites of the Kadzharan deposit. Geokhimiya 566 (1963); English Transl. Geochemistry 580 (1963).

MORACHEVSKII, D. E., NECHAEVA, A. A.: Characteristics of migration of rhenium from molybdenites. Geokhimiya 543 (1960); English Transl. Geochemistry 648 (1961).

MORGAN, J. W.: The simultaneous determination of rhenium and osmium in rocks by neutron-activation analysis. Anal. Chim. Acta **32**, 8 (1965).

MORGAN, J. W., GANAPATHY, R., LAUL, J. C., ANDERS, E.: Lunar crater Copernicus: search for debris of impacting body at Apollo 12 site. Geochim. Cosmochim. Acta **37**, 141 (1973a).

MORGAN, J. W., KRÄHENBÜHL, U., GANAPATHY, R., ANDERS, E.: Trace element abundances and petrology of separates from Apollo 12 soils. Proceeding of the Fourth Lunar Science Conference. Geochim. Cosmochim. Acta. Supplement 4, **2**, 1379 (1973b).

MORGAN, J. W., LAUL, J. C., KRÄHENBÜHL, U., GANAPATHY, R., ANDERS, E.: Major impacts on the moon: characterization from trace elements in Apollo 12 and 14 samples. Proceedings of the Third Lunar Science Conference. Geochim. Cosmochim. Acta Supplement 3, **2**, 1377 (1972).

MORGAN, J. W., LOVERING, J. F.: Rhenium and osmium abundances in stony meteorites. Science **144**, 835 (1964).

MORRIS, D. F. C., FIFIELD, F. W.: The determination of rhenium in rocks by neutron-activation analysis. Talanta **8**, 612 (1961a).

MORRIS, D. F. C., FIFIELD, F. W.: Rhenium contents of rocks. Geochim. Cosmochim. Acta **25**, 232 (1961b).

MYERS, A. T., HAMILTON, J. C.: Rhenium in plant samples from the Colorado Plateau. U.S. Geol. Surv. Profess. Papers **424-B**, 286 (1961).

MYERS, A. T., HAMILTON, J. C., WILMARTH, V. R.: A study of rhenium and molybdenum in uranium ore from the Runge Mine, Fall River County, South Dakota, by means of a spectrographic and concentration method. U.S. Geol. Surv., Profess. Papers **400-B**, B 39 (1960).

NALDRETT, S. N., LIBBY, W. F.: Natural radioactivity of rhenium. Phys. Rev. **73**, 487 (1948).

NEWTON-FRIEND, J., DRUCE, J. G. F.: Davyum a possible precursor of rhenium (element 75). Nature **165**, 819 (1950).

NODDACK, I., NODDACK, W.: Die Geochemie des Rheniums. Z. physik. Chem. **154**, 207 (1931).

NODDACK, W., NODDACK, I., BOHNSTADT, U.: Die Teilungskoeffizienten der Schwermetalle zwischen Eisensulfid und Eisen. I. Z. anorg. u. allgem. Chem. **244**, 252 (1940).

NODDACK, W., TACKE, I., BERG, O.: Zwei neue Elemente der Mangangruppe. Chemischer Teil von W. NODDACK u. I. TACKE, Röntgenspektroskopischer Teil von O. BERG u. I. TACKE. Sitzber. preuss. Akad. Wiss. **19**, 400 (1925).

PAGANELLI, L.: Rhenium content of molybdenite of Mount Mulat (Predazzo) and other Italian molybdenites. Geochim. Cosmochim. Acta **27**, 401 (1963).

PAULING, L.: The Nature of the Chemical Bond 3rd ed. Ithaca, New York: Cornell University Press 1960.

PEREZHOGIN, G. A.: Neutron-activation determination of rhenium in meteorites and molybdenites using substoichiometric separations. Zavodskaya Lab. **31**, 402 (1965) [Russian].

PETERSEN, R. G., HAMILTON, J. C., MYERS, A. T.: An occurrence of rhenium associated with uraninite in Coconino County, Arizona. Econ. Geol. **54**, 254 (1959).

POKALOV, V. T.: Forms of rhenium occurrence in molybdenite. Mineral. Syr'e, Vses. Nauchn.-Issled. Inst. Mineral'n. Syr'ya **7**, 179 (1963) [Russian].

POLYAKOVA, T. P.: X-ray investigations of a rhenium mineral in the Dzhezkazgan ores. Vestnik Akad. Nauk. Kazakh. S.S.R. **21**, 69 (1965).

PONOMAREVA, E. I., LEBEDEV, K. B.: Alkaline processing in an autoclave of molybdenum concentrates containing rhenium. Trudy Inst. Met. i Obogashch., Akad. Nauk Kazakh. S.S.R. **4**, 28 (1962) [Russian].

Poplavko, E. M.: The distribution of rhenium and its forms in the ores of Dzhezkazgan deposit. Renii, Akad. Nauk S.S.S.R. Inst. Met., Tr. 2-go Vses. Soveshch., Moscow 1962, [Russian Publ. 1964].

Poplavko, E. M.: Dzhezkazganite. In: Vlasov, K. A. (ed.) Geokhim. Mineralog. i Genet. Tipy Mestorozhd. Redkikh Elementov, Akad. Nauk S.S.S.R. Gos. Geol. Kom. S.S.S.R. Inst. Mineralog., Geokhim. i Kristallokhim. Redkikh Elementov Vol. 2, p. 739 Moscow: Izd. Nauka 1964 [Russian].

Poplavko, E. M., Marchukova, I. D., Zak, S. Sh.: Rhenium mineral in the Dzhezkazgan deposit. Doklady Akad. Nauk S.S.S.R. **146**, 433 (1962) [Russian].

Prior, G. T.: Catalogue of Meteorites. British Museum, London 1953.

Rankama, K., Sahama, Th. G.: Geochemistry. Chicago: Chicago University Press 1950.

Razenkova, N. I., Kuznetsova, V. V.: Study by electrodialysis of forms of molybdenum and rhenium occurrences in coals. Trudy Inst. Mineral. Geokhim. i Kristallokhim. Redkikh Elementov, No. 18, 20 (1963) [Russian].

Satpaeva, T. A., Kalinin, S. K., Fain, E. E.: The osmium content of ores from Dzhezkazgan deposits. Vestnik Akad. Nauk Kazakh. S.S.R. **16**, No. 8, 29 (1960) [Russian].

Satpaeva, T. A., Kalinin, S. K., Satpaeva, M. K., Marzuvanov, V. L.: Character of rhenium occurrence in ores of Dzhezkazgan. Vestnik Akad. Nauk Kazakh. S.S.R. **18**, No. 12, 58 (1962) [Russian].

Schüller, A.: Metallisation und Genese des Kupferschiefers von Mansfeld. Abhandl. deutsche Akad. Wiss. Berlin, Kl. Chem., Geol. u. Biol. **6** (1958) [Publ. 1959].

Sen, S.: Estimation of molybdenum and rhenium in Indian columbite. Sci. and Culture (Calcutta) **23**, 318 (1958).

Suess, H. E., Urey, H. C.: Abundances of the elements. Rev. Mod. Phys. **28**, 53 (1965).

Vernadsky, W.: La Géochemie. Paris 1924.

Wänke, H., Baddenhausen, H., Dreibus, G., Jagoutz, E., Kruse, H., Palme, H., Spettel, B., Teschke, F.: Multi-element analyses of Apollo 15, 16, and 17 samples and the bulk composition of the moon. Proceedings of the Fourth Lunar Science Conference Geochim. Cosmochim. Acta Supplement 4, **2**, 1461 (1973)

Wänke, H., Wlotzka, F., Jagoutz, E., Begemann, F.: Composition and structure of metallic iron particles in lunar "fines". Proceedings of the Apollo 11 Lunar Science Conference. Geochim. Cosmochim. Acta Supplement 1, **1**, 931 (1970).

Ziegler, M.: Unpublished information 1965.

Revised manuscripts received: October 1965 and September 1975

Osmium 76

and the other Platinum Group Elements 44, 45, 46, 77

see Platinum 78

Iridium 77

and the other Platinum Group Elements 44, 45, 46, 76

see Platinum 78

Platinum Group 44 - 45 - 46 76 - 77 - **78**

A E. PARTHÉ (Laboratoire de Cristallographie aux Rayons X, Université de Genève, Genève, Suisse)

B—G, K, M, O J.H. CROCKET (Department of Geology, McMaster University, Hamilton, Ontario, Canada)

78-A. Crystal Chemistry

The term "platinum metals" is generally applied to the following six elements: ruthenium (Ru), rhodium (Rh), palladium (Pd), osmium (Os), iridium (Ir) and platinum (Pt). The first three with a density $\varrho \sim 12$ g/cm³ are often called the light platinum metals, the remaining three with $\varrho \sim 22$ g/cm³ the heavy platinum metals. All these noble metals crystallize in either the cubic close-packed Cu (A1) structure type or in the hexagonal close-packed Mg (A3) structure type. Crystal structure data together with other characteristics of these elements are summarized in Table 78-A-1.

Table 78-A-1. *Crystal structure data and other important data for platinum metals*

Element	Atomic No.	Atomic wt.	Electronic ground-state configuration	Melting point (°C)	ϱ (g/cm³)	Crystal structure type	Metallic atom radius (Å)	Reference [a]
Ru	44	101.07	[Kr] $4d^7\,5s$	2,250	12.30	A3	1.32_5	1
Rh	45	102.905	[Kr] $4d^8\,5s$	1,960	12.4	A1	1.34_5	1
Pd	46	106.4	[Kr] $4d^{10}$	1,552	11.97	A1	1.37_5	2
Os	76	190.2	[Xe] $4f^{14}\,5d^6\,6s^2$	3,000	22.48[b]	A3	1.33_7	3
Ir	77	192.2	[Xe] $4f^{14}\,5d^7\,6s^2$	2,443	22.421	A1	1.35_7	2
Pt	78	195.09	[Xe] $4f^{14}\,5d^9\,6s$	1,769	21.45[b]	A1	1.38_7	4

[a] Reference for crystal structure: 1. Hull and Davey, SB 1913—1928, 69. 2. Hull and Davey, SB 1913—1928, 70. 3. Hull, SB 1913—1928, 70. 4. Hull and Davey, SB 1913—1928, 71.
[b] at 20° C

The phase diagrams of binary systems containing only platinum metals are simple. Elements with A1 structure form a complete series of solid solutions at least at high temperatures. Similarly Os and Ru with A3 structure are fully miscible in the solid state. Systems between elements with A3 structure and elements with A1 structure are of the eutectic or peritectic type, usually with wide solid solution ranges. A presentation of the known alloying behavior of the platinum metals with themselves is shown in Fig. 78-A-1. The minerals of the platinum metals found in placers may contain all six platinum metals in varying proportions. It is possible to distinguish between two groups of minerals depending on the crystal structure. *Iridosmine* (nevyanskite and sysertskite) has A3 structure. *Native platinum*, *platiniri-*

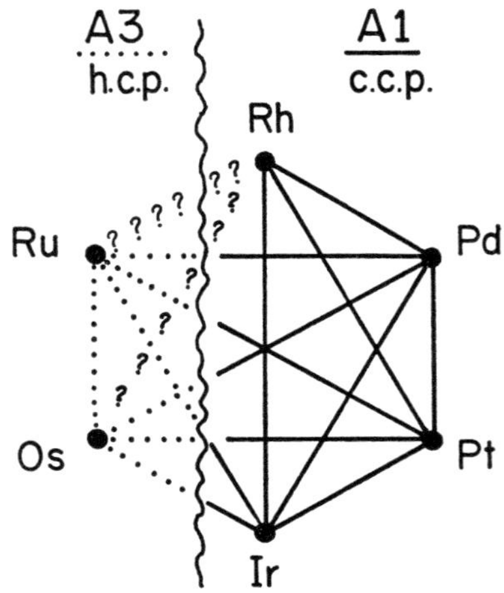

Fig. 78-A-1. The alloying behavior of the platinum metals with themselves. For the most part exact data on solubility limits in A3—A1 systems are missing. The solubility limits shown in the drawing are schematic

dium, *osmiridium* and *aurosmirid* (Au—Os—Ir) are minerals of the platinum metals with A1 structure.

The structures of the platinum metals and their alloys have been interpreted in terms of the Engel-Brewer theory (Brewer, 1965; Brewer, 1967; Hume-Rothery, 1968). This theory assumes that the crystal structures of the elements are based on definite electronic states of the atoms which are determined solely by the number of *s* and *p* electrons. The A2 (W), A3 (Mg) and A1 (Cu) structure types are regarded as resulting from the electron configurations $d^{n-1}\,s$, $d^{n-2}\,s\,p$ and $d^{n-3}\,s\,p^2$ respectively. An element will select that particular crystal structure for which the net cohesion is a maximum; the latter factor is the difference between the promotion energy (needed to create one of the specific electron configurations) and the bonding energy (which becomes available if the promoted atoms combine). All unpaired electrons of the promoted atoms, *s*, *p* and *d*, take part in the crystal bonding, but the *d* electrons play no part in determining the type of crystal structure. The electron configurations of the platinum metals are:

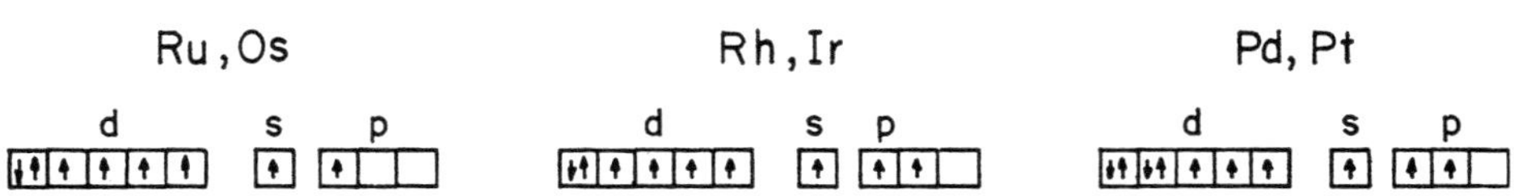

The promotion energies to create the $d^6\,s\,p^2$ and $d^7\,s\,p^2$ configurations are quite high; however, this is compensated by the increase in bonding energy due to the newly available unpaired *d* electrons.

The Engel-Brewer theory can explain the stability and structures of certain intermetallic compounds of the platinum metals. In particular, combinations between platinum metals and transition metals from the left side on the periodic table like

Table 78-A-2. *Electron configurations for the elements in intermetallic platinum metal compounds and their components in the unalloyed state according to the Engel-Brewer theory*

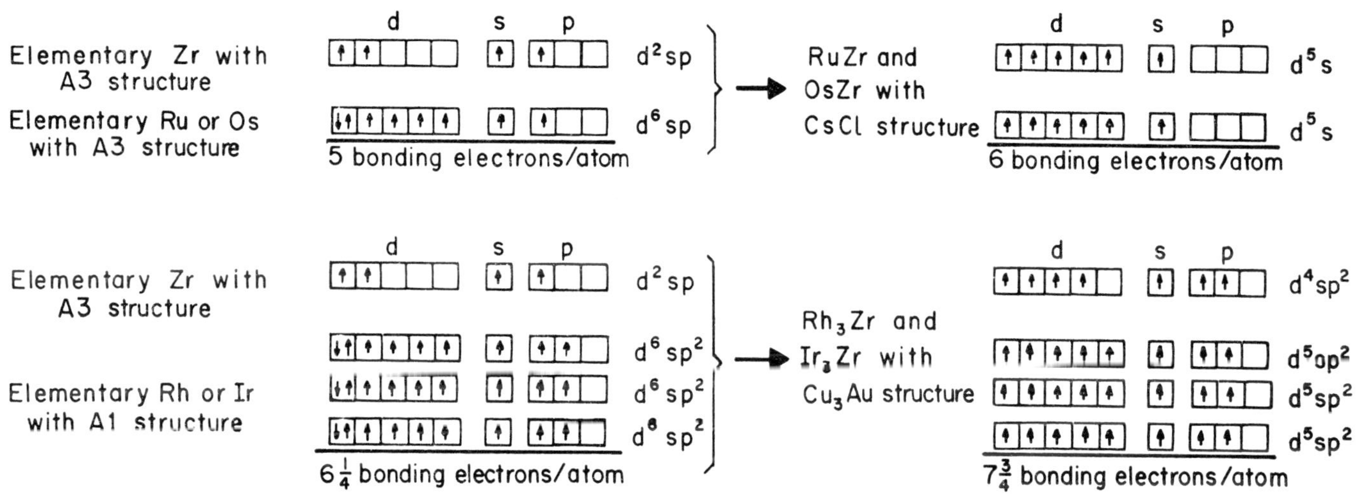

Ti, Zr and Hf, should produce very stable compounds. The differences between the transition metals from the left and right sides of the periodic table can be summarized in terms of incomplete use of available bonding orbitals and full use of available electrons on the left side and incomplete use of electrons and full use of bonding orbitals on the right side. A mixture of metals from both sides will allow electron transfer from the platinum metals with a surplus of electrons to the metals with low-lying vacant orbitals to make fullest use of all electrons and all orbitals. As an example one may consider intermetallic compounds of the platinum metals with Zr. RuZr and OsZr have the CsCl structure, an A2 derivative structure. Rh_3Zr and Ir_3Zr crystallize with the Cu_3Au structure, an A1 derivative structure. The electron configurations of the pure elements and of the elements in these intermetallic compounds are given in Table 78-A-2. The increase in the average number of bonding electrons per atom explains the unusual high stability of these compounds.

It should be mentioned that the Engel-Brewer correlation, especially the correlation between specific electron configuration and crystal structure has been the subject of heated discussions and has been attacked on theoretical grounds. However, this correlation as a whole has been surprisingly successful in explaining many observations for which no other explanations are available.

Compounds of the platinum metals which are found in nature can be grouped in metallic and non-metallic minerals as shown in Table 78-A-3.

The CuAu structure type of potarite, PdHg, corresponds to an ordered derivative of the A1 (Cu) structure, while all other metallic minerals for which the structure is known except froodite have the NiAs structure or a variant of it like the CdI_2 structure. A satisfactory crystal chemical interpretation of these metallic structures is unfortunately not possible at this time.

In the lower part of Table 78-A-3 are given the non-metallic semiconducting, diamagnetic chalcogenide and pnictide minerals of the platinum metals. These ionic-covalent compounds satisfy simultaneously the electronic conditions for covalent bonding according to the valence bond theory and the conditions for ionic bonding according to the electrostatic crystal-field theory and the general theory for valence compounds. Thus, two canonical formulae can be written for each compound, one being a covalent formula, the other an ionic formula. The valence bond theory specifies that six octahedral bonding orbitals (possibly also somewhat distorted) are formed if the transition metal atom can form a $d^2\,s\,p^3$ hybrid or four square planar orbitals with a $d\,s\,p^2$ hybrid. The theory shows further that the chalcogen and P, As, Sb atoms will form four tetrahedral orbitals with $s\,p^3$ hybridization. If all the electrons of the transition elements which participate in the hybridization are shifted momentarily to the chalcogen or P, As, Sb atoms, one obtains the limiting ionic formula. The electrostatic crystal-field theory specifies then that the diamagnetic transition metal ions must have a d^6 configuration for regular octahedral coordination, a d^4 configuration for a stretched octahedral or rhombic bi-pyramidal coordination or a d^8 configuration for a square planar coordination. Further the general valence compound equation states that the anions will use all the provided electrons to fill their octet shells either by forming isolated anions or anion dumb bells or anion chains. Using this information one can calculate as shown in Table 78-A-4 possible compositions for semiconducting platinum metal chalcogenides and pnictides assuming particular cation coordinations and anion nettings.

Table 78-A-3. *Compounds of the platinum metals found in nature*

Mineral	Formula	Structure	Reference
		Metallic minerals	
Potarite	PdHg	tetragonal, AuCu ($L1_0$) type	Terada and Cagle, SR 1960, 181; Bittner and Nowotny, SR 1952, 115
Froodite	$PdBi_2$	monoclinic, α $PdBi_2$ type	Hawley and Berry, SR 1958, 52
Niggliite	Pt(Sn, Te)	hexagonal, NiAs (B8) type	Groeneveld Meijer, SR 1955, 420
Arsenopalladinite	Pd_3As	hexagonal, does not agree with artificial Pd_3As which has tetragonal Fe_3P type	
Michenerite	$PdBi_2$	cubic pyrite, does not agree with β $PdBi_2$ which has tetragonal $MoSi_2$ (C11) type	Hawley and Berry, SR 1958, 52
Stibiopalladinite	Pd_3Sb	unknown structure	
Stannopalladinite	Pd_3Sn_2	hexagonal "filled up" NiAs (B8) type	Nowotny, Schubert and Dettinger, SR 1947—1948, 173
Kotulskite	$Pd(Te, Bi)_{1-2}$	hexagonal "partly empty" CdI_2(C6) type	
Monchëite	$(Pt, Pd)(Te, Bi)_2$	hexagonal CdI_2(C6) type	
		Semiconducting, diamagnetic minerals	
Cooperite	PtS	tetragonal PtS (B17) type	Bannister and Hey, SB 1928 bis 1932, 9, 234
Braggite	(Pt, Pd, Ni)S	tetragonal PdS (B34) type	Gaskell, SB 1937, 3, 41
Vysotskite	(Pd, Ni)S	tetragonal PdS (B34) type	
Laurite	RuS_2	cubic pyrite (C2) type	Oftedal, SB 1913—1928, 780
Sperrylithe	$PtAs_2$	cubic pyrite (C2) type	Thomassen, SB 1913—1928, 781
Geversite	$PtSb_2$	cubic pyrite (C2) type	Thomassen, SB 1913—1928, 781

In Table 78-A-5 structure types are shown of all known semiconducting and diamagnetic chalcogenides and pnictides of the platinum metals. In addition to the cation coordinations and anion nettings considered in Table 78-A-4, more complicated arrangements are also found. However, in all cases there is excellent agree-

Table 78-A-4. *Possible compositions for semiconducting platinum metal chalcogenides and pnictides with prescribed cation coordination and anion netting*

Large numerals are used for the number of valence electrons, including here in the case of the platinum metals all *d* electrons. Small numerals are conventional chemical composition parameters.

T Elements: 8: Ru or Os; 9: Rh or Ir; 10: Pd or Pt.

X Elements: 5: P, As or Sb; 6: O, S, Se or Te.

Formulae for which examples are known are enclosed in rectangles

CATION COORDINATION

d electron energy levels and their occupation with electrons for T metal ions in different coordinations.

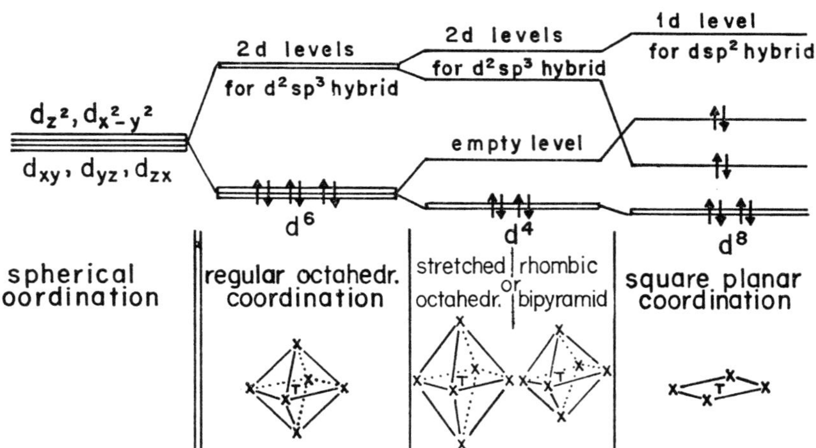

ANION NETTING	spherical coordination	regular octahedr. coordination	stretched octahedr. or rhombic bipyramid	square planar coordination
$\overline{|X|}$ isolated anions 8 electrons/anion		$9_2 6_3$	impossible	10 6
$\overline{|X}-\overline{X|}$ anions dumb bells 7 electrons/anion		86_2 $10\,5_2$ $9\,56$	85_2	impossible
$-\overline{X}-\overline{X}-\overline{X}-$ chains or rings 6 electrons/anion		95_3	impossible	86_2 $10\,5_2$ $9\,56$

Table 78-A-5. *Crystal structure types, electron configuration and ionic formulae of the semiconducting and diamagnetic chalcogenides and pnictides of the platinum metals*

Structural features	Structure type	Atoms per cell and space group	Electron distribution for covalent formula T (d s p) X (s p)	Ionic formula	Examples	References [a]
			TX Compounds			
square planar T coord., tetrahedral X coord., $\boxed{X}$ only	$Pt^{[4]}S^{[4]}$ cooperite (B17 type)	4 $P4_2/mmc$	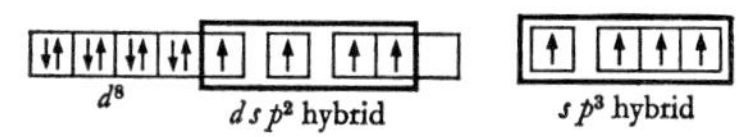	$Pt^{2+}[\boxed{S}]^{2-}$	*10 6*: PdO, PtO, PtS	1.
same; differences only in second nearest coordination	$Pd^{[4]}S^{[4]}$ (B34 type)	16 $P4_2/m$	same	$Pd^{2+}[\boxed{S}]^{2-}$	*10 6*: PdS, PdSe	2.
			T_2X_3 Compounds			
octahedral T coord., tetrahedral X coord., $\boxed{X}$ only	$Rh_2^{[6]}S_3^{[4]}$	20 Pbcn	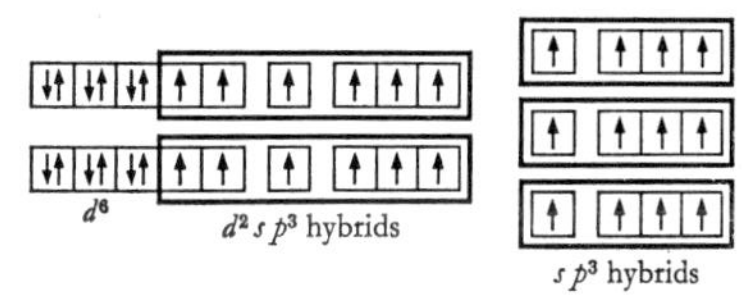	$Rh_2^{3+}[\boxed{S}]_3^{2-}$	*$9_2 6_3$*: Rh_2S_3, Rh_2Se_3, Ir_2S_3	3.
			TX_2 Compounds			
octahedral T coord., tetrahedral X coord., $\boxed{X}-\boxed{X}$ dumb bells	$Fe^{[6]}S_2^{[3+1]}$ pyrite (C2 type) or $Ni^{[3+3]}Sb^{[3+1]}S^{[3+1]}$ ullmannite ≡ ternary ordered pyrite type	12 $Pa3$ 12 $P2_13$	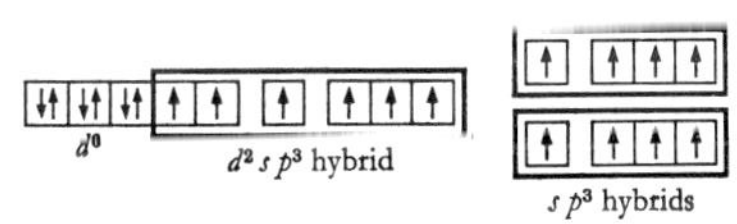	$Ru^{4+}[\boxed{S}-\boxed{S}]^{2-}$ $Rh^{3+}[\boxed{P}-\boxed{S}]^{3-}$ $Pt^{4+}[\boxed{P}-\boxed{P}]^{4-}$	*$8 6_2$*: RuS_2, $RuSe_2$, $RuTe_2$, OsS_2, $OsSe_2$, $OsTe_2$ *9 5 6*: RhPS, RhAsS, RhSbS, RhBiS, RhPSe, RhAsSe, RhSbSe, RhBiSe, RhAsTe, RhSbTe, RhBiTe, IrPS, IrAsS, IrSbS, IrBiS, IrBiSe, IrAsSe, IrSbSe, IrBiSe, IrAsTe, IrSbTe, IrBiTe *$10 5_2$*: $PdAs_2$, $PdSb_2$, PtP_2, $PtAs_2$, $PtSb_2$	4.

Table 78-A-5. (Continued)

Structural features	Structure type	Atoms per cell and space group	Electron distribution for covalent formula T (d s p) X (s p)	Ionic formula	Examples	References[a]
same; differences only in second nearest coordination	$Fe^{[6]}S^{[3+1]}$ marcasite (C18 type)	6 Pnnm	same	same		5.
rhombic bipyramidal T coordination, tetrahedral X coord., X̄—X̄ dumb bells	$Fe^{[6]}As_2^{[3+1]}$ loellingite similar to marcasite (C18 type) but with changed axial ratios ($c/a \sim 0.55$ instead of 0.74 and $c/b \sim 0.48$ instead of ~ 0.62)	6 Pnnm	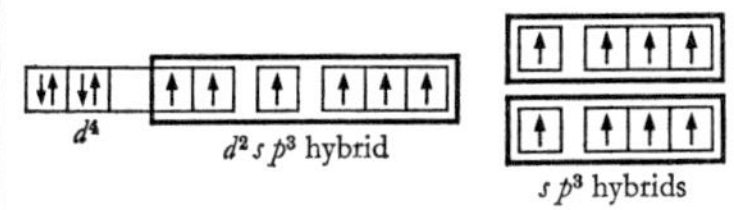	Ru^{4+}[P̄—P̄]$^{4-}$	$8\,5_2$: RuP_2, $RuAs_2$, $RuSb_2$, OsP_2, $OsAs_2$, $OsSb_2$,	6.
square planar T coord., tetrahedral X coord., —X̄—X̄—zig-zag chains	$Pd^{[4]}P_2^{[2+2]}$	12 C2/c	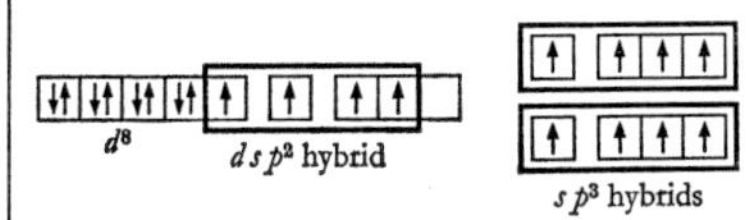	Pd^{2+}[—P̄—P̄—]$^{2-}$	$10\,5_2$: PdP_2	7.
octahedral T coord., every X atom with only 3 tetrahedral neighbors, X̄ only, layer structure	$Cd^{[6]}Cl_2^{[3]}$ (C6 type)	3 P$\bar{3}$m1	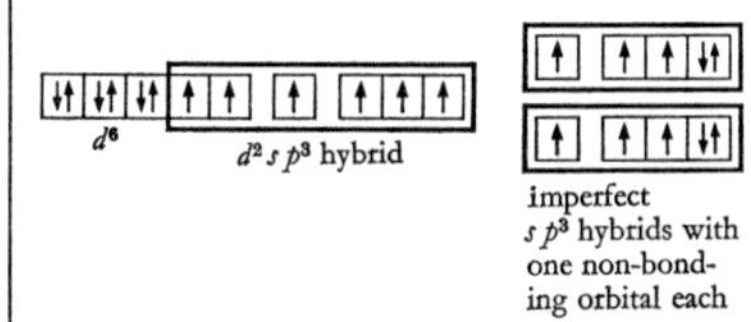	Pt^{4+}[S̄]$_2^{2-}$	$10\,6_2$: PtS_2, $PtSe_2$	8.
square planar T coordination, every X atom with only 3 tetrahedral neighbors, X̄—X̄ dumb bells, layer structure	$Pd^{[4]}Se^{[2+1]}$	12 Pbca	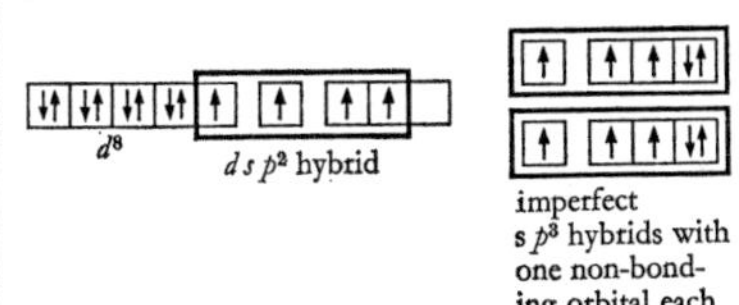	Pd^{2+}[S̄e—S̄e]$^{2-}$	$10\,6_2$: PdS_2, $PdSe_2$	9.

octahedral T coord., half of X atoms with only 3 tetrahedral neighbors, others with four; half of the X are isolated [X], half form dumb bells [X—X]	$Ir^{[6]}Se^{[3+1]}Se^{[3]}$	24 Pnma	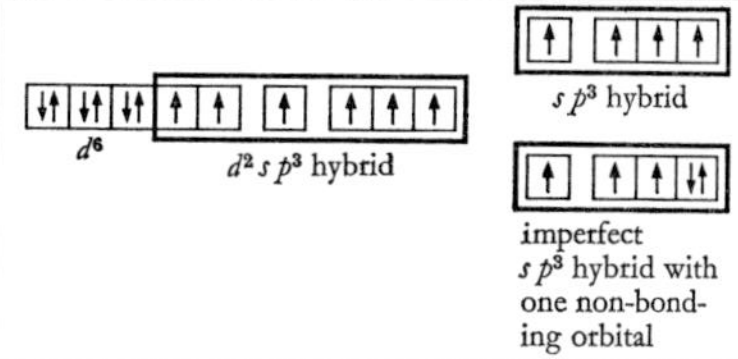	$Ir_2^{3+}[\overline{Se}]_2^{2-}[\overline{Se}—\overline{Se}]^{2-}$	96_2: RhSe$_2$, IrS$_2$, IrSe$_2$	10.
deformed octahedral T coord., tetrahedral X coordination, T—T dumb bells, [X—X] dumb bells	$Co^{[6+1]}As_2^{[3+1]}$ or $Fe^{[3+3+1]}As^{[3+1]}S^{[3+1]}$ arseno pyrite ≡ ternary ordered CoAs$_2$ type	12 P2$_1$/c 12 P2$_1$/c	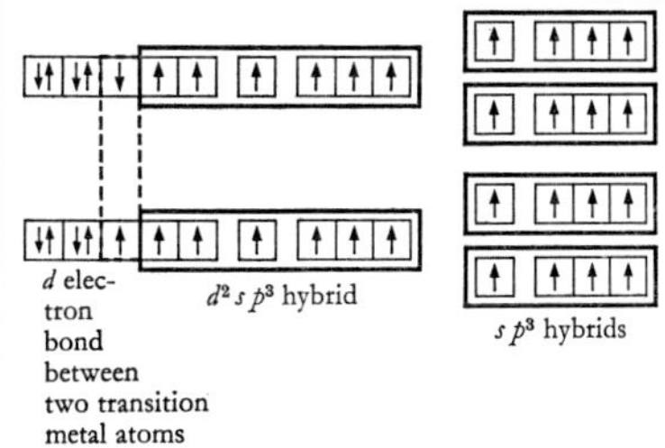	$[Rh—Rh]^{8+}[\overline{P}—\overline{P}]_2^{4-}$ $[Ru—Ru]^{6+}[\overline{P}—\overline{S}]_2^{3-}$	95_2: RhP$_2$, RhAs$_2$, RhSb$_2$, RhBi$_2$, IrP$_2$, IrAs$_2$, IrSb$_2$, IrBi$_2$ 8 56: RuPS, RuAsS, RuSbS, RuPSe, RuAsSe, RuSbSe, RuAsTe, RuSbTe, OsPSe, OsAsSe, OsSbSe, OsBiSe, OsAsTe, OsSbTe, OsPS, OsAsS, OsSbS, OsBiTe	11.
TX_3 Compounds						
octahedral T coord., deformed tetrahedral X coord., somewhat deformed ring 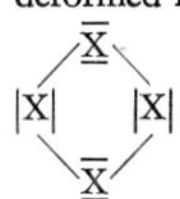	$Co^{[6]}As_3^{[2+2]}$ skutterudite (D0$_2$ type)	32 Im3	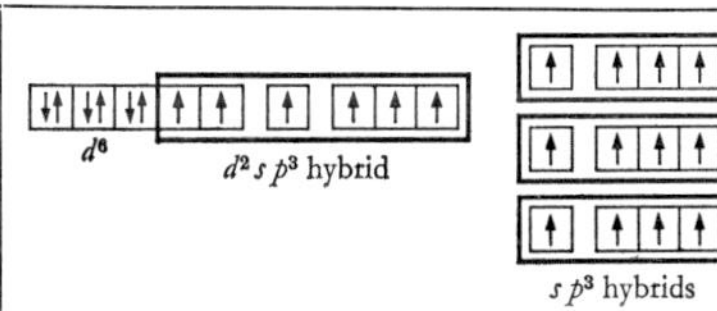	$Rh^{3+}[—\overline{P}—\overline{P}—\overline{P}—]^{3-}$	95_3: RhP$_3$, RhAs$_3$, RhSb$_3$, IrP$_3$, IrAs$_3$, IrSb$_3$	12.
octahedral T coord., on the average every X atom with only 3 tetrahedral neighbors, [X—X] dumb bells, ordered pyrite defect structure	$Ir^{[6]}Se_3^{[2+1]}$	32 P$\overline{1}$	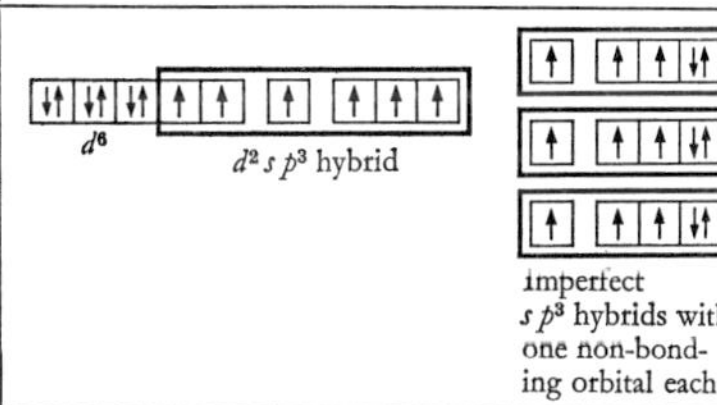	$Ir_{0.66}^{3+}\square_{0.33}[\overline{Se}—\overline{Se}]^{2-}$	96_3: IrSe$_3$	13.

[a] References for structure types: 1. BANNISTER and HEY, SB 1928—1932, 234. 2. GASKELL, SB 1937, 3, 41. 3. PARTHÉ, HOHNKE and HULLIGER (1967). 4. BRAGG, SB 1913—1928, 150, 215; TAKEUCHI, SB 1957, 34. 5. BUERGER, SB 1928—1932, 272. 6. HOLSETH and KJEKSHUS (1968); BUERGER, SB 1928—1932, 273. 7. ZACHARIASEN (1963). 8. BOZORTH, SB 1913—1928, 161, 189. 9. GRØNVOLD and RØST, SB 1957, 161. 10. BARRICELLI, SB 1958, 145. 11. DARMON and WINTENBERGER (1966). 12. OFTEDAL, SB 1913—1928, 232, 250; RUNDQVIST and ERSSON (1968). 13. PARTHÉ and HOHNKE (1970).

ment between the observed structural features and the number of valence electrons of the compounds as required by the rules given above. This can be verified by studying the limiting covalent and ionic formulae listed in Table 78-A-5. Drawings for all the different chalcogenide and pnictide structures can be found in a recent review article by HULLIGER (1968). These drawings may be examined to determine in detail the structural features of these compounds.

Revised manuscript received: March 1970

References: Section 44-45-46, 76-77-78-A

BREWER, L.: Predictions of high temperature metallic phase diagrams. In: ZACKAY, V. F. (ed.), High Strength Materials, p. 12—103. New York: John Wiley 1965.
— A most striking confirmation of the Engel metallic correlation. Acta Met. **15**, 553 (1967).
DARMON, R., WINTENBERGER, M.: Structure cristalline de $CoAs_2$. Bull. Soc. Franc. Mineral. Crist. **89**, 213 (1966).
HOLSETH, H., KJEKSHUS, A.: Compounds with the marcasite type crystal structure. I. Compositions of the binary pnictides. Acta Chem. Scand. **22**, 3273 (1968). II. On the crystal structures of the binary pnictides. Acta Chem. Scand. **22**, 3284 (1968).
HULLIGER, H.: Crystal chemistry of the chalcogenides and pnictides of the transition elements. Structure and Bonding **4**, 83 (1968).
HUME-ROTHERY, W.: The Engel-Brewer theories of metals and alloys. Progr. Mater. Sci. **13**, 229 (1968).
PARTHÉ, E., HOHNKE, D.: The ordered vacancy arrangement in pyrite defect structures. In: EYRING, L., O'KEEFE, M. (eds.), The Chemistry of Extended Defects in Non-metallic Solids, p. 220—222. Amsterdam: North-Holland Publishing Company 1970.
— — HULLIGER, F.: A new structure type with octahedron pairs for Rh_2S_3, Rh_2Se_3 and Ir_2S_3. Acta Cryst. **23**, 832 (1967).
RUNDQVIST, S., ERSSON, N. O.: Structure and bonding in skutterudite-type phosphides. Arch. Kemi **30**, 103 (1968).
ZACHARIASEN, W. H.: The crystal structure of palladium diphosphide. Acta Cryst. **16**, 1253 (1963).

Revised manuscript received: March 1970

General Introduction

The platinum metals include ruthenium, rhodium, palladium, osmium, iridium and platinum, and belong to Group VIII of the periodic table. To avoid duplication all information on the six elements has been condensed in one chapter having the order number of Pt as they are often called platinum elements.

Their abundances in terrestrial materials are very poorly known. For acid and intermediate plutonic and volcanic rocks, metamorphic rocks, limestones and sandstones, data are either extremely limited or non-existant. No data relating to behaviour during weathering, transport and solubility mechanisms in aqueous environments, concentration in natural waters or biogeochemistry are available. Basic and ultrabasic rocks have been more extensively analysed but the bulk of the data are for Pd and Pt only. Some platinum metal ores and host rocks including Sudbury, Canada, Norilsk and Inaglinsk, U.S.S.R., and the Bushveld Complex, South Africa have been studied in moderate detail, particularly for Pd, Pt and Rh.

Platinum metal concentrations as low as 10^{-1} ppb are found in some rocks. No analytical method with sufficient sensitivity for direct determination of all platinum metals in a large variety of terrestrial materials is available. This factor is primarily responsible for the paucity of platinum metal data. The sensitivity limits for various analytical methods are listed in Table 78-1.

To overcome sensitivity problems a fire assay or wet chemical concentration has been widely used with d—c arc emission spectrographic and X-ray spectrometric analysis. At very low concentrations losses by adsorption, coprecipitation, in the cupellation stage of fire assay, or alternatively, contamination by large quantities of reagents are inherent problems. The early work of I. and W. Noddack (1931), Goldschmidt and Peters (1932) and the later studies of Hagen (1954), which account for much of the available platinum metal data on rocks and minerals not associated with ore deposits, involve fire assay and wet chemical concentration steps and are subject to these considerations.

Table 78-1. *Sensitivity limits for platinum metal determination by various techniques*

Method	Reference	Sensitivity in µgm metal per gm sample (=ppm)					
		Ru	Rh	Pd	Os	Ir	Pt
d-c arc emission spectrography	1	10	10	10	50	50	50
X-ray spectrometry	2	5	8	10	10	15	15
Neutron activation	3	0.04	—	0.0005	0.005	0.00004	0.015
Colorimetric	4	0.01	0.07	0.04	0.1	0.04	0.03

References: 1. Ahrens, L., and S. Taylor (1961) (Table 6-1). 2. Kemp, J. (1956) (Table IV). 3. Winchester, J. (1960) (Table VII). The neutron activation sensitivities apply to a flux of 10^{13} neutrons/cm²-sec, a minimum detectable activity of 10 disintegrations/sec, an irradiation time of 150 hrs, and a sample weight of 100 mg. In practice Rh is seldom determined by activiation analysis due to the 4.4 min half-life of the induced radioactivity. 4. Sandell, E. (1959). Sensitivities were computed for the most sensitive reaction discussed assuming that the final volume of solution is 10 ml and that 1 gm of sample is processed. Sensitivity is defined as the number of µgms of the element giving extinction of 0.001 in a column of solution of 1 cm² cross section (µgms metal $\equiv \log I_0/I = 0.001$).

Methods with better sensitivity include colorimetry and neutron activation analysis. The latter method is not subject to post-irradiation contamination by reagents, does not require quantitative recovery of the analysis element, and can be used for direct determination of Os, Ir and Pd in a number of rock types. It is, however, critically dependent on equilibration of a carrier and the appropriate activated radionuclide.

The higher concentrations in meteorites, unlike terrestrial rocks, impose somewhat less exacting analytical requirements, and are partly responsible for the extensive platinum metal abundance data now available for meteorites.

Previous reviews of platinum metal geochemistry by GMELIN (1938), ZVYAINTSEV (Pd, 1946), ZVYAINTSEV (Ru, 1950), RANKAMA and SAHAMA (p. 688—694, 1950), GOLDSCHMIDT (p. 677—695, 1554), QUIRING (1962) and WRIGHT and FLEISCHER (1965, 1967) are available. This review includes data published to the end of 1967.

In the tables of data the number of samples analysed, the analytical method and the author are given, for example, as (2, N)[3] meaning the analysis of two samples by neutron activation by author number 3 as listed in the references for each table or group of tables. *One additional analytical symbol, FA, is used for fire assay.* Where a fire assay, FA, or wet chemical, W, concentration was used with emission spectrographic or X-ray spectrographic methods, this is indicated. Thus, FA/S means spectrographic analysis with fire assay concentration.

Revised manuscript received: March 22nd, 1968

78-B. Isotopes in Nature

The atomic percentages of the stable isotopes of the platinum group metals taken from LEDERER *et al.* (1967) are presented in Table 78-B-1. Only one platinum metal nuclide found in nature, ^{190}Pt, is radioactive. It constitutes 1.27 atom percent of plantinum and decays by alpha emission with a half-life of 4.7×10^{11} years. The only platinum metal whose natural isotopic composition is significantly affected by radioactive decay of other elements is Os. Radiogenic ^{187}Os is produced by the decay of ^{187}Re, a weak beta emitter whose half-life has been variously determined as 4.3×10^{10} years (HIRT *et al.*, 1962) and 3.0×10^{10} years (WATT and GLOVER, 1962). Both iron meteorites and some samples of molybdenite show variations in Os isotopic composition due to enrichment in radiogenic ^{187}Os (HIRT, *et al.* 1962). A study of Australian molybdenites by RILEY (1967) indicated that nine of forty-two samples contained sufficient radiogenic Os to make age determination possible (for additional data see Sect. 75-B).

Tabelle 78-B-1. *Atomic percentages of stable platinum metal isotopes.* (After LEDERER *et al.*, 1967)

Metal		Atomic percentage for corresponding mass number							References
Ru	A	96	98	99	100	101	102	104	
	%	5.46	1.868	12.63	12.53	17.02	31.6	18.87	ORDZHONIKDZE and AKIRTAVA (1960)
	%	5.57	1.86	12.7	12.6	17.1	31.6	18.5	WHITE *et al.* (1956)
	%	5.50	1.91	12.7	12.7	17.0	31.5	18.7	FRIEDMAN and IRSA (1953)
	%	5.7	2.2	12.8	—	17.0	31.3	18.3	EWALD (1944)
Rh	A				103				
	%				100				COHEN (1943)
Pd	A	102	104	105	106	108	110		
	%	0.8	9.3	22.6	27.2	26.8	13.5		SAMPSON and BLEAKNEY (1936)
	%	0.96	10.97	22.2	27.3	26.7	11.8		SITES *et al.* (1953)
Os	A	184	186	187	188	189	190	192	
	%	0.018	1.59	1.64	13.3	16.1	26.4	41.0	NIER (1937)
Ir	A				191	193			
	%				38.5	61.5			SAMPSON and BLEAKNEY (1936)
Pt	A	190	192	194	195	196	198		
	%	0.0127	0.78	32.9	33.8	25.2	7.19		WHITE *et al.* (1956)

Revised manuscript received: March 1968

78-C. Abundance in Meteorites, Tektites, Cosmos and Lunar Materials

I. Meteorites

Meteorites, as well as sulfide minerals, have been extensively analysed for platinum metals. For most chondrites and irons platinum metal concentrations of 0.1 to 10 ppm are found. These higher concentrations permit the use of a number of analytical methods including colorimetry, neutron activation and emission spectrography (in conjunction with simple wet chemical concentration procedures). The earliest significant work on platinum metal geochemistry is that of I. and W. NODDACK (1932) and GOLDSCHMIDT and PETERS (1932) whose data on iron meteorites are compared with recent determinations in Table 78-C-1. In general the early data overestimated Rh, Pd and Pt by factors of 2 to 4. Ir values may be either high or low by 2 while the limited Os data indicate that gross errors (overestimates) occur when the metal is less than 1 ppm. GOLDSCHMIDT and PETERS, however, obtained good agreement with recent data for Ru. In some cases, as for example Pd in Coahuila, the early data are high by a factor of nearly six, and for these reasons have not been included in Tables 78-C-3 to 78-C-6.

Meteorites exhibit three distinct ranges of platinum metal concentrations. Average platinum metal contents in irons vary from 1 to 10 ppm, in chondrites from 0.1 to 2 ppm, and in achondrites from 1 to 100 ppb. The data for achondrites in Table 78-C-2 indicates that their platinum metal contents are comparable with terrestrial ultramafic rocks. They are highly depleted with respect to the iron meteorites and chondrites.

Data for chondrites and iron-nickel metal separated from chondrites are given in Table 78-C-3 and 78-C-6. The concentrations in separated metal are similar to those in irons. In Forest City, Ochansk and Mocs analyses of silicate and metal phases for Os and Ru show an enrichment of at least 10 fold in the metal phase. Thus, platinum metals in chondrites are associated mostly with metallic iron-nickel.

The average atomic abundances of platinum metals in the major chondrite classes are plotted in Fig. 78-C-1. The Pd data are subject to considerable analytical uncertainty. In all but one instance where direct comparison is possible the data of GREENLAND (1967) are higher by a factor of two or more than those of REED (1963), SEN GUPTA (1965) and CROCKET *et al.* (1967). The lower values are preferred as they give better agreement with abundances derived from nucleosynthesis theory and from iron meteorites (NICHIPORUK and BROWN, 1965). The average Pd abundance data in ppm used in Fig. 78-C-1 which excludes the data of GREENLAND, are: Ce=1.11, CH=1.17, CL=0.81 and Cc=1.31.

No large fractionations are found except for Ir which is depleted in enstatite chondrites. The high iron chondrites including the carbonaceous, enstatite and bronzite classes vary by less than 52% in platinum metal content except for Ir. The hypersthene, or low iron chondrites, however, show an average depletion (excepting Ir) of 38% relative to the average of the other classes.

Table 78-C-1. *Comparison of platinum metal data for iron-meteorites and troilite, ppm*

Specimen	Reference	Ru	Rh	Pd	Os	Ir	Pt
São Julião de Moreira	Table 78-C-5			2.6			
	GOLDSCHMIDT and PETERS (1932)			10			
Cañon Diablo	Table 78-C-5	6.15	2.7	3.9	3.35	2.2	7.1
	NODDACK and NODDACK (1931)	8			3		18
	GOLDSCHMIDT and PETERS (1932)	10	5	5—10	5	5	10 to 100
Cranbourne	Table 78-C-5			4.8			
	GOLDSCHMIDT and PETERS (1932)			10			
Toluca	Table 78-C-5	5.3	1.1	4.4		2.0	6.5
	GOLDSCHMIDT and PETERS (1932)	5	0.5	10		1	1
	GOLDSCHMIDT and PETERS (1932) (unroasted)	10	5	10		1	10
	NODDACK and NODDACK (1931)	20		20			10
N'Goureyma	Table 78-C-5					<0.5	
	GOLDSCHMIDT and PETERS (1932)					1	
Mount Joy	Table 78-C-5	5.1		2.20	0.51		
	GOLDSCHMIDT and PETERS (1932)	9—10		10	<5		
Arispe	Table 78-C-5	11.85	1.7	2.7		7.6	17.2
	GOLDSCHMIDT and PETERS (1932)	10	5	10		10	10 to 100
Coahuila	Table 78-C-5	23.7	2.3	1.8		14.4	24.2
	GOLDSCHMIDT and PETERS (1932)	10	5	10		10	10 to 100
Brenham, iron phase	Table 78-C-5	4.6			0.07		
	GOLDSCHMIDT and PETERS (1932)	5—10			< 5		

Table 78-C-1 (Continued)

Specimen	Reference	Ru		Rh	Pd		Os	Ir	Pt	
		1 (1)	2 (2)	1	1 (1)	2 (2)	1.2	1	1 (1)	2 (2)
Discrepancy factor for data of GOLDSCHMIDT and PETERS (1) and NODDACK and NODDACK (2) relative to average value for Table 78-C-5; + high, — low		−1.4	+2.5	+2 excluding Toluca	+3.4	+4.6	Gross over estimate for values less than 1 ppm		+2 to −2	+2
Troilite from Cañon Diablo	Table 78-C-6	0.0324					0.0051	0.0015	0.71	
	GOLDSCHMIDT and PETERS (1932)	0.5							0.5	
	NODDACK and NODDACK (1931)	9					6	0.5	5	

Table 78-C-2. *Platinum metal content of achondritic meteorites, ppb* (for notes and references see pp. 78-C-22 and 78-C-23)

Class[a]	Specimen[b]	Ru	Os	Ir
Ae	Bishopville		4.94 (2, N)[17]	
	Cumberland Falls (white portion)			1.8 (1, N)[1]
	Norton County			2.4 (1, N)[1]
	Shallowater			140 (1, N)[1]
Ab	Johnstown	2.9 (1, N)[2]	6.1 (2, N)[2]	11 (2, N)[1] 4.6
Ap	Nuevo Laredo		7.5 (1, N)[2]	
	Pasamonte			22 (1, N)[1]

The iron meteorites contain the highest platinum metal contents of any meteorite group. The data for the irons are presented in Table 78-C-5 and the distribution of platinum metal values in irons is shown in Fig. 78-C-2. The dispersion of the Rh values is very small with more than 90% of the data in the interval 0.5 to 3.0 ppm. The Pd distribution shows some similarity to Rh in that most of the data fall within a small range, 2 to 6 ppm, but the dispersion is greater. Os and Ir distributions are characterized by much larger proportions of very high and very low values. For Os the range 1 to 4 ppm contains 54% of the data but 15% exceeds 15 ppm and 15% is below 0.2 ppm. Ruthenium and Pt are both characterized by a high dispersion of data.

Variations in platinum metal content in the different classes of iron meteorites are shown in Fig. 78-C-3 where average platinum metal content is plotted against the average Ni content (WOOD, 1963) for the various structural classes. Pd increases

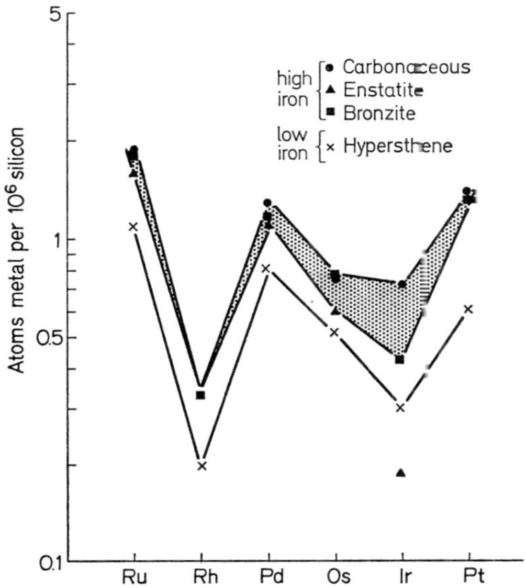

Fig. 78-C-1. Platinum metal atomic abundances in chondrites

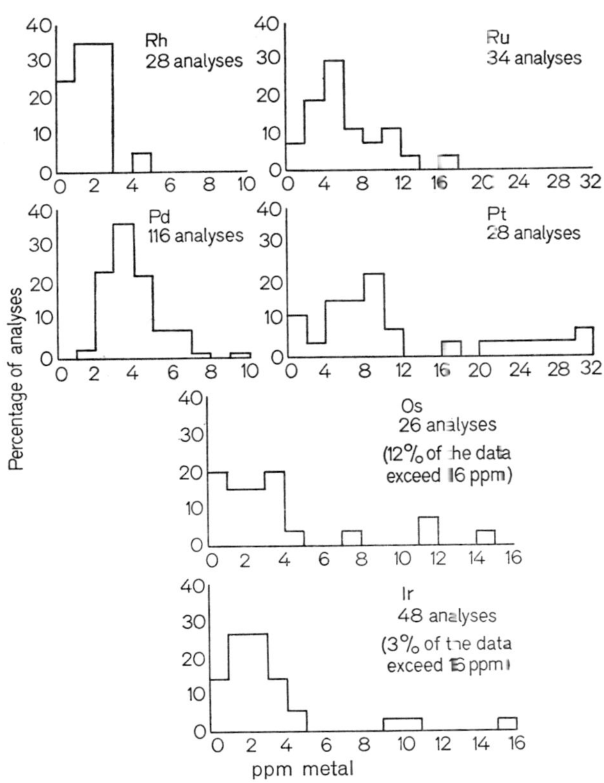

Fig. 78-C-2. Distribution of platinum metals in iron meteorites

Table 78-C-3. *Platinum metal content of chondritic meteorites, ppm and*

Class	Specimen	Ru	Rh	Pd
Ce_1	Abee	0.97, 1.52 (2, N)[5]		0.79, 1.18 (1, N)[5] 1.3, 1.9 (1, N)[7] 0.45, 0.67 (2, C)[24] 0.85, 1.3
	Indarch	1.01, 1.68 (2,N)[5]		0.80, 1.27 (1, N)[5] 2.1, 3.4 (1, N)[7] 1.4_5, 2.3
Ce_2	Altanta			
	Hvittis			1.1, 1.5 (1, N)[7]
	Daniel's Kuil			2.2, 3.2 (1, N)[7]
	Khairpur			1.4, 1.9 (1, N)[7]
	Pillistfer			1.9, 2.8 (1, N)[7]
	St. Mark's			
	Average[C]	0.99, 1.60		1.5, 2.2
Ch	Alamogordo	1.2 (1, C)[10]		
	Allegan	1.19, 1.95 (2, N)[2]		1.5, 2.3 (1, N)[7]
	Beardsley	1.12, 1.88 (1, N)[2]	0.21, 0.35 (3, N)[23]	
	Belly River			0.78_5, 1.19 (2, C)[24]
	Cope	1.1 (1, C)[10]		
	Ehole			1.5 (1, N)[7]
	Estacado	1.3, 2.2 (1, C)[10]		
	Forest City	0.95, 1.53 (1, N)[2]	0.21, 0.33 (3, N)[23]	1.0 1.5 (3, N)[9]
	Gladstone	1.1 (1, C)[10]		
	Gruver	1.1 (1, C)[10]		
	Haven	1.2 (1, C)[10]		
	Hessle		0.20, 0.32 (3, N)[23]	
	Morland	0.9, 1.5 (1, C)[10]		
	Mount Browne			
	Ochansk	1.12, 1.8 (2, N)[2]		
	Pantar			0.53, 0.82 (1, N)[20] 1.3, 2.0 (1, N)[7] 0.93, 1.42
	Penokee			
	Plainview	1.1, 1.8 (1, C)[10]		
	Pultusk			
	Richardton	0.82, 1.42 (1, N)[2]		
	Zhovtnevyi			
	Average	1.1, 1.8	0.21, 0.33	1.1, 1.6

atomic abundances. (For notes and references see pp. 78-C-22 and 78-C-23)

Os	Ir	Pt	Si (wt.-%)
0.78, 0.65, (1, N)[5]	0.083, 0.069 (1, N)[1]	1.3, 1.1 (1, N)[1]	17.66[3]
0.586, 0.490 (1, N)[18]	0.32, 0.27 (1, N)[5]	1.3, 1.1 (1, N)[5]	
0.68, 0.57		1.3, 1.1	
0.74, 0.66 (1, N)[5]	0.25, 0.22 (1, N)[5]	1.25, 1.1 (2, N)[5]	16.48[22]
0.567, 0.508 (2, N)[18]			
0.65, 0.58			
		2.2, 1.6 (1, N)[1]	19.6[21,23]
0.733, 0.660 (2, N)[18]			19.41[20]
			17.98[20]
0.851, 0.631 (2, N)[18]			19.93[20]
0.685, 0.561 (2, N)[18]			18.04[7]
0.671, 0.578 (2, N)[18]			17.14[11]
0.71, 0.60	0.22, 0.19	1.6, 1.3	
1.04, 0.904 (2, N)[2]			17.0[20,21]
0.893, 0.750 (2, N)[18]			
0.97, 0.83			
1.01, 0.899 (2, N)[2]			16.6[21]
0.791, 0.704 (2, N)[18]			
0.90, 0.80			
			17.45[17]
			16.74[15]
0.85, 0.73 (2, N)[2]	0.4, 0.3 (3, N)[9]	1.0, 0.83 (3, N)[9]	17.31[16]
	0.54, 0.46 (2, N)[1]	1.8, 1.5 (2, N)[1]	
	0.47, 0.38	1.4, 1.2	
			17.1[21]
			17.21[20]
			16.55[18]
0.784, 0.712 (2, N)[18]			16.27[20]
1.06, 0.926 (2, N)[2]	0.51, 0.44 (2, N)[1]		16.9[21,23]
0.881, 0.770 (2, N)[18]			
0.97, 0.85			
			17.03[9]
	0.39, 0.35 (1, N)[1]		16.5[21]
	0.49, 0.42 (2, N)[1]	1.6, 1.3 (1, N)[1]	17.15[21,23]
0.919, 0.833 (2, N)[18]	0.56, 0.50 (2, N)[1]		16.3[21]
1.06, 0.977 (1, N)[2]			
0.814, 0.750 (2, N)[18]		1.7, 1.5 (1, N)[1]	16.01[15]
0.94, 0.86			
0.664, 0.548 (1, N)[18]			17.9[21]
0.86, 0.77	0.48, 0.42	1.6, 0.13	

Table 78-C-3

Class	Specimen	Ru	Rh	Pd
Cl	Arriba	0.7, 1.0 (1, C)[10]		
	Bruderheim	0.72, 1.08 (1, N)[2]		1.0, 1.4 (1, N)[7]
				0.5, 0.7 (1, C)[24]
				0.75, 1.1
	Colby			
	Cynthiana			
	Elenovka			
	Harleton			
	Hayes Center	0.40, (1, S)[8]	0.26 (1, S)[8]	1.65 (1, S)[8]
	Holbrook	0.75, 1.11 (1, N)[2]	0.15, 0.22 (3, N)[23]	
		0.060, 0.089 (1, S)[15]±	0.064, 0.09 (1, S)[15]±	0.44, 0.62 (1, S)[15]
	Homestead		0.10, 0.15 (1, S)[8]	0.66, 0.95 (1, S)[8]
	Ioka			
	Khohar			
	Kingfisher	0.8 (1, C)[10]		
	Knyahinya			
	Kunashak			
	Ladder Creek	0.7 (1, C)[10]		
	Long Island		0.10, 0.15 (1, S)[8]	0.54, 0.81 (1, S)[8]
	Melrose	0.8, 1.3 (1, C)[10]		
	Mócs	0.76, 1.14 (2, N)[2]		0.84, 1.20 (1, N)[7]
	Modoc		0.16, 0.24 (3, N)[23]	0.73, 1.05 (3, N)[9]
	Ness County	0.7, 1.1 (1, C)[10]		
	Peace River			0.42, 0.60 (2, C)[24]
				0.89, 1.26 (1, N)[7]
			0.66, 0.93	
	Potter	0.8, (1, C)[10]		
	Roy	0.8, (1, C)[10]		
	Saratov	0.76, 1.10 (2, N)[2]		
	Shaw			
	Waconda		0.15 (1, S)[8]	0.95 (1, S)[8]
	Walters			1.0 (1, N)[7]
	Average	0.72, 1.1	0.15, 0.19	0.82, 0.95
CLL	Benares			
	Benton			0.60, 0.86 (1, S)[24]
	Ensisheim			0.13, 1.84 (1, N)[20]
	Lake Labyrinth			
CHL	Bencubbin inclusion			
	Chainpur			0.83, 1.54 (1, N)[7]
	Felix			2.2, 3.66 (1, N)[7]
	Karoonda			
	Lancé	1.03, 1.83 (2, N)[5]		0.81, 1.38 (2, N)[5]
				2.0, 3.40 (1, N)[7]
				1.4, 2.39

(Continued)

Os	Ir	Pt	Si (wt.-%)
			19.5[21]
0.67, 0.53, (1 N)[2]	0.46, 0.36 (3, N)[1]	0.9, 0.7 (1, N)[1]	18.58[1]
		1.4, 1.0 (1, N)[1]	19.4[21]
	0.21, 0.19 (1, N)[1]		15.9[21]
	0.33, 0.26 (1, N)[1]		18.38[7]
	0.45, 0.37 (1, N)[1]	0.9, 0.7 (2, N)[1]	17.9[21]
	0.61, (1, S)[8]	0.93, (1, S)[8]	
1.24, 0.977 (1, N)[2]	0.24, 0.19 (1, S)[15]		
0.500, 0.394 (2, N)[18]		0.44, 0.34 (1, S)[15]	18.74[13]
0.16, 0.13 (1, S)[15]±			
—			
0.87, 0.69			
0.629, 0.506 (2, N)[18]		0.16, 0.125 (1, S)[8]	18.35[20]
	0.49, 0.39 (1, N)[1]		18.5[21]
0.475 (2, N)[18]			
0.439, 0.341 (2, N)[18]			19.02[15]
	0.54, 0.44 (1, N)[1]		17.98[7]
	0.50, 0.41 (1, N)[1]	0.10, 0.88 (1, S)[8]	17.7[21]
			17.8[21]
0.89, 0.71 (2, N)[2]			18.55[12]
0.568, 0.452 (2, N)[18]			
—			
0.73, 0.58			
	0.25, 0.20 (3, N)[9]	1.0, 0.78 (3, N)[9]	18.37[17]
			18.5[21]
			18.61[2]
0.68, 0.52 (2, N)[2]			19.17[20]
	0.13, 0.096 (1, N)[1]		19.74[5]
	0.33 (1, S)[8]	0.57 (1, S)[8]	
0.64, 0.53	0.38, 0.29	0.71, 0.53	
—	—	—	—
0.503 (2, N)[18]			
			18.73[4]
			18.70[19]
	0.19, 0.15 (1, N)[1]		19.0[21]
—	—	—	—
0.831, (2, N)[18]			
0.335, 0.261 (2, N)[18]			18.99[8]
			15.85[10]
0.845, 0.803 (2, N)[18]			15.55[14]
0.856, 0.815 (2, N)[18]			
0.84, 0.80 (1, N)[5]	0.73, 0.69 (1, N)[5]	1.7, 1.6 (3, N)[5]	15.51[10]
—			
0.85, 0.81			

Table 78-C-3 (Continued)

Class	Specimen	Ru	Rh	Pd	Os	Ir	Pt	Si (wt.-%)
	Mokoia			1.3, 2.2 (1, N)[7]	0.691, 0.655 (2, N)[18]			15.59[10]
	Ngawi							
	Vigarano			1.4, 2.4 (1, N)[7]				15.37[10]
	Warrenton			1.9, 3.1 (1, N)[7]	0.799, 0.725 (1, N)[18]			16.27[22]
	Average			1.5, 2.6	0.73, 0.65			
Cc_1	Alais	0.58, 1.65 (1, N)[5]		0.33, 0.90 (1, N)[5]	0.34, 0.62 (1, N)[5]	0.38, 0.57 (1, N)[5]	0.6, 0.9 (1, N)[5]	9.71[10]
	Ivuna	0.74, 1.92 (2, N)[5]		0.58, 1.45 (2, N)[5]				
				1.8, 4.5 (1, N)[7]	0.61, 0.86 (2, N)[5]	0.41, 0.57 (2, N)[5]	1.2, 1.6 (3, N)[5]	10.60[10]
				1.2, 3.0				
	Orgueil	0.75, 1.97 (1, N)[5]		0.59, 1.48 (3, N)[5]	0.56, 0.79 (1, N)[5]			
				1.6, 4.0 (1, N)[7]	0.451, 0.633 (2, N)[17]	0.42, 0.58 (3, N)[5]	1.0, 1.4 (1, N)[5]	10.53[10]
				1.1, 2.7	0.51, 0.71			
Cc_2	Boriskino				0.536, 0.614 (1, N)[18]			12.89[10]
	Cold	0.83, 1.80 (1, N)[5]		0.57, 1.18 (1, N)[5]	0.77, 0.89 (1, N)[5]	1.25, 1.43 (1, N)[5]	1.4, 1.6 (1, N)[5]	12.76[10]
	Bokkeveld	0.70, 1.52 (1, N)[2]			0.58, 0.67 (1, N)[2]			
		0.77, 1.66			0.578, 0.699 (2, N)[18]			
					0.65, 0.75			
	Mighei	0.85, 1.81 (1, N)[5]		0.62, 1.26 (1, N)[5]	0.73, 0.83 (1, N)[5]			
				1.3, 2.6 (1, N)[7]	0.707, 0.804 (2, N)[18]	0.64, 0.72 (3, N)[5]	1.3, 1.4 (1, N)[5]	12.98[10]
				0.96, 1.9	0.72, 0.82			
	Murray			0.94, (1, N)[7]	0.685, 0.756 (2, N)[18]	0.38, 0.41 (3, N)[1]	1.5, 1.6 (1, N)[1]	13.39[10]
	Nogoya	0.86, 1.88 (1, N)[5]		0.75, 1.56 (1, N)[5]	0.75, 0.87 (1, N)[5]	0.63, 0.73 (1, N)[5]	1.0, 1.1 (1, N)[5]	12.69[10]
	Average	0.76, 1.82		0.84, 1.87	0.60, 0.75	0.59, 0.72	1.1, 1.4	

1. Silicon values determined by established wet chemical methods have been tabulated for chondritic meteorites. Fast neutron activation values (Vogt and Ehmann, 1965) are used otherwise and in case where wet chemical analyses were rejected by Urey and Craig (1953).

2. Data are given in ppm followed by atomic abundances (atoms platinum metal per 10^6 silicon atoms) for those meteorites for which silicon values are available.

smoothly with increasing Ni whereas Pt decreases except in the Ni-rich ataxites which contain relatively high Pt values. Rh is characterized by little variation in any structural class while Ru, Os and Ir show highly irregular but sympathetic trends. The highest values for these metals are in the hexahedrites, medium octahedrites and Ni-rich ataxites while general decrease in concentration with increasing Ni is shown except for the ataxites. The Pd, Pt, Ru and Rh correlations with Ni were noted by NICHIPORUK and BROWN (1965) who also drew attention to three distinct populations iron meteorites based on Ru—Rh and Pt—Ir concentrations.

Table 78-C-4. *Platinum metal content of stony-iron meteorites, ppm.*
(For notes and references see p. 78-C-22 and 78-C-23)

Class	Specimen	Ru	Pd	Os	Ir	Pt
P	Admire		4.1 (2, N)[9]		0.011 (2, N)[9]	0.75 (2, N)[9]
	metal	0.7 (1, C)[10]				
	Brenham				<0.4 (1, N)[4]	
	metal	4.6 (1, C)[10]		0.07 (1, N)[13]		
	Marjalahti				0.9 (1, N)[4]	
	Springwater, Olivine				0.003 (1, N)[1]	
M	Patwar				3.4 (1, N)[4]	

The most heterogeneous groups are the hexahedrites and the Ni-rich ataxites. The former are characterized by large variations in Ru, Os, Ir and Pt contents. Os and Ir vary from 0.025 to 50.4 ppm and 0.014 and 25.3 ppm respectively in this group. Data are lacking on Ru, Rh and Pt in the Ni-rich ataxites but the other platinum metals including Pd, Os and Ir are extremely variable. Pd, which is characterized by very uniform concentration in all other structural classes, exhibits its maximum variability in the Ni-rich ataxites.

Data for stony-irons, tektites and impact glasses are presented in Tables 78-C-4 and 78-C-8 respectively.

II. Tektites

Analyses for Ir and Os in tektites and natural glasses are given in Table 78-C-8. The low concentrations obtained for both these metals are regarded by LOVERING and MORGAN (1964) and BAEDECKER and EHMANN (1965) as supporting evidence for a terrestrial origin. Either acid igneous rocks or argillaceous sedimentary rocks were considered suitable parental material by LOVERING and MORGAN.

III. Cosmic Abundances

Meteorites are considered the most representative source of solar system abundance data for which high quality measurements can be made. Many investigators including RINGWOOD (1961), MASON (1960) and FISH *et al.* (1960) consider that the chemical composition of carbonaceous chondrites closely approximates that of primitive solar material. ANDERS (1964) and LARIMER and ANDERS (1967) suggest that

Table 78-C-5. *Platinum metal content of iron meteorites, ppm.* (For notes and references see pp. 78-C-22 and 78-C-23)

Class	Specimen	Ru	Rh	Pd	Os	Ir	Pt
Ogg	Ainsworth	3.8 (2, C)[10]					
	Arispe	13.7 (2, S)[19]	1.7 (2, S)[19]	3.1 (1, S)[19]		7.6 (2, S)[19]	17.2 (1, S)
		10.0 (2, C)[10]		2.69 (4, N)[6]			
		11.85		2.4 (1, N)[26]			
				2,7			
	Central Missouri				0.025 (1, N)[13]		
	Gladstone			3.4 (1, N)[26]			
	Helt Township					1.2 (1, N)[4]	
	Linwood			3.52 (3, N)[6]	4.52 (1, N)[13]		
	São Julião de Moreira			2.6 (1, N)[26]			
	Seeläsgen			3.4 (1, N)[26]			
	Weekeroo Station			2.9 (1, N)[26]			
	Average	7.8		3.1		4.4	
Og	Bahjoi					2.7 (1, N)[4]	
	Bendegó	11.9 (1, S)[19]	2.1 (1, S)[19]	2.7 (1, S)[19]		0.3 (1, S)[19]	9.8 (1, S)[19]
				3.1 (1, N)[26]			
				2,9			
	Billings			1.6 (1, N)[26]			
	Bischtübe			4.1 (1, N)[26]			
	Bogou					1.8 (1, N)[4]	
	Cañon Diablo	6.4 (2, S)[19]	1.5 (2, S)[19]	3.6 (1, S)[19]	2.07 (4, N)[12]	1.8 (2, S)[19]	7.4 (2, S)[19]
		5.9 (2, C)[10]	3.88 (1, C)[25]	3.98 (10, N)[6]	4.98 (1, C)[25]	2.60 (1, C)[25]	5.17 (1, C)[25]
		6.15	2.7	3.6 (1, N)[26]	3 (1, C)[22]	0.33 (3, N)[9]	5.7 (1, N)[1]
				5.1 (1, C)[25]	3.35	2.0 (2, N)[1]	9.9_5 (2, N)[9]
				4.7 (2, N)[9]		2.13	7.1
				2.58 (7, N)[3]			
				3.9			

Og	Cranbourne			4.8 (1, N)[26]			
	Deport	4.4 (1, C)[10]		4.45 (3, N)[6]			
	Loreto				3.91 (1, N)[13]		
	Mount Tabby	16.6 (2, S)[19]	2.2 (2, S)[19]	3.0 (2, S)[19]		9.1 (2, S)[19]	23.5 (2, S)[19]
	Odessa	6.0 (2, S)[19]	1.4 (2, S)[19]	3.7 (2, S)[19]	2.60 (5, N)[12]	1.8 (2, S)[19]	8.5 (2, S)[19]
				4.15 (10, N)[6]		2.0 (1, N)[1]	
				3.36 (2, N)[3]			
				3.7		1.9	
	Pan de Azucar			4.1 (1, N)[26]			
	Samelia					3.5 (1, N)[4]	
	Wichita County			3.7 (1, N)[26]			
	Yardymly					2.1 (2, N)[4]	
	Youndegin			3.9 (1, N)[26]			
	Average	9.0	2.1	3.65	3.3	2.9	12.2
Om	Albuquerque			4.46 (5, N)[6]			
	Annaheim			5.35 (2, C)[24]		3.65 (2, N)[4]	
	Bohumilitz			4.2 (1, N)[26]			
	Breece			5.0 (1, N)[26]			
	Campbellsville			3.8 (1, N)[26]			
	Cañon Diablo No. 2		2.12 (1, C)[25]	2.71 (1, C)[25]	3.90 (1, C)[25]	1.53 (1, C)[25]	1.86 (1, C)[25]
				5.30 (3, N)[6]			
				4.0			
	Canton			2.8 (1, N)[26]			
	Caperr			4.1 (1, N)[26]			
	Cape York			2.5 (1, N)[26]			
	Carbo	52.4 (1, N)[11]		3.4 (2, N)[9]	15.7 (1, N)[13]	4.3 (2, N)[9]	30.0 (2, N)[9]
	Casas Grandes		2.9 (1, C)[25]	2.6 (1, C)[25]	4.74 (1, C)[25]	1.3 (1, C)[25]	10.7 (1, C)[25]
					3.22 (5, N)[12]	4.85 (2, N)[4]	
					3.98	3.1	

Table 78-C-5 (Continued)

Class	Specimen	Ru	Rh	Pd	Os	Ir	Pt
Om	Chilkoot			3.01 (3, N)[6]			
	Chinautla			6.1 (1, N)[26]			
	Clark County			2.37 (4, N)[6]		0.79 (1, N)[1]	
				2.7 (1, N)[26]		6.95 (2, N)[4]	
				2.5		3.9?	
	Colfax			4.6 (1, N)[26]			
	Colomera				12.46 (1, N)[13]		
	Costilla Peak	13.4 (1, S)[19]	1.7 (1, S)[19]	2.5 (1, S)[19]		15.3 (1, S)[19]	17.7 (1, S)[19]
		12.4 (2, C)[10]		2.02 (3, N)[6]			
		12.9		2.3			
	Cumpas			2.9 (1, N)[26]			
	Descubridora			2.4 (1, N)[26]			
	Drum Mountains			3.23 (3, N)[6]			
	Glorieta Mountain			7.1 (1, N)[26]			
	Goose Lake	6.4 (1, S)[19]	1.2 (1, S)[19]	3.9 (1, S)[19]		2.2 (1, S)[19]	6.8 (1, S)[19]
		4.55 (2, C)[10]		3.34 (3, N)[6]			
		5.5		3.6			
	Gun Creek			4.6 (1, N)[26]			
	Henbury	6.0 (1, S)[19]	1.2 (1, S)[19]	3.9 (1, S)[19]	13.5 (2, C)[25]	2.3 (1, S)[9]	5.6 (1, S)[19]
		3.6 (2, C)[10]	2.7 (2, C)[25]	2.02 (10, N)[6]	2.25 (4, N)[12]	4.7 (2, C)[25]	9.4 (2, C)[25]
		5 (1, C)[25]	2.1	1.7 (2, C)[25]	7.9	14 (1, N)[4] +	10.0 (3, N)[9]
		4.8		2.1 (1, N)[26]		1.4 (3, N)[9]	8.3
				4.7 (3, N)[9]		2.8	
				2.9			
	Kingston			2.1 (1, N)[26]			
	La Caille			3.1 (1, N)[26]			
	Leeds			4.38 (3, N)[6]			
	Maria Elena			3.44 (3, N)[6]			

Om	Mazapil					4.75 (2, N)[4]	
	Morito			2.25 (3, N)[6]			
	Mount Edith			6.8 (1, N)[26]			
	N'Kandhla					19 (2, N)[4]	
	Norfolk					11 (1, N)[4]	
	Pará de Minas					2.4 (1, N)[4]	
	Puente del Zacate			2.57 (3, N)[6]			
	Rhine Villa			3.4 (1, N)[26]			
	Rowton					3.9 (1, N)[4]	
	Sacramento Mountains	9.8 (1, C)[10]		2.34 (3, N)[6]			
	Sams Valley			4.7 (1, N)[26]			
	San Angelo	1.1 (?, C)[10]					
	Sanderson			4.47 (3, N)[6]			
	Santa Apolonia			2.29 (4, N)[6]			
	Spearman	4.6 (1, S)[19] 4.8 (1, C)[10] 4.7	1.2 (1, S)[19]	2.9 (1, S)[19] 3.67 (3, N)[6] 3.3		0.4 (1, S)[19]	6.7 (1, S)[19]
	Toluca	5.3 (1, S)[19]	1.1 (1, S)[19]	4.0 (1, S)[19] 3.0 (1, N)[26] 5.66 (2, N)[3] 4.72 (10, N)[6] 4.4	2.55 (5, N)[12]	2.0 (1, S)[19]	6.5 (1, S)[19]
	Trenton	9.4 (1, C)[25]	4.1 (1, C)[25]	6.6 (1, C)[25]	3.52 (1, C)[25] 1.60 (1, N)[13] 2.56	3.4 (1, C)[25] 2.1 (2, N)[4] 2.75	8.4 (1, C)[25]
	Treysa				0.58 (4, N)[12]	1.5 (1, N)[4]	
	Williamstown					2.4 (1, N)[1]	
	Willow Creek	3.6 (1, C)[10]		3.70 (3, N)[6]			
	Average	12.0	2.1	3.8	6.2	4.8	10.8

Table 78-C-5 (Continued)

Class	Specimen	Ru	Rh	Pd	Os	Ir	Pt
Of	Altonah	4.7 (1, S)[19]	0.7 (1, S)[19]	4.2 (1, S)[19]		1.6 (1, S)[19]	5.1 (1, S)[19]
		3.6 (1, C)[10]		4.45 (3, N)[6]			
		4.2		4.3			
	Bear Creek	2.5 (1, S)[19]	1.5 (1, S)[19]	5.4 (1, S)[19]		<0.4 (1, S)[19]	2.4 (1, S)[19]
				5.57 (3, N)[6]			
				5.5			
	Bethany				3.20 (3, N)[12]		
	Boogaldi			6.3 (1, N)[26]			
	Bristol	5.1 (1, S)[19]	0.8 (1, S)[19]	3.9 (1, S)[19]		1.7 (1, S)[19]	5.0 (1, S)[19]
				4.36 (3, N)[6]			
				4.11 (2, N)[3]			
				4.1			
	Cambria			5.7 (1, N)[26]			
	Carlton			6.52 (4, N)[6]			
				6.7 (1, N)[26]			
				6.6			
	Charlotte					1.45 (2, N)[4]	
	Duchesne	2.1 (2, C)[10]					
	Gibeon			3.5 (1, N)[26]			
	Grand Rapids			3.5 (1, N)[26]			
	Grant	2.0 (1, C)[25]	2.55 (1, C)[25]	4.54 (2, N)[3]	2.4 (1, C)[25]		
				1.66 (1, C)[25]	0.1 (1, N)[13]		
				3.1	1.25		
	Huizopa	5.7 (1, S)[19]	0.9 (1, S)[19]	4.20 (3, N)[6]		2.5 (1, S)[19]	4.4 (1, S)[19]
				4.8 (1, S)[19]			
				3.6 (1, N)[26]			
				4.2			
	Kamkas			3.4 (1, N)[26]			
	La Grange			3.95 (3, N)[6]			

Of	Madoc			2.5 (2, C)[24]			
	Moonbi	6.3 (1, S)[19]	1.3 (1, S)[19]	3.5 (1, S)[19]		1.4 (1, S)[19]	9.2 (1, S)[19]
				3.9 (1, N)[26]			
				3.7			
	Mukerop				2.52 (3, N)[13]		
	Obernkirchen			3.1 (1, N)[26]			
	Otchinjau			3.7 (1, N)[26]			
	Puquios			3.5 (1, N)[26]			
	Average	4.0	1.3	4.2	2.3	1.3	5.2
Off	Bacubirito			3.6 (1, N)[26]			
	Ballinoo			3.8 (1, N)[29]			
	Edmonton	0.5 (1, S)[19]	0.3 (1, S)[19]	6.5 (1, S)[19]		0.6 (1, S)[19]	0.5 (1, S)[19]
		0.7_5 (2, C)[10]		5.97 (3, N)[6]			
		0.63		6.2			
	Hill City			6.23 (3, N)[6]			
	Mount Magnet			9.3 (1, N)[26]			
	Perryville			3.07 (3, N)[6]			
	Salt River			4.3 (1, N)[26]			
	Tazewell	0.5 (1, S)[19]	0.14 (1, S)[19]	5.7 (1, S)[19]		0.3 (1, S)[19]	0.5 (1, S)[19]
				7.73 (3, N)[6]			
				6.7			
	Average	0.57	0.22	5.4		0.45	0.5
Brecciated Octahedrites	Barranca Blanca			4.6 (1, N)[26]			
	Murnpeowie			2.1 (1, N)[26]			
	N'Goureyma					<0.5 (1, N)[4]	
	Kopjes Vlei			1.8 (1, N)[26]			
	Average			2.8			

Table 78-C-5 (Continued)

Class	Specimen	Ru	Rh	Pd	Os	Ir	Pt
H	Aswam				0.225 (1, N)[12]		
	Bennett County			1.7 (1, N)[26]			
	Boguslavka					23.5 (2, N)[4]	
	Braunau					10.8 (2, N)[4]	
	Cedartown			1.65 (4, N)[6]			
	Campo del Cielo			3.7 (1, N)[26]	3.71 (1, N)[13]		
	Chihuahua City			2.52 (3, N)[6]			
	Coahuila	23.7 (1, S)[19]	2.3 (1, S)[19]	2.0 (1, S)[19]		14.8 (1, S)[19]	24.2 (1, S)[19]
				1.44 (3, N)[6]		14 (1, N)[4]	
				2.0 (1, N)[26]		14.4	
				1.8			
	Coya Norte*	15.2 (1, S)[19]	2.5 (1, S)[19]	1.8 (1, S)[19]		3.3 (1, S)[19]	21.9 (1, S)[19]
		14.8 (2, C)[10]					
		15.0					
	Horse Creek			2.45 (3, N)[6]			
	Indian Valley	21.0 (1, S)[19]	2.1 (1, S)[19]	1.5 (1, S)[19]		7.9 (1, S)[19]	29.3 (1, S)[19]
				1.54 (3, N)[6]			
				1.5			
	La Primitiva			2.0 (1, N)[26]			
	Lombard				1.14 (1, N)[13]		
	Mejillones*			1.94 (3, N)[6]			
	Mount Joy	9.1 (1, C)[10]		2.20 (3, N)[6]	0.51 (1, N)[13]		
	Nedagolla			1.7 (1, N)[26]		5.9 (1, N)[4]	
	Negrillos*				50.4 (1, N)[13]		
	Okano					10 (1, N)[4]	
	Rio Loa*	21.5 (1, S)[19]	2.5 (1, S)[19]	2.2 (1, S)[19]		3.4 (1, S)[19]	26.9 (1, S)[19]
	Sandia Mountains	7.2 (1, S)[19]	1.5 (1, S)[19]	2.1 (1, S)[19]		<0.4 (1, S)[19]	9.0 (1, S)[19]
		7.45 (2, C)[10]		2.24 (3, N)[6]			
		7.3		2.46 (2, N)[3]			
				2.3			

H	San Martin*			2.0 (1, N)[26]	1.23 (1, N)[13]		
	Santa Rosa			2.4 (1, N)[26]			
	Sierra Gorda*			1.47 (3, N)[6]			
	Tocopilla*			2.1 (1, N)[26]	1.29 (4, N)[12]		
	Tombigbee River			2.4 (1, N)[26]			
	Uwet			2.1 (1, N)[26]			
	Walker County			2.0 (1, N)[26]			
	Sikhote-Alin	5.0 (1, S)[19]	1.6 (1, S)[19]	6.9 (1, S)[27]+	≪0.025 (1, N)[13]	<0.3 (1, S)[19]+	3.9 (1, S)[19]
		5.7 (1, S)[27]	0.9 (1, S)[27]			0.014 (1, N)[1]	4.1 (1, N)[1]
		5.35	1.25			⩽0.45 (2, N)[4]+	4.6 (1, S)[27]
						0.014	4.2
	Average	14.7	2.0	2.1	8.4	8.8	19.3
D	Babb's Mill			5.6 (1, N)[26]			
	Cape of Good Hope	13.3 (1, S)[19]	1.7 (1, S)[19]	4.1 (1, S)[19]		8.5 (1, S)[19]	11.8 (1, S)[19]
				7.07 (4, N)[6]			
				5.6			
	Dayton				0.047 (1, N)[13]		
	Deep Springs					12 (1, N)[4]	
	Hoba			6.6 (1, N)[26]			
	Illinois Gulch			5.6 (1, N)[26]			
	Klondike			13.2 (2, C)[24]			
	Monahans				12.7 (1, N)[13]	12 (1, N)[4]	
	Muzaffarbur					1.1 (4, N)[4]	
	Piñon			11.8 (2, N)[3]	28.3 (1, N)[13]		
				9.88 (3, N)[6]			
				10.8			
	Pitts					1.8 (1, N)[4]	
	San Cristobal			13.8 (1, N)[26]			
	Santa Catharina			19.7 (1, N)[26]	0.16 (1, N)[13]	⩽1.0 (1, N)[4]	
	Smithland			6.4 (1, N)[26]			
	Tlacotepec				39.6 (1, N)[13]		
	Washington County					<0.6 (1, N)[4]	
	Average			9.7	16.2	7.1	

Table 78-C-6. *Platinum metal content of minerals from chondritic and iron meteorites, ppm. (For notes and references see p. 78-C-22 and 78-C-23)*

Group	Class	Specimen	Mineral	Ru	Rh	Pd	Os	Ir	Pt
Chon-drites	CH	Alamogordo	Fe-Ni metal	6.1 (1, S)[19]	0.9 (1, S)[19]	4.2 (1, S)[19]		3.0 (1, S)[19]	7.9 (1, S)[19]
		Forest City	Fe-Ni metal	3.15 (2, N)[2]			1.63 (2, N)[2]		
			silicates	0.457 (2, N)[2]			0.299 (2, N)[2]		
		Gilgoin	Fe-Ni metal	5.4 (1, S)[19]	1.0 (1, S)[19]	3.7 (1, S)[19]		2.6 (1, S)[19]	8.0 (1, S)[19]
		Gladstone	Fe-Ni metal	5.9 (1, S)[19]	1.0 (1, S)[19]	3.9 (1, S)[19]		2.1 (1, S)[19]	8.5 (1, S)[19]
		Ochansk	Fe-Ni metal	6.2 (1, S)[19]	1.1 (1, S)[19]	3.7 (1, S)[19]	1.25 (1, N)[2]	2.9 (1, S)[19]	9.1 (1, S)[19]
				2.33 (1, N)[2]					
			silicates	0.222 (1, N)[2]			0.157 (1, N)[2]		
		Plainview	Fe-Ni metal	6.5 (1, S)[19]	1.0 (1, S)[19]	4.7 (1, S)[19]		3.2 (1, S)[19]	9.0 (1, S)[19]
		Average	Fe-Ni metal	5.2	1.0	4.0	1.4	2.8	8.4
	CL	Mócs	Fe-Ni metal				3.67 (1, N)[13]		
			silicates				0.35 (1, N)[13]		
		Ramsdorf	Fe-Ni metal				2.57 (1, N)[13]		
Iron	Og	Cañon Diablo	troilite	0.0324 (2, N)[2]			0.0051 (2, N)[2]	0.0015 (2, N)[1]	0.71 (1, N)[1]

Table 78-C-7. *Summary of average platinum metal contents of chondritic and iron meteorites, ppm*

Group	class	Ru	Rh	Pd	Os	Ir	Pt
Chon-drites	Ce	0.99 (2)		1.5 ± 0.2 (6)	0.71 ± 0.095 (6)	0.22 (2)	1.6 (3)
	CH	1.1 ± 0.04 (13)	0.21 (3)	1.1 ± 0.15 (5)	0.86 ± 0.038 (7)	0.48 ± 0.028 (5)	1.6 (3)
	CL	0.72 ± 0.03 (12)	0.15 ± 0.026 (5)	0.82 ± 0.12 (9)	0.64 ± 0.055 (7)	0.38 ± 0.043 (12)	0.79 ± 0.19 (8)
	CHL			1.5 ± 0.2 (6)	0.73 ± 0.06 (6)		
	Cc	0.76 ± 0.04 (6)		0.84 ± 0.12 (7)	0.60 ± 0.05 (8)	0.59 ± 0.12 (7)	1.1 ± 0.1 (7)
Irons	Ogg	7.8 (2)		3.1 ± 0.2 (6)		4.4 (2)	
	Og	9.0 ± 2.3 (5)	2.1 ± 0.3 (4)	3.65 ± 0.26 (11)	3.3 (3)	3.9 ± 0.9 (8)	12.2 ± 4.0 (4)
	Om	12.0 ± 4.6 (10)	2.1 ± 0.4 (8)	3.8 ± 0.2 (39)	6.2 ± 0.2 (8)	4.8 ± 1.2 (18)	10.8 ± 2.8 (9)
	Of	4.0 ± 0.7 (7)	1.3 ± 0.3 (6)	4.2 ± 0.3 (17)	2.3 (3)	1.5 ± 0.3 (5)	5.2 ± 1.5 (5)
	Off	0.57 (2)	0.22 (2)	5.4 ± 1.1 (7)		0.45 (2)	0.5 (2)
	Brecciated Octahedrites			2.8 (3)			
	All Octahedrites	8.05 (26)	1.65 (20)	3.82 (85)	4.75 (14)	3.53 (34)	8.45 (20)
	H	14.71 ± 2.85 (7)	2.0 ± 0.2 (6)	2.1 ± 0.1 (22)	8.4 ± 7.1 (7)	8.8 ± 2.3 (8)	19.3 ± 4.2 (6)
	D			9.7 ± 1.8 (9)	16.2 ± 7.8 (5)	7.1 ± 2.2 (5)	

Numbers in parenthese indicate the number of analyses.

Table 78-C-9. *Cosmic abundances of platinum metals (atoms metal / 10^6 silicon atoms)*

Reference	Ru	Rh	Pd	Os	Ir	Pt
CAMERON (1967)[a]	1.6	0.33	1.5	0.71	0.43	1.13
CROCKET *et al.* (1967) Type I carbonaceous chondrites[b]	1.8^5	—	1.3	0.72	0.57	1.3
NICHIPORUK and BROWN (1965) CH group chondrites	1.66	0.27	1.05	—	0.40	1.22
NCHIPORUK and BROWN (1965) (iron meteorites)[c]	1.44	0.23	0.521	—	0.31	0.89
CLAYTON and FOWLER (1961) (nucleosynthesis)	0.83	0.13	0.601	0.51	0.39	0.80

[a] The data of Cameron are based on the following determinations: Ru-chondrites; Rh-chondrites; Pd-Type I carbonaceous chondrites; Os-Type I carbonaceous chondrites, LOVERING and MORGAN (1967); Ir-Type I carbonaceous chondrites, BAEDECKER and EHMANN (1965); Pt-The abundance of Pt was adjusted relative to Os, Ir and Au to form a continuous peak in this mass range.

[b] The values of CROCKET *et al.* are the average of data for the Type I carbonaceous chondrites, Orgueil, Ivuna and Alais.

[c] For iron meteorites a contribution to mean meteoritic matter of 10% is assumed.

carbonaceous chondrites consist of a fraction, represented by matrix, which has retained a high proportion of volatile elements and a fraction, represented by chondrules, which has lost most of the volatile elements. The matrix fraction, which is of highest proportion in Type I cabonaceous chondrites, is considered most representative of primitive, undifferentiated solar material.

Table 78-C-9 presents some recent solar abundance estimates. Those of CAMERON (1967) are based largely on chondritic meteorites and those of CROCKET *et al.* (1967) on Type I carbonaceous chondrites. NICHIPORUK and BROWN'S (1965) estimates are based on CH group chondritic meteorites and iron meteorites. In view of the fractionation between the various classes of irons, data must be representative of the relative abundance of each structural class. The abundances calculated from nucleosynthesis models are approximately a factor of two lower than those obtained from meteorite data, particularly carbonaceous chondrite data, in the case of Ru, Rh and Pd. The agreement for Ir is fairly good but the nucleosynthesis figure for Pt is considerably lower than meteorite estimates.

Revised manuscript received: March 1968

IV. Lunar Materials

Concentrations of platinum group metals in lunar materials are presented in Table 78-C-10. The data are from samples collected in the Apollo 11 through Apollo 17 landings and the two U.S.S.R. unmanned spacecraft landings, Luna 16 and Luna 20. In general four categories of samples are recognized: basaltic, gabbroic and anorthositic rocks, fragmental rocks or breccias, soils which are mainly < 1 mm fines, and magnetic or metal grain separates from any of the above types of materials. Ir is the most extensively analysed platinum group metal, representing approximately 66% of the data in Table 78-C-10 compared to 14% for Os and 10% each for Pd and Ru. Other than the single Pt and Rh values reported by VINOGRADOV (1971) on Luna 16 fines, reliable data on these two metals were not found. The main geochemical characteristics of the four analysed platinum group metals are noted in the following discussion.

a) Ir in Lunar Materials

Weighted averages for Ir in lunar materials are computed from the data in Table 78-C-10 and listed below (n=number of samples):

	Mare basalts		Breccias		Soils	
	ave.	(n)	ave.	(n)	ave.	(n)
Ir (ppb)	0.055	30	8.2	63	9.2	107

There is a general consensus of opinion that the lunar breccias and soils contain a significant meteoritic Ir component. Some of the evidence for this is: (1) Discrete metal grains containing ppm levels of Ir have been separated from lunar breccias and soils. Concentrations generally range from 0.5 to 5 ppm but in one case average 82 ppm (WÄNKE *et al.*, 1970b, 1971; WLOTZKA *et al.*, 1973). The contribution of such grains to the total Ir content of lunar soils may be 50% or more as in the case of lunar soil 14163 (WLOTZKA *et al.*, 1972). Metallographic structures, silicate associations and bulk composition are cited by GOLDSTEIN *et al.* (1972) as evidence that the majority of metallic particles in lunar materials are of meteoritic origin. (2) Not only have metallic grains been separated from lunar breccias (WÄNKE *et al.*, 1971), but extremely large variations in Ir content have been observed in lithologically distinct components of lunar breccias. For example, in the black and white breccia 15455, GANAPATHY *et al.* (1973) find Ir contents of 7.0 and 4.8 ppb in two samples of the dark component compared with ≤0.002 and 0.024 ppb in the leucogabbro and anorthosite of the light fraction. This distribution suggests a localized concentration of an Ir-rich component in the dark fraction.

The most thoroughly analysed lunar rocks which appear to be largely uncontaminated with meteoritic material are the mare basalts. The weighted average from Table 78-C-10, 0.055 ppb, is probably an upper limit for the Ir content of mare basalts in that a few values are reported as upper limits only. Many other basalts such as the Apollo 16 highland rocks are characterized by much higher Ir contents. The Apollo 16 basalts listed in Table 78-C-10 average 12.0 ppb Ir. WARNER *et al.* (1973) argue that many of these rocks are derived from breccias or soils that were

Table 78-C-10. *Platinum group metal content of lunar materials*

All determinations by neutron activation analysis unless otherwise noted. Parameters given for each metal are $\bar{x}$, arithmetic mean, s, standard deviation and n, number of samples analysed. Data designated by the authors as suspect or likely representing sample contamination have not been used in computation of averages

Sample	Metal (ppb); Reference number in [ ]											
	Ru			Pd			Os			Ir		
	$\bar{x}$	s	n	$\bar{x}$	s	n	$\bar{x}$	s	n	$\bar{x}$	s	n
Apollo 17, Taurus-Littrow Site, 20° N, 30° E												
Mare basalts				<2 [37]	—	1				0.0019 [22]	0.0016	2
Dunite, troctolite, anorthosite, gabbro	1.8 [15]	1.8	2				0.8 [15]	0.4	2	5.6 [3, 17, 22]	7.0	3
Breccias	8.8 [15]	7.3	3	10.5 [37]	0.7	2	6.3 [15]	4.2	3	6.6 [6, 17, 22]	3.4	15
Mainly <1 mm fines	20.7 [15]	24	9	5 [37]	—	1	16.6 [15]	14.7	8	8.8 [6, 12, 13, 16, 22, 24]	4.7	32
Apollo 16, Descartes Mountains Cayley Plains, 8.9° S, 15.5° E												
Basalts, including poikilitic rocks, melt rocks	48 [14]	—	1				81	—	1	12.0 [12, 16, 17, 19]	7.6	9
Cataclastic anorthosite and anorthosite				25 [37]	—	1				11.9 [16, 19, 37]	11.4	3
Cataclastic anorthosite										0.0078 [16]	0.0030	2
Breccias, lithologies mainly mesostasis-rich basalt, cataclastic anorthosite, quench basalt, devitrified glass				37 [37]	21	3				12.4 [12, 16, 17, 19, 36, 37]	12.9	23
<1 mm fines	15 [14]	5	5	16.5 [36]	11	2	28 [14]	6	5	13.3 [4, 16, 19, 36]	4.7	23
Metal grains from soils										1,550 [42]	—	1
Magnetic separates from soils										45 [16]	45	2

Apollo 15, Hadley Rille-Apennine Mountains, 26.1° N, 3.7° E												
Mare basalts										0.040 [3, 13, 23]	0.039	8
Gabbro, anorthosite										0.030 [3, 13, 22]	0.037	2
Breccias										3.6 [3, 10, 13]	3.0	10
Mainly < 1 mm fines	18.7 [29]	5.5	3	6.2 [35]	—	1	22.0 [29]	7.2	3	5.6 [3, 10, 13, 23, 25, 35]	2.7	25
Apollo 14, Fra Mauro, 3.7° S, 23.4° W												
Basalt	13 [29]	—	1				11 [29]	—	1	4.6 [5, 26]	6.5	2
Breccias, both friable and coherent types, mainly dark clasts	20 [29]	—	1	8 [35]	10	2	24 [29]	—	1	6.4 [3, 5, 13, 26, 35]	3.8	7
Mainly < 1 mm fines	21 [29]	11	5	45 [8, 35, 35]	35	2	70 [29]	28	5	12.6 5, 26, 35]	1.8	12
Metal separates				2,800 [43]	—	1				879 [13, 43]		3
Apollo 12, Ocean of Storms, 3.2° S, 23.4° W												
Basalt, mare type	15 [18]	8	4	13 [31, 39]	0.8	5	0.74 [21, 28]	0.5	8	0.087 [1, 7, 9]	0.047	14
Breccias				10.2 [39]	—	1	7.5 [21]	—	1	4.6 [18, 39]	2.8	2
< 1 cm fines	53 [28]	25	2	9.7 [39]	3.5	3	5.8 [21, 28]	2.2	6	5.9 [7, 9, 18, 26, 39]	2.4	8
Metal from fines, rocks and breccias										29,370 [39]	45,740	3
Apollo 11, Sea of Tranquility, 0.67° N, 23.49° E												
Basalt, mare types				2.9 [11, 38]	3.2	5	1.3[a] [20, 27]	1.8	4	0.016 [1, 11]	0.0087	6
Breccias				8.7 [11.30]	3.8	3	4.8 [20, 21]	0.35	2	7.1 [11, 41]	4.0	6

Table 78-C-10 (continued)

Sample	Metal (ppb); Reference number in [ ]											
	Ru			Pd			Os			Ir		
	$\bar{x}$	s	n	$\bar{x}$	s	n	$\bar{x}$	s	n	$\bar{x}$	s	n
< 1 mm fines, soil core				11.1 [11, 38]	—	1	10.9[b] [20, 21, 27, 28]	7.2	2	5.6 [18, 26, 34, 41]	3.0	2
Metal grains										2,600 [40]	—	1
Luna 16, Sea of Fertility, 0° 41′ S, 56° 18′ E and Luna 20, Highlands between the Sea of Fertility and the Sea of Crises, 3° 32′ N, 56° 33′ E												
Basalt	6,300 [32][c]	—	1	27 [32]	—	1	30 [33]	—	[d]	1 [33]	—	[d]
0.07 to > 0.500 mm fines[e]	28 [32]	17	3	10 [32]	2	3	30 [33]	—	[d]	10.0 [13, 24, 26]	1.0	5

[a] Value of 220 ppb Os reported by Reed and Jovanovic (1970) omitted in calculation of the average.
[b] Value of 283 ppb Os reported by Reed and Jovanovic (1970) omitted in calculation of the average.
[c] Analytical methods used by Vinogradov (1971, 1973) are mass spectrometric and spectral procedures.
[d] Value is the average of an unreported number of analyse.
[e] Vinogradov (1971) and Vinogradov (1973) report values of 3.7 and 50 ppb for Rh and Pt respectively in these samples.

References: 1. Anders *et al.* (1971); 3. Baedecker *et al.* (1973); 4. Baedecker *et al.* (1972a); 5. Baedecker *et al.* (1972b); 6. Baedecker *et al.* (1974); 7. Baedecker *et al.* (1971); 8. Brunfelt *et al.* (1972); 9. Brunfelt *et al.* (1971); 10. Chou *et al.* (1974); 11. Ganapathy *et al.* (1970); 12. Ganapathy *et al.* (1974); 13. Ganapathy *et al.* (1973); 14. Jovanovic and Reed (1973); 15. Jovanovic and Reed (1974); 16. Krähenbühl *et al.* (1973); 17. Laul *et al.* (1974); 18. Laul *et al.* (1971); 19. Laul and Schmitt (1973); 20. Lovering and Butterfield (1970); 21. Lovering and Hughes (1971); 22. Morgan *et al.* (1974); 23. Morgan *et al.* (1972); 24. Morgan *et al.* (1973a); 25. Morgan *et al.* (1973b); 26. Morgan *et al.* (1972); 27. Reed and Jovanovic (1970); 28. Reed and Jovanovic (1971); 29. Reed *et al.* (1972); 30. Smales *et al.* (1970); 31. Smales *et al.* (1971); 32. Vinogradov (1971); 33. Vinogradov (1973); 34. Vobecky *et al.* (1971); 35. Wänke *et al.* (1972); 36. Wänke *et al.* (1973); 37. Wänke *et al.* (1974); 38. Wänke *et al.* (1970a); 39. Wänke *et al.* (1971); 40. Wänke *et al.* (1970b); 41. Wasson and Baedecker (1970); 42. Wlotzka *et al.* (1973); 43. Wlotzka *et al.* (1972).

partially melted in an impact event, and hence their high siderophile trace element contents are meteoritic in origin.

b) Os and Ru in Lunar Materials

The available date tentatively suggest that Os and Ru exhibit the same relative concentrations in lunar rocks, breccias and soils as Ir. However, the small number of basalt and breccia samples analysed for these elements limits the reliability of this generalization. Weighted averages calculated from Table 78-C-10 for Os and Ru in lunar materials are presented below (n=number of samples):

	Mare basalts		Breccias		Soils	
	ave.	(n)	ave.	(n)	ave.	(n)
Os (ppb)	0.9	(12)	9	(7)	26	(29)
Ru (ppb)	15	(4)	12	(4)	23	(27)

For Os the lowest concentrations occur in mare basalts and higher concentrations are found in breccias and soils. However, the relative depletion of Os in basalts compared to soils and breccias is approximately 6.5 times less than the comparable depletion of Ir. Similar abundances of Os and Ru are found in lunar breccias and soils but a depletion of Ru in mare basalts, comparable to that of Os, is not shown by the available data. In addition to mare basalts a dunite (72417) and a troctolite (76535) sample from the Taurus-Littrow Site were found to be relatively low in Os and Ru.

Jovanovic and Reed (1974) have found samples with meteoritic Ru/Os ratios (1.7 ± 0.3) only in Apollo 17 materials. In general, Ru/Os in soils is approximately 0.5. Apollo 17 and 16 samples exhibit several correlations attributed by Jovanovic and Reed (1974) to chemical processing of these elements on the moon. Samples characterized by meteoritic ratios are regarded as unreworked primitive materials which preserve their original complements of Os and Ru.

c) Pd in Lunar Materials

Weighted averages for Pd in lunar material are summarized below (n=number of samples):

	Mare basalt		Breccias		Soils	
	ave.	(n)	ave.	(n)	ave.	(n)
Pd (ppb)	2	(11)	17	(11)	18	(10)

In common with other platinum group metals Pd is depleted in mare basalts relative to breccias and soils, both of which have very similar average Pd contents. The average value of 2 ppb for Pd in mare basalts is an upper limit only as many of the individual determinations are reported as maximum values. The concentration of Pd in metallic grains is not as well established as is the case for Ir, but Wlotzka *et al.* (1972) report 2.8 ppm Pd in metal particles separated from lunar soil 14163.

Revised manuscript received: January 1976

78-D. Abundance in Rock-Forming Minerals (I) Platinum Metal Minerals (II)

I. Rock-Forming Minerals

Abundances of platinum metals, principally Pt and Pd, in rock-forming minerals not associated with ore deposits are given in Table 78-D-1. The bulk of these data are from LUNDE (1927) and LUNDE and JOHNSON (1928), NODDACK and NODDACK (1931) and GOLDSCHMIDT and PETERS (1932). LUNDE and JOHNSON determined total precious metals by a micro-fire assay method using lead as a collector. After separation of gold and silver the total platinum metal content was estimated by measuring the diameter of the resulting precious metal bead. GOLDSCHMIDT and PETERS (1932) also used a fire assay concentration but determined individual platinum metals by optical spectroscopy. I. and W. NODDACK separated the group by wet chemical techniques and subsequently determined individual metals by X-ray spectrographic and emission spectrographic methods. Studies of platinum metal abundances in rock-forming minerals by modern analytical methods are extremely limited and our knowledge of the geochemistry of these metals in silicate and oxide minerals is largely based on these studies.

Table 78-D-1. *Abundances of platinum metals in rock-forming minerals, ppb*

Sample description	No. of samples, methods, references[a]	Ru	Rh	Pd	Ir	Pt	Total platinum metals
Olivine[b]	(9, W/X)[5]	10		50		80	
Bronzite	(1, W/X)[5]			10		20	
Diopside	(2, W/X)[5]			20		10	
Pyroxene from eclogite, Sunmöre, Norway	(2, FA)[4]						10, 70
Garnet from eclogite, Sunnmore, Norway	(2, FA)[4]						20, 20
Enstatite, Bihar, India	(1, FA/S)[2]					6	
Serpentine, Norway	(4, FA)[4]						300, 50, 9, 9
Serpentine, Urals and Norway	(5, W/X)[5]			80		50, 100	
Garnet, Norway	(3, W/X)[5]					10	
Chromite from dunite, Urals	(1, FA)[4]						300
Chromite from pegmatitic eclogite, Sunmöre, Norway	(3, FA)[4]						240, 1,150, 300

Table 78-D-1 (Continued)

Sample description	No. of samples, methods, references[a]	Ru	Rh	Pd	Ir	Pt	Total platinum metals
Chromite from serpentinized peridotite, Quebec, Canada	(1, N)[1]			3.4			
Chromite veins in in serpentine, Fjeldtjern, Norway	(4, FA)[4]						800, 1,120, 1,286, 1,260
Chromite, Vefsen, Norway	(4, FA)[4]						830, 340, 4,700, 31,800
Chromite, Orissa, India	(2, FA/S)[2]					8,5	
Chromite (Norway)	(1, FA/S)[3]	500	200				
Chromite (Texas, U.S.A.)	(1, FA/S)[3]	500	200				
Chromite (Penn., U.S.A.)	(1, FA/S)[3]	500	500		5,000	1,000	
Columbite, Norway and	(23, W/X)[5]	10	40	1,000	100	1,000	
Madagascar				800	400	2,000 3,000	
Thortveitite, $Y_2Si_2O_7$	(2, W/X)[5]		500	3,000	800	4,000	
Ilmenite	(4, W/X)[5]	4				10	
Hornblende, Norway	(4, W/X)[5]			10		20	
Orthoclase, Harz Mtns. Germany	(8, W/X)[5]					30	
Apatite	(5, W/X)[5]					10, 40	
Zircons, Ceylon and Norway	(14, W/X)[5]					40, 100	
Zircon (alvite), Norway	(4, W/X)[5]		10	5,000	10	5,000	
Rutile	(9, W/X)[5]	10				100	
Gadolinite, Norway	(64, W/X)[5]		100	3,000, 900	300	5,000, 1,000, 100, 2,000	
Monazite, Madagascar	(6, W/X)[5]					20	
Risorite, Norway $Y(Nb,Ta)O_4 . Y_2(TiO_3)_3$	(1, W/X)[5]					800	
Samarskite, Norway	(2, W/X)[5]					60	
Tantalite, South	(18, W/X)[5]	2				30,	
Africa and Finland						100	
Yttrotantalite	(1, W/X)[5]					40	

[a] 1. Crocket, J., and G. Skippen. (1966). 2. Das Sarma, B., *et al.* (1966). 3. Goldschmidt, V., and Cl. Peters (1932). 4. Lunde, G., and M. Johnson (1928). 5. Noddack, I., and W. Noddack (1931).

[b] An average of 50 ppb Os was found in these samples.

An evaluation of the accuracy of these data is very difficult as direct comparisons with data obtained by newer methods are generally lacking. However, the troilite data of GOLDSCHMIDT and PETERS (1932) and I. and W. NODDACK (1931) given in Table 78-C-1 were obtained by the same analytical techniques. This suggests that at the ppb concentration level these methods may be subject to large errors, due possibly to poor sensitivity and reagent contamination and that much of the data in Table 78-D-1 may be high by an order of magnitude. Only the most general conclusions can, therefore, be drawn from these data. The following generalizations are suggested:

1) average Pt and Pd concentrations in common rock-forming minerals are less than 10 ppb;

2) relative to olivine, pyroxene, hornblende, garnet and orthoclase an enrichment of Pt in chromite, or in chromite-rich rocks, is found;

3) accessory minerals, particularily rare earth varieties including gadolinite and columbite, associated with granitic pegmatites are enriched in Pt and Pd relative to major rock-forming minerals, suggesting an enrichment in residual liquids produced by fractional crystallisation of acidic rocks.

It is also to be noted that no platinum metals were found in carbonates, nitrates or halogen salts in the extensive survey by I. and W. NODDACK (1931).

II. Platinum Metal Minerals

A compliation of the major platinum group metal minerals is presented in Table 78-D-2.

Table 78-D-2. *Chemical composition of the major platinum metal minerals.* [After WRIGHT and FLEISCHER (1965) and FLEISCHER (1967)]

Element	Mineral formula					
	Braggite (Pt, Pd, Ni)S	Cooperite (Pt, Pd)S	Gever-site $PtSb_2$	Holling-worthite (Rh, Pd, Pt, Ir)AsS	Merenskyite (Pd, Pt), $(Te, Bi)_2$	Irarsite (Ir, Ru, Rh, Pt) AsS
Pt	59	80.2—82.5	45	10.3—18.4	0—1.8	12.6
Ir	Present	Present	—	1.4—3.1	—	23.0
Os	—	—	—	—	—	—
Pd	18—20.9	0—4.3	—	8.7	23.1—33.2	—
Rh	Present	—	—	24.6—30.8	—	7.2
Ru	Present	—	—	12.8	—	9.4
Au	—	—	—	—	—	—
Fe	—	—	—	—	—	—
Cu	—	—	—	—	—	—
Ni	2.8—4.7	0—0.1	—	—	—	—
Pb	—	—	—	—	—	—
S	16.8—19.0	14.3—17.5	—	13.0—13.9	—	11.6
Te	—	—	—	—	50.8—56.3	—
As	—	—	—	32.6—34.0	—	34.5
Sb	—	—	51.5	—	—	—
Bi	—	—	—	—	14.2—15.1	—
Sn	—	—	—	—	—	—
Hg	—	—	—	—	—	—

Table 78-D-2 (Continued)

Element	Mineral formula				
	Kotulskite Pd (Te, Bi) $_{1\cdot 2}$	Laurite (Ru, Os)S_2	Michenerite (Pd, Pt) (Bi, Te)	Moncheite (Pt, Pd) (bi, Te)$_2$	Niggliite PtSn
Pt	0—1.1	—	8.4—9.3	18.6—42.0	58.5—63.3
Ir	—	11.4	—	—	—
Os	—	0—20.9	—	—	—
Pd	31.1—45.9	—	11.7—16.9	0—9.3	—
Rh	—	—	—	—	—
Ru	—	33.67	—	—	—
Au	—	—	—	—	—
Fe	—	—	—	—	—
Cu	—	—	—	0—8.0	—
Ni	—	—	—	—	—
Pb	—	—	—	—	—
S	—	21.4—33	—	—	—
Te	36.1—44.0	—	28.8—37.6	33.5—55.4	—
As	—	—	—	—	—
Sb	—	—	—	—	—
Bi	17.2—24.9	—	42.3—45.0	7.5—31.7	—
Sn	—	—	—	—	38.0—42
Hg	—	—	—	—	—

Table 78-D-2 (Continued)

Element	Mineral formula				
	Sperrylite $PtAs_2$	Stibiopalla-dinite Pd_3Sb	Stannopalla-dinite Pd, Pt)$_3$ Sn_2	Vysotskite (Pd, Ni, Pt)S	Zvyagint-sevite (Pd, Pt)$_3$ (Pb, Sn)
Pt	23.8—63	—	15—20	4.1—5.5	7.5
Ir	17.8	—	—	—	—
Os	—	—	—	—	—
Pd	2.1	70.4—73	40—45	57.1—61.9	55.0—55.6
Rh	11.6	—	—	—	—
Ru	—	—	—	—	—
Au	—	—	—	—	—
Fe	0—0.7	—	0.3—2.3	—	—
Cu	0—0.7	—	5—12	—	—
Ni	—	—	0.1—0.7	14.2—16.6	—
Pb	—	—	—	—	25.0—38.0
S	10.8	—	—	—	—
Te	—	—	—	—	—
As	30.7—41.9	—	—	—	—
Sb	0—0.5	25—28	—	—	—
Bi	—	—	—	—	—
Sn	0—4	—	28—33	—	12.0
Hg	—	—	—	—	—

Revised manuscript received: March 22nd, 1968

78-E. Abundance in Common Igneous Rocks

I. Introduction

Platinum metal analyses of igneous rocks are limited largely to Pt and Pd in ultrabasic rocks. Most of the data are from LUNDE (1927), LUNDE and JOHNSON (1928), HAGEN (1954) and DAS SARMA *et al.* (1966) by fire assay or fire-assay spectrographic techniques; I. and W. NODDACK (1931) who used wet chemical — X-ray spectrographic methods; HAHN-WEINHEIMER (1959) and HAHN-WEINHEIMER and ROST (1966) by combined wet chemical and spectrographic methods and CROCKET and SKIPPEN (1966) by neutron activation. The most comprehensive study is that of HAGEN which includes Pt and Pd data for 77 igneous rocks of which 35 are from the Bushveld complex, South Africa. These data, together with additional data for gabbroic rocks by VINCENT and SMALES (1956) are summarized in Table 78-E-1. HAGEN's Bushveld data are presented in Table 78-E-2 and that for the U.S.G.S. rock standards G-1 and W-1 in Table 78-E-3.

Table 78-E-1. *Platinum metal abundances in common igneous rock types, ppb*

Rock	Location and description	No. of samples, method, reference[a]	Pd	Ir	Pt	Total platinum metals
Dunite	Deneijkin Kamen, Urals	(4, FA)[8]				50
Dunite	Supreija, Urals	(3, FA)[8]				36
Dunite	Mt. Albert, Quebec, Canada 95% forsterite, 5% enstatite plus chromite	(1, N)[2]	1.3			
Peridotite nodule	Salt Lake Crater, Hawaii	(1, N)[2]	7.3			
Peridotite	Sunmöre, Norway	(7, FA)[8]				350
Peridotite	Austagder, Norway	(1, FA)[8]				120
Peridotite	Jotunheimen, Norway	(3, FA)[8]				130
Peridotite	Espedalen, Norway	(1, FA)[8]				320
Peridotite	Finnmark, Norway	(1, FA)[7]			290	
Hornblende gabbro	Associated with the peridotite sample from Finnmark, Norway	(1, FA)[7]			740	
Hornblende pegmatite	Associated with the peridotite sample from Finnmark, Norway	(1, FA)[7]			80	

Table 78-E-1 (Continued)

Rock	Location and description	No. of samples, method, reference[a]	Pd	Ir	Pt	Total platinum metals
Peridotite	Montrose Point,	(1, FA/S)[4]	8		13	
Biotite-augite peridotite	Murfreesboro, Arkansas, U.S.A.	(1, FA/S)[4]	9		11	
Peridotite	Orissa, India New York, U. S. A.	(1, FA/S)[3]			8	
Pyroxenite	Approximately equal volumes of enstatite and diallage, partially serpentinized, Thetford district, Quebec, Canada	(1, N)[2]	22.2			
Pyroxenite	Deneijkin Kamen, Urals, U.S.S.R.	(1, FA)[8]				38
Pyroxenite	Jotunheimen, Norway	(2, FA)[8]				110, 5
Olivine pyroxenite	Jotunheimen, Norway	(1, FA)[8]				13
Serpentinite	Österdal, Norway	(1, FA)[8]				300
Serpentinite	East of Rotjern, Roros, Norway	(1, FA)[8]				50
Serpentinite	Rotjern, Roros, Norway	(1, FA)[8]				9
Serpentinite	Roragen, Norway	(1, FA)[8]				9
Serpentinite	Contains chromite veins, east of Fjeldtjern, Norway	(1, FA)[8]				140
Serpentinite	Contains chromite, Norway	(1, FA)[8]				370
Serpentinite	Highly altered rocks. Contain antigorite, brucite, magnetite. Leupoldsgrun (Munchberg gneisse), Oberfranken, Germany	(6, W/S)[5]	300		85	
Serpentinite	Normal serpentinite from the Munchberg gneisse mass. Less altered than above sample	(12, W/S)[5]	12		26	
Eclogite	Munchberg gneisse mass, Germany	(2, W/S)[5]	0.50		10, 260	
Eclogite	Sunnmore, Norway	(1, FA)[8]				77
Eclogite	Norway	(1, W/X)[9]			10	
Kimberlite	Bulfontein pipe, Kimberley, South Africa	(1, FA/S)[4]	17			
Kimberlite	Wajra Karur diamond mine, Andhra Pradesh, India	(1, FA/S)[3]			19	

Table 78-E-1 (Continued)

Rock	Location and description	No. of samples, method, reference[a]	Pd	Ir	Pt	Total platinum metals
Ultrabasic rock	Andaman Islands	(1, FA/S)[3]			6	
Serpentinized ultrabasic rock	Jojohatu chromite mine, Bihar, India	(1, FA/S)[3]			11	
Gabbroic rocks:						
Quartz diorite	Sudbury, Canada	(1, N)[2]	11.5			
Norite	From the Sudbury irruptive near Levack, Ontario, Canada	(1, FA/S)[4]	9		11	
Norite	From the Sudbury irruptive transition zone near Levack, Ontario, Canada	(1, FA/S)[4]	3		10	
Gabbro	Moyie Lake, British Columbia, Canada	(1, FA/S)[4]	7.6		15	
Anorthositic gabbro	Duluth gabbro, Minnesota, U.S.A.	(1, FA/S)[4]	45		32	
Essexite	Mount Royal, Quebec, Canada	(1, FA/S)[4]	15		23	
Gabbro	Chilled marginal gabbro, Skaergaard intrusion	(1, N)[10]	17.5			
Jacupirangite	Kodal, Norway	(1, FA)[7]			<100	
Basalts:						
Basalt[c]	Kaiserstuhl, Germany	(21, W/X)[9]	50	40	100	
Basalt	Sweden	(1, W/X)[9]			30	
Basalt	Vesuvius, Italy	(3, W/X)[9]	10		40	
Tholeiitic basalt	Deccan plateau basalts, India. Weighted mean of 4 composites consisting of 20, 11, 21 and 20 samples	Composite (FA/S)[4]	111		24	
Tholeiitic basalt	Columbia River plateau. Weighted mean of 2 composites consisting of 16 and 3 samples each	Composite (FA/S)[4]	11		17	
Continental basalts	Average value of five samples Yellowstone National Park, Wyoming, U.S.A.; Keweenawan County, Michigan, U.S.A.; Jefferson County, U.S.A.; Districto Federal, Mexico; Kruger National Park, South Africa	(5, FA/S)[4]	35		27	

Table 78-E-1 (Continued)

Rock	Location and description	No. of samples, method, reference[a]	Pd	Ir	Pt	Total platinum metals
Oceanic basalts	Average of five samples: four from Hawaii, one from the Azores	(5, FA/S)[4]	17		25	
Basalts (Diabase)	Average of five samples including one from the Palisades, U.S.A.; 1 from Yellowknife, Canada, and three from Ontario, Canada	(5, FA/S)[4]	16		37	
Oceanic basalts	Average of 10 samples: Hawaii, 2; Samoa, 1; Tahiti, 1; Galapagos, 2; Reunion, 1; Azores, 1; St. Helena, 1; Ascension, 1	(10, N)[2]	1.9			
Continental plateau basaltic rocks	Average of 10 samples: Deccan, 3; Serra Geral Formation, Brazil, 3; Columbia-Snake River, 2; Karroo System, 2	(10, N)[2]	8.3			
Leucite basalt	Celebes Islands	(1, N)[2]	17.4			
Andesite	Kyosho, Japan	(1, N)[2]	1.8			
Tholeiitic basalt	Myggbukta, Greenland	(1, N)[2]	1.6			
Granitic rocks:						
Granodiorite	Dedham, Massachusetts, U.S.A.	(1, FA/S)[4]	<3	<3		
Granite	Southern Norway	(15, W/X)[9]		20		
Granite	Harz, Germany	(1, W/X)[9]		10		
Granite		(1, W/X)[9]		50		
Granite	Pikes Peak, Colorado, U.S.A.	(1, FA/S)[4]	<3		<3	
Granite	Orissa, India	(1, FA/S)[4]			8	
Granite	Mt. Abu, Rajasthan, India	(1, FA/S)[3]			12	
Granite	Jasai, Rajasthan, India	(1, FA/S)[3]			<5	
Granite	Bundelkhand, India	(1, FA/S)[3]			7	

[a] 1. BAEDECKER, P., and W. EHMANN (1965). 2. CROCKET, J., and G. SKIPPEN (1966). 3. DAS SARMA, B., *et al.* (1966). 4. HAGEN, J. (1954). 5. HAHN-WEINHEIMER, P. (1959). 6. HAHN-WEINHEIMER, P., and F. ROST (1961). 7. LUNDE, G. (1927). 8. LUNDE, G., and M. JOHNSON (1928). 9. NODDACK, I., and W. NODDACK (1931). 10. VINCENT, E., and A. SMALES (1956).

[b] 1.14 ppb Os was obtained for a dunite sample by BATE and HUIZENGA (1963).

[c] NODDACK and NODDACK (1931) report an average of 20 ppb Rh for these basalts.

Table 78-E-2. *Pd and Pt content of Bushveld Complex rocks, South Africa, ppb.* (After HAGEN, 1954)

Rock type	Description	Pd	Pt
Granitic rocks	Average of 5 samples including Rooiberg felsite, Kalkfontein granophyre, Groblersdal granite, Janefurse granite and Spitskop ijolite	~5	~3
Gabbroic rocks	Average of 4 samples including one each from the Upper zone and the chill phase and two from the Main zone	31	21
Noritic rocks	Average of 7 samples including one each from the Basal zone, Main zone and Contact zone and four from the Critical zone	15	38
Ultrabasic rocks, (Pyroxenites)	Average of 7 samples including one from the Transition zone, two from the Critical zone and four from the Basal zone	59	48
Ultrabasic rocks (Peridotites)	Average of 2 samples of harzburgite from the Basal zone	12	17

Table 78-E-3. *Platinum metal abundances in USGS standards G*-1 *and W*-1, *ppb*

Rock	Description	Pd	Os	Ir	Pt
W-1	Diabase, Centerville, Virginia, U.S.A.	18.5 (2, N)[8] 15.8 ± 1.4 (10, N)[3] 9.0 (20, C)[5]	0.46 (1, N)[2] 0.26 (2, N)[6]	0.05 (1, N)[1]	9.2 (1, FA/S)[4]
G-1	Granite, Westerly, Rhode Island, U.S.A.	< 10 (2, N)[8] 1.6 (4, N)[3] 2.5 ± 1 (2, C)[5]	0.20 (1, N)[2] 0.065 (1, N)[6] ≦0.048 (1, N)[6]	6 (1, N)[7] ≦0.07 (1, N)[1]	8.2 (1, FA/S)[4]

References: 1. BAEDECKER, P., and W. EHMANN (1965). 2. BATE, G., and J. HUIZENGA (1963). 3. CROCKET, J., and G. SKIPPEN (1966). 4. DAS SARMA, B., *et al.* (1965). 5. GRIMALDI, F., and M. SCHNEPFE (1967). 6. MORGAN, J. (1965). 7. RUSHBROOK, P., and W. EHMANN (1962). 8. VINCENT, E., and A. SMALES (1956).

Maximum values of 100 and 400 ppb for Rh and Ru respectively in both G-1 and W-1 were obtained by a spark source mass spectrometer method (BROWN and WOLSTENHOLME, 1964). These data, which represent the detection limit for Rh and Ru by this method according to BROWN and WOLSTENHOLME, are probably at least two orders of magnitude high.

II. Analytical Uncertainty of the Data

Most of the analytical methods noted have low sensitivity for direct determination of platinum metals in the ppb range and require a concentration step such as fire assay collection or wet chemical treatment. As indicated previously there is some evidence that such procedures may be subject to substantial error in this range. An evaluation of the older data of LUNDE, LUNDE and JOHNSON and the NODDACKS is particularly difficult. However recent data (HAGEN, HAHN-WEINHEIMER, CROCKET and SKIPPEN, DAS SARMA *et al.*) indicate there are very few basic or ultrabasic rocks in which Pt+Pd exceed 100 ppb suggesting that some of the high fire assay values reported by LUNDE and LUNDE and JOHNSON may be gross overestimates.

The evaluation of analytical methods by analysis of standard materials is limited by the lack of suitable standards. The U.S.G.S. rock standards G-1 and W-1 have been analysed for some platinum metals (Table 87-E-3) but very little data are yet available except for Pd. Of the contributions noted above DAS SARMA *et al.* (1965) have reported the only Pt determinations in G-1 and W-1, while CROCKET and SKIPPEN (1966), VINCENT and SMALES (1956) and GRIMALDI and SCHNEPFE (1967) determined Pd in these rocks. The agreement obtained between neutron activation and colorimetric determinations for Pd in G-1 and W-1 suggests that this metal is known to within at least a factor of two and the neutron activation data of CROCKET and SKIPPEN (1966) are accordingly regarded as accurate to at least a factor of two.

As the main body of igneous rock platinum metal data are those of HAGEN (1954), an evaluation of his analyses is particularly pertinent. HAGEN found that his fire assay method was subject to losses of platinum metals to both cupels and slags. An analysis of HAGEN's specimen 31, a sample of Mooihoek dunite from the Bushveld complex, by VINCENT and SMALES (1956) using neutron activation analysis yielded 129 ppb Pd as compared with HAGEN's corrected value of 65 ppb. This is the only direct comparison available. However, an indirect evaluation is possible by comparison with data of CROCKET and SKIPPEN for basaltic rocks. For mid-oceanic basalts the latter authors obtained an average value of 1.9 ± 1.2 ppb Pd for 10 samples whereas HAGEN found an average of 17 ± 17 ppb Pd for 5 samples of mid-oceanic basalt. An explanation of this difference by regional Pd variation is unlikely, as the samples of CROCKET and SKIPPEN are from widely separated geographic areas and demonstrate that this group of basalts is fairly homogeneous with respect to Pd content. If the activation data are accurate to a factor of two or better, HAGEN's results are probably high by a factor of at least four.

In addition to mid-oceanic basalts, both authors determined Pd in continental plateau-building basalts of which the Columbia River, U.S.A. and Deccan, India, occurrences are most significant for comparative evaluation. HAGEN obtained averages of 11 and 111 ppb Pd and CROCKET and SKIPPEN 0.5 and 20 ppb respectively for these rocks. Although both sets of data exhibit the same trend, HAGEN's values are higher by 20 for the low Pd rocks and by 5 for the high Pd rocks. Insufficient sensitivity and reagent contamination are probable sources of error in the very low concentration ranges. It is suggested that in the 1 to 20 ppb range HAGEN's data may be high by a factor of up to 4, while in the 0.1 to 1 ppb range errors by factors of 10 to 20 may occur. It is unfortunate that HAGEN's absolute values are suspect, but the trends, particularly where order of magnitude variations are involved, are probably valid, as suggested by the Columbia River and Deccan basalt data.

III. Ultrabasic and Basic Rocks

Excluding the data of LUNDE and LUNDE and JOHNSON as well as HAHN-WEINHEIMER and ROST's values for the highly serpentinized facies of the Münchberg gneiss mass, a weighted mean of 13 ppb Pd and 32 ppb Pt is obtained for the ultrabasic rock data in Table 78-E-1. The only obvious geochemical trend is the enrichment of Pd and Pt in the antigorite-brucite bearing serpentinites relative to normal serpentinites of the Münchberger gneiss mass, noted by HAHN-WEINHEIMER. The Pd enrichment, a factor of 25, was much higher than that of Pt which was approximately three times higher in the antigorite type serpentinite.

Weighted averages of 21 ppb Pd and 30 ppb Pt are obtained for basalts and gabbros. It seems, therefore, that neither element is strongly fractioned between basic and ultrabasic rocks. As discussed previously there is good evidence that the Pd average is high. The same is probably also true for Pt.

Crocket and Skippen attempted to test for regional variation in Pd in basaltic rocks. Analyses of 10 basalts from mid-oceanic, non-orogenic islands (Hawaii, Tahiti, St. Helena, the Azores and others) indicated little variation in the Pd content of these rocks. An average of 1.9±1.2 ppb Pd was obtained for this group. However, continental plateau-building tholeiites are characterized by distinctive and different Pd contents. Extreme variations were found in the Columbia-Snake River and Deccan rocks which averaged 0.5 and 19 ppb Pd respectively. These differences were attributed mainly to Pd variations in the source region.

Hagen's data for the Pd and Pt content of Bushveld rocks given in Table 78-E-2 show a significant depletion of both metals in rocks of the granitic type. In the gabbroic and ultrabasic rocks, the highest values for both metals are found in pyroxenite. The platinum-bearing Merensky Reef horizon of the Bushveld is also a pyroxenitic rock. No obvious fractionation between ultrabasic and gabbroic types is apparent.

Although data for Pt Pd in granitic rocks are extremely limited the available figures indicate a general depletion of both metals in acid relative to basic and ultrabasic rocks. The Pd, Pt and Os data for G-1 and W-1 in Table 78-E-3 are in agreement with this trend. While the dominant trend for Pt and Pd in igneous rocks appears to be removal in the basic fraction and consequent depletion in acidic rocks, there is a modest reappearance of these metals in rare earth minerals associated with acid pegmatites as previously noted. Limited solubility in silicate crystals leading to concentration in residual fluids is probably an important factor here.

Revised manuscript received: March 22nd, 1968

78-F. Behaviour during Processes Connected with Magmatism (Pegmatites, Gas Transport, Ore Deposition)

I. Introduction

The major ore deposits of the platinum metals are associated with dunitic ultrabasic rocks and gabbroic rocks containing Cu—Ni sulfides, typically pyrrhotite, pentlandite and chalcopyrite. The association of platinum metals with sulfides, tellurides, selenides, arsenides and antimonides is well established. Survey studies of platinum metal abundances, mainly in sulfide minerals, are reported by I. and W. NODDACK (1931), GOLDSCHMIDT and PETERS (1932) and HAWLEY and RIMSAITE (1953). These data are condensed and summarized in Table 78-F-1.

The analytical uncertainties noted in Sections D and E are applicable to sulfide analysis, although the higher platinum metal concentrations probably permit better accuracy because sensitivity is not such a critical factor. The main features of platinum metal geochemistry in sulfide and related minerals and ore deposits indicated by these data include the following:

a) The extensive data of I. and W. NODDACK (1931) show that Pd and Pt are the principal platinum group metals in sulfides and that Pd is more abundant than Pt. However, exceptions do occur and are noted in 3) below.

b) Selenides and tellurides contain much higher platinum metal contents than sulfides. The occurrence of both high selenium-tellurium and platinum metal values in some Cu—Ni sulfide ore deposits, and the occurrence of platinum metal tellurides such as moncheite and kotulskite (GENKIN *et al.*, 1963) indicate a strong geochemical association of these metals with the platinum group.

c) Nickeliferous sulfides, as indicated by the survey of HAWLEY and RIMSAITE (1953) are generally higher in platinum metals than non-nickeliferous sulfides. In both types of deposits chalcopyrite, rather than pyrrhotite, is the major host for Pd and probably also Pt. In Cu—Ni sulfide deposits associated with gabbroic or ultrabasic rocks such as Insizwa, South Africa or Rankin Inlet, Canada, pyrrhotite is enriched in Ru, Os, Ir and possibly Rh, relative to chalcopyrite. As chalcopyrite is usually later than the bulk of the pyrrhotite in such deposits, it seems likely that Pt and Pd are more concentrated than Ru, Ir or Os in the late residual fluids of sulfur-rich magmas.

The distribution of platinum metals in a number of ore deposits including the Sudbury district, Canada (HAWLEY, 1962; HAWLEY and RIMSAITE, 1953), the Noril'sk deposit, U.S.S.R. (GINZBURG and ROGOVER, 1961), the Inaglinsk massif, South Yakutia, U.S.S.R. (RAZIN *et al.*, 1965) and the Bushveld Complex (WAGNER, 1929; SCHNEIDERHÖHN and MORITZ, 1931; HAGEN, 1954) has been investigated and is discussed below.

II. Sudbury Complex, Canada

At Sudbury platinum metals are associated with pyrrhotite-pentlandite-chalcopyrite mineralization occuring as discrete minerals including sperrylite, michenerite and froodite and probably also in solid solution or as extremely fine disseminations in the major sulfides, particularly chalcopyrite, and in arsenides. The available data are given in Table 78-F-2 and 78-F-3, and were obtained largely by HAWLEY and RIMSAITE (1953) who used a fire assay-emission spectrographic method.

As indicated by the pyrrhotite and chalcopyrite data for the Falconbridge Mine, an extremely heterogeneous distribution of platinum metals may occasionally be found. This is probably due to sporadic inclusion of platinum minerals. Only tentative conclusions concerning relationships between major minerals and platinum metals can therefore be drawn from these data. Two generalizations, suggested particularly by the weight percentage data, are noted:

a) In pyrrhotite platinum is generally higher than palladium although the Falconbridge pyrrhotites do not appear to show this trend.

b) In general chalcopyrite and pentlandite are enriched in palladium relative to platinum.

One analysis indicated a marked enrichment of Pd in arsenides. HAWLEY (1962) regards the chalcopyrite mineralization as later than the bulk of the pyrrhotite and notes a consistent increase in palladium with the normal paragenetic order of these minerals. Ten year mine production figures suggest approximately equal Pt and Pd contents.

III. Noril'sk, U.S.S.R.

The distribution of Pt, Pd and Rh in the Noril'sk deposit was investigated by GINZBURG and ROGOVER (1961) using a spectrochemical method. Pyrrhotite, chalcopyrite and pyrrhotite-chalcopyrite concentrates were separated from the main ore-bearing rocks and from veins and encrustations. The results, summarized in Table 78-F-4 suggest that:

a) In disseminated sulfides from the main ore-bearing rocks (figures in parentheses in Table 78-F-4) Pd and Pt are concentrated in chalcopyrite whereas Rh is enriched in pyrrhotite.

b) There is an enrichment of Pd over Pt, with the Pd/Pt ratio for all data in Table 78-F-4 averaging 6.1. In the disseminated ores the enrichment of Pd relative to Pt is higher in chalcopyrite than in pyrrhotite.

c) In massive sulfides from veins and incrustations a Rh enrichment in pyrrhotite occurs as in the disseminated ores. The Pd/Pt ratio is more variable but the highest ratio, 9, is found in chalcopyrite as in the disseminated ore. However, total platinum metals are higher in pyrrhotite rather than chalcopyrite.

IV. Inaglinsk Massif, South Yakutia, U.S.S.R.

Platinum metals in the Inaglinsk Massif were studied by RAZIN *et al.* (1965). The Inaglinsk Massif is a complex pipe-like intrusive with a dunite core and a peripheral zone of alkali syenite rocks. The contact between these rocks, called apodunite, varies from peridotite at the inner contact with dunite through pyroxenite and alkalic gabbro to melanocratic alkali syenite on the outer contact.

Table 78-F-1. *Platinum metal concentrations in sulfide, selenide-telluride and arsenide minerals, ppm*

Sample description	Method and author[a]	Ru	Rh	Pd	Os	Ir	Pt
Sulfides: Mainly pyrrhotite, pyrite, chalcopyrite	(W/X)[5b]	0.01—0.04[c] 0.014 (24)	0.02—2 0.2 (23)	0.2—9 3.7 (39)	0.9—2 0.4 (26)	0.2—1.0 0.5 (26)	0.2—5 0.95 (84)
Selenide-Tellurides	(W/X)[5]	—	2—12 7.0 (2)	5—90 46 (4)	0.05—0.1 0.075 (2)	0.6—15 5.5 (4)	10—50 27.5 (4)
Molybdenite	(FA/S)[3]	—	0.05—0.2 0.08 (4)	0.02—0.5 0.09 (8)	—	0.3	0.02—1 0.17 (9)
Molybdenite: Telemark, Norway	(W/X)[5]	0.006 (60)	—	—	0.01 (60)		0.06 (60)
Co and Ni sulfides and arsenides: Mainly cobaltite, niccolite, gersdorffite, linnaiete	(FA/S)[3]	—	—	0.1—0.5 0.3 (3)	—	—	0.1—0.5 0.2 (13)
Non-nickeliferous sulfide ore deposits, mainly from eastern Canada							
Chalcopyrite	(FA/S)[4]		0.003—0.13 0.06 (6)	0.004—1.8 0.45 (13)			0.09—0.41 0.15 (4)
Pyrrhotite			0.03—0.06 0.045 (2)	0.004—0.056 0.02 (6)			
Pyrite			0.003	0.012			

Nickeliferous sulfide ore deposits							
Pyrrhotite, Rankin Inlet, Northwest Territory, Canada	(FA/S)[4]	1.6—2.1 1.9 (2)	0.3—0.6 0.5 (3)	3.8—5.2 4.5 (3)	10	1.0—1.0 1.0 (2)	1.3—2.7 2.1 (3)
Pyrrhotite, Lynn Lake, Manitoba, Canada			0.018	1.36			0.18
Pyrrhotite, Duluth Gabbro, Minnesota, U.S.A.			0.07	1.47			0.11
Pyrrhotite-Pentlandite, Insizwa, South Africa		0.41	0.38	0.38	0.34	0.51	0.48
Chalcopyrite-Cubanite, Insizwa, South Africa		0.09	0.06	13.8			2.1
Chalcopyrite, Mosalani Mines, Bihar India	(FA/S)[2]						0.025—0.070 0.050 (4)
Almalyk ore region, Central Asia, U.S.S.R.							
Pyrite	(FA/S)[1]			0.014			
Chalcopyrite				0.21			0.02
Sphalerite				0.003			
Galena				0.032			
Molybdenite					2		

[a] 1. Badalov, S., and S. Terekhovich (1966). 2. Das Sarma, B., *et al.* (1966). 3. Goldschmidt, V., and Cl. Peters (1932). 4. Hawley, J., and Y. Rimsaite (1953). 5. Noddack, I., and W. Noddack (1931).

[b] Table I, Section 6 from Reference No. 4 excluding minerals from meteorites. Averages are weighed according to the number of samples of each mineral analysed.

[c] The data include the range (top), average (middle) and number of samples (in parentheses) if more than one.

Table 78-F-2. *Platinum metals in minerals from the Sudbury complex, Canada, ppm*

Sample description		Reference[a] and method	Ru	Rh	Pd	Os	Ir	Pt
Means from South range deposits	Pyrrhotite	(FA/S)[1]	0.04 to 0.35	0.28	0.60	0.37	0.2 to 0.33	0.52
	Chalcopyrite			0.27	1.17			1.30
Falcon-bridge mine	Pyrrhotite	(FA/S)[1]			34.2 0.13			15.6 0.17
	Chalcopyrite				34.28 1.61			1.54 3.77
	Pentlandite				17.50 12.65			29.8 34.2
	Pyrite				0.31			0.37
	Arsenides				246.8			trace
Nickel-offsets	Pentlandite and Pyrrhotite	(FA/S)[1]	0.10		2.00			2.68
Creighton	Pyrrhotite	(N)[4]			0.10			0.48
	Chalcopyrite				0.013			0.72
	Pentlandite				6.3			0.2
Strathcona mine	Pyrrhotite, mafic norite	(N)[2] and (N)[3]	0.019		0.082	0.009	0.010	
	Pyrrhotite, xenolithic norite		0.054		0.025	0.023	0.053	
	Pyrrhotite, footwall		0.004		0.212	0.003	0.002	
	Chalcopyrite, footwall		0.014		0.962	0.003	0.001	

[a] HAWLEY, J., and Y. RIMSAITE (1953). 2. HSIEH, S. (1967). 3. KEAYS, R. (1968). 4. MORRIS, D., *et al.* (1963).

Platinum metals are found mainly in dunite as discrete minerals, minerals including native platinum (Pt, Fe), iridic platinum (Pt, Fe, Ir), nevyanskite (Ir, Os), sperrylite ($PtAs_2$) and cooprite (PtS), and in association with silicate and oxide minerals. The deposit is similar to the Urals type of platinum mineralization according to RAZIN *et al.*

Individual platinum metals were determined in whole rock samples by a quantitative spectrographic method following chemical enrichment. The data are presented in Table 78-F-5. High total platinum metal values occur in the dunite core carrying segregated chrome spinels. These rocks are particularily distinctive, since both Ir and Rh are higher than Pd. No other rocks noted in this survey had such high relative Rh and Ir values. The Pd/Pt ratios are low and vary from 0.13 to 0.20 in all the dunitic rocks.

Table 78-F-3. *Platinum metal ratios in common Sudbury ore minerals, percent*. (After HAWLEY, 1962)

Mineral and description	Number of samples	Rh	Pd	Pt	Total platinum metals in ppm
Pyrite	3	2.5	41.4	56.4	1.90
Pyrrhotite	37	7.8	32.7	61.6	1.36
Pentlandite	3	7.5	59.5	33.0	5.42
Chalcopyrite	7	0.9	81.4	17.7	7.69
Minerals from an offset deposit					
Pyrite		7.5	42.2	50.3	
Pyrrhotite		7.0	48.7	44.3	1.87
Pentlandite		10.3	64.6	25.1	10.52
Chalcopyrite		0.22	71.5	28.3	8.06
Mill average (6 months)		2.0	65.0	33.0	5.33
10 year production, Sudbury ores[a]		13	41	46	

[a] The Rh figure includes Ru + Ir.

Table 78-F-4. *Pt, Pd and Rh data for sulfides from the Noril'sk deposit, U.S.S.R.* (After GINZBURG and ROGOVER, 1961)

Sample Descriptions	Number of samples	Rh	Pd	Pt	Total platinum metals in ppm
		wt.-% and ppm/1% Ni (in parentheses)			
Mineral concentrates from the main ore-bearing rocks[a]					
Chalcopyrite	27	1, (0.4)	86, (32)	13, (5)	
Chalcopyrite-pyrrhotite	29	4, (0.7)	83, (14)	13, (2)	
pyrrhotite	19	21, (1)	66, (4)	13, (0.9)	
Sulfides from veins and incrustations[b]					
Pyrrhotite, veins		37%	53%	11%	19
Pyrrhotite, incrustations		15%	73%	13%	20.6
Chalcopyrite, veins		—	83%	17%	4.5
Chalcopyrite, incrustations		0.8%	90%	10%	5.1

[a] Summarized from Table 1, p. 918. For each mineral weighed averages were calculated from the original data, expressed in grams/ton/1% nickel, and converted to a weight percent basis.

[b] Summarized from Tables 4 and 5, p. 921 and 922.

Pure mineral separates were analysed for total platinum metals. The minerals were decomposed by heating with NH_4F or $HCl + H_2SO_4$ and the soluble platinum metals determined gravimetrically after purification from base metals. Chrome spinels were analysed colorimetrically after HCl decomposition. Platinum minerals occurring in intimate intergrowths with silicates did not dissolve, and were discarded after separation from host minerals. The total platinum values obtained are regarded as representing platinum metals in solid solution or adsorbed in silicates. The validity of this interpretation depends on whether the platinum minerals were

Table 78-F-5. *Platinum metal content in rocks of the Inaglinsk Massif. U.S.S.R., ppb.* (After RAZIN *et al.*, 1965)

Sampling point	Rock	No. of analyses	Pt	Pd	Ir	Rh	Os	Ru	Total average platinum metals
Core of massif	Dunite with acessory chrome-spinels	5	300 to 4,000[a] 500	40 to 600 100	2—40 17	1.7 to 80 50	7—60 8	2—30 5	680
Core of massif	Dunite with segregation chrome-spinels	6	300 to 600 400	20 to 60 50	200 to 900 400	20 to 700 270	5—9 7.5	3—6 4.5	1,132
Inner contact of the apodunite zone	Apodunite olivine-pyroxene rocks (analogues of pyroxenite)	4	30—90 80	9—20 12	0.4 to 40 15	0.4 to 5 2.5	2—8 4	0.9—4 2.2	96
Outer contact of the apodunite	Apodunite pyroxene biotite-K feldspar rock (analogue of mica shonkinite)	1	8	2	5	3	0.3	0.5	19
Periphery of massif	Pulaskites	1	2	3	2	5	traces (<0.2)	0.2	12

[a] Range above and average below.

completely insoluble in the various reagents used to attack the silicates. Proof that no dissolution of platinum minerals occurred is lacking and, as noted by WRIGHT and FLEISCHER (1967), the extreme fine grain size of these minerals would favour solubility. The data are presented in Table 78-F-6 and indicate the following:

a) chrome-spinel is the principal concentrator of platinum metals. Olivine is the second major host, but, on the average, contains about 30 times less platinum than chrome-spinel;

b) in the dunites the weakly serpentinized rocks contain higher platinum metal contents than the medium serpentinized rocks. The authors suggest that while olivine accumulates large amounts of platinum metals during magmatic crystallization there is some loss during epigenetic alteration;

c) pyroxene, mica, magnetite and apatite contain very small amounts of platinum metals.

The association of platinum metals with chrome-spinel was further investigated by analysis of morphologically similar chrome-spinel with approximately the same degree of epigenetic alteration taken from the center and periphery of the dunite core.

The average value for total platinum metals in nine samples from the core of the intrusive is 30 ppm whereas two samples from the periphery average 96 ppm. The authors consider that the lower temperature chrome-spinels at the periphery are more active concentrators of the platinum metals than the higher temperature spinels from the center of the intrusive.

Table 78-F-6. *Total platinum metals in minerals of the Inaglinsk massif. U.S.S.R., ppm.* (After RAZIN *et al.*, 1965)[a]

Rock[c] type	Description and sample number	Whole rock	Olivine	Chrysotile	Chome-spinel	% total Pt metals present as Pt minerals
Dunite with accessory chrome-spinel	Fresh, slightly serpentinized; I-135	—	0.8	not detected	20	—
	Fresh, slightly serpentinized; I-130	4.7	2	0.15	10	—
	Fresh, medium serpentinized; I-105	1.0	0.4	0.3	20	64.8
	Intensively weathered, medium serpentinized, I-101	—	0.5	0.7	50	—
Apodunite	Mica-pyroxene-olivine rock[b], fresh, slightly serpentinized, I-111	0.11	0.06	0.28 B	20	53.7

[a] Data summarized from Tables 4 and 7, p. 123 and 126 respectively, RAZIN *et al.*

[b] Diopside and phlogopite from the apodunite, sample I-111, contained 0.002 and 0.004 ppm total platinum metals respectively.

[c] Platinum metals were detected in trace amounts only in magnetite and apatite in these rocks.

V. Bushveld Complex, South Africa

In the Bushveld, platinum metals are found in high concentration in the Merensky Reef (WAGNER, 1929; COUSINS, 1964) and the Mooihoek, Driekop and Onverwacht dunite pipes. Some data of geochemical significance on the Bushveld occurrences have been presented by SCHNEIDERHÖHN and MORITZ (1931) and HAGEN (1954).

The Merensky Reef, a pegmatitic pyroxenite normally bounded on top and bottom by thin chromite seams is characterized by high platinum metal values within or adjacent to the top and bottom chromite bands, although the chromite crystals themselves do not contain platinum (COUSINS, 1964). Platinum minerals found include ferroplatinum, braggite and cooprite with smaller amounts of laurite, stibiopalladinite and sperrylite. SCHNEIDERHÖHN and MORITZ (1931) analysed sulfide and silicate minerals from the Merensky Reef for platinum metals by microdrilling of polished sections, followed by direct analysis by emission spectrography. High values for silicate minerals suggest that some of their results (Table 78-F-7) may generally be high.

Pt and Pd are the most abundant sulfides with Pt in excess of Pd and the other metals present in approximately equal but much smaller amounts. The relative proportion of Merensky Reef platinum metals, computed from Cousins data (1964) are:

Pt=63%, Pd=28%, Ru=5%, Rh=2.7%, Ir=0.7% and Os=0.6%. In the samples examined by SCHNEIDERHÖHN and MORITZ no discrete platinum minerals were observed and they concluded that the platinum metals were in solid solution in sulfides. They also noted that the highest platinum metal contents were in the earliest sulfides, pyrite and pyrrhotite. The well-established occurrence of discrete platinum metal minerals in the Merensky Reef (WAGNER, 1929; RAMDOHR, p. 759, 1960) indicates, however, that only part of the total platinum metal content can be in solid solution in sulfides.

Table 78-F-7. *Platinum metal content of minerals from the Merensky Reef, Bushveld Complex, South Africa, ppm.* (After SCHNEIDERHÖHN and MORITZ, 1931)

Location	Mineral[a]	Ru	Rh	Pd	Ir	Pt
Klipfontein-[b]	Pyrite	0.5—1	1	5—10	0.5—1	10—50
Krondaal,	Pyrrhotite					
Rustenburg	Pentlandite	0.5	0.5—1	5	0.5—1	5—10
District	Bronzite	—	0.1—0.5	0.5—1	0.1—0.5	1—5
Schildpadnest,[c]	Pyrite	0.1—1	0.1—1	10—100	0.1—1	10—100
Rustenburg	Pyrrhotite					
District	Pentlandite	0.1—1	0.1—1	1—10	0.1—1	1—10

[a] Platinum metal concentrations in chalcopyrite from both localities were below the detection limit.

[b] Olivine, diallage and plagioclase were also analysed in the Klipfontein sample. In olivine 1 and 0.5 ppm Pt and Pd respectively were found and in diallage 1—5 ppm Pt was obtained.

[c] In the Schildpadnest sample chromite, olivine, diallage and serpentine + millerite were analysed. No platinum metals were detected.

SCHNEIDERHÖHN and MORITZ also determined platinum metals in minerals from the Mooihoek and Driekop dunite pipes and a few determinations for the Mooihoek and Onverwacht pipes were reported by HAGEN (1954). The highest platinum metal concentrations are found in the iron-rich hortonolite cores of the Mooihoek and Onverwacht pipes whereas in the Driekop pipe, which lacks concentric structure and marked fractionation of iron, the distribution of high platinum metal values is more irregular. The platinum metals occur as discrete minerals including mainly ferroplatinum, sperrylite and, at Driekop, geversite, according to STUMPFL (1961). In addition small quantities of braggite, cooprite, laurite, stibiopalladinite, osmiridium and osmite are associated with the major platinum group minerals. Using an electron microanalyser STUMPFL (1961) identified eight additional new platinoid-rich minerals, principally compounds of antimony, arsenic and bismuth, in mine concentrates from Dreikop.

Data for whole-rock samples from dunite pipes are given in Table 78-F-8. An increase in platinum metal content in the hortonolite facies of the Mooihoek and Onverwacht pipes is apparent. In the Driekop mine where the degree of iron enrichment is small, little difference is found in the olivine and hyalosiderite facies. WAGNER regards the hortonolite cores of the pipes as later than the forsterite mantle because of both the lower temperature indicated by the iron-rich composition and the presence of mineralizers suggested by the association of phlogophite, hornblende and fluorapatite with the hortonolitic rocks. The platinum metals are therefore

Table 78-F-8. *Platinum metal content of dunite pipes from the Bushveld Complex, South Africa, ppm*

Pipe	Rock type and description	Author[a] and method	Ru	Rh	Pd	Ir	Pt
Mooihoek	Medium grained hortonolite dunite	S[2]	0.5	0.5—1	1—5	0.5	5—10
	Coarse grained hortonolite dunite in the vicinity of pegmatitic veins	S[2]	0.5—1	1—5	5—10	1	10—50
	Chromite band in hortonolite dunite	S[2]	0.5—1	1—5	5	1	0—50
	Hortonolite, inner core, No. 8	Fa/S[1]		0.29	0.65		12.3
	Hortonolite, inner core, No. 106	FA/S[1]			0.02		0.3
	Olivine, outer rim, No. 31[b]	FA/S[1]			0.07		0.07
Driekop	Olivine dunite, surface	S[2]		0.1—0.5	1		1—5
	Hyalosiderite dunite, surface	S[2]		0.5	1—5		1—5
	Hyalosiderite dunite, 400 foot level	S[2]		0.5	1—5		1—5
	Weathered dunite, surface	S[2]		0.5—1	1—5	0.5	5—10
Onverwacht	Olivine, outer rim, No. 16	FA/S[1]			0.004		0.02
	Hortonolite, inner core, No. 17	FA/S[1]			0.013		3.2

[a] 1. Hagen, J. (1954). 2. Schneiderhöhn, H., and H. Moritz (1931).
[b] Vincent and Smales (1956) obtained 0.123 Pd for this sample

considered to be fractionated and concentrated in late or residual fluids which gave rise to the hortonolite cores.

As in the Merensky Reef, Pt and Pd are the most abundant metals with Pt in excess of Pd. Rh is the third most abundant metal and Os and Ru minerals are known to be present.

Schneiderhöhn and Moritz also found platinum metal contents up to 5 ppm Pt in silicate minerals such as hornblende, mica and diallage from the Mooihoek and Driekop pipes. Such high platinum values were regarded as resulting from the presence of inclusions of platinum minerals. The large difference between Hagen's Pd values for two hortonolite samples from the inner core of the Mooihoek pipe suggest non-uniform distribution of this metal.

VI. Discussion

A summary of some of the properties of the platinum metal distributions and mineral associations for the ores discussed above is presented in Table 78-F-9. The platinum metals of highest concentration in these ores are Pt and Pd followed by

Table 78-F-9. *Summary of platinum metal distribution in the Sudbury, Noril'sk. Inaglinsk and Bushveld deposits*

Deposit	Pd/Pt	Sulfide, silicate or oxide minerals with highest platinum metal content		
		Rh	Pd	Pt
Sudbury	1	Pyrrhotite Pentlandite	Chalcopyrite Pentlandite	Chalcopyrite Pentlandite
Noril'sk	6	Pyrrhotite	Chalcopyrite	Chalcopyrite
Inaglinsk (whole rock samples)	0.25	Total platinum metals highest in chrome-spinels		
Bushveld Merensky Reef	0.4		Pyrite Pyrrhotite	Pyrite Pyrrhotite
Bushveld Dunite Pipes	0.2	Pt and Pd highest in hortonolitic facies		

Rh with Ru, Os and Ir occurring at considerably lower concentration levels. In Cu-Ni sulfide deposits associated with gabbroic rocks Pd equals or exceeds the Pt content on the average, whereas in deposits associated with dunitic rocks Pt is the more abundant of the two metals. The Merensky Reef ores, which are associated with significant quantities of Cu—Ni sulfides and in this sense are comparable with Sudbury and Noril'sk, are an exception. However, they do show some analogy with the dunitic deposits in that platinum metals are concentrated in the chromite layers immediately above and below the pyroxenitic facies of the Reef.

In the Cu—Ni sulfide ores of Sudbury and Noril'sk Pt and Pd are dominantly associated with chalcopyrite and pentlandite whereas Rh is distinctly higher in pyrrhotite, particularly at Noril'sk. The Merensky Reef, however, is different in that pyrite and pyrrhotite rather than chalcopyrite are richest in Pt and Pd according to SCHNEIDERHÖHN and MORITZ's data. Although distinctive platinum metal — sulfide mineral associations are apparent, it is by no means clear whether the platinum metals are in solid solution or are mechanically admixed as discrete minerals with the sulfides. STUMPFL (1962) has discussed the genesis of platinum ore deposits pointing out that discrete platinum metal minerals have been recognized in all these deposits. He concluded (p. 622) that while the "possibility that traces of Pt-metals enter the lattice of sulfides" cannot be disregarded, "the bulk of the Pt-metals in the above deposits is fixed in independent Pt-minerals".

While the presence of platinum minerals in Cu—Ni sulfide ores is well established-it is unlikely that the strong association of Pt and Pd with copper but Rh with iron sulfides is accidental. Two possible reasons for the existence of these relationships follow.

a) Because of favourable crystallochemical factors Pt and Pd form dilute solid solutions with chalcopyrite, cubanite and pentlandite. If solubility is favoured at high temperatures exsolution may occur on cooling with the formation of cooprite or braggite. Rh, again because of crystallo-chemical factors, may be capable of substitution only in the pyrrhotite lattice.

b) Pt and Pd may form complex ions, possibly with S, As or Sb and with square coplanar stoichiometry (by analogy with known low temperature S complexes of Pd) which inhibit solid solution of these metals but increase solubility in S-rich melts. These Pt metals therefore tend to occur in highest concentration in residual or late magmatic solutions as discrete minerals, and in association with the youngest major sulfide in the paragenetic sequence, usually chalcopyrite. If Rh does not form a species of high solubility in a sulfur-rich melt, it may be adsorbed, coprecipitated or otherwise enter into solid solution with the earliest major sulfide in the paragenesis, pyrrhotite.

Accurate data on the platinum metal concentrations in individual sulfide minerals from Cu—Ni sulfide ores are required to test the validity of these hypotheses.

The occurrence of platinum metals in dunitic rocks is characterized by two important features including enrichment in spinels, particularly in massive or segregation type chromite, and in the youngest dunitic rocks. High platinum metal values are found in the chromite layers marking the upper and lower boundaries of the Merensky Reef (Cousins, 1964), and in chrome -spinel of the Inaglinsk Massif. The occurrence of high platinum metal values in the youngest facies of dunitic rocks is particularly apparent in the Mooihoek and Onverwacht pipes where a core of hortonolitic olivine is present. Also at Inaglinsk the dunitic core, which presumably crystallized later than the dunite on the periphery, is highest in platinum metals. A suggested explanation is that low solubility of platinum metals in silicate lattices has led to concentration in residual magmatic fluids or melts where they have finally crystallized as native metals, sulfides, arsenides or antimonides depending on the availability of these elements. Razin *et al.* (1965) maintain that in the Inaglinsk Massif, silicates do carry platinum metals. However, the interpretation of their data is open to some doubt as previously noted. The reason for the association of platinum metals with chrome spinels is unclear. Razin *et al.* appeal to substantial solid solution or adsorption of platinum metals, whereas Cousins (1964) and Wagner (1929) maintain that Merensky Reef and other Bushveld chromite is free of platinum metals although supporting analytical data are not presented.

Revised manuscript received: March 22nd, 1968

78-G. Behaviour during Weathering and Alteration of Rocks Oxidation States of Platinum Metals

The behaviour of platinum metals during weathering, the nature of platinum metal species responsible for solubility in natural waters and the abundance of these metals in natural waters are essentially unknown.

From general chemistry we know the following:

	Ru	Rh	Pd	Os	Ir	Pt
Most common oxidation states:	III, IV	I, III	II, IV	IV, VI	III, IV	II, IV
Highest oxidation states:	+8	+6	+4	+8	+6	+6

In general terms the platinum group elements exhibit low valence states in nature as exemplified by their common occurrence as native metals or alloys. However, their oxidation states in silicate magmas, in sulfur-rich ore solutions and in weathering and transport cycles in the lithosphere are unknown.

Revised manuscript received: March 22nd, 1968

78-K. Abundances in Common Sedimentary Rock Types

Platinum metal data for sedimentary rocks and sediments are limited to a very few analyses, mostly for material of unusual composition. Normal sandstones, shales and limestones remain virtually unanalysed. The available abundance data are summarized in Table 78-K-1.

There are no data on limestones and only a single determination, 0.055 ppb Os (LOVERING and MORGAN, 1964), on a sandstone. More data are available for shales but the analytical reliability of much of it is uncertain. The Kupferschiefer data by three different analysts provide values of 0.1, 50 and 200—500 ppb Pt for this rock, but not for the same sample. The Kupferschiefer is heterogeneous with respect to different sulphides including chalcopyrite.

Little information on the sedimentary geochemical cycle of the platinum metals is available. However, the concentration of platinum metal minerals in placer deposits, often in association with ultrabasic rocks, as in the Urals, U.S.S.R., is well known. The osmiridium of the Witwatersrand conglomerates, South Africa, is often regarded as of placer origin. High specific gravity, malleability and chemical inertness are factors which promote the accumulation and preservation of platinum metal minerals in placer deposits. FISCHER (1966) has studied some aspects of the sedimentary geochemistry of Pt in the Saale River, Germany and its tributaries. He found Pt values ranging from 0.1 to 50 ppm in river sediment but did not detect the metal in water or suspended (i.e. — filterable) matter.

Table 78-K-1. *Platinum metal content of sedimentary rocks and sediments, ppb*

Rock type	Description and location	Author and method	Ru	Rh	Pd	Os	Ir	Pt	Total Pt metals
Sandstones and sands	Quartz rich sandstone, Wilpena Pound, South Astralia	(1, N)[9]				0.065 0.055			
	Sand: mixture of quartz, mica, feldspar and rutile	(3, W/X)[10]						7 40 10	
	Sand: Saale River, Germany	(FA/S)[4]						100—50,000	
Shales and clays	State Circle Shale, Etheridge Creek, A.C.T., Australia	(1, N)[9]				0.556 0.351			
	Dark shale with pyrite, Captain's Flat Railway Station, N.S.W., Australia	(1, N)[9]				0.438 0.252			
	Bituminous shale, Kupferschiefer, Mansfeld, Germany	(1, S)[2]			0.1		0.1	0.1	
	Kupferschiefer, Mansfeld, Germany	(W/X)[11]	3	0.2	20	0.3	0.4	50	
	Kupferschiefer, Mansfeld, Germany	(2,FA/S[8]			200—500 200	 500		200—500 200—500	
	Black shale	(3, S)[12]	60 80 40	20 10 10	50 50 50			50 50 10	
	Bleached shale	(3, S)[12]	40 5 20	20 10 10	50 50 50			50 50 10	
	Argillite and quartzitic slate, Katzhutter beds, Thuringia, Germany	(3, FA/S)[4]						0.5 0.5 1.0	
	Clay, Greenland	(W/X)[10]						20	

Deep sea sediments	Pelagic clay, Argentine basin	(4 N)[1]						
	100 cm					1.4		
	355 cm					0.012		
	675 cm					0.033		
	1045 cm					0.14		
	Pelagic clay, central Pacific	(2,N)[1]						
	12—16 cm					⩽0.30		
	56—60 cm					⩽0.73		
Miscel-laneous	River sediment, Orissa, India	(4, FA/S)[3]					< 5, < 5 9, 12	
	Washed river sediment, Manipur, India	(1, FA/S)[3]					98	
	Soil, India	(1, FA/S)[3]					17	
	Coal ash	(1, FA/S)[7]					200	
	Coal ash	(1, FA/S)[7]		20	50—500		100—500	
	Manganese Nodule, Challenger Station No. 286, Atlantic Ocean	(1, FA/S)[6]	200	200			500	
	Potassium-salt deposits and saline clays Germany	(S)[5]						0.9—92
	Rocks surrounding saline layers	(S)[5]						0.6 —174

References: 1. Baedecker, P., and W. Ehmann (1965). 2. Cissarz, A., and H. Moritz (1933). 3. Das Sarma, B., *et al.* (1966). 4. Fischer, K. (1966). 5. Goubeau, J., and L. Birkenbach (1936). 6. Goldschmidt, V., and Cl. Peters (1932). 7. Goldschmidt, V., and Cl. Peters (1933). 8. Goldschmidt, V., *et al.* (1948). 9. Lovering, J., and J. Morgan (1964). 10. Noddack, I., and W. Noddack (1931). 11. Noddack, I. (1936). 12. Tischendorf, G. (1959).

Revised manuscript received: March 22nd, 1968

78-M. Abundance in Common Metamorphic Rock Types

A few determinations of platinum metals in metamorphic rocks including analyses of sagvandite (a pyroxene-calcite rock) for Pt by LUNDE (1927) and LUNDE and JOHNSON (1928) are available:

		Pt (ppm)
Sample 1	fresh sagvandite	0.73
(LUNDE and JOHNSON)	amphibole-garnet contact zone	1.22
	sediment (beyond contact aureole)	0.01
Sample 2	decomposed sagvandite	0.65
(LUNDE)	hydrothermal contact zone	1.24
	chloreite schist in the contact zone	0.01

LEUTEWEIN (1939) reported Pt and Pd values of 8 to 500 ppb and 8 to 100 ppb respectively for four samples of garnetiferous amphibolite.

Revised manuscript received: March 22nd, 1968

78-O. Average Abundance, Economic Importance

I. Average Abundance

Platinum metal data are available for approximately 240 meteorites. Chondrites vary from approximately 0.1 to 2.0 ppm for individual metals. The carbonaceous, enstatite and bronzite chondrites show little variation in atomic abundance of platinum metals, except for Ir, while hypersthene chondrites are depleted by 38 % relative to averages for these three groups. The carbonaceous chondrites have the highest platinum metal abundances and appear to represent the best material for solar abundance estimates. Based on Type I carbonaceous chondrites (and CH group chondrites for Rh) the relative solar abundances are:

$$Ru > Pt > Pd > Os > Ir > Rh.$$

The irons contain the highest platinum metal contents of any meteorite group with individual platinum metal concentrations generally within the range 1 to 10 ppm. Fractionation between structural classes (that is, with varying Ni content) is characterized by distinct trends. Pd increases with Ni whereas Pt decreases, except in the Ni-rich ataxites. Rh shows very little variation. Os, Ru and Ir are more irregular but show similar relative concentrations in the same structural class. Relative platinum metal contents, as represented by medium octahedrites, are:

$$Ru > Pt > Os > Ir > Pd > Rh.$$

Pd is therefore depleted in the irons relative to other platinum metals as compared with carbonaceous chondrite abundances.

The achondrites are depleted by factors of 100 to 1,000 relative to the irons and are most comparable with terrestrial rocks.

Abundance data for platinum metals in terrestrial materials are meager, and confined mainly to Pt and Pd in mafic and ultramafic rocks. Concentrations are generally less than 100 ppb and commonly in the 1 to 30 ppb range. Average values for Pd and Pt in mafic and ultramafic rocks from Table 78-E-1 are 21 and 30 and 13 and 32 ppb respectively. Little fractionation of these elements between mafic and ultramafic rocks is indicated.

Because of the lack of data for most crustal rock types, little accuracy can be claimed for crustal abundance estimates. Many estimates (Wright and Fleischer, Table 13, 1965) are available but the primary data are essentially those of Goldschmidt and the Noddacks. Goldschmidt's (1954) crustal abundance estimates in ppb are:

Ru	Rh	Pd	Os	Ir	Pt
0.4	0.4	4	0.4	0.4	2

It is not apparent, either from terrestrial or meteoritic data, what the evidence is for an excess of Pd over Pt in the crust.

Ore deposit data demonstrate the geochemical affinity of the platinum group with S, As, Sb, Se, Te and Bi. In the gabbroic Cu-Ni sulfide ores an association of Pt and Pd with chalcopyrite and pentlandite and Rh with pyrrhotite is usually found. While these relationships may result from solid solution processes, the Pt—Pd associations are also compatible with an enrichment in residual sulfide melts and consequent association with major phases which are late in the paragenetic sequence. In dunitic rocks concentrations of platinum metals, largely as discrete minerals, particularly in hortonolite-rich facies and with segregation-type chrome spinel are noted.

Revised manuscript received: March 1968

II. Economic Importance

The major producers of platinum group metals are the USSR, the Republic of South Africa and Canada. In 1972[1] these countries accounted for 98.6 % of the world production of which 51 % came from the USSR, 39 % from South Africa and 9 % from Canada. The remainder was produced by Columbia, the USA, the Phillipines, Japan, Ethiopia and Finland.

The main producing areas in the USSR are the Norilsk district and the Kola Peninsula. About 75 % of this production is by-product recovery from the nickel-copper ores of Norilsk. Pt and Pd average 90 % of Norilsk platinoids and are in the proportion of approximately 1:2. In South Africa the platinum metals are mined as primary ores from the Merensky Reef of the Bushveld Complex. Again Pt and Pd make up approximately 90 % of the total but the Pt:Pd ratio is about 2.7:1. In Canada the platinum group metals are recovered as by-products of nickel-copper sulfides from the ores of the Sudbury and Thompson-Wabouden regions.

World production in the years 1968 to 1972 was 3.4, 3.4, 4.2, 4.1 and 4.6 million troy ounces respectively. In 1972 US consumption accounted for approximately one-third of total platinum metals produced.

The major uses of platinum group metals are based on their catalytic activity, their resistance of oxidation at high temperature, high strength and good ductility. Platinum and palladium are the major platinum group metals used. The other metals are alloyed with Pt and Pd, often to increase hardness or corrosion resistance. Important applications are found in the catalytic action of Pt and Pd, and to a large extent Rh and Ru, in the production of high octane gasolines, sulfuric and nitric acid and in the manufacture of pharmaceuticals, vitamins and antibiotics. Rh is used extensively in the plating industry and an important use of Ir is as a hardening agent in Pt-Pd alloys. Os is used mainly in the chemical industry and in dental and medical applications. As an example of usage in an industrialized country, consumption by industry in the US in 1972 was: chemical 38 %; electrical 34 %; petroleum 8 %; medical-dental 8 %; glass, jewelry and others 12 %.

Because of the very close association of platinum metals with ultramafic rocks, this class of rocks must generally be regarded as a potential resource. In particular, large layered-gabbro complexes are of special interest, particularily in view of the

[1] Minerals Yearbook, 1972. U.S. Bureau of Mines, p. 1043.

recent discovery (PAGE *et al.*, 1972)[1] of a significant platinoid horizon in the Stillwater Complex. Further, Ni-Cu sulfide ores in ultramafic rocks, such as the recently developed Western Australian nickel deposits constitute major future reserves.

Extraction and refining techniques applied to platinum group metal recovery vary somewhat for different ores but involve recovery that is coincident with that of copper and nickel in that both of these metals are electrolytically refined under condition in which platinum group metals accumulate as recoverable sludges.

The refining of such platinum group metal concentrates (WALSH and HANSMAN 1963)[2] generally involves an aqua regia extraction which separates platinum and palladium (and gold). After removal of gold from the aqua regia extract by reduction with ferrous sulfate, platinum is separated from palladium by precipitation of hexachloroplatinate with ammonium chloride. The insolubles from the aqua regia step are smelted with lead. Following cupellation the lead phase is treated with nitric acid to remove silver and the insolubles fused with sodium bisulfate which dissolves rhodium. The insoluble residues, including iridium and ruthenium, are fused with sodium peroxide and ruthenium is separated by distillation from an aqueous solution as ruthenium tetroxide. Iridium from the peroxide fusion is rendered soluble by aqua regia treatment and precipitated with ammonium chloride as ammonium hexachloroiridate.

---

[1] PAGE, N. J., RILEY, L. B., HAFFTY, J.: Vertical and lateral variation of platinum, palladium and rhodium in the Stillwater Complex, Montana. Econ. Geol. **67**, 915 (1972),

[2] WALSH, T. J., HANSMAN, E. A.: The platinum metals. In: Treatise on Analytical Chemistry **8**, 79 (1963).

Manuscript received: October 1977

References: Sections 44-45-46, 76-77-78-A to 78-O

AHRENS, L., TAYLOR, S.: Spectrochemical Analysis (2nd ed.). Reading, Mass.: Addison-Wesley 1961.

ANDERS, E.: Origin, age and composition of meteorites. Space Sci. Rev. **3**, 583 (1964).

ANDERS, E., GANAPATHY, R., KEAYS, R. R., LAUL, J. C., MORGAN, J. W.: Volatile and siderophile elements in lunar rocks: comparison with terrestrial and meteoritic basalts. Proc. Second Lunar Sci. Conf., Geochim. Cosmochim. Acta, Suppl. 2, **2**, 1021 (1971).

BAADSGAARD, H., CAMPBELL, F., FOLINSBEE, R., CUMMING, G.: The Bruderheim meteorite. J. Geophys. Res. **66**, 3574 (1961).

BAADSGAARD, H., FOLINSBEE, R., CUMMING, G.: Peace river meteorite. J. Geophys. Res. **69**, 4197 (1964).

BADALOV, S., TEREKHOVICH, S.: Geochemistry of elements of the Pt group in the Almalyk ore region. Dokl. Akad. Nauk SSSR **168**, 1397 (1966).

BAEDECKER, P., EHMANN, W.: The distribution of some noble metals in meteorites and natural materials. Geochim. Cosmochim. Acta **29**, 329 (1965).

BAEDECKER, P. A., CHOU, C.-L., GRUDEWICZ, B., WASSON, J. T.: Volatile and siderophilic trace elements in Apollo 15 samples: Geochemical implications and characterization of the long-lived and short-lived extralunar materials. Proc. Fourth Lunar Sci. Conf., Geochim. Cosmochim. Acta, Suppl. 4, **2**, 1177 (1973).

BAEDECKER, P. A., CHOU, C.-L., SUNDBERG, L. L., WASSON, J. T.: Extralunar materials in Apollo 16 soils and the decay rate of the extra lunar flux 4.0 Gy ago. Earth Planet. Sci. Lett. **17**, 79 (1972a).

BAEDECKER, P. A., CHOU, C.-L., SUNDBERG, L. L., WASSON, J. T.: Volatile and siderophilic trace elements in the soils and rocks of Taurus-Littrow. Proc. Fifth Lunar Sci. Conf., Geochim. Cosmochim. Acta, Suppl. 5, **2**, 1625 (1974).

BAEDECKER, P. A., CHOU, C.-L., WASSON, J. T.: The extralunar component in lunar soils and breccias. Proc. Third Lunar Science Conf., Geochim. Cosmochim. Acta, Suppl. 3, **2**, 1343 (1972b).

BAEDECKER, P. A., SCHAUDY, R., ELZIE, J. L., KIMBERLIN, J., WASSON, J. T.: Trace element studies of rocks and soils from Oceanus Procellarum and Mare Tranquilitatis. Proc. Second Lunar Sci. Conf., Geochim. Cosmochim. Acta, Suppl. 2, **2**, 1037 (1971).

BATE, G., HUIZENGA, J.: Abundance of ruthenium, osmium and uranium in some cosmic and terrestrial sources. Geochim. Cosmochim. Acta **27**, 345 (1963).

BREWER, L.: Predictions of high temperature metallic phase diagrams. In: ZACKAY, V. F. (ed.), High Strength Materials, p. 12—103. New York: John Wiley 1965.

BREWER, L.: A most striking confirmation of the Engel metallic correlation. Acta Met. **15**, 553 (1967).

BROWN, R., WOLSTENHOLME, W.: Analysis of geological samples by spark source mass spectrometry. Nature **201**, 598 (1964).

BRUNFELT, A. O., HEIER, K. S., NILSSEN, B., SUNDVOLL, B., STEINNES, E.: Distribution of elements between different phases of Apollo 14 rocks and soils. Proc. Third Lunar Sci. Conf., Geochim. Cosmochim. Acta, Suppl. 3, **2**, 1133 (1972).

BRUNFELT, A. O., HEIER, K. S., STEINNES, E.: Determination of 40 elements in Apollo 12 materials by neutron activation analysis. Proc. Second Lunar Sci. Conf., Geochim. Cosmochim. Acta, Suppl. 2, **2**, 1281 (1971).

CAMERON, A.: A new table of abundances of the elements in the solar system. Preprint, Belfer Graduate School of Science, Yeshiva University, N.Y. (1967).

CHAKRABURTTY, A., STEVENS, C., RUSHING, H., ANDERS, E.: Isotopic composition of silver in iron meteorites. J. Geophys. Res. **69**, 505 (1964).

Chou, C.-L., Baedecker, P. A., Bild, R. W., Wasson, J. T.: Volatile-element systematics and green glass in Apollo 15 lunar soils. Proc. Fifth Lunar Sci. Conf., Geochim. Cosmochim. Acta, Suppl. 5, **2**, 1645 (1974).

Cissarz, A., Moritz, H.: Untersuchungen über die Metallverteilung in Mansfelder Hochofenprodukten und ihre geochemische Bedeutung. Metallwirtschaft **12**, 131 (1933).

Clayton, D., Fowler, W.: Abundances of heavy nuclides. Ann. Phys. (N.Y.) **16**, 51 (1961).

Cobb, J.: A trace-element study of iron meteorites. J. Geophys. Res. **72**, 1329 (1967).

Cousins, C.: The platinum deposits of the Merensky Reef. In: Some Ore Deposits in Southern Africa, Vol. 2, Geol. Soc. of South Africa, Johannesburg 1964.

Crocket, J., Keays, R., Hsieh, S.: Precious metal abundances in some carbonaceous and enstatite chondrites. Geochim. Cosmochim. Acta **31**, 1615 (1967).

Crocket, J., Skippen, G.: Radioactivation determination of palladium in basaltic and ultrabasic rocks. Geochim. Cosmochim. Acta **30**, 129 (1966).

Darmon, R., Wintenberger, M.: Structure cristalline de $CoAs_2$. Bull. Soc. Franç. Mineral. Crist. **89**, 213 (1966).

Das Sarma, B., Sen, B., Chodwhury, A.: Platinum and gold contents of granite G-1 and diabase W-1. Econ. Geol. **60**, 373 (1965).

Das Sarma, B., Sen, B., Chowdhury, A.: Studies on geochemistry of platinum. Econ. Geol. **61**, 592 (1966).

Dawson, K., Maxwell, J., Parsons, D.: A description of the meteorite which fell near Abee, Alberta, Canada. Geochim. Cosmochim. Acta **21**, 127 (1960).

Douglas, J.: Written communication (unpublished data — Geol. Survey of Canada, (1967).

Fischer, K.: Edelmetalle in der Saale und ihrem Einzugsgebiet. Geologie (Berlin) **15**, 550 (1966).

Fish, R., Goles, G., Anders, E.: The record in the meteorites III. On the development of meteorites in asteroidal bodies. Enrico Fermi Institute Preprint 59—43 (revised, 1960). An abridged version of this paper was published in: Astrophys. J. **132**, 243 (1960).

Fredriksson, K., Mason, B.: The Shaw meteorite. Geochim. Cosmochim. Acta **31**, 1705 (1967).

Ganapathy, R., Keays, R. R., Laul, J. C., Anders, E.: Trace elements in Apollo11 lunar rocks: Implications for meteorite influx and origin of moon. Proc. Apollo 11 Lunar Sci. Conf., Geochim. Cosmochim. Acta **34**, Suppl. 1, **2**, 1117 (1970).

Ganapathy, R., Morgan, J. W., Higuchi, H., Anders, E., Anderson, A. T.: Meteoritic and volatile elements in Apollo 16 rocks and in separated phases from 14306. Proc. Fifth Lunar Sci. Conf., Geochim. Cosmochim. Acta, Suppl. 5, **2**, 1659 (1974).

Ganapathy, R., Morgan, J. W., Krähenbühl, U., Anders, E.: Ancient meteoritic components in lunar highland rocks: Clues from trace elements in Apollo 15 and 16 samples. Proc. Fourth Lunar Sci. Conf., Geochim. Cosmochim. Acta, Suppl. 4, **2**, 1239 (1973).

Genkin, A., Zhuravlev, N., Smirnova, E.: Moncheite and kotulskite — new minerals — and the composition of michenerite. Zap. Vses. Mineralog. Obshchestva **92**, 33 (1963).

Ginzburg, V., Rogover, G.: Regularities in the distribution of non-ferrous and precious metals in principal ore minerals and silicates of the Noril'sk deposits. Intern. Geol. Rev. **3**, 917 (1961).

Gmelin's Handbuch der anorganischen Chemie: System No. 68 Platin. Weinheim a. d. Bergstraße: Verlag Chemie 1938.

Goldberg, E. D., Uchiyama, A., Brown, H.: The distribution of nickel, cobalt, gallium, palladium, and gold in iron meteorites. Geochim. Cosmochim. Acta **2**, 1 (1951).

Goldschmidt, V. M.: Geochemistry. Oxford: University Press 1954.

Goldschmidt, V. M., Krejci-Graf, K., Witte, H.: Spurenelemente in Sedimenten. Nachr. Akad. Wiss. Göttingen, Math.-Phys. Kl. II. **35** (1948).

Goldschmidt, V. M., Peters, Cl.: Zur Geochemie der Edelmetalle. Nachr. Akad. Wiss. Göttingen, Math.-Phys. Kl. IV, 377 (1932).

Goldschmidt, V. M., Peters, Cl.: Über die Anreicherung seltener Elemente in Steinkohlen. Nachr. Akad. Wiss. Göttingen, Math.-Phys. Kl. **371** (1933).

Goldstein, J. I., Axon, H. J., Yen, C. F.: Metallic particles in the Apollo 14 lunar soil. Proc. Third Lunar Sci. Conf., Geochim. Cosmochim. Acta, Suppl. 3, **1**, 1037 (1972).

GOUBEAU, J., BIRKENBACH, L.: Edelmetallgehalt der Kalisalzlagerstätten. Z. Anorg. Allgem. Chem. **236**, 37 (1936).

GREENLAND, L.: The abundances of selenium, tellurium, silver, palladium, cadmium, and zinc in chondritic meteorites. Geochim. Cosmochim. Acta **31**, 849 (1967).

GRIMALDI, F., SCHNEPFE, M.: Determination of palladium in the parts-per-billion range in rocks. U.S. Geol. Surv., Profess. Papers **575-C**, C 141 (1967).

HAGEN, J.: Some aspects of the geochemistry of platinum, palladium and gold in igneous rocks with special reference to the Bushveld complex, Transvaal. Unpublished Ph. D. Thesis, M.I.T. (1954).

HAHN-WEINHEIMER, P.: Geochemische Untersuchungen an den ultrabasischen und basischen Gesteinen der Münchberger Gneismasse (Fichtelgebirge). Neues Jahrb. Mineral., Abhandl. **92**, 203 (1959).

HAHN-WEINHEIMER, P., ROST, F.: Akzessorische Mineralien und Elemente im Serpentinit von Leupoldsgrün (Münchberger Gneismasse). Ein Beitrag zur Geochemie ultrabasischer Gesteine. Geochim. Cosmochim. Acta **21**, 165 (1961).

HAMAGUCHI, H., NAKAI, T., KAMEMOTO, Y.: Determination of platinum, iridium and palladium in meteorites by neutron activation. Nippon Kagaku Zasshi **82**, 1489 (1961).

HARA, T., SANDELL, E.: Meteoritic abundance of ruthenium. Geochim. Cosmochim. Acta **21**, 145 (1960).

HAWLEY, J.: The Sudbury ores—Their mineralogy and origin. Can. Mineralogist **7**, 1 (1962).

HAWLEY, J., RIMSAITE, Y.: Platinum metals in some Canadian uranium and sulfide ores. Am. Mineralogist **38**, 463 (1953).

HERR, W., HOFFMEISTER, W., HIRT, B., GEISS, J., HOUTERMANS, F.: Versuch zur Datierung von Eisenmeteoriten nach der Rhenium-Osmium-Methode. Z. Naturforsch. **16a**, 1053 (1961).

HERR, W., HOFFMEISTER, W., LANGHOFF, J.: Die Bestimmung von Rhenium und Osmium in Eisenmeteoriten durch Neutronenaktivierung. Z. Naturforsch. **15a**, 99 (1960).

HERR, W., MERZ, E., EBERHARDT, P., GEISS, J., LANG, C., SIGNER, P.: Search for decay products of natural technetium. Geochim. Cosmochim. Acta **14**, 158 (1958).

HEY, M.: Catalogue of Meteorites, 3rd ed. London 1966.

HIRT, B., HERR, W., HOFFMEISTER, W.: Age determinations by the rhenium-osmium method. In: Radioactive Dating. International Atomic Energy Agency, Vienna 1963.

HOLSETH, H., KJEKSHUS, A.: Compounds with the marcasite type crystal structure. I. Compositions of the binary pnictides. Acta Chem. Scand. **22**, 3273 (1968). II. On the crystal structures of the binary pnictides. Acta Chem. Scand. **22**, 3284 (1968).

HULLIGER, H.: Crystal chemistry of the chalcogenides and pnictides of the transition elements. Structure and Bonding **4**, 83 (1968).

HUME-ROTHERY, W.: The Engel-Brewer theories of metals and alloys. Progr. Mater. Sci. **13**, 229 (1968).

HSIEH, S.: Analysis of ruthenium and osmium abundances in sulfide minerals from the Sudbury ores, Ontario. Unpublished M. Sc. Thesis McMaster University (1967).

JOVANOVIC, S., REED, G. W.: Volatile trace elements and the characterization of the Cayley Formation and the primitive lunar crust. Proc. Fourth Lunar Sci. Conf., Geochim. Cosmochim. Acta, Suppl. 4, **2**, 1313 (1973).

JOVANOVIC, S., REED, G. W.: Labile and nonlabile element relationships among Apollo 17 samples. Proc. Fifth Lunar Sci. Conf., Geochim. Cosmochim. Acta, Suppl. 5, **2**, 1685 (1974).

KEAYS, R.: A neutron activation analysis technique for determination of the precious metals and its application to a study of their geochemistry. Unpublished Ph. D. Thesis McMaster University (1968).

KEIL, K.: Fortschritte in der Meteoriten. Fortschr. Mineral. **38**, 202 (1960).

KEIL, K., MASON, B., WILK, H., FREDRIKSSON, K.: The Chainpur meteorite. Am. Museum. Novitates No. 2173 (1963).

KEMP, J.: Instrumentation for rapid spectrochemical analysis. Anal. Chem. **28**, 1838 (1956).

KÖNIG, H.: Die chemische Zusammenstellung der hellen und dunklen Anteile des Meteoriten Pantar. Geochim. Cosmochim. Acta **28**, 1397 (1964).

Krähenbühl, U., Ganapathy, R., Morgan, J. W., Anders, E.: Volatile elements in Apollo 16 samples: Implications for highland volcanism and accretion history of the moon. Proc. Fourth Lunar Sci. Conf., Geochim. Cosmochim. Acta, Suppl. 4, **2**, 1325 (1973).

Larimer, J., Anders, E.: Chemical fractionation in meteorites—II. Abundance patterns and their interpretation. Geochim. Cosmochim. Acta **31**, 1239 (1967).

Laul, J. C., Hill, D. W., Schmitt, R. A.: Chemical studies of Apollo 16 and 17 samples. Proc. Fifth Lunar Sci. Conf., Geochim. Cosmochim. Acta, Suppl. 5, **2**, 1047 (1974).

Laul, J. C., Morgan, J. W., Ganapathy, R., Anders, E.: Meteoritic material in lunar samples: Characterization from trace elements. Proc. Second Lunar Sci. Conf., Geochim. Cosmochim. Acta, Suppl. 2, **2**, 1139 (1971).

Laul, J. C., Schmitt, R. A.: Chemical composition of Apollo 15, 16 and 17 samples. Proc. Fourth Lunar Sci. Conf., Geochim. Cosmochim. Acta, Suppl. 4, **2**, 1349 (1973).

Lederer, C., Hollander, J., Perlman, I.: Table of Isotopes, 6th ed. New York: John Wiley & Sons, Inc. 1967.

Leutwein, F.: Über das Vorkommen einiger seltener Elemente in metamorphen Gesteinen des Schwarzwaldes, vorläufige Mitteilung. Zentr. Mineral. Geol. Abhandl. A 123 (1939).

Lovering, J. F., Butterfield, D.: Neutron activation analysis of rhenium and osmium in Apollo 11 lunar material. Proc. Apollo 11 Lunar Sci. Conf., Geochim. Cosmochim. Acta **34**, Suppl. 1, **2**, 1351 (1970).

Lovering, J. F., Hughes, T. C.: Rhenium and osmium abundance determinations and meteoritic contamination levels in Apollo 11 and Apollo 12 lunar samples. Proc. Second Lunar Sci. Conf., Geochim. Cosmochim. Acta, Suppl. 2, **2**, 1331 (1971).

Lovering, J., Morgan, J.: Rhenium and osmium abundances in tektites. Geochim. Cosmochim. Acta **28**, 761 (1964).

Lunde, G.: Über das Vorkommen des Platins in norwegischen Gesteinen und Mineralen. Z. Anorg. Allgem. Chem. **161**, 1 (1927).

Lunde, G., Johnson, M.: Vorkommen und Nachweis der Platinmetalle in norwegischen Gesteinen. II. Z. Anorg. Allgem. Chem. **172**, 167 (1928).

Mason, B.: The origin of meteorites. J. Geophys. Res. **65**, 2965 (1960).

Mason, B.: The carbonaceous chondrites. Space Sci. Rev. **1**, 621 (1963).

Mason, B.: The enstatite chondrites. Geochim. Cosmochim. Acta **30**, 23 (1966).

Mason, B., Wiik, H.: The composition of the Ottawa, Chateau-Renard, Mocs, and New Concord meteorites. Am. Museum Novitates, No. 2069 (1961a).

Mason, B., Wiik, H.: The Holbrook Arizona chondrite. Geochim. Cosmochim. Acta **21**, 276 (1961b).

Mason, B., Wiik, H.: Descriptions of two meteorites: Karoonda and Erakot. Am. Museum Novitates No. 2115 (1962).

Mason, B., Wiik, H.: The composition of the Richardton Estacado and Knyahinya meteorites. Am. Museum Novitates No. 2154 (1963).

Mason, B., Wiik, H.: The composition of the Forest City, Tennasilm, Weston and Geidam meteorites. Am. Museum Novitates No. 2220 (1965).

Mason, B., Wiik, H.: Composition of the Belly River, Bluff and Bremervoide and Modoc meteorites. Am. Museum Novitates No. 2280 (1967).

Morgan, J.: The simultaneous determination of rhenium and osmium in rocks by neutron activation analysis. Anal. Chim. Acta **32**, 8 (1965).

Morgan, J. W., Ganapathy, R., Higuchi, H., Krähenbühl, U., Anders, E.: Lunar basins: Tentative characterization of projectiles, from meteoritic elements in Apollo 17 boulders. Proc. Fifth Lunar Sci. Conf., Geochim. Cosmochim. Acta, Suppl. 5, **2**, 1703 (1974).

Morgan, J. W., Krähenbühl, U., Ganapathy, R., Anders, E.: Trace elements in Apollo 15 samples: Implications for meteorite influx and volatile depletion on the moon. Proc. Third Lunar Sci. Conf., Geochim. Cosmochim. Acta, Suppl. 3, **2**, 1361. MIT Press (1972).

Morgan, J. W., Krähenbühl, U., Ganapathy, R., Anders, E.: Luna 20 soil: abundance of 17 trace elements. Geochim. Cosmochim. Acta **37**, 953 (1973a).

Morgan, J. W., Krähenbühl, U., Ganapathy, R., Anders, E., Marvin, U. B.: Trace element abundances and petrology of separates from Apollo 15 soils. Proc. Fourth Lunar Sci. Conf., Geochim. Cosmochim. Acta, Suppl. 4, **2**, 1379 (1973b).

MORGAN, J. W., LAUL, J. C., KRÄHENBÜHL, U., GANAPATHY, R., ANDERS, E.: Major impacts on the moon: Characterization from trace elements in Apollo 12 and 14 samples. Proc. Third Lunar Sci. Conf., Geochim. Cosmochim. Acta, Suppl. 3, **2**, 1377 (1972).

MORGAN, J., LOVERING, J.: Rhenium and osmium abundances in stony meteorites. Science **144**, 835 (1964).

MORGAN, J., LOVERING, J.: Rhenium and osmium abundances in chondritic meteorites. Geochim. Cosmochim. Acta **31**, 1893 (1967).

MORRIS, D., HILL, N., SMITH, B.: The determination of traces of palladium and platinum in sulfide minerals by neutron-activation analysis. Mikrochim. Ichnoanalyt. Acta **65**, 962 (1963).

NICHIPORUK, W., BROWN, H.: The distribution of platinum and palladium metals in iron meteorites and in the metal phase of ordinary chondrites. J. Geophys. Res. **70**, 459 (1965).

NININGER, H.: Kansas meteorites since 1925. Trans. Kansas Acad. Sci. **39** 169 (1936).

NODDACK, I.: Über die Allgegenwart der chemischen Elemente. Angew. Chem. **49**, 835 (1936).

NODDACK, I., NODDACK, W.: Die Geochemie des Rheniums. Z. Physik. Chem. (Leipzig) 54, 207 (1931).

PARTHÉ, E., HOHNKE, D.: The ordered vacancy arrangement in pyrite defect structures. In: EYRING, L., O'KEEFE, M. (eds.), The Chemistry of Extended Defects in Non-metallic Solids, p. 220. Amsterdam: North-Holland Publishing Company 1970.

PARTHÉ, E., HOHNKE, D., HULLIGER, F.: A new structure type with octahedron pairs for Rh_2S_3, Rh_2Se_3 and Ir_2S_3. Acta Cryst. **23**, 832 (1967).

PRIOR, G.: On the South African meteorites Mount Ayliff and Simondium and the chemical composition of the meteorites Adare and Ensisheim. Mineral. Mag. **19**, 163 (1921).

QUIRING, H.: Die metallischen Rohstoffe, Bd. 16: Platinmetalle. Stuttgart: Ferdinand Enke 1962.

RAMDOHR, P.: Die Erzmineralien und ihre Verwachsungen. Berlin: Academie-Verlag 1960.

RANKAMA, K., SAHAMA, TH.: Geochemistry. Chicago: Chicago University Press 1950.

RAZIN, L., KHVOSTOV, V., NOVIKOV, V.: Platinum metals in the essential and accessory minerals in ultramafic rocks. Geochemistry Int. **2**, 118 (1965).

REED, G.: Heavy elements in the Pantar meteorite. J. Geophys. Res. **68**, 3531 (1963).

REED, G. W., JOVANOVIC, S.: Halogens, mercury, lithium and osmium in Apollo 11 samples. Proc. Apollo 11 Lunar Sci. Conf., Geochim. Cosmochim. Acta **34**, Suppl. 1, **2**, 1487 (1970).

REED, G. W., JR., JOVANOVIC, S.: The halogens and other trace elements in Apollo 12 samples and the implications of halides, platinum metals, and mercury surfaces. Proc. Second Lunar Sci. Conf., Geochim. Cosmochim. Acta, Suppl. 2, **2**, 1261 (1971).

REED, G. W., JOVANOVIC, S., FUCHS, L.: Trace element relations between Apollo 14 and 15 and other lunar samples, and the implications of a moon-wide C1-KREEP coherence and Pt-metal noncoherence. Proc. Third Lunar Sci. Conf., Geochim. Cosmochim. Acta, Suppl. 3, **2**, 1989 (1972).

RILEY, G.: Rhenium concentration in Australian molybdenites by stable isotope dilution. Geochim. Cosmochim. Acta **31**, 1489 (1967).

RINGWOOD, A.: Chemical and genetic relationships among meteorites. Geochim. Cosmochim. Acta **24**, 159 (1961).

RUNDQVIST, S., ERSSON, N. O.: Structure and bonding in skutterudite-type phosphides. Arkiv Kemi **30**, 103 (1968).

RUSHBROOK, P., EHMANN, W.: Iridium in stone meteorites by neutron activation analysis. Geochim. Cosmochim. Acta **26**, 649 (1962).

SANDELL, E.: Colorimetric determination of traces of osmium. Ind. Eng. Chem. **16**, 342 (1944).

SANDELL, E.: Colorimetric Determination of Traces of Metals. New York: Interscience Publ. Inc. 1959.

SCHINDEWOLF, U., WAHLGREN, M.: The rhodium, silver and indium content of some chondritic meteorites. Geochim. Cosmochim. Acta **18**, 36 (1960).

SCHNEIDERHÖHN, H., MORITZ, H.: Spektrographische Untersuchungen über die Verteilung der Platinmetalle in den Mineralien der Südafrikanischen Platinlagerstätten. Festschrift z. 50jähr. Bestehen der Platinschmelze. G. Siebert GmbH. Hanau. 1931, S. 257.

SEN GUPTA, J.: Arsenazo III as a sensitive and selective reagent for the spectrophotometric determination of palladium in iron and stony meteorites. Anal. Chem. **39**, 18 (1967).

SEN GUPTA, J., BEAMISH, F.: The determination of platinum metals in siderite meteorites. Am. Mineralogist **48**, 379 (1963).

SMALES, A., MAPPER, D., FOUCHÉ, K.: The distribution of some trace elements in iron meteorites, as determined by neutron activation. Geochim. Cosmochim. Acta **31**, 673 (1967).

SMALES, A. A., MAPPER, D., WEBB, M. S. W., WEBSTER, R. K., WILSON, J. D.: Elemental composition of lunar surface material. Proc. Apollo 11 Lunar Sci. Conf., Geochim. Cosmochim. Acta **34**, Suppl. 1, **2**, 1575 (1970).

SMALES, A. A., MAPPER, D., WEBB, M. S. W., WEBSTER, R. K., WILSON, J. D., HISLOP, J. S.: Elemental composition of lunar surface material (part 2). Proc. Second Lunar Sci. Conf., Geochim. Cosmochim. Acta, Suppl. 2, **2**, 1253 (1971).

STUMPFL, E.: Some new platinoid-rich minerals ,identified with the electron micro-analyser. Mineral. Mag. **32**, 833 (1961).

STUMPFL, E.: Some aspects of the genesis of platinum deposits. Econ. Geol. **57**, 619 (1962).

TISCHENDORF, G.: Zur Genesis einiger Selenidvorkommen, insbesondere von Tilkerode im Harz. Freiberger Forschungsh. C **69**, 1 (1959).

UREY, H., CRAIG, H.: The composition of the stony meteorites and the origin of the meteorites. Geochim. Cosmochim. Acta **4**, 36 (1953).

VINCENT, E., SMALES, A.: The determination of palladium and gold in igneous rocks by radioactivation analysis. Geochim. Cosmochim. Acta **9**, 154 (1953).

VINOGRADOV, A. P.: Preliminary data on lunar ground brought to earth by automatic probe "Luna-16". Proc. Second Lunar Sci. Conf., Geochim. Cosmochim. Acta, Suppl. 2, **1**, 1 (1971).

VINOGRADOV, A. P.: Preliminary data on lunar soil collected by the Luna 20 unmanned spacecraft. Geochim. Cosmochim. Acta **37**, 721 (1973).

VOBECKÝ, M., FRÁNA, J., BAUER, J., RANDA, Z., BENADA, J., KUNCÍŘ, J.: Radioanalytical determination of elemental compositions of lunar samples. Proc. Second Lunar Sci. Conf., Geochim. Cosmochim. Acta, Suppl. 2, **2**, 1291 (1971).

VOGT, J., EHMANN, W.: Silicon abundances in stony meteorites by fast neutron activation analysis. Geochim. Cosmochim. Acta **29**, 373 (1965).

WÄNKE, H., BADDENHAUSEN, H., BALACESCU, A., TESCHKE, F., SPETTEL, B., DREIBUS, G., PALME, H., QUIJANO-RICO, M., KRUSE, H., WLOTZKA, F., BEGEMANN, F.: Multielement analyses of lunar samples and some implications of the results. Proc. Third Lunar Sci. Conf., Geochim. Cosmochim. Acta, Suppl. 3, **2**, 1251 (1972).

WÄNKE, H., BADDENHAUSEN, H., DREIBUS, G., JAGOUTZ, E., KRUSE, H., PALME, H., SPETTEL, B., TESCHKE, F.: Multielement analyses of Apollo **15**, 16 and 17 samples and the bulk composition of the moon. Proc. Fourth Lunar Sci. Conf., Geochim. Cosmochim. Acta, Suppl. 4, **2**, 1461 (1973).

WÄNKE, H., PALME, H., BADDENHAUSEN, H., DREIBUS, G., JAGOUTZ, E., KRUSE, H., SPETTEL, B., TESCHKE, F., THACKER, R.: Chemistry of Apollo 16 and 17 samples: Bulk composition, late stage accumulation and early differentiation of the moon. Proc. Fifth Lunar Sci. Conf., Geochim. Cosmochim. Acta, Suppl. 5, **2**, 1307 (1974).

WÄNKE, H., RIEDER, R., BADDENHAUSEN, H., SPETTEL, B., TESCHKE, F., QUIJANO-RICO, M., BALACECU, A.: Major and trace elements in lunar material. Proc. Apollo 11 Lunar Sci. Conf., Geochim. Cosmochim. Acta **34**, Suppl. 1, **2**, 1719 (1970a).

WÄNKE, H., WLOTZKA, F., BADDENHAUSEN, H., BALACESCU, A., SPETTEL, B., TESCHKE, F., JAGOUTZ, E., KRUSE, H., QUIJANO-RICO, M., RIEDER, R.: Apollo 12 samples: Chemical composition and its relation to sample locations and exposure ages, the two component origin of the various soil samples and studies on lunar metallic particles. Proc. Second Lunar Sci. Conf., Geochim. Cosmochim. Acta, Suppl. 2, **2**, 1187 (1971).

WÄNKE, H., WLOTZKA, F., JAGOUTZ, E., BERGMANN, F.: Composition and structure of metallic iron particles in lunar "fines". Proc. Apollo 11 Lunar Sci. Conf., Geochim. Cosmochim. Acta **34**, Suppl. 1, **1**, 931 (1970b).

WAGNER, P.: Platinum Deposits and Mines of South Africa. London: Oliver and Boyd 1929.
WARNER, J. L., SIMONDS, C. H., PHINNEY, W. C.: Apollo 16 rocks: classification and petrogenetic model. Proc. Fourth Lunar Sci. Conf., Geochim. Cosmochim. Acta, Suppl. 4, **1**, 481 (1973).
WASSON, J. T., BAEDECKER, P. A.: Ga, Ge, In, Ir and Au in lunar, terrestrial and meteoritic basalts. Proc. Apollo 11 Lunar Sci. Conf., Geochim. Cosmochim. Acta **34**, Suppl. 1, **2**, 1741 (1970).
WATT, D., GLOVER, R.: A search for radioactivity among naturally occurring isobaric pairs. Phil. Mag. **7**, 105 (1962).
WIIK, H.: The chemical composition of some stony meteorites. Geochim. Cosmochim. Acta **9**, 279 (1956).
WINCHESTER, J.: Radioactivation analysis in inorganic geochemistry. Progr. Inorg. Chem. **2**, 1 (1960).
WING, J.: Simultaneous determination of oxygen and silicon in meteorites and rocks by nondestructive activation analysis with fast neutrons. Anal. Chem. **36**, 559 (1964).
WLOTZKA, F., SPETTEL, B., WÄNKE, H.: On the composition of metal from Apollo 16 fines and the meteoritic component. Proc. Fourth Lunar Sci. Conf., Geochim. Cosmochim. Acta, Suppl. 4, **2**, 1483 (1973).
WLOTZKA, F., JAGOUTZ, E., SPETTEL, B., BADDENHAUSEN, H., BALACECU, A., WÄNKE, H.: On lunar metallic particles and their contribution to the trace element content of Apollo 14 and 15 soils. Proc. Third Lunar Sci. Conf., Geochim. Cosmochim. Acta, Suppl. 3, **1**, 1077 (1972).
WOOD, J.: Physics and chemistry of meteorites. In: MIDDLEHURST, B., KUIPER, G. (eds). The Solar System IV. The Moon, Meteorites and Comets. Chicago: Chicago University Press 1963.
WRIGHT, T., FLEISCHER, M.: Geochemistry of the platinum metals. U.S. Geol. Survy Bull. **1214-A**, A 1 (1965) with corrections and additions (1967).
YAVNEL', A.: Spectrographic analysis of the Sikhote-Alin meteorite. Meteoritika **8**, 134 (1950).
ZACHARIASEN, W. H.: The crystal structure of palladium diphosphide. Acta Cryst. **16**, 1253 (1963).
ZVYAGINTSEV, O.: Geochemistry of palladium. Vopr. Geokhim. i Mineralog. Akad. Nauk SSSR., Otd. Geol. Geogr. Nauk 199 (1946).
ZVYAGINTSEV, O.: Ruthenium geochemistry. Izv. Sektora Platiny i Drug. Blagorodn. Metal: Inst. Obshch. i Neorgan. Khim., Acad. Nauk SSSR **25**, 129 (1950).

Revised manuscripts received: March 1968 and January 1976

Gold 79

A R. Allmann (Fachbereich Geowissenschaften der Universität, Marburg a. d. Lahn, Germany)

B—O J. H. Crocket (Department of Geology, McMaster University, Hamilton, Ontario, Canada)

Gold (× 3), Rumänien · Nugget, Rumania · Or natif, Roumanie · Oro Nativo, Rumania

79-A. Crystal Chemistry

Like copper and silver, gold belongs to group 1 b of the periodic table and has a single *s* electron outside a completed *d* shell. In spite of the similarity in electronic structures there are, however, only few resemblances between Cu, Ag and Au. Gold is strongly siderophilic and somewhat chalcophilic, whereas copper and silver are mainly chalcophilic. This means that gold occurs in nature mainly as the metal and as various alloys, especially with silver. Its chalcophilic character is limited to the formation of several gold tellurides (see Table 79-A-1).

The metallic radii of gold and silver are nearly equal, 1.44 Å, as calculated from the lattice constants of their ccp cell: $a = 4.0783$ Å[1] for Au and $a = 4.0856$ Å for Ag. The metallic radius of copper is about 11% smaller: $r = 1.28$ Å from $a = 3.6153$ Å. Because of their similar metallic radii, gold and silver form a continuous solid solution series and native gold always contains some amount of silver, commonly 10—15% by weight. Copper is mostly present too, but in smaller amounts of 0.1 to 0.5%. The purest gold was found in the Kalgoorlie district of Western Australia, having a fineness of 999.1 [= 1,000 · Au/(Au + Ag)] (PALACHE *et al.*, 1944, p. 91).

The maximum high temperature solid solubilities of some metals in gold are as follows (after WISE, 1964):

100%	Ag, Cu, Ni, Pd, Pt	10.9—7.7%	Mn, Ta, Co, In
46%	Fe	5.2—1.2%	V, Ga, Sn, Mg, Al, Ti, Ge
21.5—∼19%	Cd, Cr, Hg	< 1%	As, Bi, Ca, Mo, Pb, Pr, Rh,
13%	Zn		Sb, Th, Tl, U

At lower temperatures some of these alloys form ordered phases, e.g. the following Au—Cu phases are known: Au_3Cu (cubic, $a = 3.98$ Å, Au—Au and Au—Cu distance = 2.81 Å); CuAu (tetragonal, $a = 3.98$, $c = 3.72$ Å, Au—Au = Cu—Cu = 2.81 Å, Au—Cu = 2.73 Å); and Cu_3Au (cubic, $a = 3.71$ Å, Cu—Cu = Cu—Au = 2.65 Å). The latter two have been found in nature as auricupride (RAMDOHR, 1967).

Several gold tellurides are known, most of them belonging to the system Au—Ag—Te (CAPRI, 1965). In this system, the naturally occurring compounds montbrayite (Au_2Te_3), empressite (AgTe), and muthmannite (Ag, Au)Te could not be synthesized at the temperatures used (≧ 290° C). In the related ditellurides calaverite, krennerite and sylvanite, Au and Ag have a distorted octahedral coordination with an average Au—Te distance of 2.89 Å. These minerals may be considered as consisting of the ions Au^{3+}, Ag^{3+} and Te^{2-} with about 0.5 additional Te—Te bonds per Te atom (TUNELL and PAULING, 1952). The metallic and ionic radii of AHRENS and PAULING (from STRUNZ, 1970, p. 28/30) do not fit very well with the experimental values: $r(Au^{3+}) + r(Te^{2-}) = 0.85 + 2.21 = 3.06$ Å. According to SHANNON and PREWITT (1969), $r(Au^{3+})$ is = 0.70 Å for a square planar coordination of Au^{3+}. This value yields a better fit: $Au^{3+}—Te^{2-} = 0.70 + 2.21 = 2.91$ Å (found 2.89 Å).

[1] All cited lattice constants from STRUNZ (1970).

Table 79-A-1. *Gold-containing minerals*

Name	Formula	Wt.-% Au	Density	Comments (references, distances in Å, etc.)
a) Varieties of native gold (mostly after PALACHE *et al.*, 1944)				
Gold	(Au, Ag)	> 80	19.3 (pure Ag)	M. P. 1,062.4° C
Electrum	(Au, Ag)	50—80		argentian gold
Silver	(Ag, Au)	0—50	10.5 (pure Ag)	M. P. 960.6° C, aurian silver (küstelite)
Porpezite	(Au, Pd)	90—95		palladian gold
Rhodite	(Au, Rh)	57—66	15.5—16.8	rhodian gold, not confirmed
Auricupride	(Cu, Au) but also Cu_3Au and CuAu (RAMDOHR, 1967) = cuproauride			
Aurosmiride	(Ir, Os, Au)	19.3	20	aurian osmiridium, probably a mixture
b) Intermetallic compounds				
Gold-amalgam[c]	Au_2Hg_3(?)	34.2—41.6	15.5	(PALACHE *et al.*, 1944, p. 105)
Maldonite	Au_2Bi	64.5—65.1	15.5	(JURRIAANSE, 1935, SB **3**, 315) artificial Au_2Bi: cubic, a = 7.98 Å, Cu_2Mg-type, Dx = 15.7, Au—Au = 2.82 Å, Au—Bi = 3.30 Å, Bi—Bi = 3.45 Å
Aurostibite	$AuSb_2$	43.5—50.9	9.89 (Dx)	(GRAHAM and KAIMAN, 1952, SR **15**, 15) cubic, a = 6.66 Å, pyrit-type, Au—Sb = 2.79 Å, Sb—Sb = 2.63 Å
c) Gold tellurides, structure determined (under remarks: Au—Te distances in Å)				
Calaverite[a]	$AuTe_2$	39.2—42.8[b]	9.0—9.39[b]	2.68 (2), 2.97 (4)
Krennerite[a]	$(Au, Ag)Te_2$	30.7—43.9	8.35—8.62	2.66 (2), 2.99 (4); {2.63, 2.67, 2.87, 2.95, 3.15, 3.22; {2.63, 2.66, 2.94 (2), 3.04 (2)
Sylvanite[a]	$AuAgTe_4$	24.25—29.99	8.07—8.16	{2.67 (2), 2.75 (2), 3.25 (2)
Petzite[c]	Ag_3AuTe_2	19.0—25.2	8.7—9.02	2.53 (2) (FRUEH, SR 1959, 154)
Hessite	Ag_2Te	4.73	8.24—8.45	Ag—Te: 2.85—3.04 (see 47-A)

Table 79-A-1. (Continued)

Name	Formula	Wt.-% Au	Density	Comments(references, distances in Å, etc.)
		d) Gold tellurides, structure unknown		
Montbrayite[c]	Au_2Te_3	38.6—44.3	9.94	PEACOCK and THOMPSON (1946), SR **10**, 55
Muthmannite	(Ag, Au)Te?	22.9—31	5.6	PALACHE *et al.* (1944), p. 260
Kostovite	$CuAuTe_4$	25.2	?	TERZIEV (1966)
		e) Complex gold minerals		
Nagyagite	$AuTe_2 \cdot 6Pb(S, Te)$?	7.4—10.2	7.35—7.46	PALACHE *et al.* (1944), p. 168 (formula after STRUNZ, 1970, p. 136)
Aurobis-muthinite	$(Bi, Au, Ag)_5S_6$?	12.3	?	PALACHE *et al.* (1944), p. 278

[a] TUNELL and PAULING (1952).
[b] All gold contents and densities after PALACHE *et al.* (1944).
[c] Note added in proof: Gold-amalgam is a mixture of Au_6Hg_5 and Au_3Hg [LINDAHL, T.: Acta Chem. Scand. **24**, 946—952 (1970)]. Fischesserite, Ag_3AuSe_2, the first gold-containing selenide, is isostructural with petzite [JOHAN, Z., PICOT, P., PIERROT, R., KVAČEK, M.: Bull. Soc. Fr. Minéral. Cristallogr. **94**, 381—384 (1971)]. Montbrayite is only stable in presence of Sb [BACHECHI, F.: Amer. Mineralog. **57**, 146—154 (1972)].

Petzite, Ag_3AuTe_2, contains linear Te—Au—Te groups with a very short Au—Te distance of 2.53 Å (FRUEH, SR 1959, 154), whereas the sum of the ionic radii $r(Au^{1+}) + r(Te^{2-}) = 1.37 + 2.21 = 3.58$ Å. Therefore this Au—Te bond is more intermetallic than ionic: $r(Au) + r(Te) = 1.44 + 1.37 = 2.81$ Å (for coordination number 12, but in petzite Au has only a two-fold coordination). The silver atoms in petzite have a distorted tetrahedral surrounding as in hessite, Ag_2Te, (average Ag—Te = 2.94 Å; FRUEH, SR 1959, 240). Au may replace some Ag in hessite.

All other minerals contain only very minor amounts of Au. Older analyses tend to be significantly higher in gold than more recent analyses (including those by neutron activation) of the same minerals (for a compilation of analyses upto 1967, see JONES and FLEISCHER, 1969). Roughly one may state that the gold content of sulfides is in the ppm region and of oxides and silicates in the ppb region. Nevertheless gold commonly occurs in sulfide minerals, but largely, if not entirely, as the free metal; it is uncertain whether any gold occurs in these minerals in true isomorphous substitution.

Manuscript received: June 1971

References: Section 79-A

CABRI, L. J.: Phase relations in the Au—Ag—Te system and their mineralogical significance. Econ. Geol. **60**, 1569 (1965).

JONES, R. S., FLEISCHER, M.: Gold in minerals and the composition of native gold. U.S. Geol. Surv. Circ. **612** (1969).

PALACHE, C., BERMAN, H., FRONDEL, C.: Dana's System of Mineralogy, vol. 1 (7th ed.). New York: John Wiley & Sons, 1944.

RAMDOHR, P.: A widespread mineral association, connected with serpentinization. Neues Jahrb. Mineral., Abhandl. **107**, 241 (1967).

SHANNON, R. D., PREWITT, C. T.: Effective ionic radii in oxides and fluorides. Acta Cryst. B **25**, 925 (1969).

STRUNZ, H.: Mineralogische Tabellen (5. Aufl.). Leipzig: Akad. Verlagsges. Geest & Portig 1970.

TERZIEV, G.: Kostovite, a gold-copper telluride from Bulgaria. Am. Mineralogist **51**, 29 (1966).

TUNELL, G., PAULING, L.: The atomic arrangement and bonds in the gold-silver ditellurides. Acta Cryst. **5**, 375 (1952).

WISE, E. M.: Gold: Recovery, Properties and Applications. New York: D. Van Nostrand Co. Inc. 1964.

Manuscript received: June 1971

General Introduction

Some previous reviews of the geochemistry of gold include the following: FERSMAN (1939), ZVYAGINTSEV (1941), RANKAMA and SAHAMA (1950), GOLDSCHMIDT (1954), PHAN (1965), SHCHERBAKOV (1967), JONES (1969), JONES and FLEISCHER (1969), JONES (1970) and EHMANN (1971). Works containing large quantities of geochemical data on gold with reference to a specific topic include: RAZIN and ROZHKOV (1966; biogeochemistry, Yakutia district, U.S.S.R.), MOISEENKO *et al.* (1971; rocks of the Pacific Ocean belt, U.S.S.R.) and GOTTFRIED *et al.* (1972; igneous rocks of the U.S.). An annotated bibliography of 204 references dealing with gold geochemistry was published by COOPER (1971).

Analytical methods widely used for gold geochemical studies are briefly noted in Table 79-1.

Table 79-1. *Analytical methods frequently used in gold geochemistry studies*

Method	Sensitivity	Comment
d—c arc emission spectrography	1 ppb (occasionally to 0.2 ppb)	wet chemical preliminary concentration of gold is common; procedure of VOSKRESENSKAYA *et al.* (1965) is widely used
Atomic absorption	20 ppb	extraction of gold into methyl isobutyl ketone often used (THOMPSON *et al.*, 1968; VAN SICKLE and LAKIN, 1968; HUFFMAN *et al.*, 1967)
Neutron activation	0.05 ppb	commonly carrier-based, wet chemical methods (ROWE and SIMON, 1968, CROCKET *et al.*, 1968) are used. Substoichiometric techniques are used by ALIMARIN and PEREZHOGIN (1965), BEARDSLEY *et al.* (1965), and GILLUM and EHMANN (1972). Nondestructive methods are successful with iron meteorites (COBB, 1967; LINN and MOORE, 1967) and epithermal activation has been used for biogeochemical samples (LOBANOV *et al.*, 1966)

Revised manuscript received: June 1973

79-B. Isotopes in Nature

Gold is mono-isotopic in nature consisting of the single nuclide ^{197}Au (atom percent — 100%; atomic weight — 196.967 gm/mole).

Revised manuscript received: June 1973

79-C. Meteorites, Tektites and Lunar Materials

I. Meteorites

Reviews of the geochemistry of gold in meteorites have been presented by Jones (1968), Crocket (1970) and Ehmann (1971). The bulk of the original data are in the papers of Goldberg *et al.* (1951), Fouché and Smales (1966, 1967), Cobb (1967), Ehmann *et al.* (1970), Laul *et al.* (1972), and Ehmann and Gillum (1972). Early measurements including those of Noddack and Noddack (1930), and Goldschmidt and Peters (1932) have been summarized by Goldschmidt (1954).

Gold values for approximately 275 meteorites are included in the present survey. Averages for the main meteorite classes are given in Table 79-C-1. It is seen that gold increases from approximately 0.010 in achondrites to 1 ppm in irons. The Ca-rich and Ca-poor achondrites have similar average values, 0.012 and 0.0145 ppm respectively, although no genetic significance is attached to this due to the uneven representation of samples from the two types. A general feature of gold data for achondrites is poor analytical reproducibility due probably to inhomogeneous gold distribution arising from concentration of the metal in the small amounts of nickel-iron found in achondrites. Wänke *et al.* (1970) obtained values of 0.76, 1.64 and 1.38 ppm Au for nickel-iron grains separated from the achondrites Juvinas, Norton County 1 and Norton County 2 respectively. Such concentrations account for total gold in many achondrites if average metal is taken as 1.65 wt.-% (weighted average of metal contents for achondrites listed in Table 4—5 in Vol. I of this handbook).

Average gold for different classes of chondrites vary by only a factor of two, 0.17 to 0.34 ppm, decreasing from the metal-rich enstatite chondrites through the H and L classes to lowest values in the metal-poor carbonaceous chondrites. Thus chondrites exhibit a general correlation of gold with nickel-iron, reflecting the strong siderophile properties of gold in meteorites. Another aspect of siderophile behavior is the strong partition of gold into the metal phase of chondrites. An average metal-silicate partition of 77 is found for 26 chondrites (Table 79-C-2), a figure which is probably a minimum due to incomplete separation of metal from silicate. Further, for the enstatite chondrites Ehmann *et al.* (1970) obtained a correlation coefficient $r = +0.85$ for gold with total iron and metal phase abundances. Fouché and Smales (1967) found 2.21 ± 0.28, 1.37 ± 0.20 and 2.32 ± 0.37 ppm Au in separated nickel-iron from E, H and L-type chondrites for 4, 8 and 9 samples respectively, while Ehmann *et al.* (1970) found the gold content of separated chondrite metal followed the order $E < H < L$ (two chondrites from each class only). Fouché and Smales point out that the metal phase of chondrites appears to be much more uniform in gold than iron meteorites. Although there is some discrepancy among investigators and the number of analyses is limited, the gold content of chondritic iron is suggestive of a Prior's Law relationship with the nickel-iron of the most metal-poor class (L-type) highest in gold. Inspection of the atomic abundances in Table 79-C-1 indi-

Table 79-C-1. *Average gold content of meteorites.* (N = Number of samples)

Class	N	Gold (ppm)[a] range	Gold (ppm)[a] average	Atomic abundance[aa]	Reference
Chondrites[b]					
Carbonaceous					
C1	3	0.10—0.16	0.14		4, 6, 7, 8
C2	6	0.14—0.23	0.18		1, 4, 6, 7, 8, 15, 20, 24
C3	6	0.13—0.23	0.19		4, 6, 7, 8, 15, 20, 21
C4	1	—	0.15		7, 20
All carbonaceous chondrites, weighted average	16	0.10—0.23	0.17	0.18 (15)	
Enstatite					
E4	2	0.39—0.39	0.39		1, 4, 6, 7, 8
E5	2	0.33—0.38	0.355		1, 6, 7
E6	5	0.25—0.41	0.31		1, 6, 7, 8, 20
All enstatite chondrites, weighted average	9	0.25—0.41	0.34	0.27 (9)	
LL Group	6	0.13—0.23	0.16	0.12(5)	6, 7, 14, 16, 21, 22, 23
Ordinary chondrites					
L Group					
L3	4	0.13—0.24	0.17		6, 14, 22, 23
L4	18	0.10—0.26	0.16		6, 14, 22, 23, 24, 25
L5	7	0.13—0.21	0.16		6, 8, 14, 16, 21, 22, 23
L6	21	0.11—0.33	0.19		1, 6, 8, 11, 14, 16, 21, 22, 24, 25
All L group chondrites, weighted average	40	0.11—0.33	0.18	0.13(24)	
H Group					
H3	2	0.23—0.23	0.23		8, 20
H4	5	0.19—0.24	0.22		6, 8, 20, 24, 25
H5	13	0.17—0.30	0.23		1, 6, 8, 20, 22, 25
H6	4	0.15—0.22	0.20		6, 8, 22, 23
All H group chondrites, weighted average	24	0.15—0.30	0.22	0.19(14)	
Achondrites					
Ca-poor including 2 aubrites, 1 urelite and 1 diogenite	4	0.004—0.030	0.012		1, 7, 17, 24
Ca-rich including 1 angrite, 12 eucrites, 9 howardites, 2 nakhlites and 2 shergottites	26	0.00033—0.089	0.0145		1, 7, 16, 17, 21, 26

Table 79-C-1 (continued)

Class	N	Gold (ppm)[a] range	Gold (ppm)[a] average	Reference
Irons[c]				
Hexahedrites				
H	23	0.22—1.65	0.74	2, 3, 5, 9, 10, 12, 18, 23
Og and Ogg	28	0.28—2.28	1.27	1, 2, 3, 5, 9, 10, 12, 13, 16, 23, 24
Octahedrites				
Om	51	0.31—3.19	1.14	2, 3, 5, 9, 12, 18, 23
Of and Off	26	0.61—2.35	1.32	2, 5, 9, 10, 18, 23
Ni-rich ataxites, D	12	0.067—3.71	1.33	2, 3, 5, 9, 13, 16, 18
All irons, weighted average	140	0.067—3.71	1.15	
Stony Irons				
Pallasites and mesosiderites	7	0.27—3.19	1.63	2, 3, 9, 18

[a] All determinations by neutron activation.
[b] Classification of the chondrites according to VAN SCHMUS and WOOD (1967).
[c] Classification of the irons according to HEY (1966), Nickel-poor ataxites are included with the hexahedrites. Irons designation by HEY as metabolites or as brecciated octahedrites are not included in the compilation.
[aa] Atoms Au/10^6 Si atoms; Si data from WIIK (1956), VOGT and EHMANN (1965), and EHMANN and DURBEN (1968); figure in parentheses is the number of meteorites on which the average is calculated.

References: 1. BAEDECKER and EHMANN (1965). 2. BAUER and SCHAUDY (1970). 3. COBB (1967). 4. CROCKET *et al.* (1967). 5. CROCKET (1972). 6. EHMANN and GILLUM (1972). 7. EHMANN *et al.* (1970). 8. FOUCHÉ and SMALES (1967). 9. FOUCHÉ and SMALES (1966). 10. GOLDBERG *et al.* (1951). 11. HECHT and KIESL (1971). 12. HERMANN *et al.* (1971). 13. HÖFLER and SORANTIN (1967). 14 KEAYS *et al.* (1971). 15. KIESL (1969). 16. KIESL *et al.* (1967). 17. LAUL *et al.* (1972). 18. LINN and MOORE (1967). 19. MALISSA *et al.* (1972). 20. RIEDER and WÄNKE (1969). 21. SCHAUDY *et al.* (1967). 22. SCHAUDY *et al.* (1968). 23. SEITNER *et al.* (1971). 24. SHCHERBAKOV and PEREZHOGIN (1964). 25. VINCENT and CROCKET (1960). 26. WASSON and BAEDECKER (1970).

cates that the gold-silicon fractionation occurs in the chondrites with the Au/Si ratio decreasing as follows through the various chondrite classes: $E > H \simeq C > L \simeq LL$. LARIMER and ANDERS (1967, 1970) have considered chemical fractionations of the siderophile elements during condensation of a primitive solar nebula, and they argue that metal phase-silicate phase partition would result from the ferromagnetic properties of the metal. Such fractionation would have an important effect on gold distribution within the nebula due to strong preferential concentration of gold in the metallic phase. Based on their studies of platinum and gold in chondrites, EHMANN and GILLUM (1972) consider that such metal-silicate fractionation is adequate

Table 79-C-2. *Gold content of and fractionation between meteorite mineral phases.* (N = number of samples)

Class	Mineral phases	N	Gold (ppm) range	Gold (ppm) average	Fractionation factor	Reference[a]
Irons	Sulfide nodules	10	<0.02—0.08	<0.06	Whole meteorite / Sulfide nodules >28[b]	18
Iron	Troilite[c]	6	0.002—0.008	0.0046	Metal phase / Troilite = 335	12
Irons	Graphite[aa]	3	0.0015—0.003	0.0023	Metal phase / Graphite = 753	12
Irons	Phosphide[aa] (schreibersite)	2	0.15—0.179	0.165	Metal phase / Phosphide = 9	12
Chondrites	Magnetic (metal)	26	0.69—2.90	1.78	Magnetic / Nonmagnetic = 77	7,8
	Nonmagnetic (silicate)	26	0.0048—0.091	0.023		

[a] References are the same as for Table 79-C-1.
[b] The partition ratio is a minimum value as all gold analyses except one are reported as maximum values only. Analyzed meteorites include one pallasite.
[c] Troilite separates are from Magura, Brownfield, Cañon Diablo, Mount Joy, Odessa and Xiquipilco (Toluca). Data of Ehmann *et al.* (1970) and Shcherbakov and Perezhogin (1964) not used. See text for discussion.
[aa] Graphite separates are from Cañon Diablo, Odessa and Xiquipilco and the phosphide separates from Cañon Diablo and Odessa.

to account for the observed fractionation of Pt and Au in C, L and H group chondrites, although a more complex process is needed to explain the noble metal content of the enstatite chondrites.

Iron meteorites average 1.2 ppm gold and, as a group, show little variation in gold content among the main structural classes. The hexahedrites are slightly lower in gold than the octahedrites, while the ataxites exhibit a wider range of values than other classes. The siderophile behavior of gold is evident from the strong partition of the element into the metal phase in preference to silicates, sulfides, graphite or phosphides (see Table 79-C-2). Data on the gold content of separate mineral phases of iron meteorites have been obtained by Shcherbakov and Perezhogin (1964), Baedecker and Ehmann (1965), Linn and Moore (1967), Ehmann *et al.* (1970) and Hermann *et al.* (1971). Large differences in the gold content of troilite are reported by different laboratories, possibly due to metal contamination in the sulfide separates. The data of Hermann *et al.* (1971) average 0.0046 ppm for 6 troilite samples whereas previous determinations are much higher (0.10, 0.23 and 0.067 ppm for troilite from Cañon Diablo, Sardis and Sikhote-Alin respectively, according to Baedecker and Ehmann, 1965, Ehmann *et al.*, 1970, and Shcherbakov and Perezhogin, 1964). Linn and Moore (1967) obtained lower values averaging about 0.06 ppm for 10 sulfide nodules using a nondestructive activation method. Their

Fig. 79-C-1. Fields of iron meteorites for resolved chemical groups I, IIA ,IIIA, IIIB and IVA (WASSON and coworkers) in gold *vs.* nickel coordinates

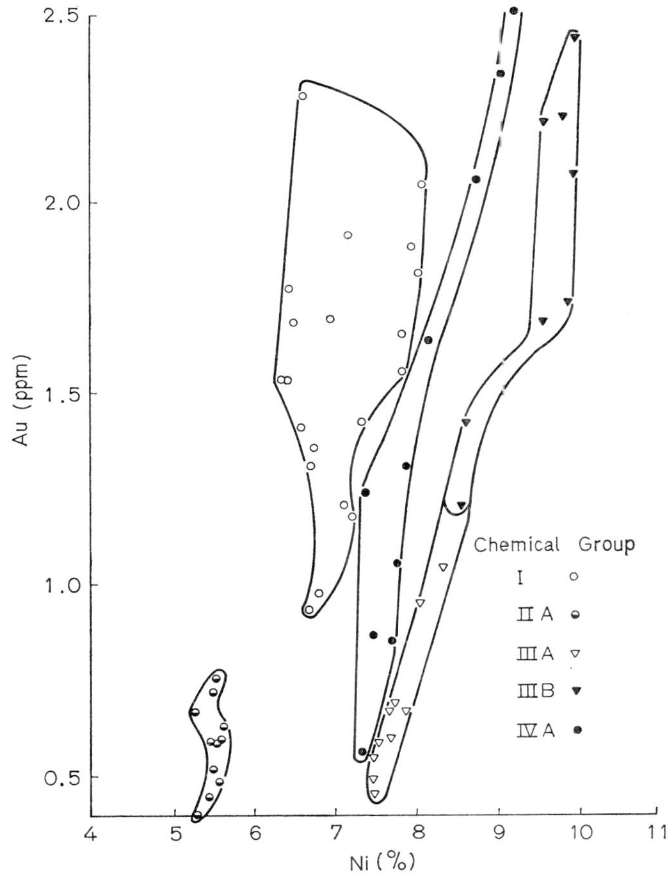

sensitivity, however, appears to be approximately 0.02 ppm and they report maximum values only for the majority of their samples. Although all troilite gold analyses indicate a depletion of the element in the sulfide with respect to co-existing metal, the recent study of HERMANN *et al.* (1971) suggests that the magnitude of the effect is much greater than indicated by previous data. They found that gold is strongly concentrated in the metal phase of iron meteorites relative to troilite, graphite and schreibersite. For troilite and graphite, very large average metal phase enrichments (335 and 753-fold respectively) are computed from their data (Table 79-C-2). Based on two schreibersite analyses, it appears that the phosphide phase is considerably higher in gold than either graphite or troilite, with only a 10-fold depletion relative to the metal phase. A study of gold distribution in the Sikhote-Alin iron by SOBOTOVICH *et al.* (1971) indicated that the metal is carried by the taenite rather than the kamacite phase.

Correlations of 17 elements in a suite of 47 iron meteorites were presented by CROCKET (1972). For the noble metals, Ga, Ge, As, Sb, Cu, Cr, Mo, In, Zn and Re, it was found that gold correlates strongly only with As and Pd ($r = +0.83$ and $+0.74$ respectively, significance level of 1%). Correlations with other noble metals, Cu and Ni were negative or very weak.

A significant geochemical feature of gold may be recognized by plotting Au *vs.* Ni for those irons included in one of the resolved chemical groups recognized by WASSON and co-workers (see WASSON and SCHAUDY, 1971, and references therein to earlier papers). It is argued that iron meteorites may be classified according to structural and chemical properties into groups of genetically-related meteorites. Important elements providing a basis for this chemical subdivision are Ga, Ge, Ir and Ni. In Fig. 79-C-1 gold-nickel values for meteorites from resolved chemical groups I, IIA, IIIA and IIIB and IVA are plotted. These meteorites are seen to occupy separate fields in Au *vs.* Ni coordinates with little mutual overlap, suggesting that the gold distribution in the resolved chemical groups of WASSON is compatible with the contention that these groups include suites of genetically-related meteorites.

II. Tektites

The gold content of 46 tektites have been determined by BAEDECKER and EHMANN (1965), EHMANN and SHOWALTER (1971), and DEGRAZIA and HASKIN (1964). These data which average 4.1 ppb are presented in Table 79-C-3. The main group of samples include 36 australites for which EHMANN and SHOWALTER (1970) obtained an average value of 3.4 ppb gold. Tektites are commonly ascribed to impact on either the lunar or terrestrial surface. Argillaceous and acid igneous rocks are often cited as possible terrestrial source rocks. Recognizing the limited significance of a single trace constituent in regard to tektite origin, it is noted that the average gold content

Table 79-C-3. *Average gold content of tektites*

Type	Number of samples	Gold (ppb)		Reference
		range	average	
Australites	36	0.1—18.3	3.4	2
Australites	3	1.5—13.2	5.7	1
Billitonite	1		24.2	1
Georgia tektite	1		6.7	1
Javaite	1		1.3	1
Moldavite	1		3.9	1
Philippinite	1		11.8	1
Bediasite	1		0.3	1
Indochinite	1		1.3	1
Bediasite	1		3.0	3
Philippinite	1		6.7	3
Weighted average, all tektites			4.1	

References: 1. BAEDECKER and EHMANN (1965). 2. EHMANN and SHOWALTER (1970). 3. DEGRAZIA and HASKIN (1964).

Table 79-C-4. *Gold content of lunar samples*

Sample[a]	LRL designation[b]	Sample No.	Au (ppb)	Reference[c]
Apollo 11 basalts	A/B ol. basalt	10017	0.72	1
	A ol. basalt	10020	0.075	4, 7
	A subophitic ol. basalt	10022	0.10	5
	B cristobalite basalt	10047	0.029	1
	B cristobalite basalt	10050	0.031	4
	A granular basalt	10057	0.015	1
	B cristobalite basalt	10058	0.72	13
	A olivine basalt	10072	0.14	1, 13[aa]
	Average Apollo 11 basalts		0.23	
Apollo 12 basalts	B porphyritic basalt	12002	0.024	3
	A porphyritic vitrophyre	12009	0.06	8
	A/B porphyritic basalt	12017	0.072	1
	A porphyritic basalt	12020	0.063	1, 8
	B ophitic basalt	12040	0.012	1
	A/B ophitic basalt	12051	0.0075	3
	A porphyritic basalt	12052	0.08	8
	A porphyritic basalt	12063	0.05	8
	A/B porphyritic basalt	12065	0.012	1
	Average Apollo 12 basalts		0.042	
Apollo 11 and 12 breccias	C	10018	5.0	12
	C	10021	2.4	13
	C	10046	2.8	13
	C	10048	2.3	4, 5, 7, 13
	C	10056	0.61	13
	C	10060	1.4	13
	C	10061	3.4	4, 7
	C	12010	0.82	10
	C	12013	0.35	1
	C	12073	2.0	10
	Weighted average, Apollo 11 and 12 breccias		1.7	
Lunar regolith, Apollo 11, Apollo 12 and Luna 16	D	10084	2.3	4, 5, 7, 12, 13
	D	12001	2.9	8
	D	12025	2.5	10
	D	12028	1.6	3
	D	12032	1.4	10
	D	12033	1.5	10
	D	12037	1.5	10
	D	12044	2.5	8
	D	12057	5.7	10
	D	12070	3.5	2, 5, 8
	—	Luna 16	2.3	6, 9, 11
	Average, all lunar regolith samples		2.5	

[a] Location of samples: Apollo 11 = Mare Tranquillitatis at 0.67° N, 23.49° E. Apollo 12 = Oceanus Procellarum at 2.45° S, 23.34° W. Luna 16 = Mare Fecunditatis at 0° 41′N, 56° 18′E.

[b] Lunar Receiving Laboratory designations are: A = fine grained crystalline rock; B = medium-grained crystalline rock; C = breccia; D = fines or lunar soil (fragments < 1 cm diameter). Apollo 11 und 12 rock names are after SCHMITT *et al.* (1970) and according

Table 79-C-4 (continued)

to Apollo 12 Lunar Sample Inventory (Proc. Second Lunar Sci. Conf., **1**, p. 983, 1971).

[c] For the crystalline rocks, Types A and B, interlaboratory comparison of Au data is poor, particularly for Apollo 11 samples. There is good evidence that the average Au content of Apollo 11 and 12 lunar basalts is less than 1 ppb. Analyses reporting greater than 1 ppb are excluded if a second laboratory has obtained a much lower result for the same rock (although not the same chip). Some powders prepared by the Lunar Receiving Laboratory are thought to be contaminated with respect to Au (Ganapathy *et al.*, 1970b) and data for these samples were also excluded. For breccias and soils where high gold contents are found, interlaboratory comparison is much better and all data have been used.

[aa] The value listed is the average of those reported by the authors noted. In general the average applies to different chips or splits from the same sample.

References: 1. Anders *et al.* (1971). 2. Brunfelt *et al.* (1971). 3. Ganapathy *et al.* (1970a). 4. Ganapathy *et al.* (1970b). 5. Haskin *et al.* (1970). 6. Jerome *et al.* (1972). 7. Keays *et al.* (1970). 8. Kharkar and Turekian (1970). 9. Laul *et al.* (1972). 10. Laul *et al.* (1971). 11. Vinogradov (1971) 12. Wänke *et al.* (1970). 13. Wasson and Baedecker (1970).

of acid igneous rocks (Table 79-E-2) is approximately half that of the australite suite analyzed by Ehmann and Showalter whereas average shale (Table 79-K-3) is virtually identical. Apollo 12 lunar basalts are depleted in gold by a factor of approximately 100 relative to tektites.

III. Lunar Materials

Table 79-C-4 presents a summary of gold analyses on rocks, breccias and regolith returned from Apollo 11 and 12 flights and Luna 16 automatic probe. According to Anders *et al.* (1971), the lunar basalts are lower in gold than their terrestrial counterparts by a factor of about 100. This is apparent from Table 79-C-4 for the Apollo 12 basalts although not for those from Apollo 11. However, the difference in average gold for these two suites is regarded as due largely to analytical error inherent in the determination of extremely low gold values (see footnote [b], Table 79-C-4). Anders *et al.* (1971), in comparing lunar and terrestrial abundance patterns for siderophile elements including gold, suggest that the best explanation for depletion in lunar rocks is a lower oxygen fugacity during metal-silicate segregation of primitive lunar material. In a more strongly reducing environment the siderophile elements should concentrate to a higher degree in the metal phase.

The lunar breccias and fines contain much higher gold contents than lunar rocks. Consideration of the abundance patterns of some 15 elements including gold and other siderophile metals by Keays *et al.* (1970), Laul *et al.* (1971) and others suggest that the lunar soils contain a meteoritic component responsible for enrichment of soil in certain elements including gold. In addition Wänke *et al.* (1970) separated metallic iron particles from a sample of lunar fines (10084—18) which contained 830 ppb Au. While they do not conclude that these particles are necessarily extralunar, it is apparent that admixture of a few percent of such a component accounts for the entire gold content of typical lunar fines.

Revised manuscript received: June 1973

79-D. Rock-Forming Minerals, Gold Minerals

I. Rock-Forming Minerals

The main sources of data on the gold content of rock-forming minerals are VINCENT and CROCKET (1960a), SHCHERBAKOV and PEREZHOGIN (1964), SHCHERBAKOV (1967), MANTEI and BROWNLOW (1967), ANOSHIN *et al.* (1970), DAVLETOV (1970), DAVLETOV and DZHAKSHIBAYEV (1970), ZVEREVA and GAVRILENKO (1971), and GOTTFRIED *et al.* (1972). Data are summarized in Table 79-D-1, and weighted averages compared in Table 79-D-2 with other averages from the literature (SHCHERBAKOV and PEREZHOGIN, 1964; SHCHERBAKOV, 1967; DAVLETOV, 1970; TILLING *et al.*, 1973). Mineral data for individual intrusions or related plutons are shown graphically in Fig. 79-D-1.

These data suggest two main generalizations: 1. average gold in felsic minerals (quartz, K-feldspar, plagioclase) is lower than in mafic minerals (biotite, hornblende, pyroxenes), and 2. the gold content of accessory minerals such as sphene and mag-

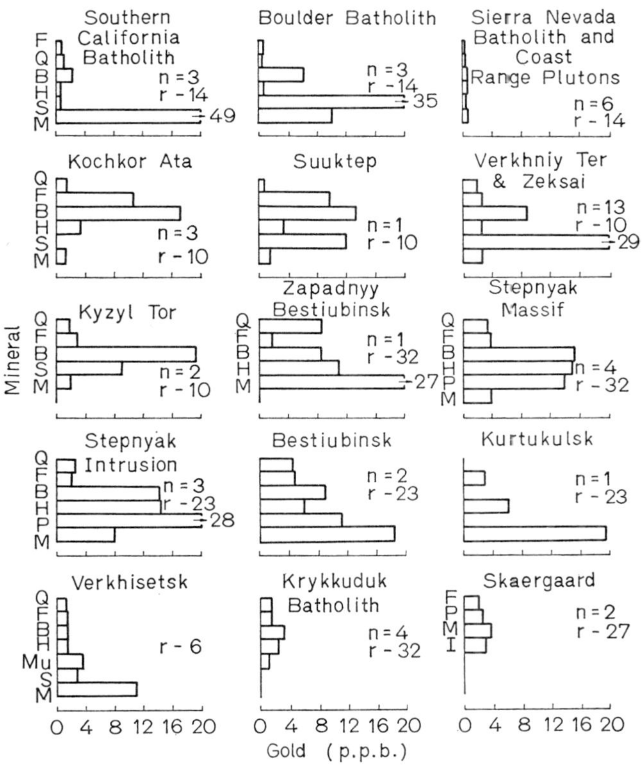

Fig. 79-D-1. Gold contents of minerals of individual plutonic intrusives. *Q* Quartz; *F* Feldspar; *B* Biotite; *H* Hornblende; *S* Sphene; *P* Pyroxene; *Mu* Muscovite; *I* Ilmenite; *M* Magnetite; *n* number of samples; *r* reference, footnotes Table 79-E-1

Table 79-D-1. *Gold content in rock-forming minerals*

Mineral	Host rock and locality	Number of samples	Gold (ppb) range	Gold (ppb) average	Method	Reference[a]
Quartz	Granite; Kolyvan' massif (USSR)	6	0.2—1.1	0.8	N	3
	Granitoids; Chatkal, Tien Shan (USSR)	36	$\sigma = \pm 0.12$[b]	1.7	N	9
	Gabbro; Stepnyak, Bestyubinsk massifs, Kazakhstan (U.S.S.R.)	8	1.5—5.4	4.0	S, N	32
	Granitoids; Verkhisetsk massif, Urals (U.S.S.R.)	6	0.7—2.8	1.7	N	6
	Granite, granodiorite, tonalite, schists; U.S.S.R.	7	1.1—8	2.6	N, S	20, 13, 32, 17
Quartz and Plagioclase	Granodiorite, quartz monozonite, quartz diorite, alaskite; California, Montana (U.S.A.)	12	0.1—3.4	0.3	N	14
Feldspar	Granite; Kolyvan' massif (U.S.S.R.)	7	0.2—0.5	0.3	N	3
	Granitoids; Chatkal, Tien Shan (U.S.S.R.)	32	$\sigma = \pm 0.14$[b]	1.7	N	9
	Granites (mainly); Altai-Sayan (U.S.S.R.)	20	0.2—7.9	3.3	N	20
	Gabbro; Stepnyak, Bestyubinsk massifs, Kazakhstan (U.S.S.R.)	8	1.9—7.9	2.8	S, N	23, 32
	Granitoids; Verkhisetsk massif, Urals (U.S.S.R.)	19	0.6—5.6	1.7	N	6, 17
	Granite and granodiorite; Iultinsk massif (U.S.S.R.)	4	2—9	4.2	S	17
	Quartz monozonite, granodiorite, alaskite; California, Montana (U.S.A.)	8	0.02—1.4	0.5	N	14
	Gabbro; Skaergaard intrusion, East Greenland	2	1.4—2.7	2.0	N	27
	Gabbro, granodiorite, granite, tonalite (U.S.S.R.)	5	0.5—3.0	1.5	S, N	32, 23, 18
Biotite	Intrusive granitoids; Chatkal, Tien Shan (U.S.S.R.)	120	$\sigma = \pm 0.04$[b]	7.1	N	9
	Gabbro, diorite; Stepnyak and Bestyubinsk massifs, Kazakhstan (U.S.S.R.)	7	8.5—17.9	12.5	S, N	32, 23
	Quartz monzonite, granodiorite, quartz diorite, alaskite; California, Montana (U.S.A.)	11	0.1—15.8	2.6	N	14
	Altai-Sayan (U.S.S.R.)	4	1.7—6.5	4.0	N	20
	Granitoids; Verkhisetsk massif, Urals (U.S.S.R.)	24	0.2—6.4	1.8	N	6
	Granodiorite, tonalite, schist, U.S.S.R.	5	1.7—3.0	1.3	N	17, 18, 32
	Granite, granodiorite; Iultinsk massif, Chukotka (U.S.S.R.)	4	2.7—9	4.8	S	17

Hornblende	Intrusive granitoids; Chatkal, Tien Shan (U.S.S.R.)	55	$\sigma = \pm 0.05$[b]	4	N	9
	Granitoids; Verkhisetsk massif, Urals (U.S.S.R.)	14	0.7—4.0	1.6	N	6
	Gabbro; Stepnyak, Bestyubinsk massif, Kazakhstan (U.S.S.R.)	4	5.0—23.8	11.3	S, N	32, 23
	Quartz monzonite, granodiorite, quartz diorite; California, Montana, (U.S.A.)	10	0.2—1.3	0.6	N	14
Amphibole	Gabbro, diorite, nepheline syenite; Altai-Sayan (U.S.S.R.)	6	1.4—24	5.9	N	20
Hornblende	Gabbro, granodiorite, granite, amphibolite, U.S.S.R.	5	1.2—8.0	4.1	N, S	19, 23, 32, 17
Pyroxene	Gabbro; Skaergaard intrusion, East Greenland	2	2.0—3.0	2.5	N	27
	Gabbro; Stepnyak, Bestyubinsk massifs, Kazakhstan (U.S.S.R.)	9	6.1—23.0	13.5	S, N	32, 23
	Gabbro, peridotite; Altai-Sayan (U.S.S.R.)	4	3.5—40	16	N	20
Hypersthene	Quartz diorite; California (U.S.A.)	1	—	0.3	N	14
Sphene	Intrusive granitoids; Chatkal, Tien Shan (U.S.S.R.)	15	$\sigma = \pm 0.72$[b]	9.3	N	9
	Tonalite, quartz monzonite; California, Montana (U.S.A.)	2	0.5—34.6	17.5	N	14
	Kuznetsk Ala-Tau (U.S.S.R.)	1	—	3.9	N	20
	Granitoids; Verkhisetsk massif, Urals (U.S.S.R.)	5	1.3—7.5	2.7	N	6
Magnetite	Intrusive granitoids; Chatkal, Tien Shan (U.S.S.R.)	40	$\sigma = 0.10$[b]	3.3	N	9
	Gabbro; Stepnyak, Bestyubinsk massifs Kazakhstan (U.S.S.R.)	8	4.0—27.2	12.0	S, N	23, 32
	Granodiorite, tonalite; Krykkuduk massif, Kazakhstan (U.S.S.R.)	3	—	<1.0	S	32
	Hybrid gabbroid and amphibolite, U.S.S.R.	2	19.6—34.0	26.7	S, N	13, 23
	Quartz monzonite, granodiorite, alaskite; California, Montana (U.S.A.)	7	0.4—49.3	11.2	N	14
	Gabbro, diorite; Siberian platform and Gornaya Shoriya	3	13—100	48.3	N	20
Magnetite, ilmenite	Granitoids; Verkhisetsk massif, Urals (U.S.S.R.)	22	0.4—8.0	3.3	N	6
	Gabbro; Skaergaard intrusion, East Greenland	4	2.9—4.2	3.3	N	27
Muscovite	Greisen, schist, U.S.S.R.	2	3.8—10	6.9	S	17, 13
	Granitoids; Verkhisetsk massif (U.S.S.R.)	3	4.2—15.5	10.9	N	6
	Pegmatite, Altai (U.S.S.R.)	4	2.3—5.5	3.8	N	20
Chromite	Alpine peridotite; Mt. Albert pluton, Quebec (Canada)	4	2.6—5.4	3.4	N	7
Tourmaline	Pegmatite; Altai (U.S.S.R.)	2	6.7—18	12.5	N	20

Table 79-D-1 (continued)

Mineral	Host rock and locality	Number of samples	Gold (ppb)		Method	Reference[a]
			range	average		
Chlorite	Granitoids; Verkhisetsk massif (U.S.S.R.)	3	1.1—5.0	2.5	N	6
Epidote	Granitoids; Verkhisetsk massif (U.S.S.R.)	15	0.3—7.0	2.5	N	6
Apatite	Granitoids; Verkhisetsk massif (U.S.S.R.)	4	1.2—3.4	2.4	N	6
Accessory minerals: sphene, zircon, apatite, sulfides	Granitoids; Chatkal, Tien Shan (U.S.S.R.)	30	$\sigma = \pm 155$[b]	536	N	9

[a] For references see footnotes to Table 79-E-1.
[b] For this entry, $\pm\sigma$ is precision in ppb at the 0.95 significance level.

Table 79-D-2. *Comparison with previous estimates of the gold content of rock-forming minerals*

Mineral	Number of samples	Average gold in ppb	Reference[a]
Quartz	9	1.7	20
	55	1.0	9
	11	0.2	TILLING *et al.* (1973)[b]
	75	1.7	This compilation
Feldspar	20	3.3	20
	65	2.0	9
	8	0.4	TILLING *et al.* (1973)
	105	2.0	This compilation
Biotite	4	4.0	20
	134	3.0	9
	12	2.4	TILLING *et al.* (1973)
	175	6.0	This compilation
Hornblende	6	5.9	20
	59	3.2	9
	10	0.6	TILLING *et al.* (1973)
	94	3.7	This compilation
Magnetite, ilmenite	3	48	20
	52	3.9	9
	8	9.8	TILLING *et al.* (1973)
	89	6.7	This compilation

[a] References given in footnotes to Table 79-E-1.
[b] Averages read directly from Fig. 6 of TILLING *et al.* (1973).

netite is extremely variable and often very high compared to the parent whole rock.

Based on studies of individual or closely-related groups of rocks, the main generalizations presented in the literature include the following: a) VINCENT and CROCKET (1960a) found little variation in the gold content of plagioclase, pyroxene, magnetite and ilmenite from Skaergaard gabbros. This observation together with inferences drawn from the overall variation of gold in Skaergaard rocks and the likely oxidation potential of the magma lead them to conclude that the metal was present as uncharged species in the magma and probably occluded in defects in crystals.

b) From a study of rocks from the Altai-Sayan region, SHCHERBAKOV and PEREZHOGIN (1964) concluded that average gold decreased from magnetite and ferromagnesian silicates to feldspars. They suggested that correlations of gold with Mg, Fe, Cu and V and the relatively small dispersion in minerals such as olivines and pyroxenes, indicate the metal is crystallochemically bound in these silicates. The larger dispersion and higher average value in magnetite and quartz were considered to indicate that gold occurs as the native metal in these two minerals.

c) In a study of granitoids from the Tien Shan, DAVLETOV and DZHAKSHIBAYEV (1970) showed that feldspar is the main carrier mineral whereas biotite is always a concentrator mineral having the highest gold content of the principal rock-forming minerals. The accessory minerals, including sphene, zircon, apatite and others, were often very high in gold but never significant contributors to total whole rock gold.

d) Gold in minerals from a large granodiorite complex (Krykkuduk) and from small gabbro-diorite intrusives (Stepnyak type) of northern Kazakhstan was determined by ZVEREVA and GAVRILENKO (1971). Pyroxene, hornblende and biotite were found to be richer in gold than quartz and feldspar. The gold content of all mafic minerals was similar, suggesting that iron and magnesium are not correlated with gold. As the gold content of minerals appeared to depend mainly on that in the whole rock, they considered it probable that gold occurs in the rock-forming minerals as neutral atoms with enrichment controlled by structural factors.

e) GOTTFRIED *et al.* (1972) determined gold in minerals from the Southern California batholith, the Boulder batholith, Montana and the Sierra Nevada batholith and Coast Range plutons, California. On average they found higher gold in mafic minerals than in salic minerals. However, in felsic rocks the quartz-plus-feldspar fraction may account for up to 90 percent of whole rock gold. The positive correlation of gold content with abundance of mafic silicates led them to suggest that for rock suites of calc-alkalic affinity, gold becomes depleted in the residual melt as magmatic differentiation proceeds.

A study not previously noted is that of MANTEI and BROWNLOW (1967) on the gold content of minerals from the Marysville quartz diorite stock, Montana (U.S.A.) an economic gold producer and possible satellitic stock of the nearby Boulder batholith. They used a neutron activation method to analyze 135 samples with the following results: biotite, 76; hornblende, 100; magnetite, 37; and quartz-feldspar, 65 ppb gold. The results are unusually high (compare Table 79-C-1) and are particularly interesting in the light of additional work on the same samples by TILLING *et al.* (1973), who removed or reduced impurities including chlorite, epidote, allanite and minor amounts of opaque minerals (magnetite and/or sulfides). TILLING *et al.* also used an activation method. Although they compare favourably with MANTEI and BROWNLOW for whole rock gold determination, the results on biotite are much lower, by approximately a factor of 55, while the results on magnetite are higher by about 40. The studies tentatively indicate very inhomogeneous gold distribution on a small scale, with accessory minerals or inclusions in the rock-forming minerals containing much higher gold than the parent whole rock.

II. Gold Minerals

The few known gold minerals are listed in Table 79-D-3.

Table 79-D-3. *Gold minerals*

Gold	Au
Electrum	(Au, Ag) (≥25% Ag)
Auricupride	$AuCu_3$; Au_2Cu_3
Rhodite	(Au, Rh)
Porpezite	(Au, Pd)
Goldamalgam	Au_2Hg_3
Montbrayite	Au_2Te_3
Nagyagite	$AuTe_2 \cdot 6\,Pb(S, Te)$ (?)
Sylvanite	$AuAgTe_4$
Krennerite	$(Au, Ag)Te_2$ (orthorhomb.)
Calaverite	$(Au, Ag)Te_2$ (monocl.)
Petzite	Ag_3AuTe_2
Aurostibite	$AuSb_2$
Maldonite	Au_2Bi

Revised manuscript received: June 1973

79-E. Abundance in Common Igneous Rocks

There are some 2500 quantitative analyses of gold in common igneous rocks, mostly by spectrochemical and neutron activation methods. Much of this data is presented in Table 79-E-1. Other data not considered representative of common igneous rocks or of doubtful analytical reliability are discussed separately in the text.

To the extent permitted by literature descriptions, rocks have been divided into very broad petrochemical types including salic, intermediate, mafic and ultramafic. In the absence of major element data, some assignments (for example, quartz diorite which may be either salic or intermediate) are probably dubious. To condense the large volume of data, it has been necessary in general to compute averages from original literature; usually for all analyses of the same rock type (i.e., granite, andesite, etc.) from the same pluton or petrographic province or suite, a weighted average is given in the table.

The compilation does not include pre-1960 data which may be found in the compilation of Jones (1969). In comparison with recent data, the early work in general over-estimates the gold content of common igneous rocks, sometimes grossly; its inclusion in averages here would erroneously bias them to high values.

Average values for the gold content of major rock types computed from Table 79-E-1 are presented in Table 79-E-2.

Some literature reporting high gold values is discussed below. Buturlintov and Latysh (1970) obtained results ranging from 10 to 220 ppb on 30 of 67 samples of andesite-trachyandesite, alkalic ultrabasics and lamprophyres from the Donets Basin, U.S.S.R.; samples from regions of widespread metasomatism contained the highest gold contents. Borisenko and Uskov (1971) and Borisenko *et al.* (1972) determined gold in ultramafic rocks in 16 differentiated massifs from the Urals and 2 from Armenia, U.S.S.R., by a spectrographic method with a sensitivity of 1 ppb. For the Gusevogorsk massif, Urals, they obtained averages of 140, 80, 80 and 120 ppb for 131 diallage and hornblende pyroxenites, 62 olivine pyroxenites, 7 wehrlites and 3 hornblendites respectively. For samples from 15 other Urals and 2 Armenian massifs, Borisenko *et al.* (1972) obtained averages of 115, 88, 66 and 95 ppb for 33 pyroxenites and hornblendites, 18 peridotites, 14 dunites and olivinites, and 10 serpentinized ultramafics respectively. Many of these samples contained separable particulate gold. The authors suggest that an average gold content of 50—100 ppb is applicable to Urals ultramafics and this distinguishes the province as one of unusually high gold content. Other investigators (Moiseenko *et al.*, 1971) also found high gold in ultramafics described as ore-bearing from the Kachkanar region, Urals. For 26 rocks, mostly pyroxenite, they obtained a range of 0.5 to 1300 ppb with an average of 17 ppb gold. These data are not tabulated in Table 79-E-1.

Table 79-E-1. *Gold abundance in common igneous rock-types*

Rock type	Location and host	Number of samples	Gold (ppb) range	Gold (ppb) average	Method	Reference
	Plutonic rocks					
	Salic rocks					
Granite	Altai and East Transbaikalia (U.S.S.R.); Kolyvan' and Khangilay-Shilinskiy massifs	27	0.5—2.3	1.2	N	3
Granite	Altai-Sayan and Kuznetsk Ala-Tau (U.S.S.R.); Sarakokshinsk, Tashtuzek and Dudetsk massifs; Turochak and Chingekat granites; Mt. Sukhaya	10	1.2—4.2	2.5	N	20
Granite	Idaho and Colorado (U.S.A.); mainly Idaho batholith, ca-prov.[a]	8	0.2—0.8	0.5	N	14
Granite	New Hampshire (U.S.A.); White Mountain plutonic series, a-prov.[b]	21	0.2—1.6	0.6	N	14
Granite	South Carolina, (U.S.A.); a-prov.	1		0.2	N	14
Granite	Tien Shan (U.S.S.R.); Verkhniy Ter, Zeksai and Kyzyl Tor intrusions	7	1.0—4.7	2.6	N	10
Granite	Georgia and Wisconsin, (U.S.A.); Algoma (Canada)	3	1.9—4.7	3.3	N	11
Granite	Northern Kazakhstan (U.S.S.R.); Krykkuduk and Aralaul' complexes	67	—	1.6	S	29
Granite	Chukotka, Magdan and Priamurya regions, Far East (U.S.S.R.); Iultinsk, Taureransk, Velitkenaysk, Emnunkenigtunsk, Magdansk, Ingaglinsk, Stoibinsk, Karauraksk and Bryus massifs	107	0.1—10	1.8	S, N	17
Granite porphyry, quartz porphyry	Amur region, Far East (U.S.S.R.)	12	0.4—3.6	1.6	S	17

Granite and granosyenite	Chukotka, Altai and Kuznetsk Ala-Tau (U.S.S.R.); Peveksk, Paliansk, Aturkolsk and Dudetsk massifs	14	0.46—6.6	2.0	N	20
Granite	Urals (U.S.S.R.); Suunduksk and Verkhisetsk massifs	22	0.5—13.0	2.3	N	1, 6
Granite	Nura-Tau, Uzbekistan (U.S.S.R.); Akchopsk, Ustuksk, Taushansk massifs	17	1.2—20	6.1	W[aa]	15
Granite	Central Kamchatka (U.S.S.R.); gabbro-granodiorite association	27	—	15	S	22
Granite	Kazakhstan (U.S.S.R.); Stepnyak intrusives, Krykkuduk complex	5	2.2—8.5	3.9	S	29
Granite, granophyre	Big Bend National Park, Texas (U.S.A.); b-a-r[c] assoc., a-prov.	6	0.2—1.2	0.55	N	14
Granodiorite	S. California, Sierra Nevada, Idaho and Boulder batholiths (U.S.A.); Coast and Transverse ranges; Tatoosh pluton	60	0.1—5.2	1.3	N	14
Granodiorite and adamellite	Chatkal, Tien Shan (U.S.S.R.); Verkhniy Ter and Zeksai intrusions	6	2.0—7.1	3.6	N	10
Granodiorite	Altai, Kuznetsk Ala-Tau, Kamchatka (U.S.S.R.)	3	3.1—4.8	4.0	N	19, 20
Granodiorite	Kazakhstan (U.S.S.R.); Krykkuduk, Aralaul' and Borovsk complexes	84	—	1.5	S	29
Granodiorite	Marysville stock, Montana (U.S.A.); unmineralized rock excluding samples within 200 meters of gold-bearing veins or mines	47	1—10	2.7	N	16
Granodiorite	Central Kamchatka (U.S.S.R.); gabbro-granite association	98	—	20	S	22
Granodiorite, granite	Urals (U.S.S.R.); Verkhisetsk and Suunduksk massifs	12	0.4—32.0	3.2	N	1, 6
Granodiorite, adamellite, granite	Chukotka, Magdan, Priamurya regions (U.S.S.R.); Iultinsk, Taurenansk, Velikenayk, Magdansk, Neorchansk, Ingaglinsk, Bryus, Peveksk, Peekeneisk and Paliansk massifs	55	0.1—15	5.3	S, N	17, 20
Granodiorite	Kamchatka (U.S.S.R.); porphyritic sills, veins and stocks	100	—	3	S	22
Porphyrites	Nura-Tau, Uzbekistan (U.S.S.R.)	13	5—40	10.7	W	15

Table 79-E-1 (continued)

Rock type	Location and host	Number of samples	Gold (ppb) range	Gold (ppb) average	Method	Reference
Granite-aplite	Kazakhstan (U.S.S.R.); Stepnyak and Kurtukulsk intrusions	5	1.9—2.6	2.3	S, N	23
Aplite and greisen	Western Siberia (U.S.S.R.); Kundatsk, Dudetsk, Buruzunsk, Aturkolsk and Belokurikhinsk massifs	6	1.9—36.0	11.4	N	20
Aplite, granite	Kazakhstan (U.S.S.R.); Krykkuduk complex	11	—	1.3	S	29
Aplite, granite	Kamchatka (U.S.S.R.)	3	8—60	29	N	19
Aplite	Boulder, Colorado (U.S.A.)	1		3.3	N	11
	Intermediate rocks					
Diorite, quartz diorite	Altai-Sayan, Tien Shan, Magadan areas (U.S.S.R.); Sarakokshinsk, Dudetsk, Kochkor Ata, Suuktepa, Magadansk batholiths; Tikhvensk and Kanatinsk diorites	16	1.0—18.0	4.4	N	10, 17, 20
Diorite, quartz diorite	Idaho batholith (U.S.A.); ca-prov.	7	0.2—6.0	2.2	N	14
Quartz diorite, granodiorite	Northern Kazakhstan (U.S.S.R.); Kurtukulsk intrusion	13	—	3.1	S, N	23
Diorite and diabase porphyrite	Priamurya region (U.S.S.R.); in part propylitized	60	0.5—50.0	6.2	S	17
Diorite	Priamurya region (U.S.S.R.)	33	1.0—110	12.7	S	17
Quartz diorite	Urals (U.S.S.R.); Verkhisetsk massif	5	0.8—3.4	1.6	N	6
Quartz diorite	Northern Kazakhstan (U.S.S.R.); Stepnyak intrusive	22	1.1—8.8	3.7	S	28
Quartz diorite, hornblende monzonite	Chatkal, Tien Shan (U.S.S.R.); Verkhniy Ter intrusion	2	3.8—4.2	4.0	N	10

Quartz diorite, quartz monzonite and tonalite	Sierra Nevada, Southern California and Idaho batholiths; Coast and Transverse Ranges, Calif.; Butte, Montana and Princeton, Colorado (U.S.A.); ca-prov.	40	0.1—11.3	1.4	N	14
Quartz diorite, quartz monzonite	Manitoba (Canada); Archean greenstone belt containing gold-quartz veins	14	0.4—3.2	1.4	N	24
Diorite and gabbro diorite	Nura-Tau, Uzbekistan (U.S.S.R.); Temirkabusk intrusive	9	5—50	17	W	15
Gabbro diorite	Central Chukotka (U.S.S.R.); East Iultinsk massif	7	0.5—5.5	2.0	S	17
Quartz diorite, tonalite and diorite porphyrite	Kazakhstan (U.S.S.R.); hybrid endocontact rocks, Krykkuduk massif; plutonic dikes, Kurtukulsk intrusion; Stepnyak intrusive	51	1.1—8.0	2.7	S, N	23, 29
Quartz monzonite	North Carolina (U.S.A.); Salisbury pluton, a-prov.	11	0.2—2.3	0.9	N	14
Quartz monzonite alaskite, aplite	Sierra Nevada and Southern California batholiths and Boulder batholith, Montana (U.S.A.); ca-prov.	13	0.3—1.9	0.75	N	14
Alkali syenite	Altai-Sayan (U.S.S.R.); Tikhvinsk, Udarninsk, Gavrilovsk massifs	4	3.6—6.2	4.4	N	20
Syenite	New Hampshire, North Carolina, Montana (U.S.A.); mainly White Mountain series, Boxelder and Shonkin Sag laccoliths, a-prov.	19	0.3—3.8	1.7	N	14
Nepheline syenite	Ontario (Canada); Wisconsin and New Hampshire (U.S.A.)	3	0.6—2.6	1.4	N	11
Syenite, syenogabbro, and syenodiorite	Big Bend National Park, Texas and Boulder batholith, Montana (U.S.A.)	7	0.2—1.0	0.6	N	14
	Mafic rocks					
Gabbro	Far East (U.S.S.R.) (Chukotka, Magadan, Kamchatka), Altai-Sayan; mainly Magdansk batholith; Sarakokshinsk, Dudetsk, Taureransk and Velitkenaysk massifs; Taskyl' and Kanatinsk gabbros	28	0.3—79.0	6.2	S. N	17, 20

Table 79-E-1 (continued)

Rock type	Location and host	Number of samples	Gold (ppb) range	Gold (ppb) average	Method	Reference
"Diabase" and gabbro	Kola Peninsula (U.S.S.R.); Kolmozero-Voronya series	5	1.8—22	6.8	S	13
Gabbro "diabase"	Siberian platform (U.S.S.R.); tholeiites of 11 magmatic complexes	82	2.0—32.0	7.4	S	5
Gabbro, qtz. gabbro-anorthosite	Northern Kazakhstan (U.S.S.R.); Stepnyak intrusion	95	2.5—7.2	5.1	S, N	23
Gabbro and basalt	Kazakhstan (U.S.S.R.); Kurtukulsk intrusion and Mt. Kul'bay	10	8—12	6.6	S, N	23
Gabbro, norite, gabbro diorite	Kazakhstan (U.S.S.R.); hybrid endocontact rocks of Krykkuduk, Arkalyk, Dologay, Otayde-Karasu massifs; Stepnyak intrusives	216	<1.0—22	3.2	S	28
Meta-gabbro, "diabase"	Manitoba (Canada)	2	17—19	18	N	24
Gabbro	Southern California batholith (U.S.A.); ca-prov.	6	0.7—11.6	4.4	N	14
Gabbro to monzo-diorite	New Hampshire and North Carolina (U.S.A.); a-provinces	3	0.3—0.7	0.4	N	14
Shonkinite and mafic phonolite	Montana (U.S.A.); Boxelder and Shonkin Sag laccoliths, a-prov., b-a-r assoc.	11	0.5—10.4	4.4	N	14
Layered gabbro	Skaergaard intrusion (East Greenland)	13	2.4—9.0	4.7	N	27
Layered gabbro	Anakita massif, Siberian platform (U.S.S.R.)	12	4.0—17	9.9	N	21
Stratified gabbro and ultramafics	Zlatogorsk intrusion (U.S.S.R.)	70	1.0—4.9	1.2	S	30
Gabbro	Romanche Depression, Mid-Atlantic Ridge	3	0.4—0.7	0.5	N	4
Gabbro, gabbro diorite	Kamchatka (U.S.S.R.) gabbro-granite and gabbro-granodiorite associations	21	—	17	S	22

Gabbro and carbonatite	Missouri (U.S.A.); Bushveld complex, SA; Tanzania	3	1.6—2.9	2.3	N	11
	Ultramafic rocks					
Peridotite, dunite and serpentinite	Mount Albert, Quebec (Canada), alpine type	12	0.6—5.9	1.8	N	7
Peridotite, dunite	North Carolina (U.S.A.)	2	2.2—2.4	2.3	N	11, 28
Peridotite	Rhum, Inner Hebrides (U.K.), olivine cumulite	1	—	11	N	28
Peridotite (dunite, harzburgite)	Far East, Caucasus, Urals (U.S.S.R.), Philippines, Indian Ocean; rocks of dunite-harzburgie clan.	23	1.4—63	20.3	S	17
Peridotites etc. (dunite, pyroxenite)	Far East, Urals, Tagil (U.S.S.R.); rocks of dunite-pyroxenite and gabbro-cortlandite clans.	30	<0.2—12	3.4	S	17
Dunite-troctolite	Stanovoy Range, far east (U.S.S.R.), layered intrusions	5	1—9	3.2	S	17
Kimberlite	Yakutai, U.S.S.R. and South Africa (Kimberley Mine)	5	1—6	2.8	S, N	12, 17
Peridotite	Northern Yakutia (U.S.S.R.), Obnazhennaya kimberlite pipe	8	0.7—2.6	1.8	S	17
Serpentinite and peridotite	Kuznetsk Ala-Tau, Eastern Sayan and Transbaikal (U.S.S.R.); Semenkovsk, Ospinsk and Dzhida massifs	3	4.3—6.0	4.7	N	20
Dunite	U.S.S.R., China, Albania	10	0.9—31	10.9	N	20
Lherzolite etc., inclusions in mafic volcanics	U.S.S.R. etc.	18	0.9—16	8.3	S, N	17, 28
Peridotite (herzolite etc.)	Romanche Depression, Mid-Atlantic Ridge	11	0.7—2.0	1.4	N	4
Garnet peridotite, pyroxenite and eclogite xenoliths	Obnazhennya kimberlite pipe Yakutia (U.S.S.R.), and Roberts Victor and Bultfontein Mines (South Africa)	21	0.6—4.5	1.7	S, N	12, 26

Table 79-E-1 (continued)

Rock type	Location and host	Number of samples	Gold (ppb) range	Gold (ppb) average	Method	Reference
	Volcanic rocks					
	Salic rocks					
Rhyodacite, rhyolite	Kazakhstan (U.S.S.R.), Krykkuduk, Aralul', Borovsk complexes	11	1.0—3.5	1.8	S	29
Rhyolite, rhyodacite, dacite and latite	Oregon, California (mainly Cascades), New Mexico (Jemez Mtns.), and Wyoming (Yellowstone) (U.S.A.)	68	0.1—1.9	0.5	N	14
Rhyolite, dacite	Eastern Kamchatka (U.S.S.R.)	48	—	1.79	S	2
Altered rhyolite, latite, rhyodacite	Jerome and Beatty, Arizona (U.S.A.), b-a-r assoc.	36	0.2—22.0	3.5	N	14
Rhyolite, rhyolite latite, dacite	Idaho, North Carolina, Virginia, (U.S.A.), b-a-r assoc.	21	0.1—2.8	0.6	N	14
Obsidian	Rotorua (New Zealand)	1	—	21	N	11
Salic volcanics	Chukotka and Altai (U.S.S.R.); Texas (U.S.A.); Australia	4	0.4—5.5	2.3	N	11, 14, 20
	Mafic and intermediate rocks					
Basalt (tholeiitic)	Hawaii; mainly Kilauea and Mauna Loa	30	0.5—6.6	2.5	N	12, 14 28
Basalt	Hawaii, mainly from Waianae	7	0.2—2.7	1.4	N	14, 31
Basalt (alkalic)	Tahiti	4	0.30—1.1	0.50	N	8
Basalt (low potassium tholeiitic)	East Pacific Rise, Mid-Atlantic Ridge	42	0.1—4.6	0.6	N	14
Basalt (alkalic)	East Pacific Rise	10	0.2—1.8	0.8	N	14
Basalt (alkalic)	Mid-Atlantic Ridge, East Pacific Rise	4	0.2—0.4	0.3	N	12
Basalt (alkalic)	Mid-Atlantic Ridge; dredge samples	5	0.5—16.0	9.5	N	4, 11

Basalt ("diabase")	Denmark Straight, Atlantic; dredge samples	2	14—48	31	N	4
Basalt (tholeiitic)	Oregon, Washington (U.S.A.), Columbia River	11	—	3.7	N	14
Basalt (tholeiitic)	Oregon (U.S.A.), Pliocene-Quaternary	4	0.4—1.8	0.9	N	14
Basalt (alkalic)	Southern Idaho (U.S.A.), Snake River Group	19	0.1—1.3	0.5	N	14
Basalt (greenstone)	Michigan (U.S.A.), Keweenawan	10	1.4—8.5	4.5	N	14
Basalt (alkalic)	Big Bend National Park, Texas (U.S.A.), b-a-r assoc.; a-prov.	8	0.3—4.3	1.1	N	14
Basalt	Kurile-Kamchatka (U.S.S.R.)	38	—	1.73	S	2
Basalt	Nevada, Wyoming, Colorado (U.S.A.); U.K.; Germany	11	0.2—8.9	2.2	N	11, 14, 20, 28
Basalt	Eastern U.S.A.	87	0.3—9.0	3.4	N	14
Basalt ("diabase")	Pennsylvania (U.S.A.), Dillsburg	6	12.2—15.5	14.3	N	14
Basalt-gabbro	Missouri (U.S.A.), two intrusions	25	0.5—7.4	2.2	N	14
Basalt ("diabase")	Tasmania; Great Lake intrusion	19	1.0—12.0	4.9	N	14
Basalt ("diabase")	Altai and Kuznetsk Ala-Tau (U.S.S.R.), Sarakokshinsk massif	3	1.6—10.0	5.4	N	20
Basalt	Central Kamchatka (U.S.S.R.), Vtulkinaya River	16	—	0.3	S	22
Basalt, dacite (metavolcanics)	Manitoba (Canada), Archean	3	0.1—0.5	0.3	N	19
Greenstones	Mother Lode California (U.S.A.)	5	1.2—3.4	2.1	N	14
Basalt, olivine basalt, andesite	Cascades, California, and Oregon (U.S.A.), b-a-r assoc.	36	0.6—19.4	5.4	N	14
Andesite	Central Idaho (U.S.A.) Challis Volcanics, b-a-r assoc.	3	0.3—1.0	0.6	N	14
Andesite (metavolcanics)	North Carolina (U.S.A.), b-a-r assoc.	3	0.5—1.9	1.1	N	14
Andesite	Kurile-Kamchatka (U.S.S.R.)	82	—	1.88	S	2

Table 79-E-1 (continued)

Rock type	Location and host	Number of samples	Gold (ppb) range	Gold (ppb) average	Method	Reference
Andesite, andesitic basalt	Altai, Kuznetsk Ala-Tau, Chukotka (U.S.S.R.)	9	1.5—10.7	6.4	N	20
Basalt, andesitic basalt	Kamchatka (U.S.S.R.)	83	—	3.7	S	22
Basalt, basaltic andesite, trachy-andesite	Boshchekul'sk region, Kazakhstan (U.S.S.R.), chert-spilite and basalt-trachyte complexes	27	1—30	9.1	S	25
Basalt, andesite, andesitic dacite	Kazakhstan (U.S.S.R.), rocks enclosing the Krykkuduk, Aralul' and Borovsk complexes	57	1.0—17.0	2.0	S	29
Trachyte	Altai (U.S.S.R.)	6	1.1—12.0	6.5	N	20
Trachyandesite trachyte	Big Bend National Park, Texas (U.S.A.), b-a-r assoc., a-prov.	6	0.2—1.9	0.65	N	14
Andesite-dacite	Central Kamchatka (U.S.S.R.), weak hydrothermal alteration	14	—	20	S	22

[a] ca-prov. = calk-alkaline province.
[b] a-prov. = alkaline province.
[c] b-a-r assoc. = basalt-andesite-rhyolite association.
[aa] Method used for most of the data given by Kushmuradov (1970) is described as a "gold chemical" method.

References: 1. Anoshin *et al.* (1970). 2. Anoshin and Kepezinskas (1972). 3. Anoshin and Potap'yev (1966). 4. Anoshin and Yemel'yanov (1969). 5. Anoshin *et al.* (1971). 6. Bushlyakov (1971). 7. Crocket and Chyi (1972). 8. Crocket *et al.* (1973). 9. Davletov (1970). 10. Davletov and Dzhakshibayev (1970). 11. DeGrazia and Haskin (1964). 12. Ehmann *et al.* (1970). 13. Gavrilenko and Garifulin (1971). 14. Gottfried *et al.* (1972). 15. Kushmuradov (1970). 16. Mantei *et al.* (1970). 17. Moiseenko *et al.* (1971). 18 Paley *et al.* (1967). 19. Rozhkov *et al.* (1970). 20. Shcherbakov (1967). 21. Shcherbakov and Perezhogin (1964). 22. Shilin (1968). 23. Spiridonov (1971). 24. Stephenson and Ehmann (1971). 25. Tolmachev *et al.* (1971). 26. Ukhanov and Pchelintseva (1972). 27. Vincent and Crocket (1960a). 28. Vincent and Crocket (1960b). 29. Voskresenskaya and Zvereva (1968). 30. Voskresenskaya *et al.* (1970). 31. Wasson and Baedecker (1970). 32. Zvereva and Gavrilenko (1971).

Table 79-E-2. *Average[a] values for igneous rocks (based on data of Table 79-E-1)*

Rock type	Number of analyses	Weighted average, all analyses
Plutonic rocks		
Salic		
Granite	310	1.7
Granodiorite	380	3.0
Aplite (including dikes and veins of granite and greisen	23	4.2
Intermediate		
Diorite, quartz diorite, monzonite, quartz monozonite, tonalite	261	3.2
Mafic		
Gabbros (including gabbro-diorite)	580	4.8
Ultramafic		
Mainly dunite and peridotite	149	6.6
Volcanic and hypabyssal rocks		
Salic		
Mainly rhyolite (including rhyodacite, dacite, felsite, latite and obsidian)	188	1.5
Mafic and intermediate		
Mainly basalt, andesitic basalt, andesite (51) trachyandesite and trachyte (2)	696	3.6

[a] A few values considered anomalously high are omitted in the calculation of average values. They include the following: DeGrazia and Haskin (1964): perlite 21 ppb; Kushmuradov (1970): granite 6.1 ppb and diorite 17 ppb; Moiseenko *et al.* (1971): small stocks of the Tokur district 28.3, 26.0, 67.2 ppb; Shilin (1968): granite 15 ppb and granodiorite 20 ppb.

I. Generalized Trends

The following generalizations concerning gold geochemisty are suggested by the averages from Table 79-E-3 and the literature:

1. The range of gold values for the majority of igneous rocks is approximately from 0.5 to 5 ppb. Some ultramafic rocks, particularly dunite-harzburgite (Moiseenko and Fatyanov, 1972; Shcheka *et al.*, 1971), may be higher in average gold content.

2. In both plutonic and volcanic rocks, gold content generally decreases from mafic to felsic types. Many authors regard this as a general geochemical trend of gold (Shcherbakov and Perezhogin, 1964; Spiridonov, 1971; Gottfried *et al.*, 1972; Moiseenko and Fatyanov, 1972; Tilling *et al.*, 1973). Voskresenskaya and Zvereva (1968), however, contend that granitoids of Northern Kazakhstan magmatic complexes show little fractionation of gold between rocks of different mafic index. Also Anoshin and Kepezinskas (1972) find no significant differences in average gold of basalt, andesite and rhyolite-dacite from Cenozoic volcanic rocks of the Kurile-Kamchatka province.

Table 79-E-3. *Comparison of average values of igneous rocks reported by different authors*

Rock type	Source						
	PHAN, (1965)	SHCHERBAKOV (1967)	VOSKRESENSKAYA and ZVEREVA (1968)	JONES (1969)	MOISEENKO and FATYANOV (1972)	GOTTFRIED *et al.* (1972)	This work
Plutonic rocks							
Granite	2.5	2.0		2.8	1.6		1.7
Quartz monzonite to granite						0.75[a]	
Granodiorite			1.5		3.7	1.2	3.0
Monzodiorite and diorite					12.7[b]		
Syenite	1.7			2.5		0.3	1.8
Diorite	4	3.4	3.5	3.5		2.7	3.2
Gabbro	5	8.2		5.4	3.4		4.8
Gabbro, calc-alkalic						3.9	
Gabbro to monzodiorite, alkalic						0.4	
Ultramafics	6	11	4.4	8.2	3.3		6.6
Ultramafics of the dunite-harzburgite clan					21.3		
Volcanic rocks							
Rhyolite		5		12.0[c]		0.7	1.5
Dacite to rhyodacite						0.8	
Trachyandesite to trachyte						0.6	
Trachyte	5			6.5			
Andesite, calc-alkalic						4.6	
Andesite, alkalic						1.5	
Andesite	6	4.9	2.2	5.2			
Basalt	3			3.2		5.4[aa]	3.6
Oceanic tholeiitic basalt						0.6	
Continental tholeiitic basalt						3.3	
Alkalic basalt						0.6	

[a] Weighted average of 29 samples of calc-alkalic affinity and 33 samples of alkalic affinity averaging 0.8 and 0.7 ppb Au respectively (author's Table 5).

[b] Includes samples of three syenite-diorite and diorite stocks from the Tokur district, Priamurya (U.S.S.R.), averaging 28.3, 26.0 and 67.2 ppb Au.

[c] This high value may result from selective sampling according to JONES (1969).

II. Plutonic Rocks

Several additional features of gold geochemistry in plutonic rocks may be noted:

1. Calc-alkalic plutonic rocks from major batholiths of Western United States show little variation in gold. They usually contain 3 ppb or less with gold decreasing from mafic to felsic members (GOTTFRIED *et al.*, 1972). By comparison alkalic plutonic rocks from New Hampshire and North and South Carolina are poorer in gold, generally containing 1.5 ppb or less (TILLING *et al.*, 1973; GOTTFRIED *et al.*, 1972).

2. In large complex granitoid massifs, gold has been noted to increase from the center outward to the peripheral contact rocks (DAVLETOV and DZHAKSHIBAYEV, 1970; SHILIN, 1968). A similar effect was demonstrated by MANTEI *et al.* (1970) for the Marysville stock. Further, in many granitoid massifs enrichment of gold in hybrid contact rocks occurs relative to the massif itself (DAVLETOV and DZHAKSHIBAYEV, 1970; SHCHERBAKOV and PEREZHOGIN, 1963; VOSKRESENSKAYA and ZVEREVA, 1968).

3. In three strongly differentiated gabbroic intrusions no compelling evidence has been found for gold concentration in late granophyric differentiates. In the Skaergaard (VINCENT and CROCKET, 1960a) and Zlatogorsk (VOSKRESENSKAYA *et al.*, 1970) intrusions, gold is very uniform over the entire range of rock types. In the Anaketa intrusion, according to SHCHERBAKOV and PEREZHOGIN (1963) there is an increase in gold content from the early gabbros up to the ferrogabbro stage. Low values, however, are found in the late granophyric differentiates. The same effect, lack of concentration of gold in late granophyres has been found in differentiated diabase sills from Tasmania (ROWE, 1969), Virginia, U.S.A. (GOTTFRIED *et al.*, 1972) and traps (tholeiite-basalt formation) from the Siberian Platform (ANOSHIN *et al.*, 1971).

4. Strong correlation of gold with copper is found in differentiated mafic rocks such as Skaergaard (VINCENT and CROCKET, 1960a) or the Great Lake intrusion, Tasmania (ROWE, 1969). Thus, depletion of gold in such magma probably occurs when saturation with respect to copper sulfides is reached.

5. TILLING *et al.* (1973) contend that available abundance data indicate that gold is not concentrated in residual silicate melts produced in differentiation of calc-alkaline magma.

III. Volcanic Rocks

With respect to geochemical behavior in volcanic rocks the following generalizations have been put forward:

1. A parallel decrease in gold from mafic to felsic volcanics is shown by average abundance data. This trend is apparent in the data of GOTTFRIED *et al.* (1972) provided only data for fresh acid volcanics are considered. Rhyolites described as "altered" (Table 7, GOTTFRIED *et al.*, 1972) are invariably higher in gold.

2. Mafic volcanic rocks, analogous to their plutonic equivalents, are generally higher in gold than other volcanic rocks. However, basalts of different petrochemical and tectonic affinities show a very wide range of average gold contents, 0.5—5 ppb, and are therefore considerably more variable in gold content than gabbros. It does not appear that there is any clear correlation of gold content with major element chemistry and in general the element appears insensitive to tholeiitic magma differen-

tiation trends. For example, GOTTFRIED *et al.* (1972) point out that the low-K_2O tholeiites from the East Pacific Rise and Mid-Atlantic ridge fracture systems have the same average gold content as alkalic basalts (see Table 79-E-3) whereas the same low-K_2O tholeiites and the chemically comparable diabase dikes of North Carolina (averaging 4.8 ppb; see GOTTFRIED and GREENLAND, 1972, for discussion) vary by approximately eight in average gold content. A possible inference from these trends is that gold abundance of basaltic rocks varies on a regional or provincial scale. On the other hand, ANOSHIN and KEPEZINSKAS (1972) suggest that Au is concentrated in rocks of high K_2O content on a basis of its variation in basalt, andesite, and rhyolite-dacite volcanics from Kurile-Kamchatka.

Revised manuscript received: June 1973

79-F. Behavior in Magmatogenic Processes

I. Sulfide Minerals

The importance of magmatic processes in the transport and deposition of gold is implicit in epithermal gold ores in which localization in volcanic calderas, vent breccias and associated lavas is common. The important recovery of the metal as a by-product of base metal mining, particularly porphyry coppers, is a further manifestation of the importance of magmatic processes in concentrating gold. In addition to important gold-bearing minerals (see Section 79-D), sulfides often carry a large fraction of total gold in some ores.

A survey of gold in sulfides and other minerals has been published by JONES and FLEISCHER (1969) and data from 1960 are presented in Table 79-F-1. The compilation shows two important features of gold distribution in sulfides: 1. the range of gold contents in various sulfide minerals species is commonly as great as 10^4; and 2. gold in sulfides appears to depend largely on environment rather than the nature of the mineral species. Thus, pyrite from gold-quartz veins generally has a higher gold content than pyrite from non-auriferous deposits.

Partition of gold between co-existing sulfides is generally small compared with the variation in the same minerals from different deposits. Thus, KEAYS and CROCKET (1970) found approximately twice as much gold in chalcopyrite as in pyrrhotite from Strathcona Mine, Sudbury district, while BADALOV (1965) noted that chalcopyrite from Cu—Mo deposits of the Almalyk district contained seven times more gold than pyrite. An analogous trend is shown in the data of SANTOS and WALTER (1971) for Philippines sulfides; no consistent trend for the gold content of pyrite and chalcopyrite from the same occurrence was found. Rather, high gold in either mineral seems dependent on whether the host deposit was gold-bearing or not. The same distributions are apparent in data (VAKHRUSHEV and TSIMBALIST, 1967) on gold in pyrite, chalcopyrite and arsenopyrite in skarns. In gold-bearing skarns, similar gold contents are found in all three minerals (14, 13 and 21 ppm respectively), whereas much lower values are found in the same minerals in gold-free skarns (0.5, 0.5 and 4 ppm respectively). Again the major differences are found not between the minerals but between the different types of deposits.

These observations suggest that differing sulfide mineral structures do not exert a major influence on the partition of gold between different minerals, and that the solubility of gold in sulfides by lattice substitution may be small. High temperature magmatic copper nickel sulfides may however constitute an exception.

II. Hydrothermal Solutions

The petrologic, tectonic and structural associations typical of epithermal gold deposits strongly suggest that igneous hydrothermal solutions played an important role in their deposition. It is less certain that a similar case can be made for gold-

Table 79-F-1. *Gold content of sulfides and sulfosalts*

Mineral	Locality and host	Number of samples	Gold (ppm) range	Gold (ppm) average	Method[a]	Reference
Pyrite	Karamazar (U.S.S.R.), magnetite skarn in carbonate rocks	7	—	0.5	—	2
	Karamazar (U.S.S.R.), molybdenum and copper-molybdenum ore, Kal'makyr deposit	69	—	3.2	—	2
	Karamazar (U.S.S.R.), quartz-gold veins in igneous rocks	131	—	211	—	2
	Uzbekistan (U.S.S.R.), quartz-gold ore in sandstones and shales	18	—	113	—	2
	Almalyk (U.S.S.R.), disseminated copper-molybdenum in syeno-diorite	—	—	3	—	1
	Middle Asia, lead-zinc deposits	14	0.19—11	1.1	—	4
	Far east (U.S.S.R.), gold-quartz veins and quartzites, Tokhur deposit	9	—	79	N, S	7
	Western Tien Shan (U.S.S.R.), ores of the Koksu River basin	13	0.02—15	4	Fa, N	5
	Philippines	21	0.003—140	14	N	8
	Altai-Sayan (U.S.S.R.), pyrite from magnetite-bearing skarns	96	0.003—9.4	0.54	S	10
	Kuznetsk Ala-Tau (U.S.S.R.), gold quartz veins	37	3.4—398	36	S	10
	Altai-Sayan (U.S.S.R.), gold-bearing skarn deposits	16	5.5—113	13.9	S	9
	Noril'sk (U.S.S.R.)	36	0.001—0.052	0.013	S	9
	New Brunswick (Canada), base metal ore, Heath Steele Mine	14	0.6—6	1.5	N	6
Pyrrhotite	Sudbury (Canada), magmatic copper-nickel sulfide ore	25	1.1—54	12.6	N	3
	Noril'sk (U.S.S.R.)	9	0.007—0.42	0.08	S	9

Chalcopyrite	Philippines	11	0.07—22	3.5	N	8
	Karamazar (U.S.S.R.), copper-bearing skarn in carbonate rocks	7	—	1	—	2
	Karamazar (U.S.S.R.), molybdenum and copper molybdenum, Kal'makyr deposit	53	—	17	—	2
	Karamazar (U.S.S.R.), quartz-gold veins in igneous rocks	6	—	8	—	2
	Uzbekistan (U.S.S.R.), scheelite skarn in carbonate rocks	2	—	8	—	2
	Almalyk (U.S.S.R.), disseminated copper-molybdenum ore	—	—	22	Fa/S	1
	Sudbury (Canada), magmatic copper-nickel sulfide ore	16	0.009—5.8	5.3	N	3
	Middle Asia, lead-zinc deposits	18	0.04—0.2	0.07		4
	Altai-Sayan (U.S.S.R.), gold-bearing skarns	5	0.9—32	13	S	10
	Kuznetsk Ala-Tau (U.S.S.R.), gold-quartz veins	2	9.5—21	15	S	10
	Altai-Sayan (U.S.S.R.), magnetite-bearing skarns	2	0.3—0.6	0.45	S	10
	Noril'sk district (U.S.S.R.)	9	0.05—13	2.8	S	9
Galena	Karamazar (U.S.S.R.), molybdenum and copper-molybdenum ore	7	—	12	—	2
	Karamazar (U.S.S.R.), quartz-gold veins in igneous rocks	11	—	11	—	2
	Tien Shan (U.S.S.R.), ores of the Koksu River basin	4	0.15—1.0	0.6	Fa, N	5
Sphalerite	Karamazar (U.S.S.R.), quartz-gold veins in igneous rocks	5	—	23	—	2
	Uzbekistan (U.S.S.R.), scheelite skarn in carbonate rocks	3	—	1.3	—	2
Pentlandite	Noril'sk (U.S.S.R.)	5	0.015—9.4	2.0	S	9

[a] The symbol "Fa" has been used for fire assay methods as described by Haber and Jaenicke (1925) and others.

References: 1. Badalov (1965). 2. Badalov and Badalova (1967). 3. Keays and Crocket (1970). 4. Koroleva (1971). 5. Maksudov (1969). 6. Mercer and Crocket (1972). 7. Moiseenko and Fatyanov (1972). 8. Santos and Walters (1971). 9. Vakhrushev *et al.* (1972). 10. Vakhrushev and Tsimbalist (1967). 11. Vincent and Crocket (1960).

quartz veins in Precambrian volcanic belts, but hot water-rich fluids, not necessarily of igneous origin, appear to have been active in the transport and deposition of such gold.

Information on gold transport and deposition by hot aqueous fluid is obtained from three main lines of research: a) thermodynamic models, b) laboratory solubility studies, and c) fluid inclusion studies.

a) Thermodynamic Studies

HELGESON and GARRELS (1968) developed a thermodynamic model for gold transport in hydrothermal fluids characterized by acidic (pH 2.7 to 3.0), chloride-rich (3 molal NaCl) waters in which $AuCl_2^-$ was responsible for solubility, and deposition occurred above 175° C. Their model is applicable to auriferous, pyritic quartz veins characterized by gangue alunite. They showed that solutions carrying no more than 20 ppb gold could potentially form a commercial ore body. Alternately BARNES and CZAMANSKE (1967) contend that gold is carried mainly as a sulfide complex, perhaps AuS^-, where formation of the complex may be governed by a reaction such as:

$$2\,Au(s) + 2\,HS^- + \tfrac{1}{2}\,O_2(g) \rightleftharpoons 2\,AuS^- + H_2O(l) \quad K = 10^{29.7}\ (25°\ C)$$

The relevance of such a reaction depends critically on the pH of the ore-transporting fluid which must be near neutral to give a significant HS^- concentration. BARNES and CZAMANSKE (1967) contend that much evidence suggests the likely range of pH is one unit on either side of neutrality up to 250° C. Detailed thermodynamic analysis of such a system at elevated temperatures has not been carried out.

b) Experimental Studies

Recent experimental determinations of gold solubility applicable to hydrothermal solutions at elevated temperatures have been made by ANDERSON and BURNHAM (1964), WEISSBERG (1970) and HENLEY (1972). High solubilities have been reported for both alkali chloride and alkaline bisulfide systems. ANDERSON and BURNHAM (1964) note solubilities of gold from 1000 to 2000 ppm in alkali chloride solutions (3.6 molar, 600° C at 3 kb and 700° C at 4 kb) saturated with respect to quartz. HENLEY (1972) found gold solubility to increase from 20 ppm at 300 °C to 900 ppm at 500° C in 2 M KCl solutions at 2 kb and at the f_{O_2} of the hematite-magnetite buffer. Both increasing HCl molality and temperature favored increasing gold solubility, but f_{O_2} did not appear to be an important variable. Because of a high degree of association of HCl and KCl under the supercritical conditions of these experiments, solubility was attributed to a species such as $Au_2Cl_6 \cdot (HCl)n$. WEISSBERG (1970) obtained solubilities of 100 and 200 ppm gold in solutions containing 0.2 and 0.3 moles NaHS per kg of solution at temperatures between 150 and 250° C and at approximately 1 kb pressure. Solubility increased with increasing temperature and concentration of NaHS. These solutions contained NaCl and were near neutral pH. WEISSBERG argued gold solubility was due to AuS^- with no significant contribution from chloride complexes. Thus, the experimental studies appear to suggest that significant gold solubility can be obtained under widely variable hydrothermal conditions from either chloride or bisulfide complexes.

c) Fluid Inclusion Studies

To the extent that fluid inclusions represent samples of hydrothermal solutions, they provide some constraints on experimental and theoretical studies. Data on inclusions pertinent to gold deposits are limited (ROEDDER, 1967) but some information on aqueous extracts from quartz of hydrothermal gold deposits of the Soviet Far East and Urals has been presented by MOISEENKO and FATYANOV (1972). Their study which includes determinations on nine different gold-quartz deposits, suggests an alkali-bicarbonate composition for the transporting fluid, in which the mole proportion of Na to K is approximately 3. These fluid inclusions are distinctive in composition in that HCO_3^- is much higher than Cl^- whereas the reverse is usually true. ROEDDER (1967) notes that epithermal mercury deposits are also characterized by this type of fluid inclusion, a particularly interesting observation in view of the association of gold and mercury in some vein deposits.

MOISEENKO and FATYANOV (1972) have also determined temperatures of quartz deposition in gold-quartz veins by the homogenization method. On a basis of data from 13 deposits (mainly Russian), they conclude that the temperature range of deposition of hydrothermal gold is 30 to 430° C with major gold deposition occurring over a relatively narrow temperature range, 130 to 200° C.

MACHAIRAS (1970) showed that in certain auriferous quartz veins high gold content is accompanied by high frequency of fluid inclusions, that particulate gold can be observed within fluid inclusions, and that gold-rich quartz (30 to 40 ppm) is characterized by high CO_2 levels in associated fluid inclusions. These observations argue for a genetic association of gold with the liquid phase of fluid inclusions in these samples.

Revised manuscript received: June 1973

79-G. Behavior of Gold during Weathering and Alteration of Rocks

I. Weathering

Early experiments on the dissolution of gold by potentially natural solvents (STOKES, 1906; MCCAUGHEY, 1909; BROKOW, 1910; OGRYZLO, 1935; EMMONS, 1940) were reviewed by KRAUSKOPF (1951, 1967) who provided an evaluation of relevant experimental results and geological observations based on consideration of standard electrode potentials. He noted that gold should be soluble at 25° C in acid, chloride-bearing waters which contain one of several moderately strong oxidizing agents such as MnO_2, O_2, Fe^{3+} and Cu^{2+}. Appreciable solubility would require higher concentrations of these constituents and greater acidity than commonly found in nature. In alkaline or neutral solutions, appreciable gold solubility occurs in the presence of HS^- but the solution mechanism is uncertain.

Anionic species forming stable complexes with gold which might be of significance in the weathering cycle, include not only Cl^- and the other halogens but also CN^-, CNS^- and $S_2O_3^{2-}$ (LAKIN *et al.*, 1970). CHUKHROV (1947) suggested that gold was carried in iodine or bromine compounds in ionic form in the zone of oxidation. LAKIN *et al.* (1970) point out that standard electrode potentials demonstrate a dependence of required oxidation potential on the nature of the gold complex formed, that is, on the anion species:

$$Au + 4\,Cl^- = AuCl_4^- + 3e^- \qquad \varepsilon^\circ = -1.00\ v$$
$$Au + 4\,Br^- = AuBr_4^- + 3e^- \qquad \varepsilon^\circ = -0.87\ v$$
$$Au + 4\,CNS^- = Au(CNS)_4^- + 3e^- \qquad \varepsilon^\circ = -0.66\ v$$
$$Au + 2\,I^- = AuI_2^- + e^- \qquad \varepsilon^\circ = -0.50\ v$$
$$Au + 2\,CN^- = Au(CN)_2^- + e^- \qquad \varepsilon^\circ = +0.60\ v$$

Thus, the ease of oxidation of gold increases from Cl^- to I^- which might be expected to favor solubility as the iodide rather than the chloride. This is probably offset by the relative abundance of halides in the earth's crust (Cl^-/I^- is approximately 1650). LAKIN *et al.* (1970) note that cyanides in soils may be produced by many plants (more than 1000 species including fungi) through hydrolysis of cyanogenetic glycosides and suggest that native gold may be mobilized as $Au(CN)_2^-$ in soils under native vegetation.

The significance of natural organic solvents in the weathering geochemistry of gold has also commanded considerable attention. ONG and SWANSON (1969) studied transportation, deposition and concentration of gold by natural organic acids and arrived at the following main conclusions: (a) organic acids do not dissolve, complex, nor oxidize gold; rather, gold chloride solutions are reduced to negative colloids of metallic gold in their presence; (b) by surrounding colloidal-size gold sols with protective coatings, organic acids inhibit their growth and precipitation.

That aspect of the weathering cycle of gold which has been of greatest interest is the behavior of the metal in the oxidation zone of sulfide ores. Of the more recent studies, those of KREUTER *et al.* (1958) and RAZIN and ROZHKOV (1963, 1966) are among the most extensive. KREUTER *et al.* (1958) argue that oxidation of sulfides is essential for supergene enrichment of gold, and that solution of dispersed gold in ferric sulfate-sulfuric acid solution is the most probable mode of gold transfer. The latter conclusion is at variance with some of the early experimental studies (OGRYSLO, 1935) and with inference from thermochemical data (KRAUSKOPF, 1951).

RAZIN and ROZHKOV (1963, 1966) studied the mobility of gold in ore deposits of the Kuranakh type (auriferous pyrite of metasomatic origin in intensely weathered carbonate rocks) from Yakutia, Aldan shield (U.S.S.R.). Primary gold occurs as microscopically fine native metal in pyrite. On oxidation of the pyrite, part of the gold remains with limonite group minerals which replace pyrite, and the remainder is apparently taken into solution and reprecipitated on clays and with iron hydroxides. RAZIN and ROZHKOV (1963) contend that such transported gold is carried by groundwaters mostly in suspension but partly in ionic form (waters are bicarbonate-type ranging in pH from 6.4 to 7.5 and are devoid of Cl^{1-} and SO_4^{2-}). Their method of distinguishing suspended and ionic (solution) gold was to allow water samples to stand for 5 months, decant the water and determine the gold content of the settled material and decanted water (ionic gold).

The oxidation of auriferous pyrite, with resultant transport of gold in iron-rich groundwaters and deposition with limonite, has been noted by KINKEL and LESURE (1968) for a number of small gold deposits in southeastern U.S.A.; such limonite may contain upto 2.9 ppm gold.

MACHAIRAS (1967) noted that cubic crystals of gold associated with goethite were lower in silver than primary xenomorphic gold disseminated in host pyrite. On a basis of these observations and experimental studies, he suggested that during oxidation of auriferous pyrite, primary gold was dissolved. Subsequent reduction caused $FeSO_4$ to precipitate soluble gold and local hydrolysis precipitated iron oxides around the newly formed secondary gold crystals.

II. Soils

Data on the gold content of soils are presented in Table 79-G-1; they are mainly for soils related to bedrock containing gold mineralization and as such represent an upper limit on the gold content of soils.

A number of studies provide information on the distribution of gold within various soil horizons. Gold in soils from Yakutia (U.S.S.R.) is associated with iron hydroxides and clays (RAZIN and ROZHKOV, 1966). ARIPOVA and TALIPOV (1966) find that in soils of the Kyzylkum area, Tamdyn territory (U.S.S.R.), the highest gold contents are formed at depths of 100 to 140 cm where the metal is in close association with secondary iron minerals. The upper soil horizon is generally impoverished. They also note that plants in arid zones are 40 to 150 times higher in gold than are soils. Thus, gold distribution in soil horizons seems highly variable and generalities should be drawn with due caution.

Slight gold enrichment in highly weathered loose rock material was found in the Tahitian basalts studied by CROCKET *et al.* (1973). Here an enrichment of

Table 79-G-1. *Gold content of soils*

Sample	Locality and description		Number of samples	Gold (ppb) range	Gold (ppb) average	Method	Reference
Soil	Tamdyn Mountains (U.S.S.R.)	ore-free zone, 0—110 cm	4	200—400	260	N	4
		ore-bearing zone, 0—200 cm	4	200—1500	600	N	4
Soil	Kazakhstan (U.S.S.R.)		10		200[a]	N	4
Forest loam etc.	Yakutia (U.S.S.R.)	A+B soil horizons.	27	100—3600	860	W[b]	5
		C soil horizon (related to carbonate rocks)	32	200—8000	1800	W	5
Soil with plant debris	South Carolina (U.S.A.)	A zone, approx. 0—25 cm (grey)	26	20—800	165	A	2
Clayey soil		B zone, approx. 25—180 cm (red)	27	20—240	100	A	2
Saprolite		C zone	4	30—140	70	A	2
Saprolite	Georgia (U.S.A.), protolithic mica schist.		12	<20—100	40	A	3
Soil	Prince Patrick and Melville Islands, Canadian Arctic		3	0.3—1.8	1.0	N	1
Intensely weathered basalt	Tahiti; brown to tan, porous, friable aggregate; groundmass completely altered to clay minerals		6	0.46—2.4	0.90	N	1

[a] All samples at or below the sensitivity limit of the method for soils.

[b] See Subsection 79-L-II for discussion of the analytical method.

References: 1. Crocket *et al.* (1973). 2. Kinkel and Lesure (1968). 3. Lesure (1971). 4. Lobanov *et al.* (1966). 5. Razin and Rozhkov (1966).

approximately 2-fold occurred in weathered basalt debris relative to fresh rock. Also LESURE (1971) found that gold was significantly enriched in quartz-mica schist saprolite at Calhoun Mine, Georgia, relative to parental unweathered quartz-mica schist.

Humus-rich forest soil from the Empire district, Colorado (U.S.A.), was not only enriched in gold but also delineated the known distribution of gold deposits in the bedrock (CURTIN *et al.*, 1968). The range of values found was <40 to >600 ppb with 62% of 962 samples containing >200 ppb. The humus-rich top soil, produced by decay of organic litter from pine and aspen, contained more gold generally than underlying soil (6 to 12-inch depth) where only 22% of the samples contained >200 ppb gold. CURTIN *et al.* (1970) studied the mobility of gold in the forest humus layer by leaching samples with demineralized water and determining gold in leachates which passed 0.05μ membrane filters. They concluded that gold is slightly mobile mainly as colloidal particles ($<0.05\,\mu$ diameter) and in minor amounts as gold cyanide or thiocyanate ions.

Enrichment of gold in the soil-plant layer (A_0 subhorizon) in the vicinity of auriferous sulfide and quartz-sulfide veins of the Darasun area (U.S.S.R.) was reported by AFEROV *et al.* (1968).

III. Bauxites

A survey of the gold content of bauxites has been presented by TENYAKOV *et al.* (1970), and included samples from 40 localities, mostly in the U.S.S.R., on which 344 determinations were made by neutron activation. Values ranged from 0.4 to 29 ppb and averaged 4.5 ppb. It was found that 26 deposits formed from igneous or irruptive mafic and ultramafic rocks averaged 8.4 ppb Au whereas 14 deposits formed from clay and shale averaged only 1.8 ppb. TENYAKOV *et al.* (1970) consider that the gold content of bauxite is clearly related to that of the parent rock. Comparison with the averages for gold in mafic-ultramafic rocks and shales found in the present survey (6.6, 4.8 and 3.6 ppb in ultramafics, gabbros and intermediate-to-mafic volcanics respectively, Table 79-E-2; 2.5 ppb for shales, Table 79-K-3) support this viewpoint.

Mechanical incorporation of native gold together with gold in individual minerals and rock fragments of the source rock is suggested by TENYAKOV *et al.* (1970) as a mechanism by which gold is incorporated into bauxite. They note that RAZIN and ROZHKOV (1966) concluded this was the principal way in which gold migrates in natural waters from Yakutia. Further, they note that if gold is dissolved as an ionic species in the weathering process, adsorption on iron or aluminum hydroxide or kaolinite may effectively inhibit its removal in ground waters.

Revised manuscript received: June 1973

79-H. Solubilities of Compounds which Control Concentrations of Gold in Natural Waters

I. Sea Water

The solubility of gold in sea water has been discussed by KRAUSKOPF (1951), SILLÉN (1959) and PESHCHEVITSKIY *et al.* (1965), and brief comment on the topic has been made by HABER (1928) and HUMMEL (1957). Haber's extensive investigations of the gold content of sea water led him to conclude that the metal occurred not only as chloroauric ion but also as colloidal particles or particles adsorbed on, or included in, suspended inorganic and organic matter. HUMMEL (1957) also concurs that much of the gold content of sea water is associated with suspended particles of organic or inorganic materials and considers that this hypothesis is supported by the decrease in gold content of samples taken progressively away from shore.

Several authors have considered the problem from the point of view of equilibrium of dissolved species with metallic gold. Using standard electrode potentials the distribution of dissolved gold between various species may be calculated and total soluble gold can be compared with the observed gold content of sea water. KRAUSKOPF (1951) presented calculated solubilities of gold in sea water at 25° C in contact with air ($p_{O_2} = 0.2$ atm.) over a pH range 7.5 to 8.4, where solubility was attributed to the species AuO_2^- and $AuCl_4^-$. Solubility was due mainly to $AuCl_4^-$ at pH 7 and AuO_2^- at pH 8.5. At pH 8, the value closest to mean sea water, a total solubility of 3×10^{-12} M (0.0006 ppb), due mainly to AuO_2^-, was calculated. This result would suggest that gold in ionic solution accounts for only a small fraction of total sea water gold which is approximately 0.011 ppb (Section 79-I). KRAUSKOPF (1951) notes that such calculations may be subject to large error (one or two orders of magnitude) due to use of concentrations rather than activities.

Other applications of this approach have yielded much higher values for total gold solubility. SILLÉN (1959) considered that solubility was due to $AuCl_4^-$, $AuCl_2^-$ and $Au(OH)_4^-$. At pH = 8.1 and taking $pE = -\log [e] = 12.5$, he computed that $[AuCl_2^-] = 6.3 \times 10^{-8}$ M and constituted more than 99.9% of total ionic gold. Thus, $[AuCl_2^-]$ is approximately 10 ppb or about 1000 times the average gold content of mid-ocean sea water. One inference from this result is that ionic species are not at equilibrium with metallic gold. SILLÉN concluded that metallic gold is probably not stable in contact with sea water and that dissolved gold is chiefly $AuCl_2^-$.

A similar result was obtained by PESHCHEVITSKIY *et al.* (1965). Based on a redox potential of sea water of 0.74 volts, they find that $AuCl_2^-$ is the dominant species in sea water followed by $AuClBr^-$. These species are present in a ratio of 10:1 and total 1.1×10^{-7} M or approximately 20 ppb Au; this estimate also greatly exceeds the gold content of sea water. PESHCHEVITSKIY *et al.* (1965) infer from their analysis that gold can accumulate in the ocean as $AuCl_2^-$.

II. Surface Waters

The minute quantities of gold found in natural surface waters (see Table 79-I-2) are presumably attributed to either solubility phenomena or to particulate gold, probably of colloidal character. GOSLING *et al.* (1971) found that filterable gold (0.1 μ membrane filter) from waters of the Colorado Front Range (U.S.A.) was generally less than solute gold, and they suggested that solubility phenomena were responsible for the observed concentrations of solute gold found in their study.

The aqueous gold species responsible for such solubility in natural waters are, of course, not amenable to direct observation at such low concentrations. It is useful, however, to determine whether observed natural solubility can be accounted for by various gold solute species known in the laboratory. ONG and SWANSON (1969) considered this problem for the species $AuCl_4^-$, $AuCl_2^-$, AuS^- and AuO_2^-. Using standard electrode potentials from MARTELL and SILLÉN (1964), they obtain electrode potentials, $\varepsilon°$, and equilibrium constants K_{eq} for the formation of these species as tabulated below:

Equation	$\varepsilon°$ (volts)	log K_{eq} (25° C)
$4Au + 4HS^- + O_2 = 4AuS^- + 2H_2O$	−0.879	60
$2Au + 8Cl^- + 6H^+ + 3/2O_2 = 2AuCl_4^- + 3H_2O$	−0.229	23
$2Au + 2H^+ + 1/2O_2 + 4Cl^- = 2AuCl_2^- + H_2O$	−0.110	4
$2Au + 2OH^- + 3/2O_2 = 2AuO_2 + H_2O$	+0.099	−10

Using the following data[1] for mean river composition: log $[Cl^-] = -3.657$ M, log $[SO_4^{2-}] = -3.921$ M, log $[HS^-] = -124$ to -119 M, pH = 4 to 9 and log P_{O_2} -0.6990 atm., the calculated solubilities of gold in fresh surface water at 25° C in contact with air are as tabulated:

	log $AuCl_2^-$	log $AuCl_4^-$	log AuO_2^-	log AuS^-	log ΣAu
pH 4	−10.5	−17	−15.5	−109	−10.5
pH 9	−15.5	−32	−10.5	−104	−10.5

This type of analysis indicates a solubility of $10^{-10.5}$ M or 0.006 ppb, a value well within the range of many determinations of surface waters and particularly close to the median value of 0.010 ppb found by GOSLING *et al.* (1971) in their study of Colorado Front Range waters.

ONG and SWANSON (1968) also investigated the effect of natural organic acids in the transportation, deposition and concentration of gold. They showed by laboratory experiments that organic concentrations in the range of 3 to 30 ppm may reduce gold chloride solutions to negative colloids of metallic gold stable for up to 8 months and of very small size, < 10 mμ. They suggest that in the pH range of most fresh surface waters (pH 4 to 9), gold can be transported as organic-protected colloids.

Thus both true solute and colloidal gold are probably important components of total gold solubility in fresh surface waters.

[1] Cl^- and SO_4^{2-} from LIVINGSTONE (1963); HS^- computed from sulfate based on the equilibrium $SO_4^- + 9H^+ + 8e^- = HS^- + 4H_2O$, $\varepsilon° = -0.252$ v according to GARRELS and CHRIST (1965); P_{O_2}, equilibrium value with the atmosphere.

III. Ground Waters

There is evidence for appreciable solubility of gold in waters of both acid and neutral to weakly alkaline character. The solubility of gold in alkaline sulfide solutions has most recently been investigated by WEISSBERG (1970) to account for gold transport in New Zealand geothermal waters. Some previous studies of alkali sulfide transport of gold include those of ORGRYZLO (1935), ZVYAGINSTEV and PAULSEN (1940), LINDNER and GRUNER (1939), SMITH (1944), KRAUSKOPF (1951), and KAKOVSKII and TYURIN (1962). WEISSBERG (1970) found experimentally that gold solubilities of 100 to 200 ppm could be obtained in neutral to weakly alkaline bisulfide solutions at temperatures of 150—200° C. In alkaline Na_2S solutions, solubilities of 1 to 6 ppm only were found and evidence is cited suggesting low solubility in H_2S solution. WEISSBERG (1970) suggests that gold solubility in bisulfide-dominated solutions is due to a complex tentatively formulated as AuS^-. Such solubility adequately accounts for gold transport in some of the near-neutral low salinity geothermal solutions of the Broadlands-Ohaki and Waiotapu areas of New Zealand where recently deposited precipitates contain up to 85 ppm Au.

Evidence for significant solubility of gold in acidic, chloride-rich waters may also be found. GOLEVA *et al.* (1970) surveyed hot and cold carbonated water of the U.S.S.R., finding the maximum gold content in strongly acid, chloride-rich fumarole waters ($pH < 1$, $Eh > 0.9$ volts, $T = 92°$ C) of Ebeko volcano (Kurile Islands), and suggested gold occurred preferentially as $AuCl_4^-$ and $AuCl_2^-$. A further example is found in Koga's (1961) study of Japanese thermal waters from Beppu, Japan, in which the highest gold contents were found in acid waters of Na_2SO_4 —NaCl type. Theoretical support for the importance of chloride complexes in gold transport and solution is provided by studies of KRAUSKOPF (1951) and KELLY and CLOKE (1964). In particular, KRAUSKOPF (1951) calculated from electrode potential and free energy data that gold would dissolve as $AuCl_4^-$ in acid solutions containing excess chloride and at a sufficiently high oxidation potential. The calculations applied to reactions at 25° C and suitable natural oxidizing agents would include MnO_2, O_2, Fe^{3+} and Cu^{2+}. More recently HELGESON and GARRELS (1968) considered gold transport in hydrothermal fluids capable of depositing gold-quartz-pyrite veins. On a basis of an estimated "geologic" solubility of gold and thermodynamic data, they concluded that gold transport in acidic concentrated NaCl solutions in equilibrium with quartz and pyrite occurs above 175° C with gold solubility due to aurous chloride complexes.

Revised manuscript received: June 1973

79-I. Abundance in Natural Waters and in the Atmosphere

I. Sea Water

A comprehensive review of gold determinations in sea water has been presented by JONES (1970) and a more limited summary may be found in SHUTZ and TUREKIAN (1965). Early attempts to measure the gold content of sea water gave average values near 5 ppb. HABER (1927; 1928) first clearly demonstrated that these results grossly over-estimated the actual gold content of sea water. His extensive research, spanning a period of ten years, suggests an average gold content of sea water of 0.004 ppb. Modern activation analyses (HUMMEL, 1957; SHUTZ and TUREKIAN, 1965) suggest an average of approximately 0.01 ppb for offshore sea water.

Table 79-I-1 presents analytical data from 1927 for which quantitative results have been reported and the analytical method described.

HABER's work (1927; 1928) is the only study of sufficient scope to bear on the question of regional variation of gold in sea water. His Arctic samples give an average value approximately 3.7 and 5.5 times higher than Pacific or South Atlantic samples respectively (Table 79-I-1), and he regarded the South Atlantic as a gold-poor region. Evidence that the gold content of near-shore sea water is higher than mid-ocean water is found particularly in the data of HABER (1927; 1928), HUMMEL (1957) and SHUTZ and TUREKIAN (1965), although all pertinent data (for example, OKA *et al.*, 1964) are not entirely consistent with this generalization. HUMMEL found approximately 10 times more gold in British coastal waters (0.139 ppb) than in samples from Bay of Biscay about 100 miles offshore (0.015 ppb). His Bay of Biscay average is similar to the average of 0.011 ppb obtained by SHUTZ and TUREKIAN (1965) for 11 mid-ocean samples. Much of HABER's mid-ocean values suggest 0.01 to 0.05 ppb Au although his San Francisco Bay samples, which are near-shore, are also in the range 0.01 to 0.02 ppb Au. In contrast, much of the other near-shore data (CALDWELL, 1938; PUTNAM, 1953; STARK, 1943) suggest values within the range 0.1 to 0.5 ppb Au. Of the 21 activation analyses of OKA *et al.* (1964), 17 are within approximately 100 miles of shore (Honshu Island) and average 0.079 ppb Au, whereas a group of 4, approximately 200 miles offshore, average 0.347 ppb Au. As a whole the group may be regarded as near-shore samples, averaging 0.130 ppb Au, and thus consistent with higher values for near-shore water. In detail, however, the trend is not consistent and there are other complicating factors. In particular, OKA *et al.* note that the highest concentrations of gold coincide with the line of convergence of the cold northern Oyashio current and the warm Kuroskio current from the southeast. Further, they point out that surface waters are higher in gold than waters collected at depth (14 surface samples average 0.167 ppb, whereas 7 samples from depths of 100 and 200 meters average 0.056 ppb Au).

Table 79-I-1. *Gold content of sea water*

Sea	Location and description	Number of samples	Gold (ppb) range	Gold (ppb) average	Method[a, b, c]	Reference[aa]
South Atlantic	186 stations	1635	—	0.004	Fa	12
South Atlantic	Lat. 42° S; 20 stations	36	0.001—0.044	0.008	Fa	11
South Atlantic	Capetown (South Africa); near-shore	1	—	0.09	S	2
North Atlantic	Near Newfoundland Banks	5	2.88—8.46	5.79	Fa	11
North Atlantic	British coast, Portland; depth 0—50 m	29	0.015—0.508	0.139	N	14
North Atlantic	Bay of Biscay; depth 0—600 m	11	0.010—0.021	0.015	N	14
North Atlantic	Gullmarfjord (Sweden); shallow water	1	—	0.008	S	17
North Atlantic		8	0.0038—0.027	0.011	N	20
North Pacific	San Francisco Bay (U.S.A.)	87	0.002—0.055	0.012	Fa	11
North Pacific	Coast of Washington and Oregon (U.S.A.)	12	0.08—1	0.27	Fa	4
North Pacific	Coast of Washington (U.S.A.)	2	5.1—44	25	E	19
North Pacific	East of Honshu Island (Japan)	21	0.006—0.43	0.13	N	18
Arctic		42	0.009—0.19	0.044	Fa	11
Mediterranean	Coast, Gibraltar to Genova, Corsica and Monaco	71	0.05—1.9	0.4	Fa	21
Adriatic	Coast, Venice to Fiume	7	0.009—0.020	0.02	Fa	21
Indian		4	0.0043—0.013	0.010	N	20

[a] Preconcentration of gold was used by all investigators except Hummel (1957).
[b] Fa — fire assay (Haber and Jaenicke, 1925 and others).
[c] E — electrolysis.
[aa] References are in the footnotes to Table 79-I-3.

Table 79-I-2. *Gold content of fresh surface waters*

Water source	Locality	Comment	Number of samples	Gold (ppb) range	Gold (ppb) average	Method[a]	Reference[b]
Rivers	Irkutsk, Transbaikalia, Caucasus and the Soviet Far East (U.S.S.R.)	Rivers: Krestovka, Zhorga, Unda, Gazimur, Baksan, Khingsn and Sutar	7	0.006—4.7	0.15[c]	C	5
Streams	Colorado (U.S.A.)	Steams from igneous, metamorphic and sedimentary terrane; areas of both extensive and minimal mining activity; average particulate contribution, 22%[aa]	29	0.001—0.081	0.022	N	8
Rhine River	Near Karlsruhe, Schopfstellen and Leverkusen (Germany)	—	24	0.0001—0.1	0.0033	Fa[cc]	13
Saale River	East Germany	Unfiltered water[bb]	1	—	0.002	S	6
Streams	Baley region, East Transbaikalia (U.S.S.R.)	Gold-Quartz veins in the area	1	—	0.06	C	7
Streams	Near Kuranakh ore area, Aldan massif (U.S.S.R.)	15 determinations; 4 quantitative results (analytical sensitivity, 0.01 ppb)	4	0.03—0.2	0.08	S	1
Lake Baikal	U.S.S.R.	—	1	—	0.12	C	5
Greek Lakes	Greece	Filtered surface samples	10	0.1—7.7	1,8	N	7
		Filtered samples, 4—15 meters depth	10	0.2—6.5	1.3	N	7

[a] Gold is usually concentrated from the water sample prior to analysis. Adsorption on charcoal, co-precipitation on PbS, co-precipitation with calomel, and ion exchange chromatography are methods widely used. Sample volumes are generally 3 to 4 liters.

[b] References are in the footnotes to Table 79-I-3.

[c] Maximum value of 4.7 ppb was not used in calculation of the average.

[aa] Samples were filtered through 0.1 μ membrane filter. Material retained on filter classified as particulate. Table gives average percentage contribution of particulate to total gold content of whole water sample.

[bb] Suspended matter in the Saale River contained 100 ppb gold.

[cc] Fa — fire assay.

For mid-ocean sea water, the average value of 0.011 ppb Au obtained by SHUTZ and TUREKIAN (1965) by activation analysis of 11 samples is regarded as the best available estimate. For near-shore samples the weighted average of HUMMEL (1957) and OKA *et al.* (1964), 0.13 ppb Au, is considered as the best estimate.

II. Fresh Surface Water

Gold contents of fresh surface waters including rivers, streams and lakes are tabulated in Table 79-I-2. They suggest an average for gold in fresh surface waters in the range of 0.001 to 1 ppb although the reliability of some of the earlier data is probably poor.

Of the more recent contributions to the hydrogeochemistry of gold, that of GOSLING *et al.* (1971) suggests a number of important features. Their values for 29 surface water samples from the Colorado Front Range average 0.022 ppb and show a very small range in total gold content. Waters did not reflect highly variable gold content of the drainage area, nor did 27 samples of water from mine tunnels or prospect pits (averaging 0.029 ppb) differ greatly from the average of non-mine waters. The gold content of suspended solids was generally less than that of solute (water) gold. GOSLING *et al.* state that the median value for solute gold was 0.010 ppb and within the range predicted from dissociation constants of the pertinent aqueous gold complexes (see Section 79-H).

Studies of lake waters are very limited. The results of GRIMANIS *et al.* (1964) are considerably higher than most analyses of river waters and are somewhat surprising in view of the analytical method employed.

The average gold content of fresh surface waters cannot be reliably estimated from the data presently available. TUREKIAN (1969) gives a value of 0.002 ppb although the source data are not indicated. The activation data of GOSLING *et al.* (1971), although from an area of gold mineralization, suggest an upper limit of 0.02 ppb for average gold content of fresh surface waters.

III. Ground Waters, Geothermal Waters etc.

Data for ground waters including geothermal waters from hot springs, fumaroles and boreholes are presented in Table 79-I-3, and the values range from 0.004 to 2.2 ppb. Reviews of the hydrogeochemistry of gold including ground waters have been published by CHERNYAYEV *et al.*, (1967) and JONES (1970).

A number of significant features of the ground water geochemistry of gold are well established. For example, geothermal waters are often very low in gold, 0.04 ppb or less, whereas their precipitates, particularly siliceous sinter, may contain ppm levels of gold. Waters and precipitates from the Taupo volcanic zone of New Zealand, Yellowstone National Park (U.S.A.), and Beppu (Japan) exhibit this relationship (BROWNE, 1969; WEISSBERG, 1969; GOTTFRIED *et al.*, 1972; TILLING *et al.*, 1973; KOGA, 1961). WEISSBERG (1969) gives values of 80, 70, 85, and 55 ppm Au respectively for precipitates, characteristically opaline silica containing As and Sb-rich red-orange amorphous sulfide, from the following localities: Champagne Pool, Waiotapu; Drill Hole 2, Rotokawa; Ohaki Pool, Broadlands, and Drill Hole 2, Broadlands. Waters from which these sulfide-bearing sinters have precipitated contain 0.04 and $<$0.04 ppb Au (Drill Hole 2, Broadlands, and Ohaki Pool respectively), a factor of

Table 79-I-3. *Gold content of ground waters*

Water	Locality and description	Number of samples	Gold (ppb) range	Gold (ppb) average	Method[a]	Reference
Hot springs	Beppu, Kyushu Island (Japan); T = 43—98° C; pH = 2.0—7.6	27	0—2.2[b]	0.53	C	15
Hot springs	Yellowstone National Park, Wyoming (U.S.A.)	—	0.004—0.1	—	N	9
Fumarole	Kuril Islands (U.S.S.R.); T = 92° C; pH = 0.65; HCl water	1	—	1.6	C	7
Hot spring	Angren area, Central Asia; T = 18° C; pH = 7.6	1	—	0.07	C	7
Hot spring	Ohaki Pool (New Zealand); T = 95° C; NaCl-$NaHCO_3$ water; pH ≃ 7	1	—	⩽0.04	S	3
Geothermal borehole discharge	Broadlands-Ohaki (New Zealand); pH = 8.3; T = max 276° C; depth = 1034 meters	1	—	0.04	S	3
Borehole water	Angren area (U.S.S.R.); T = 35° C; pH = 7.5	1	—	0.07	C	7
Carbonated spring water	Baley and Gazimur areas, East Transbaikalia; T, avg. = 6.9° C; pH avg. = 6.3	4	0.1—0.2	0.14	C	7
Near-surface ground waters	Kokchetev region, Kazakhstan	180	0.1—4	0.15	S	5
Ground water	Kazakhstan (U.S.S.R.); waters circulating in rocks averaging 3 ppb Au.	106	0.1—2.0	0.4	S	5

[a] Usually concentration of gold from 1 to 3 liter samples prior to analytical determination was required. Concentration was effected mainly by co-precipitation on tellurium, solvent extraction or absorption on activated charcoal.

[b] Lowest reported positive result is 0.1 ppb Au.

References for Tables 79-I-1, 79-I-2 and 79-I-3: 1. Borovitskiy *et al.* (1966). 2. Brooks (1960). 3. Browne (1969). 4. Caldwell (1938). 5. Chernyayev (1969). 6. Fischer (1966). 7. Goleva *et al.* (1970). 8. Gosling *et al.* (1971). 9. Gottfried *et al.* (1972). 10. Grimanis *et al.* (1964). 11. Haber (1927). 12. Haber (1928). 13. Haber and Jaenicke (1925). 14. Hummel (1957). 15. Koga (1961). 16. Konovalov (1935). 17. Noddack and Noddack (1939). 18. Oka *et al.* (1964). 19. Putnam (1953). 20. Schutz and Turekian (1965). 21. Stark (1943). 22. Weiss and Lai (1963).

approximately 1.8×10^6 less. For the thermal springs of Yellowstone National Park, Wyoming, and their siliceous precipitates, GOTTFRIED *et al.* (1972) found a similar effect with the precipitates containing up to 5 ppm Au and the waters as little as 0.004 ppb Au. Thus, an upper limit of 1.3×10^6 is found for the partition of gold between sinter and water which compares closely with the New Zealand thermal waters. The data of KOGA (1961) for 27 hot springs from Beppu, Kyushu Island, Japan, show the same water-precipitate trend although with some differences. Beppu waters average 0.53 ppb Au which is considerably higher than the New Zealand or Yellowstone waters. Precipitates from seven springs average 21.2 ppm and are in the general range given by WEISSBERG (1969) and GOTTFRIED *et al.* (1972). The ratio, gold in precipitate to gold in water, for these samples is 4×10^4.

GOTTFRIED *et al.* (1972) point out that acid volcanics, through which the Yellowstone geothermal waters circulate, are low in gold, 0.1—1.3 ppb. This suggests that high gold concentrations can be produced in the associated precipitates without circulation of such waters through sub-surface rocks enriched in gold. Factors such as discharge rate, life span of the geothermal system and a precipitation mechanism effective for small quantities of gold may be of great importance for geothermal waters.

The extensive hydrogeochemical study of CHERNYAYEV *et al.* (1967) which utilized a method having a sensitivity of 0.05 ppb Au, suggested a log-normal distribution of gold in ground waters of the Kokchetav region (U.S.S.R.). In a suite of 180 samples gold was found in 40% and varied from 0.1 to 4 ppb. The average of all positive results was 0.4 ppb and the average of all results was 0.15 ppb. CHERNYAYEV *et al.* (1967) also give values for ground waters from gold mines mainly from East Transbaikalia, Southern Yakutia and the Urals. A wide range of values, 0.001 to 50 ppb, is reported. A notable consistency appears to be high values for waters associated with Kuranakh type (dispersed gold in carbonate rocks) deposits.

IV. The Atmosphere

DUDEY *et al.* (1969) analyzed five samples of over-ocean aerosols (collected from a ship in the Atlantic Ocean) using a neutron activation method. From their results an average value of 3.3×10^{-6} μgmAu/meter^3 of air sampled is calculated. Gold was not reported by DAMS *et al.* (1970) who analyzed particulates from Chicago air by a neutron activation method with a sensitivity of 1×10^{-4} μgm Au/meter^3 for a 24 hr. sample.

Revised manuscript received: June 1974

79-K. Abundance in Common Sedimentary Rocks and Sediments

Data for sedimentary rocks and sediments are presented in Tables 79-K-1 and 79-K-2 respectively, and averages derived from these tables are compared with previous estimates in Table 79-K-3. Table 79-K-1 includes data obtained since 1960; earlier data may be found in the compilation of JONES (1969).

Only very tentative generalizations are drawn from this relatively limited data: 1. Conglomerates and sandstones tend to be higher in gold than other sedimentary rocks and are characterized by great variability in gold content; 2. Carbonate rocks are generally lower in gold than shales, sandstones and conglomerates.

Extensively studied areas include the Cretaceous and Tertiary rocks of northwest Wyoming (U.S.A.) (ANTWEILER and LOVE, 1967), and the Triassic Moenkopi Formation of the Colorado Plateau region (U.S.A.) (CADIGAN, 1971). In the Moenkopi Formation, gold enrichment is found in mudstones below the Chinle-Moenkopi contact and in claystones which tend to be higher than coarse sandstones. Significant positive geochemical association of Au is found with Fe, Cu, V and Co. In the northwest Wyoming rocks (mainly the Harebell Formation and Pinyon Conglomerate) where many of the analysed rocks are conglomerates, the gold content is largely confined to the sandy matrix and to sandstone interbeds according to ANTWEILER and LOVE (1967). The gold values reported in these studies were obtained by atomic absorption methods with sensitivities of approximately 20 ppb and they represent the highest values in Table 79-K-1. Both areas are probably anomalously high in gold. Placer gold has been known in Wyoming for more than 100 years. Further, the high values may in part reflect the limited sensitivity of the atomic absorption method. In the Moenkopi study, for example, one-third of the analyzed samples only contained detectable gold (i.e. 20 ppb). For these reasons, the data from these two areas have not been used in computing the averages presented in Table 79-K-3. The average values are compared with the previous estimates of JONES (1969) and the computer-derived geochemical balance estimates of HORN and ADAMS (1966).

Much of the data on sediments is from the studies of MOORE and SILVER (1968), CLIFTON (1968) and CLIFTON *et al.* (1967) on sand from the continental shelf of California and Oregon (U.S.A.). An average of 9.6 ppb Au was obtained from the data of MOORE and SILVER for 61 samples dredged from the coast of northern California. Approximately two-thirds of the total gold represented by these 61 samples, however, is contributed by one sample. MOORE and SILVER suggest that the background gold content of this shelf is about 0.1 ppb. They identified four anomalous tracts from 10 to 30 square kilometers in area having gold contents in excess of 10 ppb; it is suggested these are lag concentrates produced from Cenozoic strata by wave erosion during the postglacial rise in sea level.

Table 79-K-1. *Gold content of sedimentary rocks*

Rock type	Locality and description	Number of samples	Gold (ppb) range	Gold (ppb) average	Method	Reference
Quartzite conglomerate	Wyoming (U.S.A.)	407	<10—2000	87	A	1
Conglomerate	Kola Peninsula (U.S.S.R.), Kolmozero-Voronya series	8	0.7—80	11	S	9
Conglomerate	Maritime Territory (U.S.S.R.), Ogodzhinsk syncline	—	—	100	S	14
Sandstone	Wyoming (U.S.A.)	218	—	109	A	1
Greywacke	Yellowknife, Northwest Territories (Canada)	—	—	10	Fa[aa]	4
Sandstone	Colorado Plateau (U.S.A.), Moenkopi Formation (Triassic)	12	10—200	52	A	5
Sandstone	Wisconsin, Kentucky (U.S.A.); Ontario (Canada); Australia	5	0.3—12	3.6	N	3, 8
Sandstone	California (U.S.A.), Tertiary, Kettleman Hills	1	—	41	N	8
Sandstone	Martime territory (U.S.S.R.), Ogodzhinsk syncline	—	—	6.5	S	14
Sandstone	Maritime territory (U.S.S.R.), Sagursk series	—	—	2.8	S	14
Sandstone	Altai and Chukotka (U.S.S.R.), includes chert from 3 sandstones	27	0.6—6.2	2.1	N	17
Sandstone	Taskazgan, Tamdytau (U.S.S.R.), Precambrian, mainly sandstone, mudstone, schists and carbonaceous-siliceous rocks	201	?—2500	25	S[b]	11
Calcareous sandstone	Chukotka (U.S.S.R.), Iultinsk massif	1	—	5	S	15
Sandstone, siltstone, argillite	Kazakhstan (U.S.S.R.), Proterozoic to Upper Ordovician	51	1.0—7.8	2.6	S	19
Sandstone and alevro-argillite	Primurya region (U.S.S.R.)	14	0.8—4.0	1.7	S, N	15
Siltstone	Colorado Plateau (U.S.A.), Moenkopi Formation (Triassic)	6	10—30	18	A	5

Siltstone	Maritime territory (U.S.S.R.), Ogodzhinsk syncline	—	—	2.5	S	14
Siltstone	Altai (U.S.S.R.)	2	5.5—6.2	5.9	N	17
Shale	Wyoming (U.S.A.), Harebell Formation and Pinyon Conglomerate	24	—	57	A	1
Shale	Kuznetsk Ala-Tau and Chukotka (U.S.S.R.)	23	0.66—8.6	2.3	N	17
Shale	Kansas and South Dakota (U.S.A.)	2	4.7—7.2	6.0	N	8
Shale	Chukotka (U.S.S.R.), Iultinsk massif	2	1—1	1	S	15
Argillaceous rock	Maritime territory (U.S.S.R.), Sagursk series	—	—	1.2	S	14
Mudstone and claystone	Colorado Plateau (U.S.A.), Moenkopi Formation	6	10—460	142	A	1
Limestone	Wyoming (U.S.A.), Harebell Formation and Pinyon Conglomerate	13	—	38	A	1
Limestone	Kansas (U.S.A.), Paola Limestone	1	—	4.8	N	8
Limestone	Altai and Chukotka (U.S.S.R.)	3	1.3—2.6	2.1	N	17
Limestone	Kazakhstan (U.S.S.R.), Proterozoic and Upper Ordovician	11	<1.0—3.2	1.0	S	19
Limestone	Northern Nura-Tau, Uzbekistan (U.S.S.R.)	2	5—5	5	W[c]	13
Limestone, dolomite, calcareous siltstone	Colorado Plateau (U.S.A.), Moenkopi Formation (Triassic)	6	10—60	22	A	5
Carbonate	Florida (U.S.A.), Recent	2	0.8—3.9	2.4	N	8
Oolite	Cheltenham (England)	1	—	2.3	N	8

[a] See footnotes to Table 79-K-3 for references.
[b] Analyses carried out by both spectrochemical and assay methods.
[c] Analytical procedure described as a gold chemical method.
[aa] Fa — fire assay.

Table 79-K-2. *Gold content of Recent sediments*

Rock type	Locality and description	Number of samples	Gold (ppb) range	Gold (ppb) average	Method	Reference
	a) Deep sea sediments					
Pelagic clay	Brazil Basin; core, 3 samples 343 to 887 cm	1	4.2—31	16	N	8
Lutite	Argentine Basin; core, 4 samples 100 to 1045 cm	1	3.1—17.3	7.2	N	2, 8
Lutite etc.	Indian, Antarctic and Pacific Oecans	7	0.6—4.2	2.0	N	7
Clay-rich sediment	East Pacific Rise from 90° W to 139° W at 13° S	14	1.3—7.6	3.1	N	7
Sediment	Pacific; 56—60 cm core interval analyzed	1	—	9.0	N	2
Calcareous ooze	Caribbean; 15 samples 19 to 1369 cm	1	0.9—4.0	2.2	N	12
Globigerina ooze	Antarctic; core, 9 samples 0 to 1006 cm	1	0.6—2.3	1.2	N	12
Siliceous ooze	Antarctic; core, 9 samples 0 to 1600 cm	1	0.58—1.5	0.87	N	12
Pelagic clay	Antarctic; core, 4 samples 0 to 1585 cm	1	1.0—2.3	1.6	N	12
Manganese nodules	Atlantic, Pacific, Indian and Antarctic Oceans	19	0.21—8.3	2.9	N	10
	b) Near-shore, continental shelf sediments					
Sand and silt	California coast	61	<0.1—390	9.6[b]	A	16
Marine-terrace sands	Oregon coast	6	165—1100	410[b]	A	6
Beach sand	Oregon coast	1	49—56[b]	—	A	6
Near-shore clay-silt	Canadian Arctic	9	1.6—4.2	2.7	N	7

[a] See footnotes to Table 79-K-3 for references.

[b] Analysis of heavy mineral concentrates recalculated to whole sample.

The average Au content of deep-sea sediments from 28 localities is 3.4 ppb. No large differences in gold content are found in compositionally distinct types of deep sea sediments although clay-rich types appear to be slightly higher than those containing a large biogenic component. In a study of the distribution of gold across the East Pacific Rise at 13° S latitude, CROCKET *et al.* (1973) found a general enrichment of the metal in core samples taken within approximately 10° of the rise crest and suggest a volcanic source.

Table 79-K-3. *Estimates of the average gold content of sedimentary rocks*[a, b]

Rock type	Table 79-K-1		JONES (1969)		HORN and ADAMS (1966)
	average ppb Au	number of samples	average ppb Au	number of samples	average ppb Au
Sandstone[c] and siltstone	3.0	105	7.2	11	4.57
Shale	2.5	28	3.9	5	3.45
Carbonates	2.0	20	3.0	6	1.79
Deep sea sediments	3.4	28[aa]			

[a] If the number of samples analyzed is unknown, the average entered in Table 79-K-1 is considered a single value in calculation of the overall averages in Table 79-K-3.

[b] Data of ANTWEILER and LOVE (1967) and CADIGAN (1971) not used in calculation of averages for reasons noted in text.

[c] Sandstones analyzed by LOSHCHININ (1971) considered anomalously high and not used in computation of average.

[aa] Each core is treated as one sample represented by the average value of all analyses for different intervals along the core.

References for Tables 79-K-1, 79-K-2 and 79-K-3: 1. ANTWEILER and LOVE (1967). 2. BAEDECKER and EHMANN (1965). 3. BAEDECKER (1967). 4. BOYLE (1960). 5. CADIGAN (1971). 6. CLIFTON *et al.* (1968). 7. CROCKET *et al.* (1973). 8. DEGRAZIA and HASKIN (1964). 9. GAVRILENKO and GARIFULIN (1971). 10. HARRISS *et al.* (1968). 11. LOSHCHININ (1971). 12. KUO (1971). 13. KUSHMURADOV (1970). 14. MOISSENKO and MIKHAILOV (1970). 15. MOISSENKO *et al.* (1971). 16. MOORE and SILVER (1968). 17. SHCHERBAKOV (1967). 18. SHILIN (1968). 19. VOSKRESENSKAYA and ZVEREVA (1968).

Revised manuscript received: June 1973

79-L. Biogeochemistry

Gold in terrestrial plants has received considerable attention from the impetus of biogeochemical prospecting, but other organisms have not been studied in any great detail. A recent compilation by JONES (1970) presents a comprehensive tabulation of biogeochemical data on gold. Data for marine plants and animals and terrestrial plants is presented in Table 79-L-1.

I. Marine Plants and Animals

Data for marine plants and animals given in Table 79-L-1 do not include values of NODDACK and NODDACK (1939) who reported 7 to 30 ppm Au in sea urchins, sea cucumbers and starfish. By comparison with recent data, their values appear too high by approximately two orders of magnitude. The limited available data preclude a discussion of the biogeochemistry of gold in marine organisms except to note that the average value for gold in marine plants and animals is well above that for sea water, either coastal or mid-ocean (0.13 and 0.011 ppb respectively, see Section 79-I-1). Thus, the average gold content for marine organisms is approximately 345 times higher than the mean value in coastal sea water. Marine organisms therefore appear to be accumulators of gold relative to their environment (see discussion by BOWEN, 1968, with respect to sponges).

II. Terrestrial Plants

Because much of the data on terrestrial plants is for specimens growing in areas of anomalously high gold in associated bed rock, the averages presented in Table 79-L-2 probably represent upper limits with respect to the biosphere as a whole. The majority of data are from the study of RAZIN and ROZHKOV (1966) on the geochemistry of gold in the biosphere associated with gold ore deposits of Kuranakh type from the Yakutia region, U.S.S.R.

RAZIN and ROZHKOV (1966) determined gold by a dithizone titration method with a sensitivity of 100 ppb and by a spectrochemical method with sensitivity of 1 ppb on ashed samples. The spectrochemical method was applied to many samples reported as trace (below 100 ppb) on a basis of the dithizone titration method. If the ash content was reported for these samples, the ash results have been recalculated in terms of dry plant tissue. As RAZIN and ROZHKOV (1966) report slightly more than half their total analyses as trace or not found, as the area of their investigation contained ore-grade gold concentrations, and as their data accounts for the majority tabulated in Table 79-L-1, it is clear that the averages for terrestrial plants in Table 79-L-2 should be regarded as upper limits with respect to average terrestrial flora.

A second point concerning Table 79-L-1 is that the data of BABIČKA (1943) have been omitted. A comparison of BABIČKA's (1943) data with those of all other

Table 79-L-1. *Gold contents of plants and marine animals*

Organism	Locality and description	Number of samples	Gold (ppb)[a] range	Gold (ppb)[a] average	Meth-od	Refer-ence
			Marine animals			
Sponge	Plymouth (England)	1	—	126	N	3
Sponges	Hawaii	5	18—55	31	N	3
Mollusc	Tokyo Bay (Japan)	1	—	5.7(0.079)	N	6
Crustacean	Japan Isles	1	—	0.28(0.0046)	N	6
Mackerel	Japan Isles	1	—	0.12(0.0026)	N	6
			Marine plants			
Algae (Sea lettuce)	Sagami and Tokyo Bays (Japan)	2	3.5—21	12(0.054)	N	6
Algae (Kale)	Biological standard	1	—	2.1	N	2
			Land plants — nonvascular			
Fungi	Yakutia (U.S.S.R.)	3	1.0—11.5	6(1.3)	W	8
Reindeer and dog lichen	Yakutia (U.S.S.R.)	5	4.8—33.0	14.6(1.1)	W	8
Mosses	Yakutia (U.S.S.R.)	8	3.3—590	113(5.3)	W	8
			Land plants — vascular *Herbs*			
Grasses and sedges, 75% bluejoint grass	Yakutia (U.S.S.R.)	15	0.1—160	26(1.0)	W, S	8

Table 79-L-1 (continued)

Organism	Locality and description	Number of samples	Gold (ppb)[a] range	Gold (ppb)[a] average	Method	Reference
			Forbes			
Forbes	Bockhardt (Austria)	4	6.5—60	25	N	9
Forbes, wormwood	Tamdyn Mountains (U.S.S.R.)	47	3.5—63.4	30.6[b]	N	1
Forbes, Ferula sp.	Tamdytau (U.S.S.R.), stems and bark	1	—	35	N	10
Forbes, 16 genera	Yakutia (U.S.S.R.)	32	0.14—310	46(1.65)	W, S	8
Forbes, mainly genus Salsola	Central Kazakhstan (U.S.S.R.)	4	113—320	228[c](1.75)	N	7
Horsetails	British Columbia (Canada), gold placer area	4	30—75	55(0.30)	Fa[aa]	11
Horsetails	U.S.A.	16	19—66	40[c](0.15)	A	4
Horsetails	Alaska	6	25—71	50[c](0.34)	A	4
			Shrubs			
Mainly crystal tea, bramble prickly rose	Yakutia (U.S.S.R.)	83	0.005—150	12(0.86)	W, S	8
5 genera	Kazakhstan (U.S.S.R.)	5	485—400	1,359[c](10.2)	N	7
Dwarf juniper and willow	British Columbia, Canada; gold placer area	2	20—23	21(0.82)	Fa[aa]	11
Salsola, Atraphaxis, Haloxylon, Capparis and Tamarix	Tamdytau (U.S.S.R.) leaves	4	30—300	129	N	10
	stems	4	30—200	73	N	10
	bark	3		30	N	10
			Trees			
Pine, birch, aspen, larch etc.	Yakutia (U.S.S.R.), soil 1.3 ppm; leaves, branches, needles, cones, bark and wood analysed	45	0.011—53	6.5(0.5)	W, S	8

Gurjunoil tree	Kazakhstan (U.S.S.R); soil ≤0.2 ppm	1	—	48[c](0.3)	N	7
Aspen	British Columbia (Canada), gold placer area	1	—	20(0.89)	Fa[aa]	11
Lodge pole pine	British Columbia (Canada)	1	—	2.5(0.03)	Fa[aa]	11
Pine	Colorado (U.S.A.) trunk interior	4	500—1000[b]	—	A	5
	branch interior	5	200—2000	—	A	5
	needles	12	40—700	—	A	5
	root	1	—	3	A	5
Aspen	Colorado (U.S.A.) trunk interior	7	700—1600[b]		A	5
	branch interior	5	<40—1000		A	5
	leaves	9	60—100		A	5
	roots	1	—	200[b]	A	5

[a] With respect to dry tissue unless otherwise noted. If a figure in parentheses follows the average it is the gold content of ash in ppm.
[b] Gold content of ash in ppb.
[c] Value for whole plant calculated. Author gives gold content of ash and ash content of plant.
[aa] Fa — fire assay.

References: 1. Aripova and Tapilov (1966). 2. Beardsley *et al.* (1965). 3. Bowen (1968). 4. Cannon *et al.* (1968). 5. Curtin *et al.* (1971). 6. Fukai and Meinke (1962). 7. Lobanov *et al.* (1966). 8. Razin and Rozhkov (1966). 9. Schiller *et al.* (1971). 10. Tapilov *et al.* (1968). 11. Warren and Delavault (1950).

investigators (see JONES, 1970), shows that with few exceptions they are considerably higher and, as pointed out by CANNON *et al.* (1968), there are strong grounds for doubting the validity of their analytical technique.

Based on the tabulated data and other literature noted below, the following generalizations on the biogeochemistry of gold in terrestrial plants are drawn:

a) The averages in Table 79-L-2 indicate decreasing gold contents from simple to complex plants. The trend is applicable to the Yakutia region only being based largely on RAZIN and ROZHKOV's (1966) data. Its significance is difficult to assess and confirmation by study of other areas with methods of sufficient sensitivity to determine background gold in all flora is required for confirmation.

Table 79-L-2. *Average gold content of plants and marine animals (dry tissues)*

Marine plants and animals			Land plants		
Organism	Number of samples	Average ppb Au	Organism	Number of samples	Average ppb Au
Marine animals	9	57	Nonvascular (fungi, lichen, moss)	16	62
Marine plants	3	9	Herbs	82	42
All marine organisms	12	45	Shrubs	95	20[a]
			Trees	48	8

[a] Average does not include data of LOBANOV *et al.* (1966).

b) The relationship between the gold content of plants and their associated soil is obscure.

c) Some studies of the partition of gold between various plant organs suggest leaves as preferential accumulators with respect to seeds, stems, wood, bark or roots (TALIPOV *et al.*, 1968; LOBANOV *et al.*, 1966). From the data of TALIPOV *et al.* (1968) for six samples of shrubs from the Tamdytau mountains, U.S.S.R., averages of 129, 73 and 30 ppb Au are calculated for leaves, stems and bark respectively. In a more detailed study of plant organs from central Kazakhstan, LOBANOV *et al.* (1966) examined seeds, seedpods, leaves, young twigs, stems and roots of four forbs, five shrubs and one tree. From this data averages of 650, 250, 100 and 50 ppb Au (with respect to air-dried weight) are calculated for leaves, roots, stems and twigs and seeds plus seedpods respectively. Alternately the data presented by CURTIN *et al.* (1971) on pine and aspen from the Empire District, Colorado, indicate that interior wood of the trunk and branches concentrates gold more strongly than roots, leaves or needles.

d) There is some evidence for accumulator plants for gold. ARIPOVA and TALIPOV (1966) consider that wormwood is the best concentrator of gold in the dry Kyzylkum area. RAZIN and ROZHKOV (1966) cite raspberry, wild rosemary, dog-rose, poplar, mosses and fungi as the best geochemical indicators of gold based on their study of Kuranakh type deposits of the northern margin of the Aldan shield. However, the very high concentration of gold by horsetails (BABIČKA, 1943) is not substantiated by the work of WARREN and DELAVAULT (1950), RAZIN and ROZHKOV (1966), and CAN-

NON *et al.* (1968). Further the limited data of WARREN and DELAVAULT (1950), which suggested higher gold in horsetail specimens from auriferous areas relative to samples from non-auriferous areas, is not supported by the more extensive study of CANNON *et al.* (1968).

e) A study of gold and other metals in 962 mull (a forest humus layer) and soil samples from the Empire District, Colorado, showed that the distribution of gold-bearing veins in the bed rock was reflected in the gold content of the mull (CURTIN *et al.*, 1968; CURTIN *et al.*, 1971). Soil contained much less gold than mull and did not reflect gold deposits. Ash from the majority of mull samples contained 0.04 to 0.6 ppm Au compared with an upper limit of 0.7 ppm Au in ash of pine needles. CURTIN *et al.* (1968; 1971) conclude that the source of such mull gold is decay of fallen pine needles and aspen leaves, the two major trees of the area, and advocate analyses of such mull for gold as a biogeochemical prospecting tool.

III. Crude Oil

There exists almost no information on the abundance of gold in coal and little on oil. SHAH *et al.* (1970) report 0.67 to 3.01 ppb Au (average: 1.35 ppb Au) for 10 crude oil samples from the USA and Libya; the analytical method used was neutron activation.

Revised manuscript received: June 1973

79-M. Abundance in Common Metamorphic Rock Types

Abundance data for gold in metamorphic rocks is less extensive than for igneous rocks. Available data are presented in Table 79-M-1 and average values derived from this compilation given in Table 79-M-2. Previous compilations of abundance data for gold in metamorphic rocks have been given by PHAN (1965) and JONES (1969). The latter author presented a comprehensive coverage of the literature including early (pre-1954) determinations carried out mainly by fire assay. Table 79-M-1 includes data from 1960 comprised largely of spectrochemical and neutron activation determinations.

Average values for the main metamorphic rock types (argillite, schist, gneiss, metamorphosed carbonate rocks) fall in the range 1 to 5 ppb Au. The main exception is found in rocks of hornfelsic character, occurring predominantly in the contact metamorphic aureoles of intrusive igneous rocks, where an average of 8.5 ppb Au is obtained. For such rock there is sometimes evidence of introduction of gold from the igneous intrusives. Thus, the range of gold values in the majority of metamorphic rocks is very similar to the average gold content of common igneous and sedimentary rocks (see Tables 79-E-3 and 79-K-3).

A major source of data not included in calculation of the average value for schists in Table 79-M-2 is the 800 analyses of LOSHCHININ (1971) for schists from the Taskazgan rhythmic suites. The presence of gold in quartz veins in these schists together with the high gold content of associated sandstones (average of 25 ppb for 201 analyses) indicates that these rocks are probably anomalously high in gold.

Table 79-M-1. *Gold content of metamorphic rocks*

Rock type	Locality and description	Number of samples	Gold (ppb) range	Gold (ppb) average	Method	Reference
	Argillites, slates, hornfelses					
Argillite and slate	Yellowknife, Northwest Territories (Canada)	—	—	10	Fa[c]	1
Metapelite	Zhuinsk region, Patomsk highlands (U.S.S.R.), greenschist and epidote amphibolite facies	41	0.34—2.32	1.0	S	8
Regionally metamorphosed shale	Priamurya region (U.S.S.R.) epidote-actinolite facies	18	—	0.7	S	6
	greenschist facies	54	—	1.2	S	6
	zeolite facies	22	—	1.0	S	6
Hornfels	Altai (U.S.S.R.)	4	1.8—22.3	8.5	N	9
Hornfels	Primurya region (U.S.S.R.), aureole of Ingaglinsk massif	31	0.2—65.0	6.6	S	6
Hornfels	Chukotka (U.S.S.R.), aureole of Iultinsk massif	4	5—19	9.5	S	6
Hornfels	Northern Nura-Tau, Uzbekistan (U.S.S.R.)	4	5—40	22.5	S[a]	4
	Schists					
Schist	Altai, Uzbekistan, Kazakhstan, Chukotka and Taigonos (U.S.S.R.)	29	0.6—17	3.2	N, S	6, 7, 9, 12
Schist	Zhuinsk region, Patomsk highland (U.S.S.R.), greenschist, epidote-amphibolite and amphibolite facies	16	0.38—2.1	1.1	S	8
Schist	Northern Nura-Tau, Uzbekistan (U.S.S.R.)	12	5—100	29	S	4
Phyllitic schist	Taskazgan, Tamdytau (U.S.S.R.), sandstones, mudstones, schists and carbonaceous siliceous rocks with gold-quartz veins in schists	800	?—5000	35	S[b]	5
Mica schist	Yellowknife (Canada), chloritic	—	—	13.5	Fa[c]	1

Table 79-M-1 (continued)

Rock type	Locality and description	Number of samples	Gold (ppb) range	Gold (ppb) average	Method	Reference
Schist	Kola Peninsula (U.S.S.R.), Kolmozero-Voronya series	21	0.4—33	3.5	S	3
Quartz-kyanite-muscovite schist	Kola Peninsula (U.S.S.R.), Kolmozero-Voronya series (Sulfide-bearing kyanite contains 112 ppb Au)	27	3.7—65	17	S	3
Schists, partly graphitic-ferriginous	Ugakhansk region, Patomsk highland (U.S.S.R.), greenschist, epidote-amphibolite and amphibolite facies	46	0.48—1.21	0.80	S	8
	Gneisses and amphibolites					
Gneiss	Aldan shield, Taigonos Peninsula and Khankaisk massif. (U.S.S.R.); Australia. Mainly biotite, sillimanite, cordierite, garnet rocks	12	0.4—8.2	2.4	S	6
Granite gneiss	Kola Peninsula (U.S.S.R.), Kolmozero-Voronya series	7	0.2—22	4.7	S	3
Paragneiss	Manitoba (Canada), Manigotagan gneiss belt, Archean	2	0.4—0.85	0.6	N	11
Mica-garnet gneiss	New York City (U.S.A.)	1	—	1.8	N	2
Cordierite gneiss	Altai (U.S.S.R.)	3	2.0—7.7	4.3	N	12
Gneiss, amphibolite etc.	Kazakhstan (U.S.S.R.), rocks enclosing Kokchetav massif	9	1.5—5.0	2.3	S	12
Gneiss, schist and amphibolite	Central Kamchatka (U.S.S.R.)	10	—	4	S	12
Amphibolite	Kola Peninsula (U.S.S.R.), Kolmozero-Voronya series (Contains arsenopyrite and pyrrhotite with 7500 and 300 ppb Au respectively)	14	1.8—58	12	S	3
Amphibolite	Chukotka (U.S.S.R.), enclosing the Iultinsk massif	2	1.5—4.3	2.7	S	6
	Dolomite, marble and carbonate rocks					
Marble	Altai, Uzbekistan and Kazakhstan (U.S.S.R.); Vermont (U.S.A.)	4	1.0—8.6	5.1	N, S	2, 7, 9, 12

Marble	Uzbekistan (U.S.S.R.), near gold-bearing Dzhamansay intrusion	1	—	22.4	N	7
Metamorphosed marl and limestone	Ugankhansk region, Patomsk highland (U.S.S.R.), greenschist, epidote-amphibolite and amphibolite facies	45	0.25—2.3	0.71	S	8
	Others					
Metapelite and metamorphosed sandstone	Zhuinsk region, Patomsk highland (U.S.S.R.), greenschist, epidote-amphibolite and amphibolite facies	47	0.3—2.1	0.95	S	8
Quartzite	Wisconsin (U.S.A.)	2	2.4—7.3	4.9	N	2
Eclogite	Kazakhstan (U.S.S.R.), rocks enclosing Kokchetav massif	2	2.3—3.0	2.6	S	12
Regionally-metamorphosed diabase porphyrite	Priamurya region (U.S.S.R.) epidote-actinolite facies	22	—	3.6	S	6
	greenschist facies	48	—	4.1	S	6

[a] Data of Kushmuradov (1970) was determined in part (10%) spectrochemically and partly (90%) by a "gold chemical" method.
[b] Data given by Loshchinin (1971) was determined by assay and by spectrochemical methods.
[c] Fa fire assay.

References: 1. Boyle (1960). 2. DeGrazia and Haskin (1964). 3. Gavrilenko and Garifulin (1970). 4. Kushmuradov (1970). 5. Loshchinin (1971). 6. Moiseenko *et al.* (1971). 7. Paley *et al.* (1967). 8. Petrov *et al.* (1972). 9. Shcherbakov (1967). 10. Shilin (1968). 11. Stephenson and Ehmann (1971). 12. Voskresenskaya and Zvereva (1968).

Table 79-M-2. *Estimates of the average gold content of the principal metamorphic rock types (From Table 79-M-1)*

Rock type	Number of samples	Range ppb Au	Average ppb Au
Argillite and slate	135	0.34—10	1.0
Hornfels	43	0.2—40	8.5
Schists [a]	114	0.38—9	2.2
Gneisses [b]	37	0.2—22	3.9
Quartzite	2	2.4—7.3	4.9
Carbonate rocks	50	0.25—22.4	1.5

[a] Some data considered anomalously high were not used in calculation of the average. These data include: sulfide-bearing kyanite schists from Kolmozero-Voronya series averaging 17 ppb (GAVRILENKO and GARIFULIN, 1971), schists of northern Nura-Tau, Uzbekistan, averaging 29 ppb (KUSHMURADOV, 1970), and phyllitic schists of the Taskazgan rhythmic suite averaging 35 ppb (LOSHCHININ, 1971).

[b] Average does not include data for sulfide-bearing (arsenopyrite and pyrrhotite) amphibolite of Kolmozero-Voronya series, Kola Peninsula, averaging 12 ppb Au (GAVRILENKO and GARIFULIN, 1971).

Revised manuscript received: June 1973

79-N. Behavior in Metamorphic Reactions

I. Regional Metamorphism

Studies of the effect of metamorphism on gold distribution in rocks include those of Moiseenko *et al.* (1971) and Petrov *et al.* (1972).

Moiseenko *et al.* (1971) determined the average gold content of clay shale and diabase porphyrite which had undergone regional metamorphism varying from zeolite facies through greenschist to epidote-actinolite facies (see Table 79-M-1 for data). A significant gold increase in rocks of greenschist facies in comparison with rocks of higher or lower grade was observed. For metamorphosed clay shale it was found that rocks of greenschist facies were 40% higher in gold than those of the epidote-actinolite facies.

A comparable result was obtained by Petrov *et al.* (1972) for rocks of the Patomsk upland (U.S.S.R.). In metapelites subjected to greenschist, epidote-amphibolite and amphibolite facies of metamorphism, gold migrated from amphibolite facies rocks to epidote-amphibolite or upper greenschist facies rocks.

II. Contact Metamorphism

The distribution of gold in the contact metamorphic aureoles of intrusive igneous rocks of the Priamurya region (U.S.S.R.), was studied by Moiseenko *et al.* (1971; 1972). Granitoids of the East and West Iultinsk massifs average 3.0 and 1.3 ppb Au respectively (weighted average computed from data in Table 19, Moiseenko *et al.*, 1971) with even-grained granodiorite of the East Iultinsk massif containing 5.3 ppb Au as the most gold-rich facies. In the enclosing rocks of the contact metamorphic aureoles an average of 2.1 ppb Au is found in crystalline schists while amphibole, amphibole-pyroxene and biotite hornfels contain the highest gold content, 9.5 ppb, a value significantly higher than the enclosed intrusives.

A similar distribution of gold was found in contact aureole rocks of the Ingaglinsk massif. Here strong development of hornfels occurred up to 1,100 meters away from the massif with weaker hornfelsic texture observable as far as 3,300 meters from the contact. The hornfels rocks near the contact average 12.8 ppb Au, compared with 1.6 ppb for the massif itself. The highest gold contents were not at the contact but at distances of 100 to 500 meters from it.

Finally, Moiseenko *et al.* (1971) studied the distribution of gold in the contact metamorphic aureoles around small dioritic stocks and veins from the Tokur ore district (U.S.S.R.). Data from their Tables 48, 49 and 50 are presented in Table 79-N-1. In general, profiles of higher gold values in hornfels rock near the intrusive contact are found, grading away to lower gold in less intensively altered rocks away from the contact. In the case of the stock with 5.7 ppb Au, the hornfelsic rocks adjacent to it are highly enriched in gold relative to average gold in the presumed protolith (siltstone, argillite, silty argillite). This is probably best interpreted as introduction of gold into enclosing wall rock from the igneous intrusive.

Table 79-N-1. *Gold content in contact metamorphic aureoles around small dioritic stocks, Tokur (U.S.S.R.)*

Stock			Rocks of the contact aureole		
Composition	Width (meters)	Average ppb Au	Distance from contact (meters)	Main rock types (number of samples)	Average ppb Au
Quartz diorite and diorite	9	5.7	0—5[a]	Hornfels (biotite-orthoclase-quartz and orthoclase-quartz hornfels) (5)	126
			5—10[a]	Siltstone and hornfelsed siltstone (3)	4.6
			0—5[a]	Mainly metamorphosed and hornfelsed quartzite (6)	18.3
			5—17[a]	Stiltstone and silty argillite (5)	5.6
Syenite-diorite	?	5.5	0—1.6	Orthoclase-quartz hornfels (4)	24
			2.2—8.5	Hornfelsed sandstone and sandstone (4)	3.9
Syenite-diorite	?	28.3	0—2.5	Biotite-orthoclase-quartz hornfels (2)	27.3
			2.5—7.0	Hornfelsed siltstone (2)	14.8
			7.0—22.0	Silty argillite (1)	2.3

[a] Contact aureole sampled on either side of stock.

Revised manuscript received: June 1973

79-O. Economic Importance

As a mineral commodity gold is distinguished by its unique function as a common denominator of currencies used in the settlement of balance of payment accounts in the international monetary system. Gold is used in industry and in the arts as jewelry and precious metal, in electronic components and in dental equipment and supplies. However, because of its monetary function a substantial part of total gold production is hoarded and invested.

The production of primary gold is dominated by the Republic of South Africa which produces just under 70% of total world gold. Other major producers are the U.S.S.R., Canada and the U.S.A. In 1968 these four countries produced about 90% of the world's gold. The 1967 total production has been estimated to be more than 1,400 tons.

The consumption of gold in industry and art for an industrialized country is typified by the U.S.A. where in 1968 usage was: jewelry, 59%; electronics, 28%; dental, 10%, and others 3% (AGETON, 1970). Forecasts to the year 2000 for the U.S.A. suggest that gold usage in electronics will become dominant accounting for approximately 60% of consumption (AGETON, 1970).

With the exception of the Republic of South Africa where the mining of fossil placer gold is predominant, base metal mining is usually a significant source of gold in addition to primary auriferous mines and placers. For example, in Canada base metal mining accounted for approximately 22% of Canadian production in 1971 (HOGAN, 1971).

Revised manuscript received: June 1973

References: Sections 79-B to 79-O

AFEROV, YU. A., ZVYAGIN, V. G., ROSLYAKOVA, N. V., ROSLYAKOV, N. A., SHABYNIN, L. L., EPOV, I. N.: Gold in rocks, plants and waters of the Darasun deposit. Vop. Geol. Pribaikal. Zabaikal. **3**, 146 (1968) [Russian].

AGETON, R. W.: Gold. Mineral facts and problems, 1970. U.S. Bur. Mines Bull. **650**, 573 (1970).

ALIMARIN, I. P., PEREZHOGIN, G. A.: Determination of traces of gold by a radio activation method by using substoichiometric separation. Zh. Analit. Khim. **20** 48 (1965) [Russian].

ANDERS, E., GANAPATHY, R., KEAYS, R. R., LAUL, J. C., MORGAN, J. W.: Volatile and siderophile elements in lunar rocks: comparison with terrestrial and meteoritic basalts. Proc. Second Lunar Sci. Conf. Geochim. Cosmochim. Acta Suppl. 2, **2**, 1021 (1971).

ANDERSON, G. M., BURNHAM, C. W.: Solubilities of quartz, corundum, and gold in aqueous chloride and hydroxide solutions (abstr.). Geol. Soc. Am. Spec. Paper **82**, Abstracts for 1964, 4 (1964).

ANOSHIN, G. N., BUSHLYAKOV, I. N., POTAP'EV, V. V.: Raspredelenie zolota v porodakh i mineralakh granitoidnykh massivov Altaya, Zabaikal'ya i. Urala (Distribution of gold in rocks and minerals of the Altai, Transbaikalia, and Urals granitic rock massifs). Izv. Tomsk. Politekh. Inst. **239**, 39 (1970).

ANOSHIN, G. N., KEPEZINSKAS, V. V.: Raspredelenie zolota v Kainozoiskekh vulkanicheskikh porodakh Kurilo-Kamchatskoi provintsii v svyazi s ikh petrokhimicheskimi osobennostyami. (Gold distribution in Cenozoic volcanic rocks of the Kurile-Kamchatskaya province in connection with their petrochemical peculiarities). Geokhimiya **8**, 910 (1972).

ANOSHIN, G. N., POTAP'YEV, V. V.: Gold in granites of the Kolyvan' (Altay) and Khangilay-Shilinskiy (Transbaikalia) Massifs (according to radioactivation analysis data). Geochemistry Internat. **3**, 850 (1966).

ANOSHIN, G. N., YEMEL'YANOV, YE. M.: Gold in igneous rocks of the Atlantic Ocean as determined by radioactivation analysis. Doklady Acad. Sci. U.S.S.R. Earth Sci. Sect. **189**, 222 (1969).

ANOSHIN, G. N., ZOLOTAREV, B. P., TSIMBALIST, V. G.: Zoloto v porodakh toleit-bazal'tovoi formatsii. (Gold in rocks of tholeiitic basalt formations). Doklady Akad. Nauk S.S.S.R. **201**, 700 (1971).

ANTWEILER, J. C., LOVE, J. D.: Gold-bearing sedimentary rocks in northwest Wyoming—a preliminary report. U.S. Geol. Surv. Circ. **541**, 12 p. (1967).

ARIPOVA, KH., TALIPOV, R. M.: Osobennosti kontsentratsii zolota v pochvakh i rasteniyakh yuzhnoi chasti Tamdynskikh Gor. (Concentration of gold in soils and plants in the southern part of the Tamdyn Mountains). Uzbeksk. Geol. Zh. **10**, 45 (1966).

BABIČKA, J.: Gold in Lebewesen. Mikrochim. Acta **31**, 201 (1943).

BADALOV, S. T.: The role of major components in the geochemistry of minor and rare elements of ore deposits. Geochemistry Internat. **2**, 857 (1965).

BADALOV, S. T., BADALOVA, R. P.: Some regularities of distribution of gold and silver in the pricipal ore minerals of hypogene deposits of Karamazar and western Kazakhstan. Geochemistry Internat. **4**, 660 (1967).

BAEDECKER, P. A.: The distribution of gold and iridium in meteoritic and terrestrial materials. Univ. Kentucky Ph. D. dissertation, 110 p. (1967).

BAEDECKER, P. A., EHMANN, W. D.: The distribution of some noble metals in meteorites and natural materials. Geochim. Cosmochim. Acta **29**, 329 (1965).

BARNES, H. L., CZAMANSKE, G. K.: Solubilities and transport of ore minerals. In: Geochemistry of Hydrothermal Ore Deposits. BARNES, H. L. (ed.): New York: Holt, Rinehart and Winston 1967.

Bauer, R., Schaudy, R.: Activation analytical determination of elements in meteorites, 3. Determination of manganese, sodium, gallium, germanium, copper and gold in 21 iron meteorites and 2 mesosiderites. Chem. Geol. **6**, 110 (1970).

Beardsley, D. A., Briscoe, G. B., Růžička, J., Williams, M.: Substoichiometric determination of traces of gold by neutron-activation analysis. Talanta **12**, 829 (1965).

Borisenko, L. F., Miller, A. D., Fisher, E. I.: O rasprostradennosti zolota v ul'trabazitakh. (On gold abundance in ultrabasites). Geokhimiya 188 (1972).

Borovitskiy, V. P., Miller, A. D., Shemyakin, V. N.: Determination of minute amounts of gold in natural waters of the Aldan Region. Geochemistry Internat. **3**, 371 (1966).

Borisenko, L. F., Uskov, E. D.: O zolotonosnosti ul'trabazitov Gusevogorskogo masseva (Ural). (On the gold content of ultrabasites of the Gusevogorsk massif (Urals). Doklady Akad. Nauk S.S.S.R. **201**, 450 (1971).

Bowen, H. J. M.: The uptake of gold by marine sponges. J. Marine Biol. Assoc. U.K. **48**, 275 (1968).

Boyle, R. W.: The geology, geochemistry and origin of the gold deposits of the Yellowknife district. Can. Geol. Survey Mem. **310**, 193 p. (1961).

Brokaw, A. D.: The solution of gold in the surface alterations of ore bodies. J. Geol. **18**, 321 (1910).

Brooks, R. R.: The use of ion-exchange enrichment in the determination of trace elements in sea water. Analyst **85**, 745 (1960).

Browne, P. R. L.: Sulfide mineralization in a Broadlands geothermal drill hole, Taupo volcanic zone, New Zealand. Econ. Geol. **64**, 156 (1969).

Brunfelt, A. O., Heier, K. S., Steinnes, E.: Determination of 40 elements in Apollo 12 materials by neutron activation analysis. Proc. Second Lunar Sci. Conf., Geochim. Cosmochim. Acta Suppl. 2, **2**, 1281 (1971).

Bushlyakov, I. N.: Raspredelenie zolota v granitoidakh Verkhisetskogo massiva na Urale. (Distribution of gold in granitic rocks of the Verkhisetsk massif in the Urals). Geokhimiya **12**, 1442 (1971).

Buturlinov, N. V., Latysh, I. K.: Gold and silver in igneous rocks of the Donets Basin. Geol. Zh (Kiev) **30**, 69 (1970). English Translation: U.S. Geol. Survey Library.

Cadigan, R. A.: Metals in the Moenkopi Formation and related strata, Colorado Plateau region. U.S. Geol. Surv. Bull. **1344**, 56 p. (1971).

Caldwell, W. E.: Gold content of sea water. J. Chem. Educ. **15**, 507 (1938).

Cannon, H. L., Shacklette, H. T., Bastron, H.: Metal absorption by *Equisetum* (horsetail). U.S. Geol. Surv. Bull. **1278-A**, 21 p. (1968).

Chao, T. T., Jenne, E. A., Heppting, L. M.: Prevention of adsorption of trace amounts of gold by containers. U.S. Geol. Surv. Profess. Paper **600D**, D13 (1968).

Chernyayev, A. M., Chernyayev, L. E., Yeremeyeva, M. N., Andreyev, M. I.: Hydrogeochemistry of gold. Geochemistry Internat. **6**, 348 (1969).

Chukhrov, F. V.: O migratsii zolota v zone okisleniya. (Migration of gold in the oxidation zone). Akad. Nauk. S.S.S.R. Izv., Ser. Geol., no. 4, 117 (1947).

Clifton, H. E.: Gold distribution in surface sediments on the continental shelf off Southern Oregon: a preliminary report. U.S. Geol. Surv. Circ. **587**, 6 p. (1968).

Clifton, H. E., Hubert, A., Philips, R. L.: Marine sediment sample preparation for analysis for low concentrations of fine detrital gold. U.S. Geol. Surv. Circ. **545**, 11 p. (1967).

Cloke, P. L., Kelley, W. C.: Solubility of gold under inorganic supergene conditions. Econ. Geol. **59**, 259 (1964).

Cobb, J. C.: A trace-element study of iron meteorites. J. Geophys. Res. **72**, 1329 (1967).

Cooper, M.: Selected annotated bibliography on the geochemistry of gold. U.S. Geol. Surv. Bull. **1337**, 63 p. (1971).

Crocket, J. H.: Neutron activation analysis for noble metals in geochemistry. Proceedings N.A.T.O. Advanced Study Institute: Activation Analysis in Geochemistry and Cosmochemistry (eds. A. O. Brunfelt and E. Steinnes), p. 339. Oslo: Universitetsforlaget 1971.

Crocket, J. H.: Some aspects of the geochemistry of Ru, Os, Ir and Pt in iron meteorites. Geochim. Cosmochim. Acta **36**, 517 (1972).

CROCKET, J. H., CHYI, L. L.: Abundances of Pd, Ir, Os and Au in an alpine ultramafic pluton. Proc. XXIV Internatl. Geol. Congress, Section 10 (Geochemistry), 202 (1972).

CROCKET, J. H., KEAYS, R. R., HSIEH, S.: Precious metal abundances in some carbonaceous and enstatite chondrites. Geochim. Cosmochim. Acta **31**, 1615 (1967).

CROCKET, J. H., KEAYS, R. R., HSIEH, S.: Determination of some precious metals by neutron activation analysis. J. Radioanal. Chem. **1**, 487 (1968).

CROCKET, J. H., MACDOUGALL, J. D., HARRISS, R. C.: Gold, palladium and iridium in marine sediments. Geochim. Cosmochim. Acta **37**, 2547 (1973).

CURTIN, G. C., LAKIN, H. W., HUBERT, A. E.: The mobility of gold in mull (forest humus layer). U.S. Geol. Surv. Profess. Paper **700-C**, C127 (1970).

CURTIN, G. C., LAKIN, H. W., HUBERT, A. E., MOSIER, E. L., WATTS, K. C.: Utilization of mull (forest humus layer) in geochemical exploration in the Empire District, Clear Creek County, Colorado. U.S. Geol. Surv. Bull. **1278-B**, 39 p. (1971)

CURTIN, G. C., LAKIN, H. W., NEUERBURG, G. J., HUBERT, A. E.: Utilization of humus-rich forest soil (mull) in geochemical exploration of gold. U.S. Geol. Surv. Circ. **562**, 11 p. (1968).

DAMS, R., ROBBINS, J. A., RAHN, K. A., WINCHESTER, J. W.: Nondestructive neutron activation analysis of air pollution particles. Anal. Chem. **42**, 861 (1970).

DAVLETOV, I. K.: Average gold content in essential minerals of intrusive rocks. Doklady Acad. Sci. U.S.S.R. Earth Sci. Sect. **190**, 215 (1970).

DAVLETOV, I. K., DZHAKSHIBAYEV, SH.: The mineral balance and the behaviour of gold during the emplacement of an intrusive body. Geochemistry Internat. **7**, 997 (1970).

DEGRAZIA, A. R., HASKIN, L.: On the gold content of rocks. Geochim. Cosmochim. Acta **28**, 559 (1964).

DUDEY, N. D., ROSS, L. E., NOSHKIN, V. E.: Application of activation analysis and Ge(Li) detection techniques for the determination of stable elements in marine aerosols. Nat. Bur. Stand. (U.S.) Spec. Publ. 312, Vol. I (Modern Trends in Activation Analysis, ed. — J. R. DEVOE) 55 (1969).

EHMANN, W. D.: Gold (79). In: MASON, B. (Ed.): Handbook of Elemental Abundances p. 479. London: Gordon and Breach 1971.

EHMANN, W. D., BAEDECKER, P. A., MCKOWN, D. M.: Gold and iridium in meteorites and some selected rocks. Geochim. Cosmochim. Acta **34**, 493 (1970).

EHMANN, W. D., DURBEN, D. R.: Silicon abundances in some meteorites and standard rocks by activation analysis. Geochim. Cosmochim. Acta **32**, 461 (1968).

EHMANN, W. D., GILLUM, D. E.: Platinum and gold in chondritic meteorites. Chem. Geol. **9**, 1 (1972).

EHMANN, W. D., SHOWALTER, D. L.: Elemental abundance trends in the australite strewn field by non-destructive neutron activation. Proceedings N.A.T.O. Advanced Study Institute: Activation Analysis in Geochemistry and Cosmochemistry BRUNFELT, A. O. and STEINNES, E. (Eds.): p. 23. Oslo: Universitetsforlaget 1971.

EMMONS, W. H.: The Principles of Economic Geology (2nd edition). New York: McGraw-Hill 1940.

FERSMAN, A. E.: Geokhimiya Vol. IV (Geochemistry). Leningrad 1939.

FISCHER, K. W.: Edelmetalle in der Saale und in ihrem Einzugsgebiet. Geologie **15**, 550 (1966).

FOUCHÉ, K. F., SMALES, A. A.: The distribution of gold and rhenium in iron meteorites. Chem. Geol. **1**, 329 (1966).

FOUCHÉ, K. F., SMALES, A. A.: The distribution of trace elements in chondritic meteorites. 2. Antimony, arsenic, gold, palladium and rhenium. Chem. Geol. **2**, 105 (1967).

FUKAI, R., MEINKE, W. W.: Activation analyses of vandium, arsenic, molybdenum, tungsten, rhenium, and gold in marine organisms. Limnol. Oceanog. **7**, 186 (1962).

GANAPATHY, R., KEAYS, R. R., ANDERS, E.: Apollo 12 lunar samples: Trace element analysis of a core and the uniformity of the regolith. Science **170**, 533 (1970a).

GANAPATHY, R., KEAYS, R. R., LAUL, J. C., ANDERS, E.: Trace elements in Apollo 11 lunar rocks: Implications for meteorite influx and origin of moon. Proc. Apollo 11 Lunar Sci. Conf., Geochim. Cosmochim. Acta **34**, Suppl. 1, **2**, 1117 (1970b)

GARRELS, R. M., CHRIST, C. L.: Solutions, Minerals, and Equilibria. New York: Harper and Row 1965.

GAVRILENKO, B. V., GARIFULIN, L. L.: K geokhimii zolota v porodakh serii Kolmozero-Voron'ya. (Geochemistry of gold in rocks of the Kolmozero-Voron'ya series). Mater. Geol. Metallogen. Kol'sk Poluostrova **2**, 254 (1971).

GILLUM, D. E., EHMANN, W. D.: Noble metal determination in meteorites. Radiochim. Acta **16**, 123 (1971).

GOLDBERG, E. D., UCHIYAMA, A., BROWN, H.: The distribution of nickel, cobalt, gallium, palladium and gold in iron meteorites. Geochim. Cosmochim. Acta **2**, 1 (1951).

GOLDSCHMIDT, V. M.: Geochemistry. Oxford: Clarendon Press 1954.

GOLDSCHMIDT, V. M., PETERS, CL.: Zur Geochemie der Edelmetalle. Nachr. Akad. Wiss. Göttingen, Math-Physik. Kl. IV, 377 (1932).

GOLEVA, G. A., KRIVENKOV, V. A., GUTZ', Z. G.: Geochemical trends in the occurrence and migration forms of gold in natural water. Geochemistry Internat. **7**, 518 (1970).

GOSLING, A. W., JENNE, E. A., CHAO, T. T.: Gold content of natural waters in Colorado. Econ, Geol. **66**, 309 (1971).

GOTTFRIED, D., GREENLAND, L. P.: Variation of iridium and gold in oceanic and continental basalts. Proc. XXIV Internatl. Geol. Congress, Section 10 (Geochemistry), 135 (1972).

GOTTFRIED, D., ROWE, J. J., TILLING, R. I.: Distribution of gold in igneous rocks. U.S. Geol. Surv. Profess. Paper **727**, 42 p. (1972).

GRIMANIS, A. P., PANTAZIS, G., PAPADOPULOS, C., TSAMOS, N.: Determination of trace elements in the Greek lakes by neutron-activation analysis. Proc. III Internatl. Conf. Peaceful Uses At. Energy, Geneva **15**, 412 (1964).

HABER, F.: Das Gold im Meerwasser. Z. Angew. Chem. **40**, 303 (1927).

HABER, F.: Das Gold im Meere. Gesell. Erdkunde Berlin Z. ,Ergänzungsheft **3**, 3 (1928).

HABER, F., JAENICKE, J.: Beitrag zur Kenntnis des Rheinwassers. Z. Anorg. Allgem. Chem. **147**, 156 (1925).

HARRISS, R. C., CROCKET, J. H., STAINTON, M.: Palladium, iridium and gold in deep-sea manganese nodules. Geochim. Cosmochim. Acta **32**, 1049 (1968).

HASKIN, L. A., ALLEN, R. O., HELMKE, P. A., PASTER, T. P., ANDERSON, M. R., KOROTEV, R. L., ZWEIFEL, K. A.: Rare earths and other trace elements in Apollo 11 lunar samples. Proc. Apollo 11 Lunar Sci. Conf., Geochim. Cosmochim. Acta **34**, Suppl. 1, **2**, 1213 (1970).

HECHT, F., KIESL, W.: Kosmochemische Meteoritenuntersuchungen mittels Neutronen-aktivierungsanalyse. Chem. Erde **30**, 145 (1971).

HELGESON, H. C., GARRELS, R. M.: Hydrothermal transport and deposition of gold. Econ. Geol. **63**, 622 (1968).

HENLEY, R. W.: Studies in gold transport and deposition. In: Progress in Experimental Petrology. HENDERSON, C. M. B. and HAMILTON, D. L. (Eds.): Natural Environment Research Council (Great Britain), Publication Series D, No. 2, 53 (1972).

HERMANN, F., KIESL, W., KLUGER, F., HECHT, F.: Neutronenaktivierungsanalytische Bestimmung einiger Spurenelemente in meteoritischen Phasen. Mikrochim. Acta **1971**, 225.

HEY, M. H.: Catalogue of Meteorites. London: British Museum (Natural History) 1966.

HÖFLER, H., SORANTIN, H.: Application of nondestructive activation analysis to meteorites: determination of aluminum, vanadium, manganese and gold in stony and iron meteorites. Chem. Geol. **2**, 273 (1967).

HOGAN, J. J.: Gold. Canadian Minerals Yearbook 1971, preprint No. 18 (1971).

HORN, M. K., ADAMS, J. A. S.: Computer-derived geochemical balances and element abundances. Geochim. Cosmochim. Acta **30**, 279 (1966).

HUFFMAN, C., MENSIK, J. D., RILEY, L. B.: Determination of gold in geologic materials by solvent extraction and atomic-absorption spectrometry. U.S. Geol. Surv. Circ. **544**, 6 p. (1967).

HUMMEL, R. W.: Determination of gold in sea water by radioactivation analysis. Analyst **82**, 483 (1957).

IKEDA, N.: Chemical studies on the hot springs of Nasu XI. Nippon Kagaku Zasshi **76**, 1071 (1955) [Japanese].

JEROME, D. Y., PHILIPPOT, J. C., BRICHET, E.: Determination of 29 elements in Luna 16 soil by non-destructive activation analysis. Earth Planet. Sci. Lett. **13**, 436 (1972).
JONES, R. S.: Gold in meteorites and in the earth's crust. U.S. Geol. Surv. Circ. **603**, 4 p. (1968).
JONES, R. S.: Gold in igneous, sedimentary and metamorphic rocks. U.S. Geol. Surv. Circ. **610**, 27 p. (1969).
JONES, R. S.: Gold content of water, plants and animals. U.S. Geol. Surv. Circ. **625**, 15 p. (1970).
JONES, R. S., FLEISCHER, M.: Gold in minerals and the composition of native gold. U.S. Geol. Surv. Circ. **612**, 17 p. (1969).
KAKOVSKII, I. A., TYURIN, N. G.: Behaviour of gold in polysulfide solutions at elevated temperatures and pressures. Cvetnaya Metallurgiya **2**, 104 (1962) [Russian].
KEAYS, R. R., CROCKET, J. H.: A study of precious metals in the Sudbury nickel irruptive ores. Econ. Geol. **65**, 438 (1970).
KEAYS, R. R., GANAPATHY, R., ANDERS, E.: Chemical fractionations in meteorites — IV. Abundances of fourteen trace elements in L-chondrites; implications for cosmothermometry. Geochim. Cosmochim. Acta **35**, 337 (1971).
KEAYS, R. R., GANAPATHY, R., LAUL, J. C., ANDERS, E., HERZOG, G. F.: Trace elements and radioactivity in lunar rocks: Implications for meteorite infall, solar-wind flux, and formation conditions of moon. Science **167**, 490 (1970).
KEIL, K.: Meteorite composition. In: Handbook of Geochemistry, Vol. 1. WEDEPOHL, K. H. (Ed.): p. 78. Berlin-Heidelberg-New York: Springer 1969.
KHARKAR, D. P., TUREKIAN, K. K.: Neutron activation analysis of milligram quantities of Apollo 11 lunar rocks and soil. Proc. Apollo 11 Lunar Sci. Conf., Geochim. Cosmochim. Acta **34**, Suppl. 1, **2**, 1659 (1970).
KIESL, W.: Determination of trace elements in meteorites by neutron activation analysis. Nat. Bur. Stand. (U.S.) Spec. Publ. 312 (Modern Trends in Activation Analysis, ed. — J. R., DEVOE) **1**, 302 (1969).
KIESL, W., SEITNER, H., KLUGER, F., HECHT, F.: Determination of trace elements by chemical analysis and neutron activation in meteorites of the collection of the Viennese Museum of Natural History. Monatsh. Chem. **98**, 972 (1967).
KINKEL, A. R., LESURE, F. G.: Residual enrichment and supergene migration of gold, southeastern United States. U.S. Geol. Surv. Profess. Paper **600-D**, D 174 (1968).
KOGA, A.: Chemical studies on the hot springs of Beppu. XXV. Distribution of gold. Nippon Kagaku Zasshi **82**, 1476 (1961) [Japanese].
KONOVALOV, E. P.: Hydrochemical method of prospecting for gold. Sov. Geol. **2**, 114 (1941) [Russian].
KOROLEVA, N. N.: Relation between Ag:Au ratio and mineral associations. Geochemistry Internat. **8**, 200 (1971).
KRAUSKOPF, K. B.: The solubility of gold. Econ. Geol. **46**, 858 (1951).
KRAUSKOPF, K. B.: Introduction to Geochemistry. New York: McGraw-Hill 1967.
KREUTER, V. M., ARISTOV, V. V., VOLYNSKII, I. S., KRESTOVNIKOV, A. B., KUVICHINSKIY, V. V.: Povedenie zolota v zone okisleniya zoloto-sulfidnykh mestorozhdenii. (The behavior of gold in the zone of oxidation of auriferous sulfide). Moscow: Gosgeoltekhizdat 1959.
KUO, H.: Palladium, iridium and gold in deep-sea cores. McMaster University (Canada), M. Sc. Thesis, 49 p. (1971).
KUSHMURADOV, O. K.: O soderzhanii zolota v osadochno-metamorficheskikhi granitoidnykh porodakh severnogo Nuratau. (Gold content in Northern Nura-Tau sedimentary, metamorphic and granitic rocks). Uzbeksk. Geol. Zh. **14**, 8 (1970).
LAKIN, H. W., CURTIN, G. C., HUBERT, A. E.: Geochemistry of gold in the weathering cycle (abstr.) Canadian Inst. Mining and Metallurgy Special Vol. (Geochemical Exploration — BOYLE, R. W. and MCGERRIGLE, J. I. (Eds.): **11**, 196 (1971).
LARIMER, J. W., ANDERS, E.: Chemical fractionations in meteorites — II. Abundance patterns and their interpretation. Geochim. Cosmochim. Acta **31**, 1239 (1967).
LARIMER, J. W., ANDERS, E.: Chemical fractionations in meteorites — III. Major element fractionations in chondrites. Geochim. Cosmochim. Acta **34**, 367 (1970).

LAUL, J. C., GANAPATHY, R., MORGAN, J. W., ANDERS, E.: Meteoritic and non-meteoritic trace elements in Luna 16 Samples. Earth Planet. Sci. Lett. **13**, 450 (1972).

LAUL, J. C., KEAYS, R. R., GANAPATHY, R., ANDERS, E., MORGAN, J. W.: Chemical fractionation in meteorites — V. Volatile and siderophile elements in achondrites and ocean ridge basalts. Geochim. Cosmochim. Acta **36**, 329 (1972).

LAUL, J. G., MORGAN, J. W., GANAPATHY, R., ANDERS, E.: Meteoritic material in lunar samples: characterization from trace elements. Proc. Second Lunar Sci. Conf., Geochim. Cosmochim. Acta Suppl. 2, **2**, 1139 (1971).

LESURE, F. G.: Residual enrichment and supergene transport of gold, Calhoun Mine, Lumpkin County, Georgia. Econ. Geol. **66**, 178 (1971).

LINDNER, J. L., GRUNER, J. W.: Action of alkali sulfide solutions on minerals at elevated temperatures. Econ. Geol. **34**, 537 (1939).

LINN, T. A., JR., MOORE, C. B.: Neutron activation determination of gold in iron meteorites and inclusions. Earth Planet. Sci. Lett. **3**, 453 (1967).

LIVINGSTONE, D. A.: Chemical composition of rivers and lakes. U.S. Geol. Surv. Profess. Paper **440-G**, 63 p. (1963).

LOBANOV, E. M., KHATAMOV, S. H., TALIPOV, R. M.: Effektivnost primeneniya neitronoaktivatsionnogo analiza pri biogeokhimicheskikh poiskakh zolotorudnykh mestorozhdenii (na primere Tsentralnykh Kyzylkumov). (Efficiency of neutron activation logging during biogeochemical prospecting for gold ore deposits (in Central Kazakhstan)). Uzbeksk. Geol. Zh. **10**, 49 (1966).

LOSHCHININ, V. P.: Raspredelenie zolota v Dokembriiskikh otlozheniyakh Taskazganskoi ritmosvity v Ur. Taskazgan (Tamdytau). (Distribution of gold in the Precambrian formations of Taskazgan rhythmic suites in Taskazgan (Tamdytau)). Uzbeksk. Geol. Zh. **15**, 62 (1971).

MACHAIRAS, G.: MISE en évidence de différentes catégories d'or exogène par l'étude du mode de répartition de l'argent. Réflexions sur le cycle de l'or exogène. Bull. Soc. Franc. Minéral. Crist. **90**, 387 (1967).

MACHAIRAS, G.: Association des inclusions fluides et des particules d'or dans le quartz aurifère. Schweiz. Mineral. Petrogr. Mitt. **50**, 167 (1970).

MAKSUDOV, M.: Distribution characteristics of silver and gold in sulfide ores of Koksu river basin. Intern. Geol. Rev. **11**, 1427 (1969).

MALISSA, H., HERMANN, F., KLUGER, F., KIESL, W.: Chemical and microprobe investigations of the Allende-meteorite. Mikrochim. Acta (Wien) 434 (1972).

MANTEI, E., BOLTER, E., AL SHAIEB, Z.: Distribution of gold, silver, copper, lead and zinc in the productive Marysville stock, Montana. Miner. Deposita **5**, 184 (1970).

MANTEI, E. J., BROWNLOW, A. H.: Variation in gold content of minerals of the Marysville quartz diorite stock, Montana. Geochim. Cosmochim. Acta **31**, 225 (1967).

MARTELL, A. E., SILLÉN, L. G.: Stability constants of metal-ion complexes. Chem. Soc. (London), Spec. Publ. **17**, 754 p. (1964).

MCCAUGHEY, W. J.: The solvent effect of ferric and cupric salt solutions upon gold. Amer. Chem. Soc. J. **31**, 1261 (1909).

MERCER, W., CROCKET, J. H.: Gold and palladium in host rocks and ores from the Heath Steele B-1 deposit, New Brunswick. Proc. XXIV Intern. Geol. Congr. Sect. 10 (Geochemistry), 180 (1972).

MITCHELL, R. L.: Spectrographic analysis of soils, plants and related materials. Commonwealth Bur. Soil Sci., Tech. Comm. No. 44 (1948).

MOISEENKO, V. G., FAT'YANOV, I. I.: Geochemistry of gold. Proc. XXIV Intern. Geol. Congr., Sect. 10 (Geochemistry), 159 (1972).

MOISEENKO, V. G., MIKHAILOV, M. A.: Redistribution of gold during sediment accumulation and metamorphism. Vopr. Geol. Geokhimii Metallogenii Sev.-Zap. Sektora Tikhookean Poyasa, 198 (1970) [Russian].

MOISEENKO, V. G., SHCHEKA, S. H., FAT'YANOV, I. I., IVANOV, V. S.: Geokhimicheskie osobennosti raspredeleniya zolota v porodakh Tikhookeanskogo poyasa. [Geochemical aspects of the distribution of gold in rocks of the Pacific Ocean belt]. Moscow: Nauka 1971.

MOORE, G. W., SILVER, E. A.: Gold distribution on the sea floor off the Klamath Mountains, California. U.S. Geol. Surv. Circ. **605**, 9 p. (1968).

NODDACK, I., NODDACK, W.: Die Geochemie des Rheniums. Z. Physik. Chem. **54**, 207 (1931).

NODDACK, I., NODDACK, W.: Die Häufigkeiten der Schwermetalle im Meerestieren. Arkiv. Zoologie **32 A**, 1 (1939).

OKA, Y., KATO, T., SASAKI, M.: Determination of gold in sea water by neutron activation analysis. Nippon Kagaku Zasshi **85**, 643 (1964) [Japanese].

ONG, H. L., SWANSON, V. E.: Natural organic acids in the transportation, deposition, and concentration of gold. Colorado School Mines Quarterly **64**, 395 (1969).

ORGRYZLO, S. P.: Hydrothermal experiments with gold. Econ. Geol. **30**, 335 (1935).

PALEY, L. Z., MUROUTSEV, A. V., BOROZENETS, N. I.: The geochemistry of gold at Sultanuizdag. Geochemistry Internat. **4**, 1197 (1967).

PESHCHEVITSKIY, B. I., ANOSHIN, G. N., YERENBURG, A. M.: Chemical forms of gold in sea water. Doklady Acad. Sci. U.S.S.R. Earth Sci. Sect. **162**, 205 (1965).

PETROV, B. V., KRENDELEV, F. P., BOBROV, V. A., TSIMBALIST, V. G.: Provedenie radioactivnykh elementov i zolota pri metamorfizme osadochnykh porod Pamskogo nagor'ya. (Behavior of radioactive elements and gold during metamorphism of the Patmosk highland sedimentary rocks). Geokhimiya **8**, 947 (1972).

PHAN, K. D.: Enquête sur l'or dans les roches-origine de l'or des gisements. Chronique Mines Rech. Minière **33**, 175 (1965).

POHL, F. A.: Methoden zur spektrochemischen Spurenanalyse. I. Zur Spurenanalyse von Wässern. Z. Anal. Chem. **134**, 241 (1953).

PUTNAM, C. L.: The gold content of sea water. J. Chem. Educ. **30**, 576 (1953).

RANKAMA, K. K., SAHAMA, T. G.: Geochemistry. Chicago: Univ. Chicago Press 1950.

RAZIN, L. V., ROZHKOV, I. S.: K geokhimii zolota v kore vyvetrivaniya i biosfere provintsii mongoletnei merzloty na Aldanskom shchite. (Geochemistry of gold in the zone of weathering and the biosphere in the permafrost region of Aldan Shield). Akad. Nauk S.S.S.R. Yakut. Fil. Sibiroskoye Oteleniye Trudy, Geol. Ser. **16**, 5 (1963).

RAZIN, L. V., ROZHKOV, I. S.: Geokhimiya zolota v kore vyvetrivaniya i biosfere zolotorudnykh mestorozhdeniy Kuranakhskogo tipa. (Geochemistry of gold in the weathered crust and biosphere of gold deposits of the Kuranakh type). Moscow: Nauka 1966.

RIEDER, R., WÄNKE, H.: Study of trace element abundance in meteorites by neutron activation. In: MILLMAN, P. (Ed.): Meteorite Research p. 75. Reidel: Dordrecht-Holland 1969.

ROEDDER, E.: Fluid inclusions as samples of ore fluids. In: Geochemistry of Hydrothermal Ore Deposits. BARNES, H. H. (Ed.): New York: Holt, Rinehart and Winston 1967.

ROZHKOV, I. S., RAKOVSKIY, E. YE., BERENSHTEYN, L. YE., SEREBRYANYY, B. L., SHILIN, N. L.: Nonuniform distribution of gold in rocks and minerals (as determined by neutron activation analysis). Doklady Acad. Sci. U.S.S.R. Earth Sci. Sect. **191**, 200 (1970).

ROWE, J. J.: Fractionation of gold in a differentiated tholeiitic dolerite. Chem. Geol. **4**, 421 (1969).

ROWE, J. J., SIMON, F. O.: The determination of gold in geologic materials by neutron-activation analysis using fire assay for the radiochemical separations. U.S. Geol. Surv. Circ. **599**, 4 p. (1968).

SANTOS, G. G., WALTERS, L. J.: Gold provinces in the Philippines defined by activation analysis. Nucl. Tech. Miner. Explor. Exploit., Proc., 143 (1971).

SCHAUDY, R., KIESL, W., HECHT, F.: Activation analytical determination of elements in meteorites. Chem. Geol. **2**, 279 (1967).

SCHAUDY, R., KIESL, W., HECHT, F.: Activation analytical determination of elements in meteorites, 2. Chem. Geol. **3**, 307 (1968).

SCHILLER, P., COOK, G. B., BESWICK, C. K.: Contribution to geobotanical and geochemical prospecting for gold by activation analysis. Nucl. Tech. Miner. Explor. Exploit, Proc., 129 (1971).

SCHMITT, H. H., LOFGREN, G., SWANN, G. A., SIMMONS, G.: The Apollo 11 samples: Introduction. Proc. Apollo 11 Lunar Sci. Conf., Geochim. Cosmochim. Acta **34**, Suppl. 1, **1**, 1 (1970).

SCHUTZ, D. F., TUREKIAN, K. K.: The investigation of the geographical and vertical distribution of several trace elements in sea water using neutron activation analysis. Geochim. Cosmochim. Acta **29**, 259 (1965).

SEITNER, H., KIESL, W., KLUGER, F., HECHT, F.: Wet-chemical analysis and determination of trace elements by neutron activation in meteorites. J. Radioanal. Chem. **1**, 235 (1971).

SHAH, K. R., FILBY, R. H., HALLER, W. A.: Determination of trace elements in petroleum by neutron activation analysis. II. Determination of Sc, Cr, Fe, Co, Ni, Zn, As, Se, Sb, Eu, Au, Hg and U. J. Radioanal. Chem. **6**, 413 (1970).

SHCHEKA, S. A., MOISEENKO, V. G., FORMINYKH, V. G.: Osnovnye zakonomernosti raspredeleniya zolota v intruzivnykh bazitakh i giperbazitakh. (Basic relationships of the distribution of gold in intrusive basic and ultrabasic rocks). Akad. Nauk S.S.S.R. Doklady **201**, 461 (1971).

SHCHERBAKOV, YU. G.: Raspredeleniye i usloviya kontsentratsii zolota v rudnykh provintsiyakh. (Distribution and concentration conditions of gold in ore provinces). Moscow: Nauka 1967.

SHCHERBAKOV, YU. G., PEREZHOGIN, G. A.: Geochemical relation between gold mineralization, intrusives, and the enclosing rocks in western Siberia. Geochemistry **9**, 882 (1963).

SHCHERBAKOV, YU. G., PEREZHOGIN, G. A.: Geochemistry of gold. Geochemistry Internat. **3**, 489 (1964).

SHILIN, N. L.: Zoloto v porodakh Tsentral' noi Kamchatki. (Gold in rocks of Central Kamchatka). Izv. Akad. Nauk S.S.S.R. Ser. Geol. **11**, 69 (1968).

SILLÉN, L. G.: The physical chemistry of sea water. In: Oceanography (ed. M. SEARS). Amer. Assoc. for the Advancement of Science Publ. No. 67, 549 (1961).

SMITH, F. G.: The alkali sulfide theory of gold deposition. Econ. Geol. **38**, 561 (1943).

SOBOTOVICH, E. V., RUDENKO, A. K., SMIRNOVA, A. V.: Distribution of gold in kamacite of the Sikhote-Alin meteorite. Probl. Kosmokhimii Meteoritiki, Rasshir. Yubileinyi Plenum (Dokl.), 164 (ed. E. V. SOBOTOVICH). Kiev: "Naukova Dumka" 1971 [Russian].

SPIRIDONOV, E. M.: Zoloto i med v gornykh porodakh Stepnyakskogo i Bestyubinskogo zolotorudnykh polei. (Gold and copper in the rocks of the Stepnyak and Bestyube gold ore fields). Geol. Geofiz. **9**, 124 (1971).

STANTON, R. L.: Ores in vein association. In: Ore Deposits, p. 590. New York: McGraw-Hill 1972.

STARK, W.: Über die Goldführung der Meere. Helv. Chim. Acta **26**, 424 (1943).

STEELINK, C.: What is humic acid? J. Chem. Educ. **40**, 379 (1963).

STEPHENSON, J. F., EHMANN, W. D.: Neutron activation analysis of gold in Archean igneous and metamorphic rocks of the Rice Lake-Beresford Lake area, southeastern Manitoba. Econ. Geol. **66**, 933 (1971).

STOKES, H. N.: Experiments on the solution, transportation and deposition of copper, silver and gold. Econ. Geol. **1**, 644 (1906).

TALIPOV, R. M., ARIPOVA, K. H., KARABAEV, K. K., KHATAMOV, S. H., AKHUNKHODZHAEVA, N.: O vozmozhnosti ispol'zovaniya mysh'yaka pri biogeokhimicheskekh poiskakh zolotorudnykh mestorozhdenii (na primere Tamdytau). (Possible use of arsenic in biogeochemical prospecting for gold-ore deposits (as illustrated by Tamdytau)). Uzbeksk. Geol. Zh. **12**, 43 (1968).

TENYAKOV, V. A., RAKOVSKIY, E. YE., FILIPPOVA, T. P.: Abundance and some general geochemical features of gold in bauxite (as shown by neutron activation analysis). Doklady Acad. Sci. U.S.S.R. Earth Sci. Sect. **195**, 194 (1970).

THOMPSON, C. E., NAKAGAWA, H. M., VAN SICKLE, G. H.: Rapid analysis for gold in geologic materials. Geological Survey Research 1968: U.S. Geol. Surv. Profess. Paper **600-B**, B130 (1968).

TILLING, R. I., GOTTFRIED, D., ROWE, J. J.: Gold abundance in igneous rocks: bearing on gold mineralization. Econ. Geol. **68**, 168 (1973).

TOLMACHEV, I. I., KALININ, S. K., TEREKHOVICH, S. L., SYROMYATNIKOV, N. G., ZARAVNYAYEVA, V. K.: Distribution of Pd, Pt and Au in the Riphaean and Cambrian igneous rocks of the Boshchekul'sk region, northern Kazakhstan. Geochemistry Internat. **8**, 194 (1971).

TUREKIAN, K. K.: Hydrosphere distribution of the elements. 1969 Yearbook of Science and Technology. New York: McGraw-Hill 1969.

UKHANOV, A. V., PCHELINTSEVA, N. F.: Soderzhance zolota v peridotitovykh i eklogitoykh vklyucheniyakh iz kimberlitovoi trubki "Obnazhennaya". (Gold content in peridotite and eclogite xenoliths from the Obnazhennaya kimberlite pipe). Geokhimya 2, 247 (1972).

VAKHRUSHEV, V. A., RYABOV, V. V., TSIMBALIST, V. G.: O zolotonosnosti nekotorykh sul'fidnykh mineralov Noril'skogo raiona. (On the gold content of some sulfide minerals of the Noril'sk region). Geokhimiya 6, 738 (1972).

VAKHRUSHEV, V. A., TSIMBALIST, V. G.: Distribution of gold in sulfides of the Altay-Sayan skarn deposits. Geochemistry Internat. 4. 972 (1967).

VAN SCHMUS, W. R., WOOD, J. A.: A chemical-petrologic classification for the chondritic meteorites. Geochim. Cosmochim. Acta 31, 747 (1967).

VAN SICKLE, G. H., LAKIN, H. W.: An atomic-absorption method for the determination of gold in large samples of geologic materials. U.S. Geol. Surv. Circ. 561, 4 p. (1968).

VINCENT, E. A., CROCKET, J. H.: Studies in the geochemistry of gold — I. The distribution of gold in rocks and minerals of the Skaergaard Intrusion, East Greenland. Geochim. Cosmochim. Acta 18, 13 (1960a).

VINCENT, E. A., CROCKET, J. H.: Studies in the geochemistry of gold — II. The gold content of some basic and ultrabasic rocks and stone meteorites. Geochim. Cosmochim. Acta 18, 143 (1960b).

VINOGRADOV, A. P.: Preliminary data on lunar ground brought to earth by automatic probe Luna- 16. Proc. Second Lunar Sci. Conf., Geochim. Cosmochim. Acta, Suppl. 2, 1, 1 (1971).

VOGT, J. R., EHMANN, W. D.: Silicon abundances in stony meteorites by fast neutron activation analysis. Geochim. Cosmochim. Acta 29, 373 (1965).

VOSKRESENSKAYA, N. T., ZVEREVA, N. F.: Geochemistry of gold and its distribution in the magmatic complexes of northern Kazakhstan. Geochemistry Internat. 5, 373 (1968).

VOSKRESENSKAYA, N. T., ZVEREVA, N. F., KONKINA, O. M., FELDMAN, V. I.: Povedenie zolota v protsesse differentsiatsii osnovnoi magmy. (Behavior of gold in the process of basic magma differentiation). Geokhimiya 12, 1438 (1970).

VOSKRESENSKAYA, N. T., ZVEREVA, N. F., RIVKINA, L. A.: Spectrochemical determination of gold in silicate rocks and minerals. Zh. Analit. Khim. 20, 1288 (1965) [Russian].

WÄNKE, H., RIEDER, R., BADDENHAUSEN, H., SPETTEL, B., TESCHKE, F., QUIJANO-RICO, M., BALACESCU, A.: Major and trace elements in lunar material. Proc. Apollo 11 Lunar Sci. Conf., Geochim. Cosmochim. Acta 34, Suppl. 1, 2, 1719 (1970).

WÄNKE, H., WLOTZKA, F., JAGOUTZ, E., BEGEMANN, F.: Composition and structure and metallic iron particles in lunar "fines". Proc. Apollo 11 Lunar Sci. Conf., Geochim. Cosmochim. Acta 34, Suppl. 1, 1, 931 (1970).

WARREN, H. V., DELAVAULT, R. E.: Gold and silver content of some trees and horsetails (*Equisetum spp.*) in British Columbia. Geol. Soc. Am. Bull. 61, 123 (1950).

WASSON, J. T., BAEDECKER, P. A.: Ga, Ge, In, Ir and Au in lunar, terrestrial and meteoritic basalts. Proc. Apollo 11 Lunar Sci. Conf., Geochim. Cosmochim. Acta 34, Suppl. 1, 2, 1741 (1970).

WASSON, J. T., SCHAUDY, R.: The chemical classification of iron meteorites — V. Groups IIIC and IIID and other irons with germanium concentrations between 1 and 25 ppm. Icarus 14, 59 (1971).

WEISS, H. V., LAI, M.: Cocrystallisation of ultramicroquantities of elements with 2-mercaptobenzimadazole. Determination of gold in sea water. Anal. Chim. Acta 28, 242 (1963).

WEISSBERG, B. G.: Gold-silver ore-grade precipitates from New Zealand thermal waters. Econ. Geol. 64, 95 (1969).

WEISSBERG, B. C.: Solubility of gold in hydrothermal alkaline sulfide solutions. Econ. Geol. 65, 551 (1970).

WHITE, D. E.: Mercury and base-metal deposits with associated thermal and mineral waters. In: Geochemistry of Hydrothermal Ore Deposits. BARNES, H. L. (Ed.): 575. New York: Holt, Rinehart and Winston 1967.

WIIK, H. B.: The chemical composition of some stony meteorites. Geochim. Cosmochim. Acta **9**, 279 (1956).
ZVEREVA, N. F., GAVRILENKO, B. V.: Gold in the rock-forming minerals of the Krykkuduk Intrusive Complex (Northern Kazakhstan). Geochemistry Internat. **8**, 76 (1971).
ZVYAGINTSEV, O. E.: Geokhimiya zolota (Geochemistry of gold). Moscow, Akad. Nauk S.S.S.R., 114 p. (1941).
ZVYAGINTSEV, O. E., PAULSEN, I. A.: Contributions to the theory of formation of vein gold deposits. Comptes Rendus Acad. Sci. U.S.S.R. **26**, 647 (1940).

Revised manuscript received: June 1973

Mercury 80

A	D. Grdenić	(Laboratory of General and Inorganic Chemistry, Faculty of Science, University of Zagreb, Zagreb, Yugoslavia)
B—M	G. Tunell	(Department of Geological Sciences, University of California, Santa Barbara, California, U.S.A.)
O	G. Tunell	
	J. J. Rytuba	(U.S. Geological Survey, Menlo Park, California, U.S.A.)

80-A. Crystal Chemistry

The crystal chemistry of mercury differs in many respects from the crystal chemistry of the other IIb-group elements. The most striking differences are: (I) the unique structure of solid mercury and the weak interatomic forces in the metallic state, (II) mercury to mercury bond in the mercurous ion $[Hg_2]^{2+}$, (III) the rarity of ionic structures containing Hg^{2+}-ions, and (IV) the predominance of covalent mercury to non-metal bonds in simple molecular or complex polymeric crystal structures (LIPSCOMB, 1957; WELLS, 1962; BRÄNDEN, 1964; AURIVILLIUS, 1965; GRDENIĆ, 1965).

These crystallochemical characteristics of mercury originate from its $5d^{10}6s^2$ orbitals (SIDGWICK, 1950; COTTON and WILKINSON, 1966; PHILLIPS and WILLIAMS, 1966), the atomic and ionic size being less significant. Mercury occurs in nature as a metal, and in both mercurous Hg(I) and mercuric Hg(II) compounds DANA, 1944, 1951; BRAGG and CLARINGBULL, 1965; STRUNZ, 1966).

Metallic Mercury

Solid Mercury

In solid mercury (rhombohedral α-Hg) each atom is surrounded by 6 atoms at a distance of 2.999 Å. They form the apices of a trigonal antiprism having 6 equatorial atoms at 3.466 Å in a hexagon. A flattened cuboctahedron is the effective coordination polyhedron (McKEEHAN and CIOFFI, SB 1913—1928, 737; NEUBURGER, SB 1933—1935, 200; BARRETT, SB 1957, 153 cf. F.). In the structure of the tetragonal β-Hg formed under high pressure, which below 79° K is more stable than α-Hg, there are at each mercury atom two collinear bonds of 2.825 Å and eight more distant ones at 3.159 Å (ATOJI, SHIRBER, and SWENSON, SR 1959, 181; KLEMENT, JAYARAMAN, and KENNEDY, 1963, Fig. 80-A-1).

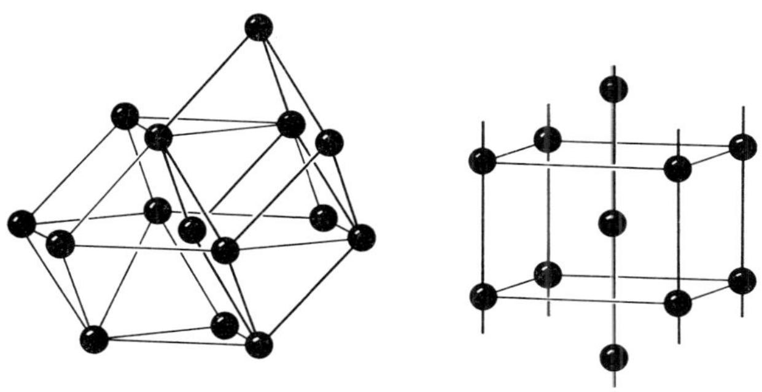

a) b)

Fig. 80-A-1. The crystal structure of metallic mercury: (*a*) rhombohedral or α-Hg, in which each atom has six neighbour atoms at 2.999 Å (heavy lines) and six more distant atoms at 3.466 Å (light lines) giving a flattened cuboctahedron as the coordination polyhedron; (*b*) tetragonal or β-Hg in which each mercury atom has two neighbour atoms at 2.825 Å (the *c*-translation) and eight more distant atoms at 3.159 Å (half the diagonal)

Liquid Mercury

In liquid mercury the ligancy is on average 10 with Hg—Hg distances of approximately 3.10 Å, the contraction on solidification being only 3.64 per cent of the liquid volume (Grosse, 1965) and the metallic radius in the liquid state not being essentially different from that obtained indirectly. A second less frequent distance of about 2.85 Å was also observed (Table 80-A-1).

Table 80-A-1. *The atomic radii of mercury*

Radius (in Å) and coordination		Source of data and reference
Metallic radius R (Hg):		
1.545	[12]	From the crystal structure of AgHg (Goldschmidt, 1928).
1.512	[12]	From the resonating valence bond theory (Pauling, 1960).
1.538	[~9.1]	Mean of the values obtained independently by X-ray scattering from liquid mercury (Black and Cundall, 1965; Wagner, Ockem, and Joshi, 1965).
1.535 1.425	[~10.0]	X-ray scattering from liquid mercury as interpreted by Rivlin, Waghorne, and Williams (1966).
Ionic radius r (Hg^{2+}):		
1.12	[8]	From the crystal structure of HgF_2 (Goldschmidt, 1926).
1.10	[6]	Theoretical or semi-empirical (Pauling, 1960; Ahrens, 1952; Stockar, 1950).
1.05	[6]	Semi-empirical (Ketelaar, 1953).
Covalent radius r (Hg):		
1.34	[2]	From Hg-*L* distances in the individual RHgX molecules based on micro-wave spectroscopy data (Aurivillius, 1965).
1.30	[2]	From C-Hg and Hg-I in individual RHgI molecules (Grdenić, 1965).
1.37	[3]	Empirical, based on Hg-I in HgI_3 as determined by Fenn, Oldham, and Phillips (1963).
1.48	[4[	From the crystal structure (cubic) of Hg*L* for $L = $ S, Se, Te (Pauling, 1960).
Van der Waals radius R_w (Hg):		
1.50	—	From the crystal structure of RHgX with $R = CH_3$, C_2H_5; X = Cl, Br (Grdenić and Kitaigorodskii, SR 1950, 449).

Amalgams

The metals having $R(M)$ within the limits of $R(\mathrm{Hg}) \pm 15\%$, the other electron factors being favourable (Hume-Rothery and Raynor, 1954), dissolve mercury to a certain extent, but never over the whole range (Pearson, 1964). In kongsbergite (Ag, Hg) with a Hg-content up to about 31 at.%, the closest Ag—Hg distance is 2.95 Å (Murphy and Preston, SB 1928—1932, 704). The metal-mercury intermetallic phases (mercurides) belong to different types of alloy structures, some of them being unique (Schubert, 1964). The minerals landsbergite Ag_3Hg_4 and gold-amalgam Au_2Hg_3 (or $AuHg_2$, Hansen and Anderko, 1958) belong to the γ-brass type and almost satisfy the Hume-Rothery rule having an electron to atom ratio of 21/13. Potarite $Pd^{[8+4]}Hg^{[8+4]}$ has a tetragonal structure of the CuAu-type with Hg—Pd and Hg—Hg bond distances of 2.82 Å and 3.03 Å respectively (Terada and Cagle, 1960).

Mercury (I) Compounds

All mercury (I) compounds are of the type L—Hg—Hg—L so that: (I) the bonds are essentially collinear, (II) L=F, Cl Br, I, O, or N, (III) the Hg—Hg bond distance varies with change of L from 2.43 to 2.69 Å (GRDENIĆ and DJORDJEVIĆ, SR 1956, 214; LEWIS and NYHOLM, 1964), while the Hg—L distance is larger than the sum of the covalent radii and less than the sum of the ionic radii, (IV) the oxonium ion $(H_2O—Hg—Hg—OH_2)^{2+}$ is formed by hydration (GRDENIĆ, SR 1956, 368; JOHANSSON, 1966). The mineral representatives are calomel Hg_2Cl_2 (HAVIGHURST SB 1913—1928, 237), eglestonite Hg_6Cl_4O (HEDLIK, SR 1950, 199), and terlinguaite $Hg_2Cl_2 \cdot 2HgO$ (ŠĆAVNIČAR, SR 1956, 242), the latter having a very close intermolecular Hg...Hg contact of 2.70 Å.

Mercury (II) Compounds

Ionic Structures

The only ionic structure is the fluorite-type structure of $Hg^{[8]}F_2^{[4]}$ (EBERT and WOITINEK, SB 1933—1935, 276).

Covalent Structures

In most of the Hg(II)-structures the Hg—L bonds are essentially covalent but, because they have a different bond order, various coordination polyhedra appear around the mercury atoms. They are conveniently systematized according to the characteristic coordination (m) and the effective coordination (n) around the mercury atom (GRDENIĆ, 1966). The characteristic coordination can have $m=2$, 3, or 4, which corresponds to the digonal sp, the trigonal sp^2, or the tetrahedral sp^3 hybrid bond orbitals of the mercury atom. The Hg—L bond distances within the characteristic coordination follow roughly the additivity rule, being obtained as the sum of the covalent radii $r^{[m]}(Hg)+r(L)$.

In crystal structures the characteristic coordination $m=2$ or 3 is completed to an effective coordination n. The $n-m$ supplementary bonds do not follow the additivity rule and are greater than the sum of the covalent radii $r^{[m]}(Hg)+r(L)$ but smaller than the expected intermolecular contact of $R_w(Hg)+R_w(L)$, the condition which determins n. The number of supplementary ligands $n-m$ depends upon the characteristic coordination m; it is 4 or 2 for $m=2$ (distorted octahedra or tetrahedra), 2 or 1 for $m=3$ (trigonal bipyramid or pyramid), and 0 for $m=4$ (regular tetrahedron) (Table 80-A-2).

The digonal characteristic bonds within an effective coordination usually depart from collinearity, e.g. in HgS (hex.) with $n=6$ the S—Hg—S bond-angle is 172° (AURIVILLIUS, SR 1950, 179) or in $(NCHg)_2O$ with $n=4$ the $C_I—Hg_I—O$ bond angle is 155.6° (ŠĆAVNIČAR, 1963; AURIVILLIUS, 1964, a). The trigonal characteristic bonds within an effective coordination with $n=4$ are usually not coplanar, e.g. the Br—Hg—Br bond angles in $[HgBr_3]^-$ are 113° and 125° instead of 120° (WHITE, 1963). The characteristic bond distances in the effective coordination polyhedron are as a rule less strongly influenced (for collected data see: AURIVILLIUS, 1965). Since the effective coordination with $n=6$ is realized from $m=2$ by adding four supplementary ligands, there is so far no known structure with the mercury atom in a regular octahedral coordination (DUNITZ and ORGEL, 1960). The digonal and

Table 80-A-2. *Bond distances (Hg-L) in crystal structures of some mercury compounds*

Compound	Hg-*L* (in Å) and references
$HgCl_2$	Hg-Cl 2.25(2); plus two pairs of Hg ... Cl contacts of 3.34(2) and 3.36(2) (Braekken and Scholten, SB 1933—1935, 277).
$HgBr_2$	Hg-Br 2.48(2); 3.32(4) (Verweel and Bijvoet, SB 1928—1932, 250).
HgI_2 (red)	Hg-I 2.80(4) (Bijvoet, Classen, and Karssen, SB 1913—1928, 189).
Hg_2F_2	Hg-Hg 2.43 Hg-F 2.31; 2.70(4) (Grdenić and Djordjević, SR 1956, 214).
Hg_2Cl_2	Hg-Hg 2.53; Hg-Cl 2.52; 3.29(4) (Havighurst, SB 1913—1928, 256).
HgO (o-rh)	Hg-O 2.04; 2.07; 2.81(2); 2.83; 2.85 (Aurivillius, 1964a).
HgS (cub)	Hg-S 2.53(4) (Kolkmeijer, Bijvoet, and Karssen, SB 1913—1928, 129; Aurivillius, 1964b).
HgS (hex)	Hg-S 2.36(2); 3.10(2); 3.30(2) (Aurivillius, SR 1950, 179).
$HgSb_4S_8$	Hg_I-S 2.37(2); 3.34(2); 3.38(2); (Hg_{II}-S) 2.33(2); 3.34(2); 3.27(2) (Niizeki and Buerger, SR 1957, 347).
$^2_\infty[Hg_3O_2]SO_4$	Hg_I-O 2.03(2); 2.4 [illegible]; 2 [illegible] 3 16 and Hg_{II}-O 2.03(2); 2.75(2); 2.78(2) (Nagorsen, Lyng, Weiss, and Weiss, 1962).
$^3_\infty[Hg_2N]Br$	Hg-N 2.06(2); Hg-Br 3.33(2); 3.39(2) (Rüdorff and Brodersen, SR 1953, 514).
$[(NCHg)_2O]$	Hg_I-O 2.09; 2.54(2) Hg_I-C 2.00; Hg_{II}-O 1.95; Hg_{II}-N 2.88(2); 2.99(2) Hg_{II}-C 1.90 (Sćavničar, 1963; Aurivillius, 1964a).

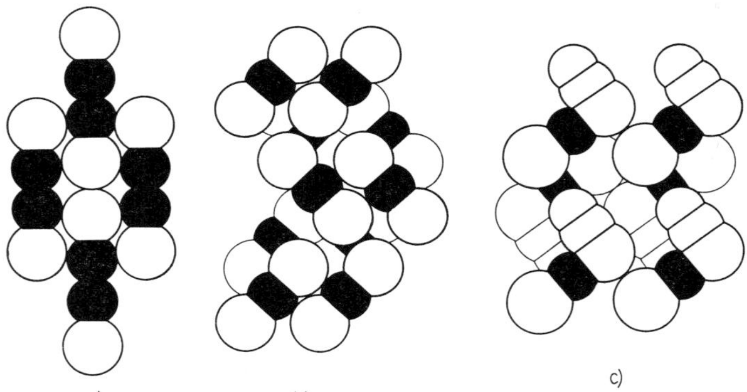

a) b) c)

Fig. 80-A-2. The layers in the crystal structure of (*a*) Hg_2Cl_2, (*b*) $HgCl_2$, and (*c*) BrHgSCN (Zvonkova and Ždanov, SR 1952, 300)

trigonal coordination appears only in free molecules like $HgCl_2$ and CH_3HgCl, or in ions like HgO_2^{2-} (Hoppe and Röhrborn, 1964) or $HgBr_3^-$, respectively. The tetrahedral coordination ($m = n$) occurs only in crystal structures containing individuai $[HgL_4]^{2-}$ ions, as in $K_2[Hg(SCN)_4]$, or layers of tetrahedra sharing corners as in $^2_\infty Hg^{[4]}I_4$ (red), as well as in the sphalerite-like structures of metacinnabar $Hg^{[4]}S^{[4]}$, tiemannite $Hg^{[4]}Se^{[4]}$, and coloradoite $Hg^{[4]}Te^{[4]}$.

Types of Connection

The crystal structure of mercury compounds containing Hg—O, Hg—S, and Hg—N bonds is best understood in terms of connections. E.g., the structure of orthorhombic $^1_\infty Hg^{[2]}O^{[2]}$ (montroydite) is built up of planar zig-zag —Hg—O—Hg—O— chains (bond angle 179.5° at Hg, 107.3° at O; Aurivillius, 1964). The structure of hexagonal HgO contains similar chains helically wound (AURIVILLIUS, 1958), i.e. isotypically with cinnabar (AURIVILLIUS, SR 1950, 179). The replacement of O by OH in the zig-zag chains gives cations $[\text{—Hg—OH—}]_n^{n+}$ found in basic chlorate and bromate (WEISS, LYNG, and WEISS, 1960) as well as in oxyfluoride Hg(OH)F (GRDENIĆ, 1966). By replacing each pair of H-atoms in neighbouring chains by one Hg atom, a slightly puckered hexagonal net ${}^{2}_{\infty}\,[Hg_3^{[20]}O_2^{[3Hg]}]_n^{2n+}$ is obtained, which is observed in basic mercury (II) sulphate ${}^{2}_{\infty}\,[Hg_3^{[2]}O_2]SO_4$, mineral schuetteite (NAGORSEN, 1962). The replacement of an O-atom in the planar —Hg—

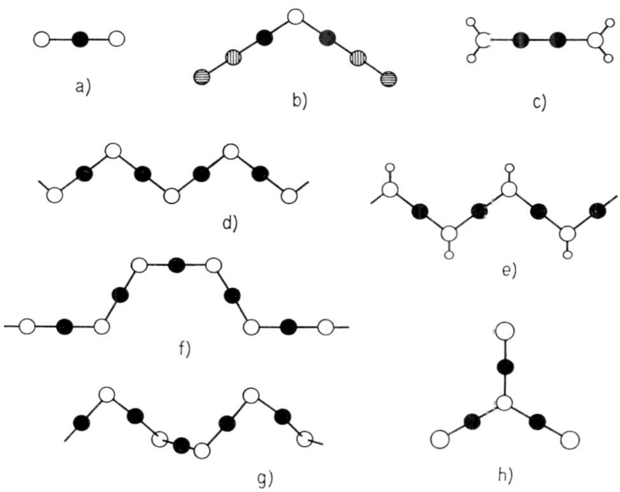

Fig. 80-A-3. The connections in the crystal structures containing the Hg—O bond: (*a*) the oxomercurate-ion HgO_2^{2-}, (*b*) the molecule of mercury (II) oxycyanide $(NCHg)_2O$, (*c*) the hydrated mercury (I)-ion $[Hg_2(OH_2)_2]^{2+}$, (*d*) the planar zig-zag chain $[\text{—Hg—O—}]_n$ orthorhombic HgO, (*e*) the planar zig-zag chain $[\text{—Hg—(OH)—}]_n^{n+}$ in mercury (II) oxychlorate, (*f*) the planar $[\text{—Hg—O—}]_n$ chain in mercury (II) oxybromide $Hg(OHg)_4Br_2$ (AURIVILLIUS, 1965), (*g*) the helically wound $[\text{—Hg—O—}]_n$ chain in hexagonal HgO, (*h*) tris-(chloromercurioxonium ion in the structure of $[(ClHg)_3O]Cl$ (WEISS *et al.*, SR 1953, 361; GRDENIĆ and ŠĆAVNIČAR, SR 1953, 361; AURIVILLIUS, 1965). The two-dimensional polymeric oxonium ion $[Hg_3O_2]_n^{2n+}$ forms an almost planar net-work of the type (*c*) in Fig. 80-A-4

O—Hg—O— zig-zag chains by $\text{—NH}_2\text{—}$ gives the ${}^{1}_{\infty}\,[HgNH_2]_n^{n+}$ cation in amidochloride $Hg^{[2N]}NH_2Cl$ (LIPSCOMB, SR 1951, 165; NIJSEN and LIPSCOMB, SR 1952, 198; BRODERSEN and RÜDORFF, SR 1952, 197; SR 1954, 397). Chains connected to layers by Hg-atoms give the cation ${}^{2}_{\infty}\,Hg^{[2N}[{}^{]}(NH)_2]_n^{2n+}$, isotypical with $[Hg_3O_2]_n^{2n+}$ in imidomercury bromide (BRODERSEN, SR 1955, 336). If layers are interconnected because mercury atoms have replaced all hydrogen atoms, a tridimensional polymeric cation ${}^{3}_{\infty}\,[Hg_3^{[2]}N^{[4]}]_n^{n+}$ of Millon's base is obtained (LIPSCOMB, SR 1951, 279; RÜDORFF and BRÖDERSEN, SR 1953, 514) whose hydrated chloride $[Hg_2N]Cl \cdot H_2O$ is the mineral mosesite (SWITZER, FOSHAG, MURATA, FAHEY, SR 1953, 440), and whose hydrated chloride sulphate $[Hg_2N](Cl, SO_4) \cdot xH_2O$ is the mineral kleinite (HERITSCH, Sr 1949, 219). The former has the cubic (cristobalite) the latter hexagonal (tridymite) type of $[Hg_2N]^+$ framework. In all the above-mentioned structures mercury atoms occur in an effective coordination $n=6$ based on digonal characteristic ligancy. The connections in the structure of chloromercurates have also been described (DAMM and WEISS, SR 1955, 331).

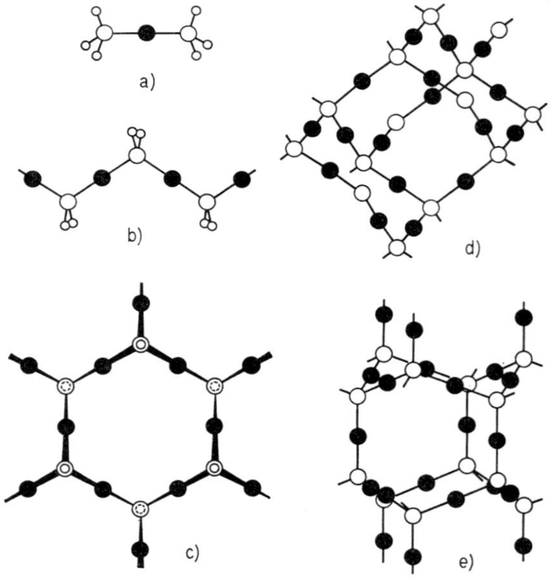

Fig. 80-A-4. The connections in the crystal structures containing the Hg—N bond: (*a*) the diaminemercury (II)-ion in $Hg(NH_3)_2Cl_2$ (MACGILLAVRY and BIJVOET, SB 1936, 147), (*b*) the planar zig-zag chain of $[—Hg(NH_2)—]_n^{n+}$ in $Hg(NH_2)Cl$, (*c*) the net-work of $[Hg_3(NH)_2]_n^{2n+}$ as in dimercury (II) imidobromide Hg_2NHBr_2, (*d*) the cubic and (*e*) the hexagonal variant of the three-dimensional framework of the $[Hg_2N]_n^{n+}$ cation in the crystal structure of Millon's base and its salts

Revised manuscript received: December 1967

80-B. Isotopes in Nature

Mercury has seven stable isotopes. The relative abundances of these isotopes in the natural state are given in Table 80-B-1. Brönsted and von Hevesy (1922) determined the densities of samples of mercury obtained from specimens of cinnabar from Gölnicz, Hungary, from Almadén, Spain, from the Pfalz, Bavaria, from Idria, Dalmatia, from California, from Santafiora, Italy, and from Ras el Mah, Tunisia; similar data were also obtained from a specimen of calomel from Terlingua, Texas, and from a specimen of mercury oxychloride from Terlingua, Texas. The densities of the samples of mercury did not differ more than 6 units in 10^6 units; the atomic weights of these samples of mercury did not differ more than 0.0012. Thus it appears that the isotopic composition of mercury in the earth's crust is very nearly invariable, if not invariable. Since a partial separation of mercury isotopes has been accomplished in the laboratory by evaporation at low pressures (Brönsted and von Hevesy, 1920); Brönsted and von Hevesy (1921); Mullikan and Harkins (1922); Harkins (1921); it is conceivable that the lighter isotopes of mercury may have become slightly more concentrated by evaporation in the air above deposits containing native mercury or above some hot springs from which mercury vapor has been given off. Thus the possibility of slight differences in the isotopic constitution of mercury from different terrestrial sources has not been completely disproved.

Table 80-B-1. *Isotopes of mercury*

Isotope mass number	Relative abundance in the natural environment (in atom percent) [a]
196	0.146 ± 0.006
198	10.02 ± 0.01
199	16.84 ± 0.02
200	23.13 ± 0.02
201	13.22 ± 0.02
202	29.80 ± 0.01
204	6.85 ± 0.01

[a] Bainbridge and Nier (1950), p. 52.

Revised manuscript received: August 1968

80-C. Abundance in Meteorites, Cosmos and Lunar Materials

I. Meteorites and Cosmos

NODDACK and NODDACK (1931) reported a mercury content of 0.2×10^{-6} gram per gram or 200 ppb in troilite from a Canyon Diablo meteorite. Subsequently by neutron activation analysis, REED *et al.* (1960) obteined a considerably smaller value, 7×10^{-9} gram per gram or 7 ppb, for the mercury content of a sample of troilite from Canyon Diablo.

More recently EHMANN and LOVERING (1967) have made an extensive series of determinations of the mercury contents of both chondritic and achondritic meteorites of various types by neutron activation analysis. The values of the mercury abundances in various types of chondrites obtained by them together with values determined previously by EHMANN and HUIZENGA (1959) and by REED *et al.* (1960) are reproduced in Table 80-C-1, and the values of the mercury abundances in various types of achondrites obtained by the same investigators are reproduced in Table 80-C-2.

Reed *et al.* (1960) concluded that the much greater mercury concentration of the carbonaceous chondrite, Orgueil, than that of the enstatite chondrite, Abee, appears to be consistent with the thermal histories of the two meteorites. They stated that the carbonaceous chondrites must have been at a low temperature throughout their history, having retained quite considerable amounts of water and easily decomposed organic compounds; on the other hand, a polished section examination of the Abee meteorite revealed an almost continuous network of metal and sulfide in a silicate matrix, which suggested high temperature exposure after accumulation.

Table 80-C-1. *Mercury contents of chondrites*

	Mercury content (ppm)	Reference [a]
Carbonaceous Chondrites		
Type I		
Orgueil I(A)	114	1
Orgueil I(B)-a [b]	22.2 [e]	1
Orgueil I(B)-b [e]	20.1 [e]	1
Orgueil I(B)-c [d]	20.8 [e]	1
Orgueil	17.26	3
Type II		
Mighei	6.82	1
Murray	1.57	1
Type III		
Karoonda	1.28	1
Mokoia I	7.3	1
Mokoia II	0.69	1

Table 80-C-1 (Continued)

	Mercury content (ppm)	Reference[a]
Olivine-hypersthene chondrites		
Bruderheim I (A)	0.066	1
Bruderheim I (B)	0.061	1
Harleton	0.16	1
Holbrook	0.17	2
Peace River	0.028	1
Olivine-bronzite chondrites		
Beardsley I (A)	0.074	2
Beardsley I (B)	0.076	1
Beardsley II	0.25	1
Forest City	0.080	2
Plainview	0.054	2
Richardton	0.36	1
Enstatite chondrites		
Type A		
Abee	0.004	3
Abee	1.52	1
Type B		
Atlanta	0.16	1
Hvittis	0.72	1
St. Marks	0.43	1

[a] References: 1. Ehmann and Lovering (1967), 357—376; 2. Ehmann and Huizenga (1959), 125—135; 3. Reed, Kigoshi, and Turkevich (1960), 122—140.

[b] Sample heated 48 hrs. at 65° C; total weight loss 7.4 percent.

[c] Sample heated 48 hrs. at 65° C and then 20 hrs. at 125° C; total weight loss 10.1 percent.

[d] Sample heated 48 hrs. at 65° C, then 20 hrs. at 125° C, and finally 8 hrs. at 190° C; total weight loss 11.4 percent.

[e] Abundance calculated on the basis of the original sample weight before heating.

For the cosmic abundance of mercury Suess and Urey (1956, p. 56) obtained a value of 0.284 atom of mercury per 10^6 atoms of silicon by the interpolation of an abundance curve based on the cosmic abundance values for the bracketing elements, gold and thallium.

Ehmann and Lovering (1967, p. 368) have stated that calculation of the cosmic abundance of mercury on the basis of various chondrite models is made difficult by the extreme variability of the mercury content between separate samples of a single meteorite and between samples of different meteorites of the same general type. In Table 80-C-3 the mean values and the ranges of values of the mercury contents per 10^6 atoms of silicon listed by Ehmann and Lovering for the different types of chondrites are reproduced. Ehmann and Lovering stated that there are reasons for thinking that the Type I carbonaceous chondrites represent the closest approach to primitive solar system material. However, they stated further that if the very high values of the mercury abundance determined in the Orgueil Type I carbonaceous chondrite approach the true value of the cosmic abundance of mercury, then the

Table 80-C-2. *Mercury contents of achondrites*

Achondrite	Mercury content (ppm)	Reference[a]
Angrite		
Angra dos Reis	2.51	1
Nakhlite		
Nakhla	0.23	1
Eucrites		
Juvinas	5.01	1
Moore County	2.74	1
Nuevo Laredo	0.078	3
Serra de Mage	0.21	1
Stannern	9.12	1
Howardite		
Binda	0.66	1
Diogenites		
Ellemeet	0.43	1
Johnstown	0.12	2
Aubrite		
Norton County	0.054	1

[a] References: 1. Ehmann and Lovering (1967), 357—376; 2. Ehmann and Huizenga (1959), 125—135; 3. Reed, Kigoshi, and Turkevich (1960), 122—140.

Table 80-C-3. *Atomic abundance of mercury in chondrites per 10^6 atoms of silicon according to* Ehmann and Lovering (1967)

	Range of values	Mean value
Carbonaceous chondrites		
Type I	22—148	29
Type II	1.6—7.2	4.4
Type III	0.62—6.6	2.8
Enstatite chondrites		
Type A	1.3	1.3
Type B	0.35—0.59	0.47
Olivine-hypersthene and olivine-bronzite chondrites	0.023—0.28	0.10

abundance curves in the vicinity of mass number 202 would need to be revised. They concluded that it is extremely important that other Type I carbonaceous chondrites be examined to establish whether or not the high mercury abundance found in Orgueil is, indeed, characteristic of meteorites of this type.

Revised manuscript received: August 1968

II. Abundance in Lunar Materials

In Apollo 11 samples, REED and JOVANOVIC (1970)[1] found concentrations of mercury of 0.6 to 13 ppb (Table 80-C-4), the largest concentration occurring in

Table 80-C-4. *Mercury concentrations in Apollo 11 samples*[a]

Sample	ppb Hg (Total)	ppb Hg (> 450° C)
Soil (10084-2)	5.3 ± 0.1	0.06 ± 0.01
Soil (10084-1)	0.95 ± 0.04	0.22 ± 0.02
Microbreccia (10061-22)	1.2 ± 0.1	0.042 ± 0.002
Micro gabbro 10044-48	0.60 ± 0.03	0.028 ± 0.003
Vuggy basalt 10072-24	5.5 ± 0.1	1.1 ± 0.1
Vesicular basalt 10017-22	13 ± 2	0.09 ± 0.01

[a] REED, G. W., JR., JOVANOVIC, S.: Proc. Apollo 11 Lunar Science Conference, **2**, 1487 (1970).

vesicular rock. These concentrations are significantly lower, in general, than those in basaltic achondrites (18 to 9,000 ppb) or in chondrites and approach more nearly the values observed in some terrestrial rocks. In contrast to meteorites and terrestrial rocks, a large mercury release below 110° C was observed, which might be explained as free mercury adsorbed on surfaces. The release trends are consistent with the following interpretation. The most degraded material, the soil 10084-2, had the largest low temperature fraction, followed by the microbreccia. The fine-grained rock, 10072-24, had a greater amount of its mercury in high temperature traps. The medium grained igneous rock occupied an intermediate position, possibly because of slower cooling to lower temperatures. The relatively large low temperature release from vesicular rock (10017-22) could be interpreted as resulting from mercury associated with the vapor phase that formed the vesicles and was trapped in the glass-like lining of the voids.

The contents of mercury determined in Apollo 12 soils by REED and JOVANOVIC (1971)[2] (Table 80-C-5) range from 2.6 to 20 ppb and in rocks from 1.6 to 12 ppb. A gradient with depth in surface-related mercury concentrations (i.e., mercury that is volatilized from samples at a temperature of 130° C or less) was observed in cores. It was concluded that some mercury is labile and can be volatilized from the surface during lunar daytime. The increasing amount of surface-related mercury with depth suggests that a cold trap exists below the surface. The total mercury extracted from core samples was also found to increase with depth. Thus, the total mercury found in rocks and soil seems to be correlated with the mercury on surfaces. The intermediate temperature fractions, 130–325° C and 130–475° C, also showed a correlation with the <130° C fraction. (Mercury was extracted from samples by stepwise heating, usually at temperatures of 75°, 130°, 175°, 225°, 275°, 325°, 475°, 625°, and 1,200° C.) These fractions constitute most of the total mercury. The 325–475° C and the 475°–1,200° C fractions do not show this correlation. The correlation is only approximate, as might be expected for material of varying composition, grain size,

[1] For reference see Table 80-C-4.
[2] For reference see Table 80-C-5.

Table 80-C-5. *Mercury Concentrations in Apollo 12 samples*[d])

Sample	ppb Hg
Soil (12070,61) (<74 μm)[a]	3.3
Soil (12070,61) (>150 μm)	0.9
Soil (12033,18)	2.6
Core (12025,74) (1.5–2.5 cm)[b]	10.0
Core (12028,215) (20–21 cm)	17.0
Core (12028,216) (38–39 cm)	20.0
Basalt (12052,18i[c])	3.0
Basalt (12052,49e)	9.0
Basalt (12053,31i)	1.6
Basalt (12053,31i)	3.8
Basalt (12053,53e)	12.0
Basalt (12022,109i)	8.0
Basalt (12022,69e)	7.6
Basalt (12021,84i)	2.0

[a] Grain size is given in parentheses.
[b] Depth in core is given in parentheses.
[c] i denotes interior and e denotes exterior of the sample.
[d] Reed, G. W. Jr., Jovanovic, S.: Proc. Second Lunar Science Conference, **2**, 1262 (1971).

and porosity. The type and abundance of sites emptied at each temperature varied from sample to sample. However, an equilibrium in the mercury distribution between the sites seems to be a reasonable assumption. Mercury released at the highest temperature >450° C, may not have equilibrated since some of this may be originally trapped mercury. In a strict sense, all the mercury below 450° C may be surface related, but only that released below 130° C is in readily emptied sites.

The concentrations of mercury in Apollo 14 soils—including trench samples, fragmental rocks, and crystalline rock 14,310, as well as a few Apollo 15 soil samples—were determined by Reed *et al.* (1972)[1] (Table 80-C-6). Sample 14305,64 consisted of a dark fraction and a light fraction. The dark fraction consisted of relatively large sub-angular fragments of plagioclase, pyroxene, and olivine, up to 0.1 mm in diameter, embedded in a fine-grained (0.01 mm) matrix of silicates and smaller (1–3 μ) opaque grains. The hard and vuggy dark clasts appeared to be partially recrystallized breccias. The light fraction, interstitial to the well-defined dark clasts, contained an assortment of less consolidated angular material and glass grains, and both recrystallized and unaltered rock fragments. The most striking aspect of the Apollo 14 mercury measurements is the large release above 450° C during stepwise heating. In samples from other sites, only a small fraction of the mercury remained above this temperature. Reed *et al.* (1972)[1] interpreted the presence of mercury in highly retentive sites as being due to high temperature equilibration. Since these samples are known to be associated with the impact that excavated the Imbrium Basin, both high temperature quenching and shock implantation, as well as pre-

[1] For reference see Table 80-C-6.

Table 80-C-6. *Mercury concentrations in Apollo 14 and 15 samples*[c])

Sample		ppb Hg (≤130° C)	ppb Hg (Total)
Apollo 14	Soil (14163,108)	0.12	2.7
	Soil (14259,96)	0.14	1.6
	Fragmental rock (14148,13 trench top)	1.2	7.3
	Fragmental rock (14156,3 trench middle)	1.4	5.9
	Fragmental rock (14149,21 trench bottom)	1.2	7.3
	Fragmental rock (14305,84-1[a])	0.83	2.4
	Fragmental rock (14305,84-2[a])	0.28	1.5
	Fragmental rock (14305,64 (light)[b])	0.80	11.0
	Fragmental rock (14305,64 (dark)[b])	0.11	0.68
	Basalt (KREEP) (14310,124-1[a])	0.35	0.76
	Basalt (KREEP) (14310,124-2[a])	0.22	0.94
	Basalt (KREEP) (14310-C)	18.0	42.0
Apollo 15	Soil (15091,32 sta 2)	0.5	10.0
	Soil (15041,32 sta 8)	1.3	3.0
	Soil (15231,29 sta 2)	7.8	11.0

[a] 14305,84 and 14310,124 are exterior samples.
[b] 14305,64 is an interior sample.
[c] REED, G. W., JR., JOVANOVIC, S., FUCHS, L.: Proc. Third Lunar Science Conference, **2**, 1989 (1972).

existing high temperatures at the impact site, may account for the mercury in retentive sites.

Of the Apollo 15 samples, the soils from station 2 were considered to be representative of the nonmare Appenine Front, while those from station 8 on the mare were thought to be a part of a ray from Aristillus or Autolycus. The results of the station 2 Apollo 15 soil samples may then be compared with the results of the Apollo 11 and Apollo 12 mare soil samples. It appears that the mercury contents of the Apollo 15 Appenine Front soil samples are a little higher than those of the Apollo 11 and Apollo 12 mare soil samples.

Manuscript received: August 1975

80-D. Mercury Minerals, Abundance in Rock-Forming Minerals

I. Mercury Minerals

Mercury occurs in nature chiefly as the red sulfide, and in lesser amounts as the black sulfide and a liquid native quicksilver. The red sulfide, cinnabar, in nature is nearly pure HgS. The black sulfide, metacinnabar, in nature is a solid solution the

Table 80-D-1. *Mercury Minerals*

Mineral	Composition
Amalgam	(Ag, Hg)
Gold-amalgam	(Au_2Hg_3) ?
Moschellandsbergite	(Ag_2Hg_3)
Potarite	(Pd_3Hg_2) ?
Schachnerite	$Ag_{1.1}Hg_{0.9}$
Paraschachnerite	Ag_3Hg_2
Metacinnabar	HgS to $(HgS)_{80}(HgSe)_{20}$
Onofrite	$(HgS)_{80}(HgSe)_{20}$ to $(HgS)_{20}(HgSe)_{80}$
Tiemannite	$(HgS)_{20}(HgSe)_{80}$ to HgSe
Guadalcazarite	(Hg, Zn)S
Leviglianite	(Hg, Zn, Fe)S
Saukovite	(Hg, Cd)S
Coloradoite	HgTe
Schwatzite	$(Cu, Fe, Hg)_{12}Sb_4S_{13}$
Hakite	$(Cu, Hg)_{12}Sb_4(S, Se)_{13}$
Cinnabar	HgS
Livingstonite	$HgSb_4S_8$
Aktashite	$Cu_6Hg_3As_5S_{12}$
Atheneite	$(Pd, Hg)_3As$
Balkanite	$Cu_9Ag_5HgS_8$
Laffittite	$AgHgAsS_3$
Routhierite	$TlHgAsS_3$
Christite	$TlHgAsS_3$
Temagamite	Pd_3HgTe_3
Vrbaite	$Tl_4Hg_3Sb_2As_8S_{20}$
Petrovicite	$Cu_3HgPbBiSe_5$
Galkhaite	$[Hg_{0.76}(Cu, Zn)_{0.24}]_{12}Tl_{0.96}(AsS_3)_8$
Tvalchrelidzeite	$Hg_{12}(Sb, As)_8S_{15}$
Corderoite	$Hg_3S_2Cl_2$
Calomel	HgCl
Eglestonite	$Hg_6Cl_3O_2H$
Terlinguaite	Hg_2OCl
Pinchite	$Hg_5O_4Cl_2$
Kleinite	Mercury nitride chloride sulfate
Gianellaite	$(Hg_2N)_2 \cdot SO_4$
Mosesite	$Hg_2N[Cl, SO_4, MoO_4, CO_3] \cdot H_2O$
Montroydite	HgO
Schuetteite	$Hg_3[SO_4]O_2$

formula of which may be written (Hg, Zn, Fe) (S, Se). Corresponding to different ranges of composition of the natural black sulfide, the various names, metacinnabar, onüfrite, guadalcazarite, and leviglianite have been applied. A number of other mercury minerals are known (see Table 80-D-1), but they have been found in only a small number of places.

Mercury is also found in a number of minerals in which it is not an essential constituent and in which it subst tutes for other elements. The small differences between the radii of mercury, copper, silver, gold, zinc, cadmium, bismuth, lead, barium, and strontium make it possible for mercury to enter the structures of various minerals containing these elements.

II. Abundance in Rock-Forming Minerals

I. Noddack and W. Noddack (1931) analyzed a number of rock minerals for mercury by methods capyble of detecting in general 100 ppb. Silicate minerals tested by them for mercury included diopside from the central Alps and gadolinite from Iveland, Norway. No mercury was found in either of these minerals. They also tested apatite from Zinnwald, columbite from Arendal (Norway), tantalite from Steinkopff (South Africa) and thorianite from Ceylon, but found no mercury in any of them.

Saukov and co-workers by spectroscopic analyses detected traces of mercury in the following minerals, microcline, oligoclase, apatite, quartz, tourmaline, muscovite, amblygonite, beryl, topaz, fluorite, orthite, and uraninite, from pegmatite vein in various parts of the Soviet Union (Saukov, 1946). Aidinyan *et al.* (1963) determined the mercury contents of numerous rock samples and mineral samples from the

Table 80-D-2. *Mercury contents of rock-forming minerals of the Khibinsky Massif of the Kola Peninsula according to* Aidinyan *et al.* (1963)

Mineral	Range of mercury contents in analyzed samples (in ppb)
Aegirine	500–6,500
Apatite	35–80
Apophyllite	180
Arfvedsonite	190
Astrophyllite	100–22,000
Calcite	30
Feldspar	100–750
Fluorite	80
Hornblende	7,000
Lamprophyllite	40–9,000
Mn-lamprophyllite	22,000
Magnetite	500–700
Pyroxene	600
Mn-pyroxene	500
Ca-rinkite	400–90,000
Rinkolite	450
Sphene	100–100,000
Wollastonite	120

Khibinsky Massif of the Kola Peninsula. The found that several of the rock-forming minerals from this area contained mercury in amounts greater than the average mercury content of the earth's crust. The result of their analyses of the minerals are summarized in Table 80-D-2. AIDINYAN *et al.* (1963) concluded that the very large contents of mercury in Mn-lamprophyllite and in some Ca-rinkite, some sphene, and some astrophyllite suggested secondary halos probably connected with pegmatic and some astrophyllite suggested secondary halos probably connected with pegmatite veins, hydrothermal manifestations, and supergene phenomena.

Revised manuscript received: August 1968 and October 1977 (Table 80-D-1)

80-E. Abundance in Igneous Rocks and Average Abundance in the Earth's Crust

Turekian and Wedepohl (1961, Table 2 opposite p. 186) gave 90 ppb as the average mercury content of basaltic rocks and 80 ppb as the average mercury content of granitic rocks, these values being averages of the determinations of Stock and Cucuel and those of Preuss. Stock and Cucuel (1934a, p. 391) reported two determinations of the mercury content of V. M. Goldschmidt's composite sample of 11 German gabbros, the values found being 76 ppb and 82 ppb. Stock and Cucuel also reported two determinations of the mercury content of Goldschmidt's composite sample of 14 German granites, the values found being 57 ppb and 58 ppb. Preuss (1941, p. 19) found 0.1 ppm of mercury in Goldschmidt's composite sample of 11 German gabbros and 0.1 ppm of mercury in Goldschmidt's composite sample of 14 German granites.

Saukov (1946, pp. 27—30) and his co-workers determined the average mercury content of acid intrusive rocks to be 64 ppb on the basis of 24 analyses of such rocks from various localities in Europe, Asia, and North America, which they carried out for this purpose, together with the one analysis of the composite sample of 14 German granites made by Stock and Cucuel. Saukov (1946, pp. 29—31) and his co-workers also determined the average mercury content of basic effusive rocks to be 90 ppb on the basis of 18 analyses of such rocks from various localities in Europe, Asia, and North America, which they carried out, together with one analysis of basalt from Kaiserstuhl, Germany, made by Stock and Cucuel.

Taylor considered that the average composition of the continental portion of the earth's crust is approximately that of a mixture of equal quantities of acidic and basic igneous rocks. Taylor (1964, p. 1281) thus calculated the average abundance of mercury in the continental crust as the mean of Turekian and Wedepohl's average abundances in acidic and basic igneous rocks, and arrived at a value of 80 ppb. In a similar way Saukov (1946, p. 31) calculated the average content of mercury in the earth's crust as the mean of his average values for acidic intrusive rocks and basic effusive rocks and thus obtained a value of 77 ppb.

More recently Ehmann and Lovering (1967, pp. 371—374) investigated the mercury contents of a number of igneous rocks by neutron activation analysis. The results of their determinations of the mercury contents of six standard rock samples prepared from igneous rocks ranging in composition from a granite to a dunite and supplied by the U.S. Geological Survey are listed in Table 80-E-1. Ehmann and Lovering also made determinations of the mercury contents of samples of tholeiitic and alkali basalts from continental (Columbia River, Washington), oceanic (Hawaiian Islands), and ocean ridge (Iceland) areas, and of samples of differentiated continental tholeiitic dolerite intrusions in Tasmania, the results of which are reproduced in Table 80-E-2. From the work of Ehmann and Lovering it appears that the average content of mercury in basaltic rocks is considerably less than the

Table 80-E-1. *Mercury contents of six U.S.G.S. standard rock samples as determined by* Ehmann and Lovering (1967) *by neutron activation analyses*

Rock type	Number of sample	Locality	Mercury content (ppb)
Granite	G 2	Westerly, Rhode Island	39
Granodiorite	GSP-1	Silver Plume, Colorado	21
Andesite	AGV-1	Guano Valley, Oregon	4
Tholeiitic basalt	BCR-1	Bridal Veil Flow Quarry, Columbia River, Washington	7
Peridotite	PCC-1	Sonoma County, California	4
Dunite	DTS-1	Twin Sisters, Washington	4

Table 80-E-2. *Mercury contents of basalts, dolerites, and granophyre as determined by* Ehmann and Lovering (1967) *by neutron activation analyses*

Rock type	Number of specimen	Mode of occurrence and locality	Mercury content (ppb)
Tholeiitic basalt	BCR-1	Flow Bridal Veil Flow Quarry, Columbia River, Washington	7
Tholeiitic basalt	GA 649	Flow Napali formation, Kauai	13
Alkali basalt	GA 810	Flow Waianae volcanics, Oahu	5
Tholeiitic basalt		Flow Reyksanes, Iceland	21
Chilled doleritic basalt	M 172	Intrusion Red Hill, Tasmania	14
Pyroxene-rich doleritic basalt	DDH 5123 (1,500 ft)	Intrusion Drill Hole 5123, 1,500 ft. depth, Great Lakes, Tasmania	11
Acid granophyre	M 162	Intrusion Red Hill, Tasmania	26

average content of mercury in granitic rocks and that the average content of mercury in all igneous rocks at the earth's surface is not as high as previously was believed.

In view of the importance that has been placed on inclusions found in deep-seated kimberlitic and alkali basalt pipes as possible samples of the earth's deep crust and upper mantle, Ehmann and Lovering (1967, pp. 373—374) determined the mercury contents of samples of peridotitic and eclogitic inclusions found in South African kimberlite pipes and samples of granulitic and eclogitic inclusions found in Australian alkali basalt pipes. The results of these analyses are given in Table 80-E-3. As Ehmann and Lovering stated, all of their analyses of inclusions from deep-seated pipes show remarkably high values for the mercury content. Ehmann and

LOVERING concluded that the average abundance of mercury in the entire earth may be very much higher than previous investigators have inferred from the low average abundance determined from samples of surface rocks.

Table 80-E-3. *Mercury contents of eclogitic and peridotitic inclusions from South African kimberlite pipes and of eclogitic and granulitic inclusions from Australian alkali basalt pipes as determined by* EHMANN *and* LOVERING *(1967) by neutron activation analyses*

Rock type	Number of specimen	Mode of occurrence and locality	Mercury content (ppb)
Garnet peridotite	R 357	Inclusion in kimberlite pipe, Bultfontein Mine, South Africa	780
Eclogite	R 269	Inclusion in kimberlite pipe, Roberts Victor Mine, Orange Free State	640
Two pyroxene granulite	R 52	Inclusion in breccia pipe, Delegate, New South Wales	1,230
Eclogite	R 11	Delegate, New South Wales	1,480

Revised manuscript received: August 1968

80-F. Behavior in Magmatogenic Processes

I. Deposits Associated with Present-Day Hot Springs

Cinnabar, metacinnabar, and native mercury are being deposited at the present time by some hot springs. The waters of several of these springs have been analyzed and the temperatures of the springs have been measured.

At Amedee Hot Springs, on the eastern shore of Honey Lake in Lassen County, California, metacinnabar and cinnabar have been precipitated from alkaline waters in and around the orifices of hot springs (Figs. 80-F-1 and 80-F-2). Native mercury

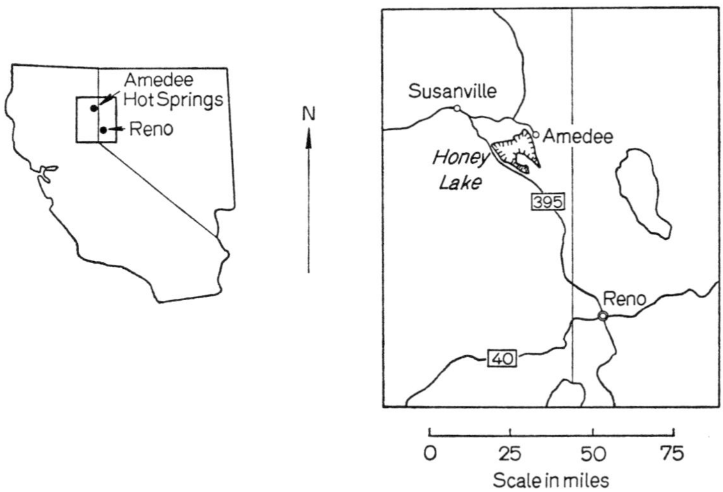

Fig. 80-F-1. Map showing location of Amedee Hot Springs, Lassen County, California

in small globules has been found in the sediment in the outflow channel below the springs[1]. The metacinnabar and cinnabar occur as thin coatings on calcareous tufa and lake sediments. The rock surfaces over which the spring waters flow are coated by layers of fine-grained cinnabar. Overlying the cinnabar is a layer of sooty metacinnabar. The total thickness of the cinnabar and metacinnabar layers together is as much as 2 millimeters in the throats of some of the springs. The mercuric sulfide forms aprons which extend as much as 75 feet (22.9 meters) down slope from the hot-spring orifices. The thickness of the mercuric sulfide layer decreases away from the hot-spring orifices. Over most of the areas of the aprons, however, the thickness of the mercuric sulfide layer is rather uniformly about 0.5 millimeter. In the throats of the hottest vents most of the mercuric sulfide is in the form of cinnabar. Away from the spring orifices the proportion of cinnabar to metacinnabar decreases. A few

[1] Personal communication from Dr. E. H. Bailey.

specimens from Amedee show wispy layers of metacinnabar alternating with thicker layers of cinnabar; in some of these specimens metacinnabar has partially altered to cinnabar. The total area of the metacinnabar and cinnabar aprons around the hot springs is at least 900 square feet (83.6 square meters) and the total mass of metacinnabar and cinnabar present in the aprons is at least 400 pounds (181 kilograms)[1]. Precipitation of cinnabar and metacinnabar appears to have continued to the present time.

The principal dissolved constituents in the hot-spring waters at Amedee are sodium sulfate, sodium chloride, and silica. An analysis of the water from Amedee

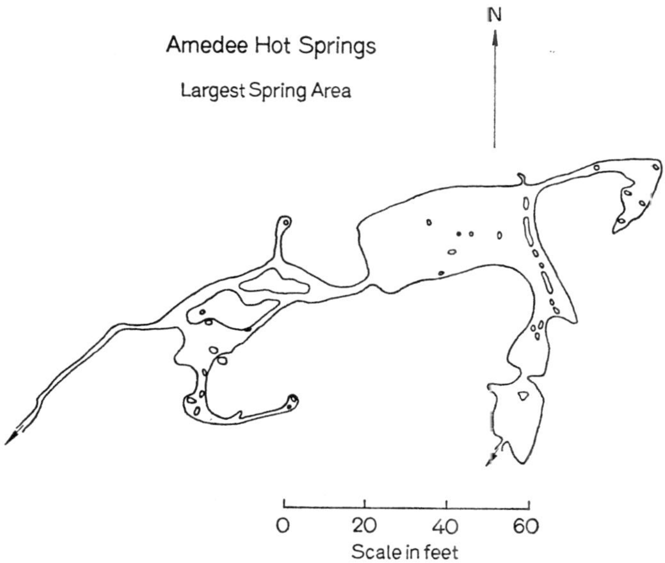

Fig. 80-F-2. Detailed map of Hot Springs Area A-B-C at Amedee Hot Springs, Lassen County, California

[1] In Hot Spring Area A-B-C at Amedee Hot Springs, which includes most of the springs at Amedee with mercuric sulfide aprons, these aprons were found to be about 0.5 millimeter thick over three subareas, one approximately 20 feet by 30 feet, one approximately 5 feet by 40 feet, and one approximately 5 feet by 20 feet. The total area made up of these three subareas was thus about 900 square feet or 83.6 square meters. Further downstream from the hot spring vents in Hot Spring Area A-B-C the mercuric sulfide coating was spotty and its mercuric sulfide content was not estimated. The product of the total area of the mercuric sulfide coating in the three subareas of Hot Spring Area A-B-C and its average thickness in the three subareas is 41,800 cubic centimeters. The density of metacinnabar given in DANA's System of Mineralogy, 7th Edition, is 7.65. The mass of mercuric sulfide in the three subareas of Hot Spring Area A-B-C calculated with this density value and the volume of 41,800 cubic centimeters is 320,000 grams or approximately 704 pounds. The mass of mercuric sulfide in the aprons at Amedee Hot Springs was originally estimated by TUNELL (1964, p. 1019) to be between 200 and 400 pounds. Since the mercuric sulfide coating was fine-grained, its density was probably somewhat less than 7.65. However, the total mass of mercuric sulfide present in the aprons at Amedee Hot Springs when the estimate was originally made by TUNELL and when the amount was redetermined by him in July, 1966, was in all probability not less than 400 pounds (181 kilograms).

Spring No. A-1 made by SCHUCH[1] in 1958 is given in Table 80-F-1. The temperature of this spring was 93° C and the pH of the spring water was 8.5. In tests of the spring water made for F. W. DICKSON and G. TUNELL by the Smith-Emery Company, no dissolved mercury was detected. The presence of sulfur in the water in a form other than sodium sulfate was proved by the formation of sulfide on the surfaces of copper and silver strips immersed in the springs for several days in tests carried out by DICKSON and TUNELL. Moreover, SCHUCH's analysis indicated that the water contained 0.2 part per million of H_2S, and, since the pH of water was 8.5, a small amount of sulfide ion was evidently present.

Table 80-F-1. *Analyses of waters. 1. from Amedee Spring No. A-1, Lassen County, California (Analyst:* J. P. SCHUCH*); 2. from the largest spring at Boiling Springs, Valley County, Idaho (Analyst:* H. KRAMER*)*

Dissolved substance in ppm	1.	2.	Dissolved substance in ppm	1.	2.
SiO_2	96.0	81.0	HCO_3	27.0	79.0
Al	0.23	0.1	CO_3	10.0	27.0
Fe^{2+}, Fe^{3+}	0.01	0.4	SO_4	288.0	12.0
			Cl	160.0	14.0
Mn	0.00	0.004	F	4.5	11.0
Ti	0.16	n.d.	Br	0.2	n.d.
As	0.02	<0.1	I	0.1	n.d.
Ca	16.0	2.2	NO_2	0.00	n.d.
Mg	0.0	0.0	NO_3	0.0	0.5
Sr	0.0	0.2	PO_4	0.40	n.d.
Ba	0.05	n.d.	B	4.1	0.1
Na	227.0	74.0	H_2S	0.2	<1.0
K	6.8	1.9	Sb	n.d.	0.1
Li	0.0	0.02	Hg	n.d.	0.0
NH_4	0.5	n.d.			

About four miles east of the town of Cedarville in Modoc County, California, there is another group of hot springs in which extensive aprons of metacinnabar and cinnabar have been deposited in recent times and in which deposition of these minerals is apparently still taking place. The principal dissolved constituents in the waters of the Cedarville Hot Springs are sodium sulfate and sodium chloride; silica, calcium bicarbonate, and boron in some form are present in lesser amounts. The composition of the hot-spring waters at Cedarville Hot Springs is thus very similar to that of the hot-spring waters at Amedee Hot Springs. The pH of the Cedarville hot-spring waters is 8.6.[2]

At Boiling Springs in Valley County, Idaho, hot springs issuing at an elevation of about 20 feet above the floor of the river valley of the Middle Fork of the Payette

[1] J. P. SCHUCH, in Data of Geochemistry, Sixth Edition, MICHAEL FLEISCHER, Technical Editor, Chapter F, Chemical Composition of Subsurface Waters, by DONALD E. WHITE, JOHN D. HEM, and G. A. WARING, U.S. Geological Survey Professional Paper **440** F, pp. F 50—F 51, 1963.

[2] Personal communication from F. W. DICKSON and W. J. RAAB.

River, flow down to the valley floor over steep walls of granitic rock. Black aprons made up chiefly of metacinnabar deposited by the springs extend as much as 25 feet down the slope from the spring orifices. Specimens collected near the hot-spring orifices showed layers of cinnabar coating the country rock, overlain in turn by thinner layers of metacinnabar. Analyses of specimens collected at one of the spring orifices and at regular intervals down to the lower edge of the dark apron were carried out by F. W. DICKSON[1] by means of the X-ray fluorescence spectrometer. The analyses showed that the dark coating near the spring orifice contained chiefly mercury with a small amount of manganese and that the ratio of mercury to manganese decreased regularly down the slope. Some mercury was found to be present, however, in all the specimens including those collected at the lower edge of the apron.

The temperature of the springs was found to be about 88° C by WHITE (1955, p. 125). This is near the boiling point at the elevation of Boiling Springs. WHITE (1955, p. 125) stated that the quantity of visible cinnabar is indeed very small, but the evidence of relation to the hot springs is conclusive. The pH of the spring waters was determined by WHITE (1955, p. 125) to be 9.2. He carried out tests for sulfide at the springs by adding cadmium chloride, but obtained a negative result. However, according to DICKSON, sufficient sulfur in a form other than sodium sulfate is present in the spring waters at Boiling Springs to cause a visible layer of silver sulfide to develop within minutes on silver metal immersed in the water. An analysis of the spring water at Boiling Springs made by KRAMER[2] is given in Table 80-F-1. The waters of the hot springs may be largely meteoric, but since Boiling Springs is situated in the interior of the Idaho Batholith, the only source for the mercuric sulfide was apparently an igneous one.

Sulphur Bank is a low, rounded hill situated on the eastern shore of Clear Lake in Lake County, California. The hill consists of lava (augite andesite), but the volcanic origin of the surface rock of the hill is not conspicuous. The bank looks like an immense oval ash-heap. The surface consists of snow-white, fine, pulverulent silica, the residue resulting from the decomposition and leaching of the lava (LE CONTE and RISING, 1882, pp. 25—26).

When the sulfur-capped ridge at Sulphur Bank was first discovered, springs discharged along the southern base of the ridge. The topography of the mine area has since been modified by the excavation of open pits and the heaping up of mine dumps.

A considerable number of analyses of the waters of the hot springs discharging at Sulphur Bank have been published through the years by various authors. Several analyses of the waters of the springs and the waters in the ponds occupying the pits were recorded by WHITE and ROBERSON (1962) and are reproduced in Tables 80-F-2 and 80-F-3. The principal constituents of the spring waters are Na, NH_4, HCO_4^{2-}, SO_4^{-}, Cl, B, and SiO_2. WHITE and ROBERSON (1962, pp. 414—415) stated that F. WANG of Buckman Laboratories, Inc., using a modified dithizone method

[1] Personal communication DICKSON.

[2] H. KRAMER, in Data of Geochemistry, Sixth Edition, MICHAEL FLEISCHER, Technical Editor, Chapter F, Chemical Composition of Subsurface Waters, by DONALD E. WHITE, JOHN D. HEM, and G. A. WARING, U.S. Geological Survey Professional Paper **440** F, pp. F 50—F 51, 1963.

Table 80-F-2. *Chemical analyses of waters from Sulphur Bank, California (in ppm)*. (From White and Roberson, 1962)

Description	Sulphur Bank spring	Canal pit stibnite-depositing spring	Herman pit "Ink" spring, north side	Herman pit spring on north side	Herman pit North side, "geyser" spring	Herman pit Southwest of Number 4	Herman pit South wall near west end	Pond in pit Filling 1947—1954	Pond Pit nearby, unwatered
Analysis number	1	2	3	4	5	6	7	8	9
Date collected	Sept. 1863	Oct. 1938	July 6, 1955	Oct. 4, 1956	Mar. 26, 1957	Mar. 26, 1957	Oct. 17, 1957	Jan. 17, 1954	Mar. 26, 1957
Temperature, °C	"hot"	80.5	57	47	69.5	77	44	cold	23
pH	—	—	6.6	6.8	6.8	7.5	6.6	2.7	6.0
Estimated discharge of sampled spring, gpm	—	—	1	1/4	0 to 15	1/2	2	—	—
Approximate altitude, feet	1,330	1,296	1,300	1,240	1,230	1,225	1,250	1,327	1,220
SiO_2	108 (?)	67	—	20	42	72	130	107	16
Al	11.4	—	—	0.4	0.6	—	—	—	—
Fe	—	Tr.	"Abundant"	<0.04	0.0	—	—	—	30
Mn	—	—	—	0.95	0.1	—	—	—	—
As	—	0	—	0.00	0.00	0.02	—	—	—
Sb	—	0	—	—	—	—	—	—	—
Hg	—	—	0.02	—	—	—	—	—	—
Ca	Tr.	56	—	118	20	7	288	—	172
Mg	Tr.	57	—	118	55	22	41	—	131
Sr	—	—	—	1.1	2.2	1.7	—	—	—
Na	1,330	1,390	1,380	1,240	1,190	1,100	184	498	616
K	Tr.	68	91	23	23	33	5.3	0.5	19
Li	—	—	4.5	3.1	4.4	4.8	0.0	1.7	1.9
NH_4	420	543	—	276[a]	464	476	59[b]	—	Present
HCO_3^-	2,380	1,890	—	2,590	3,290	2,960	1,440	—	210
CO_3^{2-}	—	—	—	0	0	0	0	—	0
SO_4^{2-}	Tr.	1,980	—	1,430	598	454	142	6,040	2,800
Cl	878	489	748	695	644	690	16	376	270

F	—	—	—	1.0	1.0	1.0	0.5	—	—
Br	Tr.	—	—	3	1.6	1.4	—	—	—
I	1.4	—	—	4.7	3.2	3.6	—	—	—
NO_3^-	—	—	—	0	0	0	0.3	—	0.1
NO_2^-	—	—	—	460[a]	0.0	0.0	4.3[b]	—	0.0
PO_4^{3-}	—	—	—	—	—	—	—	—	—
B	381	527	—	680	620	660	16	249	290
H_2S	—	—	—	1.4	12	3.6	—	—	—
CO_2	442	—	—	—	—	—	—	—	—
Total, as reported	5,957	7,067	—	7,658	6,969	6,488	2,326	—	4,556
Residue in evaporation, 180° C	—	5,010	—	6,230	5,220	5,090	—	11,300	4,600
Specific conductance in microohms, at 25° C	—	—	—	7,670	7,430	7,060	2,360	—	5,840
Analyst	G. E. Moore (recalculated)	M. Fleischer (recalculated)	C. S. Howard, Fred Wang, H. C. Whitehead	C. E. Roberson	C. E. Roberson	C. E. Roberson	H. C. Whitehead and P. Schuch	H. Kramer, S. Berman	H. C. Whitehead

[a] Nitrogen as NH_4 and NO_2^- when analyzed after bottle had been opened; all believed originally present as NH_4.
[b] Determinations made on another sample collected Sept. 17, 1958.

Table 80-F-3. *Quantitative spectrographic analyses of evaporated residues of water from Sulphur Bank (calculated to parts per million of the original water). (From* WHITE and ROBERSON, 1962)

Description	Spring	"Geyser"	Spring	Pond	Pond
Number, Table 80-F-2	4	5	6	8	9
Date collected	Oct. 4, 1956	Mar. 26, 1957	Mar. 26, 1957	Jan. 17, 1954	Mar. 26, 1957
Temperature, °C	47	$69^1/_2$	77	cold	tepid
pH	6.8	6.8	7.5	2.7	6.0
Total residue	6,320	5,200	5,090	11,300	4,600
Analyst	N. B. SHEFFEY	N. B. SHEFFEY	N. B. SHEFFEY	H. JOHN-STON	N. B. SHEFFEY
Cu	0.03	0.01	0.04	0.7	0.02
Ag	0.006	<0.005	<0.005		<0.005
Ge	0.13	0.2	0.2		<0.05
Sb	<0.6	<0.5	5.1		<0.5
Mn	1.4	0.19	0.06	19	18
Co	<0.06	<0.05	<0.05	3	0.3
Ni	0.06	<0.05	0.05	2	0.6
Fe	0.04	0.12	0.27	900	0.3
Cr	<0.01	<0.01	<0.01	0.3	<0.01
Al	0.13	0.2	0.1	1,400	2.2
Ti	<0.01	<0.01	<0.01		0.2
Zr	<0.06	<0.05	0.05		<0.05
Sr	0.25	0.63	0.5	7	2.0
Ba	0.05	0.32	0.32		0.5
Li		3.6			
Rb		<0.05			
Cs		<0.15			
Ca				300	
Mg				800	

Looked for, not found in samples 4, 5, 6, and 9, with sensitivity limits in ppm of evaporated residues: Pb (10), V (10), Ga (3), Ti (2), Mo (10), W (200), Pb (10), Sn (10), As (500), Zn (200), Cd (50), Y (10), Yb (1), and La (50).

of analysis, found 20 ppb of Hg in the water of Ink Spring (Table 80-F-2, analysis 3); 200 ppb were found in another sample collected on November 9, 1955; 60 ppb were found in pond water from Herman pit collected on August 23, 1955. According to WHITE and ROBERSON, the pH of the analyzed spring waters ranged from 6.6 to 7.5. The waters in the ponds in the pits were much more acid as a result of oxidation of H_2S to H_2SO_4. WHITE and ROBERSON (1962, p. 422) concluded that the waters at present issuing from the springs at Sulphur Bank have been derived from sedimentary rocks and are connate or metamorphic, rather than volcanic as had been previously supposed by LE CONTE and RISING (1882, p. 31), by BECKER (1888, p. 254), and by EVERHART (1946, p. 137). WHITE and ROBERSON (1962, p. 442) pointed out that the high ratio of iodine to chlorine in three of the analyses of Sulphur Bank spring waters is an especially strong indication that the waters came from sedimentary rocks.

The rocks exposed in the open pits of the Sulphur Bank mine and in the adjacent area consist of four units: 1. sandstone and shale of the Franciscan group, 2. Pleistocene brecciated rocks, conglomerate, and sandstone, 3. a Recent flow of augite andesite, and 4. post-andesite sediments consisting mostly of modern lake deposits.

The augite andesite in the mine area is leached for 15 to 25 feet below the surface to white powdery silica which is largely opal. This silica is highly porous and is saturated in places with crystalline sulfur and late sulfates (EVERHART, 1946, p. 131). According to BECKER (1888, p. 257), cinnabar did not occur in sensible quantities at or close to the surface, but was found to a considerable extent mixed with sulfur in the lower portion of the zone of oxidation.

Below the leached zone the andesite is altered to clay along joints and sheeting planes over a wide area. The white to gray clay encloses boulders of fresh andesite. Associated with the clay are large quantities of hydrous sulfates and, in places, quantities of sulfides. Glassy opal is mixed with the clay near the bottom of the leached zone, and in many places it extends down into the bouldery zone for several feet. Powdery goethite is commonly mixed with the opal. Cinnabar and pyrite form stringers and seams throughout the clay between the boulders. Near the borders of the fresh boulders the feldspar crystals are partially to completely kaolinized and other parts of the rock have been replaced by opal. The sulfides have been introduced near the joints farther from the boulders, where the original texture of the rock has been obliterated. The sulfide masses have kernels of pyrite surrounded by fringes of cinnabar (EVERHART, 1946, p. 140).

The Pleistocene sediments beneath the andesite are even more extensively kaolinized than the andesite. In structurally favorable places cinnabar and pyrite are associated with the clay formed in the sediments, and the surfaces of altered rock are coated with late acicular stibnite and various sulfates.

In the Franciscan rocks cinnabar forms finely crystalline coatings on fracture surfaces and on blocks of sandstone in fault breccia. The richest ore in the mine consists of fault gouge through which the cinnabar is disseminated as "fines" (EVERHART, 1946, p. 137).

Metacinnabar was found in the basal zone of the andesite and in lake sediments in 1956 and 1957. In most cases it was deposited on pyrite, marcasite, or cinnabar. On top of the metacinnabar there was commonly deposited stibnite. Metacinnabar was the dominant mineral in black coatings on pebbles found by WHITE on the north side of the Herman pit in August, 1955. Red coatings on the under sides of pebbles consisted, at least in part, of cinnabar. This occurrence as coatings on pebbles previously rounded by agitation in spring pools, in an area devoid of other nearby cinnabar, indicated that some mercury was still being transported and deposited at Sulphur Bank (WHITE and ROBERSON, 1962, pp. 404—405)[1].

WHITE and ROBERSON stated that the present heat flow at Sulphur Bank is on the order of 200,000 cal per second, which is about 12 times that normal for the area. The

[1] In a written communication to the author dated September 7, 1966, Dr. WHITE stated that "The rounded pebbles with coatings of HgS were obtained on August 23, 1955, from the same vent shown as Herman pit 'Ink' spring north side, on page 412 [of WHITE and ROBERSON, 1962]. During the intervening $2^1/_2$ months the level of the pond had been pumped down a few feet, 'Ink' spring water supply was draining out at lower levels and vapor was escaping under slight pressure, with a temperature of 66° C at the time. I do not know for sure whether the HgS was deposited during the earlier water stage or the later vapor-dominated stage but suspect the former. Transport and deposition of HgS after rounding of the pebbles is conclusive".

heat according to WHITE and ROBERSON (1962, p. 397) is almost certainly volcanic in origin, but, despite association with Quaternary volcanic rocks and volcanic heat, the chemical and isotopic compositions of the water and gases now being discharged indicate, in their view, that these fluids are nonvolcanic in origin. WHITE and ROBERSON (1962, p. 424) concluded that "the quicksilver of Sulphur Bank, just as the present water, may be a 'fugitive constituent' from sedimentary rocks that have been heated by volcanism and are being reconstituted by metamorphism".

At Steamboat Springs in Washoe County, Nevada, there is a group of about 50 hot springs. These springs are on the north-east edge of the Steamboat Hills, a bedrock mass having an area of about 9 square miles that lies in a structural trough between the Carson Range on the west and the Virginia Range on the east (BRANNOCK, FIX, GIANELLA, and WHITE, 1948, p. 211). The active hot springs lie along the Steamboat Springs fault zone in an area extending about 5,100 feet north from the Steamboat Post Office along the west side of Steamboat Creek (BRANNOCK, FIX, GIANELLA, and WHITE, 1948, p. 216). BRANNOCK and his co-workers have observed deposition of dark-gray to black mud containing as much as 4 percent of minute acicular stibnite crystals by some of the springs in recent years. No deposition of cinnabar at the surface by present day hot springs has been reported in the Steamboat Springs area, although fine-grained cinnabar has been found in the older, more eroded sinter west and north-west of the active springs (BRANNOCK, FIX, GIANELLA, and WHITE, 1948, p. 223).

BECKER studied the Steamboat Springs area between 1883 and 1886. He described the springs as they existed then in the following words: "The present vents of the springs lie along a series of fissures about a mile in length . . . These cracks have formed in a mass of sinter deposited from earlier vents and are in part choked up by detritus, in part covered by recent deposits, but they are nevertheless traceable for long distances. The vents in many cases retain their fissure character and the water may be seen and heard boiling in openings, evidently of great depth, from an inch to two feet in width and many yards in length. At other points the cracks are closed, with the exception of pipe-like openings, through which the water reaches the surface. At vents of the latter class there often form smooth mounds of sinter, from the summit of which water escapes continuously or fitfully. Some of the springs have a true geyser character, though on a very small scale, alternately disappearing from their basins and returning to them with noise and agitation at short intervals of time" (BECKER, 1888, p. 338).

BRANNOCK, FIX, GIANELLA, and WHITE (1948, pp. 222—223) observed that very fine-grained, dark gray to black mud was deposited from time to time in 1945 and 1946 by the hot springs of greatest discharge and highest temperature. Springs Nos. 17, 21, 22, 24, 27, 28, and 34 deposited most of the mud, at least part of which was carried up from below the visual limit which was a few feet down in the springs. The mud consisted chiefly of minute, light-colored, particles with a low index of refraction which were believed to be silica gel or opal. Minute individual acicular crystals and rosette-like clusters of crystals of stibnite made up 0.2 to 4.0 percent of the mud by weight. The color of the mud was apparently determined by the relative abundance of the stibnite crystals. Stibnite was also found floating as a thin scum on the surface of spring waters, especially in Spring No. 24. Pyrite was identified in the mud. Cinnabar, however, was not recognized in the mud. According

Table 80-F-4. *Analyses of waters from Steamboat Springs, Nevada.* (W. W. BRANNOCK, analyst)

Ion or item	Spring No. 33[a] 135.3° F (ppm)	Spring No. 16[b] 201.2° F (ppm)	Spring No. 27[c] 204. ° F (ppm)
Ca	23	d	12
Mg	2	d	<1
Na	602	666	707
K	54	81	75
Li	6	6	7
Al	<1	d	<1
Fe	<0.1	<0.1	<0.1
Ba	e	e	e
NH_4	<1	<1	<1
CO_3^{2-}	e	e	20
HCO_3^-	419	320	292
SO_4^{2-}	106	131	129
Cl	752	897	949
F	1.6	2.0	2.2
NO_3^-	e	e	e
AsO_4^{3-}	0.7	d	2.5
SbO_3^-	0.1	d	0.5
Se	e	d	e
Hg	<0.01 f	d	<0.01 f
B_2O_3	158	179	192
SiO_2	205	311	317
Total solids at 180° C	2,085	2,385	2,508
pH	7.7	d	8.2

[a] Also known as Chicken Soup Spring; southernmost of the principal hot springs; 325 feet south-west of Post Office; sample collected July 30, 1945.

[b] 4,400 ft. north of Post Office; south-east of the principal flowing springs; sample collected August 4, 1945.

[c] 460 ft. north-northwest of Spring No. 16. Deposits black siliceous mud containing minute acicular crystals of stibnite; sample collected July 30, 1945.

[d] Not determined.

[e] Looked for, but not found.

[f] Mercury determined with fluorescent lamp and screen by J. J. FAHEY in the laboratory of the U.S. Geological Survey.

to BRANNOCK and others the muds were probably precipitated chiefly in the water a few tens of feet below the surface and were then carried to the surface mechanically; in minor part the muds may have been precipitated at the surface.

The hot-spring deposits consist almost entirely of siliceous sinter, of which the total amount is large, together with very small amounts of calcium carbonate. The large accumulations around structurally favorable outlets form hot-spring terraces. The primary sinter deposits consist of opal. After burial under younger sinter and prolonged contact with hot, silica-bearing waters most of the originally porous sinter has been reworked and filled with opal, chalcedony, and some quartz (WHITE, THOMPSON, and SANDBURG, 1964, p. Bl.; BRANNOCK, FIX, GIANELLA, and WHITE, 1948, p. 223). BRANNOCK, FIX, GIANELLA, and WHITE (1948, p. 223) found radial

Table 80-F-5. *Analyses of gases from four representative hot springs, Steamboat Springs, Nevada (in mole per cent)* [a]. (According to BRANNOCK, FIX, GIANELLA, and WHITE)

Item	21[b]	24[c]	26[d]	16[e]	Limits of error
Temperature of water °F	192	204	205	202	—
pH of water [f]	6.9	7.7	8.3	7.2	—
Hydrogen	0.11	0.20	0.21	g	0.02 ±
Nitrogen	0.75	5.21	2.74	1.82	0.05 ±
Oxygen	0.11	0.10	g	0.14	0.02 ±
Hydrogen sulfide	0.38	1.06	0.45	0.77	0.04 ±
Argon	0.03	0.14	0.08	0.04	0.01 ±
Carbon dioxide	98.61	93.29	96.52	97.23	0.10 ±
Hydrocarbons[h]	g	g	g	g	—

[a] Samples collected June 20—26, 1945; determined by mass spectrometry by the National Bureau of Standards.

[b] Northernmost flowing spring in summer and autumn of 1945; 5,100 feet north of Steamboat Post Office.

[c] Intermittent spring in fissure 675 feet south of spring No. 21.

[d] Small geyser 3,000 feet south of spring No. 24; 1,400 feet north of Post Office.

[e] Perpetual spouter spring, 175 feet east of spring No. 24.

[f] Determined with a Beckman pH meter on water samples collected September 27—28, 1945.

[g] Not found.

[h] Maximum 0.03 mole per cent.

clusters of acicular crystals of stibnite on the walls of small cavities and embedded in opal in the reworked sinter. These crystals were three to four times longer than the longest crystals that were found transported to the surface of the hot springs.

BRANNOCK and his co-workers presented analyses of waters from three springs at Steamboat Springs. These three analyses are reproduced in Table 80-F-4. They also presented analyses of gases from four representative hot springs at Steamboat Springs. These four analyses are reproduced in Table 80-F-5. According to WHITE and BRANNOCK (1950, pp. 566 and 570—573) at Steamboat Springs the water discharged by the springs is mostly meteoric, but in part is derived from volcanic sources. Beginning in 1945 they made systematic measurements of spring discharges, temperatures, and mineral content. Heavy rains had marked effects on the springs. After heavy rains spring discharges increased and chloride concentrations and temperatures decreased almost immediately indicating an admixture of rainwater and thermal saline water in the uppermost part of the spring system. That a minor part of the spring water is of volcanic origin, however, was inferred by WHITE and BRANNOCK from the compositions of the spring waters. At Steamboat Springs the only available source for most of the chlorine, sulfur, boron, and excess carbon dioxide appears to be volcanic according to WHITE and BRANNOCK (1950, p. 596). WHITE and BRANNOCK (1950, p. 574) concluded that all or most of the arsenic, antimony, and fluorine in the spring waters at Steamboat Springs has come from a magmatic source. Additional evidence pointing to a volcanic source of part of the spring water was supplied by determinations of isotopic differences between the spring waters and Lake Michigan water carried out by DOLE and by UREY on the basis of which it was calculated that from 6 to 16 percent of the spring water is volcanic (WHITE and BRANNOCK, 1950, p. 573).

II. Chemical Processes Involved in the Formation of Deposits

YATES and THOMPSON (1959, pp. 79—80) stated that "The mineral products of hot springs are more like those of quicksilver deposits than are products deposited directly by volcanic gases. Spring deposits contain silica and calcium carbonate, two combinations common to quicksilver deposits and foreign to most fumarolic deposits. Sulfides of the metals, notably those of iron, antimony, and arsenic, are commonly found in small amounts. Although it is chemically possible for almost all the mineral combinations found in quicksilver deposits to be transported and deposited by gases, only a few of these combinations are typical of fumarolic deposits, whereas almost all the minerals associated with quicksilver are common to hot spring deposits. The comparisons between quicksilver deposits and the deposits of hot springs or fumaroles indicate that the environment in which the quicksilver minerals were deposited was most similar to that of hot springs. The comparisons, however, do not preclude the possibility that some of the ore elements were transported by a gas from the source to near the place of deposition."

In their comprehensive report on the geology and quicksilver deposits of the New Almaden District, California, BAILEY and EVERHART (1964, p. 125) stated that the quicksilver ores of this district resulted mainly from the replacement of silica-carbonate rock by cinnabar, but in minor part from replacement of other rocks and the filling of open space by cinnabar or quartz-dolomite veins containing cinnabar. BAILEY and EVERHART (1964, p. 123) concluded that "The fact that the ores were formed by replacement, a process that entailed the removal of large amounts of material as well as the deposition of cinnabar, indicates the presence of a liquid phase, and replacement by cinnabar just beneath the alta in structural domes indicates the absence of a vapor phase, which, if present, would fill the structural highs and keep out the ore-depositing liquid."

III. Chemical Environment of Transport and Deposition

KRAUSKOPF (1951, p. 498) stated that "Quicksilver ores, it is generally agreed, are formed by deposition from vein fluids at temperatures not as low as 80° nor as high as 250° C, and at pressures below 30 atmospheres". Cinnabar and metacinnabar are extremely insoluble in water in the range of temperature and the range of pressure under consideration, and they are only slightly less insoluble in water saturated with atmospheric carbon dioxide. According to KRAUSKOPF (1951, pp. 499, 501) the solubility of mercuric sulfide in water in equilibrium with atmospheric carbon dioxide is about 10^{-21} moles per liter at ordinary temperatures and about 100 times greater at 400° C. KRAUSKOPF (1951, p. 501) calculated from this basis that to deposit 1 Avoirdupois ton of mercuric sulfide, at least 10^{23} liters of such a solution would have to pass through a vein fissure. A volume of 10^{23} liters, he pointed out, is approximately 100 times the volume of water in all the oceans. Hence it appears that if the mercury and sulfur of cinnabar deposits were transported in liquid solutions, these solutions contained other constituents besides water and carbon dioxide that increased the solubility of mercuric sulfide. KRAUSKOPF evaluated various other chemical mechanisms for the transportation of quicksilver in vein fluids by calculations based on available thermodynamic data. KRAUSKOPF (1951, pp. 501—504) showed very clearly that chemically feasible mechanisms are transportation as HgS_2^{2-} ions in alkali sulfide solutions, as $HgCl_2$ in a vapor phase, and as the vapor of elemental quicksilver. KRAUSKOPF (1951, pp. 507—513) also showed that chemical mechanisms depending on the decrease of ionization or the decrease of solubility of H_2S at high temperatures, or on the elimination of H_2S at high temperatures by displacement of equilibrium in a gas phase, or on the formation of mercury-halogen complex ions, are inadequate to account for the transportation of the quantities of quicksilver present in ore deposits without involving unreasonably large quantities of vein fluids. Since the chloride complexes are inadequate for the transportation of sufficient quantities of mercury in the presence of sulfur, and since other complex-forming elements are probably scarce in vein solutions, KRAUSKOPF (1951, pp. 508—509) concluded that no complex ion except HgS_2^{2-} has probably played a role in the formation of quicksilver ores.

KNOX (1906, pp. 477—481) made the first quantitative studies of the solubility of cinnabar and metacinnabar in aqueous sodium sulfide solutions. He placed red mercuric sulfide in solutions of sodium sulfide of known initial sodium sulfide concentration and after the solutions became saturated with red mercuric sulfide, he analyzed them for mercuric sulfide. Similar experiments were made with black mercuric sulfide. KNOX's determinations were made at two temperatures, 25° and 33° C. His tables gave the final concentrations of mercuric sulfide in moles per liter in solutions saturated with red mercuric sulfide at 25° and 33° C, but only the initial concentrations of sodium sulfide in moles per liter. His tables likewise gave the final concentrations of mercuric sulfide in moles per liter in solutions saturated with black mercuric sulfide at 25° and 33° C, but only the initial concentrations of sodium sulfide in moles per liter.

Using the solubility data of KNOX and the equilibrium constants of KUBLI for the dissociation of H_2S to H^+ and HS^- and the dissociation of HS^- to H^+ and S^{2-}, KRAUSKOPF (1951, p. 503) calculated that at 25° C a neutral solution containing 0.1 mole per liter of total dissolved sulfur would dissolve 1.4×10^{-6} gram of mercuric sulfide per liter of solution (1.4 ppb). He calculated further that the same amount of mercuric sulfide would dissolve if the total dissolved sulfur were 0.001 mole per liter, but the pH value were increased to about 8.5. He concluded that at low temperatures sulfide solutions are adequate to transport sufficient mercuric sulfide to be considered possible ore-forming solutions. At the time KRAUSKOPF made the calculations concerning possible mercury ore-forming solutions, the only data in the literature on the effect of increased temperature on the solubility of mercuric sulfide in aqueous sodium sulfide solutions indicated that the solubility decreases with increasing temperature, but these data had been obtained from experiments carried out below 100° C.

The saturation curves of cinnabar and metacinnabar in the system HgS-Na_2S-H_2O at 25°, 50°, and 75° C and at one atmosphere were determined by DICKSON and TUNELL (1958) and subsequently DICKSON (1964) determined the solubility of cinnabar in the system HgS-Na_2S-H_2O at temperatures up to 250° C and at pressures up to 1,800 bars (cf. Section 80-H). With a constant sodium sulfide concentration and at a constant pressure, he found that the solubility of cinnabar decreases with increasing temperature up to approximately 100° C. In the vicinity of 100° C, however, the solubility of cinnabar in such a solution passes through a minimum and with increase of temperature above approximately 100° C it increases markedly. From previous experiments carried out below 100° C it did not appear probable that the falling temperature of an ore-forming solution would cause the precipitation of cinnabar or metacinnabar if the mercuric sulfide were held in solution as an alkali sulfide complex. DICKSON's experiments demonstrated, however, that falling temperature would be effective, in the range above 100° C, in causing solutions saturated with cinnabar or metacinnabar to deposit some cinnabar or metacinnabar.

It has been supposed by some authors that at constant temperature and constant sodium sulfide concentration, diminishing pressure would result in precipitation of cinnabar from a saturated solution. DICKSON's work has proved, however, that diminishing pressure would actually increase the solubility of cinnabar at constant temperature and constant sodium sulfide concentration.

The isothermal-isopiestic saturation curves of cinnabar and metacinnabar in the system HgS-Na_2S-H_2O are convex toward the field of unsaturated solution. Dilution of an aqueous sodium sulfide solution that is saturated with cinnabar or with metacinnabar consequently would cause the precipitation of cinnabar or metacinnabar until the solution became extremely dilute; with increasing dilution the precipitated cinnabar or metacinnabar would, of course, eventually redissolve if equilibrium were maintained[1]. At such high dilution, however, the precipitated cinnabar or metacinnabar would be likely to redissolve extremely slowly. Partial evaporation of a solution saturated with cinnabar or metacinnabar that is not extremely dilute, on the other hand, would cause cinnabar or metacinnabar to dissolve unless there were simultaneous oxidation of the sodium sulfide. The effect of oxidation of the sodium sulfide in an aqueous sodium sulfide solution saturated with cinnabar or metacinnabar is the practically complete precipitation of the mercuric sulfide.

An investigation of the solubility of black HgS in acidic solutions and near-neutral and neutral solutions, in which the solubility of HgS is very slight, as well as in alkaline solutions in which the solubility of HgS is appreciable, was carried out by SCHWARZENBACH and WIDMER (1963) (cf. Section 80-H). The concentrations of dissolved mercury sulfide in solutions of total sulfur concentration 0.0190 mole per liter (approximately 585 parts per million) and ionic strength 1.00 fixed by the addition of KCl at 20° C determined by SCHWARZENBACH and WIDMER as a function of pH, for pH values from 1.00 to 10.74 are given in Table 80-H-2.

Thus from the work of KNOX, DICKSON and TUNELL, DICKSON, and SCHWARZENBACH and WIDMER it appears that if the mercuric sulfide in liquid ore-forming solutions was carried chiefly in the form of an alkali sulfide complex, precipitation of cinnabar and/or metacinnabar was probably caused: by falling temperature, by reduction of alkalinity resulting from reaction with wall rocks for example, by solution of silica, and/or by admixture with less alkaline or acid ground water and, in places, to some extent, by dilution where admixture with ground water occurred, and, finally, near the surface, by oxidation of the alkali sulfide by atmospheric oxygen.

The spring waters at Amedee Hot Springs, Lassen County, California, at Cedarville Hot Springs, Modoc County, California, and at Boiling Springs, Valley County, Idaho, which have deposited appreciable quantities of metacinnabar and cinnabar

[1] A detailed graphical and numerical analysis of an analogous case, that of the dilution of a solution saturated with $Fe_2O_3 \cdot H_2O$ (goethite) in the system Fe_2O_3—SO_3—H_2O, was given by TUNELL (1930, pp. 41—50).

at the surface very recently and which appear to be still depositing these minerals are all alkaline in character (the pH values of the spring waters of these springs range from about 8.5 to about 9.2) and all contain small amounts of sulfide ion. Calculations based on the total amount of mercuric sulfide present in the spring deposits at Amedee together with the present rate of outflow of water from the springs and the estimated length of time during which the deposition of the mercuric sulfide coatings has taken place indicate that on average about 0.003 mg of mercuric sulfide has been deposited per liter (equivalent to 3 ppb) of outflowing spring water. Making use of the solubility data of SCHWARZENBACH and WIDMER, and taking into account the fact established by DICKSON and TUNELL (1958, p. 664) and DICKSON (1964, p. 628) that the changes in solubility of metacinnabar and of cinnabar in aqueous Na_2S solutions from 25° C to 100° C are very small; and also the fact established by KNOX (1906, p. 477) and by SCHWARZENBACH and WIDMER (1963, pp. 2623, 2626) that at constant temperature and constant pH value the concentration of dissolved HgS in solutions saturated with metacinnabar or cinnabar is proportional to the concentration of total sulfur (H_2S, HS^-, S^{2-}), one can calculate that the solubility of metacinnabar in the spring water emerging at the surface at Amedee Hot Springs is approximately 0.02 ppb. It thus appears that the solubility of metacinnabar in the thermal waters at some depth below the surface at Amedee Hot Springs must be many times greater than its solubility in the spring waters emerging at the surface. DICKSON's experimental work showed that the solubility of cinnabar increased markedly with increasing temperature above 100° C. The higher temperature at depth thus would account for part of the inferred greater solubility at depth. It appears probable that at least a small part of the sulfur present as SO_4^{2-} in the water emerging at the surface may be present at depth as S^{2-}, which would account for another part of the inferred greater solubility at depth. If this is the case, then the precipitation of mercuric sulfide from the rising hot, alkaline waters at Amedee is caused by the decreasing temperature of the waters as they approach the surface, by dilution by ground water, and by oxidation of S^{2-} to SO_4^{2-}. The mercuric sulfide deposited at the surface may be carried upward on the last part of its journey in supersaturated solution or in colloidal suspension. At low concentrations, solutions supersaturated with HgS or containing HgS in colloidal dispersion are not only easily made, but are difficult to avoid according to the laboratory experience of DICKSON, SCHWARZENBACH and WIDMER, and others.

The physico-chemical conditions at Cedarville Hot Springs and at Boiling Springs are quite similar to those at Amedee Hot Springs. In all three cases metacinnabar and cinnabar have been recently deposited and apparently are still being deposited at the surface from hot, alkaline waters containing small amounts of sulfide ion. Simultaneous deposition of metacinnabar and cinnabar can only occur stably if the temperature, pressure, and concentrations of the solutions are at equilibrium values, which is probably not the case at Amedee Hot Springs, Cedarville Hot Springs, or Boiling Springs. Since metacinnabar can be precipitated irreversibly from alkaline sulfide solutions at low temperatures quite readily in the laboratory, it may be depositing irreversibly in nature at Amedee Hot Springs, Cedarvile Hot Springs, and Boiling Springs.

At Sulphur Bank, California, a major mercury ore body was formed in geologically recent times by hydrothermal processes, and hot springs are still active in the mineralized area. WHITE and ROBERSON (1962, p. 423) made an estimate of the average concentration of mercury in the Sulphur Bank ore-forming fluids based on the following four assumptions: 1. The total amount of mercury in the deposits was about 200,000 flasks or 7×10^9 grams, including mining and furnace losses as well as metal still in the ground; 2. Ore deposition occurred through an interval of 10,000 years; 3. The present rate of flow of deep thermal water, 50 gpm, is also typical of the past, a total of 1.0×10^{12} liters have then been discharged[1]; 4. All mercury in liquid or vapor phase was precipitated, any that may have been discharged from the system is in addition to the calculated amount. With the foregoing assumptions, WHITE and ROBERSON (1962) calculated that the average mercury content of the solutions was 7 ppm. They considered this to be probably a maximum value unless the interval of mercury deposition was significantly less than 10,000 years. They gave as a probable minimum estimate, based on an assumed duration of 100,000 years and an average discharge of 1,000 gpm, a mercury content of 0.04 ppm. They stated that they believed the most reasonable estimate to be from 0.1 to 1.0 ppm.

Making use of the solubility data of SCHWARZENBACH and WIDMER, of DICKSON and TUNELL, of DICKSON, and of KNOX, one can infer that if the solubility of HgS in the ore-forming solution was between 0.1 and 1.0 part per million as calculated by WHITE and ROBERSON, then the solution from which the ore body was deposited would have had a pH value between 8.6 and 9.9 if cooled to room temperature and if its total sulfur content (as H_2S, HS^-, and S_2^{2-}) was about 585 ppm, and in this case the mercuric sulfide in the ore-forming solution was chiefly in the form of the complex HgS_2^{2-}. If the total sulfur content of the ore-forming solution was less than 585 ppm in order for the solubility of mercuric sulfide to be as much as 0.1 ppm, the pH value of the solution must have been greater than 8.6. The concentrations of H_2S in the analyses of spring waters from Sulphur Bank reported by WHITE and ROBERSON range from zero to 12 ppm and the pH values[2] of the analyzed spring waters reported by WHITE and ROBERSON range from 6.6 to 7.5. The solubility of mercuric sulfide in these spring waters estimated from the data of SCHWARZENBACH and WIDMER, of DICKSON and TUNELL, of DICKSON, and of KNOX is not more than 0.4 ppb.[3] Thus from the estimates of the concentration of mercuric sulfide in the ore-forming solution given by WHITE and ROBERSON together with the solubility data of SCHWARZENBACH and WIDMER, of

[1] The volume of associated gas is unknown, but its mass is unlikely to have been more than a few percent of the associated water according to WHITE and ROBERSON.

[2] WHITE stated that the pH values were determined at room temperature from samples collected in nearly full bottles, tightly stoppered immediately to prevent loss or gain of gases and measured when the bottles were first opened. (Written communication to TUNELL dated September 7, 1966.)

[3] WHITE and ROBERSON stated that F. WANG of Buckman Laboratories, Inc., using a modified dithizone method of analysis found from 0.02 to 0.2 ppm of mercury in spring and pond waters from Sulphur Bank (WHITE and ROBERSON, 1962, p. 415). SCHWARZENBACH and WIDMER found in their investigation that it was necessary to filter their solutions repeatedly through a very fine fritted glass porous plate under pressure of nitrogen gas in order to obtain true solubilities, since the precipitates of HgS formed in dilute solutions are very fine. No statement that WANG carried out such filtrations was reported by WHITE and ROBERSON. The spring and pond waters analyzed by WANG were without much doubt either suspensions containing fine particles of HgS or supersaturated solutions.

Dickson and Tunell, of Dickson, and of Knox, it appears that during the main ore-forming stage, the ore-forming solution contained more total sulfur and was probably more basic than any of the spring waters at present emerging at the surface at Sulphur Bank.

IV. Physico-Chemical Conditions of Formation of Cinnabar and Metacinnabar

The relationship of cinnabar, the red form of mercuric sulfide, to metacinnabar, the black form of mercuric sulfide containing in nature small amounts of iron and/or zinc and/or selenium in solid solution, was until recently not well-understood. Moore (1872), who first described metacinnabar, recognized that the metacinnabar in the specimens he studied had been deposited earlier than associated cinnabar, but he thought that metacinnabar was amorphous. Penfield (1885, pp. 452—453) first demonstrated that metacinnabar is crystalline and he correctly assigned it to the sphalerite group. Lindgren (1933, p. 463) regarded metacinnabar as a supergene mineral found principally in altered mercury deposits. Lindgren's opinion was widely accepted, although several workers subsequently observed that some metacinnabar was deposited before accompanying cinnabar (Ross, 1940, p. 51; Bailey and Phoenix, 1944, p. 137; Eckel and Myers, 1946, pp. 97—98).

The inversion temperature of red HgS to black HgS was determined by Dickson and Tunell (1959, pp. 475—477) to be 344° ± 2° C.[1]

Black HgS, formed by heating pure red HgS above 344° C, reverts rapidly on cooling to red HgS. The rapidity with which pure black HgS reverts to red HgS on cooling in the laboratory raises the question: How is it possible for metacinnabar to persist at ordinary temperatures, as it does in nature? In an attempt to answer this question the effect of impurities in the system was investigated.

An analysis of metacinnabar from Felsöbánya, Transylvania, made by Vavrinecz (1932, pp. 140—141) shows that it contains an appreciable quantity of iron, and an analysis of metacinnabar from Levigliani, Italy, made by Manasse (1921, p. 160) shows that it contains appreciable quantities of zinc and iron (cf. Table 80-F-6). Numerous older analyses[2] also indicate that naturally occurring metacinnabar is not pure mercuric sulfide, but is rather a solid solution containing one or more of the elements, iron, zinc, selenium. Few reliable analyses of cinnabar have been reported in the literature, but several specimens analyzed by Dickson and Tunell contained less than 0.1 percent nonvolatile impurities. By contrast, homogeneous metacinnabar from the Mt. Diablo mine, California, contained 0.5 percent iron.

[1] G. Kullerud subsequently investigated the equilibrium relations in the system mercury-sulfur. He stated that "By means of a series of heating and cooling experiments in the 340° C—350° C range, ... it was shown that the cinnabar-metacinnabar inversion is reversible and that it takes place at 345° C ± 2° C, in agreement with the value of 344° C given by Dickson and Tunell". Kullerud further stated that "The occurrence of this inversion at 345° C ± 2° C in stoichiometric HgS, as well as when excess of Hg or S is used, indicates that little or no solid solution exists in HgS at this temperature" (G. Kullerud, Carnegie Institution Yearbook 64, pp. 194—195, 1965). It should be noted that Kullerud's statement that little or no solid solution exists in HgS at the inversion temperature, applies to the system mercury-sulfur. It does not apply to the system HgS-FeS, or to the system HgS-ZnS, or to the system, HgS-HgSe.

[2] Cf. Palache, Berman, and Frondel (1944), pp. 215—217, and Doelter and Leitmeier (1926), pp. 365, 371, 830.

Table 80-F-6. *Analyses of metacinnabar*

	1	2	3
Weight % Hg	86.22	77.68	83.38
Weight % Fe		5.36	0.52
Weight % Zn			2.17
Weight % S	13.78	14.97	14.24
Remainder		1.41	
Total	100.00	99.42	100.31

1. Pure HgS.
2. Felsöbánya, Transylvania. Remainder is insoluble. Analyst, G. VAVRINECZ.
3. Levigliani, Italy (leviglianite). Analyst, E. MANASSE. (In DANA's System of Mineralogy, 7th Edition, Vol. 1, by C. PALACHE, H. BERMAN, and C. FRONDEL, MANASSE's analysis is given incorrectly on p. 216; instead of 0.52 per cent iron, the table on page 216 shows 0.52 per cent remainder, which is stated to be FeO. In MANASSE's article, however, it is clearly stated that the sample of leviglianite analyzed is a variety of metacinnabar containing zinc and iron.)

Samples of red HgS were heated in the presence of iron sulfide, zinc sulfide, and selenium, respectively, by DICKSON and TUNELL. These substances retarded the reversion of metacinnabar to cinnabar; freshly formed metacinnabar made in this way did not revert to cinnabar, but remained unchanged after being kept for months at room temperature. The metacinnabar was only metastable, however, and could be converted to cinnabar by allowing it to stand in contact with sodium sulfide solutions.

In addition to retarding the inversion of metacinnabar to cinnabar, the presence of iron or zinc actually depresses the inversion temperature. Iron disulfide lowers the inversion temperature to at least 305° C and zinc sulfide to at least 240° C. These temperatures were determined by heating intimate mixtures of red HgS and iron and zinc sulfides in glass capillary tubes in a controlled temperature-gradient furnace for one week periods. It should be emphasized that the values obtained in this way are upper bounds only. Solid-solid reactions are involved which probably become slower at low temperatures and the temperatures reported may be merely those at which the reactions became so slow as not to be detectable after one week.

Unequivocal evidence showing that some metacinnabar crystals were deposited earlier than associated cinnabar crystals was observed by DICKSON and TUNELL (1959, pp. 482, 483) in specimens from the Aetna and Knoxville mines, Napa County, California. Equant, euhedral metacinnabar crystals in cavities in altered serpentine from both localities are coated with euhedral needle-like crystals of cinnabar that radiate outward from the surfaces of the metacinnabar crystals. Some of the metacinnabar crystals in the Aetna specimens are altered to a reddish-black spongy residue about which unaltered cinnabar crystals are arranged in the same way that cinnabar crystals are arranged about unaltered metacinnabar crystals. The metacinnabar crystals evidently were less resistant to chemical attack than the cinnabar crystals. The preferential alteration and removal of metacinnabar indicates that at the time of the chemical attack cinnabar was stable and metacinnabar was unstable.

In some samples of ore from the Mt. Jackson mine, Sonoma County, California, tiny euhedral crystals of metacinnabar and cinnabar are mixed together in fractures and cavities in strongly silicified rocks. Commonly cinnabar crystals are perched on metacinnabar crystals and nearby in the same cavity metacinnabar crystals are perched on cinnabar. Such textures might have been produced by simultaneous deposition of cinnabar and metacinnabar or by repeated alternations in the deposition of cinnabar and metacinnabar.

Field and petrographic studies carried out in collaboration with F. W. DICKSON have demonstrated that: 1. metacinnabar is an earlier mineral than cinnabar as a general rule, although in some places it was deposited later; 2. in some localities mercuric sulfide appears to have been deposited continuously with early metacinnabar giving way later to cinnabar; 3. in rare cases cinnabar and metacinnabar were apparently deposited simultaneously or in repeated alternation; and 4. metacinnabar was deposited from solutions as metacinnabar and cannot be considered to be an alteration product of cinnabar.

The conclusions resulting from the petrologic studies are in accord with the determination of the stability relations of cinnabar and metacinnabar by laboratory experiments. That metacinnabar should in general be an earlier mineral than cinnabar would be expectable if deposition took place along with falling temperatures under equilibrium conditions. In this case, a high-temperature phase should be earlier than a low-temperature phase of the same composition. Where deposition has occurred under non-equilibrium conditions, however, metacinnabar may have been deposited later than cinnabar even though the temperature of the solution was falling.

The occurrence of mercuric sulfide at Amedee Hot Springs, California, and Boiling Springs, Idaho, where metacinnabar has been deposited later than cinnabar, are both unusual in that deposition has taken place at the earth's surface quite recently. The metacinnabar has probably formed irreversibly as it does very readily at ordinary temperatures and pressures in the laboratory. Some cinnabar has apparently been formed by alteration of the metacinnabar at Amedee Hot Springs, but textural evidence bearing on this point is meager. Most of the cinnabar was probably deposited as cinnabar, and, as appears likely, if the physical conditions have remained essentially the same during the period of deposition, chemical conditions must have varied to enable fluctuating deposition of cinnabar and metacinnabar.

Although metacinnabar occurs somewhat more commonly than has been reported in the literature, in most mercury deposits cinnabar is the only mercury mineral present. In general, moreover, the textural evidences indicate that the cinnabar was deposited as such, not as metacinnabar that subsequently inverted to cinnabar, although in some deposits appreciable amounts of cinnabar were formed by inversion from metacinnabar.

V. Some Tentative Conclusions Concerning the Origin of Deposits

The solutions that deposited the cinnabar and metacinnabar of mercury ores undoubtedly contained other constituents besides water and mercuric sulfide. KRAUSKOPF (1951, p. 521) concluded that the presence of a small amount of sodium sulfide in a slightly alkaline aqueous solution would permit the solution to hold enough mercuric sulfide in solution to be considered a possible ore-forming solution, and he found no other chemical mechanism by which an aqueous solution could hold sufficient mercuric sulfide in solution to be considered a possible ore-forming solution. Stibnite, and/or quartz, and/or chalcedony, and/or opal, and/or calcite, and/or dolomite, and/or barite[1] deposited in part contemporaneously with cinnabar and/or metacinnabar in some mercury ore deposits indicate that in some cases the solutions that deposited the cinnabar and/or metacinnabar contained some Sb_2S_3, and/or SiO_2, and/or $CaCO_3$, and/or $MgCO_3$, and/or $BaSO_4$. Other more soluble constituents such as NaCl, Na_2SO_4, Na_2S, and H_2S were probably present and remained in solution and were carried away by the escaping waters.

Krauskopf (1951, pp. 519—521) pointed out that transportation of mercury as elemental mercury vapor and as mercuric chloride vapor may also have taken place in sufficient quantities to be geologically significant. At 200° C the vapor pressure of elemental mercury is 0.02 atmosphere (Sidgwick, 1951, vol. 1, p. 286), and at 228° C the vapor pressure of $HgCl_2$ is 0.1 atmosphere (Kelley, 1935, p. 112). According to Krauskopf (1951, p. 520) if both Hg and $HgCl_2$ were present in a vapor phase, on cooling, calomel (Hg_2Cl_2) would be deposited if H_2S were absent and cinnabar or metacinnabar if H_2S were present. Yates and Thompson (1959, pp. 81—82) stated that the calomel and associated native mercury in the Terlingua mercury ores were probably deposited from a vapor phase in this way. The considerable quantities of native mercury observed by Bailey on the lower levels of the mines at Almadén, Spain, are believed by him to be hypogene and were probably deposited from a vapor. In the Socrates mine in Sonoma County, California, native mercury is abundant in cracks and fissures in silica-carbonate rock and is also disseminated in sandstone. It is accompanied nearly everywhere by some cinnabar. The native mercury in this deposit, as well as the cinnabar, was considered to be hypogene by Bailey (1946, p. 226).

Mercury deposits are mostly located in the belt of late Tertiary orogeny and volcanism that extends northward from the southern tip of South America along the western edge of that continent and North America to Alaska, swings westerly to northeast Asia and seemingly bifurcates north of Japan. One branch follows the island arcs through Japan, the Philippines, and Indonesia to New Zealand and eastern Australia; the other branch follows the Himalayan and Alpine belt to terminate in Spain[2]. Some mercury deposits, such as the Cordero ore body in Nevada, the veins and reefs of the Steens Mountain and Pueblo Mountains District in Oregon, and the mercury-bearing sinter at Coso Hot Springs in California, are in areas underlain entirely by volcanic and igneous rocks, and the sources of the solutions that deposited the cinnabar and other mercury minerals in these ores were apparently magmatic. Other mercury deposits, although located in volcanic belts, are in areas underlain largely or in part by sedimentary rocks. In the cases of such deposits, among which are the ore bodies at Almadén, Spain, Idria, Yugoslavia, and Sulphur Bank, California, the sources of the ore-forming solutions have not been conclusively determined.

---

[1] An occurrence at the Senator mine, Nye County, Nevada, in which there was apparently some contemporaneous deposition of barite, cinnabar, and metacinnabar was reported by Dickson and Tunell (1955, p. 58).

[2] E. H. Bailey, in J. W. Pennington: Mercury — A Materials Survey, U.S. Bureau of Mines Information Circular 7941, p. 11, 1959.

The neutron activation analyses of samples of intrusive and extrusive igneous rocks ranging from silicic to basic by EHMANN and LOVERING (1967), indicate that the average mercury content of igneous rocks in the upper part of the earth's crust is about 20 ppb. The neutron activation analyses of inclusions from deep-seated kimberlitic and alkali basalt pipes made by the same investigators, on the other hand, show a much larger average mercury content, about 1,030 ppb. These figures give support to the conclusion of GOLDSCHMIDT that the enrichment of mercury in marine hydrolyzate sediments as compared to the igneous rocks in the upper part of the earth's crust is due to the transport of much mercury to the earth's surface by magmatic fluids and to the weathering of ore deposits of magmatic origin.

Revised manuscript received: August 1968

80-G. Behavior during Weathering and Oxidation Processes

Cinnabar is very resistant to oxidation in natural weathering processes and is found in outcrops from which iron sulfides have been entirely leached away. Cinnabar is also found in stream deposits at considerable distances downstream from outcrops of mercury deposits at many places. From an investigation of the mercury deposits of the Mt. Jackson mine, Sonoma County, California, and other mercury mines, HENDERSON (1965, p. 125) concluded that only limited oxidation and leaching of cinnabar occurred in the zone of oxidation. ROSS (1942, p. 464) stated that some redistribution of quicksilver by the action of descending meteoric water had taken place in numerous localities, but generally on a scale too small to be of economic significance. SAUKOV (1953, p. 307) mentioned that leaching of cinnabar and the formation of native quicksilver and secondary cinnabar have been observed in the outcrops of numerous ore bodies. However, SAUKOV stated that unfortunately it is not always easy to distinguish secondary cinnabar from primary cinnabar. According to HENDERSON (1965, p. 116) the hypogene cinnabar remaining in the zone of oxidation at the Mt. Jackson mine is in part well-crystallized but showing some evidence of leaching, and, in part, was pulverized by the mechanical breakdown of the enclosing, leached silica-carbonate rock. Some cinnabar of probable supergene origin was observed by HENDERSON (1965, pp. 126—127) in the Mt. Jackson mine, the New Idria mine, and the Culver-Baer mine. It was pulverulent and occurred below oxidizing ore bodies. At the Mt. Jackson mine the cinnabar believed to be of supergene origin was perched on top of dolomite that was deposited in open space in a vein. At the Mt. Jackson mine dolomite appears to be invariably later than any hypogene sulfide mineral throughout the hypogene zone according to HENDERSON. At the New Idria mine HENDERSON (1965, pp. 126—127) observed cinnabar that was probably supergene in a drift underneath the Creek Pit ore body. This cinnabar was pulverulent and was the only mineral occurring on fracture surfaces in unoxidized, pyritized, and kaolinized Panoche sandstone that did not contain any disseminated cinnabar typical of the Creek Pit ore body. At the Culver-Baer mine a similar occurrence of pulverulent cinnabar of probable supergene origin was noted by HENDERSON (1965, p. 127) immediately beneath the zone of oxidation. This cinnabar was the only mineral occurring in fractures in the unaltered serpentinite.

In the Steens Mountain and Pueblo Mountains district in Oregon supergene cinnabar has been produced by the weathering of schwatzite. The schwatzite has been most readily altered where it is accompanied by abundant pyrite, which provided ferric sulfate and sulfuric acid during the oxidation of the ore. Where chalcopyrite and pyrite are not present, some of the schwatzite persists to the surface without alteration. According to WILLIAMS and COMPTON (1953, p. 44) the supergene products include dull resinous patches of limonite, rare traces of cuprite and tenorite, malachite, azurite, chrysocolla, irresolvable mixtures of antimony and arsenic minerals, and supergene cinnabar. Most of the supergene cinnabar was deposited near the parent schwatzite as red, earthy coatings mixed with limonite and peppered with specks of relic schwatzite. WILLIAMS and COMPTON stated that in many places the fresh schwatzite grades to porous, brownish-black limonite riddled with blebs and veinlets of cinnabar, malachite, azurite, and chrysocolla.

SAUKOV and AIDINYAN (1940, pp. 37—40) proved experimentally that cinnabar can be oxidized by solutions containing ferric sulfate and that this oxidation is accelerated in the presence of sodium chloride. The mercuric sulfate thus produced was partly hydrolyzed and partly dissolved. They concluded that natural waters containing ferric sulfate and chlorides have the power to dissolve mercury from cinnabar as $HgSO_4$ and $HgCl_2$. By reaction of such solutions with other sulfides, supergene cinnabar might be precipitated; under other conditions from such solutions native quicksilver, calomel, and oxychlorides might be deposited. SAUKOV (1953, pp. 307—308) stated that part of the mercury dissolved during the oxidation of cinnabar may be carried away and enter the hydrosphere, part may be absorbed by colloidal sediments, and part may evaporate.

Schuetteite, $HgSO_4 \cdot 2HgO$, has been found as a naturally occurring supergene mineral in several quicksilver mine areas in Nevada, and at single mines in Oregon, Idaho, and California (BAILEY, HILDEBRAND, CHRIST, and FAHEY, 1959, p. 1032). In all of these localities it occurs as thin, yellow films on surfaces of cinnabar-bearing rocks that are exposed to sunlight. BAILEY, HILDEBRAND, CHRIST, and FAHEY concluded that the schuetteite films were produced by a reaction involving cinnabar, a small amount of nearly neutral, surficial water containing dissolved oxygen, and sunlight, as follows:

$$HgS + 2O_2 + \text{energy from sunlight} \rightarrow HgSO_4;$$

the mercuric sulfate by hydrolysis with water then yielded the basic mercuric sulfate, which has a very limited solubility.

Montroydite, HgO, is a rare mineral that has been found in the Terlingua District, Texas, and at Huahuaxtla, Guerrero, Mexico, and a few other localities. In some places it has coated and replaced native mercury and appears to have been formed by supergene processes. In some of the Terlingua ores, on the other hand, it formed long, hair-like crystals that were enclosed in calcite. These hair-like crystals were believed by YATES and THOMPSON (1959, pp. 71, 75) to have been deposited from a vapor during the hypogene mineralization.

The calomel, eglestonite, terlinguaite, kleinite, and mosesite found in the Terlingua District were considered to be hypogene minerals by YATES and THOMPSON, but in some other localities they may have been formed by supergene processes.

Native mercury occurs as a supergene mineral in some places. In the deposits of the South Fergana mercury-antimony belt, native mercury is fairly common according to FEDORCHUK (1958, pp. 345—352) and is in part supergene and in part hypogene. FEDORCHUK stated that the supergene nature of numerous small accumulations in zones of tectonic disturbances was indicated by many leached-out cavities, some of which preserved the shapes of cinnabar crystals and contained fragments of cinnabar still adhering to the cavity walls. These cavities also contained droplets of native mercury and fine crystals of sulfur. In some places calomel was also formed by the decomposition of cinnabar. Other accumulations of native mercury in the South Fergana mercury-antimony belt found in lenses of carbonaceous shale were considered by FEDORCHUK to be hypogene because the lenses containing the native mercury were situated in shale directly above the richest cinnabar masses in underlying hornfels-jasperoid breccias and because there was no evidence of any leaching of cinnabar from the ore-bodies. FEDORCHUK stated that there were also marked differences between the chemical compositions of the supergene native mercury and the hypogene native mercury. The supergene native mercury contained only traces of copper, iron, and lead. The hypogene native mercury from the lenses of carbonaceous shale, on the other hand, contained admixed antimony, lead, traces of zinc, arsenic, copper, and silver. These same elements were also detected in the cinnabar from the main ore-bearing horizon.

Revised manuscript received: August 1968

80-H. Solubilities of Compounds that Control Concentrations of Mercury in Natural Waters

In most mercury ore deposits cinnabar is the principal ore mineral and in many mercury ore deposits metacinnabar is present in lesser amounts. Since almost all of the cinnabar and metacinnabar in mercury ore deposits is generally believed to have been deposited from aqueous solutions, knowledge of the solubility of cinnabar and metacinnabar in aqueous solutions is needed in order that an understanding of the chemical processes by which the mercury sulfide was transported and deposited may be obtained.

The saturation curves of cinnabar and metacinnabar in the system HgS-Na_2S-H_2O at 25°, 50°, and 75° C and at one atmosphere were determined by DICKSON and TUNELL (1958) and subsequently DICKSON (1964) determined the solubility of cinnabar in the system HgS-Na_2S-H_2O at temperatures up to 250° C and at pressures up to 1,800 bars. With a constant sodium sulfide concentration and at a constant pressure, he found that the solubility of cinnabar decreases with increasing temperature up to approximately 100° C. In the vicinity of 100° C, however, the solubility of cinnabar in such a solution passes through a minimum and with increase of temperature above approximately 100° C it increases markedly (cf. Table 80-H-1 and Fig. 80-H-1). DICKSON's work also proved that the solubility of cinnabar increases with diminishing pressure at constant temperature and constant sodium sulfide concentration (cf. Table 80-H-1).

The solubility of black mercuric sulfide at 20° C in aqueous solutions containing 0.0190 mole per liter (approximately 585 ppm)[1] of total sulfur, the pH values of which ranged from —0.05 to 10.74, was determined by SCHWARZENBACH and WIDMER (1963) by the use of a radiometric method. The value of the ionic strength of the solutions was fixed at 1.00 by the addition of KCl. The pH values of the solutions were fixed by the use of the following buffers: in the pH-range 0—3, HCl; in the pH-range 3—6, acetic acid + acetate; in the pH-range 6—9, imidazol-HCl+KOH; in the pH-range 8—11, ethanolamine-HCl+KOH. In the preparation of the solutions to be brought to equilibrium, mercury acetate with the radioisotope ^{203}Hg, buffer, and KCl were mixed with the necessary amount of water and, after elimination of the oxygen, a solution of NaHS was introduced. The closed vessel containing the sulfide suspension was placed in a shaking machine for several days and the suspension was brought to equilibrium at constant temperature. After equilibrium was reached the pH of the solution was determined by means of a glass

[1] The densities of the solutions were not given by SCHWARZENBACH and WIDMER. Consequently it was not possible to calculate the values in ppm exactly from their values in moles per liter. In making the approximate calculations of the values in ppm corresponding to their values in moles per liter the density value used was that of a 1 molar solution of KCl at 20° C which is 1.04. The approximate values in ppm are probably within 2 or 3 percent of the true values and are sufficiently accurate for geological purposes.

Table 80-H-1. *The solubility of cinnabar in aqueous sodium sulfide solutions as a function of temperature, pressure, and sodium sulfide concentration according to* DICKSON (1964)

Temperature (° C)	Pressure (bars)	Concentration of sodium sulfide (wt. percent)	Concentration of HgS in saturated solution (wt. percent)
50	1	3.80	3.30
50	1,822	3.80	1.81
50	10	2.04	1.07
50	783	2.04	0.80
50	1,537	2.04	0.60
50	10	1.37	0.47
50	750	1.37	0.35
50	1,537	1.37	0.26
150	10	1.37	0.43
150	750	1.37	0.35
152	1,500	1.37	0.27
149	750	0.66	0.080
200	750	0.66	0.086
250	750	0.66	0.105
50	750	2.23	0.900
100	750	2.23	0.870
150	750	2.23	0.873
199	750	2.23	0.908
250	750	2.23	0.985

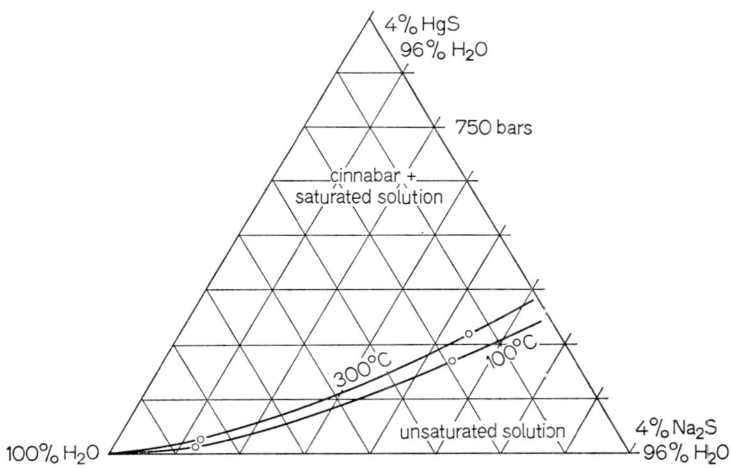

Fig. 80-H-1. Saturation curves of cinnabar in the system HgS-Na_2S-H_2O at 100° C and 300° C and 750 bars. Unpublished work of F. W. DICKSON. The curves are reproduced by kind permission of Prof. DICKSON

electrode. The separation of the homogeneous solution from the precipitate was accomplished by slow, repeated filtration through a fritted glass porous plate under pressure of nitrogen gas. The amount of radioactive mercury in the solution was determined by means of a scintillation counter.

The concentrations of total dissolved mercury, also recalculated as the concentrations of dissolved mercuric sulfide, in solutions of total sulfur concentration 0.0190 mole per liter (approximately 585 ppm) and ionic strength 1.00 fixed by the addition of KCl as a function of pH at 20° C, determined by SCHWARZENBACH and WIDMER are given in Table 80-H-2. SCHWARZENBACH and WIDMER showed that the mercuric sulfide in the solutions saturated with black HgS was present in the form of the complexes, HgS_2^{2-}, HgS_2H^-, and $Hg(SH)_2$. They calculated the concentration of each of these complexes as a function of pH in the solutions containing 0.0190 mole per liter (approximately 585 ppm) of total sulfur and ionic strength 1.00 fixed by the addition of KCl at 20° C. In Fig. 80-H-2, which is taken from the article by SCHWARZENBACH and WIDMER, the concentration of total dissolved mercury is represented as a function of pH by Curve I, the concentration of HgS_2^{2-} is represented by Curve II, the concentration of HgS_2H^- is represented by Curve III, and the concentration of $Hg(SH)_2$ is represented by Curve IV. From the curves of SCHWARZENBACH and WIDMER it appears that the complex HgS_2^{2-} is present in a greater concentration than the complex HgS_2H^- in solutions of pH greater than 8.3 at 20° C, the total sulfur concentration of which is 0.0190 mole per liter and at this pH value the concentration of the complex $Hg(SH)_2$ is zero. Further-

Table 80-H-2. *The solubility of black mercuric sulfide as a function of pH in solutions in which the total sulfur concentration is 0.0190 mole per liter (585 ppm) and the ionic strength is 1.00 (KCl) at 20° C according to* SCHWARZENBACH and WIDMER (1963)

pH values of solutions saturated with black HgS	Total dissolved mercury (in moles per liter multiplied by 10^8)	Concentrations of HgS in solutions saturated with black HgS (in ppm)[a]
1.00	2.02	0.0045
1.57	2.30	0.0051
2.00	2.20	0.0049
2.89	2.18	0.0049
3.98	2.33	0.0052
4.57	2.08	0.0047
5.32	2.13	0.0048
5.91	2.79	0.0062
6.84	6.11	0.0137
7.19	7.75	0.0173
7.63	9.82	0.0220
8.23	19.77	0.044
8.83	52.13	0.117
9.70	301.9	0.68
10.17	713.6	1.60
10.38	1,090.4	2.44
10.74	4,435.1	9.9

[a] The values of the concentrations of HgS in ppm in solutions saturated with black HgS were recalculated from SCHWARZENBACH and WIDMER's values of the concentrations of total dissolved mercury in moles per liter in solutions saturated with black mercuric sulfide. The densities of the solutions were not given by SCHWARZENBACH and WIDMER. In the recalculation of the concentrations in ppm, the value of the density used was 1.04, which is the density of a 1 molar solution of KCl at 20° C. The values in ppm are probably not in error by more than 2 or 3 percent and are sufficiently accurate for geological purposes.

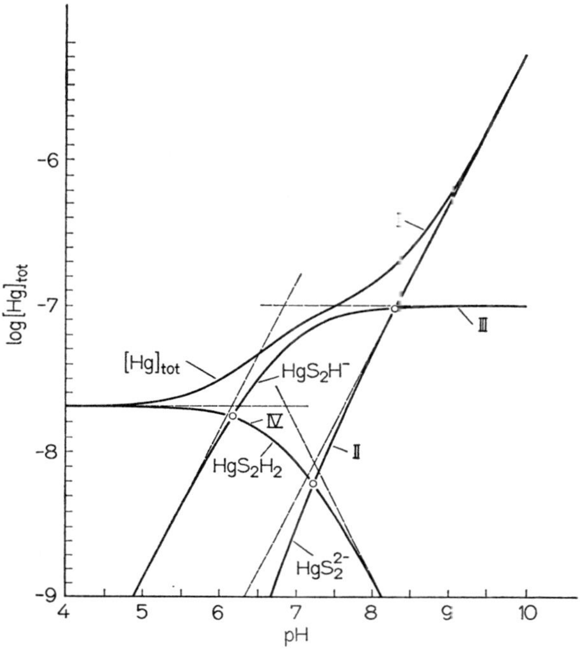

Fig. 80-H-2. The total amount of dissolved mercury in moles per liter, $[Hg]_{tot}$, in a solution saturated with black mercuric sulfide, the total sulfur concentration in moles per liter of the solution, $[S]_{tot}$, as H_2S, HS^-, and S^{2-}, being 0.0190, as a function of pH at 20° C, also the calculated concentrations of HgS_2H_2, HgS_2H^-, and HgS_2^{2-} as functions of pH under the same conditions. (After SCHWARZENBACH and WIDMER.) Curve I shows the experimentally determined concentration of total dissolved mercury in moles per liter as a function of pH. Curve II shows the calculated concentration of HgS_2^{2-} in moles per liter as a function of pH. Curve III shows the calculated concentration of HgS_2H^- in moles per liter as a function of pH. Curve IV shows the calculated concentration of $Hg(SH)_2$ in moles per liter as a function of pH

more, the concentration of the complex HgS_2^{2-} is greater than that of the complex $Hg(SH)_2$ in solutions of pH greater than 7.2 at 20° C and having the same total sulfur concentration. Finally, the concentration of the complex HgS_2H^- is greater than that of the complex $Hg(SH)_2$ in solutions of pH greater than 6.2 at 20° C and having the same total sulfur concentration. If one interpolates between the values in Table 80-H-2, one finds that the concentration of total dissolved mercury in a solution of pH 8.3 at 20° C, the total sulfur concentration of which is 0.0190 mole per liter, is 21.54×10^{-8} mole per liter (0.042 ppm). One also finds in the same way that the concentration of total dissolved mercury in a solution of pH 7.2 at 20° C and the same concentration of total sulfur is 7.76×10^{-8} mole per liter (0.015 ppm) and the concentration of total dissolved mercury in a solution of pH 6.2 at 20° C and the same concentration of total sulfur is 3.64×10^{-8} mole per liter (0.0070 ppm).

SCHWARZENBACH and WIDMER (1963, p. 2623) stated that in saturated solutions having the same pH, the concentration of mercuric sulfide is approximately proportional to the total concentration of sulfur in the form of H_2S, in the form of HS^-, and as S^{2-}.

The effect of the presence of additional constituents in the solutions on the solubility of cinnabar is of importance in attempts to elucidate the transportation of mercuric sulfide and the deposition of cinnabar and metacinnabar. For this reason LEARNED (1966) determined the concentrations of mercuric sulfide and silica in aqueous sodium sulfide solutions that were saturated with both cinnabar and quartz over the temperature range 100° to 250° C at 100 bars pressure. He found that the presence of silica in the solutions reduced the solubility of cinnabar considerably, but that the solubility of cinnabar in solutions saturated with both cinnabar and quartz is still sufficiently great so that mercuric sulfide and silica could both be carried in the same solution in geologically important amounts. LEARNED also determined the concentrations of mercuric sulfide and antimony sulfide in aqueous sodium sulfide solutions that were saturated with both cinnabar and stibnite over the temperature range 150° to 250° C at 100 bars pressure. In this case he found that the presence of the antimony sulfide in the solutions reduced the solubility of cinnabar drastically. Thus it does not appear probable that geologically important amounts of mercuric sulfide could be carried in aqueous sodium sulfide solutions that are saturated with stibnite.

KRAUSKOPF (1951, pp. 507—509) considered the effect of the presence of chlorides on the solubility of mercuric sulfide. On the basis of his calculations he concluded that in the presence of sulfur even under the most favorable conditions of high concentrations of Cl^- and H^+ and a low concentration of H_2S, the solubility of HgS would barely be high enough to be geologically significant. SCHWARZENBACH and WIDMER's determinations of the solubility of black mercuric sulfide were made in solutions containing approximately 1 mole per liter of KCl. The values of the solubility of black mercuric sulfide found by them in neutral and acid solutions containing 0.0190 mole per liter of total sulfur as H_2S, HS^-, and S^{2-}, and approximately 1 mole per liter of KCl were less than 0.02 ppm. Since SCHWARZENBACH and WIDMER pointed out furthermore that the solubility of black mercuric sulfide in solutions of the same pH is approximately proportional to the concentration of total sulfur as H_2S, HS^-, and S^{2-}, it seems clear that in the absence of sulfur as H_2S, HS^-, and S^{2-}, the presence of chlorides is insufficient to cause the solution of geologically significant quantities of HgS.

KRAUSKOPF (1951, p. 501) also calculated the effect of the presence of CO_2 dissolved in water on the solubility of mercuric sulfide and found that it was too small to be of geological importance.

Mercuric chloride is quite soluble in water, the solubility at 20° C being 6.9 grams per 100 cc of water (Handbook of Chemistry and Physics, 1967—1968). Mercurous chloride is much less soluble, but still has an appreciable solubility in water, 0.00020 gram per 100 cc of water at 25° C (Handbook of Chemistry and Physics, 1967—1968). Mercuric sulfate with water yields by hydrolysis basic mercuric sulfate, the solubility of which is very slight. Mercurous sulfate is somewhat soluble in water, a saturated solution containing 0.06 gram of mercurous sulfate per 100 cc of water at 25° C (Handbook of Chemistry and Physics, 1967—1968). Thus it is probable, as SAUKOV (1953) pointed out, that some mercury is carried in natural waters in the form of bichloride and sulfate.

Revised manuscript received: August 1968

80-I. Abundance in Natural Waters and in the Atmosphere

In a sample of sea water from the vicinity of Helgoland, STOCK and CUCUEL (1934 A, p. 392) found 0.03×10^{-6} g of mercury per 1,000 ccm of water (approximately 0.03 ppb). HAMAGUCHI, KURODA, and HOSOHARA (1961, pp. 347—349 and p. A21) reported an average content of mercury of 0.10×10^{-6} g per l (approximately 0.10 ppb) in the surface waters of the Ramapo Deep (30° N, 139—148° E). HOSOHARA (1961, pp. 1107—1108 and p. A70) determined the variation of the mercury content with depth at four stations over the Ramapo Deep (30° N, 139 to 148° E). He found that the mercury content increased slightly with depth. At a depth of 3,000 m he obtained values from 0.15×10^{-6} to 0.27×10^{-6} g per l (approximately 0.15 to 0.27 ppb). HOSOHARA, HIROTAKA, KAWASAKI, and TSURUTA (1961, pp. 1479—1480 and p. A95) reported that water from Minamata Bay, Kyushu, contained 1.6×10^{-6} to 3.6×10^{-6} g per l (approximately 1.6 to 3.6 ppb) of mercury. According to SILLÉN (1963, p. 173) mercury is present in sea water chiefly in the form of the complex, $HgCl_4^{2-}$. RANKAMA and SAHAMA (1950, p. 717) stated that the content of mercury in sea water is kept very low as a result of its adsorption by hydrolyzate and oxidate sediments.

HEIDE, LERZ, and BÖHM (1957, pp. 441—442) found that the water of the Saale River at Göschwitz contained 0.067×10^{-6} g per l (approximately 0.067 ppb) of dissolved mercury and 0.021×10^{-6} g per l (approximately 0.021 ppb) of suspended mercury. AIDINYAN (1962, pp. 9—14) reported that the content of mercury in river waters of the European portion of the Soviet Union varied from 0.4×10^{-6} to 2.8×10^{-6} g per l (approximately 0.4 to 2.8 ppb). He stated that the contents of mercury in the waters of the rivers in the European part of the Soviet Union were much higher than the mercury contents of other European rivers.

STOCK and CUCUEL (1934 a, p. 392) found that the mercury contents of samples of rain water collected at different places and different times ranged from 0.05×10^{-6} to 0.48×10^{-6} g per 1,000 ccm (approximately 0.05 to 0.48 ppb). They gave as the mean of 12 determinations, 0.2×10^{-6} g per 1,000 ccm (approximately 0.2 ppb). STOCK and CUCUEL stated that there was no appreciable difference in the mercury contents of rain water that fell near the town and rain water that fell far from dwelling places.

Mercury is present in the atmosphere in an amount slightly greater than the limit of analytical detection according to STOCK and CUCUEL (1934 b, p. 125). In two samples of air collected at a place where contamination by laboratory fumes, industrial waste gases, and other such impurities was not probable, they found that the concentration of mercury in the air was 0.008×10^{-6} g per m^3. Mercury enters the atmosphere by evaporation from the seas, from hot springs, and from native mercury in ore deposits, as well as in volcanic exhalations. From the atmosphere mercury returns to the seas and lands in rain water. An excellent diagram illustrating the different features of the geochemical cycle of mercury is to be found in the treatise of RANKAMA and SAHAMA (1950, p. 717).

Revised manuscript received: August 1968

80-K. Abundance in Sedimentary Rocks

According to TUREKIAN and WEDEPOHL (1961, Table 2 opposite p. 186) the average abundances of mercury in the various types of sedimentary rock are as follows: shales, 400 ppb; sandstones, 30 ppb; carbonate rocks, 40 ppb. The value of 400 ppb for shales was taken as the average of the determinations of STOCK and CUCUEL and of PREUSS on the same composite sample of 36 German Paleozoic marine shales. STOCK and CUCUEL (1934 a, p. 391) found the mercury content of the composite sample of 36 German shales to be 500 ppb; PREUSS (1941, p. 19) reported the mercury content of the same composite sample of German shales to be 300 ppb. STOCK and CUCUEL also determined the mercury contents of several other samples of shales and obtained the following results: composite sample of three south Swedish *Dictyonema* shales, 25 ppb; Mansfeld Kupferschiefer, 130 ppb; carboniferous marine shale with *Anthracoceras* from Recklinghausen, 1.43 ppm; kaolin-rich shale from Neurode in Schlesien, 210 ppb. The average abundance of mercury in sandstones given by TUREKIAN and WEDEPOHL was obtained from the work of STOCK and CUCUEL, who found 26 ppb and 40 ppb of mercury respectively in two samples of sandstone from Baden. PREUSS reported the mercury content of a composite sample of 23 German red Triassic sandstones to be 100 ppb. The average abundance of mercury in limestones given by TUREKIAN and WEDEPOHL was taken as the average of determinations by STOCK and CUCUEL and by HEIDE and BÖHM on the Muschelkalk. STOCK and CUCUEL determined the mercury content of a sample of the Muschelkalk from Baden to be 33 ppb. HEIDE and BÖHM (1958, pp. 199—200) analyzed samples of pure or almost pure limestone of the Muschelkalk from 12 successive horizons at Steudnitz and obtained as the mean of their determinations 48 ppb of mercury.

SAUKOV (1946, p. 44) and co-workers determined the mercury contents of samples of shales from six localities in the Soviet Union. The average value of the mercury content of these six shale samples was 130 ppb. In one sample of sandstone from the Soviet Union, SAUKOV and co-workers found 31 ppb of mercury. SAUKOV and co-workers also determined the mercury contents of five samples of limestone from the Soviet Union. The average value of the mercury content of these five samples of limestone was 38 ppb.

The average content of mercury in shales, which constitute four-fifths of the sedimentary rocks according to CLARKE and WASHINGTON (1924, p. 29), is thus higher than the average mercury content of igneous rocks exposed at the earth's surface, as RANKAMA and SAHAMA (1950, p. 717) have pointed out. According to GOLDSCHMIDT (1954, pp. 274, 278) the enrichment of mercury in the marine hydrolyzate sediments is due to the transport of much mercury to the earth's surface by processes other than the weathering of igneous rocks, such as transport by magmatic fluids and weathering of ore deposits of magmatic origin. Furthermore, mercury is readily adsorbed by the clay minerals of these sediments (RANKAMA and SAHAMA, 1950, p. 717).

Revised manuscript received: August 1968

80-L. Biogeochemistry

Mercury is absorbed by many plants and trees from the soil. In an investigation of the mercury contents of soils and plants distant from and in the vicinity of mercury, gold, silver, lead, zinc, molybdenum, and nickel ore bodies in British Columbia, WARREN, DELAVAULT, and BARAKSO (1966, p. 1025) found that where the soil contained less than 1 ppm of mercury, all vegetation growing on this soil concentrated mercury. On the other hand, where soils had a mercury content of tens or hundreds of ppm, they noted that vegetative matter contained less mercury than its related soil. RANKAMA and SAHAMA (1950, p. 334) reported that droplets of metallic mercury have been found in the seed capsules of *Holosteum umbellatum* growing on some mercury-rich soils. According to RANKAMA and SAHAMA (1950, p. 718) some species of marine algae contain more than a hundred times as much mercury as sea water does. Fish contain notable amounts of mercury. STOCK and CUCUEL (1934 a, p. 392) reported from 28 to 180 ppb of mercury in different species of fish. Little is known concerning the biological role of mercury. STOCK and CUCUEL (1934 a, p. 393) stated, however, that mercury may be one of the "catalytic elements" of living organisms.

The mercury contents of a number of samples of bituminous coal and brown coal were determined by STOCK and CUCUEL (1934 a, p. 391) with the following results: bituminous coal from the Saar, 8 ppb; bituminous coal from the Ruhr, 10 ppb; fat coal from the Ruhr, 22 ppb; English standard coal, 12 ppb; brown coal from mine, Waltershoffnung, 1.2 ppb; brown coal from Tagebau, Elise II mine, 18 ppb; brown coal from Konkordia, 25 ppb; brown coal from Deuben, 6 ppb; brown coal from von der Heydt, 5 ppb; and brown coal from Theodor, 8 ppb. Mercury has been detected in petroleum in the Cymric Field, California (BAILEY, SNAVELY, and WHITE, 1961, pp. D 307—D 308). In six samples from two wells the mercury content found ranged from 1.9 to 21.0 ppm. In samples of brine from the same oil field, mercury was also reported in amounts varying from 48 to 230 ppb. The form in which the mercury was present in the crude oil and in the brines was not determined. BAILEY, SNAVELY, and WHITE (1961, p. D309) stated that: "the differences in reported mercury contents of the waters indicate that mercury is not yet certain to be in the waters; however, the reported presence of mercury in the crude oils has been confirmed definitely".

Revised manuscript received: August 1968

80-M. Abundance in Metamorphic Rocks

JOVANOVIC and REED (1968) determined the mercury contents of samples of regionally metamorphosed pelitic schists from the Burke Area of north-eastern Vermont. The samples ranged in metamorphic grade from biotite schists, through garnet and staurolite schists, to sillimanite schists. JOVANOVIC and REED analyzed each sample for mercury given off below 450° C and for mercury given off above 450° C making use of neutron activation methods. The values which they obtained are reproduced in Table 80-M-1. The total concentrations of mercury found in the samples within a single metamorphic zone vary considerably and no systematic variation of the total concentration of mercury from one metamorphic zone to another is apparent in their data. JOVANOVIC and REED suggested that there is a tendency for the ratio of mercury released below 450° C to that released above 450° C to decrease in the samples from zones of successively higher metamorphic grade. However, they stated that many more samples need to be run.

Table 80-M-1. *Mercury contents of samples of pelitic schists from metamorphic zones ranging in grade from biotite schists, through garnet and staurolite schists, to sillimanite schists from the Burke Area of north-eastern Vermont according to* JOVANOVIC and REED (1968)

Metamorphic zone: schists containing:	Sample number	Total mercury content (ppm)	Mercury<450°C (ppm)	Mercury>450°C (ppm)	Mercury<450°C / Mercury>450°C
Biotite	Li 5125	0.376	0.36	0.016	22
Biotite	Li 5125	0.225	0.21	0.015	14
Biotite	Li 5125	0.237	0.22	0.017	13
Biotite	WB 915A	0.00988	0.0093	0.00058	16
Garnet	WB 1186	0.00686	0.0067	0.00016	41
Garnet	WB 1112	0.0062	0.0042	0.002	2.1
Staurolite	Li 5100	0.963	0.94	0.023	41
Staurolite	WB 1588	0.005283	0.0052	0.000083	63
Staurolite	WB 1517	0.01538	0.015	0.00038	39
Sillimanite	Li 5082	2.535	2.5	0.035	71
Sillimanite	Li 5121	0.537	0.52	0.017	30
Sillimanite	Li 5121	0.0440	0.040	0.0040	10
Sillimanite	WB 1302	0.00251	0.0023	0.00021	11

Revised manuscript received: August 1968

References: Sections 80-A to 80-M

AHRENS, L. H.: The use of ionization potentials, part 1. Ionic radii of the elements. Geochim. Cosmochim. Acta **2**, 155 (1952).

AIDINYAN, N. KH.: Soderzhanie rtuti v nekotorykh prirodnykh vodakh (The content of mercury in some natural waters). Tr. Inst. Geol. Rudn. Mestorozhd., Petrog., Mineralog. i Geokhim. **70**, 9—14 (1962).

— L. L. SHILIN, i G. A. BELAVSKAYA: O raspredelenii rtuti v porodakh i mineralakh Khibinskovo Massiva (Concerning the distribution of mercury in the rocks and minerals of the Khibinsky Massif). Tr. Inst. Geol. Rudn. Mestorozhd., Petrog., Mineralog. i Geokhim. **99**, 16—25 (1963).

AURIVILLIUS, K.: Least-squares refinement of the crystal structures of orthorhombic HgO and of Hg_2O_2NaI. Acta Chem. Scand. **18**, 1305 (1964a).

— An X-ray and neutron diffraction study of metacinnabarite. Acta Chem. Scand. **18**, 1552 (1964b).

— The structural chemistry of inorganic mercury (II) compounds. Arkiv Kemi **24**, 151 (1965).

—, and I. B. CARLSSON: The structure of hexagonal mercury (II) oxide. Acta Chem. Scand. **12**, 1297 (1958).

BAILEY, E. H.: Quicksilver deposits of the Western Mayacmas District, Sonoma County, California, Calif. J. Mines Geol. **42**, 199—230 (1946).

—, and D. L. EVERHART: Geology and quicksilver deposits of the New Almaden District, Santa Clara County, California. U.S. Geol. Surv., Profess. Papers 360 (1964).

—, F. A. HILDEBRAND, C. L. CHRIST, and J. J. FAHEY: Schuetteite, a new supergene mercury mineral. Am. Mineralogist **44**, 1026—1038 (1959).

—, and D. A. PHOENIX: Quicksilver deposits in Nevada. Nevada State Bureau of Mines and the Mackay School of Mines, Univ. Nevada Bull., Geology and Mining Series No. 41 (1944).

—, P. D. SNAVELY JR., and D. E. WHITE: Chemical analyses of brines and crude oil, Cymric Field, Kern County, California. In Geological survey research, 1961, Short papers in the geologic and hydrologic sciences, articles 293—435. U.S. Geol. Surv., Profess. Papers **424-D**, D 306—D 309 (1961).

BAINBRIDGE, K. T., and A. O. NIER: Relative isotopic abundances of the elements. Preliminary report No. 9, Nucl. Sci. Ser., Division of Mathematical and Physical Sciences, National Research Council, Washington, 1950.

BECKER, G. F.: Geology of the quicksilver deposits of the Pacific slope. U.S. Geol. Surv., Monograph 13 (1888).

BETHKE, P. M.: Sulfo-selenides of mercury. Bull. Geol. Soc. Am. **67**, 1671 (1956).

BLACK, P. J., and J. A. CUNDALL: The structures of liquid mercury and liquid aluminium. Acta Cryst. **19**, 807 (1965).

BRÄNDEN, C. I.: The structural chemistry of some mercury (II) complexes. Arkiv Kemi **22**, 501 (1964).

BRAGG, L., and G. F. CLARINGBULL: Crystal structure of minerals, the crystalline state, vol. 4. London: G. Bell & Sons, Ltd. 1965.

BRANNOCK, W. W., P. F. FIX, V. P. GIANELLA, and D. E. WHITE: Preliminary geochemical results at Steamboat Springs, Nevada. Trans. Am. Geophys. Union **29**, 211—226 (1948).

BRÖNSTED, J. N., and G. VON HEVESY: The separation of the isotopes of mercury. Nature **106**, 144 (1920).

— — Über die Trennung der Isotopen des Quecksilbers. Z. Physik. Chem. **99**, 189—206 (1921).

— — Das Atomgewicht von Quecksilber verschiedener Herkunft. Z. Anorg. Allgem. Chem. **124**, 22—24 (1922).

CARROBI, G., and R. PIERUCCINI: Small quantities of mercury in some celestites and strontianites. Atti soc. Toscana sci. nat. (Pisa), Mem. **57**, Ser. A, 32—37 (1950).
CLARKE, F. W., and H. S. WASHINGTON: The composition of the earth's crust. U. S. Geol. Surv., Profess. Papers **127** (1924).
COTTON, F. A., and G. WILKINSON: Advanced Inorganic Chemistry, 2. Ed. New York: Interscience Publ. 1966.
DANA's Handbook of Mineralogy, seventh ed. by C. PALACHE, H. BERMAN and C. FRONDEL, New York: John Wiley & Sons. Inc. 1944 (vol. 1), 1951 (vol. 2).
DICKSON, F. W.: Solubility of cinnabar in Na_2S solutions at 50—250° C and 1—1800 bars, with geologic applications. Econ. Geol. **59**, 625—635 (1964).
—, and G. TUNELL: Geochemical and petrographic aspects of mercury ore deposits. Final report, Office of Naval Research, Contract Nonr 233, Task Order 12, Los Angeles, 1955.
— — Equilibria of red HgS (cinnabar) and black HgS (metacinnabar) and their saturated solutions in the systems HgS-Na_2S-H_2O and HgS-Na_2S-Na_2O-H_2O from 25° C to 75° C at 1 atmosphere pressure. Am. J. Sci. **256**, 654—679 (1958).
— — The stability relations of cinnabar and metacinnabar. Am. Mineralogist **44**, 471—487 (1959).
DOELTER, C., u. H. LEITMEIER: Handbuch der Mineralchemie, Bd. 4, erste Hälfte. Dresden u. Leipzig: Theodor Steinkopff 1926.
DUNITZ, J. D., and L. E. ORGEL: Stereochemistry of ionic solids. Adv. Inorg. Chem. Radiochem. **2**, 1 (1960).
ECKEL, E. B., and W. B. MYERS: Quicksilver deposits of the New Idria District, San Benito and Fresno Counties, California. Calif. J. Mines Geol. **42**, 81—124 (1946).
EHMANN, W. D., and J. R. HUIZENGA: Bismuth, thallium and mercury in stone meteorites by activation analysis. Geochim. Cosmochim. Acta **17**, 125—135 (1959).
—, and J. F. LOVERING: The abundance of mercury in meteorites and rocks by neutron activation analysis. Geochim. Cosmochim. Acta **31**, 357—376 (1967).
EVERHART, D. L.: Quicksilver deposits at the Sulphur Bank Mine, Lake County, California. Calif. J. Mines Geol. **42**, 125—153 (1946).
FEDORCHUK, V. P.: Origin of native mercury. Geochemistry (USSR) (English Transl.) **1958**, 345—352 (1958).
FENN, R. H., J. W. H. OLDHAM, and D. C. PHILLIPS: Crystal structure of $(CH_3)_3S \cdot HgI_3$ and the configuration of the HgI_3 ion. Nature **198**, 381 (1963).
GOLDSCHMIDT, V. M.: Geochemische Verteilungsgesetze der Elemente, VII. Skrifter Norske Videnskaps-Akad. Oslo, I. Mat.-Naturv. Kl. No. 2, 24 (1926).
— Über Atomabstände in Metallen. Z. physik. Chem. (Leipzig) **133**, 397 (1928).
— Geochemistry (edit. by A. MUIR). Oxford: Clarendon Press 1954.
GRDENIĆ, D.: The structural chemistry of mercury. Quart. Rev. (London) **19**, 303 (1965).
— The crystal structure of mercuric oxyfluoride. Acta Cryst. **21**, 55 A (1966).
GRIP, E.: On the occurrence of mercury in Boliden and in some other sulphide deposits in northern Sweden. Sveriges Geol. Undersökn., Arsbok **42**, 1—12 (1948).
GROSSE, A. V.: Densities, volumes, expansion coefficients and atomic cell dimensions of metallic mercury for its entire solid and liquid temperature range, i.e., from zero absolute to its critical point (1733° K). J. Inorg. & Nuclear Chem. **27**, 733 (1965).
HAMAGUCHI, H., R. KURODA, and K. HOSOHARA: Photometric determination of mercury in sea water. Nippon Kagaku Zasshi **82**, 347—349 and A 21 (1961).
Handbook of Chemistry and Physics, 48th ed. Cleveland: The Chemical Rubber Co. 1967 ot 1968.
HANSEN, M.: Constitution of binary alloys, 2. ed. (Prepared with the cooperation of K. ANDERKO), p. 207. New York: McGraw-Hill Book Co. Inc. 1958.
HARKINS, W. D.: The separation of the elements chlorine and mercury into isotopes. Science **54**, 359—360 (1921).
HEIDE, F., u. G. BÖHM: Zur Geochemie des Quecksilbers. Chem. Erde **19**, 198—204 (1958).
—, H. LERZ u. G. BÖHM: Gehalt des Saalewassers an Blei und Quecksilber. Naturwissenschaften **44**, 441—442 (1957).
HENDERSON, F. B., III: Deposition and oxidation of "serpentinite-type" mercury deposits. Ph. D. Thesis, Harvard University, Cambridge, 1965 (Thesis HU 90.8971).

HOPPE, R., u. H. J. RÖHRBORN: Oxomercurate (II) der Alkalimetalle, M_2HgO_2. Z. anorg. u. allgem. Chem. **329**, 110 (1964).

HOSOHARA, K.: Mercury content of deep sea water. Nippon Kagaku Zasshi **82**, 1107—1108 and A 70 (1961).

—, K. HIROTAKA, K. KAWASAKI, and T. TSURUTA: Studies on the total amount of mercury in sea waters. Nippon Kagaku Zasshi **82**, 1479—1480 and A 95 (1961).

HUME-ROTHERY, W., and G. V. RAYNOR: The structure of metals and alloys, 3. ed., p. 110. London: The Institute of Metals 1954.

JANDA, I., and E. SCHROLL: Über Quecksilberspuren in Graphiten. Experientia **15**, 125—126 (1959).

JOHANSSON, G.: An X-ray investigation of the structure of the Hg_2^{2+} ion in solution. Acta Chem. Scand. **20**, 553 (1966).

JOVANOVIC, S., and G. W. REED: Hg in metamorphic rocks. Geochim. Cosmochim. Acta **32**, 341—346 (1968).

KELLEY, K. K.: Contributions to the data on theoretical metallurgy. III. The free energies of vaporization and vapor pressures of inorganic substances. U.S. Bur. Mines Bull. 383 (1935).

KETELAAR, J. A. A.: Chemical constitution, p. 29. Amsterdam: Elsevier Publ. Co. 1953.

KLEMENT JR., W., A. JAYARAMAN, and G. C. KENNEDY: Transformations in mercury at high pressures. Phys. Rev. **131**, 1 (1963).

KNOX, J.: Zur Kenntnis der Ionenbildung des Schwefels und der Komplexionen des Quecksilbers. Z. Elektrochem. **12**, 477—481 (1906).

KRAUSKOPF, K. B.: Physical chemistry of quicksilver transportation in vein fluids. Econ. Geol. **46**, 498—523 (1951).

KULLERUD, G.: The mercury-sulfur system. Carnegie Inst. Wash. Yearbook **64**, 193—195 (1965).

LEARNED, R. E.: The solubilities of quartz, quartz-cinnabar and cinnabar-stibnite in sodium sulfide solutions and their implications for ore genesis. Doctoral Diss., University of California, Riverside, 1966.

LE CONTE, J., and W. B. RISING: The phenomena of metalliferous vein formation now in progress at Sulphur Bank, California. Am. J. Sci. [3] **24**, 23—33 (1882).

LEWIS, J., and R. S. NYHOLM: Metal-metal bonds in transition metal complexes. Sci Progr. **52**, 557 (1964).

LINDGREN, W.: Mineral deposits, 4th ed. New York and London: McGraw-Hill Book Co., Inc. 1933.

LIPSCOMB, W. N.: Recent studies in the structural inorganic chemistry of mercury. Ann. N.Y. Acad. Sci. **65**, 427 (1957).

MANASSE, E.: Sulla leviglianite. Mem. Soc. Tosc. Sci. Nat. **33**, 156—163 (1921).

MOORE, G. E.: On the occurrence in nature of amorphous mercuric sulphide. Am. J. Sci. **103**, 36—43 (1872).

MULLIKAN, R. S., and W. D. HARKINS: The separation of isotopes. Theory of resolution of isotopic mixtures by diffusion and similar processes. Experimental separation of mercury by evaporation in a vacuum. J. Am. Chem. Soc. **44**, 37—65 (1922).

NAGORSEN, G., S. LYNG, A. WEISS u. A. WEISS: Zur Konstitution von $HgSO_4 \cdot 2HgO$. Angew. Chem. **74**, 119 (1962).

NIIZEKI, N., and M. J. BUERGER: The crystal structure of livingstonite, $HgSb_4S_8$. Z. Krist. **109**, 129—157 (1957).

NODDACK, I., u. W. NODDACK: Die Geochemie des Rheniums. Z. Physik. Chem. **154** A, 207—244 (1931).

PALACHE, C., H. BERMAN, and C. FRONDEL: The System of Mineralogy of James Dwight Dana and Edward Salisbury Dana, Yale University 1837—1892, 7th ed., vol. 1. New York: John Wiley & Sons, Inc.; London: Chapman & Hall, Ltd. 1944.

PAULING, L.: The Nature of the Chemical Bond, 3. ed. Ithaca (N.Y.): Cornell University Press 1960.

PEARSON, W. B.: A Handbook of Lattice Spacings and Structures of Metals and Alloys. Oxford: Pergamon Press 1964.

PENFIELD, S. L.: Crystallized tiemannite and metacinnabarite. Am. J. Sci. **127**, 449—454 (1885).

PENNINGTON, J. W.: Mercury, a materials survey. U.S. Bur. Mines Inform. Circ. **7941** (1959).
PHILLIPS, C. S. G., and R. J. P. WILLIAMS: Inorganic chemistry, vol. 1 and 2. Oxford: Clarendon Press 1966.
PREUSS, E.: Beiträge zur spektralanalytischen Methodik. II. Bestimmung von Zn, Cd, Hg, In, Tl, Ge, Sn, Pb, Sb und Bi durch fraktionierte Destillation. Z. Angew. Mineral. **3**, 8—20 (1941).
RANKAMA, K., and TH. G. SAHAMA: Geochemistry. Chicago: The University of Chicago Press 1950.
REED, G. W., JR., K. KIGOSHI, and A. TURKEVICH: Determinations of concentrations of heavy elements in meteorites by activation analysis. Geochim. Cosmochim. Acta **20**, 122—140 (1960).
RIVLIN, V. G., R. M. WAGHORNE, and G. I. WILLIAMS: The structure of liquid mercury. Phil. Mag. **13**, 1169 (1966).
ROLFE, C., and W. HUME-ROTHERY: The constitution of alloys of gold and mercury. J. Less-Common Metals **13**, 1 (1967).
ROSS, C. P.: Quicksilver deposits of the Mount Diablo District, Contra Costa County, California. U.S. Geol. Surv., Bull. **922**-B (1940).
— Some concepts on the geology of quicksilver deposits in the United States. Econ. Geol. **37**, 439—465 (1942).
SAUKOV, A. A.: Geokhimiya rtuti (Geochemistry of mercury). Tr. Inst. Geol. Nauk, Akad. Nauk SSSR, **78**, Min.-Geokhim. Ser., No. 17, 1—129 (1946).
— Geochemie. Berlin: VEB Verlag Technik 1953.
—, i N. KH. AIDINYAN: Ob okislenii kinovari (On the oxidation of cinnabar). Tr. Inst. Geol. Nauk, Akad. Nauk SSSR, **39**, Min.-Geokhim. Ser., No. 8, 37—40 (1940).
ŠĆAVNIČAR, S.: The crystal structure of mercury (II) oxycyanide. Z. Krist. **118**, 248 (1963).
SCHUBERT, K.: Kristallstrukturen zweikomponentiger Phasen. Berlin-Göttingen-Heidelberg-New York: Springer 1964.
SCHWARZENBACH, G., u. M. WIDMER: Die Löslichkeit von Metallsulfiden I. Schwarzes Quecksilbersulfid. Helv. Chim. Acta **46**, 2613—2628 (1963).
SIDGWICK, N. V.: The chemical elements and their compounds, vol. 1. Oxford: Clarendon Press 1950.
— The Chemical Elements and Their Compounds, vol. 1. Oxford: Oxford University Press 1951.
SILLÉN, L. G.: How has sea water got its present composition? Svensk Kem. Tidskr. **75**, 161—177 (1963).
STOCK, A., u. F. CUCUEL: Die Verbreitung des Quecksilbers. Naturwissenschaften H. 22/24, 390—393 (1934a).
— — Die Bestimmung des Quecksilber-Gehaltes der Luft. Ber. Deut. Chem. Ges. **67**, 122—127 (1934b).
STOCKAR, K.: Zur Kenntnis der Radien positiver Atomionen. Helv. Chim. Acta **33**, 1409 (1950).
STRUNZ, H. (unter Mitarbeit von CH. TENNYSON): Mineralogische Tabellen, 4. Aufl. Leipzig: Akademische Verlagsgesellschaft, Geest & Portig K.-G. 1966.
SUESS, H. E., and H. C. UREY: Abundances of the elements. Rev. Mod. Phys. **28**, 53—74 (1956).
SWITZER, G. S., W. F. FOSHAG, K. J. MURATA, and J. J. FAHEY: Re-examination of mosesite. Am. Mineralogist **38**, 1225—1234 (1953).
TAYLOR, S. R.: Abundance of chemical elements in the continental crust: a new table. Geochim. Cosmochim. Acta **28**, 1273—1285 (1964).
TERADA, K., and F. WM. CAGLE JR.: The crystal structure of potarite (PdHg) with some comments on allopalladium. Am. Mineralogist **45**, 1093 (1960).
TUNELL, G.: The oxidation of disseminated copper ores in altered porphyry. Doctoral Dissertation, Harvard University, Cambridge, 1930 (Thesis HU 90.2187 LN 1723).
— Chemical processes in the formation of mercury ores and ores of mercury and antimony. Geochim. Cosmochim. Acta **28**, 1019—1037 (1964).
TUREKIAN, K. K., and K. H. WEDEPOHL: Distribution of the elements in some major units of the earth's crust. Bull. Geol. Soc. Am. **72**, 175—191 (1961).

VAVRINECZ, G.: Ásványelemzések. Magy. Chem. Folyóirat **38**, 140—144 (1932).

WAGNER, C. N. J., H. OCKEN, and M. L. JOSHI: Interference and radial distribution functions of liquid copper, silver, tin, and mercury. Z. Naturforsch. **20**a, 325 (1965).

WARREN, H. V., R. E. DELAVAULT, and J. BARAKSO: Some observations on the geochemistry of mercury as applied to prospecting. Econ. Geol. **61**, 1010—1028 (1966).

WEISS, A., S. LYNG u. A. WEISS: Zur Kenntnis von basischem Quecksilberchlorat und basischem Quecksilberbromat. Z. Naturforsch. **15**b. 678 (1960).

WELLS, A. F.: Structural Inorganic Chemistry, 3. ed., p. 890. Oxford: Clarendon Press 1962.

WHITE, D. E.: Thermal springs and epithermal ore deposits. Econ. Geol., Fiftieth Anniversary Volume, 99—154 (1955).

—, and W. W. BRANNOCK: The sources of heat and water supply of thermal springs, with particular reference to Steamboat Springs, Nevada. Trans. Am. Geophys. Union **31**, 566—574 (1950).

—, J. D. HEM, and G. A. WARING: Chemical composition of subsurface waters, chapter F, Data of Geochemistry, 6th ed. U.S. Geol. Surv., Profess. Papers **440**-F (1963).

—, and C. E. ROBERSON: Sulphur Bank, California, a major hot-spring quicksilver deposit. Geol. Soc. Am. Buddington vol., 397—428 (1962).

—, G. A. THOMPSON, and C. H. SANDBURG: Rocks, structure and geologic history of Steamboat Springs thermal area, Washoe County, Nevada. U.S. Geol. Surv., Profess. Papers **458**-B (1964).

WHITE, J. G.: The crystal structure of tetramethylammonium mercury tribromide, $N(CH_3)_4HgBr_3$. Acta Cryst. **16**, 397 (1963).

WILLIAMS, H., and R. R. COMPTON: Quicksilver deposits of Steens Mountain and Pueblo Mountains, Southeast Oregon. U.S. Geol. Surv., Bull. **995**-B (1953).

YATES, R. G., and G. A. THOMPSON: Geology and quicksilver deposits of the Terlingua District, Texas. U.S. Geol. Surv., Profess. Papers **312** (1959).

Revised manuscript received: August 1968

80-O. Economic Importance

The concentration of mercury in mercury ore deposits is approximately 10,000 times greater than the concentration present in crustal rocks. The world's mercury resource is highly localized with nine countries accounting for essentially all of it (Table 80-O-1). At a price of $350 per flask (1 flask = 34.5 kg), the world's mercury resource is estimated to be 5,250,000 flasks or 181,100 t and 17 million flasks or 587,000 t if the price was $1,000 per flask (BAILEY *et al.*, 1972)[1].

Annual world production of mercury in the past ten years has remained stable, averaging about 260,000 flasks (or 8,970 t), despite large fluctuations in the average yearly price ($570 maximum in 1965—$158 minimum in 1975). Spain, U.S.S.R., China, and Italy have been the dominant producers, accounting for 62 percent of the total world production in 1975 (Table 80-O-1). A recently discovered deposit in the United States is expected to contribute about 8 percent of world's annual production for the next 20 years.

Table 80-O-1. *World reserves and annual production* (1975) *of mercury*[a]

	Reserves		Annual Production (1975)	
	Flasks	t	Flasks	t
Spain	2,000,000	65,000	47,051	1,623
U.S.S.R.	500,000	17,250	54,000	1,863
Yugoslavia	500,000	17,250	15,500	535
Peoples Republic of China	500,000	17,250	26,000	897
United States	450,000	15,525	7,366	254
Algeria	400,000	13,800	13,300	459
Italy	400,000	13,800	30,400	1,049
Mexico	250,000	8,625	14,184	489
Canada	100,000	3,450	12,000	414
Other	150,000	5,175	31,041	1,071
Total	5,250,000	181,125	250,842[b]	8,654

[a] DRAKE, H. J.: "Mercury" in Commodity Data Summaries 1976 and 1977, Bureau of Mines, United States Department of the Interior, p. 101.
[b] At an average price of 158.12 $/flask in New York (1970: 407.8 $/flask).

All major mercury deposits contain red mercuric sulfide, cinnabar, as the chief ore mineral. Liquid mercury metal and black mercuric sulfide, metacinnabar, are of some local economic importance. In one medium-size deposit mercury-antimony sulfide, livingstonite, was the main ore mineral. Mercury mines in general produce only mercury, although in a few cases antimony is also recovered.

[1] BAILEY, E. H., CLARK, A. L., SMITH, R. M.: Mercury. U.S. Geol. Surv. Prof. Paper **820**, 409 (1972).

The greatest mercury deposits in the world are located at Almadén (Spain) in the Province of Ciudad Real. Since 1499 the mines at Almadén have produced approximately 8,000,000 flasks (276,000 metric tons) of mercury according to the booklet published by the Council of Administration of the Mines of Almadén. In 1971 mining had proceeded downward to a level 504 meters below the surface. The grade of the ore mined from 1901 to 1916 was 9.3 percent mercury. The recent grade of the ore is approximately 6 percent (PARK and MACDIARMID, 1964)[1]. The mercury ore bodies are situated in three beds of Silurian quartzite interlayered with black, impervious carbonaceous slate. The chief ore mineral at Almadén is cinnabar, which is accompanied by small quantities of calomel. Considerable native mercur is also present on the lower levels.

Production from the U.S.S.R. is primarily from the Khaydarkan complex in the Kirquiz S.S.R. and the Nikitovka complex in the Ukrainian S.S.R. The Khaydarkan deposit occurs in silicified limestones where the beds are arched into anticlines. The Nikitovka deposit is similar to the Almadén deposit in that the ore bodies are stratiform and occur in quartzites. However, at Nikitovka the ore is concentrated in arched portions of dome-like folds, which are cut by faults (FEDORCHUK, 1975)[2].

The great Idrija mine, which is second only to Almadén in total mercury production, is located 38 kilometers west-southwest of Ljubljana in the extreme northwestern corner of Yugoslavia. The Idrija deposit lies in the transition zone between the east-trending Julian Alps and the southeast-trending Dinaric Mountains. The geology of the area around Idrija is complex and of the Alpine type. The geological cross-section of the Idrija region is made up of a sequence of carbonate and clastic geosynclinal sediments ranging in age from Upper Paleozoic to Eocene, with a thickness of more than 5,000 meters. The Idrija ore deposit appears in a nappe. Mercury ore occurs in all horizons of the younger Paleozoic and Lower and Middle Triassic (MILAKAR, 1974)[3]. The chief ore mineral is cinnabar, although considerable native mercury occurs locally. The grade of the ore in recent years is said to have averaged about 0.4 percent of mercury.

The mercury deposits of China (People's Republic) are closely associated with the antimony deposits. Production of both mercury and antimony has increased markedly since 1945. Although deposits of mercury are widely scattered throughout a broad zone, the most important ones are located along the Kweichow-Hunan border, among which at least six are productive, two being especially important, Wanshanchang and Tatungla. The ore bodies, in which cinnabar is the chief ore mineral, occur as irregular veins in brecciated layers of Paleozoic limestone underlying shale, especially in anticlines. Reserves are adequate to maintain the current rate of production for some time[4].

Within the United States only three mines are producing mercury in 1977 the largest being the McDermitt deposit located in the state of Nevada. The McDermitt

[1] PARK, C. F. JR., MACDIARMID, W. A.: Ore Deposits. W. H. Freeman and Co., San Francisco and London, 1964.

[2] FEDORCHUK, V. P.: Commercial types of mercury deposits. International Geology Review **17**, 989 (1975).

[3] MLAKAR, I.: I Congreso Internacional del Mercurio, **1**, 157, Barcelona, 1974

[4] The authors are indebted to Professor L. C. HSU of the Mackay School of Mines of the University of Nevada, Reno, Nevada, for information concerning mercury deposits in China.

deposit is unique in that the mercury ore consists of nearly equal amounts of cinnabar and corderoite ($Hg_3S_2Cl_2$). The ore bodies occur in tuffaceous sediments near the ring fracture of a Miocene caldera.

Development of cinnabar occurrences in the Azzaba region of Algeria in the late 1960's has substantially increased the importance of Algeria as a world producer of mercury. Three deposits of this country are reported to have substantial reserves assuring production for 20 years at the present rate of production.

In Italy mercury production has come from two major mines and several smaller ones in the Monte Amiata (Firenze) district. The mineralized area extends about 25 miles north-south and has a width of about 5 miles. The district is underlain by sedimentary rocks consisting of shale, sandstone, and limestone, ranging in age from Triassic to Quaternary. Most of the large ore bodies are in Eocene rocks. The most prominent faults trend north, and the alinement of mineral deposits, hot springs, and gas seeps (CO_2,H_2S) suggests that the major control for the entire district is one of these. Monte Amiata is a large trachyte volcano whose vent appears to lie a bit west of the fault zone near its northern end. Cinnabar is the chief ore mineral and only locally is accompanied by small quantities of native mercury. Stibnite is fairly common in the southern part of the belt. Mineralization is post-Pliocene and younger than the trachytic rocks of the Monte Amiata volcano (Bailey, 1959)[1].

Geologically, the mercury deposits of Mexico exhibit much variety, with some in acid volcanic rocks and many others in sedimentary rocks. Of unusual interest are the unique mercury-antimony ore bodies of Huitzuco, which contain the rare mineral livingstonite as the principal ore mineral.

Canada's mercury resource is primarily located along the Pinchi fault zone, a thrust fault of large displacement in Central British Columbia. The ore occurs in breccia zones near the fault and in fractured dolomitized limestone. Because of the low price of mercury production has ceased in the district (Bailey, 1977)[2].

Table 80-O-2. *United States consumption of mercury in* 1975[a]

	Flasks	t
Agricultural fungicides and bactericides	600	20.7
Catalysts	838	28.9
Dental preparations	2,340	80.7
Electrical apparatus	16,971	585.5
Electrolytic preparation of chlorine and caustic soda	15,222	525.2
Laboratory use	335	11.6
Industrial and control instruments	4,598	158.6
Paint mildew-proofing	6,928	239.0
Pharmaceuticals	445	15.3
Amalgamation and other uses	1,757	60.6
	50,034	1,726.2

[a] Drake, H. J.: Mercury in the First Quarter 1976.
Mineral Industry Surveys, Bureau of Mines, United States Department of the Interior, p. 4.

[1] Bailey, E. H. in: Pennington, J. W. (ed.): Mercury, a materials survey. Bureau of Mines, Information Circular **7941**, 22 (1959).

[2] Bailey, E. H.: (1977, unpublished).

Consumption of mercury occurs primarily in industrialized contries. The two major uses of mercury are for electrical apparatus such as batteries and as a catalyst in the manufacturing of chlorine. The consumption pattern, according to use for the United States since data on world consumption are unavailable, is given in Table 80-O-2.

Within the United States the consumption pattern of mercury has remained unchanged in the last ten years except for a 50 percent reduction in the use of mercury as an agricultural fungicide and bactericide. The present consumption pattern of mercury is likely to continue since substitutes having the unique properties of mercury are not available.

Revised manuscript received: October 1977

Thallium 81

A	K. Sahl	(Institut für Mineralogie, Ruhr-Universität, Bochum, Germany)
B—O	C. A. R. de Albuquerque and D. M. Shaw	(Department of Geology, McMaster University, Hamilton, Ontario, Canada)

81-A. Crystal Chemistry

Thallium has the electronic structure $6s^2 6p^1$ in its valency shell. In compounds, it is formally monovalent or trivalent, and the lone pair of s electrons of monovalent thallium can be expected to cause a tendency to one-sided coordinations. The accepted ionic radii are about 1.40 Å for Tl^{1+} and 0.95 Å for Tl^{3+}.

The chemical properties of Tl^{1+} salts are similar to those of alkali salts. Compounds of Tl^{3+}, however, show a tendency to complex formation and hydrolytic splitting of Tl^{3+} salts gives acid reactions in solution.

Mixed compounds of Tl^{1+} and Tl^{3+} show more intense colors than compounds in which Tl occurs in a single oxidation state only.

I. Metallic Thallium

Metallic Tl occurs in two modifications: α-$Tl^{[12]}$ with hexagonal close-packing of the atoms and Tl—Tl distances of about 3.4 Å (SEKITO, SB 1930, 173); above 232° C it converts to β-$Tl^{[8]}$ with cubic body-centered structure (PONJATOVSKIJ and ZAKHAROVA, SR 1962, 366—367) and Tl—Tl distances of 3.36 Å. Tl is also a component of numerous intermetallic phases in which it is found in a large variety of coordinations.

II. Monovalent Thallium

a) Monovalent Tl Coordinated by Oxygen

$Tl_2^{[3py]}O$ contains sheets of oxygen atoms enclosed between two sheets of Tl atoms to form layers which are packed in an alternating cubic and hexagonal sequence. The two crystallographic distinct Tl positions are coordinated by one-sided pyramids of oxygen atoms with distances Tl(1)—O = 2.52 Å, 2.53 Å (2×) and Tl(2)—O = 2.51 Å, 2.53 Å (2×). The Tl—Tl distances between layers are as short as 3.51 Å and this could indicate metallic bonding between the layers of the black compound (SABROWSKY, 1971).

According to KENNEDY and PATTERSON (SR 1961, 476—477), the orthorhombic low-temperature modification of $Tl[NO_3]$ contains Tl atoms coordinated on one side by two nearest oxygen atoms at about 2.4 Å; however, this structure is not refined in detail. Above 219° C this compound transforms to a cubic structure which can be derived from the CsCl type (FINBACK and HASSEL, SB 1937, 83).

FAWCETT *et al.* (1970) have determined and carefully refined the tetragonal structure of $Tl_2^{[2]}Tl_2^{[4py]}[P_4O_{12}]$. The $[P_4O_{12}]$-rings are connected by O—Tl—O bonds. The two crystallographically different Tl positions are coordinated on one side by two and four oxygen atoms respectively with Tl(1)—O = 2.849 Å (2×), and Tl(2)—O = 2.700 Å (2×), 2.748 Å (2×); the compound is isotypic with $Tl[AsO_3]$.

Numerous Tl(I) salts are isotypic with the corresponding K salts: $Tl_2[SO_4]$, $Tl_2[SeO_4]$ and $Tl_2[CrO_4]$ (ABBAD and RIVOIR, SR 1947, 379; GATTOW, SR 1960, 383); $Tl_2Mg[SO_4]_2 \cdot 6H_2O$ (HOFMANN, SB 1931, 436); $Tl_2Mn_2[SO_4]_3$ (GATTOW and ZEMANN, 1958); $TlAl[SO_4]_2 \cdot 12H_2O$ (CORK, SB 1927, 392) and $TlCr[SO_4]_2 \cdot 12H_2O$ (LEDSHAM and STEEPLE, 1969); $[Tl_2(Cd_5\,Fe)Fe_2[SO_4]_{12} \cdot 12H_2O] \cdot [Al \cdot 6H_2O]$, voltaite (MEREITER, 1970).

$Tl[ClO_4]$ has a baryte-type structure at low temperatures ($Tl^{[12]}$ irregular — BÜSSEM and HERRMANN, SB 1928, 372) and converts to a cubic $K[ClO_4]$ lattice above 266° C (HERRMANN and ILGE, SB 1930, 411).

$TlIO_3$ crystallizes in a rhombohedrically deformed perovskite lattice with distances Tl—O of about 3.2 Å (6×) (RIVOIR and ABBAD, SR 1947, 366).

In conclusion it can be said that one-sided coordination of monovalent Tl by oxygen atoms reduces considerably the interatomic distances (Tl—O = 2.51 to 2.53 Å in Tl_2O). In all-sided coordinations, the Tl—O distances range from 2.9 Å (Tl—Al-alum) to well above 3 Å. The shorter distances in one-sided coordination polyhedra can be considered to be due to covalent bonding or to the influence of the lone pair of 6s electrons.

b) Monovalent Tl Coordinated by Sulfur

The structure of $Tl_2^{[3py]}S$ was first investigated by X-ray methods by KETELAAR and GORTER (SB 1939, 92). They found a hexagonal layer structure of the anti-CdI_2 type with Tl in pyramidal coordination (Tl—S about 2.8 Å (3×)). The structure showed deviations from the CdI_2 type which were not analysed in detail. MAN (1970) reinvestigated the structure by electron diffraction and confirmed the anti-CdI_2 type but found distances Tl—S from 2.51 to 3.28 Å (shortest Tl—Tl distance = 3.52 Å).

ZEMANN and ZEMANN (SR 1959, 42) determined the structure type of lorandite, $Tl^{[6]}[AsS_2]$; spiralling $[AsS_2]$ chains are connected by Tl atoms which are coordinated irregularly by six sulfur atoms with a Tl—S distance between 2.9 and 3.3 Å.

OHMASA and NOWACKI (1971) have determined and refined the orthorhombic structure of vrbaite, $Hg_3Tl_4[As_8Sb_2S_{20}]$, and the Hg and Tl atoms could be distinguished by their different coordinations. The structure contains two crystallographically different Tl positions. Tl (1) is coordinated to one side by two sulfur atoms at 3.086 and 3.152 Å; five further sulfur atoms complete the irregular coordination polyhedron at distances between 3.346 and 3.433 Å. Tl (2) has a pyramidal coordination of three S atoms at 3.148 to 3.259 Å. Four additional S atoms and two As atoms at 3.322 to 3.622 Å complete the polyhedron.

In hatchite, $PbTlAg[As_2S_5]$ (MARUMO and NOWACKI, 1967), wallisite, $PbTlCu[As_2S_5]$ (TAKÉUCHI *et al.*, 1968), and hutchinsonite, $(Pb, Tl)_2[As_5S_9]$ (TAKÉUCHI *et al.*, 1965), the Pb and Tl atoms could not be distinguished from each other. Distances (Pb, Tl)—S range from 2.73 to 3.43 Å; the shorter distances are always found to one side of the (Pb, Tl) positions (Pb^{2+} also has a lone pair of 6s electrons in its valency shell, similar to Tl^{1+}).

c) Monovalent Tl Coordinated by Halides

TlF has a rhombic low-temperature modification and, above 82° C, a tetragonal high-temperature modification (BARLOW and MEREDITH, 1969) but the structures are unknown.

$Tl^{[8]}Cl$ crystallizes in the cubic CsCl lattice with Tl—Cl = 3.3 Å (8×) (DAVEY and WICK, SB 1921, 113); $Tl^{[8]}Br$ crystallizes in the same lattice with Tl—Br = 3.4 Å (8×) (VAN ARKEL, SB 1924, 113).

The yellow form of $Tl^{[1]}I$ which is stable at room temperature has a rhombic sheet structure (Fig. 81-A-1). Within each sheet of average composition TlI, the distances Tl—I are relatively large (3.5 Å). Short Tl—I bonds of about 3.36 Å (TlI molecules?) connect two sheets to a layer whereas the Tl—I distances between adjacent layers are very large (3.9 Å; HELMHOLZ, SB 1936, 96). Red $Tl^{[8]}I$, stable above 175° C or at higher pressures, crystallizes in a cubic CsCl lattice with Tl—I = 3.6 Å (8×) (DAVEY and WICK, SB 1921, 113).

When coordinated by metal-halide complexes, monovalent Tl has high coordination numbers. Low-Tl$[BF_4]$ crystallizes in the baryte lattice ($Tl^{[12]}$ irregular—PISTORIUS, 1970); $Tl_2[SiF_6]$ has a $K_2[PtCl_6]$ lattice ($Tl^{[12]}$, slightly distorted cubo-octahedron with Tl—F about 3.1 Å—TABET, SB 1933, 478); $Tl_2[SnCl_6]$, $Tl_2[PtCl_6]$ and $Tl_2[TeCl_6]$ also have a $K_2[PtCl_6]$ lattice with Tl—Cl about 3.5 Å (ENGEL, SB 1935, 478). The compounds $TlXCl_3$ (X = Mn, Fe, Co, Ni) have hexagonal or cubic close-packed arrangements of Cl and Tl atoms with $Tl^{[12]}$ and Tl—Cl from 3.43 to 3.62 Å (ZODKEVITZ *et al.*, 1970).

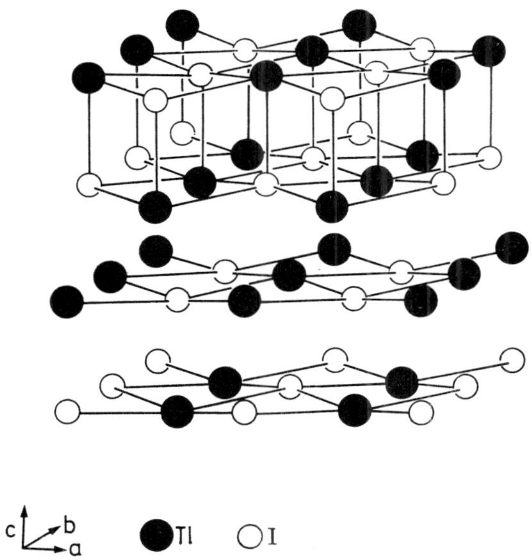

Fig. 81-A-1. Layer-structure of orthorhombic TlI. In the upper two-sheet layer, the short bonds Tl—I between sheets are stressed. In the lower two-sheet layer, the bonding within the sheets is stressed (redrawn from SB 1936, 6)

III. Mixed Compounds of Monovalent and Trivalent Thallium

The structure of TlS (= $Tl^{1+}Tl^{3+}S_2$) was determined by HAHN and KLINGNER (SR 1949, 180) and refined by SCATTURIN and FRASSON (SR 1956, 192). The tetragonal

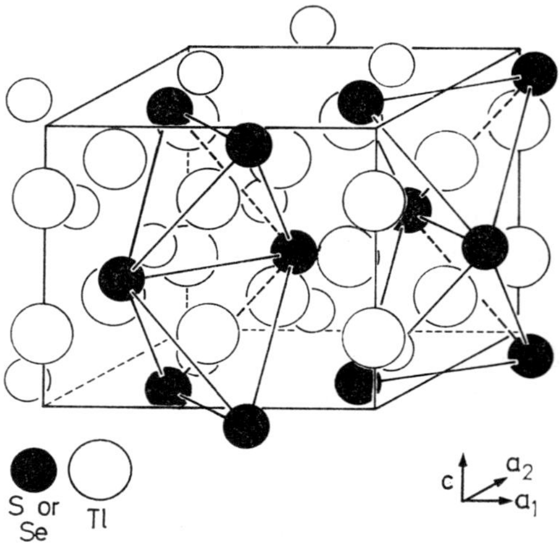

Fig. 81-A-2. Structure of tetragonal TlS and TlSe (redrawn from SB 1939, 6)

structure (Fig. 81-A-2) shows two different Tl-positions: Tl (1) is coordinated by eight sulfur atoms at the corners of a distorted tetragonal antiprism with Tl—S = 3.33 Å (8×); Tl (2) is coordinated by four sulfur atoms forming a slightly distorted tetrahedron with Tl—S = 2.59 Å (4×). From the bond distances, Tl (1) can be considered as monovalent and Tl (2) as trivalent; the short distances Tl (2)—S indicate covalent bonding.

TlSe is isotypic with TlS (Ketelaar *et al.*, SB 1939, 77) with Tl (1)—Se = 3.42 Å (8×) and Tl (2)—Se = 2.68 Å (4×).

TlTe (Burkhardt and Schubert, 1969) has a completely different structure; it is an intermetallic compound with Tl coordinated irregularly by Tl and Te (Tl—Tl from 3.52 to 3.72 Å; Tl—Te from 3.42 to 3.56 Å).

IV. Trivalent Thallium

a) Trivalent Tl Coordinated by Oxygen

The structure of $Tl_2^{[6]}O_3$, avicennite (Goldschmidt *et al.*, SB 1925, 261; Papamantellos, 1967), belongs to the C-type of the rare-earth sesqui-oxides; the Tl atoms are coordinated by six oxygen atoms at an average distance of 2.26 Å.

Low α-$LiTl^{[6]}O_2$ is orthorhombic with Tl—O = 2.13—2.59 Å. Above 560° C, this α-form transforms to an NaCl-type structure with ordered (β-form) or disordered (γ-form) distribution of Li^{1+} and Tl^{3+} ions at the Na sites; the distances Tl—O in β-$LiTlO_2$ can be calculated to be 2.28 Å from the lattice constant (Hoppe and Panek, 1971).

In $Tl^{[6]}(OH)[SO_4] \cdot 2.5\,H_2O$, Johansson (SR 1959, 452) has found chains of TlO_6 octahedra sharing corners (Tl—O = 2.16—2.26 Å), the chains being connected by hydrogen bonds.

b) Trivalent Tl Coordinated by Halides

According to HOPPE and HEBEKER (1965), $NaTlF_4$ has a structure of the CaF_2-type with a statistical distribution of Na and Tl at the Ca positions. The same authors found that $LiTlF_4$ crystallizes in a tetragonal superstructure of the CaF_2 lattice with an ordered distribution of Li and Tl atoms.

$^{2}_{\infty}Tl^{[6]}Cl_3$ is isotypic with $AlCl_3$ (TEMPLETON and CARTER, SR 1954, 352). According to HOARD and GOLDSTEIN (SB 1935, 505—506), the structure of $Cs_3Tl_2Cl_9$ contains $[Tl_2^{[6]}Cl_9]^{3-}$ groups consisting of two $TlCl_6$ octahedra sharing a common face.

Manuscript received: June 1972

References: Section 81-A

BARLOW, M., MEREDITH, C. C.: Crystallographic studies of two polymorphic forms of thallous fluoride, TlF. Z. Krist. **130**, 304 (1969).

BURKHARDT, K., SCHUBERT, K.: Kristallstruktur von TlTe. J. Less-Common Metals **18** (4), 426 (1969).

FAWCETT, J. K., KOCMAN, V., NYBURG, S. C., O'BRIEN, R. J.: The X-ray crystal structure of thallium tetrametaphosphate, $Tl_4P_4O_{12}$. J. Chem. Soc. D **18**, 1213 (1970).

GATTOW, G., ZEMANN, J.: Über Doppelsulfate vom Langbeinit-Typ, $A_2^+ B_2^{2+}(SO_4)_3$. Z. Anorg. Allgem. Chem. **293**, 233 (1958).

HOPPE, H., HEBECKER, CH.: Fluorothallate (III): $NaTlF_4$ und $LiTlF_4$. Z. Anorg. Allgem. Chem. **335**, 85 (1965).

HOPPE, R., PANEK, P.: Zur Kenntnis von α-$LiTlO_2$. Z. Anorg. Allgem. Chem. **384**, 97 (1971).

LEDSHAM, A. H. C., STEEPLE, H.: The classification of the chromium alums. Acta Cryst. **25** B, 398 (1969).

MAN, L. I.: Determination of the structure of Tl_2S by the electron diffraction method. Soviet Phys.-Cryst. **15**, (3), 399 (1970).

MARUMO, F., NOWACKI, W.: The crystal structure of hatchite, $PbTlAgAs_2S_5$. Z. Krist. **125**, 249 (1967).

MEREITER, K.: Der Strukturtyp der Voltaite. Naturwissenschaften **57**, 670 (1970).

OHMASA, M., NOWACKI, W.: The crystal structure of vrbaite, $Hg_3Tl_4As_8Sb_2S_{20}$. Z. Krist. **134**, 360 (1971).

PAPAMANTELLOS, P.: Verfeinerung der Tl_2O_3-Struktur mittels Neutronenbeugung. Z. Krist. **126**, 143 (1967).

PISTORIUS, C. W. F. T.: Crystallographic data for low-$TlBF_4$. Z. Anorg. Allgem. Chem. **376**, 308 (1970).

SABROWSKY, H.: Zur Darstellung und Kristallstruktur von Tl_2O. Z. Anorg. Allgem. Chem. **381**, 266 (1971).

TAKÉUCHI, Y., GHOSE, S., NOWACKI, W.: The crystal structure of hutchinsonite, $(Tl, Pb)_2As_5S_9$. Z. Krist. **121**, 321 (1965).

TAKÉUCHI, Y., OHMASA, M., NOWACKI, W.: The crystal structure of wallisite, $PbTlCu$-As_2S_5, the Cu analog of hatchite, $PbTlAgAs_2S_5$. Z. Krist. **127**, 349 (1968).

ZODKEVITZ, A., MAKOVSKY, J., KALMAN, Z. H.: The preparation and crystal structure of $TlMnCl_3$, $TlFeCl_3$, $TlCoCl_3$ and $TlNiCl_3$. Israel J. Chem. **8**, 755 (1970).

Revised manuscript received: June 1972

81-B. Isotopes in Nature

The two thallium stable isotopes (^{203}Tl and ^{205}Tl) were discovered by SCHÜLER and KEYSTON (1931) and ASTON (1931) and the relative abundances of these isotopes are 29.50% (^{203}Tl) and 70.50% (^{205}Tl) (MORRIS and KILLICK, 1960). OSTIC (1967) gives a value of 2.378±0.005 for the ($^{205}Tl/^{203}Tl$) ratio.

GOLUBCHINA *et al.* (1957) found that the isotopic composition of thallium does not vary with geological processes (Table 81-B-1) and this seems to be confirmed by more recent investigations (ANDERS and STEVENS, 1960; OSTIC, 1967).

Determinations of the isotopic compositions of terrestrial and meteoritic thallium agree to within 1% (ANDERS and STEVENS, 1960; OSTIC, 1967), and therefore it can be assumed that no radiogenic excess of ^{205}Tl is present in meteorites (Table 81-B-2).

The radionuclides ^{204}Tl and ^{206}Tl may be obtained on irradiation of ^{203}Tl and ^{205}Tl with thermal neutrons (MORRIS and KILLICK, 1960) and they decay to ^{204}Hg and ^{204}Pb respectively. The half-life of ^{206}Tl is 4.19 minutes and that of ^{204}Tl is 4.1 years. The latter nuclide, due to its longer life, is used for the radiometric determination of thallium.

Table 81-B-1. *Isotopic composition of thallium in rocks of different age.* (After GOLUBCHINA *et al.*, 1957)

Rock	Locality	Age (m. y.)	$^{205}Tl/^{203}Tl$ [a]
Granite	Zmeinogorsk, Altai	230	2.458 + 0.025
Granite	Zmeinogorsk, Altai	230	2.453 + 0.025
Diorite porphyry (dyke)	Zmeinogorsk, Altai	230	2.448 + 0.025
Granite	Leninogorsk, Altai	230	2.455 + 0.025
Granite	Leninogorsk, Altai	230	2.443 + 0.025
Granite	Kemenskii Massif, Eastern Siberia	1,000	2.452 ± 0.025
Granodiorite	Shakhthtama, Eastern Transbaikalia	150	2.458 + 0.025

[a] The values of this ratio are, apparently, too high, although for the study of the possible variation of the $^{205}Tl/^{203}Tl$ ratio, these figures may be considered without correction.

Table 81-B-2. *Isotopic compositions of terrestrial and meteoritic thallium*

Terrestrial thallium	$^{205}Tl/^{203}Tl$	Reference
Normal thallium	2.390	BAINBRIDGE and NIER (1950) (in ANDERS and STEVENS, 1960)
Normal thallium[a]	2.367 ± 0.020	ANDERS and STEVENS (1960)
Weighed average (7 determinations)	2.378 ± 0.005	OSTIC (1967)
Ivigtut galena	2.380 ± 0.003 (I)	OSTIC (1967)

Meteoritic thallium	Class	$^{205}Tl/^{203}Tl$	Reference
Iron meteorites			
Canyon Diablo (Metal)	Og	2.372 ± 0.003	OSTIC (1967)
Canyon Diablo (Troilite)	Og	2.375 ± 0.002	OSTIC (1967)
Canyon Diablo (Troilite)[b]	Og	2.373 ± 0.020	ANDERS and STEVENS (1960)
Toluca (Troilite)[b]	Om	2.376 ± 0.020	ANDERS and STEVENS (1960)
Chondrites			
Enstatite chondrites:			
Abee[b]	Ce_1	2.387 ± 0.020	ANDERS and STEVENS (1960)
Ordinary chondrites (high-iron group):			
Beardsley[b]	CH	2.387 ± 0.020	ANDERS and STEVENS (1960)
Plainview	CH	2.382 ± 0.003	OSTIC (1967)
Richardton[b]	CH	2.404 ± 0.040	ANDERS and STEVENS (1960)
Carbonaceous chondrites:			
Mighei[b]	Cc_2	2.381 ± 0.030 (I)	ANDERS and STEVENS (1960)

[a] These values are not corrected for mass discrimination in the instrument and therefore the ratio is systematically low.

[b] The values given in this table are the adjusted figures (OSTIC and KOHMAN, 1968) for the difference in mass spectrometers (2.378/2.367).

Revised manuscript received: March 1970

81-C. Abundances in Cosmos, Meteorites, Tektites, Earth and Lunar Samples

I. Cosmos

The estimates of the "cosmic" abundance of thallium are based on the contents of meteoritic and terrestrial matter and therefore are valid only for the solar system.

There is considerable discrepancy between the estimates given by different investigators (Table 81-C-1), due in part to the variable contents of this element in the different types of meteorite, on which these estimates are largely based.

Table 81-C-1. *Atomic abundance of thallium* $(Si = 10^6)$

Tl	^{203}Tl	Reference
0.108		Noddack and Noddack (1934)
0.170		Goldschmidt (1937) (modified by Suess and Urey, 1956)
0.110		Urey (1952)
0.108	0.03190	Suess and Urey (1956)
0.001 [a]	0.00030 [b]	Ehman and Huizenga (1959)
0.00044 [a]	0.00013	Reed *et al.* (1958), in Ehman and Huizenga (1959)
0.182 [c]	0.0537	Cameron (1968)

[a] Values obtained from the abundance of ^{203}Tl ($^{205}Tl/^{203}Tl = 2.378$).
[b] Based on the determinations on five meteorites.
[c] Based on the abundances in Type I carbonaceous chondrites.

II. Meteorites

The thallium contents of meteorites are given in Table 81-C-2 and average abundances or range of variations, according to several investigators, for different groups of meteorites are listed in Table 81-C-3.

In iron meteorites, thallium is concentrated in the troilite phase relative to the metal phase.

Enstatite chondrites and carbonaceous chondrites are enriched in thallium relative to ordinary chondrites and achondrites by factors of 120 and 170 respectively.

Values of Rb/Tl for meteorites vary widely, from about 10 for carbonaceous chondrites to more than 10,000 for some chondrites (Fig. 81-C-1).

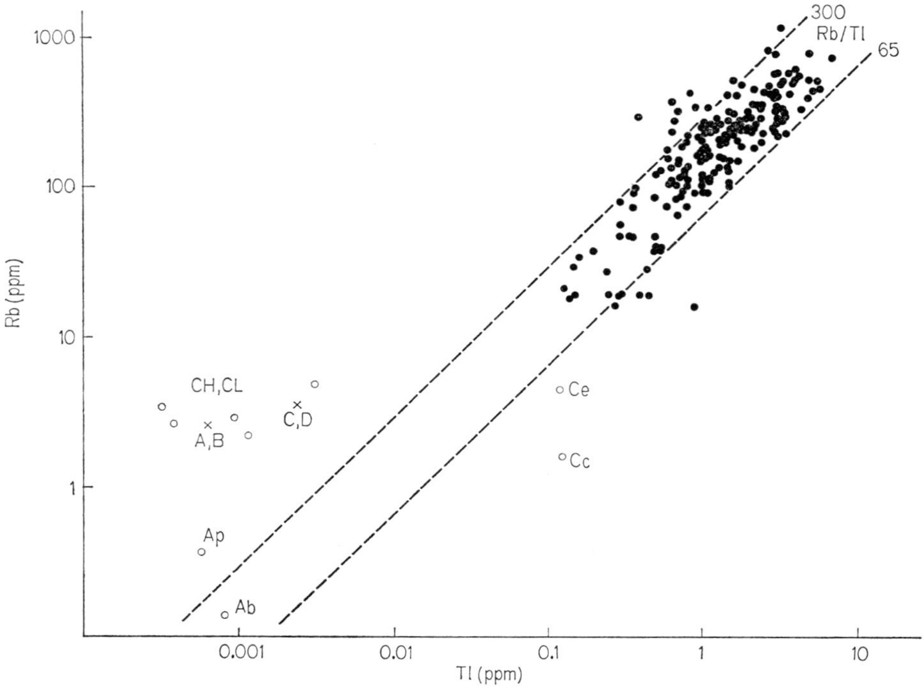

Fig. 81-C-1. *Relationship between thallium and rubidium in igneous rocks, lunar rocks and meteorites.* ○ = *meteorites*, ● = *igneous rocks*, × = *lunar rocks*

Table 81-C-2. *Thallium contents of meteorites*

Meteorite	Class	Tl (ppb)		Reference
Iron meteorites				
Canyon Diablo (Metal)	Og	0.25	(W)	EL-BADRY *et al.* (1960)
Canyon Diablo (Metal)	Og	5.8, 7.8	(N/R)	TANDON (1967)
Canyon Diablo (Troilite)	Og	1.5	(W)	EL-BADRY *et al.* (1960)
Canyon Diablo (Troilite)	Og	74, 65	(N/R)	TANDON (1967)
Canyon Diablo (Troilite)	Og	13, 7.8	(N/R)	REED *et al.* (1960)
Odessa (Metal)	Og	2.4, 0.33	(N/R)	TANDON (1967)
Toluca (Troilite)	Om	198	(N/R)	REED *et al.* (1960)
Campo del Cielo (Metal)	H	0.13, 0.21	(N/R)	TANDON (1967)
Sikhote Alin (Metal)	H	0.48, 0.26	(N/R)	TANDON (1967)
Average metal		1.35		
Average troilite		113		
Achondrites				
Nuevo Laredo	Ap	0.75, 0.57	(N/R)	REED *et al.* (1960)
Johnstown[a]	Ab	0.98, 0.61	(N/R)	EHMAN and HUIZENGA (1959)
Average		0.75		

Table 81-C-2. (Continued)

Meteorite	Class	Tl (ppb)		Reference
Chondrites				
Enstatite chondrites				
Abee	Ce_1	71, 96	(N/R)	REED *et al.* (1960)
Abee	Ce_1	150	(I)	ANDERS and STEVENS (1960)
Indarch	Ce_1	125	(N/R)	REED *et al.* (1960)
Average		120		
Ordinary chondrites				
Beardsley[b]	CH	0.36	(N/R)	REED *et al.* (1960)
Beardsley	CH	5.8	(I)	ANDERS and STEVENS (1960)
Beardsley[a]	CH	1.9, 4.6	(N/R)	EHMAN and HUIZENGA (1959)
Forest City[a]	CH	0.51, 0.35	(N/R)	EHMAN and HUIZENGA (1959)
Forest City	CH	0.51, 0.39, 0.14	(N/R)	REED *et al.* (1960)
Plainview	CH	0.7	(W)	EL-BADRY *et al.* (1960)
Richardton	CH	0.94	(I)	ANDERS and STEVENS (1960)
Average		1.3		
Holbrook	CL	0.45, 0.56, 3.8, 0.42	(N/R)	REED *et al.* (1960)
Holbrook[a]	CL	0.91, 1.18	(N/R)	EHMAN and HUIZENGA (1959)
Modoc[a]	CL	0.32	(N/R)	EHMAN and HUIZENGA (1959)
Modoc	CL	0.83, 0.14, 0.23, 0.03,	(N/R)	REED *et al.* (1960)
Average		0.75		
Carbonaceous chondrites				
Orgueil	Cc_1	141	(N/R)	REED *et al.* (1960)
Mighei	Cc_2	97	(N/R)	REED *et al.* (1960)
Mighei	Cc_2	140	(I)	ANDERS and STEVENS (1960)
Average		130		

[a] The figures given were calculated from the abundances of ^{203}Tl using the ratio $(^{205}Tl/^{203}Tl) = 2.378$.

[b] Low figure probably due to leaching of the meteorite by ground waters (ANDERS and STEVENS, 1961).

III. Tektites

Although the geochemistry of tektites is fairly well known, very few analyses are available which include thallium determinations. It is noteworthy that thallium seems to occur in very small amounts in tektites (Table 81-C-4).

Table 81-C-3. *Average thallium abundances in meteorites (ppb)*

Iron meteorites metal	Iron meteorites troilite	Enstatite chondrites	Ordinary chondrites		Carbonaceous chondrites	Meteorites	Achondrites	Feldspathic achondrites	Reference
—	—	—	—		—	1.35	—	—	EHMAN and HUIZENGA (1959)
—	10—200	70—140	0.4		70—140	—	—	—	REED *et al.* (1960)
—	—	—	1.0		140	—	—	—	TAYLOR (1964b)
—	100[a]	96[b]	1.0		97[c]	—	0.7[d]	0.8[e]	VINOGRADOV (1965)
1.35 (3)	113 (2)	120 (2)	1.3 (4) H-group	0.75 (2) L-group	130 (2)	—	0.75 (2)	—	This work

[a] Based on the value given by NICHIPORUK.
[b] Based on determinations on the Abee meteorite.
[c] Based on determinations on the Mighei meteorite.
[d] Based on determinations on the Johnstown meteorite.
[e] Based on determinations on the Nuevo Laredo meteorite.

Table 81-C-4. *Thallium contents of tektites*

Rock		Tl (ppm)	Locality	Reference
Australite	(S) (5)	0.1	Australia	TAYLOR (1966b)
Henbury impact glass	(S)	0.4	Henbury, Australia	TAYLOR (1966b)

IV. Earth

The thallium abundance in the earth's crust has been estimated by different authors and the results are summarized in Table 81-C-5.

The thallium content of the earth is, according to BROOKS and AHRENS (1961a), 0.004 ppm.

Table 81-C-5. *Thallium abundance in the Earth's crust*

Tl (ppm)	Reference
0.85×10^{-3}	CLARKE and WASHINGTON (1924)
0.1	NODDACK and NODDACK (1930)
0.3	GOLDSCHMIDT (1937)
3.0	AHRENS (1948)
0.6	RANKAMA and SAHAMA (1950)
1.3	SHAW (1952, 1957)
1.7	VINOGRADOV (1956)
1.0	MASON (1958)
0.7[a]	BROOKS and AHRENS (1961a)
1.0[a]	VINOGRADOV (1962)
0.45[b]	TAYLOR (1964a, 1966a)
0.55[b]	TAYLOR and WHITE (1967)
0.8	This paper

[a] Based on the (acid rock/basic rock) ratio of 2:1.
[b] Continental crust.

V. Lunar Samples

Thallium was detected in lunar samples collected by the U.S.A. Apollo 11 astronauts. The average Tl contents of 5 Types A and B samples and 8 Types C and D were determined to be 0.63 and 2.44 ppb (KEAYS *et al.*, 1970) with Rb/Tl ratios of 4,231 and 1,454 respectively.

Revised manuscript received: March 1970

81-D. Abundances in Rock-Forming Minerals and Thallium Minerals

I. Thallium Minerals

The thallium minerals are of very rare occurrence in nature, and are only formed during epithermal stages of hydrothermal processes or under supergene conditions (VLASOV, 1966). These minerals and their chemical composition are listed in Table 81-D-1.

Table 81-D-1. *Thallium minerals*

Mineral	Chemical composition	Paragenesis
Lorandite	$TlAsS_2$	Realgar, orpiment, colloform pyrite
Picotpaulite	$TlFe_2S_3$	
Chalkothallite	Cu_3TlS_2	
Vrbaite	$Hg_3Tl_4As_8Sb_2S_{20}$	Realgar, orpiment, colloform pyrite
Hutchinsonite	$(Pb, Tl)_2(Cu, Ag)As_5S_{10}$	Sulfides etc of As, Zn, Pb, Fe
Bukovite	$Cu_{3+a}Tl_2FeS_{4-a}$	
Wallisite	$PbTlCuAs_2S_5$	
Hatchite	$PbTlAgAs_2S_5$	
Crookesite	$(Cu, Tl, Ag)_2Se$	Selenides
Avicennite	Tl_2O_3	Hematite, calcite

II. Minerals from Ultramafic, Mafic and Intermediate Rocks

Data on the thallium contents of minerals from these rocks are scarce (Table 81-D-2).

Phlogopite from a dunite pipe from Transvaal, South Africa (AHRENS, 1945) contains 3.0 ppm Tl.

The distribution of thallium in minerals from gabbro and ferrodiorite from the Skaergaard Intrusion (Table 81-D-2) shows some irregularities. In the gabbro, thallium is concentrated in magnetite while in the ferrodiorite the highest content of this element is found in olivine. The magnetite from a lamprophyre, associated with a carbonatite complex, is richer in thallium than the augite from the same rock (Table 81-D-2).

Table 81-D-2. *Thallium contents of minerals from mafic rocks*

Mineral	Tl (ppm)				
	1	2	3	4	5
Plagioclase	0.1	0.1	—	—	0.017, 0.029
Olivine	—	0.8	—	—	
Pyroxene	0.5	0.2	0.044	0.1	
Ilmenite	1.6	—	—	—	
Magnetite	2.2	0.05	0.36	—	
Method	(N/R)	(N/R)	(S)	(S)	(S)

1 = Gabbro (MZ), Skaergaard Intrusion (East Greenland) (ADAMS, 1961, in WAGER and BROWN, 1967). 2 = Ferrodiorite (UZ), Skaergaard Intrusion (East Greenland) (ADAMS, 1961, in WAGER and BROWN, 1967). 3 = Lamprophyre, Seabrook Lake, Ontario (SHAW, 1952). 4 = Gabbro Pegmatite, Harzburg, Germany (OTTEMAN, 1940). 5 = Stillwater Complex, Montana, U.S.A. (SHAW, 1952).

III. Minerals from Granitic Rocks

Very few data are available on the thallium distribution in co-existing minerals of granitic rocks (Table 81-D-3). This element is concentrated in biotite, as can be expected from its crystal chemistry (Table 81-D-8). Relatively high amounts of thallium occur in the potassium feldspar and in amphibole and it may be noted that the amount in the latter mineral may exceed that of the co-existing potassium feldspar.

In biotite, thallium occurs generally in the range 3—15 ppm, while the amounts in potassium feldspar and possibly muscovite are mostly in the range 1—6 ppm (Tables 81-D-3; 81-D-8).

IV. Minerals from Alkalic Rocks

The available data for thallium in minerals from alkalic rocks (Table 81-D-4) show that the content of this element in both biotite and potassium feldspar is in the same range as in granitic rocks. Two leucite analyses are included in Table 81-D-8.

V. Minerals from Granite Pegmatites

Data on the thallium contents of potassium feldspar from granites and related pegmatites show a thallium enrichment in the feldspar from the pegmatites (Tables 81-D-3, 6, 7).

Large variations are observed for the thallium contents of potassium feldspars from pegmatites, where it occurs mostly in the range 0.5—50 ppm, although occasionally contents upto 150 ppm have been found. A marked enrichment of thallium is observed in strongly differentiated pegmatites, where contents upto 600 ppm have been reported in the potassium feldspar. Amazonite is also particularly rich in thallium, with amounts in the range 6—100 ppm (Table 81-D-6, 7).

Table 81-D-3. *Thallium contents of minerals from granitic rocks*

Mineral	Method	Tl (ppm) range	Tl (ppm) mean	Rock	Locality	Ref.
Quartz	(?)		0.1	Granodiorite	Almalyk area, UzbekS.S.R.	1
Potassium feldspar	(?)		1.4	Granodiorite	Almalyk area, UzbekS.S.R.	1
	(C)		2.4	Porphyritic adamellite	Susamyr batholith, U.S.S.R.	2
	(S) (2)	4.5—4.8	4.6	Granite	Dzirul Massif, U.S.S.R.	3
	(S) (3)	0.7—1.1	0.9	Granite	E. Transvaal	4
	(S) (70)	0.9—5.0		"Granite"	South Africa	5
	(S) (26)	1.3—30		Granite	S. Norway	6
	(S)		8.2	Farsundite	S. Norway	6
	(S)		1.0	Granite	Langøy, N. Norway	7
Amazonite	(C)		44	Amazonite granite	E. Transbaikalia, U.S.S.R.	8
Albite	(C)		10	Amazonite granite	E. Transbaikalia, U.S.S.R.	8
Oligoclase	(?)		0.7	Granodiorite	Almalyk area, UzbekS.S.R.	1
Biotite	(?)		8.4	Granodiorite	Almalyk area, UzbekS.S.R.	1
	(S) (4)	3.0—3.6	3.2	Biotite quartz-diorite	Dzirul Massif, U.S.S.R.	3
	(S) (3)	3.7—4.3	4.1	Granite (Hercynian)	Dzirul Massif, U.S.S.R.	3
	(S) (2)	3.0—3.8	3.4	Granite (Mesozoic)	Dzirul Massif, U.S.S.R.	3
	(C) (4)	5.0—7.2	6.0	Biotite granite	Tyrny-Auz, U.S.S.R.	9
	(C) (3)	5.2—6.0	—	"Granite"	Susamyr batholith, U.S.S.R.	2
	(N/R) (4)	2.6—12.4	—	Granite	Amo Complex, N. Nigeria	10
	(S) (22)	5—16	—	"Granite"	South Africa	5
Phlogopite	(S)		4.4	Graphic granite	Transvaal	4
Zinnwaldite	(C)		48	Amazonite granite	E. Transbaikalia, U.S.S.R.	8
Muscovite	(S) (5)	1.5—3.7		Granite	South Africa	4,11
Hornblende	(C)		0.8	Porphyritic adamellite	Susamyr batholith, U.S.S.R.	2
	(?)		2.6	Granodiorite	Almalyk area, UzbekS.S.R.	1
Magnetite	(?)		0.4	Granodiorite	Almalyk area, UzbekS.S.R.	1

References: 1. BADALOV and RABINOVICH (1966); 2. TAUSON and BUSAEV (1957); 3. ODIKADZE (1967); 4. AHRENS (1945); 5. KOLBE (1965); 6. HEIER and TAYLOR (1959); 7. HEIER (1960); 8. SITNIN (1960); 9. VOSKRESENSKAYA (1959); 10. BUTLER (1962); 11. KOLBE and TAYLOR (1966b).

Table 81-D-4. *Thallium contents of minerals from the alkalic rocks of the Mt. Sandyk Massif, U.S.S.R.* (According to ZLOBIN, 1958)

Mineral	Tl (ppm)		
	alkaline-earth syenite (Phase I)	hornblende alkalic syenite (Phase II)	nepheline syenite (Phase III)
Orthoclase	1.2	2.5	3.4
Biotite	3.2	7.6	5.3
Hornblende	—	0.7	—
Rock	1.1	2.2	2.7

Table 81-D-5. *Thallium contents of minerals from a differentiated granitic pegmatite, Mongolian Altai.* (According to SOLODOV, 1962)

Mineral	Tl (ppm) (C), pegmatite zones									
	I	II	III	IV	V	VI	VII	VIII	IX	X
Microline	18	10	25	—	—	—	—	—	61	—
Muscovite	—	15	—	21	—	—	—	—	—	—
Gilbertite	—	—	—	—	—	28	22	—	—	—
Lepidolite	—	—	—	—	—	—	—	72	—	—
Beryl	—	4	—	4	6	7	—	—	—	—
Albite	—	4	—	—	—	—	—	—	—	—
Spodumene	—	—	—	—	—	4	—	—	—	—

Zones defined in the pegmatite (from the margin to the centre): I = Graphic granite; II = Fine grained albite; III = Blocky microcline I; IV = Quartz-muscovite; V = Cleavelandite-spodumene; VI = Quartz-spodumene; VII = Fine platy albite; VIII = Scaly muscovite; IX = Blocky microcline II; X = Blocky quartz.

Muscovite is relatively rich in thallium, particularly late muscovite in replacement zones of pegmatites, which may contain upto 280 ppm. The thallium contents of this mineral vary, in general, between 2 and 60 ppm.

Lepidolite, lithium-rich biotite and pollucite contain appreciable amounts of thallium.

Other minerals from pegmatites contain very low amounts of thallium (Table 81-D-5, 7).

VI. Minerals from Syenite Pegmatites

The potassium feldspars from the Palabora pegmatites (E. Transvaal) contain thallium in the range 0.4—1.0 ppm (AHRENS, 1945), which is notably lower than in the same mineral from granitic pegmatites. Similarly, the thallium contents of phlogopite from the same pegmatite are low (<1—1.9 ppm).

VII. Minerals from Alkalic Pegmatites

GERASIMOVSKY and RASSKAZOVA (1962) investigated the distribution of thallium in the minerals from pegmatites related to the alkalic rocks of the Lovozero Massif.

Table 81-D-6. *Thallium contents*

	Tl (ppm)				
	1	2	3	4	5
Microcline	48	84	1.5–3.2	—	0.5—3.1
Amazonite	—	—	19	10—30	—
Albite	—	—	—	3.0	—
Cleavelandite	—	6.8	—	—	—
Muscovite	145	47	2.0—37	6.0—6.7	10
Lepidolite	105—210	60—95	—	95—220	—
Pollucite	—	18	—	105	—
Method	(S)	(S)	(S)	(S)	(S)

Locality: 1. Harding Mine Pegmatite, Dixon, New Mexico (AHRENS, 1948); 2. Custer Mt. Pegmatites, Black Hills, S. Dakota (AHRENS, 1948; SHAW, 1952); 3. Cape Province, South Africa (AHRENS, 1945); 4. Old granite, S. W. Africa (AHRENS, 1945; AHRENS, 1948); 5. Old granite, Natal (AHRENS, 1945); 6. Old granite, E. Transvaal (AHRENS, 1945);

Table 81-D-7. *Thallium contents of minerals from granitic pegmatites*

Mineral	Method	Tl (ppm) range	Tl (ppm) mean	Locality	Ref.
Potassium feldspar	(S) (4)	6.7—7.7	7.2	Small pegmatites, S. Norway	1
	(S) (38)	2—140	19.5	Large pegmatites, S. Norway	1
	(S) (8)	380—610	500	Differentiated vein, Kola Peninsula, U.S.S.R.	2
	(W ?) (4)	5.1—15	8.0	Karaoba Massif, U.S.S.R.	3
	(S) (8)	0.0—16	9.8	Dzirul Massif, U.S.S.R.	4
	(C)		70	Albite pegmatite, Altai, Mongolia	5
	(S) (3)	1.0—1.5	1.2	Bushveld granite	6
	(S) (10)	3.2—20	6.5	Johannesburg, S. Africa	6
Amazonite	(S)		50	Honeydew, Transvaal	6
	(S)		35	Madagascar	7
	(S) (2)		10	Labrador, Ontario	7
	(S)		77	Norway	7
	(S) (2)	6—50	—	Colorado	7
Oligoclase (An_{11})	(S)		1.0	Snarum, Norway	8
Muscovite	(W)		18	Replacement zone, Sayan Mts., U.S.S.R.	9
	(S) (2)	260—280	270	Replacement zone, Kola Peninsula, U.S.S.R.	2
Lepidolite	(S)		18	Old granite, Mozambique	6
	(S) (2)	70—95	—	Connecticut	7

of minerals from several granitic pegmatites

	Tl (ppm)				
	6	7	8	9	10
Microcline	2.6—3.7	—	9—29	5—17	12—200
Amazonite	—	15	33—99	—	—
Albite	0.4	—	0.7—1.1	3	—
Cleavelandite	—	—	1.2—1.4	—	—
Muscovite	3.3—26	2.0—58	—	4—11	82—140
Lepidolite	50—58	—	—	—	170—520
Pollucite	—	—	—	—	160—440
Method	(S)	(S)	(S)	(W)	(S)

7. Namaqualand (Ahrens, 1945); 8. Landsverk, Evje, S. Norway (Taylor *et al.*, 1960); 9. Sayan Mts., U.S.S.R. (Slepnev, 1961); 10. Kola Peninsula, U.S.S.R. (Borovik-Romanova and Sosedko, 1960).

Table 81-D-7. (Continued)

Mineral	Method	Tl (ppm)		Locality	Ref.
		range	mean		
	(S) (2)	175—200	185	Pala, California	7
	(S) (2)		155	Maine	7
	(S) (2)	230—270	250	Brown Derby, Colorado	7
	(S)		95	Copper Mt., Wyoming	7
	(S) (3)	115—165	—	S. E. Manitoba	7
	(S)		95	Lipowka, Urals	7
	(S)		125	Moravia	7
Biotite (lithium-rich)	(S)		80	Strickland Quarry, Connecticut	7
	(S)		385	King's Mt., N. Carolina	7
Zinnwaldite	(S) (2)	62—72	—	Saxony, Erzgebirge (Germany)	6
Beryl	(S)		39	Manjak, Madagascar	7
Rhodizite	(S)		95	Manjak, Madagascar	7
Pollucite	(S) (2)	13—18	—	Greenwood, Norway (Maine)	7
	(S)		77	Varuträsk, Sweden	7
Tourmaline	(W)		4	Sayan Mts., U.S.S.R.	9
Garnet	(W)		4	Sayan Mts., U.S.S.R.	9
Triphylite	(C)		4	Altai, Mongolia	5
Apatite	(S)		0.23	Black Hills, S. Dakota	10

List of References: 1. Heier and Taylor (1959); 2. Borovik-Romanova and Sosedko (1960); 3. Ganeyev and Sechina (1962); 4. Odikadze (1967); 5. Solodov (1962); 6. Ahrens (1945); 7. Ahrens (1948); 8. Heier (1962); 9. Slepnev (1961); 10. Shaw (1952).

Table 81-D-8. *Thallium contents of various rock-forming minerals*

Mineral	Method	Tl (ppm)	Locality	Reference
Zircon	(S)	0.36	Mt. Titanium, Georgia, U.S.A.	SHAW (1952)
Titanite	(S)	0.1	Eganville, Renfrew Co., Ontario	SHAW (1952)
Garnet	(S)	0.019	Adirondacks, New York	SHAW (1952)
Garnet	(S)	0.009	N. Stromfjord, Greenland	SHAW (1952)
Almandine	(S)	0.007	Koirinkoiski, Kitila, K.F.S.S.R.	SHAW (1952)
Beryl	(S)	0.40	Lyndoch Tp., Renfrew Co., Ontario	SHAW (1952)
Tremolite	(S)	0.65	Lee, Massachussets	SHAW (1952)
Biotite G-24	(S)	27.8	Australia	KOLBE (1965)
Biotite G-22	(C)	12.2	Australia	KOLBE (1965)
Biotite F-234	(C)	15.9	Australia	KOLBE (1965)
Lepidolite GMM-P	(C)	83	Australia	KOLBE (1965)
Nepheline	(S)	7.6		SHAW (1952)
Leucite	(S)	65	Civita Castellano, Italy	SHAW (1952)
Leucite	(F)	15	Italy	FORNASERI and PENTA (1963)
Opal	(S)	0.043	Grand Manan Is., New Brunswick	SHAW (1952)
Asbecasite	(C?)	0.90%	Binnatal, Kt. Wallis, Switzerland	GRAESER (1966)

Table 81-D-9. *Thallium contents of potassium feldspars from metamorphic rocks*

Rock (Number of samples)		Tl (ppm) range	Tl (ppm) mean	Locality	Reference
Monzonitic granulite	(4)	tr. – 0.9	0.7	Langøy, N. Norway	HEIER (1960)
Monzonitic and grano-dioritic granulite (banded series)	(2)	0.8—1.0	0.9	Langøy, N. Norway	HEIER (1960)
Monzonitic gneisses (retrograde meta-morphosed)	(2)	0.9—1.0	1.0	Langøy, N. Norway	HEIER (1960)
Granodioritic charnockite	(2)	0.8—0.9	0.9	Langøy, N. Norway	HEIER (1960)
"Granodioritic" gneiss (amphibolite facies)	(5)	0.6—0.9	0.8	Langøy, N. Norway	HEIER (1960)
Augen gneiss	(2)		0.8	Langøy, N. Norway	HEIER (1960)
Granitic-pegmatitic vein in gneiss			0.7	Langøy, N. Norway	HEIER (1960)
Gneiss granite			5.4	S. Norway	HEIER and TAYLOR (1959)
Granitic gneiss			5.8	S. Norway	HEIER and TAYLOR (1959)
Gneiss	(3)	5.3—12	8.2	S. Norway	HEIER and TAYLOR (1959)
Augen gneiss	(2)	1.9—7.9 (S)	4.9	S. Norway	HEIER and TAYLOR (1959)

This element is concentrated in the K—Na feldspar (2.78 ppm), nepheline contains 1.3 ppm while aegirine shows a low content of thallium —0.44 ppm.

HEIER (1966) found that the thallium concentration in potassium feldspar, nepheline and albite from the pegmatites related to the nepheline syenites of the Sternjøy Complex (Northern Norway), to be less than 0.5 ppm.

VIII. Minerals from Metamorphic Rocks

The thallium contents of potassium feldspars from gneisses from the South of Norway (Table 81-D-9) are in the same range as in the same mineral from granitic rocks.

The potassium feldspars from granulite facies rocks of different composition (Table 81-D-9), on the other hand, are particularly low in thallium, with contents of this element not exceeding 1.0 ppm.

Revised manuscript received: March 1970

81-E. Abundance in Common Igneous Rock Types

The results of thallium determinations for some international standard rocks are given in Table 81-E-1 and average thallium contents and atomic abundances in igneous rocks, according to different authors, are summarized in Table 81-E-7.

I. Ultramafic Rocks

The thallium contents of these rocks are typically low, mostly in the range 0.07—0.30 ppm (Table 81-E-2). The mineralogical composition of the rocks has, apparently, very little influence on the thallium content, as dunites, peridotites and pyroxenites show the same range of variation in the amount of this element.

II. Mafic Rocks

The thallium contents of mafic rocks are variable, mostly in the range 0.05—0.7 ppm (Table 81-E-3) and no systematic differences are found between plutonic and volcanic rocks. Dolerites also contain thallium varying in the same range. The contents of anorthosites are lower than those of other basic rocks.

Enrichment of thallium in mafic rocks is observed in the chilled margins of intrusions, e.g. the Skaergaard gabbro and the Jagersfontein Mine dolerite (Brooks and Ahrens, 1961a) (Table 81-E-3).

III. Intermediate Rocks

A definite increase in the thallium contents is observed in these rocks relative to mafic and ultramafic rocks. Most values are in the range 0.15—1.4 ppm and the higher figures are found more frequently in the more acid types (syenite, monzonite and dacite) (Table 81-E-4).

IV. Granitic-Rhyolitic Rocks

1. Granitic Rocks

Most granitic rocks contain thallium in the range 0.6—3.5 ppm (Table 81-E-5) showing that these rocks are enriched in this element relative to the more mafic rocks.

Granitic rocks of different age from South West Africa show enrichment in thallium toward the more modern types (Siedner, 1968). This would be the result of the processes of partial melting of former granitic rocks. The data available for the thallium contents of granitic rocks of different ages do not show that this element has become enriched in the more modern rocks.

Table 81-E-1. *Thallium contents of international standard rocks*

Tl (ppm)											Method	Reference
G-1	W-1	Sy-1	T-1	G-2	GSP-1	AGV-1	PCC-1	DTS-1	BCR-1	Sulphide Ore-1		
1.3	0.05	—	—	—	—	—	—	—	—	—	(W/S)	Brooks *et al.* (1960)
1.3	0.17	—	—	—	—	—	—	—	—	—	(N/R)	Morris and Killick (1960)
1.27	0.21	—	—	—	—	—	—	—	—	—	(W/S)	Brooks and Ahrens (1961b)
1.3	0.10	—	—	—	—	—	—	—	—	—	(N/R)	Vincent and Adams (1961) (*in* Fleischer and Stevens, 1962)
1.5	—	—	—	—	—	—	—	—	—	—	(N/R)	Butler (1962)
0.4	—	—	—	—	—	—	—	—	—	—	(I)	Brown and Wolstenholme (1964)
1.3	0.2	1.8	—	—	—	—	—	—	—	—	(S)	Taylor and Kolbe (1964)
—	—	1.7	—	—	—	—	—	—	—	—	(S)	Kolbe (1965)
1.3	0.14	—	—	—	—	—	—	—	—	—	(I)	Taylor (1965a)
—	0.185	—	—	—	—	—	—	—	—	—	(I)	Taylor (1965b)
1.0	0.1	—	0.5	—	—	—	—	—	—	—	(S)	Schroll and Weninger (1965)
—	—	1.8	—	—	—	—	—	—	—	—	(S)	Webber (1965)
1.2	0.12	1.3	—	—	—	—	—	—	—	1.8	(S)	Tennant (1967)
0.66	<0.05	0.86	—	0.85	0.71	0.27	<0.05	<0.05	0.36	2.90	(P)	Wahler (1968)
—	—	1.4	—	—	—	—	—	—	—	0.2	(S)	Champ (1968) (*in* Sine *et al.*, 1969)
1.3	—	—	—	—	—	—	—	—	—	—	(S)	Champ (1968) (*in* Fleischer, 1969)
—	—	—	—	1.3	1.6	1.6	—	—	—	—	(S)	Champ (1968) (*in* Flanagan, 1969)
1.3	0.13	—	—	—	—	—	—	—	—	—		Recommended value (Fleischer, 1969)

Table 81-E-2. *Thallium contents of ultramafic rocks*

Rock	Method	Tl (ppm) range	Tl (ppm) mean	Locality	Reference
Dunite	(S)		0.11	Khabozero, Kola Peninsula	SHAW (1952)
Dunite	(C)		0.13	Kola Peninsula	VOSKRESENSKAYA (1961)
Serpentinized dunite	(C) (5)		0.07	Cheremshanka, Urals	VOSKRESENSKAYA (1961)
Peridotite	(N/R)		1.04	Insch Mass, Aberdeenshire, Scotland	MORRIS and KILLICK (1960)
Peridotite	(S)		0.13	St. George, Garden Cove, Aleutian Is.	SHAW (1952)
Peridotite	(C) (10)		0.07	Kola Peninsula	VOSKRESENSKAYA (1961)
Serpentinized peridotite	(C) (5)		0.13	Kola Peninsula	VOSKRESENSKAYA (1961)
Lherzolite	(W/S)		0.12	Lac Lherz, Pyrenees, France	BROOKS *et al.* (1960)
Pyroxenite	(C) (10)		0.27	Eastern Sayan Mt., U.S.S.R.	VOSKRESENSKAYA (1961)
Feldspathic pyroxenite	(C) (10)		0.07	Kola Peninsula	VOSKRESENSKAYA (1961)
Serpentinite	(C) (4)	0.3—0.5	0.4	Urusht Complex, Greater Caucasus	VOSKRESENSKAYA (1961)

Table 81-E-3. *Thallium contents of mafic rocks and lamprophyres*

Rock	Method	Tl (ppm) range	Tl (ppm) mean	Locality	Ref.
Plutonic rocks					
Gabbro-norite	(C) (10)		0.13	Northern Karelia, U.S.S.R.	1
Gabbro (composite)	(S) (11)		0.3	Germany	2
Gabbro	(S)		0.16	Sudbury, Ontario	3
Gabbro	(S)		0.058	Duluth, Minnesota	3
Gabbro	(S)		0.060	Moskuvaara, Finland	3
Gabbro	(R)		0.6	Kohyama, Yamaguchi, Japan	4
Gabbro	(C)		0.2	Degtyarka, Urals	1
Gabbro	(W/S)		0.18	Paresis Complex, S. W. Africa	5
Hornblende gabbro	(S)		0.46	Salem, Massachusetts	3
Gabbro-diorite	(C) (5)		0.3	Central Kazakhstan, U.S.S.R.	1
Gabbro-diorite	(C) (4)		0.3	Caucasus, U.S.S.R.	1
Olivine gabbro	(N/R)		0.048	Insch Mass, Scotland	6
Syenogabbro	(N/R)		0.046	Insch Mass, Scotland	6
Olivine gabbro (LZ)	(N/R) (2)		0.03	Skaergaard Intrusion, East Greenland	7

Table 81-E-3. (Continued)

Rock	Method	Tl (ppm) range	Tl (ppm) mean	Locality	Ref.
Olivine-free gabbro (MZ)	(N/R) (2)	0.01—0.02	—	Skaergaard Intrusion, East Greenland	7
Chilled marginal gabbro	(N/R)		0.05	Skaergaard Intrusion, East Greenland	7
Ferrodiorite (UZ)	(N/R) (4)	0.01—0.04	0.025	Skaergaard Intrusion, East Greenland	7
Anorthosite	(S)		0.029	Stillwater Complex, Montana	3
Anorthosite	(S)		tr.	Egersund, Norway	3
Doleritic rocks					
Basalt ("diabase")	(C) (7)		0.22	Northern Caucasus	1
Basalt ("diabase")	(C) (4)	0.2—0.4	0.3	Greater Caucasus	1
Basalt (dolerite sill)	(W/S) (12)	0.02—0.15	0.06	Jagersfontein Mine,	8
(chilled margin)			1.10	Orange Free State,	8
Basalt (dolerite sill)	(W/S) (5)	0.03—0.45	0.12	South Africa	8
Basalt (dolerite)	(W/S)		0.54	Kloof Nek, Cape Town, S. Africa	9
Olivine basalt ("diabase porphyry")	(C) (13)		0.06	Northern Caucasus	1
Olivine basalt ("olivine diabase")	(S)		0.073	Säppi, Luvia, Finland	3
Basalt pegmatite ("diabase-pegmatite")	(S)		0.47	Säppi, Luvia, Finland	3
Volcanic rocks					
Basalt	(S) (3)	0.019—0.19	—	Aleutian Islands	3
Basalt	(R) (3)	0.6—0.9	—	Mt. Mihara, Japan	4
Basalt (Caledonian)	(C) (6)	0.2—0.3	0.2	Urusht Complex, Greater Caucasus	1
Basalt (Hercynian)	(C) (6)	0.2—0.4	0.3	Khudes, Greater Caucasus	1
Basalt (Alpine)	(C) (7)	0.1—0.4	0.22	Mt. Indyuk, Greater Caucasus	1
Basalt	(C) (10)		0.3	Northern Caucasus	1
Amygdaloidal basalt	(W/S)		0.07	Paresis, S. W. Africa	8
Spilite	(C) (9)	0.0—0.3	0.2	Khudes, Urusht, Greater Caucasus	1
Lamprophyres					
Lamprophyre	(S)		0.4	Seabrook Lake, Ontario	3
Lamprophyre	(S)		0.12	Åvo Brändö, Finland	3

List of References: 1. VOSKRESENSKAYA (1961a); 2. PREUSS (1940); 3. SHAW (1952); 4. ISHIMORI and TAKASHIMA (1955); 5. BROOKS and AHRENS (1961b); 6. MORRIS and KILLICK (1960); 7. ADAMS (1961) *in* WAGER and BROWN (1967); 8. BROOKS and AHRENS (1961a); 9. BROOKS *et al.* (1960).

Table 81-E-4. *Thallium contents of intermediate and syenitic rocks*

Rock	Method	Tl (ppm) range	Tl (ppm) mean	Locality	Ref.
Gabbro diorite and diorite	(C) (4)	0.6—0.8	0.7	Susamyr Batholith, U.S.S.R.	1
Diorite	(C) (2)	0.12—0.5	—	Greater Caucasus	2
Diorite	(?) (8)		1.1	Almalyk, UzbekS.S.R.	3
Diorite	(S)		0.3	Jersey City, N. J.	4
Diorite and monzonite	(?)		0.6	Shakhtaminski Massif, U.S.S.R.	5
Syenite-diorite	(C) (8)	0.11—0.4	—	Greater Caucasus	2
Syenodiorite	(?) (14)		1.3	Almalyk, UzbekS.S.R.	3
Monzonite	(S)		1.7	San Juan Co., Colorado	4
Syenite	(S) (2)	1.1—1.3	—	Marquis Tp., Ontario	4
Syenite	(N/R)		0.29	Insch Mass, Scotland	6
Biotite quartz-diorite	(C)		0.5	Teberda River, Greater Caucasus	2
Biotite quartz-diorite	(S) (21)	0.0—9.0	1.7	Dzirul Massif, Georgia, U.S.S.R.	7
Tonalite	(C) (7)	0.5—0.8	0.7	Greater Caucasus	2
Tonalite	(C) (9)	0.2—0.4	0.3	Malaya Laba Massif, Caucasus	8
Tonalite	(S) (2)	0.29—0.50	—	Ontario	4
Volcanic rocks					
Basaltic andesite	(C) (5)	0.2—0.6	0.4	Armenia, U.S.S.R.	2
Basaltic andesite	(C) (2)	0.4—0.5	—	S. Osetiya, Georgia, U.S.S.R.	2
Basaltic andesite	(C) (5)	0.2—0.4	0.3	Elbrus, Greater Caucasus	2
Andesite	(C)		1.1	Elbrus, Greater Caucasus	2
Andesite	(C) (8)	0.0—0.13	0.06	Mt. Indyuk, Greater Caucasus	2
Andesite	(C) (6)	0.0—0.4	0.06	Khudes, Greater Caucasus	2
Andesite	(C)		0.6	Gudzhareti, Georgia, U.S.S.R.	2
Andesite	(R) (8)	0.6—1.4	—	Mt. Asama, Mt. Yake, Japan	9
Andesite (mafic)	(S) (3)	0.14—0.90	0.43	Hahajima, Japan	9
Andesite	(S) (2)	0.16—0.50	0.33	Tongariro National Park, New Zealand	10
Andesite	(S) (2)	0.15—0.27	0.21	Asama volcano, Izu, Japan	10
Andesite-dacite	(C) (19)	0.2—1.1	0.7	Elbrus, Greater Caucasus	2
Andesite-dacite	(C) (5)	0.2—0.8	0.5	Kazbek, U.S.S.R.	2
Andesite-dacite	(C)		0.5	S. Osetiya, U.S.S.R.	2
Dacite	(S)		0.51	St. George, Aleutian Islands	4
Dacite	(R)	1.1—1.4	1.3	Japan	9
Dacite pumice tuff	(C)		1.8	Elbrus, Greater Caucasus	2
Trachyte	(C) (5)	0.13—0.33	0.23	Mt. Indyuk, Greater Caucasus	2
Keratophyres					
Keratophyre	(C) (4)		0.5	Urusht, Greater Caucasus	2
Keratophyre	(C) (2)		0.3	Buron, Northern Caucasus	2

List of References: 1. TAUSON and BUSAEV (1957); 2. VOSKRESENSKAYA (1961a); 3. BADALOV and RABINOVICH (1966); 4. SHAW (1952); 5. TAUSON (1968); 6. MORRIS and KILLICK (1960); 7. ODIKADZE (1967); 8. DEMIN and KHITAROV (1958); 9. ISHIMORI and TAKASHIMA (1955); 10. TAYLOR and WHITE (1967).

Table 81-E-5. *Thallium contents of granitic and rhyolitic rocks*

Rock	Method	Tl (ppm) range	Tl (ppm) mean	Locality	Ref.
Granitic rocks					
Granite (Precambrian)	(S) (6)	0.4—6.4	2.9	Finland	1
Rapakivi granite	(S) (3)	2.8—4.0	3.3	Finland	1
Rapakivi granite	(C) (4)	1.4—1.7	—	Ukraine, Karelia, Dresden	2
Granite	(S) (7)	0.36—0.77	0.61	Huab, S. W. Africa	3
Granite	(S) (8)	0.95—1.60	1.22	Damara, S. W. Africa	3
Granite	(S) (34)	1.1—3.1	1.8	Cape, South Africa	4
Gneissic granite	(S) (4)	1.0—1.7	1.3	Snowy Mountains, Australia	4
Granodiorite	(S) (20)	0.6—1.6	1.1	Snowy Mountains, Australia	4
Leucogranite	(S) (8)	1.5—3.2	2.1	Snowy Mountains, Australia	4
Granite	(S) (4)	1.0—2.1	1.5	Harz, Germany	5
Biotite granite	(N/R) (3)	2.7—4.0	3.2	Cornwall, S. W. England	6
Muscovite-biotite granite	(N/R)		6.9	Cornwall, S. W. England	6
Granite	(W?) (5)	3.0—5.8	—	Karaoba, C. Kazakhstan	7
Aplite	(W?)		5.9	Karaoba, C. Kazakhstan	7
Granite	(C) (2)	0.6—0.7	0.6	Greater Caucasus	2
Alaskite	(C)		0.3	Greater Caucasus	2
Microcline granite	(C) (17)	0.7—1.7	1.1	Greater Caucasus	2
Two-mica granite	(C) (38)	0.5—4.0	1.4	Greater Caucasus	2
Biotite granite	(C) (14)	0.7—1.3	0.9	Greater Caucasus	2
Porphyritic granite	(C) (87)	0.5—2.5	1.2	Greater Caucasus	2
Alaskite	(C) (3)	0.6—0.9	0.8	Greater Caucasus	2
"Granodiorite"	(R) (9)	0.4—1.1	0.9	Japan	8
"Granodiorite" (Mesozoic)	(C) (4)	0.3—1.0	—	Greater Caucasus	2
Leucocratic granite	(C) (13)	0.1—0.4	0.2	Tyrny-Auz, Greater Caucasus	2
Porphyritic biotite granite	(C) (15)	0.9—1.8	1.3	Tyrny-Auz, Greater Caucasus	2
Granite	(R) (4)	0.6—1.3	1.0	Japan	8
Granodiorite	(R) (4)	0.9—2.0	1.4	Japan	8
Mesozoic granite	(S) (13)	0.8—3.6	1.99	S. W. Africa	3
Granodiorite and granite	(C) (18)	1.2—1.8	1.3	Susamyr Batholith, U.S.S.R.	9
Leucocratic granite	(C) (8)	1.9—2.8	2.5	Susamyr Batholith, U.S.S.R.	9
Aplite	(C) (4)	4.2—6.3	4.7	Susamyr Batholith, U.S.S.R.	9
Granite	(C) (15)	0.8—2.4	1.4	Musht Complex, N. Caucasus	10
Aplite	(C) (6)	1.7—2.3	2.0	Musht Complex, N. Caucasus	10
Granite	(S) (28)	0.0—4.0	2.4	Dzirul Massif, Georgia, U.S.S.R.	11
Aplite	(S) (8)	0.0—5.0	1.1	Dzirul Massif, Georgia U.S.S.R.	11
Granite	(C) (10)		0.6	Malaya Laba Massif, Caucasus	12

Table 81-E-5. (Continued)

Rock	Method	Tl (ppm) range	Tl (ppm) mean	Locality	Ref.
Alaskite	(C) (2)		0.7	Malaya Laba Massif, Caucasus	12
Granodiorite	(?)		1.3	Shakhtaminski, E. Transbaikaliya	13
Granite	(?)		1.9	Shakhtaminski, E. Transbaikaliya	13
Leucocratic granite	(C) (6)	0.15—0.5	0.3	Turgoyak Pluton, Urals	14
Biotite granite	(C) (12)	0.5—1.0	0.8	Turgoyak Pluton, Urals	14
Granodiorite	(C) (9)	0.25—0.6	0.7	Turgoyak Pluton, Urals	14
Biotite granite	(N/R) (16)	3.0—4.9	3.9	Perak, Malaya	6
Biotite granite	(C) (2)		0.8	Baksan, U.S.S.R.	15
Muscovite-biotite granite	(C) (4)	0.9—1.3	1.1	Baksan, U.S.S.R.	15
Granophyre	(N/R) (2)	0.25—0.35	—	Skaergaard Intrusion, E. Greenland	16
Volcanic rocks					
Rhyolite	(R) (2)	1.0—1.3	—	Japan	8
Rhyolite-dacite	(C) (5)	0.10—0.25	0.16	Mt. Indyuk, Caucasus	2
Quartz porphyry	(C) (4)	0.13—0.20	0.17	Mt. Indyuk, Caucasus	2
Rhyolite-dacite	(C) (6)	0.6—1.0	0.8	Elbrus, Caucasus	2
Rhyolite	(C) (25)	0.6—1.7	1.3	Elbrus, Caucasus	2
Obsidian	(R)		1.2	Mt. Asama, Japan	8
Obsidian	(S)		1.4	Lake Co., Oregon	1
Obsidian	(S)		5.6	Millard Co., Utah	1
Vitrophyre	(C)		2.2	Elbrus, Caucasus	2
Rhyolite and ignimbrite	(S)	0.7—1.6	1.1	Taupo, New Zealand	17
Hypabyssal rocks					
Granodiorite porphyry	(?) (24)		1.2	Almalyk, UzbekS.S.R.	18
Granite porphyry	(R)		1.6	Miyazaki, Japan	8
Granite porphyry	(C) (6)	0.7—0.9	0.8	Greater Caucasus	2
Granite porphyry	(C) (38)		5.3	Northern Caucasus	2, 19
Granosyenite porphyry	(C) (96)		3.2	Northern Caucasus	2, 19
Quartz-syenite porphyry	(C) (40)		3.2	Northern Caucasus	2, 19
Quartz porphyry	(C) (6)	1.0—1.8	1.3	Malaya Laba, Greater Caucasus	2
Quartz albitophyre	(C) (4)	0.7—0.9	0.8	Malaya Laba, Greater Caucasus	2
Quartz albitophyre	(C) (2)	0.7—1.2	1.0	Kvaisa, Greater Caucasus	2
Quartz albitophyre	(C)		1.2	Kvaisinisk, U.S.S.R.	15

List of References: 1. Shaw (1952); 2. Voskresenskaya (1961a); 3. Siedner (1968); 4. Kolbe and Taylor (1966b); 5. Otteman (1940); 6. Butler (1962); 7. Ganeyev and Sechina (1962); 8. Ishimori and Takashima (1955); 9. Tauson and Busaev (1957); 10. Voskresenskaya and Su Shou-T'ien (1961); 11. Odikadze (1967); 12. Demin and Khitarov (1958); 13. Tauson (1968); 14. Kogarko (1959); 15. Voskresenskaya (1959); 16. Adams (1961) (*in* Wager and Brown, 1967); 17. Ewart *et al.* (1968); 18. Badalov and Rabinovich (1966); 19. Voskresenskaya *et al.* (1962b).

Table 81-E-6. *Thallium contents of alkalic rocks and carbonatites*

Rock	Method	Tl (ppm) range	mean	Locality	Ref.
Plutonic rocks					
Ijolite	(S)		1.4	Iivaara, Finland	1
Mariupolite	(C)		1.0	Zhdanov, Ukraine	2
Nepheline syenite (phase I)	(C) (5)	0.80—1.40	1.18	Lovozero Massif, U.S.S.R.	3, 4
Urtite, foyaite, lujavrite (phase II)	(C) (15)	0.53—1.45	0.88	Lovozero Massif, U.S.S.R.	3, 4
Lujavrite, melteijite, juvite, syenite (phase III)	(C) (14)	0.55—1.70	0.66	Lovozero Massif, U.S.S.R.	3, 4
Monchiquite (phase IV)	(C)		0.55	Lovozero Massif, U.S.S.R.	3, 4
"Monzonite" (phase I)	(C) (3)	1.0—1.3	—	Mt. Sandyk Massif, U.S.S.R.	5, 6
Syenite (phase I)	(C) (10)	1.0—2.0	—	Mt. Sandyk Massif, U.S.S.R.	5, 6
Alkali syenite (phase II)	(C) (6)	1.6—2.3	2.0	Mt. Sandyk Massif, U.S.S.R.	5, 6
Nepheline syenite (phase II)	(C) (5)	1.5—4.4	2.9	Mt. Sandyk Massif, U.S.S.R.	5, 6
Urtite	(C)	0.40—0.6	—	Khibina Massif, U.S.S.R.	3, 6
Foyaite	(C) (2)	0.45—0.8	—	Khibina Massif, U.S.S.R.	3, 6
Khibinite	(C) (4)	0.45—0.95	—	Khibina Massif, U.S.S.R.	3, 6
Rischorrite	(C)		0.66	Khibina Massif, U.S.S.R.	3, 6
Miaskite	(C)		0.5	Sakhariok Massif, U.S.S.R.	3, 6
Ijolite-urtite	(C)		1.0	Osernaya Varaka Massif, U.S.S.R.	3, 6
Nepheline syenite	(S)		2.2	Methuen Tp., Ontario	1
Nepheline syenite	(C) (5)		0.6	Vishnevyye Gory, Urals	2
Nepheline syenite	(C) (5)		0.2	Keivy, Kola Peninsula	2
Nepheline syenite	(C) (25)		1.2	Yenisei Range, U.S.S.R.	2
Nepheline syenite	(C) (10)		2.7	Central Kazakhstan, U.S.S.R.	2
Nepheline syenite	(W/S)		0.59	Paresis Complex, S.W. Africa	7
Foyaite	(W/S) (2)	0.06—0.11	0.09	Paresis Complex, S.W. Africa	8
Syenite	(W/S) (3)	0.06—0.22	—	Ondura Korume Complex, S.W. Africa	8
Ekerite	(S) (2)	1.7—1.8	—	Oslo Series, Norway	9
"Granite"	(N/R) (9)	1.3—5.0	3.0	Liruei Complex, Nigeria	10
"Granite"	(N/R) (14)	0.7—3.6	1.7	Amo Complex, Nigeria	10
"Granite"	(N/R) (6)	1.1—2.2	1.8	Shere, Northern Nigeria	10
Granite	(N/R) (4)	2.8—3.1	3.0	Ropp, Northern Nigeria	10
Granite	(N/R) (3)	1.5—1.6	1.6	Kaleri, Northern Nigeria	10

Table 81-E-6. (Continued)

Rock	Method	Tl (ppm) range	mean	Locality	Ref.
Volcanic rocks					
Leucitite	(F)		5.2	Capo di Bove, Rome	11
Phonolite	(S)		7.2	Cripple Creek, Colorado	1
Rhyolite	(N/R) (6)	1.5—2.5	2.0	Liruei Complex, Nigeria	10
Rhyolite	(W/S)		0.05	Paresis Complex, S. W. Africa	8
Riebeckite rhyolite	(W/S)		0.89	Paresis Complex, S. W. Africa	8
Na-rhyolite	(W/S)		0.34	Paresis Complex, S. W. Africa	8
Pantellerite	(S)		0.93	Mayor Island	12
Tephrite	(S)		0.16	Rheinprussen	1
Leucitophyre	(C)		0.3	Armenia	2
Leucite-basalt	(S)		0.26	Rheinpreussen	1
Carbonatites					
Micaceous carbonatite	(W/S)		0.19	Otjiwarongo, S. W. Africa	8
Dolomitic carbonatite	(W/S)		0.03	Otjiwarongo, S. W. Africa	8

List of References: 1. Shaw (1952); 2. Voskresenskaya (1961a); 3. Gerasimovskii and Rasskazova (1962); 4. Gerasimovskii *et al.* (1968). 5. Zlobin (1958); 6. Zlobin and Lebedev (1960); 7. Brooks and Ahrens (1961b); 8. Brooks and Ahrens (1961a); 9. Dietrich *et al.* (1965); 10. Butler (1962); 11. Fornaseri and Penta (1963); 12. Ewart *et al.* (1968).

2. Volcanic and Hypabyssal Rocks

Much less information exists for these rocks (Table 81-E-5). Rhyolites and obsidians show contents of thallium mostly in the range 1.0—1.6 ppm, similar to those of granitic rocks.

Quartz-syenite and granosyenite porphyries of the Caucasus show similar contents of thallium, while granite porphyries from the same region are richer in thallium. It may be added that these rocks are relatively rich in thallium, in particular those of Neogene age.

V. Alkalic Rocks

The thallium contents of alkalic rocks are generally low (Table 81-E-6), although nepheline syenites, phonolites and leucitites may contain upto 2.7, 7.2 and 5.2 ppm respectively of this element.

VI. Series of Differentiation

The evidence on the variation of the thallium contents in series of differentiation is inconclusive.

Rocks ranging from gabbro-diorite or diorite to granite may show higher thallium contents in the later differentiates (e.g. Susamyr Batholith, U.S.S.R.,

Tables 81-E-4, 81-E-5). The same can be observed in series of alkalic rocks (Mt. Sandyk Massif, Northern Kirgiziya, U.S.S.R., Table 81-E-6).

On the other hand, the alkalic rocks of the Lovozero Massif, Kola Peninsula, show a different distribution of thallium, as the amount of this element decreases towards the late differentiates (Table 81-E-6).

No trends are apparent for the variation of the thallium contents of the rocks of the Ring Complexes of the Younger Granites of Northern Nigeria (BUTLER, 1962).

An important role on the distribution of thallium may be played by the vapour phase during the crystallization of the magmas, as pointed out by DIETRICH *et al.* (1965) and TAUSON (1968). This may, in part, explain the differences on the distribution of thallium in different series of differentiation.

The geochemistry of thallium is dominated by a close coherence with potassium and rubidium. In igneous rocks the values of K/Tl and Rb/Tl fall in the ranges 4,000—112,000 and 19—590 (see Fig. 81-C-1), with averages close to 30,000 and 150 respectively.

Table 81-E-7. *Average thallium contents and atomic abundance of thallium in igneous rocks*

Ultra-mafic rocks	Mafic rocks	Inter-mediate rocks	Granitic-rhyolitic rocks	Syenite	Alkalic rocks	Reference
Thallium contents (ppm)						
0.06	0.12	0.15	3.2	1.4	—	SHAW (1952, 1957)
0.06	0.21	—	0.72 (high-Ca) 2.3 (low-Ca)	1.4	—	TUREKIAN and WEDEPOHL (1961)
—	0.11	—	0.73	—	—	BROOKS and AHRENS (1961a)
0.01	0.2	0.5	1.5	—	—	VINOGRADOV (1962)
0.05	0.1	—	0.75	0.3	—	TAYLOR (1964a, 1966a)
0.1	0.27	0.83	1.6	—	1.5	IVANOV (1966)
—	0.1	—	1.0	—	—	TAYLOR and WHITE (1967)
0.05	0.18	0.55	1.7	—	1.2	This paper
Atomic abundance (atoms /10^6 Si atoms)						
0.0072	0.114	0.264	0.636	—	—	VINOGRADOV (1962)
0.033	0.11	0.30	0.73	—	0.73	This paper

Revised manuscript received: March 1970

81-F. Behavior in Magmatogenic Processes

I. Metasomatic and Hydrothermal Processes

Enrichment of thallium is generally observed in rocks affected by metasomatic processes.

Potassium metasomatism is believed to be responsible for the high thallium content of the amazonite granite of the Etyka Massif (U.S.S.R.) (Sitnin, 1960), which contains 13 ppm of this element.

Table 81-F-1. *Thallium contents of rocks modified by metasomatic or hydrothermal processes*

Rock	Method	Tl (ppm)		Locality	Ref.
		range	mean		
Two-mica granite	(C) (7)	0.5—1.2	0.9	Kuban River, Ullu-Kam, Greater Caucasus	1
Greisenized granite	(C) (5)	1.1—2.0	1.5	Kuban River, Ullu-Kam, Greater Caucasus	1
Amazonite granite	(C)		13	Etyka Massif, U.S.S.R.	2
Greisenized contact rocks	(C) (3)	17—26		Etyka Massif, U.S.S.R.	2
Siltstone, 200 m from contact	(C)		5	Etyka Massif, U.S.S.R.	2
Leucocratic granite (unaltered)	(C) (4)	1.4—1.7	1.5	Musht, N. Caucasus	3
Leucocratic granite (fault zone)	(C) (5)	2.0—4.0	3.2	Musht, N. Caucasus	3
Silicified and microclinized granite	(C) (3)	1.2—1.8	1.4	Musht, N. Caucasus	3
Same, fault zone	(C) (2)	2.1—2.9	2.5	Musht, N. Caucasus	3
Tourmalinized coarse-grained granite	(N/R) (2)	3.7—5.2	—	Cornwall, S. W. England	4
Granite porphyry (unaltered)	(C) (6)		0.27	Tyrny-Auz, Greater Caucasus	5
Same, hydrothermally altered	(C) (2)		0.90	Tyrny-Auz, Greater Caucasus	5
Granosyenite porphyry (unaltered)	(C) (81)		3.3	Northern Caucasus	5
Same, opalized, kaolinized	(C) (5)		3.7	Northern Caucasus	5
Same, pyritized, limonitized	(C) (5)		9.8	Northern Caucasus	5
Quartz albitophyre (unaltered)	(C)		1.2	Kvaisinsk, Greater Caucasus	5
Same, albitized, sericitized	(C)		0.7	Kvaisinsk, Greater Caucasus	5

List of References: 1. Voskresenskaya (1961a); 2. Sitnin (1960); 3. Voskresenskaya and Su Shou T'ien (1961); 4. Butler (1962); 5. Voskresenskaya (1959).

The Musht granites do not show much variation in their thallium contents, even when the silicified and microclinized rocks are taken into account (Table 81-F-1), but a marked enrichment of thallium is observed in rocks occurring in the fault zones.

Hydrothermal alteration frequently leads to thallium enrichment. The kaolinized and opalized granosyenite prophyries of the Greater Caucasus (VOSKRESENSKAYA

Table 81-F-2. *Thallium contents of rocks modified by contact-metamorphic and metasomatic processes*

Rock	Method	Tl (ppm)		Locality	Ref.
		range	mean		
Limestone (beyond the boundaries of the Lachin-Khana ore deposit)	(C) (10)		n.f	Ugam Range, Western Tien Shan	1
Limestone (deposit area)	(C) (8)		3	Ugam Range, Western Tien Shan	1
Marble (near the ore bodies)	(C) (12)		6	Ugam Range, Western Tien Shan	1
Porphyritic biotite granite	(C) (14)		1.2	Tyrny-Auz, Greater Caucasus	2
Same, at contact with biotite hornfels	(C)		1.8	Tyrny-Auz, Greater Caucasus	2
Same, at limestone contact	(C) (3)	0.7—0.9	0.8	Tyrny-Auz, Greater Caucasus	2
Limestone	(C)		0.4	Tyrny-Auz, Greater Caucasus	2
Marble	(C)		0.34	Tyrny-Auz, Greater Caucasus	2
Granite porphyry	(C) (38)		5.3	Northern Caucasus	3
Same, contact with marls	(C) (4)		6.2	Northern Caucasus	3
Marl (contact with granite prophyry)	(C) (3)		8.8	Northern Caucasus	3
Granosyenite porphyry	(C) (81)		3.3	Northern Caucasus	3
Same, at contact with clay, marl, argillite	(C) (11)		5.5	Northern Caucasus	3
Same, at contact with carbonate rocks	(C) (4)		4.4	Northern Caucasus	3
Maikop clay (at contact)	(C)		5.6	Northern Caucasus	3
Same, unaltered	(C) (2)		0.7	Northern Caucasus	3
Argillaceous marl (at contact)	(C) (4)		10	Northern Caucasus	3
Same, unaltered	(C) (2)		0.8	Northern Caucasus	3
Hornfels (sandstone)	(C)		1.6	Northern Caucasus	3
Sandstone (unaltered)	(C) (2)		1.2	Northern Caucasus	3
Quartz syenite porphyry	(C) (27)		3.0	Northern Caucasus	3
Same, at contact with marls (argillaceous)	(C) (2)		4.5	Northern Caucasus	3
Same, at contact with limestone	(C) (2)		2.7	Northern Caucasus	3
Skarn (marl)	(C) (7)		7.2	Northern Caucasus	3
Skarn (limestone)	(C) (5)		2.8	Northern Caucasus	3

List of References: 1. DUNIN-BARKOVSKAYA (1961); 2. VOSKRESENSKAYA (1959); 3. VOSKRESENSKAYA *et al.* (1962b).

et al., 1962b) contain slightly higher amounts of thallium than the unaltered rocks, while a three fold enrichment of this element is observed in the pyritized and limonitized porphyries from the same region.

On the other hand, during albitization thallium may be removed. The albitized and sericitized quartz albitophyre of the Kvaisinsk Deposit region shows a decrease in thallium content relative to the unaltered rock (Table 81-F-1).

II. Contact Metamorphic and Metasomatic Processes

Enrichment of thallium in the country rocks surrounding intrusions of magmatic rocks and hydrothermal mineral deposits has been observed in different regions (Table 81-F-2 and Table 81-K-4).

The magmatic rocks themselves may also have higher thallium contents near the contact, although this enrichment is not so marked as in the country rock.

In several instances, thallium enrichment is related to the formation of greisen, and micas seem to be the main thallium carriers.

The mineralogical assemblage of the country rock is, apparently, an important factor conditioning the distribution of thallium. Limestones show a much more marked enrichment of thallium than sandstones or quartzites, if these rocks are not converted to greisen.

Table 81-F-3. *Thallium contents of the ores of the Lachin-Khana polymetallic deposit* (after DUNIN-BARKOVSKAYA, 1961) (Tl ppm, C, S)

Mineral	Stage I	Stage II	Stage III	Stage IV	Stage V
Marcasite	(5) 170	(16) 2,110	—	(4) 40	—
Pyrite	—	(3) 350	(6) 2	(5) 34	—
Sphalerite	—	(6) 55	—	—	(2) 6
Galena	—	—	—	(7) 31	—
Tetrahedrite	—	—	—	(3) 2	—

Stage I: Chalcedony-Iron Sulfide. Stage II: Iron Sulfide-Sphalerite. Stage III: Pyrite-Fluorite. Stage IV: Barite-Quartz. Stage V: Sphalerite.

Table 81-F-4. *Thallium contents of ore minerals from the Musht Region polymetallic deposit, Caucasus.* (According to VOSKRESENSKAYA, 1961b)

Mineral		Tl (ppm) (C) range	mean
Sphalerite			
1st generation	(2)	1.3—1.6	1.5
2nd generation	(2)	1.3—2.2	1.9
3rd generation	(3)	4.0—8.0	5.7
Galena	(17)	0.4—1.6	0.95
Pyrite	(3)	0.8—2.0	1.4
Marcasite			
Radiated	(2)	22—80	50
Colloform	(7)	170—300	220
From oxidised zone	(2)	10—20	15
Siderite			19.5

Table 81-F-5. *Thallium contents in ores from mineral deposits*

Mineral	Method	Tl (ppm)		Deposit	Ref.
		range	mean		
Sulfides					
Galena	(S)		0.27	Sakkijarvi, Finland	2
Galena	(?) (19)	1.0—7.8	3.2	Verkhnyaya Kvaisa, Caucasus	5
Galena	(S) (29)		53	V. Kvaisa, G. Caucasus	8
Galena					
Coarse-grained	(C) (3)	0.0—2.6	0.9	Zolotoi Kurgan polymetallic deposit, Caucasus	6
Fine-grained	(C) (4)	74—112	51	Zolotoi Kurgan polymetallic deposit, Caucasus	6
Galena	(S) (18)	< 1—150	32	Lead-zinc deposit (Replacement deposit), Kootenay, B.C.	12
Galena	(S) (23)		< 5	Lead-zinc deposit (Fissure deposit), Kootenay, B.C.	12
Galena	(W) (7)	10—5,000	—	Polymetallic deposits, E. Transbaikaliya, U.S.S.R.	10
Galena	(W) (15)		33	Polymetallic deposits (in carbonate rocks), E. Transbaikaliya, U.S.S.R.	14
Galena	(W) (30)		7	Polymetallic deposits (aluminosilicate gangue minerals), E. Transbaikaliya, U.S.S.R.	14
Galena	(C) (37)	0.0—16	—	Polymetallic deposits, Caucasus, U.S.S.R.	6
Galena	(C) (5)	14.5—38	22.5	Buron iron sulphide deposit, Caucasus, U.S.S.R.	6
Sphalerite	(C) (24)	0.0—16	—	Polymetallic deposits, Caucasus, U.S.S.R.	6
Sphalerite	(C) (2)	3.4—8.0	5.7	Urup iron sulphide deposit, Caucasus, U.S.S.R.	6
Sphalerite	(?) (5)	100—1,000	—	Belgium, Westphalia, Silesia	3
Sphalerite	(W)		700	Hanacka, Akita Pref., Japan	1
Sphalerite	(S) (38)		19	V. Kvaisa, G. Caucasus	8
Sphalerite	(S)		9.3	Sukhumi, Dusheti region, Caucasus	8
Sphalerite	(?) (10)	1.6—9.0	5.8	Verkhnyaya Kvaisa, Caucasus	5
Pyrite	(W)		70	Hanaoka, Akita Pref., Japan	1
Pyrite (Colloform)	(S)		210	Sukhumi, Dusheti region, Caucasus	8
Pyrite	(C) (6)	0.6—3.0	—	Polymetallic deposits, Caucasus, U.S.S.R.	6
Pyrite	(C) (9)	0.0—2.0	—	Iron sulfide deposits, Caucasus, U.S.S.R.	6

Table 81-F-5. (Continued)

Mineral	Method	Tl (ppm) range	mean	Deposit	Ref.
Pyrite	(C) (1)		3.0	Iron sulfide-polymetallic deposit, Khudes, Caucasus	6
Pyrite	(C) (2)		—	Iron sulfide-polymetallic deposit, Beskes, Caucasus	6
Pyrite	(C) (1)		0.6	Mercury-arsenic deposit, Tsess, Caucasus	6
Marcasite	(W)		1,250	Hanaoka, Akita Pref., Japan	1
Marcasite	(W)		850	Mori, Yamagata Pref., Japan	1
Marcasite	(S)		0.24	Veules, Normandy	2
Marcasite	(C) (3)	56—64	59	Verkhnyaya Kvaisa, Caucasus	5
Marcasite	(C) (3)	56—64	60	E. Kvaisa polymetallic deposit, Caucasus, U.S.S.R.	6
Marcasite[a]	(S) (14)		55	V. Kvaisa, G. Caucasus	8
Marcasite[a]	(S) (10)		140	Khvamli, G. Caucasus	8
Marcasite[a]	(S) (24)		450	Tkhmori, G. Caucasus	8
Marcasite	(?)	1,000—9,000	—	Daraiso lead-zinc deposit, U.S.S.R.	9
Pyrrhotite	(S)		0.074	Creighton Mine, Sudbury, Ontario	2
Pyrrhotite	(C) (5)		—	Polymetallic and iron sulfide-polymetallic deposits, Caucasus	6
Chalcopyrite	(S)		0.028	Creighton Mine, Sudbury, Ontario	2
Chalcopyrite	(C) (2)		0.8	Sadon polymetallic deposit, Caucasus, U.S.S.R.	6
Chalcopyrite	(C) (3)		—	Iron sulfide deposits, Caucasus, U.S.S.R.	6
Molybdenite	(S)		0.008	Temagami, Ontario	2
Cinnabar	(C)		—	Mercury-arsenic deposit, Akhei, Caucasus	6
Cinnabar	(C)		0.4	Mercury-arsenic deposit, Tibs, Caucasus	6
Stibnite	(C)		1.8	Mercury-arsenic deposit, Tibs, Caucasus	6
Stibnite	(C) (2)	0.2—0.3	0.25	Ferberite-stibnite deposit, Zopikhto, Caucasus	6
Ferberite	(C)		—	Ferberite-stibnite deposit, Notsaro, Caucasus	6
Berthierite	(C)		10	Ferberite-stibnite deposit, Zopikhto, Caucasus	6
Realgar	(C)		—	Mercury-arsenic deposit, Kodis-Dziri, Caucasus	6
Enargite	(W)		560	Shimokita, Aomon Pref., Japan	1
Geocrinite	(?)	100—5,000		Eastern Transbaikaliya, U.S.S.R.	10

[a] Contains pyrite as impurity.

Table 81-F-5. (Continued)

Mineral	Method	Tl (ppm) range	Tl (ppm) mean	Deposit	Ref.
Meneghinite	(?)	100—1,000		Eastern Transbaikaliya, U.S.S.R.	10
Boulangerite	(?)	10—1,000		Eastern Transbaikaliya, U.S.S.R.	10
Jamesonite	(?)	10—100		Eastern Transbaikaliya, U.S.S.R.	10
Thallium-jordanite	(?)		1.56%	Lead-zinc deposit, Upper Silesia	9
Oxides					
Braunite	(C) (3)	1.2—4.5	—	Eastern Kradzhal, U.S.S.R.	7
Braunite	(C) (8)	0.0—14	—	Dzhezda deposit, Central Kazakhstan	4
Pyrolusite	(C) (2)	2.4—5.0	—	Eastern Kradzhal, U.S.S.R.	7
Psilomelane	(C) (25)	20—300	154	Dzhesda deposit, Central Kazakhstan	7
Psilomelane	(W)	320—1,000		Eastern Transbaikaliya, U.S.S.R.	10
Manganite[b]	(C)		3.8	Dzhezda deposit, Central Kazakhstan	4
Goethite	(?)		100	Daraiso lead-zinc deposit, U.S.S.R.	9
Hydrogoethite, goethite	(?)	5—14		Sarykan deposit, Middle Asia, U.S.S.R.	11
Hydrogoethite	(?)	2—4		Kumyshkan deposit, Middle Asia, U.S.S.R.	11
Hydrous iron oxides	(?)	3—85		Kurgashinkan deposit, Middle Asia, U.S.S.R.	11
Limonite	(W)		6	Polymetallic deposits, E. Transbaikaliya, U.S.S.R.	10
Mn-rich limonite	(W)	10—50		Polymetallic deposits, E. Transbaikaliya, U.S.S.R.	10
Mn-hydroxides	(C, S) (2)	40—95	68	Lachin-Khana polymetallic deposit, U.S.S.R.	13
Iron oxides	(C, S) (18)	tr.—18	4	Lachin Khana polymetallic deposit, U.S.S.R.	13
Sulfates, carbonates etc.					
Thallium-jarosite		1.75—2.04%		Daraiso lead-zinc deposit, U.S.S.R.	9
Plumbojarosite	(?)		1.5	Sarykan lead-zinc deposit, Middle Asia	11
Plumbojarosite	(?)	3.7—4.7		Kurgashinkan lead-zinc deposit, Middle Asia	11
Plumbojarosite	(?)		—	Kurgash deposit, Middle Asia	11
Plumbojarosite	(C, S) (3)	11—44	27	Lachin-Khana polymetallic deposit, U.S.S.R.	13

[b] Contains a little braunite as impurity.

Table 81-F-5. (Continued)

Mineral	Method	Tl (ppm)		Deposit	Ref.
		range	mean		
Jarosite	(C, S) (11)	1—40	19	Lachin-Khana polymetallic deposit, U.S.S.R.	13
Melanterite	(?)		100	Daraiso lead-zinc deposit, U.S.S.R.	9
Colourless cerussite	(W) (4)	n.f.-tr.		Polymetallic deposits, E. Transbaikaliya, U.S.S.R.	10
Cerussite	(?)		tr.	Kurgash deposit, Middle Asia	11
Black cerussite	(W) (6)	0—40	—	Polymetallic deposits, E. Transbaikaliya, U.S.S.R.	10
Black cerussite	(?)		2.0	Sarykan lead-zinc deposit, Middle Asia	11
Black cerussite	(?)	10—20		Kumyshkan lead-zinc deposit, Middle Asia	11
Anglesite	(?)		tr.	Kurgash deposit, Middle Asia	11
Mimetite	(?)		41	Kurgashinkan lead-zinc deposit, Middle Asia	11
Alunite	(C, S)		2	Lachin-Khana polymetallic deposit, U.S.S.R.	13
Chalcanthite	(C, S)		2	Lachin-Khana polymetallic deposit, U.S.S.R.	13
Cuprohalloysite	(C, S)		12	Lachin-Khana polymetallic deposit, U.S.S.R.	13

List of References: 1. Murakami *et al.* (1950); 2. Shaw (1952); 3. Stoiber (1940) (*in* Shaw, 1957); 4. Voskresenskaya and Usevich (1957); 5. Voskresenskaya and Karpova (1958); 6. Voskresenskaya (1961b); 7. Voskresenskaya and Soboleva (1961); 8. Ivanitskii and Gvaramadze (1960); 9. Mogarovskii (1961); 10. Kulikova (1962); 11. Kulikova (1964); 12. Sinclair (1964); 13. Dunin-Barkovskaya (1961); 14. Meituv (1962).

III. Thallium in the Processes of Ore Deposition

The most detailed studies on the behaviour of thallium in ore deposition have been made on the hydrothermal deposits of the Caucasus region.

The ore minerals from polymetallic deposits show the highest concentrations of thallium (Table 81-F-3) and, in the Caucasus region, these deposits are related to acid intrusives (quartz albitophyres, quartz syenite and granite porphyries).

Enrichment of thallium is accompanied by concentration of volatiles such as arsenic, antimony and silver.

The critical factor governing the distribution of thallium in mineral deposits appears to be the sorption of this element in the sulfide gels, with isomorphism playing a secondary role. This explains the concentration of thallium in sphalerite rather than in galena (Tables 81-F-4, 81-F-5). Ore minerals which crystallized from colloform solutions show a strong enrichment of this element (Table 81-F-4).

Ivanov *et al.* (1956) (in Voskresenskaya and Karpova, 1958) point out that when sphalerite and galena crystallize from true solutions, thallium is concentrated in galena, its distribution being governed by the crystallochemical laws with thallium replacing lead in galena.

The mode of occurrence of thallium in sulfides is not well known. VOSKRESENSKAYA and KARPOVA (1958) found a correlation between thallium and arsenic in galena and an antipathetic behavior of arsenic and antimony in the case of a compensation replacement of lead by thallium; however, IVANOV *et al.* (1956) (in VOSKRESENSKAYA and KARPOVA, 1958) suggest the substitutions Tl—Bi and Tl—Sb for 2 Pb. Most sulfides are low in thallium (Table 81-F-5) although sulfides of lead and antimony or arsenic and lead may contain appreciable amounts of this element, in particular jordanite.

Gangue minerals do not show any appreciable enrichment in thallium (Table 81-F-6).

Table 81-F-6. *Thallium contents in gangue minerals of polymetallic ore deposits of the U.S.S.R.*

Mineral	Method	Tl (ppm)		Deposit	Ref.
		range	mean		
Garnet	(C) (7)		n.f.	Zolotoi Kurgan, Caucasus	1
Datolite	(C)		n.f.	Zolotoi Kurgan, Caucasus	1
Axinite	(W) (2)	1—7	—	Klichkinskii, E. Transbaikaliya	2
Muscovite	(W) (5)	14—20	—	Klichkinskii, E. Transbaikaliya	2
Phlogopite	(W) (4)	10—20	—	Klichkinskii, E. Transbaikaliya	2
Sericite	(W) (12)	10—27	—	Klichkinskii, E. Transbaikaliya	2
Chlorite	(C) (2)		0.6	Sadon, Caucasus	1
Serpentine	(W) (3)		n.f.	Klichkinskii, E. Transbaikaliya	2
Apophylite	(C) (2)		1.0	Zolotoi Kurgan, Caucasus	1
Potassium feldspar	(W) (3)	10—12	—	Klichkinskii, E. Transbaikaliya	2
Albite	(W)		1	Klichkinskii, E. Transbaikaliya	2
Quartz	(C, S)		1	Lachin-Khana	3
Analcime	(C)		1.5	Zolotoi Kurgan, Caucasus	1
Zeolites	(C) (3)		n.f.	Zolotoi Kurgan, Caucasus	1
Barite	(C) (7)		n.f.	Greater Caucasus	1
Barite	(C, S)		1	Lachin-Khana	3
Calcite	(C, S)		2	Lachin-Khana	3
Calcite	(C) (2)		n.f.	Sadon, Caucasus	1

List of References: 1. VOSKRESENSKAYA (1961b); 2. MEITUV (1962); 3. DUNIN-BARKOVSKAYA (1961).

IV. Discussion

It is apparent from the evidence on the thallium distribution in magmatic, contact metamorphic and metasomatic rocks that this element is highly mobile. VOSKRESENSKAYA (1959) suggested that thallium migrates from the central parts of the intrusive bodies to the endocontact zone and to the invaded rocks as halide or fluoride.

The possibility that thallium may form a monochloride in magmatic melts is advocated by ZLOBIN (1958).

Therefore, it seems that in magmatogenic processes the high volatility of the thallium halides plays a major role in the distribution of this element.

Revised manuscript received: March 1970

81-G. Behavior during Weathering and Alteration of Rocks

I. Rocks

Data on the thallium contents of unaltered and weathered rocks are given in Table 81-G-1.

The altered granites of the Musht Region, Northern Caucasus (Voskresenskaya and Su Shou T'ien, 1961), and of Perak, Malaya (Butler, 1962) do not show much variation in their thallium contents when compared to the unaltered rocks. Similarly, the clay derived from the Bosahan adamellite, Cornwall (Butler, 1953), contains the same amount of thallium as the unaltered rock.

The thallium contents of the weathered mantle of felsic and intermediate igneous rocks, in Middle Asia (Voskresenskaya *et al.*, 1962a) are similar to the contents of unaltered, comparable rocks.

Table 81-G-1. *Thallium contents of weathered rocks*

Rock	Method	Tl (ppm)		Locality	Ref.
		range	mean		
Adamellite	(S)		3.0	Bosahan, Cornwall	1
Ca-saturated clay from adamellite	(S)		3.0	Bosahan, Cornwall	1
Granite	(S) (3)	1—2	—	Cornwall, S. W. England	1
Unaltered rhyolitic welded tuffs and sericitized and kaolinized volcanic rocks	(C) (30)		1.0	Ugam Range, W. Tien Shan	2
Kaolinized welded tuffs with finely dispersed pyrite	(C) (100)	1—3	—	Ugam Range, W. Tien Shan	2
Weathered mantle of felsic and intermediate igneous rocks	(C) (4)	0.3—1.5	—	Middle Asia	3
Biotite granite (unaltered)	(C) (7)	0.8—1.3	1.1	Musht, Northern Caucasus	4
Same, altered	(C) (7)	1.1—1.5	1.4	Musht, Northern Caucasus	4
Biotite granite (unaltered)	(N/R) (6)	3.0—4.9	3.9	Perak, Malaya	5
Same, kaolinized, chloritized	(N/R) (2)	5.0—5.7	5.4	Perak, Malaya	5
Same, weathered	(N/R)		2.9	Perak, Malaya	5
Muscovite granite, kaolinized	(N/R)		4.1	Cornwall, S. W. England	5

List of References: 1. Butler (1953); 2. Dunin-Barkovskaya (1961); 3. Voskresenskaya *et al.* (1962a); 4. Voskresenskaya and Su Shou T'ien (1961); 5. Butler (1962).

Concentration of thallium in weathering products is included in KURBANAYEV'S study of the Maykain Deposit Area, Northern Kazakhstan (Table 81-G-2). A successive enrichment of thallium is observed in the weathered mantle, clay and loam and silt derived from granodiorite, andesite and shale.

It is apparent from these studies that thallium is largely retained during the weathering processes, and, probably as a result of this, enrichment of this element may be observed in the products of weathering.

Table 81-G-2. *Thallium contents in the bedrock and in weathering products of the Maykain Deposit Area, Northern Kazakhstan.* (After KURBANAYEV, 1966)

Bedrock	Weathered rock and shingle	Weathered mantle	Clays	Loam and silt
Granodiorite	0.2 (125)	0.26 (22)	0.3 (104)	0.33 (73)
Andesite	0.22 (43)	0.03 (13)	0.32 (28)	0.77 (14)
Silty shale	0.04 (121)	0.10 (115)	0.13 (122)	0.28 (79)

Thallium contents in ppm (S, W) and number of analysed samples in parentheses.

II. Oxidized Zone of Ore Deposits

Thallium tends to disperse during oxidation of sulphide ores and many oxidized minerals are very poor in this element. Therefore, the oxidized zone of ore deposits is poorer in thallium than the sulphide zone (Table 81-F-5).

In some instances, minerals from the oxidized zone of ore deposits show appreciable enrichment in thallium. Mimetite may concentrate thallium, probably due to the formation of the nearly insoluble compounds $TlAsO_3$ and $TlAsO_4$.

Of much more importance as thallium carriers are jarosite, which may form thallium-rich varieties, and manganese oxides, especially psilomelane.

Thallium seems to precipitate together with manganese, probably as Tl_2O_3, in strong oxidizing environments. According to VOSKRESENSKAYA and USEVICH (1957), thallium is preferentially associated with Mn^{4+} and, therefore, the thallium contents of the manganese minerals are not a function of the Mn-content but of the contents of the different oxides of the latter element.

It seems possible that the observed enrichment of thallium in jarosite is a consequence of the substitution of thallium for potassium of the structure of this mineral.

Revised manuscript received: March 1970

81-H. Solubilities of Compounds which Control Concentrations in Natural Waters; Adsorption Processes; Valence States in Natural Environments

The concentration of thallium in natural waters is practically unknown and the processes involved in the transport of thallium constitute a problem which needs investigation.

Thallium was detected in one out of seven samples of mine waters from the Lachin-Khana lead-zinc deposit (DUNIN-BARKOVSKAYA, 1961), in which a concentration of 0.002 ppm thallium was found.

It has been mentioned above (Section 81-G) that thallium is largely retained during the process of weathering of rocks. This element is also adsorbed in clay minerals (SHAW, 1957) owing to the fact that thallium is a large electropositive ion. It seems likely that thallium is transported in solution as a univalent cation. In strongly oxidizing conditions, thallium may be removed from solution, with formation of Tl^{3+} by precipitation with manganese and/or iron.

Adsorption processes appear to play an important role in the concentration of thallium in mineral deposits as colloform ores are particularly rich in this element. VOSKRESENSKAYA and KARPOVA (1958) explained the absence of thallium in the oxidized ores of mineral deposits as being due to the high solubility of its sulfates and carbonates.

In natural environments, thallium seems to occur in general as a univalent cation, and only in strongly oxidizing conditions will it occur as trivalent thallium as in the oxidation zones of mineral deposits. The coherence between thallium on one hand, and potassium and rubidium on the other hand, its adsorption in clays and the enrichment of thallium in sedimentary rocks formed in reducing conditions, strongly suggest that this element is present in most natural environments as univalent thallium.

Revised manuscript received: March 1970

81-I. Abundance in Natural Waters

Thallium is below the detection limits of the methods used for its determination in natural waters, and in particular in ocean water. MANHEIM (1965) (in AHRENS *et al.*, 1967) found that the thallium content of sea-water is less than 0.01 ppb.

Waters from hot springs and drillholes in thermal areas along the margin of the Taupo Volcanic Zone (New Zealand) contain upto 7 ppb of thallium (WEISSBERG, 1969). They ascent through fissures and silicic volcanic rocks and, at a depth of 1 km, the waters are slightly alkaline (pH 6—7).

The precipitates formed from these waters are amorphous sulfides containing gold, silver, arsenic, antimony and mercury. They contain thallium in variable amounts, from less than 1 ppm to 0.5%, according to the same author, and only 0.4% of the thallium in the waters is deposited as precipitate.

Revised manuscript received: March 1970

81-K. Abundance in Common Sediments and Sedimentary Rock Types

The thallium contents of sedimentary rocks are generally in the range 0.1 to 2.0 ppm. The highest values for the concentration of this element are found in sandstones (1—3 ppm) and shales; argillaceous rocks and coal may be relatively rich in thallium, with contents upto 2.2—3.0 ppm (Table 81-K-1).

Table 81-K-1. *Thallium contents of sedimentary rocks*

Rock	Method	Tl (ppm)		Locality	Ref.
		range	mean		
Sandstone (composite)	(S) (23)		2.0	Triassic, Germany	1
Sandstone	(S)		1.3	Huvitas, Finland	2
Sandstone	(C) (2)		1.2	Caucasus, U.S.S.R.	3
Siltstone (argillaceous)	(?) (5)		0.68		4
Arkose	(S)		1.0	Fair Haven, Connecticut	2
Greywacke	(S) (3)	0.13—0.21	—	North America	2
Subgreywacke	(S) (5)		0.6	Henbury, Australia	5
Clay	(C, S) (3)	0.24—0.70	—	Finland, U.S.S.R., U.S.A.	2, 3
Argillite	(S)		1.4	Michigan	2
Argillaceous rock	(S) (323)		0.36		6
Slate	(S)		0.81	Eagle Bridge, New York	2
Shale (composite)	(S) (36)		2.0	Paleozoic, Germany	1
Shale	(S, W/S) (4)	0.12—0.50	—	New York, South Africa	2, 7, 8
Shale (Dictyonema)	(S)		3.1	Tallinn, Estonia	2
Phyllite	(S)		0.79	Finland	2
Shale	(?) (7)		1.3		4
Carbonaceous schist	(S)		0.60	Manitoba	2
Schungite	(S)		0.11	Finland	2
Anthraxolite	(S)		2.8	Ontario	2
Porcellanite	(S)		0.84	Michigan	2
Carbonaceous marl	(C) (2)		0.8	Caucasus, U.S.S.R.	3
Argillaceous limestone	(C) (2)		0.7	Caucasus, U.S.S.R.	3
Argillaceous marl	(C) (2)		2.2	Caucasus, U.S.S.R.	3
Loess	(S)		0.31	Illinois	2
Laterite	(S)		0.10	Madeira	2
Ocean mud (depth 500 m)	(W/S)		0.13	Antarctica	7

Table 81-K-1. (Continued)

Rock	Method	Tl (ppm)		Locality	Ref.
		range	mean		
Eluvium-alluvium	(S, W) (230)		0.1	N. Kazakhstan	10
Clay (from parting in coal)	(C) (5)	0.3—1.0	—	Middle Asia	3
Sandstone (from parting in coal)	(C) (2)	1.0—3.0	2.0	Middle Asia	3
Coal	(C) (9)	0.3—1.5	0.85	Middle Asia	3
Coal	(C) (7)	tr. — 2.3	—	Northern Caucasus	3

List of References: 1. PREUSS (1940), in SHAW (1957); 2. SHAW (1952); 3. VOSKRESENSKAYA *et al.* (1962a); 4. ERLANK (1965), in KOLBE and TAYLOR (1966a); 5. TAYLOR (1966b); 6. CANNEY (1953); 7. BROOKS *et al.* (1960); 8. BROOKS and AHRENS (1961a); 9. BROOKS and AHRENS (1961b); 10. KURBANAYEV (1966).

Table 81-K-2. *Thallium contents of deep-sea sediments*

Rock	Method	Tl (ppm)		Reference
		range	mean	
Iron-manganese-rich ooze	(S)		0.3	SHAW (1952)
Globigerina ooze	(S)		0.16	SHAW (1952)
Siliceous ooze	(S)		—	SHAW (1952)
Clay	(S) (6)	0.16—1.5	0.66	SHAW (1952)
Manganese nodules	(S) (19)	nf. — 180	70	WILLIS and AHRENS (1962)
Manganese nodules	(S) (31)	2—164	140	AHRENS *et al.* (1967)
Manganese nodules	(?)		100	MANHEIM (1965) in AHRENS *et al.* (1967)

Deep-sea sediments contain usually low amounts of thallium (Table 81-K-2), with the exception of manganese nodules which may show concentrations of this element as high as 600 ppm and an average concentration of about 100 ppm (MANHEIM, 1965, in AHRENS *et al.*, 1967).

The average thallium contents of sedimentary rocks, according to different authors, are given in Table 81-K-5.

Argillaceous rocks with high carbonaceous content have contents of thallium about twice the value for normal pelitic rocks (SHAW, 1952, 1957). Thallium will be concentrated owing to the reducing conditions in which such rocks were formed.

Thallium occurs in the sulfide phases in coal rather than in the coal itself (Table 81-K-3, 81-K-4). According to VOSKRESENSKAYA *et al.* (1962a), thallium is not firmly held in the organic mass of the coal as it forms soluble compounds of the humate and fulvate types with organic components in coal.

Pyrite from the coal is richer in thallium than the same mineral from the enclosing sedimentary rocks. VOSKRESENSKAYA (1968) pointed out that thallium must have been delivered to the coal basins during sedimentation or early diagenesis,

Table 81-K-3. *Thallium contents of sedimentary rocks from coal basins of the U.S.S.R.* (According to VOSKRESENSKAYA, 1968)

Locality	Tl (ppm) (S, C)					
	Coal	Ash-rich coal or dull coal	Carbonaceous clay	Clay	Sandstone	Coal with sulfides
Middle Asia Basin (Jurassic)						
Soguty	0.5 (3)	—	0.8 (3)	—	n.f.	135
Kavak	n.f.	0.4 (2)	—	0.9 (2)	n.f.	—
Tashkumyr	0.1 (3)	1.0 (3)	—	—	—	—
Dzhergalan	0.5 (4)	—	—	—	0.5	—
Kara-Kiche	0.8 (4)	1.9 (2)	1.0 (3)	—	—	—
Aksay	0.6 (3)	—	1.2	—	0.5	—
Moscow Basin (Carboniferous)						
Kimovo	n.f. (2)	—	—	0.7 (2)	—	—
Dnepr Basin (Paleogene)						
Baydakov	n.f. (2)	—	—	—	n.f. (8)	—
Transcarpathian Basin (Neogene)						
Suvorovsk	—	0.4	—	—	—	4.0

Table 81-K-4. *Thallium contents of minerals from sedimentary rocks and hydrothermally altered or contact metamorphosed sedimentary rocks*

Mineral		Tl (ppm) (C) range	mean	Rock	Locality	Ref.
Pyrite	(2)		2.0	Sandstone	Northern Caucasus	2
Pyrite	(2)	2.0—2.2	2.1	Argillite	Caucasus	2
Pyrite	(9)	0.3—2.2	1.1	Sedimentary rocks (unaltered)	Greater Caucasus	1
Pyrite	(7)	2.8—270	50	Sedimentary rocks (hydrothermally alt.)	Greater Caucasus	1
Pyrite	(3)		n.f.	Marl (unaltered)	Greater Caucasus	1
Pyrite	(7)		1.0	Marl	Caucasus	2
Pyrite			18.5	Skarn	Greater Caucasus	1
Pyrite			2.0	Limestone	Suvorovsk, Moscow basin	3
Pyrite			25	Barite vein in coal	Middle Asia basin	3
Pyrite	(7)	1.0—30	10	Clay and sandstone	Middle Asia basin	3
Pyrite	(30)	1.0—200	34	Coal	Middle Asia basin	3
Pyrite	(18)	0.0—20	4.3	Clay and sandstone	Moscow basin	3

Table 81-K-4. (Continued)

Mineral		Tl (ppm) (C) range	mean	Rock	Locality	Ref.
Pyrite	(14)	0.0—13	3.3	Coal	Moscow basin	3
Pyrite	(5)	4—23	11	Coal	Transcarpathian basin	3
Pyrite	(16)	0.0—0.6	—	Clay and sandstone	Dnepr basin	3
Pyrite	(23)	0.0—0.5	0.5	Coal	Dnepr basin	3
Pyrite			1.0	Tuffite	Transcarpathian basin	3
Sphalerite			4.0	Sandstone	Northern Caucasus	2
Galena			2.6	Sandstone	Caucasus	2
Magnetite	(3)	157—300	219	Skarn	Greater Caucasus	1
Magnetite			0.6	Spotted marl	Greater Caucasus	1
Ilsemannite	(2)	12—62	37	Coal	Middle Asia basin	2
Barite		1.0—3.0	—	Clay parting in coal	Middle Asia basin	2
Hydrous iron oxides			—	Coal	Middle Asia basin	2
Sulfur			0.7	Coal	Middle Asia basin	2

List of References: 1. Voskresenskaya (1961b); 2. Voskresenskaya *et al.* (1962a); 3. Voskresenskaya (1968).

Table 81-K-5. *Average thallium contents in sedimentary rocks (ppm Tl)*

Shales	Sandstones	Carbonates	Deep-sea sediments: carbonate	Deep-sea sediments: clay	Greywackes	Manganese nodules	Reference
0.69	0.82	0.05	—	—	—	—	Shaw (1957)
1.4	0.82	0.0x	0.16	0.8	—	—	Turekian and Wedepohl (1961)
0.3	—	—	—	—	0.3	—	Taylor (1966a)
1.55	1.54	0.065	0.18	0.75	—	—	Horn and Adams (1966)
0.9	1.0 (greywacke excluded)	0.05	0.16	0.6	0.3	100	This paper

as the contents of this element do not depend on the petrographic characteristics of the coal and the contents in pyrite are uniform throughout the vertical sections of the coal beds.

According to this author the concentration of thallium in the basins is governed by the distance to the source area of the sediments and, more important, by the composition of the rock in the source area.

The leading role in the concentration of thallium in the sedimentary cycle is played by its chalcophile tendency and sorption (VOSKRESENSKAYA *et al.*, 1962a).

The possibility of thallium being adsorbed in clay minerals to a considerable extent was pointed out by SHAW (1957) and, according to CANNEY (1952), the relative degree of adsorption of the large univalent ions in clay is Cs>Rb>K>Tl.

The association of thallium with manganese in the sedimentary cycle is a consequence of the oxidizing environment precipitation (SHAW, 1957). The same seems to be valid for iron also. This explanation was offered by different authors for the presence of thallium in some ore minerals as well.

The coherence of thallium and rubidium appears to extend to sedimentary rocks. Taking the Tl values in Table 81-K-5 and Rb figures from Chapter 37, the average Rb/Tl ratios for shales and sandstones approximate to 140 and 40 respectively.

Revised manuscript received: March 1970

81-L. Biogeochemistry

The behavior of thallium in biogeochemical processes is not known in detail.

This element has been identified, according to GOLDSCHMIDT (1954), in land plants and in seaweed.

The dry matter of marine animals may contain upto 0.03 ppm Tl.

It seems possible that decaying organic matter will concentrate thallium and it has been shown that this element is enriched in sedimentary rocks containing carbonaceous matter (SHAW, 1952). Among the carbonaceous samples of argillaceous rocks analysed by SHAW (1952) was an Archean schist containing a carbonaceous aggregate believed to be a fossil. This sample contains 0.79 ppm Tl, a value above the average for such rocks (Table 81-K-1).

The thallium contents of coal from the Northern Caucasus and Middle Asia regions (VOSKRESENSKAYA *et al.*, 1962a) are mostly in the range 0.3—2.3 ppm (Table 81-K-3).

Revised manuscript received: March 1970

81-M—N. Abundance in Common Metamorphic Rock Types and Behavior in Metamorphic Reactions

The thallium distribution in metamorphic rocks has not yet been thoroughly investigated, and the variation of the contents of this element with grade of metamorphism is practically unknown.

Table 81-M—N-1. *Thallium contents of metamorphic rocks*

Rock	Method	Tl (ppm) range	Tl (ppm) mean	Locality	Ref.
Phyllite	(S) (4)	0.56—0.66	0.6	Clove Quadr., New York	1
Phyllite	(S)		0.27	Moores Mills, New York	1
Schist	(W)		2.0	Sayan Mountains, U.S.S.R.	2
Sericite-bearing quartzose rock	(S, W) (102)		0.02	Maykan, N. Kazakhstan	3
Quartz-sericite schist	(S,W) (81)		0.16	Maykan, N. Kazakhstan	3
Garnet schist	(S) (3)	0.61—1.3	1.0	Clove Quadr., New York	1
Biotite-hornblende schist	(S)		0.35	Danbury Quadr., New York	1
Sillimanite schist					
Dark layer	(S)		1.9	Clove Quadr., New York	1
Light layer	(S)		0.74	Clove Quadr., New York	1
Gneiss	(R) (3)	0.6—0.9	0.8	Oki Island , Japan	4
Gneiss	(W/S)		2.2	Lütsow-Holm Bay, Antarctica	5
Noritic gneiss	(S)		tr.	Albany Co., Wyoming	1
Dioritic gneiss	(S)		0.42	Albany Co., Wyoming	1
"Granodioritic" gneiss	(S) (3)	0.74—0.9	0.81	Albany Co., Wyoming	1
Granodioritic gneiss	(S)		0.72	Egedesminde, Greenland	1
Enderbitic gneiss	(S)		0.1	N. Isortog, Greenland	1
Garnet gneiss	(S) (2)	0.13—0.37	0.25	N. Isortog, Greenland	1
Granulite gneiss	(S)		0.026	N. Isortog, Greenland	1
Amphibolite	(S)		0.40	Impilahti, K.F.S.S.R.	1
Amphibolite	(W) (2)	2—3	—	Sayan Mountains, U.S.S.R.	2
Eclogite	(W/S) (2)		0.03	Nodule, Kimberley, S. Africa	6
Eclogite	(W/S)		0.56	Nodule, Kimberley, S. Africa	6

List of References: 1. Shaw (1952); 2. Slepnev (1961); 3. Kurbanayev (1966); 4. Ishimori and Takashima (1955); 5. Brooks *et al.* (1960); 6. Brooks and Ahrens (1961a).

Quartzose rocks contain very little thallium (0.02 ppm), according to KURBANAYEV (1966), while the quartz-sericite schists from the same region contain 0.16 ppm (Table 81-M—N-1).

Slates and phyllites contain thallium in the range 0.27—0.66 ppm and higher grade schists from the same region (New York State) show contents of this element in the range 0.35—1.3 ppm (SHAW, 1952). According to this author, the dark layers of a silimanite schist are enriched in thallium relative to the light layers by a factor of about 2.5 (Table 81-M—N-1).

The thallium contents of gneissic rocks are variable, and apparently depend mostly on the mineralogical composition of the gneiss. Granodioritic gneisses are richer in this element than dioritic or noritic gneisses by a factor of at least 2. Enderbitic gneisses are also poor in thallium (0.10 ppm) and a sample of granulite contains only 0.026 ppm thallium (Table 81-M—N-1). These figures may indicate that thallium is depleted in granulite facies rocks, confirming the observations made on the potassium feldspars from such rocks (Section 81-D, Table 81-D-9).

The scanty data on amphibolites show the thallium contents to vary in the range 0.43—3 ppm (Table 81-M—N-1). This upper limit seems to be very high for this type of rock from that predicted on theoretical grounds.

Eclogites occuring as nodules in ultrabasic rocks from South Africa (BROOKS and AHRENS, 1961 a) contain thallium in the range 0.03—0.56 ppm, and therefore no conclusions can be drawn for the abundance of thallium in these rocks until more data are available.

Revised manuscript received: March 1970

81-O. Sensitivity Limits for the Determination of Thallium by Different Methods

Method		Sensitivity	Reference
D-C arc emission spectrography	(S)	1.00 ppm	AHRENS and TAYLOR (1961)
Emission spectrography: double-arc method	(S)	0.05 ppm	SHAW (1952)
X-ray spectrometry	(X)	20.00 ppm	JENKINS and DE VRIES (1967)
Neutron activation	(N/R)	0.01 ppm	MORRIS and KILLICK (1960)
Neutron activation	(N/R)	0.03 ppb[a]	REED *et al.* (1960)
Isotope dilution	(I)	1.00 ppb[a]	ANDERS and STEVENS (1960)
Radiometric	(R)	0.30 ppm[a]	ISHIMORI and TAKASHIMA (1955)
Flame photometry	(F)	0.20 ppm	FORNASERI and PENTA (1963)
Atomic fluorescence spectrometry	(A)	0.10 ppm	MANNING and HENEAGE (1968)
Atomic absorption spectrometry	(A)	0.20 ppm	SLAVIN *et al.* (1964)
Atomic absorption: sampling boat technique	(A)	0.025 ppm	KAHN *et al.* (1968)
Chemical/spectrographic	(W/S)	0.01—0.05 ppm[a]	BROOKS *et al.* (1960)
Colorimetric	(C)	0.01 ppm[a]	VOSKRESENSKAYA (1961a)
Polarographic	(P)	0.05 ppm[a]	WAHLER (1968)

[a] Figure based on the results given, which may be taken as a limit.

Revised manuscript received: March 1970

References: Sections 81-B to 81-O

AHRENS, L.H.: The geochemical relationship between thallium and rubidium in minerals of igneous origin. Trans. Geol. Soc. S. Africa **48**, 207 (1945).
— The unique association of thallium and rubidium in minerals. J. Geol. **56**, 578 (1948).
— TAYLOR, S. R.: Spectrochemical Analysis, 2nd ed. Reading, Mass., U.S.A.: Addison-Wesely Pub. Co. Inc. 1961.
— WILLIS, J. P., OOSTHUIZEN, C. O.: Further observations on the composition of manganese nodules, with particular reference to some of the rarer elements. Geochim. Cosmochim. Acta **31**, 2169 (1967).
ANDERS, E., STEVENS, C. M.: Search for extinct lead 205 in meteorites. J. Geophys. Res. **65**, 3043 (1960).
ASTON, F. W.: Constitution of thallium and uranium. Nature **128**, 725 (1931).
BADALOV, S. T., RABINOVICH, A. V.: The geochemistry of indium and thallium in the Karamazar Ore region (the Uzbek and Tadzhik Republics). Geochem. Int. 1095 (1966).
BOROVIK-ROMANOVA, T. F., SOSEDKO, A. F.: Relation between thallium and rubidium in minerals from pegmatite veins of the Kola Peninsula. Geochem. Int. 34 (1960).
BROOKS, R. R., AHRENS, L. H.: Some observations on the distribution of thallium, cadmium and bismuth in silicate rocks and the significance of covalency on their degree of association with other elements. Geochim. Cosmochim. Acta **23**, 100 (1961a).
— — The determination of indium and thallium in G-1, W-1 and other silicate rocks by a new technique. Geochim. Cosmochim. Acta **23**, 145 (1961b).
— TAYLOR, S. R.: The determination of trace elements in silicate rocks by a combined spectrochemical-anion exchange technique. Geochim. Cosmochim. Acta **18**, 162—175 (1960).
BROWN, R., WOLSTENHOLME, W. A.: Analysis of geological samples by spark source mass spectrometry. Nature **201**, 598 (1964).
BUTLER, J. R.: The geochemistry and mineralogy of rock weathering: the Lizard Area, Cornwall. Geochim. Cosmochim. Acta **4**, 157 (1953).
— Tl in some igneous rocks. Geochem. Int. 599 (1962).
CAMERON, E. G. W.: A new table of abundances of the elements in the solar system. In: Origin and Distribution of the Elements (ed. L. H. AHRENS), **30**, 125. Pergamon Press 1968.
CANNEY, F. C.: Some aspects of the geochemistry of potassium, rubidium, caesium and thallium in sediments. Am. Mineralogist **38**, 331 (1953).
DAWSON, K. R., MAXWELL, J. A., PARSONS, D. E.: A description of the meteorite which fell near Abee, Alberta, Canada. Geochim. Cosmochim. Acta **21**, 127 (1960).
DEMIN, A. M., KHITAROV, D. N.: Geochemistry of potassium, rubidium and thallium in application to problems of petrology. Geochem. Int. 721 (1958).
DIETRICH, R. V., HEIER, K. S.: Differentiation of quartz-bearing syenite (nordmarkite) and riebeckitic-arfvedsonite granite (ekerite) of the Oslo series. Geochim. Cosmochim. Acta **31**, 275 (1967).
— TAYLOR, S. R.: Studies on the igneous rock complex of the Oslo Region. XX. Petrology and geochemistry of ekerite. Skrifter Norske Videnskaps-Akad. Oslo, I. Mat.-Naturv. Kl., Ny Ser. No. 19 (1965).
DUNIN-BARKOVSKAYA, E. A.: Thallium in the ores and minerals of Lachin Khana (Western Tien Shan, Ugam Range). Geochem. Int. 740 (1961).

EDWARDS, G.: Sodium and potassium in meteorites. Geochim. Cosmochim. Acta **8**, 285 (1955).
— UREY, H. C.: Determination of alkali metals in meteorites by a distillation process. Geochim. Cosmochim. Acta **7**, 154 (1955).
EHMANN, W. D., HUIZENGA, J. R.: Bismuth, thallium and mercury in some meteorites by activation analysis. Geochim. Cosmochim. Acta **17**, 125 (1959).
EL-BADRY, H. M., HODGE, E. S., BAER, W. K., KOHMAN T. P.: Thallium in meteorites. U.S.A.E.C. Rep. NYO-8920 (1960).
EWART, A., TAYLOR, S. R., CAPP, A. C.: Geochemistry of the pantellerites of Mayor Island, New Zealand. Contr. Mineral. and Petrol. **17**, 2, 116 (1968).
FLANAGAN, F. J.: U.S. geological survey standards. II. First compilation of data for the new U.S.G.S. rocks. Geochim. Cosmochim. Acta **33**, 80 (1969).
FLEISCHER, M.: Summary of new data on rock samples G-1 and W-1, 1962—1965. Geochim. Cosmochim. Acta **29**, 1263 (1965).
— U.S. geological survey standards. I. Additional data on rocks G-1 and W-1, 1965—1967. Geochim. Cosmochim. Acta **33**, 65 (1969).
— STEVENS, R. E.: Summary of new data on rock samples G-1 and W-1. Geochim. Cosmochim. Acta **26**, 525 (1962).
FORNASERI, M., PENTA, A.: Determinazione del tallio nei silicati per fotometria di fiamma. Metallurgia Italiana **55**, 8, 437 (1963).
GANEYEV, I. G., SECHINA, N. P.: Geochemical characteristics of albitized granites. Geochem. Int. 158 (1962).
GAST, P. W.: Alkali metals in stone meteorites. Geochim. Cosmochim. Acta **19**, 1 (1960).
— Limitations on the composition of the upper mantle. J. Geophys. Res. **65**, 1287 (1960).
GERASIMOVSKII, V. I., RASSKAZOVA, V. S.: Distribution of thallium in nepheline syenites of the Lovozero massif (Kola Peninsula). Geochem. Int. 275 (1962).
— VOLKOV, V. P., KOGARKO, L. N., POLYAKOV, A. I., SAPRYKINA, T. V., BALASHOV, YU. A.: The Geochemistry of the Lovozero Alkaline Massif. Pt. I. Geology and Petrology; Pt. 2. Geochemistry. Toronto: University of Toronto Press 1968.
GOLDSCHMIDT, V. M.: Geochemistry. Oxford: Clarendon Press 1954.
GOLUBCHINA, M. N., RABINOVICH, A. V., MURTAZINA, T. M.: Isotopic composition of thallium in igneous rocks. Geochem. Int. 231 (1957).
GRAESER, S.: Asbecasit und Cafarsit, zwei neue Mineralien aus dem Binnatal (Kt. Wallis). Schweiz. Mineral. Petrog. Mitt. **46**, 367 (1966).
HEIER, K. S.: Petrology and geochemistry of high-grade metamorphic and igneous rocks on Langøy, Northern Norway. Norg. Geol. Undersokelse 207 (1960).
— Trace elements in feldspars — a review. Norsk Geol. Tidsskr. **42**, 415 (1962).
— Some crystallochemical relations of nephelines and feldspars on Stjernoy, North Norway. J. Petrol. **7**, 95 (1966).
— TAYLOR, S. R.: Distribution of Li, Na, K, Rb, Cs, Pb and Tl in Southern Norwegian pre-Cambrian alkali feldspars. Geochim. Cosmochim. Acta **15**, 284 (1959).
HIGAZY, R. A.: Petrogenesis of perthite pegmatites in the Black Hills, South Dakota. J. Geol. **1**, 555 (1949).
— Observations on the distribution of trace elements in the perthite pegmatites of the Black Hills, South Dakota. Am. Mineralogist **38**, 172 (1953).
HORN, M. K., ADAMS, J. A. S.: Computer-derived geochemical balances and element abundances. Geochim. Cosmochim. Acta **30**, 279 (1966).
ISHIMORI, T., TAKASHIMA, Y.: Thallium content of Japanese igneous rocks. Mem. Fac. Sci. Kyushu Univ., Ser. C (Chem.) **2** (2), 65 (1955).
IVANITSKII, T. V., GVARAMADZE, N. D.: Content and distribution of certain dispersed elements in the principal sulfides of the lead-zinc and polymetallic deposits of Georgia. Geochem. Int. 167 (1960).
IVANOV, V. V.: Geochemistry of the Rare Elements Ga, Ge, Gd, In and Tl in Hydrothermal Deposits. Moscow: Nedra 1966.

JENKINS, R., DEVRIES, J. L.: Practical X-ray spectrometry. Berlin-Heidelberg-New York: Springer 1967.

KAHN, H. L., PETERSON, G. E., SCHALLIS, J. E.: Atomic absorption micro-sampling with the "sampling boat" technique. Atomic Absorption Newsletter **7**, 35 (1968).

KEAYS, R. R., GANAPATHY, R., LAUL, J. C., ANDERS, E., HERZOG, G. F., JEFFERY, P. M.: Trace elements and radioactivity in lunar rocks: implications for meteorite infall, solar-wind flux, and formation conditions of the moon. Science **167**, 490 (1970).

KOGARKO, L. N.: Distribution of alkali elements and thallium in granitic rocks of the Turgoyak massif (Central Urals). Geochem. Int. 557 (1959).

KOHMAN, T. P.: Extinct natural radioactivity: possibilities and potentialities. Ann. N.Y. Acad. Sci. **62**, 503 (1955).

KOLBE, P.: The use of an oxygen jet in the spectrochemical determination of trace amounts of Pb, Tl, Ga, Cu and Sn in some silicate rocks and minerals. Geochim. Cosmochim. Acta **29**, 153 (1965).

— TAYLOR, S. R.: Geochemical investigation of the granitic rocks of the Snowy Mountains area, New South Wales. J. Geol. Soc. Australia **13**, 1 (1966a).

— — Major and trace element relationships in granodiorites and granites from Australia and South Africa. Contr. Mineral. and Petrol. **12**, 202 (1966b).

KULIKOVA, M. F.: Trace elements in the oxidized zone of polymetallic deposits of Eastern Transbaikaliya. Geochem. Int. 176 (1962).

— Distribution of trace elements in the oxidized zone of some lead-zinc deposits of Middle Asia. Geochem. Int. 977 (1964).

KURBANAYEV, M. S.: Thallium in the dispersion aureoles of the Maykain gold-barite polymetallic deposit. Geochem. Int. 568 (1966).

LARIMER, J. W., ANDERS, E.: Chemical fractionations in meteorites. II. Abundance patterns and their interpretation. Geochim. Cosmochim. Acta **31**, 1239 (1967).

MANNING, D. C., HENEAGE, P.: Element detection limits by atomic fluorescence spectroscopy. Atomic Absorption Newsletter **7**, 80 (1968).

MASON, B.: Principles of Geochemistry (2nd. ed.). New York: John Wiley & Sons, Inc. 1958.

— WIIK, H. B.: The Holbrook, Arizona, chondrite. Geochim. Cosmochim. Acta **21**, 276 (1961).

— — The composition of the Forest City, Tennasilm, Weston, and Geidam meteorites. Am. Mus. Novitates 2220 (1965).

MEITUV, G. M.: Geochemistry of rare elements in the lead-zinc deposit of the Klichkimskii region (Eastern Transbaikaliya). Geochem. Int. 694 (1962).

MOGAROVSKII, V. V.: Geochemistry of thallium in the oxidized zone of the Daraiso sulfide deposit (Middle Asia). Geochem. Int. 848 (1961).

— Correlation between thallium and zinc content in the Daraiso sulfide deposit (Middle Asia). Geochem. Int. 734 (1962).

MORRIS, D. F. C., KILLICK, R. A.: The determination of silver and thallium in rocks neutron-activation analysis. Talanta **4**, 51 (1960).

MÜLLER, E. A.: The solar abundances. In: Origin and Distribution of the Elements (ed. L. H. AHRENS), **30**, 155. Oxford: Pergamon Press 1968.

NOCKOLDS, S. R.: Average chemical compositions of some igneous rocks. Bull. Geol. Soc. Am. **65**, 1007 (1954).

— The behaviour of some elements during fractional crystallization of magma. Geochim. Cosmochim. Acta **30**, 267 (1966).

NODDACK, I., NODDACK, W.: Die Häufigkeit der chemischen Elemente. Naturwissenschaften **18**, 35, 757 (1930).

— Die geochemischen Verteilungskoeffizienten der Elemente. Svensk Kem. Tidskr. **46**, 173 (1934).

ODIKADZE, G. L.: Some geochemical characteristics of potassium, rubidium and thallium in the granitoids of the Dzirul crystalline massif (Western Georgia). Geochem. Int. 58 (1967).

OSTIC, R. G.: Isotopic composition of thallium in meteorites. U.S.A.E.C. Rep. NYO-844-71, 45 (1967).

OSTIC, R. G., KOHMAN, T. P.: Interpretation of meteoritic thallium isotope ratios. U.S.A.E.C. Rep. NYO-844-75, 31—37 (1968).
OTTEMANN, J.: Untersuchungen zur Verteilung von Spurenelementen, insbesondere Zinn, in Tiefengesteinen und einigen gesteinsbildenden Mineralien des Harzes. Z. Angew. Mineral. **3**, 2, 142 (1940).
POLDERVAART, A.: Chemistry of the earth's crust. In: Crust of the Earth (a symposium) (A. POLDERVAART, ed.). G.S.A. Spec. Paper **62**, 119 (1955).
RANKAMA, K., SAHAMA, TH. G.: Geochemistry. Chicago: Chicago Univ. Press 1950.
REED, G. W., KIGOSHI, K., TURKEVICH, A.: Determinations of concentrations of heavy elements in meteorites by activation analysis. Geochim. Cosmochim. Acta **20**, 122 (1960).
SCHROLL, E., WENINGER, M.: Eine empfindliche spektrochemische Analysenmethode zur Bestimmung von Germanium und Zinn unter Verwendung sulfidierender thermochemischer Reagenzien. Mikrochim. Acta 378 (1965).
SCHÜLER, H., KEYSTON, J. E.: Über einen Isotopenverschiebungseffekt der Hyperfeinstrukturtherme von Thallium. Z. Phys. **69**, 1 (1931).
SHAW, D. M.: The geochemistry of thallium. Geochim. Cosmochim. Acta **2**, 118 (1952).
— The geochemistry of gallium, indium, thallium — a review. Phys. Chem. Earth **2**, 164 (1957).
SHIMA, M., HONDA, M.: Distributions of alkali, alkaline earth and rare earth elements in component minerals of chondrites. Geochim. Cosmochim. Acta **31**, 1995 (1967).
SIEDNER, G.: Distribution of alkali metals and thallium in some South-West African granites. Geochim. Cosmochim. Acta **32**, 1303 (1968).
SINCLAIR, A. J.: A lead isotope study of mineral deposits in the Kootenay Arc. Ph. D. thesis, University of British Columbia (1964).
SINE, N. M., TAYLOR, W. O., WEBBER, G. R., LEWIS, C. L.: Third report of analytical data for CAAS sulphide ore and syenite rock standards. Geochim. Cosmochim. Acta **33**, 121 (1969).
SITNIN, A. A.: Distribution of rare elements in amazonite granites of the Etyka massif, Eastern Transbaikalia. Geochem. Int. 361 (1960).
SLAVIN, W., SPRAGUE, S., MANNING, D. C.: Detection limits in analytical atomic absorption spectrophotometry. Atomic Absorption Newsletter, No. 18 (1964).
SLEPNEV, Y. S.: The ratio of Tl to Rb, Cs and K in the schists, granites and rare metal pegmatites of the Sayan Mountains. Geochem. Int. 382 (1961).
SMALES, A. A., HUGHES, T. C., MAPPER, D., MCINNES, C. A. J., WEBSTER, R. W.: The determination of rubidium and cesium in stony meteorites by neutron activation analysis and by mass spectrometry. Geochim. Cosmochim. Acta **28**, 209 (1964).
SOLODOV, N. A.: Distribution of thallium among the minerals of a zoned pegmatite. Geochem. Int. 738 (1962).
SUESS, H. E., UREY, H. C.: Abundances of the elements. Rev. Mod. Phys. **28**, 53 (1956).
TANDON, S. N.: Mercury, thallium and bismuth in metal and troilite phases of iron meteorites by neutron activation analysis. U.S.A.E.C. Rep. No. NYO-844-71, 35 (1967).
TAUSON, L. V.: Distribution regularities of trace elements in granitoid intrusions of the batholith and hypabyssal types. In: Origin and Distribution of the Elements (ed. L. H. AHRENS), **30**, 629. Oxford: Pergamon Press 1968.
— BUZAEV, N. N.: Geochemistry of thallium in granitoids of the Susamyr batholith, Central Tian-Shan. Geochem. Int. 703 (1957).
— STAVROV, O. D.: The geochemistry of rubidium in granitoids. Geochem. Int. 819 (1957).
TAYLOR, S. R.: Abundances of chemical elements in the continental crust: a new table. Geochim. Cosmochim. Acta **28**, 1273 (1964a).
— Trace element abundances and the chondritic earth model. Geochim. Cosmochim. Acta **28**, 1989 (1964b).
— Geochemical application of spark source mass spectrography. Nature **205**, 34 (1965).
— Geochemical analysis by spark source mass spectrography. Geochim. Cosmochim. Acta **29**, 1243 (1965).

Taylor, S. R.: The application of trace element data to problems in petrology. Phys. Chem. Earth **6**, 133 (1966a).
— Australites, Henbury impact glass and subgreywacke: a comparison of the abundances of 51 elements. Geochim. Cosmochim. Acta **30**, 1121 (1966b).
— Geochemistry of andesites. In: Origin and Distribution of the Elements (ed. L. H. Ahrens), **30**, 559. Oxford: Pergamon Press 1968.
— Ewart, A., Capp, A. C.: Leucogranites and rhyolites: trace element evidence for fractional crystallization and partial melting. Lithos **1**, 2, 179 (1968).
— Heier, K. S., Sverdriep, T. L.: Contributions to the mineralogy of Norway. No. 5. Trace element variations in three generations of feldspars from the Landsverk 1 pegmatite. Norsk. Geol. Tidssk. **40**, 133 (1960).
— Kolbe, P.: Geochemical standards. Geochim. Cosmochim. Acta **28**, 447 (1964).
— White, A. J. R.: Trace element abundances in andesites. Bull. Vulcan. **29**, 174 (1967).
Tennant, W. C.: General spectrographic techniques for the determination of volatile trace elements in silicates. Appl. Spectry. **21**, 282 (1967).
Turekian, K. K., Wedepohl, K. H.: Distribution of the elements in some major units of the earth's crust. Bull. Geol. Soc. Am. **72**, 175 (1961).
Urey, H. C.: A review of atomic abundances in chondrites and the origin of meteorites. Rev. Geophys. **2**, 1 (1964).
— Craig, H.: The composition of the stone meteorites and the origin of the meteorites. Geochim. Cosmochim. Acta **4**, 36 (1953).
Vinogradov, A. P.: Regularity on distribution of chemical elements in the earth's crust. Geochem. Int. 1 (1956).
— Average contents of chemical elements in the principal types of igneous rocks of the earth's crust. Geochem. Int. 641 (1962).
— Meteoritic matter. Geochem. Int. 947 (1965).
Vlasov, K. A. (ed.): Geochemistry and Mineralogy of Rare Elements and Genetic Types of Their Deposits. Vol. I: Geochemistry of Rare Elements. Israel Program for Scientific Translations 1966.
Voskresenskaya, N. T.: Geochemistry of thallium and rubidium in igneous rocks. Geochem. Int. 599 (1959).
— Thallium content in the igneous rocks of the Greater Caucasus and some other regions of the U.S.S.R. Geochem. Int. 609 (1961a).
— Thallium in some hydrothermal deposits of the Greater Caucasus. Geochem. Int. 731 (1961b).
— Thallium in coal. Geochem. Int. 158 (1968).
— Fel'dman, V. I.: Behaviour of rubidium and thallium during autometasomatism of alkalic granitoids of the Taidut massif (Central Transbaikaliya). Geochem. Int. (abstract), 56 (1964).
—Karpova, I. S.: Thallium in ore minerals of the Verkhnyaya Kvaisa. Geochem. Int. 552 (1958).
— Shou-T'ien, S.: Geochemistry of alkali metals and Tl in the Musht granites (North-ern Caucasus). Geochem. Int. 534 (1961).
— Soboleva, L. T.: Once more on thallium in manganese minerals. Geochem. Int. 309 (1961).
— Timofeyeva, N. V., Topkhana, M.: Thallium in some minerals and ores of sedimentary origin. Geochem. Int. 851 (1962a).
— Titkova, N. F., Shulyakovskaya, N. S., Ts'ui-Ying, C.: Geochemistry of thallium, rubidium and lithium in the magmatic process. Geochem. Int. 282 (1962b).
— Usevich, T. D.: The occurrence of thallium in manganese minerals. Geochem. Int. 710 (1957).
Wager, L. R., Brown, G. M.: Layered Igneous Rocks. San Francisco: W. H. Freeman & Co. 1967.
Wahler, W.: Pulse-polarographische Bestimmung der Spurenelemente Zn, Cd, In, Tl, Pb und Bi in 37 geochemischen Referenzproben nach Voranreicherung durch selektive Verdampfung. Neues Jahrb. Mineral., Abhandl. **108**, 36 (1968).

WEBBER, G. R.: Second report of analytical data for CAAS syenite and sulphide standards. Geochim. Cosmochim. Acta **29**, 229 (1965).

WEISSBERG, B. G.: Gold-silver ore-grade precipitates from New Zealand thermal waters. Econ. Geol. **64**, 95 (1969).

WIIK, H. B.: The chemical composition of some stony meteorites. Geochim. Cosmochim. Acta **9**, 279 (1956).

WILLIS, J. P., AHRENS, L. H.: Some investigations on the composition of manganese nodules with particular reference to certain trace elements. Geochim. Cosmochim. Acta **26**, 751 (1962).

ZLOBIN, B. I.: Geochemistry of Tl in alkalic rocks. Geochem. Int. 560 (1958).

— LEBEDEV, V. I.: Geochemical relationship of Li, Na, K, Rb and Tl in alkalic magma and its petrogenetic significance. Geochem. Int. 101 (1960).

Revised manuscript received: March 1970

Lead 82

A	K. Sahl	(Institut für Mineralogie, Ruhr-Universität, Bochum, Germany)
B	B. R. Doe	(U.S. Geological Survey, Denver, Colorado, U.S.A.)
C—O	K. H. Wedepohl	(Geochemisches Institut der Universität, Göttingen, Germany)

82-A. Crystal Chemistry

The main lead mineral is galena, PbS. Many varieties of other lead sulfides of minor economic importance are known. Oxidation of lead sulfides has caused the occurence of a large number of oxidic lead compounds. Lead halides are relatively rare.

Lead has the electronic structure $6s^2\,6p^2$. In compounds, it is formally divalent or tetravalent, but the more stable state is the divalent one with electronic configuration $6s^2$. The lone pair of $6s$ electrons is considered to cause the tendency of divalent lead to one-sided coordinations.

I. Metallic Lead

Metallic $Pb^{[12]}$ is sometimes found in nature. The structure is the cubic close-packed arrangement with a Pb—Pb distance of 3.49 Å (Végard, SB 1916, 55). This distance is large compared with that in other metals of high atomic number (Au—Au = 2.88 Å in $Au^{[12]}$), and it is assumed that this is to be ascribed to incomplete ionization arising from the stability of the lone pair of $6s$ electrons.

Pb is a component of numerous intermetallic phases in which Pb is found in a large variety of coordinations.

II. Formally Divalent Lead

a) Lead Coordinated with Sulfur

The most common lead sulfide is galena, $Pb^{[6]}S^{[6]}$, with NaCl structure and Pb—S $= 6 \times 2.97$ Å (Davey, SB 1921, 131). The physical properties of galena (opacity, metallic lustre, etc.) do not correspond to ionic binding. The rare minerals clausthalite, $Pb^{[6]}Se^{[6]}$ (v. Olshausen, SB 1927, 137), and altaite, $Pb^{[6]}Te^{[6]}$ (Goldschmidt, SB 1926, 138), have the same structure as galena.

In betechtinite, $Pb_2^{[7]}(Cu,Fe)_{21}^{[4]}S_{15}$, Dornberger-Schiff and Höhne (SR 1959, 140—141) found a trigonal prismatic coordination around Pb with Pb—S $= 2 \times 2.93$ Å, 2×3.17 Å and 2×3.17 Å. An additional sulfur atom is sited beyond one of the prism faces at a distance of 3.30 Å. The prisms share their bases with their neighbors and thus form columns. The copper atoms (their distribution was found to be statistical) connect the columns laterally.

A variety of lead antimony sulfide minerals of general composition xPbS. ySb_2S_3 is known. Lead arsenic sulfides of similar composition are less abundant. The structure of jamesonite, $Fe^{[6]}Pb_2^{[8]}Pb_2^{[7]}[Sb_3^{[3\,py]}S_7]_2$, has been determined with accuracy (Niizeki and Buerger, SR 1957, 346—347). Short chains of composition Sb_3S_7 (SbS_3-groups sharing corners) are connected by Fe atoms in distorted octahedral sulfur coordination, and by two different Pb atoms coordinated by 7 or 8 sulfur atoms at distances ranging from 2.85 Å to 3.39 Å. The polyhedra around the Pb are irregular. The same coordination numbers of sulfur around lead were found by Leineweber (SR 1956, 30—32) in bournonite, $Cu_2^{[4]}Pb^{[7]}Pb^{[8]}[Sb^{[3\,py]}S_3]_2$, and in the isotypic structure of seligmannite, $Cu_2^{[4]}Pb^{[7]}Pb^{[8]}[As^{[3\,py]}S_3]_2$. The coordination polyhedra around lead are irregular with Pb—S varying from 2.95 to 3.59 Å in bournonite and from 2.89 to 3.52 Å in seligmannite. In both structures, the shorter Pb—S distances (below 3.0 Å) are arranged to one side of the lead atoms within a cone of less than 180°.

In marrite, $Pb^{[4\,py+2]}Ag^{[3+1]}[As^{[3\,py]}S_3]$, the distances Pb—S are 2.799Å, 2.837 Å, 2.962 Å, 2.976 Å, 3.136 Å and 3.260 Å (WUENSCH and NOWACKI, 1967). The coordination around Pb can be described as a highly distorted octahedron. The four shorter Pb—S bonds are found at one side of the lead atom, so that Pb forms the top of a flat PbS_4-pyramid. The structure of marrite superficially resembles the arrangement in galena, according to WUENSCH and NOWACKI (1967).

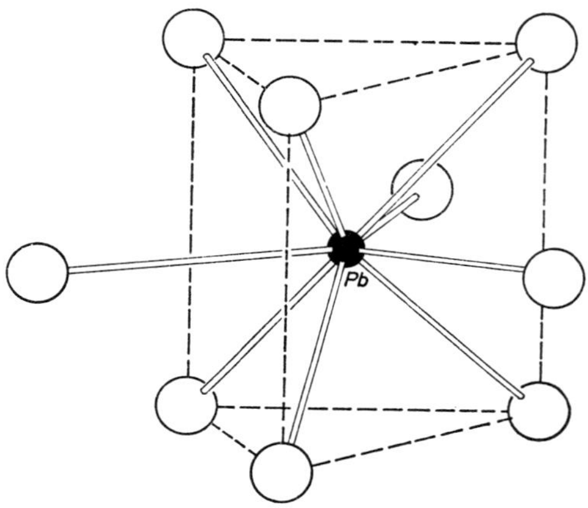

Fig. 82-A-1. Coordination polyhedron around lead as in scleroclase; nine neighbors form a trigonal prism with a further atom above each prism face

In scleroclase, $\frac{1}{\infty}Pb^{[9]}[As_2^{[3\,py]}S_4]$, IITAKA and NOWACKI (1961) have found lead atoms coordinated by nine sulfur atoms at distances between 2.95 Å and 3.42 Å. Six of the neighbors form a nearly trigonal prism, the other three form a triangle with corners located above the centre of the prism faces (Fig. 82-A-1). Similar coordination polyhedra have been found in rathite-I and in dufrenoysite. In rathite-I, $Pb_{11}Tl_1[As_{20}^{[3\,py]}S_{40}]$ (MARUMO and NOWACKI, 1965), two of the three different lead positions are coordinated by nine sulfur atoms forming a nearly trigonal prism with a further sulfur atom above each prism face. The third lead atom has seven neighbors: a trigonal prism with one further sulfur atom above one of the prism faces. The distances Pb—S range from 2.98 to 3.47 Å. The prisms form columns, sharing the bases. In dufrenoysite, $Pb_8^{[9]}Pb_8^{[7]}[As_{16}^{[3\,py]}S_{40}]$ (MARUMO and NOWACKI, 1967), the Pb atoms have the same coordination as in rathite-I with Pb—S between 2.80 and 3.50 Å.

From the described structures of lead sulfide compounds, it can be concluded that formally divalent lead is coordinated by sulfur with six to nine nearest neighbors at distances between 2.80 and 3.59 Å. The coordination polyhedra are often irregular. The regular octahedral coordination as in galena has not been found in any other compounds. The one-sided coordination in bournonite, seligmannite and marrite can possibly be explained by the presence of the inert pair of 6*s* electrons of lead. The trigonal prismatic environment with one or three further neighbors above the centre of the prism faces (betechtinite, scleroclase, rathite-I and dufrenoysite) constitutes a relatively regular polyhedron of seven or nine ligands and is frequently encountered in crystal chemistry.

b) Lead Coordinated by Oxygen

The oxide of divalent lead is known in two different modifications. The structure of litharge, the tetragonal (red) $Pb^{[4\,py]}O^{[4]}$, has been determined by Dickinson and Friauf (SB 1924, 121) and confirmed by neutron diffraction methods by Leciejewicz (1961 a). The lead atoms are located at the top of a tetragonal PbO_4-pyramid

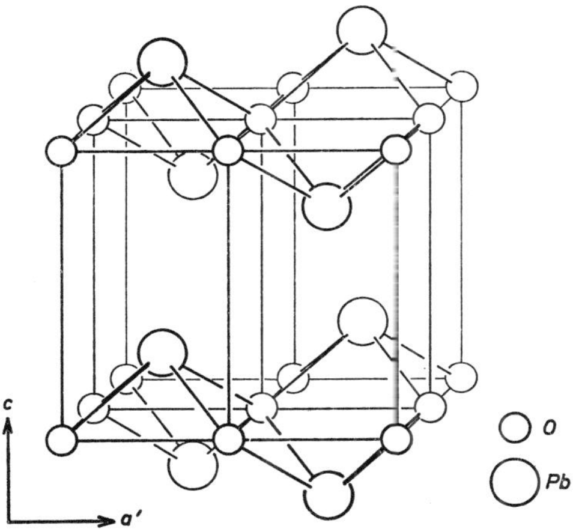

Fig. 82-A-2. Structure of tetragonal PbO (Dickinson and Friauf, SB 1924, 121)

with Pb—O $= 4 \times 2.33$ Å. The oxygen atoms are coordinated by a tetrahedron of lead atoms. The structure consists of layers of lead and layers of oxygen atoms in the sequence: Pb—2O—Pb—Pb—2O—Pb (Fig. 82-A-2). The shortest distance between lead atoms within the same layer is 3.99 Å. The shortest distance Pb—Pb between neighboring layers is 3.84 Å. However, the shortest distance between lead atoms separated by a layer of oxygen atoms is only 3.71 Å. According to Byström (SR 1947—1948, 237), the asymmetrical surrounding of the Pb atom by oxygen might be due to the inert pair of 6*s* electrons being situated at the opposite side of the oxygen atoms.

The structure of massicot, the orthorhombic (yellow) $Pb^{[4\,py]}O^{[4]}$, has been determined by Byström (SR 1947—1948, 238—239) and refined by Leciejewicz (1961 b) and by Kay (1961) with neutron diffraction methods. Here, lead is coordinated by four oxygens atoms to form a distorted PbO_4-pyramid with distances Pb—O $= 1 \times 2.214$ Å, 1×2.223 Å and 2×2.487 Å. The oxygen atoms are coordinated by a very distorted tetrahedron of lead atoms. The lead and oxygen atoms form zig-zag chains along the two shorter bonds [Leciejewicz (1961 b) assumes that the short Pb—O bonds are of a covalent nature]. The chains are connected by the two longer bonds to puckered layers in such a manner that the lead atoms form the surfaces of the layers. The shortest Pb—Pb distance is 3.47 Å within a layer and 3.63 Å between layers.

The structure of the oxides of tetravalent lead, PbO_2, and of the mixed oxide Pb_3O_4 will be discussed separately. Intermediate oxides of lead, PbO_x, have been described by BYSTRÖM (SR 1947—1948, 244—246) and by ANDERSON and STERNS (1959).

Cerussite, $Pb^{[9]}[C^{[3]}O_3]$ (COLBY and LACOSTE, SB 1933, 407—408), has a structure of the aragonite type. The coordination polyhedron around lead is irregular with distances Pb—O from 2.59 Å to 2.78 Å. The structure of hydrocerussite, $2\,Pb[CO_3]\cdot Pb(OH)_2$ (COWLEY, SR 1956, 386—388), has not been determined in detail because of extensive stacking disorder. The structure is described as consisting of layers with the following sequence of sheets within one layer: $[CO_3]$—Pb—(OH)—Pb—(OH)—Pb—$[CO_3]$.

In alamosite, $^{1}_{\infty}Pb^{[3\,py]}Pb_2^{[4\,py]}[Si_3^{[4]}O_9]$, BOUCHER and PEACOR (1968) have found three crystallographically different lead atoms with pyramidal one-sided coordination by three or four oxygen atoms. The distances are: Pb(1)—O = 2.29 Å, 2.28 Å and 2.60 Å; Pb(2)—O = 2.23 Å, 2.25 Å and 2×2.47 Å; Pb(3)—O = 2.30 Å, 2.31 Å, 2.47 Å and 2.58 Å. Chains of SiO_4-tetrahedra are connected by the lead atoms. The polyhedra around the lead atoms share edges or corners to form a continuous helical chain running parallel to the SiO_4-chains. Similar coordination polyhedra were found in larsenite, $Pb(1)^{[3\,py]}Pb(2)^{[4\,py]}Zn_2^{[4]}[Si^{[4]}O_4]_2$, with distances Pb(1)—O from 2.26 Å to 2.43 Å and Pb(2)—O from 2.35 Å to 2.65 Å (PREWITT *et al.*, 1967). The structure consists of five-membered rings of SiO_4- and ZnO_4-tetrahedra linked to form a network. The lead atoms are located in the channels of the ring system.

One-sided pyramidal coordination of lead atoms has also been described in the crystal structure of α—$K_2^{[9]}Pb_2^{[3\,py]}[Si_2O_7]$, a lead silicate usually known in the glass state (NÁRAY-SZABÓ and KÁLMÁN, 1961).

In $Pb^{[12]}[N^{[3]}O_3]_2$ (HAMILTON, SR 1957, 354), lead is coordinated by a regular icosahedron of oxygen atoms with Pb—O = 12×2.805 Å.

Pyromorphite, $Pb_2^{[9]}Pb_3^{[8]}[P^{[4]}O_4]_3Cl$ (MEHMEL, SB 1931, 457—460), and vanadinite, $Pb_2^{[9]}Pb_3^{[8]}[V^{[4]}O_4]_3Cl$ (TROTTER and BARNES, SR 1958, 438—439), have apatite structures. The eight-coordination is irregular, the nine-coordinationi s a trigonal prism with three further neighbors above the prism faces.

In carminite, $Pb^{[8]}Pb^{[6]}Fe_4^{[6]}[As^{[4\,py]}O_4]_4(OH)_4$ (FINNEY, 1963), the two crystallographically different lead atom positions are coordinated by eight, or by six to eight oxygen atoms respectively in an irregular manner. The distances Pb—O range from 2.48 to 3.06 Å. The structure consists of a network of $Fe(O,OH)_6$-octahedra joined by edges and vertices. The lead and arsenic atoms occupy vacancies in this network.

In fornacite, $Pb^{[9]}Pb^{[8\times O,\,1\times OH]}(Cu,Fe)^{[4\,pl+2]}[Cr^{[4]}O_4]\,[(As,P)^{[4]}O_4](OH)$ (COCCO *et al.*, 1967), one of the lead positions is coordinated irregularly by nine oxygen atoms with Pb—O = 2.35 Å to 3.03 Å, the shorter distances being arranged to one side of the lead atom. The coordination of the second lead atom by eight oxygen atoms and one (OH)-group is also irregular (Pb—O = 2.55 Å to 3.05 Å). The structure can be described in terms of linked polyhedra around lead forming thick sheets. Zig-zag chains of $CuO_4(OH)_2$-polyhedra and CrO_4- and $(As,P)O_4$-tetrahedra connect the sheets to a three-dimensional network. A similar structure was found in vauquelinite, $Pb^{[9]}Pb^{[9\times O,\,1\times OH]}Cu^{[4+2]}[Cr^{[4]}O_4]\,[P^{[4]}O_4](OH)$ (FANFANI and ZANAZZI, 1968) with distances Pb—O from 2.40 Å to 3.17 Å. FANFANI and ZANAZZI (1967) have pointed out the structural similarity of fornacite and vauquelinite with brackebuschite, $Pb^{[8]}Pb^{[9\times O, 1\times H_2O]}(Mn,Fe)^{[4]}[V^{[4]}O_4]_2 \cdot H_2O$ (DONALDSON and BARNES, SR 1955, 451—453), tsumebite ($Pb_2Cu[PO_4][SO_4](OH)$), and As-tsumebite ($Pb_2Cu[AsO_4][SO_4](OH)$).

Anglesite, β-$Pb^{[12]}[S^{[4]}O_4]$, is a quite common lead mineral. The structure consists of isolated SO_4-tetrahedra linked by lead atoms in an irregular 12-coordination with Pb—O = 2.63 Å to 3.05 Å (JAMES and WOOD, SB 1925, 384—385 and SAHL, 1963). The structure of the high temperature α-$PbSO_4$ is unknown.

In linarite, $Pb^{[1\times OH, 2\times O]}Cu^{[4\times OHpl]}[S^{[4]}O_4](OH)_2$, BACHMANN and ZEMANN (1961) have found the lead atoms in one-sided pyramidal coordination, with Pb—OH = 2.38 Å and Pb—O = 2×2.44 Å.

Lead titanate has a structure of the perovskite type above the Curie point: $Pb^{[12]}Ti^{[6]}O_3$. The ferroelectric phase below the Curie point has a deformed perovskite structure with tetragonal symmetry. The twelve neighbors of the lead atom are split into three groups with distances Pb—O = 4×2.53 Å, 4×2.80 Å, and 4×3.20 Å. Considering only the four nearest neighbors, the coordination polyhedron of lead is a PbO_4-pyramid (SHIRANE *et al.*, SR 1956, 288—289).

$PbZrO_3$ crystallizes in the perovskite structure above 230° C and in a pseudotetragonal, orthorhombic deformation of the perovskite structure below this temperature: antiferroelectric $Pb^{[3]}Zr^{[6]}O_3$. The lead atoms form the top of a rather flat PbO_3-pyramid with Pb—O = 1×2.53 Å and 2×2.58 Å (JONA *et al.*, SR 1957, 321—322).

The coordination of lead in pyrobelonite, $Pb^{[1\times OH, 2\times O]}Mn^{[6]}[V^{[4]}O_4](OH)$, is a trigonal $Pb(OH)O_2$-pyramid with Pb—OH = 2.3 Å and Pb—O = 2×2.5 Å (DONALDSON and BARNES, SR 1955, 451). These distances might be subject to slight changes with further refinement (QURASHI and BARNES, 1964). The coordination polyhedron around lead corresponds closely to the one in linarite (BACHMANN and ZEMANN, 1961).

In descloizite, $Pb^{[8]}(Zn,Cu)^{[6]}[V^{[4]}O_4](OH)$, QURASHI and BARNES (1964) have described a distorted square antiprism of oxygen atoms around lead with Pb—O ranging from 2.51 to 3.08 Å.

In crocoite, the monoclinic modification of $Pb^{[10]}[Cr^{[4]}O_4]$, lead is surrounded in an irregular manner by 10 oxygen atoms at distances between 2.53 Å and 3.44 Å (QUARENI and DE PIERI, 1965). The rhombic modification of lead chromate is isotypic with β-$PbSO_4$ (anglesite) according to COLLOTTI *et al.* (SR 1959, 455).

Wulfenite, $Pb^{[8]}[Mo^{[4]}O_4]$, has a scheelite-type structure. Each lead atom is surrounded by eight oxygen atoms belonging to eight different MoO_4-tetrahedra with Pb—O = 4×2.610 Å and 4×2.630 Å .The eight oxygen atoms are at the corners of two interpenetrating tetrahedra, thus forming a relatively regular coordination polyhedron (LECIEJEWICZ, 1965). Stolzite, $PbWO_4$, is isotypic with wulfenite according to an older investigation by VÉGARD and REFSUM (SB 1927, 386).

In the described structures of lead oxygen compounds, a large variety of coordination polyhedra around lead has been encountered. The coordination numbers range from three to twelve, the distances Pb—O from 2.22 Å (orthorhombic PbO) to well above 3.0 Å. Rather typical for formally divalent lead is the frequent occurrence of one-sided coordination polyhedra with short (or even very short) Pb—O distances in comparison to those in "all-sided" coordinations.

c) Lead Coordinated by Halides

Lead fluoride is known in two different modifications: a cubic β-$Pb^{[8]}F^{[4]}$ form stable above 220° C with CaF_2-structure (KOLDERUP, SB 1924, 190—191) and an orthorhombic low-temperature α-form which is isotypic with $PbCl_2$ (for description of the structure see $PbCl_2$). In a neutron diffraction reinvestigation of the structure of α-$Pb^{[7]}F^{[4]}F^{[5]}$, BOLDRINI and LOOPSTRA (1967) have found the following distances Pb—F: 2×2.41 Å, 1×2.45 Å, 2×2.53 Å, 1×2.64 Å and 1×2.69 Å (additional neighbors at 3.03 Å).

According to an older investigation by NIEUWENKAMP and BIJVOET (SB 1932, 362), matlockite, PbFCl, crystallizes in a tetragonal layer structure with layer sequence: Cl—Pb—F_2—Pb—Cl.

In cotunnite, $Pb^{[7]}Cl(1)^{[4]}Cl(2)^{[5]}$, the lead atoms are coordinated by seven chlorine atoms with distances Pb—Cl of 1×2.86 Å, 2×2.90 Å, 1×3.06 Å, 1×3.08 Å and 2×3.08 Å. Further neighbors follow at 3.64 Å (SAHL, 1963). The three closest chlorine atoms form a steep pyramid to one side of the lead atom[1]. Considering all seven neighbors, the coordination polyhedron around lead can be described as a distorted trigonal prism with one additional chlorine atom above the centre of one prism face. The coordination around Cl(1) consists of a distorted Pb-tetrahedron. The irregular 5-coordination around Cl(2) is more spacious than the tetrahedron around Cl(1).

$PbBr_2$ (NIEUWENKAMP and BIJVOET, SB 1932, 253—254), Pb(OH)Cl (BRASSEUR, SR 1940, 131) and Pb(OH)I (NÄSÄNEN *et al.*, 1964) are isotypic with $PbCl_2$. The distances in these compounds are not known with accuracy.

CALINGAERT *et al.* (SR 1949, 287—288) have investigated the isomorphous replacement of Cl by Br in the system $PbCl_2$—$PbBr_2$. They found that bromine first enters into the more spacious Cl(2) positions in the structure of $PbCl_2$.

According to VAN ARKEL (SB 1926, 191), PbI_2 forms a layer structure of the CdI_2-type with layer sequence I—Pb—I—I—Pb—I. In the CdI_2-type structure, the metal atoms are coordinated by a distorted octahedron of anions.

III. Formally Tetravalent Lead

Plattnerite, β-$Pb^{[6]}O_2$ has a rutile structure. The lead atom is coordinated by an octahedron of oxygen atoms with Pb—O = 4×2.154 Å and 2×2.163 Å (LECIEJEWICZ and PADLO, 1962). By anodic deposition of lead salt solutions, a rhombic modification is obtained: α-$Pb^{[6]}O_2$, with a structure of the niobite type, Pb replacing Fe and Nb. The distances in the Pb—O octahedron are: 2×2.16 Å, 2×2.17 Å and 2×2.22 Å (ZASLAVSKIJ *et al.*, SR 1950, 186).

In minium, $Pb_2^{[3\,py]}Pb^{[6]}O_4$, one of the lead atoms is coordinated by a nearly undistorted octahedron of oxygen atoms with Pb—O = 4×2.14 Å and 2×2.13 Å. Sharing common edges, the octahedra form chains which are connected through the other two lead atoms, located at the top of a trigonal pyramid PbO_3 with Pb—O = 2×2.18 Å and 1×2.13 Å (FAYEK and LECIEJEWICZ, 1965). The lead atoms in pyramidal environment can be considered as formally divalent. Since the sum of the ionic radii of Pb^{4+} and O^{2-} for the coordination number six is 2.24 Å, FAYEK and LECIEJEWICZ (1965) conclude that the shorter bonding in the PbO_6-octahedra is probably covalent.

[1] This pyramid of short distances is not present in the structure of $BaCl_2$, which is otherwise isotypic with $PbCl_2$ (SAHL, 1963). The difference between the two structures might be due to the presence of the lone pair of 6*s* electrons of lead.

Revised manuscript received: November 1968

References: Section 82-A

ANDERSON, J. S., and M. STERNS: The intermediate oxides of lead. J. Inorg. & Nucl. Chem. **11**, 272—285 (1959)

BACHMANN, H.-G., u. J. ZEMANN: Die Kristallstruktur von Linarit, $PbCuSO_4(OH)_2$. Acta Cryst. **14**, 747—753 (1961).

BOLDRINI, P., and B. O. LOOPSTRA: Neutron investigation of orthorhombic lead (II) fluoride. Acta Cryst. **22**, 744—745 (1967).

BOUCHER, M. L., and D. R. PEACOR: The crystal structure of alamosite, $PbSiO_3$. Z. Krist. **126**, 98—111 (1968).

COCCO, G., L. FANFANI, and P. F. ZANAZZI: The crystal structure of fornacite. Z. Krist. **124**, 385—397 (1967).

FANFANI, L., and P. F. ZANAZZI: Structural similarities of some secondary lead minerals. Mineral Mag. **36**, 522—529 (1967).

— — The crystal structure of vauquelinite and the relationships to fornacite. Z. Krist. **126**, 433—443 (1968).

FAYEK, M. K., and J. LECIEJEWICZ: Neutron diffraction study of Pb_3O_4. Z. Anorg. Allgem. Chem. **336**, 104—109 (1965).

FINNEY, J. J.: The crystal structure of carminite. Am. Mineralogist **48**, 1—13 (1963).

IITAKA, Y., and W. NOWACKI: A refinement of the pseudo crystal structure of scleroclase, $PbAs_2S_4$. Acta Cryst. **14**, 1291—1292 (1961).

KAY, I.: A neutron diffraction study of orthorhombic PbO. Acta Cryst. **14**, 80—81 (1961).

LECIEJEWICZ, J.: On the crystal structure of tetragonal (red) PbO. Acta Cryst. **14**, 1304-1305 (1961a).

— Neutron-diffraction study of orthorhombic lead monoxide. Acta Cryst. **14**, 66 (1961b).

— A neutron crystallographic investigation of lead molybdenum oxide, $PbMoO_4$. Z. Krist. **121**, 158—164 (1965).

—, and I. PADLO: Note on the oxygen parameter in tetragonal PbO_2. Naturwissenschaften **49**, 373—374 (1962).

MARUMO, F., and W. NOWACKI: The crystal structure of rathite-I. Z. Krist. **122**, 433—456 (1965).

— — The crystal structure of dufrenoysite, $Pb_{16}As_{16}S_{40}$. Z. Krist. **124**, 409—419 (1967).

NÄSÄNEN, R., P. MERILÄINEN, R. UGGLA, and M. HYLE: The crystal structure of Pb(OH)I. Suomen Kemistilehti **B 37**, No. 4, 45—46 (1964).

NÁRAY-SZABÓ, I., u. A. KÁLMÁN: Die Struktur des $K_2Pb_2Si_2O_7$ und der Bleigläser. Silikat Techn. **12**, Nr. 7 (1961).

PREWITT, C. T., E. KIRCHNER, and A. PREISINGER: Crystal structure of larsenite, $PbZnSiO_4$. Z. Krist. **124**, 115—130 (1967).

QUARENI, S., and R. DE PIERI: A 3-dimensional refinement of the structure of crocoite, $PbCrO_4$. Acta Cryst. **19**, 287—289 (1965).

QURASHI, M. M., and W. H. BARNES: The structure of the minerals of the descloizite and adelite groups: V — Descloizite and conichalcite (Part 3). The structure of descloizite. Canad. Mineralogist **8**, 23—39 (1964).

SAHL, K.: Die Verfeinerung der Kristallstrukturen von $PbCl_2$ (Cotunnit), $BaCl_2$, $PbSO_4$ (Anglesit) und $BaSO_4$ (Baryt). Beitr. Mineral. Petrog. **9**, 111—132 (1963).

WUENSCH, B. J., and W. NOWACKI: The crystal structure of marrite, $PbAgAsS_3$. Z. Krist. **125**, 459—488 (1967).

Revised manuscript received: November 1968

82-B. Isotopes in Nature[1]

I. Introduction

The geochemistry of lead isotopes is intimately associated with that of uranium and thorium and in consequence, the preparation of separate chapters on uranium, thorium and lead has necessitated a subdivision of topics involving these three elements. This section contains a discussion of U—Th—Pb dating, observed values of $^{238}U/^{204}Pb$ and $^{232}Th/^{204}Pb$ ratios in various rock types, some consideration of values of the Th/U ratio predicted from lead isotopic compositions, and some geological applications of radioactive lead that occurs as an intermediate daughter in the uranium and thorium decay chains. Chapter 92 (Uranium) includes a discussion of both uranium and thorium in connection with heat production in the earth, autoradiographic studies, fractionation of thorium and uranium during weathering and sedimentation, and the utilization of nuclear energy. Chapter 90 (Thorium) includes discussions of age determinations by means of radioactive disequilibria of short-lived uranium and thorium isotopes in the uranium and thorium decay chains, Th/U ratios in igneous rocks, and the general relationship of thorium and uranium to igneous petrology.

a) General

Lead has four stable isotopes:

Symbol	Mass (Bhanot et al., 1960)	Typical extreme abundances (%)	
		Southeast Missouri lead ore (Brown, 1962)	Cœur d'Alene district lead ore Idaho (Long *et al.*, 1960)
^{204}Pb	204.038	~ 1.28	1.38
^{206}Pb	206.040	~26.73	23.65
^{207}Pb	207.042	~20.51	22.40
^{208}Pb	208.043	~51.48	52.51

The isotopic abundances of the lead isotopes vary considerably because three of them are generated by radioactive decay: ^{206}Pb by decay of ^{238}U, ^{207}Pb by decay of ^{235}U, and ^{208}Pb by decay of ^{232}Th, ^{204}Pb has no long-lived radioactive parent. Shown above, therefore, are only the approximate average compositions of the two isotopically extreme major lead-producing deposits — the Cœur d'Alene in Idaho and the southeast Missouri lead belt — both in the U.S.A. The modes of decay

[1] Publication authorized by the Director, U.S. Geological Survey.

are complex, each parent-daughter series involving many relatively short-lived radioactive intermediate daughters before the stable lead isotope is reached.

Variations in lead isotopic composition arise directly or indirectly according to the equations in Table 82-B-1 or according to mixtures of systems that have had different values of U/Pb and Th/Pb for long periods of time. The relevant decay constants for the equations are:

$\lambda(^{238}U) = 1.54 \times 10^{-10} yr^{-1}$ (KOVARIK and ADAMS, 1955)
$\lambda(^{235}U) = 9.71 \times 10^{-10} yr^{-1}$ (FLEMING *et al.*, 1952)
$\lambda(^{232}Th) = 4.99 \times 10^{-11} yr^{-1}$ (KOVARIK and ADAMS, 1938; PICCIOTTO and WILGAIN, 1956).

Of utmost importance is the fact that the present abundance ratio of $^{238}U/^{235}U$ is nearly constant in nature:

$$\left(\frac{^{238}U}{^{235}U}\right) = 137.8 \quad \text{(INGHRAM, 1946).}$$

The values of the constants given above are those recommended in the careful evaluation by ALDRICH and WETHERILL (1958) and differ somewhat from those used by STIEFF *et al.* (1959) in compiling their tables for age calculations. Except for $\lambda(^{232}Th)$, however, the differences are within the uncertainties of the constants. For $\lambda(^{232}Th)$, STEIGER and WASSERBURG (1966) show that the constant recommended by ALDRICH and WETHERILL (1958) gives ages in better agreement with other methods than does the method used by STIEFF *et al.* (1959).

Physico-chemical variations such as those found for hydrogen-deuterium, carbon, oxygen, and sulfur are very small for heavy isotopes such as lead. There is no evidence as yet of any variations in lead isotopic composition caused by anything other than radioactive decay or mixing of leads of different isotopic compositions.

This chapter is in large part the condensation of a monograph on lead isotopes (DOE, 1970). Those who wish to see expanded discussions on the topics covered in this chapter are referred to this monograph.

b) Isotopic Measurements

Mass Spectrometry. By far the most popular method of analyzing lead isotope abundances is by mass spectrometry, preferably without an electron multiplier in the ion collector. The most precise method (0.05 percent of a ratio) appears to be that involving the use of a thermal ionization source (triple filament) (CATANZARO, 1967); however, about 500 micrograms of lead are required. Ratios are nearly absolute ($<0.1\%$ correction). This method is not yet very popular because of the large sample size required and because most mass spectrometers require modification for use in this manner.

Perhaps most data are generated using a surface emission technique where PbS and NH_4NO_3 are mixed on a single filament. This method has been evaluated by DOE *et al.* (1967) and although as little as 10 micrograms of lead may be analyzed, the reproducibility is not as good as that quoted above (about 0.05 percent per mass unit separation at best) and corrections for mass spectrometer fractionation are necessary.

Table 82-B-1. *Equations and parameters used in lead isotope calculations*

1. Basic equation: $-\frac{dN}{dt} = \lambda N$

   N = number of radioactive atoms
   t = time
   λ = constant of proportionality (decay constant)

2. Working equations (atomic ratios):

   a) *$^{206}Pb/^{238}U$ age*

   $$\left(\frac{^{206}\mathrm{Pb}}{^{204}\mathrm{Pb}}\right)_{\mathrm{observed}} - \left(\frac{^{206}\mathrm{Pb}}{^{204}\mathrm{Pb}}\right)_{\mathrm{initial}} = \left(\frac{^{238}\mathrm{U}}{^{204}\mathrm{Pb}}\right)_{\mathrm{observed}} \cdot (e^{\lambda_8 T} - 1)$$

   b) *$^{207}Pb/^{235}U$ age*

   $$\left(\frac{^{207}\mathrm{Pb}}{^{204}\mathrm{Pb}}\right)_{\mathrm{observed}} - \left(\frac{^{207}\mathrm{Pb}}{^{204}\mathrm{Pb}}\right)_{\mathrm{initial}} = \left(\frac{^{235}\mathrm{U}}{^{204}\mathrm{Pb}}\right)_{\mathrm{observed}} \cdot (e^{\lambda_5 T} - 1)$$

   c) *$^{208}Pb/^{232}Th$ age*

   $$\left(\frac{^{208}\mathrm{Pb}}{^{204}\mathrm{Pb}}\right)_{\mathrm{observed}} - \left(\frac{^{208}\mathrm{Pb}}{^{204}\mathrm{Pb}}\right)_{\mathrm{initial}} = \left(\frac{^{232}\mathrm{Th}}{^{204}\mathrm{Pb}}\right)_{\mathrm{observed}} \cdot (e^{\lambda_2 T} - 1)$$

   d) *Pb—Pb age or isochron age*

   $$\frac{\left(\frac{^{207}\mathrm{Pb}}{^{204}\mathrm{Pb}}\right)_{\mathrm{observed}} - \left(\frac{^{207}\mathrm{Pb}}{^{204}\mathrm{Pb}}\right)_{\mathrm{initial}}}{\left(\frac{^{206}\mathrm{Pb}}{^{204}\mathrm{Pb}}\right)_{\mathrm{observed}} - \left(\frac{^{206}\mathrm{Pb}}{^{204}\mathrm{Pb}}\right)_{\mathrm{initial}}} = \left(\frac{^{235}\mathrm{U}}{^{238}\mathrm{U}}\right)_{\mathrm{observed}} \cdot \frac{(e^{\lambda_5 T} - 1)}{(e^{\lambda_8 T} - 1)}$$

   e) *Explanation of abbreviated symbols*

   T = geologic age.
   λ = decay constant; subscript is the last mass number of the radioactive parent.
   Initial ratio is that ratio which the phase contained when it was formed.

3. Working equations for common lead calculations (atomic ratios)

Ratio at geologic time (t)	Ratio when earth formed	Ratio in earth today	Decay from time earth was formed (T) to the geologic time of interest (t)

Primary growth equations

a) $$\left(\frac{^{206}\mathrm{Pb}}{^{204}\mathrm{Pb}}\right)_t = \left(\frac{^{206}\mathrm{Pb}}{^{204}\mathrm{Pb}}\right)_T + \left(\frac{^{238}\mathrm{U}}{^{204}\mathrm{Pb}}\right)_n \cdot (e^{\lambda_8 T} - e^{\lambda_8 t}).$$

b) $$\left(\frac{^{207}\mathrm{Pb}}{^{204}\mathrm{Pb}}\right)_t = \left(\frac{^{207}\mathrm{Pb}}{^{204}\mathrm{Pb}}\right)_T + \left(\frac{^{235}\mathrm{U}}{^{204}\mathrm{Pb}}\right)_n \cdot (e^{\lambda_5 T} - e^{\lambda_5 t}).$$

c) $$\left(\frac{^{208}\mathrm{Pb}}{^{204}\mathrm{Pb}}\right)_t = \left(\frac{^{208}\mathrm{Pb}}{^{208}\mathrm{Pb}}\right)_T + \left(\frac{^{232}\mathrm{Th}}{^{204}\mathrm{Pb}}\right)_n \cdot (e^{\lambda_2 T} - e^{\lambda_2 t}).$$

Primary isochron equation [sometimes referred to as the Gerling-Houtermans-Holmes equation (GERLING, 1942; HOUTERMANS, 1946; HOLMES, 1946)]

d) $$\frac{\left(\frac{^{207}\mathrm{Pb}}{^{204}\mathrm{Pb}}\right)_t - \left(\frac{^{207}\mathrm{Pb}}{^{204}\mathrm{Pb}}\right)_T}{\left(\frac{^{206}\mathrm{Pb}}{^{204}\mathrm{Pb}}\right)_t - \left(\frac{^{206}\mathrm{Pb}}{^{206}\mathrm{Pb}}\right)_T} = \left(\frac{^{235}\mathrm{U}}{^{238}\mathrm{U}}\right)_n \left(\frac{e^{\lambda_5 T} - e^{\lambda_5 t}}{e^{\lambda_8 T} - e^{\lambda_8 t}}\right).$$

Table 82-B-1. (Continued)

Secondary growth equations

	Normal growth stage $\left(\frac{20\,X\,Pb}{^{204}Pb}\right)_t$ [X = 6,7,8]		Secondary growth
e)	$\left(\frac{^{206}Pb}{^{204}Pb}\right)_{t'} = \left(\frac{^{206}Pb}{^{204}Pb}\right)_T + \left(\frac{^{238}U}{^{204}Pb}\right)_n \cdot (e^{\lambda_8 T} - e^{\lambda_8 t})$	+	$\left(\frac{^{238}U}{^{204}Pb}\right)_n \cdot (e^{\lambda_8 t} - e^{\lambda_8 t'})$.
f)	$\left(\frac{^{207}Pb}{^{204}Pb}\right)_{t'} = \left(\frac{^{207}Pb}{^{204}Pb}\right)_T + \left(\frac{^{238}U}{^{204}Pb}\right)_n \cdot (e^{\lambda_5 T} - e^{\lambda_5 t})$	+	$\left(\frac{^{235}U}{^{204}Pb}\right)'_n \cdot (e^{\lambda_5 t} - e^{\lambda_5 t'})$.
g)	$\left(\frac{^{208}Pb}{^{204}Pb}\right)_{t'} = \left(\frac{^{208}Pb}{^{204}Pb}\right)_T + \left(\frac{^{238}U}{^{204}Pb}\right)_n \cdot (e^{\lambda_2 T} - e^{\lambda_5 t})$	+	$\left(\frac{^{232}Th}{^{204}Pb}\right)'_n \cdot (e^{\lambda_2 t} - e^{\lambda_2 t'})$,

where $\left(\frac{^{238}U}{^{204}Pb}\right)''_n = \left(\frac{^{235}U}{^{204}Pb}\right)'' = \left(\frac{^{232}Th}{^{204}Pb}\right)''_n = 0$.

h) Secondary isochron:

$$\frac{\left(\frac{^{207}Pb}{^{204}Pb}\right)_{t'} - \left(\frac{^{207}Pb}{^{204}Pb}\right)_t}{\left(\frac{^{206}Pb}{^{204}Pb}\right)_{t'} - \left(\frac{^{206}Pb}{^{206}Pb}\right)_t} = \left(\frac{^{235}U}{^{238}U}\right)_n \cdot \left(\frac{e^{\lambda_5 t} - e^{\lambda_5 t'}}{e^{\lambda_8 t} - e^{\lambda_8 t'}}\right).$$

i) Explanation of symbols, their equivalents in other conventions, and present estimates of key values with references to data.

	Canadian convention	Swiss convention	Value	Reference
$\left(\frac{^{206}Pb}{^{204}Pb}\right)_t$	X	α	$t=0$; 18.66	1
			18.51	2
$\left(\frac{^{206}Pb}{^{204}Pb}\right)_T$	a_0	α_0	9.56	3
			9.54	4
			9.346	5
$\left(\frac{^{238}U}{^{204}Pb}\right)_n$	137.8 V	μ_0	8.99	1
			8.7	2
			9.09	6
$\left(\frac{^{207}Pb}{^{204}Pb}\right)_t$	Y	β	$t=0$; 15.79	1
			15.72	2
$\left(\frac{^{207}Pb}{^{204}Pb}\right)_T$	b_0	β_0	10.42	3
			10.27	4
			10.218	5
$\left(\frac{^{235}U}{^{204}Pb}\right)_n$	V	$\frac{\mu_0}{137.8}$		
$\left(\frac{^{208}Pb}{^{204}Pb}\right)_t$	Z	γ	$t=0$; 39.06	1
			38.44	2
$\left(\frac{^{208}Pb}{^{204}Pb}\right)_T$	c_0	γ_0	30.0	1
			29.7	3
			29.46	4
			28.96	5
$\left(\frac{^{232}Th}{^{204}Pb}\right)_n$	W	$\mu \cdot k$	35.55	1
			34.8	2
			38.3	6
$\left(\frac{^{232}Th}{^{238}U}\right)$	$\frac{W}{V}$	k	3.92	1
			4.0	2
			3.8	3
			4.21	6

References: 1. Kanasewich (1968a); 2. Doe (1962b); 3. Murthy and Patterson (1962); 4. Cow and Patterson (1961), corrected according to Chow and Patterson (1962b); 5. Oversby (1969); 6. Stacey *et al.* (1969).

Below 10 micrograms of lead a refractory high temperature emittor is used, the most common type being silica gel with or without other high temperature additives such as $ZrSiO_4$ or H_2PO_4 (AKISHIN *et al.*, 1957; CAMERON *et al.*, 1969). The analytical uncertainties involved are not yet fully evaluated; however, TATSUMOTO and ROSHOLT (1970) have analyzed 0.2 microgram of lead.

Perhaps the greatest volume of precise published lead isotope data uses the gaseous lead tetramethyl method with electron bombardment ionization. This method has been compared with the thermal emission method by STACEY *et al.* (1969). Aside from the large sample size required, the tetramethyl method has several defects: the high toxicity of the compound being analyzed, problems of instrumental clean-up to prevent contamination between analyses, and corrections for instrumental fractionation of the data due to the gas leak. The more versatile thermal emission method now appears to be replacing the tetramethyl method.

Activation Analysis. Although it is possible to measure relative abundances of ^{204}Pb, ^{206}Pb, and ^{208}Pb by various kinds of activation analyses, the surface emission method with refractory emittors requires less speed of analysis, is probably more accurate, and measures lead quantities down to the contamination level involved in activation work. The activation method remains more of a novelty.

II. U—Th—Pb Dating

a) General

Three independent ages may be obtained in the U—Th—Pb system: $^{206}Pb/^{238}U$, $^{207}Pb/^{235}U$ or $^{207}Pb/^{206}Pb$, and $^{208}Pb/^{232}Th$. Emphasis has been placed on U—Pb dating because the value $^{238}U/^{235}U$ is a physical constant that permits internal treatment of the data not found in any other dating system. This treatment helps to eliminate the assumption that the phase being dated has remained closed to changes in the parent-daughter system. The theoretical systematics are expressed in Fig. 82-B-1. WETHERILL (1956a, b) showed that a phase, which is subject to no lead loss or uranium gain (a closed system), will have $^{207}Pb/^{235}U$ ages equal to $^{206}Pb/^{238}U$ ages and that the data will lie along a curved line called *concordia*. In addition, he showed that phases subject to lead loss or uranium gain during a period of time that is short compared with the age of the phase (episodic bulk daughter losses or parent gains), recently or in the distant past, will have data that will lie along a straight line called *discordia*. The lower intersection of discordia with concordia represents the time of the episodic event and the upper intersection represents the age of the phase. NICOLAYSEN (1957) suggested that diffusion of lead out of a phase might take place at a constant rate over the entire history of the phase (continuous diffusion), and TILTON (1960) showed that continuous diffusion with bulk lead loss or uranium gain also closely approaches a straight line over much of its length. WASSERBURG (1963) considered the possible effects of a diffusion constant that varies with time, such as might be caused by radiation damage; WETHERILL (1963) discussed most other aspects of the theory of U—Pb behavior, such as continuous diffusion with superimposed episodic loss and the theory of uranium loss and bulk radiogenic daughter gain. The U—Th—Pb system has intermediate radioactive daughters, including the gas radon. If there is radon loss over a long period of

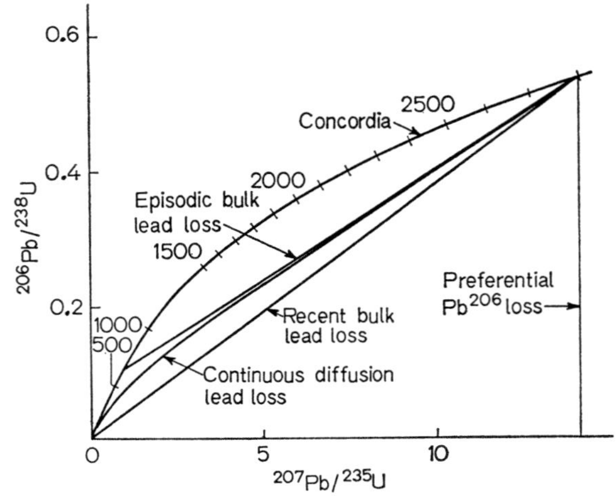

Fig. 82-B-1. Systematics of U—Pb dating. (After CATANZARO and KULP, 1964)

time, the calculated ages may be affected. If the system is subject to radon leakage, the result is reflected in preferential loss of ^{206}Pb, because the radon in the ^{238}U chain has a much longer half-life (3.8 days) than that in the ^{235}U chain (3.9 seconds). The longer half-life allows a longer time for radon escape.

Though considerable data have been obtained on Th—Pb dating, this dating has received less emphasis because of the lack of a companion radioactive parent isotope to go with ^{232}Th, so that Th—Pb dating lacks the protection against parent and daughter losses or gains inherent in the U—Pb dating method. The theoretical behavior of Th—Pb dating in terms of a U—Th—Pb concordia diagram is, however, given by STEIGER and WASSERBURG (1966).

Most U—Th—Pb dating has been done on uraninite, pitchblende, and other uranium minerals. Uraninite under some conditions gives the most concordant ages, but it is not a common accessory mineral in most rock types. Emphasis, therefore, will here be placed on the more common accessories — zircon, sphene, phosphates (particularly monazite) — and on whole-rock dating. Zircon has been subjected to the most thorough and rigorous laboratory and field investigation and is therefore chosen for first discussion.

b) Zircon ($ZrSiO_4$)

The isotopic ages found in zircon (selected data given in Table 82-B-2) are generally discordant with $^{206}Pb/^{238}U$ age $<$ $^{207}Pb/^{235}U$ age $<$ $^{207}Pb/^{206}Pb$ age. This relationship of ages is often referred to as *normal discordance*. One such example of zircons with discordant ages is the study by SILVER and DEUTSCH (1963) which was particularly useful in that the data lie along an episodic lead-loss line, and it was the first study that verified the episodic lead-loss hypothesis. The data do not lie along continuous diffusion lines and, in fact, no clear example has yet been found to support the continuous diffusion hypothesis. A good example of episodic lead loss (and uranium gain) is given in the study of DAVIS *et al.* (1968) presented in

Table 82-B-2. *U—Th—Pb data for selected minerals*

Geologic unit	Locality	Element concentration (ppm)			Atomic ratios		
		U	Th	Pb	$^{206}Pb/^{204}Pb$	$^{207}Pb/^{204}Pb$	$^{208}Pb/^{204}Pb$
		A. Zircons (from CATANZARO and KULP, 1964)					
Fresh granite gneiss	Little Belt Mountains, Montana	2,227	1,421	261	229	36.6	59.1
		549	217	138	694	90.1	126.9
		1,417	276	216	658	90.1	95.2
		553	159	115	1,042	127.2	121.3
		710	346	123	629	81.4	106.7
		1,047	338	176	303	46.4	74.2
		B. Sphenes (OOSTHUYZEN and BURGER, 1965)					
Syenite	South Africa	63.8	230	32.1	121.4	24.47	149.5
		—	—	—	28.49 [a]	16.70 [a]	50.48 [a]
		C. Apatite (OOSTHUYZEN and BURGER, 1965)					
Syenite	South Africa	20.7	—	28	38.16	17.60	83.58
		D. Monazite (CATANZARO and KULP, 1964; HEIMLICH and BANKS, 1968)					
Granite gneiss	Little Belt Mountains, U.S.A.	0.213	8.84	0.802	312.5	50.2	35.4
	Bighorn Mountains, U.S.A.	0.0830	—	0.998 [b]	206.3	53.8	4,521

[a] Acid leach of sphene.
[b] radiogenic lead only.

Figs. 82-B-2 and 82-B-3. The zircons are from the Front Range of Colorado from Precambrian rocks adjacent to Tertiary igneous stocks; in this case, the episodic event was intrusion of the Tertiary stocks. Hydrothermal experiments on a metamict Ceylon zircon in 2 molal NaCl at 500° C and 1,000 bars (PIDGEON *et al.*, 1966) showed a very rapid episodic lead loss of approximately 50 percent of the lead in 11 hours. Uranium content was little affected. Apparently weathering can also cause discordance in U—Pb ages of zircon (STERN *et al.*, 1966). The conclusion from these studies is that zircon ages are usually discordant and that this discordance may be caused rapidly by geological processes not usually considered to be strong. Also important is the fact that usually only part of the lead is lost. If the Wetherill concordia diagram is used, reliable mineral ages can be obtained.

The mechanism which causes the discordant ages in zircon is not yet known. The most promising approach seems to be consideration of the formation of domains rich in H_2O, heavy trace elements (including uranium and thorium), and yttrium, such as discussed by GRÜNENFELDER (1963), GRÜNENFELDER and EBERHARD (1969),

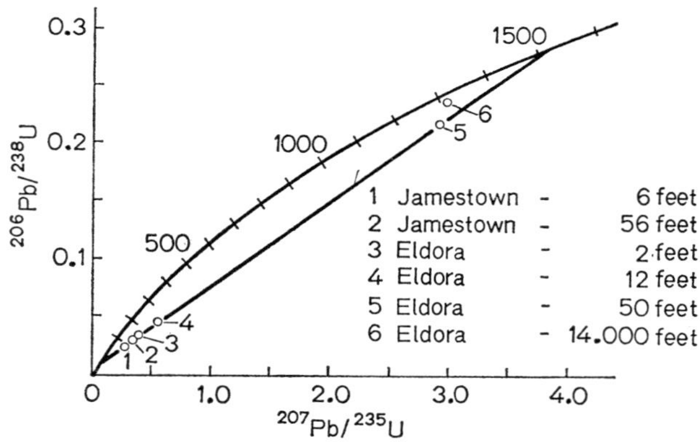

Fig. 82-B-2. Concordia diagram for zircons from the Precambrian Idaho Springs Formation adjacent to the Tertiary Eldora stock and from the Precambrian Silver Plume Granite adjacent to the Tertiary Jamestown stock, Colorado (Davis *et al.*, 1968)

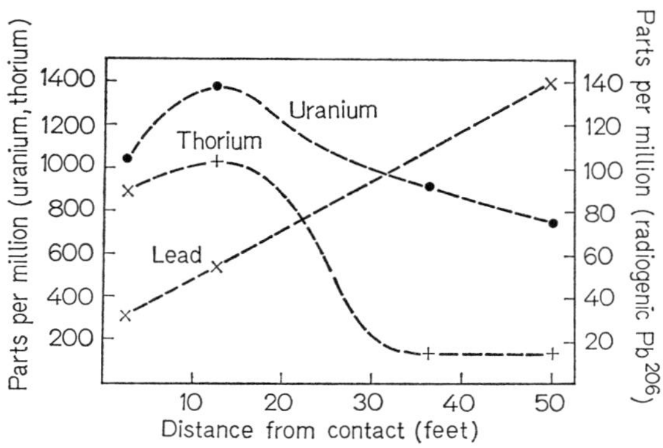

Fig. 82-B-3. Change in concentration of uranium, thorium, and lead in zircon with distance from the Eldora stock, Colorado (Davis *et al.*, 1968)

and Steiger and Wasserburg (1966). Whether this formation of domains is a continuous process or is assisted by catastrophism has not yet been determined; support for the importance of Grünenfelder's "multiphase" approach to the dating of zircons has been found by Steiger and Wasserburg (1969). They have found examples where the secondary age (lower intersection of the discordia line) determined when an episodic model is assumed, disagrees with that expected from other information. Their argument is convincing that some zircons behave like two phases — one that is nearly concordant and another that is highly discordant in a manner expected in models involving some sort of continuous diffusion.

Th—Pb ages on zircons usually, but not always, range from the $^{206}Pb/^{238}U$ age to the $^{207}Pb/^{206}Pb$ age. The Th—Pb ages are rigorously interpretable in some cases by use of the U—Th—Pb concordia diagram (Steiger and Wasserburg, 1966) but are generally unreliable for dating work.

c) Sphene ($CaTiSiO_5$)

As sphene is one of the most widespread radioactive accessory minerals in igneous and metamorphic rocks (it is particularly abundant in amphibolite), the lack of more extensive dating of this mineral is surprising (some representative data are given in Table 82-B-2). The first analyses are reported from as early as 1952; however, real interest had to await publication of the study by OOSTHUYZEN and BURGER (1965) who analyzed co-existing sphene, zircon and apatite (Fig. 82-B-4), and found the sphene to be more concordant than the zircon. In addition to the concordia

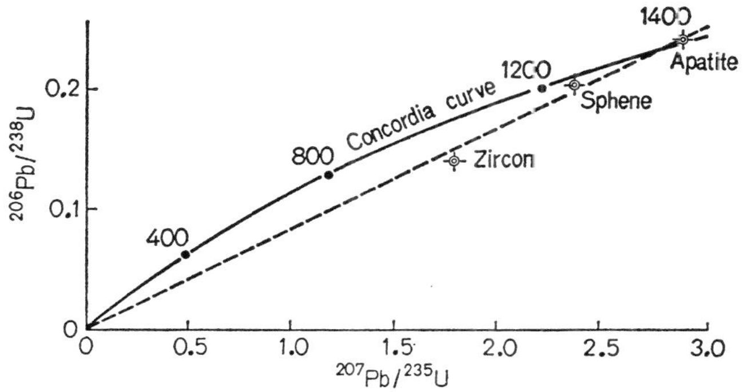

Fig. 82-B-4. Concordia diagram for Leeuwfontein Syenite, South Africa, showing isotopic relations of zircon, sphene, and apatite. (From OOSTHUYZEN and BURGER, 1965)

relationship ($t = 1{,}380$ m.y.), the Th—Pb age of the sphene was determined to be 1,190 m.y., whereas that of the zircon was only 705 m.y. The Th—Pb age on the sphene is therefore also more concordant than that of the zircon.

The major evaluation study by TILTON and GRÜNENFELDER (1968) indicates that sphenes will usually give concordant ages but they fall at either end of a discordia line. This characteristic is illustrated by the analyses of TILTON and GRÜNENFELDER on granite-gneiss from along the French River in Ontario. The whole-rock Rb—Sr isochron age of this gneiss is given by HART *et al.* (1967) as 1,725 m.y., whereas Rb—Sr mineral ages such as those for biotite reflect the age of the so-called Grenville orogeny at about 1,000 m.y. Two sphene fractions were a little discordant but gave $^{207}Pb/^{206}Pb$ ages of 1,060 and 1,100 m.y. In contrast, the zircon (KROGH, in TILTON and GRÜNENFELDER, 1968) again demonstrated the difficulty of completely resetting the zircon in nature as it has a $^{207}Pb/^{206}Pb$ age of 1,470 m.y.

Little laboratory experimental work has been done on sphene. OOSTHUYZEN and BURGER (1965) did analyze the isotopic composition of the acid leach solution of their sphene (Table 82-B-2) and found it to be somewhat radiogenic but no more so than acid leachings of zircon. The effect of this leaching on the sphene has not yet been completely evaluated.

d) Phosphates

Monazite ($CePO_4$). Next to uranium minerals and zircon, monazite receives the greatest emphasis today. Although monazite usually has normal discordance of

$^{206}Pb/^{238}U$ age $<$ $^{207}Pb/^{235}Pb$ age $<$ $^{207}Pb/^{206}Pb$ age (strongly so in some specimens), some examples are known to have strong *reverse discordance* where

$$^{206}Pb/^{238}U \text{ age} > {}^{207}Pb/^{235}U \text{ age} > {}^{207}Pb/^{206}Pb \text{ age}.$$

(For one such example see the analytical data of ZYKOV *et al.*, 1964, Table 82-B-2). Why this reverse discordance occurs in occasional monazites is not understood but the theoretical base provided by WETHERILL (1963) allows distinction of continuous diffusion of uranium from episodic loss of uranium.

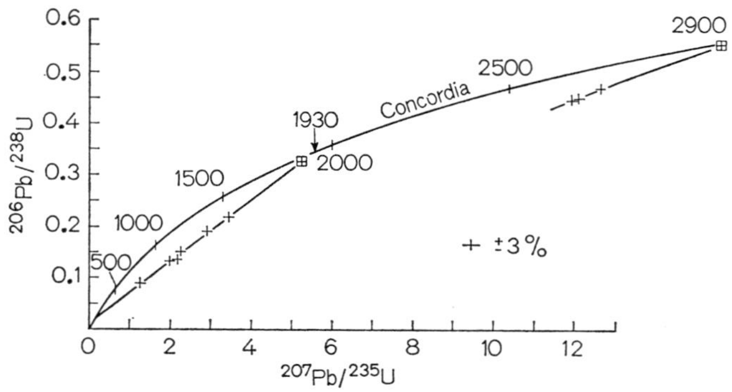

Fig. 82-B-5. Concordia diagram showing monazite and zircon relations. Discordia line intersecting concordia at 1,930 m.y. is from CATANZARO and KULP (1964), Little Belt Mountains, Montana; the discordia line intersecting concordia at 2,900 m.y. is from HEIMLICH and BANKS (1968), Bighorn Mountains, Wyoming. Crosses are plottings for zircons; boxes are plottings for monazites

Much emphasis has been placed on acid-leaching experiments (BURGER *et al.*, 1967, and papers cited therein) which have shown that some uranium and uranogenic lead are leached relatively easily in the laboratory and that Th and ^{208}Pb are less easily leached. BURGER *et al.* (1967) indicate that leaching in nature does not reflect the results of the laboratory experiments and the U—Pb system on these monazites gives the best age estimate. They also point out that it is apparent from modern detailed studies where independent evidence of age exists, that the Th—Pb age is usually too low; these studies suggest the likelihood of episodic losses in nature also. Examples of monazites with reverse discordance, which have well-established zircon isochrons, are needed to better evaluate this conclusion.

Examples are also known where monazite is nearly concordant (Fig. 82-B-5, Table 82-B-2). In the Rocky Mountain region, monazite would be a reasonable dating tool because it is relatively abundant and appears to have a tendency to be concordant.

Little rigorous evaluation has been made on the effect of metamorphism on monazite so that the ability of this mineral to retain some isotopic identity through periods of metamorphism has not yet been estimated.

Apatite $Ca_5[PO_4]_3(OH, F)$. Apatite has seldom been used or investigated for dating but is of possible interest. A comparison of apatite with zircon and sphene (data in Table 82-B-2) is given in Fig. 82-B-4 (from OOSTHUYZEN and BURGER, 1965). the apatite is nearly concordant but has slight reverse discordance which, because the radiogenic enrichment is not large, might be due to the common lead correction.

e) Whole Rock

Whole-rock dating is difficult because of the lack of radiogenic enrichment in the lead, particularly ^{207}Pb.

In a series of papers, SOBOTOVICH and his co-workers report success in dating very old rocks by the ^{207}Pb—^{206}Pb isochron method (Fig. 82-B-6). In the study of the 3,000-m.y.-old granites and gneisses of the Taromskoe Quarry in the Ukraine

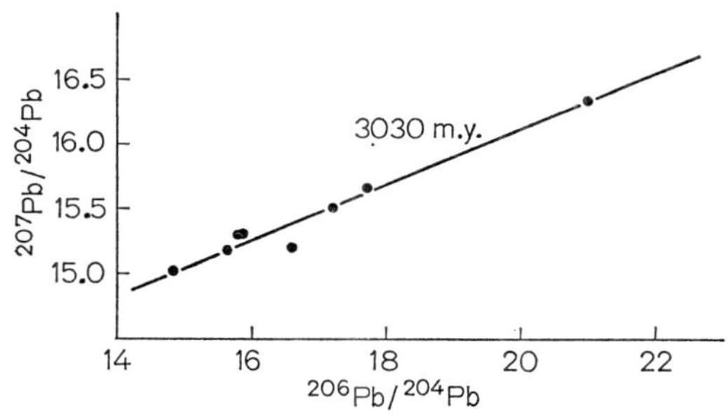

Fig. 82-B-6. Whole-rock plot of $^{207}Pb/^{204}Pb$ versus $^{206}Pb/^{204}Pb$ for rocks of granitic composition from the Taromskoe Quarry, Dnjeper Region, Ukraine (SOBOTOVICH *et al.*, 1963)

(SOBOTOVICH *et al.*, 1963), the Pb—Pb isochron was somewhat older than K—Ar ages on micas which are known to be easily affected by metamorphism. The determined values of uranium relative to lead were found to be too great for the observed lead isotopic compositions so that the $^{238}U/^{204}Pb$—$^{206}Pb/^{204}Pb$ age is much younger than the Pb—Pb age. Data on a rock sample as presented by ZARTMAN (1965) showed an opposite effect of far too little uranium. ZARTMAN's other rocks appeared to have a fairly good U—Pb balance. Apparently the disturbances in U—Pb are recent. ROSHOLT and PETERMAN (1969) have found a dramatic recent depletion of uranium relative to lead, but not thorium, in the Precambrian basement granites of the Cenozoic sandstone uranium ores of Wyoming.

III. Common Lead

a) General

"Common lead" is any lead from a phase with a low value of U/Pb and/or Th/Pb such that no significant radiogenic lead has been generated *in situ* since the phase formed. Such phases are galena and other sulfides such as pyrite, feldspars (in particular K-feldspar), micas, and most aboundant rock types of Cenozoic age. Data on common lead are used in determining ages and, more important, in the solution of genetic problems. Presented here, therefore, are data on those phases relevant to common lead applications even if sufficient radiogenic lead has been generated to markedly alter the initial isotopic composition of the lead such as in Precambrian crystalline rocks.

b) Meteorites, the Moon, and Tektites

Studies of lead isotopes in meteorites have been very fundamental to all of earth science; the first reliable age of meteorites (4,500 m.y.) was a lead isochron age (PATTERSON, 1956; MURTHY and PATTERSON, 1962) (Fig. 82-B-7). Strictly speaking, the present day lead isotopic composition of most meteorites does not fit the definition of common lead because it has been drastically changed since the meteorites formed. Lead from the troilite phase of meteorites does appear to be primordial and is the least radiogenic lead known. These values taken from the evaluations of MURTHY and PATTERSON (1962) and of OVERSBY (1970) are included in Table 82-B-1. These numbers are a very important parameter in most common lead calculations and, therefore, all meteorites are considered together here. The values determined by OVERSBY (1970) have been corrected for mass spectrometer bias, and the attempt has been made to eliminate samples that have been contaminated by terrestrial lead. STACEY *et al.* (1969) have pointed out that there is little change in derived values such as age of the earth or primary $^{238}U/^{204}Pb$ if all samples are in absolute ratios. Care should be taken to make sure that all samples are in absolute ratios when using the values of primordial lead recommended by OVERSBY (1970).

The line that best fits the data in Fig. 82-B-7 is a straight line and lead from meteorites form a physical example of a primary isochron or Gerling-Houtermans-Holmes equation (Eq. 3d in Table 82-B-1) indicating that the meteorites were formed 4,550 million years ago (PATTERSON, 1956). From the slope of the line in the plot of $^{208}Pb/^{204}Pb$ *versus* $^{206}Pb/^{204}Pb$, the value of $^{232}Th/^{238}U$ may be calculated to be 3.8 (MURTHY and PATTERSON, 1962).

The very recent results (TATSUMOTO and ROSHOLT, 1970; SILVER, 1970) on the samples returned by Apollo 11 from the Mare Tranquillitatis on the moon have values of $^{207}Pb/^{204}Pb$ of 96-590 which are much greater than those of meteorites or the earth. These large ratios for the moon materials signify depletion of lead relative to uranium a long time ago when ^{235}U was still abundant. These data on the fines and breccia from the moon lie very near the extension of the primary isochron of meteorites and the calculated U—Pb and Th—Pb ages support the age signified by the isochron. If the fines and breccia do not contain lead that has primarily been added to the moon from some foreign source, then the age of the moon must be close to 4,600 m.y., similar to that of meteorites. This age is not necessarily the age at which the observed values of U/Pb and Th/Pb were established.

The lead in the lunar volcanic rocks also give nearly concordant ages between about 3,700 and 4,200 m.y. The lead isotope ages therefore infer that the moon was tectonically active for at least a billion years after its formation.

Meteorites have large variations in values of $^{207}Pb/^{204}Pb$ — from 12 to 35 — and those of the moon are even greater — from 96 to 590 — whereas for the earth this ratio is rather uniform, ranging mainly between 15 and 16. TILTON (1958) used the difference in the nature of $^{207}Pb/^{204}Pb$ between meteorites and the earth to try to determine whether tektites are terrestrial or extraterrestrial in origin. All tektites are found to have a terrestrial type of $^{207}Pb/^{204}Pb$ value (such as included in the cross-hatched area in Fig. 83-B-7). Thus arguments based on lead isotopes suggest a derivation of tektites from a terrestrial source rather than from a meteoritic or lunar source. WAMPLER *et al.* (1969), through analyses of tektites and rocks of

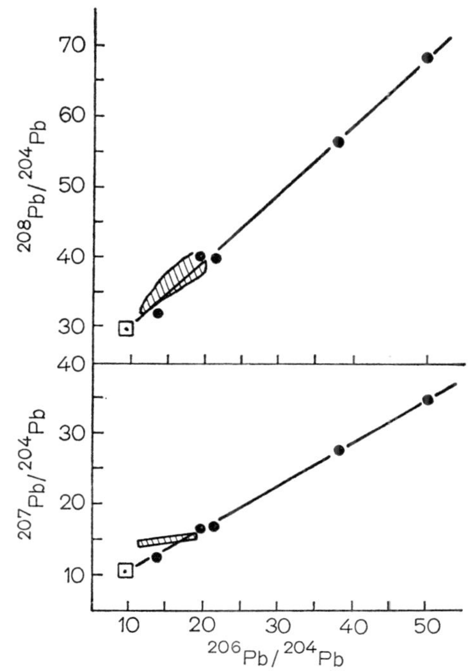

Fig. 82-B-7. A plot of $^{207}Pb/^{204}Pb$ and $^{208}Pb/^{204}Pb$ versus $^{206}Pb/^{204}Pb$ of troilite primordial lead (□) and stone meteorite leads (●) as selected by MURTHY and PATTERSON (1962). The solid line in the lower plot is a primary isochron with a slope of 0.59 and the slope of the solid line in the upper plot represents a Th/U of 3.8. Cross-hatched areas enclose terrestrial leads

possible impact crater sources, have shown that the lead isotopes support the growing body of data relating the Ivory Coast tektites to the Bosumtwi crater. They also show that moldavites are compatible with derivation from Ries crater and suggest that other tektites may be ultimately derived from pelagic sediments.

c) General Aspects for the Earth

Age of Earth. All leads from terrestrial rocks, except those from uranium and thorium ores, are enclosed in the cross-hatched areas of Fig. 82-B-7. The proximity of terrestrial lead to the primary meteorite isochron has been used by PATTERSON (PATTERSON, 1956; MURTHY and PATTERSON, 1962) in estimating the age of the earth at 4,550 m.y. as well as for estimating the isotopic composition of terrestrial primordial lead. Some studies on terrestrial materials suggest that the earth may be as much as 200 m.y. older than the estimate from meteorites (TILTON and STEIGER, 1969); however, if the earth is even that much older, either there must have also been differentiation of U relative to lead in the early history (PATTERSON and TATSUMOTO, 1964) or the values of primordial lead for the earth and for meteorites must differ (DOE *et al.*, 1965).

Table 82-B-3. *Data on selected stratiform ores whose lead approaches single stage lead isotope development*

Location	Host rock	No. of samples in average	$^{206}Pb/^{204}Pb$	$^{207}Pb/^{204}Pb$	$^{208}Pb/^{204}Pb$	Reference number
Bathurst, New Brunswick, Canada	Middle Ordovician metamorphosed rocks	3	18.291	15.781	38.526	1
	Middle Ordovician metamorphosed rocks	1	18.20	15.69	38.24	2
	Middle Ordovician metamorphosed rocks	1	18.204[a]	15.655[a]	38.122[a]	3
Balmat, New York	Marbles about 1,100 m.y. in age	1	17.019	15.677	36.857	4
	Marbles about 1,100 m.y. in age	1	16.96	15.56	36.60	2
	Marbles about 1,100 m.y. in age	1	16.935[a]	15.505[a]	36.423[a]	3
Broken Hill, Australia	Metamorphosed rocks approximately 1,600—1,700 m.y. old	6	16.12	15.54	36.06	5
	Metamorphosed rocks approximately 1,600—1,700 m.y. old.	1	16.07	15.52	36.03	2
	Metamorphosed rocks approximately 1,600—1,700 m.y. old	1	16.007[a]	15.397[a]	35.675[a]	3
Geneva Lake, Ontario, Canada	Precambrian metamorphosed rocks	2	14.08	15.02	34.10	6
	Precambrian metamorphosed rocks	1	14.002[a]	14.870[a]	33.716[a]	3
Manitouwadge Lake, Ontario, Canada	Archean metamorphosed rocks approximately 2,700—2,800 m.y. old	2	13.34	14.56	33.27	6
	Archean metamorphosed rocks approximately 2,700—2,800 m.y. old	4	13.400	14.565	33.545	1
	Archean metamorphosed rocks approximately 2,700—2,800 m.y. old	1	13.25	14.42	33.18	7
		1	13.211[a]	14.401[a]	33.069[a]	3

References: 1. Ostic *et al.* (1967); 2. Doe (1962); 3. Stacey *et al.* (1969); 4. Reynolds and Russell (1968); 5. Kollar *et al.* (1960); 6. Russell and Farquhar (1960); 7. Doe *et al.* (1965).

[a] Absolute ratio.

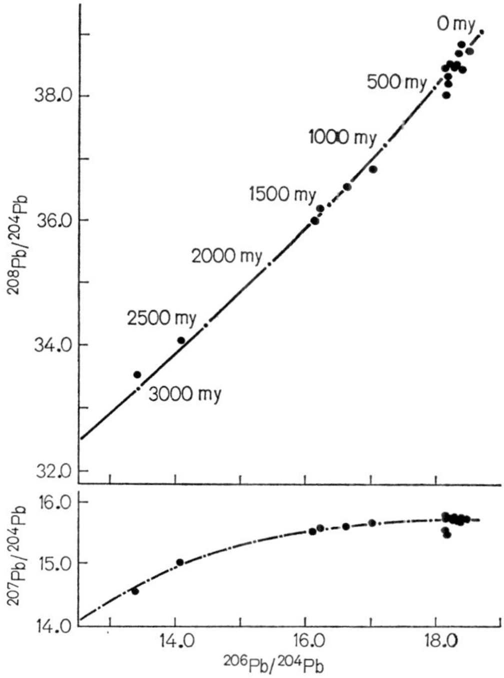

Fig. 82-B-8. $^{207}Pb/^{204}Pb$ and $^{208}Pb/^{204}Pb$ versus $^{206}Pb/^{204}Pb$ for stratiform ores which appear to have formed at approximately the same time as the enclosing sediments and approach conditions of single-stage lead isotope growth. Parameters of the growth curves are those of KANASEWICH (1968a) and are given in Table 82-B-1 and I. Introduction

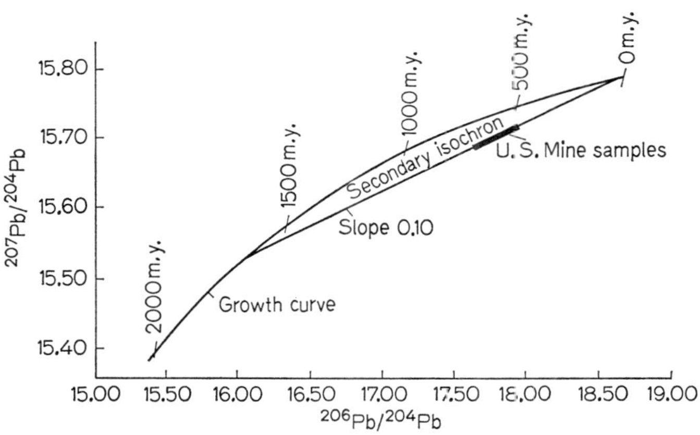

Fig. 82-B-9. Secondary isochron for galenas from the U.S. Mine (STACEY *et al.*, 1967)

Primary or single stage growth curves. A physical example where the lead isotope data forms a close approach to primary growth curves (Eqs. 3a, b and c, in Table 82-B-1) is found for the lead isotope data of *conformable ore deposits* (STANTON and RUSSELL, 1959). These ore deposits are tabular (grossly conformable with the country rock), have ill-defined margins, and are pyrite-rich; many occur in rocks bearing organic matter and most are in metamorphic terranes. STANTON has interpreted the terrane containing the deposits to be of the island-arc and related tectonic classes. The data for some of these deposits are given in Table 82-B-3 and are shown in Fig. 82-B-8. The source of the lead in such deposits is uncertain.

Secondary or two stage growth curves. One particularly good example of a secondary isochron is lead accompanying mineralization in Utah at the U.S. Mine (STACEY *et al.*, 1967), covering only a small range in $^{206}Pb/^{204}Pb$ and $^{207}Pb/^{204}Pb$ ratios (Fig. 82-B-9). Later studies (STACEY *et al.*, 1968) in Utah also showed that the lead isotopic composition has economic implications in that the largest sulfide ores in Utah have lead isotope ratios nearest those of possibly related igneous rocks.

Because the development of lead isotope ratios has occurred in two or more stages, the observed ratios may be greater than they would be in a single stage development because the lead has evolved in an integrated system with $^{238}U/^{204}Pb >$ about 9 over a period of time $\leqq 3{,}500$ m.y. Such lead is *radiogenic* (i.e., the ratios are greater) in comparison with a single stage lead of that age. Special radiogenic leads having future model lead ages are generally referred to as J-type or Joplin type. If the lead isotope ratios are less than they would be in a single stage development because of an integrated $^{238}U/^{204}Pb <$ about 9, the term *unradiogenic* lead relative to a single stage lead of that age will be used. Unradiogenic leads are sometimes called B-type or Bleiberg type. Leads from the Bleiberg deposits do not fit to the isotopic definition, however, and the use of the term B-type should be discontinued.

d) Observed Values of $^{238}U/^{204}Pb$ and $^{232}Th/^{204}Pb$ in Rocks

The lead isotopic composition of igneous rocks and ores places constraints on the observed atomic values of $^{238}U/^{204}Pb$ and $^{232}Th/^{204}Pb$ in the source rocks. Knowledge of the values of these ratios observed in various rock types is therefore desirable in radiogenic tracer studies (Fig. 82-B-10).

e) Whole-rock Studies, Precambrian and Paleozoic

Knowledge of the isotopic composition of lead in Precambrian and Paleozoic upper crustal rocks is important in answering many questions of genesis, such as the source of igneous rocks and ores. Fig. 82-B-10 shows a wide spectrum of present day isotopic compositions in these rocks (for references to data see DOE, 1970). All North American data are enclosed in the area shown and tend to be radiogenic. Lines are drawn on the diagram to show the approximate position of a terrestrial lead that lies on the meteorite geochron (a model lead of zero age). The rocks of the Altai Mountains in Asia (SOBOTOVICH and GRASHCHENKO, 1965) also appear to be radiogenic.

$^{232}Th/^{204}Pb$ observed ($\mu \cdot k$)

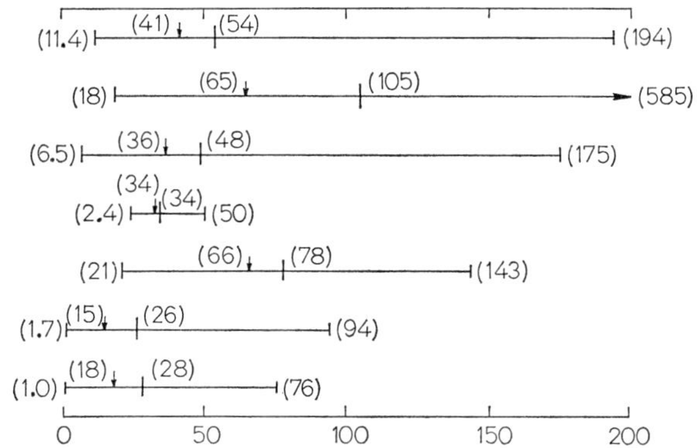

Group	General Description	No. of samples	No. with $\mu \cdot k$ above 36
7	Clastic sediments and crystalline metamorphic rocks below granulite facies	8	5
6	Silicic volcanics and intrusives	23	21
5	Intermediate volcanics and intrusives	51	26
4	Xenocryst bearing subalkaline basalts	9	3
3	Alkaline and peralkaline basalts, gabbro sills, and kimberlite	34	30
2	Subalkaline basalts	39	10
1	Mafic inclusions in pipes	5	0

$^{238}U/^{204}Pb$ observed (μ)

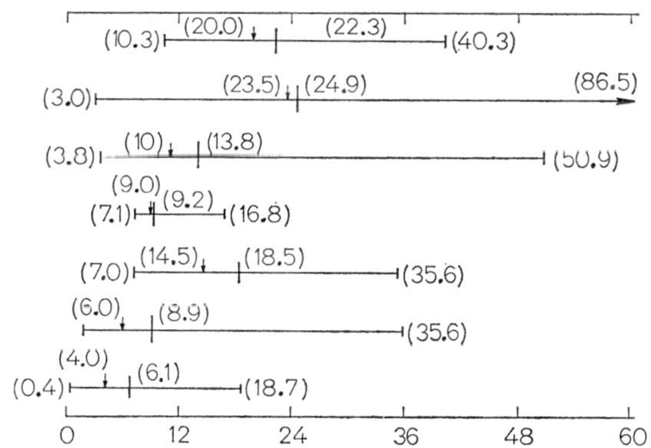

Group	General Description	No. of samples	No. with μ below 10
7	Clastic sediments and crystalline metamorphic rocks below granulite facies	8	0
6	Silicic volcanics and intrusives	43	7
5	Intermediate volcanics and intrusives	55	21
4	Xenocryst bearing subalkaline basalts	9	7
3	Alkaline and peralkaline basalts, gabbro sills, and kimberlite	34	5
2	Subalkaline basalts	44	31
1	Mafic inclusions in pipes	5	4

Fig. 82-B-10. Extremes, means (long vertical bar), and fiftieth percentiles (arrows) for various rock types for which lead isotope data are available. Numbers in parentheses give values for each indicator. (Compilation and figure are from DOE, 1970)

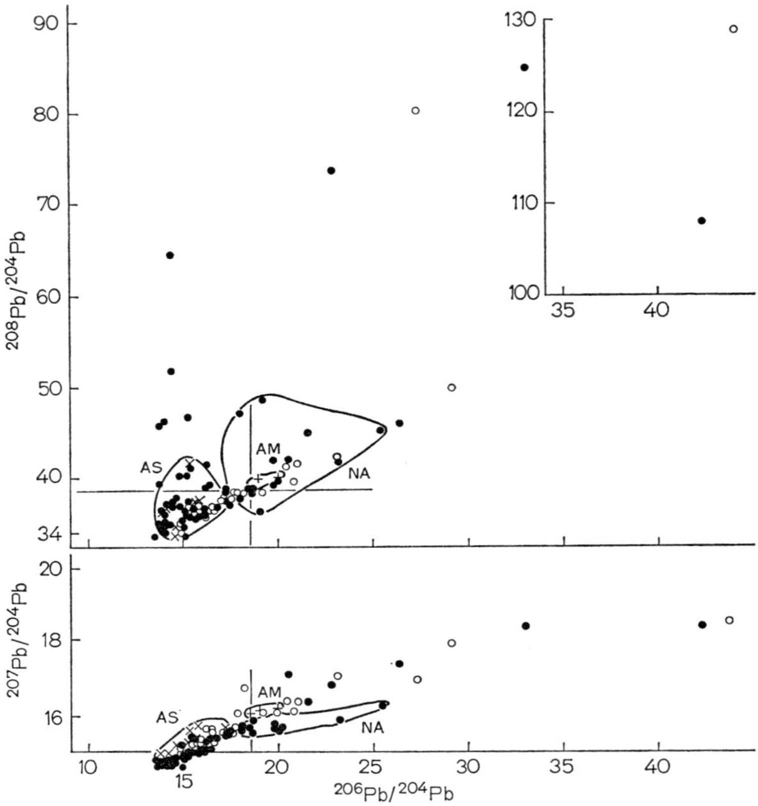

Fig. 82-B-11. $^{207}Pb/^{204}Pb$ and $^{208}Pb/^{204}Pb$ versus $^{206}Pb/^{204}Pb$ for whole-rock samples of Precambrian and Paleozoic crystalline rocks and composites: ○ Ukrainian Shield; ● Baltic Shield and Scotland; × Aldan Shield (AS); + Altai Mountains (AM); ● North America (NA). (Compilation is from DOE, 1970)

Fig. 82-B-12. $^{207}Pb/^{204}Pb$ and $^{208}Pb/^{204}Pb$ versus $^{206}Pb/^{204}Pb$ for sediments of Cenozoic age from various localities

Area or symbol	Description
A	Pelagic sediments of the Red Sea and basins of the Atlantic, Antarctic, and Indian Oceans except for the Gulf of Aqaba, HCl soluble lead
△	Sediment from the Gulf of Aqaba, HCl soluble lead
P	Pelagic sediments from the Pacific Ocean basin, HCl soluble lead
ST-GC	Calcareous clastic sediments from the Salton Sea Trough and a manganese nodule from the Gulf of California:
○	HCl soluble lead
●	residue lead
B	Sediments from the Baltic Sea:
×	HCl soluble lead

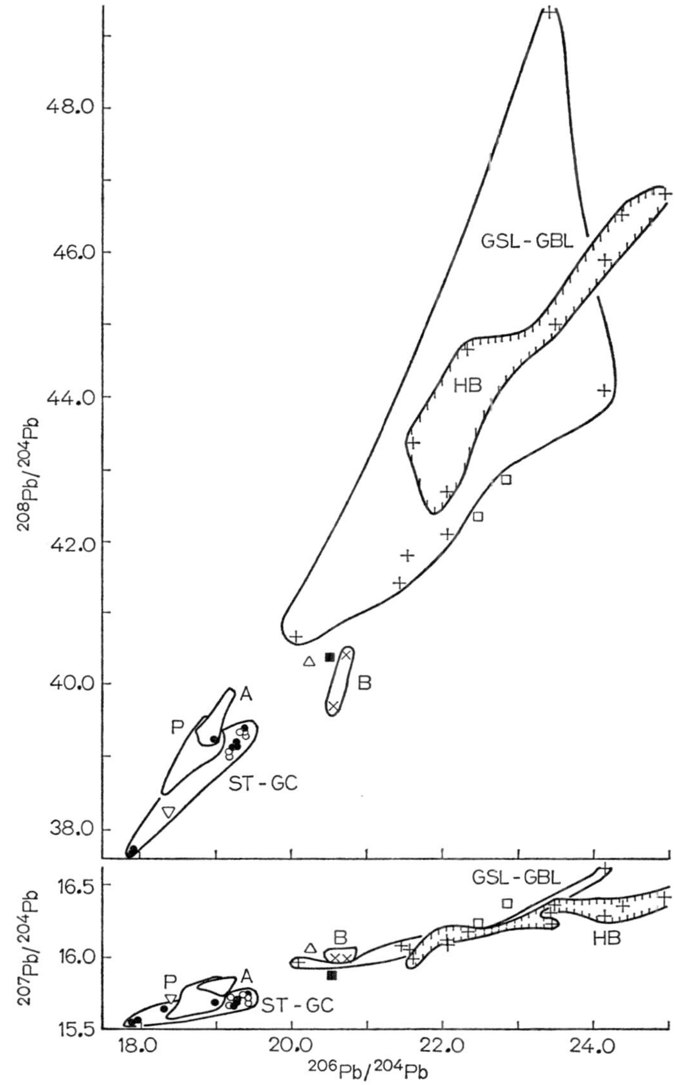

Fig. 82-B-12

	Sediments from Lake Superior in the Canadian Shield:
□	HCl and water soluble lead
■	residue lead
HB	Sediments from Hudson Bay in the Canadian Shield, HCl soluble lead
GSL-GBL	Sediments from Great Bear Lake and Great Slave Lake in the Canadian Shield:
×	HCl soluble lead
▽	Sediments from the Mediterranian Sea, HCl soluble lead

Such radiogenic rocks are not necessarily characteristic for large segments of the earth's crust as may be seen from data (Fig. 82-B-11) on the Precambrian crystalline rocks of the Baikalian block of the Aldan Shiled in Asia (SOBOTOVICH *et al.*, 1965).

Data from the Baltic and Ukrainian Shield areas range widely, which makes averages difficult to determine. The leads in rocks of the Ukrainian Shield are also probably unradiogenic today. The Baltic Shield is more problematical although many of the rocks are unradiogenic.

A detailed study of high grade Precambrian metamorphic rock from the Lewisian basement of northwest Scotland has recently become available (MOORBATH *et al.*, 1969). All present day values of $^{206}Pb/^{204}Pb$ are strongly nonradiogenic — in the range 13.5 to 16.5 — regardless of rock type. Rocks of granulite facies metamorphism about 2,600 m.y. ago and those in the transition zone to an area of amphibolite facies metamorphism about 1,400 to 1,600 m.y. ago, also have nonradiogenic values of $^{208}Pb/^{204}Pb$ — in the range 33.9 to 37.5. This range also includes many samples from the region of subsequent amphibolite facies metamorphism; however, in 40 percent of these samples, $^{208}Pb/^{204}Pb$ is greater than 40 and must be considered radiogenic if the ratios, which were determined on leads volatilized from the sample, are nearly the same as those for the whole rocks. Significant thorium but apparently little uranium may have been added during the amphibolite facies metamorphism to account for these isotopic relationships.

f) Cenozoic Sediments, Whole-rock and HCl-soluble Lead

Lead isotopic composition of the main sediment types is more a function of age of source areas for the sediment than of rock type because the source rocks contain significant quantities of U and Th relative to lead (Fig. 82-B-12). Young sediments in restricted basins draining Precambrian terranes, such as Hudson Bay, Great Slave Lake, Great Bear Lake, Lake Superior, and the Baltic Sea (CHOW, 1965; HART and TILTON, 1966), tend to be highly radiogenic and variable in lead isotope ratios.

On the other hand, sediments, well-mixed because they are far from their source, tend to be isotopically rather uniform even if the detritus is Precambrian, but even on these rocks the lead is still somewhat radiogenic (DOE *et al.*, 1966).

Pelagic sediments from the ocean basins (CHOW and PATTERSON, 1962a; CHOW and TATSUMOTO, 1964; WAMPLER and KULP, 1964; CHOW, 1968) fall in restricted ranges of lead isotopic composition (Fig. 82-B-12), although the means from the Atlantic, Indian and Antarctic Oceans and the Red Sea are somewhat more radiogenic than those of the Pacific Ocean. CHOW and PATTERSON (1962a) have concluded that this relationship reflects a younger source terrane around the Pacific than the terranes that surround the other areas.

g) Cenozoic and Mesozoic Igneous Rocks

Ocean Basins. Data on abyssal basalts from oceanic ridges and rises (TATSUMOTO, 1966) and basalts from islands in "open ocean" structures such as Hawaii [see DOE (1970) for a compilation of references] are given in Table 82-B-4 and are shown in enclosed areas along with all Hawaiian data in Fig. 82-B-13. The isotope ratios

Table 82-B-4. *Isotopic composition of lead in abyssal basalts and in some island volcanics of the East Pacific Rise* (TATSUMOTO, 1966[a])

Description	$^{206}Pb/^{204}Pb$	$^{207}Pb/^{204}Pb$	$^{208}Pb/^{204}Pb$
Abyssal basalts			
Mid-Atlantic Ridge			
High alumina tholeiite (22° 40′ S; 13° 16′ W)[b]	18.47	15.54	38.01
High alumina tholeiite (5° 47′ S; 11° 25′ W)[b]	17.82	15.54	37.52
High alumina tholeiite (9° 39′ N; 40° 27′ W)[b]	18.82	15.68	38.65
East Pacific Rise			
High alumina tholeiite (7° 47′ S; 108° 10′ W)	18.19	15.54	37.93
High alumina tholeiite (12° 52′ S; 110° 57′ W)[c]	18.24	15.53	38.03
High alumina tholeiite (18° 25′ S; 113° 20′ W)	18.50	15.58	38.34
Island volcanics, East Pacific Rise			
Easter Island			
Obsidian, Mount Ourito	19.31	15.66	39.15
Andesine andesite	19.25	15.58	38.94
Alkali basalt	19.30	15.73	39.46
Tholeiite with alkalic affinities	19.28	15.67	39.16

[a] Lead extraction by volatilization through melting. Mass spectrometric analysis uses the surface emission mode of ionization of $PbS\text{-}NH_4NO_3$ from a rhenium filament with Faraday cup collection of ions. Data are not corrected for fractionation effects in the ionization process.

[b] Average of two analyses.

[c] Average of three analyses.

are clearly not uniform, although the average of all samples would approach a modern lead within the limits of error that such models may be calculated. The age of origin for the sources of abyssal basalts may be quite old, of the order of 1,000 m.y., and there may be little differentiation of uranium relative to lead, as TATSUMOTO has demonstrated.

Island volcanics from oceanic ridges and rises have lead isotope ratios [see DOE (1970) for a compilation of references, data on Fig. 82-B-13; selected data on Table 82-B-4] which are in general more radiogenic than those of abyssal basalts and Hawaiian volcanics (DOE, 1968).

This relationship is particularly evident on the plot of $^{208}Pb/^{204}Pb$ *versus* $^{206}Pb/^{204}Pb$ where the only overlap is with Iceland Group I volcanics of WELKE *et al.* (1968).

Island arc data may be rather similar to island volcanic data from oceanic ridges and rises, although $^{207}Pb/^{204}Pb$ and $^{208}Pb/^{204}Pb$ tend to be greater in the island arc data. Values of $^{206}Pb/^{204}Pb$ and especially $^{208}Pb/^{204}Pb$ are distinctly unradiogenic for volcanics from Japan relative to most data from oceanic ridges and rises. ARMSTRONG (1968) and TATSUMOTO and KNIGHT (1969) suggest that much of the isotopic characteristics for lead in volcanics from an island arc terrane is due to contamination of the mantle in the magma-generating zone with dragged down oceanic crust. The initial data of COOPER and RICHARDS (1969b) on New Zealand andesites is rather similar to those from eastern Japan.

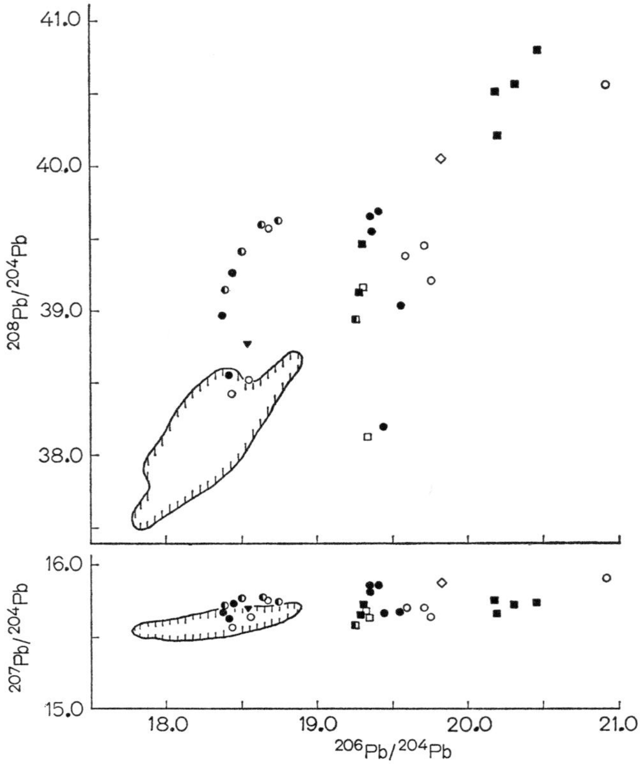

Fig. 82-B-13. $^{207}Pb/^{204}Pb$ and $^{208}Pb/^{204}Pb$ versus $^{206}Pb/^{204}Pb$ for volcanics from islands and a volcanogenic sediment from the East-Pacific Rise (■ basalt, ◧ andesite or trachyandesite, □ others), Mid-Atlantic Ridge (● basalt, ◐ andesite or trachyandesite, ○ others), Vema Seamount, S.E. Atlantic (◇ phonolite), and Reunion Islands, Indian Ocean (▼ olivine basalt). For comparison, the enclosed areas are superimposed which include all the isotopic data on abyssal basalts and volcanics of the Hawaiian Islands. (Compilation is from DOE, 1970)

Continental igneous rocks. Various kinds of igneous rocks from the continents (Fig. 82-B-14) range widely in lead isotope ratios [for a compilation of references see DOE (1970)]. Data for California, Oregon, Washington, Alaska and Nevada rocks appear similar to those for oceanic ridges and rises (DOE, 1968) and are compatible with the recent conclusion, made on the basis of other evidence, that the East Pacific Rise extends under this region (YEATS, 1968). Selected data from Oregon and Colorado are given in Table 82-B-5. In North America, the unradiogenic leads are concentrated in the Rocky Mountain region and farther east with Utah an isotopic transition zone between Nevada and the Rocky Mountain region (Fig. 82-B-14). Basalts from the Rocky Mountain region are rather similar in lead isotopic composition to silicic igneous rocks, even if obviously contaminated basalts are eliminated (Table 82-B-5). These data suggest that both rock types come from a similar age source with similar values of U/Pb and Th/Pb (DOE, 1968). Contrary

Table 82-B-5. *Selected data on continental igneous rocks from North America. (Abbreviations: prim. = primitive; cont. = contaminated)*

Sample locality	Rock type	$^{206}Pb/^{204}Pb$	$^{207}Pb/^{204}Pb$	$^{208}Pb/^{205}Pb$	Reference
		Oregon			
Northwest coast	Tholeiitic basalt[a]	18.88	15.76	39.25	2
Marys Peak	Aplite dike	19.11	15.73	39.30	2
gabbroic sill	Pegmatite	19.25	15.75	39.26	2
	Granophyric diorite	19.22	15.74	39.30	2
	Basalt, chilled border	19.31	15.75	39.34	2
Widow Creek	Alkalic basalt[a]	19.34	15.70	39.32	2
Coffin Butte	Tholeiitic basalt	18.98	15.61	38.74	2
		San Juan Mountains, Colorado			
Creede area	Silicic volcanic	18.74	15.63	38.09	1
Wagon Wheel Gap area	Silicic volcanic	18.36	15.58	37.68	1
La Jara Reservoir area	Hinsdale Formation				
	Basalt (prim.)	18.33	15.55	37.67	3
	Basalts (cont.)[a]	17.94	15.54	37.48	3
	Basalt (cont.)	18.38	15.62	38.26	1

References: 1. Doe (1967); after dissolution by HF—$HClO_4$ (and volatilization of lead in the case of the basalts) mass spectrometric analysis was made from PbS—NH_4NO_3 by the surface emission mode of ionization from rhenium filament with Faraday cup collection of ions. Data are normalized to agree with Ta-1 filament material to bring ratios within stated analytical uncertainties of Tilton's 1:1 lead isotope gravimetric standard. 2. Tatsumoto and Snavely (1969); lead extraction by volatilization below the melting point with other analytical factors as given in Table 32-B-4. No normalization of data has been made. 3. Doe *et al.* (1969); basalts were treated by volatilization extraction with one checked by HF—$HClO^4$ dissolution. All data are normalized to agree with Ta-1 filament material with is within the error limits indicated by the Tilton 1:1 lead isotope gravimetric standard.

[a] Average of 2 samples.

to the isotopic relations in the Rocky Mountain region, in Scotland the mafic igneous rocks statistically appear to be more radiogenic than the silicic igneous rocks of Cenozoic age (Moorbath and Welke (1968). The silicic igneous rocks, which are completely overlapped in their lead isotope ratios by those from whole-rock samples from the nearby Lewisian basement (Moorbath *et al.*, 1969), might be interpreted as being derived from the Lewisisan basement with the mafic igneous rocks coming from the mantle and contaminated to varying degrees with Lewisian basement.

Ultramafic rocks. Only the data for the kimberlite from South Africa (Lovering and Tatsumoto, 1968) and peridotitic portions of a layered mafic intrusion in Scotland (Moorbath and Welke, 1968) can be considered reliable whole-rock values at this time. In view of the great importance of data on these kinds of rocks and the difficult analytical problems, only those data considered reliable whole-rock or mineral values are included in Table 82-B-6.

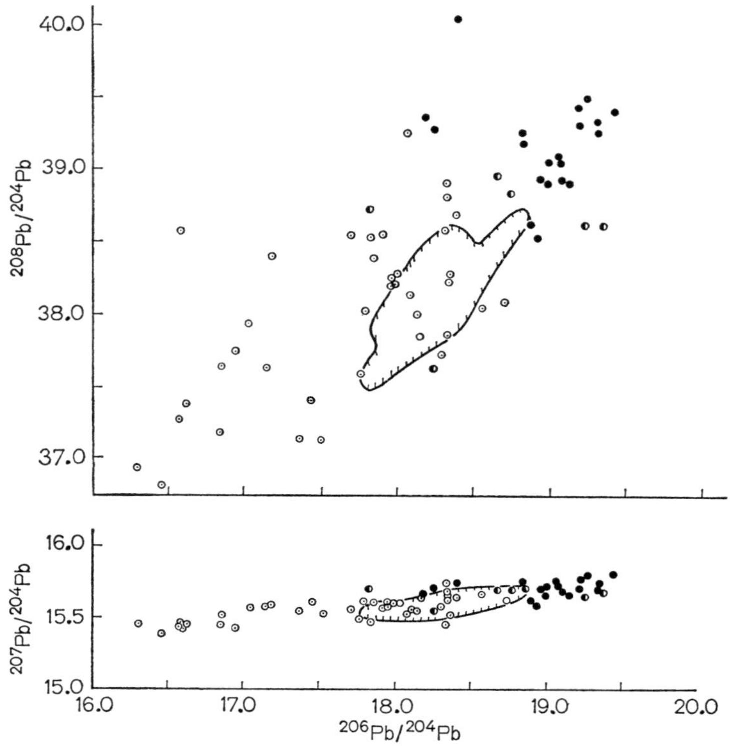

Fig. 82-B-14. $^{207}Pb/^{204}Pb$ and $^{208}Pb/^{204}Pb$ versus $^{206}Pb/^{204}Pb$ for Cenozoic and Mesozoic igneous rocks of western North America: ● California, Oregon, Washington, British Columbia, Alaska, and Nevada; ◐ Utah; ○ New Mexico, Colorado, Wyoming, Idaho Montana, South Dakota, and Texas. Enclosed areas superimposed for comparison include all the isotope data on abyssal basalts and volcanics of the Hawaiian Islands. (Compilation is from DOE, 1970)

Table 82-B-6- *Isotopic composition of lead in ultrabasic and related rocks (no normalization to data)*

Sample type	$^{206}Pb/^{204}Pb$	$^{207}Pb/^{204}Pb$	$^{208}Pb/^{204}Pb$	Reference
Kimberley pipe, South Africa				
Kimberlite (initial ratios)[a]	19.51	15.82	39.32	1
Kimberlite (present-day ratios)	19.77	15.83	39.66	
Cuillins layered intrusion, Scotland				
Gabbro, Zone I (leucocractic band)	17.84	15.69	37.97	2
Gabbro, Zone I (melanocractic band)	17.77	15.67	37.84	2
Gabbro, Zone IV	17.60	15.55	38.00	2
Peridotite, main body	18.13	15.66	38.44	2
Sligachan area, Scotland				
Peridotite dike in basalt	17.86	15.58	38.02	2

References: 1. LOVERING and TATSUMOTO (1968); 2. MOORBATH and WELKE (1968).

[a] Corrected for 174×10^6 yrs. of decay of uranium and thorium.

Lead isotope data are still valuable even when the lead is only partially extracted and its lead isotopic composition not equal to that of the whole-rock. COOPER and GREEN (1969) have volatilized lead from lherzolite inclusions and host alkali basalt from western Victoria, Australia. Although only a preliminary report is available, MANTON and TATSUMOTO (1969) find somewhat similar results to those of COOPER and GREEN between eclogite inclusions and host kimberlite from the Roberts Victor Mine in South Africa. Both the lherzolites from Australia and the eclogites from South Africa are calculated to come from sources 2,000 to 2,500 m.y. old.

h) Paleozoic and Precambrian Igneous Rocks

Igneous rocks older than Mesozoic are somewhat difficult to work on as the corrections for *in situ* radioactive decay of U and Th are large and the values of U/Pb and Th/Pb often have undergone a recent disturbance. The more successful approach is to analyze the lead in a phase with low values of U/Pb and Th/Pb such as K-feldspars. Such data for rocks about, 1,000 m.y. old are given in Fig. 82-B-15

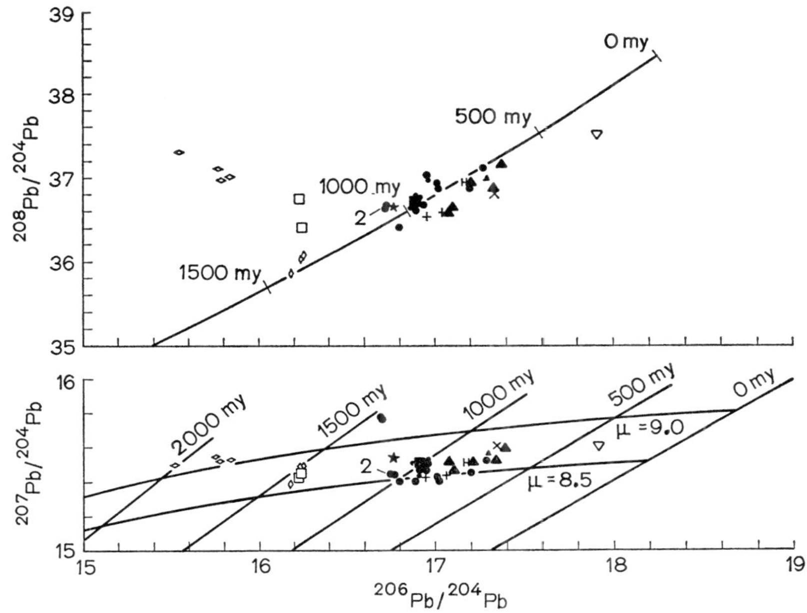

Fig. 82-B-15. Lead isotope relationships for feldspar leads from 1,000-m.y.-old igneous rocks. Curved lines on the plot of $^{207}Pb/^{204}Pb$ versus $^{206}Pb/^{204}Pb$ are the single stage growth curves of $^{238}U/^{204}Pb$ for values of 9.0 and 8.5, and the heavy sloping lines are the isochrons for the labeled ages. The normal growth curve involving $^{208}Pb/^{204}Pb$ and $^{206}Pb/^{204}Pb$ is given in that plot with positions of model leads indicated for various ages. The plot is from ZARTMAN and WASSERBURG (1969) and includes their data from: ● Texas suite, ▼ Western Adirondack Mountains, V.Y., ▽ Eastern Adirondack Mountains, N.Y., + Westport. Ontario, * Leg Pond, Newfoundland, □ ■ Port Cartier-Mt. Reed, Quebec, H Herefoss, Norway, · Pikes Peak , Colorado, × Duluth, Minnesota, ◇ Mellen.Wisconsin, ◇ Gold Butte, Nevada; as well as the following data: • Llano,Texas (ZARTMAN, 1965b), ▴ Shenandoah National Park, Viginia (DOE *et al.*, 1965), ▲ Balmat, New York (DOE, 1962), · Pikes Peak, Colorado (DOE, 1967)

Table 82-B-7. *Selected feldspar lead data on Paleozoic and Precambrian rocks (corrections for decay of U and Th are less than 2 percent)*

Sample locality	$^{206}Pb/^{204}Pb$	$^{207}Pb/^{204}Pb$	$^{208}Pb/^{204}Pb$	Ages (m.y.)			Reference
				isochron model	$^{208}Pb/^{204}Pb$	accepted	
Some K-feldspars in which model lead ages are in good agreement with accepted ages							
Pikes Peak batholith, Colorado	16.91	15.49	36.71	980	980	1,050	1
Commercial pegmatite, Wyoming	14.62	15.27	34.24	2,530	2,410	2,450	2
Microcline granite, Kola Peninsula, Baltic Shield	13.65	14.59	33.50	2,750	2,660	2,800	3
Some K-feldspars in which model lead ages are older than accepted ages							
Pegmatite, Pitkyaranta, Baltic Shield	14.53	15.09	34.43	2,450	2,220	1,900	3
Rapakivi granite, Nevada[a]	16.21	15.44	35.99	1,450	1,350	1,100	4
Granite, Wisconsin[b]	15.73	15.51	37.08	1,870	770	1,100	4
Some K-feldspars in which model lead ages are younger than accepted ages							
Pegmatite in Baltimore Gneiss, Maryland	18.60 18.50[c]	15.75 15.62[c]	38.65 38.17[c] }	−80	20	450	5
Charnockite, Sabarov, Ukraine	15.97	15.27	36.05	1,450	1,260	2,200	3
Feldspars with at least one ratio reasonably unaffected by metamorphism							
Granite gneiss, Pitkyaranta, Baltic Shield	14.35	15.13	33.51	2,600	2,320	2,600 (1,800)[d]	3
Baltimore gneiss, Maryland	17.42 18.10	15.52 15.53	40.01 37.09	610 70	−890 770 }	1,100 (300—450)[d]	5

References: 1. Doe (1967); 2. Catanzaro and Gast (1960); 3. Tugarinov *et al.* (1963a); 4. Zartman and Wasserburg (1969); 5. Doe *et al.* (1965).

[a] Average of 3 samples. [b] Average of 4 samples. [c] Muscovite.
[d] Age in ( ) is the age of metamorphism. The other age is the accepted age of the rock.

(ZARTMAN and WASSERBURG, 1969). Selected data are given in Table 82-B-7. Radiogenic leads appear to be the rule even of most granitic rocks found in terranes that are much older than the granitic rock. Leads unradiogenic for their age as in Mesozoic and Cenozoic igneous rocks have been much more difficult to find. Existence of older equivalents of the Rocky Mountain type seems likely. The ones that have been found are small, shallow intrusions (ZARTMAN and WASSERBURG, 1969) and in a pegmatite from the Baltic Shield at Pitkyaranta in Southwest Karelia (TUGARINOV *et al.*, 1963). Though many old granite whole-rock samples of the Aldan Shild are unradiogenic in their present-day ratios, TUGARINOV *et al.* (1963) give no data that demonstrates that any are very unradiogenic in their initial ratios. Apparently, most older granitic rocks form in a terrane somewhat similar to that of the present Pacific coast states of the United States or oceanic ridges or rises.

i) Ore Genesis

Magmatogenic Ores. Ores in which the isotopic composition of ore lead is the same as that in the presumably related igneous source at the time of crystallization are good candidates for a magmatogenic origin. These ores are usually isotopically uniform with $^{206}Pb/^{204}Pb < 19.5$ and $^{208}Pb/^{204}Pb < 39$. Examples are the ores at Butte, Montana (DOE *et al.*, 1968), Bingham Canyon, Utah (STACEY *et al.*, 1969), and the Nelson batholith, British Columbia (REYNOLDS and SINCLAIR, 1971). It may be, however, that in cases such as the Boulder batholith the lead was derived from the igneous rock during the wall rock alteration rather than being a differentiate of magma at depth. Both processes would give the same lead isotopic composition.

Indirect Magmatogenic Ores. Many ores thought to be derived in the igneous process have lead isotopic compositions different from the likely igneous sources of the region or of even larger areas. The lead is usually more radiogenic than the igneous rocks such as the ores of the Milford region, Utah, where $^{206}Pb/^{204}Pb$ varies between 18.6 and 20.2 for the ores, but the ratio for the igneous rock is 18.4 to 18.8 (STACEY *et al.*, 1969), although such deposits around the Nelson batholith are less radiogenic (REYNOLDS and SINCLAIR, 1971). The isotopic data demand that a large component of the ore lead is derived from sources outside the magma and is lead probably generated by radioactive decay of U and Th from rocks of a much older terrane. The facts that the igneous rock has usually the least radiogenic lead and that the ores form a continuum of more radiogenic values suggest that the igneous rock supplied lead as well as heat during ore-formation and that there was a magmatic component of lead in the ore.

Deposits Formed by Metamorphism of Much Older Rocks. Lead isotope data clearly support a metamorphic origin for some deposits of this kind, although well-documented studies are few for this little-emphasized category. Greater variability in isotope ratios may be encountered. The ratios may range from those expected for the time of formation of the source rocks to those much more radiogenic than expected for the age of metamorphism. A Pb—Pb age may be calculated to determine the age of the source terrane if the metamorphic age is known or *vice versa*; the results may be confirmed by other dating methods. The Thackaringa deposits (RUSSELL *et al.*, 1961) in Australia are the best example of this type.

Lateral Secretion Through Much Older Rocks. Many important stratiform ores, such as those of the Mississippi Valley type, do not have any close association with either

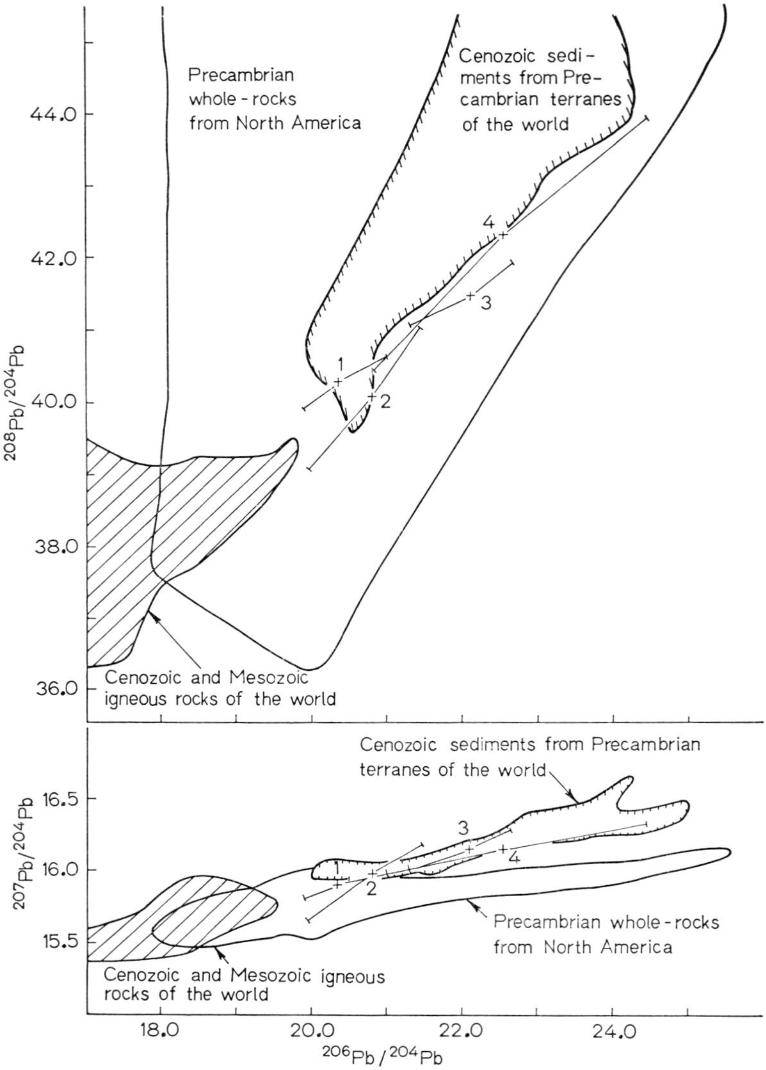

Fig. 82-B-16. $^{207}Pb/^{204}Pb$ and $^{208}Pb/^{204}Pb$ versus $^{206}Pb/^{204}Pb$ for extremes and means of four Mississippi Valley districts: **1** Illinois-Kentucky fluorspar district (HEYL *et al.*, 1966); **2** Southeast Missouri lead belt (BROWN, 1967); **3** Tri-State zinc district (RUSSEL and FARQUHAR, 1960); **4** Wisconsin-Illinois-Iowa lead-zinc district (HEYL *et al.*, 1966). For comparison, enclosed areas include the kinds of data as labeled. (Figure is from DOE, 1970)

volcanism or metamorphism. (Some lead isotope data are shown on Fig. 82-B-16.) These ores are characterized by isotopic heterogeneity, although apparently not to the extent of the indirect magmatogenic ores. The lead isotope data nicely fit a lateral secretion mechanism for these deposits, perhaps with the source of the lead being the basement rocks, as suggested by KANASEWICH (1962) and HEYL *et al.* (1966), because of the similarity of the Pb—Pb age to that accepted for the basement. These ores characteristically are highly radiogenic with isotope ratios greater than those initially present in any igneous rock of any kind, any where in the world.

Deposits of Uncertain Origin. Most other deposits carry lead isotopic compositions of a more equivocal nature. An important isotopic grouping are those ores listed in Table 82-B-3 which, in general, are a large grouping of isotopically-uniform, conformable or stratiform ores including some of the major mineral deposits of the world. The nearly "model" characteristics of the lead isotope ratios in these deposits in fact adds to the confusion as to genesis more than helping to solve this problem. The one possible source that isotopically is a good candidate is pelagic sediments, as was first noted by BROWN (1965). ARMSTRONG (1968) feels the mechanism of getting the lead from such sediments is through downwarping of oceanic crust along the Benioff zone off continental margins and island arcs. The original interpretation of STANTON and RUSSELL (1959) used a volcanic-sedimentary origin in Island arcs. In contrast, KANASEWICH (1968b) has proposed that many of these ores are related to zones of pull-apart and upwelling of fresh mantle such as, for examples, the Red Sea and California's Salton Sea basin. Heavy metal-rich chloride brines are found in these localities. OSTIC *et al.* (1967), while allowing for a mantle origin, interprets some as being derived by metamorphism of sediments soon after they are deposited (such as Bathurst, New Brunswick). WEDEPOHL (1964) has made a convincing case of syngenetic derivation from a sandstone shoreline for the Kupferschiefer Marl Slate. Many derivations may be involved, but the common tie of the noted isotopic characteristics with most conformable or stratiform deposits remains an interesting puzzle.

j) Isotopic Composition of Lead in Natural Waters

Oceans. The isotopic composition of lead has not yet been measured directly in ocean water because of the low lead content; however, the isotopic composition of HCl-soluble lead in manganese nodules is commonly accepted to represent that of lead in the oceans (Table 82-B-8). The lead isotopic compositions of manganese nodules are also not much different from those in the pelagic sediments.

Brines. The chloride-rich brines that carry great concentrations of heavy metals are a relatively new discovery and may be the ore fluid for some of the categories mentioned above. Several vein and stratiform ores near the Red Sea could well be related to the Red Sea geothermal brines for example. COOPER and RICHARDS (1969a) suggest that the variation in lead isotopic composition of acid-soluble lead in the deeper sediments may be due to contamination of the lead from the slightly acidic brine with lead from calcareous pelagic sediments. The new data available so far (Table 82-B-9) show the lead in the brines to be somewhat radiogenic (DOE *et al.*, 1966; DELEVAUX *et al.*, 1967).

IV. Radioactive Lead Isotopes

There are four radioactive isotopes of lead: ^{210}Pb and ^{214}Pb in the ^{238}U decay chain, ^{211}Pb in the ^{235}U decay chain, and ^{212}Pb in the ^{232}Th decay chain. All except ^{210}Pb have half-lives of less than 12 hours. Such short half-lives are of limited use-

Table 82-B-8. *Isotopic composition of lead in natural waters and in the atmosphere*

Locality (reference)	Material	No. samples	$^{206}Pb/^{204}Pb$	$^{207}Pb/^{204}Pb$	$^{208}Pb/^{204}Pb$
Arctic Ocean water (1)	Jurassic Mn nodule	1	18.78	15.84	39.33
Atlantic Ocean water (1)					
N.E. Atlantic, area F in (1)	Cenozoic Mn nodules	3	19.06	15.82	39.58
N.W. Atlantic, area G in (1)	Cenozoic Mn nodules	7	19.13	15.82	39.66
Pacfic Ocean water					
West Pacific (1) area A in (1)	Cenozoic Mn nodules	2	18.55	15.67	39.00
West Pacific (2), area A in (1)	Cenozoic	4	18.51	15.62	38.80
North Pacific (1), area B in (1)	Cenozoic Mn nodules	8	18.81	15.79	39.28
East Pacific (1), area C in (1)	Cenozoic Mn nodules	10	18.85	15.78	39.29
S.E. Pacific (1), area D in (1)	Cenozoic Mn nodules	4	18.99	15.88	39.47
Gulf of California (1)	Cenozoic Mn nodules	1	19.20	15.73	38.98

References: 1. Chow and Patterson (1962a); 2. Chow and Tatsumoto (1964).

Tabke 82-B-9. *Isotopic composition of lead in brines, and precipitate from brines*

Region	Lead in	No. of samples	Isotopic ratios[a]			Reference
			$^{206}Pb/^{204}Pb$	$^{207}Pb/^{204}Pb$	$^{208}Pb/^{204}Pb$	
Salton Sea Trough geothermal area, California	Chloride brine	4 (avg.)	19.13	15.67	39.07	2
Red Sea, Africa (Atlantic II deep)	Chloride brine	1	18.72	15.68	38.52	1
Red Sea, Africa	Surface precip.	1	18.68	15.75	38.82	3
	Deeper precip. from hot spots	12 (avg.)	18.92	15.77	39.08	3
Red Sea, Africa	Pelagic sediments	9 (avg.)	18.95	15.79	39.32	4

References: 1. Delevaux *et al.* (1967); 2. Doe *et al.* (1966); 3. Cooper and Richards (1969); 4. Chow (1968).

[a] All samples were analyzed by surface emission mass spectrometry. All precipitate and sediment data are on acid-soluble lead only and are not whole-rock values.

fulness in geology; therefore, ^{210}Pb with a half-life of 22.0 ± 0.5 yr. (HOUTERMANS *et al.*, 1964; RAMTHUN, 1964; GUTEN *et al.*, 1967) has received the major interest. The analytical techniques for ^{210}Pb are given by GOLDEERG (1963).

The ^{210}Pb has been found useful in studies of volcanic rocks where it is often found in disequilibrium with ^{238}U in Recent volcanics (SOMAYAJULU *et al.*, 1966; OVERSBY and GAST, 1968) and in fumarolic matter (SOMAYAJULU *et al.*, 1966; HOUTERMANS *et al.*, 1964). These studies show promise of dating Recent volcanism within the last 100 yrs. or so, where it may occur free of precursors.

The ^{210}Pb isotope is also useful as an atmospheric tracer [JAWOROWSKI, 1966; for a compilation of references see DOE (1970)]. It is of particular interest in the dating of snow (CROZAZ and LANGWAY, 1966; GOLDBERG, 1963) and other precipitation as again the ^{210}Pb occurs free of precursors. Some studies have used $^{210}Pb/^{206}Pb$ in dating of uranium minerals (KULP *et al.*, 1953).

GOLDBERG (1963) has shown that the ^{210}Pb content shows a rapid depletion between the source regions of rivers (2 to 4 dpm/l) and downstream locations (≤ 0.1 dpm/l). The reason for the rapid depletion is not known. Major rivers in regions of major enrichments of uranium have high ^{210}Pb contents (14.7 dpm/l for Colorado River water at Grand Junction, Colorado). GOLDBERG (1963) has shown that the turnover rate of ^{210}Pb in the oceans is very rapid, and he attributes this to biological causes.

Revised manuscript received: July 1970

References: Section 82-B

AKISHIN, P. A., NIKITIN, O. T., PANCHENKOV, G. M.: A new effective ion emittor for the isotopic lead analysis. Geokhimiya No. 5, 429 (1967).

ALDRICH, L. T., WETHERILL, G. W.: Geochronology by radioactive decay. In: Annual Review of Nuclear Science (Ed. E. G. SEGRÈ). Natl Research Council, Natl. Acad. Sci., Stanford, Calif., Ann. Revs. **8**, 257 (1958).

ARMSTRONG, R. L.: A model for the evolution of strontium and lead isotopes in a dynamic earth. Rev. Geophys. **6**, 175 (1968).

BHANOT, V. B., JOHNSON, W. H., JR., NIER, A. O.: Atomic masses in the heavy mass region. Phys. Rev. **120**, 235 (1960).

BROWN, J. S.: Ore leads and isotopes. Econ. Geol. **57**, 673 (1962).

— Oceanic lead isotopes and ore genesis. Econ. Geol. **60**, 47 (1965).

— Isotopic zoning of lead and sulfur in sourtheast Missouri. In: Genesis of Stratiform Lead-zinc-fluorite Deposits (Mississippi Valley Type Deposits). A symposium, New York, 1966. Econ. Geol. Mon. **3**, 410 (1967).

BURGER, A. J., NICOLAYSEN, L. O., AHRENS, L. H.: Controlled leaching of monazites. J. Geophys. Res. **72**, 3585 (1967).

CAMERON, A. E., SMITH, D. H., WALKER, R. L.: Mass spectrometry of nanogram-size samples of lead. Anal. Chem. **41**, 525 (1969).

CATANZARO, E. J.: Triple-filament method for solid-sample lead isotope analysis. J. Geophys. Res. **72**, 1325 (1967).

— GAST, P. W.: Isotopic composition of lead in pegmatitic feldspars. Geochim. Cosmochim. Acta **19**, 113 (1960).

— KULP, J. L.: Discordant zircons from the Little Belt (Montana), Bear tooth (Montana) and Santa Catalina (Arizona) Mountains. Geochim. Cosmochim. Acta **28**, 87 (1964).

— MURPHY, T. J., SHIELDS, W. H., GARNER, E. L.: Absolute isotopic abundance ratios of common, equal-atom and radiogenic lead isotope standards. J. Res. Natl. Bur. St, A. Phys. and Chem. **72**, 261 (1968).

CHOW, T. J.: Radiogenic leads of the Canadian and Baltic Shield regions. In: Symposium on Marine Geochemistry, 1964. Rhode Island Univ. Narragansett Marine Lab. Occasional Publ. **3**, 169 (1965).

— Lead isotopes of the Red Sea region. Earth Planet. Sci. Lett. **5**, 143 (1968).

— PATTERSON, C. C.: On the primordial lead of the Canyon Diablo meteorite. Geokhimiya, **12**, 1124 (1961).

— — The occurrence and significance of lead isotopes in pelagic sediments. Geochim. Cosmochim. Acta **26**, 263 (1962a).

— — Correction to "Lead isotopes in manganese nodules". Geochim. Cosmochim. Acta **26**, 973 (1962b).

— Tatsumoto, M.: Isotopic composition of lead in the sediments near Japan Trench. In: Recent Researches in the Fields of Hydrosphere, Atmosphere, and Nuclear Geochemistry, Chapt. 10. Tokyo: Maruzen Co. Ltd. 1964.

COOPER, J. A., GREEN, D. H.: Lead isotope measurements on lherzolite inclusions and host basanites from western Victoria, Australia. Earth Planet. Sci. Lett. **6**, 69 (1969).

— RICHARDS, J. R.: Lead isotope measurements on sediments from Atlantis II and Discovery deep areas. In: Hot Brines and Recent Heavy Metal Deposits in the Red Sea, 600 p. Berlin-Heidelberg-New York: Springer 1969a.

— — Lead isotope measurements on volcanics and associated galenas from Coromandel-TeAroha region, New Zealand. Geochem. Jour. (Japan) **3**, 1 (1969b).

CROZAZ, G., LANGWAY, J. R.: Dating Greenland firn-ice cores with Pb-210. Earth Planet. Sci. Lett. **1**, 194 (1966).

DAVIS, G. L., HART, S. R., TILTON, G. R.: Some effects of contact metamorphism on zircon ages. Earth Planet. Sci. Lett. **5**, 27 (1968).

DELEVAUX, M. H., DOE, B. R., BROWN, G. F.: Preliminary lead isotope investigations of brine from the Red Sea, galena from the Kingdom of Saudi Arabia, and galena from the United Arab Republic (Egypt). Earth Planet. Sci. Lett. **3**, 139 (1967).

DOE, B. R.: Relationships of lead isotopes among granites, pegmatites and sulfide ores near Balmat, New York. J. Geophys. Res. **67**, 2895 (1962).

— The bearing of lead isotopes on the source of granitic magma. J. Petrol. **8**, 51 (1967).

— Lead Isotopes. 37 p. Berlin-Heidelberg-New York: Springer 1970.

— Lead and strontium isotope studies of Cenozoic volcanic rocks in the Rocky Mountain region — a summary. Quart. Colo. School Mines **63**, 149 (1968).

— HEDGE, C. E., WHITE, D. E.: Preliminary investigation of the source of lead and strontium in deep geothermal brines underlaying the Salton Sea geothermal area. Econ. Geol. **61**, 462 (1966).

— LIPMAN, P. W., HEDGE, C. E., KURASAWA, H.: Primitive and contaminated basalts from the Southern Rocky Mountains, U.S.A. Contr. Mineral. and Petrol. **21**, 142 (1969).

— TATSUMOTO, M., DELEVAUX, M. H., PETERMAN, Z. E.: Isotope dilution determination of five elements in G-2 (granite), with a discussion of the analysis of lead. U.S. Geol. Surv. Profess. Papers **575**B, B 170 (1967)

— TILLING, R. J., HEDGE, C. E., KLEPPER, M. R.: Lead and strontium isotope studies of the Boulder batholith, southwestern Montana. Econ. Geol. **63**, 884 (1968).

— TILTON, G. R., HOPSON, C. A.: Lead isotopes in feldspars from selected granitic rocks associated with regional metamorphism. J. Geophys. Res. **70**, 1947 (1965).

GERLING, E. K.: Age of the earth according to radioactivity data. Compt. Rend. Acad. Sci. U.R.S.S. **34**, 259 (1942).

FLEMING, E. H., JR., GHIORSO, A., CUNNINGHAM, B. B.: The specific alpha-activities and half-lives of U^{234}, U^{235}, and U^{236}. Phys. Rev. **88**, 642 (1952).

GOLDBERG, E. D.: Geochronology with lead-210. In: Symposium on Radioactive Dating; Athens 1962. Proc. Vienna Internal. Atomic Energy Agency p. 121 (1963).

GRÜNENFELDER, M. H.: Heterogeneity of accessory zircons and the petrographic significance of their uranium decay age. Pt. 1: Zircons of granodiorite gneises of Acquacalda (Lukmanier Pass). [In German.] Schweiz. Mineral. Petrog. Mitt. **43**, 235 (1963).

— EBERHARD, E.: Phase unmixing in zircons: A possible cause for discordant U/Pb age patterns. In: Geochronology. Switzerland 1969 Meeting, Zürich-Bern, Aug. 29-Sept. (1969).

GUTEN, H. R., WYTTENBACK, A., DULAKAS, H.: Half-life of ^{210}Pb (RaD). J. Inorg. and Nucl. Chem. **29**, 2826 (1967).

HART, S. R., KROGH, T. E., DAVIS, G. L., ALDRICH, L. T., MUNIZAGA, F.: Rb/Sr geochronology of granitic rocks southeast of Sudbury, Ontario. Carnegie Inst. Wash. Year Book 1965—1966, **65**, 59 (1967).

— TILTON, G. R.: The isotope geochemistry of strontium and lead in Lake Superior sediments and water (p. 127—137). In: The Earth Beneath the Continents. Am. Geophys. Union Geophys. Mon., Ser. 10, 127 (1966).

HEIMLICH, R. A., BANKS, P. O.: Radiometric age determinations, Bighorn Mountains, Wyoming. Am. J. Sci. **266**, 180 (1968).

HEYL, A. V., DELEVAUX, M. H., ZARTMAN, R. E., BROCK, M. R.: Isotopic study of galenas from the upper Mississippi Valley, the Illinois-Kentucky, and some Appalachian Valley mineral districts. Econ. Geol. **61**, 933 (1966).

HOLMES, A.: An estimate of the age of the earth. Nature **157**, 680 (1946).

HOUTERMANS, F. G.: The isotope frequency in natural lead and the age of uranium. Naturwissenschaften **33**, 185 (1946).

— EBERHARDT, A., FERRARA, G.: Lead of volcanic origin. Chap. 18 (p. 233—243). In: Isotopic and Cosmic Chemistry, Chapt. 18. Amsterdam: North-Holland Pub. Co. 1964.

INGHRAM, M. G.: Manhattan Project, Tech. Ser., National Nuclear Energy Ser., Div. II, vol. 14, Chapt. 5, p. 35. New York: McGraw-Hill Book Co. 1946.

JAWOROWSKI, Z.: Temporal and geographical distribution of radium D (lead-210). Nature **212**, 886 (1966).
KANASEWICH, E. R.: Approximate age of tectonic activity using anomalous lead isotopes. Roy. Astron. Soc. Geophysics J. **7**, 158 (1962).
— The interpretation of lead isotopes and their geological significance. In: Radiometric Dating for Geologists (E. J. HAMILTON and R. M. FARQUHAR, eds.), p. 137. New York: Interscience Publ. 1968 (a).
— Precambrain rift; genesis of strata-bound ore deposits. Science **161**, 1002 (1968b).
KOLLAR, F., RUSSELL, R. D., ULRYCH, T. J.: Precision intercomparisons of lead isotope ratios, Broken Hill and Mount Isa. Nature **187**, 754 (1960).
KOVARIK, A. F., ADAMS, N. N., JR.: The disintegration constant of thorium and the branching ratio of thorium C. Phys. Rev. **54**, 413 (1938).
— — Redetermination of the disintegration constant of U^{238}. Phys. Rev. **98**, 46 (1955).
KULP, J. L., BROCKER, W. S., ECKELMAN, W. R.: Age determination of uranium minerals by the Pb^{210} method. Nucleonics **11**, 19 (1953).
LONG, A., SILVERMAN, A. J., KULP, J. L.: Isotopic composition of lead and Precambrian mineralization of the Cœur d'Alene district, Idaho. Econ. Geol. **55**, 645 (1960).
LOVERING, J. F., TATSUMOTO, M.: Lead isotopes and the origin of granulite and eclogite inclusions in deep-seated pipes. Earth Planet. Sci. Lett. **4**, 350 (1968).
MANTON, W. J., TATSUMOTO, M.: Isotopic composition of lead and strontium in eclogites and related rocks: Annual Report Geosci. Div., 1967—1968, Southwest Center Adv. Studies, p. 19—20, 22 (1969).
MOORBATH, S., WELKE, H.: Lead isotope studies on igneous rocks from the Isle of Skye, northwest Scotland. Earth Planet. Sci. Lett. **5**, 217 (1968).
— — GALE, N. H.: The significance of lead isotope studies in ancient, high-grade metamorphic basement complexes, as exemplified by the Lewisian rocks of northwest Scotland. Earth Planet. Sci. Lett. **6**, 245 (1969).
MURTHY, V. R., PATTERSON, C. C.: Primary isochron of zero age for meteorites and the earth. J. Geophys. Res. **67**, 1161 (1962).
NICOLAYSEN, L. O.: Solid diffusion in radio-active minerals and the measurement of absolute age. Geochim. Cosmochim. Acta **11**, 41 (1957).
OOSTHUYZEN, E. J., BURGER, A. J.: Radiometric dating of intrusives associated with the Waterberg system. S. Africa Geol. Survey Annals **3**, pt. 2, 87 (1965).
OSTIC, R. G., RUSSELL, R. D., STANTON, R. L.: Additional measurements of the isotopic composition of lead from stratiform deposits. Can. J. Earth Sci. **4**, 245 (1967).
OVERSBY, V. M.: The isotopic composition of lead in iron meteorites. Geochim. Cosmochim. Acta **34**, 77 (1970).
— GAST, P. W.: Lead isotopic compositions and uranium decay series disequilibrium in Recent volcanic rocks. Earth Planet. Sci. Lett. **5**, 199 (1968).
PATTERSON, C. C.: The Pb^{207}/Pb^{206} ages of some stone meteorites. Geochem. Cosmochim. Acta 10, 230 (1956).
— TATSUMOTO, M.: The significance of lead isotopes in detrital feldspar with respect to chemical differentation within the Earth's mantle. Geochim. Cosmochim. Acta **28**, 1 (1964).
PICCIOTTO, E. E., WILGAIN, S.: Confirmation of the half-lifes of thorium-232. [In French.] Nuovo Cimento **4**, 1525 (1956).
PIDGEON, R. T., O'NEIL, J. R., SILVER, L. T.: Uranium and lead isotopic stability in a metamict zircon under experimental hydrothermal conditions. Science **154**, 1538 (1966).
RAMTHUN, H.: New calorimetric determination of the half-life period of RaD (^{210}Pb). Z. Naturforsch. **19a**, 1064 (1964).
REYNOLDS, P. H., RUSSELL, R. D.: Isotopic composition of lead from Balmat, New York. Can. J. Earth Sci. **5**, 1239 (1968).
— SINCLAIR, A. J.: Rock and ore-lead isotopes from the Nelson batholith and the Kootenay Arc, British Columbia, Canada. Econ. Geol. **66**, 259 (1971).
ROSHOLT, J. N., PETERMAN, Z. E.: Uranium, thorium, lead systematics in the Granite Mountains, Wyoming. In: Geol. Soc. Am., Abstr. with Programs 1969, **1** pt. 5, 70 (1969).

RUSSELL, R. D., FARQUHAR, R. M.: Lead Isotopes in Geology. New York: Interscience Publ. 1960.
— ULRYCH, T. J., KOLLAR, F.: Anomalous leads from Broken Hill, Australia. J. Geophys. Res. **66**, 1495 (1961).
SILVER, L. T.: Uranium-thorium-lead isotope relations in some lunar materials collected by Apollo 11. Science **167**, 468 (1970).
— DEUTSCH, S.: Uranium-lead isotopic variations in zircons — a case study. J. Geol. **71**, 721 (1963).
SOBOTOVICH, E. V., GRASHENKO, S. M.: Isotopic compositions of Recent lead as age criterion for isolated igneous rock samples. [In Russian.] Akad. Nauk SSSR Izv. Serv. Geol. No. 4, 3 (1965).
— — LOVTSYUS, A. V.: Rock age of Taromskoe quarry according to data of the lead isochronous method. Akad. Nauk SSSR Kom. Opredeleniyu Absolyut. Vozrasta Geol. Formatsii Trudy, sess. 11, 353 (1963).
— — — Age of the rocks of the Sharyzhalagay series (Baikal block). [In Russian.] Akad Nauk SSSR Izv. Ser. Geol., No. 9, 38 (1965).
SOMAYAJULU, B. L. K., TATSUMOTO, M., ROSHOLT, J. N., KNIGHT, R. J.: Disequilibrium of the ^{238}U series in basalt. Earth Planet. Sci. Lett. **1**, 387 (1966).
STACEY, J. S., DELEVAUX, M. E., ULRYCH, T. J.: Some triple-filament lead isotope ratio measurements and an absolute growth curve for single stage leads. Earth Planet. Sci. Lett. **6**, 15 (1969).
— MOORE, W. J., RUBRIGHT, R. D.: Precision measurement of lead isotope ratios — Preliminary analyses from the U.S. Mine, Bingham Canyon, Utah. Earth Planet. Sci. Lett. **2**, 489 (1967).
— ZARTMAN, R. E., NKOMO, I. T.: A lead isotope study of galenas and selected feldspars from mining districts in Utah. Econ. Geol. **63**, 796 (1968).
STANTON, R. L., RUSSELL, R. D.: Anomalous leads and the emplacement of lead sulfide ores. Econ. Geol. **54**, 588 (1959).
STEIGER, R. H., WASSERBURG, G. J.: Systematics in the Pb^{208}—Th^{232}, Pb^{207}—U^{235}, and Pb^{206}—U^{238} systems. J. Geophys. Res. **71**, 6065 (1966).
— — Comparative U—Th—Pb systematics in 2.7×10^9 yr. plutons of different geologic histories. Geochim. Cosmochim. Acta **33**, 1213 (1969).
STERN, T. W., GOLDICH, S. S., NEWELL, M. F.: Effects of weathering on the U—Pb ages of zircon from the Morton Gneiss, Minnesota. Earth Planet. Sci. Lett. **1**, 369 (1966).
STIEFF, L. R., STERN, T. W., SEIKI, OSHIRO, SENFTLE, F. E.: Tables for the calculation of lead isotope ages. U.S. Geol. Surv. Profess. Papers **334**-A, 1 (1959).
TATSUMOTO, M.: Genetic relations of oceanic basalts as indicated by lead isotopes. Science **153**, 1094 (1966).
— KNIGHT, R. J.: Isotopic composition of lead in volcanic rocks from central Honshu — with regard to basalt genesis. Geochem. J. **3**, 53 (1969).
— ROSHOLT, J. N.: The age of the moon: An isotopic study of uranium-thorium-lead systematics of lunar samples. Science **167**, 461 (1970).
— SNAVELY, P. D., JR.: Isotopic composition of lead in rocks of the Coast Range, Oregon and Washington. J. Geophys. Res. **74**, 1087 (1969).
TILTON, G. R.: Volume diffusion as a mechanism for discordant ages. J. Geophys. Res. **65**, 2933 (1960).
— Isotopic composition of lead from tektites. Geochim. Cosmochim. Acta **14**, 323 (1958).
— GRÜNENFELDER, M. H.: Sphene, uranium-lead ages. Science **159**, 1458 (1968).
— STEIGER, R. H.: Mineral ages and isotopic composition of primary lead at Manitouwadge, Ontario. J. Geophys. Res. **74**, 2118 (1969).
TUGARINOV, A. I., GAVRILOVA, J. K., BEDRINOV, V. P.: Evolution of the isotopic composition of lead in Precambrain granitic rocks. Vopr. Prikladn. Radiogeol. 228 (1963).
WAMPLER, J. M., KULP, J. L.: An isotopic study of lead in sedimentary pyrite. Geochim. Cosmochim. Acta **28**, 1419 (1964).
— SMITH, D. H., CAMERON, A. E.: Isotopic comparison of lead in tektites with lead in earth materials. Geochim. Cosmochim. Acta **33**, 1045 (1969).

WASSERBURG, G. J.: Diffusion processes in lead-uranium systems. J. Geophys. Res. **68**, 4823 (1963).
WEDEPOHL, K. H.: Investigation of the Kupferschiefer in northwest Germany; a contribution to the explanation of the genesis of bituminous sediments. [In German]. Geochim. Cosmochim. Acta **28**, 305 (1964).
WELKE, H., MOORBATH, S., CUMMING, G. L., SIGURDSSON, H.: Lead isotope studies on igneous rocks from Iceland. Earth Planet. Sci. Lett. 4, 221 (1968).
WETHERILL, G. W.: An interpretation of the Rhodesia and Witwatersrand age patterns. Geochim. Cosmochim. Acta 9, 290 (1956a).
— Discordant uranium-lead ages. Am. Geophys. Union Trans. **37**, 320 (1956b).
— Discordant uranium-lead ages — Pt. 2: Discordant ages resulting from diffusion of lead and uranium. J. Geophys. Res. **68**, 2957 (1963).
YEATS, R. S.: Southern California structure, sea floor spreading, and history of the Pacific Basin. Geol. Soc. Am. Bull. **79**, 1969 (1968).
ZARTMAN, R. E.: The isotopic compositions of lead in microclines from the Llano uplift, Texas. J. Geophys. Res. **70**, 965 (1965).
— WASSERBURG, G. J.: The isotopic composition of lead in potassium feldspars from some 1.0-b.y. old North American igneous rocks. Geochim. Cosmochim. Acta **33**, 901 (1969).
ZYKOV, S. J., TUGARINOV, A. J., BEL'KOV, J. V., BIBIKOVA, E. V.: The age of the most ancient formations of the Kola Peninsula. Geokhymiya No. 4, 307 (1964).

Revised manuscript **received**: July 1970

82-C. Abundance in Cosmos, Meteorites, Tektites and Lunar Samples

I. Abundance in Cosmos

The cosmic abundance of lead relative to 10^6 atoms Si has been estimated by SUESS and UREY (1956) to be 0.47 atoms on the basis of chondrite data. The average reported in Table 82-C-3 is 0.23 atoms/10^6 Si for ordinary chondrites, half of that estimated by SUESS and UREY (1956). CAMERON (1968) based his figure of 2.9 atoms per 10^6 atoms Si on data of Type I carbonaceous chondrites. According to the latter author, lead is the most abundant element in the sequence from order number 57 to 92. In the genetic grouping of LARIMER and ANDERS (1967), which compares abundances in ordinary relative to carbonaceous (Cc1) chondrites, lead ranks with Tl, In, Bi and the halogens at the most depleted end of the strongly depleted group of elements. They explain chondrites as being composed of two types of material, a low and a high temperature fraction mixed to different degrees at certain areas of the solar nebula.

II. Abundance in Meteorites

Lead occurs in the majority of meteorites in the 0.X ppm range, but in carbonaceous and enstatite chondrites it occurs in the X ppm concentration range. Because of the low lead concentration in ordinary chondrites and irons, contamination has to be considered. A check on potential contamination is possible where data on lead isotopes and those on lead, uranium and/or thorium concentrations are available for the same sample. The concentration of uranium in iron meteorites (including troilite) is so low (0.X ppb U; see Section 92-C of this handbook, compared with 0.1 ppm Pb in metal), that decay of this uranium (and thorium) could not have significantly changed the isotopic composition of the leads listed in Table 82-C-2 within $4.6 \cdot 10^9$ years. A Pb/U ratio of 100 to 1000 has been observed for the most reliable data on iron meteorites. Numerous neutron activation uranium analyses on chondrites (listed in Table 92-C-1) average 0.01 to 0.02 ppm U so that ordinary chondrites have a Pb/U ratio of 10, and carbonaceous and enstatite chondrites a ratio of 100 (Table 82-C-1). Most authors contributing data on lead concentrations have published their isotopic compositions (compiled by OVERSBY, 1971). The least radiogenic lead isotopic composition of several troilites (PATTERSON, 1953, 1955), irons, carbonaceous and enstatite chondrites can be interpreted as the primitive lead of the time of meteorite formation. Leads close to this primitive composition have been indicated as "P" in Tables 82-C-1 and 82-C-2 and those with higher proportions of radiogenic lead have been called "R". Some proportion of radiogenic lead in irons and troilite is probably due to terrestrial contamination. Radiogenic lead of ordinary chondrites can be partly supported by their uranium data (relatively low Pb/U ratios of 10 to 40), but some should be derived from contamination (except Beardsley). Laboratory

Table 82-C-1. *Lead concentrations in chondrites and achondrites*

Name	Class	Average ppm Pb	Range ppm Pb	Method	Isotopic composition (P or R)[a]	Reference
Chondrites						
Abée	Ce 1	3.5	2.9—4.2	N/R	—	1
Indarch	Ce 1	2.2	1.8—2.6	N/R	P	1
		2.61	—	I	P	2
Bruderheim	CL	0.061	0.053—0.066	I	R	11
Elenovka	CL	0.48	—	W/C	R	3
		0.4	—	α/R	R	4
Holbrook	CL	0.30	0.036—0.51	I	R	5, 6
		0.38	0.10—0.60	N/R	—	1
Kunashak	CL	0.53	—	W/C	R	3
		0.49	—	W/C	R	7
Mezö-Madaras	CL	5.27	—	I	P	11
Mocs	CL	0.3	—	I	R	8
Modoc	CL	0.9	—	I	R	9
		0.06	0.04—0.10	N/R	—	1
		0.079	—	I	R	12
		0.3	—	I	R	8
Pervomaisky	CL	0.06	—	W/C	R	7
Saratov	CL	0.40	—	W/C	R	3
Beardsley	CH	0.13	—	I	R	6
		0.26	—	I	PÉR)	12
		0.15	0.13—0.17	N/R	—	1
Forest City	CH	0.4	—	I	R	9
		0.098	0.086—0.12	I	—	5, 6, 10
Forest City	CH	0.15	0.10—0.17	N/R	—	1
Orlovka	CH	0.38	—	W/C	R	7
Plainview	CH	0.46	—	I	R	6
Pultusk	CH	0.11	—	I	R	11
Richardton	CH	0.094	0.019—0.2	I	R	5, 6, 10
		0.075	—	I	R	12
Zhovtnevyi	CH	0.20	—	W/C	R	7
Orgueil	Cc 1	2.94	2.5—3.38	N/R	P	1
Murray	Cc 2	4.6	—	I	P	2
		1.5	1.50—1.52	I	P	12
		3.5	—	α/R	P	4
Murchison	Cc 2	1.6	—	I	P	11
Mighei	Cc 2	1.5	1.36—1.75	N/R	P	1
Mokaia	Cc 3 (CHL)	0.93	—	I	P	2
Allende	Cc 3	1.6	—	I	P	11
		1.097	—	I	P	12

Table 82-C-1 (continued)

Name	Class	Average ppm Pb	Range ppm Pb	Method	Isotopic composition (P or R)[a]	Reference
Achondrites						
Nuevo Laredo	Ap	0.7	—	I	R	9
		0.53	0.42—0.63	N/R	R	1
		0.32	0.32—0.328	I	RR	12
Pasamonte	Ap	0.4	—	I	R	8
Sioux County	Ap	0.19	—	I	RR	12
Norton County	Ae	0.57	—	W/C	R	3
		0.07	—	α/R	R	4
Pesyanoe	Ae	0.36	—	W/C	—	3
Angra dos Reis	Aa	0.55	0.541—0.550	I	RR	12
Pallasite						
Brahin (silicate fraction)	P	0.33	—	W/C	R	7

[a] P: primitive; R: "radiogenic" isotopic composition.

References: 1. Reed *et al.* (1960). 2. Marshall (1962). 3. Starik *et al.* (1958). 4. Cobb (1964). 5. Marshall and Hess (1958). 6. Hess and Marshall (1960). 7. Sobotovich *et al.* (1964). 8. Edwards and Hess (1956). 9. Patterson (1955). 10. Marshall (1968). 11. Tilton (1973). 12. Tatsumoto *et al.* (1973).

blanks ranging from 10 to 60% of the total lead concentration indicate this source of uncertainty.

Larimer (1973) has used the absolute and relative abundances of Bi, In, Pb, Cd and Tl to calculate the equilibrium accretion temperatures of ordinary and enstatite chondrites from a cooling gas of cosmic composition. These temperatures fall in the range of 420 to 540° K.

In Table 82-C-3 average abundances of lead in meteorite classes are compiled. From the above discussion on contamination, it should be expected that the average lead concentration in ordinary chondrites may be lower than the one reported. The average for irons for the total data is almost the same as that for those containing lead with primitive isotopic composition only. The example of Canyon Diablo demonstrates a case of heterogeneous distribution of lead in irons and their troilite.

For an evaluation of the isotopic data see Subsection 82-B-IIIb.

III. Tektites

Data on lead abundances in tektites are listed in Table 82-C-4. The abundance of major elements in average tektites is almost comparable with a 75 percent shale, 25 percent quartz sandstone mixture. Such a mixture is expected to contain about 20 ppm Pb. Except the set of Ivory Coast tektites analyzed by one group of authors,

Table 82-C-2. *Lead concentrations in iron meteorites, mesosiderites and troilites*

Name	Class	Average ppm Pb	Range ppm Pb	Method	Isotopic composition (P or R)[b]	Reference
Hexahedrites						
Boguslavka	H	0.02	—	W/C	R	1
Gressk	H	0.04	—	W/C	R	1
Sikhote-Alin	H (Ogg)	0.03	—	W/C	R	1
Octahedrites						
Canyon Diablo	Og	0.14	—	W/C	P	2
		0.22	—	I	—	3
Bischtübe	Og	0.18	—	W/C	P	1
Chebankol	Og	0.03	—	W/C	R	1
Yardymly	Og	0.17	—	W/C	P	1
Odessa	Og	0.099	—	I	P	3
Toluca	Om	0.16	—	W/C	P	1
		0.098	—	I	P	3
Henbury	Om	0.20	—	W/C	R	2
Toubil	Om	0.20	—	W/C	R	1
Augustinovka	Of	0.19	—	W/C	R	1
Chinga	D	0.03	—	W/C	R	1
Santa Catharina	D	0.40	—	W/C	R	1
Burgavli	iron	0.24	—	W/C	P	1
Mesosiderites						
Bondoc	M	0.071	—	I	R	4
Brahin (metal)	P	0.07	—	W/C	R	5
Troilite						
Canyon Diablo (9)	Og	(62.9) 6.5[a]	1.88—515	I, N/R	P[a] (R)	3, 6, 7, 8, 9
Odessa (2)	Og	4.6	4.2—5.0	I	P	3, 12
Baquedano (1)	Og	1.4	—	I	R	8
Bogon (1)	Og	3.68	—	I	P	8
Bischtübe (1)	Og	7.5	—	W/C	R	1
Yardymly (1)	Og	2.3	—	W/C	P	1
Toluca (7)	Om	20.0	1.66—59	I, N/R	P, R	8, 9, 10, 11
Henbury (2)	Om	5	4.9—5	I	P, R	2, 6
Stauntun (1)	Om	13.9	—	I	R	8
Gibeon (1)	Of	1.57	—	I	R	8
Grant (1)	Of	0.69	—	I	R	8
Burgavli (1)	iron	7.7	—	W/C	P	1
Gressk (1)	H	2.0	—	W/C	R	1

[a] Excluding one high value. — [b] P: primitive; R: "radiogenic" isotopic composition.

References: 1. Starik *et al.* (1960). 2. Starik *et al.* (1959). 3. Marshall and Feitknecht (1964). 4. Marshall (1968). 5. Sobotovich *et al.* (1964). 6. Patterson *et al.* (1953, 1955). 7. Chow and Patterson (1959). 8. Oversby (1970). 9. Reed *et al.* (1960). 10. Marshall and Hess (1961). 11. Ostic (1966) Tatsumoto *et al.* (1973).

Table 82-C-3. *Average Pb abundance in meteorite classes* (Number of data in brackets)

Class	Range ppm Pb	Average ppm Pb
Hexahedrites (3)	0.02—0.04	0.03
Octahedrites etc. (15)	0.098—0.40	0.17
(Troilite) (26)	1.4—515	5.9[a]
Ordinary chondrites (31)	0.019—5.3	0.45
Carbonaceous chondrites Cc 1 (2)	2.5—3.38	2.9
Carbonaceous chondrites Cc 2 (4)	1.36—4.6	2.8
Enstatite chondrites Ce 1 (5)	1.8—4.2	2.8
Achondrites (7)	0.07—0.63	0.44

[a] Excluding one high value.

Table 82-C-4. *Lead abundances in tektites and impact glasses* (Number of samples in brackets)

Type locality		Range ppm Pb	Average ppm Pb	Method	Reference
Czechoslovakia	(1)		6	W/C	1
	(2)	2.4—3.7	3	W/C	2
	(2)	4.6—6.2	5.4	I	3
Indochina	(3)	2.4—8.6	6.2	W/C	2
	(1)	—	2.8	I	3
Philippines	(1)	—	5	W/C	1
	(1)	—	1.4	W/C	2
Australia	(1)	—	2.9	I	1
Ivory Coast	(7)	<10—18	14	S	4
	(3)	≤ 1.1—≤ 1.4	≤ 1.3	I	3
Texas	(21)	<10—10	<10	S	5
Libyan Desert glass	(1)	—	5	W/C	1
	(1)	—	4.7	W/C	2
Ries glass	(3)	18—28	23	I	3
Bosumtwi glass		2.6—2.7	2.65	I	3

References: 1. TILTON (1958). 2. STARIK *et al.* (1961). 3. WAMPLER *et al.* (1969). 4. CUTTITTA *et al.* (1972). 5. CHAO (1963).

the lead concentration in tektites is about equal to or less than a quarter of the shale-sandstone mixture. The lead isotopic composition is terrestrial (TILTON, 1958; WAMPLER *et al.*, 1969) but the $^{238}U/^{204}Pb$ ratio is above 20 or even as high as 50. This can be explained by a preferential loss of lead through volatilization from terrestrial materials (probably during impact and transport) and a retention of the more refractory uranium.

Table 82-C-5. *Lead abundances in lunar materials from Mare Tranquillitatis, Oceanus Procellarum, Fra Mauro area, Mare Crisium and Descartes area.* (Number of samples in brackets). Most references are from the Proceedings of the First, Second and Third Lunar Science Conference, published as Supplement 1 (1970) indicated as P/[1], Supplement 2 (1971) indicated as P/[2], and Supplement 3 (1972) indicated as P/[3] of Geochim. Cosmochim. Acta

Rock type	Mission	Range ppm Pb	Average ppm Pb	Method	Reference
Fines (soil), lowland (6)	Apollo 11	1.3—6	2.2	I, Mass	1, 2, 3, 4, 5, 6
Fines, lowland (10)	Apollo 12	1.77—4.43	3.3	Mass, I	7, 8, 9, 10
Fines, highland (15)	Apollo 14	7.18—12	9.3	I, Mass, W/X/S	11, 12, 13, 14, 15
Fines, highland (3)	Apollo 15	1.10—1.77	1.48	I	17
Fines, highland (4)	Apollo 16	1.17—2.78	2.0	I	16
Fines, lowland (2)	Luna 20	1.26—1.33	1.3	I	16
Breccias, lowland (4)	Apollo 11	1.2—3	1.95	Mass, I	3, 4, 5
Breccias, lowland (6)	Apollo 12	3.82—16.3	9.54	I	9
Breccias, highland (23)	Apollo 14	2.9—18	8.94	I, Mass, W/X/S	11, 12, 13, 14, 15
Basalts, lowland (18)	Apollo 11	0.29—3	1.48	I, Mass	3, 4, 5
Basalts, lowland (14)	Apollo 12	0.28—1.17	0.54	I, Mass	7, 8, 9
Basalts, highland (6)	Apollo 14	1.71—13	7.0	I, W/X/S	14, 11, 15
Basalts, highland (8)	Apollo 15	0.13—0.35	0.22	I	17
Troctolites, lowland (2)	Apollo 12	0.27—0.32	0.3	I	9
Anorthosites, highland (2)	Apollo 15	0.25—0.27	0.26	I	17
Anorthosite, highland (1)	Apollo 16	—	0.0069	I	16
Plagioclase, lowland (1)	Apollo 14	—	0.7	Mass	12

References: (analytical method applied for total lead in brackets) 1. Gopalan *et al.* [P/[1]] (I). 2. Kohman *et al.* [P/[1]] (I). 3. Morrison *et al.* [P/[1]] (Mass). 4. Silver [P/[1]] (I). 5. Tatsumoto [P/[1]] (I). 6. Wanless *et al.* [P/[1]] (I). 7. Morrison *et al.* [P/[2]] (Mass). 8. Cliff *et al.* [P/[2]] (I). 9. Tatsumoto *et al.* [P/[2]] (I). 10. Huey *et al.* [P/[2]] (I). 11. Rose *et al.* [P/[3]] (W/X/S). 12. Taylor *et al.* [P/[3]] (Mass). 13. Barnes *et al.* [P/[3]] (I). 14. Tatsumoto *et al.* [P/[3]] (I). 15. Silver (1971) (I). 16. Tera and Wasserburg (1972). 17. Tatsumoto *et al.* (1972).

IV. Lunar Materials

Lead data of samples from missions Apollo 11, 12, 14, 16 and Luna 20 are listed in Table 82-C-5 to accomplish a comparison with terrestrial rocks close in composition.

Except those from the Fra Mauro site (Apollo 14), the known lunar basalts are usually much lower in common lead than terrestrial basalts. The Fra Mauro area seems to contain anomalous lead in all rock types. At the moon's surface a local depletion of readily volatile elements can partially be balanced by a local accumulation. A high fraction of radiogenic lead in all samples indicates a general depletion of primitive lead relative to the highly refractory uranium. $^{238}U/^{204}Pb$ ratios range from 50 to 2800 in contrast to those of earth materials, which are usually close to 10.

The lunar fines are to some extent accumulators of lead and some other relatively volatile elements. Highland anorthosite can be very low in lead.

Using a $^{238}U/^{204}Pb$ ratio of 100 and a value of 0.7 μcal/cm² sec for the lunar heat flux, TERA *et al.* (1974) estimate a primordial Pb concentration of 35 ppb in lunar materials.

Revised manuscript received: April 1974

82-D. Abundance in Rock-Forming Minerals, Phase Equilibria, Lead Minerals

I. Rock-Forming Minerals

Because of the oxygen fugacity in the earth's crust, divalent lead is more common than tetravalent lead. In Table 12-8 in Volume I of this handbook, values of the ionic radius of Pb^{2+} are listed as between 1.18 and 1.32 Å. Pb^{2+} is similar in ionic size to K^+ (r: 1.33 Å), Sr^{2+} (r: 1.12—1.27 Å) and Ba^{2+} (r: 1.34—1.43 Å). Even replacement of Ca^{2+} (r: 0.99—1.06 Å) and Na^+ (r: 0.95—1.0 Å) is to be observed.

A large variety of coordinations ranging from 3 to 12 nearest neighbors exists in lead oxygen compounds. Due to some tendency for covalent bonding, sometimes lower coordination numbers are observed than have been expected from the ratio of the ionic radii (Pb^{2+}/O^{2-}) (see Section 82-A). In PbO, lead is coordinated by 4 oxygen ions instead of 8, but in the carbonate (cerussite), phosphate (pyromorphite), sulfate (anglesite), chromate (crocoite) and comparable compounds, lead is surrounded by 8 and more oxygen ions which in these anions can be polarized to a minor degree compared to PbO. From these observations a more or less pronounced ionic character of Pb^{2+} is to be expected.

Because of the size of Pb^{2+}, lead will be incorporated in silicates with structural positions of large monovalent or divalent metals. It replaces K^+ in its minerals (see Table 82-D-1). Lead feldspars (low $PbAl_2Si_2O_8$ with ordered Si, Al configuration; space group I2/c; high $PbAl_2Si_2O_8$ with disordered Si, Al configuration; space group C2/m; BRUNO and FACCHINELLI, 1972) and solid solutions of the system $PbAl_2Si_2O_8$-$KAlSi_3O_8$ have been synthesized (KÖHLER, 1935; SORRELL, 1962; SCHEEL, 1971; etc.).

Because of the charge of lead and barium, the respective feldspars are expected to have a higher melting point than potassium feldspar. This is true in case of celsian (see Fig. 56-D-1). In contrast to barium, potassium feldspars of granitic rocks are on average half as low in lead (419 samples of Table 82-D-1: 53 ppm Pb) as pegmatitic microclines (219 samples of Table 82-D-1, excluding amazonites: 98 ppm Pb).

Amazonites, the green potassium feldspars of some pegmatites, are special accumulators of lead and rubidium up to the percent range. Several authors such as ZHIROV and STISHOV (1965) and PLYUSNIN (1969), concluded from a relation between color intensity and composition that the green color is caused by structural defects arising from the replacement of $2K^{2+}$ by a single Pb^{2+}. The latter author assumes charge compensation by the replacement of $2K^+O^{2-}$ through $Pb^{2+}2OH^-$.

Mica usually contains less lead than potassium feldspar. If one excludes the large series sampled by TISCHENDORF *et al.* (1969) from a limited area, 259 biotites contain 21 ppm Pb; including the Saxonian granitic biotites of these authors the average is raised up to 37 ppm Pb. The arithmetic mean of 32 muscovites is 26 ppm Pb. Lead favors the biotite lattice of co-existing biotite-muscovite; the distribution factor between

Table 82-D-1. *Lead concentrations in rock-forming potassium minerals*
(Number of samples in brackets)

Mineral	Rock, origin	Range ppm Pb	Mean ppm Pb	Method	Reference
Potassium feldspars (7)	Granites, Colorado (U.S.A.), Manitoba (Canada)	31.6—108	65.4	I	Doe (1967)
Potassium feldspars (6)	Granites, Maryland, Minnesota (U.S A), Ontario (Canada)	13.4—80	45.9	I, W/C	Doe *et al.* (1965)
Potassium feldspars (6)	Gneisses, Virginia, Maryland (U.S.A.)	34.9—55	52.7	I, W/C	Doe *et al.* (1965)
Potassium feldspars (7)	Granites, granodiorites etc., North America	9.5—79.5	42.5	I	Doe and Tilling (1967)
Potassium feldspars (87)	Quartz monzonites, Basin and Range Province, Nevada, Utah (U.S.A.)	9—126	37.5	S	Slawson and Nackowski (1959)
Potassium feldspars (microclines) (10)	Granites, rhyolite, Llano, Texas (U.S.A.)	30.4—50.2	33.3	I	Zartman (1965)
Potassium feldspars (7)	Gneisses, Llano, Texas (U.S.A.)	15.7—55.5	30.9	I	Zartman (1965)
Potassium feldspars (30)	Granites, granodiorites, rhyolites etc., Precambrian (U.S.A.)	3—79	29.5	W/C, (I)	Zartman and Wasserburg (1969)
Potassium feldspars (4)	Gneisses (U.S.A.)	23—122	53	W/C	Zartman and Wasserburg (1969)
Potassium feldspars (2)	Quartz monzonite, pegmatite, Butte Montana (U.S.A.)	45—53	49	I	Murthy and Patterson (1961)
Potassium feldspars, perthites (32 microclines; 24 orthoclases)	Granites (subvolcanic and from sedimentary and metamorphic terranes) Australia	29—222	65	X	Rhodes (1969)
Potassium feldspars (3)	Granites, granodiorite, Churchill Province (Canada), New Zealand	30.6—66.2	42.7	I	Sinha (1969)
Potassium feldspars (35)	Granites, coarse, Cape (South Africa)	23—60	41	S	Kolbe and Taylor (1966)

Table 82-D-1 (continued)

Mineral	Rock, origin	Range ppm Pb	Mean ppm Pb	Method	Reference
Potassium feldspars (18)	Granites, medium- and fine-grained, Cape (S. Africa)	<6—58	31	S	Kolbe and Taylor (1966)
Potassium feldspars (microclines) (40)	Granites, Oslo and Arendal Region (Norway)	40—150	93.1	S	Oftedal (1967)
Potassium feldspars (3)	Granites, southern Norway	34—51	41	S	Heier *et al.* (1967)
Potassium feldspars (13)	Granodiotites, adamellites, Scotland, England	10—75	25.2	S	Wedepohl (1956)
Potassium feldspars (14)	Granitic rocks, Erzgebirge, Thuringian Mountains (G.D.R.)	24—100	68	S	Bräuer (1970)
Potassium feldspars (49)	Granodiorites, quartz-monzonites, Dinarides, Balkanides (Yugoslavia)	26—74	50.5	S	Čuturič *et al.* (1968)
Potassium feldspars (18)	Granites, gneisses, Tarak, Yenisey Range (U.S.S.R.)	30—100	47	S	Zhirov and Urusova (1962)
Potassium feldspars (4)	Granites, Yenisey, M. Siberia (U.S.S.R.)	53—700	356	I	Volobuyev *et al.* (1970)
Potassium feldspars (3)	Granite, granodiorite, quartz diorite, Transbaikalia (U.S.S.R.)	15—19	16.3	W/C, P	Rabinovich and Baskova (1959)
Feldspars (3)	Granites, Susamyr batholith, Central Tienshan (U.S.S.R.)	28—40	34.3	W/C	Tauson and Kravchenko (1956)
Feldspars (19)	Granitic rocks, Central Asia (U.S.S.R.)		25.1	I	Rabinovich *et al.* (1965)
Potassium feldspars (phenocrysts) (3)	Pitchstones, Iceland, Eigg	7.5—14	9.8	W/C	Carmichael and MacDonald (1961)
Potassium feldspars (3)	Rhyolite, quartz latite, West U.S.A.	26.2—36.6	30.1	I	Doe and Tilling (1967)
Potassium feldspars (sanidines) (41)	Quartz latites, Zvecan (Yugoslavia)	18—58	37.9	S	Karamata (1969)

Table 82-D-1 (continued)

Mineral	Rock, origin	Range ppm Pb	Mean ppm Pb	Method	Reference
Potassium feldspars (sanidines (13)	Latite, Trepča (Yugoslavia)	31—45	38	S	ČUTURIČ *et al.* (1968)
Potassium feldspars (sanidines) (3)	Trachytes, Eifel (Germany)	2—46	21	S	WEDEPOHL (1956)
Potassium feldspars (adularia) (5)	Switzerland, Austria	41—100	68	S	WEDEPOHL (1956)
Potassium feldspar concentrates (6)	Beach and river sand, U.S.A.	48.1—57.6	51.9	I	PATTERSON and TATSUMOTO (1964)
Potassium feldspars (6)	Pegmatites (U.S.A.)	12—69	31.7	W/C	ZARTMAN and WASSERBURG (1969)
Potassium feldspars (8)	Pegmatites, Churchill and Superior Provinces (Canada), Maine (U.S.A.)	8.1—100	51.9	I	SINHA (1969)
Potassium feldspars (13)	Pegmatites, Maryland, Virginia, Maine (U.S.A.)	61.8—521	189	I, W/C	DOE *et al.* (1965)
Potassium feldspars (8)	Pegmatites, Front Range Colorado (U.S.A.)	28.2—88	57.4	I	DOE and HART (1963)
Potassium feldspars (microclines) (11)	Pegmatites, U.S.A. (10), Norway	20—495	177	S (I)	CATANZARO and GAST (1960)
Potassium feldspars (20)	Pegmatites, southern Norway	4.5—215.5	69.5	S/X	HEIER *et al.* (1967)
Potassium feldspars (microclines) (113)	Pegmatites, southern Norway	<10—500	101	S	OFTEDAL (1967)
Potassium feldspars (8)	Pegmatites, Yenisey, M. Siberia (U.S.S.R.)	15—1300	329	I	VOLOBUYEV *et al.* (1970)
Potassium feldspars (microclines) (13)	Pegmatites, Tarak, Yenisey Range (U.S.S.R.)	17—63	30	S	ZHIROV and URUSOVA (1962)
Potassium feldspars (microclines (19)	Pegmatites, eastern Kazakhstan (U.S.S.R.)	0.9—18	7.4	S	SRETENSKAYA (1964)
Potassium feldspars (amazonites) (9)	Pegmatites, Norway, Sweden, U.S.S.R., Č.S.R., U.S.A., Brazil, Madagascar	130—6,800	2,900	S, X	WEDEPOHL (unpublished)

Table 82-D-1 (continued)

Mineral	Rock, origin	Range ppm Pb	Mean ppm Pb	Method	Reference
Potassium feldspars (amazonites) (44)	Pegmatites, U.S.S.R. (29), Madagascar, Norway, Sweden, U.S.A., Canada, China	120—2,800	596	S	ZHIROV and STISHOV (1965)
Potassium feldspar (amazonite (1)	Pegmatite, Broken Hill (Australia)	—	11,000	W	ČECH et al. (1971)
Biotites (35)	Schists, amphibolites, granulites, U.S.A.	7—39	23.5	S	DEVORE (1955)
Biotites (86)	Quartz monzonites, Basin Range, Utah, Nevada (U.S.A.)	4—78	19.9	S	PARRY and NACKOWSKI (1963)
Biotites	Granites, granodiorites, Canada	—	50	S	RIMSAITE (1968)
Biotite-chlorite (1)	Granite, Grenville, E. Ontario (Canada)	—	23	W/C	AZZARIA (1963)
Biotites (21)	Granites, Cape (South Africa)	7—95	27	S	KOLBE and TAYLOR (1966)
Biotites (91)	Granites, gneisses, Central Europe	<10—94	≤16	S	HAACK (1969)
Biotites (348)	Granites, W. Erzgebirge, Saxony (G.D.R.)	23—89 (s range)	48.2	S	TISCHENDORF *et al.* (1969)
Biotites (14)	Granitic rocks, Erzgebirge, Thuringian Mountains (G.D.R.)	10—94	43.1	S	BRÄUER (1970)
Biotites (3)	Granites, Susamyr Batholith, Central Tienshan (U.S.S.R.)	14—20	18	W/C	TAUSON and KRAVCHENKO (1956)
Biotites (3)	Granite, granodiorite, quartz diorite, Transbaikalia (U.S.S.R.)	10—36	19.3	W/C, P	RABINOVICH and BASKOVA (1959)
Muscovites (7)	Schists, Pennsylvania, Wyoming, Colorado (U.S.A.)	12—77	49.1	S	DEVORE (1955)
Muscovites (3)	Schists, Idaho Batholith (U.S.A.)	20—70	53.3	S	HIETANEN (1969)
Muscovites (4)	Granites, Cape (South Africa)	<6—33	13	S	KOLBE and TAYLOR (1966)
Muscovites (5)	Pegmatites, U.S.A., U.S.S.R., S. Africa, Germany	9—20	12.8	S	WEDEPOHL (1956)
Muscovite (1)	Pegmatite, Baltimore (U.S.A.)		17	I	DOE et al. (1965)
Muscovites (13)	Granitic rocks, Erzgebirge, Thuringian Mountains (G.D.R.)	11—44	16	S	BRÄUER (1970)

Table 82-D-2. *Lead concentrations in rock-forming minerals* (excluding potassium minerals) (Number of samples in brackets)

Mineral	Rock, origin	Range ppm Pb	Mean ppm Pb	Method	Reference
Olivines (concentrates) (8)	Peridotite nodules in alkali olivine basalts, Germany	<0.2—7.2	<2.6	S, A	WEDEPOHL (1956), HEINRICHS (1974)
Pyroxenes (20)	Basalts, gabbro, pyroxenites, eclogites, schists, U.S.A., Italy, Norway, Germany, S. Africa	0.3—20	5.9	S	WEDEPOHL (1956)
Hornblendes (17)	Amphibolites, Emeryville, Colton N.Y. (U.S.A.)	3—14	6.7	W/C	ENGEL *et al.* (1964)
Amphiboles (40)	Schists, granulites, U.S.A. (39), Austria	6—34	13.7	S	DEVORE (1955)
Amphiboles (kaersutites) (9)	Alkaline rocks, Australia, New Zealand	1—25	11.4	X	KESSON and PRICE (1972)
Hornblendes (17)	Granitic rocks, Suhl, Thuringia (G.D.R.)	15—70	27	S	BRÄUER (1970)
Hornblendes (2)	Granodiorite, quartz diorite, Transbaikalia (U.S.S.R.)	8—14	11	W/C, P	RABINOVICH and BASKOVA (1959)
Plagioclases (12)	Granites, Granodiorites, rhyolites etc., North America	3.8—38	16.9	I	DOE and TILLING (1967)
Plagioclases (20)	Magmatic rocks, pegmatites, Scotland, Norway, U.S.A., U.S.S.R.	1—32	11.2	S	WEDEPOHL (1956)
Plagioclases (15)	Amphibolites, Emeryville, Colton N.Y. (U.S.A.)	8—36	16.2	W/C	ENGEL *et al.* (1964)
Plagioclases (14)	Granitic rocks, Erzgebirge, Thuringian Mountains (G.D.R.)	20—70	37	S	BRÄUER (1970)
Quartz (1)	Granite, Hilburton County, Ontario (Canada)	—	0.1	I	TILTON *et al.* (1955)
Quartz (2)	Quartz monzonite, Butte, Montana (U.S.A.)	0.13—0.2	0.16	I	MURTHY and PATTERSON (1961)
Quartz (2)	Alaskite, Butte, Montana (U.S.A.)	0.08—0.25	0.17	I	DOE and TILLING (1967)

Table 82-D-2 (continued)

Mineral	Rock, origin	Range ppm Pb	Mean ppm Pb	Method	Reference
Quartz (3)	Granites, Susamyr batholith, Central Tian-Shan (U.S.S.R.)	2—4	3	W/C	TAUSON and KRAVCHENKO (1956)
Quartz (3)	Granite, granodiorite, quartz diorite, Transbaikalia (U.S.S.R.)	3—4	3	W/C, P	RABINOVICH and BASKOVA (1959)
Magnetite (1)	Granite, Grenville, E. Ontario (Canada)	—	27	W/C	AZZARIA (1963)
Magnetites (8)	Gabbro, Skaergaard (Greenland)		<10	W/C	VINCENT and PHILLIPS (1954)
Magnetites (3)	Basalt, Lower Saxony (Germany)	<3—10	≤ 6.7	S	WEDEPOHL (1956)
Magnetite (1)	Quartz monzonite, Stillwater, Montana (U.S.A.)		1.5	I	NUNES and TILTON (1971)

both minerals is probably close to 2. Analcite of 3 localities ranges as high as 100 to 300 ppm Pb, whereas the only phillipsite as yet analyzed contains only 7 ppm Pb (WEDEPOHL, 1956).

Lead in rock-forming minerals excluding potassium compounds is listed in Table 82-D-2 as being lower than in potassium minerals. Plagioclases and amphiboles containing Na^+-Ca^{2+} structural positions are the only mineral groups with lead concentrations almost as high as the potassium minerals low in lead. 61 Plagioclases contain on average 19.5 ppm Pb and 85 amphiboles 15 ppm Pb. DOE and TILLING (1967) report a distribution factor of lead between co-existing K-feldspar and plagioclase of 2. Nepheline and sodium zeolites often contain lead in concentrations comparable to plagioclase.

Because of structural reasons pure olivine and quartz mostly contain less than 1 ppm Pb. Tertiary quartz sands which are composed of detrital quartz from different rock origin, analyzed by the present author, have less than 1 ppm Pb.

II. Phase Equilibria

a) Oxides. Most of the known oxides of lead have been prepared under equilibrium conditions by WHITE and ROY (1964). The authors mention PbO, Pb_3O_4, Pb_2O_3, $Pb_{12}O_{19}$ and PbO_2 as phases.

b) Silicates. In the system PbO-SiO_2, the phases $PbSiO_3$ (melting point 764° C), Pb_2SiO_4 (melting point 743 °C), and Pb_4SiO_6 (incongr. melting at about 725° C) exist according to GELLER and coworkers (for references see: LEVIN *et al.*, 1956). Of these only $PbSiO_3$ (alamosite) has been listed in Table 82-D-3 and $Pb_3Si_2O_7$ (barysilite) has not been detected in the melting experiments of the above authors. GELLER and coworkers have also investigated the systems PbO-K_2O-SiO_2, PbO-

Na_2O-SiO_2 and Pb-Al_2O_3-SiO_2, some of which can be of importance for technical glasses. The following quaternary phases exist in the system with potassium: $K_2Pb_4Si_8O_{21}$, $K_2PbSi_4O_{10}$, $K_2Pb_2Si_2O_7$, $K_4PbSi_3O_9$, and in the system with aluminum: $Pb_6Al_2Si_6O_{21}$, $Pb_8Al_2Si_4O_{19}$, $Pb_4Al_2Si_2O_{11}$. None of these are listed under the lead minerals of Table 82-D-3. The synthesis of lead feldspar has been mentioned in Subsection 82-D-I.

c) Sulfides. The literature on lead sulfosalt minerals with compositions and relations in the systems PbS-As_2S_3, PbS-Sb_2S_3 and PbS-Bi_2S_3 has been reviewed by CHANG and BEVER (1973).

One of the most important systems because of the natural abundance of galena containing appreciable concentrations of Ag and Bi, is: Ag-Bi-Pb-S, studied by VAN HOOK (1959). Ternary phases in this system are $PbBi_2S_4$ (galenobismutite, melting incongruently at $750 \pm 5°$ C) and $AgBiS_2$ ($\alpha AgBiS_2$: schapbachite; $\beta AgBiS_2$: matildite; melting point of $\alpha AgBiS_2$ $801 \pm 5°$ C). PbS melts at $1{,}115 \pm 5°$ C. VAN HOOK (1959) indicated the presence of three compounds along the join PbS-Bi_2S_3 but did not clearly identify them. According to Table 82-D-3, the phases $Pb_2Bi_2S_5$ (cosalite) and $Pb_3Bi_2S_6$ (lillianite) are expected to occur on this join.

Isothermal sections at 180°, 400° and 600° C in the Ag_2S-Bi_2S_3-PbS part of the system are drawn in Fig. 82-D-1 according to VAN HOOK (1959) to demonstrate the change of the solubility. There is a complete solid solution between $\alpha AgBiS_2$ and

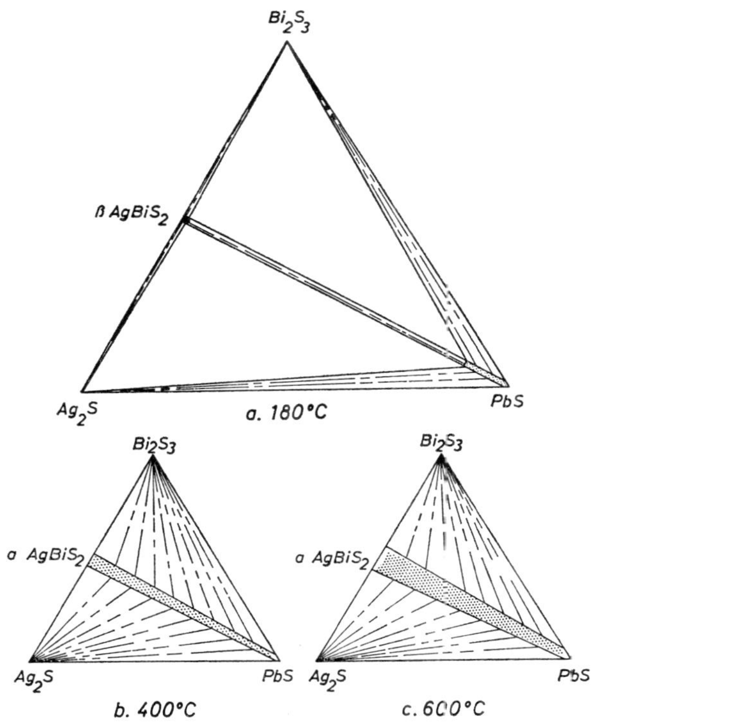

Fig. 82-D-1. Isothermal sections of the system Bi_2S_3-Ag_2S-PbS at 180, 400 and 600° C (according to VAN HOOK, 1959)

PbS above about 210° C. It decreases from this crest of the solvus to 1 mol percent PbS soluble in $AgBiS_2$ and 12 mol percent $AgBiS_2$ (matildite) soluble in PbS at 200° C (10 mol percent at 180° C). The solubility of Ag_2S in PbS was estimated as about 0.5 mol percent at only 700° C.

The solubilities of FeS in galena and of PbS in pyrite and pyrrhotite at 700° C are very small, probably less than 0.1 wt.-percent. Ternary compounds do not occur in the system Fe-Pb-S (BRETT and KULLERUD, 1967).

In a compilation on minor elements in some sulfide minerals, FLEISCHER (1955) mentions for galena those of Sb and As in numerous samples, beside the appreciable concentrations of Ag and Bi. 150 and 130 out of about 230 samples contain 200 to 5000 ppm Ag and Sb respectively. Bi occurs in a wider range of concentrations and As is lower. Mn, Ni, Tl, Sn and Cd are usually lower than 50 ppm in galena.

III. Lead Minerals

Lead minerals are listed in Table 82-D-3 according to the Mineral Tables (Mineralogische Tabellen 5th ed.; Akademische Verlagsgesellschaft, Leipzig, 1970) compiled by H. STRUNZ, the Index of New Minerals by M. FLEISCHER (Am. Miner. **51**, 1247, 1966), and recent literature. This list contains about 240 minerals of which more than one third belongs to the group of sulfides, sulfosalts etc. Sulfates, phosphates and arsenates of lead, mostly containing hydroxyl, are almost as numerous as minerals of the former group. In the large proportion of its minerals, lead occupies its specific structural positions and is not partially replacing other elements of similar properties.

Table 82-D-3. *Lead minerals*

Metals	
Lead	Pb
Zvyagintsevite	$(Pb,Pt)_3(Pb,Sn)$
Sulfides, selenides, tellurides, sulfosalts	
Galena	PbS
Clausthalite	PbSe
Altaite	PbTe
Shadlunite	$(Fe,Cu)_8(Pb,Mn,Cd)S_8$
Shandite	β-$Pb_2Ni_3S_2$
Parkerite	α-$Pb_2Ni_3S_2$
Matildite	β-$AgBiS_2$ (hex.)
Schapbachite	α-$AgBiS_2$ (cub.)
Teallite	$PbSnS_2$
Montesite	$PbSn_4S_5$
Franckeite	$Pb_5Sn_3Sb_2S_{14}$
Cylindrite	$Pb_3Sn_4Sb_2S_{14}$
Platynite	$Pb_4Bi_7Se_7S_4$
Nagyagite	$Pb_5Au(Te,Sb)_4S_{5-8}$
Seligmannite	$Pb_2Cu_2As_2S_6$
Bournonite	$Pb_2Cu_2Sb_2S_6$
Aikinite	$Pb_2Cu_2Bi_2S_6$
Hammarite	$Pb_2Cu_2Bi_4S_9$
Lindstroemite	$Pb_3Cu_3Bi_7S_{15}$
Gladite	$PbCuBi_5S_9$

Table 82-D-3 (continued)

Rezbanyite	$Pb_2Cu_2Bi_{2n}S_{3+3n}$
Larosite	$Pb(Cu,Ag)_{20}BiS_{13}$
Benjaminite	$(Cu,Ag)_2Pb_2Bi_4S_9$
Betekhtinite	$Cu_{10}(Fe,Pb)S_6$
Hutchinsonite	$(Pb,Tl)_2As_5S_9$
Hatchite	$PbTlAs_2S_7$
Wallisite	$PbTl(Cu,Ag)As_2S_5$
Fizelyite	$Pb_5Ag_2Sb_8S_{18}$
Ramdohrite	$Pb_3Ag_2Sb_6S_{13}$
Andorite	$Pb_2Ag_2Sb_6S_{12}$
Nakaseite	$Pb_4Ag_3CuSb_{12}S_{24}$
Freieslebenite	$Pb_4Ag_4Sb_4S_{12}$
Diaphorite	$Pb_2Ag_3Sb_3S_8$
Owyheeite	$Pb_5Ag_2Sb_6S_{15}$
Teremkovite	$Pb_7Ag_2Sb_8S_{20}$
Schirmerite	$PbAg_4Bi_6S_{12}$
Marrite	$PbAgAsS_3$
Sartorite	$PbAs_2S_4$
Baumhauerite	$Pb_{10}As_{17}(Ag,Tl)S_{36}$
Twinnite	$Pb(Sb,As)_2S_4$
Guettardite	$Pb_9(Sb,As)_{16}S_{33}$
Dadsonite	$Pb_{11}Sb_{12}S_{29}$
Playfairite	$Pb_{16}(Sb,As)_{18}S_{43}$
Sorbyite	$Pb_{17}(Sb,As)_{22}S_{50}$
Launayite	$Pb_{22}Sb_{26}S_{61}$
Liveingite	$Pb_4As_6S_{13}$
Rathite I	$(Pb,Tl)_3As_4(As,Ag)S_{10}$
Rathite II	$Pb_9As_{13}S_{28}$
Rathite III	$Pb_3As_5S_{10}$
Dufrenoysite	$Pb_2As_2S_5$
Veenite	$Pb_2(Sb,As)_2S_5$
Lengenbachite	$Pb_7As_4S_{13}$
Jordanite	$Pb_9As_4S_{15}$
Geokronite	$Pb_9(As,Sb)_4S_{15}$
Schulzite	$Pb_9Sb_4S_{15}$
Gratonite	$Pb_9As_4S_{15}$
Fueloeppite	$Pb_3Sb_8S_{15}$
Zinckenite	$PbSb_2S_4$ (or $Pb_6Sb_{14}S_{27}$)
Robinsonite	$Pb_7Sb_{12}S_{25}$
Plagionite	$Pb_5Sb_8S_{17}$
Heteromorphite	$Pb_7Sb_8S_{19}$
Madocite	$Pb_{17}(Sb,As)_{16}S_{41}$
Jamesonite	$Pb_4FeSb_6S_{14}$ (monocl.)
Parajamesonite	$Pb_4FeSb_6S_{14}$ (orthorhomb.?)
Sterryite	$Pb_{12}(Sb,As)_{10}S_{27}$
Semseyite	$Pb_9Sb_8S_{21}$
Boulangerite	$Pb_5Sb_4S_{11}$
Meneghinite	$Pb_{13}CuSb_7S_{24}$
Sakharovaite (bismuthjamesonite)	$(Pb,Fe)(Bi,Sb)_2S_4$
Ustarasite	$Pb(Bi,Sb)_6S_{10}$
Bonchevite (?)	$PbBi_4S_7$ (or $Pb_3Bi_2S_6$)
Galenobismutite	$PbBi_2S_4$
Cannizzarite	$Pb_3Bi_5S_{11}$ (?)

Table 82-D-3 (continued)

Wittite	$Pb_5Bi_6(S,Se)_{14}$
Cosalite	$Pb_2Bi_2S_5$
Selenocosalite	$Pb_2Bi_2(S,Se)_5$
Giessenite	$Pb_9CuBi_6Sb_{1.5}S_{30}$
Nuffieldite	$Pb_{10}Cu_4Bi_{10}S_{27}$
Lillianite	$Pb_3Bi_2S_6$
Gustavite	$Pb_6Ag_3Bi_{11}S_{24}$
Berryite	$Pb_3Bi_7(Ag_{1.5}Cu_{3.5})S_{16}$
Neyite	$Pb_7(Cu,Ag)_2Bi_6S_{17}$
Heyrovskyite	$(Pb,Bi,Ag,Cu)_6Bi_2S_9$
Kobellite	$Pb_5(Bi,Sb)_8S_{17}$
Selenokobellite	$Pb_5(Bi,Sb)_8(S,Se)_{17}$
Weibullite	$PbBi_2(S,Se)_4$
Ustarasite	$Pb(Bi,Sb)_6S_{10}$
Bursaite	$Pb_5Bi_4S_{11}$
Oxides, hydroxides, halogenides	
Cumengeite	$Pb_4Cu_4Cl_8(OH)_8 \cdot H_2O$ (or $PbCuCl_2(OH)_2$)
Pseudoboleite	$Pb_5Cu_4Cl_{10}(OH)_8 \cdot 2H_2O$
Boleite	$Pb_5Cu_4AgCl_{11}(OH)_8 \cdot 1.5H_2O$
Percylite	$PbCu(OH)_2Cl_2$
Diaboleite	$Pb_2CuCl_2(OH)_4$
Chloroxiphite	$Pb_3CuCl_2(OH)_2O_2$
Hematophanite	$Pb_5Fe_4^{3+}O_{10}(OH,Cl)_2$
Fiedlerite	$Pb_3(OH)_2Cl_4$
Paralaurionite	Pb(OH)Cl (monocl.)
Laurionite	Pb(OH)Cl (orthorhomb.)
Cotunnite	$PbCl_2$
Pseudocotunnite	K_2PbCl_4
Matlockite	PbFCl
Bideauxite	$Pb_2AgCl_3(F,OH)_2$
Mendipite	$Pb_3O_2Cl_2$
Lorettoite	$Pb_7O_6Cl_2$
Penfieldite	$Pb_2(OH)Cl_3$
Blixite	$Pb_2(O,OH)_{2-x}Cl$
Nadorite	$PbSbO_2Cl$
Perite	$PbBiO_2Cl$
Heliophyllite	$Pb_3AsO_{<4}Cl_{<2}$ (orthorhombic)
Ekdemite	$Pb_3AsO_{<4}Cl_{<2}$ (tetragon.)
Murdochite	$PbCu_6O_8$
Lithargite (litharge)	α-PbO
Massicotite (massicot)	β-PbO
Minium	Pb_3O_4
Hochschildite	$PbSnO_3 \cdot 5$—$6H_2O$ (?)
Magnetoplumbite	$(Pb,Mn)_2Fe_6^{3+}O_{11}$
Plumboferrite	$PbFe_4O_7$
Bindheimite	$Pb_{1-2}Sb_{2-1}(O,OH,H_2O)_{6-7}$
Plattnerite	β-PbO_2
Coronadite	$Pb_{\leq 2}Mn_8O_{16}$
Cesarolite	$PbH_2Mn_3O_8$
Senaite	$Pb(Ti,Fe,Mn,Mg)_{24}O_{38}$
Macedonite	$PbTiO_3$
Euxenite	$(Y,Er,Ce,U,Pb,Ca)(Nb,Ta,Ti)_2(O,OH)_6$
Quenselite	$PbMnO_2(OH)$

Table 82-D-3 (continued)

Clarkeite	$(Na,Ca,Pb)_2U_2(O,OH)_7$
Woelsendorfite	$(Pb,Ca)U_2O_7 \cdot 2H_2O$
Vandendriesscheitite	$PbU_7O_{22} \cdot 12H_2O$
Fourmarierite	$PbU_4O_{13} \cdot 7—8H_2O$
Curite	$Pb_3U_8O_{27} \cdot 4H_2O$
Arsenites, selenites, iodates	
Trigonite	$Pb_3MnH[AsO_3]_3$
Finnemanite	$Pb_5Cl[AsO_3]_3$
Molybdomenite	$Pb[SeO_3]$
Demesmaekerite	$Pb_2Cu_5[SeO_3]_6(UO_2)_2(OH)_6 \cdot 2H_2O$
Seeligerite	$Pb_3Cl_3O[IO_3]$
Schwartzembergite	$Pb_6Cl_4O_2[IO_3]_2(OH)_2$
Carbonates	
Cerussite	$Pb[CO_3]$
Beyerite	$(Ca,Pb)Bi_2[CO_3]_2O_2$
Hydrocerussite	$Pb_3(OH)[CO_3]_2$
Phosgenite	$Pb_2Cl_2[CO_3]$
Chalconatronite	$Na_2Cu[CO_3]_2 \cdot 3H_2O$
Schuilingite	$Pb_3Ca_6Cu_2[CO_3]_8(OH)_6 \cdot 6H_2O$
Dundasite	$PbAl_2(OH)_4[CO_3]_2 \cdot 2H_2O$
Sulfates, tellurates, selenates	
Palmierite	$(K,Na)_2Pb[SO_4]_2$
Anglesite	$Pb[SO_4]$
Barytoanglesite	$(Pb,Ba)[SO_4]$
Kerstenite	$Pb[SeO_4]$
Olsacherite	$Pb_2[SO_4][SeO_4]$
Schmeiderite	$(Pb,Cu)_2(OH)_2[SeO_4]$
Moctezumite	$PbU[TeO_4]_2$
Linarite	$PbCu(OH)_2[SO_4]$
Plumbojarosite	$PbFe_6^{3+}(OH)_{12}[SO_4]_4$
Beaverite	$Pb(Cu,Fe)_3(OH)_6[SO_4]_2$
Itoite	$Pb_3GeO_2[SO_4]_2(OH)_2$
Osarizawaite	$PbCuAl_2(OH)_6[SO_4]_2$
Hinsdalite	$PbAl_3(OH)_6[SO_4][PO_4]$
Hidalgoite	$PbAl_3(OH)_6[SO_4][AsO_4]$
Corkite	$PbFe_3^{3+}(OH)_6[SO_4][PO_4]$
Beudantite	$PbFe_3^{3+}(OH)_6[SO_4][AsO_4]$
Caracolite	$Pb_2Na_3(OH,Cl)[SO_4]_3$
Lanarkite	$Pb_2O[SO_4]$
Nasledovite	$PbMn_3Al_4[CO_3]_4[SO_4](OH)_{10}$
Caledonite	$Pb_5Cu_2(OH)_6[CO_3][SO_4]_3$
Leadhillite	$Pb_4(OH)_2[SO_4][CO_3]_2$
Wherryite	$Pb_4Cu(O,OH,Cl)_3[CO_3][SO_4]_2$
Fleischerite	$Pb_3Ge^{4+}(OH)_6[SO_4]_2 \cdot 3H_2O$
Arzrunite	$Pb_2Cu_4O_2Cl_6[SO_4] \cdot 4H_2O$
Chromates, wolframates, molybdates	
Krokoite	$Pb[CrO_4]$
Phoenikochroite	$Pb_3O[CrO_4]_2$
Embreyite	$Pb_5[CrO_4]_2[PO_4]_2 \cdot H_2O$
Vauquelinite	$Pb_2Cu(OH)[CrO_4][PO_4]$

Table 82-D-3 (continued)

Fornacite	$(Pb,Cu)(OH)[CrO_4][AsO_4]$
Bellite	$(Pb,Ag)_5(Cl)[(Cr,As,Si)O_4]_3$
Iranite	$Pb[CrO_4]\cdot H_2O$
Hemihedrite	$Pb_{10}Zn(F)_2[CrO_4]_6[SiO_4]_2$
Wulfenite	$Pb[MoO_4]$
Chillagite	$Pb[(Mo,W)O_4]$
Raspite	$\alpha Pb[WO_4]$
Stolzite	$\beta Pb[WO_4]$
Arsenates, phosphates, vanadates	
Schultenite	$PbH[AsO_4]$
Phosphateschultenite	$PbH[PO_4]$
Duftite	$PbCu(OH)[AsO_4]$
Gabrielsonite	$PbFe(OH)[AsO_4]$
Descloizite	$Pb(Cu,Zn)(OH)[VO_4]$
Mottramite	$Pb(Cu,Zn)(OH)[VO_4]$
Mounanaite	$PbFe_2(OH)[VO_4]_2$
Pyrobelonite	$PbMn(OH)[VO_4]$
Karyinite	$(Na,Ca)_2(Mn,Mg,Ca,Pb)_3[AsO_4]_3$
Plumbosynadelphite	$(Mn,Mg,Ca,Pb)_4(OH)_5[AsO_4]$
Tsumcorite	$PbZnFe[AsO_4]_2\cdot H_2O$
Carminite	$PbFe^{3+}_2(OH)[AsO_4]_2$
Bayldonite	$PbCu_3(OH)[AsO_4]_2$
Plumbogummite	$PbAl_3(OH)_5[PO_4]_2\cdot H_2O$
Lusungite	$(Sr,Pb)Fe^{3+}_3(OH)_5[PO_4]_2\cdot H_2O$
Ferrazite	$(Pb,Ba)_3[PO_4]_2\cdot 8H_2O$ (?)
Hydroxylpyromorphite	$Pb_5(OH)[PO_4]_3$
Pyromorphite	$Pb_5(Cl)[PO_4]_3$
Mimetesite	$Pb_5(Cl)[AsO_4]_3$
Hedyphane	$(Ca,Pb)_5(Cl)[AsO_4]_3$
Achrematite	$Pb_5(Cl_{<1})[AsO_4,WO_4]_3$
Georgiadesite	$Pb_3(Cl)_3[AsO_4]$
Sahlinite	$Pb_{14}(Cl)_4O_9[AsO_4]_2$
Vanadinite	$Pb_5(Cl)[VO_4]_3$
Brackebuschite	$Pb_2(Mn,Fe)[VO_4]_2\cdot H_2O$
Heyite	$Pb_5Fe_2[VO_4]_2O_4$
Arsentsumebite	$Pb_2Cu(OH)[SO_4][AsO_4]$
Tsumebite	$Pb_2Cu(OH)_3[PO_4]\cdot 3H_2O$
Hallimondite	$Pb_2(UO_2)[AsO_4]_2$
Renardite	$Pb(UO_2)_4(OH)_4[PO_4]_2\cdot 7H_2O$
Dewindtite	$Pb(UO_2)_2[PO_4]_2\cdot 3H_2O$
Parsonsite	$Pb_2(UO_2)[PO_4]_2\cdot 2H_2O$
Przhevalskite	$Pb(UO_2)_2[PO_4]_2\cdot 2H_2O$
Dumontite	$Pb_2(UO_2)_3(OH)_4[PO_4]_2\cdot 3H_2O$
Huegelite	$Pb_2(UO_2)_3(OH)_4[AsO_4]_2\cdot 3H_2O$
Francevillite	$(Ba,Pb)(UO_2)_2[VO_4]_2\cdot 5H_2O$
Curienite	$Pb(UO_2)_2[VO_4]_2\cdot 5H_2O$
Chervetite	$Pb_2V_2O_7$
Silicates	
Larsenite	$PbZn[SiO_4]$
Esperite	$(Ca,Pb)Zn[SiO_4]$
Roeblingite	$PbCa_3H_6[SO_4][SiO_4]_3$
Orlite	$Pb(UO_2)[SiO_4]\cdot 2H_2O$ (?)

Table8 2-D-3 (continued)

Kasolite	$Pb(UO_2)[SiO_4]\cdot H_2O$
Barysilite	$Pb_3[Si_2O_7]$
Nasonite	$Pb_6Ca_4(Cl)_2[Si_2O_7]_3$
Ganomalite	$Pb_6Ca_4(OH)_2[Si_2O_7]_3$
Molybdophyllite	$Pb_2Mg_2(OH)_2[Si_2O_7]$
Melanotekite	$Pb_2Fe_2^{3+}O_2[Si_2O_7]$
Kentrolite	$Pb_2Mn_2^{3+}O_2[Si_2O_7]$
Hancockite	$(Ca,Pb,Sr,Mn)_2(Al,Fe,Mn)_3O(OH)[SiO_4][Si_2O_7]$
Hyalotekite	$(Pb,Ca,Ba)_4B[Si_6O_{17}](F,OH)$
Jagoite	$Pb_{24}Fe_7[Si_{27}O_{84}](OH,Cl)_8$
Margarosanite	$Pb(Ca,Mn)_2[SiO_3]_3$
Alamosite	$Pb_3[SiO_3]_3$
Joesmithite (Pb amphibole)	$(Ca,Pb)Ca_2(Mg,Fe^{2+},Fe^{3+})_5[Si_6B_2O_{22}](OH)_2$
Wickenburgite	$Pb_3Al_2Ca[Si_{10}O_{24}](OH)_6$

Revised manuscript received: April 1974

82-E. Abundance in Common Magmatic Rock Types; Crustal Abundance

Potassium feldspars are the most important lead carriers of magmatic rocks. According to Table 7—11 in Volume I of this handbook, common magmatic rocks of the upper continental crust contain about 20 percent potassium feldspar (and 40 percent plagioclase). Using the means of Tables 82-D-1 and 82-D-2, the two feldspars contribute 11 and 2 ppm Pb respectively to the total lead content of the common continental rocks. On the basis of the extensive ore microscopical surveys of common magmatic rocks, published by NEWHOUSE (1936) and RAMDOHR (1940), galena is a very rare constituent of these rocks. A few authors such as BRÄUER (1970) mention that they have observed galena in heavy mineral concentrates of magmatic rocks.

I. Standard Reference Rocks (Analytical Precision and Accuracy of Lead Determinations)

Common analytical methods for the quantitative determination of lead in trace concentrations of rocks and minerals are those of optical spectrography, chemical spectrophotometry (dithizon), X-ray fluorescence, isotope dilution and atomic absorption (solid source). Sensitivities of these methods using standard sample sizes between X0 and X00 mg range from a few ppm (X-ray fluorescence and optical emission spectrography) down to less than 100 ppb (solid source atomic absorption, isotope dilution).

The largest sets of data on standard reference rocks from numerous laboratories are those by optical emission spectrography. All sets available to the present author have been averaged and listed in Table 82-E-1. Most recent averages published by FLANAGAN (1973) on a larger number of data, but not separated for the individual methods, are included in this list in square boxes. Where larger sets of data for one sample have been available, standard deviations were computed. Relative standard deviations range from 24 to 34 percent which indicates a broad scattering of data. Because of this uncertainty the averages must not always be very close to the true values; therefore some samples can only be used with caution to test accuracies of methods. This statement about errors in lead analyses has to be kept in mind when data in the following subsections are evaluated.

II. Peridotitic Rocks

Because of the low lead abundance only a few reliable analyses on rocks of this class are available.

TILTON and REED (1963) reported 19 ± 8 ppb Pb(N/R) in the Twin Sisters dunite from the same locality as reference standard DTS-1 (Table 82-E-1). Numerous authors obtained almost a thousand times this concentration in the reference rock,

Table 82-E-1- *Lead data on standard reference rocks* (symbols for samples are those introduced by the producer and reported in reviews, such as FLANAGAN and GWYN, 1967) (number of samples used for averages in brackets) (for sources of data see references in Table 29-E-1) (averages by FLANAGAN, 1973, in ☐ boxes)

Sample	Origin	Averages for common analytical methods [symbols]
Dunite DTS-1	Twin Sisters, Washington (U.S.A.)	13.6 (6) [S] 8.3 (4) [X] 16 (1) [W/C] ☐14.2☐
Peridotite PCC-1	Cazadero, Calif. (U.S.A.)	13 (6) [S] 7.9 (3) [X] ☐13.3☐
Basalt ("diabase") W-1	Centerville, Virginia (U.S.A.)	≤7.7 (26) [S] 9.4 (4) [X] 7.7 (6) [W/C] 6.0 (4) [I, Mass] ☐7.8☐
Basalt BCR-1	Bridal Veil, Oregon (U.S.A.)	17.8 (12) [S] 10.9 (5) [X] 29.5 (1) [W/C] 13.6 (2)[a] [I] ☐17.6☐[a]
Basalt JB-1	Sasebo, Nagasaki-Ken (Japan)	14 (3) [S, A] ☐14☐
Basalt BR	Essey-la-Côte, Lorraine (France)	≤11.8 (13) [S] 12.5 (2) [X] 8 (1) [W/C]
Basalt (spilite) BM	Mellenbach, Thuringia (G.D.R.)	☐12☐
Andesite AGV-1	Guano Valley, Oregon (U.S.A.)	33.1±10 (17) [S] 34.9 (6) [X] 46 (1) [W/C] 36.5 (2) [I] ☐35.1☐
Granodiorite GSP-1	Silver Plume, Colorado (U.S.A.)	51.9±15.9 (17) [S] 52.3 (6) [X] 67 (1) [W/C] 56.4 (8) [I] ☐51.3☐
Granodiorite JG-1	Sori, Gumma-Ken (Japan)	24.2 (6) [S, A] ☐24☐
Granite G-1	Westerly, Rhode Island (U.S.A.)	42.6±14.2 (42) [S] 64.8 (4) [X] 37.3 (10) [W/C] 40.6 (3) [Mass] ☐48☐
Granite G-2	Westerly, Rhode Island (U.S.A.)	30.0±6.9 (17) [S] 30.5 (3) [X] 29.5 (4) [W/C] 30.8 (1) [I] ☐31.2☐
Granite GA	Andlau, Vosges (France)	24.7±7 (17) [S] 37.8 (2) [X] 11 (1) [W/C]
Granite GH	Hoggar (Algérie)	42.2±12 (15) [S] 52 (2) [X] 15 (1) [W/C] 58 (1) [Mass]
Granite GR	Senones, Vosges (France)	34.5 (13) [S] 42.5 (2) [X]
Granite GM	Meissen, Saxonia (G.D.R.)	☐30☐
Syenite SY-1	Bancroft, Ontario (Canada)	520 (12) [S] 513 (3) [W/C] ☐495☐

[a] Due to 16 additional isotope dilution data in B. R. DOE: "Newsletter on Rock Reference Samples, Oct. 1973", the Pb content of BCR-1 must be close to 13.4 (±2%) ppm. A lower lead figure (8.6 ppm) on samples from the same quarry are suggestive of lead contamination of BRC-1.

which must be heterogeneous and/or contaminated with lead. TILTON and REED (1963) also determined lead in a dunite from St. Paul's Rock, Mid-Atlantic Ridge (110 ppb Pb) and two olivine nodules from basalts of Dreiser Weiher (Eifel, Germany; <10 ppb Pb) and Gila (Arizona, U.S.A.; 80 ± 30 ppb Pb, I). HEINRICHS (1974) reports ≤ 2.1 ppm Pb (<0.2 to 7.2 ppm Pb) in 5 lherzolite nodules from Tertiary alkali basalts and in two peridotite and harzburgite intrusions analyzed by a flameless atomic absorption procedure. His Dreiser Weiher (Eifel, Germany) sample contains less than 200 ppb Pb. COOPER and GREEN (1969) obtained <10 to 250 ppb Pb (W/C; I) (average: 93 ppb Pb) in 17 lherzolite inclusions of basanites from western Victoria (Australia). ZARTMAN and TERA (1973) mention a range of 82 to 570 ppb Pb (average: 250 ppb Pb; method: W/I) in 5 peridotite inclusions in alkali basalt from U.S.A. and Canary Islands. MOORBATH and WELKE (1969) report 1.6 and 1.77 ppm Pb (I) in two peridotite intrusives from Skye (Scotland). GRAESER (1969) lists 8 ppm and 0.8 ppm Pb (W/C) for a garnet peridotite and a peridotite from the southern Alps and the Ivrea Zone (Switzerland, Italy) respectively. Two kimberlites from Kimberley contain 4.9 and 37.3 ppm Pb (W/A, I) and two garnet peridotite inclusions in it: 1.05 and 3 ppm Pb (HEINRICHS, 1974; MANTON and TATSUMOTO, 1971).

Because of the large range of Pb concentrations in ultramafic rocks, no reasonable average from the above data can be given as yet. The mean might be comparable with ordinary chondrites containing 300 ppb Pb (Table 82-C-3).

The lead isotopic composition of closed systems of rocks is controlled by the U/Pb ratio ($^{238}U/^{204}Pb$). According to Fig. 92-E-1, abundant uranium concentrations in ultramafic rocks range from 10 to 40 ppb U (mean 25 ppb U). If the average lead of this rock class is 300 ppb (equivalent to about 4.2 ppb ^{204}Pb), the U/Pb ratio will be 0.083 ($^{238}U/^{204}Pb$: 6). This mean has been reported for mafic to ultramafic inclusions in pipes (DOE, 1970). For several ultramafic rocks and potential mantle materials, $^{238}U/^{204}Pb$ ratios have been measured and calculated as being mainly in the range of 8.2 to 9 or close to it (COOPER and GREEN, 1969; WELKE *et al.*, 1968; TATSUMOTO, 1966; etc.). This ratio fixes the average lead concentration of ultramafic rocks of the upper mantle, assuming 25 ppb average U, at about 200 ppb Pb.

III. Gabbroic and Basaltic Rocks

Because of its properties, lead is expected to be accumulated in mafic magmas from partial melting of ultramafic mantle rocks low in lead. Lead abundances in gabbroic and basaltic rocks are listed in Table 82-E-2. Excluding data of spectrographic analyses from the averaging, the arithmetic mean of high precision analyses of 10 gabbroic rocks is 2.7 ppm Pb, that of 93 tholeiitic basalts 3.7 ppm Pb (6 oceanic tholeiitic basalts contain 0.8 ppm Pb), and that of 95 alkali olivine basalts 4.3 ppm Pb. If the average lead of basalts is computed from an average U in these rocks (400 ppb; see Fig. 92-E-1), a $^{238}U/^{204}Pb$ ratio of 9 (Section 82-B, DOE, 1970, and others) and a $\frac{^{206}Pb + ^{207}Pb + ^{208}Pb}{^{204}Pb}$ ratio of 72, then we get a value of 3.2 ppm Pb, which is close to the above data.

Because of an increase in potassium, to which lead is crystal-chemically related, lead increases in the sequence oceanic tholeiitic basalts, tholeiitic basalts, alkali

Table 82-E-2. *Lead data on gabbroic and basaltic rocks* (Number of samples in brackets)

Rock	Origin	Range ppm Pb	Mean ppm Pb	Method	Reference
Gabbro (1)	Duluth, Minnesota (U.S.A.)	—	6.3	W/C	SANDELL and GOLDICH (1943)
Gabbro (1)	Duluth, Minnesota (U.S.A.)	—	5	S	FAIRBAIRN *et al.* (1953)
Gabbro (1)	Ichinome Gata, N. Honshu (Japan)	—	5.3	I	HEDGE and KNIGHT (1969)
Gabbro (3)	Cuillins layered intrusion, Skye (Scotland)	0.50—2.72	1.7	I	MOORBATH and WELKE (1969)
Gabbro (2)	Ivrea Zone (Italy)	0.35—2.5	1.4	W/C	GRAESER (1969)
Gabbro (2)	Coast Range (U.S.A.)	0.9— 6.4	3.6	I	TATSUMOTO and SNAVELY (1969)
Gabbro (composite of 11)	Germany	—	5.2	S	WEDEPOHL (1956)
Gabbro (1)	Harzburg, Harz (Germany)	—	3.3	S	WEDEPOHL (1956)
Basalt (high alumina) (12)	Cascade Mountains, California, Oregon, Washington (U.S.A.)	0.4—12.0	2.88	W/C	CHURCH and TILTON (1973)
Tholeiitic basalt	Yellowstone Nation. Park (U.S.A.)	1.5—5.5		I	DOE *et al.* (1970)
Tholeiitic basalt (5)	Coast Range, Oregon, Washington (U.S.A.)	1.2—8.5	4.9	I	TATSUMOTO and SNAVELY (1969)
Basalt (unspecified) (6)	Coast Range (U.S.A.)	0.6—3.4	2.1	I	TATSUMOTO and SNAVELY (1969)
Tholeiitic basalt ("diabase") (9)	Minnesota, Missouri, New England (U.S.A.)	5—23	9.7	W/C	SANDELL and GOLDICH (1943)
Tholeiitic basalt ("diabase") (57)	Ontario (Canada)	0.8—83	8.7	S	FAIRBAIRN *et al.* (1953)
Tholeiitic basalt, basaltic andesite (7)	Tonga-Kermadec (Pacific Ocean)	0.3—2.1	1.6	I	OVERSBY and EWART (1972)
Tholeiitic basalt (7)	Réunion (Pacific Ocean)	1.4—3.5	2.4	I	OVERSBY (1972)
Tholeiitic basalt (10)	O-shima, Hakone (Japan)	1.4—6.4	3.1	I	TATSUMOTO and KNIGHT (1969)
High alumina basalt (7)	Fuji, Omuro-Amagi (Japan)	1.9—6.8	4.1	I	TATSUMOTO and KNIGHT (1969)
Tholeiitic basalt (1)	Iwate, N. Honshu (Japan)	—	2.19	I	HEDGE and KNIGHT (1969)

Table 82-E-2 (continued)

Rock	Origin	Range ppm Pb	Mean ppm Pb	Analytical method	Reference
Tholeiitic basalt (3)	Hawaii, Japan	2.8—7.9	4.9	I	TATSUMOTO (1966a)
Tholeiitic basalt (composite of 9)	Deccan (India)	—	3.5	S	WEDEPOHL (1956)
Basalt (dolerites) (4)	Skye (Scotland)	1.14—8.44	5.32	I	MOORBATH and WELKE (1969)
Olivine basalt, tholeiitic basalt (5)	Central rift, western volcanic zone, Iceland	0.74—4.65	2.44	I	WELKE *et al.* (1968)
Tholeiitic basalt (8)	Hessia (Germany)	0.4—3.7	1.6	W/A	HEINRICHS (1974)
Oceanic tholeiitic basalt (12)	Mid-Atlantic Ridge, 25—30° N	<2—3	≤2.3	S	THOMPSON *et al.* (1972)
Oceanic tholeiitic basalt (6)	Mid-Atlantic Ridge, East Pacific Rise	0.49—1.3	0.76	I	TATSUMOTO (1966b)
Tholeiitic basalt (13)	Southern Rocky Mountains (U.S.A.)	1.9—11.1	6.2	I	DOE *et al.* (1969)
Basalt (163)	Coppermine River, NW Territories (Canada)	1.1—38	10.2	S	BARAGAR (1969)
Alkali olivine basalt (4)	Oki-dogo, Japan	—	3.35	I	TATSUMOTO (1969)
Alkali olivine basalt (2)	Coast Range (U.S.A.)	2.6—2.9	2.75	I	TATSUMOTO and SNAVELY (1969)
Alkali olivine basalt (7)	Dogo, Oki Island, Kyushu (Japan)	2.43—3.73	3.2	I	KURASAWA (1968)
Alkali olivine basalt (1)	Osima-Osima, N. Honshu (Japan)	—	2.50	I	HEDGE and KNIGHT (1969)
Alkali olivine basalt etc. (6)	Hawaii, Japan	1.3—5.4	3.4	I	TATSUMOTO (1966)
High alumina basalt (1)	Fuji (Japan)	—	4.2	I	TATSUMOTO (1966)
Basanite (5)	Western Victoria (Australia)	2—5	2.9	W/C	COOPER and GREEN (1969)
Alkali olivine basalt (5)	Guadalupe (Easter Island)	1.3—3.4	2.26	I	TATSUMOTO (1966b)
Alkali olivine basalt (5)	Réunion (Indian Ocean)	2.2—7.9	4.7	I	OVERSBY (1972)

Table 82-E-2 (continued)

Rock	Origin	Range ppm Pb	Mean ppm Pb	Analytical method	Reference
Alkali olivine basalt (7)	Tristan da Cunha, St. Helena (Atlantic Ocean)	2—6.2	3.3	I	OVERSBY and GAST (1970)
Alkali olivine basalt (4)	Faial (Azores)	4.3—5.2	4.7	I	OVERSBY and GAST (1968)
Alkali olivine basalt (10)	Tenerife (Canary Islands)	1.8—7.0	4.7	I	OVERSBY *et al.* (1971)
Alkali olivine basalts (4)	Plateau lavas, Skye (Scotland)	1.95—5.38	3.15	I	MOORBATH and WELKE (1969)
Alkali olivine basalt (11), olivine nephelinite (4)	NW Germany	0.4—10.7	3.5	W/A	HEINRICHS (1974)
Alkali olivine basalt (6)	Lower Saxony, Hessia (Germany)	5.7—15	8.8	S	WEDEPOHL (1956)
Greenstone (15)	Keweenawan, Michigan, Minnesota (U.S.A.)	<2—8	≤6.4	W/C	SANDELL and GOLDICH (1943)
Spilite (5)	Westphalia, Kellerwald, Hessia (Germany)	1.6—7.5	3.6	S, W/A	WEDEPOHL (1956) (4) HEINRICHS (1974) (1)

olivine basalts. The factor of enrichment of lead in basalts over ultramafics, from which the former originate by partial melting, is about 20.

In Fig. 82-E-1, data on lead in basaltic and andesitic rocks are plotted against those of K_2O. Only a restricted number of publications contain data on both elements: TATSUMOTO (1966a, b), OVERSBY (1972), OVERSBY and EWART (1972), HEDGE and KNIGHT (1969), CHURCH and TILTON (1973), DOE *et al.* (1969), HEINRICHS (1974; K_2O: WEDEPOHL, unpublished), and TAYLOR *et al.* (1969). The figure demonstrates some correlation between lead and potassium, but an appreciable scattering of data is observable, mainly at higher values. This can be explained by natural contamination, assimilation etc. (DOE *et al.*, 1969), and/or by the volatility of some lead compounds in fumaroles emanating from basaltic flows (see Section 82-F).

Additional information on the origin of lead in mafic magmatic rocks can be derived from lead isotope data (Section 82-B, and numerous references of Table 82-E-2). They show, for instance, that volcanic islands contain more radiogenic lead than abyssal basalts and give evidence, that the mantle can have a several stage partial melting history.

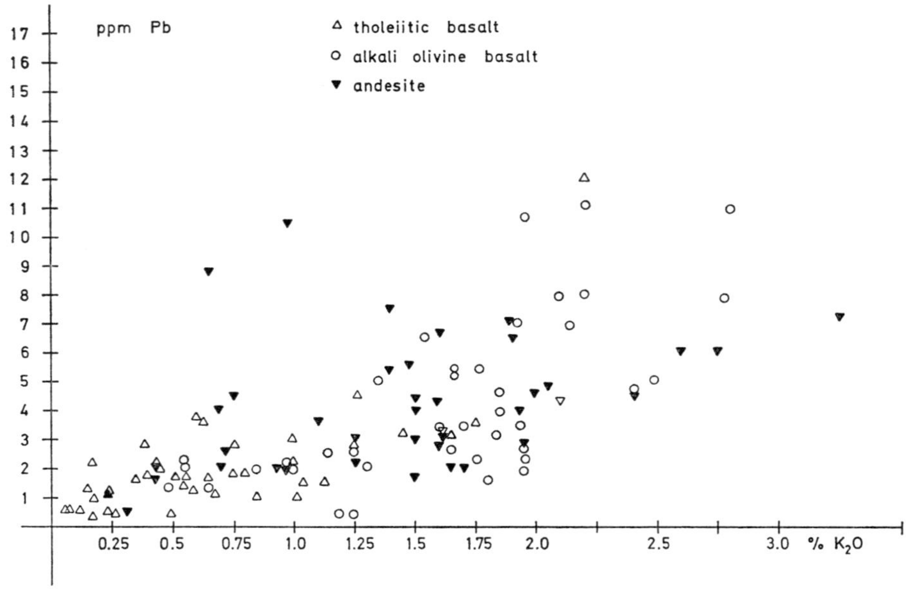

Fig. 82-E-1. Plot of selected Pb against K_2O values of basalts and andesites from those references of Table 83-E-2 which give data on both elements

IV. Diorites and Andesites

Rocks intermediate in silica content between mafic and granitic varieties contain on average lower than intermediate lead concentrations. Data from spectrographic analyses seem to be systematically high. An arithmetic mean of 79 andesites, excluding the last four sets of data of Table 82-E-3, is 5.8 ppm Pb (average of the total of 152 andesite data 10.6 ppm Pb). According to Fig. 82-E-1, some correlation between lead and potassium can be observed as in the case of basalts. A few andesites from Japan deviate from this correlation appreciably to higher lead values. This observation can be confirmed by the data published by Taylor and White (1966) not plotted in Fig. 82-E-1.

According to Table 7-3 of Volume I of this handbook, andesites contain on average 1.11% K_2O and diorites 1.33% K_2O. A potential regression line in Fig. 82-E-1 will correlate these potassium concentrations with about 3 and 4 ppm Pb only. This demonstrates that the above computed mean of andesitic lead (5.8 ppm Pb) contains some "excess" lead. The origin of a potential excess lead can only be traced by isotope work. Church and Tilton (1973) in their extensive lead and strontium isotope studies of andesites from the Cascades have ruled out mixing models of downgoing slabs of ocean sediments (low in radiogenic lead) and tholeiitic basalt magmas, or crustal contamination of magmas with radiogenic lead and strontium. The lead concentration is correlated with the addition of radiogenic lead neither in the Cascade andesites nor in those from Japan (Hedge and Knight, 1969; etc.). Some vapor transport of readily volatile lead compounds at least can explain the large range of Pb/K ratios in magmatic rocks of a fixed potassium concentration in a certain area.

Table 82-E-3. *Lead data on diorites and andesites* (Number of samples in brackets)

Rock	Origin	Range ppm Pb	Mean ppm Pb	Method	Reference
Diorite (2)	Coast Range, Oregon (U.S.A.)	9.02—10.95	10.0	I	TATSUMOTO and SNAVELY (1969)
Diorite (1)	Morrison County, Minnesota (U.S.A.)	—	8	W/C	SANDELL and GOLDICH (1943)
Andesite (16)	Cascade Mountains (U.S.A.)	2—7.5	4.66	W/C	CHURCH and TILTON (1973)
Andesite (2)	Conejos, Colorado (U.S.A.)	8.99—12	10.5	I	DOE *et al.* (1969)
Andesite (3)	Absarcka, Wyoming (U.S.A.)	11.9—16.3	14.3	I	PETERMAN *et al.* (1970)
Andesite (2)	Clear Lake area, California (U.S.A.)	7—10	8.5	W/C	SANDELL and GOLDICH (1943)
Andesite (5)	Tonga, Kermadec (Pacific Ocean)	0.55—2.63	1.8	I	OVERSBY and EWART (1972)
Andesite (13)	Bougainville, Saipan, Fiji (West Pacific)	1.7—7.2	3.49	Mass	TAYLOR *et al.* (1969)
Andesite (6)	New Zealand	1.2—10	7.1	Mass	TAYLOR and WHITE (1966)
Andesite (7)	Izu, Asama (Japan)	3.5—13	7.6	Mass	TAYLOR and WHITE (1966)
Andesite (13)	Hakone, Fuji, Omuro-Amagi (Japan)	3.06—13.34	4.57	I	TATSUMOTO and KNIGHT (1969)
Andesite (3)	Central Japan	3.08—10.31	6.53	I	TATSUMOTO (1969)
Andesite (5)	Northern Honshu (Japan)	4.8—10.58	8.0	I	HEDGE and KNIGHT (1969)
Andesite (1)	Oki Islands (Japan)	—	8.2	I	KURASAWA (1968)
Trachyandesite (5)	Tristan da Cunha (Atlantic Ocean)	10.4—13.8	12.0	I	OVERSBY and GAST (1968)
Andesite (1)	Western Iceland	—	4.31	I	WELKE *et al.* (1968)
Andesite (19)	Central Kazakhstan (U.S.S.R.)	2—23	14.2	S	ZHIROV and CHERNYSHEV (1959)
Andesite (22)	Dinarides, Serbia, Macedonia (Yugoslavia)	3—36	15.2	S	ČUTURIČ and KARAMATA (1967)
Andesite (12)	Calimanian Mountains (Roumania)	2—62	27.9	W/C	SAVUL *et al.* (1956)
Andesite (2 composites of 20)	Aegina, Methana (Greece)	—	10.6	S	WEDEPOHL (1956)

V. Alkalic Rocks (Including Monzonites, Syenites and Trachytes)

Data on lead in the alkalic rock types are compiled in Table 82-E-4. Alkalic rocks are rare and relatively variable in modal composition as well as in lead content. Therefore no arithmetic mean has been computed for the different rock classes, except for the 64 nepheline syenites containing 14.4 ppm Pb and the 37 phonolites with 14.8 ppm Pb.

Gerasimovskii and Nesmeganova (1960) have demonstrated that the lead concentration increases in the Lovozero Massif (Kola Peninsula, U.S.S.R.) in the three major stages of formation from 8 to 20.5 ppm Pb. The largest mass (80 volume percent) of stage two contains 13.4 ppm Pb (W/C). Zlobin and Gorshkova (1961) have compared lead averages of three large alkalic massifs of the U.S.S.R.: Sandyk Mountains (Northern Kirgiziya; 58 ppm Pb; W/C), Khibina (4.9 ppm Pb) and Lovozero (14.6 ppm Pb). It can be assumed that silica undersaturated alkalic melts often contain about 14 ppm Pb.

VI. Granitic Rocks and Related Effusives

The lead abundance in granitic intrusives (see Table 82-E-5) is mainly controlled by their potassium feldspar content as explained in Section 82-D (Table 82-D-1). If granites contain on average 30 percent potassium feldspar with a mean lead concentration of 50 ppm, this mineral contributes 17 ppm Pb. A less remarkable lead contribution is that of plagioclase and mica, estimated as being about 7 ppm Pb and less than 2 ppm Pb respectively. Granodiorites and quartz diorites (tonalites) contain less than 30 percent potassium feldspar; therefore a gross average of lead in granitic rocks is estimated from mineral analyses as being close to 20 ppm Pb. The arithmetic mean of the 1220 granitic rocks, which are used for plotting Fig. 82-E-2a, is 23 ppm Pb. The averages of those granitic rock species, for which the petrographic type has been indicated, differ systematically as has been expected from lead data on rock-forming minerals: 245 granodiorites contain 15 ppm Pb and 320 granites 23.6 ppm Pb. For a limited area such as Japan, Shibata *et al.* (1960) report an arithmetic mean of 11.7 ppm Pb for 160 granodiorites and of 19.6 ppm for 102 granites. 74 Dacites contain on average 15.2 ppm Pb and 145 rhyolites and obsidians 24 ppm Pb, coinciding with the means of the respective intrusives.

Lead data used for Fig. 82-E-2a and 82-E-2b are from the following publications (abbreviations: xG; xGd; xR: number of granites, granodiorites and rhyolites, respectively; rhyolites include dacites and obsidians; the symbol for the analytical method is added): Ahrens (1954) (32G;S); Bräuer (1970) (25G;S); Church and Tilton (1973) (10R;I); Clifford *et al.* (1969) (12G;S); Čuturič and Karamata (1967) (12G;9Gd;30R;S); Deleon and Ahrens (1957) (12Gd,G;S); Doe (1967) (9R;I); Doe *et al.* (1969a) (2R;I); Ewart *et al.* (1968) (6R;Mass); Fershtater *et al.* (1969) (25G;27Gd;S); Gavrilin *et al.* (1965) (6G;W/C); Gavrilin *et al.* (1972) (18G;W/C); Golubchina and Rabinovich (1957) (13G,Gd;S); Gundlach *et al.* (1967) (18G;86Gd;W/C,S); Hedge and Knight (1969) (1G,2R;I); Heinrichs (1974) (20G,Gd;A); Hügi (1956) (88G,Gd;S); Jack and Carmichael (1969) (46R;X); Kolbe and Taylor (1966) (46G,20Gd;S); Kurasawa (1968) (3R;I); Kuroda and

Table 82-E-4. *Lead data of alkalic rocks* (Number of samples in brackets)

Rock	Origin	Range ppm Pb	Mean ppm Pb	Method	Reference
Syenite (1)	Cook County, Minnesota (U.S.A.)	—	9	W/C	SANDELL and GOLDICH (1943)
Syenite (2) (reference standards NIM-S; Sy-1)	South Africa, Canada	—	13; 495	several methods	FLANAGAN (1973)
Syenite (2)	Nemuro, Hokkaido (Japan)	—	20	S	ISHIKAWA *et al.* (1971)
Syenite (8)	Matcha massif, S. Tien-Shan (U.S.S.R.)	5—37	13.6	W/C	GAVRILIN *et al.* (1965)
Syenite (32)	Kzyl-Ompul massif, Tien-Shan (U.S.S.R.)	35—90	54.1	W/C	GAVRILIN and PEVTSOVA (1963)
Trachyte (2)	Oki Islands (Japan)	15.1—18.0	16.5	I	KURASAWA (1968)
Trachyte (2)	Hawaii	5.18—8.75	7.0	I	TATSUMOTO (1966)
Trachyte (3)	Tenerife (Canary Islands)	5.73—22.7	12.16	I	OVERSBY *et al.* (1971)
Nepheline syenite (1)	Coast Range, Oregon, Washington (U.S.A.)	—	6.32	I	TATSUMOTO and SNAVELY (1969)
Nepheline syenite (composite of 22)	Brazil, U.S.A., Canada, Norway, Portugal, Roumania	—	9.6	W/A	HEINRICHS (1974)
Nepheline syenite and related rocks (33)	Lovozero massif, Kola Peninsula (U.S.S.R.)	2—102	19.3	W/C	GERASIMOVSKII and NESMEYANOVA (1960)
Nepheline syenite (8)	Matcha massif, S. Tien-Shan (U.S.S.R.)	5—10	8.4	W/C	GAVRILIN *et al.* (1965)
Phonolite (7)	Tenerife (Canary Islands)	9.10—39.4	18.05	I	OVERSBY *et al.* (1971)
Phonolite (3)	St. Helena (Atlantic Ocean)	9.6—16.2	12.7	I	OVERSBY and GAST (1970)
Phonolite (1)	Collins Peak, Vema seamount (Atlantic Ocean)	—	48	W/C	COOPER and RICHARDS (1966)
Phonolite (2 composites of 24)	Bohemia (Č.S.S.R.), Germany	—	12.8	W/A	HEINRICHS (1974)
Phonolite (2)	Westerwald, Kaiserstuhl (Germany)	9.4—20.2	14.8	W/A	HEINRICHS (1974)
Leucitite (2)	Vesuvius, Naples (Italy)	28.7—29.3	29.0	I	OVERSBY and GAST (1968)

Table 82-E-5. *Lead concentrations of selected sample sets and composites of granitic intrusives, rhyolites, obsidians and dacites* (Number of samples in brackets)

Rock type	Origin	Range ppm Pb	Average ppm Pb	Method	Reference
Granite (14)	Minnesota, Missouri, Texas, New England (U.S.A.)	<5—50	19	W/C	SANDELL and GOLDICH (1943)
Granite, granodiorite (26)	U.S.A. (24), Canada, Norway	4—79	31.1	I	ZARTMAN and WASSERBURG (1969)
Granite, granodiorite (993)	Canadian Shield	—	23	S	EADE and FAHRIG (1973)
Granite (32)	Ontario (Canada)	2—68	15.0	S	AHRENS (1954)
Granite, granodiorite (86)	Japan	<10—49	15.4	S	OKADA (1955)
Granite (102)	Japan	<10—60	19.6	S	SHIBATA *et al.* (1960)
Granite, granodiorite (239)	Japan	10—150	33.3	S(?)	KURODA and GORAI (1956)
Granite (46)	New South Wales (Australia), Cape (South Africa) (34)	13—54	29.5	S	KOLBE and TAYLOR (1966)
Granite (12)	Franzfontein (SW Africa)	8—40	21	S	CLIFFORD *et al.* (1969)
Granite, granodiorite (88)	Alps (Switzerland)	2—90	25	S	HÜGI (1956)
Granite (18)	Eastern and southern Alps	20—75	30.3	W/C/S	GUNDLACH *et al.* (1967)
Granite (composite of 14)	Germany	—	28.8	A	HEINRICHS (1974)
Granite, granodiorite (38)	G.D.R.	<7—30	15.5	S	MOENKE (1960)
Granite (20)	Ruhla, Thuringia (G.D.R.)	5—34	19.9	S	WERNER (1970)
Granite (25)	Erzgebirge (G.D.R.)	<1—39	16.4	S	BRÄUER (1970)
Granite, granophyre (7)	Skye, Scotland (U.K.)	10.8—19.7	13.6	I	MOORBATH and WELKE (1969)
Granophyre (13)	Iceland	4.14—8.98	5.9	I	WELKE *et al.* (1968)
Granite, quartz, monzonite (12)	Yugoslavia	13—44	27.9	S	ČUTURIČ and KARAMATA (1967)
Granite (5)	Susamyr, C. Tien-Shan (U.S.S.R.)	23—35	29	W/C	TAUSON and PEVTSOVA (1955)
Granite, granodiorite (55)	North Tien-Shan (U.S.S.R.)	10—44	28.9	W/C	ZLOBIN *et al.* (1965)

Table 82-E-5 (continued)

Rock type	Origin	Range ppm Pb	Average ppm Pb	Method	Reference
Granite (6)	South Tien-Shan (U.S.S.R.)	37—69	56.1	W/C	GAVRILIN *et al.* (1965)
Granite (25)	Urals (U.S.S.R.)	10—54	22.7	S	FERSHTATER *et al.* (1969)
Granite (18)	Raumid, Pamir	18—38	27.2	W/C	GAVRILIN *et al.* (1972)
Granite (26)	Tarak, Yenisey (U.S.S.R.)	5—29	14.9	S	ZHIROV and URUSOVA (1962)
Granite, granodiorite (13)	R. Altai, E. Transbaikalia (U.S.S.R.)	17—59	34.1	S	GOLUBCHINA and RABINOVICH (1957)
Granodiorite (23)	Nevada (U.S.A.)	7—22	13.6	S	WODZICKI (1971)
Granodiorite, tonalite (17)	California (U.S.A.)	<10—20	≤12.4	S	NOCKOLDS and ALLEN (1953, 1954)
Granodiorite (160)	Japan	<10—48	11.7	S	SHIBATA *et al.* (1960)
Granodiorite, quartzdiorite (5)	Central Japan	10.5—20.8	15.2	I	SHIMIZU (1970)
Granodiorite (20)	New South Wales (Australia)	15—69	29	S	KOLBE and TAYLOR (1966)
Granodiorite (86)	Eastern and southern Alps	<20—155	27.2	W/C/S	GUNDLACH *et al.* (1967)
Granodiorite (9)	Yugoslavia	7—32	21.7	S	ČUTURIČ and KARAMATA (1967)
Granodiorite, granite (12)	Yugoslavia	12—32	21.7	S	DELEON and AHRENS (1957)
Granodiorite (27)	Urals (U.S.S.R.)	<2—37	12.5	S	FERSHTATER *et al.* (1969)
Granodiorite (20)	Susamyr, Tien-Shan (U.S.S.R.)	14—35	24.3	W/C	TAUSON and PEVTSOVA (1955)
Rhyolite, obsidian (6)	Minnesota, Missouri, California (U.S.A.)	12—20	17.3	W/C	SANDELL and GOLDICH (1943)
Rhyolite, obsidian (8)	Western U.S.A.	10—50	26.9	S	NOCKOLDS and ALLEN (1953, 1954)
Rhyolite, obsidian (9)	Western U.S.A., Texas	8.7—46	24.3	I	DOE (1967)

Table 82-E-5 (continued)

Rock type	Origin	Range ppm Pb	Average ppm Pb	Method	Reference
Rhyolite (3)	Texas (U.S.A.)	3—38	15.3	I	ZARTMAN and WASSERBURG (1969)
Obsidian, rhyolite (33)	Western U.S.A.	10—55	36.7	X	JACK and CARMICHAEL (1969)
Rhyolite (5)	New Zealand	16—22.5	17.8	Mass	EWART *et al.* (1968)
Rhyolite (3)	Dogo, Oki Island (Japan)	16.6—21.7	18.9	I	KURASAWA (1968)
Felsite (8)	Skye, Scotland (U.K.)	12.1—19.2	15.6	I	MOORBATH and WELKE (1969)
Obsidian (2), felsite	Iceland	5.9—12.8	8.8	I	WELKE *et al.* (1968)
Obsidian (composite of 8)	Iceland, Italy, U.S.A.	—	13	S	WEDEPOHL (1956)
Rhyolite, quartz latite (11)	Yugoslavia	22—110	40.1	S	ČUTURIČ and KARAMATA (1967)
Rhyolite (38)	Kazakhstan (U.S.S.R.)	4—50	14.5	S	ZHIROV and CHERNYSHEV (1959)
Dacite (5)	California (U.S.A.)	11—16	14	W/C	SANDELL and GOLDICH (1943)
Dacite (8)	Cascades (U.S.A.)	1.5—8	4.5	I	CHURCH and TILTON (1973)
Dacite (3)	Western U.S.A.	10—20	15	X	JACK and CARMICHAEL (1969)
Dacite (19)	Western U.S.A., Antilles	<10—25	≤ 12.6	S	NOCKOLDS and ALLEN (1953, 1954)
Dacite (4)	Tonga, Kermadec (Pacific Ocean)	2.8—4.1	3.6	I	OVERSBY and EWART (1972)
Dacite (3)	Honshu (Japan)	5.7—7.1	6.4	I	TATSUMOTO and KNIGHT (1969)
Dacite (2)	Iceland	4.4—4.6	4.5	I	WELKE *et al.* (1968)
Dacite (7)	Munti Calimani (Roumania)	—	28.8	W/C	SAVUL *et al.* (1956)
Dacite (19)	Yugoslavia	15—42	25.6	S	ČUTURIČ and KARAMATA (1967)
Dacite (21)	Kazakhstan (U.S.S.R.)	40—>120	10.1	S	ZHIROV and CHERNYSHEV (1959)

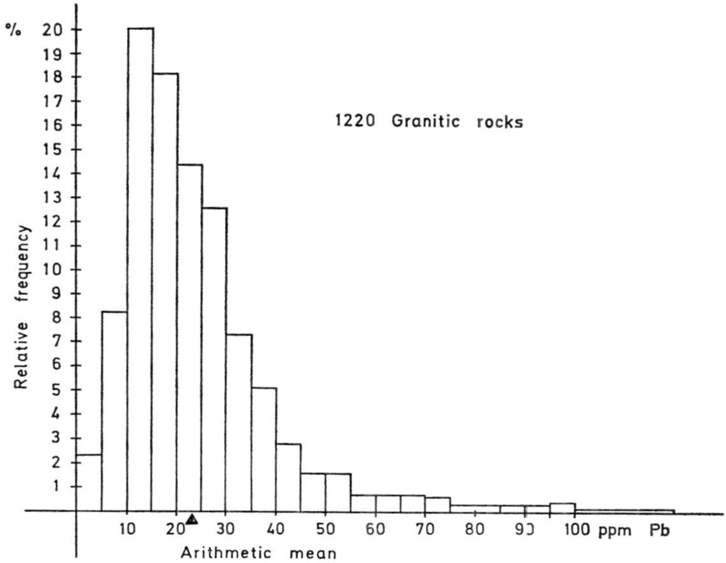

Fig. 82-E-2. a Frequency distribution of lead values in 1,220 granitic rocks (for list of references see text)

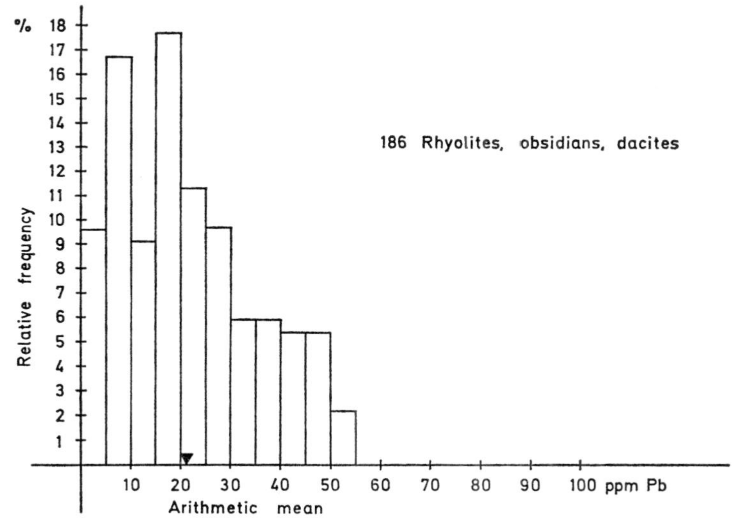

Fig. 82-E-2. b Frequency distribution of lead values in 186 rhyolites, obsidians and dacites (for list of references see text)

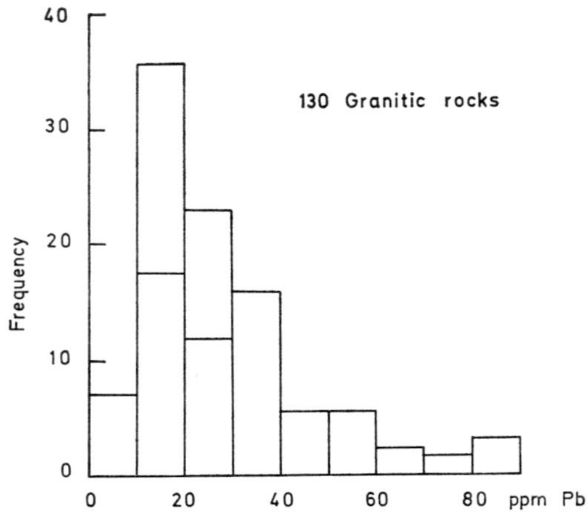

Fig. 82-E-2. c Frequency distribution of lead values in 130 granitic rocks from Africa (according to ROOKE, 1964)

GORAI (1956) (239G,Gd;S?); MOENKE (1960) (38G,Gd;S); MOORBATH and WELKE (1969) (7G,8R;I); NOCKOLDS and ALLEN (1953; 1954) (4G,17Gd,GR;S); OKADA (1955) (86G,Gd;S); OVERSBY and EWART (1972) (4R;I); RABINOVICH and BASKOVA (1959) (3G,Gd;W/C,P); ROSHOLT and BARTEL (1969) (7G;I); SANDELL and GOLDICH (1943) (14G,2Gd,17R;W/C); SAVUL *et al.* (1956) (7R;W/C); SHIBATA *et al.* (1960) (260G,Gd;S); SHIMIZU (1970) (5Gd;I); TATSUMOTO (1966) (1R;I); TATSUMOTO and KNIGHT (1969) (3R;I); TAUSON and PEVTSOVA (1955) (5G,20Gd;W/C); TAYLOR *et al.* (1969) (2R;Mass); WEDEPOHL (1956) (18G,Gd,8R;S); WELKE *et al.* (1968) (13G,5R;I); WERNER (1970) (20G;S); WODZICKI (1971) (23Gd;S); ZARTMAN and WASSERBURG (1969) (26G,Gd,3R;I); ZHIROV and CHERNYSHEV (1959) (59R;S); ZHIROV and UROSOVA (1962) (26G;S); ZLOBIN *et al.* (1965) (55G,Gd;W/C).

Except that the frequency curve for the larger number of samples is more smooth, there is no appreciable difference between the distribution of Pb concentrations in granitic intrusives (Fig. 82-E-2a) and in rhyolites, dacites and obsidians (Fig. 82-E-2b). In Fig. 82-E-2c, the frequency distribution of lead data on 130 granitic intrusives from the African continent only is drawn according to ROOKE (1964).

Most of the data used for the figures are from spectrographic analyses; because this method has a limit of sensitivity of between one and ten ppm, probably a slight shift to systematically higher lead data on granitic rocks has occurred. This assumption is confirmed by a separate plot of the isotope dilution values. The respective curve has a pronounced maximum between 5 and 15 ppm Pb for granitic intrusives.

In Table 82-E-5 a selection of lead data in form of averages and ranges of variation of sample sets and composites from worldwide sampling of granitic rocks and related effusives is listed.

The frequency distribution curves (Fig. 82-E-2a to c) as well as the single sets of Pb data, listed for certain areas, indicate an appreciable range of variation of lead values in granitic rocks. These are often due to regional differences as reported by SHIBATA *et al.* (1960). These authors list the following areal means for 14 petrographic

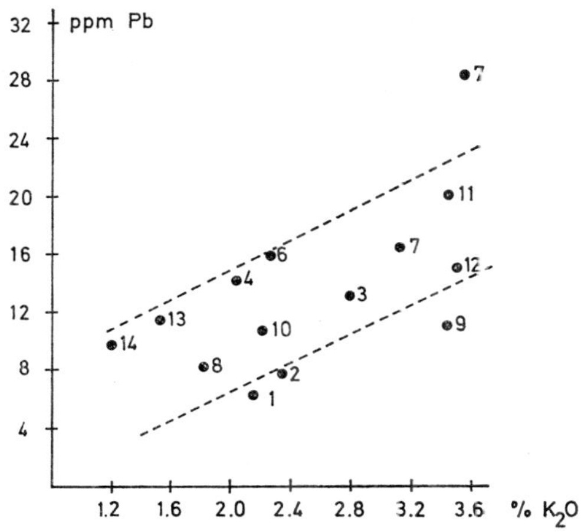

Fig. 82-E-3. Plot of mean Pb against K_2O values of granitic rocks from 14 provinces in Japan: 1. Hida; 2. Permian; 3. Fore-Ryoke; 4. Younger Ryoke; 5. Middle Upper Cretaceous; 6. Late Cretaceous; 7. Southwestern Outer Zone; 8. Shimonomoto; 9. Older Abukuma; 10. Older Ryoke; 11. Sanin; 12. Sanyo; 13. Northeastern Miocene; 14. Kansai Tertiary (according to SHIBATA *et al.*, 1960)

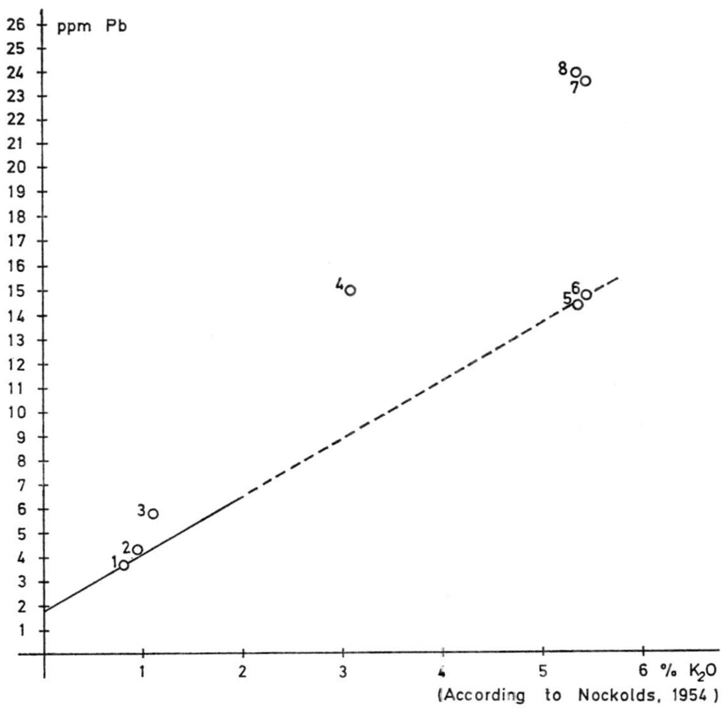

Fig. 82-E-4. Mean Pb values of rock types according to this compilation plotted against mean K_2O as compiled by NOCKOLDS (1954): 1. tholeiitic basalts; 2. olivine alkali basalts; 3. andesites; 4. granodiorites; 5. nepheline syenites; 6. phonolites; 7. granites; 8. rhyolites. Slope of solid line computed as a regression line of Fig. 82-E-1.

provinces in Japan (from a total of 262 analyses) in the sequence of decreasing age from Pre-Gotlandian to Tertiary: 6.3; 8.1; 7.9; 13.4; 11.2; 10.8; 14.6; 16.4; 20.0; 14.8; 16.2; 28.4; 10.2; 11.6 ppm Pb. There exist two maxima, one in Cretaceous and one in Tertiary time. Some correlation between K and Pb can be observed (Fig. 82-E-3), but a remarkable scattering of areal mean Pb data is not related to the respective potassium concentrations.

It can be observed that the granitic differentiation products of mafic magmas are especially low in lead: 2 granophyres from Duluth gabbro: 8.5 ppm Pb (ZARTMAN and WASSERBURG, 1969); 13 granophyres from Iceland: 5.9 ppm Pb; 2 obsidians from Iceland: 10 ppm Pb (WELKE *et al.*, 1968). Mixing of mantle and crustal lead in magmatic melts has to be investigated by isotope studies (Section 82-B).

In Fig. 82-E-4 the means of lead values for common types of magmatic rocks, as computed in this publication, are plotted against their average K_2O values as compiled by NOCKOLDS (1954). If one extrapolates a potential regression line for the "uncontaminated" basalts and andesites of Fig. 82-E-1, its slope will connect basalts and phonolites-nepheline syenites as drawn in Fig. 82-E-4. A line connecting basalts, granodiorites and granites-rhyolites has a steeper slope and indicates a different process of lead accumulation in the latter rocks.

VII. Composition of the Earth's Crust

The upper continental crust mainly consists of magmatic intrusives and metamorphic rocks. If we combine metamorphic rocks with their magmatic chemical equivalents we can use Table 7—8 of Volume I of this handbook, in which tentative abundances of the common rock species are compiled. In Table 82-E-6, those abundance figures are listed. If they are used for estimating the lead concentration of the upper continental crust one gets a value of 17 ppm Pb.

The increase of lead concentration from ultramafic to granitic rocks is almost as steep as that of uranium. Therefore the average uranium -to- lead ratio of the common magmatic rock units is rather uniform, although it decreases slightly from granitic to ultramafic rocks. From this ratio, the ratio μ for the common uranium to the non-radiogenic lead isotope can be derived $\left(\frac{^{206}Pb + {}^{207}Pb + {}^{208}Pb}{^{204}Pb} \approx 72\right)$.

The uranium lead ratio controls the increase of radiogenic lead in the closed system of a rock unit. The $^{238}U/^{204}Pb$ ratio can be computed from the isotopic com-

Table 82-E-6. *Lead and uranium abundances* (the latter according to CLARK *et al.*, 1966, and Section 92-E) *in major rock units of the earth*

Rock type	Vol.-% in the upper continental crust	Average abundance		$^{238}U/^{204}Pb$ (μ)
		ppm Pb	ppm U	
Granite, quartz monzonite	44	23.6	3.9	11.9
Granodiorite	42	15	2.6	12.5
Diorite, andesite	1	5.8	(1)	12.4
Gabbro, basalt (tholeiitic)	13	3.2	0.4	9
Peridotite	0.5	0.3	0.025	6

position of a rock to indicate its uranium lead system as has been demonstrated in detail in Section 82-B.

The lower continental crust mainly consists of high grade metamorphic rocks of granitic to intermediate chemical composition. These rocks are relatively depleted in U and Cs (LAMBERT and HEIER, 1967; HEIER and BRUNFELT, 1970), but probably not in Pb (Section 82-M).

TAYLOR (1964) lists a crustal abundance of 12.5 ppm Pb which seems to be confirmed by the present study (Sections 82-E, M and N).

There are only a few elements which are accumulated in the upper continental crust from the mantle to such a degree as Pb and U (by a factor of 30 and 300 respectively).

Revised manuscript received: April 1974

82-F. Behavior in Magmatogenic Processes

I. Pegmatites

Lead is accumulated in potassium feldspars. In Table 82-D-1, data on 219 pegmatitic microclines are listed; excluding amazonites they range from 4.5 to 1 300 ppm Pb and contain on average about 100 ppm Pb, twice as much as granitic potassium feldspars. Amazonites are even higher in lead; amazonite from Broken Hill (Australia) contains more than one percent lead (Čech *et al.*, 1971). Pegmatitic muscovites usually do not contain much lead (about 10 ppm Pb).

As the major mass of pegmatites consists of feldspar and quartz (usually in a ratio 65 to 30), the former mineral mainly contributes to an average lead content of pegmatites of 65 ppm.

II. Gas Transport

Appreciable lead concentrations in fumarolic sublimates of recent volcanoes (Vesuvius, Stromboli and Volcano, Italy; Valley of Ten Thousand Smokes, Alaska; Paricutin, Mexico; Japan, etc.—for references see Wedepohl, 1960) indicate magmatic gas transport of lead compounds. Locally they occur as cotunnite ($PbCl_2$).

The vapor pressure of $PbCl_2$ is higher than that of the nickel, iron, manganese, zinc and copper chlorides; it increases from 20 mm Hg at 600° C to 800 mm Hg at 900° C. Krauskopf (1957) has computed vapor pressures of metal compounds equilibrated with the most stable solid in a model magmatic gas of 10^3 atm H_2O, 50 atm CO_2, 30 atm H_2S, 10 atm HCl and 30 atm HF at 600° C. $PbCl_2$ is the most stable lead compound in this vapor having a partial pressure of $10^{-2.3}$ atm. For comparison it should be mentioned that at the same temperature PbS has a vapor pressure of $10^{-6.3}$ atm.

It should be expected that because of the volatility of $PbCl_2$, some lead is lost from basaltic and andesitic melts. Most consolidated magmatic rocks still contain more than 100 ppm Cl (see Section 17-E). According to Krauskopf (1957), 100 ppm Cl in magmatic vapor contains 0.05 ppm Pb at 600° C. If a rhyolite has lost 1 000 ppm Cl at this temperature, the lead loss is negligible (0.5 ppm Pb); but if the vapor pressure of $PbCl_2$ increases by a factor of 40 from 600 to 900° C, the lead loss (related to a chlorine loss of several hundred ppm Cl from a consolidating magmatic melt) can be in the concentration range of its primary lead content.

III. Magmatic Fluids — Ore Deposition

In this subsection as well as in Sections 82-G, 82-H and 82-I, lead in hydrothermal fluids and deposits will be discussed notwithstanding the origin of their metals from cooling magmatic melts or from country rocks of sedimentary, metamorphic or igneous origin.

Ore-depositing hydrothermal solutions probably range at and above 1 ppm Pb to form lead deposits. At lower concentrations water masses for transport are extremely large and the high dilution of lead sulfides by common gangue minerals with reasonable solubility will limit the economic value of the deposit. The above estimate of lead concentrations in ore-depositing fluids is confirmed by analyses of hot brines, fluid inclusions (Section 82-I), and by complexing experiments (Section 82-H). Several authors listed in Section 82-H have demonstrated experimentally the importance of chloride-rich brines to transport lead.

The lead concentration in ore-depositing brines can originate either from partitioning between consolidating granitic melts and their aqueous vapor phase, or from leaching of country rocks by hydrothermal solutions. HOLLAND (1972) has proved experimentally at 850° C and 2 kb that zinc, manganese and probably lead are strongly partitioned from silicate melts of granitic composition into the aqueous phase in relation to the chloride concentration of the latter. DURASOVA (1969; 1970) has published results on Pb and Sn partitioning experiments in the system NaCl-Na_2O-Al_2O_3-CaO-SiO_2 at 800° C and 1100° C. The distribution ratio of Pb between chloride and silicate is 0.01 to 0.02, implying that lead is retained by the magma at the above temperatures; however, this was observed in a dry system. ELLIS (1968) has investigated the behavior of lead during reactions of water and NaCl solutions (2 and 4 m) with andesite and pyrite-bearing shale at 350 to 500° C. Higher lead concentrations in the fluid have been achieved from the andesite (up to 4 ppm Pb) than from the sulfide-containing shale (up to 1 ppm; this in pure H_2O). The lead concentration in the solution extracted from andesite increases with NaCl concentration. It has a maximum of 2 ppm at 400° C and 2 m NaCl, and of 4 ppm at 400 to 500° C and 4 m NaCl. Through this extraction, the original lead content of the andesite decreases from 10 to 2 ppm Pb. Propylitization, albitization and sericitization are indicators of hydrothermal rock alteration. GAVRILIN *et al.* (1967) have investigated those processes in syenite and granite massifs of northern Tien-Shan (U.S.S.R.). They observed that during propylitization of these rocks, up to 50 percent of their original lead content has been removed (in rare cases up to 90 percent). Albitization and sericitization seem to be connected with a lead loss of 60 to almost 100 percent.

It is important to detect geochemical differences between those granitic intrusions which have produced economic galena deposits and those which have not. PUTMAN and BURNHAM (1963) studied the lead distribution in 227 biotite concentrates, mainly from granitic intrusions in Arizona. They do not report a correlation between abnormal lead concentrations in intrusives and lead mineralization. The same observation has been reported for large granitic intrusions in the U.S.S.R. by TAUSON (1964; 1967). PARRY and NACKOWSKI (1963) claim that abnormally low Pb and Zn concentrations in 86 and 95 biotites respectively (20 ppm Pb; 300 ppm Zn) from monzonitic intrusives in the Basin and Range stocks of Utah and Nevada, indicate economic galena-sphalerite deposits accompanying the stocks. However, there exist exceptions to this rule with low lead in biotite from districts that have produced no lead. These authors confirm the above mentioned observations by GAVRILIN *et al.* (1967) that hydrothermally altered intrusives contain less lead and zinc than unaltered stocks. SLAWSON and NACKOWSKY (1959) report from quartz monzonites of the same Basin and Range Province that trace lead in potassium feldspars can be correlated with galena deposits associated with igneous rocks. Again potassium feldspars from hydrothermally altered intrusions contain less lead than related unaltered rocks.

In Section 82-E no significant difference between average lead concentration in granitic intrusives and rhyolitic effusives could be observed, and this is evidence against major and general extraction of lead from granitic rocks by hydrothermal solutions.

The isotopic composition of lead is a more specific indicator than lead concentrations for detecting genetic relations (consanguinity) between source rocks and galena deposits. Examples of ore lead in which the isotopic composition is the same as that in the presumably related magmatic source during consolidation are reported on page 82-B-27. Beside Butte, Montana, Bingham, Utah (U.S.A.) and Nelson, British Columbia (Canada), ore deposits of the Yenisey Range, East Sayan and Kuznetskiy Alatan in Central Siberia could be directly related to granitoids (Volobuyev *et al.*, 1970). The other extreme, where lead isotope data fit a lateral secretion model related to observable country rocks, is exemplified in the Mississippi Valley type of ore deposits and many others. Tracing of ore sources by lead isotopes has even been tried on Recent brines such as those detected in the Salton Sea area and in the Red Sea (see Section 82-B).

Classifications into syngenetic and epigenetic types of lead deposits occurring as beds and veins and descriptions of their parageneses are a matter of special textbooks. By far the largest amount of the world lead production originates from epigenetic sources.

Revised manuscript received: April 1974

82-G. Behavior during Weathering and Rock Alteration

Lead primarily occurs in the structure of potassium feldspars and micas of magmatic and metamorphic rocks. These minerals are fairly resistant against weathering except biotite. Some mobilized lead is adsorbed on newly-formed clay minerals as kaolinite. The low rate of mobilization of lead during chemical weathering is observable from the low concentration of this element in surface run-off (see Section 82-I).

I. Weathering of Rocks

The behavior of lead during *basalt* weathering has been investigated by HEINRICHS (1974) at different stages of natural decomposition of Tertiary alkali olivine basalts in the Göttingen area. The change of the lead concentration in the products of more or less advanced chemical weathering has been compared with that of titanium, one of the least mobilized elements. A maximum lead loss relative to titanium of 40 percent has been observed in the decomposed basalts. In the case of Hoher Hagen, an appreciable gain has occurred in some of the analyzed clay fractions. The first two lines of Table 82-G-1 demonstrate the primary heterogeneity of some of the basalt complexes which limits further conclusions. SHORT (1961) has analyzed lead (Method: S) in sand, silt and clay of residual soil on basalt near Denver, Colorado (U.S.A.), and found a depletion of this element relative to the primary rock in the coarse fractions. The lead concentration in clay is 2 to 6 times that of the fresh basalt, however.

ZEISSINK (1971) has observed a relative accumulation of lead in nickeliferous laterite profiles on *serpentinites* in Queensland (Australia). Fresh serpentinite there contains 7.5 ppm Pb (method: A). The Pb accumulation is only slightly related to the iron accumulation in laterite: stage one, 14 ppm Pb and 10.9% Fe_2O_3; stage two, 20 ppm Pb and 19.8% Fe_2O_3; stage three, 17 ppm Pb and 46.7% Fe_2O_3.

Chemical weathering of *granite* affects in the first stages mainly biotite. Mechanically decomposed granites with leached biotite from the Harz Mountains (Germany) have been analyzed on lead (WEDEPOHL, 1956; method: S). The fresh Brocken granite contains 22 ppm Pb and the slightly weathered sandy material 6 to 11 ppm Pb. This indicates Pb losses of 60 to 70 percent, if the primary granite of the weathered sample was identical with the fresh rock analyzed in this investigation. The loss is too large to be explained by leaching of biotite only. More advanced decomposition of a granite during lateritization (test rock from Eklara, India; WEDEPOHL, 1956) resulted in a maximum loss of 80 percent of the primary lead (14 ppm Pb) and a pronounced accumulation of the mobilized lead in the horizon of ferruginous concretions. Several publications, which present data on lead in granite weathering, confirm the above observation mentioned under basalts, that this element is mainly accumulated in the clay fraction of decomposition products (SHORT, 1961, granodiorite from Wyoming: accumulation in clay by a factor of 3; BUTLER, 1953, grano-

Table 82-G-1. *Lead concentration (in ppm) in Tertiary alkali olivine basalt from the Göttingen area (North West Germany)*, Ratio ppm Pb·10,000/percent TiO_2 in brackets (Method: A; HEINRICHS, 1974)

	Locality			
	Bramburg	Backenberg	Hoher Hagen, Dransfeld	Steinberg, Meensen
Basalt, fresh	7.0 (3.02)	1.6 (0.65)	2.3 (0.96)	2.7 (1.32)
Basalt, almost fresh	0.8 (0.32)	2.9—3.2 (1.26—1.45)	—	2.4—3.2 (1.04—1.40)
Basalt, weathered, stage 1, olivine partly decomposed	—	4.1 (1.46)	2.5 (1.04)	2.2 (0.96)
Basalt, weathered, stage 2, pyroxene almost undecomposed	3.9 (1.05)	0.9 (0.41)	6.1 (2.26)	—
Basalt, weathered, stage 3, pyroxene partly decomposed	—	—	4.0 (1.43)	—
Basalt, weathered, stage 4, pyroxene almost completely, plagioclase partly decomposed	1.8—2.9 (0.58—0.80)	—	8.7 (1.89)	3.4 (1.06)
Basalt, weathered, stage 5, plagioclase almost decomposed	1.6 (0.48)	3.2 (0.82)	—	1.9—3.9 (0.67—0.81)

diorite from Cornwall: accumulation in clay by a factor of 7). KÖSTER (1966) reports 400 to 1500 ppm Pb (Method: W/C) in kaolinite from SE Germany which is derived from arkoses. According to his balance computation, all lead from the primary rock has been accumulated in the kaolinite fraction. According to KÖSTER (1969), the above mentioned deposit of Hirschau-Schnaittenbach (Germany), with an average of 460 ppm Pb, is extraordinarily high in lead compared with 12 other commercial kaolinite deposits of the world, ranging from 9 to 130 ppm Pb, average 83 ppm Pb (Method: W/C).

II. Weathering of Ore Deposits

By oxidation of galena under near-surface conditions, lead can be mobilized. Its concentration in mine waters (pH 2 to 9, Eh 0 to 0.7) or ground and stream waters related to sulfide deposits is usually in the ppb range and depends on anions available and adsorption processes (see Section 82-H). The solubility of secondary minerals such as anglesite ($PbSO_4$), cerussite ($PbCO_3$), pyromorphite ($Pb_5[PO_4]_3$ (Cl,OH)) etc., controls lead transport in secondary dispersion haloes around deposits. The distance of lead penetration from the deposit into the country rock is usually smaller than that of copper and zinc. Data on typical cases are reported by GINSBURG (1963),

Morris and Lovering (1952), etc. Secondary dispersions and higher concentrations of metals in streams and vegetation are the base of geochemical prospection on deposits.

III. Soils

Soils contain in their clay fraction the residual lead of weathered rocks on which they have formed. Vinogradov (1959) reports data from Russia, U.S.A., Norway and Spain, of which a mean of 14 ppm Pb can be computed. Shacklette *et al.* (1971) have published a recent survey on the chemical composition of 863 soils, sampled at about 20 centimeters depth in a regular network over the United States. The analyses (Method: S) gave a geometric mean of 16 ppm Pb. The frequency distribution is the following: 130 samples — 10 ppm Pb; 120 samples — ~10 ppm Pb; 220 samples — ~15 ppm Pb; 160 samples — ~20 ppm Pb; and 160 samples — ~30 ppm Pb; less than 30 samples — ~50 ppm Pb etc. Organic residues in soils have some influence on the lead accumulation in them.

IV. Bauxites

Bauxites are products of intensive rock alteration. The investigated specimens are derived either from nepheline syenites (Arkansas, U.S.A.) or from clay residues of limestones (southern Europe). The bauxites from Arkansas range up to 70 ppm Pb (average: 6.7 ppm Pb), whereas the primary nepheline syenite contains up to 20 ppm Pb (average 6.7 ppm Pb) (Gordon and Murata, 1952). The following lead data have been published by Wedepohl (1956) (Method: S) on composite samples representing large masses of economic bauxite: Greece, 100 ppm Pb; Yugoslavia (Montenegro, Mostar, Dalmatia, Istria), 80—140 ppm Pb; Hungary (Ujbarok, Eplény), 100—140 ppm Pb; southern France, 120 ppm Pb. A gross average of this group is 110 ppm Pb. Relative to average clay, these bauxites have accumulated Al and Ti by a factor of 4 and Pb by a factor of 5. The reason for higher lead can be some additional lead from the leached limestones.

Revised manuscript received: April 1974

82-H. Solubilities of Compounds which Control Concentrations of Lead in Natural Waters; Valence States in Natural Environments; Adsorption Processes

I. Solution Chemistry

Lead occurs as divalent and tetravalent ion in nature. Pb^{2+} predominates over Pb^{4+}. Most lead salts of Pb^{2+} with naturally common anions are only slightly soluble. The redox relationships between lead, Pb^{2+} and Pb^{4+} compounds in a system with water, carbonate and sulfur as a function of Eh and pH at 25° C and 1 atm pressure have been plotted by GARRELS and CHRIST (1965) according to MCINTYRE. In this system the minimum Eh at which a Pb^{4+} oxide (Pb_3O_4 minium) is formed is +0.2, but at the naturally uncommon pH of about 12. At pH 10 the minimum Eh of the formation of this compound is +0.4 and at pH 8 plattnerite (PbO_2) is formed above Eh +0.65. The rare environment of lead oxidation is characterized by contact with the atmosphere and absence of any reducing agent.

Standard potentials of the single ions are (according to LATIMER, 1953):

$$Pb = Pb^{2+} + 2e^- \qquad \varepsilon^0 = 0.126$$

$$Pb^{2+} = Pb^{4+} + 2e^- \qquad \varepsilon^0 \approx -1.7$$

Of the naturally abundant inorganic ions, which control solubilities of lead compounds, phosphate, carbonate and hydroxide form lead compounds of low solubility.

a) Sea Water

The large ionic strength of sea water has to be considered in the computation of lead concentrations from solubility products. An activity coefficient of 0.34 ($\log \gamma$: —0.47) has been used by ZIRINO and YAMAMOTO (1972) for divalent ions in solutions with the effective ionic strength of sea water (0.67). SILLÉN (1961) mentions an activity coefficient of 0.2 ($\log \gamma$: —0.8) for divalent ions in sea water. In the computations of Table 82-H-1, the former value of $\log \gamma$: —0.47 has been applied. The solubility of lead sulfide has been included in this table despite the fact that the occurrence of sulfide ions in ocean water is restricted to anoxic basins and interstitial waters in sediments. According to the information in Table 82-H-1, the concentration of ionic lead in the normal aerated sea water seems to be controlled by the solubility of pyromorphite (lead phosphate)[1].

The carbonate being an order of magnitude higher in solubility is another only slightly soluble compound. HEM and DURUM (1973) report hydrocerussite ($Pb_3(OH)_2CO_3$) as being the least soluble compound in a small area close to pH 8.

[1] *Note added in proof:* In a recently published investigation NRIAGU (1794) confirms this assumption and demonstrates, that the stability fields of pyromorphites and plumbogummite $PbAl_3[PO_4]_2(OH)_5 \cdot H_2O$, predominate strongly over those of other „secondary" lead minerals with in a large range of pH and Eh.

Table 82-H-1. *Solubility of some common lead compounds in sea water (25° C)*

Lead compound of low solubility	Solubility product at zero activity (reference in brackets)	Concentration (activity) of anion in sea water, moles per liter	Activity of lead in saturated solution	
			moles per liter	ppb
PbS (galena)	10^{-28} (SILLÉN and MARTELL, 1964)	10^{-4} ΣH_2S, HS^-, S^{2-} [a] $10^{-8.5}$ $(10^{-8.97})S^{2-}$	$2.8 \cdot 10^{-19}$	$5.7 \cdot 10^{-11}$
$PbCO_3$ (cerussite)	$10^{-13.2}$ (SILLÉN and MARTELL, 1964)	$10^{-4.9}$ $(10^{-5.3})CO_3^{2-}$	$3.7 \cdot 10^{-8}$	7.7
$Pb(OH)_2$ $\rightleftharpoons PbO \cdot H_2O$ or related stable compound	$10^{-14.9}$ to $10^{-19.9}$ (SILLÉN and MARTELL, 1964)	$(10^{-6})OH^-$	$3.7 \cdot 10^{-3}$ to $3.7 \cdot 10^{-8}$	$7.7 \cdot 10^5$ to 7.7
$Pb_3[PO_4]_2$	$10^{-42.1}$ (SILLÉN and MARTELL, 1964)	$10^{-9.3}$ $(10^{-10.1})PO_4^{3-}$	$1.5 \cdot 10^{-7}$	31
$Pb_5[PO_4]_3(Cl)$ (pyromorphite)	$\sim 10^{-78}$ [b] (NRIAGU, 1973)	$(10^{-10.1})PO_4^{3-}$ $10^{-0.26}$ $(10^{-0.45})Cl^-$	10^{-9}	0.21
$PbH[PO]_4$ (phosphate schultenite)	$10^{-11.1}$ (LATIMER, 1953)	$10^{-5.5}$ $(10^{-6.2})HPO_4^{2-}$	$3.6 \cdot 10^{-5}$	$7.5 \cdot 10^3$
$PbSO_4$ (anglesite)	$10^{-7.8}$ (SILLÉN and MARTELL, 1964)	$10^{-1.55}$ $(10^{-2.47})SO_4^{2-}$	$1.4 \cdot 1 10^{-5}$	$2.8 \cdot 10^3$

[a] Anoxic deep water of the Black Sea (order of magnitude).

[b] Solubility product of bromopyromorphite has been used, which should be close to that of chloropyromorphite.

According to ZIRINO and YAMAMOTO (1972) lead in sea water is mainly complexed with CO_3^{2-} and to a lesser extent with Cl^-. At pH 8.2 and 25° C, about 85% of the total lead occurs as $PbCO_3^0$, 10% as $PbCl^+$, 4% as $PbCl_2^0$ and 1.5% as Pb^{2+}.

b) Fresh Water and Ground Water

Continental waters are usually higher in lead than sea water (more than a factor of 10) (for details see Section 82-I). Streams often contain less than a quarter of the phosphate content of sea water and their pH, and their hydroxide concentration, is lower (OH: 10^{-7}). Their chloride is much less than in sea water. Therefore the slightly more soluble hydroxide pyromorphite could control the lead concentration of continental waters. With decreasing pH from 8 to 5, the solubility of lead carbonate or hydroxide increases from less than 10 ppb Pb to more than 50 ppm at $\leq 10^{-3}$ M of total dissolved CO_2 species, according to calculations by HEM and DURUM (1973). Another control mechanism is adsorption.

c) Brines and Ore-Depositing Fluids

A potential large difference in ionic strength, complexing anions, and temperature between groundwaters and deep connate or hydrothermal waters of magmatic origin, can be responsible for an appreciable lead concentration in the latter. Metal contents of brines close to and above 1 ppm Pb are needed to form ore deposits.

ANDERSON (1962) has measured the concentration of lead in aqueous solutions saturated with PbS and H_2S in the range of 30 to 90° C at 1 atm pressure as being as low as 1 to 5 ppb (the formation of a complex $PbS \cdot 2H_2S$ has been assumed). It increases up to 20 ppb by the addition of 0.2 m NaCl and 1 m NaOH.

The effect of chloride complexing of Pb has been calculated and measured by several authors: HEMLEY *et al.* (1967) (measured solubility of PbS in 2 m KCl and silicate buffer at 1,000 bar, 300° C: 620 ppm Pb; at 500° C: 16,600 ppm Pb); HELGESON (1969) (calculated solubility of PbS in 3 m NaCl in the range 50 to 300° C up to 13 ppm Pb, pH 5; see Fig. 29-H-1); NRIAGU (1971a) (measured solubility of PbS in 3 m CaCl-HCl solutions in the ranges from 28 to 200° C and from pH 1 to 5.5; at pH 4 the solubility increases from 0.5 to 300 ppm Pb with temperatures of 28 to 200° C; at 200° C it increases from 30 to 3,000 ppm Pb reciprocally with pH from 5 to 2.7; formation constants are highest for $PbCl^{+}$, $PbCl_2$ and $PbCl_4^{2-}$ complexes); NRIAGU and ANDERSON (1971) (measured and calculated saturation of $PbCl_2$ in 3 molal mixed salt media of Na (Cl,ClO_4) and degrees of formation of $PbCl_n^{2-n}$ ($n = 1$—4) species as a function of temperature from 20° to 200° C); NRIAGU (1971b) (measured solubility of PbS in 1 m and 3 m NaCl-HCl solutions saturated with H_2S at 90° C). In the pH <2.5 range solubility is mainly controlled by the formation of chloride complexes; between pH 2.5 and 6 the equilibrium solubulity can be attributed to a neutral complex $PbS \cdot SH_2S$; increase of solubility above pH 6 has been interpreted as due to $Pb(HS)_3^-$ complex formation (BARNES and CZAMANSKE, 1967); 0.2 m Σ H_2S, HS^-, S^{2-} solutions with ≥ 1 m NaCl of neutral to slightly acid pH, transport more than 10 ppm Pb at 90° C. ANDERSON (1973) has evaluated the above listed results on lead solubility and presents a convenient diagram on mineral and aqueous species stabilities to explain hydrothermal transport and deposition of galena as a function of oxygen fugacity and pH in 3 m NaCl solutions containing SO_4, Ca, Mg at 100° C.

BARANOVA and BARSUKOV (1965) present data on the solubility of $PbCO_3$ in Na-K-Cl-CO_3-HCO_3 solutions at 20, 250 and 300° C.

II. Adsorption Processes

Adsorption of lead from aqueous solutions on clay minerals has been studied by CORRENS (1924), KRAUSKOPF (1956) and LAGERWERFF and BROWER (1973), on hydrated iron oxide by KRAUSKOPF (1956) and on organic substances by KRAUSKOPF (1956) and SWANSON *et al.* (1966). CORRENS (1924) has used suspensions of kaolinite (5 grams per liter) to get 13 to 50% of a 83 to 21 ppm Pb-containing solution at pH 6.6 to 6.9 on the clay mineral surface. KRAUSKOPF (1956) reports the removal of the following fractions in suspensions of 2% (clay) to 30 ppm (Fe_2O_3) adsorbent containing initially 0.6 to 2.5 ppm Pb (in percentage of initial dissolved lead concentration): montmorillonite: >96%; freshly precipitated $Fe_2O_3 \cdot nH_2O$: 86%; peat moss: >96%. The maximum adsorbed Pb in these investigations is 25 ppm on montmorillonite and 2,500 ppm on kaolinite, mainly depending on the lead -to- clay mineral ratio of the suspension. According to SWANSON *et al.* (1966), extracted humate is able to adsorb about 10 percent (of its dry weight) of lead. The sequence of preferential sorption of metals from a mixed solution by humate is Cu>Ni>Ag>Fe^{3+}>Pb>Zn.

A comparison of the solubility of $Pb(OH)_2$ and the equilibrium solution concentration of Pb in soil clay suspensions is possible from the results of LAGERWERFF and BROWER (1973) at different NaCl concentrations and pH values. At the low NaCl concentration of freshwaters and at pH 7, about 43 ppm Pb are soluble in clay-free and 0.24 to 2.5 ppm Pb in clay-containing solutions. At pH 8, the respective concentrations are 7.5 ppm Pb and 0.13 to 1.7 ppm Pb. These data demonstrate the importance of adsorption on clay minerals (and organic substances) in the control of freshwater lead concentrations. Lead is assumed to be adsorbed as $Pb(OH)_2^{aqua}$, $PbCO_3^{aqua}$ or as Pb^{2+}.

Revised manuscript received: April 1974

82-I. Abundance in Natural Waters and in the Atmosphere

I. Rivers, Lakes and Groundwaters

Chemical weathering of common rocks can only contribute minor quantities of lead to surface waters (see Sections 82-G and 82-H). The low lead concentration of streams and lakes is probably controlled by lead adsorption on clay minerals and organic residues, as well as by precipitation of lead hydroxide, phosphate and/or carbonate. Under reducing conditions lead sulfide can be the least soluble compound.

a) Pollution of Surface Waters

The penetration of sulfide deposits or mine dumps containing galena by aerated groundwater can cause local anomalies of lead in run-off, soils and vegetation. Additional anthropogenic sources of lead pollution of surface waters are the exhaust of the automobile traffic (lead alkyl decomposition), burning of coal and firing of brick etc. The amount of lead used in the yearly world production of gasoline additives is ~500000 tons. If this lead was completely accumulated in the surface run-off, river water should contain 13 ppb instead of the lower concentration listed in Table 82-I-1. A complete accumulation of lead from coal burning in rivers contributes a tenth of the lead mass from gasoline. A reasonable fraction of the airborne lead pollution is probably fixed in soils. The World Health Organization emphasizes that drinking water should contain less than 100 ppb Pb to avoid poisoning.

The U.S. Public Health Service Drinking Water Standards quote a mandatory upper limit for lead of 50 ppb.

KONSTANTINOV (1954) mentions mine waters and creeks from mining areas in Kazakhsthan, Caucasus and Altai (U.S.S.R.) ranging from 7 to 9 ppm Pb. The lead concentration decreases from 80 to 0.4 ppm Pb in a pH gradient of a mine water from 3.4 to 7.5. Another detailed investigation of lead concentrations in the 0.0X to X ppm range in different types of mine waters is contributed by GOLEVA *et al.* (1970).

b) Continental Waters with Minor Pollution

Reliable data on lead concentrations in common types of surface and subsurface waters (except brines) are listed in Table 82-I-1. From the appreciable high lead value of rain and snow waters collected in a rural country and the low lead concentration in some springs, creeks and remote Sierra Nevada lakes, it can be concluded that an average lead concentration of 3 ppb in continental surface waters is already influenced by some pollution. Such an assumption is confirmed by an accumulation of lead in the clay fraction of all major streams in western Germany. The ratio of lead in this clay fraction to average lead in shales ranges from 2.5 to 7.5 (BANAT *et al.*, 1972). LELAND *et al.* (1973) observed a decrease of lead in Lake Michigan waters from 2 ppb Pb at off-shore Chicago to 0.4 ppb Pb at Grand Haven (Michigan),

Table 82-I-1. *Lead concentration in rain, rivers, lakes, groundwater and sea water with minor pollution* (Number of samples in brackets)

Samples	Origin	Range ppb Pb	Average ppb Pb	Method	Reference
Rain water, summer, 1966 (4)	Western Michigan (U.S.A.)	6.2—29.2	17.4	W/C	TER HAAR *et al.* (1967)
Rain water, January to December, 1971	Wraymires, Lake Windermere (England)		39	X (N/R ?)	PEIRSON *et al.* (1973)
Rain water, January to December, 1968—69 (85)	N, NW and Central European U.S.S.R.	1—48	4.7	S	DROZDOVA and MAKHONKO (1970)
Snow, rural areas (8)	Canada	2—15	8.2	A	JONASSON (1973)
Snow, 1904	77° N, 62° W Greenland	—	0.16	I	MUROZUMI *et al.* (1965)
Snow, 1964	Greenland	—	2.6	I	MUROZUMI *et al.* (1965)
Surface waters (439)	Maine (U.S.A.)		2.6	S	TUREKIAN and KLEINKOPF (1956)
Surface waters (794)	U.S.A.	<1—55 (890)	3.9[a]	W/A	DURUM *et al.* (1971)
River waters (20)	Saale, Elbe (G.D.R.)	1.4—9.2	3.9	W/C	HEIDE *et al.* (1957)
Surface and groundwaters (820)	NW Siberian platform (U.S.S.R.)		0.48	S	KONTOROVICH *et al.* (1963)
Surface and groundwaters (4376)	Siberia	0.1—2,071	1.5	S	UDODOV and PARILOV (1961)
Lake waters, summer, 1967 (96)	Erie (Canada)		4	W/A	CHAWLA and CHAN (1969)
Lake waters	High Sierra Nevada, Calif. (U.S.A.)		0.5	W/S	BRADFORD *et al.* (1968)
Creek and lake waters (16)	Harz Mountains, Göttingen area Germany	0.5—1.2	0.7	A	HEINRICHS (1974)
Spring waters (128)	Neckar tributaries Germany	0.1—1	~0.5	A	BUKENBERGER *et al.* (1972)
Sea water (nearshore, shelf, open sea), surface water (9)	Pacific	0.08—0.36[a]	0.19	I	TATSUMOTO and PATTERSON (1963)
Sea water, deep water (17)	Pacific	0.02—0.40[a]	0.09	I	TATSUMOTO and PATTERSON (1963)

Table 82-I-1 (continued)

Samples	Origin	Range ppb Pb	Average ppb Pb	Method	Reference
Sea water, deep water (>2000 m depth) (6)	Pacific	0.02—0.13[a]	0.04[b]	I	Tatsumoto and Patterson (1963)
Sea water (uncontaminated)			0.002[c]	I	Doe (personal comm. 1974)

[a] Filtered samples.
[b] The authors prefer a value of 0.03 ppb Pb for deep water.
[c] Summary of NSF Workshop on lead pollution to be published in J. Mar. Chem. (1974).

demonstrating the influence of pollution from a large city. Murozumi *et al.* (1965) have compared lead concentrations in Greenland snow increasing from 0.16 ppb to 2.6 ppb Pb from 1904 to 1964.

II. Sea Water

Because of the very low concentration of lead in sea water (0.002—0.03 ppb), there exist only few reliable data on filtered samples (Tatsumoto and Patterson, 1963; Doe, personal communication, 1974). These results are even three times lower than those published by Chow and Patterson (1962). Most of the pre-1960 data on lead in sea water range above 1 ppb and cannot be considered here.

If we assume that unpolluted river water contains only 0.5 ppb Pb or less and sea water 0.03 ppb Pb, the mean residence time in the oceans is as low as 3,500 years which is comparable with the residence time of manganese (3,000 years, at a mean concentration of 0.4 ppb Mn in sea water and 5 ppb Mn in river water). A major proportion of the lead of Recent ocean sediments occurs in manganese nodules and probably in the manganese oxide phase of clays.

In coastal waters the influence of continental pollution can be observed. Preston (1973) has conducted a survey of British coastal waters in 1971 revealing that lead concentration means range from <0.05 ppb in the northern Irish Sea to 0.15 in the Irish Sea and to 0.21 in the North Sea (47 samples). Means from Liverpool Bay, Cardigan Bay and the Bristol Channel (71 samples; polarographic method) are even as high as 1.2 to 2.2 ppb Pb (Abdullah *et al.*, 1972).

III. Hydrothermal Water

a) Thermal Springs

Minami *et al.* (1957) report that lead in hot spring water from Tamagawa (Akita Prefecture, Japan) ranges from below sensitivity to 1.3 ppm Pb (11 to 98° C; pH 1.2 to 8.1). Ellis (1968) lists 2 ppb, 2 ppb and 10 ppb Pb as lead concentrations in hydrothermal waters (230 to 275° C) from three wells (Taupo, Weirakei and Auckland) in New Zealand and gives a reference for hydrothermal water (340° C) from Cerro Prieto Mexicali (Mexico) containing 5 ppb Pb. From these data it can be concluded that thermal waters, low in NaCl and acidity, contain low lead concentrations comparable to continental waters at low temperature.

b) Hot Brines

The increase in solubility of PbS by chloride complexing has been experimentally proved as mentioned in Subsection 82-H-I c. A similar range of lead concentrations, as measured in these experiments, has been observed in natural brines: Salton Sea, California (U.S.A.): 80 ppm Pb (HELGESON, 1968); ATLANTIS II Deep of the Red Sea: 0.6 ppm Pb (EMERY *et al.*, 1969); Cheleken, Caspian Sea (U.S.S.R.): 10.7 ppm Pb (LEBEDEV, 1967). The sodium chloride concentrations of the reported brines range from 208 to 249 grams per liter.

IV. Atmosphere

The anthropogenic volatilization of lead compounds from burning of leaded automotive fuels (lead tetraethyl additive since 1922; lead tetramethyl since 1960) from industry and coal burnings, are so extensive ($4.4 \cdot 10^5$ t per year) that measurable pollution of the atmosphere and the surface waters on a hemispheric scale (see Greenland snow in Table 82-I-1) and of urban populations in industrialized countries has been measured. Several of the toxic trace elements, including Pb, were found to be most concentrated in the smallest respirable particles emitted from coal. Atmospheric concentrations in the northern hemisphere are probably 2 to 3 orders of magnitude above natural level. Some citizens of industrialized areas and users of busy highways ingest lead at rates approaching those which cause classic lead poisoning (lead blocks enzyme functions by inactivating SH-groups). In several countries the administrative safety limit is 2 μg Pb/m^3 air.

The atmosphere, measured in U.S. and British cities, contains 0.2 to 3 μg Pb per cubic meter. As an example sampling sites at Cambridge, Massachusetts (U.S.A.) with 0.1 to 2.2 μg Pb/m^3 are mentioned (LININGER *et al.*, 1966), contrasting with a rural area like Lake Windermere (England) with 0.087 μg Pb/m^3 (sampling January—December, 1971) (PEIRSON *et al.*, 1973). The current arithmetic mean U.S. aerial concentration of Pb is 0.79, the mean urban concentration is 1.1 and the average maximum U.S. urban concentration is 8.6 μg Pb/m^3 as deduced from biweekly sampling by 104 stations in 1964—65 (MORGAN *et al.*, 1970) (for lead fallout rates, see HEM and DURUM, 1973).

The bulk precipitation consists of rain and dry fallout. In the oceanic areas, air contains 0.003 to about 10 μg dust per cubic meter (CHESTER and ELDERFIELD, 1970). Particulates collected from the atmosphere ~15 m above the sea surface over large tracks of the world ocean contain on average as much as 465 ppm Pb (range: 29 to 2,750 ppm Pb)(CHESTER and STONER, 1973). This is 23 times the mean lead concentration in soil dust (average clay, shale) converted into atmospheric concentration by using a figure of 2 μg dust/m^3. This equals about 1 ng Pb/m^3. On a 20 meter high tower on the windward coast of Oahu, Hawaii, particles have been sampled for one year; 111 samples contain on average 3 ng Pb/m^3 air (range: 0.3 to 13; method: A) (HOFFMAN *et al.*, 1972). CHOW *et al.* (1969) reported that atmospheric Pb concentration over the North Pacific is about 1 ng/m^3. Even at the South Pole air contains still 0.63 ng Pb/m^3 (0.19 to 1.2; method: A) (ZOLLER *et al.*, 1974). Here it is 2,500 times the normal soil dust composition of average crustal material.

Revised manuscript received: April 1974

82-K. Abundance in Common Sediments and Sedimentary Rocks

Lead from weathering of magmatic and metamorphic rocks is expected to be mainly accumulated in detrital sediments in their potassium-bearing minerals mica and feldspar (see Table 82-D-1). Some lead will be transported into the sedimentary environment being adsorbed on clay minerals and ferric iron oxides etc.

I. Sands, Sandstones and Greywackes

Lead in psammitic rocks usually increases with decreasing quartz content. In a recent sand deposit from Buzzards Bay at the American east coast, lead concentrations are correlated with the clay fraction (MOORE, 1963). HARTMANN (1963) and others mention diagenetic lead accumulations in the light-bleached aureoles and zones of red sandstones and arkoses, sometimes correlated with local concentrates of V, Cu and U. The gross average of 924 sands and sandstones listed in Table 82-K-1 is: 10 ppm Pb. Greywackes are often higher in lead.

II. Pelites

a) Shales and Non-Bituminous, Near-Shore Deposited Clays

This most abundant sedimentary rock class contains the highest proportion of the lead mass in sediments. Lead analyses are compiled in Table 82-K-2.

The listed averages of Pb in clays and shales from different continents and of different age are close to a grand average of 179 samples of 21.6 ppm Pb (excluding the large set of South African samples). If one includes the 184 South African samples, the mean is 23.3 ppm Pb. The small scattering confirms that marine argillaceous sediments have averaged large masses of decomposed rocks and that the different analytical methods applied for this investigation are reliable.

A frequency distribution of lead values for 316 South African shales including carbonaceous is plotted as Fig. 82-K-1 (a total mean is 28.5 ppm Pb). Except the very old Fig Tree and Witwaters Rand shales (11 ppm Pb), no systematic increase of lead in shales with decreasing age could be observed.

Monomineralic concentrates of clay minerals often contain lead in the same range as clays and shales. WEDEPOHL (1956) mentions 3 montmorillonites with 15 to 34 ppm Pb, 2 illites with 15 to 25 ppm Pb and 4 kaolinites with 23 to 53 ppm Pb. The kaolinites from large deposits, investigated by KÖSTER (1969), are usually higher than 40 ppm Pb (see Section 82-G).

b) Bituminous Argillaceous Rocks (Black Shales etc.)

Data on lead in bituminous shales have been compiled by WEDEPOHL (1964) with the following average lead concentrations (analytical methods in brackets) for sets of samples from different areas listed in the sequence of decreasing age (number of

Table 82-K-1. *Lead abundances in sandstones and greywackes* (Number of samples in brackets)

Rock type	Origin	Range ppm Pb	Average ppm Pb	Method	Reference
Sands, marine (63)	Recent, Buzzards Bay, Massachusetts (U.S.A.)	2.1—82.3	25.4[a]	S	MOORE (1963)
Sands, marine (90)	Recent, Japan, Korea	<1—48	11.5	W/C	ISHIBASHI *et al.* (1960)
Sands, marine (15)	Gulf of Paria (Venezuela, Trinidad)	3.6—81	20.3	S	HIRST (1962)
Sandstones (composite of 11)	Carboniferous, Germany	—	5	A	HEINRICHS (1974)
Sandstones, arkoses, red (composite of 23)	Lower Triassic, Germany	—	9.5	A	HEINRICHS (1974)
Sandstones, arkoses, red (97)	Lower Triassic, Hessia, Lower Saxony (Germany)	2.9—37	9.6[a]	S	HARTMANN (1963)
Sandstones, red brown (85)	Jurassic, Colorado Plateau (U.S.A.)	—	5	S	MIESCH (1963)
Sandstones, red (5)	Lower Permian, Hessia (Germany)	2—50	25	S	HARTMANN (1963)
Sandstones, red (323)	Triassic, Moenkopi, Colorado Plateau (U.S.A.)	—	11	S	CADIGAN (1971)
Sandstones (composite of 12)	Cretaceous, Germany	—	5	S	WEDEPOHL (1956)
Sandstones, red, orange, yellow (205)	Tertiary, Rocky Mountains (U.S.A.)	6.9—13[b]	7.5	S	VINE and TOURTELOT (1973)
Sandstones, green (11)	Tertiary, Rocky Mountains (U.S.A.)		11	S	VINE and TOURTELOT (1973)
Greywackes (composite of 17)	Devonian, Carboniferous, Germany	—	47	A	HEINRICHS (1974)
Greywackes (4)	Lower Carboniferous, Harz Mountains (Germany)	0.95—8.7	4.4	A	HEINRICHS (1974)
Greywackes (2), sandstone (1)	Oregon (U.S.A.)	1.8—16	8.6	I	TATSUMOTO and SNAVELY (1969)

[a] Excluding one sample with more than 150 ppm Pb.
[b] Areal means.

Table 82-K-2. *Lead in shales and clays low in bituminous matter* (Number of samples in brackets)

Sediment type	Origin	Range ppm Pb	Average ppm Pb	Method	Reference
Clay muds (48)	Sea coast, Japan, Korea	5—89	23	W/C	Ishibashi *et al.* (1960)
Clays (12)	Gulf of Paria (Venezuela, Trinidad)	13—32	22	S	Hirst (1962)
Shales, marine (composite of 36)	Paleozoic, Europe	—	21	A	Heinrichs (1974)
Shales, marls, marine (4)	Lower Triassic, Jena (G.D.R.)	15—33	18.3	W/C	Heide and Lerz (1955)
Argillites (8)	Tertiary, eastern Carpathian Mountains (Roumania)	12—68	24	W/C	Savul *et al.* (1958)
Shales, marine (composite of 14)	Paleozoic, Honshu (Japan)	—	27	S	Wedepohl (1956)
Shales, marine (40)	Paleozoic, Ryoke Belt, Sanbagawa Belt (Japan)	7—46	21.7	X	Banno and Chappell (1969)
Shales, marine (composite of 10)	Mesozoic, Honshu (9), Tsushima (Japan)	—	28	S	Wedepohl (1956)
Shales, marine (7)	Devonian, New Hampshire (U.S.A.)	1—33	17.6	S	Shaw (1954)
Shales, marine (184)	Precambrian to Triassic, South Africa	see Fig. 82-K-1	24.9	X	Hofmeyer (1971)

samples in brackets): Cambrian, Sweden (7): 39 ppm (S); Ordovician-Cambrian, Sweden (>2): 36 ppm (S); Devonian, U.S.A. (21): 33 ppm (S); Carboniferous, U.S.A. (29): 23 ppm (S); Permian, U.S.A. (23): 20 ppm (S); Jurassic, Germany (36): 28 ppm (S); Cretaceous, U.S.A. (67): 14 ppm (S); Recent, Baltic Sea and Norway (24): 36 ppm (S). A gross average of these 209 samples is 23.8 ppm Pb. If one computes an average of the 8 sets giving each an equal statistical weight, this is 28.6 ppm Pb. The average of 132 carbonaceous argillaceous rocks of different age from South Africa is 33.5 ppm Pb. Vine and Tourtelot (1970) in their investigation of 20 sets of black shale samples (comprising 779 individual samples representing a wide variety of environments from the United States and Canada) report a medium of 20 ppm Pb. The frequency distribution of their data is plotted in Fig. 82-K-2.

In some Recent reducing coastal sediments such as in the Santa Monica, San Pedro and Santa Barbara basins of southern California, lead deposition has increased from pollution by factors 2 to 6 relative to pre-pollution rates. However, these basins are extreme as being situated near an area of substantially leaded gasoline combustion (Chow *et al.*, 1973).

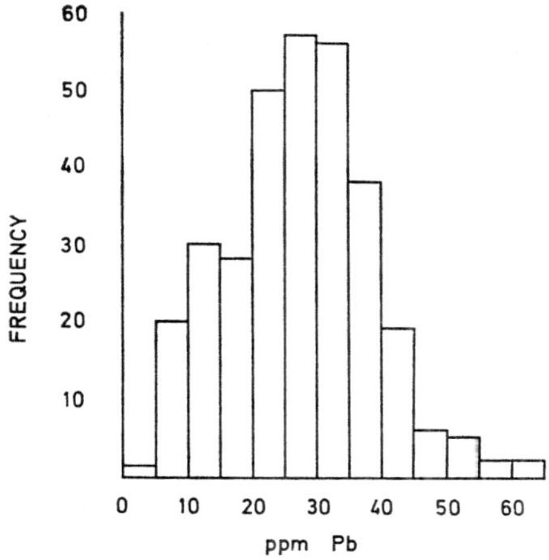

Fig. 82-K-1. Frequency distribution of lead concentrations in 316 South African shales ranging from Precambrian Fig Tree series to the Triassic Beaufort of the Karroo system (after HOFMEYR, 1971)

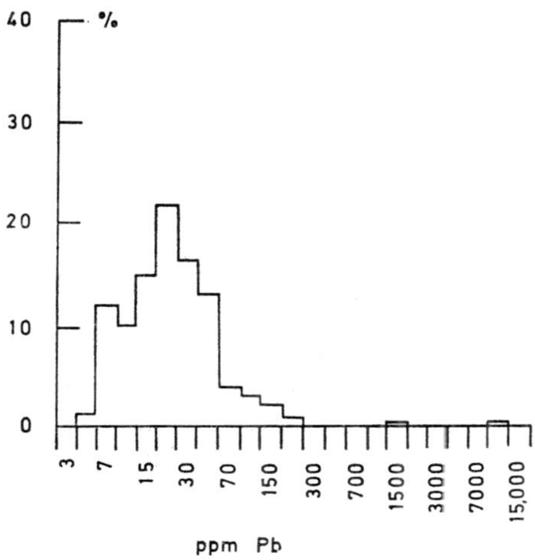

Fig. 82-K-2. Frequency distribution of lead concentrations in 20 sets of 779 black shale samples from North America according to VINE and TOURTELOT (1970). The histogram is drawn from the sum of the percent frequency distribution of individual sets giving equal weight to each set

A comparison of the above listed data with those of non-bituminous shales gives evidence for a slight enrichment of Pb in black shales. This extra lead might be connected with the organic residues (see Section 82-L) and/or the pyrite fraction. Black shales often contain one percent pyrite and more. Sedimentary pyrite ranges from <5 to 700 ppm Pb (WAMPLER and KULP, 1964). If one assumes a mean of 80 ppm Pb in pyrite in a bed containing 5% pyrite, this mineral contributes 4 ppm Pb to the shale.

Black shales containing more than 100 ppm Pb are rare. In the bituminous sediments of the Lower Jurassic of Central Europe a small fraction of galena can occur. Better known is the belt of the Upper Permian Kupferschiefer marl exposed from England to Poland, containing galena with isotopically light sulfur from bacterial reduction in the range of 100 ppm to several percent. The coverage of this bed where it is known as containing more than 0.5% Pb, exceeds 10^4 km^2. The average thickness is about one foot. The Kupferschiefer type metal deposition is rare in the earth's history because it requires a special environment of paleogeography (flat topography, slow rate of detrital deposition), composition of exposed rocks (prevalence of red beds), climate and sea (stagnant deep water with potential PbS precipitation) (for details and references see WEDEPOHL, 1964, 1971).

Beside sulfides and the organic fraction, some lead of bituminous rocks might occur in phosphorites. The range of lead in Mesozoic to Tertiary phosphorites from an area north of the Harz Mountains (Germany) is <8 to 260 ppm Pb with abundant values close to 10 ppm Pb (BENDA, 1963).

c) Pelagic Sediments

Relative to nearshore deposited argillaceous sediments, pelagic clays have accumulated several elements such as Cu, Mn, Ni, Co, Pb, Sb, Mo, Zn, etc. Syngenetically and diagenetically formed MnO_{2-x} might contain some of these additional element concentrations. In large areas of the Pacific Ocean and lesser areas in the Indian and Atlantic Ocean, ferromanganese nodules occur on the sediment surface and in the upper subsurface sediment (for distribution maps, see HORN *et al.*, 1972, etc.). MANHEIM (1965) reports an average of 1,000 ppm Pb in deep sea nodules, MERO (1962) lists 700 ppm Pb from 54 samples (Method: S) of Pacific nodules and 1,200 ppm in Atlantic nodules. AHRENS *et al.* (1967) investigated 9 samples from the Atlantic, 7 from the Indian Ocean and 14 from the Pacific Ocean ranging from 106 to 2,097 ppm Pb with an average of 1,017 ppm Pb (Method: X). Here again Atlantic nodules are higher in lead (1,300 ppm Pb) than those from the Pacific (880 ppm Pb), and the Indian Ocean is intermediate (970 ppm Pb). The results on Pacific materials are confirmed by LANGE (1974, personal communication) with an average of 730 ppm Pb on 32 nodules from the North Pacific (Method: A). From the latter investigation it seems probable that the major lead is related to the ferric iron oxide fraction of ferromanganese nodules.

If nodules from the Pacific contain on average 700 ppm Pb and 20% Mn, and if this proportion is equally typical for the manganese oxide fraction of clays, the latter might contribute about 20 ppm Pb to the total lead content of these pelagic sediments.

Lead concentrations in pelagic sediments are listed in Table 82-K-3. For the clays from the Atlantic (439 samples), an average of 47 ppm Pb can be computed and for those from the Pacific (558 samples, excluding ferruginous clays) one of 61 ppm Pb. These figures must be taken as tentative because they might be influenced by the pattern of sampling.

There are indications that ferruginous sediments, especially close to the deep sea rises, have accumulated higher lead concentrations than normal clays. HOROWITZ (1970) and PIPER (1973) give semiquantitative evidence for Pb accumulations in deposits on the East Pacific Rise. BENDER *et al.* (1971) report a value of 370 ppm Pb

Table 82-K-3. *Lead concentrations in pelagic sediments* (Number of samples in brackets)

Sediment type	Origin	Range ppm Pb	average ppm Pb	Method	Reference
Clays (14)	Atlantic[a]	19—96	45	S	WEDEPOHL (1960)
Clays (13)	Atlantic[a]	30—66	41	I	CHOW and PATTERSON (1962)
Clays (252)	Atlantic[a]	—	52	S	TUREKIAN and IMBRIE (1966)
Clays (~100)	Atlantic[a]	—	35	S	LANDERGREN and MANHEIM (1963)
Clays (2)	Atlantic[a]	61—70	65	A	BRUTY *et al.* (1973)
Clays, partly zeolitic (27)	Tertiary, Cretaceous, Atlantic	10—105	63	A	BRUTY *et al.* (1973)
Clays (<1% $CaCO_3$) (28)	Atlantic[c]	9—71	34	A	LANGE (1974)
Clays (10—30% $CaCO_3$) (3)	Atlantic[c]	20—100	51	A	LANGE (1974)
Clays (4 box cores)	Pacific[a]	<10—50	21	S	SWANSON *et al.* (1967)
Clays (5)	North Pacific	20—35	27.6	W/C	CHESTER and HUGHES (1969)
Clays (<1% $CaCO_3$) (17)	North Pacific	20—91	51	A	LANGE (1974)
Clays (26)	Pacific[a]	13—130	34	S	CRONAN (1969)
Clays (22)	North Pacific[a]	20—120	60	S	TATSUMOTO (1957)
Clays (39)	Pacific[a]	29—139	69.7	I	CHOW and PATTERSON (1962)
Clays (450)	Atlantic and Pacific[a]	—	63	S	LANDERGREN and MANHEIM (1963)
Ferruginous clays (>30% Fe_2O_3) (2)	Pacific[c]	105—138	121	A	LANGE (1974)
Carbonate ooze, chalk (4)	Atlantic[a]	55—219	105(?)	A	BRUTY *et al.* (1973)
Carbonate ooze, (>60% $CaCO_3$) (13)	Atlantic[c]	4—35	12	A	LANGE (1974)
Marl ooze (30—60% $CaCO_3$) (6)	Atlantic[c]	11—37	21	A	LANGE (1974)
Foraminiferal ooze (one core)	Mid-Atlantic Ridge[a]	—	4.4 (0.5)[b]	S	THOMPSON (1972)
Carbonate ooze (~200)	Pacific[a]	—	18	S	LANDERGREN and MANHEIM (1963)

Table 82-K-3 (continued)

Sediment type	Origin	Range ppm Pb	Average ppm Pb	Method	Reference
Carbonate ooze (7)	Pacific[a]	<10—50	19	S	TATSUMOTO (1957)
Carbonate ooze (1)	NE Pacific[a]	—	6.9	I	CHURCH (1973)
Carbonate ooze, chalk (>60% $CaCO_3$) (12)	North Pacific[c]	6—28	12	A	LANGE (1974)
Radiolarian ooze (5)	Pacific and Atlantic[c]	10—38	25	A	LANGE (1974)

[a] Surface samples of Quaternary age.
[b] Carbonate fraction.
[c] Deep Sea Drilling Project (Campanian to Recent age).

in a carbonate-free sediment from the East Pacific Rise (about 150 ppm in the original carbonate-containing sediment); DASCH *et al.* (1971) got 180 ppm Pb in similar material from the same area. Data in the same range contributed by LANGE (1974) are listed in Table 82-K-3. CHURCH (1973) reports 240 ppm Pb in an alteration crust on basalt from the Juan de Fuca Rise.

Lead isotopes in manganese nodules and pelagic clays from the Pacific are less radiogenic than those from the Atlantic (CHOW and PATTERSON, 1962; REYNOLDS and DASH, 1971). Even less radiogenic than the normal Pacific sediment seems to be the ferruginous matter (BENDER *et al.*, 1971; DASCH *et al.*, 1971). From this result one has to assume some proportion of volcanogenic lead in the deep sea accumulation.

Carbonate and siliceous oozes are usually low in lead; their major lead proportion is contributed by the clay fraction.

III. Carbonate Rocks and Evaporites

a) Carbonate Rocks

Limestones and dolomites are low in lead. Calcite and dolomite cannot incorporate appreciable concentrations because sea water and interstitial waters usually contain very little lead (see Section 82-I).

An average of the carbonate rocks, listed in Table 82-K-4, is 5 ppm Pb. Limestones very often contain between 10 and 15 percent clay and silt. This fraction will contribute about 3 ppm on average.

b) Evaporites

Several halites from Permian marine deposits in Germany have been analyzed by BORN (1934; method: active tracer) and HERRMANN (1958; method: S), and from Russian evaporite series by GALACHOWSKAJA (1953), with the following results:

Table 82-K-4. *Lead concentrations in carbonates and carbonate rocks* (Number of samples in brackets)

Sediment type	Origin	Range ppm Pb	Average ppm Pb	Method	Reference
Clam shells, marine, 17 species (313)	Recent	<1—80	1.6	S	LEUTWEIN and WASKOWIAK (1962)
Carbonate shells and skeletons, marine, 15 species (324)	Recent	<1—40[a]	3	S	LEUTWEIN and WASKOWIAK (1962)
Consolidated carbonate oozes (7), limestones (2), dolostones (3)	Tertiary, Quaternary, Atlantic (11), Pacific (1)	<2—16	≤4.6	S	THOMPSON (1972)
Limestones, marine (composite of 32)	Devonian, Germany	—	4.3	A	HEINRICHS (1974)
Limestones, marine (composite of 45)	Jurassic, Germany	—	0.4	A	HEINRICHS (1974)
Limestones, marine (composite of 16)	Cretaceous, Germany	—	10.1	A	HEINRICHS (1974)
Limestones, marine (13)	Lower and Middle Triassic, Jena (G.D.R.)	5.7—17	8.6	W/C	HEIDE and LERZ (1955)
Limestones (55)	Gorno (Italy)	—	3.4	S	HAHN-WEINHEIMER (1961)
Carbonate rocks (183)	Scotland	3—11.4	7.2	S(?)	Geological Survey of Great Britain (1956)
Limestones and dolomites (30)	Triassic and Tertiary, Carpathian Mountains (Roumania)	<1—41[b]	≤13.0	W/C	SAVUL and ABABI (1956)
Limestones and dolomites (228)	Triassic to Cretaceous, Tatra Mountains (Č.S.S.R.)	—	1.2 ± 0.7	S	DEMOVIČ (1973)
Limestones (159)	Eastern Kansas (U.S.A.)	1—200	9.1	S	RUNNELS and SCHLEICHER (1956)
Limestones, dolomites (6)	Cambrian, SE Missouri (U.S.A.)	4.2—13.4	6.6	I	DOE and DELEVAUX (1972)
Limestones and dolomites (104)	Ordovician, Silurian, Devonian, S. Ontario (Canada)	1.2—36	5.9[c]	W/C	WARREN and DELAVAULT (1961)

[a] Excluding a few Pb rich samples from the Red Sea.
[b] Excluding two samples with Pb > 50 ppm.
[c] Excluding Eramosa Goat Island, Gasport samples; all values are for aqua regia-extractable lead only.

6 samples, Salzdetfurth-Hannover (Germany): 80—490 ppb Pb (Average: 180 ppb); 2 samples, Stassfurt (Germany): $<$10 ppb Pb; 54 samples, southern Harz Mountains: 250 ppb Pb ($<$20—860 ppb Pb); 108 samples, Magdeburg-Halberstadt (Germany): 25 ppb Pb; 2 samples, Indersee (U.S.S.R.): 41.2—49 ppb Pb. The data on 5 halite rocks of different age from Cambrian to Quaternary published by MOORE (1971) are even higher (0.65—3.4 ppm Pb). If sea water with its 0.04 ppb Pb (Table 82-I-1) is completely evaporated, this corresponds to a concentration as low as 2 ppb Pb in NaCl.

Revised manuscript received: April 1974

82-L. Biogeochemistry

Lead is poisonous to many plants and several groups of animals such as mammals (blocking enzymes by inactivating SH-groups). Some varieties of plants have a remarkable tolerance for lead, growing on soils which have accumulated this element (CANNON, 1960). Several trees such as pine, fir, spruce and birch have been used for geochemical prospecting. As has been mentioned in former sections, the atmosphere, waters and soils are polluted almost everywhere from gasoline exhaust. Therefore it is hardly possible to observe the natural background of lead in organic tissues. Another problem is the sample preparation to avoid losses of this volatile element during ashing etc.

I. Marine Organisms

Table 82-L-1 contains data on lead abundances in marine organisms. Because of some scattering of the data no systematic difference between lead in phytoplankton and animals is obvious. Data on samples from the Black Sea are abnormally high to be used for averaging. A mean of the organic primary production is expected to be close to 15 ppm Pb. Because of the uncertainty about contaminations and/or losses during sample preparation, this mean for lower marine organisms can only be used with some caution.

II. Terrestrial Organisms

a) Plants

Lead can enter plants by the roots and by the foliage. Within the system Pb seems to be retained by cell membranes, mitochondria and chloroplasts (SABNIS *et al.*, 1969). Nutrition experiments have demonstrated that many plants responded to increasing lead availability to a very limited extent if aerial Pb was not a factor. Even on a soil of pH 6 with relatively high Pb mobility, a 10-fold increase in extractable lead content resulted in Pb uptake by radishes by a factor of less than 2 (for details and additional information, see LAGERWERFF, 1972). Regarding the foliar absorption of Pb from the air, rough hairy surfaces of foliar tissues were shown to retain more lead than smooth ones. In general lead in aerially-exposed plants diminishes as the distance from traffic increases and time of exposure decreases. KLOKE and RIEBARTSCH (1964) report a lead increase from 7 to 15 or 16 ppm Pb in grass approaching an urban highway with ~1,000 cars per hour, from more than 300 m distance. With increasing car frequency from 1,000/h to 2,000/h and 2,700/h, the lead content of the grass has been reported as being: 16 ppm, 48 ppm and 57 ppm, respectively. TER HAAR *et al.* (1969) have measured the uptake of lead by ryegrass and radishes from air, water and soil under controlled conditions. The lead content of grass grown in filtered air (0.03 μg Pb/m³) was 2.5 ppm while that of grass grown in unfiltered air (0.96 μg Pb/m³) was 5.2 ppm. Trees growing in the city of New

Table 82-L-1. *Lead abundances in dry substance (or ash) of organisms*
(Number of samples in brackets)

Samples	Origin	Range ppm Pb	Average ppm Pb	Method	Reference
Marine Organisms:					
Brown algae, 7 species (24)	English Coast	2—26	8.4	W/S	BLACK and MITCHELL (1952)
Planktonic diatoms	Black Sea	(2,000—3,000)[a]	—	S	VINOGRADOVA and KOVAL'SKIY (1962)
Plankton (12)	North Atlantic Ocean	(15—530)[a]	85[b] (213)[a]	S	TUREKIAN (Table 10-4 this handbook)
Phytoplankton (19)	Monterey Bay, Calif. (U.S.A.)	2—46.6	9.8	A	MARTIN and KNAUER (1973)
Microplankton (4)	N. Pacific Ocean	16.9—38.8	29.5	A	MARTIN and KNAUER (1973)
Zooplankton (11)	Atlantic Shelf (U.S.A.)	0.5—299	60	S	NICHOLLS *et al.* (1959)
Animals, 9 species	Gullmarfjord (Sweden)	0.2—43	13	S	NODDACK and NODDACK (1939)
Bivalves, soft parts, 3 species (18)	New Zealand	3—25	13	S	BROOKS and RUMSBY (1965)
Zooplankton	Black Sea	(25—1,500)[a]	169[b]	S	VINOGRADOVA and KOVAL'SKIY (1962)
Zooplankton (22)	Monterey Bay, Calif. (U.S.A.)	1—12.2	4.6[c]	A	MARTIN and KNAUER (1973)
Zooplankton (14)	N. Pacific Ocean	1.5—14.4	7.2	A	MARTIN and KNAUER (1973)
Echinodermata	—	5—21	~13	S	VINOGRADOV (1953)
Fish	North Sea, Irish Sea, English Coast, Barents Sea, etc.	<0.5—0.66[bb]	—	(?)	PRESTON (1973)
Fish	—	—	0.5	(?)	BOWEN (1966)
Terrestrial Organisms:					
Trees (691)	U.S.A.	—	2.3 (64.7)[a]	W(?)	CANNON (1960)
Trees, lesser plants, 78 species (45)	British Columbia (Canada)	0.2—6.5	1.5	W/C	WARREN (1962)
Trees (leaves) (6), ferns (1), bushes (6), far off traffic	Bavarian Alps (Germany), Austrian Alps	3—16	9.9	A	KRACKE *et al.* (1973)

Table 82-L-1 (continued)

Samples	Origin	Range ppm Pb	Average ppm Pb	Method	Reference
Bryophytes, ferns, gymnosperms, angiosperms	Finland	—	2.5	S	LOUNAMAA in: BOWEN (1966)
Equisetum, unmineralized areas (9)	U.S.A.	1.9—3.8	3.1 (11.6)[a]	S	CANNON *et al.* (1968)
Cereals, vegetables (within 300 yards of well-travelled highway) (64)	British Columbia (Canada)	0.3—18	3.3	W/C	WARREN and DELAVAULT (1962)
Grass, off traffic	Berlin (Germany)	6—8.5	7	(?)	KLOKE and RIEBARTSCH (1964)
Mammalian muscle		—	<0.2	(?)	TIPTON and COOK in: BOWEN (1966)
Mammalian heart, brain		0.2—0.24	0.22	(?)	TIPTON and COOK in: BOWEN (1966)
Mammalian kidney, liver		4.5—4.8	4.7	(?)	TIPTON and COOK in: BOWEN (1966)
Cattle, muscle (48)	Bavaria (Germany)	—	<0.02	A	KRACKE *et al.* (1973)
Cattle, kidney (52)	Bavaria (Germany)	0.1—0.57	0.33	A	KRACKE *et al.* (1973)
Cattle, liver (48)	Bavaria (Germany)	0.02—0.27	0.15	A	KRACKE *et al.* (1973)

[a] In ash.
[b] Computed, assuming an ash content of 40% in dry substance.
[c] Potential contamination of metal particles can contribute $^2/_3$ of this value.
[aa] Computed, assuming an ash content of 3.5% in dry substance.
[bb] Wet weight.

Haven with a few meters street distance contain 100 to 150 ppm Pb in dry weight of leaves (oak, maple, hemlock and spruce) (SMITH, 1972) in contrast to a natural background of 1.5 to 10 ppm Pb (Table 82-L-1).

Most terrestrial plants which are not exposed to easily observable lead pollution, contain lead in the range 0.2 to 20 ppm with abundant values close to 3 ppm Pb (Table 82-L-1).

Table 82-L-2. *Lead concentrations in coal (or coal ash) and crude oil* (Number of samples in brackets)

Samples	Origin	Range ppm Pb	Average ppm Pb	Method	Reference
Brown coal (47)	Tertiary, East Germany (G.D.R.)	<1—175	≤8.4	S	LEUTWEIN (1956)
Brown coal[b] (6)	Tertiary, Victoria (Australia)	<0.7—30	6	S	SWAINE (1971)
Hard brown coal[b] (43)	Triassic, South Australia	<2—50	10	S	SWAINE (1971)
Hard coal[b] (124)	Permian, Triassic, Jurassic, Queensland (Australia)	1.5—40	9	S	SWAINE (1971)
Hard coal[b] (151)	Permian, New South Wales (Australia)	1.5—60	10	S	SWAINE (1971)
Bituminous coal (143) (ashed at 400° C)	Pennsylvanian, Mesozoic, U.S.A.	3.8—170	11 (96)[a]	W/C, I	CHOW and EARL (1972)
Coal (378)	Carboniferous, East Germany (G.D.R.)	—	99 (1030)[a]	S	LEUTWEIN and RÖSLER (1956)
Coal (~ 150)	Cretaceous, Handlová Nováky (Č.S.S.R.)	(5—310)[a]	(~ 40)[a]	S	MECHAČEK (1972)
Crude oil (96)	Midcontinent, Rocky Mts., West Coast, U.S.A.	(<10 ppm to 30%)[a] (abundant values: 50—500)[a]	0.04 (735)[a, c]	S	HYDEN (1961)

[a] In ash.
[b] Clean coal samples.
[c] Excluding 6 values higher than 1%.

b) Animals

Lead is taken up by animals by ingestion and inhalation. Absorbed and carried by the blood, mammals accumulate Pb in liver, kidney (Table 82-L-1) and bones (12 ppm Pb; BOWEN, 1966). Animal muscles are usually low in lead.

Under average conditions man takes up lead at approximate daily rates of 300 to 400 μg. About 10% of the ingested lead is absorbed gastro-intestinally (LAGERWERFF, 1972). The major Pb proportion in the diet comes from vegetables and meat. Except for fish, aerial precipitates are a prime cause of dietary lead. The threshold concentration of Pb in diet is assumed to be near 600 μg/day. The degree of absorption of inhaled Pb by the blood depends largely on particle size. Particles <0.6 μ reach the alveoli and are deposited (about 75% particulate Pb from automobile combustion is $<1\ \mu$). The lead level in human blood and urine responds to exposure to extraordinary aerial Pb. A "normal" lead content seems to range between 10 and 20 µg Pb per 100 g blood (tentative threshold concentration: 30 µg Pb/100 g blood often approached by parking and gasoline station workers, policemen, etc.). A reasonable limit for aerial concentration of Pb is believed to be 2 to 10 $\mu g/m^3$ based on inhalation of 15 m^3 of air per day; under these conditions intake is expected to match excretion of Pb (GOLDSMITH and HEXTER, 1967).

Ingestion of toxic amounts of lead from sources beside those already mentioned can originate from plumbing, ceramics with lead pigments in kitchenware, leaded paint etc.

Lead poisoning, starting with convulsions and anemia, may proceed to peripheral nerve disease, gout, chronic nephritis, encephalophathy and death.

III. Coal and Oil

After degradation of plants and plankton some proportion of these residues can be accumulated in sediments and preserve some of their original lead. Secondary lead from solutions may be trapped by absorption, chelating reactions and sulfide precipitation.

Most coals (Table 82-L-2) contain about 10 ppm Pb; appreciably higher values are reported by LEUTWEIN and RÖSLER (1956). These authors mention a decrease of Pb with increasing ash content. YUDOVICH *et al.* (1972) in their compilation on trace elements in world coal, mention a weighted mean of 59 ppm Pb. If the factor of carbon concentration from wood to coal (2 to 3) is used to extrapolate a lead value for coal from that in Recent wood and plants, the estimate is about 10 ppm Pb. A yearly combustion of $3 \cdot 10^9$ tons coal (containing 10 ppm Pb on average) will contribute $3 \cdot 10^4$ tons Pb to soils, waters, etc.; this is less than a tenth of the pollution by automobile exhaust.

Revised manuscript received: April 1974

82-M. Abundance in Common Metamorphic Rock Types

Because of its crystal chemical properties, lead in metamorphic rocks is mainly incorporated in mica and potassium feldspar (see Section 82-D). No appreciable difference can be observed between the mean lead of metamorphic and granitic potassium-feldspars and biotites, listed in Table 82-D-1. Muscovite from schists may be higher in lead than that from granites, but the number of samples used for this observation is still small. 50 scapolites, mainly from Precambrian Grenville gneisses and skarns, contain on average 45 ppm Pb (<9—110 ppm Pb; method: S) (SHAW, 1960).

Metamorphic minerals low in potassium, such as staurolite, garnet and chlorite, usually contain less or much less than 10 ppm Pb (WEDEPOHL, 1956).

The most abundant metamorphic rocks are *gneisses* and *schists*. As can be seen in Table 82-M-1 their lead values and lead means for certain complexes scatter over a large range. An average of 3846 samples of gneisses and schists is 17 ppm Pb. Averaging seems to be influenced by sets with large numbers of samples, such as those from the Canadian Shield and Scotland. However, if we give all sets listed in Table 82-M-1 equal statistical weight, their overall mean is 19.6 ppm Pb, which is close to the former average. An average of all rocks of the Canadian Shield, weighted according to size of structural provinces and sampled by several thousand specimens during reconnaissance geological mapping, is 18 ppm Pb (EADE and FAHRIG, 1973). It seems that migmatites are systematically lower in lead than normal gneisses. 142 granulites contain on average 18.7 ppm Pb. In areas where amphibolite facies

Table 82-M-1. *Lead concentrations in metamorphic rocks* (Number of samples in brackets)

Rock type	Origin	Range ppm Pb	Average ppm Pb	Method	Reference
a) Serpentinite (2)	Queensland (Australia)	—	6.9	A	ZEISSING (1971)
b) Eclogite, (mantle nodules) (5)	Roberts Victor Mine, South Africa	0.79—5.5	2.5	I	MANTON and TATSUMOTO (1971)
Eclogite (9)	Kokchetav Massif	0.1—3.0	1.05	I (?)	NAYDENOV and YEFIMOV (1968)
Eclogite (7)	Münchberg Massif, Bavaria (Germany)	0.56—1.62	0.98	I	MARSHALL (1960)

Table 82-M-1 (continued)

Rock type	Origin	Range ppm Pb	Average ppm Pb	Method	Reference
c) Amphibolite (16)	Adirondack Mts., New York (U.S.A.)	—	7.6	W/C	ENGEL *et al.* (1964)
Amphibolite (32)	Coll, Tiree (NW Scotland)	—	16	X	DRURY (1973)
Amphibolite (41)	Dalradian (Scotland)	7—82	20.4	X	VAN DE KAMP (1970)
d) Hornfels (44)	Nevada (U.S.A.)	6—25	17.9	S	WODZICKI (1971)
e) Gneiss (amphibolite facies) (72)	Australian Shield	—	29.6	X	LAMBERT and HEIER (1968)
Gneiss, schist (para) (404)	Canadian Shield	—	20	S	EADE and FAHRIG (1973)
Gneiss, schist (158)	Canadian Shield	—	11	S	EADE and FAHRIG (1973)
Gneiss (banded), migmatite, granite gneiss (2621)	Canadian Shield	—	17	S	EADE and FAHRIG (1973)
Gneiss, schist (24)	Devonian, New Hampshire (U.S.A.)	25—65	44.5	A	DETHIER and SCHLESINGER (1973)
Gneiss, schist (30)	Devonian, New Hampshire (U.S.A.)	5—61	27.3	S	SHAW (1954)
Gneiss (73)	Adirondack Mts., New York (U.S.A.)	6—25	12	S	ENGEL and ENGEL (1958)
Granitic gneiss (5)	Adirondack Mts. (U.S.A.)	23—41	32.2	S	ENGEL and ENGEL (1958)
Gneiss (5)	Llano Uplift, Texas (U.S.A.)	9.83—2.7	19.9	I	ZARTMAN (1965)
Gneiss (2)	Hida Complex, Honshu (Japan)	18.9—20	19.5	A	MIYAZAKI *et al.* (1973)
Schist (pelitic) (7)	Sanbagawa Belt, Nagano Pref. (Japan)	12—21	16	X	BANNO and CHAPPELL (1969)
Schist (16)	Damara (SW Africa)	0.4—25	6.6	A	HEINRICHS (1974)
Gneiss, schist (23)	Ruhla, Thuringian Mts. (G.D.R.)	<10—150	≤31.7	S	WERNER (1971b)
Gneiss, migmatitic (23)	Ruhla, Thuringian Mts. (G.D.R.)	5—35	14.9	S	WERNER (1971a)
Augengneiss (13)	Ruhla, Thuringian Mts. (G.D.R.)	<3—30	≤10.8	S	WERNER (1971a)

Table 82-M-1 (continued)

Rock type	Origin	Range ppm Pb	Average ppm Pb	Method	Reference
Gneiss, migmatitic (14)	Ruhla, Thuringian Mts. (G.D.R.)	<5—36	≤19.2	S	Werner (1970)
Gneiss, metablastic (11)	Ruhla, Thuringian Mts. (G.D.R.)	<3—38	≤8.5	S	Werner (1970)
Gneiss (garnet) (14)	Coll, Tiree (NW Scotland)	—	21	X	Drury (1973)
Gneiss, migmatitic (26)	Coll, Tiree (NW Scotland)	—	15	X	Drury (1973)
Gneiss (254)	Lewisian, Sutherland (NW Scotland)	<4—37	13	X	Sheraton (1970)
Gneiss (28)	Lewisian (NW Scotland)	0.15—16.1	5.5	I	Moorbath *et al.* (1969)
Schist (chlorite, biotite, garnet, staurolite) (17)	Dalradian (Scotland)	6—41	23.9	X	Van de Kamp (1970)
Gneiss (amphibolite facies) (22)	Lofoten-Vesterålen (N Norway)	—	19.7	X	Heier and Thoresen (1971)
f) Granulite (20)	Lofoten-Vesterålen (N Norway)	—	15.1	X	Heier and Thoresen (1971)
Granulite (8)	Tiree (NW Scotland)	—	15	X	Drury (1973)
Granulite (7)	Lewisian (NW Scotland)	0.5—15.8	4.8	I	Moorbath *et al.* (1969)
Granulite (99)	Australian Shield	—	23	X	Lambert and Heier (1968)
Granulite (8)	Tomkinson Ranges (Australia)	21.2—53.2	42.3	I	Gray and Oversby (1972)
g) Marble (5)	Precambrian, S Rhodesia, Finland	0.6—1.4	0.96	I	Wampler and Kulp (1962)
Marble (3)	Nevada (U.S.A.)	5—8	6.9	S	Wodzicki (1971)

gneisses grade into granulites, the former rocks are usually higher in lead than the latter (Moorbath *et al.*, 1969; Heier and Thoresen, 1971; Eade and Fahrig, 1973; Drury, 1973).

The average lead content of shales (23 ppm Pb) and granitic rocks (23 ppm) is only slightly higher than that of gneisses and schists.

Eclogites are as low in lead as basaltic rocks.

Revised manuscript received: April 1974

82-N. Behavior in Metamorphic Reactions

The behavior of lead during contact metamorphism has been investigated at eight granodioritic stocks and their surrounding hornfels in Nevada (U.S.A.) by WODZICKI (1971). The metapelites are higher in lead than the intrusives. Within a contact aureole of a few meters some lead is lost from the pelite which might have penetrated the granodiorite.

The evidence for lead mobilization during low and medium grade regional metamorphism is controversial.

In the metasedimentary greenschists and amphibolites of the Scottish Dalradian Series, VAN DE KAMP (1970) could not detect a systematic change of the lead concentration with increasing metamorphic grade from the biotite to the staurolite isograde. SHAW (1954) observes 16 ppm Pb in low-grade, 23 ppm Pb in medium-grade and 27 ppm Pb in high-grade rocks of the Devonian Littleton Formation. HEINRICHS (1974) reports lead data (Method: A) of mica schists formed from Triassic marls in the Swiss Alps (Lukmanier Road) in a temperature gradient. Lead increases from greenschist to high amphibolite facies from 4.2 ppm Pb (3 samples) to 17.5 ppm Pb (5 samples). Another series of 6 mica schists and 3 gneisses representing progressive metamorphism in Central France (Cevennes) from greenschist to amphibolite facies shows a comparable increase of lead from 3.3 ppm Pb (3 samples) to 14.6 ppm Pb (6 samples). ENGEL and ENGEL (1960) in their study of potential migration of elements in the NW Adirondack Mountains (U.S.A.), state that they have no doubt about the release of large volumes of Ba, B, Pb, Zn and Mn during regional metamorphism (in the case of lead about 10^4 tons per cubic kilometer rock, which equals 3 ppm Pb).

Lead isotopes can indicate mobilization of this element during regional metamorphism as in the case of a Paleozoic metamorphism of Precambrian gneiss in Maryland (U.S.A.) (DOE *et al.*, 1965). GRAY and OVERSBY (1972) assume that they observed a single-stage lead system in granulites from the Australian Shield which excludes premetamorphic fractionation of U from Pb. This is followed by a metamorphic depletion of uranium. Reduction of U/Pb ratios during granulite facies metamorphism and occurrence of unradiogenic lead in high grade gneisses of Precambrian shields, seems to be a general observation (see Section 82-B, MOORBATH *et al.*, 1969, etc.).

In Section 82-M it has been mentioned that in areas where amphibolite facies rocks grade into granulites, the former rocks are usually higher in lead than the latter (Scotland: MOORBATH *et al.*, 1969; DRURY, 1973; Norway: HEIER and THORESEN, 1971). The results from New Quebec in the Canadian Shield (EADE and FAHRIG, 1973) suggest that Pb is mobile under conditions of temperature and pressure accompanying high-grade regional metamorphism. Lead seems to be lost from granulite rocks connected with the mobilization of some potassium. Pb and Zn are significantly lower in the Archean than in the Proterozoic rocks from the Canadian Shield.

Revised manuscript received: April 1974

82-O. Economic Importance

Lead is one of the oldest metals used in human culture because of its properties and the low energy consumption to reduce its oxides. The oldest lead products date from 7,000 B.C. and have been discovered at Çatal Hüyük.

World production is $3.35 \cdot 10^6$ tons per year (1970). Until 1970 $1.4 \cdot 10^8$ tons of Pb have been produced. A large increase of the yearly production occurred during industrialization. Before 1800 the total lead consumption was less than twice the current yearly production. The 4 largest producers of lead ore are: U.S.A. ($5.2 \cdot 10^5$ tons Pb per year), U.S.S.R. ($4.5 \cdot 10^5$ tons Pb), Australia ($4.0 \cdot 10^5$ tons Pb), and Canada ($3.9 \cdot 10^5$ tons Pb). In the 10^5 tons Pb per year range run Mexico, Peru, Yugoslavia, China and Bulgaria (KUBACH and MÜLLER, 1972).

About 50 percent of the current lead mining production originates from vein (and replacement) type deposits, 30 percent from Mississippi-Valley type stratiform ores in carbonate rocks, and the remainder mainly from hydrothermal bedded deposits. The average crustal lead dispersed in rocks is close to 15 ppm Pb; this is a total of $4 \cdot 10^{14}$ tons in comparison with $X \cdot 10^8$ tons economically available lead. If the economic lead would be dispersed in crustal rocks, its concentration is expected to be 0.01 ppb.

Seven large industrial countries (U.S.A., Japan, Fed. Rep. Germany, Canada, U.K. and Italy) consume 60% of the world lead production. The present use is the following: about 36% in batteries, 16% in cast materials, 13% in gasoline additives, 11% in cable mantles, 11% in chemical pigments, and 8% in alloy metals.

Revised manuscript received: April 1974

References: Section 82-C to 82-O

ABDULLAH, M. I., ROYLE, L. G., MORRIS, A. W.: Heavy metal concentration in coastal waters. Nature **235**, 158 (1972).

AHRENS, L. H.: The lognormal distribution of the elements. Geochim. Cosmochim. Acta **5**, 49 (1954).

AHRENS, L. H., WILLIS, J. P., OOSTHUIZEN, C. O.: Further observations on the composition of manganese nodules, with particular reference to some of the rarer elements. Geochim. Cosmochim. Acta **31**, 2169 (1967).

ANDERSON, G. M.: The solubility of PbS in H_2S-water solutions. Econ. Geol. **57**, 809 (1962).

ANDERSON, G. M.: The hydrothermal transport and deposition of galena and sphalerite near 100° C. Econ. Geol. **68**, 480 (1973).

AZZARIA, L. M.: A study of the distribution of traces of copper, lead and zinc in the minerals of a Precambrian granite. Can. Mineralogist **7**, 617 (1963).

BANAT, K., FOERSTNER, U., MÜLLER, G.: Schwermetalle in Sedimenten von Donau, Rhein, Ems, Weser und Elbe im Bereich der Bundesrepublik Deutschland. Naturwissenschaften **59**, 525 (1972).

BANNO, S., CHAPPELL, B. W.: X-ray fluorescent analysis of Rb, Sr, Y, Pb and Th in Japanese Paleozoic slates. Geochem. J. **3**, 127 (1969).

BARAGAR, W. R. A.: The geochemistry of Coppermine River basalts. Geol. Surv. Canada Paper **69**, **44**, 1 (1969).

BARANOVA, N. N., BARSUKOV, V. L.: Transport of lead by hydrothermal solutions in the form of carbonate complexes. Geochemistry **9**, 1093 (1965).

BARNES, H. L., CZAMANSKE, G. K.: Solubilities and transport of ore minerals. Chap. 8. In: BARNES, H. L. (Ed.): Geochemistry of Hydrothermal Ore Deposits. New York: Holt, Rinehart and Winston 1967.

BENDA, L.: Über die Anreicherung von Uran und Thorium in Phosphoriten und Bonebeds des nördlichen Harzvorlandes. Geol. Jahrb. **80**, 313 (1963).

BENDER, M., BROECKER, W., GORNITZ, V., MIDDEL, U., KAY, R., SUN, S.-S., BISCAYE, P.: Geochemistry of three cores from the East Pacific Rise. Earth Planet. Sci. Letters **12**, 425 (1971).

BLACK, W. A. P., MITCHELL, R. L.: Trace elements in the common brown algae and in sea water. J. Mar. Biol. Assoc. U.K. **30**, 575 (1952).

BORN, H. J.: Der Bleigehalt der norddeutschen Salzlager und seine Beziehungen zu radioaktiven Fragen. Chem. Erde **9**, 66 (1934).

BOWEN, H. J. M.: Trace Elements in Biochemistry. London: Academic Press 1966.

BRADFORD, G. R., BAIR, F. L., HUNSKER, V.: Trace and major element content of 170 High Sierra lakes in California. Limnol. Oceanogr. **13**, 526 (1968).

BRÄUER, H.: Spurenelementgehalte in granitischen Gesteinen des Thüringer Waldes und des Erzgebirges. Freiberger Forschungsh. C **259**, 83 (1970).

BRETT, R., KULLERUD, G.: The Fe-Pb-S system. Econ. Geol. **62**, 354 (1967).

BROOKS, R. R., RUMSBY, M. G.: The biogeochemistry of trace element uptake by some New Zealand bivalves. Limnol. Oceanogr. **10**, 521 (1965).

BROWN, H. R., SWAINE, D. J.: Inorganic constituents of Australian coals. J. Inst. Fuel **37**, 422 (1964).

BRUNO, E., FACCHINELLI, A.: Al, Si configurations in lead feldspar. Z. Kristallogr. **136**, 296 (1972).

BRUTY, D., CHESTER, R., ASTON, S. R.: Trace elements in ancient Atlantic deep-sea sediments. Nature (Phys. Sci.) **245**, 73 (1973).

Bukenberger, U., Lodemann, C. K. W., Loeschke, J.: Die Verteilung der Schwermetalle in ober- und unterirdischen Wässern sowie in den Böden des Neckartales oberhalb Tübingen. Oberrhein. Geol. Abh. **21**, 43 (1972).

Butler, J. R.: The geochemistry and mineralogy of rock weathering. 1. The Lizard area, Cornwall. Geochim. Cosmochim. Acta **4**, 157 (1953).

Cadigan, R. A.: Geochemical distribution of some metals in the Moenkopi formation and related strata, Colorado Plateau region. U.S. Geol. Surv. Bull. **1344**, 1 (1971).

Cameron, A. G. W.: A new table of abundances of the elements in the Solar System. In: Ahrens, L. H. (Ed.): Origin and Distribution of the Elements. Oxford: Pergamon Press 1968.

Cannon, H. L.: Botanical prospecting for ore deposits. Science **132**, 591 (1960).

Cannon, H. L., Shacklette, H. T., Bastron, H.: Metal absorption by Equisetum (horsetail). U.S. Geol. Surv. Bull. **1278**-A, 1 (1968).

Carmichael, J., McDonald, A.: The geochemistry of some natural acid glasses from the North Atlantic Tertiary volcanic province. Geochim. Cosmochim. Acta **25**, 189 (1961).

Catanzaro, E. J., Gast, P. W.: Isotopic composition of lead in pegmatitic feldspars. Geochim. Cosmochim. Acta **19**, 113 (1960).

Čech, F., Mísař, Z., Povondra, P.: A green lead-containing orthoclase. Tschermaks Mineral. Petrog. Mitt. **15**, 213 (1971).

Chang, L. L. Y., Bever, J. E.: Lead sulfosalt minerals: crystal structures, stability relations and paragenesis. Miner. Sci. Eng. **5**, 181 (1973).

Chao, E. C. T.: The petrographic and chemical characteristics of tektites. In: O'Keefe, J. A. (Ed.): Tektites. Chicago: Univ. Chicago Press 1963.

Chawla, V. K., Chan, Y. K.: Trace elements in lake Erie. Proc. 12th Conf. Great Lakes Res. 760 (1969).

Chester, R., Elderfield, H.: Dust over the oceans. New Scientist **27**, 432 (1970).

Chester, R., Hughes, M. J.: The trace element geochemistry of a North Pacific pelagic clay core. Deep Sea Res. **16**, 639 (1969).

Chester, R., Stoner, J. H.: Average trace element composition of low level marine atmospheric particulates. Nature **246**, 138 (1973).

Chow, T. J., Bruland, K. W., Bertine, K., Soutar, A., Koide, M., Goldberg, E. D.: Lead pollution: Records in southern California coastal sediments. Science **181**, 551 (1973).

Chow, T. J., Earl, J. L.: Lead isotopes in American coals. Science **176**, 510 (1972).

Chow, T. J., Earl, J. L., Bennett, C. F.: Lead aerosols in the marine atmosphere. Envir. Sci. Technol. **3**, 737 (1969).

Chow, T. J., Patterson, C. C.: Lead isotopes in manganese nodules. Geochim. Cosmochim. Acta **17**, 21 (1959).

Chow, T. J., Patterson, C. C.: The occurrence and significance of lead isotopes in pelagic sediments. Geochim. Cosmochim. Acta **26**, 263 (1962).

Church, S. E.: Limits of sediment involvement in the genesis of orogenic volcanic rocks. Contr. Mineral. and Petrol. **39**, 17 (1973).

Church, S. E., Tilton, G. R.: Lead and strontium isotopic studies in the Cascade Mountains: bearing on andesite genesis. Geol. Soc. Am. Bull. **84**, 431 (1973).

Clark, S. P., Jr., Peterman, Z. E., Heier, K. S.: Abundances of uranium, thorium and potassium. In: Clark, S. P., Jr. (Ed.): Handbook of Physical Constants. Geol. Soc. Am. Memoir **97**, 521 (1966).

Clifford, T. N., Rooke, J. M., Allsopp, H. L.: Petrochemistry and age of the Franzfontein granitic rocks of northern South-West Africa. Geochim. Cosmochim. Acta **33**, 973 (1969).

Cobb, J. C.: Determination of lead in meteorites by alpha activation analysis. J. Geophys. Res. **69**, 1895 (1964).

Cooper, J. A., Green, D. H.: Lead isotope measurements on lherzolite inclusions and host basanites from western Victoria, Australia. Earth Planet. Sci. Letters **6**, 69 (1969).

Cooper, J. A., Richards, J. R.: Isotopic and alkali measurements from the Vema Seamount of the South Atlantic Ocean. Nature **210**, 1245 (1966).

CORRENS, C. W.: Adsorptionsversuche mit sehr verdünnten Kupfer- und Bleilösungen und ihre Bedeutung für die Erzlagerstättenkunde. Kolloid.-Z. **34**, 341 (1924).

CRONAN, D. S.: Average abundances of Mn, Fe, Ni, Co, Cu, Pb, Mo, V, Cr, Ti and P in Pacific pelagic clays. Geochim. Cosmochim. Acta **33**, 1562 (1969).

ČUTURIČ, N., KAFOL, N., KARAMATA, S.: Lead contents in K-feldspars of young igneous rocks of the Dinarides and neighbouring areas. In: AHRENS, L. H. (ed.), Origin and Distribution of the Elements. Oxford: Pergamon Press 1966.

ČUTURIČ, N., KARAMATA, S.: Die Bleigehalte in tertiären magmatischen Gesteinen Jugoslawiens. Geol. Sbornik **18**, 27 (1967).

CUTTITTA, F., CARRON, M. K., ANNELL, C. S.: New data on selected Ivory Coast tektites. Geochim. Cosmochim. Acta **36**, 1297 (1972).

DASCH, E. J., HEATH, G. R., DYMOND, J. R.: Isotopic analysis of metalliferous sediment from the East Pacific Rise. Abstracts with Programs G.S.A. **3**, 7 (1971).

DELEON, G., AHRENS, L. H.: The distribution of Li, Rb, Cs and Pb in some Yugoslav granites. Geochim. Cosmochim. Acta **12**, 94 (1957).

DEMOVIČ, R.: Die Geochemie sedimentärer Karbonatgesteine. Die Distribution von Ba, Ti, Mn, Cu, Ag, Pb, Ni, Cr, B, V, Y. Geol. Zbornik Geol. Carpath. **24**, 339 (1973).

DETHIER, D. P., SCHLESINGER, W. H.: The copper, lead and zinc content of metamorphic rocks on Mt. Moosilauke, New Hampshire. Ohio J. Sci. **73**, 58 (1973).

DE VORE, G. W.: Crystal growth and the distribution of elements. J. Geol. **63**, 47 (1955).

DOE, B. R.: The bearing of lead isotopes on the source of granitic magma. J. Petrol. **8**, 51 (1967).

DOE, B. R.: Lead Isotopes. Berlin-Heidelberg-New York: Springer 1970.

DOE, B. R., CHRISTIANSEN, R. L., HEDGE, C. E.: Radiogenic tracers and the basalt-rhyolite association, Yellowstone National Park and vicinity. Abstracts with Progr. **2**, 538 (1970).

DOE, B. R., DELEVAUX, M. H.: Source lead in Southeast Missouri galena ores. Econ. Geol. **67**, 409 (1972).

DOE, B. R., HART, S. R.: The effect of contact metamorphism on lead in potassium feldspars near the Eldorado Stock, Colorado. J. Geophys. Res. **68**, 3521 (1963).

DOE, B. R., LIPMAN, P. W., HEDGE, C. E.: Primitive and contaminated basalts from the southern Rocky Mountains, USA. Contr. Mineral. and Petrol. **21**, 142 (1969a).

DOE, B. R., LIPMAN, P. W., HEDGE, C. E.: Radiogenic tracers and the source of continental andesites: a beginning at the San Juan volcanic field, Colorado. State Oregon Dept. Geol. Mineral. Industries Bull. **65**, 143 (1969b).

DOE, B. R., TILTON, G. R., HOPSON, C. H.: Lead isotopes in feldspars from selected granitic rocks associated with regional metamorphism. J. Geophys. Res. **70**, 1947 (1965).

DOE, B. R., TILLING, R. I.: The distribution of lead between coexisting K-feldspar and plagioclase. Am. Mineralogist **52**, 805 (1967).

DROZDOVA, V. M., MAKHONKO, E. P.: Content of trace elements in precipitation. J. Geophys. Res. **75**, 3610 (1970).

DRURY, S. A.: The geochemistry of Precambrian granulite facies rocks from the Lewisian complex of Tiree, Inner Hebrides, Scotland. Chem. Geol. **11**, 167 (1973).

DURASOVA, N. A.: Partition coefficients of tin and lead in the system $NaCl$-Na_2O-Al_2O_3-CaO-SiO_2. Geochem. Intern. **9**, 595 (1969/1970).

DURUM, W. H., HEM, J. D., HEIDEL, S. G.: Reconnaissance of selected minor elements in surface waters of the United States, October 1970. U.S. Geol. Surv. Circ. **643**, 1 (1971).

EADE, K. E., FAHRIG, W. F.: Regional, lithological and temporal variation in the abundances of some trace elements in the Canadian Shield. Geol. Surv. Canada Paper **72—46**, 1 (1973).

EDWARDS, G., HESS, D. C.: Isolation and isotopic analysis of lead in meteorites and rocks. Nuclear Processes in Geologic Settings. Proc. Second Conf. Nat. Acad. Sci. Nat. Res. Counc. **400**, 100 (1956).

ELLIS, A. J.: Natural hydrothermal systems and experimental hot-water/rock interaction: Reaction with NaCl solutions and trace metal extraction. Geochim. Cosmochim. Acta **32**, 1356 (1968).

EMERY, K. O., HUNT, J. M., HAYS, E. E.: Summary of hot brines and heavy metal deposits in the Red Sea. In: DEGENS, E. T., ROSS, D. A. (Eds.): Hot Brines and Recent Heavy Metal Deposits. New York: Springer Verlag 1969.

ENGEL, A. E. J., ENGEL, C. G.: Progressive metamorphism and granitization of the major paragneiss, Northwest Adirondack Mountains, New York. Bull. Geol. Soc. Am. **69**, 1369 (1958).

ENGEL, A. E. J., ENGEL, C. G.: Migration of elements during metamorphism in the NW Adirondack Mountains, New York. U.S. Geol. Surv. Prof. Papers **400 B**, 465 (1960).

ENGEL, A. E. J., ENGEL, C. G., HAVENS, R. G.: Mineralogy of amphibolite interlayers in the gneiss complex, Northwest Adirondack Mountains, New York. J. Geol. **72**, 131 (1964).

EWART, A., TAYLOR, S. R., CAPP, A. C.: Trace and minor element geochemistry of the rhyolitic volcanic rocks, Central North Island, New Zealand. Contr. Mineral. and Petrol. **18**, 76 (1968).

FAIRBAIRN, H. W., AHRENS, L. H., GORFINKLE, L. G.: Minor element content of Ontario diabase. Geochim. Cosmochim. Acta **3**, 34 (1953).

FERSHTATER, G. B., BORODINA, N. S., TRAJANOVA, M. V.: Lithium, rubidium, strontium and lead in the granitoids of the Urals. Geochem. Intern. **6**, 44 (1969).

FLANAGAN, F. J.: 1972 values for international geochemical reference samples. Geochim. Cosmochim. Acta **37**, 1189 (1973).

FLANAGAN, F. J., GWYN, M. E.: Sources of geochemical standards. Geochim. Cosmochim. Acta **31**, 1211 (1967).

FLEISCHER, M.: Minor elements in some sulfide minerals. Econ. Geol. **50**, 970 (1955).

GALACHOWSKAJA, T. W.: On the spectroanalytical method for small amounts of Mn, Cu, Pb, Fe and Ni in the soluble fraction of halite, sylvinite salts. Papers of the Research Institute for Halurgy of the USSR **27**, 230 (1953).

GARRELS, R. M., CHRIST, C. L.: Solutions, Minerals and Equilibria. New York: Harper's Geoscience Series, Harper and Row 1965.

GAVRILIN, R. D., PEVTSOVA, L. A.: Behavior of lead and zinc in phase and facies differentiation of magma. Geochemistry **8**, 764 (1963).

GAVRILIN, R. D., PEVTSOVA, L. A., KLASSOVA, N. S.: Lead and zinc in the differentiates of sodic granitic magma. Geochem. Intern. **2**, 783 (1965).

GAVRILIN, R. D., PEVTSOVA, L. A., KLASSOVA, N. S.: Behavior of lead and zinc in the hydrothermal alteration of intrusive rocks. Geochem. Intern. **8**, 790 (1967).

GAVRILIN, R. D., VOLKOV, V. N., NEGREY, YE. V., PEVTSOVA, L. A.: The primary vertical distribution of lead and zinc in the upper part of a granite intrusion. Geochem. Intern. **9**, 630 (1972).

GEOLOGICAL SURVEY OF GREAT BRITAIN: The limestones of Scotland, chemical analyses and petrography. Spec. Rept. Miner. Resources Great Brit. **37**, 1 (1965).

GERASIMOVSKII, V. I., NESMEYANOVA, L. I.: Lead and zinc distribution in the rocks of the Lovozero Massif. Geochemistry **7**, 704 (1960).

GINSBURG, I. I.: Grundlagen und Verfahren geochemischer Sucharbeiten auf Lagerstätten der Buntmetalle und seltenen Metalle. Berlin: Akademie Verlag 1963.

GOLDBERG, E. D., ARRHENIUS, G. O. S.: Chemistry of Pacific pelagic sediments. Geochim. Cosmochim. Acta **13**, 153 (1958).

GOLDSMITH, J. R., HEXTER, A. C.: Respiratory exposure to lead: epidemiological and experimental dose-response relationship. Science **158**, 132 (1967).

GOLEVA, G. A., POLYAKOY, V. A., NECHAYEVA, T. P.: Distribution and migration of lead in ground waters. Geochem. Intern. **7**, 256 (1970).

GOLUBCHINA, M. N., RABINOVICH, A. V.: Criteria of relationship between mineralization and magmatism based on the isotopic analysis of lead in the country rock and in the ore. Geochemistry **3**, 238 (1957).

GORDON, M., MURATA, K. J.: Minor elements in Arkansas bauxite. Econ. Geol. **47**, 169 (1952).

GRAESER, S.: Isotopic composition of lead in some basic and ultrabasic rocks from the alps. Earth Planet. Sci. Letters **6**, 491 (1969).

GRAY, C. M., OVERSBY, V. M.: The behavior of lead during granulite facies metamorphism. Geochim. Cosmochim. Acta **36**, 938 (1972).

GRUNDLACH, H., KARL, F., MÜLLER, G.: Vergleichende geochemische Untersuchungen an ost- und südalpinen Graniten, Granodioriten und Tonaliten. Contr. Mineral. and Petrol. **16**, 285 (1967).
HAACK, U. K.: Spurenelemente in Biotiten aus Graniten und Gneisen. Contr. Mineral. and Petrol. **22**, 83 (1969).
HAHN-WEINHEIMER, P.: Quantitative spektrochemische Bestimmung von Spurenelementen in Karbonatgesteinen nach der Doppelbogenmethode (SHAW, JOENSUU und AHRENS) und der Verdampfung aus der Ringrinne (HOLDT). Proc. Colloqu. Spectr. Internat. **9**, 594 (1961).
HARTMANN, M.: Einige geochemische Untersuchungen an Sandsteinen aus Perm und Trias. Geochim. Cosmochim. Acta **27**, 459 (1963).
HEDGE, C. E., KNIGHT, R. J.: Lead and strontium isotopes in volcanic rocks from northern Honshu, Japan. Geochem. J. **3**, 15 (1969).
HEIDE, F., LERZ, H.: Zur Geochemie des Bleies. Chem. Erde **17**, 217 (1955).
HEIDE, F., LERZ, H., BÖHM, G.: Gehalt des Saalewassers an Blei und Quecksilber. Naturwissenschaften **44**, 441 (1957).
HEIER, K. S., BRUNFELT, A. O.: Concentration of Cs in high grade metamorphic rocks. Earth Planet. Sci. Letters **9**, 416 (1970).
HEIER, K. S., PALMER, P. D., TAYLOR, S. R.: Comment on the Pb distribution in southern Norwegian Precambrian alkali feldspars. Norsk Geol. Tidskr. **47**, 185 (1967).
HEIER, K. S., THORESEN, K.: Geochemistry of high grade metamorphic rocks, Lofoten-Vesterålen, North Norway. Geochim. Cosmochim. Acta **35**, 89 (1971).
HEINRICHS, H.: Die Untersuchung von Gesteinen und Gewässern auf Cd, Sb, Hg, Tl, Pb, Bi mit der flammenlosen Atomabsorption. Dr. rer. nat. Thesis, Göttingen, 1974.
HELGESON, H. C.: Geologic and thermodynamic characteristics of the Salton Sea geothermal system. Am. J. Sci. **226**, 129 (1965).
HELGESON, H. C.: Thermodynamics of hydrothermal systems at elevated temperatures and pressures. Am. J. Sci. **267**, 729 (1969).
HEM, J. D., DURUM, W. H.: Solubility and occurrence of lead in surface water. J. Am. Water Works Ass. **65**, 562 (1973).
HEMLEY, J. J., MEYER, C., HODGSON, C. J., THATCHER, A. B.: Sulfide solubilities in alteration-controlled systems. Science **158**, 1580 (1967).
HERRMANN, A. G.: Geochemische Untersuchungen an Kalisalzlagerstätten im Südharz. Freiberger Forschungsh. **C 43**, 1 (1958).
HESS, D. C., MARSHALL, R. R.: The isotopic compositions and concentrations of lead in some chondritic stone meteorites. Geochim. Cosmochim. Acta **20**, 284 (1960).
HIETANEN, A.: Distribution of Fe and Mg between garnet, staurolite, and biotite in aluminium rich schist in various metamorphic zones north of the Idaho batholith. Am. J. Sci. **267**, 422 (1969).
HIRST, D. M.: The geochemistry of modern sediments from the Gulf of Paria — II. The location and distribution of trace elements. Geochim. Cosmochim. Acta **26**, 1147 (1962).
HOFFMAN, G. L., DUCE, R. A., HOFFMAN, E. J.: Trace metals in the Hawaiian atmosphere. J. Geoph. Res. **77**, 5322 (1972).
HOFMEYR, P. K.: The abundances and distribution of some trace elements in some selected South African shales. Ph. D. Thesis, Capetown, 1971.
HOLLAND, H. D.: Granites, solutions and base metal deposits. Econ. Geol. **67**, 281 (1972).
HORN, D. R., HORN, B., DELACH, M. N.: Worldwide distribution and metal content of deep-sea manganese deposits. In: CHAVE, K. E., *et al.* (Eds.): Manganese Nodule Deposits in the Pacific. Honolulu, Hawaii: Workshop Proceedings 1972.
HOROWITZ, A.: The distribution of Pb, Ag, Sn, Tl and Zn in sediments on active ocean ridges. Mar. Geol. **9**, 241 (1970).
HÜGI, T.: Vergleichende petrologische und geochemische Untersuchungen an Graniten des Aarmassivs. Beitr. Geol. Karte Schweiz. N. Folge **94**, 1 (1956).
HYDEN, H. J.: Distribution of uranium and other metals in crude oils. U.S. Geol. Surv. Bull. **1100 B**, 17 (1961).

ISHIBASHI, M., UEDA, S., YAMAMOTO, Y.: Studies on the utilization of the shallow-water deposits. On the lead content of the shallow-water deposits. Rec. Ocean. Works Japan **4**, 111 (1960).

ISHIKAWA, H., BERMAN, S., YAGI, K.: Geochemical study of trace elements in the alkalic rocks of Nemuro Peninsula, Hokkaido, Japan. Geochem. J. **5**, 187 (1971).

JACK, R. N., CARMICHAEL, I. S. E.: The chemical "fingerprinting" of acid volcanic rocks. In: Short Contributions to California Geology. California Div. Mines and Geology Spec. Rept. **100**, 17 (1969).

JONASSON, I. R.: Migration of trace metals in snow. Nature **241**, 447 (1973).

KARAMATA, S.: Lead in sanidines from quartz-latites from the Zvečan area, Yugoslavia. Zavod za Geološka i Geofizička Istraživanja Ser. A. Bull. **27**, 267 (1969).

KESSON, S., PRICE, R. C.: The major and trace element chemistry of kaersutite and its bearing on the petrogenesis of alkaline rocks. Contr. Mineral. and Petrol. **35**, 119 (1972).

KLOKE, A., RIEBARTSCH, K.: Verunreinigung von Kulturpflanzen mit Blei aus Kraftfahrzeugabgasen. Naturwissenschaften **51**, 367 (1964).

KÖHLER, A.: Optische Untersuchungen an synthetischen Mischgliedern der Feldspatgruppe. Akad. Wiss. Wien, Math. Nat. Kl. **72**, 123 (1935).

KÖSTER, H. M.: Geochemische Vorgänge bei der Verwitterung von Silikatgesteinen. Acta Albertina Ratisbonensia **26**, 55 (1966).

KÖSTER, H. M.: Beitrag zur Geochemie der Kaoline. Proc. Intern. Clay Confer. Tokyo **1**, 273 (1969).

KOLBE, P., TAYLOR, S. R.: Major and trace element relationships in granodiorites and granites from Australia and South Africa. Contr. Mineral. and Petrol. **12**, 202 (1966).

KONSTANTINOV, M. M.: The problem of sedimentary lead zinc deposits. Dokl. Akad. Nauk SSSR **2**, 65 (1954).

KONTOROVICH, A. E., SADIKOV, M. A., SHVARTSEV, S. L.: Abundances of certain elements in the surface and groundwaters of the northwestern part of the Siberian platform. Dokl. Akad. Nau SSSR **149**, 168 (1963).

KRACKE, W., DIETL, F., SANSONI, B.: Blei- und Cadmiumgehalte von Bewuchsproben in verkehrsnaher und verkehrsferner Lage. Jahresbericht 1972 Gesellschaft für Strahlen- und Umweltforschung, München, 131 (1973).

KRACKE, W., DIETL, F., TSCHÖPP, V., SANSONI, B.: Blei- und Cadmiumgehalt von Fleisch und Organen von Rindern 1972. Jahresbericht 1972 Gesellschaft für Strahlen- und Umweltforschung, München, 135 (1973).

KRAUSKOPF, K. B.: Factors controlling the concentrations of thirteen rare metals in sea-water. Geochim. Cosmochim. Acta **9**, 1 (1956).

KRAUSKOPF, K. B.: The heavy metal content of magmatic vapor at 600° C. Econ. Geol. **52**, 786 (1957).

KUBACH, I., MÜLLER, W.: Blei (47), Teil A 3 (Mineralien, Lagerstätten). In: Gmelin's Handbuch der anorganischen Chemie. (8. ed.) Weinheim: Verlag Chemie 1972.

KURASAWA, H.: Isotopic composition of lead and concentrations of uranium, thorium and lead in volcanic rocks from Dogo of the Oki Islands, Japan. Geochem. J. **2**, 11 (1968).

KURODA, R., GORAI, M.: Geochemical investigations of granites. Distribution of Pb and Ni. J. Chem. Soc. Japan **77**, 1129 (1956) [in Japanese].

LAGERWERFF, J. V.: Lead, mercury and cadmium as environmental contaminants. Micronutrients in Agriculture. Madison: Soil Science Society America 1972.

LAGERWERFF, J. V., BROWER, D. L.: Exchange adsorption or precipitation of lead in soils treated with chlorides of aluminum, calcium and sodium. Soil Sci. Soc. Am. Proc. **37**, 12 (1973).

LAMBERT, I. B., HEIER, K. S.: The vertical distribution of uranium, thorium and potassium in the continental crust. Geochim. Cosmochim. Acta **31**, 377 (1967).

LAMBERT, I. B., HEIER, K. S.: Geochemical investigations of deep-seated rocks in the Australian Shield. Lithos **1**, 30 (1968).

LANDERGREN, S., MANHEIM, F. T.: Über die Abhängigkeit der Verteilung von Schwermetallen von der Fazies. Fortschr. Geol. Rheinld. Westf. **10**, 173 (1963).

LANGE, J.: Geochemische Untersuchungen an pelagischen Sedimenten des Atlantischen und Pazifischen Ozeans (DSDP Leg I—VII). Dr. rer. nat. Thesis, Göttingen, 1974.

Larimer, J. W.: Chemical fractionations in meteorites — VII. Cosmothermometry and cosmobarometry. Geochim. Cosmochim. Acta **37**, 1603 (1973).
Larimer, J. W., Anders, E.: Chemical fractionations in meteorites — II. Abundance patterns and their interpretation. Geochim. Cosmochim. Acta **31**, 1239 (1967).
Latimer, W. M.: Oxidation Potentials (2nd ed.). New York: Prentice Hall 1953.
Lebedev, L. M.: Contemporary deposition of native lead from hot Cheleken thermal brines. Dokl. Acad. Sci. USSR Earth Sci. Sect. **174**, 173 (1967).
Leland, H. V., Shukla, S. S., Shimp, N. F.: Factors effecting distribution of lead and other trace elements in sediments of southern Lake Michigan. In: Singer, P. C. (Ed.): Trace Metals and Metal-organic Interactions in Natural Waters. Ann Arbor: Ann Arbor Science 1973.
Leutwein, F.: Untersuchungen über das Vorkommen von Spurenmetallen in Torfen und Braunkohlen. Freiberger Forschungsh. **C 30**, 28 (1956).
Leutwein, F., Rösler, H. J.: Geochemische Untersuchungen an paläozoischen und mesozoischen Kohlen Mittel- und Ostdeutschlands. Freiberger Forschungsh. **C 19**, 1 (1956).
Leutwein, F., Waskowiak, R.: Geochemische Untersuchungen an rezenten marinen Molluskenschalen. Neues Jahrb. Mineral. Abhandl. **99**, 45 (1962).
Levin, E. M., McMurdie, H. F., Hall, F. P.: Phase Diagrams for Ceramists. American Ceramic Soc. Inc. 1956.
Lininger, R. L., Duce, R. A., Winchester, J. W., Matson, W. R.: Chlorine, bromine, iodine, and lead in aerosols from Cambridge, Massachusetts. J. Geophys. Res. **71**, 2457 (1966).
Manheim, F. T.: Manganese-iron accumulations in the shallow marine environment. Narraganset Marine Lab. Occ. Publ. **3—1965**, 217 (1965).
Manton, W. I., Tatsumoto, M.: Some Pb and Sr isotopic measurements on eclogites from the Roberts Victor mine, South Africa. Earth Planet. Sci. Letters **10**, 217 (1971).
Marshall, R. R.: The amounts and isotopic compositions of lead in eclogites from the Münchberg gneiss massif (Fichtelgebirge). 21. Intern. Geol. Congr. **13**, 404 (1960).
Marshall, R. R.: Lead from troilite of the Toluca iron meteorite. Geochim. Cosmochim. Acta **21**, 161 (1961).
Marshall, R. R.: Mass spectrometric study of the lead in carbonaceous chondrites. J. Geophys. Res. **67**, 2005 (1962).
Marshall, R. R., Lead-lead age of the Bondoc meteorite. Geochim. Cosmochim. Acta **32**, 1013 (1968).
Marshall, R. R., Feitknecht, J.: Primitive lead from iron meteorites. Geochim. Cosmochim. Acta **28**, 365 (1964).
Marshall, R. R., Hess, D. C.: Lead from some stone meteorites. J. Chem. Phys. **28**, 1258 (1958).
Martin, J. H., Knauer, G. A.: The elemental composition of plankton. Geochim. Cosmochim. Acta **37**, 1639 (1973).
Mecháček, E.: Mikroelemente in Flözen des Kohlenbeckens von Handlová-Nováky. Geol. Zbornik, Geol. Carpathica **23**, 311 (1972).
Mero, J. L.: Ocean-floor manganese nodules. Econ. Geol. **57**, 747 (1962).
Miesch, A. T.: Distribution of elements in Colorado Plateau uranium deposits — a preliminary report. U.S. Geol. Surv. Bull. **1147 E**, 1 (1963).
Minami, E., Sato, G., Watanuki, K.: Contents of As and Pb in hot springs from Tamagawa (Akita Prefecture). J. Chem. Soc. Japan, Pure Chem. Sect. **78**, 1096 (1957).
Miyazaki, A., Sato, K., Saito, N.: Lead isotopic studies of metamorphic and sedimentary rocks from the Hida metamorphic terrain, Japan. Geochem. J. **6**, 116 (1973).
Moenke, H.: Spurenelemente in variskischen und praevariskischen deutschen Graniten. Chem. Erde **20**, 227 (1960).
Moorbath, S., Welke, H.: Lead isotope studies on igneous rocks from the Isle of Skye, Northwest Scotland. Earth Planet. Sci. Letters **5**, 217 (1969).
Moorbath, S., Welke, H., Gale, N. H.: The significance of lead isotope studies in ancient, high-grade metamorphic basement complexes, as exemplified by the Lewisian rocks of Northwest Scotland. Earth Planet. Sci. Letters **6**, 245 (1969).

MOORE, G. W.: Geologic significance of the minor element composition of marine salt deposits. Econ. Geol. **66**, 187 (1971).
MOORE, J. R., III: Bottom sediment studies, Buzzards Bay, Massachusetts. J. Sediment. Petrol. **33**, 511 (1963).
MORGAN, G. B., OZOLINS, G., TABOR, E. C.: Air pollution surveillance systems. Science **170**, 288 (1970).
MORRIS, H. T., LOVERING, T. S.: Supergene and hydrothermal dispersion of heavy metals in wall rocks near ore bodies, Tinctic District, Utah. Econ. Geol. **47**, 685 (1952).
MUROZUMI, M., CHOW, T. J., PATTERSON, C.: Concentrations of common lead in Greenland snows. Narragansett Marine Laboratory Occ. Publ. **3**, 213 (1965).
MURTHY, V. R., PATTERSON, C.: Lead isotopes in ores and rocks of Butte, Montana. Econ. Geol. **56**, 59 (1961).
NATUSCH, D. F. S., WALLACE, J. R., EVANS, JR., C. A.: Toxic trace elements: preferential concentration in respirable particles. Science **183**, 202 (1974).
NAYDENOV, B. M., YEFIMOV, I. A.: Lead-isotope composition data on the origin of eclogites and the time of differentiation of the primary material of the earth's mantle. Geochem. Intern. **5**, 504 (1968).
NEWHOUSE, W. H.: Opaque oxides and sulphides in common igneous rocks. Bull. Geol. Soc. Am. **47**, 1 (1936).
NICHOLLS, G. D., CURL, JR., H., BOWEN, V. T.: Spectrographic analyses of marine plankton. Limnol. Oceanogr. **4**, 472 (1959).
NOCKOLDS, S. R.: Average chemical compositions of some igneous rocks. Bull. Geol. Soc. Am. **65**, 1007 (1954).
NOCKOLDS, S. R., ALLEN, R.: The geochemistry of some igneous rock series. Parts 1 and 2. Geochim. Cosmochim. Acta **4**, 105 (1953) and **5**, 245 (1954).
NODDACK, I., NODDACK, W.: Die Häufigkeiten der Schwermetalle in Meerestieren. Arkiv Zool. **32 A**, 1 (1939).
NRIAGU, J. O.: Experimental investigation of a portion of the system PbS-NaCl-HCl-H_2O at elevated temperatures. Am. J. Sci. **271**, 157 (1971a).
NRIAGU, J. O.: Studies in the system PbS-NaCl-H_2S-H_2O: stability of lead (II) thiocomplexes at 90° C. Chem. Geol. **8**, 299 (1971b).
NRIAGU, J. O.: Lead orthophosphates — III. Stabilities of fluoropyromorphite and bromopyromorphite at 25° C. Geochim. Cosmochim. Acta **37**, 1735 (1973).
NRIAGU, J. O.: Lead orthophosphates — IV. Formation and stability in the environment. Geochim. Cosmochim. Acta **38**, 887 (1974).
NRIAGU, J. O., ANDERSON, G. M.: Stability of the lead (II) chloride complexes at elevated temperatures. Chem. Geol. **7**, 171 (1971).
NUNES, P. D.,TILTON, G. R.: Uranium-lead ages of minerals from the Stillwater igneous complex and associated rocks. Bull. Geol. Soc. Am. **82**, 2231 (1971).
OFTEDAL, I.: Lead contents in microcline from some granites and pegmatites. Norsk Geol. Tidskr. **47**, 191 (1967).
OKADA, S.: Chemical composition of Japanese granitic rocks in regard to petrographic provinces. Part III. Trace elements. Science Rep. Tokyo Kyoiku Daigaku, Sect. **C 4**, 32, 163 (1955).
OSTIC, R. G.: The concentration and isotopic composition of lead in Toluca iron meteorite. J. Geophys. Res. **71**, 4060 (1966).
OVERSBY, V. M.: The isotopic composition of lead in iron meteorites. Geochim. Cosmochim. Acta **34**, 65 (1970).
OVERSBY, V. M.: Lead (82). In: MASON, B. (Ed.): Handbook of Elemental Abundances in Meteorites. New York: Gordon and Breach 1971.
OVERSBY, V. M.: Genetic relations among the volcanic rocks of Réunion: chemical and lead isotopic evidence. Geochim. Cosmochim. Acta **36**, 1167 (1972).
OVERSBY, V. M., EWART, A.: Lead isotopic compositions of Tonga-Kermadec volcanics and their petrogenic significance. Contr. Mineral. and Petrol. **37**, 181 (1972).
OVERSBY, V. M., GAST, P. W.: Lead isotope compositions and uranium decay series disequilibrium in Recent volcanic rocks. Earth Planet. Sci. Letters **5**, 199 (1968).

OVERSBY, V. M., GAST, P. W.: Isotopic composition of lead from oceanic islands. J. Geophys. Res. **75**, 2097 (1970).
OVERSBY, V. M., LANCELOT, J., GAST, P. W.: Isotopic composition of lead in volcanic rocks from Tenerife, Canary Islands. J. Geophys. Res. **76**, 3402 (1971).
PARRY, W. T., NACKOWSKI, M. P.: Copper, lead and zinc in biotites from Basin and Range quartz monzonites. Econ. Geol. **58**, 1126 (1963).
PATTERSON, C. C.: The $^{207}Pb/^{206}Pb$ ages of some stone meteorites. Geochim. Cosmochim. Acta **7**, 151 (1955).
PATTERSON, C., BROWN, H., TILTON, G., INGHRAM, M.: Concentration of uranium and lead and the isotopic composition of lead in meteoritic material. Phys. Rev. **92**, 1234 (1953).
PATTERSON, C., TATSUMOTO, M.: The significance of lead isotopes in detrital feldspar with respect to chemical differentiation within the earth's mantle. Geochim. Cosmochim. Acta **28**, 1 (1964).
PATTERSON, C., TILTON, G., INGHRAM, M.: Age of the earth. Science **121**, 3134 (1955).
PEIRSON, D. H., CAWSE, P. A., SALIMON, L., CAMBRAY, R. S.: Trace elements in the atmospheric environment. Nature **241**, 252 (1973).
PETERMAN, Z. E., DOE, B. R., PROSTKA, H. J.: Lead and strontium isotopes in rocks of the Absaroka volcanic field, Wyoming. Contr. Mineral. and Petrol. **27**, 121 (1970).
PIPER, D. Z.: Origin of metalliferous sediments from the East Pacific Rise. Earth Planet. Sci. Letters **19**, 75 (1973).
PLYUSNIN, G. S.: On the colouring of amazonites (in Russian). Zap. Vses. Mineral. Obshch. **97**, 1 (1969).
PRESTON, A.: Heavy metals in British waters. Nature **242**, 95 (1973).
PUTMAN, G. W., BURNHAM, C. W.: Trace elements in igneous rocks, northwestern and central Arizona. Geochim. Cosmochim. Acta **27**, 53 (1963).
RABINOVICH, A. V., BASKOVA, Z. A.: The distribution of lead in some granitoids of Eastern Transbaikalia. Geochemistry **6**, 663 (1959).
RABINOVICH, A. V., GOLUBCHINA, M. N., MURTAZINA, T. M.: Isotopic composition of lead in the intrusive rocks of various metallogenetic zones of Central Asia. Geochem. Intern. **2**, 407 (1965).
RAMDOHR, P.: Die Erzmineralien in gewöhnlichen magmatischen Gesteinen. Abhandl. Preuss. Akad. Wiss. Math. Naturw. Kl. **2**, 1 (1940)
REED, G. W., KIGOSHI, K., TURKEVICH, A.: Determinations of concentrations of heavy elements in meteorites by activation analysis. Geochim. Cosmochim. Acta **20**, 122 (1960).
REYNOLDS, P. H., DASCH, E. J.: Lead isotopes in marine manganese nodules and the ore-lead growth curve. J. Geophys. Res. **76**, 5124 (1971).
RHODES, J. M.: On the chemistry of potassium feldspars in granitic rocks. Chem. Geol. **4**, 373 (1969).
RIMSAITE, J.: Geochemistry, mineralogy and petrology of poly-mica rocks. Proceed. XXIII. Internat. Geol. Congr. **6**, 45 (1968).
ROOKE, J. M.: Element distribution in some acidic igneous rocks of Africa. Geochim. Cosmochim. Acta **28**, 1187 (1964).
ROSHOLT, J. N., BARTEL, A. J.: Uranium, thorium, and lead systematics in Granite Mountains, Wyoming. Earth Planet. Sci. Letters **7**, 141 (1969).
RUNNELS, R. T., SCHLEICHER, J. A.: Chemical composition of eastern Kansas limestones. Geol. Surv. Kansas Bull. **119**, 81 (1956).
SABNIS, D. D., GORDON, M., GALSTON, A. W.: A site with an affinity for heavy metals on the thylakoid membranes of chloroplasts. Plant Physiol. **44**, 1355 (1969).
SANDELL, E. B., GOLDICH, S. S.: The rarer metallic constituents of some American igneous rocks. Part I and II. J. Geol. **51**, 99 (1943) and **51** 167 (1943).
SAVUL, M., ABABI, V.: Cuprul, zincul si plumbul ca elemente minore in calcarele si dolomitele din Carpatii Orientalii. Acad. Re. Populare Romine, Studii Cercetari Chim. **7**, 181 (1956).
SAVUL, M., ABABI, V., NICHITA, O.: Zincul, plumbul si cuprul ca elemente minore in rocile vulcanice din Muntii Calimani. Acad. Rep. Populare Romine, Studii Cercetari Chim. **2**, 89 (1956).

SAVUL, M., BOTEZ, C., JEANRENAUD, P., IONESI, L.: Geochimia elementelor minore Cu, Pb, Zn din flisul Paleogen in facies de tarcau si Fusarn-Valea Tarcanlui (Carpatii Orientali). Acad. Rep. Populare Romine, Studii Cercetari Chim. **9**, 57 (1958).

SCHEEL, H. J.: Lead feldspar. Z. Kristallogr. **133**, 264 (1971).

SHACKLETTE, H. T., HAMILTON, J. C., BOERNGEN, J. G., BOWLES, J. M.: Elemental composition of surficial materials in the conterminous United States. U.S. Geol. Surv. Profess. Papers **574 D**, 1 (1971).

SHAW, D. M.: Trace elements in pelitic rocks. Parts I and II. Bull. Geol. Soc. Am. **65**, 1151 (1954).

SHAW, D. M.: The geochemistry of scapolite. Part II Trace elements, petrology and general geochemistry. J. Petrol. **1**, 261 (1960).

SHERATON, J. W.: The origin of the Lewisian gneisses of Northwest Scotland, with particular reference to the Drumbeg area, Sutherland. Earth Planet. Sci. Letters **8**, 301 (1970).

SHIBATA, H., OKADA, S., ISHIKAWA, H.: Chemical composition of Japanese granitic rocks in regard to petrographic provinces. Part VIII, Trace elements. Sci. Rep. Tokyo Kyoiku Daigaku Sect. **C 7**, 67, 217 (1960).

SHIMIZU, N.: Lead isotopic studies on granitic rocks of the Abukuma and Sidara areas in the Ryoke-Abukuma metamorphic belt, Central Japan. J. Fac. Sci. Univ. Tokyo, Sect. **II**, **17**, 445 (1970).

SHORT, N. M.: Geochemical variations in four residual soils. J. Geol. **69**, 534 (1961).

SILLÉN, L. G.: The physical chemistry of sea water. In: SEARS, M. (Ed.): Oceanography. Washington: Am. Assoc. Adv. Sci. 1961.

SILLÉN, L. G., MARTELL, A. E.: Stability constants of metal-ion complexes. Chem. Soc. London Spec. Publ. **17** (1964).

SILVER, L. T.: U-Th-Pb isotope relations in some Apollo 14 (and 15?) materials. G.S.A. Abstracts with Programs **3**, 7 (1971).

SINHA, A. K.: Removal of radiogenic lead from potassium feldspars by volatilization. Earth Planet. Sci. Letters **7**, 109 (1969).

SLAWSON, W. F., NACKOWSKI, M. P.: Trace lead in potash feldspars associated with ore deposits. Econ. Geol. **54**, 1543 (1959).

SMITH, W. H.: Lead and mercury burden of urban woody plants. Science **176**, 1237 (1972).

SOBOTOVICH, E. V., LOVTSYUS, G. P., LOVTSYUS, A. V.: New data on the contents and isotopic compositions of lead in stone meteorites. Meteoritika **24**, 29 (1964).

SORRELL, C. A.: Solid state formation of barium, strontium and lead feldspars in clay-sulfate mixtures. Am. Mineralogist **47**, 291 (1962).

SRETENSKAYA, N. G.: Distribution of trace elements in various generations of microcline from the pegmatites of eastern Kazakhstan. Dokl. Akad. Nauk SSSR **154**, 155 (1964) (Engl. Transl.).

STARIK, I. E., SHATS, M. M., SOBOTOVICH, E. V.: The age of meteorites. Dokl. Akad. Nauk SSSR **123**, 424 (1958).

STARIK, I. E., SOBOTOVICH, E. V., LOVTSYUS, G. P., SHATS, M. M., LOVTSYUS, A. V.: The isotopic composition of lead in iron meteorites. Dokl. Akad. Nauk SSSR **128**, 688 (1959).

STARIK, I. E., SOBOTOVICH, E. V., LOVTSYUS, G. P., SHATS, M. M., LOVTSYUS, A. V.: Lead and its isotopic composition in iron meteorites. Dokl. Akad. Nauk SSSR **134**, 555 (1960).

STARIK, I. E., SOBOTOVICH, E. V., SHATS, M. M., LOVTSYUS, G. P.: Investigation of lead and uranium in meteorites. Meteoritika **20**, 206 (1961).

SUESS, H. E., UREY, H. C.: Abundances of the elements. Rev. Modern Phys. **28**, 53 (1956).

SWANSON, V. E., FROST, T. C., RADER, L. F., JR., HUFFMANN, C., JR.: Metal sorption by Northwest Florida humate. U.S. Geol. Surv. Profess. Papers **550 C**, 174 (1966).

SWANSON, V. E., PALACAS, J. G., LOVE ALONZA, H.: Geochemistry of deep-sea sediment along the 160° W meridian in the North Pacific Ocean. U.S. Geol. Surv. Profess. Papers **575-B**, 137 (1967).

SWAINE, D. J.: Personal communication (1971), mainly on the basis of the paper by: BBOWN, H. R., SWAINE, D. J. (1964) (see: BROWN and SWAINE, 1964).

TATSUMOTO, M.: Chemical investigations of deep-sea deposits (No 23). Tin and lead in sea sediments. J. Chem. Soc. Japan, Pure Chem. Sect. **78**, 42 (1957) [in Japanese].

TATSUMOTO, M.: Isotopic composition of lead in volcanic rocks from Hawaii, Iwo Jima and Japan. J. Geophys. Res. **71**, 1721 (1966a).

TATSUMOTO, M.: Genetic relations of oceanic basalts as indicated by lead isotopes. Science **153**, 1094 (1966b).

TATSUMOTO, M.: Lead isotopes in volcanic rocks and possible ocean floor thrusting beneath island arcs. Earth Planet. Sci. Letters **6**, 369 (1969).

TATSUMOTO, M., HEDGE, C. E., KNIGHT, R. J., UNRUH, D. M., DOE, B. R.: U-Th-Pb, Rb-Sr and K measurements on some Apollo 15 and Apollo 16 samples. In: CHAMBERLAIN, J. W., WATKINS, C. (Eds.): The Apollo 15 Lunar Samples 391 (1972).

TATSUMOTO, M., KNIGHT, R. J.: Isotopic composition of lead in volcanic rocks from Central Honshu — with regard to basalt genesis. Geochem. J. **3**, 53 (1969).

TATSUMOTO, M., KNIGHT, R. J., ALLEGRE, C. J.: Time difference in the formation of meteorites as determined from the ratio of lead-207 to lead-206. Science **180**, 1279 (1973).

TATSUMOTO, M., PATTERSON, C. C.: The concentration of common lead in sea water. In: GEISS, J., GOLDBERG, E. D. (Eds.): Earth Science and Meteoritics. Amsterdam: North Holland Publ. Comp. 1963.

TATSUMOTO, M., SNAVELY, P. D.: Isotopic composition of lead in rocks of the Coast Range, Oregon and Washington. J. Geophys. Res. **74**, 1087 (1969).

TAUSON, L. V.: Geochemistry of rare elements in igneous rocks and metallogenetic specialization of magmas. In: VINOGRADOV, A. P. (Ed.): Chemistry of the Earth's Crust **2**, 248 (1964) (Israel Program for Scientific Transl. 1967).

TAUSON, L. V., KRAVCHENKO, L. A.: Characteristics of lead and zinc distribution in minerals of Caledonian granitoids of the Susamyr batholith in Central Tian-Shan. Geochemistry **1**, 78 (1956).

TAUSON, L. V., PEVTSOVA, L. A.: On the relationships of the distribution of lead and zinc in the rocks of the Caledonian granite complex of Susamyr (Central Tien-Shan). Dokl. Akad. Nauk SSSR **103**, 1069 (1955).

TAYLOR, S. R.: Abundance of chemical elements in the continental crust: a new table. Geochim. Cosmochim. Acta **28**, 1273 (1964).

TAYLOR, S. R., CAPP, A. C., GRAHAM, A. L.: Trace element abundances in andesites. II. Saipan, Bougainville and Fiji. Contr. Mineral. and Petrol. **23**, 1 (1969).

TAYLOR, S. R., WHITE, A. J. R.: Trace element abundances in andesites. Bull. Volcanol. **29**, 177 (1966).

TERA, F., WASSERBURG, G. J.: U-Th-Pb systematics in Lunar highland samples from the Luna 20 and Apollo 16 missions. Earth Planet. Sci. Letters **17**, 36 (1972).

TERA, F., PAPANASTASSIOU, D. A., WASSERBURG, G. J.: Isotopic evidence for a terminal lunar cataclysm. Earth Planet. Sci. Letters **22**, 1 (1974).

TER HAAR, G. L., DEDOLPH, R. R., HOLTZMAN, R. B., LUCAS, H. F., JR.: The lead uptake by perennial ryegrass and radishes from air, water and soil. Envir. Res. **2**, 267 (1969).

TER HAAR, G. L., HOLTZMAN, R. B., LUCAS, H. F., JR.: Lead and lead-210 in rainwater. Nature **216**, 253 (1967).

THOMPSON, G.: A geochemical study of some lithified carbonate sediments from the deep-sea. Geochim. Cosmochim. Acta **36**, 1237 (1972).

THOMPSON, G., SHIDO, F., MIYASHIRO, A.: Trace element distributions in fractionated oceanic basalts. Chem. Geol. **9**, 89 (1972).

TILTON, G. R.: Isotopic composition of lead from tektites. Geochim. Cosmochim. Acta **14**, 323 (1958).

TILTON, G. R.: Isotopic lead ages of chondritic meteorites. Earth Planet. Sci. Letters **19**, 321 (1973).

TILTON, G. R., PATTERSON, C., BBOWN, H., INGHRAM, M., HAYDEN, R., HESS, D., LARSEN, E., JR.: Isotopic composition and distribution of lead, uranium and thorium in a Precambrian granite. Bull. Geol. Soc. Am. **66**, 1131 (1955).

TILTON, G. R., REED, G. W.: Radioactive heat production in eclogite and some ultramafic rocks. In: GEISS, J., GOLDBERG, E. D. (Eds.): Earth Science and Meteoritics. Amsterdam: North-Holland Publishing Co. 1963.

TISCHENDORF, G., FRIESE, G., SCHINDLER, R.: Die Dunkelglimmer der westerzgebirgisch-vogtländischen Granite und ihre Bedeutung als petrogenetische und metallogenetische Indikatoren (I). Geologie **18**, 384 (1969).
TUREKIAN, K. K., IMBRIE, J.: The distribution of trace elements in deep-sea sediments of the Atlantic Ocean. Earth Planet. Sci. Letters **1**, 161 (1966).
TUREKIAN, K. K., KLEINKOPF, M. D.: Estimates of the average abundance of Cu, Mn, Pb, Ti, Ni and Cr in surface waters of Maine. Bull. Geol. Soc. Am. **67**, 1129 (1956).
UDODOV, P. A., PARILOV, Y. S.: Certain regularities of migration of metals in natural water. Geochemistry **8**, 763 (1961) [Engl. Transl.].
VAN DE KAMP, P. C.: The green beds of the Scottish Dalradian series: geochemistry, origin, and metamorphism of mafic sediments. J. Geol. **78**, 281 (1970).
VAN HOOK, H. J.: The ternary system Ag_2S-Bi_2S_3-PbS. Ph.D Thesis Pennsylv. State University, University Park (USA) (1959) (see also: Econ. Geol. **55**, 759, 1960).
VINCENT, E. A., PHILLIPS, R.: Iron-titanium oxide minerals in layered gabbros of the Skaergaard intrusion, East Greenland. I. Chemistry and oremicroscopy. Geochim. Cosmochim. Acta **6**, 1 (1954).
VINE, J. D., TOURTELOT, E. B.: Geochemistry of black shale deposits. A summary report. Econ. Geol. **65**, 253 (1970).
VINE, J. D., TOURTELOT, E. B.: Geochemistry of Lower Eocene sandstones in the Rocky Mountain region. U.S. Geol. Surv. Profess. Papers **789**, 1 (1973).
VINOGRADOVA, Z. A., KOVALSKIY, V. V.: Elemental composition of the Black Sea plankton. Dokl. Acad. Nauk SSSR **147**, 1458 (1962).
VINOGRADOV, A. P.: The Elementary Chemical Composition of Marine Organisms. New Haven, Conn.: Sears Foundation 1953.
VINOGRADOV, A. P.: The Geochemistry of Rare and Dispersed Chemical Elements in Soils (2nd ed.). New York: Consultants Bur. Enterpr. 1959.
VOLOBUYEV, M. I., ZYKOV, S. I., STUPNIKOVA, N. I.: The use of lead isotopes in metallogenetic analysis of the source of ore. Geochem. Intern. **7**, 19 (1970).
WAMPLER, J. M., KULP, J. L.: Isotopic composition and concentration of lead in some carbonate rocks. Petrologic Studies. Buddington Volume 105 (1962).
WAMPLER, J. M., KULP, J. L.: An isotopic study in sedimentary pyrite. Geochim. Cosmochim. Acta **28**, 1419 (1964).
WAMPLER, J. M., SMITH, D. H., CAMERON, A. E.: Isotopic comparison of lead in tektites with lead in earth materials. Geochim. Cosmochim. Acta **33**, 1045 (1969).
WARREN, H. V.: Background data for biogeochemical prospecting in British Columbia. Trans. Roy. Soc. Can., Sect. III **56**, 21 (1962).
WARREN, H. V., DELAVAULT, R. E.: The lead, copper, zinc and molybdenum content of some limestones and related rocks in southern Ontario. Econ. Geol. **56**, 1265 (1961).
WARREN, H. V., DELAVAULT, R. E.: Lead in some food crops and trees. J. Sci. Food Agric. 13. Febr. (1962).
WEDEPOHL, K. H.: Untersuchungen zur Geochemie des Bleis. Geochim. Cosmochim. Acta **10**, 69 (1956).
WEDEPOHL, K. H.: Spurenanalytische Untersuchungen an Tiefseetonen aus dem Atlantik. Ein Beitrag zur Deutung der geochemischen Sonderstellung von pelagischen Tonen. Geochim. Cosmochim. Acta **18**, 200 (1960).
WEDEPOHL, K. H.: Untersuchungen am Kupferschiefer in Nordwestdeutschland (Ein Beitrag zur Deutung der Genese bituminöser Sedimente). Geochim. Cosmochim. Acta **28**, 305 (1964).
WEDEPOHL, K. H.: „Kupferschiefer" as a prototype of syngenetic sedimentary ore deposits. Soc. Mining Geol. Japan Spec. Issue **3**, 268 (1971) (Proc. IMA IAGOD Meetings 70).
WELKE, H., MOORBATH, S., CUMMING, G. L., SIGURDSSON, H.: Lead isotope studies on igneous rocks from Iceland. Earth Planet. Sci. Letters **4**, 221 (1968).
WERNER, C. D.: Geochemie der Brotteröder Serie im Ruhlaer Kristallin. Freiberg. Forschungsh. **C 259**, 1 (1970).
WERNER, C. D.: Geochemische Untersuchungen an Gesteinen der alten Baustufe des Ruhlaer Kristallins (Thüringer Wald). Ber. deutsche Ges. geol. Wiss. B. Miner. Lagerstättenf. **16**, 41 (1971a).

WERNER, C. D.: Geochemische Untersuchungen an Metamorphiten der Seimberg-Serie im Ruhlaer Kristallin (Thüringer Wald). Ber. deutsche Ges. geol. Wiss. B. Miner. Lagerstättenf. **16**, 491 (1971b).
WHITE, W. B., ROY, R.: Phase relations in the system lead-oxygen. J. Am. Ceram. Soc. **47**, 242 (1964).
WODZICKI, A.: Migration of trace elements during contact metamorphism in the Santa Rosa Range, Nevada, and its bearing on the origin of ore deposits associated with granitic intrusions. Mineral. Deposita **6**, 49 (1971).
YUDOVICH, Ya. E., KORYCHEVA, A. A., OBRUCHNIKOV, A. S., STEPANOV, Yu. V.: Mean trace-element content in coals. Geochem. Intern. **9**, 712 (1972).
ZARTMAN, R. E.: The isotopic composition of lead in microclines from the Llano Upluft Texas. J. Geophys. Res. **70**, 965 (1965).
ZARTMAN, R. E., TERA, F.: Lead concentration and isotopic composition in five peridotite inclusions of probable mantle origin. Earth Planet. Sci. Letters **20**, 54 (1973).
ZARTMAN, R. E., WASSERBURG, G. J.: The isotopic composition of lead in potassium feldspars from some 1.0-b.y. old North American igneous rocks. Geochim. Cosmochim. Acta **33**, 901 (1969).
ZEISSINK, H. E.: Trace element behavior in two nickeliferous laterite profiles. Chem. Geol. **7**, 25 (1971).
ZHIROV, K. K., CHERNYSHEV, I. V.: On the geochemistry of lead in Devonian extrusives of Central Kasakhstan. Geochemistry **2**, 141 (1959).
ZHIROV, K. K., STISHOV, S. M.: Geochemistry of amazonitization. Geochem. Intern. **2**, 16 (1965).
ZHIROV, K. K., URUSOVA, M. A.: Geochemistry of the alkalies in granites of the Tarak Massif in the Yenisei Range. Geochemistry **2**, 116 (1962).
ZIRINO, A., YAMAMOTO, S.: A pH-dependent model for the chemical specification of copper, zinc, cadmium, and lead in seawater. Limnol. Oceanogr. **17**, 661 (1972).
ZLOBIN, B. I., GORSHKOVA, M. S.: Lead and zinc in alkalic rocks and their bearing on some petrological problems. Geochemistry **4**, 317 (1961).
ZLOBIN, B. I., PEVTSOVA, L. A., KLASSOVA, N. S.: Distribution of lead and zinc and metallogenic specialization of the more calcic varieties of Variscan granitoids in the central part of the northern Tien-Shan. Geochem. Intern. **2**, 660 (1965).
ZOLLER, W. H., GLADNEY, E. S., DUCE, R. A.: Atmospheric concentrations and sources of trace metals at the South Pole. Science **183**, 198 (1974).

Revised manuscript received: April 1974

Bismuth 83

A	V. Kupčik	(Mineralogisch-Kristallographisches Institut der Universität Göttingen, Germany)
B—C I—III	L. H. Ahrens and A. J. Erlank	(Department of Geochemistry, University of Cape Town, Rondebosch, Cape, South Africa)
C IV	K. H. Wedepohl	(Geochemisches Institut der Universität Göttingen, Germany)
D—M O	L. H. Ahrens and A. J. Erlank	

83-A. Crystal Chemistry

Bismuth is crystallo-chemically very closely related to arsenic and antimony, as well as to lead. This follows from its position in the periodic table (group Vb), from the common electronic structure of Bi^{+3} and Pb^{+2}, and from the common geochemical occurrence. Bismuth is found mainly in sulfidic ore deposits and their oxidation zones. It occurs in rocks only as a trace element. The main groups of bismuth minerals include:

a) metallic bismuth and alloys (rare),
b) sulfides, selenides, tellurides, and sulfosalts,
c) oxides, oxosalts and halogen-oxosalts.

Bismuth, in contrast to arsenic and antimony, is trivalent in all its minerals. The ionic radius of Bi^{3+} is 1.20 Å according to WYCKOFF, and 0.96 Å according to AHRENS. The covalent radius (nowhere given precisely) is about 1.40 Å; the electron structure is 6 s^2p^3. The electro-positive character of Bi is the strongest within group Vb, but is only slightly stronger than that of antimony.

I. Metallic Bi and Alloys

Metallic bismuth occurs frequently in sulfidic ores. It is, like antimony and arsenic, trigonal of space group R3m. In the crystal structure (CUCKA and BARRET, 1962), each bismuth atom is coordinated by six neighbors arranged in a trigonal antiprism ($Bi^{[3+3]}$, Bi—Bi distances: 3×3.07 Å and 3×3.53 Å).

In alloy-like compounds, bismuth can, e.g., replace upto ~10% of the Ag atoms in chilenite (DANA, 1868), or it can form ordered structures, e.g., maldonite, $Au_2^{[6\,Au+6\,Bi]}\,Bi^{[12\,Au+4\,Bi]}{}_k$ (JURRIAANSE, SB 1933—1935, 315).

II. Sulfides, Selenides, Tellurides, and Sulfosalts

The structures of the S, Se, and Te compounds of Bi may be divided into two groups:

a) Bi_2S_3, with a small isomorphic addition of Se (upto a ratio of S:Se = 2:1). This has a chain structure comparable with antimonite, $^1_\infty[Sb_2^{[3+2]}S_3]$ (cf. ŠĆAVNIČAR, SB 1960, 51).[1]

b) A trigonal layer structure, which occurs with higher Se content in the presence of Te. In this structure Bi is coordinated by S, Se, or Te in the form of a trigonal antiprism. The structures of other compounds of the type $Bi_n(S, Se, Te)_m$ are most probably also layer structures (see Fig. 83-A-1 for the structure of tetradymite, $^2_\infty[Bi^{[3+3]}]Te_2S$; HARKER, SB 1933—1935, 287).

[1] A very flat trigonal antiprism has been found in artificial $Bi^{[3+3]}(Bi_2S_3)_9I_3$, MIEHE and KUPČÍK (1971).

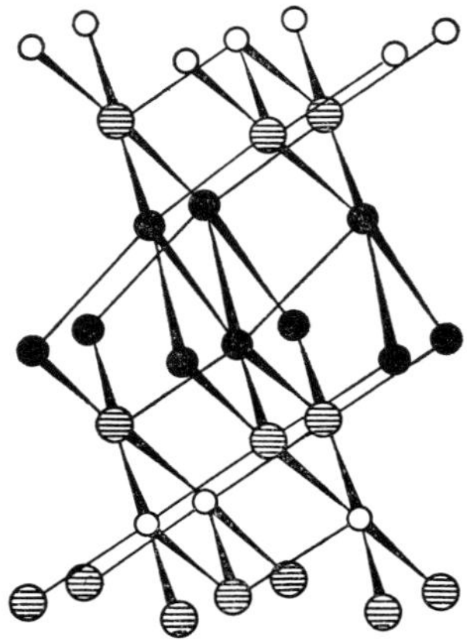

Fig. 83-A-1. Crystal structure of tetradymite, Bi_2Te_2S. Large hatched circles: Bi; medium black circles: Te; small white circles: S

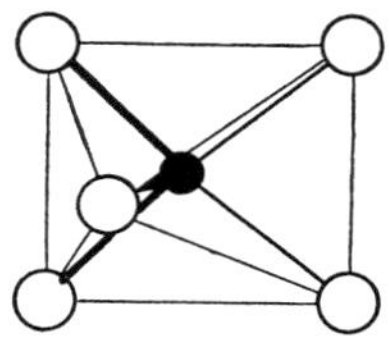

Fig. 83-A-2. Typical $[BiS_5]$ coordination pyramid

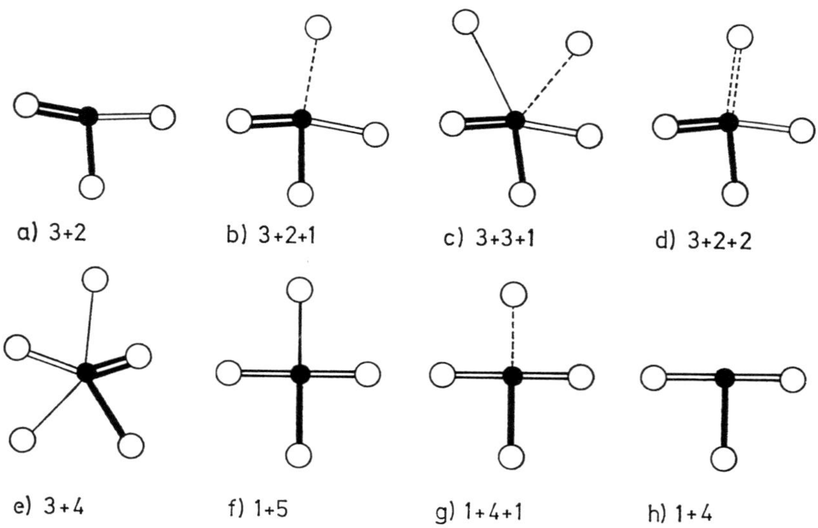

Fig. 83-A-3a—h. Different $[BiS_n]$ coordination polyhedra. See text for explanation. Small black circles: Bi; large white circles: S

Table 83-A-1

Chemical formula Total	Structural	Name	Shape of the coordination polyhedra and some interatomic distances (Å)	References
Bi_2S_3	$^{2}_{\infty}[Bi^{3+3}Bi^{[3+2+1]}S_3]$	bismuthinite	$c\begin{cases}1\times2.69\\2\times2.67\\1\times3.05\\2\times3.07\end{cases} + d\begin{cases}1\times2.54\\2\times2.76\\2\times2.97\end{cases}$	Kupčík- and Veselá-Nováková (1970)
$CuBiS_2$	$Cu^{[4]}_{2}\,\infty[Bi^{[3+3]}_{2}S_4]$	emplectite	$a\begin{cases}1\times2.55\\2\times2.66\\2\times3.16\end{cases}$	Hofmann SB 1933—35, 393; Kupčík (1965)
$Cu_4Bi_6S_{11}$	[a]	hodrushite	$4\times d + 1\times b$ + 1 octahedron	Kupčík and Makovický (1968)
$PbBi_2S_4$	$Pb^{[5]}\,^{3}_{\infty}[Bi^{[3+2+1]}_{2}S_4]$	galenobismuthite	$b\begin{cases}1\times2.63\\2\times2.73\\2\times2.99\\1\times3.12\end{cases} + e\begin{cases}2\times2.76\\1\times2.79\\1\times3.00\\2\times3.02\\1\times3.10\end{cases}$	Iitaka and Nowacki (1962)
$Pb_2Bi_2S_5$	[a]	cosalite	$3\times b + 1\times g$	Weitz and Hellner, SR 1960, 63
$CuPbBiS_3$	$Cu^{[4]}Pb^{[5]}\,\infty[Bi^{[3+2]}S_3]$	aikinite	$b\begin{cases}1\times2.66\\2\times2.76\\2\times2.95\\1\times3.16\end{cases}$	Ohmasa and Nowacki (1970)
$Pb_3Bi_2S_6$	$Pb^{[6+2]}\,^{2}_{\infty}[Pb^{[6]}_{2}Bi^{[1+5]}_{2}S_6]$	phase III	$f\begin{cases}1\times2.80\\4\times2.90\\1\times2.92\end{cases}$	Otto and Strunz (1968)
$(Cu, Fe)Pb_6(Bi, Sb)_7S_{17.5}$	[a]	kobellite	$2\times a$; $2\times b$; b; g; b	Miehe (1971)

[a] Structural formula too complicated, therefore omitted.

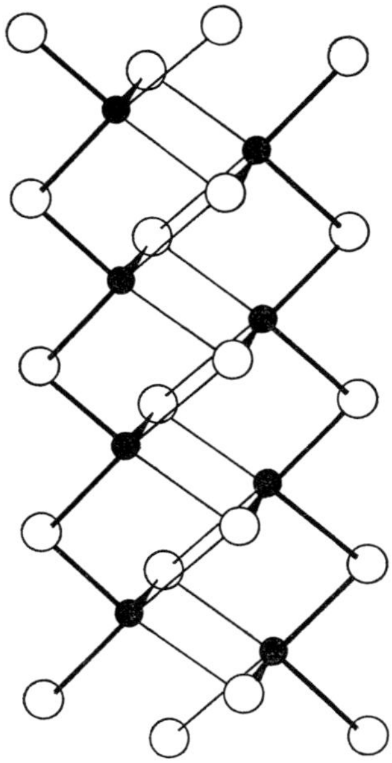

Fig. 83-A-4. Typical $[Bi_2S_4]$ double chain in emplectite, $CuBiS_2$

Bismuthinite, Bi_2S_3 (Kupčík, 1970), with its irregular coordination of sulfur atoms around bismuth, is representative of the whole group of Bi sulfosalts. The common coordination polyhedron is a tetragonal pyramid (Fig. 83-A-2) where the bismuth atom is located slightly above or below the base. One or two additional atoms are often located at the back side of the pyramid. There is only one known exception: galenobismuthite, $PbBi_2S_4$ (Iitaka and Nowacki, 1962), see Fig. 83-A-3e.

The Bi—S valences can be divided according to bond character and strength into the following groups:

a) Bi—S distances from 2.52 to 2.67 Å. These are p^3 bonds approximately rectangular to each other. The vector to the top of the coordination pyramid belongs always to this group of relatively short bonds (see Fig. 83-A-3; heavy lines).

b) Bi—S distances above 2.80 Å. These are hybrid *p-d* bonds (which mostly lie within the base of the pyramid; medium lines).

c) Bi—S distances from 2.94 to 3.17 Å. These are pure *d-γ* bonds lying within the base of the pyramid (light lines).

d) Bi—S distances above 3.27 Å. These are weak *d-γ* bonds (effect of the inner *s*-electron pair; broken lines).

In every structure there are several slightly differently coordinated Bi atoms. The coordination polyhedra polymerize by edge-sharing infinite single chains $\infty[BiS_2]$, or further to double or multiple chains, e.g., $\infty[Bi_2S_4]$, $\infty[Bi_4S_6]$ (see also Fig. 83-A-4).

Isomorphic replacement of As by Bi is unknown and of Sb by Bi possible but, for geochemical reasons, rare. A typical example for a wide-range substitution is the structure of kobellite $(Cu, Fe)Pb_6(Bi, Sb)_7S_{17.5}$ (Miehe, 1971).

The occurrence of a regular coordination (octahedron with Bi atom in the center) must be interpreted statistically. In such a case Bi may be isomorphically replaced by Pb as in hodrushite (Pb, Bi, Ag)$Cu_4Bi_5S_{11}$ (Kupčík and Makovický, 1968).

Replacement of Bi by Pb may also occur when each Bi atom replaced by a Pb results in an additional Cu atom occupying tetrahedral interstices of the bismuthinite structure. This explains the isomorphism of the bismuthinite-aikinite series, (Bi_{2-x}, Pb_x)Cu_xS_3 with $x=0$ to 1 (Ohmasa and Nowacki, 1970). It is assumed that the intermediate members have an ordered Bi—Pb, Cu distribution, which leads to multiple lattice constants and therefore to different mineral types. Data are given in Table 83-A-1.

III. Oxidic Compounds

In contrast to Bi sulfides and sulfosalts, which are very similar to the corresponding Sb compounds and in which Bi always has the same sulfur coordination, the oxidic compounds of Bi are much more complicated. E.g., α- and β-Bi_2O_3 are not related crystallo-chemically to the corresponding Sb compounds. The coordination around

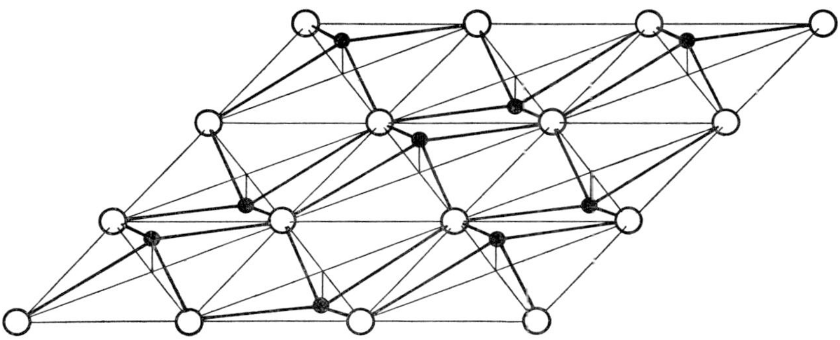

Fig. 83-A-5. $^2_\infty$ [BiO] nets formed from linked [BiO_4] pyramids in kettnerite, CaBi[OF|CO_3]. Small black circles: Bi; large white circles: O

Bi is very irregular (except in bismuthite and β-Bi_2O_3) and is difficult to describe in terms of common polyhedra types, though two frequently occurring basic arrangements may be distinguished:

a) $p^3\,d^{2-3}$ bonding (similar to sulfides) in a deformed and more or less incomplete octahedral coordination with 3 short bonds (tetragonal pyramid with Bi occupying the basis in α-Bi_2O_3 or trigonal antiprism in eulytite, $Bi_4Si_3O_{12}$);

b) extremely flat tetragonal pyramid with Bi occupying the apex.

Schematically, bond lengths can be related to the kind of hybridization:

a) ~2.09—2.30 Å pure p^3 bonding

b) ~2.30—2.40 Å $p^3\,d^n$ hybrid bonding

c) ~2.50—2.65 Å pure d bonding

d) >2.70 Å weak d bonding with repulsion effects from inert electron pair

Table 83-A-2

Chemical formula	Name	Bi—O interatomic distances [Å]	Shape of the coordination polyhedra	Reference
α-Bi_2O_3	bismite	Bi_1 2.08; 2.17; 2.21; 2.54; 2.63 B_2 2.14; 2.22; 2.29; 2.48; 2.54; 2.80	Tetragonal pyramid (3 + 2) Trigonal antiprism	MALMROS (1970)
β-$Bi_2^{[6]}O_3$	—	6×2.39	Trigonal antiprism	SILLÉN SR 1937, 69
$Bi_2^{[2]}O_2[CO_3]$	bismutite	4×2.29	Tetragonal pyramid	LAGERKRANTZ and SILLÉN, SR 1947—1948, 308
$Bi^{[4+4]}[PO_4]\cdot 0.5\,H_2O$	—	2×2.31; 2×2.33; 2×2.56;	Quadr.-planar + 2×2	MOONEY and SLATER (1962)
$Bi^{[2+2+1]}Mo^{[2+4]}O_6$	koechlinite	2×2.25; 2×2.39; 2.53	Tetragonal pyramid + 1	ZEMANN SR 1956, 449
$Bi^{[4+4]}V^{[4]}O_4$	pucherite	2×2.20; 2×2.31; 2×2.54; 2×2.73	Tetragonal prisma	QURASHI and BARNES SR 1953, 503
$CaBi_2^{[4]}O_2[CO_3]_2$	beyerite	4×2.29	Tetragonal pyramid	LAGERKRANTZ and SILLÉN, SR 1947—1948, 308
$Bi_4^{[3+3]}[SiO_4]_3$	eulytite	3×2.15; 3×2.62	Trigonal antiprism	SEGAL *et al.* (1966)
$Bi^{[4]}OCl$	bismoclite	4×2.32	Tetragonal pyramid	SILLÉN, SR 1947—1948, 312
$CaBi^{[4]}[OF/CO_3]$	kettnerite	4×2.29 + 4 from CO_3 undefined	Tetragonal pyramid	SYNEČEK and ŽÁK, SR 1960, 426

Data are given in Table 83-A-2.

There are several different ways of linking the coordination polyhedra, and the following types have been found:

a) isolated $[BiO_6]$ groups in eulytite, $Bi_4[SiO_4]_3$;

b) $^2_\infty$ $[BiO]$ nets built of BiO_4 pyramids sharing edges (see Fig. 83-A-5) in bismoclite, BiOCl, bismutite, $Bi_2O_2[CO_3]$, and kettnerite, $CaBi[OF|CO_3]$;

c) probably $^1_\infty$ $[BiO_2]$ chains with edge-sharing in koechlinite, $BiMoO_6$;

d) three-dimensional arrays $^3_\infty$ $[Bi_2O_3]$, resulting from polyhedra sharing edges and corners in bismite, α-Bi_2O_3.

Isomorphous replacement of Bi by other elements is not known in oxidic minerals.

Manuscript received: June 1972

References: Section 83-A

CUCKA, P., BARRET, C. S.: The crystal structure of Bi and solid solutions of Pb, Sn, Sb and Te in Bi. Acta Cryst. **15**, 865 (1962).

DANA in: STRUNZ, H. Mineralogische Tabellen (4. ed.), p. 100. Leipzig: Akad. Verlagsgesellschaft 1966.

IITAKA, Y., NOWACKI, W.: A redetermination of the crystal structure of galenobismutite. $PbBi_2S_4$. Acta Cryst. **15**, 691 (1962).

KUPČÍK, V.: Struktur des Emplektit $CuBiS_2$. Referate der 8. Diskussionstagung der Sektion für Kristallkunde der DMG, Marburg, 16 (1965).

— MAKOVICKÝ, E.: Die Kristallstruktur des Minerals $(Pb, Ag, Bi)Cu_4Bi_5S_{11}$. Neues Jahrb. Mineral., Monatsh. **236** (1968).

— VESELÁ-NOVÁKOVÁ, L.: Zur Kristallstruktur des Bismuthinits, Bi_2S_3. Mineral. Petrog. Mitt. **14**, 55 (1970).

MALMROS, G.: The crystal structure of α-Bi_2O_3. Acta Cryst. **24**, 384 (1970).

MIEHE, G.: Crystal structure of kobellite. Nature Physical Science **231**, 133 (1971).

— KUPČÍK, V.: Die Kristallstruktur des $Bi(Bi_2S_3)_9J_3$. Naturwissenschaften **58**, 219 (1971).

MOONEY-SLATER, R. C. L.: Polymorphic forms of bismuth phosphate. Z. Krist. **117**, 317 (1962).

OHMASA, M., NOWACKI, W.: Note on the space group and on the structure of aikinite derivatives. Neues Jahrb. Mineral., Monatsh. **158** (1970).

— — A redetermination of the crystal structure of aikinite $[BiS_2|S|Cu^{IV}Pb^{VII}]$. Z. Krist. **132**, 7 (1970).

OTTO, H. H., STRUNZ, H.: Zur Kristallchemie synthetischer Blei-Wismut-Spießglanze. Neues Jahrb. Mineral., Abhandl. **108**, 1 (1968).

SEGAL, D. J., SANTORO, R. P., NEWNHAM, R. E.: Neutron-diffraction study of $Bi_4Si_3O_{12}$. Z. Krist. **123**, 73 (1966).

Manuscript received June 1972

83-B. Isotopes in Nature

Only one bismuth isotope, ^{209}Bi, is found in nature. The 1961 atomic weight of bismuth is 208.980.

Revised manuscript received: April 1968

83-C. Abundance in Cosmos, Meteorites, Tektites and Lunar Materials

I. Cosmos

In the absence of abundance data for the solar atmosphere, two methods of assigning a cosmic or solar system abundance value for Bi have been proposed. The first involves the use of abundances in the chondritic meteorites. The large difference in the various chondrite groups (Table 83-C-1) introduces uncertainty in this respect.

Table 83-C-1. *Bismuth contents of meteorites*

Name	Class symbol	References	ppb Bi[a]
Chondrites			
Beardsley	CH	1	3.3
		2	3.5
Forest City	CH	1	0.28
		2	1.0
Plainview	CH	1	3.6
Holbrook	CL	1	2.1
		2	7.5
		2	2.0
Modoc	CL	1	1.1
		2	0–6
Mighei	Cc_2	2	180
Orgueil	Cc_1	2	130
Abec	Ce	2	80
Achondrites			
Johnstown	Ab	1	2.8
Nuevo Laredo	Ap	2	<1
Meteoritic troilite from Iron Meteorites			
Cañon Diablo troilite		2	40
Toluca troilite		2	180

References: 1.=EHMANN and HUIZENGA (1959). 2.=REED *et al.* (1960).
[a] All results by neutron activation analysis, 1 ppb $= 10^{-3}$ ppm.

The consensus view is that particular significance should be attached to the carbonaceous chondrites. In the most recent compilation of this type (CAMERON, 1968), the atomic abundance of Bi is given as 0.164 (relative to Si=10^6 atoms).

The second approach attempts to predict cosmic abundances using modern theories of nucleosynthesis and other nuclear considerations. The latest calculations by SEEGER *et al.* (1965) give a Bi value of 0.75 (relative to Si=10^6 atoms). The disparity between this values and that given by CAMERON (1967) requires further attention.

II. Meteorites

Table 83-C-1 lists all available neutron activation analyses for Bi in meteorites. It may be noted that the Bi concentration in the "ordinary" chondrites (class symbols CH and CL) varies rather widely, averaging about 3 ppb (0.003 ppm).

However, the Bi magnitude increases to 100–200 ppb (0.1–0.2 ppm) in the carbonaceous and enstatite chondrites (class symbols Cc and Ce), showing that Bi, in common with other heavy chalcophile elements, Hg and Tl in particular, is markedly fractionated in the principal chondrite groups (REED *et al.*, 1960). The contents of Bi in the troilite phases of iron meteorites bears testimony to the chalcophile behaviour of Bi.

III. Tektites

No quantitative data for Bi in tektites are available. TAYLOR (1966) analysed five australites by means of a spark source mass spectrographic technique but could not detect Bi. However, he indicates Bi to be <10 ppb (0.01 ppm). If so, this probably implies a depletion of Bi in australites as compared with suggested parent materials, as two greywackes and an impact glass from Henbury, Australia, have 0.59, 0.65 and 0.37 ppm Bi respectively.

Revised manuscript received: April 1968

IV. Lunar Materials

The average bismuth concentration in lunar surface rocks is close to 1 ppb (see Table 83-C-2). GANAPATHY and ANDERS (Proc. of the 5th Lunar Sci. Conf. **2**, 1189, 1974) assume in their model calculations that the total moon contains 0.1 ppb Bi which is about 40 times lower than the estimate for the earth.

The following averages can be computed for lunar basalts from Apollo 11 to Apollo 17 (expect Apollo 16): 0.43 ppb, 0.33 ppb, 0.34 ppb, 0.26 ppb and 0.18 ppb Bi. Common terrestrial basalt types are about 100 times higher in bismuth. Bi data for anorthositic rocks from the lunar highlands range from 0.1 to 3.6 ppb Bi, but KREEP type material does not seem to be enriched in bismuth.

Breccias and fines usually contain more bismuth than the basalts from the same sampling area. For 54 samples of fines from Apollo 11 to 17 an average of 1.5 ppb Bi can be computed. The fines from Apollo 12 and 17 are often lower than 1 ppb Bi. Those from Apollo 14 and 15 (and Luna 16 and 20) are commonly higher than 2 ppb Bi. Several investigators assume that the enhanced level of bismuth concentrations in fines and breccias is influenced by meteoritic matter. MORGAN *et al.* (Proc. of the 3rd Lunar Sci. Conf. **2**, 1361, 1972) estimate the contribution of carbonaceous chondrites (Type 1) to Apollo 11, 12 and 15 fines as 1.4, 1.8 and 2.6 percent of the total bismuth respectively. Table 83-C-2 contains information on a metal sample from Apollo 14 breccia and fines with 32 ppb Bi.

Table 83-C-2. *Bismuth in lunar materials* (Analytical method: N/R) (Number of samples in brackets). Most references are from Volume 2 of the 7 Proceedings of Lunar Science Conferences, which are Supplements 1 to 7 Geochimica Cosmochimica Acta 1970 to 1976. These are marked as [P1] to [P7] in the column "Reference"

Rock type	Range ppb Bi	Average ppb Bi	Reference
Apollo 11			
Basalt (A, B) (4)	0.15–0.73	0.33	GANAPATHY *et al.* [P1]
Basalt (A, B) (4)	0.16–1.15	0.58	ANDERS *et al.* [P2]
Low-K basalt	—	0.19	GANAPATHY and ANDERS [P5]
Breccia (2)	1.62–2.79	2.2	GANAPATHY *et al.* [P1]
Fines (1)	—	1.5	MORGAN *et al.* [P3]
Fines (3)	1.37–1.97	1.63	GANAPATHY *et al.* [P1]
Fines (1)	—	1.7	ANDERS *et al.* [P2]
Anorthosite from fines (2)	0.42–0.59	0.51	LAUL *et al.* [P2]
Apollo 12			
Basalt (A, B) (8)	0.15–0.76	0.33	ANDERS *et al.* [P2]
Breccia (7)	0.46–1.04	0.60	ANDERS *et al.* [P2]
Fines (4)	0.35–1.6	0.80	LAUL *et al.* [P7]
Fines (1)	—	0.70	MORGAN *et al.* [P3]
KREEP fragments (2)	0.4–0.7	0.55	MORGAN *et al.* [P3]
Apollo 14			
Basalt (2)	0.29–0.39	0.34	MORGAN *et al.* [P3]
Norite (from fines) (5)	0.36–$\leq$0.7	$\leq$0.56	MORGAN *et al.* [P3]
Microbreccia (4)	0.24—0.34	0.29	MORGAN *et al.* [P3]
Breccia (KREEP) (1)	—	0.28	GANAPATHY *et al.* [P5]
Breccia (2)	0.23–0.26	0.25	GROS *et al.* [P7]
Fines (3)	1.5–5.9	3.2	ALLEN *et al.* [P3]
Fines (8)	1.3–2.2	1.7	MORGAN *et al.* [P3]
Metal from breccia and fines (2)	0.37–32	16.2	GANAPATHY *et al.* [P5]
Apollo 15			
KREEP basalt (1)	—	1.8	MORGAN *et al.* [P4]
KREEP basalt (1)	—	0.29	GROS *et al.* [P7]
Basalt (1)	—	0.08	GROS *et al.* [P7]
Basalt (3)	0.089–0.41	0.24	MORGAN *et al.* [P3]
Basalt (4)	0.089–0.41	0.20	GANAPATHY *et al.* [P4]
Basalt ("diabase") (5)	0.052–7.3	2.2	KRÄHENBÜHL *et al.* [P4]
Norite, troctolite (2)	0.64–0.7	0.67	MORGAN *et al.* [P4]
Norite (1)	—	1.65	KRÄHENBÜHL *et al.* [P4]
Gabbro (3)	0.11–0.14	0.12	GANAPATHY *et al.* [P4]
Breccia (1)	—	0.20	GANAPATHY *et al.* [P5]
Breccia (2)	0.45–1.56	1.0	KRÄHENBÜHL *et al.* [P4]
Breccia (2)	0.97–1.0	0.99	MORGAN *et al.* [P4]
Breccia (1)	—	0.78	GROS *et al.* [P7]
Fines (5)	0.87–2.92	2.20	MORGAN *et al.* [P3]
Fines (8)	0.87–2.9	2.0	GANAPATHY *et al.* [P4]
Anorthosite (2)	0.097–0.14	0.12	GANAPATHY *et al.* [P4]
Anorthosite (2)	0.41–0.48	0.44	MORGAN *et al.* [P4]
Anorthosite (1)	—	0.097	MORGAN *et al.* [P3]
Anorthosite (2)	0.22–3.58	1.9	KRÄHENBÜHL *et al.* [P4]

Table 83-C-2 (continued)

Rock type	Range ppb Bi	Average ppb Bi	References
Apollo 16			
Basalt (1)	—	0.08	GANAPATHY *et al.* [P5]
Breccia (10)	0.12–2.02	0.45	GANAPATHY *et al.* [P5]
Breccia (2)	12.2–12.6	12.4	ALLEN *et al.* [P5]
Fines (4)	0.7–1.72	1.27	ALLEN *et al.* [P5]
Fines (9)	0.32–1.10	0.83	KRÄHENBÜHL *et al.* [P4]
Orange soil (1)	—	0.72	ALLEN *et al.* [P5]
Apollo 17			
Dunite (2)	0.41–1.24	0.82	HIGUCHI and MORGAN [P6]
Basalt (1)	—	0.40	HIGUCHI and MORGAN [P6]
Norite (1)	—	0.05	HIGUCHI and MORGAN [P6]
Basalt (2)	0.043–0.099	0.07	MORGAN *et al.* [P5]
Troctolite (2)	0.037–0.14	0.09	MORGAN *et al.* [P5]
Orange glass (1)	—	1.53	MORGAN *et al.* [P5]
Breccia (11)	$\leq$0.12–0.67	0.33	HIGUCHI and MORGAN [P6]
Breccia (13)	0.098–0.70	0.32	MORGAN *et al.* [P5]
Breccia (6)	0.16–0.75	0.49	MORGAN *et al.* [P7]
Fines (7)	0.54–1.43	0.76	MORGAN *et al.* [P5]
Gabbroic anorthosite (1)	—	1.76	GROS *et al.* [P7]
Gabbroic anorthosite (1)	—	1.3	MORGAN *et al.* [P‘]
Anorthosite (2)	0.17–0.29	0.23	MORGAN *et al.* [P5]
Luna 16			
Fines (2)	4.6–5.1	4.9	MORGAN *et al.* [P3]
Luna 20			
Breccia (1)	—	2.8	MORGAN *et al.* (1973)[a]
Fines (1)	—	2.63	MORGAN *et al.* (1973)[a]
Anorthosite (1)	—	$\leq$4.8	GANAPATHY *et al.* [P4]

[a] MORGAN, J.W., KRÄHENBÜHL, U., GANAPATHY, R., ANDERS, E.: Luna 20 soil: abundance of 17 trace elements. Geochim. Cosmochim. Acta **37**, 953 (1973).

Manuscript received: June 1977

83-D. Abundance in Sulfides and in Rock-Forming Minerals; Bismuth Minerals

Because of its chalcophile tendency, Bi occurs most commonly in sulfide minerals and hence most data are available on sulfides. Data for normal silicate minerals are virtually non-existent, as the abundance levels likely to be encountered can only be estimated by sensitive techniques such as neutron activation analysis or other techniques employing pre-enrichment procedures. Hence the geochemistry of Bi is dominated by its presence in sulfides, and it is necessary to consider this aspect in some detail. It should of course be borne in mind that sulfides are not normally rock forming minerals.

Ionic Radius of Bi^{5+}

V. M. GOLDSCHMIDT (1926) estimated the radius of Bi^{+3} as 1.20 Å. Subsequently, AHRENS (1952) recommended a value of 0.96 Å. According to AHRENS and KABLE (in press) this value is too low and they recommend a radius of 1.16 Å.

I. Chalcophile Tendency of Bismuth

A distinct indication based on meteorite evidence, that Bi is mainly chalcophile was provided by I. and W. NODDACK (1930 and 1934), who reported the Bi contents in metal, sulfide (troilite) and silicate phases of meteorites as 0.5, 2 and 0.02 ppm, respectively.

The other two Group V-b elements, As and Sb, also tend to be chalcophile, but have distinctly greater siderophile tendencies than Bi. Although the techniques used by the NODDACKS do not appear to be highly accurate, the relationships they observed between the three principal meteorite phases probably reflect the relative intensities of the respective tentencies. The chalcophile behaviour of bismuth is also borne out by the results reported by REED *et al.* (1960) who report 0.04 and 0.18 ppm Bi in two samples of troilite from iron meteorites, as compared with a level of 0.003 ppm Bi in ordinary chondrites (see Table 83-C-1).

II. Chalcophile Behavior of Bismuth

Bi forms several sulfide minerals of its own (Table 83-A-1). The major proportion of "chalcophile" Bi in the earth's crust occurs, however, in trace quantities in various common sulfide minerals, notably galena which appears to be the principal host mineral for this element. In the discussions which follow on the chalcophile behaviour of Bi, galena will be considered first. For convenience, much of the available data for common sulfides is summarized in Table 83-D-1. It is not possible to calculate realistic averages for the individual minerals as many of the analysed samples are below the detection limits of the methods employed. Note also the variation in sensitivity given by the various authors, hence in FLEISCHER's compilation the category "not found" in some cases indicates a concentration $>$10 ppm.

Table 83-D-1. *Summary of bismuth contents of common sulphide minerals*

Max. conc. % or ppm	No. of samples in each concentration range: 1 % or more	5,000 to 10,000 ppm	1,000 to 5,000 ppm	500 to 1,000 ppm	200 to 500 ppm	100 to 200 ppm	50 to 100 ppm	10 to 50 ppm	< 10 ppm	not found	Total no. of samples	Sensi-tivity ppm	Refer-ence	Method
Galena														
5 %	23	1	28	18	26	19	8	43	36	125	327	1—100	1	S
3.9 %	1	2		3		1		3	10	12	32	10	2	C
Sphalerite														
1,000						7	1	13	2	143	166	1—10	1	S
4,000			4	3	3	7	12	16	57	120	222	1	3	S
900				3	7	7	11	35	64	59	186	1—3	4	S
Chalcopyrite														
2,000			6	7	5	4	9	30	31	137	229	1	3	S
2,000			2	1	8	3	7	14	10	82	127	1—3	4	S
Pyrite														
100						1	—	4	1	11	17	10—100	1	S
780				4	9	21	27	63	49	422	605	10	5	S

References: 1 = Fleischer (1955), 2 = Nesterova (1958), 3 = Burnham (1959), 4 = Rose (1967), 5 = Cambel and Jarkovský (1967).

In a detailed review on the occurrence of Bi in sulphide minerals FLEISCHER (1955) refers to several pre-1954 investigations on galena (PINA DE RUBIES and DOETSCH, LOPEZ DE AZCONA, WASSERSTEIN, OFTEDAL, WAHLSTROM, LEUTWEIN and HERRMANN: for reference details, see FLEISCHER's review), and on the basis of their observations it is quite clear that the Bi content varies enormously from < 10 ppm up to 5%. Subsequent observations (NESTEROVA, 1958; HERTEL, 1966, for example) corroborate this conclusion. In 40 specimens of U.S.S.R. galena, Bi was found to vary from < 10 ppm to almost 4%.

The mode of occurrence of Bi in galena is problematical. Some Bi, together with Ag, may occur structurally because of the possibility of the coupled substitution

$$Ag^{I}Bi^{III} \rightleftharpoons 2\,Pb^{II}.$$

If, however, the Bi and Ag concentrations are high, schapbachite (matildite) ($AgBiS_2$) may tend to exsolve (RAMDOHR, 1938 etc. — References in RAMDOHR, 1960; OFTEDAL, 1942). HOEHNE (1934) had earlier concluded that a substantial proportion of Bi occurred in galena as inclusion impurities. This view was also held by BETEKHTIN (1950) and later by NESTEROVA (1958) who maintained that the inclusions were mainly bismuthinite. The more complex Bi sulphides (benjaminite, tetradymite and matildite) are relatively rare.

The possibility that the Bi content of galena might be temparature dependent was investigated first by OFTEDAL (1941). American sphalerite data are given by SIMS and BARTON (1961).

CAMBEL and JARKOVSKY (1967) carried out a very detailed study of the abundances and distribution of several elements, including Bi, in pyrites from Czechoslovakia. Samples from various modes of origin and from a wide variety of localities were analyzed. The Bi concentrations in the great majority of specimens are below the detection limit (10 ppm) of the spectro-chemical method that was used. Several specimens, but nevertheless a minority, contain 10—40 ppm, and occasionally the level rises above 100 ppm reaching a maximum of $\sim$1,000 ppm. (one specimen only).

The concentrations of most elements investigated by CAMBEL and JARKOVSKY appear to follow a lognormal-type distribution in the pyrite specimens they studied. Data for Bi were, however, not sufficient to establish the statistical nature of the distribution of the concentration of this element in pyrite. The data of NESTEROVA (1958), BURNHAM (1959) and ROSE (1967), as indicated in Table 83-D-1, would, however, suggest that Bi does follow a lognormal distribution in galena, sphalerite and chalcopyrite.

In summary, two features are notable, (1) the preference of Bi for galenas, and (2) the large spread of concentration for all the sulphides discussed. In the latter respect, it is worth re-emphasizing that many of the higher concentrations encountered in the various sulphides may be explained by admixture of small amounts of foreign bismuth minerals. Despite considerable variations in the Bi concentration in galena from different modes of occurrence, there is a distinct indication that the Bi content tended to increase with increase in temperature. This possible temperature effect has been discussed by GOLDSCHMIDT (1954). SCHROLL (1955) is also of the opinion that relatively high Bi and Ag concentrations tended to occur in galenas which had been formed at high temperatures. Further discussion may be found in the section dealing with phase relationships in the Pb—Bi—S system (subsection D-VI).

The Bi concentration in other common sulphide minerals is usually much less than in galena. In his review of data up to 1954 FLEISCHER (1955) concluded that in the zinc sulphides (sphalerite and wurtzite), most specimens contain <1—10 ppm Bi; concentrations up to a few hundred ppm are occasionally found. According to the pre-1955 information, the Bi content of chalcopyrite is usually less (maximum, a few ppm) than in sphalerite and in pyrite, concentrations are still lower. The investigations by BURNHAM (1959) and ROSE (1967) on the composition of chalcopyrite and sphalerite, mainly from metallogenic provinces of the south western U.S.A. and northern New Mexico, show that where as the Bi levels are often low in these minerals, occasional high concentrations (up to $\sim$1,000 ppm) may be found in both sphalerite and chalcopyrite. It is quite clear from the detailed work of ROSE in particular that not only does the Bi content vary enormously — together with several other chalcophile elements — from specimen to specimen, whether from one area or not, but variation can be quite extreme in single specimens. Further data on the Bi contents of amounts of foreign bismuth minerals.

Detailed knowledge regarding the mechanism of substitution of Bi in sulphide minerals is lacking. Although bonding in sulphides is typically covalent, ionic size is apparently a consideration, as the ionic radius of Bi^{3+} (1.16 A) is closer to that of Pb^{2+} (1.20 A) than it is to Fe^{2+} (0.74 A), Cu^{2+} (0.72 A) or Zn^{2+} (0.74 A), hence providing some evidence for the higher concentration of Bi in galena. Covalent radii show the same relationships. Further discussion may be found in FLEISCHER (1955).

III. Lithophile Behavior of Bismuth

Although Bi is a distinctly chalcophile element (noted above), Bi^{III} exhibits some lithophile tendency. It will be recalled that Bi^{3+} is a relatively large cation ($r=1.16$ A) and silicates the possibility of a Bi^{3+}—Ca^{2+} ($r=1.01$ A) substitution exists, as mentioned by GOLDSCHMIDT (1954). If such substitution does take place to a significant extent, Bi would be expected to be enriched in plagioclase rich (basaltic) rocks, as compared with acidic rocks. BROOKS *et al.* (1960) and particularly BROOKS and AHRENS (1961) found no such relationship.

The possibility of a substitution of Bi^{3+} for Ca^{2+} in the mineral apatite has been stressed by GOLDSCHMIDT (1954) who stated that Bi is "... known to be captured in apatite." Neither the source of information nor any data are given. GURNEY and AHRENS (in press) have recently carried out a reconnaissance survey on 13 specimens of apatite from various localities. They estimate 5 of the samples to have Bi contents within the range 5—100 ppm with the highest Bi contents occurring in apatites from pegmatites. The other samples have Bi contents below the detection limit ($\sim$1 ppm) of the semi-quantitative spectrographic method employed.

Further data are presented by CLARK (1965) who reports the presence of Bi in several apatites from hydrothermal veins in Finland. Concentrations of up to 1,150 ppm Bi are recorded. Although the veins contain bismuth minerals, CLARK suggests that Bi occurs in the apatite structure, substituting for Ca.

Table 83-D-2 lists the meagre data available for silicate minerals. From this, and from the data available for silicate rocks (see next section), it appears that a bismuth content of <1 ppm can normally be expected in the common silicate minerals.

IV. Inter-Element Relationship Between Bismuth and the Rare Earth Elements

As the radius of Bi^{3+} is close to the radii of the trebly charged Rare Earth (RE) cations (0.88 A to 1.15 A) and as the crystal chemistry of several Bi^{III} compounds is similar to that of Rare Earth compounds (some are isostructural) some Bi—RE association might be expected.

Work on the Bi contents of RE minerals, notably gadolinite, has been carried out by I. and W. NODDACK (1931) and by GURNEY and AHRENS (in press); results are summarized in Table 83-D-3. Both studies indicate that of the minerals investigated gadolinite ($Y_2FeBe_2[O|SiO_4]_2$) contains the highest amounts of Bi. This suggests that the Bi^{3+}—Y^{3+} association may be closer than that existing between Bi and the other RE elements. In common with other authorities, we class Y as a Rare Earth element.

Table 83-D-2. *Bismuth contents of silicate minerals*

Mineral	No.	Method	ppm Bi average
Perthite, REED *et al.* (1960)	1	N/R	0.011
Biotite, BROOKS and AHRENS (1961)	1	S	0.21
Glauconite, BROOKS and AHRENS (1961)	1	S	0.13
Crocidolite, BROOKS and AHRENS (1961)	1	S	0.01

Table 83-D-3. *Bismuth contents of rare earth minerals*

Mineral	No.	Method	ppm Bi	
			range	average
Gadolinite, I. and W. NODDACK (1931)	4	X	~1—20	
Gadolinite, GURNEY and AHRENS (in press)	29	S	12—262	63
Fergusonite, GURNEY and AHRENS (in press)	5	S[a]	~20—40	
Allanite, GURNEY and AHRENS (in press)	6	S[a]	<1—10	
Monazite, GURNEY and AHRENS (in press)	3	S[a]	≈ 5	
Euxenite, GURNEY and AHRENS (in press)	2	S[a]	≈ 40	

[a] Semi-quantitative method.

V. Bismuth Minerals

Some bismuth minerals are listed in Table 83-D-4. Many of these are relatively rare and are mainly found in hydrothermal ore deposits where bismuth usually occurs in combination with either sulphur or other metals such as Au, Se and Te. The commonest bismuth minerals are bismuthinite (Bi_2S_3) and bismite (Bi_2O_3), the latter usually occuring as an alteration product. The only bismuth silicate minerals known are eulytite ($Bi_4(SiO_4)_3$) and bismutoferrite ($BiFe_2(SiO_4)_2(OH)$). However, bismuth oxides, carbonates, vanadates, arsenates, molybdates are also found. It should be noted that Table 83-D-4 does not contain a complete list of bismuth minerals, nor does it list those which are considered mixtures or which are discredited in the recent review by FLEISCHER (1966). Further details may be found in the references listed in Table 83-D-4.

Table 83-D-4. *Bismuth minerals*

Aikinite	$Cu_2S \cdot 2PbS \cdot Bi_2S_3$	Kettnerite	$CaBi(CO_3)OF$
Aramayoite	$Ag(Sb,Bi)S_2$	Kotulskite	$Pd(Te,Bi)_{1-2}$
Arsenobismutite	$Bi_2(AsO_4)(OH)_3$	Koechlinite	$(BiO_2)(MoO_4)$
Atelestite	$Bi_3(AsO_4)O_2(OH)_2$?	Laitakarite	Bi_4Se_2S
Benjaminite	$(Cu, Ag)_2Pb_2Bi_4S_9$	Lindstromite	$PbCuBi_3S_6$
Beyerite	$(Ca, Pb)Bi_2(CO_3)_2O_2$	Maldonite	Au_2Bi
Bismite	Bi_2O_3	Michenerite	$(Pd, Pt)BiTe$
Bismoclite	$BiOCl$	Mixite	$Cu_{11}Bi(AsO_4)_5(OH)_{10} \cdot 6H_2O$
(Daubreéite)	$BiO(OH,Cl)$	Moncheite	$(Pt, Pd)(Te, Bi)_2$
Bismuthinite	Bi_2S_3	Montanite	$(BiO)_2(TeO_4) \cdot 2H_2O$?
Bismutite	$Bi_2(CO_3)O_2$	Paraguanajuatite	$Bi_2(S,Se)_3$
Bismutoferrite	$BiFe_2(SiO_4)_2(OH)$	Parkerite	$Ni_3(Bi, Pb)_2S_2$
Bismutotantalite	$Bi(Ta,Nb)O_4$	Pavonite	$AgBi_3S_5$
Bonchevite	$PbBi_4S_7$	Perite	$PbBiO_2Cl$
Bursaite	$Pb_5Bi_4S_{11}$	Pucherite	$BiVO_4$
Cosalite	$2PbS \cdot Bi_2S_3$	Rooseveltite	$BiAsO_4$
Csiklovaite	$Bi_2Te(S,Se)_2$	Russellite	$(Bi_2, W)O_3$
Cuprobismutite	$CuBiS_2$	Schapbachite (Matildite)	$AgBiS_2$
Emplectite	$Cu_2S \cdot Bi_2S_3$		
Eulytite	$Bi_4(SiO_4)_3$	Sakharovaite	$(Pb, Fe) (Bi, Sb)_2S_4$
Froodite	$PdBi_2$	Sillenite	Bi_2O_3
Galenobismutite	$PbS \cdot Bi_2S_3$	Telluro-bismuthite	Bi_2Te_3
(Cannizzarite)	$Pb_3Bi_5S_{11}$?		
(Bismutoplagionite)	$5PbS \cdot 4Bi_2S_3$	Tetradymite	Bi_2Te_2S
Giessenite	$Pb_9CuBi_6Sb_{1.5}S_{30}$	Ustarasite	$Pb(Bi, Sb)_6S_{10}$
Gladite	$PbCuBi_5S_9$	Walpurgite	$Bi_{10}(UO_2)(AsO_4)_2O_4 \cdot 3H_2O$?
Grunlingite	Bi_4TeS_3	Waylandite	$(Bi, Ca)Al_3(PO_4,SiO_4)_2(OH)_6$
Guanajuatite	$BiSe_3$	Wehrlite	Bi_3Te_2?
Hammarite	$Pb_2Cu_2Bi_4S_9$?	Westgrenite	$(Bi, Ca)(Ta, Nb)_2O_6(OH)$
Healeyite	Bi_7Te_3	Wittichenite	Cu_3BiS_3
Ikunolite	$Bi_4(S, Se)_3$	Wittite	$Pb_5Bi_6(S, Se)_{14}$
Joseite	Bi_4TeS_2	Zavaritskite	$BiOF$

References: Palache, Berman and Frondel (1958), Short (1940), American Society for Testing and Materials, ASTM powder diffraction file (1966), Fleischer (1966), Strunz (1963).

Note: Minerals listed in brackets and not in alphabetical order are possibly varieties of the minerals listed immediately before them.

Sulphide minerals of bismuth are found widely distributed in hydrothermal deposits, in pyrometasomatic deposits alongside granite contacts with limestones, and in other pneumatolytic deposits. In the famous hydrothermal deposits of Tazna, Bolivia, bismuth is associated with tin deposits. Bismuth minerals are also found in granite pegmatites. Goldschmidt (1954) refers to crystals up to 1 inch wide and 6 inches long, in the Precambrian granite pegmatites from Southern Norway. Bismutotantalite is present in granite pegmatites of East Africa. Goldschmidt also mentions that he found major amounts of Bi in other tantalates and niobates in Norwegian pegmatites (no data given).

Metallic bismuth often occurs as a secondary mineral after bismuthinite, or in secondary zones of cementation. Further discussion on the occurrence of bismuth minerals and on the general geochemistry of Bi may be found in Goldschmidt (1954), Rankama and Sahama (1950), and Ramdohr (1960).

VI. Phase Equilibria

A useful summary of the phase equilibriu mrelationships in bismuth systems: (Bi_2O_3—MoO_2—PbO; Bi_2O_3—PbO; Bi_2O_3—PbO—WO_3) may be found in LEVIN, ROBINS and MCMURDIE (1964).

VAN HOOK (1960), SALANCI (1965) and CRAIG (1967) have carried out extensive experimental work on the Bi_2S_3—PbS system. The latest data by CRAIG indicate the maximum solubility of Bi_2S_3 in galena to be 10 mol-% (19 weight-%) at 829 ± 6° C,

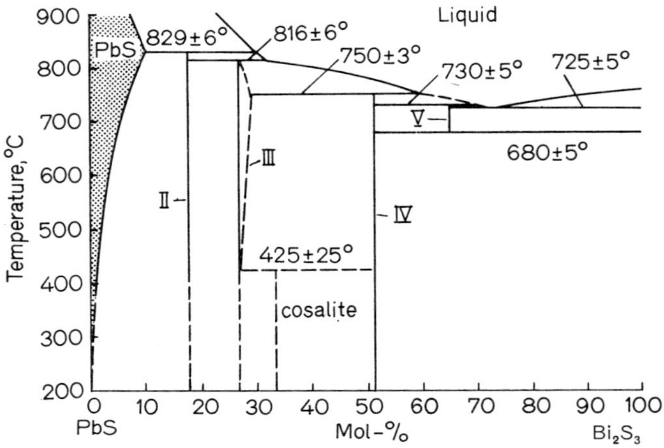

Fig. 83-D-1. System PbS-Bi_2S_3 after CRAIG (1967)

decreasing at lower temperatures (Fig 83-D-1). CRAIG suggests that the solid solution may vary sufficiently with temperature to provide a potential geologic thermometer.

VAN HOOK (1960) and CRAIG (1967) have also investigated the $AgBiS_2$—PbS system. The latter author's data show that solid solution is complete above 215 ± 15° C; below this temperature characteristic Widmanstätten structure-like textures are formed through exsolution. Even so, about 8% $AgBiS_2$ may remain dissolved in PbS at 200° C. Hence the solubility of bismuth in galena may be extensive, even at low temperatures, when silver is present.

Revised manuscript received: April 1968

83-E. Abundance in Common Igneous Rock Types

I. Bismuth in Standard Rocks, Granite G-1 and Diabase (tholeiitic basalt) W-1

A summary of the available Bi data (in ppm) in standard rocks G-1 and W-1 is given in Table 83-E-1.

Table 83-E-1. *Bismuth in standard rocks, granite G-1 and diabase (basalt) W-1*

G-1 ppm Bi	W-1 ppm Bi	Method	Source
0.09	—	ion exchange enrichment — spectrochemical	BROOKS *et al.* (1960)
0.25	—	polarographic	CARMICHAEL and MCDONALD (1961)
0.1	—	spectrographic	SCHROLL and WENINGER (1965)
< 0.9	< 0.9	spark source — mass spectrometer	BROWN and WOLSTENHOLME (1964)
0.10	0.25	spark source — mass spectrometer	TAYLOR (1965)

Some of the data on G-1 agree reasonably well at a magnitude of 0.1 ppm Bi. Data on W-1 are much less satisfactory. These data emphasize the need to improve the analytical techniques for determining Bi in rocks.

II. Bismuth in Igneous Rocks

Table 83-E-2 lists all other available data for Bi in igneous rocks. In mafic rocks, mainly in Karroo doleritic basalt, the Bi content varies enormously (< 0.01 ppm to 0.80 ppm) and averages at 0.15 ppm (BROOKS *et al.*, 1960; BROOKS and AHRENS, 1961). According to the data of the same workers, variations of the Bi content in acid rocks is also quite large. They report a range of 0.02 to 0.9 ppm Bi and an average of 0.18 ppm, close to the average for basaltic rocks. Alkaline rocks (syenite and foyaite) appear to contain less Bi (ave. ~0.03 ppm) but as only five specimens were analysed, an accurate abundance estimate cannot be made.

Table 83-E-2. *Bismuth contents of igneous rocks*

Rock type	No.	Method	ppm Bi range	ppm Bi ave.
Ultramafic rocks				
Peridotite, Norway, BROOKS *et al.* (1960)	1	S[a]		1.20
Lherzolite, France, BROOKS *et al.* (1960)	1	S		0.03
Eclogites (see Table 83-M-1)				

Table 83-E-2 (Continued)

Rock type	No.	Method	ppm Bi range	ave.
Mafic rocks				
Norite, South Africa, BROOKS *et al.* (1960)	1	S		0.05
Dolerite, South Africa, BROOKS *et al.* (1960)	1	S		0.15
Karroo dolerite, South Africa, BROOKS and AHRENS (1961)	16	S	0.01—0.80	0.22
Basalt, S.W. Africa, BROOKS and AHRENS (1961)	1	S		0.02
Basalt, U.S.A., REED *et al.* (1960)	1	N/R		0.008
Intermediate rocks				
Trachyte, New Zealand, BROOKS *et al.* (1960)	1	S		0.09
Syenite, S.W. Africa, BROOKS and AHRENS (1961)	3	S	0.02—0.03	0.03
Syenite, Canada, INGAMELLS and SUHR (1963)	1	S		0.25
Foyaite, S.W. Africa, BROOKS and AHRENS (1961)	2	S	0.04, 0.04	0.04
Andesite, New Zealand, TAYLOR and WHITE (1966)	1	I		0.12
Carbonatites				
Soevite, Ondurakorume Complex, S.W. Africa, BROOKS and AHRENS (1961)	1	S		0.04
Mocaceous, Ondurakorume Complex, S.W. Africa, BROOKS and AHRENS (1961)	1	S		0.06
Dolomitic, Ondurakorume Complex, S.W. Africa, BROOKS and AHRENS (1961)	1	S		0.06
Acidic rocks				
Granites composite 14, Germany, PREUSS (1940)		S		2
Granite, S. Africa, BROOKS *et al.* (1960)	1	S		0.9
Granite, France, INGAMELLS and SUHR (1963)	1	S		0.15
Tonalite, Tanganyika, SCHROLL and WENINGER (1965)	1	S		0.5
Rhyolite, S.W. Africa, BROOKS and AHRENS (1961)	4	S	0.02—0.22	0.12
Acid porphyry, S.W. Africa, BROOKS and AHRENS (1961)	2	S	0.04, 0.04	0.04

[a] Data obtained by an ion-exchange enrichment spectrochemical procedure.

Hence, the present data allow no firm conclusions to be drawn as to the variation of Bi in the various igneous rock types and with magmatic diferentiation.

III. Crustal Abundance of Bismuth

Some estimated crustal abundances of Bi are given in Table 83-E-3.

Table 83-E-3. *Estimated crustal abundances bismuth (in ppm)*

Bi	Source
0.07	NODDACK and NODDACK (1934)
0.20	GOLDSCHMIDT (1937), RANKAMA and SAHAMA (1950)
0.009	VINOGRADOV (1962)
0.17	BROOKS and AHRENS (1961), TAYLOR (1964)

There is little doubt that many more data are required before the crustal abundance of Bi can be established accurately. The most recent estimate (TAYLOR, 1964) is based on the work of BROOKS, AHRENS and TAYLOR (1960) and BROOKS and AHRENS (1961). As TAYLOR points out, his estimate refers particularly to the *continental* crust, but as the Bi content of acid rocks and of basalts appears to be of the same general magnitude, the continental crust value might be similar to the value for the entire crust (continental and oceanic). Data on the Bi content of oceanic basalts are unfortunately not available.

Revised manuscript received: August, 15th, 1968

83-F. Behavior in Magmatogenic Processes

In the previous sections it was noted that Bi is not easily accepted into common silicate minerals, the Bi abundance level usually being <1 ppm. Furthermore, as the average igneous rock has only 260 ppm S (TAYLOR, 1964), mainly in the form of pyrite and chalcopyrite where Bi is usually <10 ppm, a negligible proportion of Bi in igneous rocks is present in sulphides. These features suggest that Bi will tend to accumulate in residual melts during the magmatic differentiation process. Support for this suggestion is provided by the observation that Bi minerals are typically found in granite pegmatites and hydrothermal ore deposits. Quantitative data regarding the paragenesis of Bi minerals are apparently lacking. RANKAMA and SAHAMA (1950) mention that bismuth minerals separate and crystallize mainly from hot hydrothermal solutions, after arsenic minerals (pegmatitic stage), but before the crystallisation of antimony minerals.

Revised manuscript received: April 1968

83-G. Behavior during Weathering and Alteration of Rocks

Few data are available concerning the behaviour of Bi during weathering processes and hence it is not possible to discuss quantitatively the relationship existing between igneous and sedimentary rocks in terms of the geochemical cycle of Bi. To a certain extent, however, the geochemical cycle of Bi will be influenced by the stability of sulphide minerals such as bismuthinite and galena. Sulphides are readily attacked by acid solutions and hence Bi may be released for re-distribution in the secondary geochemical environment. Bismite (Bi_2O_3) is in fact often found as a secondary alteration product. There is a suggestion that some sedimentary rock types may be relatively enriched in Bi (see Table 83-K-1). GOLDSCHMIDT (1954) mentions that Bi concentrates in ferruginous bauxites and sedimentary oxidic iron ores; no data are given.

Revised manuscript received: April 1968

83-H. Behavior in Natural Waters

The characteristic oxidation state of Bi in the geological environment is Bi^{3+}. VINOGRADOV (1967) expects that Bi occurs in sea water in the form of the species BiO^+, $BiCl^-$ and BiOCl, while the residence time of Bi in the oceans is given by GOLDBERG (1965) as 4.5×10^4 years. The solubility of Bi_2S_3 in water is 0.18 ppm at 18° C; that of BiO(OH) is 1.44 ppm at 20° C (STEPHEN and STEPHEN, 1963).

RANKAMA and SAHAMA (1950) state that Bi is comparatively readily precipitated as basic carbonate which is formed during the hydrolytic decomposition of bismuth salts in aqueous solution. Hence, Bi should be enriched in hydrolyzate sediments. Bismutite $[Bi_2(CO_3)O_3]$ is an important carbonate mineral of bismuth.

Revised manuscript received: April 1968

83-I. Abundance in Ocean and Lake Water

A few data on the concentration of Bi in ocean water are available. Most of this information, summarized in Table 83-I-1, refers to surface water.

The data indicate strongly that the magnitude of Bi in surface waters of the oceans is ~0.02—0.04 ppb (μg/l) and there is an indication also that the concentration is lower in deep water. Distinctly higher concentrations, several above 2 ppb (μg/l) (maximum 7.1 ppb (μg/l)), have been reported by BOSWELL, BROOKS and WILSON (1967) in Lakes of McMurdo oasis, Antarctica.

Table 83-I-1. *Bismuth concentration in ppb Bi in ocean water*

ppb Bi	Method	Location	Source
0.2	Spectrochemical after co-precipitation	Gulmarsfjord, Norway	NODDACK and NODDACK (1940)
0.02	Spectrochemical after ion exchange enrichment	Surface water, south Atlantic	BROOKS (1960)
0.024—0.026	Spectrophotometric, after dithizone extraction	Surface water, English Channel	PORTMANN and RILEY (1966)
0.035—0.042	Spectrophotometric, after dithizone extraction	Surface water, Irish Sea	PORTMANN and RILEY (1966)
0.033	Spectrophotometric, after dithizone extraction	Surface water, North Atlantic	PORTMANN and RILEY (1966)
0.015	Spectrophotometric, after dithizone extraction	Water at 2000 m, North Atlantic	PORTMANN and RILEY (1966)

Revised manuscript received: April 1968

83-K. Abundance in Common Sedimentary Rock Types

Bi has been estimated in very few sediments and little is known about its abundance in them. The available data is summarized in Table 83-K-1. The normal clastic types of sediment show a large spread in concentration (0.06—2 ppm), but even so, there is the suggestion that the abundance level in some sediments may be higher than in igneous rocks. More data are obviously required.

Table 83-K-1. *Bismuth contents of sedimentary rocks*

Rock Type	No.	Method	ppm Bi	
			range	average
A. *Clastic Sediments*				
Shales composite p. 36, Germany, Preuss (1940)		S		1
Sandstones, composite of 23, Germany, Preuss (1940)		S		0.3
Ocean Mud, 68° S, 32° E, Brooks *et al.* (1960)	1	S		0.06
Shale, Table Mountain, S. Africa, Brooks *et al.* (1960)	1	S		2.1
Shale, Malmesbury, S. Africa, Brooks and Ahrens (1961)	1	S		0.10
Subgreywacke, Australia, Taylor (1966)	2	I	0.59, 0.65	0.62
Conglomerate, Eastern Alps, Weninger (1965)	4	S	0.5—3.0	1.3
Kupferschiefer, Mansfeld, Goldschmidt (1954)		S		100
B. "*Organic*" *Sediments*				
Anthracite, Eastern Alps, Weninger (1965)	2	S	1, 10	5.5
Coal and bitumen bearing sediments, Eastern Austria, Brandenstein *et al.* (1960)				
(1) Total rock samples	25	S	0.02—4.1	1.1
(11) Volatile, HF acid insoluble fraction	25	S	0.1—20	3.5
C. *Manganese Nodules*				
Atlantic, Indian, Pacific Oceans, Ahrens *et al.* (1967)	26	S	0.5—24	7
	6	S	<0.5	

With regard to the types classed as "organic", a distinct enrichment of Bi is noted, especially in the acid insoluble fractions of the coal and bitumen samples analysed by Brandenstein *et al.* (1960). However, lack of sample detail precludes further discussion. Goldscmhidt (1954) also mentioned an enrichment of Bi in the ashes of coal seams; however, no data are given.

Both GOLDSCHMIDT (1954) and RANKAMA and SAHAMA (1950) mention the possibility of enrichment of Bi in hydrolyzate sediments, associated with ferric hydroxide fractions. GOLDSCHMIDT (1954) quotes 100 ppm Bi in the sulphide bearing Permian Kupferschiefer from Mansfeld and considers this to be due to precipitation of Bi_2S_3 by hydrogen sulphide from stagnant bottom waters. This figure is way in excess of others indicated in Table 83-K-1.

The available data are sufficiently interesting to merit further investigations of the distribution of Bi in sedimentary materials.

The Bi concentration in 26 *manganese nodules* has been estimated by AHRENS, WILLIS and OOSTHUIZEN (1967). An ion exchange pre-enrichment — spectrochemical procedure was used. In common with several other elements, the Bi content was found to be much higher (Ave. = 7 ppm) in the nodules relative to common rocks and sediments. Variation of concentration is quite extreme (from <0.5 ppm to 24 ppm), a feature which is characteristic for several elements in manganese nodules.

Revised manuscript received: April 1968

83-L. Biogeochemistry

Gurney and Ahrens (in press), using a spectrochemical technique (detection limit $\sim$1 ppm) were unable to detect Bi in 10 samples of calcium phosphate (shark's teeth, whale bones and phosphate nodules) dredged from the ocean floor.

I. and W. Noddack (1940), found from 0.025 to 0.6 ppm Bi in the dry substance of various marine animals, highest amounts being present in a sponge and a jellyfish.

Bismuth in coals has been mentioned already under 83-K and in table 83-K-1.

Revised manuscript received: April 1968

83-M. Abundance in Common Metamorphic Rock Types

The few available data are summarized in Table 83-M-1. The Bi levels quoted by WENINGER (1965) for graphite bearing rocks are distinctly higher than found in common igneous rocks and sediments, but are similar to the Bi contents of "organic" sediments (Table 83-K-1, B). However, their composition is unusual and hence too much significance should not be attached to the results. The few results cannot be used to study the distribution of Bi with increasing grades of metamorphism.

Table 83-M-1. *Bismuth contents of metamorphic rocks*

Rock type	No.	Method	ppm Bi	
			range	average
Graphite and talc bearing rocks, Eastern Alps, WENINGER (1965)				
Graphite schists	2	S	3, 10	6.5
Graphite phyllites	5	S	0.5—10	3.1
Quartz phyllites	2	S	0.5, 1.0	0.75
Talc phyllites	3	S	0.5—5	2.1
Grey Gneiss, Antarctica, BROOKS *et al.* (1960)	1	S		0.04
Eclogite, Almeklovdalen, Norway, BROOKS *et al.* (1960)	1	S		0.07
Eclogite, South Africa, BROOKS and AHRENS (1961)	3	S[a]	0.02—0.10	0.05

[a] Data obtained by ion exchange enrichment and spectrochemical procedure.

Revised manuscript received: April 1968

83-O Relations to Other Elements; Economic Importance etc.

I. Inter-Element Relationships

Possible associations existing between Bi and other elements are discussed in section D. However, for the sake of completeness, a short summary is provided here. The chalcophile behaviour of Bi dominates its geochemistry and, therefore, apart from bismuth minerals themselves, highest concentrations of Bi are found in common sulphides. This is particularly so for galena where a strong Bi^{3+}—Pb^{2+} association is observed.

In Rare Earth (RE) minerals, some Bi—RE coherence appears to exist, notably between Bi^{3+} and Y^{3+} in gadolinites. In common silicate minerals, the most likely association is between Bi^{3+} and Ca^{2+}; however, the available data do not support this suggestion and it appears that such an association might be non-existent. The abundance levels of Bi in normal silicate rocks are, moreover, extremely low.

II. Technical and Economic Importance

Rich deposits of bismuth are uncommon, and most of the world's supply is obtained from the treatment of lead and copper refinery slimes, and as a by-product from the mining of other ores. An exception is the Tasna mine, Bolivia, where quartz veins contain native bismuth, bismuthinite and bismutite. The ores carry from 20—30% bismuth.

Bismuth has been extensively used in pharmaceutical products where bismuth compounds are used in medicinal remedies, cosmetics, ointments and for internal X-ray examinations. Bismuth compounds are also used to give a glaze to porcelain, for enamelling, and making optical glass.

Bismuth forms various alloys which melt far below the melting points of the separate metals. For example, bismuth melts at 271° C, but Woods metal (Bi—Pb—Sn—Cd) melts at 72° C. These so-called fusible alloys play an important part in the manufacture of fire alarms, automatic sprinklers, boiler safety plugs, hand grenades and electrical apparatus. There are also innumerable other minor uses.

Bismuth is stated to play an important part in nuclear energy projects.

Further details may be found in Bateman (1942) and Johnstone (1954).

Acknowledgment

We are grateful to Miss C. C. Fullard for assistance with the compilation of data used in this study.

Revised manuscript received: April 1968

References: Sections 83-B to 83-M, 83-O

AHRENS, L. H.: The use of ionization potentials. Part I. Ionic radii of the elements. Geochim. Cosmochim. Acta 2, 155 (1952).

—, and E. J. D. KABLE: A revised set of cationic radii. (In press.)

— J. P. WILLIS, and C. O. OOSTHUIZEN: Further observations on the composition of manganese nodules, with particular reference to some of the rarer elements. Geochim. Cosmochim. Acta **31**, 2169 (1967).

American Society for Testing and Materials. Index to the Powder Diffraction File, ASTM Publication PD1S—16i, Philadelphia (1966).

BATEMAN, A. M.: Economic mineral deposits. London: Chapman & Hall Ltd., 1942.

BETEKHIN, A. G.: Mineralogiya (Mineralogy). Moscow: Gosgeolizdat 1950.

BOSWELL, C. R., R. R. BROOKS, and A. T. WILSON: Some trace elements in lakes of McMurdo Oasis, Antarctica. Geochim. Cosmochim. Acta **31**, 731 (1967).

BRANDENSTEIN, M., I. JANDA u. E. SCHROLL: Seltene Elemente in österreichischen Kohlen- und Bitumengesteinen. Tschermaks Mineral. Petrog. Mitt. **7**, 260 (1960).

BROOKS, R. R.: The use of ion-exchange enrichment in the determination of trace elements in sea water. Analyst **85**, 745 (1960).

—, and L. H. AHRENS: Some observations on the distribution of Tl, Cd, and Bi in silicate rocks and the significance of covalency on their degree of association with other elements. Geochim. Cosmochim. Acta **23**, 100 (1961).

— —, and S. R. TAYLOR: The determination of trace elements in silicate rocks by a combined spectrochemical — anion exchange technique. Geochim. Cosmochim. Acta **18**, 162 (1960).

BROWN, R., and W. A. WOLSTENHOLME: Analysis of geological samples by spark source mass spectrometry. Nature **201**, 598 (1964).

BURNHAM, C. W.: Metallogenic Provinces of the south-western United States and Northern Mexico. Bull. 65, State Bureau of Mines and Mineral Resources, New Mexico Inst. Mining and Technology, Socorro, New Mexico (1959).

CAMBEL, B., u. J. JARKOVSKÝ: Geochemie der Pyrite einiger Lagerstätten der Tschechoslowakei. Vydavatefstvo Slovenskej akadémie vied Bratislava (1967).

CAMERON, A. G. W.: A new table of abundances of the elements in the solar system. Intern. Ser. of Monographs in Earth Sciences, Origin and distribution of the elements" (Paris, 1967). Oxford: Pergamon Press 1968.

CARMICHAEL, I., and A. McDONALD: The colorimetric and polarographic determination of some trace elements in the standard rocks G-1 and W-1. Geochim. Cosmochim. Acta **22**, 87 (1961).

CLARK, A. H.: The mineralogy and geochemistry of the Ylöjärvi Cu—W deposit, southwest Finland. Bull. Comm. Géol. Finlande **218**, 195 (1965).

CRAIG, J. R.: Phase relations and mineral assemblages in the Ag—Bi—Pb—S system. Mineral. Deposita **1**, 278 (1967.)

EHMANN, W. D., and J. R. HUIZENGA: Bismuth, thallium and mercury in stone meteorites by activation analysis. Geochim. Cosmochim. Acta **17**, 125 (1959).

FLEISCHER, M.: Minor elements in some sulphide minerals. Econ. Geol. Fiftieth Anniversary Volume, 970 (1955).

— Index of new mineral names, discredited minerals, and changes of mineralogical nomenclature in volumes 1—50 of The American Mineralogist. Am. Mineralogist **51**, 1247 (1966).

GOLDBERG, E. D.: Minor elements in sea water. In: Chemical oceanography (Edit. J. P. RILEY and G. SKIRROW). London and New York: Academic Press 1965.

GOLDSCHMIDT, V. M.: Geochemische Verteilungsgesetze der Elemente VII. Skrifter Norske Videnskaps-Akad. Oslo, Mat.-Naturv. Kl. No. 2 (1926).

GOLDSCHMIDT, V. M.: Geochemische Verteilungsgesetze der Elemente IX. Skrifter Norske Videnskaps-Akad. Oslo, Mat.-Naturv. Kl. No. 4 (1937).
— Geochemistry (edited by ALEX MUIR). Oxford: Clarendon Press 1954.
GURNEY, J. J., and L. H. AHRENS: The Bi contents of some rare earth minerals, notably gadolinite. (In press.)
HERTEL, G.: Die Fremdelementführung der Bleiglanze als Hilfe zur Bestimmung der Bildungstemperatur. Erzmetall **19**, 632 (1966).
HOEHNE, K.: Quantitative chemische und erzmikroskopische Bestimmung von As, Sb, Zn und Bi in vorwiegend schlesischen Bleiglanzen. Chem. Erde **9**, 219 (1934/35).
HOOK, H. J. VAN: The ternary system Ag_2S—Bi_2S_3—PbS. Econ. Geol. **55**, 759 (1960).
INGAMELLS, C. O., and N. H. SUHR: Chemical and spectrochemical analysis standard silicate samples. Geochim. Cosmochim. Acta **27**, 897 1963).
JOHNSTONE, S. J.: Minerals for the chemical and allied industries. London: Chapman & Hall Ltd. 1954.
LEVIN, E. M., C. R. ROBINS, and H. F. MCMURDIE: Phase diagrams for ceramists. Ohio: The American Ceramic Society 1964.
NESTEROVA, YU. S.: The chemical composition of galena. Geochemistry No. 7, 835 (1958).
NODDACK, I., u. W. NODDACK: Die Häufigkeit der chemischen Elemente. Naturwissenschaften **18**, 757 (1930).
— — Die Geochemie des Rheniums. Z. Physik. Chem. A **154**, 207 (1931).
— — Die geochemischen Verteilungskoefficienten der Elemente. Svensk. Kem. Tidskr. **46**, 173 (1934).
— — Die Häufigkeiten der Schwermetalle in Meerestieren. Arkiv Zool. No.4, 32 A (1940).
OFTEDAL, I.: Untersuchungen über die Nebenbestandteile von Erzmineralien Norwegischer zinkblendeführender Vorkommen. Skrifter Norske Videnskaps-Akad. Oslo, Mat.-Naturv. Kl. No. 8 (1941).
— Om betingelsane for oktaedrisk delbarheit hos vismutrik blyglans. Norsk Geol. Tidsskr. **22**, 61 (1942).
PALACHE, C., H. BERMAN, and C. FRONDEL: The system of mineralogy of James Dwight Dana and Edward Salisbury Dana, 7th edit. New York: John Wiley & Sons 1958.
PORTMANN, J. E., and J. P. RILEY: The determination of Bi in sea and natural waters. Anal. Chim. Acta **34**, 201 (1966).
PREUSS, E.: Beiträge zur spektralanalytischen Methodik II. Bestimmung von Zn, Cd, Hg, In, Tl, Gn, Sn, Pb, Sb, und Bi durch fraktionierte Distillation. Z. Angew. Mineral. **3**, 8 (1940).
RAMDOHR, P.: Die Erzmineralien und ihre Verwachsungen, 3. Aufl. Berlin: Akademie-Verlag 1960.
RANKAMA, K., and TH. G. SAHAMA: Geochemistry. Chicago: Chicago University Press 1950.
REED, G. W., K. KIGOSHI, and A. TURKEVICH: Determination of concentrations of heavy elements in meteorites by activation analysis. Geochim. Cosmochim. Acta **20**, 122 (1960).
ROSE, A. W.: Trace elements in sulfide minerals from the Central district New Mexico and the Bingham district, Utah. Geochim. Cosmochim. Acta **31**, 547 (1967).
SALANCI, B.: Untersuchungen am system Bi_2S_3—PbS. Neues Jahrb. Mineral. Monatsh. **12**, 384 (1965).
SCHROLL, E.: Über das Vorkommen einiger Spurenmetalle Blei-Zink-Erzen der ostalpinen Metallprovinz. Tschermaks Mineral. Petrog. Mitt. **5**, 183 (1965).
—, u. M. WENINGER: Eine empfindliche spektrochemische Analysenmethode zur Bestimmung von Germanium und Zinn unter Verwendung sulfidierender thermochemischer Reagenzien. Mikrochim. Acta **2**, 378 (1965).
SEEGER, P. A., W. A. FOWLER, and D. D. CLAYTON: Nucleosynthesis of heavy elements by neutron capture. Astrophys. J., Suppl. II, No. 97 121, (1965).
SHORT, M. N.: Microscopic determination of the ore minerals. U.S.G.S. Bull. 914, 2nd edit. (1940).
SIMS, P. K., and P. B. BARTON: Some aspects of the geochemistry of sphalerite, Central City District, Colorado. Econ. Geol. **56**, 1211 (1961).

STEPHEN, H., and T. STEPHEN: Solubilities of inorganic and organic compounds, vol. 1. London: Pergamon Press 1963.
STRUNZ, H.: Homöotypie Bi_2Se_2—Bi_2Se_3—Bi_3Se_4—Bi_4Se_5 usw. (Platynit, Ikunolith, Laitakarit.) Neues Jahrb. Mineral. Monatsh. 7, 154 (1963).
TAYLOR, S. R.: Abundance of chemical elements in the continental crust: a new table. Geochim. Cosmochim. Acta **28**, 1273 (1964).
— Geochemical application of spark source mass spectrometry. Nature **205**, 34 (1965).
— Australites, Henbury impact glass and subgreywacke: a comparison of the abundance of 51 elements. Geochim. Cosmochim. Acta **30**, 1121 (1966).
—, and A. J. R. WHITE: Trace element abundances in Andesites. Bull. Volcanol. **29**, 177 (1966).
VINOGRADOV, A. P.: Average contents of chemical elements in the principal types of igneous rocks of the earth's crust. Geochemistry No. 7, 641 (1962).
— Introduction to geochemistry of the oceans [in Russian]. Moscow: Science Publishing House 1967.
WENINGER, M.: Über Gehalte an Germanium, Zinn und einigen anderen Spurenelementen in ostalpinen Graphit- und Talkgesteinen. Tschermaks Mineral. Petrog. Mitt. **10**, 475 (1965).

Revised manuscript received: April 1968

Radon 86

and the other Noble Gases 10, 18, 36, 54

see Helium 2

Thorium 90

A	G. Bayer	(Institut für Kristallographie und Petrographie, Eidgen. Techn. Hochschule, Zürich, Switzerland)
B, C I	J. J. W. Rogers and J. A. S. Adams	(Department of Geology, Rice University, Houston, Texas, U.S.A.)
C II	U. K. Haack	(Geochemisches Institut der Universität, Göttingen, Germany)
D—O	J. J. W. Rogers and J. A. S. Adams	

90-A. Crystal Chemistry

I. Introduction

Thorium belongs to the series of actinide elements which are characterized by an incomplete 5f shell, analogous to the rare earth elements with incomplete 4f shells; such elements, therefore, may occur in a variety of valence states. Thorium is predominantly stable in the tetravalent state but is sometimes 3-valent and 2-valent in some halides and chalcogenides. It is the largest of the tetravalent cations with an ionic radius comparable to that of U^{4+} and Ce^{4+} (see Table 12-8 of Volume I of this handbook; also SHANNON and PREWITT, 1969).

The similarities in ionic size, outer electron configuration and bond character are the main reasons for the close relationship between the crystal chemistry of thorium, cerium, uranium and zirconium. There are many examples for isostructural compounds, e.g. ThS, US, CeS, ZrS(NaCl-type); ThO_2, UO_2, CeO_2, ZrO_2 (fluorite-type); $ThSiO_4$, $USiO_4$, $ZrSiO_4$ (zircon-type); $ThGeO_4$, $UGeO_4$, $CeGeO_4$, $ZrGeO_4$ (scheelite-type); and $BaThO_3$, $BaUO_3$, $BaCeO_3$, $BaZrO_3$ (perovskite-type). Although there is a definite preference by Th for 8- and 6-fold coordination in nonmetallic compounds, higher coordination numbers such as 9, 10 and 11 are also found. Pure thorium minerals, like thorianite (ThO_2) or thorite ($ThSiO_4$), are rather rare, but this element occurs as a minor constituent of a large number of minerals containing uranium and rare earths as monazite (see Section 90-D-2; also FRONDEL, 1958).

II. Metallic Compounds, Borides, Carbides, Nitrides, etc.

Elemental thorium exists in two modifications. The face-centered cubic α-form is stable from room temperature upto 1,400° C. Above this temperature it transforms reversibly into the body-centered cubic β-modification. The existence of many intermetallic compounds of thorium, varying greatly in composition and structure, has been reported by MAKAROV (1959) and NEVITT (1963); most of these structures are close-packed with high coordination numbers for thorium.

The stable carbides of thorium are ThC (NaCl-type like other IV-B group carbides) and ThC_2 (structurally closely related to NaCl). The C_2-groups instead of Cl-atoms change their orientation with temperature and ThC_2 is therefore known in three modifications: monoclinic from room temperature to 1,430° C, tetragonal from 1,430—1,480° C, and cubic above 1,480° C (BOWMAN *et al.*, 1968). The silicides of thorium have compositions ThSi, $ThSi_2$ and Th_3Si_2 (JACOBSON, FREEMAN, THARP and SEARCY, SR 1956, 191).

The nitride ThN crystallizes in the NaCl-structure, Th_2N_3 in the hexagonal A-type structure of the rare earth sesquioxides (ZACHARIASEN, SR 1949, 172), and Th_3N_4 in a special close-packed configuration of thorium atoms where nitrogen occupies octahedral and tetrahedral holes (BENZ, 1966). This structure is closely related to that of Th_2N_2O where the thorium atoms are in hexagonal close-packing. The compounds $Th_2(N, O)_2X$ (X = P, S, As or Se) crystallize in the Ce_2O_2S structure which is related to Th_2N_3 (BENZ and ZACHARIASEN, 1969).

III. Oxides and Oxosalts

The very stable oxide ThO_2 (thorianite) crystallizes in the fluorite structure; natural thorianite contains varying amounts of UO_2 and also rare earth oxides (ROBINSON and SABINA, 1955). On the other hand thorium is usually present in the minerals uraninite (UO_2) and cerianite (CeO_2). A continuous series of solid solutions exists between the isostructural oxides ThO_2 and UO_2. Thorium oxide may also form defect fluorite-type phases with Y_2O_3 (SCHUSTERIUS and PADUROV, SR 1955, 377) and with rare earth oxides (GINGERICH, SR 1961, 493). At higher sesquioxide concentrations the structure changes to the C-type of the rare earth oxides.

Various compounds between ThO_2 and other oxides, and also oxosalts of tetravalent thorium, are known. A compilation of such compounds is given in Table 90-A-1; in a few cases the crystal structure is not known. Oxide and silicate minerals containing thorium as a major constituent are listed in Subsection 90-D-3, Table 90-D-2; many of such minerals occur in the metamict state.

Thorium germanate crystallizes in the scheelite structure below 1,100° C and in the zircon structure above this temperature. The high temperature form of $ThTi_2O_6$ is isostructural with UTi_2O_6 and with the mineral brannerite. The linking of the (TiO_6) octahedra is similar to that in anatase. Thoro-aeschynite and thorbastnaesite, which are listed in Subsection 90-D-3, contain upto 30% ThO_2 and 47% ThO_2

Table 90-A-1. *Oxide compounds and oxosalts of thorium*

Compound	Coordination number of Th	Reference
ThO_2	8	GOLDSCHMIDT and THOMASSEN (SB 1913—28, 208)
α-$Th[GeO_4]$	8	BERTAUT and DURIF (SR 1954, 451)
β-$Th[GeO_4]$	8	BERTAUT and DURIF (SR 1954, 451)
α-$ThTi_2O_6$	6+2	HARARI and THERY (1969)
β-$ThTi_2O_6$	6	RUH and WADSLEY (1966)
α-$ThMo_2O_8$	8	THORET *et al.* (1970)
β-$ThMo_2O_8$	8	THORET *et al.* (1968)
$SrThO_3$, $BaThO_3$, $PbThO_3$	6	NARAY-SZABO (SR 1947/48, 456)
$Th_2(OH)_2[NO_3]_6\cdot(H_2O)_8$	11	JOHANSSON (1968)
$Th[NO_3]_4\cdot 5\,H_2O$	11	TAYLOR *et al.* (1966)
$Th[NO_3]_4\cdot 6\,H_2O$	8	TEMPLETON and DAUBEN, (SR 1951, 278)
$Th[P_2O_7]$	6	BURDESE and BORLERA (SR 1960, 405)
$NaTh_2[PO_4]_3$, $KTh_2[PO_4]_3$	9	MATKOVIC *et al.* (1970)
$CaTh[PO_4]_2$, (Ca, Ce, La, Th)$[PO_4]$	9	BOWIC and HORNE (SR 1953, 540)
$Me^{2+}Th[PO_4]_2$	9	SCHWARZ (1965)
$Me^{2+}Th[AsO_4]_2$	9	SCHWARZ (1965)
$Me^{2+}Th[VO_4]_2$ (Me^{2+} = Cd, Ca, Sr, Ba, Pb)	8	SCHWARZ (1965)
$Th_3[VO_4]_4$	8	LE FLEM *et al.* (1965)
$Th[V_2O_7]$	8	QUARTON, *et al.* (1970)
$Th(OH)_2[SO_4]$	8	LUNDGREN (SR 1950, 327)
$Th(OH)_2[CrO_4]\cdot H_2O$	8	LUNDGREN and SILLÉN (SR 1949, 254)

respectively (ES'KOVA *et al.*, 1964; PAVLENKO *et al.*, 1965). Cheralite, (Ca, Ce, La, Th) $[PO_4]$, and its synthetic analogue $CaTh[PO_4]_2$ are isostructural with huttonite, $ThSiO_4$. Table 90-A-1 demonstrates the importance of thorium-containing phosphates and their corresponding vanadates. $Th_3[VO_4]_4$ crystallizes in the zircon structure with partly unoccupied Zr positions; $PbTh[VO_4]_2$ occurs in three modifications, scheelite, huttonite and zircon.

IV. Silicates

In spite of the close crystal chemical relationship between thorium and zirconium, only a few thorium silicates are known so far as compared to the large number of Zr-containing silicates (see Subsection 40-A-IV). By far the most important silicate is the orthosilicate $ThSiO_4$ and its substituted derivatives. It occurs in two modifications, in the tetragonal structure of zircon as thorite, and in the monoclinic structure of monazite as huttonite. Contrary to expectation the denser huttonite structure is stable at high temperature and is formed irreversibly from thorite above 1,225° C (FINCH *et al.*, 1964). The more abundant thorite is usually found in the metamict state and frequently contains U, RE, Pb or Zr e.g. uranothorite, calciothorite, eucrasite and thorogummite (PABST, SR 1951, 307). A continuous series of solid solutions probably exists between $USiO_4$ and $ThSiO_4$ in the zircon structure (FUCHS and GEBERT, 1958); also partial replacement of $[SiO_4]^{4-}$ by $[OH_4]^{4-}$ is found in the thorite-variety thorogummite, as in zircon (FRONDEL, SR 1953, 565; FRONDEL and COLETTE, 1957). The monoclinic huttonite form seems to be inherently much more stable and is not so easily decomposed to a metamict state as thorite.

The metamict silicate, thorosteenstrupine, contains thorium instead of the rare earth elements in steenstrupine (see Table 90-D-2). The rare mineral ekanite, (Th, U) $(Ca, Fe, Pb)_2Si_8O_{20}$, is also found in the metamict state; it recrystallizes above 650° C to some unknown phases and to huttonite (ANDERSON *et al.*, SR 1961, 538). A new thorium-beryllium-silicate, $ThBe_2Si_2O_8$, is probably structurally related to anorthite (GREBENSHCHIKOV, 1964).

V. Chalkogenides, Phosphides and Halides

A large number of thorium sulfides, selenides and tellurides is known. ThS and ThSe crystallize in the NaCl-structure and ThTe in the CsCl-structure (ZACHARIASEN, SR 1949, 139; D'EYE *et al.*, SR 1952, 135; FERRO, SR 1955, 289); ThS_2 occurs in the orthorhombic $PbCl_2$ structure (ZACHARIASEN, SR 1949, 155) as is $ThSe_2$ (D'EYE, SR 1953, 257). A typical layer structure (PbFCl-type) is present in the compounds ThOS (Fig. 90-A-1), ThOSe, and ThOTe (ZACHARIASEN, SR 1949, 217; FERRO, SR 1955, 289). Th_2S_3 and Th_2Se_3 are isostructural with Sb_2S_3 (ZACHARIASEN, SR 1949, 381), and Th_7S_{12} and Th_7Se_{12} have a special structure type with 8- and 9-fold coordinated thorium (ZACHARIASEN, SR 1949, 184; D'EYE, SR 1953, 257).

Most of the phosphides, arsenides and antimonides of thorium are also isostructural (FERRO, SR 1955, 57 and SR 1956, 37), e.g. ThP, ThAs, ThSb (NaCl-type), $ThAs_2$, $ThSb_2$, $ThBi_2$ (anti-Fe_2As-type), and Th_3P_4, Th_3As_4, Th_3Sb_4, Th_3Bi_4 (cubic Th_3P_4-type).

The simple halides of thorium are listed in Table 90-A-2.

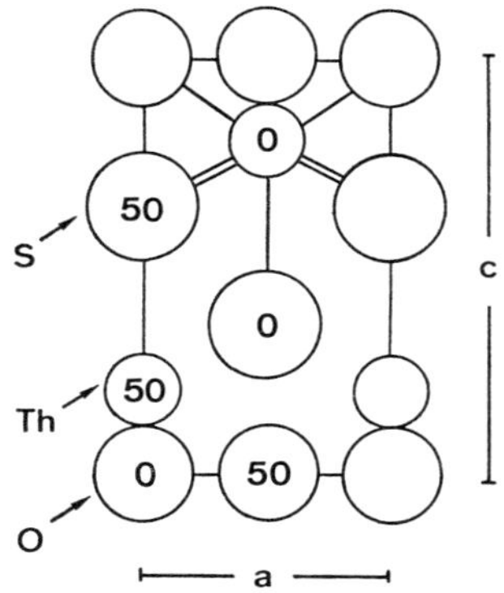

Fig. 90-A-1. The atomic arrangement in ThOS (PbFCl-structure) projected along one of the a-axes (FERRO, SR 1955, 289)

Table 90-A-2. *Thorium halides*

Compound	Structure type	Coordination number of Th	Reference
ThF_4	UF_4	8	ZACHARIASEN (SR 1949, 168)
$ThCl_4$	UCl_4	8	MOONEY (SR 1949, 167) MUCKER *et al.* (1969)
$ThBr_4$	UCl_4	8	D'EYE (SR 1950, 188)
ThI_2	CdI_2	6	ANDERSON and D'EYE (SR 1949, 159)
ThOF	CaF_2	8	RANNON and LUCAS (1969)
$ThO_{0.5}F_{2.5}$	LaF_3	11	RANNON and LUCAS (1969)
$ThOF_2$	LaF_3	11	ZACHARIASEN (SR 1949, 165)

Complex thorium fluorides have been reported representing a wide variety of compositions and structures: $CaThF_6$, $SrThF_6$, $BaThF_6$, $PbThF_6$ (LaF_3-type; ZACHARIASEN, SR 1949, 165); $(NH_4)_4ThF_8$ (RYAN *et al.*, 1969); $(NH_4)_3ThF_7$ (PENNEMANN *et al.*, 1971); $(Na, Li)_7Th_6F_{31}$ (BRUNTON and SEARS, 1969); α-KTh_6F_{25} (BRUNTON, 1972); $RbTh_3F_{13}$ (BRUNTON, 1971a); and $K_7Th_6F_{31}$ (BRUNTON, 1971b).

Revised manuscript received: March 1973

References: Section 90-A

Benz, R.: Th_3N_4 crystal structure and comparison with that of Th_2N_2O. Acta Cryst. **21**, 838 (1966).

Benz, R., Zachariasen, W. H.: Crystal structure of the compounds U_2N_2X and $Th_2(N,O)_2X$ with X = P, S, As and Se. Acta Cryst. B **25**, 294 (1969).

Bowman, A. L., Krikorian, W. H., Arnold, G. P., Wallace, T. C., Nereson, N. G.: The crystal structures of ThC_2. Acta Cryst. B **24**, 1121 (1968).

Brunton, C.: The crystal structure of $RbTh_3F_{13}$. Acta Cryst. B **27**, 1823 (1971a).

Brunton, C.: The crystal structure of $K_7Th_6F_{31}$. Acta Cryst. B **27**, 2290 (1971b).

Brunton, C.: The crystal structure of α-KTh_6F_{25}, a polymorph of CsU_6F_{25}. Acta Cryst. B **28**, 144 (1972).

Brunton, C., Sears, D. R.: The neutron and x-ray crystal structure of $(Na, Li)_7Th_6F_{31}$ with more than one least-squares minimum. Acta Cryst. B **25**, 2519 (1969).

Es'Kova, E. M., Zhabin, A. G., Mukhitdinov, G. N.: Mineralogy and geochemistry of the rare elements of Vishnevye Gor. Inst. Mineral Geokhim., Kristallokhim. Redk. Elementov 150 (1964).

Finch, C. B., Harris, L. A., Clark, G. W.: The thorite-huttonite phase transformation as determined by growth of synthetic thorite and huttonite single crystals. Am. Mineralogist **49**, 782 (1964).

Frondel, C.: Systematic mineralogy of uranium and thorium. U.S. Geol. Surv. Bull. **1064** (1958).

Frondel, C., Collette, R. L.: Hydrothermal synthesis of zircon, thorite and huttonite. Am. Mineralogist **42**, 759 (1957).

Fuchs, L. H., Gebert, E.: X-ray studies of synthetic coffinite, thorite and uranothorites. Am. Mineralogist **43**, 243 (1958).

Grebenshchikov, R. G.: Silicate de Th et Be, $ThBe_2Si_2O_8$ et ses rapports isomorphes avec les minéraux feldspathiques. Zh. Prikl. Khim. S.S.S.R **37**, 2044 (1964).

Harari, A., Thery, J.: Etude comparée des oxydes mixtes MO_2,nM'O_2. Rev. Int. Hautes Temp. Réfract .**6**, 29 (1969).

LeFlem, G., Hardy, A., Hagenmuller, P.: L'orthovandate de thorium. Compt. Rend. **260**, 1663 (1965).

Makarov, E. S.: Crystal Chemistry of Simple Compounds of Uranium, Thorium, Plutonium, Neptunium (transl. from Russian). New York: Consultants Bureau Inc. 1959.

Matkovic, B., Kojic-Prodic, B., Šliukič, M., Topič, M., Willett, R. D., Pullen, F.: Crystal structure of a new ferroelectric compound, sodium triphosphate-dithorate. Inorg. Chim. Acta **4**, 571 (1970).

Mucker, K., Smith, G. S., Johnson, Q., Elson, R. E.: Refinement of the crystal structure of $ThCl_4$. Acta Cryst. B **25**, 2362 (1969).

Nevitt, M. V.: In: Beck, P. A. (ed.): Alloy Chemistry of Transition Elements: Electronic Structure and Alloy Chemistry of the Transition Elements. New York: Interscience Publ., John Wiley 1963.

Pavlenko, A. S., Orlova, L. P., Akhmanova, M. V., Tobelko, K. I.: A thorium fluocarbonate, thorbastnaesite. Zap. Vses. Mineralog. Obshchestva **94**, 105 (1965).

Pennemann, R. A., Ryan, R. R., Kressin, I. K.: The crystal structure of ammonium heptafluorothorate. Acta Cryst. B **27**, 2279 (1971).

Quarton, M., Rimsky, A., Freundlich, W.: Crystal structure of the ThV_2O_7 phase. Compt. Rend. C **271**, 1439 (1970).

Rannov, J. P., Lucas, J.: L'oxyfluorure de thorium trivalent ThOF. Etude du diagramme ThOF—ThO_2—ThF_4. Mater. Res. Bull. **4**, 443 (1969).

ROBINSON, S. C., SABINA, A. P.: Uraninite and thorianite from Ontario and Quebec. Am. Mineralogist **40**, 62 (1955).

RUM, R., WADSLEY, A. D.: The crystal structure of $ThTi_2O_6$ (brannerite). Acta Cryst. **21**, 974 (1966).

RYAN, R. R., PENNEMAN, R. A., ROSENZWEIG, A.: The crystal structure of ammonium thiorium-(IV) octafluoride, $(NH_4)_4ThF_8$. Acta Cryst. B **25**, 1958 (1969).

SHANNON, R. D., PREWITT, C. T.: Effective ionic radii in oxides and fluorides. Acta Cryst. B **25**, 925 (1969).

TAYLOR, J. C., MUELLER, M. H., HITTERMAN, R. L.: Crystal structure of thorium nitrate pentahydrate by neutron diffraction. Acta Cryst. **20**, 842 (1966).

THORET, J., RIMSKY, A., FREUNDLICH, W.: Structure cristalline de là phase hexagonale $ThMo_2O_8$ haute temperature. Comptes Rend. C **267**, 1682 (1968).

THORET, J., RIMSKY, A., FREUNDLICH, W.: Structure cristalline de là phase orthorhombique $ThMo_2O_8$ basse temperature. Compt. Rend. C **270**, 2045 (1970).

Revised manuscript received: March 1973

Introduction

Owing to the geochemical similarity of thorium and uranium, it is necessary in the chapters on each element to refer to the geochemisty of the other. The chapter on uranium (92) specifically considers a number of topics pertaining to both elements, namely: the contribution of radioactive elements to heat flow; autoradiography; fractionation of thorium and uranium during weathering and the use of Th/U ratios as indices of sedimentary depositional environments; and the utilization of nuclear fuels as energy resources. This chapter on thorium includes a discussion of age dating by disequilibrium methods involving both the thorium and uranium series; the Th/U ratio in igneous rocks; and the regional variation in the abundances of radioactive elements in igneous rocks and the petrologic conclusions which can be drawn from these variations. A discussion of Th/Pb and U/Pb dating can be found in Chapter 82 (lead), Section B.

Reviews of the geochemistry of thorium have been published in a number of places. The present writers presented a review of the geochemistry of thorium and uranium (ADAMS *et al.*, 1959), and PETERMAN (1963) prepared a preliminary version of a very detailed description of the abundances of thorium and uranium in verious materials. A bibliography of the geology of uranium and thorium has been published by EL SHAZLY (1962). The present sections 90-B to 90-O are based primarily on world-wide literature published prior to 1965 with a few references to later papers. In the case of some papers which are difficult to obtain or are in a language other than English, German, or French, the data reported in this chapter have been taken from PETERMAN's 1963 summary.

Revised manuscript received: July 18th, 1967

90-B. Isotopes in Nature

Only two long-lived isotopes of thorium occur in nature. The major one is Th^{232}, which is the parent nuclide of a long decay series terminating with Pb^{208}. The half-life of Th^{232} is 1.39×10^{10} years, which is considerably longer than that of either U^{238} or U^{235}. The other important natural isotope of thorium, Th ^{230}, is generally present in minerals which contain uranium. Thorium-230 is the decay product of U^{234}, an intermediate in the decay of U^{238}. The half-life of Th^{230} is 80,000 years.

The fact that thorium and uranium are geochemically separable during sedimentation and other surficial processes permits disequilibrium dating of sediments in which some of the isotopes occur. The various techniques are summarized in a number of papers, including ROSHOLT *et al.* (1961, 1962), ROSHOLT and ANTAL (1963), SACKETT and POTRATZ (1963), SACKETT (1965), and THURBER (1965). In outline form, the various dating techniques are as follows:

(1) Uranium is precipitated from sea water in calcareous organisms or other forms of calcium carbonate. The calcium carbonate does not incorporate the thorium daughter products, and thus the rate of accumulation of Th^{230} is a measure of the age of the precipitated carbonate.

(2) The limited amounts of Th^{230} and Pa^{231} in sea water are absorbed in clay sediments. Because the thorium and protoactinium are unsupported by their parent uranium, which is not incorporated in the sediments, the rate of disappearance of thorium and protoactinium can be used to date the formation of the sediment. The ratio Pa^{231}/Th^{230} has a half-life of 60,100 years.

(3) The excess of U^{234} commonly found in sea water above the amount required for secular equilibrium permits a check on other methods of dating. The 15 percent excess of U^{234} gradually disappears in deposited sediments which incorporate uranium in any form, such as in the calcium carbonate lattice. The rate of establishment of secular equilibrium permits dating the sediment.

In addition to the two thorium isotopes discussed, four other short-lived isotopes also occur naturally:

Th^{234} — a member of the U^{238} decay series
Th^{231} — a member of the U^{235} decay series
Th^{228} — a member of the Th^{232} decay series
Th^{227} — a member of the U^{235} decay series.

Revised manuscript received: July 18th, 1967

90-C. Abundance in Meteorites, Tektites, and Lunar Materials

I. Meteorites and Tektites

The abundance of thorium and the Th/U ratios in meteorites are shown in Table 90-C-1. The uranium values from which the Th/U ratios are calculated are given in Table 92-C-1. Owing to the greater difficulty in analyzing for thorium than uranium, there are many more analyses for uranium than for thorium.

The average concentration of thorium in chondritic meteorites is in the range of 0.04 ppm. The average for achondrites is higher and the range is much greater than

Table 90-C-1. *Contents of thorium and Th/U ratios in meteorites*

Name	Class symbol	No. of samples	Th $10^2 \times$ ppm	Th/U	Method	Reference
Stones						
Achondrites						
Angra dos Reis	Aa	2	96.7	4.88	N	Morgan and Lovering (1964)
Binda	Ap	2	6.31	2.78	N	Morgan and Lovering (1964)
Bishopville	Ae	2	4.46	8.61	N	Morgan and Lovering (1964)
Ellemeet	Ab	2	0.379	2.51	N	Morgan and Lovering (1964)
Johnstown	Ab	2	0.57		N	Bate *et al.* (1957, 1959)
Juvinas		1	60		R	Rowe *et al.* (1963)
Moore Co.	Ap	2	6.20	3.16	N	Morgan and Lovering (1964)
Nakhla	Ado	2	19.0	3.86	N	Morgan and Lovering (1964)
Nuevo Laredo	Ap	1	54		N	Bate *et al.* (1957)
		2	47.6		N	Bate *et al.* (1959)
		1	47		R	Rowe *et al.* (1963)
fragment II		2	45.9	3.5	N	Morgan and Lovering (1965)
fragment III		2	49.2	3.5	N	Morgan and Lovering (1965)
Pasamonte	Aor	1	52		R	Rowe *et al.* (1963)
Sioux Co.	Ap	1	35		R	Rowe *et al.* (1963)
Stannern		1	50		R	Rowe *et al.* (1963)
Chondrites						
Beardsley	CH	1	4.3		N	Bate *et al.* (1957)
		1	4.77		N	Bate *et al.* (1959)
Forest City	CH	2	4.35		N	Bate *et al.* (1957)
		1	3.87		N	Bate *et al.* (1959)
Holbrook	CL	2	3.80		N	Bate *et al.* (1959)
Hvittis	Ce	2	3.11	5.02	N	Lovering and Morgan (1964)
Karoonda	CHL	2	5.66	4.13	N	Lovering and Morgan (1964)
Modoc	CL	1	4.5		N	Bate *et al.* (1957)
		2	3.92		N	Bate *et al.* (1959)
Mokoia	CHL	2	6.05	4.32	N	Lovering and Morgan (1964)
Orgueil	Cc	2	6.48	2.68	N	Lovering and Morgan (1964)
Richardton	CH	2	3.80		N	Bate *et al.* (1959)

Table 90-C-1 (continued)

Name	Class symbol	No. of samples	Th $10^2 \times$ ppm	Th/U	Method	Reference
Stony irons						
Brenham	P					
(metal)		1	<0.07		R	Reasbeck and Mayne (1955)
(stone)		1	5.3		R	Reasbeck and Mayne (1955)
(olivine)		1	1.1		N	Bate *et al.* 1958)
Irons						
Arispe I		1	0.0015		N	Bate *et al.* (1958)
Arispe II		1	0.00058		N	Bate *et al.* (1958)
Sandia Mts.		1	0.001		N	Bate*et al.* (1958)

Table 90-C-2. *Contents of thorium and Th/U ratios in tektites and glasses*

Name	No. of samples	Th ppm	Th/U	Method	Reference
Australite	2	11.7	4.8	R	Dubey (1933)
	1	9.19	5.28	I	Tilton (1958)
	6	9.4	5.0	R	Cherry (1962)
	1	12.9	4.8	R	Cherry and Adams (1963)
	1	12.2		R	Rowe *et al.* (1963)
	8	12.7	6.1	R	Balacek and Adams (1966)
Bediasite	3	4	3.3	R	Adams *et al.* (1959)
	2	8.6	5.9	R	Cherry and Adams (1963)
	2	9.0		R	Rowe *et al.* (1963)
	10	7.2	3.9	R	Balacek and Adams (1966)
Billitonite	1	9.6	3.6	R	Dubey (1933)
Indochinite	2	10.5	6.5	R	Adams (1956)
	4	13.0	6.4	R	Cherry and Adams (1963)
	1	11.9		R	Rowe *et al.* (1963)
	5	12.7	5.9	R	Balacek and Adams (1966)
Moldavite	4	15.4	5.8	R	Dubey (1933)
	1	11	6	R	Adams (1956)
	1	9.8	4.7	R	Cherry and Adams (1963)
Philippinite	1	15	9	R	Adams (1956)
	1	15.7	8.5	R	Cherry and Adams (1963)
	4	14.7		R	Rowe *et al.* (1963)
	12	14.6	5.8	R	Balacek and Adams (1966)
Darwin glass	1	11.3	8.1	R	Dubey (1933)
Libyan Desert glass	1	3.3		R	Rowe *et al.* (1963)
Peruvian glass	2	3.6	0.2	R	Cherry and Adams (1963)

for the chondrites. The thorium content of iron meteorites is extremely small. The average Th/U ratio for all of the stony meteorites is in the range of 3–4, with a few exceptional ones outside of this range. The average Th/U ratio is presumed to be the fundamental ratio for the solar system, and calculations of the age of the earth based on lead isotope data are generally based on the assumption of a primordial Th/U ratio of 3.5. The reported values of Th/U ratio are, in some cases, based on uranium measurements by one investigator and thorium measurements on other pieces of the same meteorite by another investigator. The combination of separate values clearly yields less precision than determination of the Th/U ratio by measurements by the same investigator on the same sample; Table 90-C-1 reports Th/U ratios only for measurements on the same sample.

Thorium concentrations in tektites are reported in Table 90-C-2, and comparable uranium values are given in Table 92-C-2. The thorium concentrations in tektites are much higher than in meteorites and average about 8 to 10 ppm, which is very close to the average concentration in the continental crust. The Th/U ratios for tektites are also reported in Table 90-C-2, and the uranium values on which they are based are given in Table 92-C-2. The Th/U ratios in tektites are on the order of 4 to 5 and seem to be somewhat higher than in meteorites or in most crustal materials. The origin of these slightly higher ratios is unknown. As discussed in Chapter 92, the high thorium and uranium contents of tektites are difficult to explain in terms of an extraterrestrial origin.

Revised manuscript received: July 1967

II. Lunar Materials

More than 500 analyses of thorium in lunar rocks obtained by neutron activation, γ-ray spectroscopy of natural radioactivity, and mass spectrometry have been summarized in Table 90-C-3 (spark source mass spectrometry data were not included). As the distribution of thorium in lunar rocks is intimately associated with that of uranium, the same rocks as in Table 92-C-3 are grouped together. The same problems, that existed with uranium with proper classification of rock types and with the inadequacy of the sample sizes also exist with thorium. The rock groups of Table 90-C-3 could certainly be arranged in many other ways. The variation within a rock group at one particular landing site is, in general, smaller than between similar rock groups of different landing sites. If the rocks are classified as in Table 90-C-3, the relative standard deviations from the means in 67% of the cases are less than 30% and exceed 50% only in 19% of the cases. Although hardly any sample can be claimed to be representative of a rock type, the large number of analyses and the remarkable agreement of the results by different analytical methods on rocks of the same group permits some generalizations to be made.

The lowest Th content, 0.0032 ppm, was found in an anorthosite, the highest, 34.2 ppm, in a granitic portion of a breccia. The average anorthosite contains 0.09 ppm Th and the average low-K mare basalt 0.7 ppm Th. The Th content of the other rocks is dominated by their KREEP component, the hypothetical rock which besides K, REE and P is enriched in other incompatible elements like U and Th. The different proportion of KREEP in the soils, breccias and melt rocks (anorthositic, troctolitic, very high alumina basalts and Apollo 14, 15 and 17 KREEP basalts) is reflected by their varying Th content.

The geochemical coherence of Th and U in lunar samples is even more pronounced than in terrestrial rocks. This is demonstrated by the constancy of the Th/U

Table 90-C-3. *Thorium in lunar rocks*

	Mean ppm Th	Range ppm Th	s	Number of analyses	References
Apollo 11					
Soils	2.05	1.31–2.3	0.30	12	16, 20, 39, 44, 46, 53, 60, 64, 66
Breccias	2.50	1.90–3.72	0.65	6	38, 44, 60
High-K basalts	3.33	2.80–4.03	0.38	17	21, 38, 39, 44, 46, 60, 61
Low-K basalts	0.82	0.53–1.08	0.19	12	38, 39, 46, 60, 61
Apollo 12					
Soils	6.73	3.2–13.5	1.98	39	2, 9, 17, 36, 40, 49, 53, 55, 61, 63, 66
Breccias	16.85	8.45–34.2	8.45	13	17, 19, 31, 36, 40, 49, 62, 63
Basalts	0.90	0.24–2.19	0.31	58	2, 9, 19, 24, 31, 36, 40, 43, 49, 55, 61, 63
Apollo 14					
Soils	13.30	11.3–15.9	1.38	23	6, 10, 15, 27, 28, 41, 48, 50, 56, 65
Breccias	14.14	4.2–21	3.65	21	3, 10, 15, 24, 56, 65
Melt rocks	11.37	8.6–13.3	1.29	11	3, 10, 24, 27, 41, 48, 51
Basalts	2.19	1.98–2.38	0.18	4	24, 48, 51
Apollo 15					
Soils	4.07	1.8–7.35	1.05	44	1, 14, 18, 24, 26, 27, 37, 41, 56, 58
Breccias	3.39	2.0–5.9	1.19	11	24, 37
KREEP-rich breccias	12.25	12.0–12.6	0.30	4	24, 41
Basalts	0.52	0.40–0.70	0.10	11	1, 24, 37, 41, 58
Anorthosites	0.065	0.028–0.102	—	2	24
Apollo 16					
Soils	1.91	0.51–2.77	0.50	65	1, 4, 8, 11, 13, 26, 30, 32, 52, 57, 58, 67
Breccias	2.10	0.22–7.8	1.68	50	4, 8, 11, 13, 22, 25, 30, 32, 58, 59
KREEP-rich breccias	9.15	8.1–10.34	0.82	8	4, 8, 11, 25, 29, 32
Melt rocks	2.23	0.63–3.55	1.16	7	1, 8, 11, 13, 22, 32
Anorthosites	0.096	0.0032–0.49	0.18	7	8, 22, 32, 52, 67
Apollo 17					
Soils	1.69	0.3–3.0	0.87	53	4, 7, 12, 13, 18, 25, 33, 42, 58, 59
Breccias	4.48	1.86–6.05	1.29	26	5, 12, 18, 22, 25, 34, 58, 59
Melt rocks	1.03	0.4–1.49	0.29	11	22, 25, 35
Basalts	0.38	0.17–0.79	0.11	32	5, 12, 13, 18, 25, 33, 58
Luna 16					
Soils	0.94	0.47–1.23	0.33	4	23, 50, 54
Luna 20					
Soils	1.09	0.73–1.54	0.29	5	26, 45, 47, 52

References: Abbreviations P I to P VI = Proceedings of the Lunar Science Conferences, Geochim. et Cosmochim. Acta, Supplements 1 to 6; GCA = Geochim. et Cosmochim. Acta; EPSL = Earth Planet. Sci. Letters.

1. Barnes, I. L. *et al.* (1973), P IV, 1197. 2. Brunfelt, A. O. *et al.* (1971), P II, 1281. 3. Brunfelt, A. O. *et al.* (1972), P III, 1133. 4. Brunfelt, A. O. *et al.* (1973), P IV, 1209. 5. Brunfelt, A. O. *et al.* (1974), P V, 981. 6. Brunfelt, A. O. *et al.* (1971), EPSL, **11**, 351. 7. Church, S. E. *et al.* (1974), P V, 1389. 8. Clark, R. S. *et al.* (1973), P IV, 2105. 9. Cliff, R. A. *et al.* (1971), P II, 1493. 10. Eldridge, J. S. *et al.* (1972), P III, 1651. 11. Eldridge, J. S. *et al.* (1973), P IV, 2115. 12. Eldridge, J. S. *et al.* (1974), P V, 1025. 13. Eldridge, J. S. *et al.* (1975), P VI, 1407. 14. Fields, P. R. *et al.* (1973), P IV, 2123. 15. Fields, P. R. *et al.* (1972), P III, 1637. 16. Fields, P. R. *et al.* (1970), P I, 1097. 17. Fields, P. R. *et al.* (1971), P II, 1571. 18. Fruchter, J. S. *et al.* (1975), P VI, 1399. 19. Goles, G. G. *et al.* (1971), P I, 1063. 20. Gopalan, K. *et al.* (1970), P I, 1194. 21. Herzog, G. F. *et al.* (1970), P I, 1239. 22. Hubbard, N. J. *et al.* (1974), P V, 1227. 23. Jerome, D. Y. *et al.* (1972), EPSL 13, 436. 24. Keith, J. E. *et al.* (1971), P III, 1671. 25. Laul, J. C. *et al.* (1974), P V, 1047. 26. Laul, J. C. *et al.* (1973), GCA **37**, 927. 27. Laul, J. C. *et al.* (1972), P III, 1181. 28. Lindstrom, M. M. *et al.* (1972), P III, 1201. 29. Lovering, J. F. *et al.* (1975), P VI, 1203. 30. Morrison, G. H. *et al.* (1973), P IV, 1399. 31. Morrison, G. H. *et al.* (1971), P II, 1169. 32. Nunes, P. D. *et al.* (1973), P IV, 1797. 33. Nunes, P. D. *et al.* (1974), P V, 1487. 34. Nunes, P. D. *et al.* (1974), EPSL **23**, 445. 35. Nunes, P. D. *et al.* (1975), P VI, 1431. 36. O'Kelley, G. D. *et al.* (1971), P II, 1159. 37. O'Kelley, G. D. *et al.* (1972), P III, 1659. 38. O'Kelley, G. D. *et al.* (1970), P I, 1407. 39. Perkins, R. W. *et al.* (1970), P I, 1455. 40. Rancitelli, L. A. *et al.* (1971), P II, 1757. 41. Rancitelli, L. A. *et al.* (1972), P III, 1681. 42. Rhodes, J. M. *et al.* (1974), P V, 1097. 43. Rosholt, J. N. *et al.* (1971), P II, 1577. 44. Silver, L. T. (1970), P I, 1533. 45. Surkov, Yu. A. *et al.* (1973), P IV, 1437. 46. Tatsumoto, M. (1970), P I, 1595. 47. Tatsumoto, M. (1973), GCA **37**, 1079. 48. Tatsumoto, M. *et al.* (1972), P III, 1531. 49. Tatsumoto, M. *et al.* (1971), P II, 1521. 50. Tera, F. *et al.* (1972), EPSL **13**, 457. 51. Tera, F. *et al.* (1972), EPSL **14**, 281. 52. Tera, F. *et al.* (1972), EPSL **17**, 36. 53. Travesi, A. *et al.* (1971), P II, 1277. 54. Vinogradov, A. P. (1971), P II, 1. 55. Vobecky, M. *et al.* (1971), P II, 1291. 56. Wänke, H. *et al.* (1972), P III, 1251. 57. Wänke, H. *et al.* (1973), P IV, 1461. 58. Wänke, H. *et al.* (1975), P VI, 1313. 59. Wänke, H. *et al.* (1974), P V, 1307. 60. Wänke, H. *et al.* (1970), P I, 1719. 61. Wänke, H. *et al.* (1971), P II, 1187. 62. Wakita, H. *et al.* (1970), EPSL **9**, 169. 63. Wakita, H. *et al.* (1971), P II, 1319. 64. Wanless, R. K. *et al.* (1970), P I, 1729. 65. Wiik, H. B. *et al.* (1973), EPSL **17**, 365. 66. Wrigley, R. C. (1971), P II, 1791. 67. Wrigley, R. C. (1973), P IV, 2203.

ratio for whole rock samples. The mean and median Th/U ratio of 477 determinations on the same rock samples (for references see Table 90-C-3) is 3.65, with 67% within ± 0.3 of this value. The smallest ratio reported is 1.8 in an anorthosite of Apollo 16 and the largest 6.1 in a breccia of Apollo 11. However, on a small scale there exist differences of two orders of magnitude: apparently U and Th concentrate during crystallization differentiation and are enriched in interstitial residual liquids. They enter discrete minerals only in the very last stage of crystallization; it is then that the two elements segregate. Burnett *et al.* (1971, P II, 1503)[1] report Th/U ratios for zircon of between 0.2 and 3.6, for a Zr-Ti-mineral of 2.9, for apatite $\sim 5 \pm 1$ and for whitlockite, of between 6.8 and 20. Andersen *et al.* (1972, EPSL **14**, 195) found Th/U ratios of 0.5–0.8 in zircon, 0.5 in baddeleyite, 1.6–1.7 in zirkelite, 2.2 in a Ti-Zr

[1] The abbreviations P I to P VI refer to the Proceedings of the First to Sixth Lunar Science Conferences, Geochim. Cosmochim. Acta Supplements 1 to 6. EPSL means Earth and Planetary Science Letters.

silicate, and 5.5 in REE phosphate and 20 in a whitlockite. BUSCHE *et al.* (1972, EPSL **14**, 313) report Th/U ratios of between 1.8 and 2.6 for a mineral which is probably zirkelite and HAINES *et al.* (1971, EPSL **12**, 145 and 1975, PVI, 3527) report ratios of 8–14 in whitlockite and 1.5 in a REE-rich Zr-Ti phase (in this the highest Th content of 4.1% was found). From these data it is obvious that Th is preferably incorporated in phosphates and U in Zr-minerals. These differences probably explain part of the spread for whole rock Th/U ratios because many of the samples analyzed were extremely small.

The orbital γ-ray mapping experiment (METZGER *et al.*, 1974, P V, 1067) yielded average Th concentrations of 0.9 ppm for the highlands, 3.2 ppm for the maria, and 2.2 ppm as an average for the surface. Although these numbers are very different from those that may be calculated from the soil data of Table 90-C-3 they are probably more representative. The reason for this is that the orbital data for the least heterogeneous landing sites (Apollo 11 and 16) agree well with the averages of the soils. As is the case with uranium, these concentrations are due to strong enrichment in the lunar crust. Many models for the evolution of the moon (for references see Sect. 92-C-III) arrive at $\sim$0.06 ppm U as the average for the entire moon. Under the assumption that the ratio Th/U = 3.65 is constant this implies that the average Th content of the moon is $\sim$0.2 ppm.

The isotopic composition of lunar thorium was studied by ROSHOLT *et al.* (1970, P I, 1499 and 1595; 1971, P II, 1577). They measured the ^{230}Th/^{232}Th ratio by α-particle spectrometry. Comparison of the measured ratio with the ratio which has to be expected from the ^{232}Th and ^{238}U contents revealed excesses at the α-particle energy of the ^{230}Th 4.68 MeV peak in some crystalline rocks. This excess seems to increase with the KREEP component and may reach 80%. As the effect cannot be explained by disequilibrium in the ^{238}U-^{234}U-^{230}Th decay sequence, the existence of an isomer of ^{232}Th (^{232m}Th) is suggested which emits α-particles of $\sim$4.7 MeV and which is accompanied by a decay series collateral to the regular decay series for ^{232}Th. It is possible that a variable cosmogenic history for ^{232}Th and ^{232m}Th may exist.

Manuscript received: January 1977

90-D. Abundance in Rock Forming Minerals (I); Thorium Minerals (II); Phase Equilibria (III)

I. Rock Forming Minerals

The concentrations of thorium in various rock-forming minerals are given in Table 90-D-1. The Th/U ratios shown in the table are calculated with the aid of the uranium data given in Table 92-D-1. As discussed in Chapter 92, Section D, the precise location of radioactive elements in rock-forming minerals is uncertain. In minerals such as allanite, huttonite, and monazite, the thorium clearly occupies one or more of the cation sites in the lattice as a major constituent. In zircon it may be presumed that thorium substitutes for zirconium, and in fact the Th/Zr ratio is relatively constant for a variety of crustal materials (for discussion see ADAMS *et al.*, 1959). In minerals such as epidote or potassium feldspar, it is uncertain whether thorium occurs in lattice sites, interstitial positions, submicroscopic defects, or perhaps in the form of individual small grains of thorium minerals.

The Th/U ratios of the various minerals reflect the ability of crystal lattices to discriminate between thorium and uranium. Thus allanite, huttonite, and monazite are primarily thorium minerals and have high Th/U ratios, whereas xenotime is primarily a uranium mineral. Zircon and sphene, common accessory minerals in granite, have Th/U ratios of 1–2, which contrasts with the general 3.5–4 ratios for whole igneous rocks. Whether these low ratios for zircon and sphene reflect differential inclusion of uranium in the mineral or possibly precipitation from an early magma of lower Th/U ratio than the magma from which the final rock crystallized is uncertain.

II. Thorium Minerals

Thorium is a major cation in a variety of minerals, none of which are at all common. Table 90-D-2, based largely on FRONDEL (1956), lists the various thorium minerals. The scarcity of thorium, its ability to substitute for other elements in crystal lattices, and the absence of a geochemical method of concentration of the element all combine to render thorium a highly dispersed material. Dispersal contrasts markedly with the tendency of uranium to become concentrated by precipitation from waters carrying the soluble, oxidized uranyl ion (see Chapter 92). Thorium occurs naturally only as the comparatively insoluble tetravalent ion.

Mineral formulas are those given by FRONDEL (1956) except in the case of minerals discovered since 1956 or minerals whose formulas have been modified by later work. References are given in each case of more recent discovery or new description. Papers consulted only secondarily through summaries in the American Mineralogist are shown in the reference list with a parenthetical reference to volume and page in the American Mineralogist.

Table 90-D-1. *Contents of thorium and Th/U ratios in minerals in igneous rocks*

	Th in ppm	Th/U	References (no. of samples)
Major minerals			
Quartz	0.5—10	1—5	A (3); C (1); J (6); K (5); M (1)
Feldspar (including potassium feldspar and plagioclase)	0.5—10	1—6	A (5); C (1); J (6); K (5); M (1)
Biotite	0.5—50	0.5—3	A (1); J (6); K (5); M (1)
Hornblende	5—50	2—4	A (1); J (4); M (1)
Olivine (from dunite)	0.02	low	L (2)
Accessory minerals			
Allanite	9100	high	G (10)
	1×10^3 to 2×10^4	high	B (1); F (4); G (10); J (6); K (5); M (1); N (1)
Apatite	70	1.3	E (29)
	50—250	~1	C (1); E (29); J (6); K (5); M (1)
Epidote	200	4.8	E (9)
	50—500	2—10	E (9)
Huttonite	nearly pure $ThSiO_4$		N (individual grains in one rock)
Magnetite (and other opaque minerals)	0.3—20		A (12); C (1); J (6); K (5); M (1)
Monazite	125,000	high	F (15)
	49,700	~25	E (11)
	2×10^4 to 2×10^5	high	E (11); F (15); H (4)
Sphene	510	1.7	E (25)
	100—1,000	1—3	C (1); E (25); J (6); K (5)
Thorianite and Uraninite	varies from ThO_2 to UO_2		
Thorite and Uranothorite	varies from $ThSiO_4$ to $USiO_4$ (?) (solid solution may not be complete)		
Zircon	560	0.4	E (43)
	$100—10^4$	0.2—2	C (1); D (21); E (43); J (6); K (5)

Ranges shown in the table are those within which thorium concentrations and Th/U ratios commonly occur rather than the extreme ranges for suites of samples. Averages are given only for: suites of minerals separated from North American rocks and reported by reference E, using gamma-radiometric methods; monazites reported by reference F, using dilution; and allanites reported by reference G, using radiometric methods.

A: Keevil (1944); B: Hutton (1951); C: Tilton *et al.* (1955); D: Webber *et al.* (1956); E: Hurley and Fairbairn (1957); F: Murata *et al.* (1957); G: W. L. Smith *et al.* (1957); H: Tilton and Nicolaysen (1957); J: Leonova and Tauson (1958); K: Leonova (1962); L: Heier and Carter (1964); M: Leonova and Renne (1964); N: Richardson (1964).

Data on Th/U ratios are given by Hurley and Fairbairn (1955), and preliminary data on accessory allanite is provided by Phair (1955).

III. Phase Equilibria

Mumpton and Roy (1961) have investigated the system ThO_2—UO_2—SiO_2 both dry and in the presence of water. A preliminary diagram of the system is shown in Fig. 90-D-1. Their conclusions may be summarized as follows:

1. Huttonite (monoclinic) is the stable form of $ThSiO_4$ at all temperatures, and thorite (tetragonal) is metastable.

2. Thorite dissolves not more than about 35 mol percent "$USiO_4$" (which does not exist as a separate phase), although other workers have reported considerable isomorphism between $ThSiO_4$ and the hypothetical "$USiO_4$".

3. ThO_2 and UO_2 are completely isomorphous at high temperatures; (experiments were performed at 1,350 degrees C.).

4. $ThSiO_4$ melts incongruently near 2,000 degrees C. to ThO_2 and SiO_2; (this conclusion is a citation of the work of another author).

5. Zircon dissolves less than 1 mol percent $ThSiO_4$ and 5 mol percent "$USiO_4$".

The frequency distribution of thorium concentrations in zircons has been studied by AHRENS (1965). The presumed lognormality of this distribution is also characteristic of uranium.

Table 90-D-2. *Minerals containing thorium as a major constituent*

Brockite	$Ca_{0.43}Sr_{0.03}Ba_{0.01}Th_{0.41}RE_{0.11}(PO_4)_{0.83}(CO_3)_{0.17} \cdot 0.9\ H_2O$ (FISHER and MEYROWITZ, 1962)
Cheralite	(Th, Ca, Ce) (PO_4, SiO_4)
Ekanite	(Th, U) (Ca, Fe, Pb)$_2$$Si_8O_{20}$ (ANDERSON *et al.*, 1961)
Huttonite	$ThSiO_4$
Monazite	(Ce, Y, La, Th) PO_4
Thorbastnaesite	$Th(Ca_{0.8}RE_{0.2})(CO_3)_2F_2 \cdot 3\ H_2O$ (PAVLENKO *et al.*, 1965)
Thorianite	ThO_2
Thorite	$ThSiO_4$
Thorogummite	$Th(SiO_4)_{1-x}(OH)_{4x}$
Thorosteenstrupine	$(Ca, Th, Mn)_3Si_4(O_{11.24}F_{0.95})_{12.2} \cdot 5\ H_2O$ (KUPRIYANOVA *et al.*, 1962)
Thorutite (Smirnovite)	$ThTi_2O_6$ (GOTMAN and KHAPAEV, 1958)
Uranothorianite	(Th, U) O_2
Uranothorite	(Th, U) SiO_4 solid solution may be incomplete

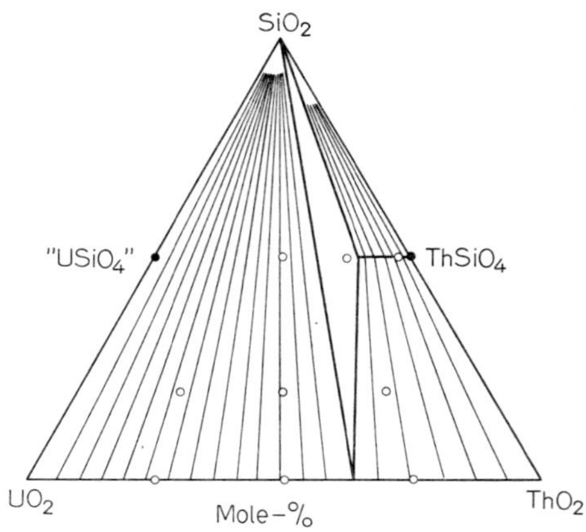

Fig. 90-D-1. Possible, though not completely verified, phase diagram of the ThO_2—UO_2—SiO_2 system in the neighborhood of 1350° C. (From MUMPTON and ROY, 1961)

Revised manuscript received: July 18th, 1967

90-E. Abundances in Common Igneous Rocks

Thorium contents and the Th/U ratios in various types of igneous rocks (Table 90-E-1) are discussed under several subheadings. This section includes a description of the average thorium content and Th/U ratios of various major rock groups, the variation in thorium content and Th/U ratios in differentiation sequences, the recognition of provinces of high thorium and uranium abundance, and various problems concerning the homogeneity of the distribution of thorium and uranium within the earth. The Th/U ratios shown in Table 90-E-1 are calculated using uranium data presented in Table 92-E-1.

Table 90-E-1. *Thorium contents and Th/U ratios of selected igneous rock suites*

Suite	No. of samples	Th ppm	Th/U	Method	Reference
Granitic rocks					
Russia:					
Precambrian mica granites of Ukranian shield (middle Dnepr area)	100	33	5.8	emanation	Filippov and Komlev (1959)
Tertiary Megrinsk intrusions of Southern Armenia; alkalic granodiorites to alkalic granites of 2nd intrusive phase	7	32	6.3		Meliksetyan (1961)
Northern Tien Shan; weighted av. of all crystalline rocks	—	15	3.7		Krylov (1963)
Paleozoic granites and syenites of Central and Western Tuva	47	16.0	4.1	radiochemical	Abramovich (1959)
United States:					
Conway granite, New Hampshire, Triassic, alkalic	1,265	56	3.7	R	Adams *et al.* (1962); Rogers *et al.* (1965)
New England granites	123	26	4.5	R	Rogers (1964)
Enchanted Rock batholith, Texas, granite	81	20	6	R	Rogers (1964)
Precambrian granites of Front Range, Colorado	38	25.5	5.1	C	Phair and Gottfried (1964)
Laramide stocks of Front Range, Colorado	23	30	4.0	C	Phair and Gottfried (1964)
Laramide stocks west of Front Range, Colorado	25	9.7	4.5	C	Phair and Gottfried (1964)

Table 90-E-1 (Continued)

Suite	No. of samples	Th ppm	Th/U	Method	References
Precambrian granites of middle Rocky Mts.	24	16	—	C	Phair and Gottfried (1964)
Idaho batholith; Laramide:					
tonalites and granodiorites	17	11	5.0	C	Larsen Jr. and Gottfried (1961)
qtz. monzonites and granites	9	17.8	4.7	C	Larsen Jr. and Gottfried (1961)
Sierra Nevada batholith, Cal.; granodiorites to granites; Jurassic	20	18	3.5	various	Rogers (1964)
Laramide granites of Idaho batholith, Boulder batholith, and northern Washington	38	10	4.7	various	Rogers (1964)
Granites and granodiorites of Southern Cal. batholith; Cretaceous	54	8	3.8	various	Rogers (1964)
Gabbroic and ultramafic rocks					
General average for "gabbroid" rocks	24	3.84	4.3	various	Heier and Carter (1964)
Dunite (North America)	3	0.05	low	various	Heier and Carter (1964)
Eclogites					
Pipe eclogite (Africa)	10	0.44	2.3	R	Heier (1963)
Pipe eclogite (Africa-1; Australia-3)	4	0.22	4.2	N	Lovering and Morgan (1963)
Average eclogite from metamorphic terraines (Norway and central Europe)	9	0.46	2.2	various	Heier and Carter (1964)
Alkaline intrusive rocks					
Nepheline syenites (northern Norway)	16	0.55	6	R	Heier (1962a)
Nepheline syenites of Lovozero massif, Kola Peninsula; weighted av.	—	35	—	C	Polyakov and Kot (1964)
Variscan alkalic syenites and nepheline syenites of southern Tien Shan	26	17	5.6	C	Leonova (1964)
Various suites from Kola Peninsula, Russia:					
Lesnaya Varaka (olivinite)	2	0.65		C	Polyakov and Volynets (1961)
Afrikanda (pyroxenite)	10	21.5		C	Polyakov and Volynets (1961)
Lake Varaka (alkalic)	5	27.8		C	Polyakov and Volynets (1961)

Table 90-E-1 (Continued)

Suite	No. of samples	Th ppm	Th/U	Method	References
Basaltic rocks					
Oceanic tholeiitic basalts	22	0.18	1.8	I	Tatsumoto *et al.* (1965)
Basalts and basaltic andesites of Mariana Islands	20	1.1	2.2	C	Gottfried *et al.* (1963)
Tholeiitic basalts of Hawaii	6	0.69	4.0	R	Heier *et al.* (1964)
	23	0.95	3.0	C	Larsen, 3d, and Gottfried (1960)
Alkali basalts of Hawaii	2	3.9	3.7	R	Heier *et al.* (1964)
	7	5.4	3.1	C	Larsen, 3d, and Gottfried (1960)
Japan					
tholeiitic basalts	2	0.19	1.6	R	Heier and Rogers (1963)
high-alumina basalts	2	0.78	3.7	R	Heier and Rogers (1963)
alkali-olivine basalts	2	3.9	7.6	R	Heier and Rogers (1963)
Caribbean Island arc					
spilites	6	0.26	1	R	Rogers and Donnelly (1966)
intrusive keratophyres	5	0.84	1	R	Rogers and Donnelly (1966)
extrusive keratophyres	4	0.39	1.4	R	Rogers and Donnelly (1966)
basaltic andesites	12	1.90	2.4	R	Rogers and Donnelly (1966)
Plateau basalts (av.)	21	1.96	4.2	various	Heier and Carter (1964)
Columbia River	2	1.0	3.3	R	Heier and Rogers (1963)
Palisades (New York)	5	1.8	5.3	R	Heier and Rogers (1963)

I. Abundances in Various Rock Groups

Table 90-E-1 shows the thorium contents and Th/U ratios for a variety of groups of igneous rocks. These groups represent suites of rocks or large assemblages of similar rock types which may be considered as individual populations. No effort has been made in Table 90-E-1 to summarize all of the measurements of thorium in igneous rocks.

Although an average thorium concentration for granites, for example, is essentially unobtainable owing to the extreme variation in thorium content from one type of granite to another, it is possible, nevertheless, to indicate the general range of thorium concentrations for major rock types. As shown in Table 90-E-1 the thorium content of granite is in the range of 10—20 ppm and decreases steadily through the intermediate rock types to a concentration in the range of 0.5—2 ppm for various types of basalt and gabbro. The thorium concentration thus shows the same increase in abundance as the uranium concentration toward the more petrologically differentiated rock types. In fact, the K/Th ratio has been found to have a nearly constant value of approximately 3×10^3 in a large variety of igneous rocks (HEIER and ROGERS, 1963). This constancy of the K/Th ratio has been noted in sediments as well as in igneous rocks by PLILER and ADAMS (1962a), and the K/Th ratio is, in general, even more constant than the K/U ratio in surficial materials. The extremely high concentration of thorium in alkaline rocks is noteworthy in view of the general tendency for thorium and potassium to occur together. Table 90-E-1 shows no significant differences between extrusive and intrusive rocks in regard to their thorium contents or their Th/U ratios.

The average thorium content of the continental crust may be estimated from the data given in Table 90-E-1 providing that the proportion of various rock types composing the crust is known. This problem has been discussed by a number of authors including PHAIR and GOTTFRIED (1964) for thorium and uranium and ROGERS (1966) for the general problem of geochemical balances. The abundance of granite in the earth's crust is almost certainly less than 50 percent, and most of the other rock types contain much less thorium and uranium than granite. Thus the average thorium content of the continental crust may be estimated in the range of 6—10 ppm, approximately one half the concentration in the highly differentiated granites. As discussed in Section 92-E, the uranium content of the continental crust is approximately 2 ppm, and thus the Th/U ratio for the continental crust should be about 3—4.

The average Th/U ratio for igneous rocks is shown by Table 90-E-1 to be in the range of 3.5 to 4. As discussed in Section 90-C, a primordial Th/U ratio of 3.5 is commonly used for calculations of the age of the earth by lead isotope methods. The similarity of crustal and cosmic ratios, however, does not indicate that the crust is a comprehensive sample of the cosmos. As pointed out, for example, by GAST (1960) and a number of other workers, the ratio of K/U and K/Th in the average chondritic meteorite is not the same as in crustal rocks. Furthermore, as discussed below, the Th/U ratio in rocks derived directly from the upper mantle is generally in the range of 1 to 2, thus indicating some type of preferential enrichment of thorium in crustal materials during igneous evolution.

The thorium and uranium contents and Th/U ratios in the upper mantle are of considerable interest in connection with heat flow studies, isotopic studies of lead, and also in regard to understanding the separation of the continental crust. The subject of heat flow is discussed more completely in the chapter on uranium (Section 92-B).

Obviously no direct measurements of the upper mantle are possible, but the study of primitive basalts derived from the mantle gives some information concerning its composition. TATSUMOTO *et al.* (1965) have determined the thorium and uranium contents of some mid-oceanic tholeiitic basalts. As discussed in that paper and references cited by it, these oceanic basalts appear to be the most primitive direct derivatives from the upper mantle. The values obtained by TATSUMOTO *et al.* of approximately 0.1 ppm uranium and 0.18 ppm thorium in the tholeiitic basalts correlate rather well with the thorium and uranium contents of the spilites reported by ROGERS and DONNELLY (1966) from the Caribbean area and some similar early-orogenic basalts from western Oregon. Apparently, as discussed by ROGERS and DONNELLY, the early stages of eugeosynclinal activity represent a tapping of relatively undifferentiated upper mantle material. The thorium and uranium content of the Carribbean and other eugeosynclinal materials gradually increases from the low values found in the spilites characteristic of early volcanic stages, and the orogenic sequence presumably terminates with the development of the typical silicic, highly radioactive rocks of a continental craton.

The thorium and uranium contents of the primitive oceanic tholeiitic basalts and eugeosynclinal spilites are essentially identical to the abundances of the elements estimated by heat flow studies for the upper mantle (see, for example, BIRCH, 1965).

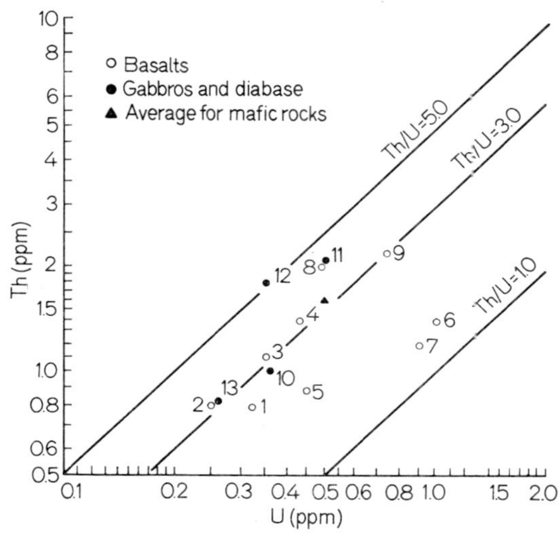

Fig. 90-E-1. Thorium and uranium contents of selected groups of mafic igneous rocks. Note the general tendency toward Th/U ratios of 3.0 or less. [From PETERMANN, Z. E.: unpublished, personal communication (1963)]

The similarity cannot, of course, result from bulk melting in the upper mantle, and it must be concluded that the fractional melting of the basaltic lava in the mantle takes place without fractionation of thorium and uranium between the solid and liquid phases. If there is no fractionation during partial melting, then the Th/U ratio of the upper mantle is the same as in the oceanic tholeiitic basalts (approximately 2). In view of a cosmic Th/U ratio of about 3.5, this ratio of 2 is probably not representative of the mantle as a whole. Additional data are needed, however, before this conclusion can be verified.

The fact that the Th/U ratio of primitive basalts is low (Fig. 90-E-1), plus the dissimilarity of the K/U ratio between rocks of the continental crust and the presumed bulk composition of the earth (e.g., see GAST, 1960) indicate segregation of radioactive constituents during continental evolution. It may be that the continental masses represent an ultimate stable segregate from the remainder of the earth, and thus their thorium and uranium concentrations and Th/U ratio of 3.5 to 4 give a comprehensive estimate of the radioactivity of terrestrial material. It is also possible, however, that different elements such as thorium and uranium diffuse upward out of the mantle at different rates, and the continents may not represent the ultimate differentiate. The fact that the Th/U ratio in continental crusts is similar to that assumed for the universe as a whole probably indicates that the craton is a stable ultimate differentiate from the mantle, although considerable doubt must be attached to this conclusion.

II. Variation within Igneous Rock Series

As mentioned in the preceding section, the thorium concentration of igneous rocks generally increases toward the silicic, more differentiated members of a series.

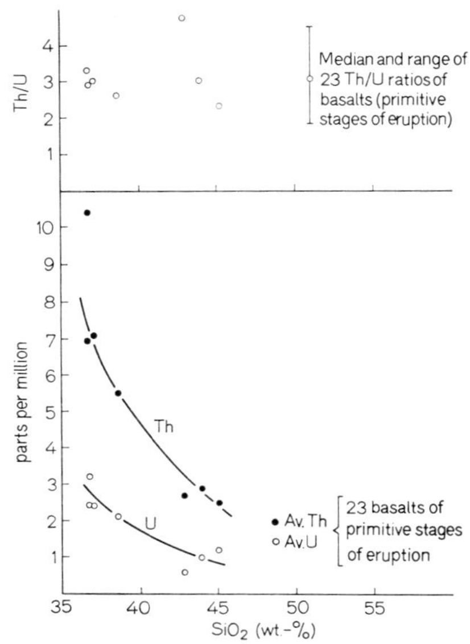

Fig. 90-E-2

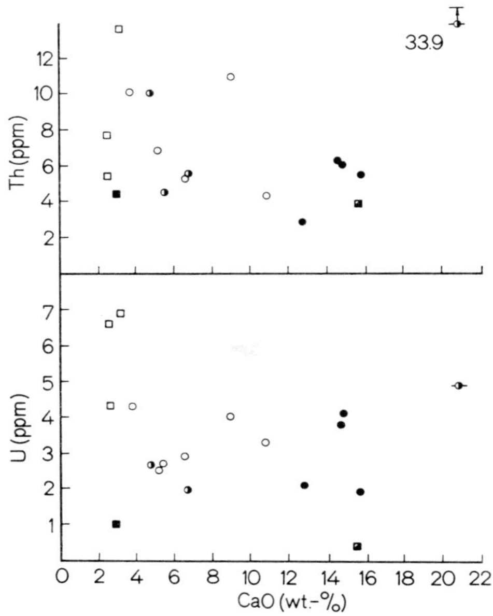

Fig. 90-E-3

Fig. 90-E-2. Variation of thorium and uranium concentrations and Th/U ratios in rocks of the Honolulu Series, Hawaiian Islands. The Honolulu Series is a group of alkalic basalts younger than the main tholeiitic basalts. [From PETERMAN, Z. E.: unpublished, personal communication (1963); based on data from LARSEN, 3rd, and GOTTFRIED, 1960]

Fig. 90-E-3. Distribution of uranium and thorium in rocks of the Magnet Cove igneous complex. Concentrations are plotted against CaO content. Plots of thorium and uranium concentrations against other major element abundances show no better correlation. [From PETERMAN, Z. E.: unpublished, personal communication (1963); based on data from ERICKSON and BLADE, 1963]

A large number of differentiation sequences is described by PETERMAN (1963), and the only one in which thorium content decreases towards the younger rocks is the Honolulu sequence of Hawaii (Fig. 90-E-2). As mentioned in section 92-E, the Honolulu series is also the only one in which uranium content decreases with differentiation. Plots of thorium content against silica content in differentiation sequences are generally more scattered (Figs. 90-E-2 to 90-E-7) than plots of thorium against other differentiation indexes such as the Larsen index or the potassium

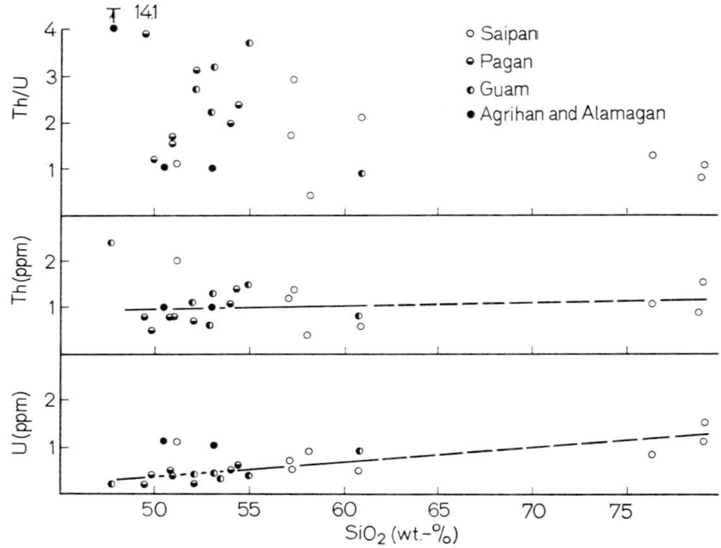

Fig. 90-E-4. Variation of thorium and uranium concentrations and Th/U ratios in volcanic rocks of the Marianas Islands. [From PETERMAN, Z. E.: unpublished, personal communication (1963); based on data from GOTTFRIED *et al.*, 1963]

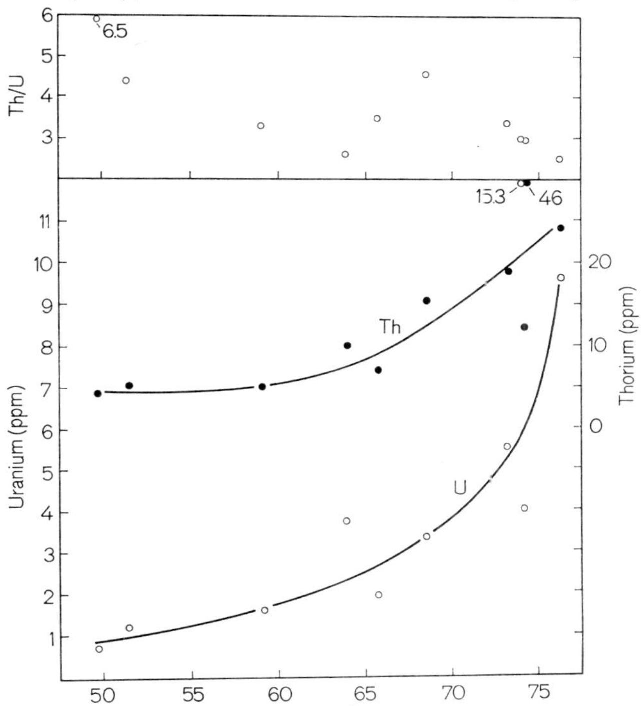

Fig. 90-E-5. Variation of thorium and uranium concentrations and Th/U ratios in rocks of the Valles Mountains, New Mexico, U.S.A. [From PETERMAN, Z. E.: unpublished, personal communication (1963); based on data from LARSEN, 3rd, and GOTTFRIED, 1960]

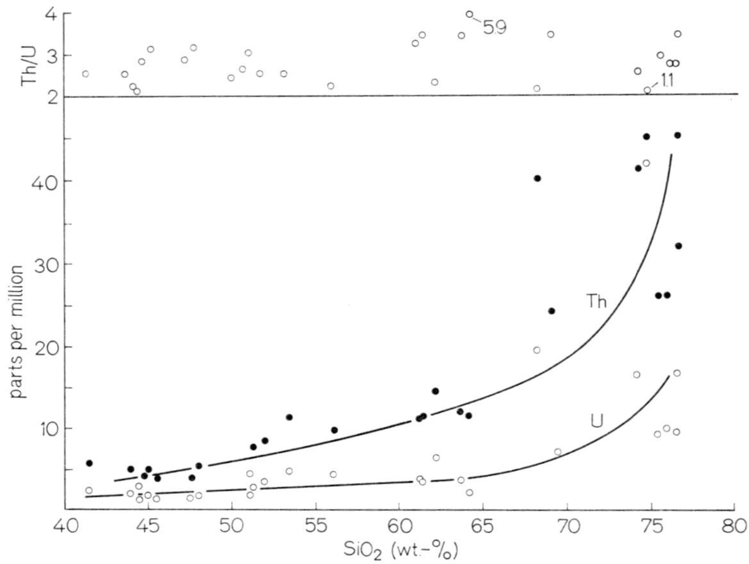

Fig. 90-E-6. Variation of thorium and uranium concentrations and Th/U ratios in alkalic volcanic and hypabyssal rocks of Big Bend National Park, Texas, U.S.A. [From PETERMAN, Z. E.: unpublished, personal communication (1963); based on data from LARSEN and GOTTFRIED, 1960]

content. Plots of thorium against potassium are almost invariably quite linear. In PETERMAN'S (1963) summary, the number of differentiation sequences in which thorium has been studied is less than the number in which uranium has been studied. The sequences in which uranium content increases with differentiation are listed in Table 92-E-2. Among those sequences are several in which the thorium content has been found to increase toward the younger members of the sequence; these are listed in Table 90-E-2 (Figs. 90-E-5 to 90-E-8).

Variation in the Th/U ratios in igneous rock series has been a topic of considerable discussion. LARSEN, *3rd* and GOTTFRIED (1960) found no general relationship between the ratios and position in several differentiation sequences. Among the sequences summarized by PETERMAN (1963), however, some increase in the Th/U ratio toward the younger members of the series is shown by the White Mountain and Oliverian intrusive sequences, whereas the extrusive sequences in the Mariana Islands, Lassen Park, and the Valles Mountains show a decrease in the Th/U ratio with increasing differentiation. A general increase in the Th/U ratio with differentiation has been proposed in several papers. ROGERS and RAGLAND (1961) found a weak, though statistically valid, relationship between Th/U ratios and potassium content for a large suite of silicic intrusive rocks (Fig. 90-E-9). The increase in ratio with differentiation is attributed to the oxidation proposed by OSBORN (1959) as necessary for the evolution of granitic magmas. HEIER and ROGERS (1963) found a linear relationship between Th/U ratio and potassium content for various apparently primary basalts (Fig. 90-E-10). As shown in Table 90-E-1, the Th/U ratio for many basic rocks is somewhat lower than for silicic ones.

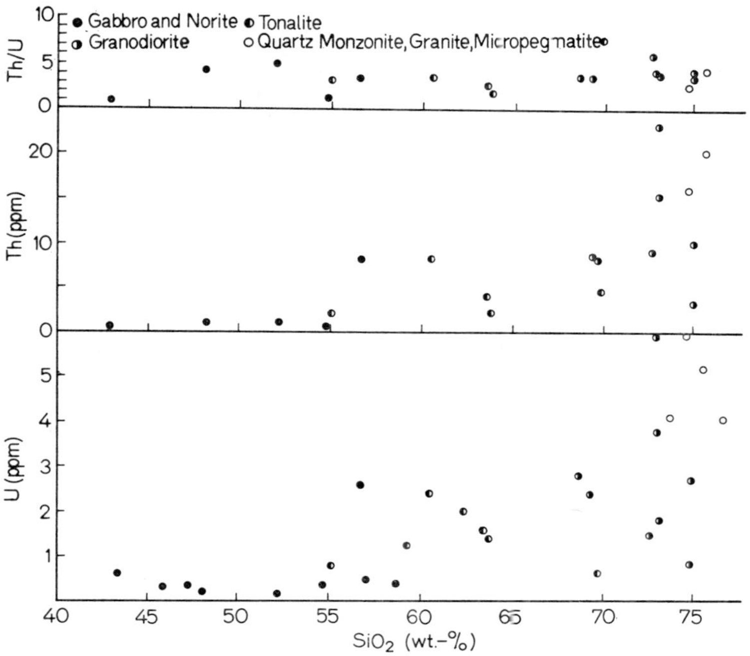

Fig. 90-E-7. Variation of thorium and uranium concentrations and Th/U ratios for the Southern California Batholith, U.S.A. [From PETERMAN, Z. E.: unpublished, personal communication (1963); based on data from LARSEN, 3d, and GOTTFRIED, 1960; LARSEN JR., and GOTTFRIED, 1961; WHITFIELD *et al.*, 1959; ROGERS and RAGLAND, 1961; HEIER and ROGERS, 1963; and personal communication from D. GOTTFRIED]

Table 90-E-2. *Sequences of igneous rocks showing increase of thorium content toward the younger rocks*

Extrusive series

Lassen Volcanic National Park (ADAMS, 1955; LARSEN, 3rd, and GOTTFRIED, 1960)
Modoc area, California (GOTTFRIED and LARSEN JR., 1958; LARSEN, 3rd, and GOTTFRIED, 1960)
Valles Mountains, New Mexico (LARSEN, 3rd, and GOTTFRIED, 1960) (Fig. 90-E-5)
Big Bend National Park, Texas (GOTTFRIED *et al.*, 1962) (part intrusive) (Fig. 90-E-6)
Mt. Garibaldi area, British Columbia (GOTTFRIED *et al.*, 1963)
Strawberry Mountains, Oregon (GOTTFRIED *et al.*, 1963)

Sills and shallow intrusives

Augusta County, Virginia (GOTTFRIED *et al.*, 1962)
Duluth gabbro, Minnesota (HEIER and ROGERS, 1963)

Plutonic sequences

Southern California batholith (LARSEN, 3rd, and GOTTFRIED, 1960; LARSEN JR. and GOTTFRIED, 1961; WHITFIELD *et al.*, 1959; ROGERS and RAGLAND, 1961; HEIER and ROGERS, 1963) (Fig. 90-E-7)
White Mountain magma series, New Hampshire (BUTLER, 1961; ROGERS and RAGLAND, 1961; ADAMS *et al.*, 1962)
Oliverian series, New Hampshire (ROGERS and RAGLAND, 1961; LYONS, 1961) (Fig. 90-E-8,

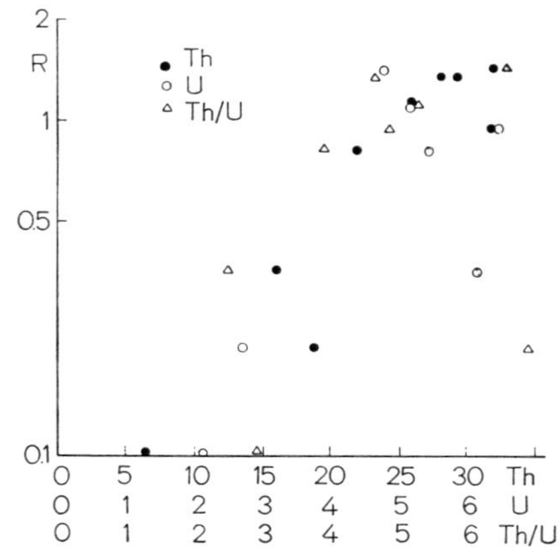

Fig. 90-E-8. Variation of thorium and uranium concentrations and Th/U ratios in the Oliverian series of mantled gneiss domes, New Hampshire, U.S.A. with R (ratio of potassium feldspar to plagioclase). (From ROGERS and RAGLAND, 1961)

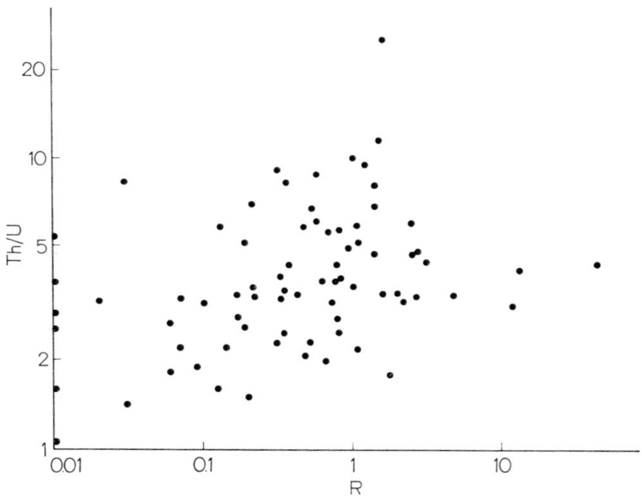

Fig. 90-E-9. Th/U ratio vs. the ratio of potassium feldspar to plagioclase (R) for 77 granitic rocks. Rocks with R values of less than 0.01 are plotted along the 0.01 line. The correlation coefficient of the logarithms of the ratios is 0.31, which is significant at the 99% confidence level. (From ROGERS and RAGLAND, 1961)

The preceding paragraph indicates a discrepancy in the variation of Th/U ratios which has not been satisfactorily explained. Where rocks from a variety of plutonic and/or extrusive sequences are compared, the Th/U ratio is generally higher in the more differentiated rocks, i.e., in those containing more silica or potassium. The few intrusive sequences which show systematic variation in Th/U ratios also exhibit an increase in the ratio with differentiation. Where there is variation in the Th/U ratio in extrusive sequences, however, the ratio decreases toward the younger rocks, despite the pronounced correlation between Th/U and potassium content found for apparently primary basalts. No comprehensive explanation of these observations is yet available.

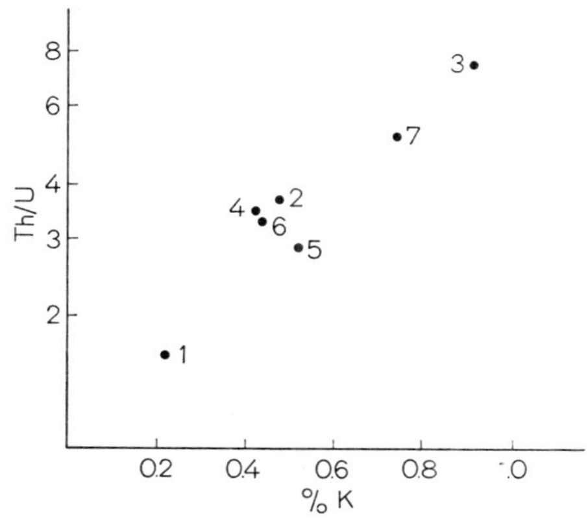

Fig. 90-E-10. Relationship between Th/U ratio and potassium content in various mafic igneous rocks. (From HEIER and ROGERS, 1963)

III. Provinces of High Thorium and Uranium Contents

High thorium and uranium concentrations appear to characterize the igneous rocks formed over considerable periods of time within certain areas. One excellent example is the Colorado Front Range, studied by PHAIR and GOTTFRIED (1964). Here thorium concentrations at least twice as large as the average for cratonic rocks are found in igneous rocks ranging in age from Precambrian through Laramide. Thorium concentrations of rocks within the Front Range area contrast markedly with the concentrations in rocks of equivalent age outside of the province.

Another area of high thorium concentration is the general New England region. Here the dominantly alkaline granites ranging in age from Paleozoic through lower Mesozoic have concentrations of thorium as high as 50 or more ppm and equivalent high abundances of uranium. The major example is the Triassic Conway granite of New Hampshire (ADAMS *et al.*, 1962). Although the data are not adequate, there seems to be some tendency for New England granites to have a slightly higher Th/U ratio than those elsewhere in the United States (ROGERS, 1964). High contents of thorium and uranium had not been found in the rocks of the Pacific Coast area until the comparatively recent work of WOLLENBERG and SMITH (1964). It is possible that the general tendency for granites from the western United States to be high in sodium and comparatively deficient in potassium, whereas those of the eastern United States are more potassic, is reflected in the tendency for higher thorium and uranium contents in the eastern granites. The concept of thorium and uranium provinces is discussed in more detail by PETERMAN (1963).

A possible relationship between age and Th/U ratios has been suggested by WHITFIELD *et al.* (1959) for granites in North America. In general, Precambrian granites in this area have a slightly higher Th/U ratio than granites of other ages. This higher ratio is primarily the result of slightly lower uranium content, and it is possible that Precambrian rocks have suffered depletion of uranium owing to upward migration of the element throughout geologic time.

The homogeneity of thorium and uranium distributions in the silicic intrusive rocks has been discussed by ROGERS (1964) and ROGERS *et al.* (1965) based on an interpretation of frequency distribution of the elemental abundances. The distribution of thorium within homogeneous bodies such as the Conway granite of New Hampshire was found to be homogeneous (lognormal), although the distribution of uranium was non-homogeneous. Presumably the thorium distribution is related to primary crystallization processes, whereas the uranium has undergone surficial redistribution which has altered the primary homogeneous distribution (Section 92-G). The thorium contents of New England rocks are not homogeneous for populations which include several plutons owing to differences in the abundances of thorium in the separate bodies. The secondary surficial control of uranium, however, yields a homogeneous distribution from the surficial rocks of a population which includes a large number of New England plutons (Fig. 92-E-2). The problem of distribution of thorium and uranium within relatively small areas of rock has also been discussed in the Russian literature; (see, for example, LEONOVA and RENNE, 1964).

Revised manuscript received: July 18th, 1967

90-F. Behavior during Processes Connected with Magmatism

The tendency of thorium to be incorporated in primary minerals in igneous rocks reduces its ability to concentrate in late-stage fluids. Thorium minerals, primarily allanite, are found in some pegmatites, but high concentrations are not common. The isomorphism between uraninite and thorianite permits small concentrations of thorium to accompany uranium in hydrothermal deposits, but the Th/U ratio in most such ores is extremely low. In North America only one major vein deposit has been discovered, the recently developed thorianite area at Lemhi Pass on the Idaho-Montana border. The geology of this deposit is discussed in detail by ANDERSON (1961).

Revised manuscript received: July 18th, 1967

90-G. Behavior during Weathering and Rock Alteration

As discussed more fully in the chapter on uranium (Section 92-G), thorium and uranium are commonly fractionated during surficial processes owing to oxidation of uranium to the soluble uranyl ion. The relatively immobile thorium becomes concentrated in residual materials such as soil and weathered rocks. Two examples may be cited of the residual accumulation of thorium. One is the high concentration in bauxite as reported by ADAMS and RICHARDSON (1960), who found concentrations in the general range of 50 ppm or more. Thorium in bauxites is contained largely in resistate minerals such as zircon but also partly in the aluminum hydroxide minerals, possibly by adsorption. Residual accumulations in clays are also noted by ADAMS and WEAVER (1958).

A second example of retention of thorium during weathering and alteration is the Pennsylvanian weathering profile on the granite of Flagstaff Mountain, Colorado (PLILER and ADAMS, 1962b). Thorium concentrations in this subtropically weathered material are as high as 34 ppm in the most thoroughly kaolinized rock. The thorium is partly in resistate minerals but in large part attached to the clay (Fig. 92-G-2).

Revised manuscript received: July 18th, 1967

90-H. Solubilities in Waters; Valence States

Thorium occurs in nature only as the tetravalent ion. The close association of thorium and potassium in a wide variety of rocks, however, together with the concentration of all radioactive materials toward the outer part of the earth (see Section 92-C), indicates that thorium is probably closely coordinated with oxygen ions. This coordination complex provides the low density necessary for upward migration which the individual thorium ions would not show.

Thorium forms a large number of salts such as the sulfate, chloride, etc. which are soluble in acid solutions. Slightly basic solutions however, hydrolyse the thorium ion to an oxide or hydroxide precipitate. The solubility is commonly given in terms of the solubility product of $Th(OH)_4$, which is approximately 10^{-42}.

Revised manuscript received: July 18th, 1967

90-I. Abundance in Natural Waters

The low solubility of thorium in natural waters is demonstrated in Table 90-I-1 for sea water and in Table 90-I-2 for fresh water. Solubilities are in the range of 10^{-5} to 10^{-6} ppm (10^{-2} to 10^{-3} ppb). Comparison with Tables 92-I-1 and 92-I-2, which show comparable values for uranium, indicate a Th/U ratio of about 10^{-2} to 10^{-3} for most waters. The determination of thorium concentrations in waters is extremely difficult and in some cases is inferential; the problem is discussed in various papers by F. KOCZY and co-workers.

Table 90-I-1. *Contents of thorium in sea water*

Location	Th ppb	Method	Reference
Scandinavian waters	<0.5	W	FØYN *et al.* (1939)
Scandinavian waters	0.00004		F. KOCZY (1956)
Scandinavian waters	<0.02	autoradio-graphic	F. KOCZY *et al.* (1957)
Pacific (deep ocean)	<0.05	R	SACKETT *et al.* (1958)
Indian Ocean	0.01		BARANOV and KHRISTIANOVA (1959)
Black Sea	0.0022	W	STARIK *et al.* (1959)
Sea of Azov (central part)	0.004	R	NIKOLAEV *et al.* (1961)

Table 90-I-2. *Contents of thorium in continental waters*

Location	No. of samples	Th ppb	Method	Reference
River water		0.01	calcu-lation	KOCZY (1956)
River waters of Japan	10	0.027	C	MIYAKE *et al.* (1964)
Waters of Florida	7	0.4	R	OSMOND (1964)

Revised manuscript received: July 18th, 1967

90-K. Abundance in Common Sedimentary Rocks

As discussed in the chapter on uranium (Section 92-K), the major factors controlling the distribution of thorium and uranium in sedimentary rocks are:

1. The tendency of thorium and uranium to fractionate because of oxidation of the uranium to the soluble uranyl ion.

2. The precipitation of uranium out of solution in reducing environments, carbonates, etc.

3. The selective adsorption of thorium in clays and its retention in heavy resistate minerals.

A more complete discussion of the comparative geochemistry of thorium and uranium in sedimentary rocks is given in Section 92-K. The present section considers only the concentrations of thorium and the Th/U ratios in various sedimentary rocks. Data are given in Table 90-K-1, and comparable data for uranium are given in Table 92-K-1.

The concentrations in sandstones shown in Table 90-K-1 are based on reasonably complete sampling of orthoquartzites, fair sampling of graywackes, and very poor sampling of arkoses. The problem of sampling sandstones and also of establishing valid sandstone populations is discussed by ROGERS and RICHARDSON (1964). The volcanic graywackes shown in Table 90-K-1 are important because they represent the average composition of rocks of the Caribbean orogen and thus give an integrated average of the amount of radioactive material that has been derived from igneous sources during eugeosynclinal activity.

Ordinary shales with colors of gray, green, etc. have a thorium content of 12 ppm and a Th/U ratio in the range of 3.5 to 4 (Fig. 90-K-1). As discussed in Section 92-K, these values are similar to those for the average granite. Black shales have a variable thorium content but a low Th/U ratio owing to their high content of uranium. The high concentration of thorium in bauxites (Fig. 90-K-2) is discussed in Section 90-G. The concentration of thorium in bentonites is high for the same reason as in bauxites.

Thorium concentrations in limestone (Fig. 90-K-3) are low because thorium does not enter the carbonate lattice readily. A close relationship has been established between thorium concentration and insoluble residue content of limestones (ADAMS and WEAVER, 1958), thus demonstrating that most of the thorium in limestones is in the clay or heavy mineral fraction. The Th/U ratio of limestones is obviously very low. Thorium is almost completely absent from evaporites.

The thorium content and Th/U ratios are given in Table 90-K-1 for a variety of modern oceanic sediments. These sediments are also discussed in section 92-K. It should be noted here that the low Th/U ratios reported by BARANOV and KHRISTIANOVA (1963) are not easily explained in terms of sedimentation processes. In particular, the finding of high thorium concentrations in pelagic clays and managanese-rich sediments (see, for example, GOLDBERG and KOIDE, 1962) seems incompatible with the low Th/U ratios reported in the sediments summarized by BARANOV and KHRISTIANOVA.

Table 90-K-1. *Contents of thorium and Th/U ratios in sedimentary rocks*

Rock type	No. of samples	Th ppm	Th/U	Method	Reference
Sandstones					
Orthoquartzite					
North American av.	16	1.7	3.8	R	MURRAY and ADAMS (1958)
Mesa Verde	8	2.7	1.6	R	PLILER and ADAMS (1962a)
Gulf Coast beach sand	29	1.49	2.53	R	MAHDAVI (1964)
Atlantic Coast beach sand	54	9.05	2.82	R	MAHDAVI (1964)
Volcanic graywacke					
Caribbean	14	1—2	~2	R	ROGERS and DONNELLY (1966)
Umpqua (Oregon)	4	2.8	~2	R	ROGERS (1966)
Other graywackes (estimated average)		6.7			ROGERS and RICHARDSON (1964)
Arkose (estimated average)		5			ROGERS and RICHARDSON (1964)
Shales					
North American gray and green	52	13.1	4.9	R	ADAMS and WEAVER (1958)
Russian platform	178 (comp)	11.0	2.7	emanation	BARANOV *et al.* (1956)
Mancos	102	10.2	3.1	R	PLILER and ADAMS (1962a)
red and yellow	10	12—13	6—7	R	ADAMS and WEAVER (1958)
black	numerous		low	various	SWANSON (1961); BATES and STRAHL (1958)
Bauxites	29	48.9	5.1	R	ADAMS and RICHARDSON (1960)
Bentonites	64	24.0	5.8	various	ADAMS and WEAVER (1958)
Limestones					
North American av.	25 (comp)	1.1	0.7	R	ADAMS and WEAVER (1958)
Russian platform	128 (comp)	2.4	1.1	emanation	BARANOV *et al.* (1956)
Eniwetok core E-1	21	0.05		0.02	SACKETT and POTRATZ (1963)

Table 90-K-1 (Continued)

Rock type	No. of samples	Th ppm	Th/U	Method	Reference
Phosphate rocks		1—5	<0.1	various	References summarized by McKelvey (1956)
Pacific ocean sediments					
Sands	5	1.2	0.4	various	Summary by Baranov and Khristianova (1963)
Muds	3	2.7	1.2	various	Summary by Baranov and Khristianova (1963)
Clayey muds	6	4.8	1.8	various	Summary by Baranov and Khristianova (1963)
Siliceous and diatomaceous muds and clayey muds	9	2.2	low	various	Summary by Baranov and Khristianova (1963)
Pelagic clays	13	6.8	1.7	various	Summary by Baranov and Khristianova (1963)
Pelagic clays	5	3.1—8.5	2—4	radiochemical	Picciotto and Wilgain (1954)
Pelagic clays	3	9.3—11.2		R	Goldberg and Koide (1962)
Manganese nodules	4	24—124		C	Goldberg and Picciotto (1955)
Indian ocean sediments					
Various muds		1—2	3—4.5	various	Summary by Baranov and Khristianova (1960)
Atlantic ocean sediments					
Pelagic clays	2	30			El Wakeel and Riley (1961)
Globigerina oozes	5	5.1	9.9	R	Rosholt *et al.* (1961)
Caribbean sediments					
Globigerina oozes	17	5.5	5.5	R	Rosholt *et al.* (1961, 1962)
Black Sea sediments					
Muds	44	7.3	3.3 (41 spls.)		Starik *et al.* (1961)

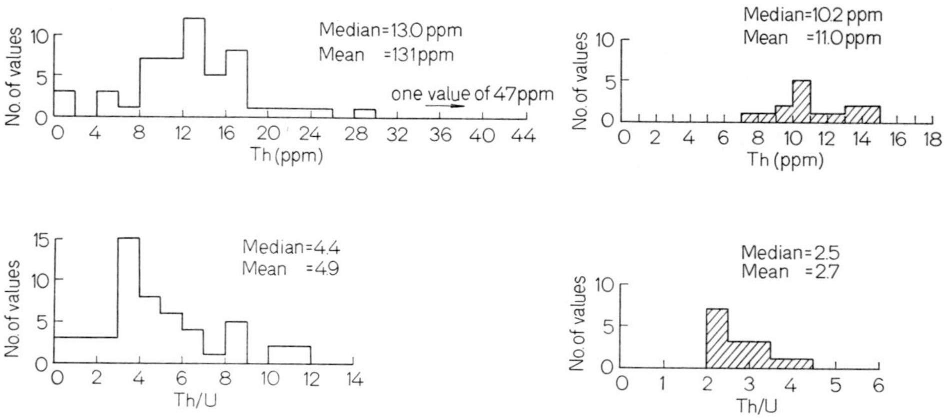

Fig. 90-K-1. Histogram of thorium concentrations and Th/U ratios in North American gray and green shales and Russian Platform shales. (From ADAMS and WEAVER, 1958; Russian Platform data from BARANOV, RONOV, and KUNASHOVA, 1956)

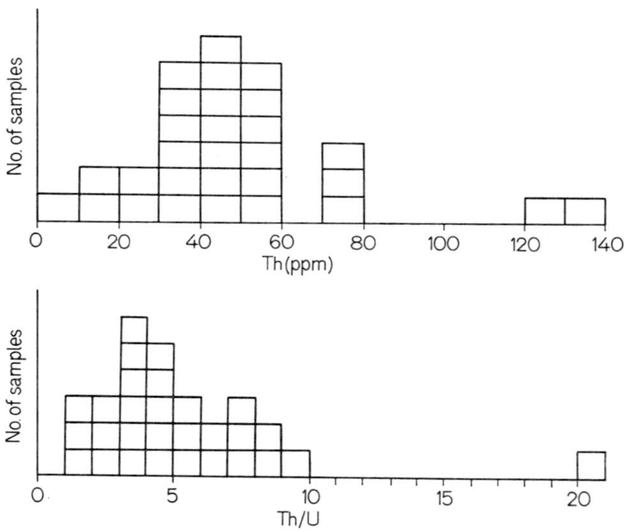

Fig. 90-K-2. Histograms of thorium concentrations and Th/U ratios in a world-wide sampling of bauxites. (From ADAMS and RICHARDSON, 1960)

It is possible, however, that some major sedimentary fractionation occurs on the sea floor, yielding high thorium concentrations in some varieties of clay and comparatively high uranium concentrations in other materials.

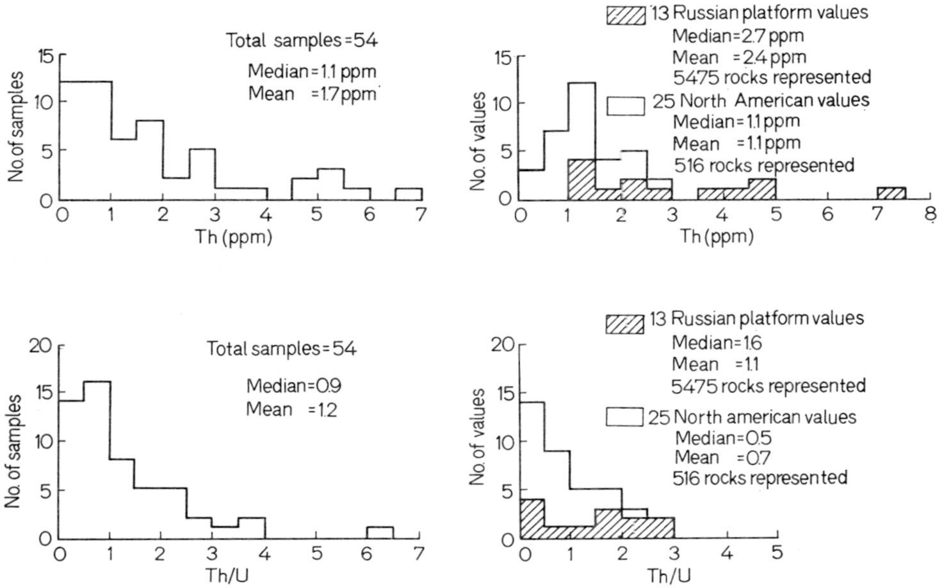

Fig. 90-K-3. Histograms of thorium concentrations and Th/U ratios in aggregate samples of carbonate rocks. (From ADAMS and WEAVER, 1958; Russian Platform data from BARANOV, RONOV, and KUNASHOVA, 1956)

Revised manuscript received: July 18th, 1967

90-L. Biogeochemistry

Thorium is not known to be an essential material in any organic process. Bowie and Atkin (1956) report a thorium concentration of as much as 4,500 ppm in a fossil fish skeleton, but most organisms and organically derived materials contain essentially no thorium. The problem of dealing with the exceptionally low thorium concentrations which occur in organic material and the possible biological significance of thorium is discussed briefly by Edgington (1965).

Revised manuscript received: July 18th, 1967

90-M. Abundance in Common Metamorphic Rock Types

The concentration of thorium in a variety of metamorphic rocks is shown in Table 90-M-1. The Th/U ratios are also shown in this table and are based on uranium data given in Table 92-M-1. Concentrations of thorium and the Th/U ratio are highly variable from one metamorphic rock to another and clearly depend on a variety of factors, including parent rock composition, metasomatic effects, and the possible sampling difficulties caused by segregation during metamorphism. HEIER and ADAMS (1965) have suggested that high-rank metamorphic rocks contain less thorium and uranium than low-rank rocks as a result of selective removal of these materials from the deeper parts of the crust. The data for this conclusion, however, are not complete. Other generalizations concerning the thorium and uranium contents of metamorphic rocks cannot be made from the small number of measurements currently available.

Table 90-M-1. *Contents of thorium and Th/U ratios in metamorphic rocks*

Rock type	No. of samples	Th ppm	Th/U	Method	Reference
Amphibolite (Schwarzwald, Germany)	3	5	1.4	R	HUSMANN (1956)
Biotite-Hornbl. Paragneiss (Schwarzwald, Germany)	2	27	13.5	R	HUSMANN (1956)
Paragneiss (Schwarzwald, Germany)	4	6.4	0.9	R	HUSMANN (1956)
Granulite (Schwarzwald, Germany)	2	21	4.3	R	HUSMANN (1956)
Orthogneiss (Schwarzwald, Germany)	4	5.5	1.5	R	HUSMANN (1956)
Metatectite (Schwarzwald, Germany)	4	22	4.6	R	HUSMANN (1956)
Diatectite (Schwarzwald, Germany)	4	3	0.3	R	HUSMANN (1956)
Cordierite Gneiss (Schwarzwald, Germany)	1	21.8	3.8	R	HUSMANN (1956)
Orthoclase metacrysts (Schwarzwald, Germany)	1	67.7	1.5	R	HUSMANN (1956)
Kinzigite (Schwarzwald, Germany)	1	7		R	HUSMANN (1956)
Marble	2	0.03	0.2	R	PLILER (1956)
Slate (mainly Michigan)	14	7.5	2.8	R	PLILER (1956)
Phyllite (mainly Arizona-New Mexico)	7	5.5	2.8	R	PLILER (1956)
Schist (New Mexico)	4	7.5	3.0	R	PLILER (1956)
Mafic rocks (Terskei Ala Tau Mts.)	82	10	3.1	W	KRYLOV (1958)

Table 90-M-1 (Continued)

Rock type	No. of samples	Th ppm	Th/U	Method	Reference
Gneiss (Texas) (Almandine-Amph. Facies)	13	13.1	4.75	R	BILLINGS (1962)
Amphibolite (Texas) (Almandine-Amph. Facies)	14	0.31	0.79	R	BILLINGS (1962)
Graphite Schist (Texas) (Almandine-Amph. Facies)	3	6.16	2.66	R	BILLINGS (1962)
Light Gneiss (Langøy, Norway) (Amphibolite Facies)	5	15.08	> 7.36	R	HEIER (1962b)
Amphibolite (Langøy) (Amphibolite Facies)	7	4.2	~ 3.8	R	HEIER (1962b)
Intermediate Gneiss (Langøy) (Granulite Facies)	2	9.25		R	HEIER (1962b)
Monzonite (Langøy) (Retrograde Gneiss)	2	3.3	> 2.8	R	HEIER (1962b)
Granite (Langøy) (Retrograde Gneiss)	8	<14.4	>12.89	R	HEIER (1962b)
Biotite Schist (Front Range, Colorado)	11	18.8		W	PHAIR and GOTTFRIED (1964)
Biotite-Hornbl. Schist and Amphibolite (Front Range)	4	10.6		W	PHAIR and GOTTFRIED (1964)
Monzonite (Langøy) (High Granulite Facies)	3	1.18	1.7	R	HEIER and ADAMS (1965)
Banded Gneiss (Langøy) (High Granulite Facies)	4	0.85	3.07	R	HEIER and ADAMS (1965)
Gneiss (Langøy) (Low Granulite Facies)	5	4.09	30	R	HEIER and ADAMS (1965)
Gneiss (Langøy) (High Amphibolite Facies)	3	9.39	13.4	R	HEIER and ADAMS (1965)

Revised manuscript received: July 18th, 1967

90-N. Behavior in Metamorphic Reactions

As mentioned in Section 90-M, there is some possibility that thorium and uranium are depleted from high-rank rocks owing to upward movement through the crust during major metamorphism. No data are available on the behavior of thorium during metamorphism either from laboratory studies or from complete sampling of metamorphic terrains.

Revised manuscript received: July 18th, 1967

90-O. Economic Utilization

At the present time thorium is of minor economic interest with annual production rarely exceeding a thousand tons of metal. The metal is used in magnesium alloys and incandescent gas mantles. Current research on the breeding of Th^{232} to make fissionable U^{233} for power generation may lead to a greatly expanded market.

Acknowledgments

Preparation of this review was greatly assisted by Grant C-009 from the ROBERT A. WELCH Foundation to JOHN A. S. ADAMS and JOHN J. W. ROGERS.

Revised manuscript received: July 18th, 1967

References: Sections 90-B to 90-O

ABRAMOVICH, I. I.: Uranium and thorium in the intrusive rocks of Central and Western Tuva. Geochemistry (USSR) (English Transl.) 442 (1959).

ADAMS, J. A. S.: The uranium geochemistry of Lassen Volcanic National Park, Calif. Geochim. Cosmochim. Acta **8**, 74 (1955).

— The uranium contents and alpha particle activities of tektites. Proc. 20th Int. Geol. Congr., Mexico City. In press (1956).

— M.-C. KLINE, K. A. RICHARDSON, and J. J. W. ROGERS: The Conway granite of New Hampshire as a major low-grade thorium resource. Proc. Natl. Acad. Sci. U. S. **48**, 1898 (1962).

— J. K. OSMOND, and J. J. W. ROGERS: The geochemistry of uranium and thorium. Phys. Chem. Earth **3**, 298 (1959).

—, and K. A. RICHARDSON: Thorium, uranium, and zirconium concentrations in bauxite. Econ. Geol. **55**, 1653 (1960).

—, and C. E. WEAVER: Thorium-to-uranium ratios as indicators of sedimentary processes — an example of geochemical facies. Bull. Am. Assoc. Petrol. Geologists **42**, 387 (1958).

AHRENS, L. H.: Some observations on the uranium and thorium distributions in accessory zircon from granitic rocks. Geochim. Cosmochim. Acta **29**, 711 (1965).

ANDERSON, A. L.: Thorium mineralization in the Lemhi Pass Area, Lemhi Co., Idaho. Econ. Geol. **56**, 177 (1961).

ANDERSON, B. W., G. F. CLARINGBULL, R. J. DAVIS, and D. K. HULL: Ekanite, a new metamict mineral from Ceylon. Nature **190**, 997 (1961). [Am. Mineralogist **46**, 1516 (1961).]

BALACEK, K. J., and J. A. S. ADAMS: Radioactivity in tektites. In preparation (1966).

BARANOV, V. I., and L. A. KHRISTIANOVA: Radioactivity of the waters of the Indian Ocean. Geochemistry (USSR) (English Transl.) 765 (1959).

— — New data on radioactivity of the Indian Ocean. Akad. Nauk SSSR Otd. Geol.-Geograf. Nauk 392 (1960) [in Russian].

— — Radioactivity of ocean sediments. Akad. Nauk. SSSR, Khim. Zemnoi. Kory, Tr. Geokhim. Konf. **1**, 401 (1963).

— A. B. RONOV, and K. G. KUNASHOVA: Geochemistry of dispersed thorium and uranium in clays and carbonate rock of the Russian Platform. Geochemistry (USSR) (English Transl.) 227 (1956).

BATE, G. L., J. R. HUIZENGA, and H. A. POTRATZ: Thorium content of stone meteorites. Science **126**, 612 (1957).

— — — Thorium in stone meteorites by neutron activation analysis. Geochim. Cosmochim. Acta **16**, 88 (1959).

— H. A. POTRATZ, and S. R. HUIZENGA: Thorium in iron meteorites — a preliminary investigation. Geochim. Cosmochim. Acta **14**, 118 (1958).

BATES, T. F., and E. O. STRAHL: Mineralogy and chemistry of uranium-bearing black shales. Proc. of Second Int. United Nations Conf. on Peaceful Uses of Atomic Energy, Geneva **2**, 407 (1958).

BILLINGS, G. K.: A geochemical investigation of the Valley Spring gneiss and Packsaddle schist, Llano uplift, Tex. Texas J. Sci. **9**, 328 (1962).

BIRCH, F.: Speculations on the earth's thermal history. Bull. Geol. Soc. Am. **76** 133 (1965).

BOWIE, S. H. U., and D. ATKIN: An unusually radioactive fossil fish from Thurso, Scotland. Nature **177**, 487 (1956).

BUTLER JR., A. P.: Ratio of thorium to uranium in some plutonic rocks of the White Mountain plutonic-volcanic series, New Hampshire. U.S. Geol. Surv. Profess. Papers **424-B**, 67 (1961).

CHERRY, R. D.: Thorium and uranium contents of Australites. Nature **195**, 1184 (1962).
—, and J. A. S. ADAMS: Gamma-spectrometric determinations of thorium, uranium, and potassium in tektites. Geochim. Cosmochim. Acta **27**, 1089 (1963).
DUBEY, V. S.: The origin of tektites. Nature **132**, 678 (1933).
EDGINGTON, D. N.: Thorium and uranium in the natural environment — a review of present knowledge. Annual Report of Radiological Physics Division, July 1964 through June 1965, Argonne National Laboratory, ANL-7060, 73 (1966).
EL SHAZLY, E. M.: Geology of uranium and thorium. Internat. Atomic Energy Agency, Vienna, Bibliog. Ser. No. 4, 130 p. (1962).
EL WAKEEL, S. K., and J. P. RILEY: Chemical and mineralogical studies of deep sea sediments. Geochim. Cosmochim. Acta **25**, 110 (1961).
FILIPPOV, M. S., and L. V. KOMLEV: Uranium and thorium in granitoids of the middle Dnepr region. Geochemistry (USSR) (English Transl.) 535 (1959).
FISHER, F. G., and R. MEYROWITZ: Brockite, a new calcium thorium phosphate from the Wet Mountains, Colorado. Am. Mineralogist **47**, 1346 (1962).
FØYN, E. E., B. KARLIK, H. PETTERSON, and E. RONA: Radioactivity of sea water. Nature **143**, 275 (1939).
FRONDEL, C.: Mineralogy of thorium. U.S. Geol. Surv. Profess. Papers **300**, 567 (1956).
GAST, P. W.: Limitations on the composition of the upper mantle. J. Geophys. Res. **65**, 1287 (1960).
GOLDBERG, E. D., and M. KOIDE: Geochronological studies of deep sea sediments by the ionium/thorium method. Geochim. Cosmochim. Acta **26**, 417 (1962).
—, and E. PICCIOTTO: Thorium determinations in manganese nodules. Science **121**, 613 (1955).
GOTMAN, YA. D., and I. A. KHAPAEV: Thorutite — a new mineral of the group of titanates of thorium. Zap. Vses. Mineralog. Obshchestva **87**, 201 (1958) [in Russian]. [Am. Mineralogist **43**, 1007 (1958).]
GOTTFRIED, D., and E. S. LARSEN 3rd.: Distribution of uranium in the volcanic rocks of the Modoc Area. Calif. U.S. Geol. Surv. TEI **740**, 282 (1958).
— R. MOORE, and E. CAMPBELL: Thorium and uranium in some volcanic rocks from the Circum-Pacific Province. U.S. Geol. Surv. Profess. Papers **450-E**, 85 (1963).
— —, and A. CAEMERER: Thorium and uranium in some alkalic igneous rocks from Virginia and Texas. U.S. Geol. Surv. Profess. Papers **450-B**, 70 (1962).
HEIER, K. S.: A note of the U, Th, and K contents in the nepheline syenite and carbonatite on Stjernøy, North Norway. Norsk Geol. Tidsskr. **42**, 287 (1962a).
— Spectrometric uranium and thorium determinations on some high-grade metamorphic rocks on Langöy, Northern Norway. Norsk Geol. Tidsskr. **42**, 143 (1962b).
— Uranium, thorium, and potassium in eclogitic rocks. Geochim. Cosmochim. Acta **27**, 849 (1963).
—, and J. A. S. ADAMS: Concentration of radioactive elements in deep crustal material. Geochim. Cosmochim. Acta **29**, 53 (1965).
—, and J. L. CARTER: Uranium, thorium, and potassium contents in basic rocks and their bearing on the nature of the upper mantle. In: J. A. S. ADAMS and W. M. LOWDER (eds.), The natural radiation environment, p. 75. Chicago: Chicago University Press 1964.
— I. MCDOUGALL, and J. A. S. ADAMS: Th, U, and K in Hawaiian lavas. Nature **201**, 254 (1964).
—, and J. J. W. ROGERS: Radiometric determination of thorium, uranium, and potassium in basalts and in two magmatic differentiation series. Geochim. Cosmochim. Acta **27**, 137 (1963).
HURLEY, P. M., and H. W. FAIRBAIRN: Ratio of thorium to uranium in zircon, sphene, and apatite. Bull. Geol. Soc. Am. **66**, 1578 (1955).
— — Abundance and distribution of uranium and thorium in zircon, sphene, apatite, epidote, and monazite in granitic rocks. Trans. Am. Geophys. Union **38**, 939 (1957).
HUSMANN, O.: Bestimmungen des Thorium- und Uranium-Gehaltes an Gneisen und magmatischen Gesteinen des mittleren Schwarzwaldes mittels der Koinzidenz-Methode. Neues Jahrb. Mineral., Monatsh. 108 (1956).
HUTTON, C. O.: Uranoan thorite and thorian monazite from blacksand paystreaks, San Mateo County, California. Bull. Geol. Soc. Am. **62**, 1518 (1951).

KEEVIL, N. B.: Thorium-uranium ratios in rocks and minerals. Am. J. Sci. **242**, 309 (1944).

KOCZY, F. F.: Geochemistry of the radioactive elements in the ocean. Deep-Sea Res. **3**, 93 (1956).

— E. PICCIOTTO, G. POULAERT et S. WILGAIN: Mesures des isotopes du thorium dans l'eau de mer. Geochim. Cosmochim. Acta **11**, 103 (1957).

KRYLOV, A. YA.: Radioactivity in various rock complexes of the Terskei Ala Tau Mountains. Geochemistry (USSR) (English Transl.) 240 (1958).

— The evolution of radioactivity and general chemical composition, North Tien Shan rocks. Radiokhimiya 173 (1963). [in Russian.]

KUPRIYANOVA, I. I., T. I. STOLYAROVA, and G. A. SIDORENKO: A new thorium silicate-thorosteenstrupine. Zap. Vses. Mineral. Obshchestva **91**, 325 (1962) [in Russian]. [Am. Mineralogist **48**, 433 (1963).]

LARSEN, E. S., 3rd., and D. GOTTFRIED: Uranium and thorium in selected suites of igneous rocks. Am. J. Sci. **258-A**, 151 (1960).

LARSEN, E. S., JR., and D. GOTTFRIED: Distribution of uranium in rocks and minerals of Mesozoic batholiths in western United States. U.S. Geol. Surv. Bull **1070**-C, 63 (1961).

LEONOVA, L. L.: Distribution of thorium in minerals of the granites of the Susamyr batholith (central Tien Shan). Geochemistry (USSR) (English Transl.) 571 (1962).

— Uranium and thorium in the variscan alkaline intrusives of Southern Tien Shan (Matcha and Turpi Massifs). Geochemistry Intern. (English Transl.) 925 (1964).

—, and O. S. RENNE: Distribution of uranium, thorium, and potassium in homogeneous granites. Geochemistry Intern. (English Transl.) 775 (1964).

—, and L. V. TAUSON: The distribution of uranium in the minerals of the caledonian granitoids of the Susamyr batholith (central Tien Shan). Geochemistry (USSR) (English Transl.) 815 (1958).

LOVERING, J. F., and J. W. MORGAN: Uranium and thorium in possible upper mantle materials. Nature **197**, 138 (1963).

— — Uranium and thorium abundances in stony meteorites, 1. The chondritic meteorites. J. Geophys. Res. **69**, 1979 (1964).

LYONS, J. B.: Uranium and thorium in the older plutonic rocks of New Hampshire. Art 32. U.S. Geol. Surv., Profess. Papers **424-B**, B 69 (1961).

MAHDAVI, A.: Thorium, uranium, and potassium contents of Atlantic and Gulf Coast beach sand. In: J. A. S. ADAMS and W. M. LOWDER (eds.), The natural radiation environment, p. 87. Chicago: Chicago University Press 1964.

MCKELVEY, V. E., D. L. EVERHART, and R. M. GARRELS: Origin of uranium deposits. Econ. Geol. 50th Anniversary vol., 464 (1955).

MELIKSETYAN, B. M.: The geochemistry of uranium and thorium in some granitic rocks of southern Armenian SSR. Izv. Akad. Nauk Arm. SSR, Ser. Geol. i Geogr. Nauk **14**, 21 (1961). [in Russian].

MIYAKE, Y., Y. SUGIMURA, and H. TSUBOTA: Content of uranium, radium, and thorium in river waters of Japan. In: J. A. S. ADAMS and W. M. LOWDER (eds), The natural radiation environment, p. 221. Chicago: Chicago University Press 1964.

MORGAN, J. W., and J. F. LOVERING: Uranium and thorium abundances in stony meteorites, 2. The achondritic meteorites. J. Geophys. Res. **69**, 1989 (1964).

— —Uranium and thorium in the Nuevo Laredo achondrite. J. Geophys. Res. **70**, 2005 (1965).

MUMPTON, F. A., and R. ROY: Hydrothermal studies of the zircon-thorite group. Geochim. Cosmochim. Acta **21**, 217 (1961).

MURATA, K. J., H. J. ROSE, M. K. CARRON, and J. J. GLASS: Systematic variation of rare-earth elements in cerium-earth minerals. Geochim. Cosmochim. Acta **11**, 141 (1957).

MURRAY, E. G., and J. A. S. ADAMS: Amount and distribution of thorium, uranium, and potassium in sandstones. Geochim. Cosmochim. Acta **13**, 260 (1958).

NIKOLAEV, D. S., K. F. LAZAREV, and S. M. GRASHENKO: The concentration of thorium isotopes in the water of the Sea of Azov. Dokl. Akad. Nauk SSSR (English Transl.) **138**, 674 (1961).

Osborn, E. F.: Role of oxygen pressure in the crystallization and differentiation of basaltic magma. Am. J. Sci. **257**, 609 (1959).
Osmond, J. K.: The distribution of the heavy radioelements in the rocks and waters of Florida. In: J. A. S. Adams and W. M. Lowder (eds.), The natural radiation environment, p. 158. Chicago: Chicago University Press 1964.
Pavlenko, A. S., L. P. Orlova, M. V. Akmanova, and K. I. Tobelko: A thorium fluocarbonate, thorbastnaesite. Zap. Vses. Mineral. Obshchestva **94**, 105 (1965) [in Russian]. [Am. Mineralogist **50**, 1505 (1965).]
Peterman, Z. E.: Unpublished, personal communication (1963).
Phair, G.: Geologic investigations of radioactive deposits. Semiannual Progress Report, Dec., 1954 to May, 1955. U.S. Geol. Surv. TEI **540**, 237 (1955).
—, and D. Gottfried: The Colorado Front Range, Colorado, U.S.A., as a uranium and thorium province. In: J. A. S. Adams and W. M. Lowder (eds.), The natural radiation environment, p. 7. Chicago: Chicago University Press 1964.
Picciotto, E., and S. Wilgain: Thorium determination in deep-sea sediments. Nature **173**, 632 (1954).
Pliler, R.: The distribution of thorium and uranium in sedimentary rocks and the oxygen content of the pre-cambrian atmosphere. M. A. Thesis Rice Institute 1956.
—, and J. A. S. Adams: The distribution of thorium, uranium, and potassium in the Mancos Shale. Geochim. Cosmochim. Acta **26**, 1115 (1962a).
— — The distribution of thorium and uranium in a Pennsylvanian weathering profile. Geochim. Cosmochim. Acta **26**, 1137 (1962b).
Polyakov, A. I., and G. A. Kot: Geochemistry of thorium in the nepheline syenites of Kola Peninsula. Geochemistry Intern. (USSR) (English Transl.) 479 (1964).
—, and M. P. Volynets: Distribution of thorium in a series of ultrabasic alkalic rocks of the Kola Peninsula. Geochemistry (USSR) (English Transl.) 446 (1961).
Reasbeck, P., and K. I. Mayne: The ages and origin of meteorites. Nature **176**, 186 (1955).
Richardson, K. A.: Thorium, uranium, and potassium in the Conway granite, New Hampshire, U.S.A. In: J. A. S. Adams and W. M. Lowder (eds.), The natural radiation environment, p. 39. Chicago: Chicago University Press 1964.
Rogers, J. J. W.: Statistical test of the homogeneity of the radioactive components of granitic rocks. In: J. A. S. Adams and W. M. Lowder (eds.), The natural radiation environment, p. 51. Chicago: Chicago University Press 1964.
— Geochemical significance of the source rocks of some graywackes from Western Oregon and Washington. Texas J. Sci. **18**, 5 (1966).
— J. A. S. Adams, and B. Gatlin: Distribution of thorium, uranium and potassium concentration in three cores from the Conway granite, New Hampshire, U.S.A. Am. J. Sci. **263**, 817 (1965).
—, and T. W. Donnelly: Radiometric evidence for the origin of eugeosynclinal materials. Tulane Studies in Geology **4**, 133 (1966).
—, and P. C. Ragland: Variation of thorium and uranium in selected granitic rocks. Geochim. Cosmochim. Acta **25**, 99 (1961).
—, and K. A. Richardson: Thorium and uranium contents of some sandstones. Geochim. Cosmochim. Acta **28**, 2005 (1964).
Rosholt, J. N., and P. S. Antal: Evaluation of the $Pa^{231}/U—Th^{230}/U$ method for dating Pleistocene carbonate rock. U.S. Geol. Surv., Profess. Papers **450-E**, E 108 (1963).
— C. Emiliani, J. Geiss, F. F. Koczy, and P. J. Wangersky: Absolute dating of deep-sea cores by the Pa^{231}/Th^{230} Method. J. Geol. **69**, 162 (1961).
— — — — — Pa^{231}/Th^{230} dating and O^{18}/O^{16} temperature analysis of core A 254-BR-C. J. Geophys. Res. **67**, 2907 (1962).
Rowe, M. W., M. A. van Dilla, and E. C. Anderson: On the radioactivity of stone meteorites. Geochim. Cosmochim. Acta **27**, 983 (1963).
Sackett, W. M.: Deposition rates by the protoactinium method. Univ. of Rhode Island, Grad. School of Oceanog., Occasional Publication 3-1965, 29 (1965).
—, and H. A. Potratz: Dating of carbonate rocks by ionium-uranium ratios. In: Subsurface geology of Eniwetok Atoll. U.S. Geol. Surv. Profess. Papers **260-BB**, 1053 (1953)
— —, and E. D. Goldberg: Thorium content of ocean water. Science **128**, 204 (1958).

Shavrova, N. N.: The content of radium and thorium in lavas of the volcanos of the Semyachinskii Group. Akad. Nauk SSSR, Lab. Volkanol. Byull. Volkanol. Stantsii, No. 31, 28 (1958) [in Russian].
Smith, W. L., M. L. Franck, and A. M. Sherwood: Uranium and thorium in the accessory allanite of igneous rocks. Am. Mineralogist **42**, 367 (1957)
Starik, I. E., K. F. Lazarev, D. S. Nikolaev, S. M. Grashchenko, L. B. Kolyadin, and Yu. V. Kuznetsov: Concentration of thorium isotopes in the water of the Black Sea. Dokl. Akad. Nauk SSSR (English Transl.) **129**, 1045 (1959).
— D. S. Nikolaev, Yu. V. Kuznetsov, and V. K. Legin: The radioactivity of Black Sea sediments. Dokl. Akad. Nauk SSSR (English Transl.) **139**, 704 (1961).
Swanson, V. E.: Geology and geochemistry of uranium in marine black shales, a review. U.S. Geol. Surv, Profess. Papers **356-C**, 67 (1961).
Tatsumoto, M., C. E. Hedge, and A. E. J. Engel: Potassium, rubidium, strontium, thorium, uranium, and the ratio of strontium-87 to strontium-86 in oceanic tholeiitic basalt. Science **150**, 886 (1965).
Thurber, D. L.: The dating of molluscs from raised marine terraces. Univ. of Rhode Island, Grad. School of Oceanog., Occasional Publication 3-1965, 1 (1965).
Tilton, G. R.: Isotopic composition of lead from tektites. Geochim. Cosmochim. Acta **14**, 323 (1958).
—, and L. B. Nicolayson: The use of monazites for age determination. Geochim. Cosmochim. Acta **11**, 28 (1957).
— C. Patterson, H. Brown, M. G. Inghram, R. Hayden, D. Hess, and E. S. Larsen Jr.: The isotopic composition and distribution of lead, uranium, and thorium in a Precambrian granite. Bull. Geol. Soc. Am. **66**, 1131 (1955).
Webber, G. R., P. M. Hurley, and H.W. Fairbairn: Relative ages of eastern Massachusetts granites by total lead ratios in zircon. Am. J. Sci. **254**, 574 (1956).
Whitfield, J. M., J. J. W. Rogers, and J. A. S. Adams: The relationship between petrology and the thorium and uranium contents of some granitic rocks. Geochim. Cosmochim. Acta **17**, 248 (1959).
Wollenberg, H. A., and A. R. Smith: Radioactivity and radiogenic heat in Sierra Nevada plutons. J. Geophys. Res. **69**, 3471 (1964).

Revised manuscript received: July 18th, 1967

Uranium 92

A	F. Pertlik	(Institut für Mineralogie und Kristallographie, Universität Wien, Austria)
B, C I, II	J. J. W. Rogers and J. A. S. Adams	(Department of Geology, Rice University, Houston, Texas, U.S.A.)
C III	U. K. Haack	(Geochemisches Institut der Universität, Göttingen, Germany)
D—O	J. J. W. Rogers and J. A. S. Adams	

92-A. Crystal Chemistry

The 5-f element uranium in the ground state has the probable electronic structure $5f^3\,6d^1\,7s^2$ and the formal valence states vary between 2+ and 6+. In minerals, however, only the valences 4+, 5+ and 6+ are known. With many elements (e.g. C, N, Si, P, S etc.), binary compounds of uranium are semimetallic and often non-stoichiometric.

I. Metallic Uranium

Metallic uranium (melting point 1,132° C) has three crystalline modifications; α-uranium (below 660° C) crystallizes orthorhombic, and each uranium atom is coordinated by twelve neighbors at distances between 2.75 and 3.34 Å (STURCKEN and POST, SR 1960, 240). β-uranium (between 660° and 760° C) is probably tetragonal and non-centrosymmetric, and the coordination number of the six structurally different uranium atoms is between 12 and 15 (THEWLIS and STEEPLE, SR 1954, 304; STEEPLE and ASHWORTH, 1966). Because of some uncertainties in the indexing of the powder data (DONOHUE and EINSPAHR, 1971), it is possible that the structure determination is not quite correct. The phase γ-uranium (from 760° C upto the melting point) is cubic, body-centered (THEWLIS, SR 1951, 123) with U—U distances of 3.0 Å (8×).

II. Formally Di-, Tri- and Penta-valent Uranium

One example of a compound with divalent uranium is UO, crystallizing in the NaCl-type (RUNDLE *et al.*, SR 1947—1948, 220).

For trivalent uranium the structures of some artificial halides are the best known. The layer structure of UF_3 is isotypic with CeF_3; in the hexagonal arrangement the uranium atoms have a [2+3+6] coordination of fluorine. The fluorides AUF_6 (A = Sr, Ba, Pb) with formally tetravalent uranium are isotypic with UF_3 and the atoms A and U seem to be randomly distributed over the metal sites (ZACHARIASEN, SR 1949, 164). UCl_3 and UBr_3 crystallize in the $Y(OH)_3$ structure type. Uranium is surrounded by nine halogen neighbors at nearly equal distances, e.g. U—Cl ~ 2.9 Å (ZACHARIASEN, SR 1947—1948, 277). The fourth halogenide, UI_3 crystallizes orthorhombic and each U is coordinated by 8 iodine neighbors (ZACHARIASEN, SR 1947—1948, 282).

Pentavalent uranium occurs in two tetragonal modifications of UF_5. In the α-form each uranium is coordinated by six fluorine neighbors in the form of an octahedron with distances U—F ~ 2.2 Å. The β-form shows a seven-fold coordination to uranium, with U—F distances between 2.18 and 2.29 Å (ZACHARIASEN, SR 1949, 169). In UCl_5 the uranium atoms are coordinated octahedrally by six chlorine atoms (U—Cl distances between 2.43 and 2.70 Å); two such octahedra share an edge to form U_2Cl_{10} units (SMITH *et al.*, 1967).

In the system involving UO_2 and U_3O_8 (which is only tentatively known — GRONVOLD, 1955), the crystal structures of only a few compounds have been determined. UO_2 crystallizes in the CaF_2-type (see below) and the compound U_4O_9 has been reported to crystallize in a cubic superstructure of UO_2 with 4 times the UO_2 cell dimensions (MASAKI and DOI, 1972). In U_3O_8 there are two crystallographically different uranium atoms: U(1) is coordinated by six oxygens with U—O distances of 2.07 (2×) and 2.18 Å (4×) (after ANDRESEN, SR 1958, 296) and is pentavalent; U(2) has the typical [2+5] coordination of hexavalent uranium with U—O distances of 2.07 (2×), 2.17 (2×), 2.21 and 2.42 (2×). The system U—O is very complex and deviations from stoichiometry are the rule, but only the UO_2 to U_3O_8 region seems to be significant for the natural occurrence of uranium oxides.

III. Formally Tetravalent Uranium

The structures of two halogenides are well-known. UF_4 crystallizes in a monoclinic structure, isotypic with ZrF_4, and uranium is coordinated by 8 fluorine atoms (ZACHARIASEN, SR 1949, 168). UCl_4 is tetragonal with four chlorine atoms at distances of 2.41 Å, and four at 2.09 Å (MOONEY, SR 1949, 166).

The coordination number of oxygen around U(IV) is six or eight and examples are given in Table 92-A-1. In minerals with structures similar to columbite, (Fe, Mn) $(Nb, Ta)_2O_6$ (STURDIVANT, SB 1928—1932, 337), the atoms (Fe, Mn) with six-fold coordination (octahedral) can probably be partially substituted by uranium. In the

Table 92-A-1. *Structural data of some uranium (IV) compounds*

Mineral name	Chemical formula	Coordination number and form of the polyhedron about uranium	U-O distances (Å)	References
Coffinite	$U[SiO_4]$	[4+4] two-fold disphenoide	2.32 (4 ×) 2.52 (4 ×)	FUCHS and GEBERT (SR 1958, 504)
Uraninite	UO_2	[8] cube	2.37 (8 ×)	several authors (SB 1913—1928, 148 and 212)
	UTi_2O_6[a]	[6] octahedron	~2.3 (6 ×)	RUH and WADSLEY (1966)
	$UGeO_4$	[8] two-fold disphenoide	~2.3 to 2.5 (8 ×)	DURIF (SR 1956, 283)
	$U[SO_4](OH)_2$	[8] square Archimedian antiprism	~2.3 (8 ×)	LUNDGREN (SR 1952, 287)
	$U_6O_4[SO_4]_6(OH)_4$	[8] square Archimedian antiprism	~2.3 (8 ×)	LUNDGREN (SR 1952, 287)
	$U[SO_4]_2.4\ H_2O$	[8] square Archimedian antiprism	~2.4 (8 ×)	KIERKEGAARD (SR 1956, 354)

[a] UTi_2O_6 is isotypic to brannerite, (U, Ca, Th, Y) $(Ti, Fe)_2O_6$.

pyrochlore-like structures, (Ca, Na)$_2$(Nb, Ta)$_2$O$_6$ (O, OH, F) (several authors; SB 1928—1932, 340), uranium can probably occupy the position of (Ca, Na) with a [2+6] coordination of oxygens. In both these minerals, the valency state of the substituting uranium is not known exactly.

IV. Formally Hexavalent Uranium

The compound UF_6 is orthorhombic and each uranium is octahedrally coordinated by six fluorines at distances between 2.01 and 2.13 Å (HOARD and STROUPE, 1958). The uranium atoms are arranged in the octahedral holes of the double-hexagonal close-packing of fluorine atoms. In UCl_6 the chlorine atoms form a nearly perfect octahedron about uranium with U—Cl distances of 2.4 Å (ZACHARIASEN, SR 1947—1948, 479); the chlorine atoms are arranged in a hexagonal close-packing.

The coordination number of hexavalent uranium against oxygen is six, seven or eight. Two oxygens having significantly shorter distances to the central atom (2.0 Å

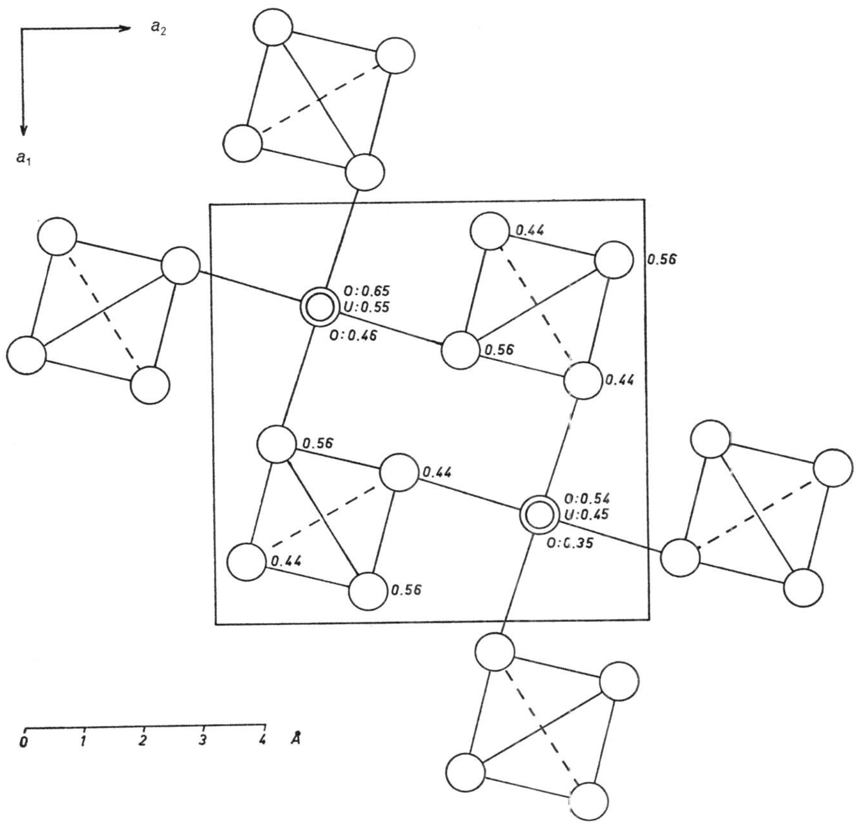

Fig. 92-A-1. Abernathyite, projection of a single [UO$_2$]—[AsO$_4$] sheet of (001). The numbers are the z parameters of the atoms in the tetragonal arrangement (c_0 = 18.13 Å). The As-atoms in the center of the tetrahedra with z = $^1/_2$ are not drawn

Table 92-A-2. *Structural data of some uranium(VI) compounds*

Mineral name	Chemical formula	U-O distances in the uranyl group (Å)		U-O distances in the plane perpendicular to the uranyl group (Å)		References
A) Coordination number [2+4]						
Abernathyite	$K[UO_2][AsO_4] \cdot 3\ H_2O$	1.81	1.70	2.35 (4×)		Ross and Evans (1964)
Meta-autunite	$Ca[UO_2]_2[PO_4]_2 \cdot 6\ H_2O$	1.99	1.79	2.32 (4×)		Makarov and Ivanov (SR 1960, 412)
Meta-torber-nite	$Cu[UO_2]_2[PO_4]_2 \cdot 8\ H_2O^a$	1.82	1.77	2.33 (4×)		Ross *et al.* (1964)
		1.80	1.94	2.31 (4×)		
Meta-zeunerite	$Cu[UO_2]_2[AsO_4]_2 \cdot 8\ H_2O$	1.94	1.78	2.18 (4×)		Hanic (SR 1960, 415)
	$Ba[UO_2]O_2$	1.90	1.90	2.12 (2×)	2.22 (2×)	Samson and Sillén (SR 1947—1948, 441)
	$Mg[UO_2]O_2$	1.92	1.92	2.16 (2×)	2.20 (2×)	Zachariasen (SR 1954, 463)
	$K_2[UO_2]O_2$	1.90	1.90	2.17 (4×)		Kovba *et al.* (SR 1958, 320)
	$Rb_2[UO_2]O_2$	1.91	1.91	2.17 (4×)		Kovba *et al.* (SR 1958, 320)
	$Cs_2[UO_2]O_2$	1.91	1.91	2.19 (4×)		Kovba *et al.* (SR 1958, 320)

	β-$Na_2[UO_2]O_2$	1.93	1.93	2.12 (4×)					Kovba et al. (SR 1958, 321)
	α-$Li_2[UO_2]O_2$	1.89	1.89	1.98 (4×)					Kovba et al. (SR 1958, 321)
	α-$Na_2[UO_2]O_2$	1.90	1.90	2.24 (4×)					Kovba et al. (SR 1958, 321)
	$NH_4[UO_2][AsO_4]\cdot 3\ H_2O$	1.81	1.73	2.37 (4×)					Ross and Evans (1964)
	$K(H_3O)[UO_2]_2[AsO_4]_2\cdot 6\ H_2O$	1.77	1.70	2.33 (4×)					Ross and Evans (1964)
	$Cu[UO_2]O_2$	1.90	1.90	2.15 (2×)	2.24 (2×)				Siegel and Hoekstra (1968)
	β-$[UO_2](OH)_2$	1.81	1.81	2.27 (2×)	2.32 (2×)				Bannister and Taylor (1970)
	$[UO_2]_3O_2(OH)_2$ [b]	1.83	1.83	2.07 (2×)	2.42 (2×)				Siegel et al. (1972)
	$Ca_3[UO_2]O_4$ [c]	2.02	1.91	2.26	2.16 (2×)	2.24			Rietveld (1966)
	$Sr_3[UO_2]O_4$ [c]	1.86	1.85	2.39	2.11	2.42	2.36		Rietveld (1966)
B) Coordination number [2+5]									
"Anhydrous carnotite"	$K_2[UO_2]_2[V_2O_8]$	1.8	1.8	2.2 to	2.4 (5×)				Appleman and Evans (1965)
Curienite	$Pb[UO_2]_2[V_2O_8]\cdot 5\ H_2O$	1.67	1.66	2.39	2.37	2.30	2.29	2.27	Borène and Cesbron (1971)
Dumontite	$Pb_2[UO_2]_3[PO_4]_2(OH)_4\cdot 3\ H_2O$ [b]	1.76	1.76	2.33 (2×)	2.37 (2×)	2.32			Piret-Meunier, Leonard and van Meersche (SR 1962, 583)

Table 92-A-2 (continued)

Mineral name	Chemical formula	U-O distances in the uranyl group (Å)		U-O distances in the plane perpendicular to the uranyl group (Å)					References
Jachimovite	$CuH_2[UO_2]_2[SiO_4]_2 \cdot 5\ H_2O$ [aa]	1.90	1.87	2.27	2.63	2.36	2.26	2.45	PIRET-MEUNIER and van MEERSCHE (1963)
Kasolite	$Pb[UO_2][SiO_4] \cdot H_2O$	1.71	1.72	2.22	2.15	2.50	2.21	2.47	MOKEEVA (1964a)
Sklodowskite	$MgH_2[UO_2]_2[SiO_4]_2 \cdot 6\ H_2O$ [aa]	1.7	1.7	2.2	2.3	2.5 (2 ×)	2.7		MOKEEVA (1964b)
Uranophane	$Ca(H_3O)_2[UO_2]_2[SiO_4]_2 \cdot 3\ H_2O$	1.91	1.91	2.5 (2 ×)	2.3 (2 ×)	2.2			SMITH, GRUNER and LIPSKOMB (SR 1957, 338)
β-Uranophane	$Ca[UO_2][UOOH][SiO_4][SiO_3OH] \cdot 4\ H_2O$ [a]	1.77	1.81	2.27	2.35	2.38 (2 ×)	2.41		SMITH and STOHL (1972)
		1.76	1.81	2.26	2.34	2.35	2.38	2.48	
	$Cs_2[UO_2]_2[SO_4]_3$	1.74	1.74	2.37 (2 ×)	2.47 (2 ×)	2.46			ROSS and EVANS (SR 1960, 389)
	$Cs_2[UO_2]_2[V_2O_8]$	1.77	1.77	2.40	2.37	2.38	2.25	2.29	APPLEMAN and EVANS (1965)
	$Ni[UO_2]_2[V_2O_8] \cdot 4\ H_2O$	1.79	1.84	2.33	2.29	2.34	2.35	2.21	BORÈNE and CESBRON (1970)
	UVO_5 [bb]	2.05	2.07	2.34	2.18 (2 ×)	2.25 (2 ×)			CHEVALIER and GASPERIN (1970)
	$[UO_2]_3O_2(OH)_2$ [b]	1.74	1.78	2.48	2.33	2.27	2.31	2.40	SIEGEL *et al.* (1972)
C) Coordination number [2+6]									
Cliffordite	$[UO_2]Te_3O_7$ [a]	2.04	2.04	2.30 (6 ×)					FISCHER *et al.* (1969)
		1.96	1.96	2.51 (6 ×)					GALY and MEUNIER (1971)

Dumontite	$Pb_2[UO_2]_3[PO_4]_2(OH)_4 \cdot 3\,H_2O$[b]	1.76	1.76	2.11 (2 ×)	2.73 (4 ×)		Piret-Meunier, Leonard and van Meersche (SR 1962, 583)
Rutherfordin	$[UO_2][CO_3]$	~1.9	~1.9	~2.5 (4 ×)	~2.4 (2 ×)		Christ, Clark and Evans (resp. Cromer and Harper; SR 1955, 425)
	$Ca[UO_2]O_2$	1.91	1.91	2.29 (6 ×)			Zachariasen (SR 1948, 323)
	$Sr[UO_2]O_2$	1.91	1.91	2.33 (6 ×)			Zachariasen (SR 1948, 323)
	$[UO_2][NO_3]_2 \cdot 6\,H_2O$	1.77	1.75	2.50 (2 ×)	2.55 (2 ×)	2.40 (2 ×)	Taylor and Mueller (1965)
	$Rb[UO_2][NO_3]_3$	1.78	1.78	2.48 (6 ×)			Barcley *et al.* (1965)
	$[UO_2]Nb_3O_8$[bb]	1.95	1.95	2.42 (4 ×)	2.57 (2 ×)		Chevalier and Gasperin (1968)
	$[UO_2]TiNb_2O_8$	1.79	1.79	2.34 (4 ×)	2.50 (2 ×)		Chevalier and Gasperin (1969)
	α-$[UO_2](OH)_2$	1.71	1.71	2.46 (4 ×)	2.51 (2 ×)		Taylor (1971)

[a] Two structurally different U-positions with essentially identical coordination.
[b] For one of the two structurally different U-positions.
[c] Uranium (VI) compounds with an insignificant "uranyl-group,"
[aa] The bond of the two H atoms is unknown.
[bb] Valency of V and Nb with respect to U not clearly defined by composition.

and less) compared to the other atoms (between 2.2 and 2.5 Å) are characteristic for the so-called "uranyl-group". The O—U—O angle within this group is 180° (metazeunerite) or nearly this value ($Ni[UO_2]_2[V_2O_8] \cdot 4H_2O$). Equal distances of the two oxygens lead to a symmetric UO_2-group as in rutherfordine. An asymmetric UO_2-group has been found in meta-zeunerite, abernathyite and meta-torbernite where the other oxygens of the uranium polyhedron are arranged in a plane, perpendicular to the axial direction of the linear uranyl group.

To summarize, EVANS (1963) has made an attempt to explain the different relations in the coordination of hexavalent uranium. This author also tabulated the crystal structures with uranyl groups according to the varying coordination within these groups; similarly, the coordination and U—O distances of some uranium(VI) compounds are listed in Table 92-A-2.

In the minerals of the torbernite group, isolated [2+4] uranium polyhedra, and $[PO_4]$ or $[AsO_4]$ tetrahedra are linked together in two dimensions (Fig. 92-A-1). Interlayered water and partially hydrated metal-atoms connect these sheets in the third direction. For the first time, BEINTEMA (SB 1938, 24) has described these structural elements in the proposed crystal structure of autunite.

A large number of synthesized "uranates" show the same [2+4] coordination principle. In $Mg[UO_2]O_2$, the uranium polyhedra linked by two shared edges form endless chains in one direction. $Ba[UO_2]O_2$ has a layer structure built up by uranium octahedra with four shared corners.

The [2+5] coordination plays an important role in uranyl vanadates and silicates. In curienite and also in $Cs_2[UO_2]_2[V_2O_8]$, the Cs-analogue of carnotite (cf. Fig. 23-A-5, this Handbook), the uranium polyhedra share edges, forming U_2O_{12} units. Uranophane and β-uranophane show typical sheet-structures, in which uranium polyhedra and silicon tetrahedra are connected in two dimensions.

$[UO_2]Te_3O_7$ contains UO_2 groups in a $(Te_3O_7)^{2-}$ framework; cliffordite crystallizes in the same structure type. Rutherfordine has a sheet-structure in which the carbonate groups and [2+6] coordinated uranium atoms share edges. Synthetic $[UO_2][NO_3]_2 \cdot 6H_2O$ is characterized by two O—O edges of NO_3 groups belonging to the coordination polyhedron around U^{6+}; a similar coordination is found in $Rb[UO_2][NO_3]_3$ where three edges of nitrate groups coordinate one uranyl group.

Several forms of uranium trioxide are known. α-UO_3 crystallizes hexagonally, and each uranium is coordinated by eight oxygens (U—O: 2.08 (2×), 2.39 (6×); ZACHARIASEN, SR 1947—1948, 224), but the structure now seems to be more complicated (LOOPSTRA and CORDFUNKE, 1966) than described here. The form β-UO_3 has five structurally different uranium atoms, with [6] and [7] coordination of oxygens and U—O distances between 1.52 and 2.77 Å (DEBETS, 1966). In γ-UO_3 the oxygens form distorted octahedra around uranium (ENGMANN and WOLFF, 1963) and δ-UO_3 is said to be isostructural with ReO_3 (WAIT, SR 1955, 356). Finally, in a high pressure form of UO_3 uranium has a seven-fold coordination with U—O distances from 1.80 to 2.56 Å (SIEGEL *et al.*, 1966).

Manuscript received: August 1972

References: Section 92-A

APPLEMAN, D. E., EVANS, H. T., JR.: The crystal structures of synthetic anhydrous carnotite, $K_2(UO_2)_2V_2O_8$, and its caesium analogue, $Cs_2(UO_2)_2V_2O_8$. Am. Mineralogist **50**, 825 (1965).

BANNISTER, M. J., TAYLOR, J. C.: The crystal structure and anisotropic thermal expansion of β-uranyl dihydroxide, $UO_2(OH)_2$. Acta Cryst. B **26**, 1775 (1970).

BARCLEY, G. A., SABINE, T. M., TAYLOR, J. C.: The crystal structure of rubidium uranyl nitrate: a neutron-diffraction study. Acta Cryst. **19**, 205 (1965).

BORÈNE, J., CESBRON, F.: Structure cristalline de l'uranyl-vanadate de nickel tetrahydraté, $Ni(UO_2)_2(VO_4)_2 \cdot 4H_2O$. Bull. Soc. Franç. Minéral. Crist. **93**, 426 (1970).

BORÈNE, J., CESBRON, F.: Structure cristalline de la curiénite, $Pb(UO_2)_2(VO_4)_2 \cdot 5H_2O$. Bull. Soc. Franç. Minéral. Crist. **94**, 8 (1971).

CHEVALIER, R., GASPERIN, M.: Structure cristalline de l'oxyde double UNb_3O_{10}. Compt. Rend. C **267**, 481 (1968).

CHEVALIER, R., GASPERIN, M.: Synthèse en monocristaux et structure cristalline de l'oxyde $UTiNb_2O_{10}$. Compt. Rend. C **268**, 1426 (1969).

CHEVALIER, R., GASPERIN, M.: Structure cristalline de UVO_5. Bull. Soc. Franç. Minéral. Crist. **93**, 18 (1970).

DEBETS, P. A.: The structure of β-UO_3. Acta Cryst. **21**, 589 (1966).

DONOHUE, J., EINSPAHR, H.: The structure of β-uranium. Acta Cryst. B **27**, 1740 (1971).

ENGMANN, R., DE WOLFF, P. M.: The crystal structure of γ-UO_3. Acta Cryst. **16**, 993 (1963).

EVANS, H. T., JR.: Uranyl ion coordination. Science **141**, 154 (1963).

FISCHER, R., SCHLATTI, M., ZEMANN, J.: Der Strukturtyp von Cliffordit, UTe_3O_8. Österr. Akad. Wiss., Math.-Naturw. Kl., Anz. 93 (1969).

GALY, J., MEUNIER, G.: A propos de la cliffordite, UTe_3O_8. Le système UO_3—TeO_2 à 700° C. Structure cristalline de UTe_3O_9. Acta Cryst. B **27**, 608 (1971).

GRONVOLD, F.: High-temperature x-ray study of uranium oxides in the UO_2—U_3O_8 region. J. Inorg. Nucl. Chem. **1**, 357 (1955).

HOARD, J. L., STROUPE, J. D.: The structure of crystalline uranium hexafluoride. U.S. At. Energy Comm. Rept. TID-5290, Bk. **1**, 325 (1958).

LOOPSTRA, B. O., CORDFUNKE, E. H. P.: On the structure of α-UO_3. Rec. Trav. Chim. **85**, 135 (1966).

MASAKI, N., DOI, K.: Analysis of the superstructure of U_4O_9 by neutron diffraction. Acta Cryst. B **28**, 785 (1972).

MOKEEVA, V. I.: The crystal structure of kasolite [in Russian]. Kristallografiya **9**, 738 (1964a).

MOKEEVA, V. I.: The structure of sklodowskite [in Russian]. Kristallografiya **9**, 277 (1964b).

PIRET-MEUNIER, J., VAN MEERSCHE, M.: Structure de la jachimovite, $CuH_2(UO_2 \cdot SiO_4)_2 \cdot 5H_2O$. Acad. Roy. Belg., Bull. Classe Sci. **49**, 181 (1963).

RIETVELD, H. M.: The crystal structure of some alkaline earth metal uranates of the type M_3UO_6. Acta Cryst. **20**, 508 (1966).

ROSS, M., EVANS, H. T., JR.: Studies of torbernite minerals (I): The crystal structure of abernathyite and the structurally related compounds $NH_4(UO_2AsO_4) \cdot 3H_2O$ and $K(H_3O)\ (UO_2AsO_4)_2 \cdot 6H_2O$. Am. Mineralogist **49**, 1578 (1964).

ROSS, M., EVANS, H. T., JR., APPLEMAN, D. E.: Studies of the torbernite minerals (II): The crystal structure of meta-torbernite. Am. Mineralogist **49**, 1603 (1964).

RUH, R., WADSLEY, A. D.: The crystal structure of $ThTi_2O_6$ (brannerite). Acta Cryst. **21**, 974 (1966).

SIEGEL, S., HOEKSTRA, H. R.: The crystal structure of copper uranium tetroxide. Acta Cryst. B **24**, 967 (1968).

Siegel, S., Hoekstra, H., Sherry, E.: The crystal structure of high-pressure UO_3. Acta Cryst. **20**, 292 (1966).
Siegel, S., Viste, A., Hoekstra, H. R., Tani, B.: The structure of hydrogen triuranate. Acta Cryst. B **28**, 117 (1972).
Smith, D. K., Stohl, F. V.: The crystal structures of uranophane and beta-uranophane. Am. Mineralogist **37**, 321 (1972), and personal communication.
Smith, G. S., Johnson, Q., Elson, R. E.: The crystal structure of uranium pentachloride. Acta Cryst. **22**, 300 (1967).
Steeple, H., Ashworth, T.: The minimum-residual refinement of β-uranium from polycrystalline data. Acta Cryst. **21**, 995 (1966).
Taylor, J. C.: The structure of the α-form of uranyl hydroxide. Acta Cryst. B **27**, 1088 (1971).
Taylor, J. C., Mueller, M. H.: A neutron diffraction study of uranyl nitrate hexahydrate. Acta Cryst. **19**, 536 (1965).

Manuscript received: August 1972

Introduction

The geochemistry of uranium is intimately associated with that of thorium. Consequently, the preparation of separate chapters on thorium and uranium has necessitated a subdivision of topics involving these two elements. In this chapter on uranium, the writers have included a discussion of both thorium and uranium in connection with heat production in the earth, autoradiographic studies, fractionation of thorium and uranium during weathering and sedimentation, and the utilization of nuclear energy. The chapter on thorium (90) includes a discussion of age dating by means of radioactive disequilibria, Th/U ratios in igneous rocks, and the general problem of the relationship of thorium and uranium to igneous petrology. The chapter on lead (82) Section B includes a discussion of U/Pb and Th/Pb age dating.

A number of reviews of the geochemistry of thorium and uranium have been published. ADAMS, OSMOND, and ROGERS prepared one in 1959. DYBEK (1962) reviews the geochemistry of uranium and EL SHAZLY (1962) provides a detailed bibliography. An extensive review now in preparation by PETERMAN was available in preliminary form in 1963. Owing to the enormous volume of published material, no effort is made in this chapter to refer to all papers dealing with the geochemistry of uranium. Reference is made to those papers from which numerical data have been obtained and to recent summary papers which will introduce readers to specific sub-topics.

Revised manuscript received: July 18th, 1967

92-B. Isotopes in Nature; Heat Production in Common Rocks

I. Varieties of Isotopes

Naturally-occurring uranium consists of three different isotopes: U^{238}, U^{235}, and U^{234}. The U^{238} and U^{235} are parent isotopes for two separate decay series which ultimately yield Pb^{206} and Pb^{207} respectively. No natural isotopic fractionation of U^{238} and U^{235} has been observed, and today all natural materials have a ratio of U^{238}/U^{235} equal to 137.5 ± 0.5. In the past this ratio has systematically increased to the present value, owing to the fact that U^{235} has a half-life of 7.1×10^8 years, which is much less than that of U^{238} (4.5×10^9 years). The U^{238}/U^{235} ratio 4.5×10^9 years ago, the presumed age of the earth, was 3.45. The third isotope of uranium, U^{234}, is an intermediate in the decay series of U^{238} and is formed from U^{238} by emission of an alpha followed by two beta particles; consequently, the abundance ratio of U^{234}/U^{238} is proportional to the ratio of their half-lives. The half-life of U^{234} is 2.5×10^5 years, making U^{234} equal to 0.0056 percent of total natural uranium.

Uranium, but generally not its daughter products, is incorporated from sea water into calcium carbonate shells and other precipitates. As a result of this selective fixation, the daughter of U^{234} (Th^{230} by alpha decay) is largely absent in freshly deposited carbonate, and thus the rate of building of Th^{230} can be used to date the carbonate materials; (see THURBER, 1965, for a review). For reasons not wholly understood, sea water contains about 15 percent more U^{234} than would be expected to be in secular radioactive equilibrium with the U^{238} in the water. Apparently the intermediate daughters formed in the decay process permit chemical fractionation of the U^{238} and U^{234}. Consequently, the rate at which the U^{234}/U^{238} ratio approaches that of secular radioactive equilibrium in carbonate shells may be used to check U^{234}/Th^{230} dating (THURBER, 1965). RICHARDSON (1964) detected a 15 percent deficiency in U^{234} in weathered samples of the Conway granite, New Hampshire. This deficiency in U^{234} correlates well with the apparent excess of U^{234} in sea water.

II. Heat Production in the Earth

The study of the flow of heat through the earth's surface has recently assumed great importance[1]. As examples, two of the questions involved are: 1) determination of whether or not the earth has the bulk composition of chondritic meteorites (the "chondritic hypothesis"); and 2) comparison of oceanic and continental heat flows, possibly for the purpose of establishing compositional differences in the upper mantle underlying continental and oceanic areas. Studies of heat flow in the earth have recently been summarized by a number of writers, including BIRCH (1965) and LEE and UYEDA (1965).

One major line of evidence for the chondritic hypothesis is the fact that the total terrestrial heat flow can be accounted for by the heat generated by radioactive elements

[1] See Chapter 6, Section V in Volume I of this handbook.

in an earth having the composition of an average chondrite. Taking into account regional variations in heat flow, the average flow through the earth's surface each year is about 2×10^{20} calories for the entire earth. The total amount of heat produced by radioactive decay in a chondritic earth can be calculated from the average concentrations of radioactive elements in chondrites. Tables 92-C-1 and 90-C-1 show the average uranium and thorium concentrations in chondrites as approximately 10^{-2} and 4×10^{-2} ppm respectively. (The data of STARIK and SHATS (1956) are not included in the averages because their values are so markedly different from those of other workers.) The average concentration of potassium in chondrites is 0.1 percent. The amount of heat generated by one gram of the metal of each of these three elements plus the daughter products in secular radioactive equilibrium with the uranium and thorium is:

potassium	2.7×10^{-5} cal/gm/yr
thorium	0.20 cal/gm/yr
uranium	0.73 cal/gm/yr.

From these data plus the mass of the earth, the total heat produced by radioactivity in a chondritic earth would be approximately 2×10^{20} calories per year, a figure which matches that of the observed heat flow.

Most studies seem to indicate an approximate uniformity of heat flow in oceanic and continental areas. Consequently, any models proposed for the crust and upper mantle must provide roughly the same total amount of radioactive material underlying equal surface areas of continents and ocean basins. The heat produced by rocks commonly used in such models is shown below in terms of the amount of heat generated in one year in a column of rock one kilometer long and having a cross-sectional area of one square centimeter. Thorium and uranium concentrations are taken from Tables 90-E-1 and 92-E-1 respectively, and potassium concentrations are from Chapter 19 (courtesy of Dr. K. S. HEIER).

	Th ppm	U ppm	K ppm metal	Heat cal/cm²/yr for 1 km of rock
Granite (> 70% SiO_2)	18	4	3.6×10^4	2.0
Oceanic tholeiitic basalt	0.18	0.10	0.17×10^4	0.05
Plateau basalt	1.96	0.53	0.75×10^4	0.3
Dunite and peridotite	< 0.05	0.01	10—20	0.01
Pipe eclogite	0.22	0.05	$0.1—0.2 \times 10^4$	0.03
Eclogite in metamorphic terraines	0.46	0.20	$0.1—0.2 \times 10^4$	0.1

The table clearly shows the extreme concentration of radioactive elements in continental-type rocks. Thus, an approximate equality of heat flow in continental and oceanic areas presumably indicates that the upper mantle under the oceans contains a higher concentration of radioactive elements than it does under the continents. This fact has been used to argue against a continental drifting by means of some simple shifting of continental masses over a more plastic mantle (see discussion by BIRCH, 1965, and MACDONALD, 1965).

Comparison of average concentrations of thorium and uranium in chondrites (Tables 90-C-1 and 92-C-1) with the concentrations in crustal rocks shown above clearly indicates extreme outward migration of thorium and uranium in the earth. TILTON and REED (1963) point out that the abundances of radioactive elements in chondrites are much lower than those required to account for the observed oceanic heat flow, which must come largely from the mantle, unless the oceanic upper mantle has undergone considerable fractionation from a chondritic earth. A mantle with the composition of the average dunite also would not be sufficiently radioactive to account for oceanic heat flow, and TILTON and REED conclude that eclogites may be important constituents of the oceanic upper mantle.

Revised manuscript received: July 18th, 1967

92-C. Abundance in Meteorites, Tektites, and Lunar Materials

I. Meteorites

The exceptionally low concentrations of radioactive elements in meteorites have made their determination difficult. The refinement of neutron activation techniques, however, has made it possible to estimate the small concentrations present even in the iron meteorites. The concentrations of uranium in a variety of meteorites are given in Table 92-C-1. Measurements of the concentration of thorium in meteorites are far less abundant than those of uranium. The thorium concentrations and Th/U ratios where available are given in Table 90-C-1.

Table 92-C-1. *Contents of uranium in meteorites*

Name	Class symbol	No. of samples	U $10^2 \times$ ppm	Method	Reference
Stones					
Achondrites					
Angra dos Reis	Aa	2	19.85	N	MORGAN and LOVERING (1964)
Binda	Ap	2	2.27	N	MORGAN and LOVERING (1964)
Bishopville	Ae	2	0.517	N	MORGAN and LOVERING (1964)
Cumberland Falls	Ac	1	1.0	R	DAVIS (1950)
Ellemeet	Ab	2	0.151	N	MORGAN and LOVERING (1964)
Johnstown	Ab	2	0.22	N	KÖNIG and WÄNKE (1959)
		1	1.1	N	BATE and HUIZENGA (1963)
		2	1.15	N	MORGAN and LOVERING (1964)
Moore Co.	Ap.	2	1.96	N	MORGAN and LOVERING (1964)
Nakhla	Ado	2	4.92	N	MORGAN and LOVERING (1964)
Norton Co.	Ae	1	1.00	I	PATTERSON *et al.* (1953)
Nuevo Laredo	Ap	1	12.6	N	HAMAGUCHI *et al.* (1957)
		2	15.5	N	REED *et al.* (1960)
fragment II		2	13.2	N	MORGAN and LOVERING (1965a)
fragment III		2	13.9	N	MORGAN and- LOVERING (1965a)
Pasamonte	Aor	1	5.42	N	KÖNIG and WÄNKE (1959)
Sioux Co.	Ap	1	6.30	N	KÖNIG and WÄNKE (1959)
Staroe Pas' Yanoe	Ae		17–18	L and N	STARIK and SHATS (1960)

Table 92-C-1 (continued)

Name	Class symbol	No. of samples	U $10^2 \times$ ppm	Method	Reference
Chondrites					
Abee	Ce	3	1.10	N	Reed *et al.* (1960)
		2	1.50	N	Goles and Anders (1962)
Akaba		1	0.9	N	Hernegger and Wänke (1957)
		1	0.8	N	Ebert *et al.* (1957)
Beardsley	CH	3	1.20	N	Reed *et al.* (1960)
		1	1.13	N	König and Wänke (1959)
Breitscheid		1	1.5	N	Hernegger and Wänke (1957)
		1	1.23	N	Ebert *et al.* (1957)
		1	1.24	N	König and Wänke (1959)
Bruderheim	CL	3	1.50	N	Goles and Anders (1962)
Elenovka (total)	CL	1	11.0	L and N	Starik and Shats (1960)
chondrules		2	21.4	L and N	Starik and Shats (1956)
matrix		2	28.6	L and N	Starik and Shats (1956)
Ergheo	CL	1	2.1	N	Goles and Anders (1962)
Forest City	CH	1	1.02	N	Hamaguchi *et al.* (1957)
		2	1.50	N	Reed *et al.* (1960)
Holbrook	CL	2	1.26	N	Hamaguchi *et al.* (1957)
		4	1.60	N	Reed *et al.* (1960)
Hvittis	Ce	2	0.619	N	Lovering and Morgan (1964)
Indarch	Ce	1	1.6	N	Reed *et al.* (1960)
		3	0.9	N	Goles and Anders (1962)
Karoonda	CHL	2	1.37	N	Lovering and Morgan (1964)
Kunashak		4	22.7	L and N	Starik and Shats (1956)
Mighei	Cc_2	1	1.6	N	Reed *et al.* (1960)
		2	1.4	N	Goles and Anders (1962)
Mocs	CL	1	1.1	N	Goles and Anders (1962)
Modoc	CL	1	1.1	I	Patterson *et al.* (1953)
		1	1.08	N	Hamaguchi *et al.* (1957)
		4	1.40	N	Reed *et al.* (1960)
		4	2.7	I	Patterson in Hamaguchi *et al.* (1957)
		1	1.04	N	König and Wänke (1959)
Mokoia	CHL	2	1.40	N	Lovering and Morgan (1964)
Murray	Cc_2	3	2.00	N	Goles and Anders (1962)
Orgueil	Cc	2	0.8	N	Reed *et al.* (1960)
		2	2.42	N	Lovering and Morgan (1964)
Pultusk	CH or CL	1	1.20	N	Ebert *et al.* (1957)
		1	1.23	N	König and Wänke (1959)

Table 92-C-1 (Continued)

Name	Class symbol	No. of samples	U $10^2 \times$ ppm	Method	Reference
Richardton	CR	2	1.22	N	HAMAGUCHI *et al.* (1957)
		1	1.3	N	GOLES and ANDERS (1962)
Saratov (total)	CL		8.5—9.0	L and N	STARIK and SHATS (1960)
Chondrules		1	20.0	L and N	STARIK and SHATS (1956)
Matrix		1	33.2	L and N	STARIK and SHATS (1956)
St. Marks	Ce	1	0.9	N	GOLES and ANDERS (1962)
Ställdalen	CH or CL	1	1.1	N	GOLES and ANDERS (1962)
Stony irons					
Antofagasta (metal)	P	1	1.0	R	DAVIS (1950)
Brenham	P	1	0.79	R	DAVIS (1950)
(olivine)		1	<0.09	R	DAVIS (1950)
(metal)		1	0.06	R	REASBECK and MAYNE (1955)
(stone)		1	0.64	R	REASBECK and MAYNE (1955)
(metal)		1	0.0005	N	REED *et al.* (1958)
		1	0.012	N	REED *et al.* (1958)
Salta	P				
(metal)		1	1.9	R	DAVIS (1950)
(olivine)		1	0.12	R	DAVIS (1950)
Irons					
Cañon Diablo	Og		0.0005 to 0.0014	N	REED *et al.* (1958)
(metal)		1	<0.06	N	GOLES and ANDERS (1962)
(troilite)		1	0.35	N	GOLES and ANDERS (1962)
		1	0.4	N	REED *et al.* (1960)
		1	0.9	I	PATTERSON *et al.* (1953)
Carbo	Om		0.0003 to 0.007	N	REED *et al.* (1958)
Chebankol	Og	1	1.68	L and N	STARIK and SHATS (1956)
Chinghe	D	1	1.73	L and N	STARIK and SHATS (1956)
Grant	Of				
(metal)		1	<0.05	N	GOLES and ANDERS (1962)
(troilite)		1	0.65	N	GOLES and ANDERS (1962)
Sardis (troilite)		1	0.65	N	GOLES and ANDERS (1962)
Sikhote-Alin	H	2	2.19	L and N	STARIK and SHATS (1956)
		1	<0.002	N	FIREMAN and FISHER (1961)
Soroti (troilite)	uncl.	1	1.7	N	GOLES and ANDERS (1962)
Thunda			0.0005 to 0.032	N	REED *et al.* (1958)
Toluca	Om				
(metal)		2	<0.015	N	GOLES and ANDERS (1962)
(troilite)		1	1.0	N	GOLES and ANDERS (1962)
(troilite)		1	<1.3	N	REED *et al.* (1960)

With regard to the various types of meteorites, the abundance of uranium is particularly low in the irons. Early determinations in the range of hundredths of a part per million uranium in the irons are probably not accurate, and the more recent data are in the range of tenths of a part per billion. The abundances of uranium in the stony meteorites are higher than in the irons, with the chondrites averaging 0.01 to 0.02 part per million of the metal, roughly two orders of magnitude greater than in the irons. The achondrites may contain slightly more uranium than the chondrites.

II. Tektites

The abundances of uranium in various tektites are shown in Table 92-C-2. Comparable values for thorium and the Th/U ratio are given in Table 90-C-2. Tektites are interesting in view of the possibility that they might be of extraterrestrial origin. The concentrations of radioactive elements in tektites, however, are at least two orders of magnitude greater than in meteorites. In fact, concentrations of radioactive materials in tektites are rather similar to those found in surficial materials on the earth. This similarity, of course, does not prove terrestrial origin but is at least

Table 92-C-2. *Contents of uranium in tektites and glasses*

Name	No. of samples	U ppm	Method	Reference
Australite	2	2.6	R	Dubey (1933)
	4	1.6	L	Adams (1956)
	1	1.74	I	Tilton (1958)
	6	2.0	R	Cherry (1962)
	1	2.7	R	Cherry and Adams (1963)
	8	2.1	R	Balacek and Adams (1966)
Bediasite	4	1.3	L	Adams *et al.* (1959)
	2	1.5	R	Cherry and Adams (1963)
	10	1.9	R	Balacek and Adams (1966)
Billitonite	1	2.7	R	Dubey (1933)
Ivory Coast Tektite	1	0.91	L	Friedman (1958)
Indochinite	2	1.7	L	Adams (1956)
	1	2.24	L	Heide (1961)
	3	2.2	L	Starik *et al.* (1961b)
	4	2.1	R	Cherry and Adams (1963)
	5	2.2	R	Balacek and Adams (1966)
Moldavite	4	2.7	R	Dubey (1933)
	1	1.8	L	Adams *et al.* (1959)
	2	1.9	L	Heide (1961)
	2	2.2	L	Starik *et al.* (1961b)
	1	2.1	R	Cherry and Adams (1963)
Philippinite	1	1.6	L	Adams (1956)
	1	2.5	L	Starik *et al.* (1961b)
	1	1.85	R	Cherry and Adams (1963)
	12	2.6	R	Balacek and Adams (1966)
Darwin glass	1	1.4	R	Dubey (1933)
	1	1.9	L	Adams (1956)
Libyan glass	1	0.8	L	Starik *et al.* (1961b)
Peruvian Glass	2	18.4	R	Cherry and Adams (1963)
Tektites Ave.		1—2		Adams (1956)

a strong indication of that possibility. Regional differences in the abundances of thorium and uranium in tektites have recently been summarized by BALACEK and ADAMS (1966), who also report a large number of new analyses. The origin of these regional differences has not been established.

Revised manuscript received: July 1967

III. Lunar Materials

The results from approximately 700 uranium analyses of lunar rocks can be found in the literature, but it is very difficult to summarize them in a meaningful way. The reason for this is that rock types are often ill defined or form only small portions of larger rocks, e.g. breccias. An additional problem with many analyses is that the samples analyzed were very small and far from being representative for a particular hand specimen or rock type. On the other hand, it seems that at each landing site the U concentration of a particular rock type keeps close to a specific concentration level with rather small variation. The large number of analyses therefore, probably provides a good estimate of the U concentration of the various petrographic units. Agreement between average concentrations obtained by neutron activation, mass spectrometry, γ-ray spectrometry of natural radioactivity and occasionally fission track counting is good. Spark source mass spectrometry in many cases yields clearly deviating values and therefore these data were not used.

Table 92-C-3 lists the average uranium concentrations, their range and standard deviations for the various rock types encountered at each landing site. The grouping of the rocks is to a very large degree arbitrary and an example of this is the melt rocks. They comprise the so-called anorthositic, troctolitic, very high aluminia and Apollo 14 KREEP basalts which contain clasts or xenolithic materials and frequently show evidence of shock or reaction with the melt matrix containing them (IRVING, 1975, P VI, 363)[1]; the Apollo 15 and 17 KREEP basalts were also included with this group. It is certainly possible to subdivide a particular group on petrographic grounds, but in most cases there is either no consensus on the classification or the differences are not borne out in the uranium concentrations, e.g. for the mare basalts.

Table 92-C-3. *Uranium in lunar rocks*

	Mean ppm U	Range ppm U	s	Number of analyses	References
Apollo 11					
Soils	0.56	0.29–1.2	0.17	22	15, 22, 26, 28, 49, 60, 67, 69, 77, 78, 85, 88, 90
Breccias	0.56	0.36–0.69	0.08	31	26, 27, 49, 59, 67, 69, 77, 85
High-K basalts	0.75	0.47–0.88	0.12	29	26, 27, 31, 49, 59, 60, 66 67, 69, 77, 85, 86
Low-K basalts	0.23	0.14–0.41	0.06	27	26, 27, 49, 59, 60, 67, 69, 77, 85

[1] The abbreviations P I to P VI refer to the Proceedings of the First to Sixth Lunar Science Conferences, Geochim. Cosmochim. Acta Supplements 1 to 6 (1970–1975).

Table 92-C-3 (continued)

	Mean ppm U	Range ppm U	s	Number of analyses	References
Apollo 12					
Soils	1.85	0.9–3.27	0.54	42	4, 13, 23, 38, 57, 61, 63, 65, 72, 80, 86, 90
Breccias	5.25	2.19–10.3	3.02	11	23, 51, 57, 61, 63, 72, 87
Basalts	0.23	0.11–0.37	0.05	52	4, 9, 13, 37, 38, 51, 57, 61, 63, 65, 72, 86
Apollo 14					
Soils	3.34	2.1–4.07	0.39	24	8, 16, 21, 41, 42, 62, 71, 73, 81, 89
Breccias	3.52	1.24–5.4	0.93	28	5, 16, 21, 29, 37, 81, 89
Melt rocks	3.15	2.6–3.9	0.37	11	5, 16, 37, 41, 62, 71, 74
Basalt 14053	0.58	0.52–0.64	0.05	4	37, 71, 74
1 Anorthositic fragment 14161	0.58	—	—	1	33
Apollo 15					
Soils	1.11	0.51–1.89	0.26	42	3, 20, 25, 37, 41, 58, 62, 81
Breccias	0.83	0.35–1.46	0.32	12	35, 37, 58
KREEP-rich breccias	3.15	2.9–3.28	0.17	4	37, 62
Melt rocks	1.51	1.17–1.84	—	2	35
Basalts	0.15	0.10–0.20	0.03	12	3, 29, 37, 58, 62, 83
Anorthosites	0.0076	0.0015–0.043	0.0065	5	33, 37, 83
Apollo 16					
Soils	1.11	0.51–1.89	0.26	42	29, 40, 50, 52, 53, 75, 82, 91
Breccias	0.57	0.062–1.31	0.37	49	6, 12, 17, 19, 29, 35, 39, 50, 53, 83, 84
KREEP-rich breccias	2.35	1.7–3.2	0.45	8	6, 12, 17, 39, 45, 53
Melt rocks	0.61	0.28–1.10	0.34	11	2, 3, 12, 17, 19, 35, 53
Anorthosites	0.026	<0.0006–0.101	0.31	11	12, 35, 52, 53, 75, 83, 84, 91
Apollo 17					
Soils	0.46	0.16–1.0	0.26	66	6, 11, 18, 19, 25, 39, 52, 54, 64, 84
Breccias	1.27	0.51–2.06	0.41	27	7, 18, 25, 35, 39, 55, 83, 84
Melt rocks	0.29	0.20–0.41	0.06	12	35, 56
Basalts	0.13	0.07–0.22	0.03	33	7, 18, 25, 54, 83, 84
Luna 16					
Soils	0.29	0.1–0.36	0.10	6	1, 34, 73, 79
Luna 20					
Soils	0.32	0.22–0.41	0.07	6	2, 48, 68, 70, 75

References: Abbreviations P I to P VI = Proceedings of the Lunar Science Conferences, Geochim. et Cosmochim. Acta, Supplements 1 to 6; GCA = Geochim. et Cosmochim. Acta; EPSL = Earth Planet. Sci. Letters.

1. Albee, A. L. *et al.* (1972), EPSL **13**, 353. 2. Bansal, B. M. *et al.* (1972), EPSL **17**, 29. 3. Barnes, I. L. *et al.* (1973), P IV, 1197. 4. Brunfelt, A. O. *et al.* (1971), P II, 1281. 5. Brunfelt, A. O. *et al.* (1972), P III, 1133. 6. Brunfelt, A. O. *et al.* (1973), P IV, 1209. 7. Brunfelt, A. O. *et al.* (1974), P V, 981. 8. Brunfelt, A. O. *et al.* (1971), EPSL **11**, 351; 9. Burnett, D. *et al.* (1971), P II, 1503. 10. Busche, F. D. *et al.* (1972), EPSL **14**, 313. 11. Church, S. E. *et al.* (1974), P V, 1389. 12. Clark, R. S. *et al.* (1973), P IV, 2105. 13. Cliff, R. A. *et al.* (1971), P II, 1493. 14. Crosaz, G. *et al.* (1972), P III, 1623. 15. Crosaz, G. *et al.* (1970), P I, 2051. 16. Eldridge, J. S. *et al.* (1972), P III, 1651. 17. Eldridge, J. S. *et al.* (1973), P IV, 2115. 18. Eldridge, J. S. *et al.* (1974), P V, 1025. 19. Eldridge, J. S. *et al.* (1975), P VI, 1407. 20. Fields, P. R. *et al.* (1973), P IV, 2123. 21. Fields, P. R. *et al.* (1972), P III, 1637. 22. Fields, P. R. *et al.* (1970), P I, 1097. 23. Fields, P. R. *et al.* (1971), P II, 1571. 24. Fleischer, R. L. *et al.* (1971), P II, 2559. 25. Fruchter, J. S. *et al.* (1975), P VI, 1399. 26. Goles, G. G. *et al.* (1970), P I, 1177. 27. Goles, G. G. *et al.* (1970), P I, 1165; 28. Gopalan, K. *et al.* (1970), P I, 1194. 29. Graf, H. *et al.* (1973), P IV, 2145. 30. Haines, E. L. *et al.* (1971), EPSL **12**, 145. 31. Herzog, G. F. *et al.* (1970), P I, 1239. 32. Hubbard, N. J. *et al.* (1972), P III, 1161. 33. Hubbard, N. J. *et al.* (1971), EPSL **13**, 71. 34. Hubbard, N. J. *et al.* (1972), EPSL **13**, 423. 35. Hubbard, N. J. *et al.* (1974), P V, 1227. 36. Hutcheon, I. D. *et al.* (1974), P V, 2597. 37. Keith, J. E. *et al.* (1971), P III, 1671. 38. Kharkar, D. P. *et al.* (1971), P II, 1301. 39. Laul, J. C. *et al.* (1974), P V, 1047. 40. Laul, J. C. *et al.* (1973), GCA **37**, 927. 41. Laul, J. C. *et al.* (1972), P III, 1181. 42. Lindstrom, M. M. *et al.* (1972), P III, 1201. 43. Lovering, J. F. *et al.* (1970), P I, 627. 44. Lovering, J. F. *et al.* (1971), P II, 151. 45. Lovering, J. F. *et al.* (1975), P VI, 1203. 46. Lovering, J. F. *et al.* (1974), EPSL **21**, 164. 47. Lovering, J. F. *et al.* (1971), P II, 39. 48. Morgan, J. W. *et al.* (1973), GCA **37**, 953. 49. Morrison, G. H. *et al.* (1970), P I, 1383. 50. Morrsion, G. H. *et al.* (1973), P IV, 1399. 51. Morrison, G. H. *et al.* (1971), P II, 1169. 52. Müller, O. (1975), P VI, 1303. 53. Nunes, P. D. *et al.* (1973), P IV, 1797. 54. Nunes, P. D. *et al.* (1974), P V, 1487. 55. Nunes, P. D. *et al.* (1974), EPSL **23**, 445. 56. Nunes, P. D. *et al.* (1975), P VI, 1431. 57. O'Kelley, G. D. *et al.* (1971), P II, 1159. 58. O'Kelley, G. D. *et al.* (1972), P III, 1659. 59. O'Kelley, G. D. *et al.* (1970), P I, 1407. 60. Perkins, R. W. *et al.* (1970), P I, 1455. 61. Rancitelli, L. A. *et al.* (1971), P II, 1757. 62. Rancitelli, L. A. *et al.* (1972), P III, 1681. 63. Reed, G. W. *et al.* (1971), P II, 1261. 64. Rhodes, J. M. *et al.* (1974), P V, 1097. 65. Rosholt, J. N. *et al.* (1971), P II, 1577. 66. Schonfield, E. *et al.* (1972), P III, 1379. 67. Silver, L. T. (1970), P I, 1533. 68. Surkov, Yu. A. *et al.* (1973), P IV, 1437. 69. Tatsumoto, M. (1970), P I, 1595. 70. Tatsumoto, M. (1973), GCA 37, 1079. 71. Tatsumoto, M. *et al.* (1972), P III, 1531. 72. Tatsumoto, M. *et al.* (1971), P II, 1521. 73. Tera, F. *et al.* (1972), EPSL **13**, 457. 74. Tera, F. *et al.* (1972), EPSL **14**, 281. 75. Tera, F. *et al.* (1972), EPSL **17**, 36. 76. Thiel, K. *et al.* (1972), EPSL **16**, 31. 77. Turekian, K. K. *et al.* (1970), P I, 1659. 78. Turkevich, A. *et al.* (1971), P II, 1565. 79. Vinogradov, A. P. (1971), P II, 1. 80. Vobecký, M. *et al.* (1971), P II, 1291. 81. Wänke, H. *et al.* (1972), P III, 1251. 82. Wänke, H. *et al.* (1973), P IV, 1461. 83. Wänke, H. *et al.* (1975), P VI, 1313. 84. Wänke, H. *et al.* (1974), P V, 1307. 85. Wänke, H. *et al.* (1970), P I, 1719. 86. Wänke, H. *et al.* (1971), P II, 1187. 87. Wakita, H. *et al.* (1970), EPSL **9**, 169. 88. Wanless, R. K. *et al.* (1970), P I, 1729 89. Wiik, H. B. *et al.* (1973), EPSL **17**, 365. 90. Wrigley, R. C. (1971), P II, 1791. 91. Wrigley, R. C. (1973), P IV, 2203.

The variation within each group at one particular landing site is, in general, smaller than between analogous rock groups of different landing sites.

With respect to many incompatible elements, e.g. K and U, the composition of the great majority of the soils, breccias and most other rocks, except the anorthositic ones, can be approximated by a two-component mixing model with varying proportions of the end-members (e.g. SCHONFIELD, 1974, P V, 1135). For Apollo 11 these are low-K ($\lesssim 0.06\%$ K, ~ 0.2 ppm U) and high-K ($\sim 0.25\%$ K, 0.75 ppm U) mare basalt. For the other lunar landing sites the end-members are low-K mare basalt and the hypothetical rock KREEP. The latter is very rich in K, REE, P and other incompatible elements like Th and U which are present due to concentration by differentiation processes. In general, the U content of a rock is dominated by its KREPP component which provides up to 3–4 ppm U in some breccias. No pure KREEP, however, has ever been found. The highest U concentration of 6–10 ppm was encountered in a granitic differentiate (white portion of breccia 12013), the lowest of < 0.0006 ppm U in an almost monomineralic anorthosite (rock 65315). The average for anorthosites is 0.02 ppm U. The emerald green glass of Apollo 15, which is considered a representative of the least differentiated mare basalt contains 0.045 ppm U (average of 445 fission track analyses, FLEISCHER *et al.*, 1974, P V, 2251; STORZER *et al.*, 1973, P IV, 2363), average low-K mare basalts ~ 0.2 ppm U and high-K mare basalts ~ 0.75 ppm U. The melt rocks are all uranium-rich types with uranium contents around 1 ppm.

Geochemically the uranium in lunar rocks is characterized by its stronger coherence with Th than in terrestrial rocks. This is probably due to the absence of hexavalent uranium on the moon, which prevents a possible separation of the two elements. Fission track studies have shown, that uranium in lunar rocks is very inhomogeneously distributed and that the uranium content of the rock-forming minerals is extremely low: Pyroxenes, feldspars, olivines, cristobalites and ilmenites contain < 5 ppb U (BURNETT *et al.*, 1971, P II, 1503; CROSAZ *et al.*, 1970, P I, 2051; LOVERING *et al.*, 1970, P I, 627). Most of the uranium in volcanic rocks is localized in indeterminate interstitial areas of possibly glassy material enriched in Si, K, P. It probably represents a quenched residual magmatic liquid which concentrates uranium until the very last stage of crystallization. Only then does the uranium enter discrete phases leaving the glassy matrix essentially devoid of uranium (BURNETT *et al.*, 1971, P II, 1503; LOVERING *et al.*, 1970, P I, 627; 1971, P II, 151, and 1972, P III, 281; RICE *et al.*, 1971, P II, 159). These phases are baddeleyite, zircon, tranquillityite, a mineral which perhaps is zirconolite, whitlockite and apatite. In lunar breccias the uranium is less strictly confined to certain phases of interstitial material (LOVERING *et al.*, 1970, P I, 627) and the distributed uranium may amount up to 95% (GRAF *et al.*, 1973, P IV, 2145) of the total. Table 92-C-4 gives an indication of the uranium content in the uranium concentrating minerals and Table 92-C-5 is an example of the uranium balance in two lunar magmatic rocks.

The orbital γ-ray mapping experiment (METZGER *et al.*, 1974, P V, 1067) yielded average Th concentrations of 0.9 ppm for the highlands, 3.2 ppm for the maria and 2.2 ppm as an average for the surface. Since the Th/U ratio on the moon is very constant at 3.65 ± 0.3, these data permit the conclusion to be drawn that on average the highlands contain ~ 0.25 ppm U, the maria 0.9 ppm U and the average crustal rock on the lunar surface 0.6 ppm U. Although these numbers deviate from averages that may be calculated from the values for the soils of Table 92-C-3 they are probably representative, since for the least heterogeneous landing sites (Apollo 11 and 16) the orbital data are close to the averages of the soils. The U and Th concentrations in the crust are due to strong enrichment. Based on element correlations, geochronology and heatflow measurements (LANGSETH *et al.*, 1972, The Moon **4**, 390) and depending

Table 92-C-4. *U content of U concentrating phases*[a]

Mineral	Typical values ppm U	Range ppm U	No. of grains analyzed	Rock numbers	References[b]
Apatite	40–60, 100–200	31–430	27	1084, 12013, 14305, 14321, 72315, 72395	9, 30, 36, 76
Whitlockite	20–80	10–880	34	12013, 12040, 14301, 14318, 7225	9, 29, 30, 36
Monazite	270	—	1	10047	46
Zircon	100–400	<10–1,500	24	10084, 12013, 14301, 14311	9, 14, 24, 30
Baddeleyite	50–150, 300–450	~10–1,500	>20	10047, 10057, 10069, 12002, 12020, 12022, 12028, 12040, 12051, 12053, 14301, 14306, 14310, 14311	9, 14, 24, 43, 44
Tranquillityite[c])	50–300	40–~1,000	>13	10047, 10057, 10069, 12002, 12020, 12022, 12051, 12053	44, 47
Zirkelite[d])	2,000	1,900–2,200	5	12036	10

[a] Besides the minerals listed, there are other major U carriers which were not yet identified. They are all Zr-Ti or phosphate minerals which probably are tranquillityite, zirkelite or zirconolite and unresolved intergrowths of apatite and whitlockite. The highest U concentration (3.6%) was reported for a Zr-Ti mineral in rock 12013 containing also much Fe, Y, Nb, REE, Th and Pb (Reference 30).
[b] For references see footnote of Table 92-C-3.
[c] $Fe_8^{2+}(Zr+Y)_2Ti_3Si_3O_{24}$.
[d] $CaZrTiO_5$.

Table 92-C-5. *Uranium balance in the lunar rocks 12013 and 12040* (Taken from reference 9 of Table 92-C-3)

Phase	12013		12040	
	U (ppm)	% of total U	U (ppm)	% of total U
Apatite	170	25	a	a
Whitlockite	190	25	104	78
Zircon	330	17	a	a
Baddeleyite	a	a	840	10
Zr-Ti phase	2,000	10	1,300	5
Distributed	6[b])	20		7[c])
		97		100

[a] Not present.
[b] Dark areas rich in P.
[c] From major minerals, both from buried sources and distributed.

on the model of lunar evolution, overall uranium concentrations of between 30 and 86 ppb were estimated, with values of around 60 ppb being the most probable ones (Papanastassiou *et al.*, 1971, Earth Planet. Sci. Letters **11**, 37; Solomon *et al.*, 1973,

Phys. Earth Planet. Interiors **7**, 15; Toksöz *et al.*, 1973, The Moon **7**, 251; Wänke *et al.*, 1973, P IV, 1461; Schonfield *et al.*, 1974, P V, 1135; Ganapathy *et al.*, 1974, P V, 1181; Hubbard *et al.*, 1975, P VI, 1057; Conel *et al.*, 1975, The Moon **14**, 263).

The $^{238}U/^{235}U$ ratio in lunar rocks is the same as in terrestrial samples: 137.8 ± 0.3 (Barnes *et al.*, 1972, P III, 1465 and 1973, P IV, 1197; Fields *et al.*, 1970, P I, 1729 and 1972, P III, 1637; Rosholt *et al.*, 1970, P I, 1499 and 1971, P II, 1577; Wanless *et al.*, 1970, P I, 1729). As a 10^7 year difference in the time of nucleosynthesis of the U isotopes would be reflected in a 1% difference in the $^{238}U/^{235}U$ ratios of lunar and terrestrial uranium, this implies that the U in the two bodies has a common origin within ± 5 million years. The ^{234}U daughter is in radioactive equilibrium with parent ^{238}U in the lunar rocks (Rosholt *et al.*, 1971, P II, 1577).

Manuscript received: January 1977

92-D. Abundances in Rock Forming Minerals (I), Uranium Minerals (II)

I.

Uranium occurs in a variety of minerals but characteristically is concentrated in a few species of minor abundance. Where uranium is present as trace quantities in such major minerals as quartz and the feldspars, its mode of occurrence is uncertain, though the following possibilities may be listed: (1) isomorphous substitution in the lattice, (2) concentration in lattice defects, (3) adsorption along crystal imperfections and grain borders, and (4) inclusion as microcrystals of uranium minerals. Table 92-D-1 summarizes the abundances of uranium in rock-forming minerals and uraniferous accessories in igneous rocks. Comparable concentrations of thorium and Th/U ratios are given in Table 90-D-1.

Table 92-D-1. *Contents of uranium in minerals in igneous rocks*

	U in ppm	References (No. of samples)
Major minerals		
Quartz	1.7	P (20)
	0.1–10	A (3); C (3); D (1); F (2); L (6); M (3); P (20); Q (2); R (2); U (1)
Feldspar (including potassium feldspar and plagioclase)	2.7	P (42)
	0.1–10	A (5); C (3); D (1); F (2); L (6); M (3); N (2); P 42); Q (7); R (2); U (1)
Biotite	8.1	P (23)
	1–60	A (1); C (3); F (2); L (6); M (3); N (2); P (23); Q (5); R (2); U (1)
Muscovite	2–8	C (1); M (2); P (4)
Hornblende	7.9	P (19)
	0.2–60	A (1); C (2); L (4); P (19); Q (5); R (2); U (1)
Pyroxene	3.6	P (11)
	0.1–50	A (1); P (11); Q (1); R (2)
Olivine (from dunite)	~0.05	T (20)
Accessory minerals		
Allanite	200	J (10)
	30–1,000	B (1); C (1); J (10); L (6); M (3); N (2); P (10); R (2); S (5); U (1)
Apatite	65	H (29)
	10–100	C (2); D (1); F (2); H (29); L (6); M (3); N (2); P (14); Q (7); U (1)
Epidote	43	H (9)
	20–200	H (9); P (3)
Garnet	6–30	C (2); P (4)

Table 92-D-1 (Continued)

	U in ppm	References (No. of samples)
Huttonite	3×10^3—7×10^4	V (individual grains from one rock)
Magnetite (and other opaque minerals)	1—30	A (12); C (2); D (1); F (2); L (6); M (3); N (2); P (23); Q (6); R (1); U (1)
Monazite	3000	H (11)
	500—3000	C (2); H (11); K (4); M (3); N (2); P (4); R (1)
Sphene	280	H (25)
	10—700	D (1); F (2); H (25); L (6); M (3); Q (4)
Thorianite and Uraninite	varies from ThO_2 to UO_2	
Thorite and Uranothorite	varies from $ThSiO_4$ to $USiO_4$ (?) (solid solution may not be complete)	
Xenotime	300—4×10^4	C (2); P (3)
Zircon	1330	H (43)
	100—6000	C (3); D (1); G (21); H (43); L (6); M (3); N (2); P (14); Q (7); U (1); V (1)

Ranges of concentrations shown are those within which uranium values commonly lie, rather than the extreme values of the various mineral suites. Average values are reported for: major minerals in several batholiths of the western U.S.A. analyzed by reference P, using fluorimetric methods; accessory minerals from a sampling of North American granites by reference H, using gamma-spectrometric methods; and a suite of allanite separates by reference J, using radiometric methods. References listed below are cited for each mineral, and the numbers in parentheses in the table by each reference indicate the number of samples analyzed.

A. Keevil (1944); B. Hutton (1951); C. Larsen Jr. and Phair (1954); D. Tilton *et al.* (1955); E. Larsen Jr. *et al.* (1956); F. Tauson (1956); G. Webber *et al.* (1956); H. Hurley and Fairbairn (1957); J. W. L. Smith *et al.* (1957); K. Tilton and Nicolaysen (1957); L. Leonova and Tauson (1958); M. Chentsov (1959); N. Baranov and Du (1961); P. Larsen Jr. and Gottfried (1961); Q. Leonova and Pogiblova (1961); R. Meliksetyan (1961); S. Leonova (1962); T. Heier and Carter (1964); U. Leonova and Renne (1964); V. Richardson (1964).

Four papers which give preliminary data on mineral compositions but which are not cited in the table are Gottfried and Larsen Jr. (1954); Larsen Jr. (1954); Phair (1955); and Phair (1956).

A major method of investigation of alpha-emitting elements in rocks is autoradiography. It is possible to construct autoradiographic techniques which discriminate between uranium and thorium, but the common and easily used methods involving thin films or plates of alpha-sensitive emulsions do not permit such a distinction. Modern techniques have been summarized by Ragland (1964), who also gives an extensive bibliography of autoradiographic investigations. The principal conclusions derived from autoradiographic studies of igneous rocks may be summarized as follows: (1) Most granites have a major portion of their alpha activity concentrated in small grains of accessory minerals. In many granites, however, there is also appreciable concentration on grain borders or disseminated through the major minerals. (2) Gabbros and other basic igneous rocks have their alpha activity distributed fairly evenly among the various major constituents. (3) Volcanic rocks have a uniform distribution of radioactivity among their constituent phases. These facts lead to the conclusion that most minerals containing appreciable concentrations of thorium and uranium form quite late in the igneous crystallization sequence. In addition to the summary by RAGLAND (1964) cited above, other papers which might be consulted for an introduction to the field are those of MERLIN *et al.* (1957), ROGERS and ADAMS (1957), SPEARS (1961), and COPPENS (1950).

Variation of thorium and uranium abundances within individual crystals of accessory minerals has recently been demonstrated by SILVER and DEUTSCH (1963). By selectively dissolving various layers of zircon grains, they showed variations in abundances of thorium and uranium from place to place in single grains by as much as an order of magnitude. AHRENS (1965) has recently proposed a lognormal frequency distribution of uranium concentrations in accessory zircons of granites.

II. Uranium Minerals

The most abundant uranium mineral is uraninite, with a formula varying from UO_2 to U_3O_8. The variability of chemical composition apparently results largely from oxidation of UO_2 to U_3O_8 after deposition, and all compositions between the end members are found. Some geologists use the term "pitchblende" as synonomous with uraninite, whereas others restrict it to the more oxidized form. Uraninite is the common black ore of uranium and occurs in hydrothermal vein and disseminated deposits and also in sedimentary rocks throughout the world. A more complete mineralogic description of uraninite is given by PALACHE *et al.* (1944).

A characteristic feature of uranium is its ability to be oxidized to the highly soluble uranyl ion (UO_2^{++}). This ion is easily mobilized in surface and near-surface waters; consequently, regardless of their origin, most uranium deposits contain an assemblage of supergene minerals which are invariably bright yellow, green, or orange and are generally fluorescent. In some cases these minerals are widely dispersed and not obviously associated with the original, primary mineralization. The various minerals in this category consist of carbonates, phosphates, vanadates, silicates, sulfates, etc., commonly in combination with uranium plus some other cation. A complete list of uranium minerals is given in Table 92-D-2; the list represents an addition of some 25 new species to those given by FRONDEL (1956).

Table 92-D-2. *Minerals containing uranium as an essential constituent*

Oxides	
Becquerelite	$7UO_3 \cdot H_2O$
Billietite	$BaO \cdot 6UO_3 \cdot 11H_2O$
Clarkeite	$(Na, K, Ca, Pb)_2U_2O_7 \cdot nH_2O$
Compreignacite	$K_2O \cdot 6UO_3 \cdot 11H_2O$ (PROTAS, 1964)
Curite	$3PbO \cdot 8UO_3 \cdot 4H_2O$
Fourmarierite	$PbO \cdot 4UO_3 \cdot 7H_2O$
Ianthinite	$UO_2 \cdot 5UO_3 \cdot 10.56H_2O$ (GUILLEMIN and PROTAS, 1959)
Masuyite	$\approx UO_3 \cdot 2H_2O$
Richetite	Hydrated oxide of U and Pb (?)
Schoepite	$2UO_3 \cdot 5H_2O$
Uraninite	UO_2
Uranospherite	$(BiO)(UO_2)(OH)_3$ (?)
Vandenbrandeite	$Cu(UO_2)O_2 \cdot 2H_2O$
Vandendriesscheite	$PbO \cdot 7UO_3 \cdot 12H_2O$

Table 92-D-2 (Continued)

Carbonates	
Andersonite	$Na_2Ca(UO_2)(CO_3)_3 \cdot 6H_2O$
Bayleyite	$Mg_2(UO_2)(CO_3)_3 \cdot 18H_2O$
Diderichite	Identical with rutherfordine
Liebigite	$Ca_2(UO_2)(CO_3)_3 \cdot 10-11H_2O$
Rabbittite	$Ca_3Mg_3(UO_2)_2(CO_3)_6(OH)_4 \cdot 18H_2O$
Rutherfordine	$(UO_2)(CO_3)$
Schroeckingerite	$NaCa_3(UO_2)(CO_3)_3(SO_4)F \cdot 10H_2O$
Sharpite	$(UO_2)(CO_3) \cdot H_2O$(?)
Studtite	Pb, uranyl carbonate (?)
Swartzite	$CaMg(UO_2)(CO_3)_3 \cdot 12H_2O$
Voglite	Ca, Cu, uranyl carbonate (?)
Wyartite	$3CaO \cdot UO_2 \cdot 6UO_3 \cdot 2CO_2 \cdot 12-14H_2O$ (GUILLEMIN and PROTAS, 1959)
Sulfates	
Johannite	$Cu(UO_2)_2(SO_4)_2(OH)_2 \cdot 6H_2O$
Meta-uranopilite	Doubtful basic uranyl sulfate
Peligotite	Identical with johannite
Uranopilite	$(UO_2)_6(SO_4)(OH)_{10} \cdot 12H_2O$
Zippeite	Near $2UO_3 \cdot SO_3 \cdot 5H_2O$
Phosphates-Arsenates	
Abernathyite	$K_2(UO_2)_2(AsO_4)_2 \cdot 8H_2O$
Arsenuranylite	$Ca(UO_2)_4(AsO_4)_2(OH)_4 \cdot 6H_2O$ (BELOVA, 1958)
Autunite	$Ca(UO_2)_2(PO_4)_2 \cdot 8-12H_2O$
Bassetite	$Fe(UO_2)_2(PO_4)_2 \cdot 8H_2O$
Bergenite	$Ba(UO_2)_4(PO_4)_2(OH)_4 \cdot 8H_2O$ (BÜLTEMANN and MOH, 1959)
Dewindtite	Doubtful Pb, uranyl phosphate
Dumontite	$Pb_2(UO_2)_3(PO_4)_2(OH)_4 \cdot 3H_2O$
Fritzscheite	Doubtful Mn, uranyl phosphate-vanadate
Hallimondite	$Pb_2(UO_2)(AsO_4)_2 \cdot nH_2O$ (WALENTA and WIMMENHAUER, 1961)
Heinrichite	$Ba(UO_2)_2(AsO_4)_2 \cdot 10-12H_2O$ (GROSS *et al.*, 1958)
Hügelite	$Pb_2(UO_2)_3(AsO_4)_2(OH)_4 \cdot 3H_2O$ (WALENTA and WIMMENHAUER, 1961)
Kahlerite	$Fe(UO_2)_2(AsO_4)_2 \cdot 8H_2O$
Lermontovite	$(U, Ca, TR)_3(PO_4)_4 \cdot 6H_2O$ (G. and S., p. 199)
Meta-autunite	$Ca(UO_2)_2(PO_4)_2 \cdot 6-8H_2O$
Metaheinrichite	$Ba(UO_2)_2(AsO_4)_2 \cdot 8H_2O$ (GROSS *et al.*, 1958)
Metakahlerite	$Fe(UO_2)_2(AsO_4)_2 \cdot 8H_2O$ (WALENTA, 1958)
Meta-kircheimerite	$Co(UO_2)_2(AsO_4)_2 \cdot 8H_2O$ (WALENTA, 1958)
Metanovacekite	$Mg(UO_2)_2(AsO_4)_2 \cdot 8H_2O$
Metatorbernite	$Cu(UO_2)_2(PO_4)_2 \cdot 6-8H_2O$
Meta-uranocircite	$Ba(UO_2)_2(PO_4)_2 \cdot 8H_2O$
Meta-uranospinite	$Ca(UO_2)_2(AsO_4)_2 \cdot 8H_2O$ (WALENTA, 1958)
Metazeunerite	$Cu(UO_2)_2(AsO_4)_2 \cdot 8H_2O$
Novacekite	$Mg(UO_2)_2(AsO_4)_2 \cdot 8-10H_2O$
Parsonite	$Pb_2(UO_2)_3(PO_4)_2(OH)_4 \cdot 7H_2O$
Phosphuranylite	$Ca(UO_2)_4(PO_4)_2(OH)_4 \cdot 7H_2O$
Przhevalskite	$Pb(UO_2)_2(PO_4)_2 \cdot 2H_2O$ (G. and S., p. 213)
Renardite	$Pb(UO_2)_4(PO_4)_2(OH)_4 \cdot 7H_2O$
Sabugalite	$HAl(UO_2)_4(PO_4)_4 \cdot 16H_2O$
Saleeite	$Mg(UO_2)_2(PO_4)_2 \cdot 8-10H_2O$
Torbernite	$Cu(UO_2)_2(PO_4)_2 \cdot 12H_2O$
Troegerite	$H_2(UO_2)_2(AsO_4)_2 \cdot 8H_2O$

Table 92-D-2 (Continued)

Uramphite	$NH_4(UO_2)(PO_4) \cdot 3H_2O$ (NEKRASOVA, 1957)
Uranocircite	$Ba(UO_2)_2(PO_4)_2 \cdot 8—10H_2O$
Uranospathite	$Cu(UO_2)_2(AsO_4, PO_4)_2 \cdot 16H_2O$ (?)
Uranospinite	$Ca(UO_2)_2(AsO_4)_2 \cdot 10H_2O$
Walpurgite	$Bi_4(UO_2)(AsO_4)_2 \cdot 3H_2O$
Zeunerite	$Cu(UO_2)_2(AsO_4)_2 \cdot 8—10H_2O$
Vanadates	
Carnotite	$K_2(UO_2)_2(VO_4)_2 \cdot 1—3H_2O$
Ferghanite	$U_3(VO_4)_2 \cdot 6H_2O$
Francevillite	$(Ba, Pb)(UO_2)_2(VO_4)_2 \cdot 5H_2O$ (BRANCHE *et al.*, 1957)
Metatyuyamunite	$Ca(UO_2)_2(VO_4)_2 \cdot 5—7H_2O$
Rauvite	$CaO \cdot 2UO_3 \cdot 2V_2O_5 \cdot 16H_2O$ (?)
Sengierite	$Cu(UO_2)(VO_4)(OH) \cdot 4—5H_2O$
Tyuyamunite	$Ca(UO_2)_2(VO_4)_2 \cdot 7—11H_2O$
Uvanite	$U_2V_6O_{21} \cdot 15H_2O$ (?)
Vanuralite	$Al_2O_3 \cdot 2V_2O_5 \cdot 4UO_3 \cdot 17H_2O$ (BRANCHE *et al.*, 1963)
Silicates	
Beta-uranophane	$Ca(UO_2)_2(SiO_3)_2(OH)_2 \cdot 5H_2O$
Boltwoodite	$K_2(UO_2)_2(SiO_3)_2(OH)_2 \cdot 5H_2O$ (FRONDEL and ITO, 1956)
Coffinite	$U(SiO_4)_{1-x}(OH)_{4x}$
Cuprosklodowskite	$Cu(UO_3)_2(SiO_3)_2(OH)_2 \cdot 5H_2O$
Gastunite	Ca, Pb, uranyl silicate
Kasolite	$Pb(UO_2)(SiO_3)(OH)_2$
Orlite	$3PbO \cdot 3UO_3 \cdot 4SiO_2 \cdot 6H_2O$ (G. and S., p. 239)
Pilbarite	$UO_3 \cdot PbO \cdot ThO_2 \cdot 2SiO_2 \cdot 4H_2O$ (?)
Ranquilite	$1.5CaO \cdot 2UO_3 \cdot 5SiO_2 \cdot 12H_2O$ (DE ABELEDO, 1960)
Sklodowskite	$Mg(UO_2)_2(SiO_3)_2(OH)_2 \cdot 5H_2O$
Soddyite	$(UO_2)_5(SiO_4)_2(OH)_2 \cdot 5H_2O$
Uranophane	$Ca(UO_2)_2(SiO_3)_2(OH)_2 \cdot 5H_2O$
Ursilite	$2(Ca, Mg) \cdot O.2UO_3 \cdot 5SiO_2 \cdot 9—10H_2O$ (CHERNIKOV *et al.*, 1957)
Weeksite	$K_2(UO_2)_2(Si_2O_5)_3 \cdot 4H_2O$
Niobates — Tantalates — Titanates	
Betafite	$(U, Ca)(Nb, Ta, Ti)_3O_9 \cdot nH_2O$
Brannerite	UTi_2O_6
Lodochnikovite (Lodochnikite)	$2(U, Th)O_2 \cdot 3UO_3 \cdot 14TiO_2$ (GERASIMOVSKY, 1956)
Pisekite	U, Ti, Th, rare-earth niobate-tantalate
Molybdates	
Calcurmolite	$Ca(UO_2)_3(MoO_4)_3(OH)_2 \cdot 11H_2O$ (FEDEROV, 1963)
Iriginite	$UO_3 \cdot 2MoO_3 \cdot 4H_2O$ (G. and S., p. 197)
Moluranite	$UO_2 \cdot 3UO_3 \cdot 7MoO_2 \cdot 20H_2O$ (G. and S., p. 196)
Mourite	formula uncertain; contains U and Mo (KOPCHENOVA *et al.*, 1962)
Umohoite	$(UO_2)(MoO_4) \cdot 4H_2O$

Mineral formulas are those given by FRONDEL (1956) except in the case of minerals discovered since 1956 or minerals whose formulas have been modified by later work. References are given in each case of more recent discovery or new description. G. and S. refer-to GETSEVA and SAVAL'EVA (1957). Papers consulted only secondarily through summaries in the American Mineralogist are shown in the reference list with a parenthetical reference to volume and page in the American Mineralogist.

Revised manuscript received: July 18th, 1967

92-E. Abundances in Common Igneous Rocks

The abundances of uranium in igneous rocks are discussed under three subheadings. The first is a generalized discussion of the content of uranium in rocks broadly classified as granites, basalts, etc. The second category for discussion concerns the variation in abundances of uranium in cogenetic suites of rocks, and the third category discusses the variation of abundance of uranium within individual plutons.

I. Abundances in Major Rock Types

The fact that the uranium and thorium concentrations of igneous rocks are both so closely related to the compositions of those rocks makes it difficult to subdivide igneous rocks into meaningful radiometric groups. For example, the phrase "uranium content of basalt" is a meaningless term because the basalts clearly do not represent one petrological population, and sufficient work on contents of radioactive materials has been done to indicate radiometric differences between different varieties of basalt. The data given in Table 92-E-1, therefore, represent only average compositions of those groups of rocks which might possibly be considered single populations. Indeed, many of the groups shown in Table 92-E-1 probably do not actually represent one homogeneous population. Similar groups are used in Table 90-E-1, which gives the contents of thorium and the Th/U ratios in igneous rocks.

Table 92-E-1 shows the general tendency for uranium to increase in abundance toward the more silicic varieties of igneous rocks. The average uranium content of various igneous rocks is summarized graphically in Figs. 92-E-1 to 92-E-2. In view of the marked concentration of uranium toward the outer parts of the earth (see Section 92-B), the fact that uranium increases in abundance toward the later members of igneous rock series further demonstrates the tendency of uranium to become segregated from other rock-forming constituents during general petrologic evolution.

The average concentration of uranium in the earth's continental crust can be variously estimated according to the proportions of rock types assumed to constitute the crust. As discussed by Rogers (1966), there is a general tendency to overestimate the abundance of granite in cratonic areas. Assuming that granites constitute about half of the continental crust, and that the remainder of the crust contains little radioactive material, the average concentration of uranium in continental masses must lie in the range of 1—4 ppm and is probably somewhere near the middle of this range. Phair and Gottfried (1964) estimate an average uranium content of 2.8 ppm for continental areas.

The concentration of uranium in dunites and other ultrabasic rocks is extremely low. The average of 0.014 ppm from the data of Heier and Carter (1964) is about one order of magnitude below that estimated for the upper mantle from heat flow studies (Birch, 1965). The dunites and peridotites included in this average come both from major intrusive bodies and from inclusions in basalts. Most of them, though not all, are from continental areas, and it is possible that they represent the upper mantle

Tables 92-E-1. *Uranium contents of selected igneous rock suites*

Suite	No. of samples	U ppm	Method	Reference
Granitic rocks				
France				
Luzy granites (Paleozoic)	167	4.2	L	Coulomb (1959)
Paleozoic granites of western France	814	6.1	L	Hussein (1961) and Coulomb (1959)
Russia				
Precambrian mica granites of Ukrainian shield (Middle Dnepr area)	100	5.7	emanation	Filippov and Komlev (1959)
Tertiary Megrinsk intrusions of Southern Armenia; alkalic granodiorites to alkalic granites of 2nd intrusive phase	16	5.1		Meliksetyan (1961)
Paleozoic Susamyr batholith of Tien Shan; granodiorites to granites	152	3.4	emanation	Tauson *et al.* (1958)
Northern Tien Shan; weighted average of all crystalline rocks		4.1		Krylov (1963)
Paleozoic granites and syenites of Central and western Tuva	47	3.9	radiochemical	Abramovich (1959)
United States				
Conway granite, New Hampshire; Triassic, alkalic	1051	15	R	Rogers *et al.* (1965)
New England granites	123	6.2	R	Rogers (1964)
Enchanted Rock batholith, Texas; granite	81	3.5	R	Rogers (1964)
Precambrian granites of Front Range, Colo.	38	5.0 (weighted)	L	Phair and Gottfried (1964)
Laramide stocks of Front Range, Colorado	27	7.6	L	Phair and Gottfried (1964)
Laramide stocks west of Front Range, Colo.	25	2.2	L	Phair and Gottfried (1964)
Precambrian granites of Middle Rocky Mountains	25	2.2	L	Phair and Gottfried (1964)
Idaho batholith; Laramide;				
tonalites and granodiorites	23	2.3	L	Larsen Jr. and Gottfried (1961)
qtz. monzonites and granites	9	3.7	L	
Sierra Nevada batholith, Cal.: granodiorites to granites; Jurrassic	20	5.4	various	Rogers (1964)
Laramide granites of Idaho and Boulder batholiths and northern Washington	38	2.5	various	Rogers (1964)

Table 92-E-1 (Continued)

Suite	No. of samples	U ppm	Method	Reference
Granites and granodiorites of Southern California batholith; Cretaceous	54	2.4	various	ROGERS (1964)
Gabbroic and ultramafic rocks				
General average for "gabbroid" rocks	34	0.84	various	HEIER and CARTER (1964)
Dunite (intrusive and nodules)	24	0.014	various	HEIER and CARTER (1964)
Dunite nodules (Hawaii, Germany, Arizona)	3	0.003 to 0.009	I and N	TILTON and REED (1963)
Eclogites				
Pipe eclogite (Africa)	10	0.26	R	HEIER (1963)
Pipe eclogite (Africa-1; Australia-3)	4	0.052	N	LOVERING and MORGAN (1963)
Eclogite nodules (Hawaii)	2	0.030 to 0.043	I	TILTON and REED (1963)
Average eclogite from metamorphic terraines (Norway and central Europe)	9	0.20	various	HEIER and CARTER (1964)
Alkaline intrusive rocks				
Nepheline syenites from northern Norway	16	0.04	R	HEIER (1962a)
Variscan alkalic syenites and nepheline syenites of southern Tien Shan	26	1.4	L	LEONOVA (1964)
Lovozero massif, Kola Peninsula, Russia:				SAPRYKINA (1959)
first intrusive phase (neph. syenites)	14	9.9	L	
second intrusive phase (lujavrites, urtites, and foyaites)	25	15.3	L	
third intrusive phase (mainly lujavrites)	19	19.7	L	
fourth intrusive phase (monchiquites)	2	2.8	L	
Silicic extrusive rocks				
Rhyolites and dacites of western U.S.A.	116	5.0	—	Coats (1956)
Basaltic rocks				
Oceanic tholeiitic basalts	22	0.10	I	TATSUMOTO *et al.* (1965)
Basalts and basaltic andesites of Mariana Islands	20	0.56	L	GOTTFRIED *et al.* (1963)
Tholeiitic basalts of Hawaii	6	0.18	R L	HEIER *et al.* (1964) LARSEN, 3d, and GOTTFRIED (1960)

Table 92-E-1 (Continued)

Suite	No. of samples	U ppm	Method	Reference
Alkali basalts of Hawaii	2	0.99	R	HEIER *et al.* (1964)
			L	LARSEN, 3d, and GOTTFRIED (1960)
Japan:				
tholeiitic basalts	2	0.15	R	HEIER and ROGERS (1963)
high-alumina basalts	2	0.22	R	HEIER and ROGERS (1963)
alkali-olivine basalts	2	0.53	R	HEIER and ROGERS (1963)
Caribbean Island arc:				
spilites	6	0.22	R	ROGERS and DONNELLY (1966)
intrusive keratophyres	5	0.77	R	ROGERS and DONNELLY (1966)
extrusive keratophyres	4	0.29	R	ROGERS and DONNELLY (1966)
basaltic andesites	12	0.79	R	ROGERS and DONNELLY (1966)
Plateau basalts (ave.)	26	0.53	various	HEIER and CARTER (1964)
Columbia River	2	0.36	R	HEIER and ROGERS (1963)
Palisades diabase (New York)	5	0.35	R	HEIER and ROGERS (1963)

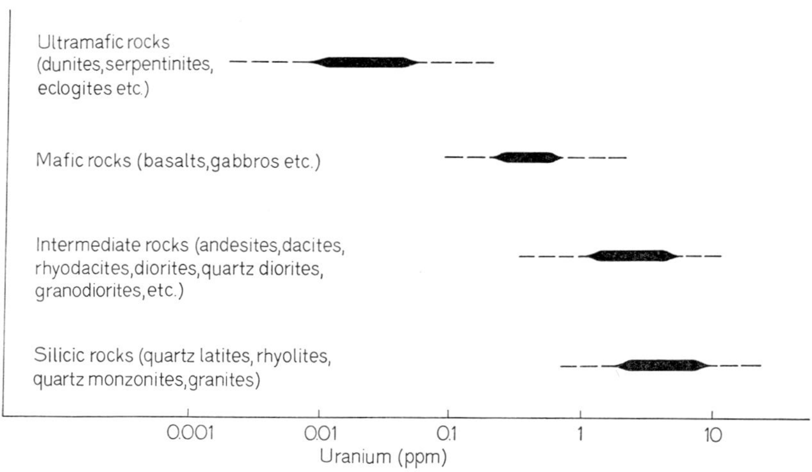

Fig. 92-E-1. Generalized distribution of uranium in various broad groups of igneous rocks. [From PETERMAN, Z. E.: unpublished communication (1963)]

under the continent that has been depleted in thorium and uranium during separation of the cratonic materials. Data are not yet adequate to determine whether or not there are differences between oceanic and continental dunites and peridotites. It is interesting to note that the concentration of 0.2 ppm found by HEIER and CARTER for pipe eclogites is comparable to that presumed for the upper mantle (also see TILTON and

Reed, 1963). With regard to upper mantle studies, Morgan and Lovering (1965b) have recently reported the uranium and thorium abundances in the basalt of the Guadalupe experimental Mohole. This basalt contained 0.195 ppm uranium.

The alkaline rocks reported in Table 92-E-1 are merely a few representatives of a broad sampling of such rocks that have been studied radiometrically. The wide variability of uranium concentrations in such alkaline rocks is discussed in more detail by Peterman (1963). The areas which have been primarily studied include the Kola Peninsula of Russia (Table 92-E-1 summarizes data on 60 samples by

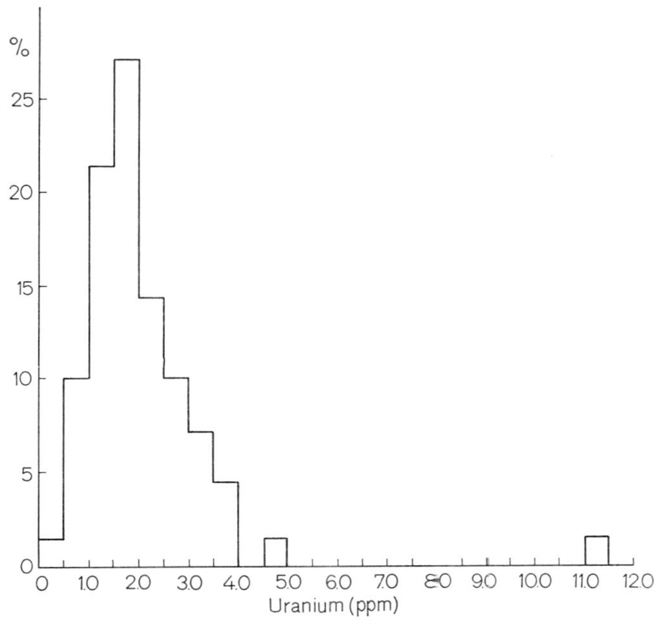

Fig. 92-E-2. Histogram of uranium concentrations in silicic igneous rocks. Plot represents 242 individual analyses and 3 average values for volcanic rocks and 194 individual analyses and 96 average values for plutonic and hypabyssal rocks. [From Peterman, Z.E.: unpublished communication (1963); data are summarized variously in Table 92-E-1]

Saprykina, 1959), several suites of rocks from the southern Urals and Tien Shan area (e.g., see Yes'hova *et al.*, 1962), the rocks of Stjernoy in northern Norway (Heier, 1962a), a suite from Magnet Cove, Arkansas (Erickson and Blade, 1963) (Fig. 90-E-3), and rocks from Virginia and Texas (Gottfried *et al.*, 1962) (Fig. 90-E-6). The extreme variability between and within groups makes averaging almost useless, and the only valid generalization concerning the uranium content of alkaline rocks is that it is generally high. It should be noted in this regard that the K/U ratio varies only slightly between igneous rocks of wide petrologic differences (Heier and Rogers, 1963).

The averages and general ranges of uranium concentrations shown in Table 92-E-1 for various types of rocks are similar for intrusive and extrusive rocks of compositional equivalence. The chances of occurrence of very large concentrations of uranium are, of course, higher for intrusive than for extrusive rocks.

II. Variations within Rock Suites

The almost universal tendency for the concentration of uranium to increase toward the later members of an igneous differentiation series has been documented repeatedly. Many of these investigations are thoroughly summarized by PETERMAN (1963). In his manuscript, the sequences in Table 92-E-2 show a tendency toward increasing uranium content during igneous differentiation (Figs. 92-E-3; 90-E-3 to 90-E-6). Note that some differentiation sequences do not show the tendency to

Table 92-E-2. *Sequences of igneous rocks showing increase in uranium content toward the younger rocks*

Extrusive series
Lassen Volcanic National Park (ADAMS, 1955; LARSEN, 3d, and GOTTFRIED, 1960)
Modoc area, California (GOTTFRIED and LARSEN JR., 1958; LARSEN, 3rd, and GOTTFRIED, 1960)
San Juan Mountains, Colorado (LARSEN JR. *et al.*, 1958)
Valles Mountains, New Mexico (LARSEN, 3d, and GOTTFRIED, 1960) (Fig. 90-E-5)
Chaine des Puys, France (GOLDSZTEIN and GUILLERMO, 1961)
Big Bend National Park, Texas (GOTTFRIED *et al.*, 1962) (Fig. 90-E-6) (part intrusive)
Marianas Islands (GOTTFRIED *et al.*, 1963) (Fig. 90-E-4)
Mt. Garibaldi area, British Columbia (GOTTFRIED *et al.*, 1963)
Strawberry Mountains, Oregon (GOTTFRIED *et al.*, 1963)
Sills and shallow intrusives (Fig. 92-E-3)
Dillsburg sill, Pennsylvania (GOTTFRIED, 1959)
Skaergaard, Greenland (HAMILTON, 1959)
Duluth gabbro, Minnesota (HEIER and ROGERS, 1963)
Augusta County, Virginia (GOTTFRIED *et al.*, 1962)
Plutonic sequences
Southern California Batholith (LARSEN, 3d, and GOTTFRIED, 1960; LARSEN JR. and GOTTFRIED, 1961; WHITFIELD *et al.*, 1959; ROGERS and RAGLAND, 1961; HEIER and ROGERS, 1963) (Fig. 90-E-7)
White Mountain magma series, New Hampshire (BUTLER, 1961; ROGERS and RAGLAND, 1961; ADAMS *et al.*, 1962)
Oliverian series, New Hampshire (ROGERS and RAGLAND, 1961; LYONS, 1961)

increase in uranium content toward younger rocks. Most such series merely show no detectable trends in composition, but the Honolulu series of Hawaii, studied by LARSEN, *3rd*, and GOTTFRIED (1960) shows a decrease in uranium with increasing silica content (Fig. 90-E-2).

Plots of uranium content against some type of differentiation index are much smoother in extrusive sequences than in intrusive ones. Most extrusive series show consistent trends when plotted against silica content, the Larsen differentiation index, and potassium content. In intrusive rocks the scatter of values is quite high in plots of uranium abundance versus silica content. The most consistent variation of uranium in intrusive rocks is shown by a comparison of uranium and potassium abundances. In fact the K/U ratio is about 10^4 in all crustal materials (HEIER and ROGERS, 1963). The use of this ratio for the earth's mantle, however, creates difficulties in explaining the earth's thermal history; (see calculations of MACDONALD, 1965). Discrepancies between the K/U ratio of chondrites and the apparent ratio for the whole earth are discussed by GAST (1960).

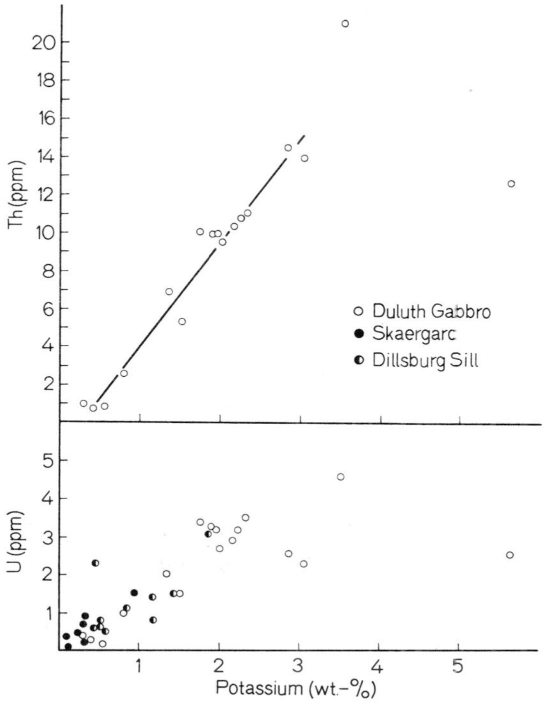

Fig. 92-E-3. Variation of thorium and uranium concentrations in three gabbro-granophyre sequences. [From Peterman, Z. E.: unpublished, personal communication (1963); data on the Skaergaard intrusion, Greenland, are from Hamilton, 1959; data on the Dillsburg sill, Pennsylvania, U.S.A. are from Gottfried, 1959; data on the Duluth complex, Minnesota, U.S.A. are from Heier and Rogers, 1963]

III. Variation within Batholithic Complexes

Economic interest in uranium has led to considerable investigation of the distribution of uranium within individual plutons and associated wallrocks and also within complex batholiths. In many complexes such as that of the Colorado Front Range (Phair and Gottfried, 1964), variability in uranium content within plutons and between members of the plutonic complex is not simply related to some differentiation index. In some studied plutonic complexes or individual plutons there are progressive changes in uranium concentration toward wallrocks. Most of the work on these problems has been done by Russian workers; their studies are summarized by Peterman (1963). Surficial redistribution of uranium makes detection of areal trends difficult (e.g., see Fig. 92-E-4).

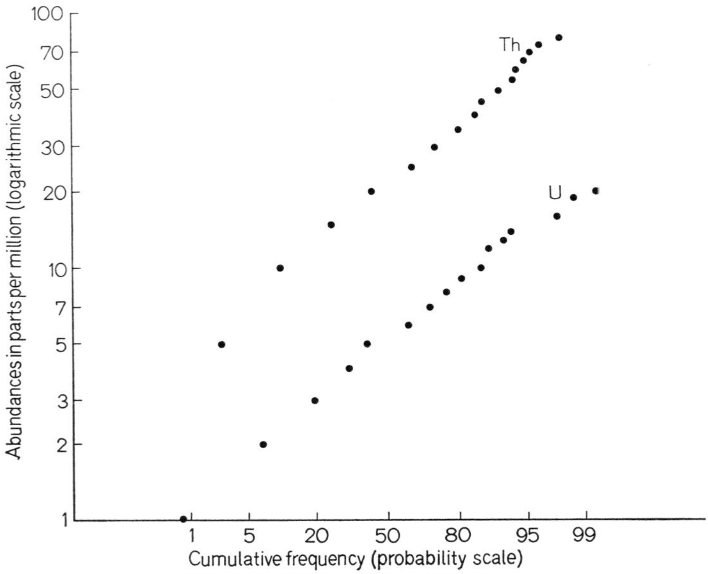

Fig. 92-E-4. Cumulative percentage plot of uranium concentrations in granitic rocks of New England, U.S.A. The lognormality shown on this logarithmic-probability diagram may indicate the importance of surficial processes in redistributing uranium. (From ROGERS, 1964)

Revised manuscript received: July 18th, 1967

92-F. Behavior during Processes Connected with Magmatism

As mentioned above in Section 92-D, uranium commonly occurs in veins as the mineral uraninite or pitchblende. Uraninite has also been found in pegmatites along with uranium-bearing varieties of such rare earth minerals as samarskite and euxenite. The economic geology of uranium in magmatically-derived ore deposits is summarized by a number of workers. A discussion in English is provided by McKelvey *et al.* (1955) along with a description of the general economic geology of uranium. Kotlyar (1961) has published a summary in Russian of the various types of uranium deposits.

Revised manuscript received: July 18th, 1967

92-G. Behavior during Weathering and Rock Alteration

Few studies have been oriented directly toward an understanding of the behavior of uranium during weathering and alteration. The tendency of uranium to be oxidized to the comparatively soluble uranyl ion permits uranium to be mobilized easily in surficial processes. This oxidation is primarily responsible for the wide variations in Th/U ratios shown by surficial material. The comparatively insoluble thorium is concentrated in resistate minerals or is adsorbed on clays, whereas uranium is redistributed in surface and ground waters. The solubilities and oxidation potentials pertinent to these processes are given in Section 92-H.

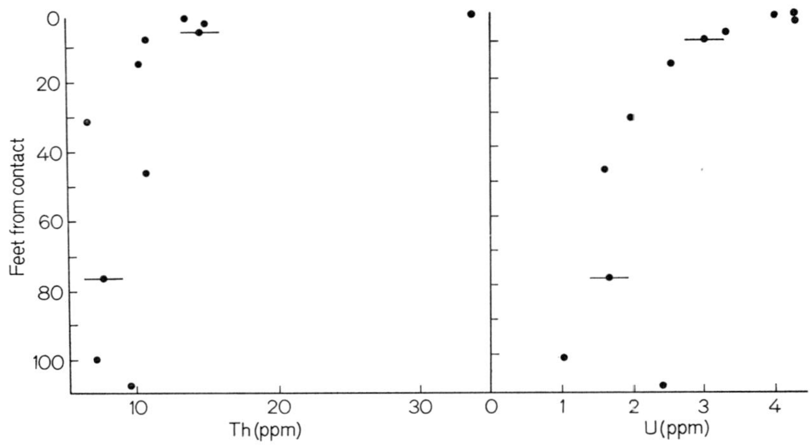

Fig. 92-G-1. Variation in thorium and uranium concentrations in the weathering zone in the granodiorite on Flagstaff Mountain, Colorado, U.S.A. (From PLILER and ADAMS, 1962b)

One study of the behavior of uranium during weathering was made by PLILER and ADAMS (1962b) (Fig. 92-G-1). In the Pennsylvanian weathering profile on a granodiorite at Flagstaff Mountain, Colorado, the variation in both thorium and uranium contents can be investigated throughout the soil profile. The profile is somewhat lateritic and clearly results from humid, subtropical weathering. The uranium which is retained in the soil is held primarily in resistate minerals such as the zircons, whereas thorium is incorporated both in resistate minerals and also in clays formed during the weathering process. The highest concentrations of both elements are found in the residual clays at the top of the soil profile. Artificial leaching experiments on the granodiorite with dilute acid indicate that the vast bulk of the uranium was removed by natural leaching whereas the thorium was less affected (Fig. 92-G-2). These results correlate well with the enrichment of thorium relative to uranium in the soil zone. A generalized study of weathering by HARRISS and ADAMS

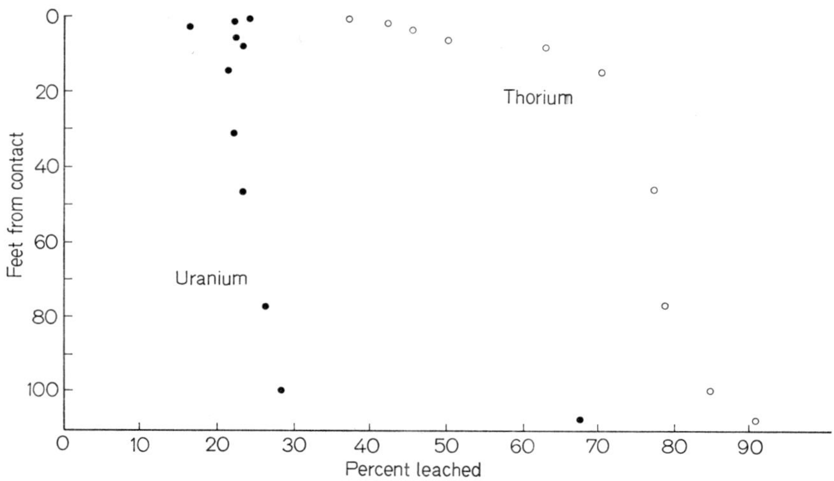

Fig. 92-G-2. Variation in the percentage of leachable thorium and uranium in the weathering zone in the granodiorite on Flagstaff Mountain, Colorado, U.S.A. (From PLILER and ADAMS, 1962b)

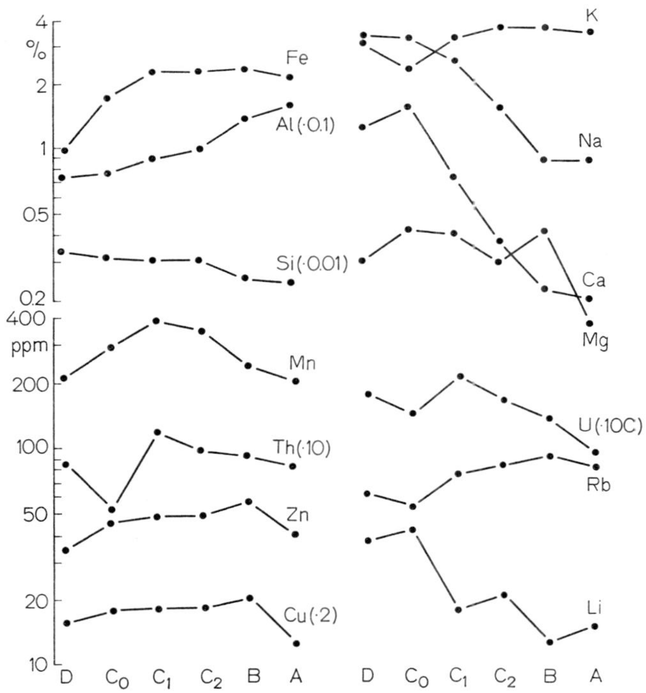

Fig. 92-G-3. Variation of thorium and uranium concentrations in a soil zone compared with variations in abundances of other elements. Tishomingo granite, Oklahoma, U.S.A. (From HARRISS and ADAMS, 1966)

(1966) (Fig. 92-G-3 and 92-G-4) provided information about the mobilization of thorium and uranium, and the results of this study are roughly similar to those found by PLILER and ADAMS (1962b) and indicate the generally greater ease of mobilization of uranium than of thorium.

Investigations of the radioactivity of the Conway granite, New Hamsphire, have involved the drilling of cores into unaltered rock below surficially weathered

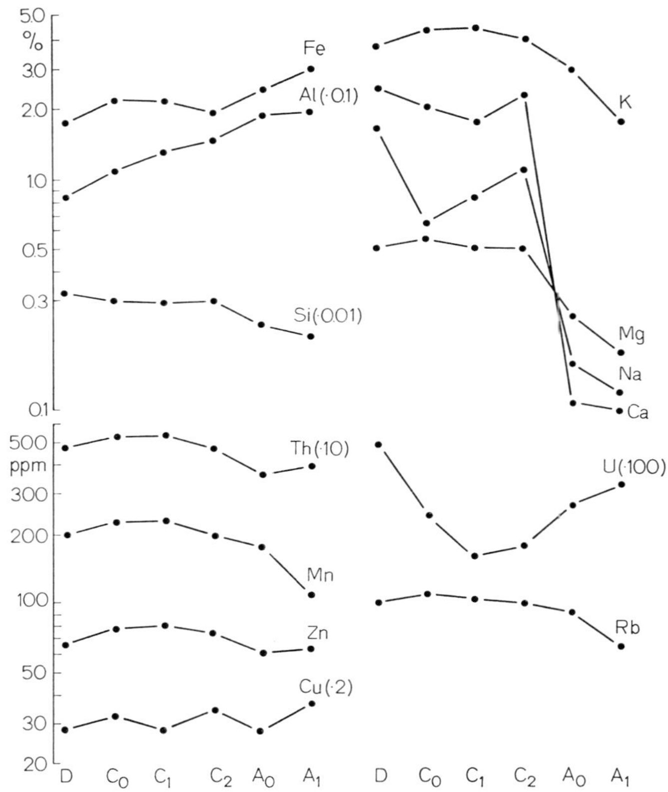

Fig. 92-G-4. Variation of thorium and uranium concentrations in a soil zone compared with variations in abundances of other elements. Elberton granodiorite, Georgia, U.S.A. (From HARRISS and ADAMS, 1966)

material. Data reported by RICHARDSON (1964) and ROGERS *et al.*, (1965) indicate a near-surface depletion of uranium except in a few zones where the mobilized uranium is redeposited by circulating ground water. In some areas, the remobilization of uranium extends to depths of several hundred feet, generally in zones of fracturing and ground water movement. As reported by RICHARDSON (1964), in some of the weathered rocks the U^{234} isotope is about 15 percent less abundant than it should be for secular equilibrium with U^{238}.

Surficially deposited uranium minerals consist of the complex uranyl salts listed in Section 92-D.

Revised manuscript received: July 18th, 1967

92-H. Solubilities in Waters; Valence States

The solubility and redox data pertinent to the distribution of uranium under surficial conditions are:

$$UO_2^{++} + 4H^+ + 2e^- \rightleftharpoons U^{+4} + 2H_2O \qquad \varepsilon^0 = +0.62\,v,$$
$$U(OH)_4 \rightleftharpoons U^{+4} + 4OH^- \qquad K_{sp} = 10^{-52}.$$

The uranyl (UO_2^{++}) ion is almost completely soluble, whereas the uranous (U^{+4}) ion can be precipitated easily in alkaline solutions. The uranyl ion precipitates in nature only in the uranyl salts listed in Table 92-E-1 and in certain laboratory preparations. The carbonate ion strongly complexes the uranyl ion and inhibits precipitation; a dissociation constant of 10^{-10} (approx.) has been proposed for the uranyl carbonate complex $(UO_2)(CO_3)_2^{--}$. Uranyl salts coprecipitate with calcium phosphates, but the true solubility products of naturally-occurring uranyl phosphates and other salts are unknown.

Uranium is known to be adsorbed from solution by organic material (see SWANSON, 1963) and also by clays (GOLDSZTAUB and WEY, 1955). The amount adsorbed by clays is proportional to the normal base exchange capacity of the clay, and GOLDSZTAUB and WEY report the following grams of uranium adsorbed from a 1-percent solution of uranyl nitrate:

montmorillonite	7.5 g/100 g calcined clay
kaolinite	2.0 g/100 g calcined clay.

Revised manuscript received: July 18th, 1967

92-I. Abundance in Natural Waters

Uranium concentrations are constant in sea water and are in the range of 1—4 × 10^{-3} ppm (1—4 ppb) (Table 92-I-1). These values may be compared with an average of 10^{-5} to 10^{-6} ppm for thorium (Table 90-I-1); it is clear that the Th/U ratio in sea water is extremely small. The geochemistry of uranium in sea water is discussed in various papers by F. Koczy and coworkers and also by Tatsumoto and Goldberg (1959).

Concentrations of uranium in fresh water are considerably more variable than in the oceans. The concentrations are clearly controlled by a variety of factors, such as rock type leached by the water, rate of flow and rate of evaporation, etc. The concentrations of a number of samples of continental waters are shown in Table 92-I-2. Variability in concentrations of uranium in continental ground and underground waters has been used for commercial purposes such as prospecting for uranium (see, for example, Cannon and Kleinhampl, 1956; Illsley, 1957). Some efforts have also been made to provide mass balances of uranium in source areas and in surface and ground waters (for example, Agamirov, 1965; Lopatkina, 1964). The general concentration of uranium in most fresh waters may be somewhat less than in sea water.

Table 92-I-1. *Contents of uranium in sea water*

Location	U ppb	Method	Reference
Atlantic, Pacific	1.5	L	Føyn *et al.* (1939)
Atlantic, Pacific	1.1—1.5	L	G. Koczy (1950)
Pacific (deep ocean)	2.82	L	Nakanishi (1951)
Pacific (mouth of Ito Bay)	1.18	L	Nakanishi (1951)
Pacific (surface, 1 km off Aomori Pref.)	0.46	L	Nakanishi (1951)
North Atlantic	0.3—1	L	Rona and Urry (1952)
Atlantic	1.15	L	Smith and Grimaldi (1954)
Gulf of Mexico	3.15	L	Smith and Grimaldi (1954)
Pacific	2.5	L	Stewart and Bentley (1954)
Atlantic, Pacific, Gulf of Mexico	3.3	I	Rona *et al.* (1956)
Baltic Sea	0.8—5.9	L	F. Koczy *et al.* (1957b)
Bering Sea	2.2	L	Starik *et al.* (1958)
Barents Sea	1.6	L	Starik *et al.* (1958)
White Sea (eastern)	1.4	L	Starik *et al.* (1958)
White Sea (Onezh Bay)	1.8	L	Starik *et al.* (1958)
Sea of Azov	3.3	L	Starik *et al.* (1958)
Caspian Sea	3—10	L	Starik *et al.* (1958)
Indian Ocean (SW part)	2.7	L	Baranov and Khristianova (1959)
Indian Ocean (central)	1.8	L	Baranov and Khristianova (1959)
Black Sea	2	L	Nikolaev *et al.* (1960)
English Channel and Bay of Biscay	3.3	various	Wilson *et al.* (1960)

Table 92-I-2. *Contents of uranium in continental waters*

Location	U ppb	Method	Reference
Colorado ground water	0.4—4	L	Bruce and Ferguson (1951)
Illinois ground water	<0.3	L	Bruce and Ferguson (1951)
Florida ground water	<1.0	L	Bruce and Ferguson (1951)
Three N. American Rivers	0.026	L	Rona and Urry (1952)
River water, by calculation	0.09—1.1	L	Holland and Kulp (1954)
Great Salt Lake, Utah	4.8	L	Stewart and Bentley (1954)
Texas ground water	2.3—10	L	Judson and Osmond (1955)
Wisconsin and Illinois ground waters	0.8	L	Judson and Osmond (1955)
Mineralized Aquifer	100—460	L	Judson and Osmond (1955)
Mineralized Aquifer	10—250	L	Denson *et al.* (1956)
Mineralized Aquifer	20—200	L	Fix (1956)
Thermal Springs	<7	L	Fix (1956)
U.S.A., Alaska, regional average	0.1	L	Fix (1956)
Colorado Plateau	0.5—3	L	Fix (1956)
France, Springs	0.35—7.68	L	Jurain (1957)
Baltic region, igneous drainage	0.5	L	Koczy *et al.* (1957b)
Baltic region, sedimentary drainage	1—13	L	Koczy et al. (1957b)
Llano Estacado, Texas ground water	0.9—46	L	Barker and Scott (1958)
New Mexico ground water	1.2—4.7	L	Barker and Scott (1958)
Austria, Vienna Basin; Springs	1.4—12	L	Hecht *et al.* (1958)
Austria, Tyrol; Springs	0.25—40	L	Hecht *et al.* (1958)
Austria, Danube River	0.5—2.3	L	Hecht *et al.* (1958)
Austria, Tyrol; Rivers	0.2—2.8	L	Hecht *et al.* (1958)
Wales, Spring	1.3	L	Smith and Chandler (1958)
Jamaica, Rivers	1.5—5.9	L	Smith and Chandler (1958)
Uganda, Spring	1.3	L	Smith and Chandler (1958)
England, Cornish Waters	0.6—40	L	Smith and Chandler (1958)
Germany, Saale River	0.34—4.27 dissolved	L	Heide and Proft (1959)
Germany, Saale River	0.03—0.28 suspended		Heide and Proft (1959)
Great Central Plains, Wells (ground water)	9.5	L	Landis (1960)
Great Central Plains, Springs (ground water)	11.3	L	Landis (1960)
Pacific NW, ground water in silicic, subsilicic rocks	<0.1—10	L	Barker and Scott (1961)
Pacific NW, ground water in Idaho batholith	<0.1—13	L	Barker and Scott (1961)
Pacific NW, ground water in Snake River basalt	<0.1—2.6	L	Barker and Scott (1961)
Pacific NW, ground water in Columbia River basalt	<0.1—0.6	L	Barker and Scott (1961)
Nevada (Truckee Meadows)	0.1—43	L	Cohen (1961)
England, surface waters	≦1.1	L	Peacock (1961)
U.S.A. (general)	<0.1—120	L	Scott and Barker (1962)
India, surface water	0.03—4.2	L	Kamath *et al.* (1964)
River water of Japan	0.34—1.23	L	Miyake *et al.* (1964)
Waters of Florida	0.9	R	Osmond (1964)

Revised manuscript received: July 18th, 1967

92-K. Abundance in Common Sedimentary Rocks

The concentrations of uranium in sedimentary rocks are difficult to summarize because of the various designations used by different authors to describe the rocks which they study. Thus, for example, the term "graywacke" means different things to different geologists. (For definition see Chapter 8 of Volume I of this Handbook.) The names used for various rocks in this section are thought to be reasonably standard, and the abundances of uranium in different types are listed in Table 92-K-1. A correlative summary of thorium values is given in Table 90-K-1.

Table 92-K-1. *Contents of uranium in sedimentary rocks*

Rock type	No. of samples	U ppm	Method	Reference
Sandstones				
Orthoquartzites				
North American av.	16	0.45	R	MURRAY and ADAMS (1958)
Mesa Verde	8	1.7	R	PLILER and ADAMS (1962a)
Gulf Coast beach sands	29	0.59	R	MAHDAVI (1964)
Atlantic Coast beach sands	54	3.21	R	MAHDAVI (1964)
"Volcanic" graywackes				
Caribbean	14	0.5—1	R	ROGERS and DONNELLY (1966)
Umpqua (Oregon)	4	1.3	R	ROGERS (1966)
Other graywackes (estimated average)		2.1		ROGERS and RICHARDSON (1964)
Arkoses (estimated average)		1.5		ROGERS and RICHARDSON (1964)
Shales				
North American gray and green	52	3.2	L	ADAMS and WEAVER (1958)
Russian platform (comp)	178	4.1	emanation	BARANOV *et al.* (1956)
Mancos	102	3.7	R	PLILER and ADAMS (1962a)
Red and yellow	10	~2	L	ADAMS and WEAVER (1958)
Black (average)	numerous	8	various	SWANSON (1961)
Black Shales				
Marine black shales (excluding geosynclinal)		20	L	SWANSON (1961)
All black shales		8	L	SWANSON (1961)
Chattanooga		79	L	SWANSON (1961)
Alum shale; Cambrian; Sweden	309	168	L	from summary by ,, BATES and STRAHL (1958)

Table 92-K-1 (Continued)

Rock type	No. of samples	U ppm	Method	References
Reedsville; Ord; Pennsylvania	4	4	L	„ Bates and Strahl (1958)
Ohio shale; Devonian; Ohio	220	10	L	„ Bates and Strahl (1958)
Woodford; Devonian; New Mexico	116	9	L	„ Bates and Strahl (1958)
Hamilton; Devonian; Pennsylvania	6	6	L	„ Bates and Strahl (1958)
Chemung; Devonian; Pennsylvania	4	3	L	„ Bates and Strahl (1958)
St. Hippolyte; Carb.; France	150	1244	L	„ Bates and Strahl (1958)
Allegheny; Penn.; Pennsylvania	34	4	L	„ Bates and Strahl (1958)
Kansas shale; Penn.; Kansas	21	16	L	„ Bates and Strahl (1958)
Kupferschiefer; Perm.; Germany	4	39	L	„ Bates and Strahl (1958)
Bauxites	29	11.4	R	Adams and Richardson (1960)
Bentonites	64	5.0	various	Adams and Weaver (1958)
Limestones				
North American av.	25 (comp)	2.2	L and R	Adams and Weaver (1958)
Russian platform	128 (comp)	2.1	emanation	Baranov *et al.* (1956)
Eniwetok cores				
calcite		0.35		Sackett and Potratz (1963)
aragonite		2.3		Sackett and Potratz (1963)
Eniwetok core E-1	21	2.34		Sackett and Potratz (1963)
Miscellaneous	49	2		Bell (1963)
Dolomites				
Dolomites and dolomitic limestones	100	0.03—2.0	L	Deininger (1964)
Phosphate rock			mostly L	from summary by McKelvey (1956), based on a variety of published and unpublished measurements
Marine phosphorites		50—300		
Residual phosphorites		50		
River-pebble deposits		unknown; probably low		
Guano		negligible		
Phosphatized rock		negligible (?)		
Fossil bone		50—300		
Evaporites				
Estimate of pure evaporite minerals		<0.1		Bell (1960)

Table 92-K-1 (Continued)

Rock type	No. of samples	U ppm	Method	Reference
Pacific ocean sediments			mostly L	summary by Baranov and Khristianova (1963)
Sands	5	3.0		
Muds	3	2.3		
Clayey muds	6	2.7		
Siliceous and diatomaceous muds and clayey muds	7	3.6		
Pelagic clays	13	4.0	various	summary by Baranov and Khristianova (1963)
Pelagic clays	5	1.5—2.5	L	Hahofer and Hecht (1954)
Indian ocean sediments				
Various muds		0.2—0.5	various	summary by Baranov and Khristianova (1960)
Atlantic ocean sediments				
Globigerina oozes	5	0.74	L	Rosholt *et al.* (1961)
Caribbean sediments				
Globigerina oozes	17	1.0	L	Rosholt *et al.* (1961)
Black sea sediments				
Muds	41	3.7		Starik *et al.* (1961a)
Organic matter				
Lignite; Great Plains, U.S.A.		80	L	Vine (1956)
Coal; Eocene; Wyoming		30	L	Vine (1956)
Coaly shale; Cret.; Idaho		20	L	Vine (1956)
Coal and coaly shale; Cret.; New Mexico		<6000	L	Vine (1956)
Lignite; Tertiary; Nevada		<50	L	Vine (1956)
Coal; Upper Paleozoic eastern U.S.A.		mostly <10	L	Vine (1956)
Oil; ranges of regional averages:			L	Hyden (1956)
Mid-continent, U.S.A.	36	5—77 in ash (0.17—10) $\times 10^{-3}$ in oil		
Rocky Mts., U.S.A.	61	4—46 in ash (0.4—3.1) $\times 10^{-3}$ in oil; omitting one very large value		
California	10	4.5 in ash 7.0 $\times 10^{-3}$ in oil		
Asphalts from western U.S.A.	45	10—3760; normal range of 100—500	L	Hail *et al.* (1956)
Black muds; Norwegian Fjords	10	36 (13—60 range)	L	Swanson (1963)
Black and gray muds; Baltic Sea	10	48 (22—103 range)	L	Swanson (1963)

I. Sandstones

Sandstones are subdivided into three categories in Table 92-K-1. The average concentration of uranium in the orthoquartzites is reasonably well known because the range of values among various orthoquartzitic suites is comparatively small. In orthoquartzites, uranium may occur both disseminated in the quartz or as heavy resistate grains. The sites of radioactivity have been investigated by both MURRAY and ADAMS (1958) and ROGERS and RICHARDSON (1964). Both investigations showed that most of the uranium is held within the quartz grains themselves.

Two special groups of graywackes are reported in Table 92-K-1 in addition to an estimated "average" for all graywackes. The Caribbean "volcanic" graywackes are important because they represent the average composition of rocks of the Caribbean orogen. As ROGERS and DONNELLY (1966) have pointed out, the low thorium and uranium contents of eugeosynclinal materials indicates derivation from nearly undifferentiated material of the upper mantle. The Umpqua graywackes of Oregon are similar geologically and also radiometrically to the Caribbean orogenic wackes. The average graywacke estimate in Table 92-I-1 includes both the highly-volcanic, generally uranium-poor, graywackes of the Caribbean and Oregon areas and also some more-highly radioactive rocks of the type which are generally classified as graywacke simply because of their high clay content. Arkoses have not been well studied, and the estimate in Table 92-K-1 for their uranium content is highly tentative.

II. Shales

The abundance of uranium in common *shales* (Fig. 92-K-1) shown in Table 92-K-1 is well established by several independent studies. Both ADAMS and WEAVER (1958), working with North American shales, and BARANOV *et al.* (1956), working with shales of the Russian platform arrived at almost identical values of approximately 3.7 ppm for the average uranium content. It is interesting to note that both the average thorium contents (Table 90-K-1) and uranium contents of ordinary shales are very close to the values commonly cited for the average concentration in the continental crust.

Both black and red shales are distinct from the ordinary green and gray shales in thorium and uranium contents. The black shales have been studied by numerous workers, and only summary reference is made in Table 92-K-1. A discussion of the economic significance of the black shales is given by BATES and STRAHL (1958). The black shales characteristically have a very high uranium concentration and comparatively low thorium content with a consequently low Th/U ratio. The high concentration of uranium results presumably from absorption of uranium by organic material in the shale and/or reduction of uranyl ion in sea water to the less soluble uranous form by the decay of organic matter and consequent precipitation of uranium oxide. The red and yellow shales are comparatively enriched in thorium, and their relatively low uranium content (1—2 ppm) gives them a high Th/U ratio. Depletion of uranium in the red shales presumably results from the solubility of the oxidized uranyl ion. This discussion of shales shows the value of the Th/U ratio for the interpretation of the environment of formation of sedimentary rocks (Fig. 92-K-2).

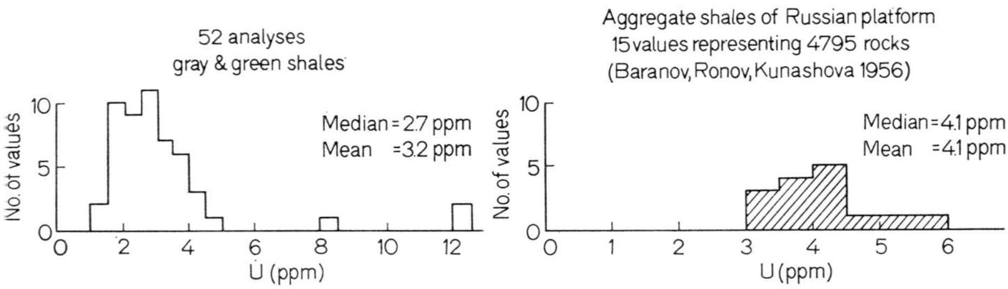

Fig. 92-K-1. Histogram of uranium concentrations in North American gray and green shales and Russian platform shales. (From Adams and Weaver, 1958; Russian platform data from Baranov *et al.*, 1956)

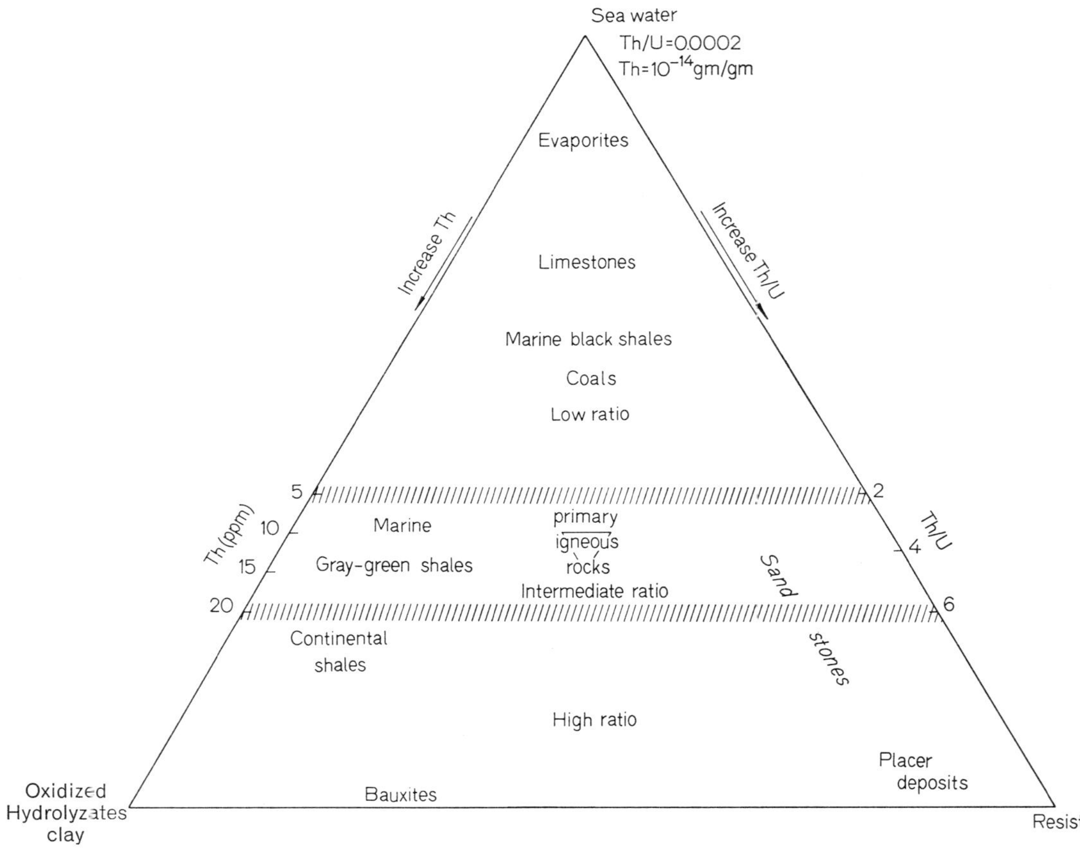

Fig. 92-K-2. Generalized distribution of thorium and uranium in sedimentary rocks. (From Adams and Weaver, 1958)

III. Bauxites, Bentonites

Bauxites and bentonites are examples of the concentration of thorium and uranium in clays. In bauxites, the average uranium content (11.4 ppm) and thorium content (45.6 ppm; Table 90-K-1) are both abnormally high. Presumably during weathering both the thorium and uranium have been concentrated in resistate residues and adsorbed on clays. Some tendency for greater retention of thorium than of uranium is shown by the generally high Th/U ratios.

IV. Carbonate Rocks

Carbonate rocks (Fig. 92-K-3) have uranium concentrations in the range found for typical calcareous organisms. The uranium in calcium carbonate apparently substitutes in the lattice in place of calcium, although some may accompany thorium in the insoluble residue fraction of limestones; (see Section 90-K). Several studies have been made of the concentration of uranium in calcareous organic material, and reference may be made to TATSUMOTO and GOLDBERG (1959), SACKETT and POTRATZ

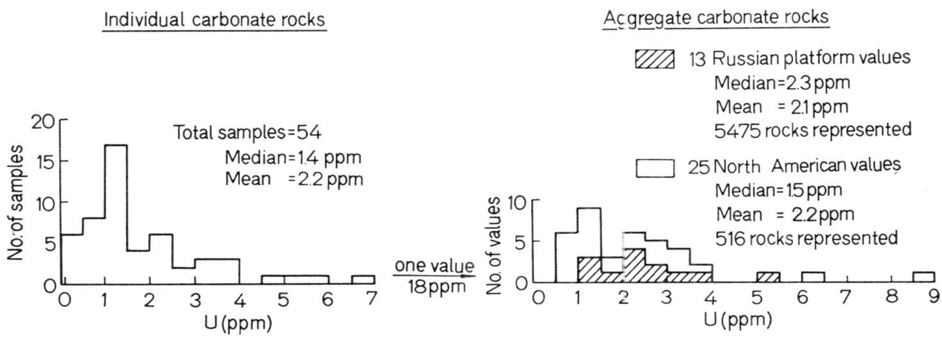

Fig. 92-K-3. Histogram of uranium concentrations in aggregate samples of carbonate rocks. (From ADAMS and WEAVER, 1958; Russian platform data from BARANOV, RONOV, and KUNASHOVA, 1956)

(1963), ROSHOLT and ANTAL (1963), and BELL (1963). Estimates of the average uranium content of carbonate rocks are quite uniform at approximately 2 ppm. These estimates include the North American limestones studied by ADAMS and WEAVER (1958), the carbonate rocks of the Russian platform studied by BARANOV *et al.* (1956), and the highly assorted suite of limestones investigated by BELL (1963). DEININGER (1964) shows, in general, little variation between limestones and dolomites with respect to uranium content.

V. Phosphates

Phosphate rocks, listed in Table 92-K-1, generally have a high concentration of uranium. A large number of studies have been made on the uranium in phosphate rocks and in apatite minerals because of the potential economic importance of uranium recovery from such rocks. The references are summarized in a paper by MCKELVEY (1956) cited in the table.

VI. Oceanic Sediments

Uranium contents are given in Table 92-K-1 for a variety of modern oceanic sediments. With the exception of studies made primarily for the purpose of dating cores by means of various disequilibria methods, not a great deal of atten-

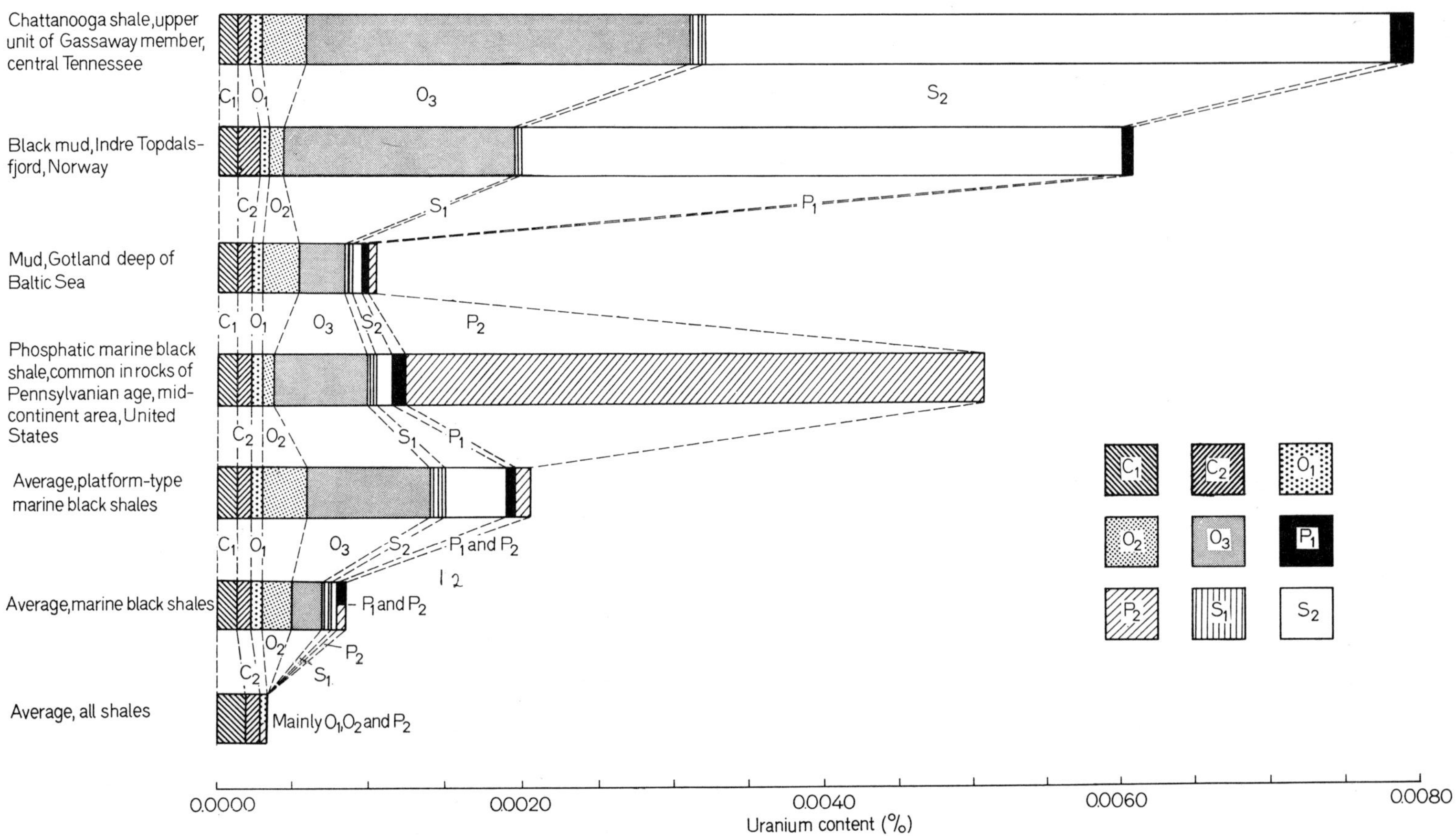

Fig. 92-K-4. Estimated average amount contributed to total uranium content by nine uranium-bearing constituents of shales and muds. (From SWANSON, 1963). For list of symbols see p. 92-K-8

Fig. 92-K-4 (Continued). List of symbols:

C_1: Uranium in main mass of clastic mineral sediment, particularly clay minerals, but exclusive of resistates.
C_2: Uranium in resistates or heavy minerals, such as zircon, apatite, and sphene.
O_1: Uranium inherent to living plants or animals, either land or marine types.
O_2: Uranium sorbed from sea water by solid or colloidal carbonaceous remains of plankton settling from surface waters to bottom.
O_3: Uranium sorbed from stream water and sea water by either solid or colloidal organic matter derived from decaying land plants of the humic type.
P_1: Uranium in phosphate precipitated on and within sediment as oolites and nodules where hydrogen sulfide environment extends into overlying water; also uranium in phosphatic fossil shells and bones.
P_2: Uranium in phosphate precipitated from, settling through and exposed to sea water above sediment-water interface, and formed where hydrogen sulfide environment confined to sediment.
S_1: Uranium precipitated as colloidal-size uraninite(?) from connate water by hydrogen sulfide formed and confined within sediment.
S_2: Uranium precipitated as colloidal-size uraninite(?) from sea water by hydrogen sulfide generated within and above sediment by anaerobic bacteria, and diffused into overlying water.

tion has been paid to the radioactive contents of oceanic materials. The low Th/U ratios reported by BARANOV and KHRISTIANOVA (1963) certainly do not match the ratios obtained for similar sediments by other investigators.

In many sedimentary rocks uranium is commonly associated with organic matter. A typical example of this occurrence is the Colorado Plateau Province of the western United States, where uranium is found absorbed in petrified logs and also intimately admixed with asphaltic bands in sandstones. The nature of the uranium mineral or minerals in these occurrences has been investigated in detail (KELLEY, 1963). On a broader scale, the general association of uranium with organic material is well documented by the high uranium concentrations in black shales and also in coals, asphalts, and oil (Table 92-K-1). The site of uranium in these organic materials is uncertain. Some asphaltic deposits contain admixed uraninite, but the organic phases themselves also seem to be rich in uranium, which may be in the form of some organic complex. The term "thucholite" has been proposed for mineraloids containing thorium, uranium, carbon, hydrogen, and oxygen.

The reasons for the association of thorium and uranium with organic material have never been fully ascertained. Experimental work has clearly demonstrated that wood and other organic material are capable of absorbing uranium from solution. It has also been proposed that the radioactivity of the thorium and uranium may aid in polymerization of organic materials around the two metals. Organic uranium geochemistry is discussed more fully by BREGER and DEUL (1956). SWANSON (1963) shows quantitatively the amounts of uranium contributed to black muds and shales by various processes (Fig. 92-K-4).

Revised manuscript received: July 18th, 1968

92-L. Biogeochemistry

Uranium is not known to be an essential element for the life process of any organism. The several parts per million concentration of uranium in carbonate shells is probably explained by substitution of uranium in the calcium carbonate lattice. The concentration of uranium in woody and other plant residues is not easily understood. Possibly the organic matter simply acts as a reducing agent which causes precipitation of the comparatively insoluble uranous form. It is also possible, however, that the uranium occurs in some particularly stable organo-metallic complexes. Sulfate-reducing bacteria may play a role in the reduction and fixation of uranium (JENSEN, 1963). The association of uranium with organic and woody material is discussed in a number of papers summarized by BREGER and DEUL (1956). The biochemical importance of uranium is summarized by F. KOCZY (1963) and EDGINGTON (1966). Some analyses of animal and plant materials are given by HOFFMANN (1941, 1943). BOWIE and ATKIN (1956) report 2,700 ppm uranium in an unusual fossil fish skeleton.

Revised manuscript received: July 18th, 1967

92-M. Abundance in Common Metamorphic Rock Types

The abundances of uranium have been determined in a number of different metamorphic rocks, and the data are given in Table 92-M-1. The names used for rock designations are those given by the original authors. As might be expected for rocks derived from a variety of parent materials under a variety of conditions, the range of compositions is quite broad.

HEIER and CARTER (1964) have summarized studies of the radiometry of various eclogites from metamorphic environments. They find an average of 0.20 ppm uranium and 0.38 ppm thorium in a total of 14 samples analyzed by various methods.

Table 92-M-1. *Contents of uranium in metamorphic rocks*

Rock type	No. of samples	U ppm	Method	References
Gneiss, Japan	1	2.3	R	EVANS and GOODMAN (1941)
Augen Gneiss, Finland	1	2.1	R	EVANS and GOODMAN (1941)
Eclogite, Switzerland	1	0.2	R	EVANS and GOODMAN (1941)
Amphibolite (Schwarzwald, Germany)	3	3.5	R	HUSMANN (1956)
Biotite-Hornbl. Paragneiss (Schwarzwald, Germany)	2	2.0	R	HUSMANN (1956)
Paragneiss (Schwarzwald, Germany)	4	7.0	R	HUSMANN (1956)
Granulite (Schwarzwald, Germany)	2	4.9	R	HUSMANN (1956)
Orthogneiss (Schwarzwald, Germany)	4	3.6	R	HUSMANN (1956)
Metatectite (Schwarzwald, Germany)	4	4.8	R	HUSMANN (1956)
Diatectite (Schwarzwald, Germany)	4	11.2	R	HUSMANN (1956)
Cordierite Gneiss (Schwarzwald, Germany)	1	5.8	R	HUSMANN (1956)
Orthoclase metacrysts (Schwarzwald, Germany)	1	45	R	HUSMANN (1956)
Marble	2	0.17	L	PLILER (1956)
Slate (mainly Michigan)	14	2.7	L	PLILER (1956)
Phyllite (mainly Arizona — New Mexico)	7	1.9	L	PLILER (1956)
Schist (New Mexico)	4	2.5	L	PLILER (1956)
Mafic Rocks (Terskei Ala Tau Mts.)	82 (comp)	3.2	W	KRYLOV (1958)
Idaho Batholith Gneisses	12	2.2	L	LARSEN JR. and GOTTFRIED (1961)
(Almandine-Amph. Facies) Gneiss (Texas)	13	2.82	R	BILLINGS (1962)

Table 92-M-1 (Continued)

Rock type	No. of samples	U ppm	Method	Reference
(Almandine-Amph. Facies) Amphibolite (Texas)	14	0.33	R	Billings (1962)
(Almandine-Amph. Facies) Graphite Schist (Texas)	3	3.49	R	Billings (1962)
Marble (Grenville Province)	2	0.36	L	Doe (1962)
(Amphibolite Facies) Light Gneiss (Langøy, Norway)	5	< 2.5	R	Heier (1962b)
(Amphibolite Facies) Amphibolite (Langøy)	7	< 1.06	R	Heier (1962b)
(Retrograde Gneiss) Monzonite (Langøy)	2	< 1.25	R	Heier (1962b)
(Retrograde Gneiss) Granite (Langøy)	8	< 0.94	R	Heier (1962b)
Biotite Schist (Front Range, Colorado)	9	4.7	L	Phair and Gottfried (1964)
Biotite Hornblende Schist and Amphibolite (Front Range)	4	4.7	L	Phair and Gottfried (1964)
(High Granulite Facies) Monzonite (Langøy, Norway)	3	0.61	R	Heier and Adams (1965)
(High Granulite Facies) Banded Gneiss (Langøy)	4	0.22	R	Heier and Adams (1965)
(Low Granulite Facies) Gneiss (Langøy)	5	0.88	R	Heier and Adams (1965)
(High Amphibolite Facies) Gneiss (Langøy)	3	1.22	R	Heier and Adams (1965)

Revised manuscript received: July 18th, 1967

92-N. Behavior in Metamorphic Reactions

The mobility of the uranyl ion in surface waters plus the tendency for potassium and uranium to occur together lead to the presumption that uranium is highly involved in metasomatic activity. Unfortunately, data for metamorphic rocks are not yet adequate to test the hypothesis. HEIER and ADAMS (1965) have proposed that high-rank rocks in one area in Norway are impoverished in uranium in comparison with lower-rank rocks owing to upward movement of uranium. This evidence of mobility, however, has not been confirmed in other areas owing to lack of adequate samples.

Revised manuscript received: July 18th, 1967

92-O. Economic Utilization

Aside from very minor uses as a pigment and chemical reagent, the economic utilization of uranium has been dominated in the first 40 years of this century by a demand for radium and since 1942 by a demand for fissionable materials. The demand for radium for medical and industrial purposes led to widespread exploration for the parent uranium in the first quarter of this century, but this exploration was drastically cut back after the discovery in 1923 of the rich uranium deposits in Katanga (then the Belgian Congo). The Katanga source and the Great Bear Lake, Canada, sources of rich pitchblende discovered in 1930 were more than adequate to meet the demand for radium.

Beginning in 1943 a second widespread exploration effort was initiated to develop uranium supplies for military purposes. This military demand for uranium was substantially oversaturated in the U.S.A. by 1958, when guaranteed prices were restricted and new import arrangements ceased to be made. A steady exploration effort continued in France, Sweden, and other countries.

As of this writing (1966) uranium-fueled nuclear reactors are competitive in many areas with conventional power plants for the generation of electricity for civilian uses. To the extent that this competition continues to shift in favor of the nuclear reactors, a new and still larger demand for uranium will be created, leading to a third widespread exploration effort.

The last year for which world production figures of uranium are reasonably available is 1963. Among the countries reporting production, the leader is the United States, with 14,218 short tons of U_3O_8 during the year; the U.S. is followed by Canada, with 8,141 short tons, and South Africa, with 4,532 short tons. Other countries reporting significant production are the Congo, France, and Portugal.

Most of the uranium production in the United States is from mineralized sandstones in the Colorado Plateau province centering around Wyoming, Utah, and New Mexico. Canadian production comes primarily from vein deposits in the Northwest Territories, with large reserves in the Blind River conglomerate. South African production is almost exclusively a by-product of gold mining in the Witwatersrand area, where pitchblende mineralization has apparently accompanied gold deposition. The largest reserves are disseminated in sediments, usually sandstones. The majority of smaller deposits are pitchblende veins such as those of Great Bear Lake, Canada; Erzgebirge, Germany and Czechoslovakia; Katanga, Congo; and Portugal.

Acknowledgments

Preparation of this review was greatly assisted by Grant C-009 from the Robert A. Welch Foundation to John A. S. Adams and John J. W. Rogers.

Revised manuscript received: July 18th, 1967

References: Sections 92-B to 92-O

ABELEDO, M. V. DE, M. R. DE BENYACAR, and E. E. GALLONI: Ranquilite, a calcium uranyl silicate. Am. Mineralogist **45**, 1078 (1960).

ABRAMOVICH, I. I.: Uranium and thorium in the intrusive rocks of Central and Western Tuva. Geochemistry (USSR) (English Transl.) 442 (1959).

ADAMS, J. A. S.: The uranium geochemistry of Lassen Volcanic National Park, Calif. Geochim. Cosmochim. Acta **8**, 74 (1955).

— The uranium contents and alpha particle activities of tektites. Proc. 20th Int. Geol. Congr., Mexico City (1956).

— M. C. KLINE, K. A. RICHARDSON, and J. J. W. ROGERS: The Conway granite of New Hampshire as a major low-grade thorium resource. Proc. Natl. Acad. Sci. U.S. **48**, 1898 (1962).

— J. K. OSMOND, and J. J. W. ROGERS: The geochemistry of uranium and thorium. Phys. Chem. Earth **3**, 298 (1959).

—, and K. A. RICHARDSON: Thorium, uranium, and zirconium concentrations in bauxite. Econ. Geol. **55**, 1653 (1960).

—, and C. E. WEAVER: Thorium-to-uranium ratios as indicators of sedimentary processes — an example of geochemical facies. Bull. Am. Assoc. Petrol. Geologists **42**, 387 (1958).

AGAMIROV, S. SH.: Geochemical balance of radioactive elements in the Black Sea Basin. Geochemistry (USSR) (English Transl.) 630 (1963).

AHRENS, L. H.: Some observations on the uranium and thorium distributions in accessory zircon from granitic rocks. Geochim. Cosmochim. Acta **29**, 711 (1965).

BALACEK, K. J., and J. A. S. ADAMS: Radioactivity in tektites. In preparation (1966).

BARANOV, V. I., and L. A. KHRISTIANOVA: Radioactivity of the waters of the Indian Ocean. Geochemistry (USSR) (English Transl.) 765 (1959).

— — New data on radioactivity of the Indian Ocean. Akad. Nauk SSSR, Otdel. Geol.-Geograf. Nauk 392 (1960) [in Russian].

— — Radioactivity of ocean sediments. Akad. Nauk SSSR, Khim. Zemnoi Kory, Tr. Geokhim. Konf. **1**, 401 (1963) [in Russian.].

—, and LIEH-TIEN DU: Geochemistry of uranium and thorium in the granites of the Kyzyltau Massif (central Kazakhstan). Geochemistry (USSR) (English Transl.) 1180 (1961).

— A. B. RONOV, and K. G. KUNASHOVA: Geochemistry of dispersed thorium and uranium in clays and carbonate rock of the Russian platform. Geochemistry (USSR) (English Transl.) 227 (1956).

BARKER, F. B., and R. C. SCOTT: Uranium and radium in the ground water of the Llano Estacado, Texas and New Mexico. Trans. Am. Geophys. Union **39**, 459 (1958).

— — Uranium and radium in ground water from igneous terranes of the Pacific North West. U.S. Geol. Surv. Profess. Papers **424-B**, 298 (1961).

BATE, G. L., and J. R. HUIZENGA: Abundance of ruthenium, osmium, and uranium in some cosmic and terrestrial sources. Geochim. Cosmochim. Acta **27**, 345 (1963).

— —, and H. A. POTRATZ: Thorium in stone meteorites by neutron activation analysis. Geochim. Cosmochim. Acta **16**, 88 (1959).

BATES, T. F., and E. O. STRAHL: Mineralogy and chemistry of uranium-bearing black shales. Proc. of Second Int. United Nations Conf. on Peaceful Uses of Atomic Energy, Geneva, **2**, 407 (1958).

BELL, K. G.: Deposition of uranium in salt-pan basins. U.S. Geol. Surv. Profess. Papers **354-G**, 161 (1960).

— Uranium in carbonate rocks. U.S. Geol. Surv. Profess. Papers **474-A**, 29 p. (1963).

BELOVA, L. N.: Arsenuranylite, the arsenic analogue of phosphuranylite. Zap. Vses. Mineralog. Obshchestva **87**, 589 (1958) [in Russian]. [Am. Mineralogist **44**, 208 (1959).]

BILLINGS, G. K.: A geochemical investigation of the Valley Spring gneiss and Packsaddle schist, Llano uplift, Texas. Texas J. Sci. **9**, 328 (1962).

BIRCH, F.: Speculations on the earth's thermal history. Bull. Geol. Soc. Am. **76**, 133 (1965).

BOWIE, S. H. U., and D. ATKIN: An unusually radioactive fossil fish from Thurso, Scotland. Nature **177**, 487 (1956).

BRANCHE, G., P. BARIAND, F. CHANTRET, R. POUGET, and A. RIMSKY: La vanuralite, nouveau minéral uranifère. Compt. Rend. **256**, 5374 (1963). [Am. Mineralogist **48**, 1415 (1963)].

— M.-E. ROPERT, F. CHANTRET, B. MORIGNAT: La francevillite, nouveau minéral uranifère. Compt. Rend. **245**, 89 (1957). [Am. Mineralogist **43**, 180 (1958).]

BREGER, I. A., and M. DEUL: The organic geochemistry of uranium. U.S. Geol. Surv. Profess. Papers **300**, 505 (1956).

BRUCE, R. F., and D. E. FERGUSON: The recovery of uranium from natural waters of the United States. Cf-2-102, declassified 1955. (1951.)

BÜLTEMANN, H. W., u. G. H. MOH: Bergenit, ein neues Mineral der Phosphuranylite-Gruppe. Neues Jahrb. Mineral., Monatsh. 232 (1959). [Am. Mineralogist **45**, 907 (1960).]

BUTLER, A. P., JR.: Ratio of thorium to uranium in some plutonic rocks of the White Mountain Plutonic-Volcanic Series, New Hampshire. U.S. Geol. Surv. Profess. Papers **424-B**, 67 (1961).

CANNON, H. L., and F. KLEINHAMPL: Botanical methods of prospecting for uranium. U.S. Geol. Surv. Profess. Papers **300**, 681 (1956).

CHENTSOV, I. G.: The entry of uranium into some rock-forming minerals. Akad. Nauk SSSR, Inst. Geo. Rudnykh Mestorzhdenii; Petrog. Mineral. i. Geokhim, Trudy **28**, 142 (1959) [in Russian].

CHERNIKOV, A. A., O. V. KRUMETSKAYA, and V. D. SIEDEL'NIKOVA: Ursilite — a new silicate of uranium. Vopr. Geol. Urana, Atomizdat, 73 (1957): Consultants Bureau Transl. 61 (1958). [Am. Mineralogist **44**, 464 (1959).]

CHERRY, R. D.: Thorium and uranium contents of Australites. Nature **195**, 1184 (1962).

—, and J. A. S. ADAMS: Gamma-spectrometric determinations of thorium, uranium, and potassium in tektites. Geochim. Cosmochim. Acta **27**, 1089 (1963).

COATS, R. R.: Uranium and certain other trace elements in felsic volcanic rocks of Cenozoic Age in Western United States. U.S. Geol. Surv. Profess. Papers **300**, 75 (1956).

COHEN, P.: An evaluation of uranium as a tool for studying the hydrogeochemistry of the Truckee Meadows Area, Nevada. J. Geophys. Res. **66**, 4199 (1961).

COPPENS, R.: Sur l'emploi de l'emulsion photographique pour la determination de la radioactivite des roches par l'examen des trajectoires des rayons alpha. J. Phys. Radium **11**, 21 (1950).

COULOMB, R.: Contribution a la geochemie de l'uranium dans les granites intrusifs. Comm. Énergie Atomique (France), Rappt No. 1173, 85 p (1959).

DAVIS, G. L.: Radium content of ultramafic igneous rocks, III-meteorites. Am. J. Sci. **248**, 107 (1950).

DEININGER, R. W.: Ferrous ion and uranium concentrations and distributions in 100 selected limestones and dolomites. Ph. D. Thesis, Rice University, Houston, Texas 1964, 85 p.

DENSON, N., H. ZELLER, and J. STEPHENS: Water sampling as a guide in the search for uranium deposits and its use in evaluating widespread volcanic units as potential source beds for uranium. U.S. Geol. Surv. Profess. Papers **300**, 673 (1956).

DOE, B. R.: Distribution and composition of sulfide minerals at Balmat, New York. Bull. Geol. Soc. Am. **73**, 833 (1962).

DUBEY, V. S.: The origin of tektites. Nature **132**, 678 (1933).

DYBEK, I. J.: Zur Geochemie und Lagerstättenkunde des Urans. Clausthaler Hefte Lagerstättenkunde und Geochemie der Mineralogischen Rohstoffe, H. 1, 163 p. (1962).

EBERT, K. H., H. KÖNIG u. H. WÄNKE: Eine neue Methode zur Bestimmung kleinster Uranmengen und ihre Anwendung auf die Urananalyse von Steinmeteoriten. Z. Naturforsch. **12** A, 763 (1957).

EDGINGTON, D. N.: Thorium and uranium in the natural environment — a review of present knowledge. Annual Report of Radiological Physics Division, July 1964 through June 1965, Argonne National Laboratory, ANL-7060, 73 (1966).

EL SHAZLY, E. M.: Geology of uranium and thorium. Internat. Atomic Energy Agency, Vienna, Bibliog. Ser. No. 4, 130 p (1962).
ERICKSON, R. L., and L. V. BLADE: Geochemistry and petrology of the alkalic igneous complex at Magnet Cove, Arkansas. U.S. Geol. Surv. Profess. Papers **425**, 95 p. (1963).
EVANS, R. D., and C. GOODMAN: Radioactivity of rocks. Bull. Geol. Soc. Am. **52**, 459 (1941).
FEDOROV, O. V.: Second find of Ca-U molybdate in USSR. Zap. Vses. Mineralog. Obshchestva **92**, 464 (1963) [in Russian]. [Am. Mineralogist **49**, 1152 (1964).]
FILIPPOV, M. S., and L. V. KOMLEV: Uranium and thorium in granitoids of the middle Dnepr region. Geochemistry (USSR) (English Transl.) 535 (1959).
FIREMAN, E. L., and D. E. FISHER: Uranium in Sikhote-Alin meteorite and its relation to the lead method of age determinations. Nature **192**, 644 (1961).
FIX, P. F.: Hydrogeochemical exploration for uranium. U.S. Geol. Surv. Profess. Papers **300**, 667 (1956).
FØYN, E. E., B. KARLIK, H. PETTERSON, and E. RONA: Radioactivity of sea water. Nature **143**, 275 (1939).
FRIEDMAN, I.: The water, deuterium, gas, and uranium content of tektites. Geochim. Cosmochim. Acta **14**, 316, (1958).
FRONDEL, C.: Mineralogy of thorium. U.S. Geol. Surv. Profess. Papers **300**, 567 (1956).
—, and J. ITO: Boltwoodite, a new uranium silicate. Science **124**, 931 (1956). [Am. Mineralogist **42**, 307 (1957).]
GAST, P. W.: Limitations on the composition of the upper mantle. J. Geophys. Res. **65**, 1287 (1960).
GERASIMOVSKY, V. I.: Uranium minerals. Atomnaya Energ. No. 4, 118 (1956) [in Russian]. [Am. Mineralogist **42**, 307 (1957).]
GETSEVA, R. V., and K. T. SAVEL'EVA: Handbook for the determination of uranium minerals. Moscow 1957 [in Russian]. [Am. Mineralogist **43**, 379—381 (1958).]
GOLDSZTAUB, S., et R. WEY: Absorption des ions uranyles par les argiles. Bull. Soc. Franc. Mineral. Crist. **78**, 242 (1955).
GOLDSZTEIN, M., et S. GUILLERMO: Radioactivite des lavas de la Chaine des Puys. Bull. Soc. Franc. Mineral. Crist. **84**, 6 (1961).
GOLES, G. G., and E. ANDERS: Abundance of iodine, tellurium and uranium in meteorites. Geochim. Cosmochim. Acta **26**, 723 (1962).
GOTTFRIED, D.: Distribution of uranium in igneous complexes. U.S. Geol. Surv. TEI **751**, 102 (1959).
—, and E. S. LARSEN JR.: Geologic investigation of radioactive deposits. Semi-annual progress report, Dec. 1953 to May 1954. U.S. Geol. Surv. TEI **440**, 202 (1954).
— — Distribution of uranium in the volcanic rocks of the Modoc Area, Calif. U.S. Geol. Surv. TEI **740**, 282 (1958).
— R. MOORE, and E. CAMPBELL: Thorium and uranium in some volcanic rocks from the circum-pacific province. U.S. Geol. Surv. Profess. Papers **450-E**, 85 (1963).
— —, and A. COEMERER: Thorium and uranium in some alkalic igneous rocks from Virginia and Texas. U.S. Geol. Surv. Profess. Papers **450-B**, 70 (1962).
GROSS, E. B., A. S. COREY, R. S. MITCHELL, and K. WALENTA: Heinrichite and metaheinrichite, hydrated barium uranyl arsenate minerals. Am. Mineralogist **43**, 1134 (1958).
GUILLEMIN, C., et J. PROTAS: Ianthinite et wyartite. Bull. Soc. Franc. Mineral. Crist. **82**, 80 (1959). [Am. Mineralogist **44**, 908, 1003 (1959).]
HAHOFER, E., u. F. HECHT: Uranbestimmung in Tiefseeproben. Microchim. Acta **41**, 417 (1954).
HAIL, W. J., JR., A. T. MYERS, and C. A. HORR: Uranium in asphalt-bearing rocks of Western United States. U.S. Geol. Surv. Profess. Papers **300**, 521 (1956).
HAMAGUCHI, H., G. W. REED, and A. TURKEVICH: Uranium and barium in stone meteorites. Geochim. Cosmochim. Acta **12**, 337 (1957).
HAMILTON, E.: The uranium content of the differentiated Skaergaard intrusion. Medd. Groenland **162**, 34 p. (1959).
HARRISS, R. C., and J. A. S. ADAMS: Geochemical and mineralogical studies on the weathering of granitic rocks. Am. J. Sci. **264**, 146 (1966).

HECHT, F., H. KEUPPER, and W. E. PETRASCHECK: Preliminary remarks on the determination of uranium in Austrian springs and rocks. Proc. of Second Int. United Nations Conf. on Peaceful Uses of Atomic Energy, Geneva **2**, 158 (1958).
HEIDE, F.: Über den Urangehalt der Tektite. Acta Geol. Acad. Sci. Hung. **7**, 1 (1961)
—, u. G. PROFT: Der Urangehalt des Saalewassers. Naturwissenschaften **46**, 352 (1959).
HEIER, K. S.: A note on the U, Th, and K contents in the nepheline syenite and carbonatite on Stjernøy, North Norway. Norsk Geol. Tidsskr. **42**, 287 (1962a).
— Spectrometric uranium and thorium determinations on some high-grade metamorphic rocks on Langöy, Northern Norway. Norsk Geol. Tidsskr. **42**, 143 (1962b).
— Uranium, thorium, and potassium in eglogitic rocks. Geochim. Cosmochim. Acta **27**, 849 (1963).
—, and J. A. S. ADAMS: Concentration of radioactive elements in deep crustal material. Geochim. Cosmochim. Acta **29**, 53 (1965).
—, and J. L. CARTER: Uranium, thorium, and potassium contents in basic rocks and their bearing on the nature of the upper mantle. In: J. A. S. ADAMS and W. M. LOWDER (eds.), The natural radiation environment, p. 75, Chicago: Chicago University Press 1964.
— I. McDOUGALL, and J. A. S. ADAMS: Th, U and K in Hawaiian lavas. Nature **201**, 254 (1964).
—, and J. J. W. ROGERS: Radiometric determination of thorium, uranium and potassium in basalts and in two magmatic differentiation series. Geochim. Cosmochim. Acta **27**, 137 (1963).
HERNEGGER, F., u. H. WÄNKE: Über den Urangehalt der Steinmeteoriten und deren „Alter". Z. Naturforsch. **12** A, 759 (1957).
HOFFMANN, J.: Experimentelle Erfassung von Uran in lebenden Süßwasseralgen. Naturwissenschaften **29**, 403 (1941).
— Bioelement Uran im Pflanzen- und Tierreich sowie in menschlichen Organismus. Biochem. Z. **313**, 377 (1943).
HOLLAND, H. D., and L. KULP: The transport and deposition of uranium, ionium and radium in rivers, oceans, and ocean sediments. Geochim. Cosmochim. Acta **5**, 197 (1954).
HURLEY, P. M., and H. W. FAIRBAIRN: Abundance and distribution of uranium and thorium in zircon, sphene, apatite, epidote, and monazite in granitic rocks. Trans. Am. Geophys. Union **38**, 939 (1957).
HUSMANN, O.: Bestimmungen des Thorium- und Uran-Gehaltes an Gneissen und Magmatischen Gesteinen des mittleren Schwarzwaldes mittels der Koinzidenz-Methode. Neues Jahrb. Mineral., Monatsh. 108 (1956).
HUSSEIN, A.-M. H.: Étude radiogeologique de la region située au Nord de Nantes. Sci. Terre **7**, 7 (1961).
HUTTON, C. O.: Uranoan thorite and thorian monazite from blacksand paystreaks, San Mateo County, California. Bull. Geol. Soc. Am. **62**, 1518 (1951).
HYDEN, H. J.: Uranium and other trace metals in crude oils of Western United States. U.S. Geol. Surv. Profess. Papers **300**, 511 (1956).
ILLSLEY, C. T.: Hydrogeochemical exploration for uranium in the Mt. Spokane Area, Washington. Bull. Geol. Soc. Am. **68**, 1750 (1957).
JENSEN, M. L.: Sulfur isotopes and biogenic origin of uraniferous deposits of the Grants and Laguna districts. New Mexico, Bur. Mines Mineral Resources, Mem. **15**, 182 (1963).
JUDSON, S., and J. K. OSMOND: Radioactivity in ground and surface water. Am. J. Sci. **253**, 104 (1955).
JURAIN, G.: Remarques sur la teneur a uranium des eaux des vosges meridionale. Compt. Rend. **245**, 1071 (1957).
KAMATH, P. R., A. A. KHAN, S. R. RAO, T. N. V. PILLAI, M. L. BORKAR, and S. GANAPATHI: Environmental natural radioactivity measurements at Trombay Establishment. In: J. A. S. ADAMS and W. M. LOWDER (eds.), The natural radiation environment, p. 221. Chicago: Chicago University Press 1964.
KEEVIL, N. B.: Thorium-uranium ratios in rocks and minerals. Am. J. Sci. **242**, 309 (1944).
KELLEY, V. C. (ed.): Geology and technology of the Grants uranium region. New Mexico, Bur. Mines Mineral Resources, Mem. **15**, 277 p. (1963).

KOCZY, F. F.: The natural radioactive series in organic material. In :V. SCHULTZ and A. W. KLEMENT (eds.), Radioecology, p. 611. New York: Am. Inst. Biological Sciences and Reinholt Publ. Co. (1963).

— E. TOMIC, and F. HECHT: Zur Geochemie des Urans im Ostseebecken. Geochim. Cosmochim. Acta **11**, 86 (1957).

KOCZY, G.: Weitere Uranbestimmungen an Meerwasserproben. Oesterr. Akad. Wiss., Math-. Naturw. Kl., Sitzber., Abt. IIa, **158**, 113 (1950).

KÖNIG, H., u. H. WÄNKE: Uranbestimmungen an Steinmeteoriten mittels neutronenaktivierung über die Xenon-Isotopen 133 und 135. Z. Naturforsch. **14** A, 866 (1959).

KOPCHENOVA, E. V., K. V. SKVORTSOVA, N. I. SILANTIEVA, G. A. SIDORENKO, and L. V. MIKHAILOVA: Mourite, a new supergene uranium-molybdenum mineral. Zap. Vses. Mineralog. Obshchestva **91**, 67 (1962) [in Russian]. [Am. Mineralogist **47**, 1217 (1962).]

KOTLYAR, V. N.: Geology and genetic types of commercial uranium deposits. Gosudar. Nauk.-Tekh. Izd. Lit. Geol. i Okhrane Nedr., Moscow 1961, 246 p. [in Russian].

KRYLOV, A. YA.: Radioactivity in various rock complexes of the Terskei Ala-Tau Mountains. Geochemistry (USSR) (English Transl.) 240 (1958).

— The evolution of radioactivity and general chemical composition, North Tien-Shan Rocks. Radiokhimiya **5**, 173 (1963) [in Russian].

LANDIS, E. R.: Uranium content of ground and surface waters in a part of the central great plains. U.S. Geol. Surv. Bull. **1087-G**, 223 (1960).

LARSEN, E. S., JR.: Geologic investigation of radioactive deposits. Semi-Annual Progress Report, June through November, 1954. U.S. Geol. Surv. TEI **490**, 255 (1954).

—, and D. GOTTFRIED: Distribution of uranium in rocks and minerals of Mesozoic batholiths in Western United States. U.S. Geol. Surv. Bull. **1070**-C, 63 (1961).

— —, and M. MOLLOY: Distribution of uranium in the volcanic rocks of the San Juan Mountains, Southwestern Colorado. Proc. of Second Int. United Nations Conf. on Paeceful Uses of Atomic Energy, Geneva, **2**, 509 (1958).

—, and G. PHAIR: The distribution of uranium and thorium in igneous rocks. In: H. FAUL (ed.), Nuclear geology, p. 75. New York: John Wiley & Sons 1954.

— — D. GOTTFRIED, and W. L. SMITH: Uranium in magmatic differentiation. U.S. Geol. Surv. Profess. Papers **300**, 65 (1956).

LARSEN, E. S. 3rd, and D. GOTTFRIED: Uranium and thorium in selected suites of igneous rocks. Am. J. Sci. **258**-A, 151 (1960).

LEE, W. H. K., and S. UYEDA: Review of heat flow data. In: W. H. K. LEE (ed.), Terrestrial heat flow. Am. Geophys. Union, Geophys. Monograph **8**, 87 (1965).

LEONOVA, L. L.: Distribution of thorium in minerals of the granites of the Susamyr Batholith (Central Tien Shan). Geochemistry (USSR) (English Transl.) 571 (1962).

— Uranium and thorium in the Variscan alkaline intrusives of Southern Tien Shan (Matcha and Turpi Massifs). Geochemistry Int. (USSR) (English Transl.) 925 (1964).

—, and L. S. POGIBLOVA: Uranium in minerals of the intrusive rocks of the Kyzyl-Ompul Mountains (northern Kirgiziya). Geochemistry (USSR) (English Transl.) 999 (1961).

—, and O. S. RENNE: Distribution of uranium, thorium, and potassium in homogeneous granites. Geochemistry Int. (USSR) (English Transl.) 775 (1964).

—, and L. V. TAUSON: The distribution of uranium in the minerals of the Caledonian granitoids of the Susamyr Batholith (central Tien Shan). Geochemistry (USSR) (English Transl.) 815 (1958).

LOPATKINA, A. P.: Characteristics of migration of uranium in the natural waters of humid regions and their use in the determination of the geochemical background for uranium. Geochemistry Int. (USSR) (English Transl.) 788 (1964).

LOVERING, J. F., and J. W. MORGAN: Uranium and thorium in possible upper mantle materials. Nature **197**, 138 (1963).

— — Uranium and thorium abundances in stony meteorites, 1. The chondritic meteorites. J. Geophys. Res. **69**, 1979 (1964).

LYONS, J. B.: Uranium and thorium in the older plutonic rocks of New Hampshire. Art 32. U.S. Geol. Surv. Profess. Papers **424-B**, B 69 (1961).

MacDonald, G. J. F.: Geophysical deductions from observations of heat flow. In: W. H. K. Lee (ed.), Terrestrial heat flow. Am. Geophys. Union, Geophys. Monograph **8**, 191 (1965).
Mahdavi, A.: Thorium, uranium, and potassium contents of Atlantic and Gulf Coast beach sands. In: J. A. S. Adams, and W. M. Lowder (eds.), The natural radiation environment, p. 87. Chicago: Chicago University Press 1964.
McKelvey, V. E.: Uranium in phosphate rock. U.S. Geol. Surv. Profess. Papers **300**, 477 (1956).
— D. L. Everhart, and R. M. Garrels: Origin of uranium deposits. Econ. Geol., 50th Anniversary vol., 464 (1955).
Meliksetyan, B. M.: The geochemistry of uranium and thorium in some granitic rocks of Southern Armenian SSR. Akad. Nauk Armyan. SSR, Geol.-Geograf. Nauk **14**, 21 (1961) [in Russian].
Merlin, O. H., E. E. Picciotto, et S. Wilgain: Étude photographique de la distribution de la radioactive dans la granodiorite de l'Adamello. Geochim. Cosmochim. Acta **11**, 171 (1957).
Miyake, Y., Y. Sugimura, and H. Tsubota: Content of uranium, radium, and thorium in river waters of Japan. In: J. A. S. Adams and W. M. Lowder (eds.), The natural radiation environment. p. 221. Chicago: Chicago University Press 1964.
Morgan, J. W., and J. F. Lovering: Uranium and thorium abundances in stony meteorites, 2. The achondritic meteorites. J. Geophys. Res. **69**, 1989 (1964).
— — Uranium and thorium in the Nuevo Laredo achondrite. J. Geophys. Res. **70**, 2002 (1965a).
— — Uranium and thorium abundances in the basalt cored in Mohole Project (Guadalupe Site). J. Geophys. Res. **70**, 4724 (1965b).
Murray, E. G., and J. A. S. Adams: Amount and distribution of thorium, uranium, and potassium in sandstones. Geochim. Cosmochim. Acta **13**, 260 (1958).
Nakanishi, M.: Fluorometric microdetermination of uranium. V. The uranium content of sea water. Bull. Chem. Soc. Japan **24**, 36 (1951).
Nekrasova, Z. A.: A hydrous uranyl ammonium phosphate (uramphite), $NH_4(UO_2)(PO_4) \cdot 3H_2O$. Voprosy Geologii Urana, Atomizdat, 67 (1957): Consultants Bureau Transl. 56 (1958). [Am. Mineralogist **44**, 464 (1959).]
Nikolaev, D. S., O. P. Korn, K. F. Lazarev, L. B. Kolyadin, Yu. V. Kuznetsov, and S. M. Grashchenko: Concentration of uranium in waters of the Black Sea. Dokl. Akad. Nauk SSSR (English Transl.) **132**, 1411 (1960).
Osmond, J. K.: The distribution of the heavy radioelements in the rocks and waters of Florida. In: J. A. S. Adams and W. M. Lowder (eds.), The natural radiation environment, p. 158. Chicago: Chicago University Press 1964.
Palache, C., H. Berman, and C. Frondel: The system of mineralogy, 7th ed., vol. 1, p. 611—620. New York: John Wiley & Sons 1944.
Patterson, C., H. Brown, G. Tilton, and M. Ingram: Concentration of uranium and lead and the isotopic composition of lead in meteoritic material. Phys. Rev. **92**, 1234 (1953).
Peacock, J. D.: Uranium in British surface and underground waters. Nature **191**, 1189 (1961).
Peterman, Z. E.: Unpublished, personal communication (1963).
Phair, G.: Geologic investigations of radioactive deposits. Semiannual Progress Report, Dec., 1954 to May, 1955. U.S. Geol. Surv. TEI **540**, 237 (1955).
— Geologic investigations of radioactive deposits. Semi-annual Progress Report, Dec., 1955 to May, 1956. U.S. Geol. Surv. TEI **620**, 310 (1956).
—, and D. Gottfried: The Colorado Front Range, Colorado, U.S.A., as a uranium and thorium province. In: J. A. S. Adams and W. M. Lowder (eds.), The natural radiation environment, p. 7. Chicago: Chicago University Press 1964.
Pliler, R.: The distribution of thorium and uranium in sedimentary rocks and the oxygen content of the pre-Cambrian atmosphere. M. A. Thesis Rice Institute 1956.
—, and J. A. S. Adams: The distribution of thorium, uranium, and potassium in the Mancos shale. Geochim. Cosmochim. Acta **26**, 1115 (1962a).
— — The distribution of thorium and uranium in a Pennsylvanian weathering profile. Geochim. Cosmochim. Acta **26**, 1137 (1962b).

PROTAS, J.: Une nouvelle espece minérale: la compreignacite, $K_2O \cdot 6UO_3 \cdot 11H_2O$. Bull. Soc. Franc. Mineral Crist. **87**, 365 (1964). [Am. Mineralogist **50**, 807 (1965).]
RAGLAND, P. C.: Autoradiographic investigation of naturally occurring materials. In: J. A. S. ADAMS and W. M. LOWDER (eds.), The natural radiation environment, p. 129. Chicago: Chicago University Press 1964.
REASBECK, P., and K. I. MAYNE: The ages and origin of meteorites. Nature **176**, 186 (1955).
REED, G. W., H. HAMAGUCHI, and A. TURKEVICH: The uranium contents of iron meteorites. Geochim. Cosmochim. Acta **13**, 248 (1958).
— K. KIGOSHI, and A. TURKEVICH: Determinations of concentrations of heavy elements in meteorites by activation analysis. Geochim. Cosmochim. Acta **20**, 122 (1960).
RICHARDSON, K. A.: Thorium, uranium, and potassium in the Conway granite, New Hampshire, U.S.A. In: J. A. S. ADAMS and W. M. LOWDER (eds.), The natural radiation environment, p. 47. Chicago: Chicago University Press 1964.
ROGERS, J. J. W.: Statistical tests of the homogeneity of the radioactive components of granitic rocks. In: J. A. S. ADAMS and W. M. LOWDER (eds.), The natural radiation environment, p. 51. Chicago: Chicago University Press 1964.
— Geochemical significance of the source rocks of some graywackes from Western Oregon and Washington. Texas J. Sci. **18**, 5 (1966).
—, and J. A. S. ADAMS: Autoradiography of volcanic rocks of Mount Lassen. Science **125**, 1150 (1957).
— —, and B. GATLIN: Distribution of thorium, uranium, and potassium concentrations in three cores from the Conway granite, New Hampshire, U.S.A. Am. J. Sci. **263**, 817 (1965).
—, and T. W. DONNELLY: Radiometric evidence for the origin of eugeosynclinal materials. Tulane Studies in Geology **4**, 133 (1966).
—, and P. C. RAGLAND: Variation of thorium and uranium in selected granitic rocks. Geochim. Cosmochim. Acta **25**, 99 (1961).
—, and K. A. RICHARDSON: Thorium and uranium contents of some sandstones. Geochim. Cosmochim. Acta **28**, 2005 (1964).
RONA, E., L. O. GILPATRICK, and L. M. JEFFREY: Uranium determination in sea water. Trans. Am. Geophys. Union **37**, 697 (1956).
—, and W. D. URRY: Radioactivity of ocean sediments. VIII. Radium and uranium contents of ocean and river waters. Am. J. Sci. **250**, 241 (1952).
ROSHOLT, J. N., and P. S. ANTAL: Evaluation of the $Pa^{231}/U—Th^{230}/U$ method for dating Pleistocene carbonate rocks. U.S. Geol. Surv. Profess. Papers **450-E**, E 108 (1963).
— C. EMILIANI, J. GEISS, F. F. KOCZY, and P. J. WANGERSKY: Absolute dating of deep-sea cores by the Pa^{231}/Th^{230} method. J. Geol. **69**, 162 (1961).
SACKETT, W. M., and H. A. POTRATZ: Dating of carbonate rocks by ionium-uranium ratios. In: Subsurface geology of Eniwetok Atoll. U.S. Geol. Surv. Profess. Papers **260-BB**, 1053 (1963).
SAPRYKINA, T. V.: Distribution of uranium in the rocks of the Lovozero Alkali-Massif. Geochemistry (USSR) (English Transl.) 565 (1959).
SCOTT, R. C., and F. B. BARKER: Data on uranium and radium in groundwater in the United States, 1954—1957. U.S. Geol. Surv. Profess. Papers **426**, 115 p (1962).
SILVER, L. T., and S. DEUTSCH: Uranium-lead isotopic variations in zircons — a case study. J. Geol. **71**, 721 (1963).
SMITH, A. P., and F. S. GRIMALDI: Fluorometric determination of uranium in nonsaline and saline waters. U.S. Geol. Surv. Bull. **1006**, 125 (1954).
SMITH, G. H., and T. R. D. CHANDLER: A field method for the determination of uranium in natural waters. Proc. Second United Nations Int. Conf. on Peaceful Uses of Atomic Energy, Geneva **2**, 148 (1958).
SMITH, W. L., M. L. FRANCK, and A. M. SHERWOOD: Uranium and thorium in the accessory allanite of igneous rocks. Am. Mineralogist **42**, 367 (1957).
SPEARS, D. A.: The distribution of alpha radioactivity in a specimen of Shap granite. Geol. Mag. **48**, 483 (1961).
STARIK, I. E., YU. V. KUZNETSOV, S. M. GRASHCHENKO, and M. S. FRENKLIKH: The ionium method of determination of age of marine sediments. Geochemistry (USSR) (English Transl.) 1 (1958).

Starik, I. E., D. S. Nikolaev, Yu. V. Kuznetsov, and V. K. Legin: The radioactivity of Black Sea sediments. Dok. Akad. Nauk SSSR (English Transl.) **139**, 704 (1961 a).

—, and M. M. Shats: Determination of uranium in stony and iron meteorites. Geochemistry (USSR) (English Transl.) 140 (1956).

— — The uranium content of meteorites. Meteoritika 83 (1960) [in Russian].

— E. V. Sobotovich, M. M. Shats, and G. P. Lovtsyus: Uranium and lead in tektites. Meteoritika, No. 20, 204 (1961 b) [in Russian].

Stewart, D. C., and W. C. Bentley: Analysis of uranium in sea water. Science **120**, 50 (1954).

Swanson, V. E.: Geology and geochemistry of uranium in marine black shales, a review. U.S. Geol. Surv. Profess. Papers **356-C**, 67 (1961).

Tatsumoto, M., and E. D. Goldberg: Some aspects of the marine geochemistry of uranium. Geochim. Cosmochim. Acta **17**, 201 (1959).

— C. E. Hedge, and A. E. J. Engel: Potassium, rubidium, strontium, thorium, uranium, and the ratio of strontium-87 to strontium-86 in oceanic tholeiitic basalt. Science **150**, 886 (1965).

Tauson, L. V.: Geochemistry of uranium in granitoids of the Chernovinsky massif. Geochemistry (USSR) (English Transl.) 236 (1956).

— B. I. Zlobin, and L. L. Leonova: Uranium distribution in the granitoid complex of the Susamyr Batholith, Central Tien Shan. Geochemistry (USSR) (English Transl.) 650 (1958).

Thurber, D. L.: The dating of molluscs from raised marine terraces. Univ. of Rhode Island, Grad. School of Oceanog., Occasional Publication 3-1965, 1 (1965).

Tilton, G. R.: Isotopic composition of lead from tektites. Geochim. Cosmochim. Acta **14**, 323 (1958).

—, and L. O. Nicolayson: The use of monazites for age determination. Geochim. Cosmochim. Acta **11**, 28 (1957).

— C. Patterson, H. Brown, M. G. Inghram, R. Hayden, D. Hess, and E. S. Larsen Jr.: The isotopic composition and distribution of lead, uranium, and thorium in a pre-Cambrian granite. Bull. Geol. Soc. Am. **66**, 1131 (1955).

—, and G. W. Reed: Radioactive heat production in eclogite and some ultramafic rocks. In: J. Geiss and E. D. Goldberg (eds.), Earth science and meteoritics, p. 31. New York: Interscience Publ. (1963).

Vine, J. D.: Uranium-bearing coal in the United States. U.S. Geol. Surv. Profess. Papers **300**, 405 (1956).

Walenta, K.: Die sekundären Uranmineralien des Schwarzwaldes. Jahresh. Geol. Landesamtes Baden-Württemberg **3**, 17 (1958). [Am. Mineralogist **44**, 466 (1959); **45**, 254 (1960).]

—, u. W. Wimmenhauer: Der Mineralbestand des Michaelganges im Weiler bei Lahr (Schwarzwald). Jahresh. Geol. Landesamtes Baden-Württemberg **4**, 7 (1961). [Am. Mineralogist **47**, 414, 418 (1962).]

Webber, G. R., P. M. Hurley, and H. W. Fairbairn: Relative ages of eastern Massachusetts granites by total lead ratios in zircon. Am. J. Sci. **254**, 574 (1956).

Whitfield, J. M., J. J. W. Rogers, and J. A. S. Adams: The relationship between petrology and the thorium and uranium contents of some granitic rocks. Geochim. Cosmochim. Acta **17**, 248 (1959).

Wilson, J. D., R. K. Webster, G. W. C. Milner, G. A. Barnett, and A. A. Smales: A comparison of three methods of determining the concentration of uranium in sea water. Anal. Chim. Acta **23**, 505 (1960).

Yes'hova, E. M., D. A. Mineyev, and I. G. Mineyeva: Uranium and thorium in the alkalic rocks of the Urals. Geochemistry (USSR) (English Transl.) 885 (1962).

Revised manuscript received: July 18th, 1967

Notice to Users (Vol. II/5)

The Chapters of the handbook are arranged in the order of the atomic numbers of the chemical elements (see "List of Chapters" in the title pages). There is a standard sequence of Sections within the Chapters. This is explained in "Standard Sections of Chapters" on the title pages and the book marks.

The handbook is now complete with the present instalment. The pages of this instalment are approximately half in a binder and half in a package.

It has been tested that the complete material, including the present instalment, can be put into the five binders (Vol. II/1-Vol. II/5) in the following order:

Vol. II/1: Elements H (**1**) to Al (**13**)
Vol. II/2: Elements Si (**14**) to V (**23**)
Vol. II/3: Elements Cr (**24**) to Br (**35**)
Vol. II/4: Elements Kr (**36**) to Ba (**56**)
Vol. II/5: Elements La (**57**) to U (**92**)

Before beginning to place the present material, be sure that the instalments Vols. II/1 to II/4 have been sorted in accordance with the grey cards "Notice to Users". If you have not yet sorted Vols. II/1-II/4, then do this first. Place the pages from Vol. II/2 in Vol. II/1, Vol. II/3 in Vols. II/1 and II/2 etc.

For easy use of the final handbook a label can be stuck on the spine of the five binders, giving the elements in the individual binders. The label for Vol. II/5 has already been set in the correct place. Please do the same with remaining 4 labels which are to be found in a plastic cover at the beginning of binder Vol. II/5.

When the material has been thus far sorted out, please continue as follows:

Distribute first the sorted material of Vols. II/1 to II/4 according to the divisions (see above and title page) amongst the binders which you now have.

Begin at the end, that is, take the material of the elements La (**57**) to U (**92**) out of binder II/4 and place it in binder II/5. Then fill binder II/4 with the material Kr (**36**) to Ba (**56**) etc.

Now arrange the title pages in the binders of the appropriate instalments. The title part of Vol. II/1 comprise 16 connected pages, the title pages of Vol. II/2 to II/5 comprise 8 connected pages each. The title pages are to be found at the beginning of the present instalment II/5, each separate for Vols. II/1, II/2, II/3, II/4, and II/5. Include these title pages as first pages in each binder. Please remove the old title pages in the instalments Vol. II/1 to II/4. The title pages of Vol. II/1 contain additionally a new Preface and a "List of Authors".

Included moreover is a "List of Sections" printed on a yellow card. This provides an overview of all the Sections contained in the Handbook and indicates in what form the References appear in the Handbook (see footnote on this list).

The following Chapters are completed in instalment II/5:

Hydrogen **1**, Helium **2**, Aluminum **13**, Phosphorus **15**, Sulfur **16**, Calcium **20**, Titanium **22**, Chromium **24**, Manganese **25**, Cobalt **27**, Nickel **28**, Arsenic **33**, Strontium **38**, Zirconium **40**, Niobium **41**, Molybdenum **42**, Yttrium and Lanthanides 39, 57-71, Hafnium **72**, Tantalum **73**.

Notice to Users

For the elements Neon **10**, Argon **18**, Krypton **36**, Yttrium **39**, Ruthenium **44**, Rhodium **45**, Palladium **46**, Xenon **54**, Osmium **76**, Iridium **77**, and Radon **86** pink cards have been prepared for easier handling.

The following pink cards from the instalments II/1 to II/4 should be exchanged for the new cards which are in instalment II/5:

Hydrogen **1**, Helium **2**, Lithium **3**, Carbon **6**, Sodium **11**, Magnesium **12**, Aluminum **13**, Potassium **19**, Chromium **24**, Iron **26**, Arsenic **33**, Rubidium **37**, Strontium **38**, Zirconium **40**, Niobium **41**, Molybdenum **42**, Indium **49**, Tin **50**, Cesium **55**, Yttrium and Lanthanides 39, 57-71, Hafnium **72**, Tantalum **73**, Rhenium **75**, Platinum **78**, Mercury **80**, Bismuth **83**, Thorium **90**, and Uranium **92**.

For the Chapters

Lithium **3**, Beryllium **4**, Sodium **11**, Potassium **19**, Scandium **21**, Iron **26**, Germanium **32**, Rubidium **37**, Cadmium **48**, Indium **49**, Tin **50**, Cesium **55**, Rhenium **75**, Platinum **78**, Mercury **80**, Bismuth **83**, Thorium **90** and Uranium **92**

Subsections on "Lunar Materials" are additionally supplied.

Please insert these pages as follows:

Lithium 3:	old pages	3-C-1/3-C-2, 3-C-3/3-D-1 replace by
	new pages	3-C-1/3-C-2, 3-C-3/3-C-4, and 3-D-1;
Beryllium 4:	old pages	4-C-1/4-D-1 replace by
	new pages	4-C-1/4-C-2, 4-C-3/4-C-4, 4-C-5/4-D-1;
Sodium 11:	old pages	11-B-1/11-C-1, 11-C-4/11-D-1, 11-D-2 to 11-D-5 replace by
	new pages	11-B-1/11-C-1, 11-C-4 to 11-C-6, 11-D-1 to 11-D-5;
Potassium 19:	old pages	19-C-1/19-C-2, 19-C-5/19-D-1 replace by
	new pages	19-C-1/19-C-2, 19-C-5/19-C-6, and 19-D-1;
Scandium 21:	old pages	21-B-1/21-C-1, 21-C-2/21-D-1 replace by
	new pages	21-B-1/21-C-1, 21-C-2/21-C-3, and 21-D-1;
Iron 26:	old pages	26-C-1/26-C-2 replace by
	new pages	26-C-1/26-C-2;
Germanium 32:	old pages	32-B-1/32-C-1, 32-C-8/32-C-9 replace by
	new pages	32-B-1/32-C-1, 32-C-8 to 32-C-12;
Rubidium 37:	old pages	37-B-1/37-C-1, 37-C-4/37-D-1 replace by
	new pages	37-B-1/37-C-1, 37-C-4/37-C-5, and 37-D-1;
Cadmium 48:	old pages	48-B-1/48-C-1, 48-C-2/48-D-1 replace by
	new pages	48-B-1/48-C-1, 48-C-2/48-C-3, 48-C-4/48-C-5, and 48-D-1;
Indium 49:	old pages	49-B-1/49-C-1, 49-C-2/49-C-3 replace by
	new pages	49-B-1/49-C-1, 49-C-2 to 49-C-5;
Tin 50:	old pages	50-B-1/50-C-1, 50-C-2 to 50-C-5 replace by
	new pages	50-B-1/50-C-1, 50-C-2 to 50-C-6;
Cesium 55:	old pages	55-B-1/55-C-1 replace by
	new pages	55-B-1/55-C-1, 55-C-4/55-C-5;

Notice to Users

Rhenium 75:	old pages	75-C-1/75-C-2 replace by
	new pages	75-C-1 to 75-C-5;

Platinum Group 44-45-46, 76-77-78:

	old pages	44-45-46, 76-77-78-B-1/44-45-46, 76-77-78-C-1, 44-45-46, 76-77-78-C-22/44-45-46, 76-77-78-C-23 replace by
	new pages	44-45-46, 76-77-78-B-1/44-45-46, 76-77-78-C-1, 44-45-46, 76-77-78-C-22 to 44-45-46, 76-77-78-C-28;
Mercury 80:	old pages	80-B-1/80-C-1 replace by
	new pages	80-B-1/80-C-1, 80-C-4 to 80-C-6;
Bismuth 83:	old pages	83-B-1/83-C-1, 83-C-2/83-D-1 replace by
	new pages	83-B-1/83-C-1, 83-C-2/83-C-3, 83-C-4/83-D-1;
Thorium 90:	old pages	90-C-1/90-C-2, 90-C-3/90-D-1 replace by
	new pages	90-C-1 to 90-C-6, 90-D-1;
Uranium 92:	old pages	92-C-1/92-C-2, 92-C-5/92-D-1 replace by
	new pages	92-C-1/92-C-2, 92-C-5 to 92-C-10, and 92-D-1.

For Section O "Relation to other Elements" new pages are included for the following Chapters to be exchanged against the respective old pages:

Carbon **6**, Arsenic **33**, Tin **50**, Antimony **51**, Tungsten **74**, Platinum **78**, and Mercury **80**.

In detail the following pages in the old instalments are to be exchanged against the following new pages:

Carbon 6:	old pages	6-N-1/6-O-1, 6-O-2/6-O-3 replace by
	new pages	6-N-1/6-O-1, 6-O-2 to 6-O-4;
Arsenic 33:	old pages	33-O-1 replace by
	new pages	33-O-1;
Tin 50:	include new page 50-O-1;	
Antimony 51:	old pages	51-O-1 replace by
	new pages	51-O-1;
Tungsten 74:	old pages	74-O-1/74-O-2 replace by
	new pages	74-O-1 to 74-O-4;

Platinum Group 44-45-46, 76-77-78:

	old pages	44-45-46, 76-77-78-O-1/44-45-46, 76-77-78-O-2 replace by
	new pages	44-45-46, 76-77-78-O-1/44-45-46, 76-77-78-O-2, 44-45-46, 76-77-78-O-3;
Mercury 80:	include new pages 80-O-1 to 80-O-4.	

The following pages from earlier instalments include errors and must be exchanged:

Carbon 6:	pages	6-B-16/6-C-1;
Magnesium 12:	pages	12-C-1/12-C-2;
Scandium 21:	pages	21-I-1/21-K-1;
Zinc 30:	pages	30-H-2/30-H-3;
Arsenic 33:	pages	33-D-2 to 33-D-7, and 33-E-1;

Notice to Users

Tin 50:	pages	50-D-7/50-E-1;
Antimony 51:	pages	51-D-1 to 51-D-3;
Mercury 80:	pages	80-D-1 to 80-D-3.

Additionally old pages 2, 10, 18, 36, 54, 86-A-3 must be removed and replaced by the new pages 2, 10, 18, 36, 54, 86-A-3/2, 10, 18, 36, 54, 86-B-1.

By mistake the References of various Chapters were printed on the left side and were therefore impossible to be placed correctly. The following pages should be exchanged for these erroneous pages:

Oxygen 8:	old pages	8-B-27/References 8-B-1, References 8-B-2 to References 8-B-5 replace by
	new pages	8-B-27 and References 8-B-1 to References 8-B-5 (after page References 8-A-3);
Vanadium 23:	old pages	23-A-7/References 23-A-1, References 23-A-2 replace by
	new pages	23-A-7 and References 23-A-1/References 23-A-2 (after page 23-O-1);
Silver 47:	old pages	47-A-3/References 47-A-1 replace by
	new pages	47-A-3 and References 47-A-1 (after page 47-O-2).

References to Sections in formerly published Chapters have been combined with References to Sections included on this instalment. This occurs on the following Chapters:

Aluminum 13:	old pages	References 13-G to 13-L-1/13-G to 13-L-2 replace
	new pages	References 13-B to 13-O-1 to References 13-B to 13-O-4;
Sulfur 16:	old pages	References 16-A-1 to References 16-A-4, References 16-L-1 to References 16-L-4 replace by
	new pages	References 16-A to 16-O-1 to References 16-A to 16-O-26;
Strontium 38:	old pages	References 38-A-1, References 38-B-1 to References 38-B-5 replace by
	new pages	References 38-A to 38-N-1 to References 38-A to 38-N-32;
Rhenium 75:	old pages	References 75-A to 75-O-1 to References 75-A to 75-O-4 replace by
	new pages	References 75-A to 75-C, 75-E to 75-G, 75-I to 75-L, 75-O-1 to References 75-A to 75-C, 75-E to 75-G, 75-I to 75-L, 75-O-5;
Platinum Group 44-45-46, 76-77-78:		
	old pages	References 44-45-46, 76-77-78-1, References 44-45-46, 76-77-78-B to 78-O-1 to References 44-45-46, 76-77-78-B to 78-O-4 replace by
	new pages	References 44-45-46, 76-77-78-A to 78-O-1 to References 44-45-46, 76-77-78-A to 78-O-7.